4.21.98

D0208656

HANDBOOK of
APPLIED
CRYPTOGRAPHY

The CRC Press Series on

DISCRETE MATHEMATICS

AND

ITS APPLICATIONS

Series Editor

Kenneth H. Rosen, Ph.D.

AT&T Bell Laboratories

Charles J. Colbourn and Jeffery H. Dinitz, The CRC Handbook of Combinatorial Designs

Steven Furino, Ying Miao, and Jianxing Yin, Frames and Resolvable Designs: Uses, Constructions, and Existence

Daryl D. Harms, Miroslav Kraetzl, Charles J. Colbourn, and Stanley J. Devitt, Network Reliability: Experiments with A Symbolic Algebra Environment

Richard A. Mollin, Quadratics

Douglas R. Stinson, Cryptography: Theory and Practice

HANDBOOK of
APPLIED
CRYPTOGRAPHY

Alfred J. Menezes
Paul C. van Oorschot
Scott A. Vanstone

CRC Press
Boca Raton   New York   London   Tokyo

Library of Congress Cataloging-in-Publication Data

Menezes, A. J. (Alfred J.), 1965–
    Handbook of applied cryptography / Alfred Menezes, Paul van Oorschot,
Scott Vanstone.
       p. cm. -- (CRC Press series on discrete mathematics and its
applications)
    Includes bibliographical references and index.
    ISBN 0-8493-8523-7 (alk. paper)
    1. Computers--Access control--Handbooks, manuals, etc.
2. Cryptography--Handbooks, manuals, etc.   I. Van Oorschot, Paul C.
II. Vanstone, Scott A.   III. Title.   IV. Series: Discrete
mathematics and its applications.
QA76.9.A25M463   1996
005.8¢2--dc20
                                           96-27609
                                              CIP

No claim to original U.S. Government works
International Standard Book Number 0-8493-8523-7
Library of Congress Card Number 96-27609
Printed in the United States of America 2 3 4 5 6 7 8 9 0
Printed on acid-free paper

To Archie and Lida Menezes

To Cornelis Henricus van Oorschot
and Maria Anna Buys van Vugt

To Margret and Gordon Vanstone

Contents in Brief

Table of Contents

List of Tables

List of Figures

Foreword

by R.L. Rivest

As we draw near to closing out the twentieth century, we see quite clearly that the information-processing and telecommunications revolutions now underway will continue vigorously into the twenty-first. We interact and transact by directing flocks of digital packets towards each other through cyberspace, carrying love notes, digital cash, and secret corporate documents. Our personal and economic lives rely more and more on our ability to let such ethereal carrier pigeons mediate at a distance what we used to do with face-to-face meetings, paper documents, and a firm handshake. Unfortunately, the technical wizardry enabling remote collaborations is founded on broadcasting everything as sequences of zeros and ones that one's own dog wouldn't recognize. What is to distinguish a digital dollar when it is as easily reproducible as the spoken word? How do we converse privately when every syllable is bounced off a satellite and smeared over an entire continent? How should a bank know that it really *is* Bill Gates requesting from his laptop in Fiji a transfer of $10,000,000,000 to another bank? Fortunately, the magical mathematics of cryptography can help. Cryptography provides techniques for keeping information secret, for determining that information has not been tampered with, and for determining who authored pieces of information.

Cryptography is fascinating because of the close ties it forges between theory and practice, and because today's practical applications of cryptography are pervasive and critical components of our information-based society. Information-protection protocols designed on theoretical foundations one year appear in products and standards documents the next. Conversely, new theoretical developments sometimes mean that last year's proposal has a previously unsuspected weakness. While the theory is advancing vigorously, there are as yet few true guarantees; the security of many proposals depends on unproven (if plausible) assumptions. The theoretical work refines and improves the practice, while the practice challenges and inspires the theoretical work. When a system is "broken," our knowledge improves, and next year's system is improved to repair the defect. (One is reminded of the long and intriguing battle between the designers of bank vaults and their opponents.)

Cryptography is also fascinating because of its game-like adversarial nature. A good cryptographer rapidly changes sides back and forth in his or her thinking, from attacker to defender and back. Just as in a game of chess, sequences of moves and counter-moves must be considered until the current situation is understood. Unlike chess players, cryptographers must also consider all the ways an adversary might try to gain by breaking the rules or violating expectations. (Does it matter if she measures how long I am computing? Does it matter if her "random" number isn't one?)

The current volume is a major contribution to the field of cryptography. It is a rigorous encyclopedia of known techniques, with an emphasis on those that are both (believed to be) secure and practically useful. It presents in a coherent manner most of the important cryptographic tools one needs to implement secure cryptographic systems, and explains many of the cryptographic principles and protocols of existing systems. The topics covered range from low-level considerations such as random-number generation and efficient modular exponentiation algorithms and medium-level items such as public-key signature techniques, to higher-level topics such as zero-knowledge protocols. This book's excellent organization and style allow it to serve well as both a self-contained tutorial and an indispensable desk reference.

In documenting the state of a fast-moving field, the authors have done incredibly well at providing error-free comprehensive content that is up-to-date. Indeed, many of the chapters, such as those on hash functions or key-establishment protocols, break new ground in both their content and their unified presentations. In the trade-off between comprehensive coverage and exhaustive treatment of individual items, the authors have chosen to write simply and directly, and thus efficiently, allowing each element to be explained together with their important details, caveats, and comparisons.

While motivated by practical applications, the authors have clearly written a book that will be of as much interest to researchers and students as it is to practitioners, by including ample discussion of the underlying mathematics and associated theoretical considerations. The essential mathematical techniques and requisite notions are presented crisply and clearly, with illustrative examples. The insightful historical notes and extensive bibliography make this book a superb stepping-stone to the literature. (I was very pleasantly surprised to find an appendix with complete programs for the CRYPTO and EUROCRYPT conferences!)

It is a pleasure to have been asked to provide the foreword for this book. I am happy to congratulate the authors on their accomplishment, and to inform the reader that he/she is looking at a landmark in the development of the field.

Ronald L. Rivest
Webster Professor of Electrical Engineering and Computer Science
Massachusetts Institute of Technology

Preface

This book is intended as a reference for professional cryptographers, presenting the techniques and algorithms of greatest interest to the current practitioner, along with the supporting motivation and background material. It also provides a comprehensive source from which to learn cryptography, serving both students and instructors. In addition, the rigorous treatment, breadth, and extensive bibliographic material should make it an important reference for research professionals.

Our goal was to assimilate the existing cryptographic knowledge of industrial interest into one consistent, self-contained volume accessible to engineers in practice, to computer scientists and mathematicians in academia, and to motivated non-specialists with a strong desire to learn cryptography. Such a task is beyond the scope of each of the following: research papers, which by nature focus on narrow topics using very specialized (and often non-standard) terminology; survey papers, which typically address, at most, a small number of major topics at a high level; and (regretably also) most books, due to the fact that many book authors lack either practical experience or familiarity with the research literature or both. Our intent was to provide a detailed presentation of those areas of cryptography which we have found to be of greatest practical utility in our own industrial experience, while maintaining a sufficiently formal approach to be suitable both as a trustworthy reference for those whose primary interest is further research, and to provide a solid foundation for students and others first learning the subject.

Throughout each chapter, we emphasize the relationship between various aspects of cryptography. Background sections commence most chapters, providing a framework and perspective for the techniques which follow. Computer source code (e.g. C code) for algorithms has been intentionally omitted, in favor of algorithms specified in sufficient detail to allow direct implementation without consulting secondary references. We believe this style of presentation allows a better understanding of how algorithms actually work, while at the same time avoiding low-level implementation-specific constructs (which some readers will invariably be unfamiliar with) of various currently-popular programming languages.

The presentation also strongly delineates what has been established as fact (by mathematical arguments) from what is simply current conjecture. To avoid obscuring the very applied nature of the subject, rigorous proofs of correctness are in most cases omitted; however, references given in the Notes section at the end of each chapter indicate the original or recommended sources for these results. The trailing Notes sections also provide information (quite detailed in places) on various additional techniques not addressed in the main text, and provide a survey of research activities and theoretical results; references again indicate where readers may pursue particular aspects in greater depth. Needless to say, many results, and indeed some entire research areas, have been given far less attention than they warrant, or have been omitted entirely due to lack of space; we apologize in advance for such major omissions, and hope that the most significant of these are brought to our attention.

To provide an integrated treatment of cryptography spanning foundational motivation through concrete implementation, it is useful to consider a hierarchy of thought ranging from conceptual ideas and end-user services, down to the tools necessary to complete actual implementations. Table 1 depicts the hierarchical structure around which this book is organized. Corresponding to this, Figure 1 illustrates how these hierarchical levels map

Information Security Objectives	
Confidentiality	
Data integrity	
Authentication (entity and data origin)	
Non-repudiation	
Cryptographic functions	
Encryption	Chapters 6, 7, 8
Message authentication and data integrity techniques	Chapter 9
Identification/entity authentication techniques	Chapter 10
Digital signatures	Chapter 11
Cryptographic building blocks	
Stream ciphers	Chapter 6
Block ciphers (symmetric-key)	Chapter 7
Public-key encryption	Chapter 8
One-way hash functions (unkeyed)	Chapter 9
Message authentication codes	Chapter 9
Signature schemes (public-key, symmetric-key)	Chapter 11
Utilities	
Public-key parameter generation	Chapter 4
Pseudorandom bit generation	Chapter 5
Efficient algorithms for discrete arithmetic	Chapter 14
Foundations	
Introduction to cryptography	Chapter 1
Mathematical background	Chapter 2
Complexity and analysis of underlying problems	Chapter 3
Infrastructure techniques and commercial aspects	
Key establishment protocols	Chapter 12
Key installation and key management	Chapter 13
Cryptographic patents	Chapter 15
Cryptographic standards	Chapter 15

Table 1: *Hierarchical levels of applied cryptography.*

onto the various chapters, and their inter-dependence.

Table 2 lists the chapters of the book, along with the primary author(s) of each who should be contacted by readers with comments on specific chapters. Each chapter was written to provide a self-contained treatment of one major topic. Collectively, however, the chapters have been designed and carefully integrated to be entirely complementary with respect to definitions, terminology, and notation. Furthermore, there is essentially no duplication of material across chapters; instead, appropriate cross-chapter references are provided where relevant.

While it is not intended that this book be read linearly from front to back, the material has been arranged so that doing so has some merit. Two primary goals motivated by the "handbook" nature of this project were to allow easy access to stand-alone results, and to allow results and algorithms to be easily referenced (e.g., for discussion or subsequent cross-reference). To facilitate the ease of accessing and referencing results, items have been categorized and numbered to a large extent, with the following classes of items jointly numbered consecutively in each chapter: *Definitions*, *Examples*, *Facts*, *Notes*, *Remarks*, *Algorithms*, *Protocols*, and *Mechanisms*. In more traditional treatments, *Facts* are usually identified as propositions, lemmas, or theorems. We use numbered *Notes* for additional technical points,

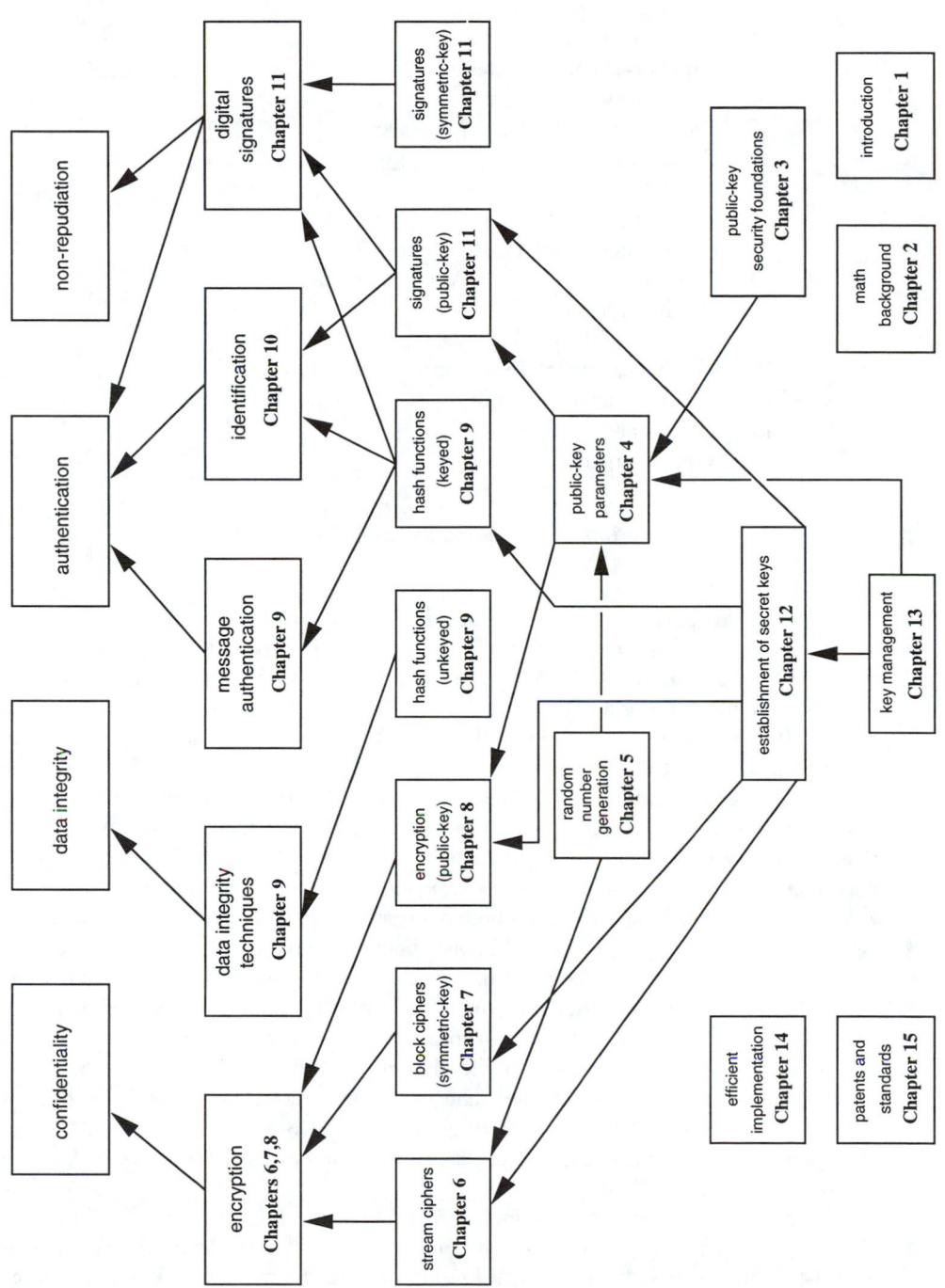

Figure 1: *Roadmap of the book.*

	Chapter	Primary Author		
		AJM	PVO	SAV
1.	Overview of Cryptography	*	*	*
2.	Mathematical Background	*		
3.	Number-Theoretic Reference Problems	*		
4.	Public-Key Parameters	*	*	
5.	Pseudorandom Bits and Sequences	*		
6.	Stream Ciphers	*		
7.	Block Ciphers		*	
8.	Public-Key Encryption	*		
9.	Hash Functions and Data Integrity		*	
10.	Identification and Entity Authentication		*	
11.	Digital Signatures			*
12.	Key Establishment Protocols		*	
13.	Key Management Techniques		*	
14.	Efficient Implementation			*
15.	Patents and Standards		*	
—	Overall organization	*	*	

Table 2: *Primary authors of each chapter.*

while numbered *Remarks* identify non-technical (often non-rigorous) comments, observations, and opinions. *Algorithms*, *Protocols* and *Mechanisms* refer to techniques involving a series of steps. *Examples*, *Notes*, and *Remarks* generally begin with parenthetical summary titles to allow faster access, by indicating the nature of the content so that the entire item itself need not be read in order to determine this. The use of a large number of small subsections is also intended to enhance the handbook nature and accessibility to results.

Regarding the partitioning of subject areas into chapters, we have used what we call a *functional organization* (based on functions of interest to end-users). For example, all items related to entity authentication are addressed in one chapter. An alternative would have been what may be called an *academic organization*, under which perhaps, all protocols based on zero-knowledge concepts (including both a subset of entity authentication protocols and signature schemes) might be covered in one chapter. We believe that a functional organization is more convenient to the practitioner, who is more likely to be interested in options available for an entity authentication protocol (Chapter 10) or a signature scheme (Chapter 11), than to be seeking a zero-knowledge protocol with unspecified end-purpose.

In the front matter, a top-level Table of Contents (giving chapter numbers and titles only) is provided, as well as a detailed Table of Contents (down to the level of subsections, e.g., §5.1.1). This is followed by a List of Figures, and a List of Tables. At the start of each chapter, a brief Table of Contents (specifying section number and titles only, e.g., §5.1, §5.2) is also given for convenience.

At the end of the book, we have included a list of papers presented at each of the Crypto, Eurocrypt, Asiacrypt/Auscrypt and Fast Software Encryption conferences to date, as well as a list of all papers published in the *Journal of Cryptology* up to Volume 9. These are in addition to the *References* section, each entry of which is cited at least once in the body of the handbook. Almost all of these references have been verified for correctness in their exact titles, volume and page numbers, etc. Finally, an extensive Index prepared by the authors is included. The Index begins with a List of Symbols.

Our intention was not to introduce a collection of new techniques and protocols, but rather to selectively present techniques from those currently available in the public domain. Such a consolidation of the literature is necessary from time to time. The fact that many good books in this field include essentially no more than what is covered here in Chapters 7, 8 and 11 (indeed, these might serve as an introductory course along with Chapter 1) illustrates that the field has grown tremendously in the past 15 years. The mathematical foundation presented in Chapters 2 and 3 is hard to find in one volume, and missing from most cryptography texts. The material in Chapter 4 on generation of public-key parameters, and in Chapter 14 on efficient implementations, while well-known to a small body of specialists and available in the scattered literature, has previously not been available in general texts. The material in Chapters 5 and 6 on pseudorandom number generation and stream ciphers is also often absent (many texts focus entirely on block ciphers), or approached only from a theoretical viewpoint. Hash functions (Chapter 9) and identification protocols (Chapter 10) have only recently been studied in depth as specialized topics on their own, and along with Chapter 12 on key establishment protocols, it is hard to find consolidated treatments of these now-mainstream topics. Key management techniques as presented in Chapter 13 have traditionally not been given much attention by cryptographers, but are of great importance in practice. A focused treatment of cryptographic patents and a concise summary of cryptographic standards, as presented in Chapter 15, are also long overdue.

In most cases (with some historical exceptions), where algorithms are known to be insecure, we have chosen to leave out specification of their details, because most such techniques are of little practical interest. Essentially all of the algorithms included have been verified for correctness by independent implementation, confirming the test vectors specified.

Acknowledgements

This project would not have been possible without the tremendous efforts put forth by our peers who have taken the time to read endless drafts and provide us with technical corrections, constructive feedback, and countless suggestions. In particular, the advice of our Advisory Editors has been invaluable, and it is impossible to attribute individual credit for their many suggestions throughout this book. Among our Advisory Editors, we would particularly like to thank:

Mihir Bellare	Don Coppersmith	Dorothy Denning	Walter Fumy
Burt Kaliski	Peter Landrock	Arjen Lenstra	Ueli Maurer
Chris Mitchell	Tatsuaki Okamoto	Bart Preneel	Ron Rivest
Gus Simmons	Miles Smid	Jacques Stern	Mike Wiener
Yacov Yacobi			

In addition, we gratefully acknowledge the exceptionally large number of additional individuals who have helped improve the quality of this volume, by providing highly appreciated feedback and guidance on various matters. These individuals include:

Carlisle Adams	Rich Ankney	Tom Berson
Simon Blackburn	Ian Blake	Antoon Bosselaers
Colin Boyd	Jörgen Brandt	Mike Burmester
Ed Dawson	Peter de Rooij	Yvo Desmedt
Whit Diffie	Hans Dobbertin	Carl Ellison
Luis Encinas	Warwick Ford	Amparo Fuster
Shuhong Gao	Will Gilbert	Marc Girault
Jovan Golić	Dieter Gollmann	Li Gong

Carrie Grant   Blake Greenlee   Helen Gustafson
Darrel Hankerson   Anwar Hasan   Don Johnson
Mike Just   Andy Klapper   Lars Knudsen
Neal Koblitz   Çetin Koç   Judy Koeller
Evangelos Kranakis   David Kravitz   Hugo Krawczyk
Xuejia Lai   Charles Lam   Alan Ling
S. Mike Matyas   Willi Meier   Peter Montgomery
Mike Mosca   Tim Moses   Serge Mister
Volker Müeller   David Naccache   James Nechvatal
Kaisa Nyberg   Andrew Odlyzko   Richard Outerbridge
Walter Penzhorn   Birgit Pfitzmann   Kevin Phelps
Leon Pintsov   Fred Piper   Carl Pomerance
Matt Robshaw   Peter Rodney   Phil Rogaway
Rainer Rueppel   Mahmoud Salmasizadeh   Roger Schlafly
Jeff Shallit   Jon Sorenson   Doug Stinson
Andrea Vanstone   Serge Vaudenay   Klaus Vedder
Jerry Veeh   Fausto Vitini   Lisa Yin
Robert Zuccherato

We apologize to those whose names have inadvertently escaped this list. Special thanks are due to Carrie Grant, Darrel Hankerson, Judy Koeller, Charles Lam, and Andrea Vanstone. Their hard work contributed greatly to the quality of this book, and it was truly a pleasure working with them. Thanks also to the folks at CRC Press, including Tia Atchison, Gary Bennett, Susie Carlisle, Nora Konopka, Mary Kugler, Amy Morrell, Tim Pletscher, Bob Stern, and Wayne Yuhasz. The second author would also like to thank his colleagues past and present at Nortel Secure Networks (Bell-Northern Research), many of whom are mentioned above, for their contributions on this project, and in particular Brian O'Higgins for his encouragement and support; all views expressed, however, are entirely that of the author.

Any errors that remain are, of course, entirely our own. We would be grateful if readers who spot errors, missing references or credits, or incorrectly attributed results would contact us with details. It is our hope that this volume facilitates further advancement of the field, and that we have helped play a small part in this.

Alfred J. Menezes
Paul C. van Oorschot
Scott A. Vanstone

Chapter 1

Overview of Cryptography

Contents in Brief

1.1 Introduction

Cryptography has a long and fascinating history. The most complete non-technical account of the subject is Kahn's *The Codebreakers*. This book traces cryptography from its initial and limited use by the Egyptians some 4000 years ago, to the twentieth century where it played a crucial role in the outcome of both world wars. Completed in 1963, Kahn's book covers those aspects of the history which were most significant (up to that time) to the development of the subject. The predominant practitioners of the art were those associated with the military, the diplomatic service and government in general. Cryptography was used as a tool to protect national secrets and strategies.

The proliferation of computers and communications systems in the 1960s brought with it a demand from the private sector for means to protect information in digital form and to provide security services. Beginning with the work of Feistel at IBM in the early 1970s and culminating in 1977 with the adoption as a U.S. Federal Information Processing Standard for encrypting unclassified information, DES, the Data Encryption Standard, is the most well-known cryptographic mechanism in history. It remains the standard means for securing electronic commerce for many financial institutions around the world.

The most striking development in the history of cryptography came in 1976 when Diffie and Hellman published *New Directions in Cryptography*. This paper introduced the revolutionary concept of public-key cryptography and also provided a new and ingenious method for key exchange, the security of which is based on the intractability of the discrete logarithm problem. Although the authors had no practical realization of a public-key encryption scheme at the time, the idea was clear and it generated extensive interest and activity in the cryptographic community. In 1978 Rivest, Shamir, and Adleman discovered the first practical public-key encryption and signature scheme, now referred to as RSA. The RSA scheme is based on another hard mathematical problem, the intractability of factoring large integers. This application of a hard mathematical problem to cryptography revitalized efforts to find more efficient methods to factor. The 1980s saw major advances in this area but none which rendered the RSA system insecure. Another class of powerful and practical public-key schemes was found by ElGamal in 1985. These are also based on the discrete logarithm problem.

One of the most significant contributions provided by public-key cryptography is the digital signature. In 1991 the first international standard for digital signatures (ISO/IEC 9796) was adopted. It is based on the RSA public-key scheme. In 1994 the U.S. Government adopted the Digital Signature Standard, a mechanism based on the ElGamal public-key scheme.

The search for new public-key schemes, improvements to existing cryptographic mechanisms, and proofs of security continues at a rapid pace. Various standards and infrastructures involving cryptography are being put in place. Security products are being developed to address the security needs of an information intensive society.

The purpose of this book is to give an up-to-date treatise of the principles, techniques, and algorithms of interest in cryptographic practice. Emphasis has been placed on those aspects which are most practical and applied. The reader will be made aware of the basic issues and pointed to specific related research in the literature where more indepth discussions can be found. Due to the volume of material which is covered, most results will be stated without proofs. This also serves the purpose of not obscuring the very applied nature of the subject. This book is intended for both implementers and researchers. It describes algorithms, systems, and their interactions.

Chapter 1 is a tutorial on the many and various aspects of cryptography. It does not attempt to convey all of the details and subtleties inherent to the subject. Its purpose is to introduce the basic issues and principles and to point the reader to appropriate chapters in the book for more comprehensive treatments. Specific techniques are avoided in this chapter.

1.2 Information security and cryptography

The concept of *information* will be taken to be an understood quantity. To introduce cryptography, an understanding of issues related to information security in general is necessary. Information security manifests itself in many ways according to the situation and requirement. Regardless of who is involved, to one degree or another, all parties to a transaction must have confidence that certain objectives associated with information security have been met. Some of these objectives are listed in Table 1.1.

Over the centuries, an elaborate set of protocols and mechanisms has been created to deal with information security issues when the information is conveyed by physical documents. Often the objectives of information security cannot solely be achieved through mathematical algorithms and protocols alone, but require procedural techniques and abidance of laws to achieve the desired result. For example, privacy of letters is provided by sealed envelopes delivered by an accepted mail service. The physical security of the envelope is, for practical necessity, limited and so laws are enacted which make it a criminal

privacy or confidentiality	keeping information secret from all but those who are authorized to see it.
data integrity	ensuring information has not been altered by unauthorized or unknown means.
entity authentication or identification	corroboration of the identity of an entity (e.g., a person, a computer terminal, a credit card, etc.).
message authentication	corroborating the source of information; also known as data origin authentication.
signature	a means to bind information to an entity.
authorization	conveyance, to another entity, of official sanction to do or be something.
validation	a means to provide timeliness of authorization to use or manipulate information or resources.
access control	restricting access to resources to privileged entities.
certification	endorsement of information by a trusted entity.
timestamping	recording the time of creation or existence of information.
witnessing	verifying the creation or existence of information by an entity other than the creator.
receipt	acknowledgement that information has been received.
confirmation	acknowledgement that services have been provided.
ownership	a means to provide an entity with the legal right to use or transfer a resource to others.
anonymity	concealing the identity of an entity involved in some process.
non-repudiation	preventing the denial of previous commitments or actions.
revocation	retraction of certification or authorization.

Table 1.1: *Some information security objectives.*

offense to open mail for which one is not authorized. It is sometimes the case that security is achieved not through the information itself but through the physical document recording it. For example, paper currency requires special inks and material to prevent counterfeiting.

Conceptually, the way information is recorded has not changed dramatically over time. Whereas information was typically stored and transmitted on paper, much of it now resides on magnetic media and is transmitted via telecommunications systems, some wireless. What has changed dramatically is the ability to copy and alter information. One can make thousands of identical copies of a piece of information stored electronically and each is indistinguishable from the original. With information on paper, this is much more difficult. What is needed then for a society where information is mostly stored and transmitted in electronic form is a means to ensure information security which is independent of the physical medium recording or conveying it and such that the objectives of information security rely solely on digital information itself.

One of the fundamental tools used in information security is the signature. It is a building block for many other services such as non-repudiation, data origin authentication, identification, and witnessing, to mention a few. Having learned the basics in writing, an individual is taught how to produce a handwritten signature for the purpose of identification. At contract age the signature evolves to take on a very integral part of the person's identity. This signature is intended to be unique to the individual and serve as a means to identify, authorize, and validate. With electronic information the concept of a signature needs to be redressed; it cannot simply be something unique to the signer and independent of the information signed. Electronic replication of it is so simple that appending a signature to a document not signed by the originator of the signature is almost a triviality.

Analogues of the "paper protocols" currently in use are required. Hopefully these new electronic based protocols are at least as good as those they replace. There is a unique opportunity for society to introduce new and more efficient ways of ensuring information security. Much can be learned from the evolution of the paper based system, mimicking those aspects which have served us well and removing the inefficiencies.

Achieving information security in an electronic society requires a vast array of technical and legal skills. There is, however, no guarantee that all of the information security objectives deemed necessary can be adequately met. The technical means is provided through cryptography.

1.1 Definition *Cryptography* is the study of mathematical techniques related to aspects of information security such as confidentiality, data integrity, entity authentication, and data origin authentication.

Cryptography is not the only means of providing information security, but rather one set of techniques.

Cryptographic goals

Of all the information security objectives listed in Table 1.1, the following four form a framework upon which the others will be derived: (1) privacy or confidentiality (§1.5, §1.8); (2) data integrity (§1.9); (3) authentication (§1.7); and (4) non-repudiation (§1.6).

1. *Confidentiality* is a service used to keep the content of information from all but those authorized to have it. *Secrecy* is a term synonymous with confidentiality and privacy. There are numerous approaches to providing confidentiality, ranging from physical protection to mathematical algorithms which render data unintelligible.

2. *Data integrity* is a service which addresses the unauthorized alteration of data. To assure data integrity, one must have the ability to detect data manipulation by unauthorized parties. Data manipulation includes such things as insertion, deletion, and substitution.

3. *Authentication* is a service related to identification. This function applies to both entities and information itself. Two parties entering into a communication should identify each other. Information delivered over a channel should be authenticated as to origin, date of origin, data content, time sent, etc. For these reasons this aspect of cryptography is usually subdivided into two major classes: *entity authentication* and *data origin authentication*. Data origin authentication implicitly provides data integrity (for if a message is modified, the source has changed).

4. *Non-repudiation* is a service which prevents an entity from denying previous commitments or actions. When disputes arise due to an entity denying that certain actions were taken, a means to resolve the situation is necessary. For example, one entity may authorize the purchase of property by another entity and later deny such authorization was granted. A procedure involving a trusted third party is needed to resolve the dispute.

A fundamental goal of cryptography is to adequately address these four areas in both theory and practice. Cryptography is about the prevention and detection of cheating and other malicious activities.

This book describes a number of basic *cryptographic tools* (*primitives*) used to provide information security. Examples of primitives include encryption schemes (§1.5 and §1.8), hash functions (§1.9), and digital signature schemes (§1.6). Figure 1.1 provides a schematic listing of the primitives considered and how they relate. Many of these will be briefly introduced in this chapter, with detailed discussion left to later chapters. These primitives should

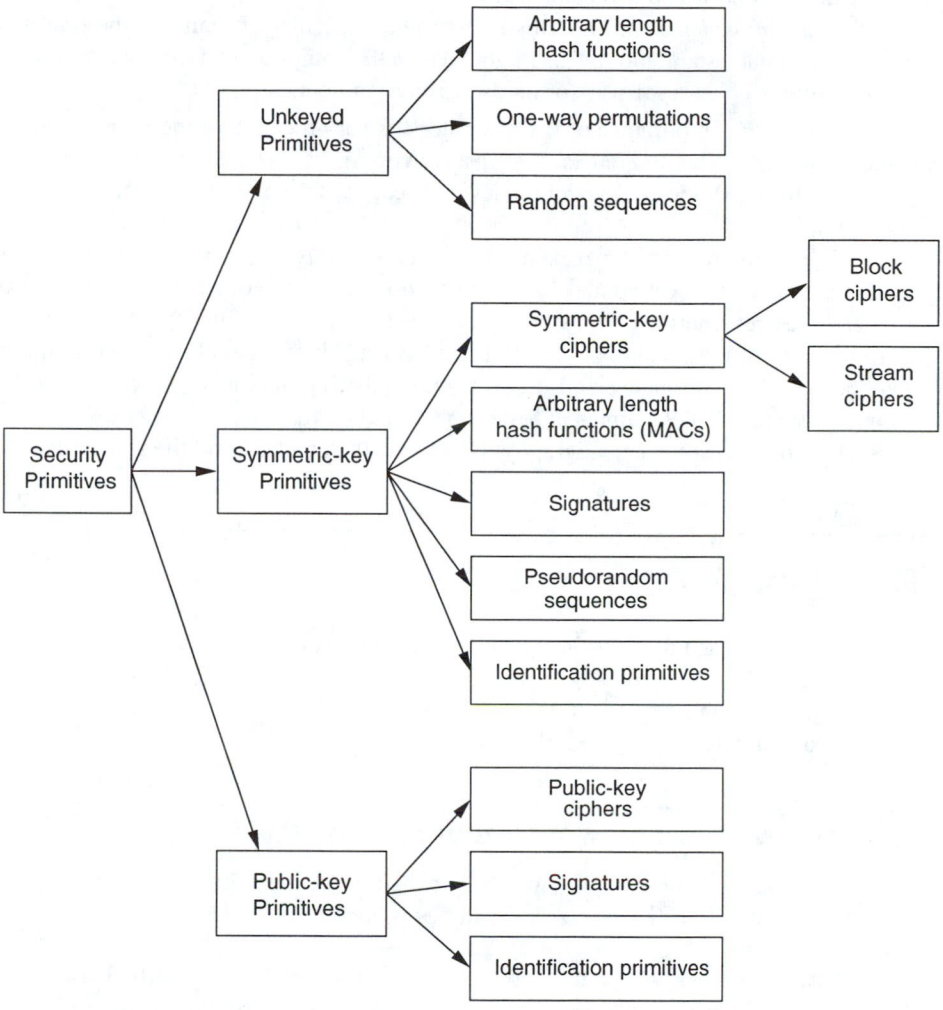

Figure 1.1: *A taxonomy of cryptographic primitives.*

be evaluated with respect to various criteria such as:

1. *level of security.* This is usually difficult to quantify. Often it is given in terms of the number of operations required (using the best methods currently known) to defeat the intended objective. Typically the level of security is defined by an upper bound on the amount of work necessary to defeat the objective. This is sometimes called the work factor (see §1.13.4).

2. *functionality.* Primitives will need to be combined to meet various information security objectives. Which primitives are most effective for a given objective will be determined by the basic properties of the primitives.

3. *methods of operation.* Primitives, when applied in various ways and with various inputs, will typically exhibit different characteristics; thus, one primitive could provide

very different functionality depending on its mode of operation or usage.

4. *performance.* This refers to the efficiency of a primitive in a particular mode of operation. (For example, an encryption algorithm may be rated by the number of bits per second which it can encrypt.)

5. *ease of implementation.* This refers to the difficulty of realizing the primitive in a practical instantiation. This might include the complexity of implementing the primitive in either a software or hardware environment.

The relative importance of various criteria is very much dependent on the application and resources available. For example, in an environment where computing power is limited one may have to trade off a very high level of security for better performance of the system as a whole.

Cryptography, over the ages, has been an art practised by many who have devised ad hoc techniques to meet some of the information security requirements. The last twenty years have been a period of transition as the discipline moved from an art to a science. There are now several international scientific conferences devoted exclusively to cryptography and also an international scientific organization, the International Association for Cryptologic Research (IACR), aimed at fostering research in the area.

This book is about cryptography: the theory, the practice, and the standards.

1.3 Background on functions

While this book is not a treatise on abstract mathematics, a familiarity with basic mathematical concepts will prove to be useful. One concept which is absolutely fundamental to cryptography is that of a *function* in the mathematical sense. A function is alternately referred to as a *mapping* or a *transformation*.

1.3.1 Functions (1-1, one-way, trapdoor one-way)

A *set* consists of distinct objects which are called *elements* of the set. For example, a set X might consist of the elements a, b, c, and this is denoted $X = \{a, b, c\}$.

1.2 Definition A *function* is defined by two sets X and Y and a *rule* f which assigns to each element in X precisely one element in Y. The set X is called the *domain* of the function and Y the *codomain*. If x is an element of X (usually written $x \in X$) the *image* of x is the element in Y which the rule f associates with x; the image y of x is denoted by $y = f(x)$. Standard notation for a function f from set X to set Y is $f \colon X \longrightarrow Y$. If $y \in Y$, then a *preimage* of y is an element $x \in X$ for which $f(x) = y$. The set of all elements in Y which have at least one preimage is called the *image* of f, denoted $\mathrm{Im}(f)$.

1.3 Example (*function*) Consider the sets $X = \{a, b, c\}$, $Y = \{1, 2, 3, 4\}$, and the rule f from X to Y defined as $f(a) = 2$, $f(b) = 4$, $f(c) = 1$. Figure 1.2 shows a schematic of the sets X, Y and the function f. The preimage of the element 2 is a. The image of f is $\{1, 2, 4\}$. $\square$

Thinking of a function in terms of the schematic (sometimes called a *functional diagram*) given in Figure 1.2, each element in the domain X has precisely one arrowed line originating from it. Each element in the codomain Y can have any number of arrowed lines incident to it (including zero lines).

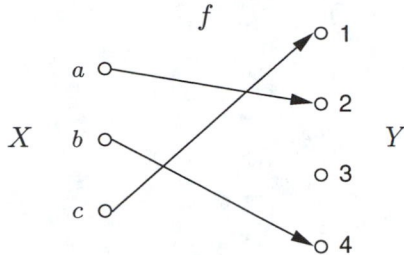

Figure 1.2: *A function f from a set X of three elements to a set Y of four elements.*

Often only the domain X and the rule f are given and the codomain is assumed to be the image of f. This point is illustrated with two examples.

1.4 Example *(function)* Take $X = \{1, 2, 3, \ldots, 10\}$ and let f be the rule that for each $x \in X$, $f(x) = r_x$, where r_x is the remainder when x^2 is divided by 11. Explicitly then

$$f(1) = 1 \quad f(2) = 4 \quad f(3) = 9 \quad f(4) = 5 \quad f(5) = 3$$
$$f(6) = 3 \quad f(7) = 5 \quad f(8) = 9 \quad f(9) = 4 \quad f(10) = 1.$$

The image of f is the set $Y = \{1, 3, 4, 5, 9\}$. □

1.5 Example *(function)* Take $X = \{1, 2, 3, \ldots, 10^{50}\}$ and let f be the rule $f(x) = r_x$, where r_x is the remainder when x^2 is divided by $10^{50} + 1$ for all $x \in X$. Here it is not feasible to write down f explicitly as in Example 1.4, but nonetheless the function is completely specified by the domain and the mathematical description of the rule f. □

(i) 1-1 functions

1.6 Definition A function (or transformation) is $1 - 1$ *(one-to-one)* if each element in the codomain Y is the image of at most one element in the domain X.

1.7 Definition A function (or transformation) is *onto* if each element in the codomain Y is the image of at least one element in the domain. Equivalently, a function $f \colon X \longrightarrow Y$ is onto if $\mathrm{Im}(f) = Y$.

1.8 Definition If a function $f \colon X \longrightarrow Y$ is $1 - 1$ and $\mathrm{Im}(f) = Y$, then f is called a *bijection*.

1.9 Fact If $f \colon X \longrightarrow Y$ is $1 - 1$ then $f \colon X \longrightarrow \mathrm{Im}(f)$ is a bijection. In particular, if $f \colon X \longrightarrow Y$ is $1 - 1$, and X and Y are finite sets of the same size, then f is a bijection.

In terms of the schematic representation, if f is a bijection, then each element in Y has exactly one arrowed line incident with it. The functions described in Examples 1.3 and 1.4 are not bijections. In Example 1.3 the element 3 is not the image of any element in the domain. In Example 1.4 each element in the codomain has two preimages.

1.10 Definition If f is a bijection from X to Y then it is a simple matter to define a bijection g from Y to X as follows: for each $y \in Y$ define $g(y) = x$ where $x \in X$ and $f(x) = y$. This function g obtained from f is called the *inverse function* of f and is denoted by $g = f^{-1}$.

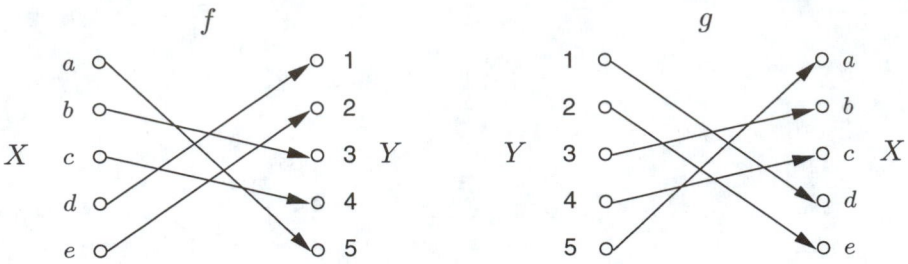

Figure 1.3: *A bijection f and its inverse $g = f^{-1}$.*

1.11 Example (*inverse function*) Let $X = \{a, b, c, d, e\}$, and $Y = \{1, 2, 3, 4, 5\}$, and consider the rule f given by the arrowed edges in Figure 1.3. f is a bijection and its inverse g is formed simply by reversing the arrows on the edges. The domain of g is Y and the codomain is X.   □

Note that if f is a bijection, then so is f^{-1}. In cryptography bijections are used as the tool for encrypting messages and the inverse transformations are used to decrypt. This will be made clearer in §1.4 when some basic terminology is introduced. Notice that if the transformations were not bijections then it would not be possible to always decrypt to a unique message.

(ii) One-way functions

There are certain types of functions which play significant roles in cryptography. At the expense of rigor, an intuitive definition of a one-way function is given.

1.12 Definition A function f from a set X to a set Y is called a *one-way function* if $f(x)$ is "easy" to compute for all $x \in X$ but for "most" elements $y \in \mathrm{Im}(f)$ it is "computationally infeasible" to find any $x \in X$ such that $f(x) = y$.

1.13 Note (*clarification of terms in Definition 1.12*)

    (i) A rigorous definition of the terms "easy" and "computationally infeasible" is necessary but would detract from the simple idea that is being conveyed. For the purpose of this chapter, the intuitive meaning will suffice.

    (ii) The phrase "for most elements in Y" refers to the fact that there are a few values $y \in Y$ for which it is easy to find an $x \in X$ such that $y = f(x)$. For example, one may compute $y = f(x)$ for a small number of x values and then for these, the inverse is known by table look-up. An alternate way to describe this property of a one-way function is the following: for a random $y \in \mathrm{Im}(f)$ it is computationally infeasible to find any $x \in X$ such that $f(x) = y$.

The concept of a one-way function is illustrated through the following examples.

1.14 Example (*one-way function*) Take $X = \{1, 2, 3, \ldots, 16\}$ and define $f(x) = r_x$ for all $x \in X$ where r_x is the remainder when 3^x is divided by 17. Explicitly,

x	1	2	3	4	5	6	7	8	9	10	11	12	13	14	15	16
$f(x)$	3	9	10	13	5	15	11	16	14	8	7	4	12	2	6	1

Given a number between 1 and 16, it is relatively easy to find the image of it under f. However, given a number such as 7, without having the table in front of you, it is harder to find

x given that $f(x) = 7$. Of course, if the number you are given is 3 then it is clear that $x = 1$ is what you need; but for most of the elements in the codomain it is not that easy. □

One must keep in mind that this is an example which uses very small numbers; the important point here is that there is a difference in the amount of work to compute $f(x)$ and the amount of work to find x given $f(x)$. Even for very large numbers, $f(x)$ can be computed efficiently using the repeated square-and-multiply algorithm (Algorithm 2.143), whereas the process of finding x from $f(x)$ is much harder.

1.15 Example (*one-way function*) A *prime number* is a positive integer greater than 1 whose only positive integer divisors are 1 and itself. Select primes $p = 48611$, $q = 53993$, form $n = pq = 2624653723$, and let $X = \{1, 2, 3, \ldots, n - 1\}$. Define a function f on X by $f(x) = r_x$ for each $x \in X$, where r_x is the remainder when x^3 is divided by n. For instance, $f(2489991) = 1981394214$ since $2489991^3 = 5881949859 \cdot n + 1981394214$. Computing $f(x)$ is a relatively simple thing to do, but to reverse the procedure is much more difficult; that is, given a remainder to find the value x which was originally cubed (raised to the third power). This procedure is referred to as the computation of a modular cube root with modulus n. If the factors of n are unknown and large, this is a difficult problem; however, if the factors p and q of n are known then there is an efficient algorithm for computing modular cube roots. (See §3.3 for details.) □

Example 1.15 leads one to consider another type of function which will prove to be fundamental in later developments.

(iii) Trapdoor one-way functions

1.16 Definition A *trapdoor one-way function* is a one-way function $f\colon X \longrightarrow Y$ with the additional property that given some extra information (called the *trapdoor information*) it becomes feasible to find for any given $y \in \text{Im}(f)$, an $x \in X$ such that $f(x) = y$.

Example 1.15 illustrates the concept of a trapdoor one-way function. With the additional information of the factors of $n = 2624653723$ (namely, $p = 48611$ and $q = 53993$, each of which is five decimal digits long) it becomes much easier to invert the function. The factors of 2624653723 are large enough that finding them by hand computation would be difficult. Of course, any reasonable computer program could find the factors relatively quickly. If, on the other hand, one selects p and q to be very large distinct prime numbers (each having about 100 decimal digits) then, by today's standards, it is a difficult problem, even with the most powerful computers, to deduce p and q simply from n. This is the well-known *integer factorization problem* (see §3.2) and a source of many trapdoor one-way functions.

It remains to be rigorously established whether there actually are any (true) one-way functions. That is to say, no one has yet definitively proved the existence of such functions under reasonable (and rigorous) definitions of "easy" and "computationally infeasible". Since the existence of one-way functions is still unknown, the existence of trapdoor one-way functions is also unknown. However, there are a number of good candidates for one-way and trapdoor one-way functions. Many of these are discussed in this book, with emphasis given to those which are practical.

One-way and trapdoor one-way functions are the basis for public-key cryptography (discussed in §1.8). The importance of these concepts will become clearer when their application to cryptographic techniques is considered. It will be worthwhile to keep the abstract concepts of this section in mind as concrete methods are presented.

1.3.2 Permutations

Permutations are functions which are often used in various cryptographic constructs.

1.17 Definition Let S be a finite set of elements. A *permutation* p on S is a bijection (Definition 1.8) from S to itself (i.e., $p\colon S \longrightarrow S$).

1.18 Example *(permutation)* Let $S = \{1, 2, 3, 4, 5\}$. A permutation $p\colon S \longrightarrow S$ is defined as follows:

$$p(1) = 3, \; p(2) = 5, \; p(3) = 4, \; p(4) = 2, \; p(5) = 1.$$

A permutation can be described in various ways. It can be displayed as above or as an array:

$$p = \begin{pmatrix} 1 & 2 & 3 & 4 & 5 \\ 3 & 5 & 4 & 2 & 1 \end{pmatrix}, \tag{1.1}$$

where the top row in the array is the domain and the bottom row is the image under the mapping p. Of course, other representations are possible.   □

Since permutations are bijections, they have inverses. If a permutation is written as an array (see 1.1), its inverse is easily found by interchanging the rows in the array and reordering the elements in the new top row if desired (the bottom row would have to be reordered correspondingly). The inverse of p in Example 1.18 is $p^{-1} = \begin{pmatrix} 1 & 2 & 3 & 4 & 5 \\ 5 & 4 & 1 & 3 & 2 \end{pmatrix}$.

1.19 Example *(permutation)* Let X be the set of integers $\{0, 1, 2, \ldots, pq - 1\}$ where p and q are distinct *large* primes (for example, p and q are each about 100 decimal digits long), and suppose that neither $p - 1$ nor $q - 1$ is divisible by 3. Then the function $p(x) = r_x$, where r_x is the remainder when x^3 is divided by pq, can be shown to be a permutation. Determining the inverse permutation is computationally infeasible by today's standards unless p and q are known (cf. Example 1.15).   □

1.3.3 Involutions

Another type of function which will be referred to in §1.5.3 is an involution. Involutions have the property that they are their own inverses.

1.20 Definition Let S be a finite set and let f be a bijection from S to S (i.e., $f\colon S \longrightarrow S$). The function f is called an *involution* if $f = f^{-1}$. An equivalent way of stating this is $f(f(x)) = x$ for all $x \in S$.

1.21 Example *(involution)* Figure 1.4 is an example of an involution. In the diagram of an involution, note that if j is the image of i then i is the image of j.   □

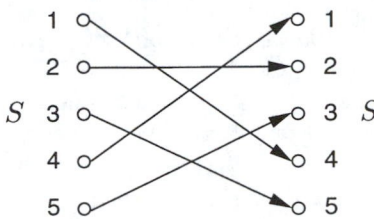

Figure 1.4: *An involution on a set S of 5 elements.*

1.4 Basic terminology and concepts

The scientific study of any discipline must be built upon rigorous definitions arising from fundamental concepts. What follows is a list of terms and basic concepts used throughout this book. Where appropriate, rigor has been sacrificed (here in Chapter 1) for the sake of clarity.

Encryption domains and codomains

- $\mathcal{A}$ denotes a finite set called the *alphabet of definition*. For example, $\mathcal{A} = \{0, 1\}$, the *binary alphabet*, is a frequently used alphabet of definition. Note that any alphabet can be encoded in terms of the binary alphabet. For example, since there are 32 binary strings of length five, each letter of the English alphabet can be assigned a unique binary string of length five.
- $\mathcal{M}$ denotes a set called the *message space*. $\mathcal{M}$ consists of strings of symbols from an alphabet of definition. An element of $\mathcal{M}$ is called a *plaintext message* or simply a *plaintext*. For example, $\mathcal{M}$ may consist of binary strings, English text, computer code, etc.
- $\mathcal{C}$ denotes a set called the *ciphertext space*. $\mathcal{C}$ consists of strings of symbols from an alphabet of definition, which may differ from the alphabet of definition for $\mathcal{M}$. An element of $\mathcal{C}$ is called a *ciphertext*.

Encryption and decryption transformations

- $\mathcal{K}$ denotes a set called the *key space*. An element of $\mathcal{K}$ is called a *key*.
- Each element $e \in \mathcal{K}$ uniquely determines a bijection from $\mathcal{M}$ to $\mathcal{C}$, denoted by E_e. E_e is called an *encryption function* or an *encryption transformation*. Note that E_e must be a bijection if the process is to be reversed and a unique plaintext message recovered for each distinct ciphertext.[1]
- For each $d \in \mathcal{K}$, D_d denotes a bijection from $\mathcal{C}$ to $\mathcal{M}$ (i.e., $D_d \colon \mathcal{C} \longrightarrow \mathcal{M}$). D_d is called a *decryption function* or *decryption transformation*.
- The process of applying the transformation E_e to a message $m \in \mathcal{M}$ is usually referred to as *encrypting m* or the *encryption* of m.
- The process of applying the transformation D_d to a ciphertext c is usually referred to as *decrypting c* or the *decryption* of c.

---

[1] More generality is obtained if E_e is simply defined as a $1 - 1$ transformation from $\mathcal{M}$ to $\mathcal{C}$. That is to say, E_e is a bijection from $\mathcal{M}$ to $\mathrm{Im}(E_e)$ where $\mathrm{Im}(E_e)$ is a subset of $\mathcal{C}$.

- An *encryption scheme* consists of a set $\{E_e : e \in \mathcal{K}\}$ of encryption transformations and a corresponding set $\{D_d : d \in \mathcal{K}\}$ of decryption transformations with the property that for each $e \in \mathcal{K}$ there is a unique key $d \in \mathcal{K}$ such that $D_d = E_e^{-1}$; that is, $D_d(E_e(m)) = m$ for all $m \in \mathcal{M}$. An encryption scheme is sometimes referred to as a *cipher*.
- The keys e and d in the preceding definition are referred to as a *key pair* and sometimes denoted by (e, d). Note that e and d could be the same.
- To *construct* an encryption scheme requires one to select a message space $\mathcal{M}$, a ciphertext space $\mathcal{C}$, a key space $\mathcal{K}$, a set of encryption transformations $\{E_e : e \in \mathcal{K}\}$, and a corresponding set of decryption transformations $\{D_d : d \in \mathcal{K}\}$.

Achieving confidentiality

An encryption scheme may be used as follows for the purpose of achieving confidentiality. Two parties Alice and Bob first secretly choose or secretly exchange a key pair (e, d). At a subsequent point in time, if Alice wishes to send a message $m \in \mathcal{M}$ to Bob, she computes $c = E_e(m)$ and transmits this to Bob. Upon receiving c, Bob computes $D_d(c) = m$ and hence recovers the original message m.

The question arises as to why keys are necessary. (Why not just choose one encryption function and its corresponding decryption function?) Having transformations which are very similar but characterized by keys means that if some particular encryption/decryption transformation is revealed then one does not have to redesign the entire scheme but simply change the key. It is sound cryptographic practice to change the key (encryption/decryption transformation) frequently. As a physical analogue, consider an ordinary resettable combination lock. The structure of the lock is available to anyone who wishes to purchase one but the combination is chosen and set by the owner. If the owner suspects that the combination has been revealed he can easily reset it without replacing the physical mechanism.

1.22 Example (*encryption scheme*) Let $\mathcal{M} = \{m_1, m_2, m_3\}$ and $\mathcal{C} = \{c_1, c_2, c_3\}$. There are precisely $3! = 6$ bijections from $\mathcal{M}$ to $\mathcal{C}$. The key space $\mathcal{K} = \{1, 2, 3, 4, 5, 6\}$ has six elements in it, each specifying one of the transformations. Figure 1.5 illustrates the six encryption functions which are denoted by $E_i, 1 \leq i \leq 6$. Alice and Bob agree on a trans-

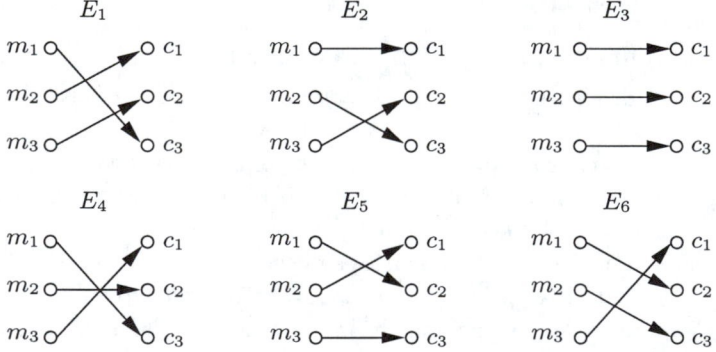

Figure 1.5: *Schematic of a simple encryption scheme.*

formation, say E_1. To encrypt the message m_1, Alice computes $E_1(m_1) = c_3$ and sends c_3 to Bob. Bob decrypts c_3 by reversing the arrows on the diagram for E_1 and observing that c_3 points to m_1.

When $\mathcal{M}$ is a small set, the functional diagram is a simple visual means to describe the mapping. In cryptography, the set $\mathcal{M}$ is typically of astronomical proportions and, as such, the visual description is infeasible. What is required, in these cases, is some other simple means to describe the encryption and decryption transformations, such as mathematical algorithms. □

Figure 1.6 provides a simple model of a two-party communication using encryption.

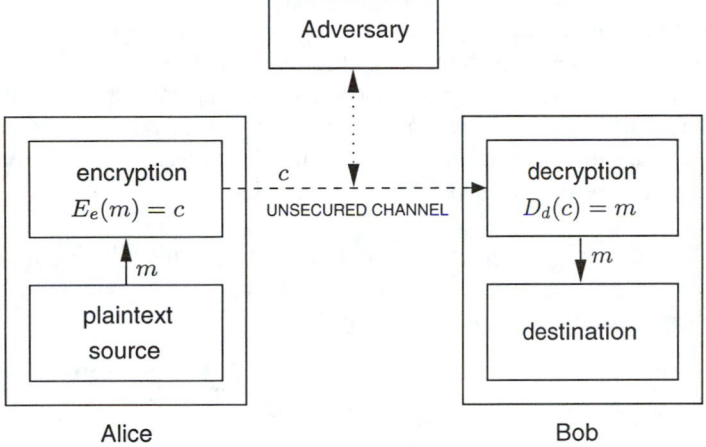

Figure 1.6: *Schematic of a two-party communication using encryption.*

Communication participants

Referring to Figure 1.6, the following terminology is defined.

- An *entity* or *party* is someone or something which sends, receives, or manipulates information. Alice and Bob are entities in Example 1.22. An entity may be a person, a computer terminal, etc.
- A *sender* is an entity in a two-party communication which is the legitimate transmitter of information. In Figure 1.6, the sender is Alice.
- A *receiver* is an entity in a two-party communication which is the intended recipient of information. In Figure 1.6, the receiver is Bob.
- An *adversary* is an entity in a two-party communication which is neither the sender nor receiver, and which tries to defeat the information security service being provided between the sender and receiver. Various other names are synonymous with adversary such as enemy, attacker, opponent, tapper, eavesdropper, intruder, and interloper. An adversary will often attempt to play the role of either the legitimate sender or the legitimate receiver.

Channels

- A *channel* is a means of conveying information from one entity to another.
- A *physically secure channel* or *secure channel* is one which is not physically accessible to the adversary.
- An *unsecured channel* is one from which parties other than those for which the information is intended can reorder, delete, insert, or read.
- A *secured channel* is one from which an adversary does not have the ability to reorder, delete, insert, or read.

One should note the subtle difference between a physically secure channel and a secured channel – a secured channel may be secured by physical or cryptographic techniques, the latter being the topic of this book. Certain channels are assumed to be physically secure. These include trusted couriers, personal contact between communicating parties, and a dedicated communication link, to name a few.

Security

A fundamental premise in cryptography is that the sets $\mathcal{M}, \mathcal{C}, \mathcal{K}, \{E_e : e \in \mathcal{K}\}, \{D_d : d \in \mathcal{K}\}$ are public knowledge. When two parties wish to communicate securely using an encryption scheme, the only thing that they keep secret is the particular key pair (e, d) which they are using, and which they must select. One can gain additional security by keeping the class of encryption and decryption transformations secret but one should not base the security of the entire scheme on this approach. History has shown that maintaining the secrecy of the transformations is very difficult indeed.

1.23 Definition An encryption scheme is said to be *breakable* if a third party, without prior knowledge of the key pair (e, d), can systematically recover plaintext from corresponding ciphertext within some appropriate time frame.

An appropriate time frame will be a function of the useful lifespan of the data being protected. For example, an instruction to buy a certain stock may only need to be kept secret for a few minutes whereas state secrets may need to remain confidential indefinitely.

An encryption scheme can be broken by trying all possible keys to see which one the communicating parties are using (assuming that the class of encryption functions is public knowledge). This is called an *exhaustive search* of the key space. It follows then that the number of keys (i.e., the size of the key space) should be large enough to make this approach computationally infeasible. It is the objective of a designer of an encryption scheme that this be the best approach to break the system.

Frequently cited in the literature are *Kerckhoffs' desiderata*, a set of requirements for cipher systems. They are given here essentially as Kerckhoffs originally stated them:

1. the system should be, if not theoretically unbreakable, unbreakable in practice;
2. compromise of the system details should not inconvenience the correspondents;
3. the key should be rememberable without notes and easily changed;
4. the cryptogram should be transmissible by telegraph;
5. the encryption apparatus should be portable and operable by a single person; and
6. the system should be easy, requiring neither the knowledge of a long list of rules nor mental strain.

This list of requirements was articulated in 1883 and, for the most part, remains useful today. Point 2 allows that the class of encryption transformations being used be publicly known and that the security of the system should reside only in the key chosen.

Information security in general

So far the terminology has been restricted to encryption and decryption with the goal of privacy in mind. Information security is much broader, encompassing such things as authentication and data integrity. A few more general definitions, pertinent to discussions later in the book, are given next.

- An *information security service* is a method to provide some specific aspect of security. For example, integrity of transmitted data is a security objective, and a method to ensure this aspect is an information security service.

- *Breaking* an information security service (which often involves more than simply encryption) implies defeating the objective of the intended service.
- A *passive adversary* is an adversary who is capable only of reading information from an unsecured channel.
- An *active adversary* is an adversary who may also transmit, alter, or delete information on an unsecured channel.

Cryptology

- *Cryptanalysis* is the study of mathematical techniques for attempting to defeat cryptographic techniques, and, more generally, information security services.
- A *cryptanalyst* is someone who engages in cryptanalysis.
- *Cryptology* is the study of cryptography (Definition 1.1) and cryptanalysis.
- A *cryptosystem* is a general term referring to a set of cryptographic primitives used to provide information security services. Most often the term is used in conjunction with primitives providing confidentiality, i.e., encryption.

Cryptographic techniques are typically divided into two generic types: *symmetric-key* and *public-key*. Encryption methods of these types will be discussed separately in §1.5 and §1.8. Other definitions and terminology will be introduced as required.

1.5 Symmetric-key encryption

§1.5 considers symmetric-key encryption. Public-key encryption is the topic of §1.8.

1.5.1 Overview of block ciphers and stream ciphers

1.24 Definition Consider an encryption scheme consisting of the sets of encryption and decryption transformations $\{E_e : e \in \mathcal{K}\}$ and $\{D_d : d \in \mathcal{K}\}$, respectively, where $\mathcal{K}$ is the key space. The encryption scheme is said to be *symmetric-key* if for each associated encryption/decryption key pair (e, d), it is computationally "easy" to determine d knowing only e, and to determine e from d.

Since $e = d$ in most practical symmetric-key encryption schemes, the term symmetric-key becomes appropriate. Other terms used in the literature are *single-key*, *one-key*, *private-key*,[2] and *conventional* encryption. Example 1.25 illustrates the idea of symmetric-key encryption.

1.25 Example (*symmetric-key encryption*) Let $\mathcal{A} = \{A, B, C, \ldots, X, Y, Z\}$ be the English alphabet. Let $\mathcal{M}$ and $\mathcal{C}$ be the set of all strings of length five over $\mathcal{A}$. The key e is chosen to be a permutation on $\mathcal{A}$. To encrypt, an English message is broken up into groups each having five letters (with appropriate padding if the length of the message is not a multiple of five) and a permutation e is applied to each letter one at a time. To decrypt, the inverse permutation $d = e^{-1}$ is applied to each letter of the ciphertext. For instance, suppose that the key e is chosen to be the permutation which maps each letter to the one which is three positions to its right, as shown below

$$e = \begin{pmatrix} A\ B\ C\ D\ E\ F\ G\ H\ I\ J\ K\ L\ M\ N\ O\ P\ Q\ R\ S\ T\ U\ V\ W\ X\ Y\ Z \\ D\ E\ F\ G\ H\ I\ J\ K\ L\ M\ N\ O\ P\ Q\ R\ S\ T\ U\ V\ W\ X\ Y\ Z\ A\ B\ C \end{pmatrix}$$

---

[2] Private key is a term also used in quite a different context (see §1.8). The term will be reserved for the latter usage in this book.

A message

$$m = \text{THISC IPHER ISCER TAINL YNOTS ECURE}$$

is encrypted to

$$c = E_e(m) = \text{WKLVF LSKHU LVFHU WDLQO BQRWV HFXUH.} \qquad \square$$

A two-party communication using symmetric-key encryption can be described by the block diagram of Figure 1.7, which is Figure 1.6 with the addition of the secure (both con-

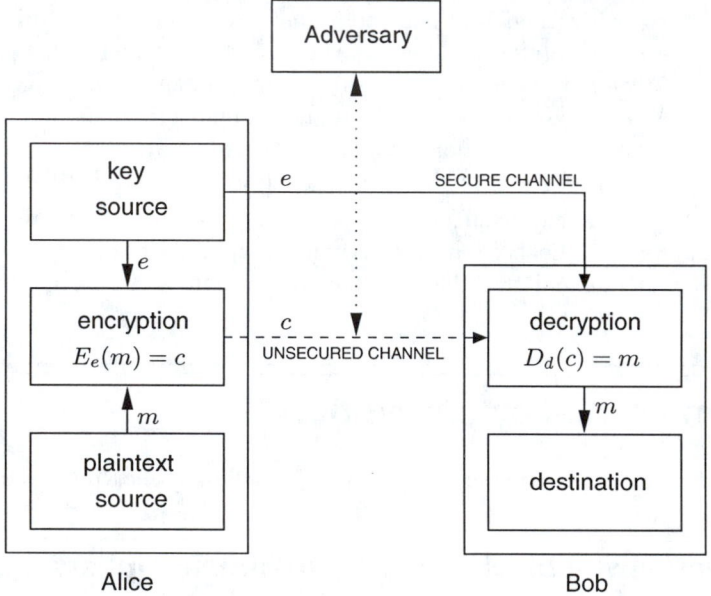

Figure 1.7: *Two-party communication using encryption, with a secure channel for key exchange. The decryption key d can be efficiently computed from the encryption key e.*

fidential and authentic) channel. One of the major issues with symmetric-key systems is to find an efficient method to agree upon and exchange keys securely. This problem is referred to as the *key distribution problem* (see Chapters 12 and 13).

It is assumed that all parties know the set of encryption/decryption transformations (i.e., they all know the encryption scheme). As has been emphasized several times the only information which should be required to be kept secret is the key d. However, in symmetric-key encryption, this means that the key e must also be kept secret, as d can be deduced from e. In Figure 1.7 the encryption key e is transported from one entity to the other with the understanding that both can construct the decryption key d.

There are two classes of symmetric-key encryption schemes which are commonly distinguished: *block ciphers* and *stream ciphers*.

1.26 Definition A *block cipher* is an encryption scheme which breaks up the plaintext messages to be transmitted into strings (called *blocks*) of a fixed length t over an alphabet $\mathcal{A}$, and encrypts one block at a time.

Most well-known symmetric-key encryption techniques are block ciphers. A number of examples of these are given in Chapter 7. Two important classes of block ciphers are *substitution ciphers* and *transposition ciphers* (§1.5.2). Product ciphers (§1.5.3) combine these. Stream ciphers are considered in §1.5.4, while comments on the key space follow in §1.5.5.

1.5.2 Substitution ciphers and transposition ciphers

Substitution ciphers are block ciphers which replace symbols (or groups of symbols) by other symbols or groups of symbols.

Simple substitution ciphers

1.27 Definition Let $\mathcal{A}$ be an alphabet of q symbols and $\mathcal{M}$ be the set of all strings of length t over $\mathcal{A}$. Let $\mathcal{K}$ be the set of all permutations on the set $\mathcal{A}$. Define for each $e \in \mathcal{K}$ an encryption transformation E_e as:

$$E_e(m) = (e(m_1)e(m_2) \cdots e(m_t)) = (c_1 c_2 \cdots c_t) = c,$$

where $m = (m_1 m_2 \cdots m_t) \in \mathcal{M}$. In other words, for each symbol in a t-tuple, replace (substitute) it by another symbol from $\mathcal{A}$ according to some fixed permutation e. To decrypt $c = (c_1 c_2 \cdots c_t)$ compute the inverse permutation $d = e^{-1}$ and

$$D_d(c) = (d(c_1)d(c_2) \cdots d(c_t)) = (m_1 m_2 \cdots m_t) = m.$$

E_e is called a *simple substitution cipher* or a *mono-alphabetic substitution cipher*.

The number of distinct substitution ciphers is $q!$ and is independent of the block size in the cipher. Example 1.25 is an example of a simple substitution cipher of block length five.

Simple substitution ciphers over small block sizes provide inadequate security even when the key space is extremely large. If the alphabet is the English alphabet as in Example 1.25, then the size of the key space is $26! \approx 4 \times 10^{26}$, yet the key being used can be determined quite easily by examining a modest amount of ciphertext. This follows from the simple observation that the distribution of letter frequencies is preserved in the ciphertext. For example, the letter E occurs more frequently than the other letters in ordinary English text. Hence the letter occurring most frequently in a sequence of ciphertext blocks is most likely to correspond to the letter E in the plaintext. By observing a modest quantity of ciphertext blocks, a cryptanalyst can determine the key.

Homophonic substitution ciphers

1.28 Definition To each symbol $a \in \mathcal{A}$, associate a set $H(a)$ of strings of t symbols, with the restriction that the sets $H(a)$, $a \in \mathcal{A}$, be pairwise disjoint. A *homophonic substitution cipher* replaces each symbol a in a plaintext message block with a randomly chosen string from $H(a)$. To decrypt a string c of t symbols, one must determine an $a \in \mathcal{A}$ such that $c \in H(a)$. The key for the cipher consists of the sets $H(a)$.

1.29 Example (*homophonic substitution cipher*) Consider $\mathcal{A} = \{a, b\}$, $H(a) = \{00, 10\}$, and $H(b) = \{01, 11\}$. The plaintext message block ab encrypts to one of the following: 0001, 0011, 1001, 1011. Observe that the codomain of the encryption function (for messages of length two) consists of the following pairwise disjoint sets of four-element bitstrings:

$$
\begin{aligned}
aa &\longrightarrow \{0000, 0010, 1000, 1010\} \\
ab &\longrightarrow \{0001, 0011, 1001, 1011\} \\
ba &\longrightarrow \{0100, 0110, 1100, 1110\} \\
bb &\longrightarrow \{0101, 0111, 1101, 1111\}
\end{aligned}
$$

Any 4-bitstring uniquely identifies a codomain element, and hence a plaintext message. $\square$

Often the symbols do not occur with equal frequency in plaintext messages. With a simple substitution cipher this non-uniform frequency property is reflected in the ciphertext as illustrated in Example 1.25. A homophonic cipher can be used to make the frequency of occurrence of ciphertext symbols more uniform, at the expense of data expansion. Decryption is not as easily performed as it is for simple substitution ciphers.

Polyalphabetic substitution ciphers

1.30 Definition A *polyalphabetic substitution cipher* is a block cipher with block length t over an alphabet $\mathcal{A}$ having the following properties:

(i) the key space $\mathcal{K}$ consists of all ordered sets of t permutations $(p_1, p_2, \ldots, p_t)$, where each permutation p_i is defined on the set $\mathcal{A}$;

(ii) encryption of the message $m = (m_1 m_2 \cdots m_t)$ under the key $e = (p_1, p_2, \ldots, p_t)$ is given by $E_e(m) = (p_1(m_1)p_2(m_2) \cdots p_t(m_t))$; and

(iii) the decryption key associated with $e = (p_1, p_2, \ldots, p_t)$ is $d = (p_1^{-1}, p_2^{-1}, \ldots, p_t^{-1})$.

1.31 Example (*Vigenère cipher*) Let $\mathcal{A} = \{A, B, C, \ldots, X, Y, Z\}$ and $t = 3$. Choose $e = (p_1, p_2, p_3)$, where p_1 maps each letter to the letter three positions to its right in the alphabet, p_2 to the one seven positions to its right, and p_3 ten positions to its right. If

$$m = \text{THI SCI PHE RIS CER TAI NLY NOT SEC URE}$$

then

$$c = E_e(m) = \text{WOS VJS SOO UPC FLB WHS QSI QVD VLM XYO.} \qquad \square$$

Polyalphabetic ciphers have the advantage over simple substitution ciphers that symbol frequencies are not preserved. In the example above, the letter E is encrypted to both O and L. However, polyalphabetic ciphers are not significantly more difficult to cryptanalyze, the approach being similar to the simple substitution cipher. In fact, once the block length t is determined, the ciphertext letters can be divided into t groups (where group i, $1 \leq i \leq t$, consists of those ciphertext letters derived using permutation p_i), and a frequency analysis can be done on each group.

Transposition ciphers

Another class of symmetric-key ciphers is the simple transposition cipher, which simply permutes the symbols in a block.

1.32 Definition Consider a symmetric-key block encryption scheme with block length t. Let $\mathcal{K}$ be the set of all permutations on the set $\{1, 2, \ldots, t\}$. For each $e \in \mathcal{K}$ define the encryption function

$$E_e(m) = (m_{e(1)} m_{e(2)} \cdots m_{e(t)})$$

where $m = (m_1 m_2 \cdots m_t) \in \mathcal{M}$, the message space. The set of all such transformations is called a *simple transposition cipher.* The decryption key corresponding to e is the inverse permutation $d = e^{-1}$. To decrypt $c = (c_1 c_2 \cdots c_t)$, compute $D_d(c) = (c_{d(1)} c_{d(2)} \cdots c_{d(t)})$.

A simple transposition cipher preserves the number of symbols of a given type within a block, and thus is easily cryptanalyzed.

1.5.3 Composition of ciphers

In order to describe product ciphers, the concept of composition of functions is introduced. Compositions are a convenient way of constructing more complicated functions from simpler ones.

Composition of functions

1.33 Definition Let $\mathcal{S}$, $\mathcal{T}$, and $\mathcal{U}$ be finite sets and let $f: \mathcal{S} \longrightarrow \mathcal{T}$ and $g: \mathcal{T} \longrightarrow \mathcal{U}$ be functions. The *composition* of g with f, denoted $g \circ f$ (or simply gf), is a function from $\mathcal{S}$ to $\mathcal{U}$ as illustrated in Figure 1.8 and defined by $(g \circ f)(x) = g(f(x))$ for all $x \in \mathcal{S}$.

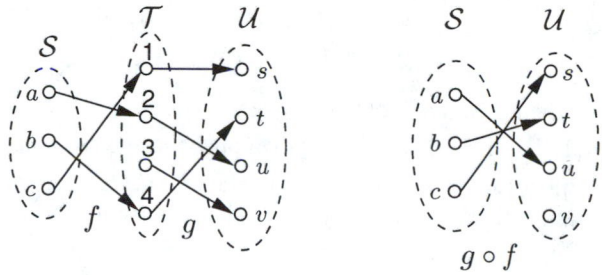

Figure 1.8: *The composition $g \circ f$ of functions g and f.*

Composition can be easily extended to more than two functions. For functions f_1, f_2, $\dots$, f_t, one can define $f_t \circ \cdots \circ f_2 \circ f_1$, provided that the domain of f_t equals the codomain of f_{t-1} and so on.

Compositions and involutions

Involutions were introduced in §1.3.3 as a simple class of functions with an interesting property: $E_k(E_k(x)) = x$ for all x in the domain of E_k; that is, $E_k \circ E_k$ is the identity function.

1.34 Remark (*composition of involutions*) The composition of two involutions is not necessarily an involution, as illustrated in Figure 1.9. However, involutions may be composed to get somewhat more complicated functions whose inverses are easy to find. This is an important feature for decryption. For example if $E_{k_1}, E_{k_2}, \dots, E_{k_t}$ are involutions then the inverse of $E_k = E_{k_1} E_{k_2} \cdots E_{k_t}$ is $E_k^{-1} = E_{k_t} E_{k_{t-1}} \cdots E_{k_1}$, the composition of the involutions in the reverse order.

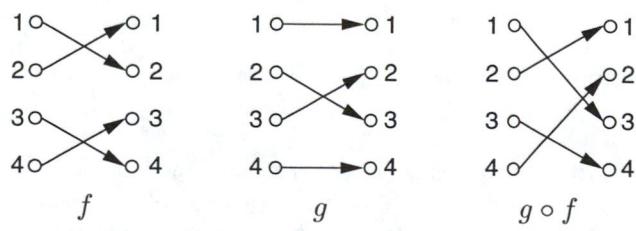

Figure 1.9: *The composition $g \circ f$ of involutions g and f is not an involution.*

Product ciphers

Simple substitution and transposition ciphers individually do not provide a very high level of security. However, by combining these transformations it is possible to obtain strong ciphers. As will be seen in Chapter 7 some of the most practical and effective symmetric-key systems are product ciphers. One example of a *product cipher* is a composition of $t \geq 2$ transformations $E_{k_1} E_{k_2} \cdots E_{k_t}$ where each E_{k_i}, $1 \leq i \leq t$, is either a substitution or a transposition cipher. For the purpose of this introduction, let the composition of a substitution and a transposition be called a *round*.

1.35 Example (*product cipher*) Let $\mathcal{M} = \mathcal{C} = \mathcal{K}$ be the set of all binary strings of length six. The number of elements in $\mathcal{M}$ is $2^6 = 64$. Let $m = (m_1 m_2 \cdots m_6)$ and define

$$E_k^{(1)}(m) = m \oplus k, \text{ where } k \in \mathcal{K},$$
$$E^{(2)}(m) = (m_4 m_5 m_6 m_1 m_2 m_3).$$

Here, $\oplus$ is the *exclusive-OR* (XOR) operation defined as follows: $0 \oplus 0 = 0, 0 \oplus 1 = 1, 1 \oplus 0 = 1, 1 \oplus 1 = 0$. $E_k^{(1)}$ is a polyalphabetic substitution cipher and $E^{(2)}$ is a transposition cipher (not involving the key). The product $E_k^{(1)} E^{(2)}$ is a round. While here the transposition cipher is very simple and is not determined by the key, this need not be the case.   □

1.36 Remark (*confusion and diffusion*) A substitution in a round is said to add *confusion* to the encryption process whereas a transposition is said to add *diffusion*. Confusion is intended to make the relationship between the key and ciphertext as complex as possible. Diffusion refers to rearranging or spreading out the bits in the message so that any redundancy in the plaintext is spread out over the ciphertext. A round then can be said to add both confusion and diffusion to the encryption. Most modern block cipher systems apply a number of rounds in succession to encrypt plaintext.

1.5.4 Stream ciphers

Stream ciphers form an important class of symmetric-key encryption schemes. They are, in one sense, very simple block ciphers having block length equal to one. What makes them useful is the fact that the encryption transformation can change for each symbol of plaintext being encrypted. In situations where transmission errors are highly probable, stream ciphers are advantageous because they have no error propagation. They can also be used when the data must be processed one symbol at a time (e.g., if the equipment has no memory or buffering of data is limited).

1.37 Definition Let $\mathcal{K}$ be the key space for a set of encryption transformations. A sequence of symbols $e_1 e_2 e_3 \cdots e_i \in \mathcal{K}$, is called a *keystream*.

1.38 Definition Let $\mathcal{A}$ be an alphabet of q symbols and let E_e be a simple substitution cipher with block length 1 where $e \in \mathcal{K}$. Let $m_1 m_2 m_3 \cdots$ be a plaintext string and let $e_1 e_2 e_3 \cdots$ be a keystream from $\mathcal{K}$. A *stream cipher* takes the plaintext string and produces a ciphertext string $c_1 c_2 c_3 \cdots$ where $c_i = E_{e_i}(m_i)$. If d_i denotes the inverse of e_i, then $D_{d_i}(c_i) = m_i$ decrypts the ciphertext string.

A stream cipher applies simple encryption transformations according to the keystream being used. The keystream could be generated at random, or by an algorithm which generates the keystream from an initial small keystream (called a *seed*), or from a seed and previous ciphertext symbols. Such an algorithm is called a *keystream generator*.

The Vernam cipher

A motivating factor for the Vernam cipher was its simplicity and ease of implementation.

1.39 Definition The *Vernam Cipher* is a stream cipher defined on the alphabet $\mathcal{A} = \{0, 1\}$. A binary message $m_1 m_2 \cdots m_t$ is operated on by a binary key string $k_1 k_2 \cdots k_t$ of the same length to produce a ciphertext string $c_1 c_2 \cdots c_t$ where

$$c_i = m_i \oplus k_i, \ 1 \le i \le t.$$

If the key string is randomly chosen and never used again, the Vernam cipher is called a *one-time system* or a *one-time pad*.

To see how the Vernam cipher corresponds to Definition 1.38, observe that there are precisely two substitution ciphers on the set $\mathcal{A}$. One is simply the identity map E_0 which sends 0 to 0 and 1 to 1; the other E_1 sends 0 to 1 and 1 to 0. When the keystream contains a 0, apply E_0 to the corresponding plaintext symbol; otherwise, apply E_1.

If the key string is reused there are ways to attack the system. For example, if $c_1 c_2 \cdots c_t$ and $c'_1 c'_2 \cdots c'_t$ are two ciphertext strings produced by the same keystream $k_1 k_2 \cdots k_t$ then

$$c_i = m_i \oplus k_i, \quad c'_i = m'_i \oplus k_i$$

and $c_i \oplus c'_i = m_i \oplus m'_i$. The redundancy in the latter may permit cryptanalysis.

The one-time pad can be shown to be theoretically unbreakable. That is, if a cryptanalyst has a ciphertext string $c_1 c_2 \cdots c_t$ encrypted using a random key string which has been used only once, the cryptanalyst can do no better than guess at the plaintext being any binary string of length t (i.e., t-bit binary strings are equally likely as plaintext). It has been proven that to realize an unbreakable system requires a random key of the same length as the message. This reduces the practicality of the system in all but a few specialized situations. Reportedly until very recently the communication line between Moscow and Washington was secured by a one-time pad. Transport of the key was done by trusted courier.

1.5.5 The key space

The size of the key space is the number of encryption/decryption key pairs that are available in the cipher system. A key is typically a compact way to specify the encryption transformation (from the set of all encryption transformations) to be used. For example, a transposition cipher of block length t has $t!$ encryption functions from which to select. Each can be simply described by a permutation which is called the key.

It is a great temptation to relate the security of the encryption scheme to the size of the key space. The following statement is important to remember.

1.40 Fact A necessary, but usually not sufficient, condition for an encryption scheme to be secure is that the key space be large enough to preclude exhaustive search.

For instance, the simple substitution cipher in Example 1.25 has a key space of size $26! \approx 4 \times 10^{26}$. The polyalphabetic substitution cipher of Example 1.31 has a key space of size $(26!)^3 \approx 7 \times 10^{79}$. Exhaustive search of either key space is completely infeasible, yet both ciphers are relatively weak and provide little security.

1.6 Digital signatures

A cryptographic primitive which is fundamental in authentication, authorization, and non-repudiation is the *digital signature*. The purpose of a digital signature is to provide a means for an entity to bind its identity to a piece of information. The process of *signing* entails transforming the message and some secret information held by the entity into a tag called a *signature*. A generic description follows.

Nomenclature and set-up

- $\mathcal{M}$ is the set of messages which can be signed.
- $\mathcal{S}$ is a set of elements called *signatures*, possibly binary strings of a fixed length.
- S_A is a transformation from the message set $\mathcal{M}$ to the signature set $\mathcal{S}$, and is called a *signing transformation* for entity A.[3] The transformation S_A is kept secret by A, and will be used to create signatures for messages from $\mathcal{M}$.
- V_A is a transformation from the set $\mathcal{M} \times \mathcal{S}$ to the set $\{true, false\}$.[4] V_A is called a *verification transformation* for A's signatures, is publicly known, and is used by other entities to verify signatures created by A.

1.41 Definition The transformations S_A and V_A provide a *digital signature scheme* for A. Occasionally the term *digital signature mechanism* is used.

1.42 Example (*digital signature scheme*) $\mathcal{M} = \{m_1, m_2, m_3\}$ and $\mathcal{S} = \{s_1, s_2, s_3\}$. The left side of Figure 1.10 displays a signing function S_A from the set $\mathcal{M}$ and, the right side, the corresponding verification function V_A. □

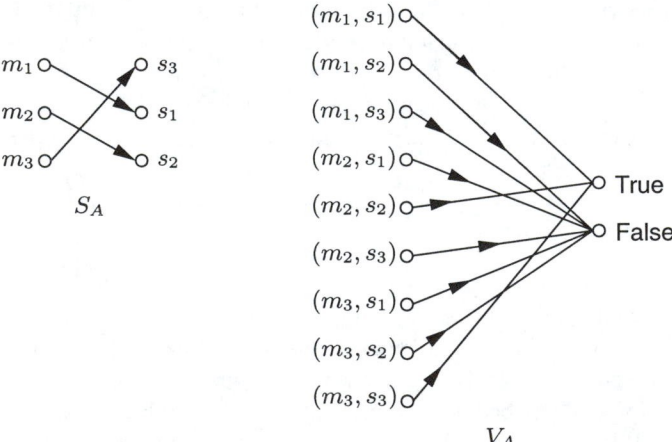

Figure 1.10: *A signing and verification function for a digital signature scheme.*

---

[3] The names of Alice and Bob are usually abbreviated to A and B, respectively.
[4] $\mathcal{M} \times \mathcal{S}$ consists of all pairs (m, s) where $m \in \mathcal{M}$, $s \in \mathcal{S}$, called the *Cartesian product* of $\mathcal{M}$ and $\mathcal{S}$.

Signing procedure

Entity A (the *signer*) creates a signature for a message $m \in \mathcal{M}$ by doing the following:

1. Compute $s = S_A(m)$.
2. Transmit the pair (m, s). s is called the *signature* for message m.

Verification procedure

To verify that a signature s on a message m was created by A, an entity B (the *verifier*) performs the following steps:

1. Obtain the verification function V_A of A.
2. Compute $u = V_A(m, s)$.
3. Accept the signature as having been created by A if $u = true$, and reject the signature if $u = false$.

1.43 Remark (*concise representation*) The transformations S_A and V_A are typically characterized more compactly by a key; that is, there is a class of signing and verification algorithms publicly known, and each algorithm is identified by a key. Thus the signing algorithm S_A of A is determined by a key k_A and A is only required to keep k_A secret. Similarly, the verification algorithm V_A of A is determined by a key l_A which is made public.

1.44 Remark (*handwritten signatures*) Handwritten signatures could be interpreted as a special class of digital signatures. To see this, take the set of signatures S to contain only one element which is the handwritten signature of A, denoted by s_A. The verification function simply checks if the signature on a message purportedly signed by A is s_A.

An undesirable feature in Remark 1.44 is that the signature is not message-dependent. Hence, further constraints are imposed on digital signature mechanisms as next discussed.

Properties required for signing and verification functions

There are several properties which the signing and verification transformations must satisfy.

(a) s is a valid signature of A on message m if and only if $V_A(m, s) = true$.
(b) It is computationally infeasible for any entity other than A to find, for any $m \in \mathcal{M}$, an $s \in S$ such that $V_A(m, s) = true$.

Figure 1.10 graphically displays property (a). There is an arrowed line in the diagram for V_A from (m_i, s_j) to *true* provided there is an arrowed line from m_i to s_j in the diagram for S_A. Property (b) provides the security for the method – the signature uniquely binds A to the message which is signed.

No one has yet formally proved that digital signature schemes satisfying (b) exist (although existence is widely believed to be true); however, there are some very good candidates. §1.8.3 introduces a particular class of digital signatures which arise from public-key encryption techniques. Chapter 11 describes a number of digital signature mechanisms which are believed to satisfy the two properties cited above. Although the description of a digital signature given in this section is quite general, it can be broadened further, as presented in §11.2.

1.7 Authentication and identification

Authentication is a term which is used (and often abused) in a very broad sense. By itself it has little meaning other than to convey the idea that some means has been provided to guarantee that entities are who they claim to be, or that information has not been manipulated by unauthorized parties. Authentication is specific to the security objective which one is trying to achieve. Examples of specific objectives include access control, entity authentication, message authentication, data integrity, non-repudiation, and key authentication. These instances of authentication are dealt with at length in Chapters 9 through 13. For the purposes of this chapter, it suffices to give a brief introduction to authentication by describing several of the most obvious applications.

Authentication is one of the most important of all information security objectives. Until the mid 1970s it was generally believed that secrecy and authentication were intrinsically connected. With the discovery of hash functions (§1.9) and digital signatures (§1.6), it was realized that secrecy and authentication were truly separate and independent information security objectives. It may at first not seem important to separate the two but there are situations where it is not only useful but essential. For example, if a two-party communication between Alice and Bob is to take place where Alice is in one country and Bob in another, the host countries might not permit secrecy on the channel; one or both countries might want the ability to monitor all communications. Alice and Bob, however, would like to be assured of the identity of each other, and of the integrity and origin of the information they send and receive.

The preceding scenario illustrates several independent aspects of authentication. If Alice and Bob desire assurance of each other's identity, there are two possibilities to consider.

1. Alice and Bob could be communicating with no appreciable time delay. That is, they are both active in the communication in "real time".

2. Alice or Bob could be exchanging messages with some delay. That is, messages might be routed through various networks, stored, and forwarded at some later time.

In the first instance Alice and Bob would want to verify identities in real time. This might be accomplished by Alice sending Bob some challenge, to which Bob is the only entity which can respond correctly. Bob could perform a similar action to identify Alice. This type of authentication is commonly referred to as *entity authentication* or more simply *identification*.

For the second possibility, it is not convenient to challenge and await response, and moreover the communication path may be only in one direction. Different techniques are now required to authenticate the originator of the message. This form of authentication is called *data origin authentication*.

1.7.1 Identification

1.45 Definition An *identification* or *entity authentication* technique assures one party (through acquisition of corroborative evidence) of both the identity of a second party involved, and that the second was active at the time the evidence was created or acquired.

Typically the only data transmitted is that necessary to identify the communicating parties. The entities are both active in the communication, giving a timeliness guarantee.

1.46 Example (*identification*) A calls B on the telephone. If A and B know each other then entity authentication is provided through voice recognition. Although not foolproof, this works effectively in practice. □

1.47 Example (*identification*) Person A provides to a banking machine a personal identification number (PIN) along with a magnetic stripe card containing information about A. The banking machine uses the information on the card and the PIN to verify the identity of the card holder. If verification succeeds, A is given access to various services offered by the machine. □

Example 1.46 is an instance of *mutual authentication* whereas Example 1.47 only provides *unilateral authentication*. Numerous mechanisms and protocols devised to provide mutual or unilateral authentication are discussed in Chapter 10.

1.7.2 Data origin authentication

1.48 Definition *Data origin authentication* or *message authentication* techniques provide to one party which receives a message assurance (through corroborative evidence) of the identity of the party which originated the message.

Often a message is provided to B along with additional information so that B can determine the identity of the entity who originated the message. This form of authentication typically provides no guarantee of timeliness, but is useful in situations where one of the parties is not active in the communication.

1.49 Example (*need for data origin authentication*) A sends to B an electronic mail message (e-mail). The message may travel through various network communications systems and be stored for B to retrieve at some later time. A and B are usually not in direct communication. B would like some means to verify that the message received and purportedly created by A did indeed originate from A. □

Data origin authentication implicitly provides data integrity since, if the message was modified during transmission, A would no longer be the originator.

1.8 Public-key cryptography

The concept of public-key encryption is simple and elegant, but has far-reaching consequences.

1.8.1 Public-key encryption

Let $\{E_e : e \in \mathcal{K}\}$ be a set of encryption transformations, and let $\{D_d : d \in \mathcal{K}\}$ be the set of corresponding decryption transformations, where $\mathcal{K}$ is the key space. Consider any pair of associated encryption/decryption transformations (E_e, D_d) and suppose that each pair has the property that knowing E_e it is computationally infeasible, given a random ciphertext $c \in \mathcal{C}$, to find the message $m \in \mathcal{M}$ such that $E_e(m) = c$. This property implies that given e it is infeasible to determine the corresponding decryption key d. (Of course e and d are simply means to describe the encryption and decryption functions, respectively.) E_e is being viewed here as a trapdoor one-way function (Definition 1.16) with d being the trapdoor information necessary to compute the inverse function and hence allow decryption. This is unlike symmetric-key ciphers where e and d are essentially the same.

Under these assumptions, consider the two-party communication between Alice and Bob illustrated in Figure 1.11. Bob selects the key pair (e, d). Bob sends the encryption key e (called the *public key*) to Alice over any channel but keeps the decryption key d (called the *private key*) secure and secret. Alice may subsequently send a message m to Bob by applying the encryption transformation determined by Bob's public key to get $c = E_e(m)$. Bob decrypts the ciphertext c by applying the inverse transformation D_d uniquely determined by d.

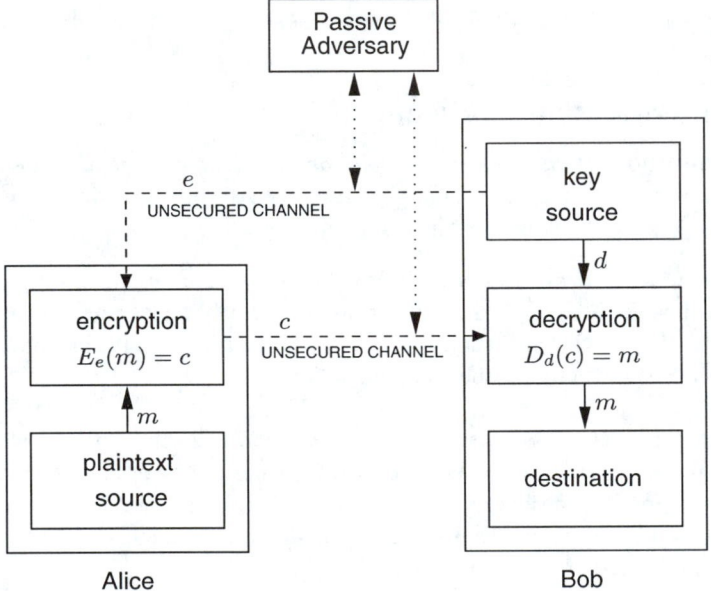

Figure 1.11: *Encryption using public-key techniques.*

Notice how Figure 1.11 differs from Figure 1.7 for a symmetric-key cipher. Here the encryption key is transmitted to Alice over an unsecured channel. This unsecured channel may be the same channel on which the ciphertext is being transmitted (but see §1.8.2).

Since the encryption key e need not be kept secret, it may be made public. Any entity can subsequently send encrypted messages to Bob which only Bob can decrypt. Figure 1.12 illustrates this idea, where A_1, A_2, and A_3 are distinct entities. Note that if A_1 destroys message m_1 after encrypting it to c_1, then even A_1 cannot recover m_1 from c_1.

As a physical analogue, consider a metal box with the lid secured by a combination lock. The combination is known only to Bob. If the lock is left open and made publicly available then anyone can place a message inside and lock the lid. Only Bob can retrieve the message. Even the entity which placed the message into the box is unable to retrieve it.

Public-key encryption, as described here, assumes that knowledge of the public key e does not allow computation of the private key d. In other words, this assumes the existence of trapdoor one-way functions (§1.3.1(iii)).

1.50 Definition Consider an encryption scheme consisting of the sets of encryption and decryp-

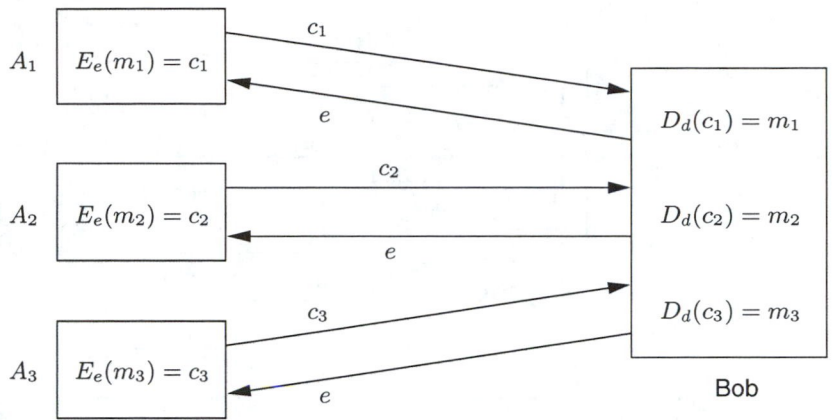

Figure 1.12: *Schematic use of public-key encryption.*

tion transformations $\{E_e : e \in \mathcal{K}\}$ and $\{D_d : d \in \mathcal{K}\}$, respectively. The encryption method is said to be a *public-key encryption scheme* if for each associated encryption/decryption pair (e, d), one key e (the *public key*) is made publicly available, while the other d (the *private key*) is kept secret. For the scheme to be *secure*, it must be computationally infeasible to compute d from e.

1.51 Remark (*private key vs. secret key*) To avoid ambiguity, a common convention is to use the term *private key* in association with public-key cryptosystems, and *secret key* in association with symmetric-key cryptosystems. This may be motivated by the following line of thought: it takes two or more parties to *share* a secret, but a key is truly *private* only when one party alone knows it.

There are many schemes known which are widely believed to be secure public-key encryption methods, but none have been mathematically proven to be secure independent of qualifying assumptions. This is not unlike the symmetric-key case where the only system which has been proven secure is the one-time pad (§1.5.4).

1.8.2 The necessity of authentication in public-key systems

It would appear that public-key cryptography is an ideal system, not requiring a secure channel to pass the encryption key. This would imply that two entities could communicate over an unsecured channel without ever having met to exchange keys. Unfortunately, this is not the case. Figure 1.13 illustrates how an active adversary can defeat the system (decrypt messages intended for a second entity) without breaking the encryption system. This is a type of *impersonation* and is an example of *protocol failure* (see §1.10). In this scenario the adversary impersonates entity B by sending entity A a public key e' which A assumes (incorrectly) to be the public key of B. The adversary intercepts encrypted messages from A to B, decrypts with its own private key d', re-encrypts the message under B's public key e, and sends it on to B. This highlights the necessity to *authenticate* public keys to achieve data origin authentication of the public keys themselves. A must be convinced that she is encrypting under the legitimate public key of B. Fortunately, public-key techniques also allow an elegant solution to this problem (see §1.11).

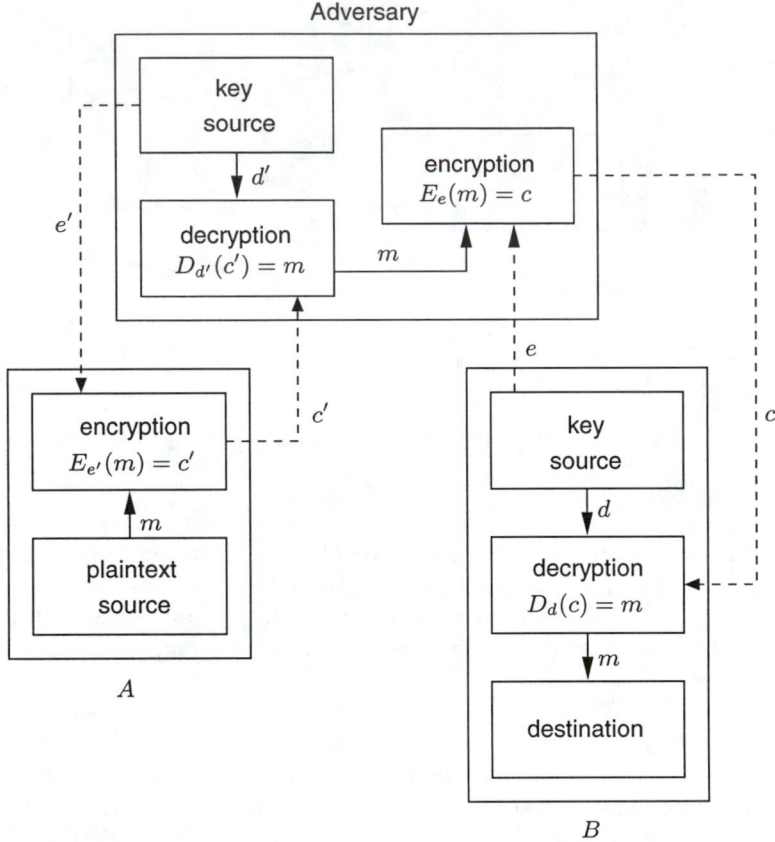

Figure 1.13: *An impersonation attack on a two-party communication.*

1.8.3 Digital signatures from reversible public-key encryption

This section considers a class of digital signature schemes which is based on public-key encryption systems of a particular type.

Suppose E_e is a public-key encryption transformation with message space $\mathcal{M}$ and ciphertext space $\mathcal{C}$. Suppose further that $\mathcal{M} = \mathcal{C}$. If D_d is the decryption transformation corresponding to E_e then since E_e and D_d are both permutations, one has

$$D_d(E_e(m)) = E_e(D_d(m)) = m, \quad \text{for all } m \in \mathcal{M}.$$

A public-key encryption scheme of this type is called *reversible*.[5] Note that it is essential that $\mathcal{M} = \mathcal{C}$ for this to be a valid equality for all $m \in \mathcal{M}$; otherwise, $D_d(m)$ will be meaningless for $m \notin \mathcal{C}$.

---

[5]There is a broader class of digital signatures which can be informally described as arising from *irreversible* cryptographic algorithms. These are described in §11.2.

Construction for a digital signature scheme

1. Let $\mathcal{M}$ be the message space for the signature scheme.
2. Let $\mathcal{C} = \mathcal{M}$ be the signature space $\mathcal{S}$.
3. Let (e, d) be a key pair for the public-key encryption scheme.
4. Define the signing function S_A to be D_d. That is, the signature for a message $m \in \mathcal{M}$ is $s = D_d(m)$.
5. Define the verification function V_A by

$$V_A(m, s) = \begin{cases} true, & \text{if } E_e(s) = m, \\ false, & \text{otherwise.} \end{cases}$$

The signature scheme can be simplified further if A only signs messages having a special structure, and this structure is publicly known. Let $\mathcal{M}'$ be a subset of $\mathcal{M}$ where elements of $\mathcal{M}'$ have a well-defined special structure, such that $\mathcal{M}'$ contains only a negligible fraction of messages from the set. For example, suppose that $\mathcal{M}$ consists of all binary strings of length $2t$ for some positive integer t. Let $\mathcal{M}'$ be the subset of $\mathcal{M}$ consisting of all strings where the first t bits are replicated in the last t positions (e.g., 101101 would be in $\mathcal{M}'$ for $t = 3$). If A only signs messages within the subset $\mathcal{M}'$, these are easily recognized by a verifier.

Redefine the verification function V_A as

$$V_A(s) = \begin{cases} true, & \text{if } E_e(s) \in \mathcal{M}', \\ false, & \text{otherwise.} \end{cases}$$

Under this new scenario A only needs to transmit the signature s since the message $m = E_e(s)$ can be recovered by applying the verification function. Such a scheme is called a *digital signature scheme with message recovery*. Figure 1.14 illustrates how this signature function is used. The feature of selecting messages of special structure is referred to as selecting messages with *redundancy*.

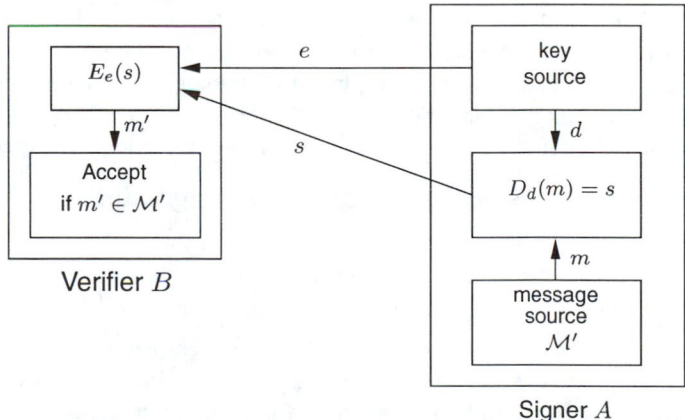

Figure 1.14: *A digital signature scheme with message recovery.*

The modification presented above is more than a simplification; it is absolutely crucial if one hopes to meet the requirement of property (b) of signing and verification functions (see page 23). To see why this is the case, note that any entity B can select a random element $s \in \mathcal{S}$ as a signature and apply E_e to get $u = E_e(s)$, since $\mathcal{S} = \mathcal{M}$ and E_e is public knowledge. B may then take the message $m = u$ and the signature on m to be s and transmits (m, s). It is easy to check that s will verify as a signature created by A for m but in which A has had no part. In this case B has *forged* a signature of A. This is an example of what is called *existential forgery*. (B has produced A's signature on some message likely not of B's choosing.)

If $\mathcal{M}'$ contains only a negligible fraction of messages from $\mathcal{M}$, then the probability of some entity forging a signature of A in this manner is negligibly small.

1.52 Remark (*digital signatures vs. confidentiality*) Although digital signature schemes based on reversible public-key encryption are attractive, they require an encryption method as a primitive. There are situations where a digital signature mechanism is required but encryption is forbidden. In such cases these digital signature schemes are inappropriate.

Digital signatures in practice

For digital signatures to be useful in practice, concrete realizations of the preceding concepts should have certain additional properties. A digital signature must

1. be easy to compute by the signer (the signing function should be easy to apply);
2. be easy to verify by anyone (the verification function should be easy to apply); and
3. have an appropriate lifespan, i.e., be computationally secure from forgery until the signature is no longer necessary for its original purpose.

Resolution of disputes

The purpose of a digital signature (or any signature method) is to permit the resolution of disputes. For example, an entity A could at some point deny having signed a message or some other entity B could falsely claim that a signature on a message was produced by A. In order to overcome such problems a *trusted third party* (TTP) or *judge* is required. The TTP must be some entity which all parties involved agree upon in advance.

If A denies that a message m held by B was signed by A, then B should be able to present the signature s_A for m to the TTP along with m. The TTP rules in favor of B if $V_A(m, s_A) = \textit{true}$ and in favor of A otherwise. B will accept the decision if B is confident that the TTP has the same verifying transformation V_A as A does. A will accept the decision if A is confident that the TTP used V_A and that S_A has not been compromised. Therefore, fair resolution of disputes requires that the following criteria are met.

Requirements for resolution of disputed signatures

1. S_A and V_A have properties (a) and (b) of page 23.
2. The TTP has an authentic copy of V_A.
3. The signing transformation S_A has been kept secret and remains secure.

These properties are necessary but in practice it might not be possible to guarantee them. For example, the assumption that S_A and V_A have the desired characteristics given in property 1 might turn out to be false for a particular signature scheme. Another possibility is that A claims falsely that S_A was compromised. To overcome these problems requires an agreed method to validate the time period for which A will accept responsibility for the verification transformation. An analogue of this situation can be made with credit card revocation. The holder of a card is responsible until the holder notifies the card issuing company that the card has been lost or stolen. §13.8.2 gives a more indepth discussion of these problems and possible solutions.

1.8.4 Symmetric-key vs. public-key cryptography

Symmetric-key and public-key encryption schemes have various advantages and disadvantages, some of which are common to both. This section highlights a number of these and summarizes features pointed out in previous sections.

(i) Advantages of symmetric-key cryptography

1. Symmetric-key ciphers can be designed to have high rates of data throughput. Some hardware implementations achieve encrypt rates of hundreds of megabytes per second, while software implementations may attain throughput rates in the megabytes per second range.
2. Keys for symmetric-key ciphers are relatively short.
3. Symmetric-key ciphers can be employed as primitives to construct various cryptographic mechanisms including pseudorandom number generators (see Chapter 5), hash functions (see Chapter 9), and computationally efficient digital signature schemes (see Chapter 11), to name just a few.
4. Symmetric-key ciphers can be composed to produce stronger ciphers. Simple transformations which are easy to analyze, but on their own weak, can be used to construct strong product ciphers.
5. Symmetric-key encryption is perceived to have an extensive history, although it must be acknowledged that, notwithstanding the invention of rotor machines earlier, much of the knowledge in this area has been acquired subsequent to the invention of the digital computer, and, in particular, the design of the Data Encryption Standard (see Chapter 7) in the early 1970s.

(ii) Disadvantages of symmetric-key cryptography

1. In a two-party communication, the key must remain secret at both ends.
2. In a large network, there are many key pairs to be managed. Consequently, effective key management requires the use of an unconditionally trusted TTP (Definition 1.65).
3. In a two-party communication between entities A and B, sound cryptographic practice dictates that the key be changed frequently, and perhaps for each communication session.
4. Digital signature mechanisms arising from symmetric-key encryption typically require either large keys for the public verification function or the use of a TTP (see Chapter 11).

(iii) Advantages of public-key cryptography

1. Only the private key must be kept secret (authenticity of public keys must, however, be guaranteed).
2. The administration of keys on a network requires the presence of only a functionally trusted TTP (Definition 1.66) as opposed to an unconditionally trusted TTP. Depending on the mode of usage, the TTP might only be required in an "off-line" manner, as opposed to in real time.
3. Depending on the mode of usage, a private key/public key pair may remain unchanged for considerable periods of time, e.g., many sessions (even several years).
4. Many public-key schemes yield relatively efficient digital signature mechanisms. The key used to describe the public verification function is typically much smaller than for the symmetric-key counterpart.

5. In a large network, the number of keys necessary may be considerably smaller than in the symmetric-key scenario.

(iv) Disadvantages of public-key encryption

1. Throughput rates for the most popular public-key encryption methods are several orders of magnitude slower than the best known symmetric-key schemes.
2. Key sizes are typically much larger than those required for symmetric-key encryption (see Remark 1.53), and the size of public-key signatures is larger than that of tags providing data origin authentication from symmetric-key techniques.
3. No public-key scheme has been proven to be secure (the same can be said for block ciphers). The most effective public-key encryption schemes found to date have their security based on the presumed difficulty of a small set of number-theoretic problems.
4. Public-key cryptography does not have as extensive a history as symmetric-key encryption, being discovered only in the mid 1970s.[6]

Summary of comparison

Symmetric-key and public-key encryption have a number of complementary advantages. Current cryptographic systems exploit the strengths of each. An example will serve to illustrate.

Public-key encryption techniques may be used to establish a key for a symmetric-key system being used by communicating entities A and B. In this scenario A and B can take advantage of the long term nature of the public/private keys of the public-key scheme and the performance efficiencies of the symmetric-key scheme. Since data encryption is frequently the most time consuming part of the encryption process, the public-key scheme for key establishment is a small fraction of the total encryption process between A and B.

To date, the computational performance of public-key encryption is inferior to that of symmetric-key encryption. There is, however, no proof that this must be the case. The important points in practice are:

1. public-key cryptography facilitates efficient signatures (particularly non-repudiation) and key mangement; and
2. symmetric-key cryptography is efficient for encryption and some data integrity applications.

1.53 Remark (*key sizes: symmetric key vs. private key*) Private keys in public-key systems must be larger (e.g., 1024 bits for RSA) than secret keys in symmetric-key systems (e.g., 64 or 128 bits) because whereas (for secure algorithms) the most efficient attack on symmetric-key systems is an exhaustive key search, all known public-key systems are subject to "short-cut" attacks (e.g., factoring) more efficient than exhaustive search. Consequently, for equivalent security, symmetric keys have bitlengths considerably smaller than that of private keys in public-key systems, e.g., by a factor of 10 or more.

---

[6] It is, of course, arguable that some public-key schemes which are based on hard mathematical problems have a long history since these problems have been studied for many years. Although this may be true, one must be wary that the mathematics was not studied with this application in mind.

1.9 Hash functions

One of the fundamental primitives in modern cryptography is the cryptographic hash function, often informally called a one-way hash function. A simplified definition for the present discussion follows.

1.54 Definition A *hash function* is a computationally efficient function mapping binary strings of arbitrary length to binary strings of some fixed length, called *hash-values*.

For a hash function which outputs n-bit hash-values (e.g., $n = 128$ or 160) and has desirable properties, the probability that a randomly chosen string gets mapped to a particular n-bit hash-value (image) is 2^{-n}. The basic idea is that a hash-value serves as a compact representative of an input string. To be of cryptographic use, a hash function h is typically chosen such that it is computationally infeasible to find two distinct inputs which hash to a common value (i.e., two *colliding* inputs x and y such that $h(x) = h(y)$), and that given a specific hash-value y, it is computationally infeasible to find an input (pre-image) x such that $h(x) = y$.

The most common cryptographic uses of hash functions are with digital signatures and for data integrity. With digital signatures, a long message is usually hashed (using a publicly available hash function) and only the hash-value is signed. The party receiving the message then hashes the received message, and verifies that the received signature is correct for this hash-value. This saves both time and space compared to signing the message directly, which would typically involve splitting the message into appropriate-sized blocks and signing each block individually. Note here that the inability to find two messages with the same hash-value is a security requirement, since otherwise, the signature on one message hash-value would be the same as that on another, allowing a signer to sign one message and at a later point in time claim to have signed another.

Hash functions may be used for data integrity as follows. The hash-value corresponding to a particular input is computed at some point in time. The integrity of this hash-value is protected in some manner. At a subsequent point in time, to verify that the input data has not been altered, the hash-value is recomputed using the input at hand, and compared for equality with the original hash-value. Specific applications include virus protection and software distribution.

A third application of hash functions is their use in protocols involving a priori commitments, including some digital signature schemes and identification protocols (e.g., see Chapter 10).

Hash functions as discussed above are typically publicly known and involve no secret keys. When used to detect whether the message input has been altered, they are called *modification detection codes* (MDCs). Related to these are hash functions which involve a secret key, and provide data origin authentication (§9.76) as well as data integrity; these are called *message authentication codes* (MACs).

1.10 Protocols and mechanisms

1.55 Definition A *cryptographic protocol* (*protocol*) is a distributed algorithm defined by a sequence of steps precisely specifying the actions required of two or more entities to achieve a specific security objective.

1.56 Remark (*protocol vs. mechanism*) As opposed to a protocol, a *mechanism* is a more general term encompassing protocols, algorithms (specifying the steps followed by a single entity), and non-cryptographic techniques (e.g., hardware protection and procedural controls) to achieve specific security objectives.

Protocols play a major role in cryptography and are essential in meeting cryptographic goals as discussed in §1.2. Encryption schemes, digital signatures, hash functions, and random number generation are among the primitives which may be utilized to build a protocol.

1.57 Example (*a simple key agreement protocol*) Alice and Bob have chosen a symmetric-key encryption scheme to use in communicating over an unsecured channel. To encrypt information they require a key. The communication protocol is the following:

1. Bob constructs a public-key encryption scheme and sends his public key to Alice over the channel.
2. Alice generates a key for the symmetric-key encryption scheme.
3. Alice encrypts the key using Bob's public key and sends the encrypted key to Bob.
4. Bob decrypts using his private key and recovers the symmetric (secret) key.
5. Alice and Bob begin communicating with privacy by using the symmetric-key system and the common secret key.

This protocol uses basic functions to attempt to realize private communications on an unsecured channel. The basic primitives are the symmetric-key and the public-key encryption schemes. The protocol has shortcomings including the impersonation attack of §1.8.2, but it does convey the idea of a protocol.   □

Often the role of public-key encryption in privacy communications is exactly the one suggested by this protocol – public-key encryption is used as a means to exchange keys for subsequent use in symmetric-key encryption, motivated by performance differences between symmetric-key and public-key encryption.

Protocol and mechanism failure

1.58 Definition A *protocol failure* or *mechanism failure* occurs when a mechanism fails to meet the goals for which it was intended, in a manner whereby an adversary gains advantage not by breaking an underlying primitive such as an encryption algorithm directly, but by manipulating the protocol or mechanism itself.

1.59 Example (*mechanism failure*) Alice and Bob are communicating using a stream cipher. Messages which they encrypt are known to have a special form: the first twenty bits carry information which represents a monetary amount. An active adversary can simply XOR an appropriate bitstring into the first twenty bits of ciphertext and change the amount. While the adversary has not been able to read the underlying message, she has been able to alter the transmission. The encryption has not been compromised but the protocol has failed to perform adequately; the inherent assumption that encryption provides data integrity is incorrect.   □

1.60 Example (*forward search attack*) Suppose that in an electronic bank transaction the 32-bit field which records the value of the transaction is to be encrypted using a public-key scheme. This simple protocol is intended to provide privacy of the value field – but does it? An adversary could easily take all 2^{32} possible entries that could be plaintext in this field and encrypt them using the public encryption function. (Remember that by the very nature of public-key encryption this function must be available to the adversary.) By comparing each of the 2^{32} ciphertexts with the one which is actually encrypted in the transaction, the adversary can determine the plaintext. Here the public-key encryption function is not compromised, but rather the way it is used. A closely related attack which applies directly to authentication for access control purposes is the dictionary attack (see §10.2.2). □

1.61 Remark (*causes of protocol failure*) Protocols and mechanisms may fail for a number of reasons, including:

1. weaknesses in a particular cryptographic primitive which may be amplified by the protocol or mechanism;
2. claimed or assumed security guarantees which are overstated or not clearly understood; and
3. the oversight of some principle applicable to a broad class of primitives such as encryption.

Example 1.59 illustrates item 2 if the stream cipher is the one-time pad, and also item 1. Example 1.60 illustrates item 3. See also §1.8.2.

1.62 Remark (*protocol design*) When designing cryptographic protocols and mechanisms, the following two steps are essential:

1. identify *all* assumptions in the protocol or mechanism design; and
2. for each assumption, determine the effect on the security objective if that assumption is violated.

1.11 Key establishment, management, and certification

This section gives a brief introduction to methodology for ensuring the secure distribution of keys for cryptographic purposes.

1.63 Definition *Key establishment* is any process whereby a shared secret key becomes available to two or more parties, for subsequent cryptographic use.

1.64 Definition *Key management* is the set of processes and mechanisms which support key establishment and the maintenance of ongoing keying relationships between parties, including replacing older keys with new keys as necessary.

Key establishment can be broadly subdivided into *key agreement* and *key transport*. Many and various protocols have been proposed to provide key establishment. Chapter 12 describes a number of these in detail. For the purpose of this chapter only a brief overview of issues related to key management will be given. Simple architectures based on symmetric-key and public-key cryptography along with the concept of certification will be addressed.

As noted in §1.5, a major issue when using symmetric-key techniques is the establishment of pairwise secret keys. This becomes more evident when considering a network of entities, any two of which may wish to communicate. Figure 1.15 illustrates a network consisting of 6 entities. The arrowed edges indicate the 15 possible two-party communications which could take place. Since each pair of entities wish to communicate, this small network requires the secure exchange of $\binom{6}{2} = 15$ key pairs. In a network with n entities, the number of secure key exchanges required is $\binom{n}{2} = \frac{n(n-1)}{2}$.

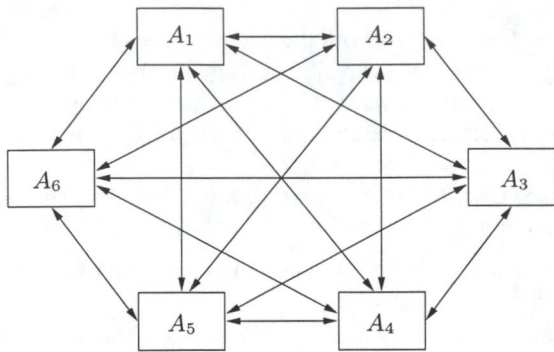

Figure 1.15: *Keying relationships in a simple 6-party network.*

The network diagram depicted in Figure 1.15 is simply the amalgamation of 15 two-party communications as depicted in Figure 1.7. In practice, networks are very large and the key management problem is a crucial issue. There are a number of ways to handle this problem. Two simplistic methods are discussed; one based on symmetric-key and the other on public-key techniques.

1.11.1 Key management through symmetric-key techniques

One solution which employs symmetric-key techniques involves an entity in the network which is trusted by all other entities. As in §1.8.3, this entity is referred to as a *trusted third party* (TTP). Each entity A_i shares a distinct symmetric key k_i with the TTP. These keys are assumed to have been distributed over a secured channel. If two entities subsequently wish to communicate, the TTP generates a key k (sometimes called a *session key*) and sends it encrypted under each of the fixed keys as depicted in Figure 1.16 for entities A_1 and A_5.

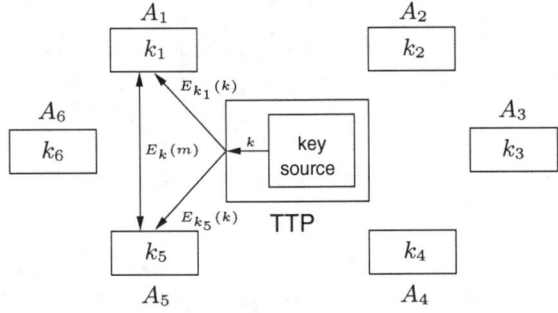

Figure 1.16: *Key management using a trusted third party (TTP).*

Advantages of this approach include:

1. It is easy to add and remove entities from the network.
2. Each entity needs to store only one long-term secret key.

Disadvantages include:

1. All communications require initial interaction with the TTP.
2. The TTP must store n long-term secret keys.

3. The TTP has the ability to read all messages.

4. If the TTP is compromised, all communications are insecure.

1.11.2 Key management through public-key techniques

There are a number of ways to address the key management problem through public-key techniques. Chapter 13 describes many of these in detail. For the purpose of this chapter a very simple model is considered.

Each entity in the network has a public/private encryption key pair. The public key along with the identity of the entity is stored in a central repository called a *public file*. If an entity A_1 wishes to send encrypted messages to entity A_6, A_1 retrieves the public key e_6 of A_6 from the public file, encrypts the message using this key, and sends the ciphertext to A_6. Figure 1.17 depicts such a network.

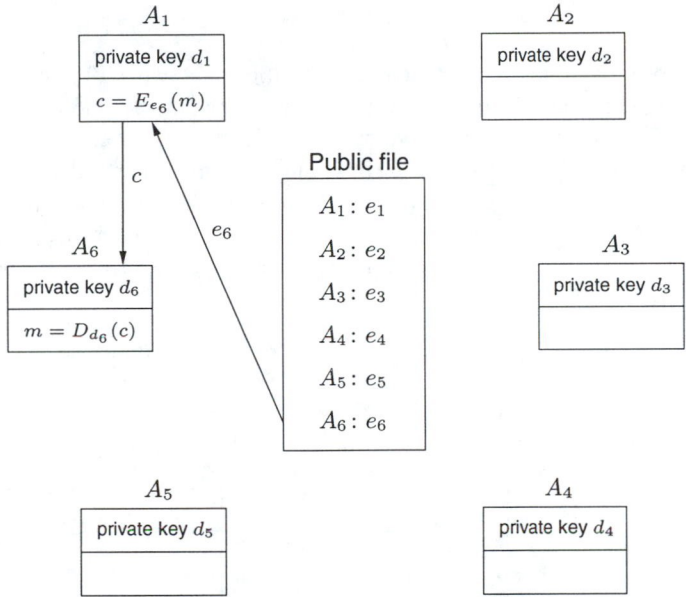

Figure 1.17: *Key management using public-key techniques.*

Advantages of this approach include:

1. No trusted third party is required.

2. The public file could reside with each entity.

3. Only n public keys need to be stored to allow secure communications between any pair of entities, assuming the only attack is that by a passive adversary.

The key management problem becomes more difficult when one must take into account an adversary who is *active* (i.e. an adversary who can alter the public file containing public keys). Figure 1.18 illustrates how an active adversary could compromise the key management scheme given above. (This is directly analogous to the attack in §1.8.2.) In the figure, the adversary alters the public file by replacing the public key e_6 of entity A_6 by the adversary's public key e^*. Any message encrypted for A_6 using the public key from the public file can be decrypted by only the adversary. Having decrypted and read the message, the adversary can now encrypt it using the public key of A_6 and forward the ciphertext to A_6. A_1 however believes that only A_6 can decrypt the ciphertext c.

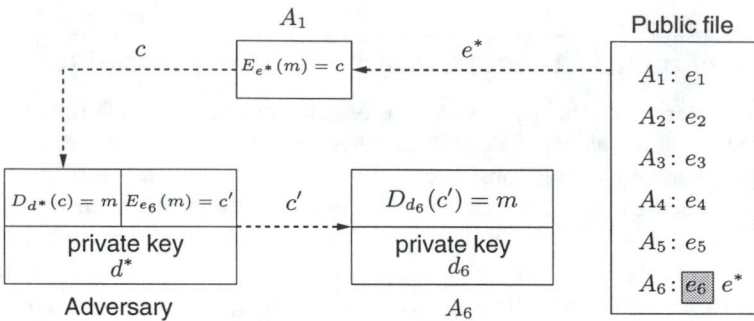

Figure 1.18: *An impersonation of A_6 by an active adversary with public key e^*.*

To prevent this type of attack, the entities may use a TTP to *certify* the public key of each entity. The TTP has a private signing algorithm S_T and a verification algorithm V_T (see §1.6) assumed to be known by all entities. The TTP carefully verifies the identity of each entity, and signs a message consisting of an identifier and the entity's authentic public key. This is a simple example of a *certificate*, binding the identity of an entity to its public key (see §1.11.3). Figure 1.19 illustrates the network under these conditions. A_1 uses the public key of A_6 only if the certificate signature verifies successfully.

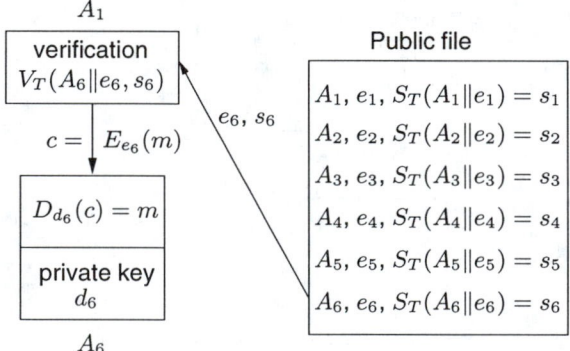

Figure 1.19: *Authentication of public keys by a TTP. $\|$ denotes concatenation.*

Advantages of using a TTP to maintain the integrity of the public file include:

1. It prevents an active adversary from impersonation on the network.
2. The TTP cannot monitor communications. Entities need trust the TTP only to bind identities to public keys properly.
3. Per-communication interaction with the public file can be eliminated if entities store certificates locally.

Even with a TTP, some concerns still remain:

1. If the signing key of the TTP is compromised, all communications become insecure.
2. All trust is placed with one entity.

1.11.3 Trusted third parties and public-key certificates

A trusted third party has been used in §1.8.3 and again here in §1.11. The trust placed on this entity varies with the way it is used, and hence motivates the following classification.

1.65 Definition A TTP is said to be *unconditionally trusted* if it is trusted on all matters. For example, it may have access to the secret and private keys of users, as well as be charged with the association of public keys to identifiers.

1.66 Definition A TTP is said to be *functionally trusted* if the entity is assumed to be honest and fair but it does not have access to the secret or private keys of users.

§1.11.1 provides a scenario which employs an unconditionally trusted TTP. §1.11.2 uses a functionally trusted TTP to maintain the integrity of the public file. A functionally trusted TTP could be used to register or certify users and contents of documents or, as in §1.8.3, as a judge.

Public-key certificates

The distribution of public keys is generally easier than that of symmetric keys, since secrecy is not required. However, the integrity (authenticity) of public keys is critical (recall §1.8.2).

A *public-key certificate* consists of a *data part* and a *signature part*. The data part consists of the name of an entity, the public key corresponding to that entity, possibly additional relevant information (e.g., the entity's street or network address, a validity period for the public key, and various other attributes). The signature part consists of the signature of a TTP over the data part.

In order for an entity B to verify the authenticity of the public key of an entity A, B must have an authentic copy of the public signature verification function of the TTP. For simplicity, assume that the authenticity of this verification function is provided to B by non-cryptographic means, for example by B obtaining it from the TTP in person. B can then carry out the following steps:

1. Acquire the public-key certificate of A over some unsecured channel, either from a central database of certificates, from A directly, or otherwise.
2. Use the TTP's verification function to verify the TTP's signature on A's certificate.
3. If this signature verifies correctly, accept the public key in the certificate as A's authentic public key; otherwise, assume the public key is invalid.

Before creating a public-key certificate for A, the TTP must take appropriate measures to verify the identity of A and the fact that the public key to be certificated actually belongs to A. One method is to require that A appear before the TTP with a conventional passport as proof of identity, and obtain A's public key from A in person along with evidence that A knows the corresponding private key. Once the TTP creates a certificate for a party, the trust that all other entities have in the authenticity of the TTP's public key can be used transitively to gain trust in the authenticity of that party's public key, through acquisition and verification of the certificate.

1.12 Pseudorandom numbers and sequences

Random number generation is an important primitive in many cryptographic mechanisms. For example, keys for encryption transformations need to be generated in a manner which is unpredictable to an adversary. Generating a random key typically involves the selection of random numbers or bit sequences. Random number generation presents challenging issues. A brief introduction is given here with details left to Chapter 5.

Often in cryptographic applications, one of the following steps must be performed:

(i) From a finite set of n elements (e.g., $\{1, 2, \ldots, n\}$), select an element at random.

(ii) From the set of all sequences (strings) of length m over some finite alphabet $\mathcal{A}$ of n symbols, select a sequence at random.

(iii) Generate a random sequence (string) of symbols of length m over a set of n symbols.

It is not clear what exactly it means to *select at random* or *generate at random*. Calling a number random without a context makes little sense. Is the number 23 a random number? No, but if 49 identical balls labeled with a number from 1 to 49 are in a container, and this container mixes the balls uniformly, drops one ball out, and this ball happens to be labeled with the number 23, then one would say that 23 was generated randomly from a uniform distribution. The *probability* that 23 drops out is 1 in 49 or $\frac{1}{49}$.

If the number on the ball which was dropped from the container is recorded and the ball is placed back in the container and the process repeated 6 times, then a random sequence of length 6 defined on the alphabet $\mathcal{A} = \{1, 2, \ldots, 49\}$ will have been generated. What is the chance that the sequence $17, 45, 1, 7, 23, 35$ occurs? Since each element in the sequence has probability $\frac{1}{49}$ of occuring, the probability of the sequence $17, 45, 1, 7, 23, 35$ occurring is

$$\frac{1}{49} \times \frac{1}{49} \times \frac{1}{49} \times \frac{1}{49} \times \frac{1}{49} \times \frac{1}{49} = \frac{1}{13841287201}.$$

There are precisely 13841287201 sequences of length 6 over the alphabet $\mathcal{A}$. If each of these sequences is written on one of 13841287201 balls and they are placed in the container (first removing the original 49 balls) then the chance that the sequence given above drops out is the same as if it were generated one ball at a time. Hence, (ii) and (iii) above are essentially the same statements.

Finding good methods to generate random sequences is difficult.

1.67 Example (*random sequence generator*) To generate a random sequence of 0's and 1's, a coin could be tossed with a head landing up recorded as a 1 and a tail as a 0. It is assumed that the coin is *unbiased*, which means that the probability of a 1 on a given toss is exactly $\frac{1}{2}$. This will depend on how well the coin is made and how the toss is performed. This method would be of little value in a system where random sequences must be generated quickly and often. It has no practical value other than to serve as an example of the idea of random number generation.   □

1.68 Example (*random sequence generator*) A *noise diode* may be used to produce random binary sequences. This is reasonable if one has some way to be convinced that the probability that a 1 will be produced on any given trial is $\frac{1}{2}$. Should this assumption be false, the sequence generated would not have been selected from a uniform distribution and so not all sequences of a given length would be equally likely. The only way to get some feeling for the reliability of this type of random source is to carry out statistical tests on its output. These are considered in Chapter 5. If the diode is a source of a uniform distribution on the set of all binary sequences of a given length, it provides an effective way to generate random sequences.   □

Since most *true sources* of random sequences (if there is such a thing) come from *physical means*, they tend to be either costly or slow in their generation. To overcome these problems, methods have been devised to construct *pseudorandom sequences* in a deterministic manner from a shorter random sequence called a *seed*. The pseudorandom sequences appear to be generated by a truly random source to anyone not knowing the method of generation. Often the generation algorithm is known to all, but the seed is unknown except by the entity generating the sequence. A plethora of algorithms has been developed to generate pseudorandom bit sequences of various types. Many of these are completely unsuitable for cryptographic purposes and one must be cautious of claims by creators of such algorithms as to the random nature of the output.

1.13 Classes of attacks and security models

Over the years, many different types of attacks on cryptographic primitives and protocols have been identified. The discussion here limits consideration to attacks on encryption and protocols. Attacks on other cryptographic primitives will be given in appropriate chapters.

In §1.11 the roles of an active and a passive adversary were discussed. The attacks these adversaries can mount may be classified as follows:.

1. A *passive attack* is one where the adversary only monitors the communication channel. A passive attacker only threatens confidentiality of data.
2. An *active attack* is one where the adversary attempts to delete, add, or in some other way alter the transmission on the channel. An active attacker threatens data integrity and authentication as well as confidentiality.

A passive attack can be further subdivided into more specialized attacks for deducing plaintext from ciphertext, as outlined in §1.13.1.

1.13.1 Attacks on encryption schemes

The objective of the following attacks is to systematically recover plaintext from ciphertext, or even more drastically, to deduce the decryption key.

1. A *ciphertext-only attack* is one where the adversary (or cryptanalyst) tries to deduce the decryption key or plaintext by only observing ciphertext. Any encryption scheme vulnerable to this type of attack is considered to be completely insecure.
2. A *known-plaintext attack* is one where the adversary has a quantity of plaintext and corresponding ciphertext. This type of attack is typically only marginally more difficult to mount.
3. A *chosen-plaintext attack* is one where the adversary chooses plaintext and is then given corresponding ciphertext. Subsequently, the adversary uses any information deduced in order to recover plaintext corresponding to previously unseen ciphertext.
4. An *adaptive chosen-plaintext attack* is a chosen-plaintext attack wherein the choice of plaintext may depend on the ciphertext received from previous requests.
5. A *chosen-ciphertext attack* is one where the adversary selects the ciphertext and is then given the corresponding plaintext. One way to mount such an attack is for the adversary to gain access to the equipment used for decryption (but not the decryption key, which may be securely embedded in the equipment). The objective is then to be able, without access to such equipment, to deduce the plaintext from (different) ciphertext.

6. An *adaptive chosen-ciphertext attack* is a chosen-ciphertext attack where the choice of ciphertext may depend on the plaintext received from previous requests.

Most of these attacks also apply to digital signature schemes and message authentication codes. In this case, the objective of the attacker is to forge messages or MACs, as discussed in Chapters 11 and 9, respectively.

1.13.2 Attacks on protocols

The following is a partial list of attacks which might be mounted on various protocols. Until a protocol is proven to provide the service intended, the list of possible attacks can never be said to be complete.

1. *known-key attack.* In this attack an adversary obtains some keys used previously and then uses this information to determine new keys.
2. *replay.* In this attack an adversary records a communication session and replays the entire session, or a portion thereof, at some later point in time.
3. *impersonation.* Here an adversary assumes the identity of one of the legitimate parties in a network.
4. *dictionary.* This is usually an attack against passwords. Typically, a password is stored in a computer file as the image of an unkeyed hash function. When a user logs on and enters a password, it is hashed and the image is compared to the stored value. An adversary can take a list of probable passwords, hash all entries in this list, and then compare this to the list of true encrypted passwords with the hope of finding matches.
5. *forward search.* This attack is similar in spirit to the dictionary attack and is used to decrypt messages. An example of this method was cited in Example 1.60.
6. *interleaving attack.* This type of attack usually involves some form of impersonation in an authentication protocol (see §12.9.1).

1.13.3 Models for evaluating security

The security of cryptographic primitives and protocols can be evaluated under several different models. The most practical security metrics are computational, provable, and ad hoc methodology, although the latter is often dangerous. The confidence level in the amount of security provided by a primitive or protocol based on computational or ad hoc security increases with time and investigation of the scheme. However, time is not enough if few people have given the method careful analysis.

(i) Unconditional security

The most stringent measure is an information-theoretic measure – whether or not a system has *unconditional security*. An adversary is assumed to have unlimited computational resources, and the question is whether or not there is enough information available to defeat the system. Unconditional security for encryption systems is called *perfect secrecy*. For perfect secrecy, the uncertainty in the plaintext, after observing the ciphertext, must be equal to the a priori uncertainty about the plaintext – observation of the ciphertext provides no information whatsoever to an adversary.

A necessary condition for a symmetric-key encryption scheme to be unconditionally secure is that the key be at least as long as the message. The one-time pad (§1.5.4) is an example of an unconditionally secure encryption algorithm. In general, encryption schemes do not offer perfect secrecy, and each ciphertext character observed decreases the theoretical uncertainty in the plaintext and the encryption key. Public-key encryption schemes cannot be unconditionally secure since, given a ciphertext c, the plaintext can in principle be recovered by encrypting all possible plaintexts until c is obtained.

(ii) Complexity-theoretic security

An appropriate model of computation is defined and adversaries are modeled as having polynomial computational power. (They mount attacks involving time and space polynomial in the size of appropriate security parameters.) A proof of security relative to the model is then constructed. An objective is to design a cryptographic method based on the weakest assumptions possible anticipating a powerful adversary. Asymptotic analysis and usually also worst-case analysis is used and so care must be exercised to determine when proofs have practical significance. In contrast, polynomial attacks which are feasible under the model might, in practice, still be computationally infeasible.

Security analysis of this type, although not of practical value in all cases, may nonetheless pave the way to a better overall understanding of security. Complexity-theoretic analysis is invaluable for formulating fundamental principles and confirming intuition. This is like many other sciences, whose practical techniques are discovered early in the development, well before a theoretical basis and understanding is attained.

(iii) Provable security

A cryptographic method is said to be *provably secure* if the difficulty of defeating it can be shown to be essentially as difficult as solving a well-known and *supposedly* difficult (typically number-theoretic) problem, such as integer factorization or the computation of discrete logarithms. Thus, "provable" here means provable subject to assumptions.

This approach is considered by some to be as good a practical analysis technique as exists. Provable security may be considered part of a special sub-class of the larger class of computational security considered next.

(iv) Computational security

This measures the amount of computational effort required, by the best currently-known methods, to defeat a system; it must be assumed here that the system has been well-studied to determine which attacks are relevant. A proposed technique is said to be *computationally secure* if the perceived level of computation required to defeat it (using the best attack known) exceeds, by a comfortable margin, the computational resources of the hypothesized adversary.

Often methods in this class are related to hard problems but, unlike for provable security, no proof of equivalence is known. Most of the best known public-key and symmetric-key schemes in current use are in this class. This class is sometimes also called *practical security*.

(v) Ad hoc security

This approach consists of any variety of convincing arguments that every successful attack requires a resource level (e.g., time and space) greater than the fixed resources of a perceived adversary. Cryptographic primitives and protocols which survive such analysis are said to have *heuristic security*, with security here typically in the computational sense.

Primitives and protocols are usually designed to counter standard attacks such as those given in §1.13. While perhaps the most commonly used approach (especially for protocols), it is, in some ways, the least satisfying. Claims of security generally remain questionable and unforeseen attacks remain a threat.

1.13.4 Perspective for computational security

To evaluate the security of cryptographic schemes, certain quantities are often considered.

1.69 Definition The *work factor* W_d is the minimum amount of work (measured in appropriate units such as elementary operations or clock cycles) required to compute the private key d given the public key e, or, in the case of symmetric-key schemes, to determine the secret key k. More specifically, one may consider the work required under a ciphertext-only attack given n ciphertexts, denoted $W_d(n)$.

If W_d is t years, then for sufficiently large t the cryptographic scheme is, for all practical purposes, a secure system. To date no public-key system has been found where one can prove a sufficiently large lower bound on the work factor W_d. The best that is possible to date is to rely on the following as a basis for security.

1.70 Definition The *historical work factor* $\overline{W_d}$ is the minimum amount of work required to compute the private key d from the public key e using the best known algorithms at a given point in time.

The historical work factor $\overline{W_d}$ varies with time as algorithms and technology improve. It corresponds to computational security, whereas W_d corresponds to the true security level, although this typically cannot be determined.

How large is large?

§1.4 described how the designer of an encryption system tries to create a scheme for which the best approach to breaking it is through exhaustive search of the key space. The key space must then be large enough to make an exhaustive search completely infeasible. An important question then is "How large is large?". In order to gain some perspective on the magnitude of numbers, Table 1.2 lists various items along with an associated magnitude.

Reference	Magnitude
Seconds in a year	$\approx 3 \times 10^7$
Age of our solar system (years)	$\approx 6 \times 10^9$
Seconds since creation of solar system	$\approx 2 \times 10^{17}$
Clock cycles per year, 50 MHz computer	$\approx 1.6 \times 10^{15}$
Binary strings of length 64	$2^{64} \approx 1.8 \times 10^{19}$
Binary strings of length 128	$2^{128} \approx 3.4 \times 10^{38}$
Binary strings of length 256	$2^{256} \approx 1.2 \times 10^{77}$
Number of 75-digit prime numbers	$\approx 5.2 \times 10^{72}$
Electrons in the universe	$\approx 8.37 \times 10^{77}$

Table 1.2: *Reference numbers comparing relative magnitudes.*

Some powers of 10 are referred to by prefixes. For example, high-speed modern computers are now being rated in terms of *teraflops* where a teraflop is 10^{12} floating point operations per second. Table 1.3 provides a list of commonly used prefixes.

Prefix	Symbol	Magnitude
exa	E	10^{18}
peta	P	10^{15}
tera	T	10^{12}
giga	G	10^{9}
mega	M	10^{6}
kilo	k	10^{3}
hecto	h	10^{2}
deca	da	10

Prefix	Symbol	Magnitude
deci	d	10^{-1}
centi	c	10^{-2}
milli	m	10^{-3}
micro	μ	10^{-6}
nano	n	10^{-9}
pico	p	10^{-12}
femto	f	10^{-15}
atto	a	10^{-18}

Table 1.3: *Prefixes used for various powers of 10.*

1.14 Notes and further references

§1.1

Kahn [648] gives a thorough, comprehensive, and non-technical history of cryptography, published in 1967. Feistel [387] provides an early exposition of block cipher ideas. The original specification of DES is the 1977 U.S. Federal Information Processing Standards Publication 46 [396]. Public-key cryptography was introduced by Diffie and Hellman [345]. The first concrete realization of a public-key encryption scheme was the knapsack scheme by Merkle and Hellman [857]. The RSA public-key encryption and signature scheme is due to Rivest, Shamir, and Adleman [1060], while the ElGamal public-key encryption and signature schemes are due to ElGamal [368]. The two digital signature standards, ISO/IEC 9796 [596] and the Digital Signature Standard [406], are discussed extensively in Chapter 11.

Cryptography has used specialized areas of mathematics such as number theory to realize very practical mechanisms such as public-key encryption and digital signatures. Such usage was not conceived as possible a mere twenty years ago. The famous mathematician, Hardy [539], went as far as to apologize for its lack of utility:

> " ... both Gauss and lesser mathematicians may be justified in rejoicing that there is one science at any rate, and that their own, whose very remoteness from ordinary human activities should keep it gentle and clean."

§1.2

This section was inspired by the foreword to the book *Contemporary Cryptology, The Science of Information Integrity*, edited by Simmons [1143]. The handwritten signature came into the British legal system in the seventeenth century as a means to provide various functions associated with information security. See Chapter 9 of Meyer and Matyas [859] for details.

This book only considers cryptography as it applies to information in digital form. Chapter 9 of Beker and Piper [84] provides an introduction to the encryption of analogue signals, in particular, speech. Although in many cases physical means are employed to facilitate privacy, cryptography plays the major role. Physical means of providing privacy include fiber optic communication links, spread spectrum technology, TEMPEST techniques, and tamper-resistant hardware. *Steganography* is that branch of information privacy which attempts to obscure the existence of data through such devices as invisible inks, secret compartments, the use of subliminal channels, and the like. Kahn [648] provides an historical account of various steganographic techniques.

Excellent introductions to cryptography can be found in the articles by Diffie and Hellman [347], Massey [786], and Rivest [1054]. A concise and elegant way to describe cryptography was given by Rivest [1054]: *Cryptography is about communications in the presence of adversaries.* The taxonomy of cryptographic primitives (Figure 1.1) was derived from the classification given by Bosselaers, Govaerts, and Vandewalle [175].

§1.3

The theory of functions is fundamental in modern mathematics. The term *range* is often used in place of image of a function. The latter, being more descriptive, is preferred. An alternate term for one-to-one is *injective*; an alternate term for onto is *surjective*.

One-way functions were introduced by Diffie and Hellman [345]. A more extensive history is given on page 377. Trapdoor one-way functions were first postulated by Diffie and Hellman [345] and independently by Merkle [850] as a means to obtain public-key encryption schemes; several candidates are given in Chapter 8.

§1.4

The basic concepts of cryptography are treated quite differently by various authors, some being more technical than others. Brassard [192] provides a concise, lucid, and technically accurate account. Schneier [1094] gives a less technical but very accessible introduction. Salomaa [1089], Stinson [1178], and Rivest [1054] present more mathematical approaches. Davies and Price [308] provide a very readable presentation suitable for the practitioner.

The comparison of an encryption scheme to a resettable combination lock is from Diffie and Hellman [347]. Kerckhoffs' desiderata [668] were originally stated in French. The translation stated here is given in Kahn [648]. Shannon [1121] also gives desiderata for encryption schemes.

§1.5

Symmetric-key encryption has a very long history, as recorded by Kahn [648]. Most systems invented prior to the 1970s are now of historical interest only. Chapter 2 of Denning [326] is also a good source for many of the more well known schemes such as the Caesar cipher, Vigenère and Beaufort ciphers, rotor machines (Enigma and Hagelin), running key ciphers, and so on; see also Davies and Price [308] and Konheim [705]. Beker and Piper [84] give an indepth treatment, including cryptanalysis of several of the classical systems used in World War II. Shannon's paper [1121] is considered the seminal work on secure communications. It is also an excellent source for descriptions of various well-known historical symmetric-key ciphers.

Simple substitution and transposition ciphers are the focus of §1.5. Hill ciphers [557], a class of substitution ciphers which substitute blocks using matrix methods, are covered in Example 7.52. The idea of confusion and diffusion (Remark 1.36) was introduced by Shannon [1121].

Kahn [648] gives 1917 as the date when Vernam discovered the cipher which bears Vernam's name, however, Vernam did not publish the result until 1926 [1222]; see page 274 for further discussion. Massey [786] states that reliable sources have suggested that the Moscow-Washington hot-line (channel for very high level communications) is no longer secured with a one-time pad, which has been replaced by a symmetric-key cipher requiring a much shorter key. This change would indicate that confidence and understanding in the ability to construct very strong symmetric-key encryption schemes exists. The one-time pad seems to have been used extensively by Russian agents operating in foreign countries. The highest ranking Russian agent ever captured in the United States was Rudolph Abel. When apprehended in 1957 he had in his possession a booklet the size of a postage stamp ($1\frac{7}{8} \times \frac{7}{8} \times \frac{7}{8}$ inches) containing a one-time key; see Kahn [648, p.664].

§1.6

The concept of a digital signature was introduced by Diffie and Hellman [345] and independently by Merkle [850]. The first practical realization of a digital signature scheme appeared in the paper by Rivest, Shamir, and Adleman [1060]. Rabin [1022] (see also [1023]) also claims to have independently discovered RSA but did not publish the result.

Most introductory sources for digital signatures stress digital signatures with message recovery coming from a public-key encryption system. Mitchell, Piper, and Wild [882] give a good general treatment of the subject. Stinson [1178] provides a similar elementary but general introduction. Chapter 11 generalizes the definition of a digital signature by allowing randomization. The scheme described in §1.8 is referred to as *deterministic*. Many other types of digital signatures with specific properties have been created, such as blind signatures, undeniable signatures, and failstop signatures (see Chapter 11).

§1.7

Much effort has been devoted to developing a theory of authentication. At the forefront of this is Simmons [1144], whose contributions are nicely summarized by Massey [786]. For a more concrete example of the necessity for authentication without secrecy, see the article by Simmons [1146].

§1.8

1976 marked a major turning point in the history of cryptography. In several papers that year, Diffie and Hellman introduced the idea of public-key cryptography and gave concrete examples of how such a scheme might be realized. The first paper on public-key cryptography was "Multiuser cryptographic techniques" by Diffie and Hellman [344], presented at the National Computer Conference in June of 1976. Although the authors were not satisfied with the examples they cited, the concept was made clear. In their landmark paper, Diffie and Hellman [345] provided a more comprehensive account of public-key cryptography and described the first viable method to realize this elegant concept. Another good source for the early history and development of the subject is Diffie [343]. Nechvatal [922] also provides a broad survey of public-key cryptography.

Merkle [849, 850] independently discovered public-key cryptography, illustrating how this concept could be realized by giving an elegant and ingenious example now commonly referred to as the *Merkle puzzle scheme*. Simmons [1144, p.412] notes the first reported application of public-key cryptography was fielded by Sandia National Laboratories (U.S.) in 1978.

§1.9

Much of the early work on cryptographic hash functions was done by Merkle [850]. The most comprehensive current treatment of the subject is by Preneel [1004].

§1.10

A large number of successful cryptanalytic attacks on systems claiming security are due to protocol failure. An overview of this area is given by Moore [899], including classifications of protocol failures and design principles.

§1.11

One approach to distributing public-keys is the so-called *Merkle channel* (see Simmons [1144, p.387]). Merkle proposed that public keys be distributed over so many independent public channels (newspaper, radio, television, etc.) that it would be improbable for an adversary to compromise all of them.

In 1979 Kohnfelder [702] suggested the idea of using *public-key certificates* to facilitate the distribution of public keys over unsecured channels, such that their authenticity can be verified. Essentially the same idea, but by on-line requests, was proposed by Needham and Schroeder (ses Wilkes [1244]).

A provably secure key agreement protocol has been proposed whose security is based on the Heisenberg uncertainty principle of quantum physics. The security of so-called *quantum cryptography* does not rely upon any complexity-theoretic assumptions. For further details on quantum cryptography, consult Chapter 6 of Brassard [192], and Bennett, Brassard, and Ekert [115].

§1.12

For an introduction and detailed treatment of many pseudorandom sequence generators, see Knuth [692]. Knuth cites an example of a complex scheme to generate random numbers which on closer analysis is shown to produce numbers which are far from random, and concludes: *...random numbers should not be generated with a method chosen at random.*

§1.13

The seminal work of Shannon [1121] on secure communications, published in 1949, remains as one of the best introductions to both practice and theory, clearly presenting many of the fundamental ideas including redundancy, entropy, and unicity distance. Various models under which security may be examined are considered by Rueppel [1081], Simmons [1144], and Preneel [1003], among others; see also Goldwasser [476].

Chapter 4

Mathematical Background

Contents in Brief

This chapter is a collection of basic material on probability theory, information theory, complexity theory, number theory, abstract algebra, and finite fields that will be used throughout this book. Further background and proofs of the facts presented here can be found in the references given in §2.7. The following standard notation will be used throughout:

1. $\mathbb{Z}$ denotes the set of *integers*; that is, the set $\{\ldots, -2, -1, 0, 1, 2, \ldots\}$.
2. $\mathbb{Q}$ denotes the set of *rational numbers*; that is, the set $\{\frac{a}{b} \mid a, b \in \mathbb{Z}, b \neq 0\}$.
3. $\mathbb{R}$ denotes the set of *real numbers*.
4. π is the mathematical constant; $\pi \approx 3.14159$.
5. e is the base of the natural logarithm; $e \approx 2.71828$.
6. $\lfloor x \rfloor$ is the largest integer less than or equal to x. For example, $\lfloor 5.2 \rfloor = 5$ and $\lfloor -5.2 \rfloor = -6$.
7. $\lceil x \rceil$ is the smallest integer greater than or equal to x. For example, $\lceil 5.2 \rceil = 6$ and $\lceil -5.2 \rceil = -5$.
8. If A is a finite set, then $|A|$ denotes the number of elements in A, called the *cardinality* of A.
9. $a \in A$ means that element a is a member of the set A.
10. $A \subseteq B$ means that A is a subset of B.
11. $A \subset B$ means that A is a proper subset of B; that is $A \subseteq B$ and $A \neq B$.
12. The *intersection* of sets A and B is the set $A \cap B = \{x \mid x \in A \text{ and } x \in B\}$.
13. The *union* of sets A and B is the set $A \cup B = \{x \mid x \in A \text{ or } x \in B\}$.
14. The *difference* of sets A and B is the set $A - B = \{x \mid x \in A \text{ and } x \notin B\}$.
15. The *Cartesian product* of sets A and B is the set $A \times B = \{(a, b) \mid a \in A \text{ and } b \in B\}$. For example, $\{a_1, a_2\} \times \{b_1, b_2, b_3\} = \{(a_1, b_1), (a_1, b_2), (a_1, b_3), (a_2, b_1), (a_2, b_2), (a_2, b_3)\}$.

16. A *function* or *mapping* $f : A \longrightarrow B$ is a rule which assigns to each element a in A precisely one element b in B. If $a \in A$ is mapped to $b \in B$ then b is called the *image* of a, a is called a *preimage* of b, and this is written $f(a) = b$. The set A is called the *domain* of f, and the set B is called the *codomain* of f.

17. A function $f : A \longrightarrow B$ is $1 - 1$ (*one-to-one*) or *injective* if each element in B is the image of at most one element in A. Hence $f(a_1) = f(a_2)$ implies $a_1 = a_2$.

18. A function $f : A \longrightarrow B$ is *onto* or *surjective* if each $b \in B$ is the image of at least one $a \in A$.

19. A function $f : A \longrightarrow B$ is a *bijection* if it is both one-to-one and onto. If f is a bijection between finite sets A and B, then $|A| = |B|$. If f is a bijection between a set A and itself, then f is called a *permutation* on A.

20. $\ln x$ is the natural logarithm of x; that is, the logarithm of x to the base e.

21. $\lg x$ is the logarithm of x to the base 2.

22. $\exp(x)$ is the exponential function e^x.

23. $\sum_{i=1}^{n} a_i$ denotes the sum $a_1 + a_2 + \cdots + a_n$.

24. $\prod_{i=1}^{n} a_i$ denotes the product $a_1 \cdot a_2 \cdots \cdot a_n$.

25. For a positive integer n, the factorial function is $n! = n(n-1)(n-2)\cdots 1$. By convention, $0! = 1$.

2.1 Probability theory

2.1.1 Basic definitions

2.1 Definition An *experiment* is a procedure that yields one of a given set of outcomes. The individual possible outcomes are called *simple events*. The set of all possible outcomes is called the *sample space*.

This chapter only considers *discrete* sample spaces; that is, sample spaces with only finitely many possible outcomes. Let the simple events of a sample space S be labeled $s_1, s_2, \ldots, s_n$.

2.2 Definition A *probability distribution* P on S is a sequence of numbers $p_1, p_2, \ldots, p_n$ that are all non-negative and sum to 1. The number p_i is interpreted as the *probability* of s_i being the outcome of the experiment.

2.3 Definition An *event* E is a subset of the sample space S. The *probability* that event E occurs, denoted $P(E)$, is the sum of the probabilities p_i of all simple events s_i which belong to E. If $s_i \in S$, $P(\{s_i\})$ is simply denoted by $P(s_i)$.

2.4 Definition If E is an event, the *complementary event* is the set of simple events not belonging to E, denoted $\overline{E}$.

2.5 Fact Let $E \subseteq S$ be an event.

   (i) $0 \le P(E) \le 1$. Furthermore, $P(S) = 1$ and $P(\emptyset) = 0$. ($\emptyset$ is the empty set.)

   (ii) $P(\overline{E}) = 1 - P(E)$.

(iii) If the outcomes in S are equally likely, then $P(E) = \frac{|E|}{|S|}$.

2.6 Definition Two events E_1 and E_2 are called *mutually exclusive* if $P(E_1 \cap E_2) = 0$. That is, the occurrence of one of the two events excludes the possibility that the other occurs.

2.7 Fact Let E_1 and E_2 be two events.
   (i) If $E_1 \subseteq E_2$, then $P(E_1) \le P(E_2)$.
   (ii) $P(E_1 \cup E_2) + P(E_1 \cap E_2) = P(E_1) + P(E_2)$. Hence, if E_1 and E_2 are mutually exclusive, then $P(E_1 \cup E_2) = P(E_1) + P(E_2)$.

2.1.2 Conditional probability

2.8 Definition Let E_1 and E_2 be two events with $P(E_2) > 0$. The *conditional probability of E_1 given E_2*, denoted $P(E_1|E_2)$, is

$$P(E_1|E_2) = \frac{P(E_1 \cap E_2)}{P(E_2)}.$$

$P(E_1|E_2)$ measures the probability of event E_1 occurring, given that E_2 has occurred.

2.9 Definition Events E_1 and E_2 are said to be *independent* if $P(E_1 \cap E_2) = P(E_1)P(E_2)$.

    Observe that if E_1 and E_2 are independent, then $P(E_1|E_2) = P(E_1)$ and $P(E_2|E_1) = P(E_2)$. That is, the occurrence of one event does not influence the likelihood of occurrence of the other.

2.10 Fact (*Bayes' theorem*) If E_1 and E_2 are events with $P(E_2) > 0$, then

$$P(E_1|E_2) = \frac{P(E_1)P(E_2|E_1)}{P(E_2)}.$$

2.1.3 Random variables

Let S be a sample space with probability distribution P.

2.11 Definition A *random variable* X is a function from the sample space S to the set of real numbers; to each simple event $s_i \in S$, X assigns a real number $X(s_i)$.

    Since S is assumed to be finite, X can only take on a finite number of values.

2.12 Definition Let X be a random variable on S. The *expected value* or *mean* of X is $E(X) = \sum_{s_i \in S} X(s_i)P(s_i)$.

2.13 Fact Let X be a random variable on S. Then $E(X) = \sum_{x \in \mathbb{R}} x \cdot P(X = x)$.

2.14 Fact If $X_1, X_2, \ldots, X_m$ are random variables on S, and $a_1, a_2, \ldots, a_m$ are real numbers, then $E(\sum_{i=1}^m a_i X_i) = \sum_{i=1}^m a_i E(X_i)$.

2.15 Definition The *variance* of a random variable X of mean μ is a non-negative number defined by

$$\mathrm{Var}(X) = E((X - \mu)^2).$$

The *standard deviation* of X is the non-negative square root of $\mathrm{Var}(X)$.

If a random variable has small variance then large deviations from the mean are unlikely to be observed. This statement is made more precise below.

2.16 Fact (*Chebyshev's inequality*) Let X be a random variable with mean $\mu = E(X)$ and variance $\sigma^2 = \text{Var}(X)$. Then for any $t > 0$,

$$P(|X - \mu| \geq t) \leq \frac{\sigma^2}{t^2}.$$

2.1.4 Binomial distribution

2.17 Definition Let n and k be non-negative integers. The *binomial coefficient* $\binom{n}{k}$ is the number of different ways of choosing k distinct objects from a set of n distinct objects, where the order of choice is not important.

2.18 Fact (*properties of binomial coefficients*) Let n and k be non-negative integers.

(i) $\binom{n}{k} = \frac{n!}{k!(n-k)!}$.

(ii) $\binom{n}{k} = \binom{n}{n-k}$.

(iii) $\binom{n+1}{k+1} = \binom{n}{k} + \binom{n}{k+1}$.

2.19 Fact (*binomial theorem*) For any real numbers a, b, and non-negative integer n, $(a+b)^n = \sum_{k=0}^{n} \binom{n}{k} a^k b^{n-k}$.

2.20 Definition A *Bernoulli trial* is an experiment with exactly two possible outcomes, called *success* and *failure*.

2.21 Fact Suppose that the probability of success on a particular Bernoulli trial is p. Then the probability of exactly k successes in a sequence of n such independent trials is

$$\binom{n}{k} p^k (1-p)^{n-k}, \text{ for each } 0 \leq k \leq n. \tag{2.1}$$

2.22 Definition The probability distribution (2.1) is called the *binomial distribution*.

2.23 Fact The expected number of successes in a sequence of n independent Bernoulli trials, with probability p of success in each trial, is np. The variance of the number of successes is $np(1-p)$.

2.24 Fact (*law of large numbers*) Let X be the random variable denoting the fraction of successes in n independent Bernoulli trials, with probability p of success in each trial. Then for any $\epsilon > 0$,

$$P(|X - p| > \epsilon) \longrightarrow 0, \text{ as } n \longrightarrow \infty.$$

In other words, as n gets larger, the proportion of successes should be close to p, the probability of success in each trial.

2.1.5 Birthday attacks

2.25 Definition

(i) For positive integers m, n with $m \geq n$, the number $m^{(n)}$ is defined as follows:

$$m^{(n)} = m(m-1)(m-2)\cdots(m-n+1).$$

(ii) Let m, n be non-negative integers with $m \geq n$. The *Stirling number of the second kind*, denoted $\left\{ {m \atop n} \right\}$, is

$$\left\{ {m \atop n} \right\} = \frac{1}{n!} \sum_{k=0}^{n} (-1)^{n-k} \binom{n}{k} k^m,$$

with the exception that $\left\{ {0 \atop 0} \right\} = 1$.

The symbol $\left\{ {m \atop n} \right\}$ counts the number of ways of partitioning a set of m objects into n non-empty subsets.

2.26 Fact (*classical occupancy problem*) An urn has m balls numbered 1 to m. Suppose that n balls are drawn from the urn one at a time, with replacement, and their numbers are listed. The probability that exactly t different balls have been drawn is

$$P_1(m, n, t) = \left\{ {n \atop t} \right\} \frac{m^{(t)}}{m^n}, \quad 1 \leq t \leq n.$$

The birthday problem is a special case of the classical occupancy problem.

2.27 Fact (*birthday problem*) An urn has m balls numbered 1 to m. Suppose that n balls are drawn from the urn one at a time, with replacement, and their numbers are listed.

(i) The probability of at least one coincidence (i.e., a ball drawn at least twice) is

$$P_2(m, n) = 1 - P_1(m, n, n) = 1 - \frac{m^{(n)}}{m^n}, \quad 1 \leq n \leq m. \tag{2.2}$$

If $n = O(\sqrt{m})$ (see Definition 2.55) and $m \longrightarrow \infty$, then

$$P_2(m, n) \longrightarrow 1 - \exp\left(-\frac{n(n-1)}{2m} + O\left(\frac{1}{\sqrt{m}}\right)\right) \approx 1 - \exp\left(-\frac{n^2}{2m}\right).$$

(ii) As $m \longrightarrow \infty$, the expected number of draws before a coincidence is $\sqrt{\frac{\pi m}{2}}$.

The following explains why probability distribution (2.2) is referred to as the *birthday surprise* or *birthday paradox*. The probability that at least 2 people in a room of 23 people have the same birthday is $P_2(365, 23) \approx 0.507$, which is surprisingly large. The quantity $P_2(365, n)$ also increases rapidly as n increases; for example, $P_2(365, 30) \approx 0.706$.

A different kind of problem is considered in Facts 2.28, 2.29, and 2.30 below. Suppose that there are two urns, one containing m white balls numbered 1 to m, and the other containing m red balls numbered 1 to m. First, n_1 balls are selected from the first urn and their numbers listed. Then n_2 balls are selected from the second urn and their numbers listed. Finally, the number of coincidences between the two lists is counted.

2.28 Fact (*model A*) If the balls from both urns are drawn one at a time, with replacement, then the probability of at least one coincidence is

$$P_3(m, n_1, n_2) = 1 - \frac{1}{m^{n_1+n_2}} \sum_{t_1, t_2} m^{(t_1+t_2)} \left\{ {n_1 \atop t_1} \right\} \left\{ {n_2 \atop t_2} \right\},$$

where the summation is over all $0 \leq t_1 \leq n_1, 0 \leq t_2 \leq n_2$. If $n = n_1 = n_2, n = O(\sqrt{m})$ and $m \longrightarrow \infty$, then

$$P_3(m, n_1, n_2) \longrightarrow 1 - \exp\left(-\frac{n^2}{m}\left[1 + O\left(\frac{1}{\sqrt{m}}\right)\right]\right) \approx 1 - \exp\left(-\frac{n^2}{m}\right).$$

2.29 Fact (*model B*) If the balls from both urns are drawn without replacement, then the probability of at least one coincidence is

$$P_4(m, n_1, n_2) = 1 - \frac{m^{(n_1+n_2)}}{m^{(n_1)}m^{(n_2)}}.$$

If $n_1 = O(\sqrt{m})$, $n_2 = O(\sqrt{m})$, and $m \longrightarrow \infty$, then

$$P_4(m, n_1, n_2) \longrightarrow 1 - \exp\left(-\frac{n_1 n_2}{m}\left[1 + \frac{n_1 + n_2 - 1}{2m} + O\left(\frac{1}{m}\right)\right]\right).$$

2.30 Fact (*model C*) If the n_1 white balls are drawn one at a time, with replacement, and the n_2 red balls are drawn without replacement, then the probability of at least one coincidence is

$$P_5(m, n_1, n_2) = 1 - \left(1 - \frac{n_2}{m}\right)^{n_1}.$$

If $n_1 = O(\sqrt{m})$, $n_2 = O(\sqrt{m})$, and $m \longrightarrow \infty$, then

$$P_5(m, n_1, n_2) \longrightarrow 1 - \exp\left(-\frac{n_1 n_2}{m}\left[1 + O\left(\frac{1}{\sqrt{m}}\right)\right]\right) \approx 1 - \exp\left(-\frac{n_1 n_2}{m}\right).$$

2.1.6 Random mappings

2.31 Definition Let $\mathcal{F}_n$ denote the collection of all functions (mappings) from a finite domain of size n to a finite codomain of size n.

Models where random elements of $\mathcal{F}_n$ are considered are called *random mappings models*. In this section the only random mappings model considered is where every function from $\mathcal{F}_n$ is equally likely to be chosen; such models arise frequently in cryptography and algorithmic number theory. Note that $|\mathcal{F}_n| = n^n$, whence the probability that a particular function from $\mathcal{F}_n$ is chosen is $1/n^n$.

2.32 Definition Let f be a function in $\mathcal{F}_n$ with domain and codomain equal to $\{1, 2, \ldots, n\}$. The *functional graph* of f is a directed graph whose *points* (or *vertices*) are the elements $\{1, 2, \ldots, n\}$ and whose *edges* are the ordered pairs $(x, f(x))$ for all $x \in \{1, 2, \ldots, n\}$.

2.33 Example (*functional graph*) Consider the function $f : \{1, 2, \ldots, 13\} \longrightarrow \{1, 2, \ldots, 13\}$ defined by $f(1) = 4$, $f(2) = 11$, $f(3) = 1$, $f(4) = 6$, $f(5) = 3$, $f(6) = 9$, $f(7) = 3$, $f(8) = 11$, $f(9) = 1$, $f(10) = 2$, $f(11) = 10$, $f(12) = 4$, $f(13) = 7$. The functional graph of f is shown in Figure 2.1. $\qquad\qquad\square$

As Figure 2.1 illustrates, a functional graph may have several *components* (maximal connected subgraphs), each component consisting of a directed *cycle* and some directed *trees* attached to the cycle.

2.34 Fact As n tends to infinity, the following statements regarding the functional digraph of a random function f from $\mathcal{F}_n$ are true:

(i) The expected number of components is $\frac{1}{2}\ln n$.

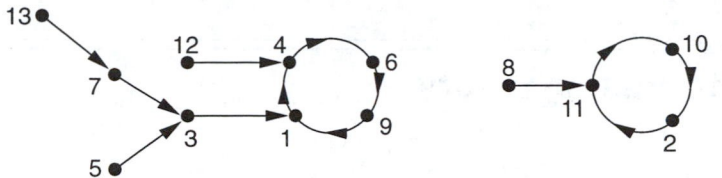

Figure 2.1: *A functional graph (see Example 2.33).*

(ii) The expected number of points which are on the cycles is $\sqrt{\pi n/2}$.

(iii) The expected number of *terminal points* (points which have no preimages) is n/e.

(iv) The expected number of *k-th iterate image points* (x is a k-th iterate image point if $x = \underbrace{f(f(\cdots f(y)\cdots))}_{k \text{ times}}$ for some y) is $(1 - \tau_k)n$, where the τ_k satisfy the recurrence $\tau_0 = 0$, $\tau_{k+1} = e^{-1+\tau_k}$ for $k \geq 0$.

2.35 Definition Let f be a random function from $\{1, 2, \ldots, n\}$ to $\{1, 2, \ldots, n\}$ and let $u \in \{1, 2, \ldots, n\}$. Consider the sequence of points $u_0, u_1, u_2, \ldots$ defined by $u_0 = u$, $u_i = f(u_{i-1})$ for $i \geq 1$. In terms of the functional graph of f, this sequence describes a path that connects to a cycle.

(i) The number of edges in the path is called the *tail length* of u, denoted $\lambda(u)$.

(ii) The number of edges in the cycle is called the *cycle length* of u, denoted $\mu(u)$.

(iii) The *rho-length* of u is the quantity $\rho(u) = \lambda(u) + \mu(u)$.

(iv) The *tree size* of u is the number of edges in the maximal tree rooted on a cycle in the component that contains u.

(v) The *component size* of u is the number of edges in the component that contains u.

(vi) The *predecessors size* of u is the number of iterated preimages of u.

2.36 Example The functional graph in Figure 2.1 has 2 components and 4 terminal points. The point $u = 3$ has parameters $\lambda(u) = 1$, $\mu(u) = 4$, $\rho(u) = 5$. The tree, component, and predecessors sizes of $u = 3$ are 4, 9, and 3, respectively. □

2.37 Fact As n tends to infinity, the following are the expectations of some parameters associated with a random point in $\{1, 2, \ldots, n\}$ and a random function from $\mathcal{F}_n$: (i) tail length: $\sqrt{\pi n/8}$ (ii) cycle length: $\sqrt{\pi n/8}$ (iii) rho-length: $\sqrt{\pi n/2}$ (iv) tree size: $n/3$ (v) component size: $2n/3$ (vi) predecessors size: $\sqrt{\pi n/8}$.

2.38 Fact As n tends to infinity, the expectations of the maximum tail, cycle, and rho lengths in a random function from $\mathcal{F}_n$ are $c_1\sqrt{n}$, $c_2\sqrt{n}$, and $c_3\sqrt{n}$, respectively, where $c_1 \approx 0.78248$, $c_2 \approx 1.73746$, and $c_3 \approx 2.4149$.

Facts 2.37 and 2.38 indicate that in the functional graph of a random function, most points are grouped together in one giant component, and there is a small number of large trees. Also, almost unavoidably, a cycle of length about $\sqrt{n}$ arises after following a path of length $\sqrt{n}$ edges.

2.2 Information theory

2.2.1 Entropy

Let X be a random variable which takes on a finite set of values $x_1, x_2, \ldots, x_n$, with probability $P(X = x_i) = p_i$, where $0 \leq p_i \leq 1$ for each i, $1 \leq i \leq n$, and where $\sum_{i=1}^{n} p_i = 1$. Also, let Y and Z be random variables which take on finite sets of values.

The entropy of X is a mathematical measure of the amount of information provided by an observation of X. Equivalently, it is the uncertainity about the outcome before an observation of X. Entropy is also useful for approximating the average number of bits required to encode the elements of X.

2.39 Definition The *entropy* or *uncertainty* of X is defined to be $H(X) = -\sum_{i=1}^{n} p_i \lg p_i = \sum_{i=1}^{n} p_i \lg \left(\frac{1}{p_i}\right)$ where, by convention, $p_i \cdot \lg p_i = p_i \cdot \lg \left(\frac{1}{p_i}\right) = 0$ if $p_i = 0$.

2.40 Fact (*properties of entropy*) Let X be a random variable which takes on n values.

(i) $0 \leq H(X) \leq \lg n$.
(ii) $H(X) = 0$ if and only if $p_i = 1$ for some i, and $p_j = 0$ for all $j \neq i$ (that is, there is no uncertainty of the outcome).
(iii) $H(X) = \lg n$ if and only if $p_i = 1/n$ for each i, $1 \leq i \leq n$ (that is, all outcomes are equally likely).

2.41 Definition The *joint entropy* of X and Y is defined to be

$$H(X, Y) = -\sum_{x,y} P(X = x, Y = y) \lg(P(X = x, Y = y)),$$

where the summation indices x and y range over all values of X and Y, respectively. The definition can be extended to any number of random variables.

2.42 Fact If X and Y are random variables, then $H(X, Y) \leq H(X) + H(Y)$, with equality if and only if X and Y are independent.

2.43 Definition If X, Y are random variables, the *conditional entropy of X given $Y = y$* is

$$H(X|Y = y) = -\sum_{x} P(X = x|Y = y) \lg(P(X = x|Y = y)),$$

where the summation index x ranges over all values of X. The *conditional entropy of X given Y*, also called the *equivocation of Y about X*, is

$$H(X|Y) = \sum_{y} P(Y = y)H(X|Y = y),$$

where the summation index y ranges over all values of Y.

2.44 Fact (*properties of conditional entropy*) Let X and Y be random variables.

(i) The quantity $H(X|Y)$ measures the amount of uncertainty remaining about X after Y has been observed.

(ii) $H(X|Y) \geq 0$ and $H(X|X) = 0$.

(iii) $H(X,Y) = H(X) + H(Y|X) = H(Y) + H(X|Y)$.

(iv) $H(X|Y) \leq H(X)$, with equality if and only if X and Y are independent.

2.2.2 Mutual information

2.45 Definition The *mutual information* or *transinformation* of random variables X and Y is $I(X;Y) = H(X) - H(X|Y)$. Similarly, the transinformation of X and the pair Y, Z is defined to be $I(X;Y,Z) = H(X) - H(X|Y,Z)$.

2.46 Fact (*properties of mutual transinformation*)

(i) The quantity $I(X;Y)$ can be thought of as the amount of information that Y reveals about X. Similarly, the quantity $I(X;Y,Z)$ can be thought of as the amount of information that Y and Z together reveal about X.

(ii) $I(X;Y) \geq 0$.

(iii) $I(X;Y) = 0$ if and only if X and Y are independent (that is, Y contributes no information about X).

(iv) $I(X;Y) = I(Y;X)$.

2.47 Definition The *conditional transinformation* of the pair X, Y given Z is defined to be $I_Z(X;Y) = H(X|Z) - H(X|Y,Z)$.

2.48 Fact (*properties of conditional transinformation*)

(i) The quantity $I_Z(X;Y)$ can be interpreted as the amount of information that Y provides about X, given that Z has already been observed.

(ii) $I(X;Y,Z) = I(X;Y) + I_Y(X;Z)$.

(iii) $I_Z(X;Y) = I_Z(Y;X)$.

2.3 Complexity theory

2.3.1 Basic definitions

The main goal of complexity theory is to provide mechanisms for classifying computational problems according to the resources needed to solve them. The classification should not depend on a particular computational model, but rather should measure the intrinsic difficulty of the problem. The resources measured may include time, storage space, random bits, number of processors, etc., but typically the main focus is time, and sometimes space.

2.49 Definition An *algorithm* is a well-defined computational procedure that takes a variable input and halts with an output.

Of course, the term "well-defined computational procedure" is not mathematically precise. It can be made so by using formal computational models such as Turing machines, random-access machines, or boolean circuits. Rather than get involved with the technical intricacies of these models, it is simpler to think of an algorithm as a computer program written in some specific programming language for a specific computer that takes a variable input and halts with an output.

It is usually of interest to find the most efficient (i.e., fastest) algorithm for solving a given computational problem. The time that an algorithm takes to halt depends on the "size" of the problem instance. Also, the unit of time used should be made precise, especially when comparing the performance of two algorithms.

2.50 Definition The *size* of the input is the total number of bits needed to represent the input in ordinary binary notation using an appropriate encoding scheme. Occasionally, the size of the input will be the number of items in the input.

2.51 Example (*sizes of some objects*)

(i) The number of bits in the binary representation of a positive integer n is $1 + \lfloor \lg n \rfloor$ bits. For simplicity, the size of n will be approximated by $\lg n$.

(ii) If f is a polynomial of degree k, each coefficient being a non-negative integer at most n, then the size of f is $(k+1) \lg n$ bits.

(iii) If A is a matrix with r rows, s columns, and with non-negative integer entries each at most n, then the size of A is $rs \lg n$ bits. □

2.52 Definition The *running time* of an algorithm on a particular input is the number of primitive operations or "steps" executed.

Often a step is taken to mean a bit operation. For some algorithms it will be more convenient to take step to mean something else such as a comparison, a machine instruction, a machine clock cycle, a modular multiplication, etc.

2.53 Definition The *worst-case running time* of an algorithm is an upper bound on the running time for any input, expressed as a function of the input size.

2.54 Definition The *average-case running time* of an algorithm is the average running time over all inputs of a fixed size, expressed as a function of the input size.

2.3.2 Asymptotic notation

It is often difficult to derive the exact running time of an algorithm. In such situations one is forced to settle for approximations of the running time, and usually may only derive the *asymptotic* running time. That is, one studies how the running time of the algorithm increases as the size of the input increases without bound.

In what follows, the only functions considered are those which are defined on the positive integers and take on real values that are always positive from some point onwards. Let f and g be two such functions.

2.55 Definition (*order notation*)

(i) (*asymptotic upper bound*) $f(n) = O(g(n))$ if there exists a positive constant c and a positive integer n_0 such that $0 \le f(n) \le cg(n)$ for all $n \ge n_0$.

(ii) (*asymptotic lower bound*) $f(n) = \Omega(g(n))$ if there exists a positive constant c and a positive integer n_0 such that $0 \leq cg(n) \leq f(n)$ for all $n \geq n_0$.

(iii) (*asymptotic tight bound*) $f(n) = \Theta(g(n))$ if there exist positive constants c_1 and c_2, and a positive integer n_0 such that $c_1 g(n) \leq f(n) \leq c_2 g(n)$ for all $n \geq n_0$.

(iv) (*o-notation*) $f(n) = o(g(n))$ if for any positive constant $c > 0$ there exists a constant $n_0 > 0$ such that $0 \leq f(n) < cg(n)$ for all $n \geq n_0$.

Intuitively, $f(n) = O(g(n))$ means that f grows no faster asymptotically than $g(n)$ to within a constant multiple, while $f(n) = \Omega(g(n))$ means that $f(n)$ grows at least as fast asymptotically as $g(n)$ to within a constant multiple. $f(n) = o(g(n))$ means that $g(n)$ is an upper bound for $f(n)$ that is not asymptotically tight, or in other words, the function $f(n)$ becomes insignificant relative to $g(n)$ as n gets larger. The expression $o(1)$ is often used to signify a function $f(n)$ whose limit as n approaches ∞ is 0.

2.56 Fact (*properties of order notation*) For any functions $f(n)$, $g(n)$, $h(n)$, and $l(n)$, the following are true.

(i) $f(n) = O(g(n))$ if and only if $g(n) = \Omega(f(n))$.

(ii) $f(n) = \Theta(g(n))$ if and only if $f(n) = O(g(n))$ and $f(n) = \Omega(g(n))$.

(iii) If $f(n) = O(h(n))$ and $g(n) = O(h(n))$, then $(f + g)(n) = O(h(n))$.

(iv) If $f(n) = O(h(n))$ and $g(n) = O(l(n))$, then $(f \cdot g)(n) = O(h(n)l(n))$.

(v) (*reflexivity*) $f(n) = O(f(n))$.

(vi) (*transitivity*) If $f(n) = O(g(n))$ and $g(n) = O(h(n))$, then $f(n) = O(h(n))$.

2.57 Fact (*approximations of some commonly occurring functions*)

(i) (*polynomial function*) If $f(n)$ is a polynomial of degree k with positive leading term, then $f(n) = \Theta(n^k)$.

(ii) For any constant $c > 0$, $\log_c n = \Theta(\lg n)$.

(iii) (*Stirling's formula*) For all integers $n \geq 1$,

$$\sqrt{2\pi n}\left(\frac{n}{e}\right)^n \leq n! \leq \sqrt{2\pi n}\left(\frac{n}{e}\right)^{n+(1/(12n))}.$$

Thus $n! = \sqrt{2\pi n}\left(\frac{n}{e}\right)^n \left(1 + \Theta(\frac{1}{n})\right)$. Also, $n! = o(n^n)$ and $n! = \Omega(2^n)$.

(iv) $\lg(n!) = \Theta(n \lg n)$.

2.58 Example (*comparative growth rates of some functions*) Let ϵ and c be arbitrary constants with $0 < \epsilon < 1 < c$. The following functions are listed in increasing order of their asymptotic growth rates:

$$1 < \ln\ln n < \ln n < \exp(\sqrt{\ln n \ln\ln n}) < n^\epsilon < n^c < n^{\ln n} < c^n < n^n < c^{c^n}. \qquad \square$$

2.3.3 Complexity classes

2.59 Definition A *polynomial-time algorithm* is an algorithm whose worst-case running time function is of the form $O(n^k)$, where n is the input size and k is a constant. Any algorithm whose running time cannot be so bounded is called an *exponential-time algorithm*.

Roughly speaking, polynomial-time algorithms can be equated with *good* or *efficient* algorithms, while exponential-time algorithms are considered *inefficient*. There are, however, some practical situations when this distinction is not appropriate. When considering polynomial-time complexity, the degree of the polynomial is significant. For example, even though an algorithm with a running time of $O(n^{\ln \ln n})$, n being the input size, is asymptotically slower that an algorithm with a running time of $O(n^{100})$, the former algorithm may be faster in practice for smaller values of n, especially if the constants hidden by the big-O notation are smaller. Furthermore, in cryptography, average-case complexity is more important than worst-case complexity — a necessary condition for an encryption scheme to be considered secure is that the corresponding cryptanalysis problem is difficult on average (or better yet, always difficult), and not just for some isolated cases.

2.60 Definition A *subexponential-time algorithm* is an algorithm whose worst-case running time function is of the form $e^{o(n)}$, where n is the input size.

A subexponential-time algorithm is asymptotically faster than an algorithm whose running time is fully exponential in the input size, while it is asymptotically slower than a polynomial-time algorithm.

2.61 Example (*subexponential running time*) Let A be an algorithm whose inputs are either elements of a finite field $\mathbb{F}_q$ (see §2.6), or an integer q. If the expected running time of A is of the form

$$L_q[\alpha, c] = O\left(\exp\left((c + o(1))(\ln q)^\alpha (\ln \ln q)^{1-\alpha}\right)\right), \qquad (2.3)$$

where c is a positive constant, and α is a constant satisfying $0 < \alpha < 1$, then A is a subexponential-time algorithm. Observe that for $\alpha = 0$, $L_q[0, c]$ is a polynomial in $\ln q$, while for $\alpha = 1$, $L_q[1, c]$ is a polynomial in q, and thus fully exponential in $\ln q$. □

For simplicity, the theory of computational complexity restricts its attention to *decision problems*, i.e., problems which have either YES or NO as an answer. This is not too restrictive in practice, as all the computational problems that will be encountered here can be phrased as decision problems in such a way that an efficient algorithm for the decision problem yields an efficient algorithm for the computational problem, and vice versa.

2.62 Definition The complexity class **P** is the set of all decision problems that are solvable in polynomial time.

2.63 Definition The complexity class **NP** is the set of all decision problems for which a YES answer can be verified in polynomial time using some extra information, called a *certificate*.

2.64 Definition The complexity class **co-NP** is the set of all decision problems for which a NO answer can be verified in polynomial time using an appropriate certificate.

It must be emphasized that if a decision problem is in **NP**, it may not be the case that the certificate of a YES answer can be easily obtained; what is asserted is that such a certificate does exist, and, if known, can be used to efficiently verify the YES answer. The same is true of the NO answers for problems in **co-NP**.

2.65 Example (*problem in* **NP**) Consider the following decision problem:
COMPOSITES
INSTANCE: A positive integer n.
QUESTION: Is n composite? That is, are there integers $a, b > 1$ such that $n = ab$?
COMPOSITES belongs to **NP** because if an integer n is composite, then this fact can be verified in polynomial time if one is given a divisor a of n, where $1 < a < n$ (the certificate in this case consists of the divisor a). It is in fact also the case that COMPOSITES belongs to **co-NP**. It is still unknown whether or not COMPOSITES belongs to **P**. □

2.66 Fact $\mathbf{P} \subseteq \mathbf{NP}$ and $\mathbf{P} \subseteq \mathbf{co\text{-}NP}$.

The following are among the outstanding unresolved questions in the subject of complexity theory:

1. Is $\mathbf{P} = \mathbf{NP}$?
2. Is $\mathbf{NP} = \mathbf{co\text{-}NP}$?
3. Is $\mathbf{P} = \mathbf{NP} \cap \mathbf{co\text{-}NP}$?

Most experts are of the opinion that the answer to each of the three questions is NO, although nothing along these lines has been proven.

The notion of reducibility is useful when comparing the relative difficulties of problems.

2.67 Definition Let L_1 and L_2 be two decision problems. L_1 is said to *polytime reduce* to L_2, written $L_1 \leq_P L_2$, if there is an algorithm that solves L_1 which uses, as a subroutine, an algorithm for solving L_2, and which runs in polynomial time if the algorithm for L_2 does.

Informally, if $L_1 \leq_P L_2$, then L_2 is at least as difficult as L_1, or, equivalently, L_1 is no harder than L_2.

2.68 Definition Let L_1 and L_2 be two decision problems. If $L_1 \leq_P L_2$ and $L_2 \leq_P L_1$, then L_1 and L_2 are said to be *computationally equivalent*.

2.69 Fact Let L_1, L_2, and L_3 be three decision problems.

(i) *(transitivity)* If $L_1 \leq_P L_2$ and $L_2 \leq_P L_3$, then $L_1 \leq_P L_3$.
(ii) If $L_1 \leq_P L_2$ and $L_2 \in \mathbf{P}$, then $L_1 \in \mathbf{P}$.

2.70 Definition A decision problem L is said to be $\mathbf{NP}$-*complete* if

(i) $L \in \mathbf{NP}$, and
(ii) $L_1 \leq_P L$ for every $L_1 \in \mathbf{NP}$.

The class of all $\mathbf{NP}$-complete problems is denoted by $\mathbf{NPC}$.

$\mathbf{NP}$-complete problems are the hardest problems in $\mathbf{NP}$ in the sense that they are at least as difficult as every other problem in $\mathbf{NP}$. There are thousands of problems drawn from diverse fields such as combinatorics, number theory, and logic, that are known to be $\mathbf{NP}$-complete.

2.71 Example (*subset sum problem*) The *subset sum problem* is the following: given a set of positive integers $\{a_1, a_2, \dots, a_n\}$ and a positive integer s, determine whether or not there is a subset of the a_i that sum to s. The subset sum problem is $\mathbf{NP}$-complete. $\square$

2.72 Fact Let L_1 and L_2 be two decision problems.

(i) If L_1 is $\mathbf{NP}$-complete and $L_1 \in \mathbf{P}$, then $\mathbf{P} = \mathbf{NP}$.
(ii) If $L_1 \in \mathbf{NP}$, L_2 is $\mathbf{NP}$-complete, and $L_2 \leq_P L_1$, then L_1 is also $\mathbf{NP}$-complete.
(iii) If L_1 is $\mathbf{NP}$-complete and $L_1 \in \mathbf{co\text{-}NP}$, then $\mathbf{NP} = \mathbf{co\text{-}NP}$.

By Fact 2.72(i), if a polynomial-time algorithm is found for any single $\mathbf{NP}$-complete problem, then it is the case that $\mathbf{P} = \mathbf{NP}$, a result that would be extremely surprising. Hence, a proof that a problem is $\mathbf{NP}$-complete provides strong evidence for its intractability. Figure 2.2 illustrates what is widely believed to be the relationship between the complexity classes $\mathbf{P}$, $\mathbf{NP}$, $\mathbf{co\text{-}NP}$, and $\mathbf{NPC}$.

Fact 2.72(ii) suggests the following procedure for proving that a decision problem L_1 is $\mathbf{NP}$-complete:

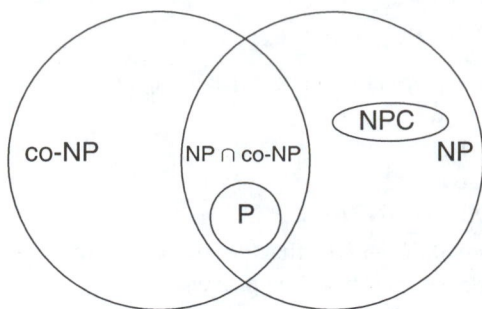

Figure 2.2: *Conjectured relationship between the complexity classes* **P**, **NP**, **co-NP**, *and* **NPC**.

1. Prove that $L_1 \in$ **NP**.
2. Select a problem L_2 that is known to be **NP**-complete.
3. Prove that $L_2 \leq_P L_1$.

2.73 Definition A problem is **NP**-*hard* if the existence of a polynomial-time algorithm for its solution implies that **P** = **NP**.

Note that the **NP**-hard classification is not restricted to only decision problems. Observe also (using Fact 2.72(i)) that an **NP**-complete problem is also **NP**-hard.

2.74 Example (**NP**-*hard problem*) Given positive integers $a_1, a_2, \dots, a_n$ and a positive integer s, the computational version of the subset sum problem would ask to actually find a subset of the a_i which sums to s, provided that such a subset exists. This problem is **NP**-hard.   □

2.3.4 Randomized algorithms

The algorithms studied so far in this section have been *deterministic*; such algorithms follow the same execution path (sequence of operations) each time they execute with the same input. By contrast, a *randomized* algorithm makes random decisions at certain points in the execution; hence their execution paths may differ each time they are invoked with the same input. The random decisions are based upon the outcome of a random number generator. Remarkably, there are many problems for which randomized algorithms are known that are more efficient, both in terms of time and space, than the best known deterministic algorithms.

Randomized algorithms for decision problems can be classified according to the probability that they return the correct answer.

2.75 Definition Let A be a randomized algorithm for a decision problem L, and let I denote an arbitrary instance of L.

   (i) A has *0-sided error* if $P(A$ outputs YES $\mid I$'s answer is YES $) = 1$, and
      $P(A$ outputs YES $\mid I$'s answer is NO $) = 0$.

  (ii) A has *1-sided error* if $P(A$ outputs YES $\mid I$'s answer is YES $) \geq \frac{1}{2}$, and
      $P(A$ outputs YES $\mid I$'s answer is NO $) = 0$.

(iii) A has *2-sided error* if $P(A$ outputs YES $\mid I$'s answer is YES $) \geq \frac{2}{3}$, and $P(A$ outputs YES $\mid I$'s answer is NO $) \leq \frac{1}{3}$.

The number $\frac{1}{2}$ in the definition of 1-sided error is somewhat arbitrary and can be replaced by any positive constant. Similarly, the numbers $\frac{2}{3}$ and $\frac{1}{3}$ in the definition of 2-sided error, can be replaced by $\frac{1}{2} + \epsilon$ and $\frac{1}{2} - \epsilon$, respectively, for any constant ϵ, $0 < \epsilon < \frac{1}{2}$.

2.76 Definition The *expected running time* of a randomized algorithm is an upper bound on the expected running time for each input (the expectation being over all outputs of the random number generator used by the algorithm), expressed as a function of the input size.

The important randomized complexity classes are defined next.

2.77 Definition (*randomized complexity classes*)

(i) The complexity class **ZPP** ("zero-sided probabilistic polynomial time") is the set of all decision problems for which there is a randomized algorithm with 0-sided error which runs in expected polynomial time.

(ii) The complexity class **RP** ("randomized polynomial time") is the set of all decision problems for which there is a randomized algorithm with 1-sided error which runs in (worst-case) polynomial time.

(iii) The complexity class **BPP** ("bounded error probabilistic polynomial time") is the set of all decision problems for which there is a randomized algorithm with 2-sided error which runs in (worst-case) polynomial time.

2.78 Fact $\mathbf{P} \subseteq \mathbf{ZPP} \subseteq \mathbf{RP} \subseteq \mathbf{BPP}$ and $\mathbf{RP} \subseteq \mathbf{NP}$.

2.4 Number theory

2.4.1 The integers

The set of integers $\{\ldots, -3, -2, -1, 0, 1, 2, 3, \ldots\}$ is denoted by the symbol $\mathbb{Z}$.

2.79 Definition Let a, b be integers. Then a *divides* b (equivalently: a is a *divisor* of b, or a is a *factor* of b) if there exists an integer c such that $b = ac$. If a divides b, then this is denoted by $a|b$.

2.80 Example (i) $-3|18$, since $18 = (-3)(-6)$. (ii) $173|0$, since $0 = (173)(0)$. $\qquad\square$

The following are some elementary properties of divisibility.

2.81 Fact (*properties of divisibility*) For all a, b, $c \in \mathbb{Z}$, the following are true:

(i) $a|a$.

(ii) If $a|b$ and $b|c$, then $a|c$.

(iii) If $a|b$ and $a|c$, then $a|(bx + cy)$ for all $x, y \in \mathbb{Z}$.

(iv) If $a|b$ and $b|a$, then $a = \pm b$.

2.82 Definition (*division algorithm for integers*) If a and b are integers with $b \geq 1$, then ordinary long division of a by b yields integers q (the *quotient*) and r (the *remainder*) such that

$$a = qb + r, \quad \text{where } 0 \leq r < b.$$

Moreover, q and r are unique. The remainder of the division is denoted $a \bmod b$, and the quotient is denoted $a \operatorname{div} b$.

2.83 Fact Let $a, b \in \mathbb{Z}$ with $b \neq 0$. Then $a \operatorname{div} b = \lfloor a/b \rfloor$ and $a \bmod b = a - b\lfloor a/b \rfloor$.

2.84 Example If $a = 73$, $b = 17$, then $q = 4$ and $r = 5$. Hence $73 \bmod 17 = 5$ and $73 \operatorname{div} 17 = 4$. $\square$

2.85 Definition An integer c is a *common divisor* of a and b if $c|a$ and $c|b$.

2.86 Definition A non-negative integer d is the *greatest common divisor* of integers a and b, denoted $d = \gcd(a, b)$, if

    (i) d is a common divisor of a and b; and
    (ii) whenever $c|a$ and $c|b$, then $c|d$.

Equivalently, $\gcd(a, b)$ is the largest positive integer that divides both a and b, with the exception that $\gcd(0, 0) = 0$.

2.87 Example The common divisors of 12 and 18 are $\{\pm 1, \pm 2, \pm 3, \pm 6\}$, and $\gcd(12, 18) = 6$. $\square$

2.88 Definition A non-negative integer d is the *least common multiple* of integers a and b, denoted $d = \operatorname{lcm}(a, b)$, if

    (i) $a|d$ and $b|d$; and
    (ii) whenever $a|c$ and $b|c$, then $d|c$.

Equivalently, $\operatorname{lcm}(a, b)$ is the smallest non-negative integer divisible by both a and b.

2.89 Fact If a and b are positive integers, then $\operatorname{lcm}(a, b) = a \cdot b / \gcd(a, b)$.

2.90 Example Since $\gcd(12, 18) = 6$, it follows that $\operatorname{lcm}(12, 18) = 12 \cdot 18/6 = 36$. $\square$

2.91 Definition Two integers a and b are said to be *relatively prime* or *coprime* if $\gcd(a, b) = 1$.

2.92 Definition An integer $p \geq 2$ is said to be *prime* if its only positive divisors are 1 and p. Otherwise, p is called *composite*.

The following are some well known facts about prime numbers.

2.93 Fact If p is prime and $p|ab$, then either $p|a$ or $p|b$ (or both).

2.94 Fact There are an infinite number of prime numbers.

2.95 Fact (*prime number theorem*) Let $\pi(x)$ denote the number of prime numbers $\leq x$. Then

$$\lim_{x \to \infty} \frac{\pi(x)}{x/\ln x} = 1.$$

This means that for large values of x, $\pi(x)$ is closely approximated by the expression $x/\ln x$. For instance, when $x = 10^{10}$, $\pi(x) = 455,052,511$, whereas $\lfloor x/\ln x \rfloor = 434,294,481$. A more explicit estimate for $\pi(x)$ is given below.

2.96 Fact Let $\pi(x)$ denote the number of primes $\leq x$. Then for $x \geq 17$

$$\pi(x) > \frac{x}{\ln x}$$

and for $x > 1$

$$\pi(x) < 1.25506\frac{x}{\ln x}.$$

2.97 Fact (*fundamental theorem of arithmetic*) Every integer $n \geq 2$ has a factorization as a product of prime powers:

$$n = p_1^{e_1} p_2^{e_2} \cdots p_k^{e_k},$$

where the p_i are distinct primes, and the e_i are positive integers. Furthermore, the factorization is unique up to rearrangement of factors.

2.98 Fact If $a = p_1^{e_1} p_2^{e_2} \cdots p_k^{e_k}$, $b = p_1^{f_1} p_2^{f_2} \cdots p_k^{f_k}$, where each $e_i \geq 0$ and $f_i \geq 0$, then

$$\gcd(a, b) = p_1^{\min(e_1,f_1)} p_2^{\min(e_2,f_2)} \cdots p_k^{\min(e_k,f_k)}$$

and

$$\mathrm{lcm}(a, b) = p_1^{\max(e_1,f_1)} p_2^{\max(e_2,f_2)} \cdots p_k^{\max(e_k,f_k)}.$$

2.99 Example Let $a = 4864 = 2^8 \cdot 19$, $b = 3458 = 2 \cdot 7 \cdot 13 \cdot 19$. Then $\gcd(4864, 3458) = 2 \cdot 19 = 38$ and $\mathrm{lcm}(4864, 3458) = 2^8 \cdot 7 \cdot 13 \cdot 19 = 442624$. $\square$

2.100 Definition For $n \geq 1$, let $\phi(n)$ denote the number of integers in the interval $[1, n]$ which are relatively prime to n. The function ϕ is called the *Euler phi function* (or the *Euler totient function*).

2.101 Fact (*properties of Euler phi function*)
  (i) If p is a prime, then $\phi(p) = p - 1$.
  (ii) The Euler phi function is *multiplicative*. That is, if $\gcd(m, n) = 1$, then $\phi(mn) = \phi(m) \cdot \phi(n)$.
  (iii) If $n = p_1^{e_1} p_2^{e_2} \cdots p_k^{e_k}$ is the prime factorization of n, then

$$\phi(n) = n \left(1 - \frac{1}{p_1}\right) \left(1 - \frac{1}{p_2}\right) \cdots \left(1 - \frac{1}{p_k}\right).$$

Fact 2.102 gives an explicit lower bound for $\phi(n)$.

2.102 Fact For all integers $n \geq 5$,

$$\phi(n) > \frac{n}{6 \ln \ln n}.$$

2.4.2 Algorithms in $\mathbb{Z}$

Let a and b be non-negative integers, each less than or equal to n. Recall (Example 2.51) that the number of bits in the binary representation of n is $\lfloor \lg n \rfloor + 1$, and this number is approximated by $\lg n$. The number of bit operations for the four basic integer operations of addition, subtraction, multiplication, and division using the classical algorithms is summarized in Table 2.1. These algorithms are studied in more detail in §14.2. More sophisticated techniques for multiplication and division have smaller complexities.

Operation		Bit complexity
Addition	$a + b$	$O(\lg a + \lg b) = O(\lg n)$
Subtraction	$a - b$	$O(\lg a + \lg b) = O(\lg n)$
Multiplication	$a \cdot b$	$O((\lg a)(\lg b)) = O((\lg n)^2)$
Division	$a = qb + r$	$O((\lg q)(\lg b)) = O((\lg n)^2)$

Table 2.1: *Bit complexity of basic operations in $\mathbb{Z}$.*

The greatest common divisor of two integers a and b can be computed via Fact 2.98. However, computing a gcd by first obtaining prime-power factorizations does not result in an efficient algorithm, as the problem of factoring integers appears to be relatively difficult. The Euclidean algorithm (Algorithm 2.104) is an efficient algorithm for computing the greatest common divisor of two integers that does not require the factorization of the integers. It is based on the following simple fact.

2.103 Fact If a and b are positive integers with $a > b$, then $\gcd(a, b) = \gcd(b, a \bmod b)$.

2.104 Algorithm Euclidean algorithm for computing the greatest common divisor of two integers

INPUT: two non-negative integers a and b with $a \geq b$.
OUTPUT: the greatest common divisor of a and b.

    1. While $b \neq 0$ do the following:

        1.1 Set $r \leftarrow a \bmod b$, $a \leftarrow b$, $b \leftarrow r$.

    2. Return(a).

2.105 Fact Algorithm 2.104 has a running time of $O((\lg n)^2)$ bit operations.

2.106 Example (*Euclidean algorithm*) The following are the division steps of Algorithm 2.104 for computing $\gcd(4864, 3458) = 38$:

$$
\begin{aligned}
4864 &= 1 \cdot 3458 + 1406 \\
3458 &= 2 \cdot 1406 + 646 \\
1406 &= 2 \cdot 646 + 114 \\
646 &= 5 \cdot 114 + 76 \\
114 &= 1 \cdot 76 + 38 \\
76 &= 2 \cdot 38 + 0.
\end{aligned}
$$
$\square$

The Euclidean algorithm can be extended so that it not only yields the greatest common divisor d of two integers a and b, but also integers x and y satisfying $ax + by = d$.

2.107 Algorithm Extended Euclidean algorithm

INPUT: two non-negative integers a and b with $a \geq b$.
OUTPUT: $d = \gcd(a, b)$ and integers x, y satisfying $ax + by = d$.

1. If $b = 0$ then set $d \leftarrow a$, $x \leftarrow 1$, $y \leftarrow 0$, and return(d,x,y).
2. Set $x_2 \leftarrow 1$, $x_1 \leftarrow 0$, $y_2 \leftarrow 0$, $y_1 \leftarrow 1$.
3. While $b > 0$ do the following:
   3.1 $q \leftarrow \lfloor a/b \rfloor$, $r \leftarrow a - qb$, $x \leftarrow x_2 - qx_1$, $y \leftarrow y_2 - qy_1$.
   3.2 $a \leftarrow b$, $b \leftarrow r$, $x_2 \leftarrow x_1$, $x_1 \leftarrow x$, $y_2 \leftarrow y_1$, and $y_1 \leftarrow y$.
4. Set $d \leftarrow a$, $x \leftarrow x_2$, $y \leftarrow y_2$, and return(d,x,y).

2.108 Fact Algorithm 2.107 has a running time of $O((\lg n)^2)$ bit operations.

2.109 Example (*extended Euclidean algorithm*) Table 2.2 shows the steps of Algorithm 2.107 with inputs $a = 4864$ and $b = 3458$. Hence $\gcd(4864, 3458) = 38$ and $(4864)(32) + (3458)(-45) = 38$. □

q	r	x	y	a	b	x_2	x_1	y_2	y_1
—	—	—	—	4864	3458	1	0	0	1
1	1406	1	−1	3458	1406	0	1	1	−1
2	646	−2	3	1406	646	1	−2	−1	3
2	114	5	−7	646	114	−2	5	3	−7
5	76	−27	38	114	76	5	−27	−7	38
1	38	32	−45	76	38	−27	32	38	−45
2	0	−91	128	38	0	32	−91	−45	128

Table 2.2: *Extended Euclidean algorithm (Algorithm 2.107) with inputs $a = 4864$, $b = 3458$.*

Efficient algorithms for gcd and extended gcd computations are further studied in §14.4.

2.4.3 The integers modulo n

Let n be a positive integer.

2.110 Definition If a and b are integers, then a is said to be *congruent to b modulo n*, written $a \equiv b \pmod{n}$, if n divides $(a-b)$. The integer n is called the *modulus* of the congruence.

2.111 Example (i) $24 \equiv 9 \pmod 5$ since $24 - 9 = 3 \cdot 5$.
(ii) $-11 \equiv 17 \pmod 7$ since $-11 - 17 = -4 \cdot 7$. □

2.112 Fact (*properties of congruences*) For all $a, a_1, b, b_1, c \in \mathbb{Z}$, the following are true.
(i) $a \equiv b \pmod n$ if and only if a and b leave the same remainder when divided by n.
(ii) (*reflexivity*) $a \equiv a \pmod n$.
(iii) (*symmetry*) If $a \equiv b \pmod n$ then $b \equiv a \pmod n$.

(iv) (*transitivity*) If $a \equiv b \pmod{n}$ and $b \equiv c \pmod{n}$, then $a \equiv c \pmod{n}$.

(v) If $a \equiv a_1 \pmod{n}$ and $b \equiv b_1 \pmod{n}$, then $a + b \equiv a_1 + b_1 \pmod{n}$ and $ab \equiv a_1 b_1 \pmod{n}$.

The *equivalence class* of an integer a is the set of all integers congruent to a modulo n. From properties (ii), (iii), and (iv) above, it can be seen that for a fixed n the relation of congruence modulo n partitions $\mathbb{Z}$ into equivalence classes. Now, if $a = qn + r$, where $0 \leq r < n$, then $a \equiv r \pmod{n}$. Hence each integer a is congruent modulo n to a unique integer between 0 and $n - 1$, called the *least residue* of a modulo n. Thus a and r are in the same equivalence class, and so r may simply be used to represent this equivalence class.

2.113 Definition The *integers modulo* n, denoted $\mathbb{Z}_n$, is the set of (equivalence classes of) integers $\{0, 1, 2, \ldots, n - 1\}$. Addition, subtraction, and multiplication in $\mathbb{Z}_n$ are performed modulo n.

2.114 Example $\mathbb{Z}_{25} = \{0, 1, 2, \ldots, 24\}$. In $\mathbb{Z}_{25}$, $13 + 16 = 4$, since $13 + 16 = 29 \equiv 4 \pmod{25}$. Similarly, $13 \cdot 16 = 8$ in $\mathbb{Z}_{25}$. $\qquad\square$

2.115 Definition Let $a \in \mathbb{Z}_n$. The *multiplicative inverse* of a modulo n is an integer $x \in \mathbb{Z}_n$ such that $ax \equiv 1 \pmod{n}$. If such an x exists, then it is unique, and a is said to be *invertible*, or a *unit*; the inverse of a is denoted by a^{-1}.

2.116 Definition Let $a, b \in \mathbb{Z}_n$. *Division* of a by b modulo n is the product of a and b^{-1} modulo n, and is only defined if b is invertible modulo n.

2.117 Fact Let $a \in \mathbb{Z}_n$. Then a is invertible if and only if $\gcd(a, n) = 1$.

2.118 Example The invertible elements in $\mathbb{Z}_9$ are 1, 2, 4, 5, 7, and 8. For example, $4^{-1} = 7$ because $4 \cdot 7 \equiv 1 \pmod{9}$. $\qquad\square$

The following is a generalization of Fact 2.117.

2.119 Fact Let $d = \gcd(a, n)$. The congruence equation $ax \equiv b \pmod{n}$ has a solution x if and only if d divides b, in which case there are exactly d solutions between 0 and $n - 1$; these solutions are all congruent modulo n/d.

2.120 Fact (*Chinese remainder theorem*) If the integers $n_1, n_2, \ldots, n_k$ are pairwise relatively prime, then the system of simultaneous congruences

$$
\begin{aligned}
x &\equiv a_1 \pmod{n_1} \\
x &\equiv a_2 \pmod{n_2} \\
&\vdots \\
x &\equiv a_k \pmod{n_k}
\end{aligned}
$$

has a unique solution modulo $n = n_1 n_2 \cdots n_k$.

2.121 Algorithm (*Gauss's algorithm*) The solution x to the simultaneous congruences in the Chinese remainder theorem (Fact 2.120) may be computed as $x = \sum_{i=1}^{k} a_i N_i M_i \bmod n$, where $N_i = n/n_i$ and $M_i = N_i^{-1} \bmod n_i$. These computations can be performed in $O((\lg n)^2)$ bit operations.

Another efficient practical algorithm for solving simultaneous congruences in the Chinese remainder theorem is presented in §14.5.

2.122 Example The pair of congruences $x \equiv 3 \pmod 7$, $x \equiv 7 \pmod{13}$ has a unique solution $x \equiv 59 \pmod{91}$. □

2.123 Fact If $\gcd(n_1, n_2) = 1$, then the pair of congruences $x \equiv a \pmod{n_1}$, $x \equiv a \pmod{n_2}$ has a unique solution $x \equiv a \pmod{n_1 n_2}$.

2.124 Definition The *multiplicative group* of $\mathbb{Z}_n$ is $\mathbb{Z}_n^* = \{a \in \mathbb{Z}_n \mid \gcd(a, n) = 1\}$. In particular, if n is a prime, then $\mathbb{Z}_n^* = \{a \mid 1 \le a \le n - 1\}$.

2.125 Definition The *order* of $\mathbb{Z}_n^*$ is defined to be the number of elements in $\mathbb{Z}_n^*$, namely $|\mathbb{Z}_n^*|$.

It follows from the definition of the Euler phi function (Definition 2.100) that $|\mathbb{Z}_n^*| = \phi(n)$. Note also that if $a \in \mathbb{Z}_n^*$ and $b \in \mathbb{Z}_n^*$, then $a \cdot b \in \mathbb{Z}_n^*$, and so $\mathbb{Z}_n^*$ is closed under multiplication.

2.126 Fact Let $n \ge 2$ be an integer.
   (i) (*Euler's theorem*) If $a \in \mathbb{Z}_n^*$, then $a^{\phi(n)} \equiv 1 \pmod n$.
   (ii) If n is a product of distinct primes, and if $r \equiv s \pmod{\phi(n)}$, then $a^r \equiv a^s \pmod n$ for all integers a. In other words, when working modulo such an n, exponents can be reduced modulo $\phi(n)$.

A special case of Euler's theorem is Fermat's (little) theorem.

2.127 Fact Let p be a prime.
   (i) (*Fermat's theorem*) If $\gcd(a, p) = 1$, then $a^{p-1} \equiv 1 \pmod p$.
   (ii) If $r \equiv s \pmod{p - 1}$, then $a^r \equiv a^s \pmod p$ for all integers a. In other words, when working modulo a prime p, exponents can be reduced modulo $p - 1$.
   (iii) In particular, $a^p \equiv a \pmod p$ for all integers a.

2.128 Definition Let $a \in \mathbb{Z}_n^*$. The *order* of a, denoted $\operatorname{ord}(a)$, is the least positive integer t such that $a^t \equiv 1 \pmod n$.

2.129 Fact If the order of $a \in \mathbb{Z}_n^*$ is t, and $a^s \equiv 1 \pmod n$, then t divides s. In particular, $t \mid \phi(n)$.

2.130 Example Let $n = 21$. Then $\mathbb{Z}_{21}^* = \{1, 2, 4, 5, 8, 10, 11, 13, 16, 17, 19, 20\}$. Note that $\phi(21) = \phi(7)\phi(3) = 12 = |\mathbb{Z}_{21}^*|$. The orders of elements in $\mathbb{Z}_{21}^*$ are listed in Table 2.3. □

$a \in \mathbb{Z}_{21}^*$	1	2	4	5	8	10	11	13	16	17	19	20
order of a	1	6	3	6	2	6	6	2	3	6	6	2

Table 2.3: Orders of elements in $\mathbb{Z}_{21}^*$.

2.131 Definition Let $\alpha \in \mathbb{Z}_n^*$. If the order of α is $\phi(n)$, then α is said to be a *generator* or a *primitive element* of $\mathbb{Z}_n^*$. If $\mathbb{Z}_n^*$ has a generator, then $\mathbb{Z}_n^*$ is said to be *cyclic*.

2.132 Fact (*properties of generators of* $\mathbb{Z}_n^*$)

    (i) $\mathbb{Z}_n^*$ has a generator if and only if $n = 2, 4, p^k$ or $2p^k$, where p is an odd prime and $k \geq 1$. In particular, if p is a prime, then $\mathbb{Z}_p^*$ has a generator.

    (ii) If α is a generator of $\mathbb{Z}_n^*$, then $\mathbb{Z}_n^* = \{\alpha^i \bmod n \mid 0 \leq i \leq \phi(n) - 1\}$.

    (iii) Suppose that α is a generator of $\mathbb{Z}_n^*$. Then $b = \alpha^i \bmod n$ is also a generator of $\mathbb{Z}_n^*$ if and only if $\gcd(i, \phi(n)) = 1$. It follows that if $\mathbb{Z}_n^*$ is cyclic, then the number of generators is $\phi(\phi(n))$.

    (iv) $\alpha \in \mathbb{Z}_n^*$ is a generator of $\mathbb{Z}_n^*$ if and only if $\alpha^{\phi(n)/p} \not\equiv 1 \pmod{n}$ for each prime divisor p of $\phi(n)$.

2.133 Example $\mathbb{Z}_{21}^*$ is not cyclic since it does not contain an element of order $\phi(21) = 12$ (see Table 2.3); note that 21 does not satisfy the condition of Fact 2.132(i). On the other hand, $\mathbb{Z}_{25}^*$ is cyclic, and has a generator $\alpha = 2$. $\qquad\square$

2.134 Definition Let $a \in \mathbb{Z}_n^*$. a is said to be a *quadratic residue* modulo n, or a *square* modulo n, if there exists an $x \in \mathbb{Z}_n^*$ such that $x^2 \equiv a \pmod{n}$. If no such x exists, then a is called a *quadratic non-residue* modulo n. The set of all quadratic residues modulo n is denoted by Q_n and the set of all quadratic non-residues is denoted by $\overline{Q}_n$.

    Note that by definition $0 \notin \mathbb{Z}_n^*$, whence $0 \notin Q_n$ and $0 \notin \overline{Q}_n$.

2.135 Fact Let p be an odd prime and let α be a generator of $\mathbb{Z}_p^*$. Then $a \in \mathbb{Z}_p^*$ is a quadratic residue modulo p if and only if $a = \alpha^i \bmod p$, where i is an even integer. It follows that $|Q_p| = (p-1)/2$ and $|\overline{Q}_p| = (p-1)/2$; that is, half of the elements in $\mathbb{Z}_p^*$ are quadratic residues and the other half are quadratic non-residues.

2.136 Example $\alpha = 6$ is a generator of $\mathbb{Z}_{13}^*$. The powers of α are listed in the following table.

i	0	1	2	3	4	5	6	7	8	9	10	11
$\alpha^i \bmod 13$	1	6	10	8	9	2	12	7	3	5	4	11

Hence $Q_{13} = \{1, 3, 4, 9, 10, 12\}$ and $\overline{Q}_{13} = \{2, 5, 6, 7, 8, 11\}$. $\qquad\square$

2.137 Fact Let n be a product of two distinct odd primes p and q, $n = pq$. Then $a \in \mathbb{Z}_n^*$ is a quadratic residue modulo n if and only if $a \in Q_p$ and $a \in Q_q$. It follows that $|Q_n| = |Q_p| \cdot |Q_q| = (p-1)(q-1)/4$ and $|\overline{Q}_n| = 3(p-1)(q-1)/4$.

2.138 Example Let $n = 21$. Then $Q_{21} = \{1, 4, 16\}$ and $\overline{Q}_{21} = \{2, 5, 8, 10, 11, 13, 17, 19, 20\}$. $\qquad\square$

2.139 Definition Let $a \in Q_n$. If $x \in \mathbb{Z}_n^*$ satisfies $x^2 \equiv a \pmod{n}$, then x is called a *square root* of a modulo n.

2.140 Fact (*number of square roots*)

    (i) If p is an odd prime and $a \in Q_p$, then a has exactly two square roots modulo p.

    (ii) More generally, let $n = p_1^{e_1} p_2^{e_2} \cdots p_k^{e_k}$ where the p_i are distinct odd primes and $e_i \geq 1$. If $a \in Q_n$, then a has precisely 2^k distinct square roots modulo n.

2.141 Example The square roots of 12 modulo 37 are 7 and 30. The square roots of 121 modulo 315 are 11, 74, 101, 151, 164, 214, 241, and 304. $\qquad\square$

2.4.4 Algorithms in $\mathbb{Z}_n$

Let n be a positive integer. As before, the elements of $\mathbb{Z}_n$ will be represented by the integers $\{0, 1, 2, \ldots, n-1\}$.

Observe that if $a, b \in \mathbb{Z}_n$, then

$$(a+b) \bmod n = \begin{cases} a+b, & \text{if } a+b < n, \\ a+b-n, & \text{if } a+b \geq n. \end{cases}$$

Hence modular addition (and subtraction) can be performed without the need of a long division. Modular multiplication of a and b may be accomplished by simply multiplying a and b as integers, and then taking the remainder of the result after division by n. Inverses in $\mathbb{Z}_n$ can be computed using the extended Euclidean algorithm as next described.

2.142 Algorithm Computing multiplicative inverses in $\mathbb{Z}_n$

INPUT: $a \in \mathbb{Z}_n$.
OUTPUT: $a^{-1} \bmod n$, provided that it exists.

1. Use the extended Euclidean algorithm (Algorithm 2.107) to find integers x and y such that $ax + ny = d$, where $d = \gcd(a, n)$.
2. If $d > 1$, then $a^{-1} \bmod n$ does not exist. Otherwise, return(x).

Modular exponentiation can be performed efficiently with the repeated square-and-multiply algorithm (Algorithm 2.143), which is crucial for many cryptographic protocols. One version of this algorithm is based on the following observation. Let the binary representation of k be $\sum_{i=0}^{t} k_i 2^i$, where each $k_i \in \{0, 1\}$. Then

$$a^k = \prod_{i=0}^{t} a^{k_i 2^i} = (a^{2^0})^{k_0} (a^{2^1})^{k_1} \cdots (a^{2^t})^{k_t}.$$

2.143 Algorithm Repeated square-and-multiply algorithm for exponentiation in $\mathbb{Z}_n$

INPUT: $a \in \mathbb{Z}_n$, and integer $0 \leq k < n$ whose binary representation is $k = \sum_{i=0}^{t} k_i 2^i$.
OUTPUT: $a^k \bmod n$.

1. Set $b \leftarrow 1$. If $k = 0$ then return(b).
2. Set $A \leftarrow a$.
3. If $k_0 = 1$ then set $b \leftarrow a$.
4. For i from 1 to t do the following:
   4.1 Set $A \leftarrow A^2 \bmod n$.
   4.2 If $k_i = 1$ then set $b \leftarrow A \cdot b \bmod n$.
5. Return(b).

2.144 Example (*modular exponentiation*) Table 2.4 shows the steps involved in the computation of $5^{596} \bmod 1234 = 1013$. $\square$

The number of bit operations for the basic operations in $\mathbb{Z}_n$ is summarized in Table 2.5. Efficient algorithms for performing modular multiplication and exponentiation are further examined in §14.3 and §14.6.

i	0	1	2	3	4	5	6	7	8	9
k_i	0	0	1	0	1	0	1	0	0	1
A	5	25	625	681	1011	369	421	779	947	925
b	1	1	625	625	67	67	1059	1059	1059	1013

Table 2.4: *Computation of 5^{596} mod 1234.*

Operation		Bit complexity
Modular addition	$(a + b) \bmod n$	$O(\lg n)$
Modular subtraction	$(a - b) \bmod n$	$O(\lg n)$
Modular multiplication	$(a \cdot b) \bmod n$	$O((\lg n)^2)$
Modular inversion	$a^{-1} \bmod n$	$O((\lg n)^2)$
Modular exponentiation	$a^k \bmod n, k < n$	$O((\lg n)^3)$

Table 2.5: *Bit complexity of basic operations in $\mathbb{Z}_n$.*

2.4.5 The Legendre and Jacobi symbols

The Legendre symbol is a useful tool for keeping track of whether or not an integer a is a quadratic residue modulo a prime p.

2.145 Definition Let p be an odd prime and a an integer. The *Legendre symbol* $\left(\frac{a}{p}\right)$ is defined to be

$$\left(\frac{a}{p}\right) = \begin{cases} 0, & \text{if } p|a, \\ 1, & \text{if } a \in Q_p, \\ -1, & \text{if } a \in \overline{Q}_p. \end{cases}$$

2.146 Fact (*properties of Legendre symbol*) Let p be an odd prime and $a, b \in \mathbb{Z}$. Then the Legendre symbol has the following properties:

(i) $\left(\frac{a}{p}\right) \equiv a^{(p-1)/2} \pmod{p}$. In particular, $\left(\frac{1}{p}\right) = 1$ and $\left(\frac{-1}{p}\right) = (-1)^{(p-1)/2}$. Hence $-1 \in Q_p$ if $p \equiv 1 \pmod 4$, and $-1 \in \overline{Q}_p$ if $p \equiv 3 \pmod 4$.

(ii) $\left(\frac{ab}{p}\right) = \left(\frac{a}{p}\right)\left(\frac{b}{p}\right)$. Hence if $a \in \mathbb{Z}_p^*$, then $\left(\frac{a^2}{p}\right) = 1$.

(iii) If $a \equiv b \pmod{p}$, then $\left(\frac{a}{p}\right) = \left(\frac{b}{p}\right)$.

(iv) $\left(\frac{2}{p}\right) = (-1)^{(p^2-1)/8}$. Hence $\left(\frac{2}{p}\right) = 1$ if $p \equiv 1$ or $7 \pmod 8$, and $\left(\frac{2}{p}\right) = -1$ if $p \equiv 3$ or $5 \pmod 8$.

(v) (*law of quadratic reciprocity*) If q is an odd prime distinct from p, then

$$\left(\frac{p}{q}\right) = \left(\frac{q}{p}\right)(-1)^{(p-1)(q-1)/4}.$$

In other words, $\left(\frac{p}{q}\right) = \left(\frac{q}{p}\right)$ unless both p and q are congruent to 3 modulo 4, in which case $\left(\frac{p}{q}\right) = -\left(\frac{q}{p}\right)$.

The Jacobi symbol is a generalization of the Legendre symbol to integers n which are odd but not necessarily prime.

2.147 Definition Let $n \geq 3$ be odd with prime factorization $n = p_1^{e_1} p_2^{e_2} \cdots p_k^{e_k}$. Then the *Jacobi symbol* $\left(\frac{a}{n}\right)$ is defined to be

$$\left(\frac{a}{n}\right) = \left(\frac{a}{p_1}\right)^{e_1} \left(\frac{a}{p_2}\right)^{e_2} \cdots \left(\frac{a}{p_k}\right)^{e_k}.$$

Observe that if n is prime, then the Jacobi symbol is just the Legendre symbol.

2.148 Fact (*properties of Jacobi symbol*) Let $m \geq 3$, $n \geq 3$ be odd integers, and $a, b \in \mathbb{Z}$. Then the Jacobi symbol has the following properties:

(i) $\left(\frac{a}{n}\right) = 0, 1,$ or -1. Moreover, $\left(\frac{a}{n}\right) = 0$ if and only if $\gcd(a, n) \neq 1$.

(ii) $\left(\frac{ab}{n}\right) = \left(\frac{a}{n}\right)\left(\frac{b}{n}\right)$. Hence if $a \in \mathbb{Z}_n^*$, then $\left(\frac{a^2}{n}\right) = 1$.

(iii) $\left(\frac{a}{mn}\right) = \left(\frac{a}{m}\right)\left(\frac{a}{n}\right)$.

(iv) If $a \equiv b \pmod{n}$, then $\left(\frac{a}{n}\right) = \left(\frac{b}{n}\right)$.

(v) $\left(\frac{1}{n}\right) = 1$.

(vi) $\left(\frac{-1}{n}\right) = (-1)^{(n-1)/2}$. Hence $\left(\frac{-1}{n}\right) = 1$ if $n \equiv 1 \pmod 4$, and $\left(\frac{-1}{n}\right) = -1$ if $n \equiv 3 \pmod 4$.

(vii) $\left(\frac{2}{n}\right) = (-1)^{(n^2-1)/8}$. Hence $\left(\frac{2}{n}\right) = 1$ if $n \equiv 1$ or $7 \pmod 8$, and $\left(\frac{2}{n}\right) = -1$ if $n \equiv 3$ or $5 \pmod 8$.

(viii) $\left(\frac{m}{n}\right) = \left(\frac{n}{m}\right)(-1)^{(m-1)(n-1)/4}$. In other words, $\left(\frac{m}{n}\right) = \left(\frac{n}{m}\right)$ unless both m and n are congruent to 3 modulo 4, in which case $\left(\frac{m}{n}\right) = -\left(\frac{n}{m}\right)$.

By properties of the Jacobi symbol it follows that if n is odd and $a = 2^e a_1$ where a_1 is odd, then

$$\left(\frac{a}{n}\right) = \left(\frac{2^e}{n}\right)\left(\frac{a_1}{n}\right) = \left(\frac{2}{n}\right)^e \left(\frac{n \bmod a_1}{a_1}\right)(-1)^{(a_1-1)(n-1)/4}.$$

This observation yields the following recursive algorithm for computing $\left(\frac{a}{n}\right)$, which does not require the prime factorization of n.

2.149 Algorithm Jacobi symbol (and Legendre symbol) computation

---

JACOBI(a,n)
INPUT: an odd integer $n \geq 3$, and an integer a, $0 \leq a < n$.
OUTPUT: the Jacobi symbol $\left(\frac{a}{n}\right)$ (and hence the Legendre symbol when n is prime).

1. If $a = 0$ then return(0).
2. If $a = 1$ then return(1).
3. Write $a = 2^e a_1$, where a_1 is odd.
4. If e is even then set $s \leftarrow 1$. Otherwise set $s \leftarrow 1$ if $n \equiv 1$ or $7 \pmod 8$, or set $s \leftarrow -1$ if $n \equiv 3$ or $5 \pmod 8$.
5. If $n \equiv 3 \pmod 4$ and $a_1 \equiv 3 \pmod 4$ then set $s \leftarrow -s$.
6. Set $n_1 \leftarrow n \bmod a_1$.
7. Return($s \cdot$ JACOBI(n_1, a_1)).

2.150 Fact Algorithm 2.149 has a running time of $O((\lg n)^2)$ bit operations.

2.151 Remark (*finding quadratic non-residues modulo a prime p*) Let p denote an odd prime. Even though it is known that half of the elements in $\mathbb{Z}_p^*$ are quadratic non-residues modulo p (see Fact 2.135), there is no *deterministic* polynomial-time algorithm known for finding one. A *randomized* algorithm for finding a quadratic non-residue is to simply select random integers $a \in \mathbb{Z}_p^*$ until one is found satisfying $\left(\frac{a}{p}\right) = -1$. The expected number iterations before a non-residue is found is 2, and hence the procedure takes expected polynomial-time.

2.152 Example (*Jacobi symbol computation*) For $a = 158$ and $n = 235$, Algorithm 2.149 computes the Jacobi symbol $\left(\frac{158}{235}\right)$ as follows:

$$\left(\frac{158}{235}\right) = \left(\frac{2}{235}\right)\left(\frac{79}{235}\right) = (-1)\left(\frac{235}{79}\right)(-1)^{78 \cdot 234/4} = \left(\frac{77}{79}\right)$$

$$= \left(\frac{79}{77}\right)(-1)^{76 \cdot 78/4} = \left(\frac{2}{77}\right) = -1. \qquad \square$$

Unlike the Legendre symbol, the Jacobi symbol $\left(\frac{a}{n}\right)$ does not reveal whether or not a is a quadratic residue modulo n. It is indeed true that if $a \in Q_n$, then $\left(\frac{a}{n}\right) = 1$. However, $\left(\frac{a}{n}\right) = 1$ does not imply that $a \in Q_n$.

2.153 Example (*quadratic residues and non-residues*) Table 2.6 lists the elements in $\mathbb{Z}_{21}^*$ and their Jacobi symbols. Recall from Example 2.138 that $Q_{21} = \{1, 4, 16\}$. Observe that $\left(\frac{5}{21}\right) = 1$ but $5 \notin Q_{21}$. $\qquad \square$

$a \in \mathbb{Z}_{21}^*$	1	2	4	5	8	10	11	13	16	17	19	20
$a^2 \bmod n$	1	4	16	4	1	16	16	1	4	16	4	1
$\left(\frac{a}{3}\right)$	1	-1	1	-1	-1	1	-1	1	1	-1	1	-1
$\left(\frac{a}{7}\right)$	1	1	1	-1	1	-1	1	-1	1	-1	-1	-1
$\left(\frac{a}{21}\right)$	1	-1	1	1	-1	-1	-1	-1	1	1	-1	1

Table 2.6: *Jacobi symbols of elements in $\mathbb{Z}_{21}^*$.*

2.154 Definition Let $n \geq 3$ be an odd integer, and let $J_n = \{a \in \mathbb{Z}_n^* \mid \left(\frac{a}{n}\right) = 1\}$. The set of *pseudosquares* modulo n, denoted $\widetilde{Q}_n$, is defined to be the set $J_n - Q_n$.

2.155 Fact Let $n = pq$ be a product of two distinct odd primes. Then $|Q_n| = |\widetilde{Q}_n| = (p-1)(q-1)/4$; that is, half of the elements in J_n are quadratic residues and the other half are pseudosquares.

2.4.6 Blum integers

2.156 Definition A *Blum integer* is a composite integer of the form $n = pq$, where p and q are distinct primes each congruent to 3 modulo 4.

2.157 Fact Let $n = pq$ be a Blum integer, and let $a \in Q_n$. Then a has precisely four square roots modulo n, exactly one of which is also in Q_n.

2.158 Definition Let n be a Blum integer and let $a \in Q_n$. The unique square root of a in Q_n is called the *principal square root* of a modulo n.

2.159 Example (*Blum integer*) For the Blum integer $n = 21$, $J_n = \{1, 4, 5, 16, 17, 20\}$ and $\widetilde{Q}_n = \{5, 17, 20\}$. The four square roots of $a = 4$ are 2, 5, 16, and 19, of which only 16 is also in Q_{21}. Thus 16 is the principal square root of 4 modulo 21. $\square$

2.160 Fact If $n = pq$ is a Blum integer, then the function $f : Q_n \longrightarrow Q_n$ defined by $f(x) = x^2 \bmod n$ is a permutation. The inverse function of f is:

$$f^{-1}(x) = x^{((p-1)(q-1)+4)/8} \bmod n.$$

2.5 Abstract algebra

This section provides an overview of basic algebraic objects and their properties, for reference in the remainder of this handbook. Several of the definitions in §2.5.1 and §2.5.2 were presented earlier in §2.4.3 in the more concrete setting of the algebraic structure $\mathbb{Z}_n^*$.

2.161 Definition A *binary operation* $*$ on a set S is a mapping from $S \times S$ to S. That is, $*$ is a rule which assigns to each ordered pair of elements from S an element of S.

2.5.1 Groups

2.162 Definition A *group* $(G, *)$ consists of a set G with a binary operation $*$ on G satisfying the following three axioms.

(i) The group operation is *associative*. That is, $a * (b * c) = (a * b) * c$ for all $a, b, c \in G$.

(ii) There is an element $1 \in G$, called the *identity element*, such that $a * 1 = 1 * a = a$ for all $a \in G$.

(iii) For each $a \in G$ there exists an element $a^{-1} \in G$, called the *inverse* of a, such that $a * a^{-1} = a^{-1} * a = 1$.

A group G is *abelian* (or *commutative*) if, furthermore,

(iv) $a * b = b * a$ for all $a, b \in G$.

Note that multiplicative group notation has been used for the group operation. If the group operation is addition, then the group is said to be an *additive* group, the identity element is denoted by 0, and the inverse of a is denoted $-a$.

Henceforth, unless otherwise stated, the symbol $*$ will be omitted and the group operation will simply be denoted by juxtaposition.

2.163 Definition A group G is *finite* if $|G|$ is finite. The number of elements in a finite group is called its *order*.

2.164 Example The set of integers $\mathbb{Z}$ with the operation of addition forms a group. The identity element is 0 and the inverse of an integer a is the integer $-a$. $\square$

2.165 Example The set $\mathbb{Z}_n$, with the operation of addition modulo n, forms a group of order n. The set $\mathbb{Z}_n$ with the operation of multiplication modulo n is not a group, since not all elements have multiplicative inverses. However, the set $\mathbb{Z}_n^*$ (see Definition 2.124) is a group of order $\phi(n)$ under the operation of multiplication modulo n, with identity element 1. $\square$

2.166 Definition A non-empty subset H of a group G is a *subgroup* of G if H is itself a group with respect to the operation of G. If H is a subgroup of G and $H \neq G$, then H is called a *proper* subgroup of G.

2.167 Definition A group G is *cyclic* if there is an element $\alpha \in G$ such that for each $b \in G$ there is an integer i with $b = \alpha^i$. Such an element α is called a *generator* of G.

2.168 Fact If G is a group and $a \in G$, then the set of all powers of a forms a cyclic subgroup of G, called the subgroup *generated by* a, and denoted by $\langle a \rangle$.

2.169 Definition Let G be a group and $a \in G$. The *order* of a is defined to be the least positive integer t such that $a^t = 1$, provided that such an integer exists. If such a t does not exist, then the order of a is defined to be ∞.

2.170 Fact Let G be a group, and let $a \in G$ be an element of finite order t. Then $|\langle a \rangle|$, the size of the subgroup generated by a, is equal to t.

2.171 Fact (*Lagrange's theorem*) If G is a finite group and H is a subgroup of G, then $|H|$ divides $|G|$. Hence, if $a \in G$, the order of a divides $|G|$.

2.172 Fact Every subgroup of a cyclic group G is also cyclic. In fact, if G is a cyclic group of order n, then for each positive divisor d of n, G contains exactly one subgroup of order d.

2.173 Fact Let G be a group.

   (i) If the order of $a \in G$ is t, then the order of a^k is $t/\gcd(t,k)$.

  (ii) If G is a cyclic group of order n and $d|n$, then G has exactly $\phi(d)$ elements of order d. In particular, G has $\phi(n)$ generators.

2.174 Example Consider the multiplicative group $\mathbb{Z}_{19}^* = \{1, 2, \ldots, 18\}$ of order 18. The group is cyclic (Fact 2.132(i)), and a generator is $\alpha = 2$. The subgroups of $\mathbb{Z}_{19}^*$, and their generators, are listed in Table 2.7. $\square$

Subgroup	Generators	Order
$\{1\}$	1	1
$\{1, 18\}$	18	2
$\{1, 7, 11\}$	7, 11	3
$\{1, 7, 8, 11, 12, 18\}$	8, 12	6
$\{1, 4, 5, 6, 7, 9, 11, 16, 17\}$	4, 5, 6, 9, 16, 17	9
$\{1, 2, 3, \ldots, 18\}$	2, 3, 10, 13, 14, 15	18

Table 2.7: *The subgroups of $\mathbb{Z}_{19}^*$.*

2.5.2 Rings

2.175 Definition A *ring* $(R, +, \times)$ consists of a set R with two binary operations arbitrarily denoted $+$ (addition) and $\times$ (multiplication) on R, satisfying the following axioms.

   (i) $(R, +)$ is an abelian group with identity denoted 0.

(ii) The operation $\times$ is associative. That is, $a \times (b \times c) = (a \times b) \times c$ for all $a, b, c \in R$.

(iii) There is a multiplicative identity denoted 1, with $1 \neq 0$, such that $1 \times a = a \times 1 = a$ for all $a \in R$.

(iv) The operation $\times$ is *distributive* over $+$. That is, $a \times (b+c) = (a \times b) + (a \times c)$ and $(b+c) \times a = (b \times a) + (c \times a)$ for all $a, b, c \in R$.

The ring is a *commutative ring* if $a \times b = b \times a$ for all $a, b \in R$.

2.176 Example The set of integers $\mathbb{Z}$ with the usual operations of addition and multiplication is a commutative ring. $\square$

2.177 Example The set $\mathbb{Z}_n$ with addition and multiplication performed modulo n is a commutative ring. $\square$

2.178 Definition An element a of a ring R is called a *unit* or an *invertible element* if there is an element $b \in R$ such that $a \times b = 1$.

2.179 Fact The set of units in a ring R forms a group under multiplication, called the *group of units* of R.

2.180 Example The group of units of the ring $\mathbb{Z}_n$ is $\mathbb{Z}_n^*$ (see Definition 2.124). $\square$

2.5.3 Fields

2.181 Definition A *field* is a commutative ring in which all non-zero elements have multiplicative inverses.

2.182 Definition The *characteristic* of a field is 0 if $\overbrace{1 + 1 + \cdots + 1}^{m \text{ times}}$ is never equal to 0 for any $m \geq 1$. Otherwise, the characteristic of the field is the least positive integer m such that $\sum_{i=1}^{m} 1$ equals 0.

2.183 Example The set of integers under the usual operations of addition and multiplication is not a field, since the only non-zero integers with multiplicative inverses are 1 and -1. However, the rational numbers $\mathbb{Q}$, the real numbers $\mathbb{R}$, and the complex numbers $\mathbb{C}$ form fields of characteristic 0 under the usual operations. $\square$

2.184 Fact $\mathbb{Z}_n$ is a field (under the usual operations of addition and multiplication modulo n) if and only if n is a prime number. If n is prime, then $\mathbb{Z}_n$ has characteristic n.

2.185 Fact If the characteristic m of a field is not 0, then m is a prime number.

2.186 Definition A subset F of a field E is a *subfield* of E if F is itself a field with respect to the operations of E. If this is the case, E is said to be an *extension field* of F.

2.5.4 Polynomial rings

2.187 Definition If R is a commutative ring, then a *polynomial* in the indeterminate x over the ring R is an expression of the form

$$f(x) = a_n x^n + \cdots + a_2 x^2 + a_1 x + a_0$$

where each $a_i \in R$ and $n \geq 0$. The element a_i is called the *coefficient* of x^i in $f(x)$. The largest integer m for which $a_m \neq 0$ is called the *degree* of $f(x)$, denoted $\deg f(x)$; a_m is called the *leading coefficient* of $f(x)$. If $f(x) = a_0$ (a *constant polynomial*) and $a_0 \neq 0$, then $f(x)$ has degree 0. If all the coefficients of $f(x)$ are 0, then $f(x)$ is called the *zero polynomial* and its degree, for mathematical convenience, is defined to be $-\infty$. The polynomial $f(x)$ is said to be *monic* if its leading coefficient is equal to 1.

2.188 Definition If R is a commutative ring, the *polynomial ring* $R[x]$ is the ring formed by the set of all polynomials in the indeterminate x having coefficients from R. The two operations are the standard polynomial addition and multiplication, with coefficient arithmetic performed in the ring R.

2.189 Example (*polynomial ring*) Let $f(x) = x^3 + x + 1$ and $g(x) = x^2 + x$ be elements of the polynomial ring $\mathbb{Z}_2[x]$. Working in $\mathbb{Z}_2[x]$,

$$f(x) + g(x) = x^3 + x^2 + 1$$

and

$$f(x) \cdot g(x) = x^5 + x^4 + x^3 + x. \qquad \square$$

For the remainder of this section, F will denote an arbitrary field. The polynomial ring $F[x]$ has many properties in common with the integers (more precisely, $F[x]$ and $\mathbb{Z}$ are both *Euclidean domains*, however, this generalization will not be pursued here). These similarities are investigated further.

2.190 Definition Let $f(x) \in F[x]$ be a polynomial of degree at least 1. Then $f(x)$ is said to be *irreducible over* F if it cannot be written as the product of two polynomials in $F[x]$, each of positive degree.

2.191 Definition (*division algorithm for polynomials*) If $g(x), h(x) \in F[x]$, with $h(x) \neq 0$, then ordinary polynomial long division of $g(x)$ by $h(x)$ yields polynomials $q(x)$ and $r(x) \in F[x]$ such that

$$g(x) = q(x)h(x) + r(x), \text{ where } \deg r(x) < \deg h(x).$$

Moreover, $q(x)$ and $r(x)$ are unique. The polynomial $q(x)$ is called the *quotient*, while $r(x)$ is called the *remainder*. The remainder of the division is sometimes denoted $g(x) \bmod h(x)$, and the quotient is sometimes denoted $g(x) \operatorname{div} h(x)$ (cf. Definition 2.82).

2.192 Example (*polynomial division*) Consider the polynomials $g(x) = x^6 + x^5 + x^3 + x^2 + x + 1$ and $h(x) = x^4 + x^3 + 1$ in $\mathbb{Z}_2[x]$. Polynomial long division of $g(x)$ by $h(x)$ yields

$$g(x) = x^2 h(x) + (x^3 + x + 1).$$

Hence $g(x) \bmod h(x) = x^3 + x + 1$ and $g(x) \operatorname{div} h(x) = x^2$. $\qquad \square$

2.193 Definition If $g(x), h(x) \in F[x]$ then $h(x)$ *divides* $g(x)$, written $h(x)|g(x)$, if $g(x) \bmod h(x) = 0$.

Let $f(x)$ be a fixed polynomial in $F[x]$. As with the integers (Definition 2.110), one can define congruences of polynomials in $F[x]$ based on division by $f(x)$.

2.194 Definition If $g(x), h(x) \in F[x]$, then $g(x)$ is said to be *congruent to $h(x)$ modulo $f(x)$* if $f(x)$ divides $g(x) - h(x)$. This is denoted by $g(x) \equiv h(x) \pmod{f(x)}$.

2.195 Fact (*properties of congruences*) For all $g(x), h(x), g_1(x), h_1(x), s(x) \in F[x]$, the following are true.

(i) $g(x) \equiv h(x) \pmod{f(x)}$ if and only if $g(x)$ and $h(x)$ leave the same remainder upon division by $f(x)$.

(ii) (*reflexivity*) $g(x) \equiv g(x) \pmod{f(x)}$.

(iii) (*symmetry*) If $g(x) \equiv h(x) \pmod{f(x)}$, then $h(x) \equiv g(x) \pmod{f(x)}$.

(iv) (*transitivity*) If $g(x) \equiv h(x) \pmod{f(x)}$ and $h(x) \equiv s(x) \pmod{f(x)}$, then $g(x) \equiv s(x) \pmod{f(x)}$.

(v) If $g(x) \equiv g_1(x) \pmod{f(x)}$ and $h(x) \equiv h_1(x) \pmod{f(x)}$, then $g(x) + h(x) \equiv g_1(x) + h_1(x) \pmod{f(x)}$ and $g(x)h(x) \equiv g_1(x)h_1(x) \pmod{f(x)}$.

Let $f(x)$ be a fixed polynomial in $F[x]$. The *equivalence class* of a polynomial $g(x) \in F[x]$ is the set of all polynomials in $F[x]$ congruent to $g(x)$ modulo $f(x)$. From properties (ii), (iii), and (iv) above, it can be seen that the relation of congruence modulo $f(x)$ partitions $F[x]$ into equivalence classes. If $g(x) \in F[x]$, then long division by $f(x)$ yields unique polynomials $q(x), r(x) \in F[x]$ such that $g(x) = q(x)f(x) + r(x)$, where $\deg r(x) < \deg f(x)$. Hence every polynomial $g(x)$ is congruent modulo $f(x)$ to a unique polynomial of degree less than $\deg f(x)$. The polynomial $r(x)$ will be used as representative of the equivalence class of polynomials containing $g(x)$.

2.196 Definition $F[x]/(f(x))$ denotes the set of (equivalence classes of) polynomials in $F[x]$ of degree less than $n = \deg f(x)$. Addition and multiplication are performed modulo $f(x)$.

2.197 Fact $F[x]/(f(x))$ is a commutative ring.

2.198 Fact If $f(x)$ is irreducible over F, then $F[x]/(f(x))$ is a field.

2.5.5 Vector spaces

2.199 Definition A *vector space* V over a field F is an abelian group $(V, +)$, together with a multiplication operation $\bullet : F \times V \longrightarrow V$ (usually denoted by juxtaposition) such that for all $a, b \in F$ and $v, w \in V$, the following axioms are satisfied.

(i) $a(v + w) = av + aw$.

(ii) $(a + b)v = av + bv$.

(iii) $(ab)v = a(bv)$.

(iv) $1v = v$.

The elements of V are called *vectors*, while the elements of F are called *scalars*. The group operation $+$ is called *vector addition*, while the multiplication operation is called *scalar multiplication*.

2.200 Definition Let V be a vector space over a field F. A *subspace* of V is an additive subgroup U of V which is closed under scalar multiplication, i.e., $av \in U$ for all $a \in F$ and $v \in U$.

2.201 Fact A subspace of a vector space is also a vector space.

2.202 Definition Let $S = \{v_1, v_2, \ldots, v_n\}$ be a finite subset of a vector space V over a field F.

   (i) A *linear combination* of S is an expression of the form $a_1v_1 + a_2v_2 + \cdots + a_nv_n$, where each $a_i \in F$.

   (ii) The *span* of S, denoted $\langle S \rangle$, is the set of all linear combinations of S. The span of S is a subspace of V.

   (iii) If U is a subspace of V, then S is said to *span* U if $\langle S \rangle = U$.

   (iv) The set S is *linearly dependent* over F if there exist scalars $a_1, a_2, \ldots, a_n$, not all zero, such that $a_1v_1 + a_2v_2 + \cdots + a_nv_n = 0$. If no such scalars exist, then S is *linearly independent* over F.

   (v) A linearly independent set of vectors that spans V is called a *basis* for V.

2.203 Fact Let V be a vector space.

   (i) If V has a finite spanning set, then it has a basis.

   (ii) If V has a basis, then in fact all bases have the same number of elements.

2.204 Definition If a vector space V has a basis, then the number of elements in a basis is called the *dimension* of V, denoted $\dim V$.

2.205 Example If F is any field, then the n-fold Cartesian product $V = F \times F \times \cdots \times F$ is a vector space over F of dimension n. The *standard basis* for V is $\{e_1, e_2, \ldots, e_n\}$, where e_i is a vector with a 1 in the ith coordinate and 0's elsewhere. $\quad\square$

2.206 Definition Let E be an extension field of F. Then E can be viewed as a vector space over the subfield F, where vector addition and scalar multiplication are simply the field operations of addition and multiplication in E. The dimension of this vector space is called the *degree* of E over F, and denoted by $[E : F]$. If this degree is finite, then E is called a *finite extension* of F.

2.207 Fact Let F, E, and L be fields. If L is a finite extension of E and E is a finite extension of F, then L is also a finite extension of F and

$$[L : F] = [L : E][E : F].$$

2.6 Finite fields

2.6.1 Basic properties

2.208 Definition A *finite field* is a field F which contains a finite number of elements. The *order* of F is the number of elements in F.

2.209 Fact (*existence and uniqueness of finite fields*)

    (i) If F is a finite field, then F contains p^m elements for some prime p and integer $m \geq 1$.

    (ii) For every prime power order p^m, there is a unique (up to isomorphism) finite field of order p^m. This field is denoted by $\mathbb{F}_{p^m}$, or sometimes by $GF(p^m)$.

Informally speaking, two fields are *isomorphic* if they are structurally the same, although the representation of their field elements may be different. Note that if p is a prime then $\mathbb{Z}_p$ is a field, and hence every field of order p is isomorphic to $\mathbb{Z}_p$. Unless otherwise stated, the finite field $\mathbb{F}_p$ will henceforth be identified with $\mathbb{Z}_p$.

2.210 Fact If $\mathbb{F}_q$ is a finite field of order $q = p^m$, p a prime, then the characteristic of $\mathbb{F}_q$ is p. Moreover, $\mathbb{F}_q$ contains a copy of $\mathbb{Z}_p$ as a subfield. Hence $\mathbb{F}_q$ can be viewed as an extension field of $\mathbb{Z}_p$ of degree m.

2.211 Fact (*subfields of a finite field*) Let $\mathbb{F}_q$ be a finite field of order $q = p^m$. Then every subfield of $\mathbb{F}_q$ has order p^n, for some n that is a positive divisor of m. Conversely, if n is a positive divisor of m, then there is exactly one subfield of $\mathbb{F}_q$ of order p^n; an element $a \in \mathbb{F}_q$ is in the subfield $\mathbb{F}_{p^n}$ if and only if $a^{p^n} = a$.

2.212 Definition The non-zero elements of $\mathbb{F}_q$ form a group under multiplication called the *multiplicative group* of $\mathbb{F}_q$, denoted by $\mathbb{F}_q^*$.

2.213 Fact $\mathbb{F}_q^*$ is a cyclic group of order $q - 1$. Hence $a^q = a$ for all $a \in \mathbb{F}_q$.

2.214 Definition A generator of the cyclic group $\mathbb{F}_q^*$ is called a *primitive element* or *generator* of $\mathbb{F}_q$.

2.215 Fact If $a, b \in \mathbb{F}_q$, a finite field of characteristic p, then

$$(a + b)^{p^t} = a^{p^t} + b^{p^t} \text{ for all } t \geq 0.$$

2.6.2 The Euclidean algorithm for polynomials

Let $\mathbb{Z}_p$ be the finite field of order p. The theory of greatest common divisors and the Euclidean algorithm for integers carries over in a straightforward manner to the polynomial ring $\mathbb{Z}_p[x]$ (and more generally to the polynomial ring $F[x]$, where F is any field).

2.216 Definition Let $g(x), h(x) \in \mathbb{Z}_p[x]$, where not both are 0. Then the *greatest common divisor* of $g(x)$ and $h(x)$, denoted $\gcd(g(x), h(x))$, is the monic polynomial of greatest degree in $\mathbb{Z}_p[x]$ which divides both $g(x)$ and $h(x)$. By definition, $\gcd(0, 0) = 0$.

2.217 Fact $\mathbb{Z}_p[x]$ is a *unique factorization domain*. That is, every non-zero polynomial $f(x) \in \mathbb{Z}_p[x]$ has a factorization

$$f(x) = a f_1(x)^{e_1} f_2(x)^{e_2} \cdots f_k(x)^{e_k},$$

where the $f_i(x)$ are distinct monic irreducible polynomials in $\mathbb{Z}_p[x]$, the e_i are positive integers, and $a \in \mathbb{Z}_p$. Furthermore, the factorization is unique up to rearrangement of factors.

The following is the polynomial version of the Euclidean algorithm (cf. Algorithm 2.104).

2.218 Algorithm Euclidean algorithm for $\mathbb{Z}_p[x]$

INPUT: two polynomials $g(x), h(x) \in \mathbb{Z}_p[x]$.
OUTPUT: the greatest common divisor of $g(x)$ and $h(x)$.
1. While $h(x) \neq 0$ do the following:
    1.1 Set $r(x) \leftarrow g(x) \bmod h(x)$, $g(x) \leftarrow h(x)$, $h(x) \leftarrow r(x)$.
2. Return($g(x)$).

2.219 Definition A $\mathbb{Z}_p$-*operation* means either an addition, subtraction, multiplication, inversion, or division in $\mathbb{Z}_p$.

2.220 Fact Suppose that $\deg g(x) \leq m$ and $\deg h(x) \leq m$. Then Algorithm 2.218 has a running time of $O(m^2)$ $\mathbb{Z}_p$-operations, or equivalently, $O(m^2 (\lg p)^2)$ bit operations.

As with the case of the integers (cf. Algorithm 2.107), the Euclidean algorithm can be extended so that it also yields two polynomials $s(x)$ and $t(x)$ satisfying

$$s(x)g(x) + t(x)h(x) = \gcd(g(x), h(x)).$$

2.221 Algorithm Extended Euclidean algorithm for $\mathbb{Z}_p[x]$

INPUT: two polynomials $g(x), h(x) \in \mathbb{Z}_p[x]$.
OUTPUT: $d(x) = \gcd(g(x), h(x))$ and polynomials $s(x), t(x) \in \mathbb{Z}_p[x]$ which satisfy $s(x)g(x) + t(x)h(x) = d(x)$.
1. If $h(x) = 0$ then set $d(x) \leftarrow g(x)$, $s(x) \leftarrow 1$, $t(x) \leftarrow 0$, and return($d(x),s(x),t(x)$).
2. Set $s_2(x) \leftarrow 1$, $s_1(x) \leftarrow 0$, $t_2(x) \leftarrow 0$, $t_1(x) \leftarrow 1$.
3. While $h(x) \neq 0$ do the following:
    3.1 $q(x) \leftarrow g(x) \operatorname{div} h(x)$, $r(x) \leftarrow g(x) - h(x)q(x)$,
    3.2 $s(x) \leftarrow s_2(x) - q(x)s_1(x)$, $t(x) \leftarrow t_2(x) - q(x)t_1(x)$.
    3.3 $g(x) \leftarrow h(x)$, $h(x) \leftarrow r(x)$,
    3.4 $s_2(x) \leftarrow s_1(x)$, $s_1(x) \leftarrow s(x)$, $t_2(x) \leftarrow t_1(x)$, and $t_1(x) \leftarrow t(x)$.
4. Set $d(x) \leftarrow g(x)$, $s(x) \leftarrow s_2(x)$, $t(x) \leftarrow t_2(x)$.
5. Return($d(x),s(x),t(x)$).

2.222 Fact (*running time of Algorithm 2.221*)
  (i) The polynomials $s(x)$ and $t(x)$ given by Algorithm 2.221 have small degree; that is, they satisfy $\deg s(x) < \deg h(x)$ and $\deg t(x) < \deg g(x)$.
  (ii) Suppose that $\deg g(x) \leq m$ and $\deg h(x) \leq m$. Then Algorithm 2.221 has a running time of $O(m^2)$ $\mathbb{Z}_p$-operations, or equivalently, $O(m^2 (\lg p)^2)$ bit operations.

2.223 Example (*extended Euclidean algorithm for polynomials*) The following are the steps of Algorithm 2.221 with inputs $g(x) = x^{10} + x^9 + x^8 + x^6 + x^5 + x^4 + 1$ and $h(x) = x^9 + x^6 + x^5 + x^3 + x^2 + 1$ in $\mathbb{Z}_2[x]$.

Initialization
    $s_2(x) \leftarrow 1$, $s_1(x) \leftarrow 0$, $t_2(x) \leftarrow 0$, $t_1(x) \leftarrow 1$.

Iteration 1

$q(x) \leftarrow x + 1,\ r(x) \leftarrow x^8 + x^7 + x^6 + x^2 + x,$
$s(x) \leftarrow 1,\ t(x) \leftarrow x + 1,$
$g(x) \leftarrow x^9 + x^6 + x^5 + x^3 + x^2 + 1,\ h(x) \leftarrow x^8 + x^7 + x^6 + x^2 + 1,$
$s_2(x) \leftarrow 0,\ s_1(x) \leftarrow 1,\ t_2(x) \leftarrow 1,\ t_1(x) \leftarrow x + 1.$

Iteration 2

$q(x) \leftarrow x + 1,\ r(x) \leftarrow x^5 + x^2 + x + 1,$
$s(x) \leftarrow x + 1,\ t(x) \leftarrow x^2,$
$g(x) \leftarrow x^8 + x^7 + x^6 + x^2 + 1,\ h(x) \leftarrow x^5 + x^2 + x + 1,$
$s_2(x) \leftarrow 1,\ s_1(x) \leftarrow x + 1,\ t_2(x) \leftarrow x + 1,\ t_1(x) \leftarrow x^2.$

Iteration 3

$q(x) \leftarrow x^3 + x^2 + x + 1,\ r(x) \leftarrow x^3 + x + 1,$
$s(x) \leftarrow x^4,\ t(x) \leftarrow x^5 + x^4 + x^3 + x^2 + x + 1,$
$g(x) \leftarrow x^5 + x^2 + x + 1,\ h(x) \leftarrow x^3 + x + 1,$
$s_2(x) \leftarrow x + 1,\ s_1(x) \leftarrow x^4,\ t_2(x) \leftarrow x^2,\ t_1(x) \leftarrow x^5 + x^4 + x^3 + x^2 + x + 1.$

Iteration 4

$q(x) \leftarrow x^2 + 1,\ r(x) \leftarrow 0,$
$s(x) \leftarrow x^6 + x^4 + x + 1,\ t(x) \leftarrow x^7 + x^6 + x^2 + x + 1,$
$g(x) \leftarrow x^3 + x + 1,\ h(x) \leftarrow 0,$
$s_2(x) \leftarrow x^4,\ s_1(x) \leftarrow x^6 + x^4 + x + 1,$
$t_2(x) \leftarrow x^5 + x^4 + x^3 + x^2 + x + 1,\ t_1(x) \leftarrow x^7 + x^6 + x^2 + x + 1.$

Hence $\gcd(g(x), h(x)) = x^3 + x + 1$ and

$$(x^4)g(x) + (x^5 + x^4 + x^3 + x^2 + x + 1)h(x) = x^3 + x + 1. \qquad \square$$

2.6.3 Arithmetic of polynomials

A commonly used representation for the elements of a finite field $\mathbb{F}_q$, where $q = p^m$ and p is a prime, is a *polynomial basis representation*. If $m = 1$, then $\mathbb{F}_q$ is just $\mathbb{Z}_p$ and arithmetic is performed modulo p. Since these operations have already been studied in Section 2.4.2, it is henceforth assumed that $m \geq 2$. The representation is based on Fact 2.198.

2.224 Fact Let $f(x) \in \mathbb{Z}_p[x]$ be an irreducible polynomial of degree m. Then $\mathbb{Z}_p[x]/(f(x))$ is a finite field of order p^m. Addition and multiplication of polynomials is performed modulo $f(x)$.

The following fact assures that all finite fields can be represented in this manner.

2.225 Fact For each $m \geq 1$, there exists a monic irreducible polynomial of degree m over $\mathbb{Z}_p$. Hence, every finite field has a polynomial basis representation.

An efficient algorithm for finding irreducible polynomials over finite fields is presented in §4.5.1. Tables 4.6 and 4.7 list some irreducible polynomials over the finite field $\mathbb{Z}_2$.

Henceforth, the elements of the finite field $\mathbb{F}_{p^m}$ will be represented by polynomials in $\mathbb{Z}_p[x]$ of degree $< m$. If $g(x), h(x) \in \mathbb{F}_{p^m}$, then addition is the usual addition of polynomials in $\mathbb{Z}_p[x]$. The product $g(x)h(x)$ can be formed by first multiplying $g(x)$ and $h(x)$ as polynomials by the ordinary method, and then taking the remainder after polynomial division by $f(x)$. Multiplicative inverses in $\mathbb{F}_{p^m}$ can be computed by using the extended Euclidean algorithm for the polynomial ring $\mathbb{Z}_p[x]$.

2.226 Algorithm Computing multiplicative inverses in $\mathbb{F}_{p^m}$

INPUT: a non-zero polynomial $g(x) \in \mathbb{F}_{p^m}$. (The elements of the field $\mathbb{F}_{p^m}$ are represented as $\mathbb{Z}_p[x]/(f(x))$, where $f(x) \in \mathbb{Z}_p[x]$ is an irreducible polynomial of degree m over $\mathbb{Z}_p$.)
OUTPUT: $g(x)^{-1} \in \mathbb{F}_{p^m}$.

   1. Use the extended Euclidean algorithm for polynomials (Algorithm 2.221) to find two polynomials $s(x)$ and $t(x) \in \mathbb{Z}_p[x]$ such that $s(x)g(x) + t(x)f(x) = 1$.
   2. Return($s(x)$).

Exponentiation in $\mathbb{F}_{p^m}$ can be done efficiently by the repeated square-and-multiply algorithm (cf. Algorithm 2.143).

2.227 Algorithm Repeated square-and-multiply algorithm for exponentiation in $\mathbb{F}_{p^m}$

INPUT: $g(x) \in \mathbb{F}_{p^m}$ and an integer $0 \leq k < p^m - 1$ whose binary representation is $k = \sum_{i=0}^{t} k_i 2^i$. (The field $\mathbb{F}_{p^m}$ is represented as $\mathbb{Z}_p[x]/(f(x))$, where $f(x) \in \mathbb{Z}_p[x]$ is an irreducible polynomial of degree m over $\mathbb{Z}_p$.)
OUTPUT: $g(x)^k \bmod f(x)$.

   1. Set $s(x) \leftarrow 1$. If $k = 0$ then return($s(x)$).
   2. Set $G(x) \leftarrow g(x)$.
   3. If $k_0 = 1$ then set $s(x) \leftarrow g(x)$.
   4. For i from 1 to t do the following:
      4.1 Set $G(x) \leftarrow G(x)^2 \bmod f(x)$.
      4.2 If $k_i = 1$ then set $s(x) \leftarrow G(x) \cdot s(x) \bmod f(x)$.
   5. Return($s(x)$).

The number of $\mathbb{Z}_p$-operations for the basic operations in $\mathbb{F}_{p^m}$ is summarized in Table 2.8.

Operation		Number of $\mathbb{Z}_p$-operations
Addition	$g(x) + h(x)$	$O(m)$
Subtraction	$g(x) - h(x)$	$O(m)$
Multiplication	$g(x) \cdot h(x)$	$O(m^2)$
Inversion	$g(x)^{-1}$	$O(m^2)$
Exponentiation	$g(x)^k,\ k < p^m$	$O((\lg p)m^3)$

Table 2.8: *Complexity of basic operations in $\mathbb{F}_{p^m}$.*

In some applications (cf. §4.5.3), it may be preferable to use a primitive polynomial to define a finite field.

2.228 Definition An irreducible polynomial $f(x) \in \mathbb{Z}_p[x]$ of degree m is called a *primitive polynomial* if x is a generator of $\mathbb{F}_{p^m}^*$, the multiplicative group of all the non-zero elements in $\mathbb{F}_{p^m} = \mathbb{Z}_p[x]/(f(x))$.

2.229 Fact The irreducible polynomial $f(x) \in \mathbb{Z}_p[x]$ of degree m is a primitive polynomial if and only if $f(x)$ divides $x^k - 1$ for $k = p^m - 1$ and for no smaller positive integer k.

2.230 Fact For each $m \geq 1$, there exists a monic primitive polynomial of degree m over $\mathbb{Z}_p$. In fact, there are precisely $\phi(p^m - 1)/m$ such polynomials.

2.231 Example (*the finite field* $\mathbb{F}_{2^4}$ *of order* 16) It can be verified (Algorithm 4.69) that the polynomial $f(x) = x^4 + x + 1$ is irreducible over $\mathbb{Z}_2$. Hence the finite field $\mathbb{F}_{2^4}$ can be represented as the set of all polynomials over $\mathbb{F}_2$ of degree less than 4. That is,

$$\mathbb{F}_{2^4} = \{a_3 x^3 + a_2 x^2 + a_1 x + a_0 \mid a_i \in \{0, 1\}\}.$$

For convenience, the polynomial $a_3 x^3 + a_2 x^2 + a_1 x + a_0$ is represented by the vector $(a_3 a_2 a_1 a_0)$ of length 4, and

$$\mathbb{F}_{2^4} = \{(a_3 a_2 a_1 a_0) \mid a_i \in \{0, 1\}\}.$$

The following are some examples of field arithmetic.

(i) Field elements are simply added componentwise: for example, $(1011) + (1001) = (0010)$.

(ii) To multiply the field elements (1101) and (1001), multiply them as polynomials and then take the remainder when this product is divided by $f(x)$:

$$\begin{aligned}(x^3 + x^2 + 1) \cdot (x^3 + 1) &= x^6 + x^5 + x^2 + 1 \\ &\equiv x^3 + x^2 + x + 1 \pmod{f(x)}.\end{aligned}$$

Hence $(1101) \cdot (1001) = (1111)$.

(iii) The multiplicative identity of $\mathbb{F}_{2^4}$ is (0001).

(iv) The inverse of (1011) is (0101). To verify this, observe that

$$\begin{aligned}(x^3 + x + 1) \cdot (x^2 + 1) &= x^5 + x^2 + x + 1 \\ &\equiv 1 \pmod{f(x)},\end{aligned}$$

whence $(1011) \cdot (0101) = (0001)$.

$f(x)$ is a primitive polynomial, or, equivalently, the field element $x = (0010)$ is a generator of $\mathbb{F}_{2^4}^*$. This may be checked by verifying that all the non-zero elements in $\mathbb{F}_{2^4}$ can be obtained as a powers of x. The computations are summarized in Table 2.9. □

A list of some primitive polynomials over finite fields of characteristic two is given in Table 4.8.

2.7 Notes and further references

§2.1

A classic introduction to probability theory is the first volume of the book by Feller [392]. The material on the birthday attacks (§2.1.5) is summarized from Nishimura and Sibuya [931]. See also Girault, Cohen, and Campana [460]. The material on random mappings (§2.1.6) is summarized from the excellent article by Flajolet and Odlyzko [413].

§2.2

The concept of entropy was introduced in the seminal paper of Shannon [1120]. These ideas were then applied to develop a mathematical theory of secrecy systems by Shannon [1121]. Hellman [548] extended the Shannon theory approach to cryptography, and this work was further generalized by Beauchemin and Brassard [80]. For an introduction to information theory see the books by Welsh [1235] and Goldie and Pinch [464]. For more complete treatments, consult Blahut [144] and McEliece [829].

i	$x^i \bmod x^4 + x + 1$	vector notation
0	1	(0001)
1	x	(0010)
2	x^2	(0100)
3	x^3	(1000)
4	$x + 1$	(0011)
5	$x^2 + x$	(0110)
6	$x^3 + x^2$	(1100)
7	$x^3 + x + 1$	(1011)
8	$x^2 + 1$	(0101)
9	$x^3 + x$	(1010)
10	$x^2 + x + 1$	(0111)
11	$x^3 + x^2 + x$	(1110)
12	$x^3 + x^2 + x + 1$	(1111)
13	$x^3 + x^2 + 1$	(1101)
14	$x^3 + 1$	(1001)

Table 2.9: *The powers of x modulo $f(x) = x^4 + x + 1$.*

§2.3

Among the many introductory-level books on algorithms are those of Cormen, Leiserson, and Rivest [282], Rawlins [1030], and Sedgewick [1105]. A recent book on complexity theory is Papadimitriou [963]. Example 2.58 is from Graham, Knuth, and Patashnik [520, p.441]. For an extensive list of **NP**-complete problems, see Garey and Johnson [441].

§2.4

Two introductory-level books in number theory are Giblin [449] and Rosen [1069]. Good number theory books at a more advanced level include Koblitz [697], Hardy and Wright [540], Ireland and Rosen [572], and Niven and Zuckerman [932]. The most comprehensive works on the design and analysis of algorithms, including number theoretic algorithms, are the first two volumes of Knuth [691, 692]. Two more recent books exclusively devoted to this subject are Bach and Shallit [70] and Cohen [263]. Facts 2.96 and 2.102 are due to Rosser and Schoenfeld [1070]. Shallit [1108] describes and analyzes three algorithms for computing the Jacobi symbol.

§2.5

Among standard references in abstract algebra are the books by Herstein [556] and Hungerford [565].

§2.6

An excellent introduction to finite fields is provided in McEliece [830]. A comprehensive treatment of the theory and applications of finite fields is given by Lidl and Niederreitter [764]. Two books which discuss various methods of representing the elements of a finite field are those of Jungnickel [646] and Menezes et al. [841].

Chapter 3

Number-Theoretic Reference Problems

Contents in Brief

3.1 Introduction and overview

The security of many public-key cryptosystems relies on the apparent intractability of the computational problems studied in this chapter. In a cryptographic setting, it is prudent to make the assumption that the adversary is very powerful. Thus, informally speaking, a computational problem is said to be *easy* or *tractable* if it can be solved in (expected)[1] polynomial time, at least for a non-negligible fraction of all possible inputs. In other words, if there is an algorithm which can solve a non-negligible fraction of all instances of a problem in polynomial time, then any cryptosystem whose security is based on that problem must be considered insecure.

The computational problems studied in this chapter are summarized in Table 3.1. The true computational complexities of these problems are not known. That is to say, they are widely believed to be intractable,[2] although no proof of this is known. Generally, the only lower bounds known on the resources required to solve these problems are the trivial linear bounds, which do not provide any evidence of their intractability. It is, therefore, of interest to study their relative difficulties. For this reason, various techniques of reducing one

---

[1] For simplicity, the remainder of the chapter shall generally not distinguish between deterministic polynomial-time algorithms and randomized algorithms (see §2.3.4) whose *expected* running time is polynomial.

[2] More precisely, these problems are intractable if the problem parameters are carefully chosen.

Problem	Description
FACTORING	*Integer factorization problem*: given a positive integer n, find its prime factorization; that is, write $n = p_1^{e_1} p_2^{e_2} \ldots p_k^{e_k}$ where the p_i are pairwise distinct primes and each $e_i \geq 1$.
RSAP	*RSA problem* (also known as *RSA inversion*): given a positive integer n that is a product of two distinct odd primes p and q, a positive integer e such that $\gcd(e, (p-1)(q-1)) = 1$, and an integer c, find an integer m such that $m^e \equiv c \pmod{n}$.
QRP	*Quadratic residuosity problem*: given an odd composite integer n and an integer a having Jacobi symbol $\left(\frac{a}{n}\right) = 1$, decide whether or not a is a quadratic residue modulo n.
SQROOT	*Square roots modulo n*: given a composite integer n and $a \in Q_n$ (the set of quadratic residues modulo n), find a square root of a modulo n; that is, an integer x such that $x^2 \equiv a \pmod{n}$.
DLP	*Discrete logarithm problem*: given a prime p, a generator α of $\mathbb{Z}_p^*$, and an element $\beta \in \mathbb{Z}_p^*$, find the integer x, $0 \leq x \leq p-2$, such that $\alpha^x \equiv \beta \pmod{p}$.
GDLP	*Generalized discrete logarithm problem*: given a finite cyclic group G of order n, a generator α of G, and an element $\beta \in G$, find the integer x, $0 \leq x \leq n-1$, such that $\alpha^x = \beta$.
DHP	*Diffie-Hellman problem*: given a prime p, a generator α of $\mathbb{Z}_p^*$, and elements $\alpha^a \bmod p$ and $\alpha^b \bmod p$, find $\alpha^{ab} \bmod p$.
GDHP	*Generalized Diffie-Hellman problem*: given a finite cyclic group G, a generator α of G, and group elements α^a and α^b, find α^{ab}.
SUBSET-SUM	*Subset sum problem*: given a set of positive integers $\{a_1, a_2, \ldots, a_n\}$ and a positive integer s, determine whether or not there is a subset of the a_j that sums to s.

Table 3.1: *Some computational problems of cryptographic relevance.*

computational problem to another have been devised and studied in the literature. These reductions provide a means for converting any algorithm that solves the second problem into an algorithm for solving the first problem. The following intuitive notion of reducibility (cf. §2.3.3) is used in this chapter.

3.1 Definition Let A and B be two computational problems. A is said to *polytime reduce* to B, written $A \leq_P B$, if there is an algorithm that solves A which uses, as a subroutine, a hypothetical algorithm for solving B, and which runs in polynomial time if the algorithm for B does.[3]

Informally speaking, if A polytime reduces to B, then B is at least as difficult as A; equivalently, A is no harder than B. Consequently, if A is a well-studied computational problem that is widely believed to be intractable, then proving that $A \leq_P B$ provides strong evidence of the intractability of problem B.

3.2 Definition Let A and B be two computational problems. If $A \leq_P B$ and $B \leq_P A$, then A and B are said to be *computationally equivalent*, written $A \equiv_P B$.

---

[3]In the literature, the hypothetical polynomial-time subroutine for B is sometimes called an *oracle* for B.

Informally speaking, if $A \equiv_P B$ then A and B are either both tractable or both intractable, as the case may be.

Chapter outline

The remainder of the chapter is organized as follows. Algorithms for the integer factorization problem are studied in §3.2. Two problems related to factoring, the RSA problem and the quadratic residuosity problem, are briefly considered in §3.3 and §3.4. Efficient algorithms for computing square roots in $\mathbb{Z}_p$, p a prime, are presented in §3.5, and the equivalence of the problems of finding square roots modulo a composite integer n and factoring n is established. Algorithms for the discrete logarithm problem are studied in §3.6, and the related Diffie-Hellman problem is briefly considered in §3.7. The relation between the problems of factoring a composite integer n and computing discrete logarithms in (cyclic subgroups of) the group $\mathbb{Z}_n^*$ is investigated in §3.8. The tasks of finding partial solutions to the discrete logarithm problem, the RSA problem, and the problem of computing square roots modulo a composite integer n are the topics of §3.9. The L^3-lattice basis reduction algorithm is presented in §3.10, along with algorithms for the subset sum problem and for simultaneous diophantine approximation. Berlekamp's Q-matrix algorithm for factoring polynomials is presented in §3.11. Finally, §3.12 provides references and further chapter notes.

3.2 The integer factorization problem

The security of many cryptographic techniques depends upon the intractability of the integer factorization problem. A partial list of such protocols includes the RSA public-key encryption scheme (§8.2), the RSA signature scheme (§11.3.1), and the Rabin public-key encryption scheme (§8.3). This section summarizes the current knowledge on algorithms for the integer factorization problem.

3.3 Definition The *integer factorization problem* (FACTORING) is the following: given a positive integer n, find its prime factorization; that is, write $n = p_1^{e_1} p_2^{e_2} \cdots p_k^{e_k}$ where the p_i are pairwise distinct primes and each $e_i \geq 1$.

3.4 Remark (*primality testing vs. factoring*) The problem of *deciding* whether an integer is composite or prime seems to be, in general, much easier than the factoring problem. Hence, before attempting to factor an integer, the integer should be tested to make sure that it is indeed composite. Primality tests are a main topic of Chapter 4.

3.5 Remark (*splitting vs. factoring*) A *non-trivial factorization* of n is a factorization of the form $n = ab$ where $1 < a < n$ and $1 < b < n$; a and b are said to be *non-trivial factors* of n. Here a and b are not necessarily prime. To solve the integer factorization problem, it suffices to study algorithms that *split* n, that is, find a non-trivial factorization $n = ab$. Once found, the factors a and b can be tested for primality. The algorithm for splitting integers can then be recursively applied to a and/or b, if either is found to be composite. In this manner, the prime factorization of n can be obtained.

3.6 Note (*testing for perfect powers*) If $n \geq 2$, it can be efficiently checked as follows whether or not n is a *perfect power*, i.e., $n = x^k$ for some integers $x \geq 2$, $k \geq 2$. For each prime

$p \leq \lg n$, an integer approximation x of $n^{1/p}$ is computed. This can be done by performing a binary search for x satisfying $n = x^p$ in the interval $[2, 2^{\lfloor \lg n/p \rfloor + 1}]$. The entire procedure takes $O((\lg^3 n) \lg \lg \lg n)$ bit operations. For the remainder of this section, it will always be assumed that n is not a perfect power. It follows that if n is composite, then n has at least two distinct prime factors.

Some factoring algorithms are tailored to perform better when the integer n being factored is of a special form; these are called *special-purpose* factoring algorithms. The running times of such algorithms typically depend on certain properties of the factors of n. Examples of special-purpose factoring algorithms include trial division (§3.2.1), Pollard's rho algorithm (§3.2.2), Pollard's $p - 1$ algorithm (§3.2.3), the elliptic curve algorithm (§3.2.4), and the special number field sieve (§3.2.7). In contrast, the running times of the so-called *general-purpose* factoring algorithms depend solely on the size of n. Examples of general-purpose factoring algorithms include the quadratic sieve (§3.2.6) and the general number field sieve (§3.2.7).

Whenever applicable, special-purpose algorithms should be employed as they will generally be more efficient. A reasonable overall strategy is to attempt to find small factors first, capitalize on any particular special forms an integer may have, and then, if all else fails, bring out the general-purpose algorithms. As an example of a general strategy, one might consider the following.

1. Apply trial division by small primes less than some bound b_1.
2. Next, apply Pollard's rho algorithm, hoping to find any small prime factors smaller than some bound b_2, where $b_2 > b_1$.
3. Apply the elliptic curve factoring algorithm, hoping to find any small factors smaller than some bound b_3, where $b_3 > b_2$.
4. Finally, apply one of the more powerful general-purpose algorithms (quadratic sieve or general number field sieve).

3.2.1 Trial division

Once it is established that an integer n is composite, before expending vast amounts of time with more powerful techniques, the first thing that should be attempted is trial division by all "small" primes. Here, "small" is determined as a function of the size of n. As an extreme case, trial division can be attempted by all primes up to $\sqrt{n}$. If this is done, trial division will completely factor n but the procedure will take roughly $\sqrt{n}$ divisions in the worst case when n is a product of two primes of the same size. In general, if the factors found at each stage are tested for primality, then trial division to factor n completely takes $O(p + \lg n)$ divisions, where p is the second-largest prime factor of n.

Fact 3.7 indicates that if trial division is used to factor a randomly chosen large integer n, then the algorithm can be expected to find some small factors of n relatively quickly, and expend a large amount of time to find the second largest prime factor of n.

3.7 Fact Let n be chosen uniformly at random from the interval $[1, x]$.
   (i) If $\frac{1}{2} \leq \alpha \leq 1$, then the probability that the largest prime factor of n is $\leq x^\alpha$ is approximately $1 + \ln \alpha$. Thus, for example, the probability that n has a prime factor $> \sqrt{x}$ is $\ln 2 \approx 0.69$.
  (ii) The probability that the second-largest prime factor of n is $\leq x^{0.2117}$ is about $\frac{1}{2}$.
 (iii) The expected total number of prime factors of n is $\ln \ln x + O(1)$. (If $n = \prod p_i^{e_i}$, the *total* number of prime factors of n is $\sum e_i$.)

3.2.2 Pollard's rho factoring algorithm

Pollard's rho algorithm is a special-purpose factoring algorithm for finding small factors of a composite integer.

Let $f : S \longrightarrow S$ be a random function, where S is a finite set of cardinality n. Let x_0 be a random element of S, and consider the sequence $x_0, x_1, x_2, \ldots$ defined by $x_{i+1} = f(x_i)$ for $i \geq 0$. Since S is finite, the sequence must eventually cycle, and consists of a *tail* of expected length $\sqrt{\pi n/8}$ followed by an endlessly repeating *cycle* of expected length $\sqrt{\pi n/8}$ (see Fact 2.37). A problem that arises in some cryptanalytic tasks, including integer factorization (Algorithm 3.9) and the discrete logarithm problem (Algorithm 3.60), is of finding distinct indices i and j such that $x_i = x_j$ (a *collision* is then said to have occurred).

An obvious method for finding a collision is to compute and store x_i for $i = 0, 1, 2, \ldots$ and look for duplicates. The expected number of inputs that must be tried before a duplicate is detected is $\sqrt{\pi n/2}$ (Fact 2.27). This method requires $O(\sqrt{n})$ memory and $O(\sqrt{n})$ time, assuming the x_i are stored in a hash table so that new entries can be added in constant time.

3.8 Note (*Floyd's cycle-finding algorithm*) The large storage requirements in the above technique for finding a collision can be eliminated by using *Floyd's cycle-finding algorithm*. In this method, one starts with the pair (x_1, x_2), and iteratively computes (x_i, x_{2i}) from the previous pair (x_{i-1}, x_{2i-2}), until $x_m = x_{2m}$ for some m. If the tail of the sequence has length λ and the cycle has length μ, then the first time that $x_m = x_{2m}$ is when $m = \mu(1 + \lfloor \lambda/\mu \rfloor)$. Note that $\lambda < m \leq \lambda + \mu$, and consequently the expected running time of this method is $O(\sqrt{n})$.

Now, let p be a prime factor of a composite integer n. Pollard's rho algorithm for factoring n attempts to find duplicates in the sequence of integers $x_0, x_1, x_2, \ldots$ defined by $x_0 = 2$, $x_{i+1} = f(x_i) = x_i^2 + 1 \bmod p$ for $i \geq 0$. Floyd's cycle-finding algorithm is utilized to find x_m and x_{2m} such that $x_m \equiv x_{2m} \pmod{p}$. Since p divides n but is unknown, this is done by computing the terms x_i modulo n and testing if $\gcd(x_m - x_{2m}, n) > 1$. If also $\gcd(x_m - x_{2m}, n) < n$, then a non-trivial factor of n is obtained. (The situation $\gcd(x_m - x_{2m}, n) = n$ occurs with negligible probability.)

3.9 Algorithm Pollard's rho algorithm for factoring integers

INPUT: a composite integer n that is not a prime power.
OUTPUT: a non-trivial factor d of n.

1. Set $a \leftarrow 2$, $b \leftarrow 2$.
2. For $i = 1, 2, \ldots$ do the following:
   2.1 Compute $a \leftarrow a^2 + 1 \bmod n$, $b \leftarrow b^2 + 1 \bmod n$, $b \leftarrow b^2 + 1 \bmod n$.
   2.2 Compute $d = \gcd(a - b, n)$.
   2.3 If $1 < d < n$ then return(d) and terminate with success.
   2.4 If $d = n$ then terminate the algorithm with failure (see Note 3.12).

3.10 Example (*Pollard's rho algorithm for finding a non-trivial factor of $n = 455459$*) The following table lists the values of variables a, b, and d at the end of each iteration of step 2 of Algorithm 3.9.

a	b	d
5	26	1
26	2871	1
677	179685	1
2871	155260	1
44380	416250	1
179685	43670	1
121634	164403	1
155260	247944	1
44567	68343	743

Hence two non-trivial factors of 455459 are 743 and $455459/743 = 613$.   □

3.11 Fact Assuming that the function $f(x) = x^2 + 1 \bmod p$ behaves like a random function, the expected time for Pollard's rho algorithm to find a factor p of n is $O(\sqrt{p})$ modular multiplications. This implies that the expected time to find a non-trivial factor of n is $O(n^{1/4})$ modular multiplications.

3.12 Note (*options upon termination with failure*) If Pollard's rho algorithm terminates with failure, one option is to try again with a different polynomial f having integer coefficients instead of $f(x) = x^2 + 1$. For example, the polynomial $f(x) = x^2 + c$ may be used as long as $c \neq 0, -2$.

3.2.3 Pollard's $p - 1$ factoring algorithm

Pollard's $p - 1$ factoring algorithm is a special-purpose factoring algorithm that can be used to efficiently find any prime factors p of a composite integer n for which $p - 1$ is smooth (see Definition 3.13) with respect to some relatively small bound B.

3.13 Definition Let B be a positive integer. An integer n is said to be *B-smooth*, or *smooth with respect to a bound B*, if all its prime factors are $\leq B$.

The idea behind Pollard's $p - 1$ algorithm is the following. Let B be a smoothness bound. Let Q be the least common multiple of all powers of primes $\leq B$ that are $\leq n$. If $q^l \leq n$, then $l \ln q \leq \ln n$, and so $l \leq \lfloor \frac{\ln n}{\ln q} \rfloor$. Thus

$$Q = \prod_{q \leq B} q^{\lfloor \ln n / \ln q \rfloor},$$

where the product is over all distinct primes $q \leq B$. If p is a prime factor of n such that $p-1$ is B-smooth, then $p - 1 | Q$, and consequently for any a satisfying $\gcd(a, p) = 1$, Fermat's theorem (Fact 2.127) implies that $a^Q \equiv 1 \pmod{p}$. Hence if $d = \gcd(a^Q - 1, n)$, then $p|d$. It is possible that $d = n$, in which case the algorithm fails; however, this is unlikely to occur if n has at least two large distinct prime factors.

3.14 Algorithm Pollard's $p - 1$ algorithm for factoring integers

---

INPUT: a composite integer n that is not a prime power.
OUTPUT: a non-trivial factor d of n.

1. Select a smoothness bound B.
2. Select a random integer a, $2 \leq a \leq n - 1$, and compute $d = \gcd(a, n)$. If $d \geq 2$ then return(d).
3. For each prime $q \leq B$ do the following:
   3.1 Compute $l = \lfloor \frac{\ln n}{\ln q} \rfloor$.
   3.2 Compute $a \leftarrow a^{q^l} \bmod n$ (using Algorithm 2.143).
4. Compute $d = \gcd(a - 1, n)$.
5. If $d = 1$ or $d = n$, then terminate the algorithm with failure. Otherwise, return(d).

3.15 Example (*Pollard's $p - 1$ algorithm for finding a non-trivial factor of $n = 19048567$*)

1. Select the smoothness bound $B = 19$.
2. Select the integer $a = 3$ and compute $\gcd(3, n) = 1$.
3. The following table lists the intermediate values of the variables q, l, and a after each iteration of step 3 in Algorithm 3.14:

q	l	a
2	24	2293244
3	15	13555889
5	10	16937223
7	8	15214586
11	6	9685355
13	6	13271154
17	5	11406961
19	5	554506

4. Compute $d = \gcd(554506 - 1, n) = 5281$.
5. Two non-trivial factors of n are $p = 5281$ and $q = n/p = 3607$ (these factors are in fact prime).

Notice that $p - 1 = 5280 = 2^5 \times 3 \times 5 \times 11$, and $q - 1 = 3606 = 2 \times 3 \times 601$. That is, $p - 1$ is 19-smooth, while $q - 1$ is not 19-smooth. □

3.16 Fact Let n be an integer having a prime factor p such that $p - 1$ is B-smooth. The running time of Pollard's $p - 1$ algorithm for finding the factor p is $O(B \ln n / \ln B)$ modular multiplications.

3.17 Note (*improvements*) The smoothness bound B in Algorithm 3.14 is selected based on the amount of time one is willing to spend on Pollard's $p - 1$ algorithm before moving on to more general techniques. In practice, B may be between 10^5 and 10^6. If the algorithm terminates with $d = 1$, then one might try searching over prime numbers $q_1, q_2, \ldots, q_l$ larger than B by first computing $a \leftarrow a^{q_i} \bmod n$ for $1 \leq i \leq l$, and then computing $d = \gcd(a - 1, n)$. Another variant is to start with a large bound B, and repeatedly execute step 3 for a few primes q followed by the gcd computation in step 4. There are numerous other practical improvements of the algorithm (see page 125).

3.2.4 Elliptic curve factoring

The details of the *elliptic curve factoring algorithm* are beyond the scope of this book; nevertheless, a rough outline follows. The success of Pollard's $p-1$ algorithm hinges on $p-1$ being smooth for some prime divisor p of n; if no such p exists, then the algorithm fails. Observe that $p-1$ is the order of the group $\mathbb{Z}_p^*$. The elliptic curve factoring algorithm is a generalization of Pollard's $p-1$ algorithm in the sense that the group $\mathbb{Z}_p^*$ is replaced by a random elliptic curve group over $\mathbb{Z}_p$. The order of such a group is roughly uniformly distributed in the interval $[p+1-2\sqrt{p}, p+1+2\sqrt{p}]$. If the order of the group chosen is smooth with respect to some pre-selected bound, the elliptic curve algorithm will, with high probability, find a non-trivial factor of n. If the group order is not smooth, then the algorithm will likely fail, but can be repeated with a different choice of elliptic curve group.

The elliptic curve algorithm has an expected running time of $L_p[\frac{1}{2}, \sqrt{2}]$ (see Example 2.61 for definition of L_p) to find a factor p of n. Since this running time depends on the size of the prime factors of n, the algorithm tends to find small such factors first. The elliptic curve algorithm is, therefore, classified as a special-purpose factoring algorithm. It is currently the algorithm of choice for finding t-decimal digit prime factors, for $t \leq 40$, of very large composite integers.

In the hardest case, when n is a product of two primes of roughly the same size, the expected running time of the elliptic curve algorithm is $L_n[\frac{1}{2}, 1]$, which is the same as that of the quadratic sieve (§3.2.6). However, the elliptic curve algorithm is not as efficient as the quadratic sieve in practice for such integers.

3.2.5 Random square factoring methods

The basic idea behind the random square family of methods is the following. Suppose x and y are integers such that $x^2 \equiv y^2 \pmod{n}$ but $x \not\equiv \pm y \pmod{n}$. Then n divides $x^2 - y^2 = (x-y)(x+y)$ but n does not divide either $(x-y)$ or $(x+y)$. Hence, $\gcd(x-y, n)$ must be a non-trivial factor of n. This result is summarized next.

3.18 Fact Let x, y, and n be integers. If $x^2 \equiv y^2 \pmod{n}$ but $x \not\equiv \pm y \pmod{n}$, then $\gcd(x-y, n)$ is a non-trivial factor of n.

The random square methods attempt to find integers x and y at random so that $x^2 \equiv y^2 \pmod{n}$. Then, as shown in Fact 3.19, with probability at least $\frac{1}{2}$ it is the case that $x \not\equiv \pm y \pmod{n}$, whence $\gcd(x-y, n)$ will yield a non-trivial factor of n.

3.19 Fact Let n be an odd composite integer that is divisible by k distinct odd primes. If $a \in \mathbb{Z}_n^*$, then the congruence $x^2 \equiv a^2 \pmod{n}$ has exactly 2^k solutions modulo n, two of which are $x = a$ and $x = -a$.

3.20 Example Let $n = 35$. Then there are four solutions to the congruence $x^2 \equiv 4 \pmod{35}$, namely $x = 2, 12, 23$, and 33.   □

A common strategy employed by the random square algorithms for finding x and y at random satisfying $x^2 \equiv y^2 \pmod{n}$ is the following. A set consisting of the first t primes $S = \{p_1, p_2, \dots, p_t\}$ is chosen; S is called the *factor base*. Proceed to find pairs of integers (a_i, b_i) satisfying

(i) $a_i^2 \equiv b_i \pmod{n}$; and

(ii) $b_i = \prod_{j=1}^{t} p_j^{e_{ij}}$, $e_{ij} \geq 0$; that is, b_i is p_t-smooth.

Next find a subset of the b_i's whose product is a perfect square. Knowing the factorizations of the b_i's, this is possible by selecting a subset of the b_i's such that the power of each prime p_j appearing in their product is even. For this purpose, only the parity of the non-negative integer exponents e_{ij} needs to be considered. Thus, to simplify matters, for each i, associate the binary vector $v_i = (v_{i1}, v_{i2}, \ldots, v_{it})$ with the integer exponent vector $(e_{i1}, e_{i2}, \ldots, e_{it})$ such that $v_{ij} = e_{ij} \bmod 2$. If $t+1$ pairs (a_i, b_i) are obtained, then the t-dimensional vectors $v_1, v_2, \ldots, v_{t+1}$ must be linearly dependent over $\mathbb{Z}_2$. That is, there must exist a non-empty subset $T \subseteq \{1, 2, \ldots, t+1\}$ such that $\sum_{i \in T} v_i = 0$, and hence $\prod_{i \in T} b_i$ is a perfect square. The set T can be found using ordinary linear algebra over $\mathbb{Z}_2$. Clearly, $\prod_{i \in T} a_i^2$ is also a perfect square. Thus setting $x = \prod_{i \in T} a_i$ and y to be the integer square root of $\prod_{i \in T} b_i$ yields a pair of integers (x, y) satisfying $x^2 \equiv y^2 \pmod{n}$. If this pair also satisfies $x \not\equiv \pm y \pmod{n}$, then $\gcd(x - y, n)$ yields a non-trivial factor of n. Otherwise, some of the (a_i, b_i) pairs may be replaced by some new such pairs, and the process is repeated. In practice, there will be several dependencies among the vectors $v_1, v_2, \ldots, v_{t+1}$, and with high probability at least one will yield an (x, y) pair satisfying $x \not\equiv \pm y \pmod{n}$; hence, this last step of generating new (a_i, b_i) pairs does not usually occur.

This description of the random square methods is incomplete for two reasons. Firstly, the optimal choice of t, the size of the factor base, is not specified; this is addressed in Note 3.24. Secondly, a method for efficiently generating the pairs (a_i, b_i) is not specified. Several techniques have been proposed. In the simplest of these, called *Dixon's algorithm*, a_i is chosen at random, and $b_i = a_i^2 \bmod n$ is computed. Next, trial division by elements in the factor base is used to test whether b_i is p_t-smooth. If not, then another integer a_i is chosen at random, and the procedure is repeated.

The more efficient techniques strategically select an a_i such that b_i is relatively small. Since the proportion of p_t-smooth integers in the interval $[2, x]$ becomes larger as x decreases, the probability of such b_i being p_t-smooth is higher. The most efficient of such techniques is the quadratic sieve algorithm, which is described next.

3.2.6 Quadratic sieve factoring

Suppose an integer n is to be factored. Let $m = \lfloor \sqrt{n} \rfloor$, and consider the polynomial $q(x) = (x + m)^2 - n$. Note that

$$q(x) = x^2 + 2mx + m^2 - n \approx x^2 + 2mx, \tag{3.1}$$

which is small (relative to n) if x is small in absolute value. The quadratic sieve algorithm selects $a_i = (x + m)$ and tests whether $b_i = (x + m)^2 - n$ is p_t-smooth. Note that $a_i^2 = (x + m)^2 \equiv b_i \pmod{n}$. Note also that if a prime p divides b_i then $(x + m)^2 \equiv n \pmod{p}$, and hence n is a quadratic residue modulo p. Thus the factor base need only contain those primes p for which the Legendre symbol $\left(\frac{n}{p}\right)$ is 1 (Definition 2.145). Furthermore, since b_i may be negative, -1 is included in the factor base. The steps of the quadratic sieve algorithm are summarized in Algorithm 3.21.

3.21 Algorithm Quadratic sieve algorithm for factoring integers

---

INPUT: a composite integer n that is not a prime power.
OUTPUT: a non-trivial factor d of n.

1. Select the factor base $S = \{p_1, p_2, \ldots, p_t\}$, where $p_1 = -1$ and p_j ($j \geq 2$) is the $(j-1)^{\text{th}}$ prime p for which n is a quadratic residue modulo p.
2. Compute $m = \lfloor \sqrt{n} \rfloor$.
3. (*Collect $t+1$ pairs (a_i, b_i). The x values are chosen in the order $0, \pm 1, \pm 2, \ldots$.*)
   Set $i \leftarrow 1$. While $i \leq t+1$ do the following:

   3.1 Compute $b = q(x) = (x+m)^2 - n$, and test using trial division (cf. Note 3.23) by elements in S whether b is p_t-smooth. If not, pick a new x and repeat step 3.1.

   3.2 If b is p_t-smooth, say $b = \prod_{j=1}^{t} p_j^{e_{ij}}$, then set $a_i \leftarrow (x+m)$, $b_i \leftarrow b$, and $v_i = (v_{i1}, v_{i2}, \ldots, v_{it})$, where $v_{ij} = e_{ij} \bmod 2$ for $1 \leq j \leq t$.

   3.3 $i \leftarrow i+1$.

4. Use linear algebra over $\mathbb{Z}_2$ to find a non-empty subset $T \subseteq \{1, 2, \ldots, t+1\}$ such that $\sum_{i \in T} v_i = 0$.
5. Compute $x = \prod_{i \in T} a_i \bmod n$.
6. For each j, $1 \leq j \leq t$, compute $l_j = (\sum_{i \in T} e_{ij})/2$.
7. Compute $y = \prod_{j=1}^{t} p_j^{l_j} \bmod n$.
8. If $x \equiv \pm y \pmod{n}$, then find another non-empty subset $T \subseteq \{1, 2, \ldots, t+1\}$ such that $\sum_{i \in T} v_i = 0$, and go to step 5. (In the unlikely case such a subset T does not exist, replace a few of the (a_i, b_i) pairs with new pairs (step 3), and go to step 4.)
9. Compute $d = \gcd(x - y, n)$ and return(d).

3.22 Example (*quadratic sieve algorithm for finding a non-trivial factor of $n = 24961$*)

1. Select the factor base $S = \{-1, 2, 3, 5, 13, 23\}$ of size $t = 6$. (7, 11, 17 and 19 are omitted from S since $\left(\frac{n}{p}\right) = -1$ for these primes.)
2. Compute $m = \lfloor \sqrt{24961} \rfloor = 157$.
3. Following is the data collected for the first $t+1$ values of x for which $q(x)$ is 23-smooth.

i	x	$q(x)$	factorization of $q(x)$	a_i	v_i
1	0	-312	$-2^3 \cdot 3 \cdot 13$	157	$(1,1,1,0,1,0)$
2	1	3	3	158	$(0,0,1,0,0,0)$
3	-1	-625	-5^4	156	$(1,0,0,0,0,0)$
4	2	320	$2^6 \cdot 5$	159	$(0,0,0,1,0,0)$
5	-2	-936	$-2^3 \cdot 3^2 \cdot 13$	155	$(1,1,0,0,1,0)$
6	4	960	$2^6 \cdot 3 \cdot 5$	161	$(0,0,1,1,0,0)$
7	-6	-2160	$-2^4 \cdot 3^3 \cdot 5$	151	$(1,0,1,1,0,0)$

4. By inspection, $v_1 + v_2 + v_5 = 0$. (In the notation of Algorithm 3.21, $T = \{1, 2, 5\}$.)
5. Compute $x = (a_1 a_2 a_5 \bmod n) = 936$.
6. Compute $l_1 = 1, l_2 = 3, l_3 = 2, l_4 = 0, l_5 = 1, l_6 = 0$.
7. Compute $y = -2^3 \cdot 3^2 \cdot 13 \bmod n = 24025$.
8. Since $936 \equiv -24025 \pmod{n}$, another linear dependency must be found.
9. By inspection, $v_3 + v_6 + v_7 = 0$; thus $T = \{3, 6, 7\}$.
10. Compute $x = (a_3 a_6 a_7 \bmod n) = 23405$.
11. Compute $l_1 = 1, l_2 = 5, l_3 = 2, l_4 = 3, l_5 = 0, l_6 = 0$.

12. Compute $y = (-2^5 \cdot 3^2 \cdot 5^3 \bmod n) = 13922$.
13. Now, $23405 \not\equiv \pm 13922 \pmod{n}$, so compute $\gcd(x-y, n) = \gcd(9483, 24961) = 109$. Hence, two non-trivial factors of 24961 are 109 and 229. $\qquad\square$

3.23 Note (*sieving*) Instead of testing smoothness by trial division in step 3.1 of Algorithm 3.21, a more efficient technique known as *sieving* is employed in practice. Observe first that if p is an odd prime in the factor base and p divides $q(x)$, then p also divides $q(x + lp)$ for every integer l. Thus by solving the equation $q(x) \equiv 0 \pmod{p}$ for x (for example, using the algorithms in §3.5.1), one knows either one or two (depending on the number of solutions to the quadratic equation) entire sequences of other values y for which p divides $q(y)$.

The *sieving process* is the following. An array $Q[\ ]$ indexed by x, $-M \le x \le M$, is created and the xth entry is initialized to $\lfloor \lg |q(x)| \rfloor$. Let x_1, x_2 be the solutions to $q(x) \equiv 0 \pmod{p}$, where p is an odd prime in the factor base. Then the value $\lfloor \lg p \rfloor$ is subtracted from those entries $Q[x]$ in the array for which $x \equiv x_1$ or $x_2 \pmod{p}$ and $-M \le x \le M$. This is repeated for each odd prime p in the factor base. (The case of $p = 2$ and prime powers can be handled in a similar manner.) After the sieving, the array entries $Q[x]$ with values near 0 are most likely to be p_t-smooth (roundoff errors must be taken into account), and this can be verified by factoring $q(x)$ by trial division.

3.24 Note (*running time of the quadratic sieve*) To optimize the running time of the quadratic sieve, the size of the factor base should be judiciously chosen. The optimal selection of $t \approx L_n[\frac{1}{2}, \frac{1}{2}]$ is derived from knowledge concerning the distribution of smooth integers close to $\sqrt{n}$. With this choice, Algorithm 3.21 with sieving (Note 3.23) has an expected running time of $L_n[\frac{1}{2}, 1]$, independent of the size of the factors of n.

3.25 Note (*multiple polynomial variant*) In order to collect a sufficient number of (a_i, b_i) pairs, the sieving interval must be quite large. From equation (3.1) it can be seen that $|q(x)|$ increases linearly with $|x|$, and consequently the probability of smoothness decreases. To overcome this problem, a variant (the *multiple polynomial quadratic sieve*) was proposed whereby many appropriately-chosen quadratic polynomials can be used instead of just $q(x)$, each polynomial being sieved over an interval of much smaller length. This variant also has an expected running time of $L_n[\frac{1}{2}, 1]$, and is the method of choice in practice.

3.26 Note (*parallelizing the quadratic sieve*) The multiple polynomial variant of the quadratic sieve is well suited for parallelization. Each node of a parallel computer, or each computer in a network of computers, simply sieves through different collections of polynomials. Any (a_i, b_i) pair found is reported to a central processor. Once sufficient pairs have been collected, the corresponding system of linear equations is solved on a single (possibly parallel) computer.

3.27 Note (*quadratic sieve vs. elliptic curve factoring*) The elliptic curve factoring algorithm (§3.2.4) has the same[4] expected running time as the quadratic sieve factoring algorithm in the special case when n is the product of two primes of equal size. However, for such numbers, the quadratic sieve is superior in practice because the main steps in the algorithm are single precision operations, compared to the much more computationally intensive multiprecision elliptic curve operations required in the elliptic curve algorithm.

---

[4]This does not take into account the different $o(1)$ terms in the two expressions $L_n[\frac{1}{2}, 1]$.

3.2.7 Number field sieve factoring

For several years it was believed by some people that a running time of $L_n[\frac{1}{2}, 1]$ was, in fact, the best achievable by any integer factorization algorithm. This barrier was recently broken with the discovery of the *number field sieve*. Like the quadratic sieve, the number field sieve is an algorithm in the random square family of methods (§3.2.5). That is, it attempts to find integers x and y such that $x^2 \equiv y^2 \pmod{n}$ and $x \not\equiv \pm y \pmod{n}$. To achieve this goal, two factor bases are used, one consisting of all prime numbers less than some bound, and the other consisting of all prime ideals of norm less than some bound in the ring of integers of a suitably-chosen algebraic number field. The details of the algorithm are quite complicated, and are beyond the scope of this book.

A special version of the algorithm (the *special number field sieve*) applies to integers of the form $n = r^e - s$ for small r and $|s|$, and has an expected running time of $L_n[\frac{1}{3}, c]$, where $c = (32/9)^{1/3} \approx 1.526$.

The general version of the algorithm, sometimes called the *general number field sieve*, applies to all integers and has an expected running time of $L_n[\frac{1}{3}, c]$, where $c = (64/9)^{1/3} \approx 1.923$. This is, asymptotically, the fastest algorithm known for integer factorization. The primary reason why the running time of the number field sieve is smaller than that of the quadratic sieve is that the candidate smooth numbers in the former are much smaller than those in the latter.

The general number field sieve was at first believed to be slower than the quadratic sieve for factoring integers having fewer than 150 decimal digits. However, recent experiments have indicated that the general number field sieve is substantially faster than the quadratic sieve even for numbers in the 115 digit range. This implies that the crossover point between the effectiveness of the quadratic sieve vs. the general number field sieve may be 110–120 digits. For this reason, the general number field sieve is considered the current champion of all general-purpose factoring algorithms.

3.3 The RSA problem

The intractability of the RSA problem forms the basis for the security of the RSA public-key encryption scheme (§8.2) and the RSA signature scheme (§11.3.1).

3.28 Definition The *RSA problem* (RSAP) is the following: given a positive integer n that is a product of two distinct odd primes p and q, a positive integer e such that $\gcd(e, (p-1)(q-1)) = 1$, and an integer c, find an integer m such that $m^e \equiv c \pmod{n}$.

In other words, the RSA problem is that of finding e^{th} roots modulo a composite integer n. The conditions imposed on the problem parameters n and e ensure that for each integer $c \in \{0, 1, \ldots, n-1\}$ there is exactly one $m \in \{0, 1, \ldots, n-1\}$ such that $m^e \equiv c \pmod{n}$. Equivalently, the function $f : \mathbb{Z}_n \longrightarrow \mathbb{Z}_n$ defined as $f(m) = m^e \bmod n$ is a permutation.

3.29 Remark (*SQROOT vs. RSA problems*) Since $p - 1$ is even, it follows that e is odd. In particular, $e \neq 2$, and hence the SQROOT problem (Definition 3.43) is *not* a special case of the RSA problem.

As is shown in §8.2.2(i), if the factors of n are known then the RSA problem can be easily solved. This fact is stated next.

3.30 Fact RSAP $\leq_P$ FACTORING. That is, the RSA problem polytime reduces to the integer factorization problem.

It is widely believed that the RSA and the integer factorization problems are computationally equivalent, although no proof of this is known.

3.4 The quadratic residuosity problem

The security of the Goldwasser-Micali probabilistic public-key encryption scheme (§8.7) and the Blum-Blum-Shub pseudorandom bit generator (§5.5.2) are both based on the apparent intractability of the quadratic residuosity problem.

Recall from §2.4.5 that if $n \geq 3$ is an odd integer, then J_n is the set of all $a \in \mathbb{Z}_n^*$ having Jacobi symbol 1. Recall also that Q_n is the set of quadratic residues modulo n and that the set of pseudosquares modulo n is defined by $\widetilde{Q}_n = J_n - Q_n$.

3.31 Definition The *quadratic residuosity problem* (QRP) is the following: given an odd composite integer n and $a \in J_n$, decide whether or not a is a quadratic residue modulo n.

3.32 Remark (*QRP with a prime modulus*) If n is a prime, then it is easy to decide whether $a \in \mathbb{Z}_n^*$ is a quadratic residue modulo n since, by definition, $a \in Q_n$ if and only if $\left(\frac{a}{n}\right) = 1$, and the Legendre symbol $\left(\frac{a}{n}\right)$ can be efficiently calculated by Algorithm 2.149.

Assume now that n is a product of two distinct odd primes p and q. It follows from Fact 2.137 that if $a \in J_n$, then $a \in Q_n$ if and only if $\left(\frac{a}{p}\right) = 1$. Thus, if the factorization of n is known, then QRP can be solved simply by computing the Legendre symbol $\left(\frac{a}{p}\right)$. This observation can be generalized to all integers n and leads to the following fact.

3.33 Fact QRP $\leq_P$ FACTORING. That is, the QRP polytime reduces to the FACTORING problem.

On the other hand, if the factorization of n is unknown, then there is no efficient procedure known for solving QRP, other than by guessing the answer. If $n = pq$, then the probability of a correct guess is $\frac{1}{2}$ since $|Q_n| = |\widetilde{Q}_n|$ (Fact 2.155). It is believed that the QRP is as difficult as the problem of factoring integers, although no proof of this is known.

3.5 Computing square roots in $\mathbb{Z}_n$

The operations of squaring modulo an integer n and extracting square roots modulo an integer n are frequently used in cryptographic functions. The operation of computing square roots modulo n can be performed efficiently when n is a prime, but is difficult when n is a composite integer whose prime factors are unknown.

3.5.1 Case (i): n prime

Recall from Remark 3.32 that if p is a prime, then it is easy to decide if $a \in \mathbb{Z}_p^*$ is a quadratic residue modulo p. If a is, in fact, a quadratic residue modulo p, then the two square roots of a can be efficiently computed, as demonstrated by Algorithm 3.34.

3.34 Algorithm Finding square roots modulo a prime p

INPUT: an odd prime p and an integer a, $1 \leq a \leq p - 1$.
OUTPUT: the two square roots of a modulo p, provided a is a quadratic residue modulo p.

1. Compute the Legendre symbol $\left(\frac{a}{p}\right)$ using Algorithm 2.149. If $\left(\frac{a}{p}\right) = -1$ then return(a does not have a square root modulo p) and terminate.
2. Select integers b, $1 \leq b \leq p - 1$, at random until one is found with $\left(\frac{b}{p}\right) = -1$. ($b$ is a quadratic non-residue modulo p.)
3. By repeated division by 2, write $p - 1 = 2^s t$, where t is odd.
4. Compute $a^{-1} \bmod p$ by the extended Euclidean algorithm (Algorithm 2.142).
5. Set $c \leftarrow b^t \bmod p$ and $r \leftarrow a^{(t+1)/2} \bmod p$ (Algorithm 2.143).
6. For i from 1 to $s - 1$ do the following:
   6.1 Compute $d = (r^2 \cdot a^{-1})^{2^{s-i-1}} \bmod p$.
   6.2 If $d \equiv -1 \pmod{p}$ then set $r \leftarrow r \cdot c \bmod p$.
   6.3 Set $c \leftarrow c^2 \bmod p$.
7. Return(r, $-r$).

Algorithm 3.34 is a randomized algorithm because of the manner in which the quadratic non-residue b is selected in step 2. No deterministic polynomial-time algorithm for finding a quadratic non-residue modulo a prime p is known (see Remark 2.151).

3.35 Fact Algorithm 3.34 has an expected running time of $O((\lg p)^4)$ bit operations.

This running time is obtained by observing that the dominant step (step 6) is executed $s - 1$ times, each iteration involving a modular exponentiation and thus taking $O((\lg p)^3)$ bit operations (Table 2.5). Since in the worst case $s = O(\lg p)$, the running time of $O((\lg p)^4)$ follows. When s is small, the loop in step 6 is executed only a small number of times, and the running time of Algorithm 3.34 is $O((\lg p)^3)$ bit operations. This point is demonstrated next for the special cases $s = 1$ and $s = 2$.

Specializing Algorithm 3.34 to the case $s = 1$ yields the following simple deterministic algorithm for finding square roots when $p \equiv 3 \pmod{4}$.

3.36 Algorithm Finding square roots modulo a prime p where $p \equiv 3 \pmod{4}$

INPUT: an odd prime p where $p \equiv 3 \pmod{4}$, and a square $a \in Q_p$.
OUTPUT: the two square roots of a modulo p.

1. Compute $r = a^{(p+1)/4} \bmod p$ (Algorithm 2.143).
2. Return(r, $-r$).

Specializing Algorithm 3.34 to the case $s = 2$, and using the fact that 2 is a quadratic non-residue modulo p when $p \equiv 5 \pmod{8}$, yields the following simple deterministic algorithm for finding square roots when $p \equiv 5 \pmod{8}$.

3.37 Algorithm Finding square roots modulo a prime p where $p \equiv 5 \pmod 8$

INPUT: an odd prime p where $p \equiv 5 \pmod 8$, and a square $a \in Q_p$.
OUTPUT: the two square roots of a modulo p.

1. Compute $d = a^{(p-1)/4} \bmod p$ (Algorithm 2.143).
2. If $d = 1$ then compute $r = a^{(p+3)/8} \bmod p$.
3. If $d = p - 1$ then compute $r = 2a(4a)^{(p-5)/8} \bmod p$.
4. Return($r, -r$).

3.38 Fact Algorithms 3.36 and 3.37 have running times of $O((\lg p)^3)$ bit operations.

Algorithm 3.39 for finding square roots modulo p is preferable to Algorithm 3.34 when $p - 1 = 2^s t$ with s large.

3.39 Algorithm Finding square roots modulo a prime p

INPUT: an odd prime p and a square $a \in Q_p$.
OUTPUT: the two square roots of a modulo p.

1. Choose random $b \in \mathbb{Z}_p$ until $b^2 - 4a$ is a quadratic non-residue modulo p, i.e., $\left(\frac{b^2 - 4a}{p}\right) = -1$.
2. Let f be the polynomial $x^2 - bx + a$ in $\mathbb{Z}_p[x]$.
3. Compute $r = x^{(p+1)/2} \bmod f$ using Algorithm 2.227. (Note: r will be an integer.)
4. Return($r, -r$).

3.40 Fact Algorithm 3.39 has an expected running time of $O((\lg p)^3)$ bit operations.

3.41 Note (*computing square roots in a finite field*) Algorithms 3.34, 3.36, 3.37, and 3.39 can be extended in a straightforward manner to find square roots in any finite field $\mathbb{F}_q$ of odd order $q = p^m$, p prime, $m \geq 1$. Square roots in finite fields of even order can also be computed efficiently via Fact 3.42.

3.42 Fact Each element $a \in \mathbb{F}_{2^m}$ has exactly one square root, namely $a^{2^{m-1}}$.

3.5.2 Case (ii): n composite

The discussion in this subsection is restricted to the case of computing square roots modulo n, where n is a product of two distinct odd primes p and q. However, all facts presented here generalize to the case where n is an arbitrary composite integer.

Unlike the case where n is a prime, the problem of deciding whether a given $a \in \mathbb{Z}_n^*$ is a quadratic residue modulo a composite integer n, is believed to be a difficult problem. Certainly, if the Jacobi symbol $\left(\frac{a}{n}\right) = -1$, then a is a quadratic non-residue. On the other hand, if $\left(\frac{a}{n}\right) = 1$, then deciding whether or not a is a quadratic residue is precisely the quadratic residuosity problem, considered in §3.4.

3.43 Definition The *square root modulo n problem* (SQROOT) is the following: given a composite integer n and a quadratic residue a modulo n (i.e. $a \in Q_n$), find a square root of a modulo n.

If the factors p and q of n are known, then the SQROOT problem can be solved efficiently by first finding square roots of a modulo p and modulo q, and then combining them using the Chinese remainder theorem (Fact 2.120) to obtain the square roots of a modulo n. The steps are summarized in Algorithm 3.44, which, in fact, finds all of the four square roots of a modulo n.

3.44 Algorithm Finding square roots modulo n given its prime factors p and q

---

INPUT: an integer n, its prime factors p and q, and $a \in Q_n$.
OUTPUT: the four square roots of a modulo n.
1. Use Algorithm 3.39 (or Algorithm 3.36 or 3.37, if applicable) to find the two square roots r and $-r$ of a modulo p.
2. Use Algorithm 3.39 (or Algorithm 3.36 or 3.37, if applicable) to find the two square roots s and $-s$ of a modulo q.
3. Use the extended Euclidean algorithm (Algorithm 2.107) to find integers c and d such that $cp + dq = 1$.
4. Set $x \leftarrow (rdq + scp) \bmod n$ and $y \leftarrow (rdq - scp) \bmod n$.
5. Return($\pm x \bmod n, \pm y \bmod n$).

3.45 Fact Algorithm 3.44 has an expected running time of $O((\lg p)^3)$ bit operations.

Algorithm 3.44 shows that if one can factor n, then the SQROOT problem is easy. More precisely, SQROOT $\leq_P$ FACTORING. The converse of this statement is also true, as stated in Fact 3.46.

3.46 Fact FACTORING $\leq_P$ SQROOT. That is, the FACTORING problem polytime reduces to the SQROOT problem. Hence, since SQROOT $\leq_P$ FACTORING, the FACTORING and SQROOT problems are computationally equivalent.

Justification. Suppose that one has a polynomial-time algorithm A for solving the SQ-ROOT problem. This algorithm can then be used to factor a given composite integer n as follows. Select an integer x at random with $\gcd(x, n) = 1$, and compute $a = x^2 \bmod n$. Next, algorithm A is run with inputs a and n, and a square root y of a modulo n is returned. If $y \equiv \pm x \pmod{n}$, then the trial fails, and the above procedure is repeated with a new x chosen at random. Otherwise, if $y \not\equiv \pm x \pmod{n}$, then $\gcd(x - y, n)$ is guaranteed to be a non-trivial factor of n (Fact 3.18), namely, p or q. Since a has four square roots modulo n ($\pm x$ and $\pm z$ with $\pm z \not\equiv \pm x \pmod{n}$), the probability of success for each attempt is $\frac{1}{2}$. Hence, the expected number of attempts before a factor of n is obtained is two, and consequently the procedure runs in expected polynomial time. □

3.47 Note (*strengthening of Fact 3.46*) The proof of Fact 3.46 can be easily modified to establish the following stronger result. Let $c \geq 1$ be any constant. If there is an algorithm A which, given n, can find a square root modulo n in polynomial time for a $\frac{1}{(\lg n)^c}$ fraction of all quadratic residues $a \in Q_n$, then the algorithm A can be used to factor n in expected polynomial time. The implication of this statement is that if the problem of factoring n is difficult, then for *almost all* $a \in Q_n$ it is difficult to find square roots modulo n.

The computational equivalence of the SQROOT and FACTORING problems was the basis of the first provably secure public-key encryption and signature schemes, presented in §8.3.

3.6 The discrete logarithm problem

The security of many cryptographic techniques depends on the intractability of the discrete logarithm problem. A partial list of these includes Diffie-Hellman key agreement and its derivatives (§12.6), ElGamal encryption (§8.4), and the ElGamal signature scheme and its variants (§11.5). This section summarizes the current knowledge regarding algorithms for solving the discrete logarithm problem.

Unless otherwise specified, algorithms in this section are described in the general setting of a (multiplicatively written) finite cyclic group G of order n with generator α (see Definition 2.167). For a more concrete approach, the reader may find it convenient to think of G as the multiplicative group $\mathbb{Z}_p^*$ of order $p - 1$, where the group operation is simply multiplication modulo p.

3.48 Definition Let G be a finite cyclic group of order n. Let α be a generator of G, and let $\beta \in G$. The *discrete logarithm of β to the base α*, denoted $\log_\alpha \beta$, is the unique integer x, $0 \leq x \leq n - 1$, such that $\beta = \alpha^x$.

3.49 Example Let $p = 97$. Then $\mathbb{Z}_{97}^*$ is a cyclic group of order $n = 96$. A generator of $\mathbb{Z}_{97}^*$ is $\alpha = 5$. Since $5^{32} \equiv 35 \pmod{97}$, $\log_5 35 = 32$ in $\mathbb{Z}_{97}^*$. $\qquad\square$

The following are some elementary facts about logarithms.

3.50 Fact Let α be a generator of a cyclic group G of order n, and let $\beta, \gamma \in G$. Let s be an integer. Then $\log_\alpha(\beta\gamma) = (\log_\alpha \beta + \log_\alpha \gamma) \bmod n$ and $\log_\alpha(\beta^s) = s \log_\alpha \beta \bmod n$.

The groups of most interest in cryptography are the multiplicative group $\mathbb{F}_q^*$ of the finite field $\mathbb{F}_q$ (§2.6), including the particular cases of the multiplicative group $\mathbb{Z}_p^*$ of the integers modulo a prime p, and the multiplicative group $\mathbb{F}_{2^m}^*$ of the finite field $\mathbb{F}_{2^m}$ of characteristic two. Also of interest are the group of units $\mathbb{Z}_n^*$ where n is a composite integer, the group of points on an elliptic curve defined over a finite field, and the jacobian of a hyperelliptic curve defined over a finite field.

3.51 Definition The *discrete logarithm problem* (DLP) is the following: given a prime p, a generator α of $\mathbb{Z}_p^*$, and an element $\beta \in \mathbb{Z}_p^*$, find the integer x, $0 \leq x \leq p - 2$, such that $\alpha^x \equiv \beta \pmod{p}$.

3.52 Definition The *generalized discrete logarithm problem* (GDLP) is the following: given a finite cyclic group G of order n, a generator α of G, and an element $\beta \in G$, find the integer x, $0 \leq x \leq n - 1$, such that $\alpha^x = \beta$.

The discrete logarithm problem in elliptic curve groups and in the jacobians of hyperelliptic curves are not explicitly considered in this section. The discrete logarithm problem in $\mathbb{Z}_n^*$ is discussed further in §3.8.

3.53 Note (*difficulty of the DLP is independent of generator*) Let α and γ be two generators of a cyclic group G of order n, and let $\beta \in G$. Let $x = \log_\alpha \beta$, $y = \log_\gamma \beta$, and $z = \log_\alpha \gamma$. Then $\alpha^x = \beta = \gamma^y = (\alpha^z)^y$. Consequently $x = zy \bmod n$, and

$$\log_\gamma \beta = (\log_\alpha \beta)(\log_\alpha \gamma)^{-1} \bmod n.$$

This means that any algorithm which computes logarithms to the base α can be used to compute logarithms to any other base γ that is also a generator of G.

3.54 Note (*generalization of GDLP*) A more general formulation of the GDLP is the following: given a finite group G and elements $\alpha, \beta \in G$, find an integer x such that $\alpha^x = \beta$, provided that such an integer exists. In this formulation, it is not required that G be a cyclic group, and, even if it is, it is not required that α be a generator of G. This problem may be harder to solve, in general, than GDLP. However, in the case where G is a cyclic group (for example if G is the multiplicative group of a finite field) and the order of α is known, it can be easily recognized whether an integer x satisfying $\alpha^x = \beta$ exists. This is because of the following fact: if G is a cyclic group, α is an element of order n in G, and $\beta \in G$, then there exists an integer x such that $\alpha^x = \beta$ if and only if $\beta^n = 1$.

3.55 Note (*solving the DLP in a cyclic group G of order n is in essence computing an isomorphism between G and $\mathbb{Z}_n$*) Even though any two cyclic groups of the same order are *isomorphic* (that is, they have the same structure although the elements may be written in different representations), an efficient algorithm for computing logarithms in one group does not necessarily imply an efficient algorithm for the other group. To see this, consider that every cyclic group of order n is isomorphic to the additive cyclic group $\mathbb{Z}_n$, i.e., the set of integers $\{0, 1, 2, \dots, n-1\}$ where the group operation is addition modulo n. Moreover, the discrete logarithm problem in the latter group, namely, the problem of finding an integer x such that $ax \equiv b \pmod{n}$ given $a, b \in \mathbb{Z}_n$, is easy as shown in the following. First note that there does not exist a solution x if $d = \gcd(a, n)$ does not divide b (Fact 2.119). Otherwise, if d divides b, the extended Euclidean algorithm (Algorithm 2.107) can be used to find integers s and t such that $as + nt = d$. Multiplying both sides of this equation by the integer b/d gives $a(sb/d) + n(tb/d) = b$. Reducing this equation modulo n yields $a(sb/d) \equiv b \pmod{n}$ and hence $x = (sb/d) \bmod n$ is the desired (and easily obtainable) solution.

The known algorithms for the DLP can be categorized as follows:

1. algorithms which work in arbitrary groups, e.g., exhaustive search (§3.6.1), the baby-step giant-step algorithm (§3.6.2), Pollard's rho algorithm (§3.6.3);
2. algorithms which work in arbitrary groups but are especially efficient if the order of the group has only small prime factors, e.g., Pohlig-Hellman algorithm (§3.6.4); and
3. the index-calculus algorithms (§3.6.5) which are efficient only in certain groups.

3.6.1 Exhaustive search

The most obvious algorithm for GDLP (Definition 3.52) is to successively compute α^0, α^1, $\alpha^2, \dots$ until β is obtained. This method takes $O(n)$ multiplications, where n is the order of α, and is therefore inefficient if n is large (ie. in cases of cryptographic interest).

3.6.2 Baby-step giant-step algorithm

Let $m = \lceil \sqrt{n} \rceil$, where n is the order of α. The baby-step giant-step algorithm is a time-memory trade-off of the method of exhaustive search and is based on the following observation. If $\beta = \alpha^x$, then one can write $x = im + j$, where $0 \le i, j < m$. Hence, $\alpha^x = \alpha^{im} \alpha^j$, which implies $\beta(\alpha^{-m})^i = \alpha^j$. This suggests the following algorithm for computing x.

3.56 Algorithm Baby-step giant-step algorithm for computing discrete logarithms

---

INPUT: a generator α of a cyclic group G of order n, and an element $\beta \in G$.
OUTPUT: the discrete logarithm $x = \log_\alpha \beta$.

1. Set $m \leftarrow \lceil \sqrt{n} \rceil$.
2. Construct a table with entries (j, α^j) for $0 \leq j < m$. Sort this table by second component. (Alternatively, use conventional hashing on the second component to store the entries in a hash table; placing an entry, and searching for an entry in the table takes constant time.)
3. Compute α^{-m} and set $\gamma \leftarrow \beta$.
4. For i from 0 to $m-1$ do the following:
   4.1 Check if γ is the second component of some entry in the table.
   4.2 If $\gamma = \alpha^j$ then return($x = im + j$).
   4.3 Set $\gamma \leftarrow \gamma \cdot \alpha^{-m}$.

Algorithm 3.56 requires storage for $O(\sqrt{n})$ group elements. The table takes $O(\sqrt{n})$ multiplications to construct, and $O(\sqrt{n} \lg n)$ comparisons to sort. Having constructed this table, step 4 takes $O(\sqrt{n})$ multiplications and $O(\sqrt{n})$ table look-ups. Under the assumption that a group multiplication takes more time than $\lg n$ comparisons, the running time of Algorithm 3.56 can be stated more concisely as follows.

3.57 Fact The running time of the baby-step giant-step algorithm (Algorithm 3.56) is $O(\sqrt{n})$ group multiplications.

3.58 Example (*baby-step giant-step algorithm for logarithms in $\mathbb{Z}^*_{113}$*) Let $p = 113$. The element $\alpha = 3$ is a generator of $\mathbb{Z}^*_{113}$ of order $n = 112$. Consider $\beta = 57$. Then $\log_3 57$ is computed as follows.

1. Set $m \leftarrow \lceil \sqrt{112} \rceil = 11$.
2. Construct a table whose entries are $(j, \alpha^j \bmod p)$ for $0 \leq j < 11$:

j	0	1	2	3	4	5	6	7	8	9	10
$3^j \bmod 113$	1	3	9	27	81	17	51	40	7	21	63

and sort the table by second component:

j	0	1	8	2	5	9	3	7	6	10	4
$3^j \bmod 113$	1	3	7	9	17	21	27	40	51	63	81

3. Using Algorithm 2.142, compute $\alpha^{-1} = 3^{-1} \bmod 113 = 38$ and then compute $\alpha^{-m} = 38^{11} \bmod 113 = 58$.
4. Next, $\gamma = \beta \alpha^{-mi} \bmod 113$ for $i = 0, 1, 2, \ldots$ is computed until a value in the second row of the table is obtained. This yields:

i	0	1	2	3	4	5	6	7	8	9
$\gamma = 57 \cdot 58^i \bmod 113$	57	29	100	37	112	55	26	39	2	3

Finally, since $\beta \alpha^{-9m} = 3 = \alpha^1$, $\beta = \alpha^{100}$ and, therefore, $\log_3 57 = 100$. □

3.59 Note (*restricted exponents*) In order to improve performance, some cryptographic protocols which use exponentiation in $\mathbb{Z}^*_p$ select exponents of a special form, usually having small Hamming weight. (The *Hamming weight* of an integer is the number of ones in its binary representation.) Suppose that p is a k-bit prime, and only exponents of Hamming weight t are used. The number of such exponents is $\binom{k}{t}$. Algorithm 3.56 can be modified to search the exponent space in roughly $\binom{k}{t/2}$ steps. The algorithm also applies to exponents that are restricted in certain other ways, and extends to all finite groups.

3.6.3 Pollard's rho algorithm for logarithms

Pollard's rho algorithm (Algorithm 3.60) for computing discrete logarithms is a randomized algorithm with the same expected running time as the baby-step giant-step algorithm (Algorithm 3.56), but which requires a negligible amount of storage. For this reason, it is far preferable to Algorithm 3.56 for problems of practical interest. For simplicity, it is assumed in this subsection that G is a cyclic group whose order n is prime.

The group G is partitioned into three sets S_1, S_2, and S_3 of roughly equal size based on some easily testable property. Some care must be exercised in selecting the partition; for example, $1 \notin S_2$. Define a sequence of group elements $x_0, x_1, x_2, \ldots$ by $x_0 = 1$ and

$$x_{i+1} = f(x_i) \stackrel{\text{def}}{=} \begin{cases} \beta \cdot x_i, & \text{if } x_i \in S_1, \\ x_i^2, & \text{if } x_i \in S_2, \\ \alpha \cdot x_i, & \text{if } x_i \in S_3, \end{cases} \tag{3.2}$$

for $i \geq 0$. This sequence of group elements in turn defines two sequences of integers $a_0, a_1, a_2, \ldots$ and $b_0, b_1, b_2, \ldots$ satisfying $x_i = \alpha^{a_i} \beta^{b_i}$ for $i \geq 0$: $a_0 = 0, b_0 = 0$, and for $i \geq 0$,

$$a_{i+1} = \begin{cases} a_i, & \text{if } x_i \in S_1, \\ 2a_i \bmod n, & \text{if } x_i \in S_2, \\ a_i + 1 \bmod n, & \text{if } x_i \in S_3, \end{cases} \tag{3.3}$$

and

$$b_{i+1} = \begin{cases} b_i + 1 \bmod n, & \text{if } x_i \in S_1, \\ 2b_i \bmod n, & \text{if } x_i \in S_2, \\ b_i, & \text{if } x_i \in S_3. \end{cases} \tag{3.4}$$

Floyd's cycle-finding algorithm (Note 3.8) can then be utilized to find two group elements x_i and x_{2i} such that $x_i = x_{2i}$. Hence $\alpha^{a_i} \beta^{b_i} = \alpha^{a_{2i}} \beta^{b_{2i}}$, and so $\beta^{b_i - b_{2i}} = \alpha^{a_{2i} - a_i}$. Taking logarithms to the base α of both sides of this last equation yields

$$(b_i - b_{2i}) \cdot \log_\alpha \beta \equiv (a_{2i} - a_i) \pmod{n}.$$

Provided $b_i \not\equiv b_{2i} \pmod{n}$ (note: $b_i \equiv b_{2i}$ occurs with negligible probability), this equation can then be efficiently solved to determine $\log_\alpha \beta$.

3.60 Algorithm Pollard's rho algorithm for computing discrete logarithms

---

INPUT: a generator α of a cyclic group G of prime order n, and an element $\beta \in G$.
OUTPUT: the discrete logarithm $x = \log_\alpha \beta$.

    1. Set $x_0 \leftarrow 1$, $a_0 \leftarrow 0$, $b_0 \leftarrow 0$.
    2. For $i = 1, 2, \ldots$ do the following:

        2.1 Using the quantities $x_{i-1}, a_{i-1}, b_{i-1}$, and $x_{2i-2}, a_{2i-2}, b_{2i-2}$ computed previously, compute x_i, a_i, b_i and x_{2i}, a_{2i}, b_{2i} using equations (3.2), (3.3), and (3.4).
        2.2 If $x_i = x_{2i}$, then do the following:

            Set $r \leftarrow b_i - b_{2i} \bmod n$.

            If $r = 0$ then terminate the algorithm with failure; otherwise, compute $x = r^{-1}(a_{2i} - a_i) \bmod n$ and return(x).

In the rare case that Algorithm 3.60 terminates with failure, the procedure can be repeated by selecting random integers a_0, b_0 in the interval $[1, n-1]$, and starting with $x_0 = \alpha^{a_0} \beta^{b_0}$. Example 3.61 with artificially small parameters illustrates Pollard's rho algorithm.

3.61 Example (*Pollard's rho algorithm for logarithms in a subgroup of* $\mathbb{Z}_{383}^*$) The element $\alpha = 2$ is a generator of the subgroup G of $\mathbb{Z}_{383}^*$ of order $n = 191$. Suppose $\beta = 228$. Partition the elements of G into three subsets according to the rule $x \in S_1$ if $x \equiv 1 \pmod{3}$, $x \in S_2$ if $x \equiv 0 \pmod{3}$, and $x \in S_3$ if $x \equiv 2 \pmod{3}$. Table 3.2 shows the values of x_i, a_i, b_i, x_{2i}, a_{2i}, and b_{2i} at the end of each iteration of step 2 of Algorithm 3.60. Note that $x_{14} = x_{28} = 144$. Finally, compute $r = b_{14} - b_{28} \bmod 191 = 125$, $r^{-1} = 125^{-1} \bmod 191 = 136$, and $r^{-1}(a_{28} - a_{14}) \bmod 191 = 110$. Hence, $\log_2 228 = 110$. □

i	x_i	a_i	b_i	x_{2i}	a_{2i}	b_{2i}
1	228	0	1	279	0	2
2	279	0	2	184	1	4
3	92	0	4	14	1	6
4	184	1	4	256	2	7
5	205	1	5	304	3	8
6	14	1	6	121	6	18
7	28	2	6	144	12	38
8	256	2	7	235	48	152
9	152	2	8	72	48	154
10	304	3	8	14	96	118
11	372	3	9	256	97	119
12	121	6	18	304	98	120
13	12	6	19	121	5	51
14	144	12	38	144	10	104

Table 3.2: *Intermediate steps of Pollard's rho algorithm in Example 3.61.*

3.62 Fact Let G be a group of order n, a prime. Assume that the function $f : G \longrightarrow G$ defined by equation (3.2) behaves like a random function. Then the expected running time of Pollard's rho algorithm for discrete logarithms in G is $O(\sqrt{n})$ group operations. Moreover, the algorithm requires negligible storage.

3.6.4 Pohlig-Hellman algorithm

Algorithm 3.63 for computing logarithms takes advantage of the factorization of the order n of the group G. Let $n = p_1^{e_1} p_2^{e_2} \cdots p_t^{e_t}$ be the prime factorization of n. If $x = \log_\alpha \beta$, then the approach is to determine $x_i = x \bmod p_i^{e_i}$ for $1 \leq i \leq t$, and then use Gauss's algorithm (Algorithm 2.121) to recover $x \bmod n$. Each integer x_i is determined by computing the digits $l_0, l_1, \ldots, l_{e_i-1}$ in turn of its p_i-ary representation: $x_i = l_0 + l_1 p_i + \cdots + l_{e_i-1} p_i^{e_i-1}$, where $0 \leq l_j \leq p_i - 1$.

To see that the output of Algorithm 3.63 is correct, observe first that in step 2.3 the order of $\overline{\alpha}$ is q. Next, at iteration j of step 2.4, $\gamma = \alpha^{l_0 + l_1 q + \cdots + l_{j-1} q^{j-1}}$. Hence,

$$
\begin{aligned}
\overline{\beta} &= (\beta/\gamma)^{n/q^{j+1}} = (\alpha^{x - l_0 - l_1 q - \cdots - l_{j-1} q^{j-1}})^{n/q^{j+1}} \\
&= (\alpha^{n/q^{j+1}})^{x_i - l_0 - l_1 q - \cdots - l_{j-1} q^{j-1}} \\
&= (\alpha^{n/q^{j+1}})^{l_j q^j + \cdots + l_{e-1} q^{e-1}} \\
&= (\alpha^{n/q})^{l_j + \cdots + l_{e-1} q^{e-1-j}} = (\overline{\alpha})^{l_j},
\end{aligned}
$$

the last equality being true because $\overline{\alpha}$ has order q. Hence, $\log_{\overline{\alpha}} \overline{\beta}$ is indeed equal to l_j.

3.63 Algorithm Pohlig-Hellman algorithm for computing discrete logarithms

INPUT: a generator α of a cyclic group G of order n, and an element $\beta \in G$.
OUTPUT: the discrete logarithm $x = \log_\alpha \beta$.

1. Find the prime factorization of n: $n = p_1^{e_1} p_2^{e_2} \cdots p_r^{e_r}$, where $e_i \geq 1$.
2. For i from 1 to r do the following:
   (*Compute* $x_i = l_0 + l_1 p_i + \cdots + l_{e_i-1} p_i^{e_i-1}$, *where* $x_i = x \bmod p_i^{e_i}$)

   2.1 (*Simplify the notation*) Set $q \leftarrow p_i$ and $e \leftarrow e_i$.
   2.2 Set $\gamma \leftarrow 1$ and $l_{-1} \leftarrow 0$.
   2.3 Compute $\overline{\alpha} \leftarrow \alpha^{n/q}$.
   2.4 (*Compute the* l_j) For j from 0 to $e - 1$ do the following:
       Compute $\gamma \leftarrow \gamma \alpha^{l_{j-1} q^{j-1}}$ and $\overline{\beta} \leftarrow (\beta \gamma^{-1})^{n/q^{j+1}}$.
       Compute $l_j \leftarrow \log_{\overline{\alpha}} \overline{\beta}$ (e.g., using Algorithm 3.56; see Note 3.67(ii)).
   2.5 Set $x_i \leftarrow l_0 + l_1 q + \cdots + l_{e-1} q^{e-1}$.
3. Use Gauss's algorithm (Algorithm 2.121) to compute the integer x, $0 \leq x \leq n - 1$, such that $x \equiv x_i \pmod{p_i^{e_i}}$ for $1 \leq i \leq r$.
4. Return(x).

Example 3.64 illustrates Algorithm 3.63 with artificially small parameters.

3.64 Example (*Pohlig-Hellman algorithm for logarithms in* $\mathbb{Z}_{251}^*$) Let $p = 251$. The element $\alpha = 71$ is a generator of $\mathbb{Z}_{251}^*$ of order $n = 250$. Consider $\beta = 210$. Then $x = \log_{71} 210$ is computed as follows.

1. The prime factorization of n is $250 = 2 \cdot 5^3$.
2. (a) (Compute $x_1 = x \bmod 2$)
        Compute $\overline{\alpha} = \alpha^{n/2} \bmod p = 250$ and $\overline{\beta} = \beta^{n/2} \bmod p = 250$. Then $x_1 = \log_{250} 250 = 1$.
    (b) (Compute $x_2 = x \bmod 5^3 = l_0 + l_1 5 + l_2 5^2$)
        i. Compute $\overline{\alpha} = \alpha^{n/5} \bmod p = 20$.
        ii. Compute $\gamma = 1$ and $\overline{\beta} = (\beta \gamma^{-1})^{n/5} \bmod p = 149$. Using exhaustive search,[5] compute $l_0 = \log_{20} 149 = 2$.
        iii. Compute $\gamma = \gamma \alpha^2 \bmod p = 21$ and $\overline{\beta} = (\beta \gamma^{-1})^{n/25} \bmod p = 113$. Using exhaustive search, compute $l_1 = \log_{20} 113 = 4$.
        iv. Compute $\gamma = \gamma \alpha^{4 \cdot 5} \bmod p = 115$ and $\overline{\beta} = (\beta \gamma^{-1})^{(p-1)/125} \bmod p = 149$. Using exhaustive search, compute $l_2 = \log_{20} 149 = 2$.
        Hence, $x_2 = 2 + 4 \cdot 5 + 2 \cdot 5^2 = 72$.
3. Finally, solve the pair of congruences $x \equiv 1 \pmod 2$, $x \equiv 72 \pmod{125}$ to get $x = \log_{71} 210 = 197$. $\square$

3.65 Fact Given the factorization of n, the running time of the Pohlig-Hellman algorithm (Algorithm 3.63) is $O(\sum_{i=1}^{r} e_i (\lg n + \sqrt{p_i}))$ group multiplications.

3.66 Note (*effectiveness of Pohlig-Hellman*) Fact 3.65 implies that the Pohlig-Hellman algorithm is efficient only if each prime divisor p_i of n is relatively small; that is, if n is a smooth

---

[5] Exhaustive search is preferable to Algorithm 3.56 when the group is very small (here the order of $\overline{\alpha}$ is 5).

integer (Definition 3.13). An example of a group in which the Pohlig-Hellman algorithm is effective follows. Consider the multiplicative group $\mathbb{Z}_p^*$ where p is the 107-digit prime:

$$p = 227088231986781039743145181950291021585250524967592855$$
$$96453269189798311427475159776411276642277139650833937.$$

The order of $\mathbb{Z}_p^*$ is $n = p - 1 = 2^4 \cdot 104729^8 \cdot 224737^8 \cdot 350377^4$. Since the largest prime divisor of $p - 1$ is only 350377, it is relatively easy to compute logarithms in this group using the Pohlig-Hellman algorithm.

3.67 Note (*miscellaneous*)

(i) If n is a prime, then Algorithm 3.63 (Pohlig-Hellman) is the same as baby-step giant-step (Algorithm 3.56).

(ii) In step 1 of Algorithm 3.63, a factoring algorithm which finds small factors first (e.g., Algorithm 3.9) should be employed; if the order n is not a smooth integer, then Algorithm 3.63 is inefficient anyway.

(iii) The storage required for Algorithm 3.56 in step 2.4 can be eliminated by using instead Pollard's rho algorithm (Algorithm 3.60).

3.6.5 Index-calculus algorithm

The index-calculus algorithm is the most powerful method known for computing discrete logarithms. The technique employed does not apply to all groups, but when it does, it often gives a subexponential-time algorithm. The algorithm is first described in the general setting of a cyclic group G (Algorithm 3.68). Two examples are then presented to illustrate how the index-calculus algorithm works in two kinds of groups that are used in practical applications, namely $\mathbb{Z}_p^*$ (Example 3.69) and $\mathbb{F}_{2^m}^*$ (Example 3.70).

The index-calculus algorithm requires the selection of a relatively small subset S of elements of G, called the *factor base*, in such a way that a significant fraction of elements of G can be efficiently expressed as products of elements from S. Algorithm 3.68 proceeds to precompute a database containing the logarithms of all the elements in S, and then reuses this database each time the logarithm of a particular group element is required.

The description of Algorithm 3.68 is incomplete for two reasons. Firstly, a technique for selecting the factor base S is not specified. Secondly, a method for efficiently generating relations of the form (3.5) and (3.7) is not specified. The factor base S must be a subset of G that is small (so that the system of equations to be solved in step 3 is not too large), but not too small (so that the expected number of trials to generate a relation (3.5) or (3.7) is not too large). Suitable factor bases and techniques for generating relations are known for some cyclic groups including $\mathbb{Z}_p^*$ (see §3.6.5(i)) and $\mathbb{F}_{2^m}^*$ (see §3.6.5(ii)), and, moreover, the multiplicative group $\mathbb{F}_q^*$ of a general finite field $\mathbb{F}_q$.

3.68 Algorithm Index-calculus algorithm for discrete logarithms in cyclic groups

INPUT: a generator α of a cyclic group G of order n, and an element $\beta \in G$.
OUTPUT: the discrete logarithm $y = \log_\alpha \beta$.

1. (*Select a factor base S*) Choose a subset $S = \{p_1, p_2, \dots, p_t\}$ of G such that a "significant proportion" of all elements in G can be efficiently expressed as a product of elements from S.

2. (*Collect linear relations involving logarithms of elements in S*)

2.1 Select a random integer k, $0 \le k \le n - 1$, and compute α^k.

2.2 Try to write α^k as a product of elements in S:

$$\alpha^k = \prod_{i=1}^{t} p_i^{c_i}, \quad c_i \ge 0. \tag{3.5}$$

If successful, take logarithms of both sides of equation (3.5) to obtain a linear relation

$$k \equiv \sum_{i=1}^{t} c_i \log_\alpha p_i \pmod{n}. \tag{3.6}$$

2.3 Repeat steps 2.1 and 2.2 until $t + c$ relations of the form (3.6) are obtained (c is a small positive integer, e.g. $c = 10$, such that the system of equations given by the $t + c$ relations has a unique solution with high probability).

3. (*Find the logarithms of elements in* S) Working modulo n, solve the linear system of $t + c$ equations (in t unknowns) of the form (3.6) collected in step 2 to obtain the values of $\log_\alpha p_i$, $1 \le i \le t$.

4. (*Compute* y)

4.1 Select a random integer k, $0 \le k \le n - 1$, and compute $\beta \cdot \alpha^k$.

4.2 Try to write $\beta \cdot \alpha^k$ as a product of elements in S:

$$\beta \cdot \alpha^k = \prod_{i=1}^{t} p_i^{d_i}, \quad d_i \ge 0. \tag{3.7}$$

If the attempt is unsuccessful then repeat step 4.1. Otherwise, taking logarithms of both sides of equation (3.7) yields $\log_\alpha \beta = (\sum_{i=1}^{t} d_i \log_\alpha p_i - k) \bmod n$; thus, compute $y = (\sum_{i=1}^{t} d_i \log_\alpha p_i - k) \bmod n$ and return(y).

(i) Index-calculus algorithm in $\mathbb{Z}_p^*$

For the field $\mathbb{Z}_p$, p a prime, the factor base S can be chosen as the first t prime numbers. A relation (3.5) is generated by computing $\alpha^k \bmod p$ and then using trial division to check whether this integer is a product of primes in S. Example 3.69 illustrates Algorithm 3.68 in $\mathbb{Z}_p^*$ on a problem with artificially small parameters.

3.69 Example (*Algorithm 3.68 for logarithms in $\mathbb{Z}_{229}^*$*) Let $p = 229$. The element $\alpha = 6$ is a generator of $\mathbb{Z}_{229}^*$ of order $n = 228$. Consider $\beta = 13$. Then $\log_6 13$ is computed as follows, using the index-calculus technique.

1. The factor base is chosen to be the first 5 primes: $S = \{2, 3, 5, 7, 11\}$.

2. The following six relations involving elements of the factor base are obtained (unsuccessful attempts are not shown):

$$6^{100} \bmod 229 = 180 = 2^2 \cdot 3^2 \cdot 5$$
$$6^{18} \bmod 229 = 176 = 2^4 \cdot 11$$
$$6^{12} \bmod 229 = 165 = 3 \cdot 5 \cdot 11$$
$$6^{62} \bmod 229 = 154 = 2 \cdot 7 \cdot 11$$
$$6^{143} \bmod 229 = 198 = 2 \cdot 3^2 \cdot 11$$
$$6^{206} \bmod 229 = 210 = 2 \cdot 3 \cdot 5 \cdot 7.$$

These relations yield the following six equations involving the logarithms of elements in the factor base:

$$100 \equiv 2\log_6 2 + 2\log_6 3 + \log_6 5 \pmod{228}$$
$$18 \equiv 4\log_6 2 + \log_6 11 \pmod{228}$$
$$12 \equiv \log_6 3 + \log_6 5 + \log_6 11 \pmod{228}$$
$$62 \equiv \log_6 2 + \log_6 7 + \log_6 11 \pmod{228}$$
$$143 \equiv \log_6 2 + 2\log_6 3 + \log_6 11 \pmod{228}$$
$$206 \equiv \log_6 2 + \log_6 3 + \log_6 5 + \log_6 7 \pmod{228}.$$

3. Solving the linear system of six equations in five unknowns (the logarithms $x_i = \log_6 p_i$) yields the solutions $\log_6 2 = 21$, $\log_6 3 = 208$, $\log_6 5 = 98$, $\log_6 7 = 107$, and $\log_6 11 = 162$.

4. Suppose that the integer $k = 77$ is selected. Since $\beta \cdot \alpha^k = 13 \cdot 6^{77} \bmod 229 = 147 = 3 \cdot 7^2$, it follows that

$$\log_6 13 = (\log_6 3 + 2\log_6 7 - 77) \bmod 228 = 117. \qquad \square$$

(ii) Index-calculus algorithm in $\mathbb{F}_{2^m}^*$

The elements of the finite field $\mathbb{F}_{2^m}$ are represented as polynomials in $\mathbb{Z}_2[x]$ of degree at most $m-1$, where multiplication is performed modulo a fixed irreducible polynomial $f(x)$ of degree m in $\mathbb{Z}_2[x]$ (see §2.6). The factor base S can be chosen as the set of all irreducible polynomials in $\mathbb{Z}_2[x]$ of degree at most some prescribed bound b. A relation (3.5) is generated by computing $\alpha^k \bmod f(x)$ and then using trial division to check whether this polynomial is a product of polynomials in S. Example 3.70 illustrates Algorithm 3.68 in $\mathbb{F}_{2^m}^*$ on a problem with artificially small parameters.

3.70 Example (*Algorithm 3.68 for logarithms in* $\mathbb{F}_{2^7}^*$) The polynomial $f(x) = x^7 + x + 1$ is irreducible over $\mathbb{Z}_2$. Hence, the elements of the finite field $\mathbb{F}_{2^7}$ of order 128 can be represented as the set of all polynomials in $\mathbb{Z}_2[x]$ of degree at most 6, where multiplication is performed modulo $f(x)$. The order of $\mathbb{F}_{2^7}^*$ is $n = 2^7 - 1 = 127$, and $\alpha = x$ is a generator of $\mathbb{F}_{2^7}^*$. Suppose $\beta = x^4 + x^3 + x^2 + x + 1$. Then $y = \log_x \beta$ can be computed as follows, using the index-calculus technique.

1. The factor base is chosen to be the set of all irreducible polynomials in $\mathbb{Z}_2[x]$ of degree at most 3: $S = \{x, x+1, x^2+x+1, x^3+x+1, x^3+x^2+1\}$.

2. The following five relations involving elements of the factor base are obtained (unsuccessful attempts are not shown):

$$x^{18} \bmod f(x) = x^6 + x^4 \qquad\qquad = x^4(x+1)^2$$
$$x^{105} \bmod f(x) = x^6 + x^5 + x^4 + x \qquad = x(x+1)^2(x^3+x^2+1)$$
$$x^{72} \bmod f(x) = x^6 + x^5 + x^3 + x^2 \qquad = x^2(x+1)^2(x^2+x+1)$$
$$x^{45} \bmod f(x) = x^5 + x^2 + x + 1 \qquad\qquad = (x+1)^2(x^3+x+1)$$
$$x^{121} \bmod f(x) = x^6 + x^5 + x^4 + x^3 + x^2 + x + 1 = (x^3+x+1)(x^3+x^2+1).$$

These relations yield the following five equations involving the logarithms of elements in the factor base (for convenience of notation, let $p_1 = \log_x x$, $p_2 = \log_x(x+$

1), $p_3 = \log_x(x^2 + x + 1)$, $p_4 = \log_x(x^3 + x + 1)$, and $p_5 = \log_x(x^3 + x^2 + 1)$):

$$
\begin{aligned}
18 &\equiv 4p_1 + 2p_2 \pmod{127} \\
105 &\equiv p_1 + 2p_2 + p_5 \pmod{127} \\
72 &\equiv 2p_1 + 2p_2 + p_3 \pmod{127} \\
45 &\equiv 2p_2 + p_4 \pmod{127} \\
121 &\equiv p_4 + p_5 \pmod{127}.
\end{aligned}
$$

3. Solving the linear system of five equations in five unknowns yields the values $p_1 = 1$, $p_2 = 7$, $p_3 = 56$, $p_4 = 31$, and $p_5 = 90$.

4. Suppose $k = 66$ is selected. Since

$$
\beta\alpha^k = (x^4 + x^3 + x^2 + x + 1)x^{66} \bmod f(x) = x^5 + x^3 + x = x(x^2 + x + 1)^2,
$$

it follows that

$$
\log_x(x^4 + x^3 + x^2 + x + 1) = (p_1 + 2p_3 - 66) \bmod 127 = 47. \qquad \square
$$

3.71 Note (*running time of Algorithm 3.68*) To optimize the running time of the index-calculus algorithm, the size t of the factor base should be judiciously chosen. The optimal selection relies on knowledge concerning the distribution of smooth integers in the interval $[1, p-1]$ for the case of $\mathbb{Z}_p^*$, and for the case of $\mathbb{F}_{2^m}^*$ on the distribution of *smooth polynomials* (that is, polynomials all of whose irreducible factors have relatively small degrees) among polynomials in $\mathbb{F}_2[x]$ of degree less than m. With an optimal choice of t, the index-calculus algorithm as described above for $\mathbb{Z}_p^*$ and $\mathbb{F}_{2^m}^*$ has an expected running time of $L_q[\frac{1}{2}, c]$ where $q = p$ or $q = 2^m$, and $c > 0$ is a constant.

3.72 Note (*fastest algorithms known for discrete logarithms in $\mathbb{Z}_p^*$ and $\mathbb{F}_{2^m}^*$*) Currently, the best algorithm known for computing logarithms in $\mathbb{F}_{2^m}^*$ is a variation of the index-calculus algorithm called *Coppersmith's algorithm*, with an expected running time of $L_{2^m}[\frac{1}{3}, c]$ for some constant $c < 1.587$. The best algorithm known for computing logarithms in $\mathbb{Z}_p^*$ is a variation of the index-calculus algorithm called the *number field sieve*, with an expected running time of $L_p[\frac{1}{3}, 1.923]$. The latest efforts in these directions are surveyed in the Notes section (§3.12).

3.73 Note (*parallelization of the index-calculus algorithm*)

(i) For the optimal choice of parameters, the most time-consuming phase of the index-calculus algorithm is usually the generation of relations involving factor base logarithms (step 2 of Algorithm 3.68). The work for this stage can be easily distributed among a network of processors by simply having the processors search for relations independently of each other. The relations generated are collected by a central processor. When enough relations have been generated, the corresponding system of linear equations can be solved (step 3 of Algorithm 3.68) on a single (possibly parallel) computer.

(ii) The database of factor base logarithms need only be computed once for a given finite field. Relative to this, the computation of individual logarithms (step 4 of Algorithm 3.68) is considerably faster.

3.6.6 Discrete logarithm problem in subgroups of $\mathbb{Z}_p^*$

The discrete logarithm problem in subgroups of $\mathbb{Z}_p^*$ has special interest because its presumed intractability is the basis for the security of the U.S. Government NIST Digital Signature Algorithm (§11.5.1), among other cryptographic techniques.

Let p be a prime and q a prime divisor of $p-1$. Let G be the unique cyclic subgroup of $\mathbb{Z}_p^*$ of order q, and let α be a generator of G. Then the discrete logarithm problem in G is the following: given p, q, α, and $\beta \in G$, find the unique integer $x, 0 \le x \le q-1$, such that $\alpha^x \equiv \beta \pmod{p}$. The powerful index-calculus algorithms do not appear to apply directly in G. That is, one needs to apply the index-calculus algorithm in the group $\mathbb{Z}_p^*$ itself in order to compute logarithms in the smaller group G. Consequently, there are two approaches one could take to computing logarithms in G:

1. Use a "square-root" algorithm directly in G, such as Pollard's rho algorithm (Algorithm 3.60). The running time of this approach is $O(\sqrt{q})$.
2. Let γ be a generator of $\mathbb{Z}_p^*$, and let $l = (p-1)/q$. Use an index-calculus algorithm in $\mathbb{Z}_p^*$ to find integers y and z such that $\alpha = \gamma^y$ and $\beta = \gamma^z$. Then $x = \log_\alpha \beta = (z/l)(y/l)^{-1} \bmod q$. (Since y and z are both divisible by l, y/l and z/l are indeed integers.) The running time of this approach is $L_p[\frac{1}{3}, c]$ if the number field sieve is used.

Which of the two approaches is faster depends on the relative size of $\sqrt{q}$ and $L_p[\frac{1}{3}, c]$.

3.7 The Diffie-Hellman problem

The Diffie-Hellman problem is closely related to the well-studied discrete logarithm problem (DLP) of §3.6. It is of significance to public-key cryptography because its apparent intractability forms the basis for the security of many cryptographic schemes including Diffie-Hellman key agreement and its derivatives (§12.6), and ElGamal public-key encryption (§8.4).

3.74 Definition The *Diffie-Hellman problem (DHP)* is the following: given a prime p, a generator α of $\mathbb{Z}_p^*$, and elements $\alpha^a \bmod p$ and $\alpha^b \bmod p$, find $\alpha^{ab} \bmod p$.

3.75 Definition The *generalized Diffie-Hellman problem (GDHP)* is the following: given a finite cyclic group G, a generator α of G, and group elements α^a and α^b, find α^{ab}.

Suppose that the discrete logarithm problem in $\mathbb{Z}_p^*$ could be efficiently solved. Then given $\alpha, p, \alpha^a \bmod p$ and $\alpha^b \bmod p$, one could first find a from α, p, and $\alpha^a \bmod p$ by solving a discrete logarithm problem, and then compute $(\alpha^b)^a = \alpha^{ab} \bmod p$. This establishes the following relation between the Diffie-Hellman problem and the discrete logarithm problem.

3.76 Fact DHP $\le_P$ DLP. That is, DHP polytime reduces to the DLP. More generally, GDHP $\le_P$ GDLP.

The question then remains whether the GDLP and GDHP are computationally equivalent. This remains unknown; however, some recent progress in this regard is summarized in Fact 3.77. Recall that ϕ is the Euler phi function (Definition 2.100), and an integer is B-smooth if all its prime factors are $\le B$ (Definition 3.13).

3.77 Fact (*known equivalences between GDHP and GDLP*)

    (i) Let p be a prime where the factorization of $p-1$ is known. Suppose also that $\phi(p-1)$ is B-smooth, where $B = O((\ln p)^c)$ for some constant c. Then the DHP and DLP in $\mathbb{Z}_p^*$ are computationally equivalent.

    (ii) More generally, let G be a finite cyclic group of order n where the factorization of n is known. Suppose also that $\phi(n)$ is B-smooth, where $B = O((\ln n)^c)$ for some constant c. Then the GDHP and GDLP in G are computationally equivalent.

    (iii) Let G be a finite cyclic group of order n where the factorization of n is known. If for each prime divisor p of n either $p-1$ or $p+1$ is B-smooth, where $B = O((\ln n)^c)$ for some constant c, then the GDHP and GDLP in G are computationally equivalent.

3.8 Composite moduli

The group of units of $\mathbb{Z}_n$, namely $\mathbb{Z}_n^*$, has been proposed for use in several cryptographic mechanisms, including the key agreement protocols of Yacobi and McCurley (see §12.6 notes on page 536) and the identification scheme of Girault (see §10.4 notes on page 423). There are connections of cryptographic interest between the discrete logarithm and Diffie-Hellman problems in (cyclic subgroups of) $\mathbb{Z}_n^*$, and the problem of factoring n. This section summarizes the results known along these lines.

3.78 Fact Let n be a composite integer. If the discrete logarithm problem in $\mathbb{Z}_n^*$ can be solved in polynomial time, then n can be factored in expected polynomial time.

    In other words, the discrete logarithm problem in $\mathbb{Z}_n^*$ is at least as difficult as the problem of factoring n. Fact 3.79 is a partial converse to Fact 3.78 and states that the discrete logarithm in $\mathbb{Z}_n^*$ is no harder than the combination of the problems of factoring n and computing discrete logarithms in $\mathbb{Z}_p^*$ for each prime factor p of n.

3.79 Fact Let n be a composite integer. The discrete logarithm problem in $\mathbb{Z}_n^*$ polytime reduces to the combination of the integer factorization problem and the discrete logarithm problem in $\mathbb{Z}_p^*$ for each prime factor p of n.

    Fact 3.80 states that the Diffie-Hellman problem in $\mathbb{Z}_n^*$ is at least as difficult as the problem of factoring n.

3.80 Fact Let $n = pq$ where p and q are odd primes. If the Diffie-Hellman problem in $\mathbb{Z}_n^*$ can be solved in polynomial time for a non-negligible proportion of all bases $\alpha \in \mathbb{Z}_n^*$, then n can be factored in expected polynomial time.

3.9 Computing individual bits

While the discrete logarithm problem in $\mathbb{Z}_p^*$ (§3.6), the RSA problem (§3.3), and the problem of computing square roots modulo a composite integer n (§3.5.2) appear to be intractable, when the problem parameters are carefully selected, it remains possible that it is much easier to compute some partial information about the solution, for example, its least significant bit. It turns out that while some bits of the solution to these problems are indeed easy to compute, other bits are equally difficult to compute as the entire solution. This section summarizes the results known along these lines. The results have applications to the construction of probabilistic public-key encryption schemes (§8.7) and pseudorandom bit generation (§5.5).

Recall (Definition 1.12) that a function f is called a one-way function if $f(x)$ is easy to compute for all x in its domain, but for most y in the range of f, it is computationally infeasible to find any x such that $f(x) = y$.

Three (candidate) one-way functions

Although no proof is known for the existence of a one-way function, it is widely believed that one-way functions do exist. The following are candidate one-way functions (in fact, one-way permutations) since they are easy to compute, but their inversion requires the solution of the discrete logarithm problem in $\mathbb{Z}_p^*$, the RSA problem, or the problem of computing square roots modulo n, respectively:

1. *exponentiation modulo p*. Let p be a prime and let α be a generator of $\mathbb{Z}_p^*$. The function is $f : \mathbb{Z}_p^* \longrightarrow \mathbb{Z}_p^*$ defined as $f(x) = \alpha^x \bmod p$.
2. *RSA function*. Let p and q be distinct odd primes, $n = pq$, and let e be an integer such that $\gcd(e, (p-1)(q-1)) = 1$. The function is $f : \mathbb{Z}_n \longrightarrow \mathbb{Z}_n$ defined as $f(x) = x^e \bmod n$.
3. *Rabin function*. Let $n = pq$, where p and q are distinct primes each congruent to 3 modulo 4. The function is $f : Q_n \longrightarrow Q_n$ defined as $f(x) = x^2 \bmod n$. (Recall from Fact 2.160 that f is a permutation, and from Fact 3.46 that inverting f, i.e., computing principal square roots, is difficult assuming integer factorization is intractable.)

The following definitions are used in §3.9.1, 3.9.2, and 3.9.3.

3.81 Definition Let $f : S \longrightarrow S$ be a one-way function, where S is a finite set. A Boolean predicate $B : S \longrightarrow \{0, 1\}$ is said to be a *hard predicate* for f if:

   (i) $B(x)$ is easy to compute given $x \in S$; and
   (ii) an oracle which computes $B(x)$ correctly with probability[6] significantly greater than $\frac{1}{2}$ given only $f(x)$ (where $x \in S$) can be used to invert f easily.

   Informally, B is a hard predicate for the one-way function f if determining the single bit $B(x)$ of information about x, given only $f(x)$, is as difficult as inverting f itself.

3.82 Definition Let $f : S \longrightarrow S$ be a one-way function, where S is a finite set. A k-bit predicate $B^{(k)} : S \longrightarrow \{0,1\}^k$ is said to be a *hard k-bit predicate* for f if:

   (i) $B^{(k)}(x)$ is easy to compute given $x \in S$; and
   (ii) for every Boolean predicate $B : \{0,1\}^k \longrightarrow \{0,1\}$, an oracle which computes $B(B^{(k)}(x))$ correctly with probability significantly greater than $\frac{1}{2}$ given only $f(x)$ (where $x \in S$) can be used to invert f easily.

If such a $B^{(k)}$ exists, then f is said to *hide* k bits, or the k bits are said to be *simultaneously secure*.

   Informally, $B^{(k)}$ is a hard k-bit predicate for the one-way function f if determining any partial information whatsoever about $B^{(k)}(x)$, given only $f(x)$, is as difficult as inverting f itself.

---

[6]In Definitions 3.81 and 3.82, the probability is taken over all choices of $x \in S$ and random coin tosses of the oracle.

3.9.1 The discrete logarithm problem in $\mathbb{Z}_p^*$ — individual bits

Let p be an odd prime and α a generator of $\mathbb{Z}_p^*$. Assume that the discrete logarithm problem in $\mathbb{Z}_p^*$ is intractable. Let $\beta \in \mathbb{Z}_p^*$, and let $x = \log_\alpha \beta$. Recall from Fact 2.135 that β is a quadratic residue modulo p if and only if x is even. Hence, the least significant bit of x is equal to $\left(1 - \left(\frac{\beta}{p}\right)\right)/2$, where the Legendre symbol $\left(\frac{\beta}{p}\right)$ can be efficiently computed (Algorithm 2.149). More generally, the following is true.

3.83 Fact Let p be an odd prime, and let α be a generator of $\mathbb{Z}_p^*$. Suppose that $p - 1 = 2^s t$, where t is odd. Then there is an efficient algorithm which, given $\beta \in \mathbb{Z}_p^*$, computes the s least significant bits of $x = \log_\alpha \beta$.

3.84 Fact Let p be a prime and α a generator of $\mathbb{Z}_p^*$. Define the predicate $B : \mathbb{Z}_p^* \longrightarrow \{0, 1\}$ by

$$B(x) = \begin{cases} 0, & \text{if } 1 \leq x \leq (p-1)/2, \\ 1, & \text{if } (p-1)/2 < x \leq p - 1. \end{cases}$$

Then B is a hard predicate for the function of exponentiation modulo p. In other words, given p, α, and β, computing the single bit $B(x)$ of the discrete logarithm $x = \log_\alpha \beta$ is as difficult as computing the entire discrete logarithm.

3.85 Fact Let p be a prime and α a generator of $\mathbb{Z}_p^*$. Let $k = O(\lg \lg p)$ be an integer. Let the interval $[1, p-1]$ be partitioned into 2^k intervals $I_0, I_1, \ldots, I_{2^k - 1}$ of roughly equal lengths. Define the k-bit predicate $B^{(k)} : \mathbb{Z}_p^* \longrightarrow \{0, 1\}^k$ by $B^{(k)}(x) = j$ if $x \in I_j$. Then $B^{(k)}$ is a hard k-bit predicate for the function of exponentiation modulo p.

3.9.2 The RSA problem — individual bits

Let n be a product of two distinct odd primes p and q, and let e be an integer such that $\gcd(e, (p-1)(q-1)) = 1$. Given n, e, and $c = x^e \bmod n$ (for some $x \in \mathbb{Z}_n$), some information about x is easily obtainable. For example, since e is an odd integer,

$$\left(\frac{c}{n}\right) = \left(\frac{x^e}{n}\right) = \left(\frac{x}{n}\right)^e = \left(\frac{x}{n}\right),$$

and hence the single bit of information $\left(\frac{x}{n}\right)$ can be obtained simply by computing the Jacobi symbol $\left(\frac{c}{n}\right)$ (Algorithm 2.149). There are, however, other bits of information about x that are difficult to compute, as the next two results show.

3.86 Fact Define the predicate $B : \mathbb{Z}_n \longrightarrow \{0, 1\}$ by $B(x) = x \bmod 2$; that is, $B(x)$ is the least significant bit of x. Then B is a hard predicate for the RSA function (see page 115).

3.87 Fact Let $k = O(\lg \lg n)$ be an integer. Define the k-bit predicate $B^{(k)} : \mathbb{Z}_n \longrightarrow \{0, 1\}^k$ by $B^{(k)}(x) = x \bmod 2^k$. That is, $B^{(k)}(x)$ consists of the k least significant bits of x. Then $B^{(k)}$ is a hard k-bit predicate for the RSA function.

Thus the RSA function has $\lg \lg n$ simultaneously secure bits.

3.9.3 The Rabin problem — individual bits

Let $n = pq$, where p and q are distinct primes each congruent to 3 modulo 4.

3.88 Fact Define the predicate $B : Q_n \longrightarrow \{0,1\}$ by $B(x) = x \bmod 2$; that is, $B(x)$ is the least significant bit of the quadratic residue x. Then B is a hard predicate for the Rabin function (see page 115).

3.89 Fact Let $k = O(\lg \lg n)$ be an integer. Define the k-bit predicate $B^{(k)} : Q_n \longrightarrow \{0,1\}^k$ by $B^{(k)}(x) = x \bmod 2^k$. That is, $B^{(k)}(x)$ consists of the k least significant bits of the quadratic residue x. Then $B^{(k)}$ is a hard k-bit predicate for the Rabin function.

Thus the Rabin function has $\lg \lg n$ simultaneously secure bits.

3.10 The subset sum problem

The difficulty of the subset sum problem was the basis for the (presumed) security of the first public-key encryption scheme, called the Merkle-Hellman knapsack scheme (§8.6.1).

3.90 Definition The *subset sum problem* (SUBSET-SUM) is the following: given a set $\{a_1, a_2, \dots, a_n\}$ of positive integers, called a *knapsack set*, and a positive integer s, determine whether or not there is a subset of the a_j that sum to s. Equivalently, determine whether or not there exist $x_i \in \{0,1\}$, $1 \le i \le n$, such that $\sum_{i=1}^{n} a_i x_i = s$.

The subset sum problem above is stated as a decision problem. It can be shown that the problem is computationally equivalent to its computational version which is to actually determine the x_i such that $\sum_{i=1}^{n} a_i x_i = s$, provided that such x_i exist. Fact 3.91 provides evidence of the intractability of the subset sum problem.

3.91 Fact The subset sum problem is **NP**-complete. The computational version of the subset sum problem is **NP**-hard.

Algorithms 3.92 and 3.94 give two methods for solving the computational version of the subset sum problem; both are exponential-time algorithms. Algorithm 3.94 is the fastest method known for the general subset sum problem.

3.92 Algorithm Naive algorithm for subset sum problem

INPUT: a set of positive integers $\{a_1, a_2, \dots, a_n\}$ and a positive integer s.
OUTPUT: $x_i \in \{0,1\}$, $1 \le i \le n$, such that $\sum_{i=1}^{n} a_i x_i = s$, provided such x_i exist.

1. For each possible vector $(x_1, x_2, \dots, x_n) \in (\mathbb{Z}_2)^n$ do the following:
   1.1 Compute $l = \sum_{i=1}^{n} a_i x_i$.
   1.2 If $l = s$ then return(a solution is $(x_1, x_2, \dots, x_n)$).
2. Return(no solution exists).

3.93 Fact Algorithm 3.92 takes $O(2^n)$ steps and, hence, is inefficient.

3.94 Algorithm Meet-in-the-middle algorithm for subset sum problem

INPUT: a set of positive integers $\{a_1, a_2, \ldots, a_n\}$ and a positive integer s.
OUTPUT: $x_i \in \{0, 1\}$, $1 \leq i \leq n$, such that $\sum_{i=1}^{n} a_i x_i = s$, provided such x_i exist.

1. Set $t \leftarrow \lfloor n/2 \rfloor$.
2. Construct a table with entries $(\sum_{i=1}^{t} a_i x_i, (x_1, x_2, \ldots, x_t))$ for $(x_1, x_2, \ldots, x_t) \in (\mathbb{Z}_2)^t$. Sort this table by first component.
3. For each $(x_{t+1}, x_{t+2}, \ldots, x_n) \in (\mathbb{Z}_2)^{n-t}$, do the following:
    3.1 Compute $l = s - \sum_{i=t+1}^{n} a_i x_i$ and check, using a binary search, whether l is the first component of some entry in the table.
    3.2 If $l = \sum_{i=1}^{t} a_i x_i$ then return(a solution is $(x_1, x_2, \ldots, x_n)$).
4. Return(no solution exists).

3.95 Fact Algorithm 3.94 takes $O(n 2^{n/2})$ steps and, hence, is inefficient.

3.10.1 The L^3-lattice basis reduction algorithm

The L^3-lattice basis reduction algorithm is a crucial component in many number-theoretic algorithms. It is useful for solving certain subset sum problems, and has been used for cryptanalyzing public-key encryption schemes which are based on the subset sum problem.

3.96 Definition Let $x = (x_1, x_2, \ldots, x_n)$ and $y = (y_1, y_2, \ldots, y_n)$ be two vectors in $\mathbb{R}^n$. The *inner product* of x and y is the real number

$$< x, y > = x_1 y_1 + x_2 y_2 + \cdots + x_n y_n.$$

3.97 Definition Let $y = (y_1, y_2, \ldots, y_n)$ be a vector in $\mathbb{R}^n$. The *length* of y is the real number

$$\|y\| = \sqrt{< y, y >} = \sqrt{y_1^2 + y_2^2 + \cdots + y_n^2}.$$

3.98 Definition Let $B = \{b_1, b_2, \ldots, b_m\}$ be a set of linearly independent vectors in $\mathbb{R}^n$ (so that $m \leq n$). The set L of all integer linear combinations of $b_1, b_2, \ldots, b_m$ is called a *lattice* of *dimension* m; that is, $L = \mathbb{Z}b_1 + \mathbb{Z}b_2 + \cdots + \mathbb{Z}b_m$. The set B is called a *basis* for the lattice L.

A lattice can have many different bases. A basis consisting of vectors of relatively small lengths is called *reduced*. The following definition provides a useful notion of a reduced basis, and is based on the Gram-Schmidt orthogonalization process.

3.99 Definition Let $B = \{b_1, b_2, \ldots, b_n\}$ be a basis for a lattice $L \subset \mathbb{R}^n$. Define the vectors b_i^* ($1 \leq i \leq n$) and the real numbers $\mu_{i,j}$ ($1 \leq j < i \leq n$) inductively by

$$\mu_{i,j} = \frac{< b_i, b_j^* >}{< b_j^*, b_j^* >}, \quad 1 \leq j < i \leq n, \tag{3.8}$$

$$b_i^* = b_i - \sum_{j=1}^{i-1} \mu_{i,j} b_j^*, \quad 1 \leq i \leq n. \tag{3.9}$$

The basis B is said to be *reduced* (more precisely, *Lovász-reduced*) if

$$|\mu_{i,j}| \leq \frac{1}{2}, \quad \text{for } 1 \leq j < i \leq n,$$

and

$$\|b_i^*\|^2 \geq \left(\frac{3}{4} - \mu_{i,i-1}^2\right) \|b_{i-1}^*\|^2, \quad \text{for } 1 < i \leq n. \tag{3.10}$$

Fact 3.100 explains the sense in which the vectors in a reduced basis are relatively short.

3.100 Fact Let $L \subset \mathbb{R}^n$ be a lattice with a reduced basis $\{b_1, b_2, \ldots, b_n\}$.

(i) For every non-zero $x \in L$, $\|b_1\| \leq 2^{(n-1)/2} \|x\|$.

(ii) More generally, for any set $\{a_1, a_2, \ldots, a_t\}$ of linearly independent vectors in L,
$$\|b_j\| \leq 2^{(n-1)/2} \max(\|a_1\|, \|a_2\|, \ldots, \|a_t\|), \quad \text{for } 1 \leq j \leq t.$$

The L^3-lattice basis reduction algorithm (Algorithm 3.101) is a polynomial-time algorithm (Fact 3.103) for finding a reduced basis, given a basis for a lattice.

3.101 Algorithm L^3-lattice basis reduction algorithm

INPUT: a basis $(b_1, b_2, \ldots, b_n)$ for a lattice L in $\mathbb{R}^m$, $m \geq n$.
OUTPUT: a reduced basis for L.

1. $b_1^* \leftarrow b_1$, $B_1 \leftarrow\ <b_1^*, b_1^*>$.
2. For i from 2 to n do the following:
   2.1 $b_i^* \leftarrow b_i$.
   2.2 For j from 1 to $i-1$, set $\mu_{i,j} \leftarrow\ <b_i, b_j^*>/B_j$ and $b_i^* \leftarrow b_i^* - \mu_{i,j} b_j^*$.
   2.3 $B_i \leftarrow\ <b_i^*, b_i^*>$.
3. $k \leftarrow 2$.
4. Execute subroutine RED($k, k-1$) to possibly update some $\mu_{i,j}$.
5. If $B_k < (\frac{3}{4} - \mu_{k,k-1}^2) B_{k-1}$ then do the following:
   5.1 Set $\mu \leftarrow \mu_{k,k-1}$, $B \leftarrow B_k + \mu^2 B_{k-1}$, $\mu_{k,k-1} \leftarrow \mu B_{k-1}/B$, $B_k \leftarrow B_{k-1} B_k/B$, and $B_{k-1} \leftarrow B$.
   5.2 Exchange b_k and b_{k-1}.
   5.3 If $k > 2$ then exchange $\mu_{k,j}$ and $\mu_{k-1,j}$ for $j = 1, 2, \ldots, k-2$.
   5.4 For $i = k+1, k+2, \ldots, n$:
      Set $t \leftarrow \mu_{i,k}$, $\mu_{i,k} \leftarrow \mu_{i,k-1} - \mu t$, and $\mu_{i,k-1} \leftarrow t + \mu_{k,k-1} \mu_{i,k}$.
   5.5 $k \leftarrow \max(2, k-1)$.
   5.6 Go to step 4.
   Otherwise, for $l = k-2, k-3, \ldots, 1$, execute RED(k, l), and finally set $k \leftarrow k+1$.
6. If $k \leq n$ then go to step 4. Otherwise, return($b_1, b_2, \ldots, b_n$).

RED(k, l) If $|\mu_{k,l}| > \frac{1}{2}$ then do the following:

1. $r \leftarrow \lfloor 0.5 + \mu_{k,l} \rfloor$, $b_k \leftarrow b_k - r b_l$.
2. For j from 1 to $l-1$, set $\mu_{k,j} \leftarrow \mu_{k,j} - r \mu_{l,j}$.
3. $\mu_{k,l} \leftarrow \mu_{k,l} - r$.

3.102 Note (*explanation of selected steps of Algorithm 3.101*)

(i) Steps 1 and 2 initialize the algorithm by computing b_i^* ($1 \leq i \leq n$) and $\mu_{i,j}$ ($1 \leq j < i \leq n$) as defined in equations (3.9) and (3.8), and also $B_i =\ <b_i^*, b_i^*>$ ($1 \leq i \leq n$).

(ii) k is a variable such that the vectors $b_1, b_2, \ldots, b_{k-1}$ are reduced (initially $k = 2$ in step 3). The algorithm then attempts to modify b_k, so that $b_1, b_2, \ldots, b_k$ are reduced.

(iii) In step 4, the vector b_k is modified appropriately so that $|\mu_{k,k-1}| \leq \frac{1}{2}$, and the $\mu_{k,j}$ are updated for $1 \leq j < k - 1$.

(iv) In step 5, if the condition of equation (3.10) is violated for $i = k$, then vectors b_k and b_{k-1} are exchanged and their corresponding parameters are updated. Also, k is decremented by 1 since then it is only guaranteed that $b_1, b_2, \ldots, b_{k-2}$ are reduced. Otherwise, b_k is modified appropriately so that $|\mu_{k,j}| \leq \frac{1}{2}$ for $j = 1, 2, \ldots, k - 2$, while keeping (3.10) satisfied. k is then incremented because now $b_1, b_2, \ldots, b_k$ are reduced.

It can be proven that the L^3-algorithm terminates after a finite number of iterations. Note that if L is an integer lattice, i.e. $L \subset \mathbb{Z}^n$, then the L^3-algorithm only operates on rational numbers. The precise running time is given next.

3.103 Fact Let $L \subset \mathbb{Z}^n$ be a lattice with basis $\{b_1, b_2, \ldots, b_n\}$, and let $C \in \mathbb{R}$, $C \geq 2$, be such that $\|b_i\|^2 \leq C$ for $i = 1, 2, \ldots, n$. Then the number of arithmetic operations needed by Algorithm 3.101 is $O(n^4 \log C)$, on integers of size $O(n \log C)$ bits.

3.10.2 Solving subset sum problems of low density

The density of a knapsack set, as defined below, provides a measure of the size of the knapsack elements.

3.104 Definition Let $S = \{a_1, a_2, \ldots, a_n\}$ be a knapsack set. The *density* of S is defined to be

$$d = \frac{n}{\max\{\lg a_i \mid 1 \leq i \leq n\}}.$$

Algorithm 3.105 reduces the subset sum problem to one of finding a particular short vector in a lattice. By Fact 3.100, the reduced basis produced by the L^3-algorithm includes a vector of length which is guaranteed to be within a factor of $2^{(n-1)/2}$ of the shortest non-zero vector of the lattice. In practice, however, the L^3-algorithm usually finds a vector which is much shorter than what is guaranteed by Fact 3.100. Hence, the L^3-algorithm can be expected to find the short vector which yields a solution to the subset sum problem, provided that this vector is shorter than most of the non-zero vectors in the lattice.

3.105 Algorithm Solving subset sum problems using L^3-algorithm

INPUT: a set of positive integers $\{a_1, a_2, \ldots, a_n\}$ and an integer s.
OUTPUT: $x_i \in \{0, 1\}$, $1 \leq i \leq n$, such that $\sum_{i=1}^{n} a_i x_i = s$, provided such x_i exist.

1. Let $m = \lceil \frac{1}{2}\sqrt{n} \rceil$.
2. Form an $(n+1)$-dimensional lattice L with basis consisting of the rows of the matrix

$$A = \begin{pmatrix} 1 & 0 & 0 & \cdots & 0 & ma_1 \\ 0 & 1 & 0 & \cdots & 0 & ma_2 \\ 0 & 0 & 1 & \cdots & 0 & ma_3 \\ \vdots & \vdots & \vdots & \ddots & \vdots & \vdots \\ 0 & 0 & 0 & \cdots & 1 & ma_n \\ \frac{1}{2} & \frac{1}{2} & \frac{1}{2} & \cdots & \frac{1}{2} & ms \end{pmatrix}$$

3. Find a reduced basis B of L (use Algorithm 3.101).
4. For each vector $y = (y_1, y_2, \ldots, y_{n+1})$ in B, do the following:

4.1 If $y_{n+1} = 0$ and $y \in \{-\frac{1}{2}, \frac{1}{2}\}$ for all $i = 1, 2, \ldots, n$, then do the following:

For $i = 1, 2, \ldots, n$, set $x_i \leftarrow y_i + \frac{1}{2}$.

If $\sum_{i=1}^{n} a_i x_i = s$, then return(a solution is $(x_1, x_2, \ldots, x_n)$).

For $i = 1, 2, \ldots, n$, set $x_i \leftarrow - y_i + \frac{1}{2}$.

If $\sum_{i=1}^{n} a_i x_i = s$, then return(a solution is $(x_1, x_2, \ldots, x_n)$).

5. Return(FAILURE). (Either no solution exists, or the algorithm has failed to find one.)

Justification. Let the rows of the matrix A be $b_1, b_2, \ldots, b_{n+1}$, and let L be the $(n+1)$-dimensional lattice generated by these vectors. If $(x_1, x_2, \ldots, x_n)$ is a solution to the subset sum problem, the vector $y = \sum_{i=1}^{n} x_i b_i - b_{n+1}$ is in L. Note that $y_i \in \{-\frac{1}{2}, \frac{1}{2}\}$ for $i = 1, 2, \ldots, n$ and $y_{n+1} = 0$. Since $\|y\| = \sqrt{y_1^2 + y_2^2 + \cdots + y_{n+1}^2}$ the vector y is a vector of short length in L. If the density of the knapsack set is small, i.e. the a_i are large, then most vectors in L will have relatively large lengths, and hence y may be the unique shortest non-zero vector in L. If this is indeed the case, then there is good possibility of the L^3-algorithm finding a basis which includes this vector.

Algorithm 3.105 is not guaranteed to succeed. Assuming that the L^3-algorithm always produces a basis which includes the shortest non-zero lattice vector, Algorithm 3.105 succeeds with high probability if the density of the knapsack set is less than 0.9408.

3.10.3 Simultaneous diophantine approximation

Simultaneous diophantine approximation is concerned with approximating a vector $(\frac{q_1}{q}, \frac{q_2}{q}, \ldots, \frac{q_n}{q})$ of rational numbers (more generally, a vector $(\alpha_1, \alpha_2, \ldots, \alpha_n)$ of real numbers) by a vector $(\frac{p_1}{p}, \frac{p_2}{p}, \ldots, \frac{p_n}{p})$ of rational numbers with a smaller denominator p. Algorithms for finding simultaneous diophantine approximation have been used to break some knapsack public-key encryption schemes (§8.6).

3.106 Definition The vector $(\frac{p_1}{p}, \frac{p_2}{p}, \ldots, \frac{p_n}{p})$ of rational numbers is said to be a *simultaneous diophantine approximation of δ-quality* to the vector $(\frac{q_1}{q}, \frac{q_2}{q}, \ldots, \frac{q_n}{q})$ of rational numbers if $p < q$ and

$$\left| p \frac{q_i}{q} - p_i \right| \leq q^{-\delta} \text{ for } i = 1, 2, \ldots, n.$$

Furthermore, it is an *unusually good simultaneous diophantine approximation* (UGSDA) if $\delta > \frac{1}{n}$.

Fact 3.107 shows that an UGSDA is indeed unusual.

3.107 Fact For $n \geq 2$, the set

$$S_n(q) = \left\{ \left(\frac{q_1}{q}, \frac{q_2}{q}, \ldots, \frac{q_n}{q} \right) \mid 0 \leq q_i < q, \ \gcd(q_1, q_2, \ldots, q_n, q) = 1 \right\}$$

has at least $\frac{1}{2} q^n$ members. Of these, at most $O(q^{n(1-\delta)+1})$ members have at least one δ-quality simultaneous diophantine approximation. Hence, for any fixed $\delta > \frac{1}{n}$, the fraction of members of $S_n(q)$ having at least one UGSDA approaches 0 as $q \to \infty$.

Algorithm 3.108 reduces the problem of finding a δ-quality simultaneous diophantine approximation, and hence also a UGSDA, to the problem of finding a short vector in a lattice. The latter problem can (usually) be solved using the L^3-lattice basis reduction.

3.108 Algorithm Finding a δ-quality simultaneous diophantine approximation

INPUT: a vector $w = (\frac{q_1}{q}, \frac{q_2}{q}, \ldots, \frac{q_n}{q})$ of rational numbers, and a rational number δ.
OUTPUT: a δ-quality simultaneous diophantine approximation $(\frac{p_1}{p}, \frac{p_2}{p}, \ldots, \frac{p_n}{p})$ of w.

1. Choose an integer $\lambda \approx q^\delta$.
2. Use Algorithm 3.101 to find a reduced basis B for the $(n+1)$-dimensional lattice L which is generated by the rows of the matrix

$$A = \begin{pmatrix} \lambda q & 0 & 0 & \cdots & 0 & 0 \\ 0 & \lambda q & 0 & \cdots & 0 & 0 \\ 0 & 0 & \lambda q & \cdots & 0 & 0 \\ \vdots & \vdots & \vdots & \ddots & \vdots & \vdots \\ 0 & 0 & 0 & \cdots & \lambda q & 0 \\ -\lambda q_1 & -\lambda q_2 & -\lambda q_3 & \cdots & -\lambda q_n & 1 \end{pmatrix}$$

3. For each $v = (v_1, v_2, \ldots, v_n, v_{n+1})$ in B such that $v_{n+1} \neq q$, do the following:
   3.1 $p \leftarrow v_{n+1}$.
   3.2 For i from 1 to n, set $p_i \leftarrow \frac{1}{q}\left(\frac{v_i}{\lambda} + pq_i\right)$.
   3.3 If $|p\frac{q_i}{q} - p_i| \leq q^{-\delta}$ for each i, $1 \leq i \leq n$, then return$(\frac{p_1}{p}, \frac{p_2}{p}, \ldots, \frac{p_n}{p})$.
4. Return(FAILURE). (Either no δ-quality simultaneous diophantine approximation exists, or the algorithm has failed to find one.)

Justification. Let the rows of the matrix A be denoted by $b_1, b_2, \ldots, b_{n+1}$. Suppose that $(\frac{q_1}{q}, \frac{q_2}{q}, \ldots, \frac{q_n}{q})$ has a δ-quality approximation $(\frac{p_1}{p}, \frac{p_2}{p}, \ldots, \frac{p_n}{p})$. Then the vector

$$\begin{aligned} x &= p_1 b_1 + p_2 b_2 + \cdots + p_n b_n + p b_{n+1} \\ &= (\lambda(p_1 q - pq_1), \lambda(p_2 q - pq_2), \ldots, \lambda(p_n q - pq_n), p) \end{aligned}$$

is in L and has length less than approximately $(\sqrt{n+1})q$. Thus x is short compared to the original basis vectors, which are of length roughly $q^{1+\delta}$. Also, if $v = (v_1, v_2, \ldots, v_{n+1})$ is a vector in L of length less than q, then the vector $(\frac{p_1}{p}, \frac{p_2}{p}, \ldots, \frac{p_n}{p})$ defined in step 3 is a δ-quality approximation. Hence there is a good possibility that the L^3-algorithm will produce a reduced basis which includes a vector v that corresponds to a δ-quality approximation.

3.11 Factoring polynomials over finite fields

The problem considered in this section is the following: given a polynomial $f(x) \in \mathbb{F}_q[x]$, with $q = p^m$, find its factorization $f(x) = f_1(x)^{e_1} f_2(x)^{e_2} \cdots f_t(x)^{e_t}$, where each $f_i(x)$ is an irreducible polynomial in $\mathbb{F}_q[x]$ and each $e_i \geq 1$. (e_i is called the *multiplicity* of the factor $f_i(x)$.) Several situations call for the factoring of polynomials over finite fields, such as index-calculus algorithms in $\mathbb{F}_{2^m}^*$ (Example 3.70) and Chor-Rivest public-key encryption (§8.6.2). This section presents an algorithm for square-free factorization, and Berlekamp's classical deterministic algorithm for factoring polynomials which is efficient if the underlying field is small. Efficient randomized algorithms are known for the case of large q; references are provided on page 132.

3.11.1 Square-free factorization

Observe first that $f(x)$ may be divided by its leading coefficient. Thus, it may be assumed that $f(x)$ is monic (see Definition 2.187). This section shows how the problem of factoring a monic polynomial $f(x)$ may then be reduced to the problem of factoring one or more monic square-free polynomials.

3.109 Definition Let $f(x) \in \mathbb{F}_q[x]$. Then $f(x)$ is *square-free* if it has no repeated factors, i.e., there is no polynomial $g(x)$ with $\deg g(x) \geq 1$ such that $g(x)^2$ divides $f(x)$. The *square-free factorization* of $f(x)$ is $f(x) = \prod_{i=1}^{k} f_i(x)^i$, where each $f_i(x)$ is a square-free polynomial and $\gcd(f_i(x), f_j(x)) = 1$ for $i \neq j$. (Some of the $f_i(x)$ in the square-free factorization of $f(x)$ may be 1.)

Let $f(x) = \sum_{i=0}^{n} a_i x^i$ be a polynomial of degree $n \geq 1$. The (formal) *derivative* of $f(x)$ is the polynomial $f'(x) = \sum_{i=0}^{n-1} a_{i+1}(i+1)x^i$. If $f'(x) = 0$, then, because p is the characteristic of $\mathbb{F}_q$, in each term $a_i x^i$ of $f(x)$ for which $a_i \neq 0$, the exponent of x must be a multiple of p. Hence, $f(x)$ has the form $f(x) = a(x)^p$, where $a(x) = \sum_{i=0}^{n/p} a_{ip}^{q/p} x^i$, and the problem of finding the square-free factorization of $f(x)$ is reduced to finding that of $a(x)$. Now, it is possible that $a'(x) = 0$, but repeating this process as necessary, it may be assumed that $f'(x) \neq 0$.

Next, let $g(x) = \gcd(f(x), f'(x))$. Noting that an irreducible factor of multiplicity k in $f(x)$ will have multiplicity $k - 1$ in $f'(x)$ if $\gcd(k, p) = 1$, and will retain multiplicity k in $f'(x)$ otherwise, the following conclusions may be drawn. If $g(x) = 1$, then $f(x)$ has no repeated factors; and if $g(x)$ has positive degree, then $g(x)$ is a non-trivial factor of $f(x)$, and $f(x)/g(x)$ has no repeated factors. Note, however, the possibility of $g(x)$ having repeated factors, and, indeed, the possibility that $g'(x) = 0$. Nonetheless, $g(x)$ can be refined further as above. The steps are summarized in Algorithm 3.110.

3.110 Algorithm Square-free factorization

SQUARE-FREE($f(x)$)
INPUT: a monic polynomial $f(x) \in \mathbb{F}_q[x]$, where the characteristic of $\mathbb{F}_q$ is p.
OUTPUT: the square-free factorization of $f(x)$.

1. Set $i \leftarrow 1$, $F \leftarrow 1$, and compute $f'(x)$.
2. If $f'(x) = 0$ then set $f(x) \leftarrow f(x)^{1/p}$ and $F \leftarrow (\text{SQUARE-FREE}(f(x)))^p$.
   Otherwise (i.e. $f'(x) \neq 0$) do the following:
   2.1 Compute $g(x) \leftarrow \gcd(f(x), f'(x))$ and $h(x) \leftarrow f(x)/g(x)$.
   2.2 While $h(x) \neq 1$ do the following:
       Compute $\overline{h}(x) \leftarrow \gcd(h(x), g(x))$ and $l(x) \leftarrow h(x)/\overline{h}(x)$.
       Set $F \leftarrow F \cdot l(x)^i$, $i \leftarrow i+1$, $h(x) \leftarrow \overline{h}(x)$, and $g(x) \leftarrow g(x)/\overline{h}(x)$.
   2.3 If $g(x) \neq 1$ then set $g(x) \leftarrow g(x)^{1/p}$ and $F \leftarrow F \cdot (\text{SQUARE-FREE}(g(x)))^p$.
3. Return(F).

Once the square-free factorization $f(x) = \prod_{i=1}^{k} f_i(x)^i$ is found, the square-free polynomials $f_1(x)$, $f_2(x), \ldots, f_k(x)$ need to be factored in order to obtain the complete factorization of $f(x)$.

3.11.2 Berlekamp's Q-matrix algorithm

Let $f(x) = \prod_{i=1}^{t} f_i(x)$ be a monic polynomial in $\mathbb{F}_q[x]$ of degree n having distinct irreducible factors $f_i(x)$, $1 \leq i \leq t$. Berlekamp's Q-matrix algorithm (Algorithm 3.111) for factoring $f(x)$ is based on the following facts. The set of polynomials

$$\mathcal{B} = \{b(x) \in \mathbb{F}_q[x]/(f(x)) \mid b(x)^q \equiv b(x) \pmod{f(x)}\}$$

is a vector space of dimension t over $\mathbb{F}_q$. $\mathcal{B}$ consists of precisely those vectors in the null space of the matrix $Q - I_n$, where Q is the $n \times n$ matrix with (i,j)-entry q_{ij} specified by

$$x^{iq} \bmod f(x) = \sum_{j=0}^{n-1} q_{ij} x^j, \quad 0 \leq i \leq n-1,$$

and where I_n is the $n \times n$ identity matrix. A basis $B = \{v_1(x), v_2(x), \dots, v_t(x)\}$ for $\mathcal{B}$ can thus be found by standard techniques from linear algebra. Finally, for each pair of distinct factors $f_i(x)$ and $f_j(x)$ of $f(x)$ there exists some $v_k(x) \in B$ and some $\alpha \in \mathbb{F}_q$ such that $f_i(x)$ divides $v_k(x) - \alpha$ but $f_j(x)$ does not divide $v_k(x) - \alpha$; these two factors can thus be split by computing $\gcd(f(x), v_k(x) - \alpha)$. In Algorithm 3.111, a vector $w = (w_0, w_1, \dots, w_{n-1})$ is identified with the polynomial $w(x) = \sum_{i=0}^{n-1} w_i x^i$.

3.111 Algorithm Berlekamp's Q-matrix algorithm for factoring polynomials over finite fields

INPUT: a square-free monic polynomial $f(x)$ of degree n in $\mathbb{F}_q[x]$.
OUTPUT: the factorization of $f(x)$ into monic irreducible polynomials.

1. For each i, $0 \leq i \leq n-1$, compute the polynomial

$$x^{iq} \bmod f(x) = \sum_{j=0}^{n-1} q_{ij} x^j.$$

   Note that each q_{ij} is an element of $\mathbb{F}_q$.
2. Form the $n \times n$ matrix Q whose (i,j)-entry is q_{ij}.
3. Determine a basis $v_1, v_2, \dots, v_t$ for the null space of the matrix $(Q - I_n)$, where I_n is the $n \times n$ identity matrix. The number of irreducible factors of $f(x)$ is precisely t.
4. Set $F \leftarrow \{f(x)\}$. (F is the set of factors of $f(x)$ found so far; their product is equal to $f(x)$.)
5. For i from 1 to t do the following:
   5.1 For each polynomial $h(x) \in F$ such that $\deg h(x) > 1$ do the following: compute $\gcd(h(x), v_i(x) - \alpha)$ for each $\alpha \in \mathbb{F}_q$, and replace $h(x)$ in F by all those polynomials in the gcd computations whose degrees are ≥ 1.
6. Return(the polynomials in F are the irreducible factors of $f(x)$).

3.112 Fact The running time of Algorithm 3.111 for factoring a square-free polynomial of degree n over $\mathbb{F}_q$ is $O(n^3 + tqn^2)$ $\mathbb{F}_q$-operations, where t is the number of irreducible factors of $f(x)$. The method is efficient only when q is small.

3.12 Notes and further references

§3.1

Many of the topics discussed in this chapter lie in the realm of algorithmic number theory. Excellent references on this subject include the books by Bach and Shallit [70], Cohen [263], and Pomerance [993]. Adleman and McCurley [15] give an extensive survey of the important open problems in algorithmic number theory. Two other recommended surveys are by Bach [65] and Lenstra and Lenstra [748]. Woll [1253] gives an overview of the reductions among thirteen of these problems.

§3.2

A survey of the integer factorization problem is given by Pomerance [994]. See also Chapters 8 and 10 of Cohen [263], and the books by Bressoud [198] and Koblitz [697]. Brillhart et al. [211] provide extensive listings of factorizations of integers of the form $b^n \pm 1$ for "small" n and $b = 2, 3, 5, 6, 7, 10, 11, 12$.

Bach and Sorenson [71] presented some algorithms for recognizing perfect powers (cf. Note 3.6), one having a worst-case running time of $O(\lg^3 n)$ bit operations, and a second having an average-case running time of $O(\lg^2 n)$ bit operations. A more recent algorithm of Bernstein [121] runs in essentially linear time $O((\lg n)^{1+o(1)})$. Fact 3.7 is from Knuth [692]. Pages 367–369 of this reference contain explicit formulas regarding the expected sizes of the largest and second largest prime factors, and the expected total number of prime factors, of a randomly chosen positive integer. For further results, see Knuth and Trabb Pardo [694], who prove that the average number of bits in the kth largest prime factor of a random m-bit number is asymptotically equivalent to the average length of the kth longest cycle in a permutation on m objects.

Floyd's cycle-finding algorithm (Note 3.8) is described by Knuth [692, p.7]. Sedgewick, Szymanski, and Yao [1106] showed that by saving a small number of values from the x_i sequence, a collision can be found by doing roughly one-third the work as in Floyd's cycle-finding algorithm. Pollard's rho algorithm for factoring (Algorithm 3.9) is due to Pollard [985]. Regarding Note 3.12, Cohen [263, p.422] provides an explanation for the restriction $c \neq 0, -2$. Brent [196] presented a cycle-finding algorithm which is better on average than Floyd's cycle-finding algorithm, and applied it to yield a factorization algorithm which is similar to Pollard's but about 24 percent faster. Brent and Pollard [197] later modified this algorithm to factor the eighth Fermat number $F_8 = 2^{2^8} + 1$. Using techniques from algebraic geometry, Bach [67] obtained the first rigorously proven result concerning the expected running time of Pollard's rho algorithm: for fixed k, the probability that a prime factor p is discovered before step k is at least $\binom{k}{2}/p + O(p^{-3/2})$ as $p \to \infty$.

The $p - 1$ algorithm (Algorithm 3.14) is due to Pollard [984]. Several practical improvements have been proposed for the $p - 1$ algorithm, including those by Montgomery [894] and Montgomery and Silverman [895], the latter using fast Fourier transform techniques. Williams [1247] presented an algorithm for factoring n which is efficient if n has a prime factor p such that $p+1$ is smooth. These methods were generalized by Bach and Shallit [69] to techniques that factor n efficiently provided n has a prime factor p such that the kth cyclotomic polynomial $\Phi_k(p)$ is smooth. The first few cyclotomic polynomials are $\Phi_1(p) = p - 1$, $\Phi_2(p) = p + 1$, $\Phi_3(p) = p^2 + p + 1$, $\Phi_4(p) = p^2 + 1$, $\Phi_5(p) = p^4 + p^3 + p^2 + p + 1$, and $\Phi_6(p) = p^2 - p + 1$.

The elliptic curve factoring algorithm (ECA) of §3.2.4 was invented by Lenstra [756]. Montgomery [894] gave several practical improvements to the ECA. Silverman and

Wagstaff [1136] gave a practical analysis of the complexity of the ECA, and suggested optimal parameter selection and running-time guidelines. Lenstra and Manasse [753] implemented the ECA on a network of MicroVAX computers, and were successful in finding 35-decimal digit prime factors of large (at least 85 digit) composite integers. Later, Dixon and Lenstra [350] implemented the ECA on a 16K MasPar (massively parallel) SIMD (single instruction, multiple data) machine. The largest factor they found was a 40-decimal digit prime factor of an 89-digit composite integer. On November 26 1995, Peter Montgomery reported finding a 47-decimal digit prime factor of the 99-digit composite integer $5^{256} + 1$ with the ECA.

Hafner and McCurley [536] estimated the number of integers $n \leq x$ that can be factored with probability at least $\frac{1}{2}$ using at most t arithmetic operations, by trial division and the elliptic curve algorithm. Pomerance and Sorenson [997] provided the analogous estimates for Pollard's $p-1$ algorithm and Williams' $p+1$ algorithm. They conclude that for a given running time bound, both Pollard's $p-1$ and Williams' $p+1$ algorithms factor more integers than trial division, but fewer than the elliptic curve algorithm.

Pomerance [994] credits the idea of multiplying congruences to produce a solution to $x^2 \equiv y^2 \pmod{n}$ for the purpose of factoring n (§3.2.5) to some old work of Kraitchik circa 1926-1929. The *continued fraction factoring algorithm*, first introduced by Lehmer and Powers [744] in 1931, and refined more than 40 years later by Morrison and Brillhart [908], was the first realization of a random square method to result in a subexponential-time algorithm. The algorithm was later analyzed by Pomerance [989] and conjectured to have an expected running time of $L_n[\frac{1}{2}, \sqrt{2}]$. If the smoothness testing in the algorithm is done with the elliptic curve method, then the expected running time drops to $L_n[\frac{1}{2}, 1]$. Morrison and Brillhart were also the first to use the idea of a factor base to test for good (a_i, b_i) pairs. The continued fraction algorithm was the champion of factoring algorithms from the mid 1970s until the early 1980s, when it was surpassed by the quadratic sieve algorithm.

The quadratic sieve (QS) (Algorithm 3.2.6) was discovered by Pomerance [989, 990]. The multiple polynomial variant of the quadratic sieve (Note 3.25) is due to P. Montgomery, and is described by Pomerance [990]; see also Silverman [1135]. A detailed practical analysis of the QS is given by van Oorschot [1203]. Several practical improvements to the original algorithms have subsequently been proposed and successfully implemented. The first serious implementation of the QS was by Gerver [448] who factored a 47-decimal digit number. In 1984, Davis, Holdridge, and Simmons [311] factored a 71-decimal digit number with the QS. In 1988, Lenstra and Manasse [753] used the QS to factor a 106-decimal digit number by distributing the computations to hundreds of computers by electronic mail; see also Lenstra and Manasse [754]. In 1993, the QS was used by Denny et al. [333] to factor a 120-decimal digit number. In 1994, the 129-decimal digit (425 bits) RSA-129 challenge number (see Gardner [440]), was factored by Atkins et al. [59] by enlisting the help of about 1600 computers around the world. The factorization was carried out in 8 months. Table 3.3 shows the estimated time taken, in mips years, for the above factorizations. A *mips year* is equivalent to the computational power of a computer that is rated at 1 mips (million instructions per second) and utilized for one year, or, equivalently, about $3 \cdot 10^{13}$ instructions.

The number field sieve was first proposed by Pollard [987] and refined by others. Lenstra et al. [752] described the special number field sieve (SNFS) for factoring integers of the form $r^e - s$ for small positive r and $|s|$. A readable introduction to the algorithm is provided by Pomerance [995]. A detailed report of an SNFS implementation is given by Lenstra et al. [751]. This implementation was used to factor the ninth Fermat number $F_9 = 2^{512} + 1$, which is the product of three prime factors having 7, 49, and 99 decimal digits. The general number field sieve (GNFS) was introduced by Buhler, Lenstra, and Pomerance [219].

Year	Number of digits	mips years
1984	71	0.1
1988	106	140
1993	120	825
1994	129	5000

Table 3.3: *Running time estimates for numbers factored with QS.*

Coppersmith [269] proposed modifications to the GNFS which improve its running time to $L_n[\frac{1}{3}, 1.902]$, however, the method is not practical; another modification (also impractical) allows a precomputation taking $L_n[\frac{1}{3}, 2.007]$ time and $L_n[\frac{1}{3}, 1.639]$ storage, following which all integers in a large range of values can be factored in $L_n[\frac{1}{3}, 1.639]$ time. A detailed report of a GNFS implementation on a massively parallel computer with 16384 processors is given by Bernstein and Lenstra [122]. See also Buchmann, Loho, and Zayer [217], and Golliver, Lenstra, and McCurley [493]. More recently, Dodson and Lenstra [356] reported on their GNFS implementation which was successful in factoring a 119-decimal digit number using about 250 mips years of computing power. They estimated that this factorization completed about 2.5 times faster than it would with the quadratic sieve. Most recently, Lenstra [746] announced the factorization of the 130-decimal digit RSA-130 challenge number using the GNFS. This number is the product of two 65-decimal digit primes. The factorization was estimated to have taken about 500 mips years of computing power (compare with Table 3.3). The book edited by Lenstra and Lenstra [749] contains several other articles related to the number field sieve.

The ECA, continued fraction algorithm, quadratic sieve, special number field sieve, and general number field sieve have *heuristic* (or *conjectured*) rather than *proven* running times because the analyses make (reasonable) assumptions about the proportion of integers generated that are smooth. See Canfield, Erdös, and Pomerance [231] for bounds on the proportion of y-smooth integers in the interval $[2, x]$. Dixon's algorithm [351] was the first rigorously analyzed subexponential-time algorithm for factoring integers. The fastest rigorously analyzed algorithm currently known is due to Lenstra and Pomerance [759] with an expected running time of $L_n[\frac{1}{2}, 1]$. These algorithms are of theoretical interest only, as they do not appear to be practical.

§3.3

The RSA problem was introduced in the landmark 1977 paper by Rivest, Shamir, and Adleman [1060].

§3.4

The quadratic residuosity problem is of much historical interest, and was one of the main algorithmic problems discussed by Gauss [444].

§3.5

An extensive treatment of the problem of finding square roots modulo a prime p, or more generally, the problem of finding d^{th} roots in a finite field, can be found in Bach and Shallit [70]. The presentation of Algorithm 3.34 for finding square roots modulo a prime is derived from Koblitz [697, pp.48-49]; a proof of correctness can be found there. Bach and Shallit attribute the essential ideas of Algorithm 3.34 to an 1891 paper by A. Tonelli. Algorithm 3.39 is from Bach and Shallit [70], who attribute it to a 1903 paper of M. Cipolla.

The computational equivalence of computing square roots modulo a composite n and factoring n (Fact 3.46 and Note 3.47) was first discovered by Rabin [1023].

§3.6

A survey of the discrete logarithm problem is given by McCurley [827]. See also Odlyzko [942] for a survey of recent advances.

Knuth [693] attributes the baby-step giant-step algorithm (Algorithm 3.56) to D. Shanks. The baby-step giant-step algorithms for searching restricted exponent spaces (cf. Note 3.59) are described by Heiman [546]. Suppose that p is a k-bit prime, and that only exponents of Hamming weight t are used. Coppersmith (personal communication, July 1995) observed that this exponent space can be searched in $k \cdot \binom{k/2}{t/2}$ steps by dividing the exponent into two equal pieces so that the Hamming weight of each piece is $t/2$; if k is much smaller than $2^{t/2}$, this is an improvement over Note 3.59.

Pollard's rho algorithm for logarithms (Algorithm 3.60) is due to Pollard [986]. Pollard also presented a *lambda method* for computing discrete logarithms which is applicable when x, the logarithm sought, is known to lie in a certain interval. More specifically, if the interval is of width w, the method is expected to take $O(\sqrt{w})$ group operations and requires storage for only $O(\lg w)$ group elements. van Oorschot and Wiener [1207] showed how Pollard's rho algorithm can be parallelized so that using m processors results in a speedup by a factor of m. This has particular significance to cyclic groups such as elliptic curve groups, for which no subexponential-time discrete logarithm algorithm is known.

The Pohlig-Hellman algorithm (Algorithm 3.63) was discovered by Pohlig and Hellman [982]. A variation which represents the logarithm in a mixed-radix notation and does not use the Chinese remainder theorem was given by Thiong Ly [1190].

According to McCurley [827], the basic ideas behind the index-calculus algorithm (Algorithm 3.68) first appeared in the work of Kraitchik (circa 1922-1924) and of Cunningham (see Western and Miller [1236]), and was rediscovered by several authors. Adleman [8] described the method for the group $\mathbb{Z}_p^*$ and analyzed the complexity of the algorithm. Hellman and Reyneri [555] gave the first description of an index-calculus algorithm for extension fields $\mathbb{F}_{p^m}$ with p fixed.

Coppersmith, Odlyzko, and Schroeppel [280] presented three variants of the index-calculus method for computing logarithms in $\mathbb{Z}_p^*$: the *linear sieve*, the *residue list sieve*, and the *Gaussian integer method*. Each has a heuristic expected running time of $L_p[\frac{1}{2}, 1]$ (cf. Note 3.71). The Gaussian integer method, which is related to the method of ElGamal [369], was implemented in 1990 by LaMacchia and Odlyzko [736] and was successful in computing logarithms in $\mathbb{Z}_p^*$ with p a 192-bit prime. The paper concludes that it should be feasible to compute discrete logarithms modulo primes of about 332 bits (100 decimal digits) using the Gaussian integer method. Gordon [510] adapted the number field sieve for factoring integers to the problem of computing logarithms in $\mathbb{Z}_p^*$; his algorithm has a heuristic expected running time of $L_p[\frac{1}{3}, c]$, where $c = 3^{2/3} \approx 2.080$. Schirokauer [1092] subsequently presented a modification of Gordon's algorithm that has a heuristic expected running time of $L_p[\frac{1}{3}, c]$, where $c = (64/9)^{1/3} \approx 1.923$ (Note 3.72). This is the same running time as conjectured for the number field sieve for factoring integers (see §3.2.7). Recently, Weber [1232] implemented the algorithms of Gordon and Schirokauer and was successful in computing logarithms in $\mathbb{Z}_p^*$, where p is a 40-decimal digit prime such that $p-1$ is divisible by a 38-decimal digit (127-bit) prime. More recently, Weber, Denny, and Zayer (personal communication, April 1996) announced the solution of a discrete logarithm problem modulo a 75-decimal digit (248-bit) prime p with $(p-1)/2$ prime.

Blake et al. [145] made improvements to the index-calculus technique for $\mathbb{F}_{2^m}^*$ and computed logarithms in $\mathbb{F}_{2^{127}}^*$. Coppersmith [266] dramatically improved the algorithm and showed that under reasonable assumptions the expected running time of his improved algorithm is $L_{2^m}[\frac{1}{3}, c]$ for some constant $c < 1.587$ (Note 3.72). Later, Odlyzko [940] gave several refinements to Coppersmith's algorithm, and a detailed practical analysis; this paper provides the most extensive account to date of the discrete logarithm problem in $\mathbb{F}^*_{2^m}$. A similar practical analysis was also given by van Oorschot [1203]. Most recently in 1992, Gordon and McCurley [511] reported on their massively parallel implementation of Coppersmith's algorithm, combined with their own improvements. Using primarily a 1024 processor nCUBE-2 machine with 4 megabytes of memory per processor, they completed the precomputation of logarithms of factor base elements (which is the dominant step of the algorithm) required to compute logarithms in $\mathbb{F}^*_{2^{227}}$, $\mathbb{F}^*_{2^{313}}$, and $\mathbb{F}^*_{2^{401}}$. The calculations for $\mathbb{F}^*_{2^{401}}$ were estimated to take 5 days. Gordon and McCurley also completed most of the precomputations required for computing logarithms in $\mathbb{F}^*_{2^{503}}$; the amount of time to complete this task on the 1024 processor nCUBE-2 was estimated to be 44 days. They concluded that computing logarithms in the multiplicative groups of fields as large as $\mathbb{F}_{2^{593}}$ still seems to be out of their reach, but might be possible in the near future with a concerted effort.

It was not until 1992 that a subexponential-time algorithm for computing discrete logarithms over all finite fields $\mathbb{F}_q$ was discovered by Adleman and DeMarrais [11]. The expected running time of the algorithm is conjectured to be $L_q[\frac{1}{2}, c]$ for some constant c. Adleman [9] generalized the number field sieve from algebraic number fields to algebraic function fields which resulted in an algorithm, called the *function field sieve*, for computing discrete logarithms in $\mathbb{F}^*_{p^m}$; the algorithm has a heuristic expected running time of $L_{p^m}[\frac{1}{3}, c]$ for some constant $c > 0$ when $\log p \leq m^{g(m)}$, and where g is any function such that $0 < g(m) < 0.98$ and $\lim_{m \to \infty} g(m) = 0$. The practicality of the function field sieve has not yet been determined. It remains an open problem to find an algorithm with a heuristic expected running time of $L_q[\frac{1}{3}, c]$ for *all* finite fields $\mathbb{F}_q$.

The algorithms mentioned in the previous three paragraphs have *heuristic* (or *conjectured*) rather than *proven* running times because the analyses make some (reasonable) assumptions about the proportion of integers or polynomials generated that are smooth, and also because it is not clear when the system of linear equations generated has full rank, i.e., yields a unique solution. The best rigorously analyzed algorithms known for the discrete logarithm problem in $\mathbb{Z}^*_p$ and $\mathbb{F}^*_{2^m}$ are due to Pomerance [991] with expected running times of $L_p[\frac{1}{2}, \sqrt{2}]$ and $L_{2^m}[\frac{1}{2}, \sqrt{2}]$, respectively. Lovorn [773] obtained rigorously analyzed algorithms for the fields $\mathbb{F}_{p^2}$ and $\mathbb{F}_{p^m}$ with $\log p < m^{0.98}$, having expected running times of $L_{p^2}[\frac{1}{2}, \frac{3}{2}]$ and $L_{p^m}[\frac{1}{2}, \sqrt{2}]$, respectively.

The linear system of equations collected in the quadratic sieve and number field sieve factoring algorithms, and the index-calculus algorithms for computing discrete logarithms in $\mathbb{Z}^*_p$ and $\mathbb{F}^*_{2^m}$, are very large. For the problem sizes currently under consideration, these systems cannot be solved using ordinary linear algebra techniques, due to both time and space constraints. However, the equations generated are extremely *sparse*, typically with at most 50 non-zero coefficients per equation. The technique of *structured* or so-called *intelligent* Gaussian elimination (see Odlyzko [940]) can be used to reduce the original sparse system to a much smaller system that is still fairly sparse. The resulting system can be solved using either ordinary Gaussian elimination, or one of the *conjugate gradient*, *Lanczos* (Coppersmith, Odlyzko, and Schroeppel [280]), or *Wiedemann* algorithms [1239] which were also designed to handle sparse systems. LaMacchia and Odlyzko [737] have implemented some of these algorithms and concluded that the linear algebra stages arising in both integer factorization and the discrete logarithm problem are not running-time bottlenecks in practice. Recently, Coppersmith [272] proposed a modification of the Wiedemann algorithm which allows parallelization of the algorithm; for an analysis of Coppersmith's algorithm, see Kaltofen [657]. Coppersmith [270] (see also Montgomery [896]) presented a modification of the Lanczos algorithm for solving sparse linear equations over $\mathbb{F}_2$; this variant appears to be the most efficient in practice.

As an example of the numbers involved, Gordon and McCurley's [511] implementation for computing logarithms in $\mathbb{F}_{2^{401}}^*$ produced a total of 117164 equations from a factor base consisting of the 58636 irreducible polynomials in $\mathbb{F}_2[x]$ of degree at most 19. The system of equations had 2068707 non-zero entries. Structured Gaussian elimination was then applied to this system, the result being a 16139×16139 system of equations having 1203414 non-zero entries, which was then solved using the conjugate gradient method. Another example is from the recent factorization of the RSA-129 number (see [59]). The sieving step produced a sparse matrix of 569466 rows and 524339 columns. Structured Gaussian elimination was used to reduce this to a dense 188614×188160 system, which was then solved using ordinary Gaussian elimination.

There are many ways of representing a finite field, although any two finite fields of the same order are isomorphic (see also Note 3.55). Lenstra [757] showed how to compute an isomorphism between any two explicitly given representations of a finite field in deterministic polynomial time. Thus, it is sufficient to find an algorithm for computing discrete logarithms in one representation of a given field; this algorithm can then be used, together with the isomorphism obtained by Lenstra's algorithm, to compute logarithms in any other representation of the same field.

Menezes, Okamoto, and Vanstone [843] showed how the discrete logarithm problem for an elliptic curve over a finite field $\mathbb{F}_q$ can be reduced to the discrete logarithm problem in some extension field $\mathbb{F}_{q^k}$. For the special class of *supersingular curves*, k is at most 6, thus providing a subexponential-time algorithm for the former problem. This work was extended by Frey and Rück [422]. No subexponential-time algorithm is known for the discrete logarithm problem in the more general class of *non-supersingular* elliptic curves.

Adleman, DeMarrais, and Huang [12] presented a subexponential-time algorithm for finding logarithms in the jacobian of large genus hyperelliptic curves over finite fields. More precisely, there exists a number c, $0 < c \leq 2.181$, such that for all sufficiently large $g \geq 1$ and all odd primes p with $\log p \leq (2g + 1)^{0.98}$, the expected running time of the algorithm for computing logarithms in the jacobian of a genus g hyperelliptic curve over $\mathbb{Z}_p$ is conjectured to be $L_{p^{2g+1}}[\frac{1}{2}, c]$.

McCurley [826] invented a subexponential-time algorithm for the discrete logarithm problem in the class group of an imaginary quadratic number field. See also Hafner and McCurley [537] for further details, and Buchmann and Düllmann [216] for an implementation report.

In 1994, Shor [1128] conceived randomized polynomial-time algorithms for computing discrete logarithms and factoring integers on a *quantum computer*, a computational device based on quantum mechanical principles; presently it is not known how to build a quantum computer, nor if this is even possible. Also recently, Adleman [10] demonstrated the feasibility of using tools from molecular biology to solve an instance of the directed Hamiltonian path problem, which is **NP**-complete. The problem instance was encoded in molecules of DNA, and the steps of the computation were performed with standard protocols and enzymes. Adleman notes that while the currently available fastest supercomputers can execute approximately 10^{12} operations per second, it is plausible for a DNA computer to execute 10^{20} or more operations per second. Moreover such a DNA computer would be far more energy-efficient than existing supercomputers. It is not clear at present whether it is feasible to build a DNA computer with such performance. However, should either quantum computers or DNA computers ever become practical, they would have a very significant impact on public-key cryptography.

§3.7

Fact 3.77(i) is due to den Boer [323]. Fact 3.77(iii) was proven by Maurer [817], who also proved more generally that the GDHP and GDLP in a group G of order n are computationally equivalent when certain extra information of length $O(\lg n)$ bits is given. The extra information depends only on n and not on the definition of G, and consists of parameters that define cyclic elliptic curves of smooth order over the fields $\mathbb{Z}_{p_i}$ where the p_i are the prime divisors of n.

Waldvogel and Massey [1228] proved that if a and b are chosen uniformly and randomly from the interval $\{0, 1, \dots, p-1\}$, the values $\alpha^{ab} \bmod p$ are roughly uniformly distributed (see page 537).

§3.8

Facts 3.78 and 3.79 are due to Bach [62]. Fact 3.80 is due to Shmuely [1127]. McCurley [825] refined this result to prove that for specially chosen composite n, the ability to solve the Diffie-Hellman problem in $\mathbb{Z}_n^*$ for the *fixed* base $\alpha = 16$ implies the ability to factor n.

§3.9

The notion of a hard Boolean predicate (Definition 3.81) was introduced by Blum and Micali [166], who also proved Fact 3.84. The notion of a hard k-bit predicate (Definition 3.82) was introduced by Long and Wigderson [772], who also proved Fact 3.85; see also Peralta [968]. Fact 3.83 is due to Peralta [968]. The results on hard predicates and k-bit predicates for the RSA functions (Facts 3.86 and 3.87) are due to Alexi et al. [23]. Facts 3.88 and 3.89 are due to Vazirani and Vazirani [1218].

Yao [1258] showed how any one-way length-preserving permutation can be transformed into a more complicated one-way length-preserving permutation which has a hard predicate. Subsequently, Goldreich and Levin [471] showed how any one-way function f can be transformed into a one-way function g which has a hard predicate. Their construction is as follows. Define the function g by $g(p, x) = (p, f(x))$, where p is a binary string of the same length as x, say n. Then g is also a one-way function and $B(p, x) = \sum_{i=1}^{n} p_i x_i \bmod 2$ is a hard predicate for g.

Håstad, Schrift, and Shamir [543] considered the one-way function $f(x) = \alpha^x \bmod n$, where n is a Blum integer and $\alpha \in \mathbb{Z}_n^*$. Under the assumption that factoring Blum integers is intractable, they proved that all the bits of this function are individually hard. Moreover, the lower half as well as the upper half of the bits are simultaneously hard.

§3.10

The subset sum problem (Definition 3.90) is sometimes confused with the *knapsack problem* which is the following: given two sets $\{a_1, a_2, \dots, a_n\}$ and $\{b_1, b_2, \dots, b_n\}$ of positive integers, and given two positive integers s and t, determine whether or not there is a subset S of $\{1, 2, \dots, n\}$ such that $\sum_{i \in S} a_i \le s$ and $\sum_{i \in S} b_i \ge t$. The subset sum problem is actually a special case of the knapsack problem when $a_i = b_i$ for $i = 1, 2, \dots, n$ and $s = t$.

The L^3-lattice basis reduction algorithm (Algorithm 3.101) and Fact 3.103 are both due to Lenstra, Lenstra, and Lovász [750]. Improved algorithms have been given for lattice basis reduction, for example, by Schnorr and Euchner [1099]; consult also Section 2.6 of Cohen [263]. Algorithm 3.105 for solving the subset sum problem involving knapsacks sets of low density is from Coster et al. [283]. Unusually good simultaneous diophantine approximations were first introduced and studied by Lagarias [723]; Fact 3.107 and Algorithm 3.108 are from this paper.

§3.11

A readable introduction to polynomial factorization algorithms is given by Lidl and Nieder-reiter [764, Chapter 4]. Yun [1261] presented an algorithm that is more efficient than Algorithm 3.110 for finding the square-free factorization of a polynomial. The running time of the algorithm is only $O(n^2)$ $\mathbb{Z}_p$-operations when $f(x)$ is a polynomial of degree n in $\mathbb{Z}_p[x]$. A lucid presentation of Yun's algorithm is provided by Bach and Shallit [70]. Berlekamp's Q-matrix algorithm (Algorithm 3.111) was first discovered by Prange [999] for the purpose of factoring polynomials of the form $x^n - 1$ over finite fields. The algorithm was later and independently discovered by Berlekamp [117] who improved it for factoring general polynomials over finite fields.

There is no deterministic polynomial-time algorithm known for the problem of factoring polynomials over finite fields. There are, however, many efficient randomized algorithms that work well even when the underlying field is very large, such as the algorithms given by Ben-Or [109], Berlekamp [119], Cantor and Zassenhaus [232], and Rabin [1025]. For recent work along these lines, see von zur Gathen and Shoup [1224], as well as Kaltofen and Shoup [658].

Public-Key Parameters

Contents in Brief

4.1 Introduction

The efficient generation of public-key parameters is a prerequisite in public-key systems. A specific example is the requirement of a prime number p to define a finite field $\mathbb{Z}_p$ for use in the Diffie-Hellman key agreement protocol and its derivatives (§12.6). In this case, an element of high order in $\mathbb{Z}_p^*$ is also required. Another example is the requirement of primes p and q for an RSA modulus $n = pq$ (§8.2). In this case, the prime must be of sufficient size, and be "random" in the sense that the probability of any particular prime being selected must be sufficiently small to preclude an adversary from gaining advantage through optimizing a search strategy based on such probability. Prime numbers may be required to have certain additional properties, in order that they do not make the associated cryptosystems susceptible to specialized attacks. A third example is the requirement of an irreducible polynomial $f(x)$ of degree m over the finite field $\mathbb{Z}_p$ for constructing the finite field $\mathbb{F}_{p^m}$. In this case, an element of high order in $\mathbb{F}_{p^m}^*$ is also required.

Chapter outline

The remainder of §4.1 introduces basic concepts relevant to prime number generation and summarizes some results on the distribution of prime numbers. Probabilistic primality tests, the most important of which is the Miller-Rabin test, are presented in §4.2. True primality tests by which arbitrary integers can be proven to be prime are the topic of §4.3; since these tests are generally more computationally intensive than probabilistic primality tests, they are not described in detail. §4.4 presents four algorithms for generating prime numbers, strong primes, and provable primes. §4.5 describes techniques for constructing irreducible and primitive polynomials, while §4.6 considers the production of generators and elements of high orders in groups. §4.7 concludes with chapter notes and references.

4.1.1 Generating large prime numbers naively

To motivate the organization of this chapter and introduce many of the relevant concepts, the problem of generating large prime numbers is first considered. The most natural method is to generate a random number n of appropriate size, and check if it is prime. This can be done by checking whether n is divisible by any of the prime numbers $\leq \sqrt{n}$. While more efficient methods are required in practice, to motivate further discussion consider the following approach:

1. Generate as *candidate* a random odd number n of appropriate size.
2. Test n for primality.
3. If n is composite, return to the first step.

A slight modification is to consider candidates restricted to some *search sequence* starting from n; a trivial search sequence which may be used is $n, n+2, n+4, n+6, \ldots$. Using specific search sequences may allow one to increase the expectation that a candidate is prime, and to find primes possessing certain additional desirable properties *a priori*.

In step 2, the test for primality might be either a test which *proves* that the candidate is prime (in which case the outcome of the generator is called a *provable prime*), or a test which establishes a weaker result, such as that n is "probably prime" (in which case the outcome of the generator is called a *probable prime*). In the latter case, careful consideration must be given to the exact meaning of this expression. So-called *probabilistic primality tests* typically determine, with mathematical certainty, which candidates n are composite, but do not provide a mathematical proof that n is prime in the case when such a number is declared to be "probably" so. In the latter case, however, when used properly one may often be able to draw conclusions more than adequate for the purpose at hand. For this reason, such tests are more properly called *compositeness tests* than probabilistic primality tests. True primality tests, which allow one to conclude with mathematical certainty that a number is prime, also exist, but generally require considerably greater computational resources.

While (true) primality tests can determine (with mathematical certainty) whether a typically random candidate number is prime, other techniques exist whereby candidates n are specially constructed such that it can be established by mathematical reasoning whether a candidate actually is prime. These are called *constructive prime generation* techniques.

A final distinction between different techniques for prime number generation is the use of randomness. Candidates are typically generated as a function of a random input. The technique used to judge the primality of the candidate, however, may or may not itself use random numbers. If it does not, the technique is *deterministic*, and the result is reproducible; if it does, the technique is said to be *randomized*. Both deterministic and randomized probabilistic primality tests exist.

In some cases, prime numbers are required which have additional properties. For example, to make the extraction of discrete logarithms in $\mathbb{Z}_p^*$ resistant to an algorithm due to Pohlig and Hellman (§3.6.4), it is a requirement that $p-1$ have a large prime divisor. Thus techniques for generating public-key parameters, such as prime numbers, of special form need to be considered.

4.1.2 Distribution of prime numbers

Let $\pi(x)$ denote the number of primes in the interval $[2, x]$. The prime number theorem (Fact 2.95) states that $\pi(x) \sim \frac{x}{\ln x}$.[1] In other words, the number of primes in the interval

---

[1] If $f(x)$ and $g(x)$ are two functions, then $f(x) \sim g(x)$ means that $\lim_{x \to \infty} \frac{f(x)}{g(x)} = 1$.

$[2, x]$ is approximately equal to $\frac{x}{\ln x}$. The prime numbers are quite uniformly distributed, as the following three results illustrate.

4.1 Fact (*Dirichlet theorem*) If $\gcd(a, n) = 1$, then there are infinitely many primes congruent to a modulo n.

A more explicit version of Dirichlet's theorem is the following.

4.2 Fact Let $\pi(x, n, a)$ denote the number of primes in the interval $[2, x]$ which are congruent to a modulo n, where $\gcd(a, n) = 1$. Then

$$\pi(x, n, a) \sim \frac{x}{\phi(n) \ln x}.$$

In other words, the prime numbers are roughly uniformly distributed among the $\phi(n)$ congruence classes in $\mathbb{Z}_n^*$, for any value of n.

4.3 Fact (*approximation for the nth prime number*) Let p_n denote the nth prime number. Then $p_n \sim n \ln n$. More explicitly,

$$n \ln n \; < \; p_n \; < \; n(\ln n + \ln \ln n) \;\; \text{for } n \geq 6.$$

4.2 Probabilistic primality tests

The algorithms in this section are methods by which arbitrary positive integers are tested to provide partial information regarding their primality. More specifically, probabilistic primality tests have the following framework. For each odd positive integer n, a set $W(n) \subset \mathbb{Z}_n$ is defined such that the following properties hold:

(i) given $a \in \mathbb{Z}_n$, it can be checked in deterministic polynomial time whether $a \in W(n)$;
(ii) if n is prime, then $W(n) = \emptyset$ (the empty set); and
(iii) if n is composite, then $\#W(n) \geq \frac{n}{2}$.

4.4 Definition If n is composite, the elements of $W(n)$ are called *witnesses* to the compositeness of n, and the elements of the complementary set $L(n) = \mathbb{Z}_n \setminus W(n)$ are called *liars*.

A probabilistic primality test utilizes these properties of the sets $W(n)$ in the following manner. Suppose that n is an integer whose primality is to be determined. An integer $a \in \mathbb{Z}_n$ is chosen at random, and it is checked if $a \in W(n)$. The test outputs "composite" if $a \in W(n)$, and outputs "prime" if $a \notin W(n)$. If indeed $a \in W(n)$, then n is said to *fail the primality test for the base a*; in this case, n is surely composite. If $a \notin W(n)$, then n is said to *pass the primality test for the base a*; in this case, no conclusion with absolute certainty can be drawn about the primality of n, and the declaration "prime" may be incorrect.[2]

Any single execution of this test which declares "composite" establishes this with certainty. On the other hand, successive independent runs of the test all of which return the answer "prime" allow the confidence that the input is indeed prime to be increased to whatever level is desired — the cumulative probability of error is multiplicative over independent trials. If the test is run t times independently on the composite number n, the probability that n is declared "prime" all t times (i.e., the probability of error) is at most $(\frac{1}{2})^t$.

---

[2]This discussion illustrates why a probabilistic primality test is more properly called a *compositeness* test.

4.5 Definition An integer n which is believed to be prime on the basis of a probabilistic primality test is called a *probable prime*.

Two probabilistic primality tests are covered in this section: the Solovay-Strassen test (§4.2.2) and the Miller-Rabin test (§4.2.3). For historical reasons, the Fermat test is first discussed in §4.2.1; this test is not truly a probabilistic primality test since it usually fails to distinguish between prime numbers and special composite integers called Carmichael numbers.

4.2.1 Fermat's test

Fermat's theorem (Fact 2.127) asserts that if n is a prime and a is any integer, $1 \le a \le n-1$, then $a^{n-1} \equiv 1 \pmod{n}$. Therefore, given an integer n whose primality is under question, finding any integer a in this interval such that this equivalence is not true suffices to prove that n is composite.

4.6 Definition Let n be an odd composite integer. An integer a, $1 \le a \le n - 1$, such that $a^{n-1} \not\equiv 1 \pmod{n}$ is called a *Fermat witness* (to compositeness) for n.

Conversely, finding an integer a between 1 and $n - 1$ such that $a^{n-1} \equiv 1 \pmod{n}$ makes n appear to be a prime in the sense that it satisfies Fermat's theorem for the base a. This motivates the following definition and Algorithm 4.9.

4.7 Definition Let n be an odd composite integer and let a be an integer, $1 \le a \le n - 1$. Then n is said to be a *pseudoprime to the base* a if $a^{n-1} \equiv 1 \pmod{n}$. The integer a is called a *Fermat liar* (to primality) for n.

4.8 Example (*pseudoprime*) The composite integer $n = 341 (= 11 \times 31)$ is a pseudoprime to the base 2 since $2^{340} \equiv 1 \pmod{341}$. $\qquad\square$

4.9 Algorithm Fermat primality test

FERMAT(n,t)
INPUT: an odd integer $n \ge 3$ and security parameter $t \ge 1$.
OUTPUT: an answer "prime" or "composite" to the question: "Is n prime?"
1. For i from 1 to t do the following:
    1.1 Choose a random integer a, $2 \le a \le n - 2$.
    1.2 Compute $r = a^{n-1} \bmod n$ using Algorithm 2.143.
    1.3 If $r \ne 1$ then return("composite").
2. Return("prime").

If Algorithm 4.9 declares "composite", then n is certainly composite. On the other hand, if the algorithm declares "prime" then no proof is provided that n is indeed prime. Nonetheless, since pseudoprimes for a given base a are known to be rare, Fermat's test provides a correct answer on *most* inputs; this, however, is quite distinct from providing a correct answer most of the time (e.g., if run with different bases) on *every* input. In fact, it does not do the latter because there are (even rarer) composite numbers which are pseudoprimes to *every* base a for which $\gcd(a, n) = 1$.

4.10 Definition A *Carmichael number n* is a composite integer such that $a^{n-1} \equiv 1 \pmod{n}$ for all integers a which satisfy $\gcd(a, n) = 1$.

If n is a Carmichael number, then the only Fermat witnesses for n are those integers $a, 1 \leq a \leq n - 1$, for which $\gcd(a, n) > 1$. Thus, if the prime factors of n are all large, then with high probability the Fermat test declares that n is "prime", even if the number of iterations t is large. This deficiency in the Fermat test is removed in the Solovay-Strassen and Miller-Rabin probabilistic primality tests by relying on criteria which are stronger than Fermat's theorem.

This subsection is concluded with some facts about Carmichael numbers. If the prime factorization of n is known, then Fact 4.11 can be used to easily determine whether n is a Carmichael number.

4.11 Fact (*necessary and sufficient conditions for Carmichael numbers*) A composite integer n is a Carmichael number if and only if the following two conditions are satisfied:

    (i) n is square-free, i.e., n is not divisible by the square of any prime; and

    (ii) $p - 1$ divides $n - 1$ for every prime divisor p of n.

A consequence of Fact 4.11 is the following.

4.12 Fact Every Carmichael number is the product of at least three distinct primes.

4.13 Fact (*bounds for the number of Carmichael numbers*)

    (i) There are an infinite number of Carmichael numbers. In fact, there are more than $n^{2/7}$ Carmichael numbers in the interval $[2, n]$, once n is sufficiently large.

    (ii) The best upper bound known for $C(n)$, the number of Carmichael numbers $\leq n$, is:

$$C(n) \leq n^{1 - \{1 + o(1)\} \ln \ln \ln n / \ln \ln n} \quad \text{for } n \to \infty.$$

The smallest Carmichael number is $n = 561 = 3 \times 11 \times 17$. Carmichael numbers are relatively scarce; there are only 105212 Carmichael numbers $\leq 10^{15}$.

4.2.2 Solovay-Strassen test

The Solovay-Strassen probabilistic primality test was the first such test popularized by the advent of public-key cryptography, in particular the RSA cryptosystem. There is no longer any reason to use this test, because an alternative is available (the Miller-Rabin test) which is both more efficient and always at least as correct (see Note 4.33). Discussion is nonetheless included for historical completeness and to clarify this exact point, since many people continue to reference this test.

Recall (§2.4.5) that $\left(\frac{a}{n}\right)$ denotes the Jacobi symbol, and is equivalent to the Legendre symbol if n is prime. The Solovay-Strassen test is based on the following fact.

4.14 Fact (*Euler's criterion*) Let n be an odd prime. Then $a^{(n-1)/2} \equiv \left(\frac{a}{n}\right) \pmod{n}$ for all integers a which satisfy $\gcd(a, n) = 1$.

Fact 4.14 motivates the following definitions.

4.15 Definition Let n be an odd composite integer and let a be an integer, $1 \leq a \leq n - 1$.

    (i) If either $\gcd(a, n) > 1$ or $a^{(n-1)/2} \not\equiv \left(\frac{a}{n}\right) \pmod{n}$, then a is called an *Euler witness* (to compositeness) for n.

(ii) Otherwise, i.e., if $\gcd(a, n) = 1$ and $a^{(n-1)/2} \equiv \left(\frac{a}{n}\right)$ (mod n), then n is said to be an *Euler pseudoprime to the base* a. (That is, n acts like a prime in that it satisfies Euler's criterion for the particular base a.) The integer a is called an *Euler liar* (to primality) for n.

4.16 Example (*Euler pseudoprime*) The composite integer 91 ($= 7 \times 13$) is an Euler pseudoprime to the base 9 since $9^{45} \equiv 1$ (mod 91) and $\left(\frac{9}{91}\right) = 1$.   □

Euler's criterion (Fact 4.14) can be used as a basis for a probabilistic primality test because of the following result.

4.17 Fact Let n be an odd composite integer. Then at most $\phi(n)/2$ of all the numbers a, $1 \leq a \leq n - 1$, are Euler liars for n (Definition 4.15). Here, ϕ is the Euler phi function (Definition 2.100).

4.18 Algorithm Solovay-Strassen probabilistic primality test

SOLOVAY-STRASSEN(n,t)
INPUT: an odd integer $n \geq 3$ and security parameter $t \geq 1$.
OUTPUT: an answer "prime" or "composite" to the question: "Is n prime?"
  1. For i from 1 to t do the following:
      1.1 Choose a random integer a, $2 \leq a \leq n - 2$.
      1.2 Compute $r = a^{(n-1)/2} \bmod n$ using Algorithm 2.143.
      1.3 If $r \neq 1$ and $r \neq n - 1$ then return("composite").
      1.4 Compute the Jacobi symbol $s = \left(\frac{a}{n}\right)$ using Algorithm 2.149.
      1.5 If $r \not\equiv s$ (mod n) then return ("composite").
  2. Return("prime").

If $\gcd(a, n) = d$, then d is a divisor of $r = a^{(n-1)/2} \bmod n$. Hence, testing whether $r \neq 1$ is step 1.3, eliminates the necessity of testing whether $\gcd(a, n) \neq 1$. If Algorithm 4.18 declares "composite", then n is certainly composite because prime numbers do not violate Euler's criterion (Fact 4.14). Equivalently, if n is actually prime, then the algorithm always declares "prime". On the other hand, if n is actually composite, then since the bases a in step 1.1 are chosen independently during each iteration of step 1, Fact 4.17 can be used to deduce the following probability of the algorithm erroneously declaring "prime".

4.19 Fact (*Solovay-Strassen error-probability bound*) Let n be an odd composite integer. The probability that SOLOVAY-STRASSEN(n,t) declares n to be "prime" is less than $\left(\frac{1}{2}\right)^t$.

4.2.3 Miller-Rabin test

The probabilistic primality test used most in practice is the Miller-Rabin test, also known as the *strong pseudoprime test*. The test is based on the following fact.

4.20 Fact Let n be an odd prime, and let $n - 1 = 2^s r$ where r is odd. Let a be any integer such that $\gcd(a, n) = 1$. Then either $a^r \equiv 1$ (mod n) or $a^{2^j r} \equiv -1$ (mod n) for some j, $0 \leq j \leq s - 1$.

Fact 4.20 motivates the following definitions.

4.21 Definition Let n be an odd composite integer and let $n - 1 = 2^s r$ where r is odd. Let a be an integer in the interval $[1, n - 1]$.

(i) If $a^r \not\equiv 1 \pmod{n}$ and if $a^{2^j r} \not\equiv -1 \pmod{n}$ for all j, $0 \leq j \leq s - 1$, then a is called a *strong witness* (to compositeness) for n.

(ii) Otherwise, i.e., if either $a^r \equiv 1 \pmod{n}$ or $a^{2^j r} \equiv -1 \pmod{n}$ for some j, $0 \leq j \leq s - 1$, then n is said to be a *strong pseudoprime to the base a*. (That is, n acts like a prime in that it satisfies Fact 4.20 for the particular base a.) The integer a is called a *strong liar* (to primality) for n.

4.22 Example (*strong pseudoprime*) Consider the composite integer $n = 91$ ($= 7 \times 13$). Since $91 - 1 = 90 = 2 \times 45$, $s = 1$ and $r = 45$. Since $9^r = 9^{45} \equiv 1 \pmod{91}$, 91 is a strong pseudoprime to the base 9. The set of all strong liars for 91 is:

$$\{1, 9, 10, 12, 16, 17, 22, 29, 38, 53, 62, 69, 74, 75, 79, 81, 82, 90\}.$$

Notice that the number of strong liars for 91 is $18 = \phi(91)/4$, where ϕ is the Euler phi function (cf. Fact 4.23). □

Fact 4.20 can be used as a basis for a probabilistic primality test due to the following result.

4.23 Fact If n is an odd composite integer, then at most $\frac{1}{4}$ of all the numbers a, $1 \leq a \leq n - 1$, are strong liars for n. In fact, if $n \neq 9$, the number of strong liars for n is at most $\phi(n)/4$, where ϕ is the Euler phi function (Definition 2.100).

4.24 Algorithm Miller-Rabin probabilistic primality test

MILLER-RABIN(n,t)
INPUT: an odd integer $n \geq 3$ and security parameter $t \geq 1$.
OUTPUT: an answer "prime" or "composite" to the question: "Is n prime?"
1. Write $n - 1 = 2^s r$ such that r is odd.
2. For i from 1 to t do the following:
    2.1 Choose a random integer a, $2 \leq a \leq n - 2$.
    2.2 Compute $y = a^r \bmod n$ using Algorithm 2.143.
    2.3 If $y \neq 1$ and $y \neq n - 1$ then do the following:
        $j \leftarrow 1$.
        While $j \leq s - 1$ and $y \neq n - 1$ do the following:
            Compute $y \leftarrow y^2 \bmod n$.
            If $y = 1$ then return("composite").
            $j \leftarrow j + 1$.
        If $y \neq n - 1$ then return ("composite").
3. Return("prime").

Algorithm 4.24 tests whether each base a satisfies the conditions of Definition 4.21(i). In the fifth line of step 2.3, if $y = 1$, then $a^{2^j r} \equiv 1 \pmod{n}$. Since it is also the case that $a^{2^{j-1} r} \not\equiv \pm 1 \pmod{n}$, it follows from Fact 3.18 that n is composite (in fact $\gcd(a^{2^{j-1} r} - 1, n)$ is a non-trivial factor of n). In the seventh line of step 2.3, if $y \neq n - 1$, then a is a strong witness for n. If Algorithm 4.24 declares "composite", then n is certainly composite because prime numbers do not violate Fact 4.20. Equivalently, if n is actually prime, then the algorithm always declares "prime". On the other hand, if n is actually composite, then Fact 4.23 can be used to deduce the following probability of the algorithm erroneously declaring "prime".

4.25 Fact (*Miller-Rabin error-probability bound*) For any odd composite integer n, the probability that MILLER-RABIN(n,t) declares n to be "prime" is less than $\left(\frac{1}{4}\right)^t$.

4.26 Remark (*number of strong liars*) For most composite integers n, the number of strong liars for n is actually much smaller than the upper bound of $\phi(n)/4$ given in Fact 4.23. Consequently, the Miller-Rabin error-probability bound is much smaller than $\left(\frac{1}{4}\right)^t$ for most positive integers n.

4.27 Example (*some composite integers have very few strong liars*) The only strong liars for the composite integer $n = 105 \, (= 3 \times 5 \times 7)$ are 1 and 104. More generally, if $k \geq 2$ and n is the product of the first k odd primes, there are only 2 strong liars for n, namely 1 and $n - 1$.   □

4.28 Remark (*fixed bases in Miller-Rabin*) If a_1 and a_2 are strong liars for n, their product $a_1 a_2$ is very likely, but not certain, to also be a strong liar for n. A strategy that is sometimes employed is to fix the bases a in the Miller-Rabin algorithm to be the first few primes (composite bases are ignored because of the preceding statement), instead of choosing them at random.

4.29 Definition Let $p_1, p_2, \ldots, p_t$ denote the first t primes. Then ψ_t is defined to be the smallest positive composite integer which is a strong pseudoprime to all the bases $p_1, p_2, \ldots, p_t$.

The numbers ψ_t can be interpreted as follows: to determine the primality of any integer $n < \psi_t$, it is sufficient to apply the Miller-Rabin algorithm to n with the bases a being the first t prime numbers. With this choice of bases, the answer returned by Miller-Rabin is always correct. Table 4.1 gives the value of ψ_t for $1 \leq t \leq 8$.

t	ψ_t
1	2047
2	1373653
3	25326001
4	3215031751
5	2152302898747
6	3474749660383
7	341550071728321
8	341550071728321

Table 4.1: *Smallest strong pseudoprimes. The table lists values of ψ_t, the smallest positive composite integer that is a strong pseudoprime to each of the first t prime bases, for $1 \leq t \leq 8$.*

4.2.4 Comparison: Fermat, Solovay-Strassen, and Miller-Rabin

Fact 4.30 describes the relationships between Fermat liars, Euler liars, and strong liars (see Definitions 4.7, 4.15, and 4.21).

4.30 Fact Let n be an odd composite integer.

    (i) If a is an Euler liar for n, then it is also a Fermat liar for n.

    (ii) If a is a strong liar for n, then it is also an Euler liar for n.

4.31 Example (*Fermat, Euler, strong liars*) Consider the composite integer $n = 65 (= 5 \times 13)$. The Fermat liars for 65 are $\{1, 8, 12, 14, 18, 21, 27, 31, 34, 38, 44, 47, 51, 53, 57, 64\}$. The Euler liars for 65 are $\{1, 8, 14, 18, 47, 51, 57, 64\}$, while the strong liars for 65 are $\{1, 8, 18, 47, 57, 64\}$. □

For a fixed composite candidate n, the situation is depicted in Figure 4.1. This set-

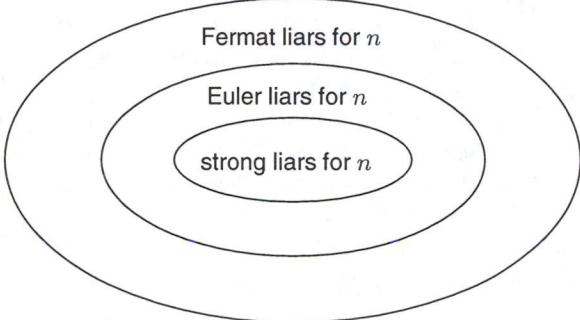

Figure 4.1: *Relationships between Fermat, Euler, and strong liars for a composite integer n.*

tles the question of the relative accuracy of the Fermat, Solovay-Strassen, and Miller-Rabin tests, not only in the sense of the relative correctness of each test on a fixed candidate n, but also in the sense that given n, the specified containments hold for *each* randomly chosen base a. Thus, from a correctness point of view, the Miller-Rabin test is never worse than the Solovay-Strassen test, which in turn is never worse than the Fermat test. As the following result shows, there are, however, some composite integers n for which the Solovay-Strassen and Miller-Rabin tests are equally good.

4.32 Fact If $n \equiv 3 \pmod 4$, then a is an Euler liar for n if and only if it is a strong liar for n.

What remains is a comparison of the computational costs. While the Miller-Rabin test may appear more complex, it actually requires, at worst, the same amount of computation as Fermat's test in terms of modular multiplications; thus the Miller-Rabin test is better than Fermat's test in all regards. At worst, the sequence of computations defined in MILLER-RABIN(n,1) requires the equivalent of computing $a^{(n-1)/2} \bmod n$. It is also the case that MILLER-RABIN(n,1) requires less computation than SOLOVAY-STRASSEN(n,1), the latter requiring the computation of $a^{(n-1)/2} \bmod n$ and possibly a further Jacobi symbol computation. For this reason, the Solovay-Strassen is both computationally and conceptually more complex.

4.33 Note (*Miller-Rabin is better than Solovay-Strassen*) In summary, both the Miller-Rabin and Solovay-Strassen tests are correct in the event that either their input is actually prime, or that they declare their input composite. There is, however, no reason to use the Solovay-Strassen test (nor the Fermat test) over the Miller-Rabin test. The reasons for this are summarized below.

(i) The Solovay-Strassen test is computationally more expensive.
(ii) The Solovay-Strassen test is harder to implement since it also involves Jacobi symbol computations.
(iii) The error probability for Solovay-Strassen is bounded above by $(\frac{1}{2})^t$, while the error probability for Miller-Rabin is bounded above by $(\frac{1}{4})^t$.

(iv) Any strong liar for n is also an Euler liar for n. Hence, from a correctness point of view, the Miller-Rabin test is never worse than the Solovay-Strassen test.

4.3 (True) Primality tests

The primality tests in this section are methods by which positive integers can be *proven* to be prime, and are often referred to as *primality proving algorithms*. These primality tests are generally more computationally intensive than the probabilistic primality tests of §4.2. Consequently, before applying one of these tests to a candidate prime n, the candidate should be subjected to a probabilistic primality test such as Miller-Rabin (Algorithm 4.24).

4.34 Definition An integer n which is believed to be prime on the basis of a primality proving algorithm is called a *provable prime*.

4.3.1 Testing Mersenne numbers

Efficient algorithms are known for testing primality of some special classes of numbers, such as Mersenne numbers and Fermat numbers. Mersenne primes n are useful because the arithmetic in the field $\mathbb{Z}_n$ for such n can be implemented very efficiently (see §14.3.4). The Lucas-Lehmer test for Mersenne numbers (Algorithm 4.37) is such an algorithm.

4.35 Definition Let $s \geq 2$ be an integer. A *Mersenne number* is an integer of the form $2^s - 1$. If $2^s - 1$ is prime, then it is called a *Mersenne prime*.

The following are necessary and sufficient conditions for a Mersenne number to be prime.

4.36 Fact Let $s \geq 3$. The Mersenne number $n = 2^s - 1$ is prime if and only if the following two conditions are satisfied:

    (i) s is prime; and

    (ii) the sequence of integers defined by $u_0 = 4$ and $u_{k+1} = (u_k^2 - 2) \bmod n$ for $k \geq 0$ satisfies $u_{s-2} = 0$.

Fact 4.36 leads to the following deterministic polynomial-time algorithm for determining (with certainty) whether a Mersenne number is prime.

4.37 Algorithm Lucas-Lehmer primality test for Mersenne numbers

INPUT: a Mersenne number $n = 2^s - 1$ with $s \geq 3$.
OUTPUT: an answer "prime" or "composite" to the question: "Is n prime?"

    1. Use trial division to check if s has any factors between 2 and $\lfloor \sqrt{s} \rfloor$. If it does, then return("composite").
    2. Set $u \leftarrow 4$.
    3. For k from 1 to $s - 2$ do the following: compute $u \leftarrow (u^2 - 2) \bmod n$.
    4. If $u = 0$ then return("prime"). Otherwise, return("composite").

It is unknown whether there are infinitely many Mersenne primes. Table 4.2 lists the 33 known Mersenne primes.

Index j	M_j	decimal digits	Index j	M_j	decimal digits
1	2	1	18	3217	969
2	3	1	19	4253	1281
3	5	2	20	4423	1332
4	7	3	21	9689	2917
5	13	4	22	9941	2993
6	17	6	23	11213	3376
7	19	6	24	19937	6002
8	31	10	25	21701	6533
9	61	19	26	23209	6987
10	89	27	27	44497	13395
11	107	33	28	86243	25962
12	127	39	29	110503	33265
13	521	157	30	132049	39751
14	607	183	31	216091	65050
15	1279	386	32?	756839	227832
16	2203	664	33?	859433	258716
17	2281	687			

Table 4.2: *Known Mersenne primes. The table shows the 33 known exponents M_j, $1 \leq j \leq 33$, for which $2^{M_j} - 1$ is a Mersenne prime, and also the number of decimal digits in $2^{M_j} - 1$. The question marks after $j = 32$ and $j = 33$ indicate that it is not known whether there are any other exponents s between M_{31} and these numbers for which $2^s - 1$ is prime.*

4.3.2 Primality testing using the factorization of $n - 1$

This section presents results which can be used to prove that an integer n is prime, provided that the factorization or a partial factorization of $n-1$ is known. It may seem odd to consider a technique which requires the factorization of $n - 1$ as a subproblem — if integers of this size can be factored, the primality of n itself could be determined by factoring n. However, the factorization of $n-1$ may be easier to compute if n has a special form, such as a *Fermat number* $n = 2^{2^k} + 1$. Another situation where the factorization of $n - 1$ may be easy to compute is when the candidate n is "constructed" by specific methods (see §4.4.4).

4.38 Fact Let $n \geq 3$ be an integer. Then n is prime if and only if there exists an integer a satisfying:

(i) $a^{n-1} \equiv 1 \pmod{n}$; and
(ii) $a^{(n-1)/q} \not\equiv 1 \pmod{n}$ for each prime divisor q of $n - 1$.

This result follows from the fact that $\mathbb{Z}_n^*$ has an element of order $n - 1$ (Definition 2.128) if and only if n is prime; an element a satisfying conditions (i) and (ii) has order $n - 1$.

4.39 Note (*primality test based on Fact 4.38*) If n is a prime, the number of elements of order $n - 1$ is precisely $\phi(n - 1)$. Hence, to prove a candidate n prime, one simply chooses an integer $a \in \mathbb{Z}_n$ at random and uses Fact 4.38 to check if a has order $n - 1$. If this is the case, then n is certainly prime. Otherwise, another $a \in \mathbb{Z}_n$ is selected and the test is repeated. If n is indeed prime, the expected number of iterations before an element a of order $n - 1$ is selected is $O(\ln \ln n)$; this follows since $(n - 1)/\phi(n - 1) < 6 \ln \ln n$ for

$n \geq 5$ (Fact 2.102). Thus, if such an a is not found after a "reasonable" number (for example, $12 \ln \ln n$) of iterations, then n is probably composite and should again be subjected to a probabilistic primality test such as Miller-Rabin (Algorithm 4.24).[3] This method is, in effect, a probabilistic compositeness test.

The next result gives a method for proving primality which requires knowledge of only a *partial* factorization of $n - 1$.

4.40 Fact (*Pocklington's theorem*) Let $n \geq 3$ be an integer, and let $n = RF + 1$ (i.e. F divides $n - 1$) where the prime factorization of F is $F = \prod_{j=1}^{t} q_j^{e_j}$. If there exists an integer a satisfying:

  (i) $a^{n-1} \equiv 1 \pmod{n}$; and
  (ii) $\gcd(a^{(n-1)/q_j} - 1, n) = 1$ for each j, $1 \leq j \leq t$,

then every prime divisor p of n is congruent to 1 modulo F. It follows that if $F > \sqrt{n} - 1$, then n is prime.

If n is indeed prime, then the following result establishes that most integers a satisfy conditions (i) and (ii) of Fact 4.40, provided that the prime divisors of $F > \sqrt{n} - 1$ are sufficiently large.

4.41 Fact Let $n = RF + 1$ be an odd prime with $F > \sqrt{n} - 1$ and $\gcd(R, F) = 1$. Let the distinct prime factors of F be $q_1, q_2, \ldots, q_t$. Then the probability that a randomly selected base a, $1 \leq a \leq n - 1$, satisfies both: (i) $a^{n-1} \equiv 1 \pmod{n}$; and (ii) $\gcd(a^{(n-1)/q_j} - 1, n) = 1$ for each j, $1 \leq j \leq t$, is $\prod_{j=1}^{t}(1 - 1/q_j) \geq 1 - \sum_{j=1}^{t} 1/q_j$.

Thus, if the factorization of a divisor $F > \sqrt{n} - 1$ of $n - 1$ is known, one simply chooses random integers a in the interval $[2, n - 2]$ until one is found satisfying conditions (i) and (ii) of Fact 4.40, implying that n is prime. If such an a is not found after a "reasonable" number of iterations,[4] then n is probably composite and should be subjected to a probabilistic primality test (footnote 3 also applies here). This method is, in effect, a probabilistic compositeness test.

The next result gives a method for proving primality which only requires the factorization of a divisor F of $n - 1$ that is greater than $\sqrt[3]{n}$. For an example of the use of Fact 4.42, see Note 4.63.

4.42 Fact Let $n \geq 3$ be an odd integer. Let $n = 2RF + 1$, and suppose that there exists an integer a satisfying both: (i) $a^{n-1} \equiv 1 \pmod{n}$; and (ii) $\gcd(a^{(n-1)/q} - 1, n) = 1$ for each prime divisor q of F. Let $x \geq 0$ and y be defined by $2R = xF + y$ and $0 \leq y < F$. If $F \geq \sqrt[3]{n}$ and if $y^2 - 4x$ is neither 0 nor a perfect square, then n is prime.

4.3.3 Jacobi sum test

The *Jacobi sum test* is another true primality test. The basic idea is to test a set of congruences which are analogues of Fermat's theorem (Fact 2.127(i)) in certain *cyclotomic rings*. The running time of the Jacobi sum test for determining the primality of an integer n is $O((\ln n)^{c \ln \ln \ln n})$ bit operations for some constant c. This is "almost" a polynomial-time algorithm since the exponent $\ln \ln \ln n$ acts like a constant for the range of values for

---

[3] Another approach is to run both algorithms in parallel (with an unlimited number of iterations), until one of them stops with a definite conclusion "prime" or "composite".

[4] The number of iterations may be taken to be T where $P^T \leq (\frac{1}{2})^{100}$, and where $P = 1 - \prod_{j=1}^{t}(1 - 1/q_j)$.

n of interest. For example, if $n \leq 2^{512}$, then $\ln \ln \ln n < 1.78$. The version of the Jacobi sum primality test used in practice is a randomized algorithm which terminates within $O(k(\ln n)^{c \ln \ln \ln n})$ steps with probability at least $1 - (\frac{1}{2})^k$ for every $k \geq 1$, and always gives a correct answer. One drawback of the algorithm is that it does not produce a "certificate" which would enable the answer to be verified in much shorter time than running the algorithm itself.

The Jacobi sum test is, indeed, practical in the sense that the primality of numbers that are several hundred decimal digits long can be handled in just a few minutes on a computer. However, the test is not as easy to program as the probabilistic Miller-Rabin test (Algorithm 4.24), and the resulting code is not as compact. The details of the algorithm are complicated and are not given here; pointers to the literature are given in the chapter notes on page 166.

4.3.4 Tests using elliptic curves

Elliptic curve primality proving algorithms are based on an elliptic curve analogue of Pocklington's theorem (Fact 4.40). The version of the algorithm used in practice is usually referred to as *Atkin's test* or the *Elliptic Curve Primality Proving algorithm* (ECPP). Under heuristic arguments, the expected running time of this algorithm for proving the primality of an integer n has been shown to be $O((\ln n)^{6+\epsilon})$ bit operations for any $\epsilon > 0$. Atkin's test has the advantage over the Jacobi sum test (§4.3.3) that it produces a short *certificate of primality* which can be used to efficiently verify the primality of the number. Atkin's test has been used to prove the primality of numbers more than 1000 decimal digits long.

The details of the algorithm are complicated and are not presented here; pointers to the literature are given in the chapter notes on page 166.

4.4 Prime number generation

This section considers algorithms for the generation of prime numbers for cryptographic purposes. Four algorithms are presented: Algorithm 4.44 for generating *probable* primes (see Definition 4.5), Algorithm 4.53 for generating *strong* primes (see Definition 4.52), Algorithm 4.56 for generating *probable* primes p and q suitable for use in the Digital Signature Algorithm (DSA), and Algorithm 4.62 for generating *provable* primes (see Definition 4.34).

4.43 Note (*prime generation vs. primality testing*) Prime number *generation* differs from primality *testing* as described in §4.2 and §4.3, but may and typically does involve the latter. The former allows the construction of candidates of a fixed form which may lead to more efficient testing than possible for random candidates.

4.4.1 Random search for probable primes

By the prime number theorem (Fact 2.95), the proportion of (positive) integers $\leq x$ that are prime is approximately $1/\ln x$. Since half of all integers $\leq x$ are even, the proportion of *odd* integers $\leq x$ that are prime is approximately $2/\ln x$. For instance, the proportion of all odd integers $\leq 2^{512}$ that are prime is approximately $2/(512 \cdot \ln(2)) \approx 1/177$. This suggests that a reasonable strategy for selecting a random k-bit (probable) prime is to repeatedly pick random k-bit odd integers n until one is found that is declared to be "prime" by MILLER-RABIN(n,t) (Algorithm 4.24) for an appropriate value of the security parameter t (discussed below).

If a random k-bit odd integer n is divisible by a small prime, it is less computationally expensive to rule out the candidate n by trial division than by using the Miller-Rabin test. Since the probability that a random integer n has a small prime divisor is relatively large, before applying the Miller-Rabin test, the candidate n should be tested for small divisors below a pre-determined bound B. This can be done by dividing n by all the primes below B, or by computing greatest common divisors of n and (pre-computed) products of several of the primes $\leq B$. The proportion of candidate odd integers n not ruled out by this trial division is $\prod_{3 \leq p \leq B}(1 - \frac{1}{p})$ which, by Merten's theorem, is approximately $1.12/\ln B$ (here p ranges over prime values). For example, if $B = 256$, then only 20% of candidate odd integers n pass the trial division stage, i.e., 80% are discarded before the more costly Miller-Rabin test is performed.

4.44 Algorithm Random search for a prime using the Miller-Rabin test

---

RANDOM-SEARCH(k,t)
INPUT: the required bitlength k of the prime, and a security parameter t. (See Note 4.49 for guidance on selecting t.)
OUTPUT: a random k-bit probable prime.
1. Generate an odd k-bit integer n at random.
2. Use trial division to determine whether n is divisible by any odd prime $\leq B$ (see Note 4.45 for guidance on selecting B). If it is then go to step 1.
3. If MILLER-RABIN(n,t) (Algorithm 4.24) outputs "prime" then return(n). Otherwise, go to step 1.

4.45 Note (*optimal trial division bound B*) Let E denote the time for a full k-bit modular exponentiation, and let D denote the time required for ruling out one small prime as divisor of a k-bit integer. (The values E and D depend on the particular implementation of long-integer arithmetic.) Then the trial division bound B that minimizes the expected running time of Algorithm 4.44 for generating a k-bit prime is roughly $B = E/D$. A more accurate estimate of the optimum choice for B can be obtained experimentally. The odd primes up to B can be precomputed and stored in a table. If memory is scarce, a value of B that is smaller than the optimum value may be used.

Since the Miller-Rabin test does not provide a mathematical proof that a number is indeed prime, the number n returned by Algorithm 4.44 is a probable prime (Definition 4.5). It is important, therefore, to have an estimate of the probability that n is in fact composite.

4.46 Definition The probability that RANDOM-SEARCH(k,t) (Algorithm 4.44) returns a composite number is denoted by $p_{k,t}$.

4.47 Note (*remarks on estimating $p_{k,t}$*) It is tempting to conclude directly from Fact 4.25 that $p_{k,t} \leq (\frac{1}{4})^t$. This reasoning is flawed (although typically the conclusion will be correct in practice) since it does not take into account the distribution of the primes. (For example, if the candidate n were chosen from a set S of composite numbers, the probability of error is 1.) The following discussion elaborates on this point. Let X represent the event that n is composite, and let Y_t denote the event than MILLER-RABIN(n,t) declares n to be prime. Then Fact 4.25 states that $P(Y_t|X) \leq (\frac{1}{4})^t$. What is relevant, however, to the estimation of $p_{k,t}$ is the quantity $P(X|Y_t)$. Suppose that candidates n are drawn uniformly and randomly from a set S of odd numbers, and suppose p is the probability that n is prime (this depends on the candidate set S). Assume also that $0 < p < 1$. Then by Bayes' theorem (Fact 2.10):

$$P(X|Y_t) = \frac{P(X)P(Y_t|X)}{P(Y_t)} \leq \frac{P(Y_t|X)}{P(Y_t)} \leq \frac{1}{p}\left(\frac{1}{4}\right)^t,$$

since $P(Y_t) \geq p$. Thus the probability $P(X|Y_t)$ may be considerably larger than $(\frac{1}{4})^t$ if p is small. However, the error-probability of Miller-Rabin is usually far smaller than $(\frac{1}{4})^t$ (see Remark 4.26). Using better estimates for $P(Y_t|X)$ and estimates on the number of k-bit prime numbers, it has been shown that $p_{k,t}$ is, in fact, smaller than $(\frac{1}{4})^t$ for all sufficiently large k. A more concrete result is the following: if candidates n are chosen at random from the set of odd numbers in the interval $[3, x]$, then $P(X|Y_t) \leq (\frac{1}{4})^t$ for all $x \geq 10^{60}$.

Further refinements for $P(Y_t|X)$ allow the following explicit upper bounds on $p_{k,t}$ for various values of k and t. [5]

4.48 Fact (*some upper bounds on $p_{k,t}$ in Algorithm 4.44*)

(i) $p_{k,1} < k^2 4^{2-\sqrt{k}}$ for $k \geq 2$.

(ii) $p_{k,t} < k^{3/2} 2^t t^{-1/2} 4^{2-\sqrt{tk}}$ for ($t = 2$, $k \geq 88$) or ($3 \leq t \leq k/9$, $k \geq 21$).

(iii) $p_{k,t} < \frac{7}{20}k2^{-5t} + \frac{1}{7}k^{15/4}2^{-k/2-2t} + 12k2^{-k/4-3t}$ for $k/9 \leq t \leq k/4$, $k \geq 21$.

(iv) $p_{k,t} < \frac{1}{7}k^{15/4}2^{-k/2-2t}$ for $t \geq k/4$, $k \geq 21$.

For example, if $k = 512$ and $t = 6$, then Fact 4.48(ii) gives $p_{512,6} \leq (\frac{1}{2})^{88}$. In other words, the probability that RANDOM-SEARCH(512,6) returns a 512-bit composite integer is less than $(\frac{1}{2})^{88}$. Using more advanced techniques, the upper bounds on $p_{k,t}$ given by Fact 4.48 have been improved. These upper bounds arise from complicated formulae which are not given here. Table 4.3 lists some improved upper bounds on $p_{k,t}$ for some sample values of k and t. As an example, the probability that RANDOM-SEARCH(500,6) returns a composite number is $\leq (\frac{1}{2})^{92}$. Notice that the values of $p_{k,t}$ implied by the table are considerably smaller than $(\frac{1}{4})^t = (\frac{1}{2})^{2t}$.

k	t									
	1	2	3	4	5	6	7	8	9	10
100	5	14	20	25	29	33	36	39	41	44
150	8	20	28	34	39	43	47	51	54	57
200	11	25	34	41	47	52	57	61	65	69
250	14	29	39	47	54	60	65	70	75	79
300	19	33	44	53	60	67	73	78	83	88
350	28	38	48	58	66	73	80	86	91	97
400	37	46	55	63	72	80	87	93	99	105
450	46	54	62	70	78	85	93	100	106	112
500	56	63	70	78	85	92	99	106	113	119
550	65	72	79	86	93	100	107	113	119	126
600	75	82	88	95	102	108	115	121	127	133

Table 4.3: *Upper bounds on $p_{k,t}$ for sample values of k and t. An entry j corresponding to k and t implies $p_{k,t} \leq (\frac{1}{2})^j$.*

---

[5]The estimates of $p_{k,t}$ presented in the remainder of this subsection were derived for the situation where Algorithm 4.44 does not use trial division by small primes to rule out some candidates n. Since trial division never rules out a prime, it can only give a better chance of rejecting composites. Thus the error probability $p_{k,t}$ might actually be even smaller than the estimates given here.

4.49 Note (*controlling the error probability*) In practice, one is usually willing to tolerate an error probability of $(\frac{1}{2})^{80}$ when using Algorithm 4.44 to generate probable primes. For sample values of k, Table 4.4 lists the smallest value of t that can be derived from Fact 4.48 for which $p_{k,t} \leq (\frac{1}{2})^{80}$. For example, when generating 1000-bit probable primes, Miller-Rabin with $t = 3$ repetitions suffices. Algorithm 4.44 rules out most candidates n either by trial division (in step 2) or by performing just one iteration of the Miller-Rabin test (in step 3). For this reason, the only effect of selecting a larger security parameter t on the running time of the algorithm will likely be to increase the time required in the final stage when the (probable) prime is chosen.

k	t	k	t	k	t	k	t	k	t
100	27	500	6	900	3	1300	2	1700	2
150	18	550	5	950	3	1350	2	1750	2
200	15	600	5	1000	3	1400	2	1800	2
250	12	650	4	1050	3	1450	2	1850	2
300	9	700	4	1100	3	1500	2	1900	2
350	8	750	4	1150	3	1550	2	1950	2
400	7	800	4	1200	3	1600	2	2000	2
450	6	850	3	1250	3	1650	2	2050	2

Table 4.4: *For sample k, the smallest t from Fact 4.48 is given for which $p_{k,t} \leq (\frac{1}{2})^{80}$.*

4.50 Remark (*Miller-Rabin test with base $a = 2$*) The Miller-Rabin test involves exponentiating the base a; this may be performed using the repeated square-and-multiply algorithm (Algorithm 2.143). If $a = 2$, then multiplication by a is a simple procedure relative to multiplying by a in general. One optimization of Algorithm 4.44 is, therefore, to fix the base $a = 2$ when first performing the Miller-Rabin test in step 3. Since most composite numbers will fail the Miller-Rabin test with base $a = 2$, this modification will lower the expected running time of Algorithm 4.44.

4.51 Note (*incremental search*)

(i) An alternative technique to generating candidates n at random in step 1 of Algorithm 4.44 is to first select a random k-bit odd number n_0, and then test the s numbers $n = n_0, n_0 + 2, n_0 + 4, \ldots, n_0 + 2(s - 1)$ for primality. If all these s candidates are found to be composite, the algorithm is said to have *failed*. If $s = c \cdot \ln 2^k$ where c is a constant, the probability $q_{k,t,s}$ that this incremental search variant of Algorithm 4.44 returns a composite number has been shown to be less than $\delta k^3 2^{-\sqrt{k}}$ for some constant δ. Table 4.5 gives some explicit bounds on this error probability for $k = 500$ and $t \leq 10$. Under reasonable number-theoretic assumptions, the probability of the algorithm failing has been shown to be less than $2e^{-2c}$ for large k (here, $e \approx 2.71828$).

(ii) Incremental search has the advantage that fewer random bits are required. Furthermore, the trial division by small primes in step 2 of Algorithm 4.44 can be accomplished very efficiently as follows. First the values $R[p] = n_0 \bmod p$ are computed for each odd prime $p \leq B$. Each time 2 is added to the current candidate, the values in the table R are updated as $R[p] \leftarrow (R[p] + 2) \bmod p$. The candidate passes the trial division stage if and only if none of the $R[p]$ values equal 0.

(iii) If B is large, an alternative method for doing the trial division is to initialize a table $S[i] \leftarrow 0$ for $0 \leq i \leq (s - 1)$; the entry $S[i]$ corresponds to the candidate $n_0 + 2i$. For each odd prime $p \leq B$, $n_0 \bmod p$ is computed. Let j be the smallest index for

c	t = 1	2	3	4	5	6	7	8	9	10
1	17	37	51	63	72	81	89	96	103	110
5	13	32	46	58	68	77	85	92	99	105
10	11	30	44	56	66	75	83	90	97	103

Table 4.5: *Upper bounds on the error probability of incremental search (Note 4.51) for* $k = 500$ *and sample values of* c *and* t. *An entry* j *corresponding to* c *and* t *implies* $q_{500,t,s} \leq (\frac{1}{2})^j$, *where* $s = c \cdot \ln 2^{500}$.

which $(n_0 + 2j) \equiv 0 \pmod{p}$. Then $S[j]$ and each pth entry after it are set to 1. A candidate $n_0 + 2i$ then passes the trial division stage if and only if $S[i] = 0$. Note that the estimate for the optimal trial division bound B given in Note 4.45 does not apply here (nor in (ii)) since the cost of division is amortized over all candidates.

4.4.2 Strong primes

The RSA cryptosystem (§8.2) uses a modulus of the form $n = pq$, where p and q are distinct odd primes. The primes p and q must be of sufficient size that factorization of their product is beyond computational reach. Moreover, they should be random primes in the sense that they be chosen as a function of a random input through a process defining a pool of candidates of sufficient cardinality that an exhaustive attack is infeasible. In practice, the resulting primes must also be of a pre-determined bitlength, to meet system specifications. The discovery of the RSA cryptosystem led to the consideration of several additional constraints on the choice of p and q which are necessary to ensure the resulting RSA system safe from cryptanalytic attack, and the notion of a strong prime (Definition 4.52) was defined. These attacks are described at length in Note 8.8(iii); as noted there, it is now believed that strong primes offer little protection beyond that offered by random primes, since randomly selected primes of the sizes typically used in RSA moduli today will satisfy the constraints with high probability. On the other hand, they are no less secure, and require only minimal additional running time to compute; thus, there is little real additional cost in using them.

4.52 Definition A prime number p is said to be a *strong prime* if integers r, s, and t exist such that the following three conditions are satisfied:

(i) $p - 1$ has a large prime factor, denoted r;
(ii) $p + 1$ has a large prime factor, denoted s; and
(iii) $r - 1$ has a large prime factor, denoted t.

In Definition 4.52, a precise qualification of "large" depends on specific attacks that should be guarded against; for further details, see Note 8.8(iii).

4.53 Algorithm Gordon's algorithm for generating a strong prime

---

SUMMARY: a strong prime p is generated.

1. Generate two large random primes s and t of roughly equal bitlength (see Note 4.54).
2. Select an integer i_0. Find the first prime in the sequence $2it + 1$, for $i = i_0, i_0 + 1, i_0 + 2, \ldots$ (see Note 4.54). Denote this prime by $r = 2it + 1$.
3. Compute $p_0 = (2s^{r-2} \bmod r)s - 1$.
4. Select an integer j_0. Find the first prime in the sequence $p_0 + 2jrs$, for $j = j_0, j_0 + 1, j_0 + 2, \ldots$ (see Note 4.54). Denote this prime by $p = p_0 + 2jrs$.
5. Return(p).

Justification. To see that the prime p returned by Gordon's algorithm is indeed a strong prime, observe first (assuming $r \neq s$) that $s^{r-1} \equiv 1 \pmod{r}$; this follows from Fermat's theorem (Fact 2.127). Hence, $p_0 \equiv 1 \pmod{r}$ and $p_0 \equiv -1 \pmod{s}$. Finally (cf. Definition 4.52),

  (i) $p - 1 = p_0 + 2jrs - 1 \equiv 0 \pmod{r}$, and hence $p - 1$ has the prime factor r;
  (ii) $p + 1 = p_0 + 2jrs + 1 \equiv 0 \pmod{s}$, and hence $p + 1$ has the prime factor s; and
  (iii) $r - 1 = 2it \equiv 0 \pmod{t}$, and hence $r - 1$ has the prime factor t.

4.54 Note (*implementing Gordon's algorithm*)

  (i) The primes s and t required in step 1 can be probable primes generated by Algorithm 4.44. The Miller-Rabin test (Algorithm 4.24) can be used to test each candidate for primality in steps 2 and 4, after ruling out candidates that are divisible by a small prime less than some bound B. See Note 4.45 for guidance on selecting B. Since the Miller-Rabin test is a probabilistic primality test, the output of this implementation of Gordon's algorithm is a probable prime.

  (ii) By carefully choosing the sizes of primes s, t and parameters i_0, j_0, one can control the exact bitlength of the resulting prime p. Note that the bitlengths of r and s will be about half that of p, while the bitlength of t will be slightly less than that of r.

4.55 Fact (*running time of Gordon's algorithm*) If the Miller-Rabin test is the primality test used in steps 1, 2, and 4, the expected time Gordon's algorithm takes to find a strong prime is only about 19% more than the expected time Algorithm 4.44 takes to find a random prime.

4.4.3 NIST method for generating DSA primes

Some public-key schemes require primes satisfying various specific conditions. For example, the NIST Digital Signature Algorithm (DSA of §11.5.1) requires two primes p and q satisfying the following three conditions:

  (i) $2^{159} < q < 2^{160}$; that is, q is a 160-bit prime;
  (ii) $2^{L-1} < p < 2^L$ for a specified L, where $L = 512 + 64l$ for some $0 \leq l \leq 8$; and
  (iii) q divides $p - 1$.

This section presents an algorithm for generating such primes p and q. In the following, H denotes the SHA-1 hash function (§9.53) which maps bitstrings of bitlength $< 2^{64}$ to 160-bit hash-codes. Where required, an integer x in the range $0 \leq x < 2^g$ whose binary representation is $x = x_{g-1}2^{g-1} + x_{g-2}2^{g-2} + \cdots + x_2 2^2 + x_1 2 + x_0$ should be converted to the g-bit sequence $(x_{g-1}x_{g-2} \cdots x_2 x_1 x_0)$, and vice versa.

4.56 Algorithm NIST method for generating DSA primes

INPUT: an integer l, $0 \leq l \leq 8$.
OUTPUT: a 160-bit prime q and an L-bit prime p, where $L = 512 + 64l$ and $q|(p-1)$.

1. Compute $L = 512 + 64l$. Using long division of $(L-1)$ by 160, find n, b such that $L - 1 = 160n + b$, where $0 \leq b < 160$.
2. Repeat the following:
   2.1 Choose a random seed s (not necessarily secret) of bitlength $g \geq 160$.
   2.2 Compute $U = H(s) \oplus H((s+1) \bmod 2^g)$.
   2.3 Form q from U by setting to 1 the most significant and least significant bits of U. (Note that q is a 160-bit odd integer.)
   2.4 Test q for primality using MILLER-RABIN(q,t) for $t \geq 18$ (see Note 4.57).

   Until q is found to be a (probable) prime.
3. Set $i \leftarrow 0$, $j \leftarrow 2$.
4. While $i < 4096$ do the following:
   4.1 For k from 0 to n do the following: set $V_k \leftarrow H((s + j + k) \bmod 2^g)$.
   4.2 For the integer W defined below, let $X = W + 2^{L-1}$. (X is an L-bit integer.)
   $$W = V_0 + V_1 2^{160} + V_2 2^{320} + \cdots + V_{n-1} 2^{160(n-1)} + (V_n \bmod 2^b) 2^{160n}.$$
   4.3 Compute $c = X \bmod 2q$ and set $p = X - (c - 1)$. (Note that $p \equiv 1 \pmod{2q}$.)
   4.4 If $p \geq 2^{L-1}$ then do the following:
       Test p for primality using MILLER-RABIN(p,t) for $t \geq 5$ (see Note 4.57).
       If p is a (probable) prime then return(q,p).
   4.5 Set $i \leftarrow i + 1$, $j \leftarrow j + n + 1$.
5. Go to step 2.

4.57 Note (*choice of primality test in Algorithm 4.56*)

(i) The FIPS 186 document where Algorithm 4.56 was originally described only specifies that a *robust* primality test be used in steps 2.4 and 4.4, i.e., a primality test where the probability of a composite integer being declared prime is at most $(\frac{1}{2})^{80}$. If the heuristic assumption is made that q is a randomly chosen 160-bit integer then, by Table 4.4, MILLER-RABIN(q,18) is a robust test for the primality of q. If p is assumed to be a randomly chosen L-bit integer, then by Table 4.4, MILLER-RABIN(p,5) is a robust test for the primality of p. Since the Miller-Rabin test is a probabilistic primality test, the output of Algorithm 4.56 is a probable prime.

(ii) To improve performance, candidate primes q and p should be subjected to trial division by all odd primes less than some bound B before invoking the Miller-Rabin test. See Note 4.45 for guidance on selecting B.

4.58 Note (*"weak" primes cannot be intentionally constructed*) Algorithm 4.56 has the feature that the random seed s is not input to the prime number generation portion of the algorithm itself, but rather to an unpredictable and uncontrollable randomization process (steps 2.2 and 4.1), the output of which is used as the actual random seed. This precludes manipulation of the input seed to the prime number generation. If the seed s and counter i are made public, then anyone can verify that q and p were generated using the approved method. This feature prevents a central authority who generates p and q as system-wide parameters for use in the DSA from intentionally constructing "weak" primes q and p which it could subsequently exploit to recover other entities' private keys.

4.4.4 Constructive techniques for provable primes

Maurer's algorithm (Algorithm 4.62) generates random *provable* primes that are almost uniformly distributed over the set of all primes of a specified size. The expected time for generating a prime is only slightly greater than that for generating a probable prime of equal size using Algorithm 4.44 with security parameter $t = 1$. (In practice, one may wish to choose $t > 1$ in Algorithm 4.44; cf. Note 4.49.)

The main idea behind Algorithm 4.62 is Fact 4.59, which is a slight modification of Pocklington's theorem (Fact 4.40) and Fact 4.41.

4.59 Fact Let $n \geq 3$ be an odd integer, and suppose that $n = 1 + 2Rq$ where q is an odd prime. Suppose further that $q > R$.

    (i) If there exists an integer a satisfying $a^{n-1} \equiv 1 \pmod{n}$ and $\gcd(a^{2R} - 1, n) = 1$, then n is prime.

    (ii) If n is prime, the probability that a randomly selected base a, $1 \leq a \leq n-1$, satisfies $a^{n-1} \equiv 1 \pmod{n}$ and $\gcd(a^{2R} - 1, n) = 1$ is $(1 - 1/q)$.

Algorithm 4.62 recursively generates an odd prime q, and then chooses random integers R, $R < q$, until $n = 2Rq + 1$ can be proven prime using Fact 4.59(i) for some base a. By Fact 4.59(ii) the proportion of such bases is $1 - 1/q$ for prime n. On the other hand, if n is composite, then most bases a will fail to satisfy the condition $a^{n-1} \equiv 1 \pmod{n}$.

4.60 Note (*description of constants c and m in Algorithm 4.62*)

    (i) The optimal value of the constant c defining the trial division bound $B = ck^2$ in step 2 depends on the implementation of long-integer arithmetic, and is best determined experimentally (cf. Note 4.45).

    (ii) The constant $m = 20$ ensures that I is at least 20 bits long and hence the interval from which R is selected, namely $[I + 1, 2I]$, is sufficiently large (for the values of k of practical interest) that it most likely contains at least one value R for which $n = 2Rq + 1$ is prime.

4.61 Note (*relative size r of q with respect to n*) The *relative size r* of q with respect to n is defined to be $r = \lg q / \lg n$. In order to assure that the generated prime n is chosen randomly with essentially uniform distribution from the set of all k-bit primes, the size of the prime factor q of $n - 1$ must be chosen according to the probability distribution of the largest prime factor of a randomly selected k-bit integer. Since q must be greater than R in order for Fact 4.59 to apply, the relative size r of q is restricted to being in the interval $[\frac{1}{2}, 1]$. It can be deduced from Fact 3.7(i) that the cumulative probability distribution of the relative size r of the largest prime factor of a large random integer, given that r is at least $\frac{1}{2}$, is $(1 + \lg r)$ for $\frac{1}{2} \leq r \leq 1$. In step 4 of Algorithm 4.62, the relative size r is generated according to this distribution by selecting a random number $s \in [0, 1]$ and then setting $r = 2^{s-1}$. If $k \leq 2m$ then r is chosen to be the smallest permissible value, namely $\frac{1}{2}$, in order to ensure that the interval from which R is selected is sufficiently large (cf. Note 4.60(ii)).

4.62 Algorithm Maurer's algorithm for generating provable primes

PROVABLE_PRIME(k)
INPUT: a positive integer k.
OUTPUT: a k-bit prime number n.

1. (*If k is small, then test random integers by trial division. A table of small primes may be precomputed for this purpose.*)
   If $k \leq 20$ then repeatedly do the following:
   1.1 Select a random k-bit odd integer n.
   1.2 Use trial division by all primes less than $\sqrt{n}$ to determine whether n is prime.
   1.3 If n is prime then return(n).
2. Set $c \leftarrow 0.1$ and $m \leftarrow 20$ (see Note 4.60).
3. (*Trial division bound*) Set $B \leftarrow c \cdot k^2$ (see Note 4.60).
4. (*Generate r, the size of q relative to n — see Note 4.61*) If $k > 2m$ then repeatedly do the following: select a random number s in the interval $[0, 1]$, set $r \leftarrow 2^{s-1}$, until $(k - rk) > m$. Otherwise (i.e. $k \leq 2m$), set $r \leftarrow 0.5$.
5. Compute $q \leftarrow$ PROVABLE_PRIME($\lfloor r \cdot k \rfloor + 1$).
6. Set $I \leftarrow \lfloor 2^{k-1}/(2q) \rfloor$.
7. success $\leftarrow 0$.
8. While (success $= 0$) do the following:
   8.1 (*select a candidate integer n*) Select a random integer R in the interval $[I + 1, 2I]$ and set $n \leftarrow 2Rq + 1$.
   8.2 Use trial division to determine whether n is divisible by any prime number $< B$. If it is not then do the following:
   > Select a random integer a in the interval $[2, n - 2]$.
   > Compute $b \leftarrow a^{n-1} \bmod n$.
   > If $b = 1$ then do the following:
   > > Compute $b \leftarrow a^{2R} \bmod n$ and $d \leftarrow \gcd(b - 1, n)$.
   > > If $d = 1$ then success $\leftarrow 1$.
9. Return(n).

4.63 Note (*improvements to Algorithm 4.62*)

(i) A speedup can be achieved by using Fact 4.42 instead of Fact 4.59(i) for proving $n = 2Rq + 1$ prime in step 8.2 of Maurer's algorithm — Fact 4.42 only requires that q be greater than $\sqrt[3]{n}$.

(ii) If a candidate n passes the trial division (in step 8.2), then a Miller-Rabin test (Algorithm 4.24) with the single base $a = 2$ should be performed on n; only if n passes this test should the attempt to prove its primality (the remainder of step 8.2) be undertaken. This leads to a faster implementation due to the efficiency of the Miller-Rabin test with a single base $a = 2$ (cf. Remark 4.50).

(iii) Step 4 requires the use of real number arithmetic when computing 2^{s-1}. To avoid these computations, one can precompute and store a list of such values for a selection of random numbers $s \in [0, 1]$.

4.64 Note (*provable primes vs. probable primes*) Probable primes are advantageous over provable primes in that Algorithm 4.44 for generating probable primes with $t = 1$ is slightly faster than Maurer's algorithm. Moreover, the latter requires more run-time memory due to its recursive nature. Provable primes are preferable to probable primes in the sense that the former have zero error probability. In any cryptographic application, however, there is always a non-zero error probability of some catastrophic failure, such as the adversary guessing a secret key or hardware failure. Since the error probability of probable primes can be efficiently brought down to acceptably low levels (see Note 4.49 but note the dependence on t), there appears to be no reason for mandating the use of provable primes over probable primes.

4.5 Irreducible polynomials over $\mathbb{Z}_p$

Recall (Definition 2.190) that a polynomial $f(x) \in \mathbb{Z}_p[x]$ of degree $m \geq 1$ is said to be *irreducible over* $\mathbb{Z}_p$ if it cannot be written as a product of two polynomials in $\mathbb{Z}_p[x]$ each having degree less than m. Such a polynomial $f(x)$ can be used to represent the elements of the finite field $\mathbb{F}_{p^m}$ as $\mathbb{F}_{p^m} = \mathbb{Z}_p[x]/(f(x))$, the set of all polynomials in $\mathbb{Z}_p[x]$ of degree less than m where the addition and multiplication of polynomials is performed modulo $f(x)$ (see §2.6.3). This section presents techniques for constructing irreducible polynomials over $\mathbb{Z}_p$, where p is a prime. The characteristic two finite fields $\mathbb{F}_{2^m}$ are of particular interest for cryptographic applications because the arithmetic in these fields can be efficiently performed both in software and in hardware. Thus, attention in the section will be focused on irreducible polynomials over $\mathbb{Z}_2$.

The arithmetic in finite fields can usually be implemented more efficiently if the irreducible polynomial chosen has few non-zero terms. Irreducible *trinomials*, i.e., irreducible polynomials having exactly three non-zero terms, are considered in §4.5.2. *Primitive* polynomials, i.e., irreducible polynomials $f(x)$ of degree m in $\mathbb{Z}_p[x]$ for which x is a generator of $\mathbb{F}_{p^m}^*$, the multiplicative group of the finite field $\mathbb{F}_{p^m} = \mathbb{Z}_p[x]/(f(x))$ (Definition 2.228), are the topic of §4.5.3. Primitive polynomials are also used in the generation of linear feedback shift register sequences having the maximum possible period (Fact 6.12).

4.5.1 Irreducible polynomials

If $f(x) \in \mathbb{Z}_p[x]$ is irreducible over $\mathbb{Z}_p$ and a is a non-zero element in $\mathbb{Z}_p$, then $a \cdot f(x)$ is also irreducible over $\mathbb{Z}_p$. Hence it suffices to restrict attention to *monic* polynomials in $\mathbb{Z}_p[x]$, i.e., polynomials whose leading coefficient is 1. Observe also that if $f(x)$ is an irreducible polynomial, then its constant term must be non-zero. In particular, if $f(x) \in \mathbb{Z}_2[x]$, then its constant term must be 1.

There is a formula for computing exactly the number of monic irreducible polynomials in $\mathbb{Z}_p[x]$ of a fixed degree. The Möbius function, which is defined next, is used in this formula.

4.65 Definition Let m be a positive integer. The *Möbius function* μ is defined by

$$
\mu(m) = \begin{cases}
1, & \text{if } m = 1, \\
0, & \text{if } m \text{ is divisible by the square of a prime}, \\
(-1)^k, & \text{if } m \text{ is the product of } k \text{ distinct primes}.
\end{cases}
$$

4.66 Example (*Möbius function*) The following table gives the values of the Möbius function $\mu(m)$ for the first 10 values of m:

m	1	2	3	4	5	6	7	8	9	10
$\mu(m)$	1	-1	-1	0	-1	1	-1	0	0	1

$\square$

4.67 Fact (*number of monic irreducible polynomials*) Let p be a prime and m a positive integer.

(i) The number $N_p(m)$ of monic irreducible polynomials of degree m in $\mathbb{Z}_p[x]$ is given by the following formula:

$$N_p(m) = \frac{1}{m} \sum_{d|m} \mu(d) p^{m/d},$$

where the summation ranges over all positive divisors d of m.

(ii) The probability of a random monic polynomial of degree m in $\mathbb{Z}_p[x]$ being irreducible over $\mathbb{Z}_p$ is roughly $\frac{1}{m}$. More specifically, the number $N_p(m)$ satisfies

$$\frac{1}{2m} \leq \frac{N_p(m)}{p^m} \approx \frac{1}{m}.$$

Testing irreducibility of polynomials in $\mathbb{Z}_p[x]$ is significantly simpler than testing primality of integers. A polynomial can be tested for irreducibility by verifying that it has no irreducible factors of degree $\leq \lfloor \frac{m}{2} \rfloor$. The following result leads to an efficient method (Algorithm 4.69) for accomplishing this.

4.68 Fact Let p be a prime and let k be a positive integer.

(i) The product of all monic irreducible polynomials in $\mathbb{Z}_p[x]$ of degree dividing k is equal to $x^{p^k} - x$.

(ii) Let $f(x)$ be a polynomial of degree m in $\mathbb{Z}_p[x]$. Then $f(x)$ is irreducible over $\mathbb{Z}_p$ if and only if $\gcd(f(x), x^{p^i} - x) = 1$ for each i, $1 \leq i \leq \lfloor \frac{m}{2} \rfloor$.

4.69 Algorithm Testing a polynomial for irreducibility

INPUT: a prime p and a monic polynomial $f(x)$ of degree m in $\mathbb{Z}_p[x]$.
OUTPUT: an answer to the question: "Is $f(x)$ irreducible over $\mathbb{Z}_p$?"

1. Set $u(x) \leftarrow x$.
2. For i from 1 to $\lfloor \frac{m}{2} \rfloor$ do the following:

   2.1 Compute $u(x) \leftarrow u(x)^p \bmod f(x)$ using Algorithm 2.227. (Note that $u(x)$ is a polynomial in $\mathbb{Z}_p[x]$ of degree less than m.)
   2.2 Compute $d(x) = \gcd(f(x), u(x) - x)$ (using Algorithm 2.218).
   2.3 If $d(x) \neq 1$ then return("reducible").
3. Return("irreducible").

Fact 4.67 suggests that one method for finding an irreducible polynomial of degree m in $\mathbb{Z}_p[x]$ is to generate a random monic polynomial of degree m in $\mathbb{Z}_p[x]$, test it for irreducibility, and continue until an irreducible one is found (Algorithm 4.70). The expected number of polynomials to be tried before an irreducible one is found is approximately m.

4.70 Algorithm Generating a random monic irreducible polynomial over $\mathbb{Z}_p$

INPUT: a prime p and a positive integer m.
OUTPUT: a monic irreducible polynomial $f(x)$ of degree m in $\mathbb{Z}_p[x]$.

1. Repeat the following:
   1.1 *(Generate a random monic polynomial of degree m in $\mathbb{Z}_p[x]$)*
       Randomly select integers $a_0, a_1, a_2, \ldots, a_{m-1}$ between 0 and $p-1$ with $a_0 \neq 0$. Let $f(x)$ be the polynomial $f(x) = x^m + a_{m-1}x^{m-1} + \cdots + a_2x^2 + a_1x + a_0$.
   1.2 Use Algorithm 4.69 to test whether $f(x)$ is irreducible over $\mathbb{Z}_p$.

   Until $f(x)$ is irreducible.
2. Return($f(x)$).

It is known that the expected degree of the irreducible factor of least degree of a random polynomial of degree m in $\mathbb{Z}_p[x]$ is $O(\lg m)$. Hence for each choice of $f(x)$, the expected number of times steps 2.1 – 2.3 of Algorithm 4.69 are iterated is $O(\lg m)$. Each iteration takes $O((\lg p)m^2)$ $\mathbb{Z}_p$-operations. These observations, together with Fact 4.67(ii), determine the running time for Algorithm 4.70.

4.71 Fact Algorithm 4.70 has an expected running time of $O(m^3(\lg m)(\lg p))$ $\mathbb{Z}_p$-operations.

Given one irreducible polynomial of degree m over $\mathbb{Z}_p$, Note 4.74 describes a method, which is more efficient than Algorithm 4.70, for randomly generating additional such polynomials,

4.72 Definition Let $\mathbb{F}_q$ be a finite field of characteristic p, and let $\alpha \in \mathbb{F}_q$. A *minimum polynomial* of α over $\mathbb{Z}_p$ is a monic polynomial of least degree in $\mathbb{Z}_p[x]$ having α as a root.

4.73 Fact Let $\mathbb{F}_q$ be a finite field of order $q = p^m$, and let $\alpha \in \mathbb{F}_q$.
   (i) The minimum polynomial of α over $\mathbb{Z}_p$, denoted $m_\alpha(x)$, is unique.
   (ii) $m_\alpha(x)$ is irreducible over $\mathbb{Z}_p$.
   (iii) The degree of $m_\alpha(x)$ is a divisor of m.
   (iv) Let t be the smallest positive integer such that $\alpha^{p^t} = \alpha$. (Note that such a t exists since, by Fact 2.213, $\alpha^{p^m} = \alpha$.) Then

$$m_\alpha(x) = \prod_{i=0}^{t-1}(x - \alpha^{p^i}). \tag{4.1}$$

4.74 Note *(generating new irreducible polynomials from a given one)* Suppose that $f(y)$ is a given irreducible polynomial of degree m over $\mathbb{Z}_p$. The finite field $\mathbb{F}_{p^m}$ can then be represented as $\mathbb{F}_{p^m} = \mathbb{Z}_p[y]/(f(y))$. A random monic irreducible polynomial of degree m over $\mathbb{Z}_p$ can be efficiently generated as follows. First generate a random element $\alpha \in \mathbb{F}_{p^m}$ and then, by repeated exponentiation by p, determine the smallest positive integer t for which $\alpha^{p^t} = \alpha$. If $t < m$, then generate a new random element $\alpha \in \mathbb{F}_{p^m}$ and repeat; the probability that $t < m$ is known to be at most $(\lg m)/q^{m/2}$. If indeed $t = m$, then compute $m_\alpha(x)$ using the formula (4.1). Then $m_\alpha(x)$ is a random monic irreducible polynomial of degree m in $\mathbb{Z}_p[x]$. This method has an expected running time of $O(m^3(\lg p))$ $\mathbb{Z}_p$-operations (compare with Fact 4.71).

4.5.2 Irreducible trinomials

If a polynomial $f(x)$ in $\mathbb{Z}_2[x]$ has an even number of non-zero terms, then $f(1) = 0$, whence $(x + 1)$ is a factor of $f(x)$. Hence, the smallest number of non-zero terms an irreducible polynomial of degree ≥ 2 in $\mathbb{Z}_2[x]$ can have is three. An irreducible *trinomial* of degree m in $\mathbb{Z}_2[x]$ must be of the form $x^m + x^k + 1$, where $1 \leq k \leq m - 1$. Choosing an irreducible trinomial $f(x) \in \mathbb{Z}_2[x]$ of degree m to represent the elements of the finite field $\mathbb{F}_{2^m} = \mathbb{Z}_2[x]/(f(x))$ can lead to a faster implementation of the field arithmetic. The following facts are sometimes of use when searching for irreducible trinomials.

4.75 Fact Let m be a positive integer, and let k denote an integer in the interval $[1, m - 1]$.

(i) If the trinomial $x^m + x^k + 1$ is irreducible over $\mathbb{Z}_2$ then so is $x^m + x^{m-k} + 1$.

(ii) If $m \equiv 0 \pmod 8$, there is no irreducible trinomial of degree m in $\mathbb{Z}_2[x]$.

(iii) Suppose that either $m \equiv 3 \pmod 8$ or $m \equiv 5 \pmod 8$. Then a necessary condition for $x^m + x^k + 1$ to be irreducible over $\mathbb{Z}_2$ is that either k or $m - k$ must be of the form $2d$ for some positive divisor d of m.

Tables 4.6 and 4.7 list an irreducible trinomial of degree m over $\mathbb{Z}_2$ for each $m \leq 1478$ for which such a trinomial exists.

4.5.3 Primitive polynomials

Primitive polynomials were introduced at the beginning of §4.5. Let $f(x) \in \mathbb{Z}_p[x]$ be an irreducible polynomial of degree m. If the factorization of the integer $p^m - 1$ is known, then Fact 4.76 yields an efficient algorithm (Algorithm 4.77) for testing whether or not $f(x)$ is a primitive polynomial. If the factorization of $p^m - 1$ is unknown, there is no efficient algorithm known for performing this test.

4.76 Fact Let p be a prime and let the distinct prime factors of $p^m - 1$ be $r_1, r_2, \ldots, r_t$. Then an irreducible polynomial $f(x) \in \mathbb{Z}_p[x]$ is primitive if and only if for each i, $1 \leq i \leq t$:

$$x^{(p^m - 1)/r_i} \not\equiv 1 \pmod{f(x)}.$$

(That is, x is an element of order $p^m - 1$ in the field $\mathbb{Z}_p[x]/(f(x))$.)

4.77 Algorithm Testing whether an irreducible polynomial is primitive

---

INPUT: a prime p, a positive integer m, the distinct prime factors $r_1, r_2, \ldots, r_t$ of $p^m - 1$, and a monic irreducible polynomial $f(x)$ of degree m in $\mathbb{Z}_p[x]$.
OUTPUT: an answer to the question: "Is $f(x)$ a primitive polynomial?"

1. For i from 1 to t do the following:
   1.1 Compute $l(x) = x^{(p^m - 1)/r_i} \bmod f(x)$ (using Algorithm 2.227).
   1.2 If $l(x) = 1$ then return("not primitive").
2. Return("primitive").

There are precisely $\phi(p^m - 1)/m$ monic primitive polynomials of degree m in $\mathbb{Z}_p[x]$ (Fact 2.230), where ϕ is the Euler phi function (Definition 2.100). Since the number of monic irreducible polynomials of degree m in $\mathbb{Z}_p[x]$ is roughly p^m/m (Fact 4.67(ii)), it follows that the probability of a random monic irreducible polynomial of degree m in $\mathbb{Z}_p[x]$

m	k	m	k	m	k	m	k	m	k	m	k	m	k
2	1	93	2	193	15	295	48	402	171	508	9	618	295
3	1	94	21	194	87	297	5	404	65	510	69	620	9
4	1	95	11	196	3	300	5	406	141	511	10	622	297
5	2	97	6	198	9	302	41	407	71	513	26	623	68
6	1	98	11	199	34	303	1	409	87	514	67	625	133
7	1	100	15	201	14	305	102	412	147	516	21	626	251
9	1	102	29	202	55	308	15	414	13	518	33	628	223
10	3	103	9	204	27	310	93	415	102	519	79	631	307
11	2	105	4	207	43	313	79	417	107	521	32	633	101
12	3	106	15	209	6	314	15	418	199	522	39	634	39
14	5	108	17	210	7	316	63	420	7	524	167	636	217
15	1	110	33	212	105	318	45	422	149	526	97	639	16
17	3	111	10	214	73	319	36	423	25	527	47	641	11
18	3	113	9	215	23	321	31	425	12	529	42	642	119
20	3	118	33	217	45	322	67	426	63	532	1	646	249
21	2	119	8	218	11	324	51	428	105	534	161	647	5
22	1	121	18	220	7	327	34	431	120	537	94	649	37
23	5	123	2	223	33	329	50	433	33	538	195	650	3
25	3	124	19	225	32	330	99	436	165	540	9	651	14
28	1	126	21	228	113	332	89	438	65	543	16	652	93
29	2	127	1	231	26	333	2	439	49	545	122	654	33
30	1	129	5	233	74	337	55	441	7	550	193	655	88
31	3	130	3	234	31	340	45	444	81	551	135	657	38
33	10	132	17	236	5	342	125	446	105	553	39	658	55
34	7	134	57	238	73	343	75	447	73	556	153	660	11
35	2	135	11	239	36	345	22	449	134	558	73	662	21
36	9	137	21	241	70	346	63	450	47	559	34	663	107
39	4	140	15	242	95	348	103	455	38	561	71	665	33
41	3	142	21	244	111	350	53	457	16	564	163	668	147
42	7	145	52	247	82	351	34	458	203	566	153	670	153
44	5	146	71	249	35	353	69	460	19	567	28	671	15
46	1	147	14	250	103	354	99	462	73	569	77	673	28
47	5	148	27	252	15	358	57	463	93	570	67	676	31
49	9	150	53	253	46	359	68	465	31	574	13	679	66
52	3	151	3	255	52	362	63	468	27	575	146	682	171
54	9	153	1	257	12	364	9	470	9	577	25	684	209
55	7	154	15	258	71	366	29	471	1	580	237	686	197
57	4	155	62	260	15	367	21	473	200	582	85	687	13
58	19	156	9	263	93	369	91	474	191	583	130	689	14
60	1	159	31	265	42	370	139	476	9	585	88	690	79
62	29	161	18	266	47	372	111	478	121	588	35	692	299
63	1	162	27	268	25	375	16	479	104	590	93	694	169
65	18	166	37	270	53	377	41	481	138	593	86	695	177
66	3	167	6	271	58	378	43	484	105	594	19	697	267
68	9	169	34	273	23	380	47	486	81	596	273	698	215
71	6	170	11	274	67	382	81	487	94	599	30	700	75
73	25	172	1	276	63	383	90	489	83	601	201	702	37
74	35	174	13	278	5	385	6	490	219	602	215	705	17
76	21	175	6	279	5	386	83	492	7	604	105	708	15
79	9	177	8	281	93	388	159	494	17	606	165	711	92
81	4	178	31	282	35	390	9	495	76	607	105	713	41
84	5	180	3	284	53	391	28	497	78	609	31	714	23
86	21	182	81	286	69	393	7	498	155	610	127	716	183
87	13	183	56	287	71	394	135	500	27	612	81	718	165
89	38	185	24	289	21	396	25	503	3	614	45	719	150
90	27	186	11	292	37	399	26	505	156	615	211	721	9
92	21	191	9	294	33	401	152	506	23	617	200	722	231

Table 4.6: *Irreducible trinomials $x^m + x^k + 1$ over $\mathbb{Z}_2$. For each m, $1 \le m \le 722$, for which an irreducible trinomial of degree m in $\mathbb{Z}_2[x]$ exists, the table lists the smallest k for which $x^m + x^k + 1$ is irreducible over $\mathbb{Z}_2$.*

m	k	m	k	m	k	m	k	m	k	m	k	m	k
724	207	831	49	937	217	1050	159	1159	66	1265	119	1374	609
726	5	833	149	938	207	1052	291	1161	365	1266	7	1375	52
727	180	834	15	942	45	1054	105	1164	19	1268	345	1377	100
729	58	838	61	943	24	1055	24	1166	189	1270	333	1380	183
730	147	839	54	945	77	1057	198	1167	133	1271	17	1383	130
732	343	841	144	948	189	1058	27	1169	114	1273	168	1385	12
735	44	842	47	951	260	1060	439	1170	27	1276	217	1386	219
737	5	844	105	953	168	1062	49	1174	133	1278	189	1388	11
738	347	845	2	954	131	1063	168	1175	476	1279	216	1390	129
740	135	846	105	956	305	1065	463	1177	16	1281	229	1391	3
742	85	847	136	959	143	1071	7	1178	375	1282	231	1393	300
743	90	849	253	961	18	1078	361	1180	25	1284	223	1396	97
745	258	850	111	964	103	1079	230	1182	77	1286	153	1398	601
746	351	852	159	966	201	1081	24	1183	87	1287	470	1399	55
748	19	855	29	967	36	1082	407	1185	134	1289	99	1401	92
750	309	857	119	969	31	1084	189	1186	171	1294	201	1402	127
751	18	858	207	972	7	1085	62	1188	75	1295	38	1404	81
753	158	860	35	975	19	1086	189	1190	233	1297	198	1407	47
754	19	861	14	977	15	1087	112	1191	196	1298	399	1409	194
756	45	862	349	979	178	1089	91	1193	173	1300	75	1410	383
758	233	865	1	982	177	1090	79	1196	281	1302	77	1412	125
759	98	866	75	983	230	1092	23	1198	405	1305	326	1414	429
761	3	868	145	985	222	1094	57	1199	114	1306	39	1415	282
762	83	870	301	986	3	1095	139	1201	171	1308	495	1417	342
767	168	871	378	988	121	1097	14	1202	287	1310	333	1420	33
769	120	873	352	990	161	1098	83	1204	43	1311	476	1422	49
772	7	876	149	991	39	1100	35	1206	513	1313	164	1423	15
774	185	879	11	993	62	1102	117	1207	273	1314	19	1425	28
775	93	881	78	994	223	1103	65	1209	118	1319	129	1426	103
777	29	882	99	996	65	1105	21	1210	243	1321	52	1428	27
778	375	884	173	998	101	1106	195	1212	203	1324	337	1430	33
780	13	887	147	999	59	1108	327	1214	257	1326	397	1431	17
782	329	889	127	1001	17	1110	417	1215	302	1327	277	1433	387
783	68	890	183	1007	75	1111	13	1217	393	1329	73	1434	363
785	92	892	31	1009	55	1113	107	1218	91	1332	95	1436	83
791	30	894	173	1010	99	1116	59	1220	413	1334	617	1438	357
793	253	895	12	1012	115	1119	283	1223	255	1335	392	1441	322
794	143	897	113	1014	385	1121	62	1225	234	1337	75	1442	395
798	53	898	207	1015	186	1122	427	1226	167	1338	315	1444	595
799	25	900	1	1020	135	1126	105	1228	27	1340	125	1446	421
801	217	902	21	1022	317	1127	27	1230	433	1343	348	1447	195
804	75	903	35	1023	7	1129	103	1231	105	1345	553	1449	13
806	21	905	117	1025	294	1130	551	1233	151	1348	553	1452	315
807	7	906	123	1026	35	1134	129	1234	427	1350	237	1454	297
809	15	908	143	1028	119	1135	9	1236	49	1351	39	1455	52
810	159	911	204	1029	98	1137	277	1238	153	1353	371	1457	314
812	29	913	91	1030	93	1138	31	1239	4	1354	255	1458	243
814	21	916	183	1031	68	1140	141	1241	54	1356	131	1460	185
815	333	918	77	1033	108	1142	357	1242	203	1358	117	1463	575
817	52	919	36	1034	75	1145	227	1246	25	1359	98	1465	39
818	119	921	221	1036	411	1146	131	1247	14	1361	56	1466	311
820	123	924	31	1039	21	1148	23	1249	187	1362	655	1468	181
822	17	926	365	1041	412	1151	90	1252	97	1364	239	1470	49
823	9	927	403	1042	439	1153	241	1255	589	1366	1	1471	25
825	38	930	31	1044	41	1154	75	1257	289	1367	134	1473	77
826	255	932	177	1047	10	1156	307	1260	21	1369	88	1476	21
828	189	935	417	1049	141	1158	245	1263	77	1372	181	1478	69

Table 4.7: *Irreducible trinomials $x^m + x^k + 1$ over $\mathbb{Z}_2$. For each m, $723 \leq m \leq 1478$, for which an irreducible trinomial of degree m in $\mathbb{Z}_2[x]$ exists, the table gives the smallest k for which $x^m + x^k + 1$ is irreducible over $\mathbb{Z}_2$.*

being primitive is approximately $\phi(p^m - 1)/p^m$. Using the lower bound for the Euler phi function (Fact 2.102), this probability can be seen to be at least $1/(6 \ln \ln p^m)$. This suggests the following algorithm for generating primitive polynomials.

4.78 Algorithm Generating a random monic primitive polynomial over $\mathbb{Z}_p$

INPUT: a prime p, integer $m \geq 1$, and the distinct prime factors $r_1, r_2, \ldots, r_t$ of $p^m - 1$.
OUTPUT: a monic primitive polynomial $f(x)$ of degree m in $\mathbb{Z}_p[x]$.

1. Repeat the following:

    1.1 Use Algorithm 4.70 to generate a random monic irreducible polynomial $f(x)$ of degree m in $\mathbb{Z}_p[x]$.

    1.2 Use Algorithm 4.77 to test whether $f(x)$ is primitive.

    Until $f(x)$ is primitive.
2. Return($f(x)$).

For each m, $1 \leq m \leq 229$, Table 4.8 lists a polynomial of degree m that is primitive over $\mathbb{Z}_2$. If there exists a primitive trinomial $f(x) = x^m + x^k + 1$, then the trinomial with the smallest k is listed. If no primitive trinomial exists, then a primitive pentanomial of the form $f(x) = x^m + x^{k_1} + x^{k_2} + x^{k_3} + 1$ is listed.

If $p^m - 1$ is prime, then Fact 4.76 implies that every irreducible polynomial of degree m in $\mathbb{Z}_p[x]$ is also primitive. Table 4.9 gives either a primitive trinomial or a primitive pentanomial of degree m over $\mathbb{Z}_2$ where m is an exponent of one of the first 27 Mersenne primes (Definition 4.35).

4.6 Generators and elements of high order

Recall (Definition 2.169) that if G is a (multiplicative) finite group, the *order* of an element $a \in G$ is the least positive integer t such that $a^t = 1$. If there are n elements in G, and if $a \in G$ is an element of order n, then G is said to be *cyclic* and a is called a *generator* or a *primitive element* of G (Definition 2.167). Of special interest for cryptographic applications are the multiplicative group $\mathbb{Z}_p^*$ of the integers modulo a prime p, and the multiplicative group $\mathbb{F}_{2^m}^*$ of the finite field $\mathbb{F}_{2^m}$ of characteristic two; these groups are cyclic (Fact 2.213). Also of interest is the group $\mathbb{Z}_n^*$ (Definition 2.124), where n is the product of two distinct odd primes. This section deals with the problem of finding generators and other elements of high order in $\mathbb{Z}_p^*$, $\mathbb{F}_{2^m}^*$, and $\mathbb{Z}_n^*$. See §2.5.1 for background in group theory and §2.6 for background in finite fields.

Algorithm 4.79 is an efficient method for determining the order of a group element, given the prime factorization of the group order n. The correctness of the algorithm follows from the fact that the order of an element must divide n (Fact 2.171).

m	k or (k_1, k_2, k_3)	m	k or (k_1, k_2, k_3)	m	k or (k_1, k_2, k_3)	m	k or (k_1, k_2, k_3)
2	1	59	22, 21, 1	116	71, 70, 1	173	100, 99, 1
3	1	60	1	117	20, 18, 2	174	13
4	1	61	16, 15, 1	118	33	175	6
5	2	62	57, 56, 1	119	8	176	119, 118, 1
6	1	63	1	120	118, 111, 7	177	8
7	1	64	4, 3, 1	121	18	178	87
8	6, 5, 1	65	18	122	60, 59, 1	179	34, 33, 1
9	4	66	10, 9, 1	123	2	180	37, 36, 1
10	3	67	10, 9, 1	124	37	181	7, 6, 1
11	2	68	9	125	108, 107, 1	182	128, 127, 1
12	7, 4, 3	69	29, 27, 2	126	37, 36, 1	183	56
13	4, 3, 1	70	16, 15, 1	127	1	184	102, 101, 1
14	12, 11, 1	71	6	128	29, 27, 2	185	24
15	1	72	53, 47, 6	129	5	186	23, 22, 1
16	5, 3, 2	73	25	130	3	187	58, 57, 1
17	3	74	16, 15, 1	131	48, 47, 1	188	74, 73, 1
18	7	75	11, 10, 1	132	29	189	127, 126, 1
19	6, 5, 1	76	36, 35, 1	133	52, 51, 1	190	18, 17, 1
20	3	77	31, 30, 1	134	57	191	9
21	2	78	20, 19, 1	135	11	192	28, 27, 1
22	1	79	9	136	126, 125, 1	193	15
23	5	80	38, 37, 1	137	21	194	87
24	4, 3, 1	81	4	138	8, 7, 1	195	10, 9, 1
25	3	82	38, 35, 3	139	8, 5, 3	196	66, 65, 1
26	8, 7, 1	83	46, 45, 1	140	29	197	62, 61, 1
27	8, 7, 1	84	13	141	32, 31, 1	198	65
28	3	85	28, 27, 1	142	21	199	34
29	2	86	13, 12, 1	143	21, 20, 1	200	42, 41, 1
30	16, 15, 1	87	13	144	70, 69, 1	201	14
31	3	88	72, 71, 1	145	52	202	55
32	28, 27, 1	89	38	146	60, 59, 1	203	8, 7, 1
33	13	90	19, 18, 1	147	38, 37, 1	204	74, 73, 1
34	15, 14, 1	91	84, 83, 1	148	27	205	30, 29, 1
35	2	92	13, 12, 1	149	110, 109, 1	206	29, 28, 1
36	11	93	2	150	53	207	43
37	12, 10, 2	94	21	151	3	208	62, 59, 3
38	6, 5, 1	95	11	152	66, 65, 1	209	6
39	4	96	49, 47, 2	153	1	210	35, 32, 3
40	21, 19, 2	97	6	154	129, 127, 2	211	46, 45, 1
41	3	98	11	155	32, 31, 1	212	105
42	23, 22, 1	99	47, 45, 2	156	116, 115, 1	213	8, 7, 1
43	6, 5, 1	100	37	157	27, 26, 1	214	49, 48, 1
44	27, 26, 1	101	7, 6, 1	158	27, 26, 1	215	23
45	4, 3, 1	102	77, 76, 1	159	31	216	196, 195, 1
46	21, 20, 1	103	9	160	19, 18, 1	217	45
47	5	104	11, 10, 1	161	18	218	11
48	28, 27, 1	105	16	162	88, 87, 1	219	19, 18, 1
49	9	106	15	163	60, 59, 1	220	15, 14, 1
50	27, 26, 1	107	65, 63, 2	164	14, 13, 1	221	35, 34, 1
51	16, 15, 1	108	31	165	31, 30, 1	222	92, 91, 1
52	3	109	7, 6, 1	166	39, 38, 1	223	33
53	16, 15, 1	110	13, 12, 1	167	6	224	31, 30, 1
54	37, 36, 1	111	10	168	17, 15, 2	225	32
55	24	112	45, 43, 2	169	34	226	58, 57, 1
56	22, 21, 1	113	9	170	23	227	46, 45, 1
57	7	114	82, 81, 1	171	19, 18, 1	228	148, 147, 1
58	19	115	15, 14, 1	172	7	229	64, 63, 1

Table 4.8: *Primitive polynomials over $\mathbb{Z}_2$. For each m, $1 \leq m \leq 229$, an exponent k is given for which the trinomial $x^m + x^k + 1$ is primitive over $\mathbb{Z}_2$. If no such trinomial exists, a triple of exponents (k_1, k_2, k_3) is given for which the pentanomial $x^m + x^{k_1} + x^{k_2} + x^{k_3} + 1$ is primitive over $\mathbb{Z}_2$.*

j	m	k (k_1, k_2, k_3)
1	2	1
2	3	1
3	5	2
4	7	1, 3
5	13	none (4,3,1)
6	17	3, 5, 6
7	19	none (5,2,1)
8	31	3, 6, 7, 13
9	61	none (43,26,14)
10	89	38
11	107	none (82,57,31)
12	127	1, 7, 15, 30, 63
13	521	32, 48, 158, 168
14	607	105, 147, 273
15	1279	216, 418
16	2203	none (1656,1197,585)
17	2281	715, 915, 1029
18	3217	67, 576
19	4253	none (3297,2254,1093)
20	4423	271, 369, 370, 649, 1393, 1419, 2098
21	9689	84, 471, 1836, 2444, 4187
22	9941	none (7449,4964,2475)
23	11213	none (8218,6181,2304)
24	19937	881, 7083, 9842
25	21701	none (15986,11393,5073)
26	23209	1530, 6619, 9739
27	44497	8575, 21034

Table 4.9: *Primitive polynomials of degree m over $\mathbb{Z}_2$, $2^m - 1$ a Mersenne prime. For each exponent $m = M_j$ of the first 27 Mersenne primes, the table lists all values of k, $1 \le k \le m/2$, for which the trinomial $x^m + x^k + 1$ is irreducible over $\mathbb{Z}_2$. If no such trinomial exists, a triple of exponents (k_1, k_2, k_3) is listed such that the pentanomial $x^m + x^{k_1} + x^{k_2} + x^{k_3} + 1$ is irreducible over $\mathbb{Z}_2$.*

4.79 Algorithm Determining the order of a group element

INPUT: a (multiplicative) finite group G of order n, an element $a \in G$, and the prime factorization $n = p_1^{e_1} p_2^{e_2} \cdots p_k^{e_k}$.
OUTPUT: the order t of a.

1. Set $t \leftarrow n$.
2. For i from 1 to k do the following:
   2.1 Set $t \leftarrow t/p_i^{e_i}$.
   2.2 Compute $a_1 \leftarrow a^t$.
   2.3 While $a_1 \ne 1$ do the following: compute $a_1 \leftarrow a_1^{p_i}$ and set $t \leftarrow t \cdot p_i$.
3. Return(t).

Suppose now that G is a cyclic group of order n. Then for any divisor d of n the number of elements of order d in G is exactly $\phi(d)$ (Fact 2.173(ii)), where ϕ is the Euler phi function (Definition 2.100). In particular, G has exactly $\phi(n)$ generators, and hence the probability of a random element in G being a generator is $\phi(n)/n$. Using the lower bound for the Euler phi function (Fact 2.102), this probability can be seen to be at least $1/(6 \ln \ln n)$. This suggests the following efficient randomized algorithm for finding a generator of a cyclic group.

4.80 Algorithm Finding a generator of a cyclic group

INPUT: a cyclic group G of order n, and the prime factorization $n = p_1^{e_1} p_2^{e_2} \cdots p_k^{e_k}$.
OUTPUT: a generator α of G.

1. Choose a random element α in G.
2. For i from 1 to k do the following:
   2.1 Compute $b \leftarrow \alpha^{n/p_i}$.
   2.2 If $b = 1$ then go to step 1.
3. Return(α).

4.81 Note (*group elements of high order*) In some situations it may be desirable to have an element of high order, and not a generator. Given a generator α in a cyclic group G of order n, and given a divisor d of n, an element β of order d in G can be efficiently obtained as follows: $\beta = \alpha^{n/d}$. If q is a prime divisor of the order n of a cyclic group G, then the following method finds an element $\beta \in G$ of order q without first having to find a generator of G: select a random element $g \in G$ and compute $\beta = g^{n/q}$; repeat until $\beta \neq 1$.

4.82 Note (*generators of* $\mathbb{F}_{2^m}^*$) There are two basic approaches to finding a generator of $\mathbb{F}_{2^m}^*$. Both techniques require the factorization of the order of $\mathbb{F}_{2^m}^*$, namely $2^m - 1$.

(i) Generate a monic primitive polynomial $f(x)$ of degree m over $\mathbb{Z}_2$ (Algorithm 4.78). The finite field $\mathbb{F}_{2^m}$ can then be represented as $\mathbb{Z}_2[x]/(f(x))$, the set of all polynomials over $\mathbb{Z}_2$ modulo $f(x)$, and the element $\alpha = x$ is a generator.

(ii) Select the method for representing elements of $\mathbb{F}_{2^m}$ first. Then use Algorithm 4.80 with $G = \mathbb{F}_{2^m}^*$ and $n = 2^m - 1$ to find a generator α of $\mathbb{F}_{2^m}^*$.

If $n = pq$, where p and q are distinct odd primes, then $\mathbb{Z}_n^*$ is a non-cyclic group of order $\phi(n) = (p-1)(q-1)$. The maximum order of an element in $\mathbb{Z}_n^*$ is $\text{lcm}(p-1, q-1)$. Algorithm 4.83 is a method for generating such an element which requires the factorizations of $p-1$ and $q-1$.

4.83 Algorithm Selecting an element of maximum order in $\mathbb{Z}_n^*$, where $n = pq$

INPUT: two distinct odd primes, p, q, and the factorizations of $p-1$ and $q-1$.
OUTPUT: an element α of maximum order $\text{lcm}(p-1, q-1)$ in $\mathbb{Z}_n^*$, where $n = pq$.

1. Use Algorithm 4.80 with $G = \mathbb{Z}_p^*$ and $n = p-1$ to find a generator a of $\mathbb{Z}_p^*$.
2. Use Algorithm 4.80 with $G = \mathbb{Z}_q^*$ and $n = q-1$ to find a generator b of $\mathbb{Z}_q^*$.
3. Use Gauss's algorithm (Algorithm 2.121) to find an integer α, $1 \leq \alpha \leq n - 1$, satisfying $\alpha \equiv a \pmod{p}$ and $\alpha \equiv b \pmod{q}$.
4. Return(α).

4.6.1 Selecting a prime p and generator of $\mathbb{Z}_p^*$

In cryptographic applications for which a generator of $\mathbb{Z}_p^*$ is required, one usually has the flexibility of selecting the prime p. To guard against the Pohlig-Hellman algorithm for computing discrete logarithms (Algorithm 3.63), a security requirement is that $p-1$ should contain a "large" prime factor q. In this context, "large" means that the quantity $\sqrt{q}$ represents an infeasible amount of computation; for example, $q \geq 2^{160}$. This suggests the following algorithm for selecting appropriate parameters (p, α).

4.84 Algorithm Selecting a k-bit prime p and a generator α of $\mathbb{Z}_p^*$

INPUT: the required bitlength k of the prime and a security parameter t.
OUTPUT: a k-bit prime p such that $p - 1$ has a prime factor $\geq t$, and a generator α of $\mathbb{Z}_p^*$.

    1. Repeat the following:

        1.1 Select a random k-bit prime p (for example, using Algorithm 4.44).
        1.2 Factor $p - 1$.

        Until $p - 1$ has a prime factor $\geq t$.
    2. Use Algorithm 4.80 with $G = \mathbb{Z}_p^*$ and $n = p - 1$ to find a generator α of $\mathbb{Z}_p^*$.
    3. Return(p,α).

    Algorithm 4.84 is relatively inefficient as it requires the use of an integer factorization algorithm in step 1.2. An alternative approach is to generate the prime p by first choosing a large prime q and then selecting relatively small integers R at random until $p = 2Rq + 1$ is prime. Since $p - 1 = 2Rq$, the factorization of $p - 1$ can be obtained by factoring R. A particularly convenient situation occurs by imposing the condition $R = 1$. In this case the factorization of $p - 1$ is simply $2q$. Furthermore, since $\phi(p - 1) = \phi(2q) = \phi(2)\phi(q) = q - 1$, the probability that a randomly selected element $\alpha \in \mathbb{Z}_p^*$ is a generator is $\frac{q-1}{2q} \approx \frac{1}{2}$.

4.85 Definition A *safe prime* p is a prime of the form $p = 2q + 1$ where q is prime.

    Algorithm 4.86 generates a safe (probable) prime p and a generator of $\mathbb{Z}_p^*$.

4.86 Algorithm Selecting a k-bit safe prime p and a generator α of $\mathbb{Z}_p^*$

INPUT: the required bitlength k of the prime.
OUTPUT: a k-bit safe prime p and a generator α of $\mathbb{Z}_p^*$.

    1. Do the following:

        1.1 Select a random $(k - 1)$-bit prime q (for example, using Algorithm 4.44).
        1.2 Compute $p \leftarrow 2q + 1$, and test whether p is prime (for example, using trial division by small primes and Algorithm 4.24).

        Until p is prime.
    2. Use Algorithm 4.80 to find a generator α of $\mathbb{Z}_p^*$.
    3. Return(p,α).

4.7 Notes and further references

§4.1

Several books provide extensive treatments of primality testing including those by Bressoud [198], Bach and Shallit [70], and Koblitz [697]. The book by Kranakis [710] offers a more theoretical approach. Cohen [263] gives a comprehensive treatment of modern primality tests. See also the survey articles by A. Lenstra [747] and A. Lenstra and H. Lenstra [748]. Facts 4.1 and 4.2 were proven in 1837 by Dirichlet. For proofs of these results, see Chapter 16 of Ireland and Rosen [572]. Fact 4.3 is due to Rosser and Schoenfeld [1070]. Bach and Shallit [70] have further results on the distribution of prime numbers.

§4.2

Fact 4.13(i) was proven by Alford, Granville, and Pomerance [24]; see also Granville [521]. Fact 4.13(ii) is due to Pomerance, Selfridge, and Wagstaff [996]. Pinch [974] showed that there are 105212 Carmichael numbers up to 10^{15}.

The Solovay-Strassen probabilistic primality test (Algorithm 4.18) is due to Solovay and Strassen [1163], as modified by Atkin and Larson [57].

Fact 4.23 was proven independently by Monier [892] and Rabin [1024]. The Miller-Rabin test (Algorithm 4.24) originated in the work of Miller [876] who presented it as a non-probabilistic polynomial-time algorithm assuming the correctness of the Extended Riemann Hypothesis (ERH). Rabin [1021, 1024] rephrased Miller's algorithm as a probabilistic primality test. Rabin's algorithm required a small number of gcd computations. The Miller-Rabin test (Algorithm 4.24) is a simplification of Rabin's algorithm which does not require any gcd computations, and is due to Knuth [692, p.379]. Arazi [55], making use of Montgomery modular multiplication (§14.3.2), showed how the Miller-Rabin test can be implemented by "divisionless modular exponentiations" only, yielding a probabilistic primality test which does not use any division operations.

Miller [876], appealing to the work of Ankeny [32], proved under assumption of the Extended Riemann Hypothesis that, if n is an odd composite integer, then its least strong witness is less than $c(\ln n)^2$, where c is some constant. Bach [63] proved that this constant may be taken to be $c = 2$; see also Bach [64]. As a consequence, one can test n for primality in $O((\lg n)^5)$ bit operations by executing the Miller-Rabin algorithm for all bases $a \le 2(\ln n)^2$. This gives a deterministic polynomial-time algorithm for primality testing, under the assumption that the ERH is true.

Table 4.1 is from Jaeschke [630], building on earlier work of Pomerance, Selfridge, and Wagstaff [996]. Arnault [56] found the following 46-digit composite integer

$$n = 1195068768795265792518361315725116351898245581$$

that is a strong pseudoprime to all the 11 prime bases up to 31. Arnault also found a 337-digit composite integer which is a strong pseudoprime to all 46 prime bases up to 199.

The Miller-Rabin test (Algorithm 4.24) randomly generates t independent bases a and tests to see if each is a strong witness for n. Let n be an odd composite integer and let $t = \lceil \frac{1}{2} \lg n \rceil$. In situations where random bits are scarce, one may choose instead to generate a single random base a and use the bases $a, a + 1, \ldots, a + t - 1$. Bach [66] proved that for a randomly chosen integer a, the probability that $a, a + 1, \ldots, a + t - 1$ are all strong liars for n is bounded above by $n^{-1/4+o(1)}$; in other words, the probability that the Miller-Rabin algorithm using these bases mistakenly declares an odd composite integer "prime" is at most $n^{-1/4+o(1)}$. Peralta and Shoup [969] later improved this bound to $n^{-1/2+o(1)}$.

Monier [892] gave exact formulas for the number of Fermat liars, Euler liars, and strong liars for composite integers. One consequence of Monier's formulas is the following improvement (in the case where n is not a prime power) of Fact 4.17 (see Kranakis [710, p.68]). If $n \geq 3$ is an odd composite integer having r distinct prime factors, and if $n \equiv 3 \pmod{4}$, then there are at most $\phi(n)/2^{r-1}$ Euler liars for n. Another consequence is the following improvement (in the case where n has at least three distinct prime factors) of Fact 4.23. If $n \geq 3$ is an odd composite integer having r distinct prime factors, then there are at most $\phi(n)/2^{r-1}$ strong liars for n. Erdös and Pomerance [373] estimated the average number of Fermat liars, Euler liars, and strong liars for composite integers. Fact 4.30(ii) was proven independently by Atkin and Larson [57], Monier [892], and Pomerance, Selfridge, and Wagstaff [996].

Pinch [975] reviewed the probabilistic primality tests used in the *Mathematica*, *Maple V*, *Axiom*, and *Pari/GP* computer algebra systems. Some of these systems use a probabilistic primality test known as the *Lucas test*; a description of this test is provided by Pomerance, Selfridge, and Wagstaff [996].

§4.3

If a number n is composite, providing a non-trivial divisor of n is evidence of its compositeness that can be verified in polynomial time (by long division). In other words, the decision problem "is n composite?" belongs to the complexity class **NP** (cf. Example 2.65). Pratt [1000] used Fact 4.38 to show that this decision problem is also in **co-NP**. That is, if n is prime there exists some evidence of this (called a *certificate of primality*) that can be verified in polynomial time. Note that the issue here is not in finding such evidence, but rather in determining whether such evidence exists which, if found, allows efficient verification. Pomerance [992] improved Pratt's results and showed that every prime n has a certificate of primality which requires $O(\ln n)$ multiplications modulo n for its verification.

Primality of the *Fermat number* $F_k = 2^{2^k} + 1$ can be determined in deterministic polynomial time by *Pepin's test*: for $k \geq 2$, F_k is prime if and only if $5^{(F_k-1)/2} \equiv -1 \pmod{F_k}$. For the history behind Pepin's test and the Lucas-Lehmer test (Algorithm 4.37), see Bach and Shallit [70].

In Fact 4.38, the integer a does not have to be the same for all q. More precisely, Brillhart and Selfridge [212] showed that Fact 4.38 can be refined as follows: an integer $n \geq 3$ is prime if and only if for each prime divisor q of $n - 1$, there exists an integer a_q such that $a_q^{n-1} \equiv 1 \pmod{n}$ and $a_q^{(n-1)/q} \not\equiv 1 \pmod{n}$. The same is true of Fact 4.40, which is due to Pocklington [981]. For a proof of Fact 4.41, see Maurer [818]. Fact 4.42 is due to Brillhart, Lehmer, and Selfridge [210]; a simplified proof is given by Maurer [818].

The original Jacobi sum test was discovered by Adleman, Pomerance, and Rumely [16]. The algorithm was simplified, both theoretically and algorithmically, by Cohen and H. Lenstra [265]. Cohen and A. Lenstra [264] give an implementation report of the Cohen-Lenstra Jacobi sum test; see also Chapter 9 of Cohen [263]. Further improvements of the Jacobi sum test are reported by Bosma and van der Hulst [174].

Elliptic curves were first used for primality proving by Goldwasser and Kilian [477], who presented a randomized algorithm which has an expected running time of $O((\ln n)^8)$ bit operations for *most* inputs n. Subsequently, Adleman and Huang [13] designed a primality proving algorithm using hyperelliptic curves of genus two whose expected running time is polynomial for *all* inputs n. This established that the decision problem "is n prime?" is in the complexity class **RP** (Definition 2.77(ii)). The Goldwasser-Kilian and Adleman-Huang algorithms are inefficient in practice. Atkin's test, and an implementation of it, is extensively described by Atkin and Morain [58]; see also Chapter 9 of Cohen [263]. The largest number proven prime as of 1996 by a general purpose primality proving algorithm is a 1505-decimal digit number, accomplished by Morain [903] using Atkin's test. The total time for the computation was estimated to be 4 years of CPU time distributed among 21 SUN 3/60 workstations. See also Morain [902] for an implementation report on Atkin's test which was used to prove the primality of the 1065-decimal digit number $(2^{3539} + 1)/3$.

§4.4

A proof of Merten's theorem can be found in Hardy and Wright [540]. The optimal trial division bound (Note 4.45) was derived by Maurer [818]. The discussion (Note 4.47) on the probability $P(X|Y_t)$ is from Beauchemin et al. [81]; the result mentioned in the last sentence of this note is due to Kim and Pomerance [673]. Fact 4.48 was derived by Damgård, Landrock, and Pomerance [300], building on earlier work of Erdös and Pomerance [373], Kim and Pomerance [673], and Damgård and Landrock [299]. Table 4.3 is Table 2 of Damgård, Landrock, and Pomerance [300]. The suggestions to first do a Miller-Rabin test with base $a = 2$ (Remark 4.50) and to do an incremental search (Note 4.51) in Algorithm 4.44 were made by Brandt, Damgård, and Landrock [187]. The error and failure probabilities for incremental search (Note 4.51(i)) were obtained by Brandt and Damgård [186]; consult this paper for more concrete estimates of these probabilities.

Algorithm 4.53 for generating strong primes is due to Gordon [514, 513]. Gordon originally proposed computing $p_0 = (s^{r-1} - r^{s-1}) \bmod rs$ in step 3. Kaliski (personal communication, April 1996) proposed the modified formula $p_0 = (2s^{r-2} \bmod r)s - 1$ which can be computed more efficiently. Williams and Schmid [1249] proposed an algorithm for generating strong primes p with the additional constraint that $p - 1 = 2q$ where q is prime; this algorithm is not as efficient as Gordon's algorithm. Hellman and Bach [550] recommended an additional constraint on strong primes, specifying that $s - 1$ (where s is a large prime factor of $p + 1$) must have a large prime factor (see §15.2.3(v)); this thwarts cycling attacks based on Lucas sequences.

The NIST method for prime generation (Algorithm 4.56) is that recommended by the NIST Federal Information Processing Standards Publication (FIPS) 186 [406].

Fact 4.59 and Algorithm 4.62 for provable prime generation are derived from Maurer [818]. Algorithm 4.62 is based on that of Shawe-Taylor [1123]. Maurer notes that the total diversity of reachable primes using the original version of his algorithm is roughly 10% of all primes. Maurer also presents a more complicated algorithm for generating provable primes with a better diversity than Algorithm 4.62, and provides extensive implementation details and analysis of the expected running time. Maurer [812] provides heuristic justification that Algorithm 4.62 generates primes with virtually uniform distribution. Mihailescu [870] observed that Maurer's algorithm can be improved by using the Eratosthenes sieve method for trial division (in step 8.2 of Algorithm 4.62) and by searching for a prime n in an appropriate interval of the arithmetic progression $2q + 1, 4q + 1, 6q + 1, \ldots$ instead of generating R's at random until $n = 2Rq + 1$ is prime. The second improvement comes at the expense of a reduction of the set of primes which may be produced by the algorithm. Mihailescu's paper includes extensive analysis and an implementation report.

§4.5

Lidl and Niederreiter [764] provide a comprehensive treatment of irreducible polynomials; proofs of Facts 4.67 and 4.68 can be found there.

Algorithm 4.69 for testing a polynomial for irreducibility is due to Ben-Or [109]. The fastest algorithm known for generating irreducible polynomials is due to Shoup [1131] and has an expected running time of $O(m^3 \lg m + m^2 \lg p)$ $\mathbb{Z}_p$-operations. There is no *deterministic* polynomial-time algorithm known for finding an irreducible polynomial of a specified degree m in $\mathbb{Z}_p[x]$. Adleman and Lenstra [14] give a deterministic algorithm that runs in polynomial time under the assumption that the ERH is true. The best deterministic algorithm known is due to Shoup [1129] and takes $O(m^4 \sqrt{p})$ $\mathbb{Z}_p$-operations, ignoring powers of $\log m$ and $\log p$. Gordon [512] presents an improved method for computing minimum polynomials of elements in $\mathbb{F}_{2^m}$.

Zierler and Brillhart [1271] provide a table of all irreducible trinomials of degree ≤ 1000 in $\mathbb{Z}_2[x]$. Blake, Gao, and Lambert [146] extended this list to all irreducible trinomials of degree ≤ 2000 in $\mathbb{Z}_2[x]$. Fact 4.75 is from their paper.

Table 4.8 extends a similar table by Stahnke [1168]. The primitive pentanomials $x^m + x^{k_1} + x^{k_2} + x^{k_3} + 1$ listed in Table 4.8 have the following properties: (i) $k_1 = k_2 + k_3$; (ii) $k_2 > k_3$; and (iii) k_3 is as small as possible, and for this particular value of k_3, k_2 is as small as possible. The rational behind this form is explained in Stahnke's paper. For each $m < 5000$ for which the factorization of $2^m - 1$ is known, Živković [1275, 1276] gives a primitive trinomial in $\mathbb{Z}_2[x]$, one primitive polynomial in $\mathbb{Z}_2[x]$ having five non-zero terms, and one primitive polynomial in $\mathbb{Z}_2[x]$ having seven non-zero terms, provided that such polynomials exist. The factorizations of $2^m - 1$ are known for all $m \leq 510$ and for some additional $m \leq 5000$. A list of such factorizations can be found in Brillhart et al. [211] and updates of the list are available by anonymous ftp from `sable.ox.ac.uk` in the `/pub/math/cunningham/` directory. Hansen and Mullen [538] describe some improvements to Algorithm 4.78 for generating primitive polynomials. They also give tables of primitive polynomials of degree m in $\mathbb{Z}_p[x]$ for each prime power $p^m \leq 10^{50}$ with $p \leq 97$. Moreover, for each such p and m, the primitive polynomial of degree m over $\mathbb{Z}_p$ listed has the smallest number of non-zero coefficients among all such polynomials.

The entries of Table 4.9 were obtained from Zierler [1270] for Mersenne exponents M_j, $1 \leq j \leq 23$, and from Kurita and Matsumoto [719] for Mersenne exponents M_j, $24 \leq j \leq 27$.

Let $f(x) \in \mathbb{Z}_p[x]$ be an irreducible polynomial of degree m, and consider the finite field $\mathbb{F}_{p^m} = \mathbb{Z}_p[x]/(f(x))$. Then $f(x)$ is called a *normal polynomial* if the set $\{x, x^p, x^{p^2}, \ldots, x^{p^{m-1}}\}$ forms a basis for $\mathbb{F}_{p^m}$ over $\mathbb{Z}_p$; such a basis is called a *normal basis*. Mullin et al. [911] introduced the concept of an *optimal normal basis* in order to reduce the hardware complexity of multiplying field elements in the finite field $\mathbb{F}_{2^m}$. A VLSI implementation of the arithmetic in $\mathbb{F}_{2^m}$ which uses optimal normal bases is described by Agnew et al. [18]. A normal polynomial which is also primitive is called a *primitive normal polynomial*. Davenport [301] proved that for any prime p and positive integer m there exists a primitive normal polynomial of degree m in $\mathbb{Z}_p[x]$. See also Lenstra and Schoof [760] who generalized this result from prime fields $\mathbb{Z}_p$ to prime power fields $\mathbb{F}_q$. Morgan and Mullen [905] give a primitive normal polynomial of degree m over $\mathbb{Z}_p$ for each prime power $p^m \leq 10^{50}$ with $p \leq 97$. Moreover, each polynomial has the smallest number of non-zero coefficients among all primitive normal polynomials of degree m over $\mathbb{Z}_p$; in fact, each polynomial has at most five non-zero terms.

§4.6

No polynomial-time algorithm is known for finding generators, or even for testing whether an element is a generator, of a finite field $\mathbb{F}_q$ if the factorization of $q - 1$ is unknown. Shoup [1130] considered the problem of deterministically generating in polynomial time a subset of $\mathbb{F}_q$ that contains a generator, and presented a solution to the problem for the case where the characteristic p of $\mathbb{F}_q$ is small (e.g. $p = 2$). Maurer [818] discusses how his algorithm (Algorithm 4.62) can be used to generate the parameters (p, α), where p is a *provable* prime and α is a generator of $\mathbb{Z}_p^*$.

Chapter 5

Pseudorandom Bits and Sequences

Contents in Brief

5.1 Introduction

The security of many cryptographic systems depends upon the generation of unpredictable quantities. Examples include the keystream in the one-time pad (§1.5.4), the secret key in the DES encryption algorithm (§11.5.1), the primes p, q in the RSA encryption (§8.2) and digital signature (§11.3.1) schemes, the private key a in the DSA (§11.5.1), and the challenges used in challenge-response identification systems (§10.3). In all these cases, the quantities generated must be of sufficient size and be "random" in the sense that the probability of any particular value being selected must be sufficiently small to preclude an adversary from gaining advantage through optimizing a search strategy based on such probability. For example, the key space for DES has size 2^{56}. If a secret key k were selected using a true random generator, an adversary would on average have to try 2^{55} possible keys before guessing the correct key k. If, on the other hand, a key k were selected by first choosing a 16-bit random secret s, and then expanding it into a 56-bit key k using a complicated but publicly known function f, the adversary would on average only need to try 2^{15} possible keys (obtained by running every possible value for s through the function f).

This chapter considers techniques for the generation of random and pseudorandom bits and numbers. Related techniques for pseudorandom bit generation that are generally discussed in the literature in the context of stream ciphers, including linear and nonlinear feedback shift registers (Chapter 6) and the output feedback mode (OFB) of block ciphers (Chapter 7), are addressed elsewhere in this book.

Chapter outline

The remainder of §5.1 introduces basic concepts relevant to random and pseudorandom bit generation. §5.2 considers techniques for random bit generation, while §5.3 considers some techniques for pseudorandom bit generation. §5.4 describes statistical tests designed to measure the quality of a random bit generator. Cryptographically secure pseudorandom bit generators are the topic of §5.5. §5.6 concludes with references and further chapter notes.

5.1.1 Background and Classification

5.1 Definition A *random bit generator* is a device or algorithm which outputs a sequence of statistically independent and unbiased binary digits.

5.2 Remark (*random bits vs. random numbers*) A random bit generator can be used to generate (uniformly distributed) random numbers. For example, a random integer in the interval $[0, n]$ can be obtained by generating a random bit sequence of length $\lfloor \lg n \rfloor + 1$, and converting it to an integer; if the resulting integer exceeds n, one option is to discard it and generate a new random bit sequence.

§5.2 outlines some physical sources of random bits that are used in practice. Ideally, secrets required in cryptographic algorithms and protocols should be generated with a (true) random bit generator. However, the generation of random bits is an inefficient procedure in most practical environments. Moreover, it may be impractical to securely store and transmit a large number of random bits if these are required in applications such as the one-time pad (§6.1.1). In such situations, the problem can be ameliorated by substituting a random bit generator with a pseudorandom bit generator.

5.3 Definition A *pseudorandom bit generator* (PRBG) is a deterministic[1] algorithm which, given a truly random binary sequence of length k, outputs a binary sequence of length $l \gg k$ which "appears" to be random. The input to the PRBG is called the *seed*, while the output of the PRBG is called a *pseudorandom bit sequence*.

The output of a PRBG is *not* random; in fact, the number of possible output sequences is at most a small fraction, namely $2^k/2^l$, of all possible binary sequences of length l. The intent is to take a small truly random sequence and expand it to a sequence of much larger length, in such a way that an adversary cannot efficiently distinguish between output sequences of the PRBG and truly random sequences of length l. §5.3 discusses ad-hoc techniques for pseudorandom bit generation. In order to gain confidence that such generators are secure, they should be subjected to a variety of statistical tests designed to detect the specific characteristics expected of random sequences. A collection of such tests is given in §5.4. As the following example demonstrates, passing these statistical tests is a *necessary* but not *sufficient* condition for a generator to be secure.

5.4 Example (*linear congruential generators*) A *linear congruential generator* produces a pseudorandom sequence of numbers $x_1, x_2, x_3, \ldots$ according to the linear recurrence

$$x_n = a x_{n-1} + b \bmod m, \quad n \geq 1;$$

a, b, and m are *parameters* which characterize the generator, while x_0 is the (secret) *seed*. While such generators are commonly used for simulation purposes and probabilistic algorithms, and pass the statistical tests of §5.4, they are *predictable* and hence entirely insecure for cryptographic purposes: given a partial output sequence, the remainder of the sequence can be reconstructed even if the parameters a, b, and m are unknown. □

---

[1] *Deterministic* here means that given the same initial seed, the generator will always produce the same output sequence.

A minimum security requirement for a pseudorandom bit generator is that the length k of the random seed should be sufficiently large so that a search over 2^k elements (the total number of possible seeds) is infeasible for the adversary. Two general requirements are that the output sequences of a PRBG should be statistically indistinguishable from truly random sequences, and the output bits should be unpredictable to an adversary with limited computational resources; these requirements are captured in Definitions 5.5 and 5.6.

5.5 Definition A pseudorandom bit generator is said to pass all *polynomial-time*[2] *statistical tests* if no polynomial-time algorithm can distinguish between an output sequence of the generator and a truly random sequence of the same length with probability significantly greater that $\frac{1}{2}$.

5.6 Definition A pseudorandom bit generator is said to pass the *next-bit test* if there is no polynomial-time algorithm which, on input of the first l bits of an output sequence s, can predict the $(l+1)^{\text{st}}$ bit of s with probability significantly greater than $\frac{1}{2}$.

Although Definition 5.5 appears to impose a more stringent security requirement on pseudorandom bit generators than Definition 5.6 does, the next result asserts that they are, in fact, equivalent.

5.7 Fact (*universality of the next-bit test*) A pseudorandom bit generator passes the next-bit test if and only if it passes all polynomial-time statistical tests.

5.8 Definition A PRBG that passes the next-bit test (possibly under some plausible but unproved mathematical assumption such as the intractability of factoring integers) is called a *cryptographically secure pseudorandom bit generator* (CSPRBG).

5.9 Remark (*asymptotic nature of Definitions 5.5, 5.6, and 5.8.*) Each of the three definitions above are given in complexity-theoretic terms and are asymptotic in nature because the notion of "polynomial-time" is meaningful for asymptotically large inputs only; the resulting notions of security are relative in the same sense. In Definitions 5.5, 5.6, 5.8, and Fact 5.7, a pseudorandom bit generator is actually a *family* of such PRBGs. Thus the theoretical security results for a family of PRBGs are only an indirect indication about the security of individual members.

Two cryptographically secure pseudorandom bit generators are presented in §5.5.

5.2 Random bit generation

A (true) random bit generator requires a naturally occurring source of randomness. Designing a hardware device or software program to exploit this randomness and produce a bit sequence that is free of biases and correlations is a difficult task. Additionally, for cryptographic applications, the generator must not be subject to observation or manipulation by an adversary. This section surveys some potential sources of random bits.

Random bit generators based on natural sources of randomness are subject to influence by external factors, and also to malfunction. It is imperative that such devices be tested periodically, for example by using the statistical tests of §5.4.

---

[2] The running time of the test is bounded by a polynomial in the length l of the output sequence.

(i) Hardware-based generators

Hardware-based random bit generators exploit the randomness which occurs in some physical phenomena. Such physical processes may produce bits that are biased or correlated, in which case they should be subjected to de-skewing techniques mentioned in (iii) below. Examples of such physical phenomena include:

1. elapsed time between emission of particles during radioactive decay;
2. thermal noise from a semiconductor diode or resistor;
3. the frequency instability of a free running oscillator;
4. the amount a metal insulator semiconductor capacitor is charged during a fixed period of time;
5. air turbulence within a sealed disk drive which causes random fluctuations in disk drive sector read latency times; and
6. sound from a microphone or video input from a camera.

Generators based on the first two phenomena would, in general, have to be built externally to the device using the random bits, and hence may be subject to observation or manipulation by an adversary. Generators based on oscillators and capacitors can be built on VLSI devices; they can be enclosed in tamper-resistant hardware, and hence shielded from active adversaries.

(ii) Software-based generators

Designing a random bit generator in software is even more difficult than doing so in hardware. Processes upon which software random bit generators may be based include:

1. the system clock;
2. elapsed time between keystrokes or mouse movement;
3. content of input/output buffers;
4. user input; and
5. operating system values such as system load and network statistics.

The behavior of such processes can vary considerably depending on various factors, such as the computer platform. It may also be difficult to prevent an adversary from observing or manipulating these processes. For instance, if the adversary has a rough idea of when a random sequence was generated, she can guess the content of the system clock at that time with a high degree of accuracy. A well-designed software random bit generator should utilize as many good sources of randomness as are available. Using many sources guards against the possibility of a few of the sources failing, or being observed or manipulated by an adversary. Each source should be sampled, and the sampled sequences should be combined using a complex *mixing function*; one recommended technique for accomplishing this is to apply a cryptographic hash function such as SHA-1 (Algorithm 9.53) or MD5 (Algorithm 9.51) to a concatenation of the sampled sequences. The purpose of the mixing function is to distill the (true) random bits from the sampled sequences.

(iii) De-skewing

A natural source of random bits may be defective in that the output bits may be *biased* (the probability of the source emitting a 1 is not equal to $\frac{1}{2}$) or *correlated* (the probability of the source emitting a 1 depends on previous bits emitted). There are various techniques for generating truly random bit sequences from the output bits of such a defective generator; such techniques are called *de-skewing techniques*.

5.10 Example (*removing biases in output bits*) Suppose that a generator produces biased but uncorrelated bits. Suppose that the probability of a 1 is p, and the probability of a 0 is $1 - p$, where p is unknown but fixed, $0 < p < 1$. If the output sequence of such a generator is grouped into pairs of bits, with a 10 pair transformed to a 1, a 01 pair transformed to a 0, and 00 and 11 pairs discarded, then the resulting sequence is both unbiased and uncorrelated. □

A practical (although not provable) de-skewing technique is to pass sequences whose bits are biased or correlated through a cryptographic hash function such as SHA-1 or MD5.

5.3 Pseudorandom bit generation

A one-way function f (Definition 1.12) can be utilized to generate pseudorandom bit sequences (Definition 5.3) by first selecting a random seed s, and then applying the function to the sequence of values $s, s+1, s+2, \ldots$; the output sequence is $f(s), f(s+1), f(s+2), \ldots$. Depending on the properties of the one-way function used, it may be necessary to only keep a few bits of the output values $f(s + i)$ in order to remove possible correlations between successive values. Examples of suitable one-way functions f include a cryptographic hash function such as SHA-1 (Algorithm 9.53), or a block cipher such as DES (§7.4) with secret key k.

Although such ad-hoc methods have not been proven to be cryptographically secure, they appear sufficient for most applications. Two such methods for pseudorandom bit and number generation which have been standardized are presented in §5.3.1 and §5.3.2. Techniques for the cryptographically secure generation of pseudorandom bits are given in §5.5.

5.3.1 ANSI X9.17 generator

Algorithm 5.11 is a U.S. Federal Information Processing Standard (FIPS) approved method from the ANSI X9.17 standard for the purpose of pseudorandomly generating keys and initialization vectors for use with DES. E_k denotes DES E-D-E two-key triple-encryption (Definition 7.32) under a key k; the key k should be reserved exclusively for use in this algorithm.

5.11 Algorithm ANSI X9.17 pseudorandom bit generator

INPUT: a random (and secret) 64-bit seed s, integer m, and DES E-D-E encryption key k.
OUTPUT: m pseudorandom 64-bit strings $x_1, x_2, \ldots, x_m$.
  1. Compute the intermediate value $I = E_k(D)$, where D is a 64-bit representation of the date/time to as fine a resolution as is available.
  2. For i from 1 to m do the following:
     2.1 $x_i \leftarrow E_k(I \oplus s)$.
     2.2 $s \leftarrow E_k(x_i \oplus I)$.
  3. Return($x_1, x_2, \ldots, x_m$).

Each output bitstring x_i may be used as an initialization vector (IV) for one of the DES modes of operation (§7.2.2). To obtain a DES key from x_i, every eighth bit of x_i should be reset to odd parity (cf. §7.4.2).

5.3.2 FIPS 186 generator

The algorithms presented in this subsection are FIPS-approved methods for pseudorandomly generating the secret parameters for the DSA (Algorithm 11.5.1). Algorithm 5.12 generates a DSA private key a, while Algorithm 5.14 generates the per-message secrets k to be used in signing messages. Both algorithms use a secret seed s which should be randomly generated, and utilize a one-way function constructed by using either SHA-1 (Algorithm 9.53) or DES (Algorithm 7.82), respectively described in Algorithms 5.15 and 5.16.

5.12 Algorithm FIPS 186 pseudorandom number generator for DSA private keys

---

INPUT: an integer m and a 160-bit prime number q.
OUTPUT: m pseudorandom numbers $a_1, a_2, \ldots, a_m$ in the interval $[0, q-1]$ which may be used as DSA private keys.

1. If Algorithm 5.15 is to be used in step 4.3 then select an arbitrary integer b, $160 \leq b \leq 512$; if Algorithm 5.16 is to be used then set $b \leftarrow 160$.
2. Generate a random (and secret) b-bit seed s.
3. Define the 160-bit string $t = 67452301$ efcdab89 98badcfe 10325476 c3d2e1f0 (in hexadecimal).
4. For i from 1 to m do the following:
    4.1 (optional user input) Either select a b-bit string y_i, or set $y_i \leftarrow 0$.
    4.2 $z_i \leftarrow (s + y_i) \bmod 2^b$.
    4.3 $a_i \leftarrow G(t, z_i) \bmod q$. ($G$ is either that defined in Algorithm 5.15 or 5.16.)
    4.4 $s \leftarrow (1 + s + a_i) \bmod 2^b$.
5. Return($a_1, a_2, \ldots, a_m$).

5.13 Note (*optional user input*) Algorithm 5.12 permits a user to augment the seed s with random or pseudorandom strings derived from alternate sources. The user may desire to do this if she does not trust the quality or integrity of the random bit generator which may be built into a cryptographic module implementing the algorithm.

5.14 Algorithm FIPS 186 pseudorandom number generator for DSA per-message secrets

---

INPUT: an integer m and a 160-bit prime number q.
OUTPUT: m pseudorandom numbers $k_1, k_2, \ldots, k_m$ in the interval $[0, q-1]$ which may be used as the per-message secret numbers k in the DSA.

1. If Algorithm 5.15 is to be used in step 4.1 then select an integer b, $160 \leq b \leq 512$; if Algorithm 5.16 is to be used then set $b \leftarrow 160$.
2. Generate a random (and secret) b-bit seed s.
3. Define the 160-bit string $t = $ efcdab89 98badcfe 10325476 c3d2e1f0 67452301 (in hexadecimal).
4. For i from 1 to m do the following:
    4.1 $k_i \leftarrow G(t, s) \bmod q$. ($G$ is either that defined in Algorithm 5.15 or 5.16.)
    4.2 $s \leftarrow (1 + s + k_i) \bmod 2^b$.
5. Return($k_1, k_2, \ldots, k_m$).

5.15 Algorithm FIPS 186 one-way function using SHA-1

INPUT: a 160-bit string t and a b-bit string c, $160 \leq b \leq 512$.
OUTPUT: a 160-bit string denoted $G(t, c)$.

1. Break up t into five 32-bit blocks: $t = H_1\|H_2\|H_3\|H_4\|H_5$.
2. Pad c with 0's to obtain a 512-bit message block: $X \leftarrow c\|0^{512-b}$.
3. Divide X into 16 32-bit words: $x_0 x_1 \ldots x_{15}$, and set $m \leftarrow 1$.
4. Execute step 4 of SHA-1 (Algorithm 9.53). (This alters the H_i's.)
5. The output is the concatenation: $G(t, c) = H_1\|H_2\|H_3\|H_4\|H_5$.

5.16 Algorithm FIPS 186 one-way function using DES

INPUT: two 160-bit strings t and c.
OUTPUT: a 160-bit string denoted $G(t, c)$.

1. Break up t into five 32-bit blocks: $t = t_0\|t_1\|t_2\|t_3\|t_4$.
2. Break up c into five 32-bit blocks: $c = c_0\|c_1\|c_2\|c_3\|c_4$.
3. For i from 0 to 4 do the following: $x_i \leftarrow t_i \oplus c_i$.
4. For i from 0 to 4 do the following:
   4.1 $b_1 \leftarrow c_{(i+4)\bmod 5}$, $b_2 \leftarrow c_{(i+3)\bmod 5}$.
   4.2 $a_1 \leftarrow x_i$, $a_2 \leftarrow x_{(i+1)\bmod 5} \oplus x_{(i+4)\bmod 5}$.
   4.3 $A \leftarrow a_1\|a_2$, $B \leftarrow b_1'\|b_2$, where b_1' denotes the 24 least significant bits of b_1.
   4.4 Use DES with key B to encrypt A: $y_i \leftarrow \mathrm{DES}_B(A)$.
   4.5 Break up y_i into two 32-bit blocks: $y_i = L_i\|R_i$.
5. For i from 0 to 4 do the following: $z_i \leftarrow L_i \oplus R_{(i+2)\bmod 5} \oplus L_{(i+3)\bmod 5}$.
6. The output is the concatenation: $G(t, c) = z_0\|z_1\|z_2\|z_3\|z_4$.

5.4 Statistical tests

This section presents some tests designed to measure the quality of a generator purported to be a random bit generator (Definition 5.1). While it is impossible to give a mathematical proof that a generator is indeed a random bit generator, the tests described here help detect certain kinds of weaknesses the generator may have. This is accomplished by taking a sample output sequence of the generator and subjecting it to various statistical tests. Each statistical test determines whether the sequence possesses a certain attribute that a truly random sequence would be likely to exhibit; the conclusion of each test is not definite, but rather *probabilistic*. An example of such an attribute is that the sequence should have roughly the same number of 0's as 1's. If the sequence is deemed to have failed any one of the statistical tests, the generator may be *rejected* as being non-random; alternatively, the generator may be subjected to further testing. On the other hand, if the sequence passes all of the statistical tests, the generator is *accepted* as being random. More precisely, the term "accepted" should be replaced by "not rejected", since passing the tests merely provides probabilistic evidence that the generator produces sequences which have certain characteristics of random sequences.

§5.4.1 and §5.4.2 provide some relevant background in statistics. §5.4.3 establishes some notation and lists Golomb's randomness postulates. Specific statistical tests for randomness are described in §5.4.4 and §5.4.5.

5.4.1 The normal and chi-square distributions

The normal and χ^2 distributions are widely used in statistical applications.

5.17 Definition If the result X of an experiment can be any real number, then X is said to be a *continuous* random variable.

5.18 Definition A *probability density function* of a continuous random variable X is a function $f(x)$ which can be integrated and satisfies:

   (i) $f(x) \geq 0$ for all $x \in \mathbb{R}$;

   (ii) $\int_{-\infty}^{\infty} f(x)\, dx = 1$; and

   (iii) for all $a, b \in \mathbb{R}$, $P(a < X \leq b) = \int_{a}^{b} f(x)\, dx$.

(i) The normal distribution

The normal distribution arises in practice when a large number of independent random variables having the same mean and variance are summed.

5.19 Definition A (continuous) random variable X has a *normal distribution* with *mean* μ and *variance* σ^2 if its probability density function is defined by

$$f(x) \; = \; \frac{1}{\sigma\sqrt{2\pi}} \exp\left\{ \frac{-(x-\mu)^2}{2\sigma^2} \right\}, \quad -\infty < x < \infty.$$

Notation: X is said to be $N(\mu, \sigma^2)$. If X is $N(0,1)$, then X is said to have a *standard normal distribution*.

A graph of the $N(0,1)$ distribution is given in Figure 5.1. The graph is symmetric

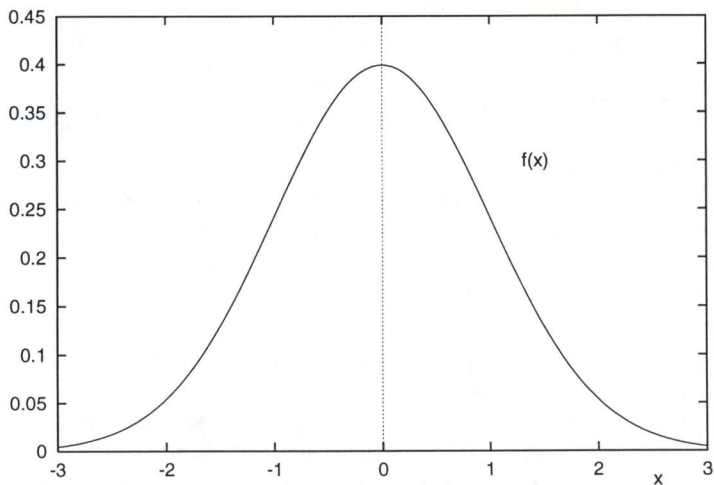

Figure 5.1: *The normal distribution $N(0,1)$.*

about the vertical axis, and hence $P(X > x) = P(X < -x)$ for any x. Table 5.1 gives some percentiles for the standard normal distribution. For example, the entry ($\alpha = 0.05$, $x = 1.6449$) means that if X is $N(0,1)$, then X exceeds 1.6449 about 5% of the time.

Fact 5.20 can be used to reduce questions about a normal distribution to questions about the standard normal distribution.

α	0.1	0.05	0.025	0.01	0.005	0.0025	0.001	0.0005
x	1.2816	1.6449	1.9600	2.3263	2.5758	2.8070	3.0902	3.2905

Table 5.1: *Selected percentiles of the standard normal distribution. If X is a random variable having a standard normal distribution, then $P(X > x) = \alpha$.*

5.20 Fact If the random variable X is $N(\mu, \sigma^2)$, then the random variable $Z = (X - \mu)/\sigma$ is $N(0, 1)$.

(ii) The χ^2 distribution

The χ^2 distribution can be used to compare the *goodness-of-fit* of the observed frequencies of events to their expected frequencies under a hypothesized distribution. The χ^2 distribution with v degrees of freedom arises in practice when the squares of v independent random variables having standard normal distributions are summed.

5.21 Definition A (continuous) random variable X has a χ^2 *(chi-square) distribution* with v *degrees of freedom* if its probability density function is defined by

$$f(x) = \begin{cases} \dfrac{1}{\Gamma(v/2)2^{v/2}} x^{(v/2)-1} e^{-x/2}, & 0 \le x < \infty, \\ 0, & x < 0, \end{cases}$$

where Γ is the gamma function. [3] The *mean* and *variance* of this distribution are $\mu = v$, and $\sigma^2 = 2v$.

A graph of the χ^2 distribution with $v = 7$ degrees of freedom is given in Figure 5.2. Table 5.2 gives some percentiles of the χ^2 distribution for various degrees of freedom. For

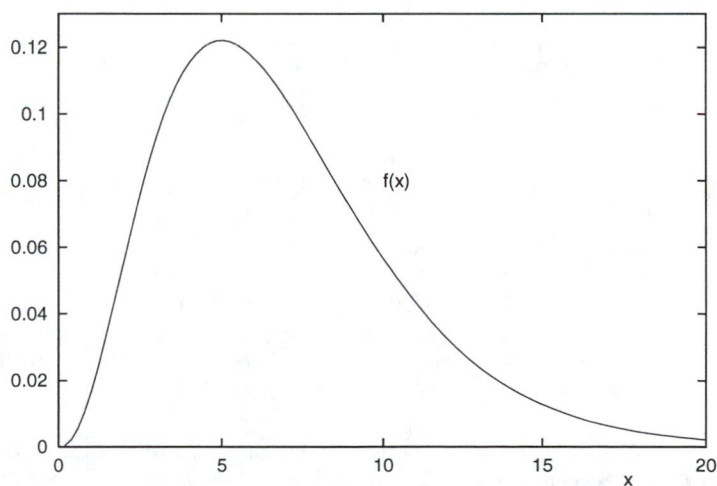

Figure 5.2: *The χ^2 (chi-square) distribution with $v = 7$ degrees of freedom.*

example, the entry in row $v = 5$ and column $\alpha = 0.05$ is $x = 11.0705$; this means that if X has a χ^2 distribution with 5 degrees of freedom, then X exceeds 11.0705 about 5% of the time.

---

[3] The *gamma function* is defined by $\Gamma(t) = \int_0^\infty x^{t-1} e^{-x} dx$, for $t > 0$.

v	α					
	0.100	0.050	0.025	0.010	0.005	0.001
1	2.7055	3.8415	5.0239	6.6349	7.8794	10.8276
2	4.6052	5.9915	7.3778	9.2103	10.5966	13.8155
3	6.2514	7.8147	9.3484	11.3449	12.8382	16.2662
4	7.7794	9.4877	11.1433	13.2767	14.8603	18.4668
5	9.2364	11.0705	12.8325	15.0863	16.7496	20.5150
6	10.6446	12.5916	14.4494	16.8119	18.5476	22.4577
7	12.0170	14.0671	16.0128	18.4753	20.2777	24.3219
8	13.3616	15.5073	17.5345	20.0902	21.9550	26.1245
9	14.6837	16.9190	19.0228	21.6660	23.5894	27.8772
10	15.9872	18.3070	20.4832	23.2093	25.1882	29.5883
11	17.2750	19.6751	21.9200	24.7250	26.7568	31.2641
12	18.5493	21.0261	23.3367	26.2170	28.2995	32.9095
13	19.8119	22.3620	24.7356	27.6882	29.8195	34.5282
14	21.0641	23.6848	26.1189	29.1412	31.3193	36.1233
15	22.3071	24.9958	27.4884	30.5779	32.8013	37.6973
16	23.5418	26.2962	28.8454	31.9999	34.2672	39.2524
17	24.7690	27.5871	30.1910	33.4087	35.7185	40.7902
18	25.9894	28.8693	31.5264	34.8053	37.1565	42.3124
19	27.2036	30.1435	32.8523	36.1909	38.5823	43.8202
20	28.4120	31.4104	34.1696	37.5662	39.9968	45.3147
21	29.6151	32.6706	35.4789	38.9322	41.4011	46.7970
22	30.8133	33.9244	36.7807	40.2894	42.7957	48.2679
23	32.0069	35.1725	38.0756	41.6384	44.1813	49.7282
24	33.1962	36.4150	39.3641	42.9798	45.5585	51.1786
25	34.3816	37.6525	40.6465	44.3141	46.9279	52.6197
26	35.5632	38.8851	41.9232	45.6417	48.2899	54.0520
27	36.7412	40.1133	43.1945	46.9629	49.6449	55.4760
28	37.9159	41.3371	44.4608	48.2782	50.9934	56.8923
29	39.0875	42.5570	45.7223	49.5879	52.3356	58.3012
30	40.2560	43.7730	46.9792	50.8922	53.6720	59.7031
31	41.4217	44.9853	48.2319	52.1914	55.0027	61.0983
63	77.7454	82.5287	86.8296	92.0100	95.6493	103.4424
127	147.8048	154.3015	160.0858	166.9874	171.7961	181.9930
255	284.3359	293.2478	301.1250	310.4574	316.9194	330.5197
511	552.3739	564.6961	575.5298	588.2978	597.0978	615.5149
1023	1081.3794	1098.5208	1113.5334	1131.1587	1143.2653	1168.4972

Table 5.2: *Selected percentiles of the χ^2 (chi-square) distribution. A (v, α)-entry of x in the table has the following meaning: if X is a random variable having a χ^2 distribution with v degrees of freedom, then $P(X > x) = \alpha$.*

Fact 5.22 relates the normal distribution to the χ^2 distribution.

5.22 Fact If the random variable X is $N(\mu, \sigma^2)$, $\sigma^2 > 0$, then the random variable $Z = (X - \mu)^2/\sigma^2$ has a χ^2 distribution with 1 degree of freedom. In particular, if X is $N(0,1)$, then $Z = X^2$ has a χ^2 distribution with 1 degree of freedom.

5.4.2 Hypothesis testing

A *statistical hypothesis*, denoted H_0, is an assertion about a distribution of one or more random variables. A *test* of a statistical hypothesis is a procedure, based upon observed values of the random variables, that leads to the acceptance or rejection of the hypothesis H_0. The test only provides a measure of the strength of the evidence provided by the data against the hypothesis; hence, the conclusion of the test is not definite, but rather probabilistic.

5.23 Definition The *significance level* α of the test of a statistical hypothesis H_0 is the probability of rejecting H_0 when it is true.

In this section, H_0 will be the hypothesis that a given binary sequence was produced by a random bit generator. If the significance level α of a test of H_0 is too high, then the test may reject sequences that were, in fact, produced by a random bit generator (such an error is called a *Type I error*). On the other hand, if the significance level of a test of H_0 is too low, then there is the danger that the test may accept sequences even though they were not produced by a random bit generator (such an error is called a *Type II error*).[4] It is, therefore, important that the test be carefully designed to have a significance level that is appropriate for the purpose at hand; a significance level α between 0.001 and 0.05 might be employed in practice.

A statistical test is implemented by specifying a *statistic* on the random sample.[5] Statistics are generally chosen so that they can be efficiently computed, and so that they (approximately) follow an $N(0,1)$ or a χ^2 distribution (see §5.4.1). The value of the statistic for the sample output sequence is computed and compared with the value expected for a random sequence as described below.

1. Suppose that a statistic X for a random sequence follows a χ^2 distribution with v degrees of freedom, and suppose that the statistic can be expected to take on larger values for nonrandom sequences. To achieve a significance level of α, a *threshold* value x_α is chosen (using Table 5.2) so that $P(X > x_\alpha) = \alpha$. If the value X_s of the statistic for the sample output sequence satisfies $X_s > x_\alpha$, then the sequence *fails* the test; otherwise, it *passes* the test. Such a test is called a *one-sided* test. For example, if $v = 5$ and $\alpha = 0.025$, then $x_\alpha = 12.8325$, and one expects a random sequence to fail the test only 2.5% of the time.

2. Suppose that a statistic X for a random sequence follows an $N(0,1)$ distribution, and suppose that the statistic can be expected to take on both larger and smaller values for nonrandom sequences. To achieve a significance level of α, a *threshold* value x_α is chosen (using Table 5.1) so that $P(X > x_\alpha) = P(X < -x_\alpha) = \alpha/2$. If the value

---

[4]Actually, the probability β of a Type II error may be completely independent of α. If the generator is not a random bit generator, the probability β depends on the nature of the defects of the generator, and is usually difficult to determine in practice. For this reason, assuming that the probability of a Type II error is proportional to α is a useful intuitive guide when selecting an appropriate significance level for a test.

[5]A *statistic* is a function of the elements of a random sample; for example, the number of 0's in a binary sequence is a statistic.

X_s of the statistic for the sample output sequence satisfies $X_s > x_\alpha$ or $X_s < -x_\alpha$, then the sequence *fails* the test; otherwise, it *passes* the test. Such a test is called a *two-sided* test. For example, if $\alpha = 0.05$, then $x_\alpha = 1.96$, and one expects a random sequence to fail the test only 5% of the time.

5.4.3 Golomb's randomness postulates

Golomb's randomness postulates (Definition 5.28) are presented here for historical reasons – they were one of the first attempts to establish some *necessary* conditions for a periodic pseudorandom sequence to look random. It is emphasized that these conditions are far from being *sufficient* for such sequences to be considered random. Unless otherwise stated, all sequences are binary sequences.

5.24 Definition Let $s = s_0, s_1, s_2, \ldots$ be an infinite sequence. The subsequence consisting of the first n terms of s is denoted by $s^n = s_0, s_1, \ldots, s_{n-1}$.

5.25 Definition The sequence $s = s_0, s_1, s_2, \ldots$ is said to be *N-periodic* if $s_i = s_{i+N}$ for all $i \geq 0$. The sequence s is *periodic* if it is N-periodic for some positive integer N. The *period* of a periodic sequence s is the smallest positive integer N for which s is N-periodic. If s is a periodic sequence of period N, then the *cycle* of s is the subsequence s^N.

5.26 Definition Let s be a sequence. A *run* of s is a subsequence of s consisting of consecutive 0's or consecutive 1's which is neither preceded nor succeeded by the same symbol. A run of 0's is called a *gap*, while a run of 1's is called a *block*.

5.27 Definition Let $s = s_0, s_1, s_2, \ldots$ be a periodic sequence of period N. The *autocorrelation function* of s is the integer-valued function $C(t)$ defined as

$$C(t) = \frac{1}{N} \sum_{i=0}^{N-1} (2s_i - 1) \cdot (2s_{i+t} - 1), \quad \text{for } 0 \leq t \leq N - 1.$$

The autocorrelation function $C(t)$ measures the amount of similarity between the sequence s and a shift of s by t positions. If s is a random periodic sequence of period N, then $|N \cdot C(t)|$ can be expected to be quite small for all values of t, $0 < t < N$.

5.28 Definition Let s be a periodic sequence of period N. *Golomb's randomness postulates* are the following.

R1: In the cycle s^N of s, the number of 1's differs from the number of 0's by at most 1.

R2: In the cycle s^N, at least half the runs have length 1, at least one-fourth have length 2, at least one-eighth have length 3, etc., as long as the number of runs so indicated exceeds 1. Moreover, for each of these lengths, there are (almost) equally many gaps and blocks.[6]

R3: The autocorrelation function $C(t)$ is two-valued. That is for some integer K,

$$N \cdot C(t) = \sum_{i=0}^{N-1} (2s_i - 1) \cdot (2s_{i+t} - 1) = \begin{cases} N, & \text{if } t = 0, \\ K, & \text{if } 1 \leq t \leq N - 1. \end{cases}$$

---

[6] Postulate R2 implies postulate R1.

5.29 Definition A binary sequence which satisfies Golomb's randomness postulates is called a *pseudo-noise sequence* or a *pn-sequence*.

Pseudo-noise sequences arise in practice as output sequences of maximum-length linear feedback shift registers (cf. Fact 6.14).

5.30 Example (*pn-sequence*) Consider the periodic sequence s of period $N = 15$ with cycle

$$s^{15} = 0, 1, 1, 0, 0, 1, 0, 0, 0, 1, 1, 1, 1, 0, 1.$$

The following shows that the sequence s satisfies Golomb's randomness postulates.

R1: The number of 0's in s^{15} is 7, while the number of 1's is 8.

R2: s^{15} has 8 runs. There are 4 runs of length 1 (2 gaps and 2 blocks), 2 runs of length 2 (1 gap and 1 block), 1 run of length 3 (1 gap), and 1 run of length 4 (1 block).

R3: The autocorrelation function $C(t)$ takes on two values: $C(0) = 1$ and $C(t) = \frac{-1}{15}$ for $1 \le t \le 14$.

Hence, s is a pn-sequence. □

5.4.4 Five basic tests

Let $s = s_0, s_1, s_2, \ldots, s_{n-1}$ be a binary sequence of length n. This subsection presents five statistical tests that are commonly used for determining whether the binary sequence s possesses some specific characteristics that a truly random sequence would be likely to exhibit. It is emphasized again that the outcome of each test is not definite, but rather probabilistic. If a sequence passes all five tests, there is no guarantee that it was indeed produced by a random bit generator (cf. Example 5.4).

(i) Frequency test (monobit test)

The purpose of this test is to determine whether the number of 0's and 1's in s are approximately the same, as would be expected for a random sequence. Let n_0, n_1 denote the number of 0's and 1's in s, respectively. The statistic used is

$$X_1 = \frac{(n_0 - n_1)^2}{n} \tag{5.1}$$

which approximately follows a χ^2 distribution with 1 degree of freedom if $n \ge 10$. [7]

(ii) Serial test (two-bit test)

The purpose of this test is to determine whether the number of occurrences of 00, 01, 10, and 11 as subsequences of s are approximately the same, as would be expected for a random sequence. Let n_0, n_1 denote the number of 0's and 1's in s, respectively, and let n_{00}, n_{01}, n_{10}, n_{11} denote the number of occurrences of 00, 01, 10, 11 in s, respectively. Note that $n_{00} + n_{01} + n_{10} + n_{11} = (n - 1)$ since the subsequences are allowed to overlap. The statistic used is

$$X_2 = \frac{4}{n-1} \left(n_{00}^2 + n_{01}^2 + n_{10}^2 + n_{11}^2 \right) - \frac{2}{n} \left(n_0^2 + n_1^2 \right) + 1 \tag{5.2}$$

which approximately follows a χ^2 distribution with 2 degrees of freedom if $n \ge 21$.

---

[7] In practice, it is recommended that the length n of the sample output sequence be much larger (for example, $n \gg 10000$) than the minimum specified for each test in this subsection.

(iii) Poker test

Let m be a positive integer such that $\lfloor \frac{n}{m} \rfloor \geq 5 \cdot (2^m)$, and let $k = \lfloor \frac{n}{m} \rfloor$. Divide the sequence s into k non-overlapping parts each of length m, and let n_i be the number of occurrences of the ith type of sequence of length m, $1 \leq i \leq 2^m$. The poker test determines whether the sequences of length m each appear approximately the same number of times in s, as would be expected for a random sequence. The statistic used is

$$X_3 = \frac{2^m}{k} \left(\sum_{i=1}^{2^m} n_i^2 \right) - k \tag{5.3}$$

which approximately follows a χ^2 distribution with $2^m - 1$ degrees of freedom. Note that the poker test is a generalization of the frequency test: setting $m = 1$ in the poker test yields the frequency test.

(iv) Runs test

The purpose of the runs test is to determine whether the number of runs (of either zeros or ones; see Definition 5.26) of various lengths in the sequence s is as expected for a random sequence. The expected number of gaps (or blocks) of length i in a random sequence of length n is $e_i = (n-i+3)/2^{i+2}$. Let k be equal to the largest integer i for which $e_i \geq 5$. Let B_i, G_i be the number of blocks and gaps, respectively, of length i in s for each i, $1 \leq i \leq k$. The statistic used is

$$X_4 = \sum_{i=1}^{k} \frac{(B_i - e_i)^2}{e_i} + \sum_{i=1}^{k} \frac{(G_i - e_i)^2}{e_i} \tag{5.4}$$

which approximately follows a χ^2 distribution with $2k - 2$ degrees of freedom.

(v) Autocorrelation test

The purpose of this test is to check for correlations between the sequence s and (non-cyclic) shifted versions of it. Let d be a fixed integer, $1 \leq d \leq \lfloor n/2 \rfloor$. The number of bits in s not equal to their d-shifts is $A(d) = \sum_{i=0}^{n-d-1} s_i \oplus s_{i+d}$, where $\oplus$ denotes the XOR operator. The statistic used is

$$X_5 = 2 \left(A(d) - \frac{n-d}{2} \right) / \sqrt{n-d} \tag{5.5}$$

which approximately follows an $N(0,1)$ distribution if $n - d \geq 10$. Since small values of $A(d)$ are as unexpected as large values of $A(d)$, a two-sided test should be used.

5.31 Example (*basic statistical tests*) Consider the (non-random) sequence s of length $n = 160$ obtained by replicating the following sequence four times:

$$11100 \ \ 01100 \ \ 01000 \ \ 10100 \ \ 11101 \ \ 11100 \ \ 10010 \ \ 01001.$$

  (i) (*frequency test*) $n_0 = 84$, $n_1 = 76$, and the value of the statistic X_1 is 0.4.
  (ii) (*serial test*) $n_{00} = 44$, $n_{01} = 40$, $n_{10} = 40$, $n_{11} = 35$, and the value of the statistic X_2 is 0.6252.
  (iii) (*poker test*) Here $m = 3$ and $k = 53$. The blocks $000, 001, 010, 011, 100, 101, 110,$ 111 appear 5, 10, 6, 4, 12, 3, 6, and 7 times, respectively, and the value of the statistic X_3 is 9.6415.
  (iv) (*runs test*) Here $e_1 = 20.25$, $e_2 = 10.0625$, $e_3 = 5$, and $k = 3$. There are 25, 4, 5 blocks of lengths 1, 2, 3, respectively, and 8, 20, 12 gaps of lengths 1, 2, 3, respectively. The value of the statistic X_4 is 31.7913.

(v) (*autocorrelation test*) If $d = 3$, then $A(3) = 35$. The value of the statistic X_5 is -6.9434.

For a significance level of $\alpha = 0.05$, the threshold values for X_1, X_2, X_3, X_4, and X_5 are 3.8415, 5.9915, 14.0671, 9.4877, and 1.96, respectively (see Tables 5.1 and 5.2). Hence, the given sequence s passes the frequency, serial, and poker tests, but fails the runs and autocorrelation tests. □

5.32 Note (*FIPS 140-1 statistical tests for randomness*) FIPS 140-1 specifies four statistical tests for randomness. Instead of making the user select appropriate significance levels for these tests, explicit bounds are provided that the computed value of a statistic must satisfy. A single bitstring s of length 20000 bits, output from a generator, is subjected to each of the following tests. If any of the tests fail, then the generator fails the test.

(i) *monobit test*. The number n_1 of 1's in s should satisfy $9654 < n_1 < 10346$.

(ii) *poker test*. The statistic X_3 defined by equation (5.3) is computed for $m = 4$. The poker test is passed if $1.03 < X_3 < 57.4$.

(iii) *runs test*. The number B_i and G_i of blocks and gaps, respectively, of length i in s are counted for each i, $1 \le i \le 6$. (For the purpose of this test, runs of length greater than 6 are considered to be of length 6.) The runs test is passed if the 12 counts B_i, G_i, $1 \le i \le 6$, are each within the corresponding interval specified by the following table.

Length of run	Required interval
1	$2267 - 2733$
2	$1079 - 1421$
3	$502 - 748$
4	$223 - 402$
5	$90 - 223$
6	$90 - 223$

(iv) *long run test*. The long run test is passed if there are no runs of length 34 or more.

For high security applications, FIPS 140-1 mandates that the four tests be performed each time the random bit generator is powered up. FIPS 140-1 allows these tests to be substituted by alternative tests which provide equivalent or superior randomness checking.

5.4.5 Maurer's universal statistical test

The basic idea behind Maurer's universal statistical test is that it should not be possible to significantly compress (without loss of information) the output sequence of a random bit generator. Thus, if a sample output sequence s of a bit generator can be significantly compressed, the generator should be rejected as being defective. Instead of actually compressing the sequence s, the universal statistical test computes a quantity that is related to the length of the compressed sequence.

The *universality* of Maurer's universal statistical test arises because it is able to detect any one of a very general class of possible defects a bit generator might have. This class includes the five defects that are detectable by the basic tests of §5.4.4. A drawback of the universal statistical test over the five basic tests is that it requires a much longer sample output sequence in order to be effective. Provided that the required output sequence can be efficiently generated, this drawback is not a practical concern since the universal statistical test itself is very efficient.

Algorithm 5.33 computes the statistic X_u for a sample output sequence $s = s_0, s_1, \ldots,$ s_{n-1} to be used in the universal statistical test. The parameter L is first chosen from the

L	μ	σ_1^2
1	0.7326495	0.690
2	1.5374383	1.338
3	2.4016068	1.901
4	3.3112247	2.358
5	4.2534266	2.705
6	5.2177052	2.954
7	6.1962507	3.125
8	7.1836656	3.238

L	μ	σ_1^2
9	8.1764248	3.311
10	9.1723243	3.356
11	10.170032	3.384
12	11.168765	3.401
13	12.168070	3.410
14	13.167693	3.416
15	14.167488	3.419
16	15.167379	3.421

Table 5.3: *Mean μ and variance σ^2 of the statistic X_u for random sequences, with parameters L, K as $Q \to \infty$. The variance of X_u is $\sigma^2 = c(L, K)^2 \cdot \sigma_1^2/K$, where $c(L, K) \approx 0.7 - (0.8/L) + (1.6 + (12.8/L)) \cdot K^{-4/L}$ for $K \geq 2^L$.*

interval $[6, 16]$. The sequence s is then partitioned into non-overlapping L-bit blocks, with any leftover bits discarded; the total number of blocks is $Q+K$, where Q and K are defined below. For each i, $1 \leq i \leq Q+K$, let b_i be the integer whose binary representation is the ith block. The blocks are scanned in order. A table T is maintained so that at each stage $T[j]$ is the position of the last occurrence of the block corresponding to integer j, $0 \leq j \leq 2^L - 1$. The first Q blocks of s are used to initialize table T; Q should be chosen to be at least $10 \cdot 2^L$ in order to have a high likelihood that each of the 2^L L-bit blocks occurs at least once in the first Q blocks. The remaining K blocks are used to define the statistic X_u as follows. For each i, $Q + 1 \leq i \leq Q + K$, let $A_i = i - T[b_i]$; A_i is the number of positions since the last occurrence of block b_i. Then

$$X_u = \frac{1}{K} \sum_{i=Q+1}^{Q+K} \lg A_i. \qquad (5.6)$$

K should be at least $1000 \cdot 2^L$ (and, hence, the sample sequence s should be at least $(1010 \cdot 2^L \cdot L)$ bits in length). Table 5.3 lists the mean μ and variance σ^2 of X_u for random sequences for some sample choices of L as $Q \to \infty$.

5.33 Algorithm Computing the statistic X_u for Maurer's universal statistical test

---

INPUT: a binary sequence $s = s_0, s_1, \ldots, s_{n-1}$ of length n, and parameters L, Q, K.
OUTPUT: the value of the statistic X_u for the sequence s.

1. *Zero the table T.* For j from 0 to $2^L - 1$ do the following: $T[j] \leftarrow 0$.
2. *Initialize the table T.* For i from 1 to Q do the following: $T[b_i] \leftarrow i$.
3. sum$\leftarrow 0$.
4. For i from $Q + 1$ to $Q + K$ do the following:
   4.1 sum$\leftarrow$sum $+ \lg(i - T[b_i])$.
   4.2 $T[b_i] \leftarrow i$.
5. $X_u \leftarrow$sum$/K$.
6. Return(X_u).

Maurer's universal statistical test uses the computed value of X_u for the sample output sequence s in the manner prescribed by Fact 5.34. To test the sequence s, a two-sided test should be used with a significance level α between 0.001 and 0.01 (see §5.4.2).

5.34 Fact Let X_u be the statistic defined in (5.6) having mean μ and variance σ^2 as given in Table 5.3. Then, for random sequences, the statistic $Z_u = (X_u - \mu)/\sigma$ approximately follows an $N(0, 1)$ distribution.

5.5 Cryptographically secure pseudorandom bit generation

Two cryptographically secure pseudorandom bit generators (CSPRBG – see Definition 5.8) are presented in this section. The security of each generator relies on the presumed intractability of an underlying number-theoretic problem. The modular multiplications that these generators use make them relatively slow compared to the (ad-hoc) pseudorandom bit generators of §5.3. Nevertheless they may be useful in some circumstances, for example, generating pseudorandom bits on hardware devices which already have the circuitry for performing modular multiplications. Efficient techniques for implementing modular multiplication are presented in §14.3.

5.5.1 RSA pseudorandom bit generator

The RSA pseudorandom bit generator is a CSPRBG under the assumption that the RSA problem is intractable (§3.3; see also §3.9.2).

5.35 Algorithm RSA pseudorandom bit generator

SUMMARY: a pseudorandom bit sequence $z_1, z_2, \ldots, z_l$ of length l is generated.

1. *Setup*. Generate two secret RSA-like primes p and q (cf. Note 8.8), and compute $n = pq$ and $\phi = (p-1)(q-1)$. Select a random integer e, $1 < e < \phi$, such that $\gcd(e, \phi) = 1$.
2. Select a random integer x_0 (the *seed*) in the interval $[1, n-1]$.
3. For i from 1 to l do the following:
   3.1 $x_i \leftarrow x_{i-1}^e \bmod n$.
   3.2 $z_i \leftarrow$ the least significant bit of x_i.
4. The output sequence is $z_1, z_2, \ldots, z_l$.

5.36 Note (*efficiency of the RSA PRBG*) If $e = 3$ is chosen (cf. Note 8.9(ii)), then generating each pseudorandom bit z_i requires one modular multiplication and one modular squaring. The efficiency of the generator can be improved by extracting the j least significant bits of x_i in step 3.2, where $j = c \lg \lg n$ and c is a constant. Provided that n is sufficiently large, this modified generator is also cryptographically secure (cf. Fact 3.87). For a modulus n of a fixed bitlength (e.g., 1024 bits), an explicit range of values of c for which the resulting generator remains cryptographically secure (cf. Remark 5.9) under the intractability assumption of the RSA problem has not been determined.

The following modification improves the efficiency of the RSA PRBG.

5.37 Algorithm Micali-Schnorr pseudorandom bit generator

SUMMARY: a pseudorandom bit sequence is generated.

1. *Setup.* Generate two secret RSA-like primes p and q (cf. Note 8.8), and compute $n = pq$ and $\phi = (p-1)(q-1)$. Let $N = \lfloor \lg n \rfloor + 1$ (the bitlength of n). Select an integer e, $1 < e < \phi$, such that $\gcd(e, \phi) = 1$ and $80e \leq N$. Let $k = \lfloor N(1 - \frac{2}{e}) \rfloor$ and $r = N - k$.
2. Select a random sequence x_0 (the *seed*) of bitlength r.
3. *Generate a pseudorandom sequence of length $k \cdot l$.* For i from 1 to l do the following:

    3.1 $y_i \leftarrow x_{i-1}^e \bmod n$.
    3.2 $x_i \leftarrow$ the r most significant bits of y_i.
    3.3 $z_i \leftarrow$ the k least significant bits of y_i.
4. The output sequence is $z_1 \parallel z_2 \parallel \cdots \parallel z_l$, where $\parallel$ denotes concatenation.

5.38 Note (*efficiency of the Micali-Schnorr PRBG*) Algorithm 5.37 is more efficient than the RSA PRBG since $\lfloor N(1 - \frac{2}{e}) \rfloor$ bits are generated per exponentiation by e. For example, if $e = 3$ and $N = 1024$, then $k = 341$ bits are generated per exponentiation. Moreover, each exponentiation requires only one modular squaring of an $r = 683$-bit number, and one modular multiplication.

5.39 Note (*security of the Micali-Schnorr PRBG*) Algorithm 5.37 is cryptographically secure under the assumption that the following is true: the distribution $x^e \bmod n$ for random r-bit sequences x is indistinguishable by all polynomial-time statistical tests from the uniform distribution of integers in the interval $[0, n-1]$. This assumption is stronger than requiring that the RSA problem be intractable.

5.5.2 Blum-Blum-Shub pseudorandom bit generator

The Blum-Blum-Shub pseudorandom bit generator (also known as the $x^2 \bmod n$ generator or the *BBS* generator) is a CSPRBG under the assumption that integer factorization is intractable (§3.2). It forms the basis for the Blum-Goldwasser probabilistic public-key encryption scheme (Algorithm 8.55).

5.40 Algorithm Blum-Blum-Shub pseudorandom bit generator

SUMMARY: a pseudorandom bit sequence $z_1, z_2, \ldots, z_l$ of length l is generated.

1. *Setup.* Generate two large secret random (and distinct) primes p and q (cf. Note 8.8), each congruent to 3 modulo 4, and compute $n = pq$.
2. Select a random integer s (the *seed*) in the interval $[1, n-1]$ such that $\gcd(s, n) = 1$, and compute $x_0 \leftarrow s^2 \bmod n$.
3. For i from 1 to l do the following:

    3.1 $x_i \leftarrow x_{i-1}^2 \bmod n$.
    3.2 $z_i \leftarrow$ the least significant bit of x_i.
4. The output sequence is $z_1, z_2, \ldots, z_l$.

5.41 Note (*efficiency of the Blum-Blum-Shub PRBG*) Generating each pseudorandom bit z_i requires one modular squaring. The efficiency of the generator can be improved by extracting the j least significant bits of x_i in step 3.2, where $j = c \lg \lg n$ and c is a constant. Provided that n is sufficiently large, this modified generator is also cryptographically secure. For a modulus n of a fixed bitlength (eg. 1024 bits), an explicit range of values of c for which the resulting generator is cryptographically secure (cf. Remark 5.9) under the intractability assumption of the integer factorization problem has not been determined.

5.6 Notes and further references

§5.1

Chapter 3 of Knuth [692] is the definitive reference for the classic (non-cryptographic) generation of pseudorandom numbers. Knuth [692, pp.142-166] contains an extensive discussion of what it means for a sequence to be random. Lagarias [724] gives a survey of theoretical results on pseudorandom number generators. Luby [774] provides a comprehensive and rigorous overview of pseudorandom generators.

For a study of linear congruential generators (Example 5.4), see Knuth [692, pp.9-25]. Plumstead/Boyar [979, 980] showed how to predict the output of a linear congruential generator given only a few elements of the output sequence, and when the parameters a, b, and m of the generator are unknown. Boyar [180] extended her method and showed that linear *multivariate* congruential generators (having recurrence equation $x_n = a_1 x_{n-1} + a_2 x_{n-2} + \cdots + a_l x_{n-l} + b \bmod m$), and *quadratic* congruential generators (having recurrence equation $x_n = a x_{n-1}^2 + b x_{n-1} + c \bmod m$) are cryptographically insecure. Finally, Krawczyk [713] generalized these results and showed how the output of any *multivariate polynomial* congruential generator can be efficiently predicted. A *truncated* linear congruential generator is one where a fraction of the least significant bits of the x_i are discarded. Frieze et al. [427] showed that these generators can be efficiently predicted if the generator parameters a, b, and m are known. Stern [1173] extended this method to the case where only m is known. Boyar [179] presented an efficient algorithm for predicting linear congruential generators when $O(\log \log m)$ bits are discarded, and when the parameters a, b, and m are unknown. No efficient prediction algorithms are known for truncated multivariate polynomial congruential generators. For a summary of cryptanalytic attacks on congruential generators, see Brickell and Odlyzko [209, pp.523-526].

For a formal definition of a statistical test (Definition 5.5), see Yao [1258]. Fact 5.7 on the universality of the next-bit test is due to Yao [1258]. For a proof of Yao's result, see Kranakis [710] and §12.2 of Stinson [1178]. A proof of a generalization of Yao's result is given by Goldreich, Goldwasser, and Micali [468]. The notion of a cryptographically secure pseudorandom bit generator (Definition 5.8) was introduced by Blum and Micali [166]. Blum and Micali also gave a formal description of the next-bit test (Definition 5.6), and presented the first cryptographically secure pseudorandom bit generator whose security is based on the discrete logarithm problem (see page 189). Universal tests were presented by Schrift and Shamir [1103] for verifying the assumed properties of a pseudorandom generator whose output sequences are not necessarily uniformly distributed.

The first provably secure pseudorandom *number* generator was proposed by Shamir [1112]. Shamir proved that predicting the next number of an output sequence of this generator is equivalent to inverting the RSA function. However, even though the numbers as a whole may be unpredictable, certain parts of the number (for example, its least significant bit) may be biased or predictable. Hence, Shamir's generator is not cryptographically secure in the sense of Definition 5.8.

§5.2

Agnew [17] proposed a VLSI implementation of a random bit generator consisting of two identical metal insulator semiconductor capacitors close to each other. The cells are charged over the same period of time, and then a 1 or 0 is assigned depending on which cell has a greater charge. Fairfield, Mortenson, and Coulthart [382] described an LSI random bit generator based on the frequency instability of a free running oscillator. Davis, Ihaka, and Fenstermacher [309] used the unpredictability of air turbulence occurring in a sealed disk drive as a random bit generator. The bits are extracted by measuring the variations in the time to access disk blocks. Fast Fourier Transform (FFT) techniques are then used to remove possible biases and correlations. A sample implementation generated 100 random bits per minute. For further guidance on hardware and software-based techniques for generating random bits, see RFC 1750 [1043].

The de-skewing technique of Example 5.10 is due to von Neumann [1223]. Elias [370] generalized von Neumann's technique to a more efficient scheme (one where fewer bits are discarded). Fast Fourier Transform techniques for removing biases and correlations are described by Brillinger [213]. For further ways of removing correlations, see Blum [161], Santha and Vazirani [1091], Vazirani [1217], and Chor and Goldreich [258].

§5.3

The idea of using a one-way function f for generating pseudorandom bit sequences is due to Shamir [1112]. Shamir illustrated why it is difficult to prove that such ad-hoc generators are cryptographically secure without imposing some further assumptions on f. Algorithm 5.11 is from Appendix C of the ANSI X9.17 standard [37]; it is one of the approved methods for pseudorandom bit generation listed in FIPS 186 [406]. Meyer and Matyas [859, pp.316-317] describe another DES-based pseudorandom bit generator whose output is intended for use as data-encrypting keys. The four algorithms of §5.3.2 for generating DSA parameters are from FIPS 186.

§5.4

Standard references on statistics include Hogg and Tanis [559] and Wackerly, Mendenhall, and Scheaffer [1226]. Tables 5.1 and 5.2 were generated using the Maple symbolic algebra system [240]. Golomb's randomness postulates (§5.4.3) were proposed by Golomb [498].

The five statistical tests for local randomness outlined in §5.4.4 are from Beker and Piper [84]. The serial test (§5.4.4(ii)) is due to Good [508]. It was generalized to subsequences of length greater than 2 by Marsaglia [782] who called it the *overlapping m-tuple test*, and later by Kimberley [674] who called it the *generalized serial test*. The underlying distribution theories of the serial test and the runs test (§5.4.4(iv)) were analyzed by Good [507] and Mood [897], respectively. Gustafson [531] considered alternative statistics for the runs test and the autocorrelation test (§5.4.4(v)).

There are numerous other statistical tests of local randomness. Many of these tests, including the gap test, coupon collector's test, permutation test, run test, maximum-of-t test, collision test, serial test, correlation test, and spectral test are described by Knuth [692]. The poker test as formulated by Knuth [692, p.62] is quite different from that of §5.4.4(iii). In the former, a sample sequence is divided into m-bit blocks, each of which is further subdivided into l-bit sub-blocks (for some divisor l of m). The number of m-bit blocks having r distinct l-bit sub-blocks ($1 \leq r \leq m/l$) is counted and compared to the corresponding expected numbers for random sequences. Erdmann [372] gives a detailed exposition of many of these tests, and applies them to sample output sequences of six pseudorandom bit generators. Gustafson et al. [533] describe a computer package which implements various statistical tests for assessing the strength of a pseudorandom bit generator. Gustafson, Dawson, and Golić [532] proposed a new *repetition test* which measures the number of repetitions of l-bit blocks. The test requires a count of the number of patterns repeated, but does not require the frequency of each pattern. For this reason, it is feasible to apply this test for larger values of l (e.g. $l = 64$) than would be permissible by the poker test or Maurer's universal statistical test (Algorithm 5.33). Two spectral tests have been developed, one based on the discrete Fourier transform by Gait [437], and one based on the Walsh transform by Yuen [1260]. For extensions of these spectral tests, see Erdmann [372] and Feldman [389].

FIPS 140-1 [401] specifies security requirements for the design and implementation of cryptographic modules, including random and pseudorandom bit generators, for protecting (U.S. government) unclassified information.

The universal statistical test (Algorithm 5.33) is due to Maurer [813] and was motivated by source coding algorithms of Elias [371] and Willems [1245]. The class of defects that the test is able to detect consists of those that can be modeled by an ergodic stationary source with limited memory; Maurer argues that this class includes the possible defects that could occur in a practical implementation of a random bit generator. Table 5.3 is due to Maurer [813], who provides derivations of formulae for the mean and variance of the statistic X_u.

§5.5

Blum and Micali [166] presented the following general construction for CSPRBGs. Let D be a finite set, and let $f\colon D \to D$ be a permutation that can be efficiently computed. Let $B\colon D \to \{0,1\}$ be a Boolean predicate with the property that $B(x)$ is hard to compute given only $x \in D$, however, $B(x)$ can be efficiently computed given $y = f^{-1}(x)$. The output sequence $z_1, z_2, \dots, z_l$ corresponding to a seed $x_0 \in D$ is obtained by computing $x_i = f(x_{i-1})$, $z_i = B(x_i)$, for $1 \leq i \leq l$. This generator can be shown to pass the next-bit test (Definition 5.6). Blum and Micali [166] proposed the first concrete instance of a CSPRBG, called the *Blum-Micali generator*. Using the notation introduced above, their method can be described as follows. Let p be a large prime, and α a generator of $\mathbb{Z}_p^*$. Define $D = \mathbb{Z}_p^* = \{1, 2, \dots, p-1\}$. The function $f : D \to D$ is defined by $f(x) = \alpha^x \bmod p$. The function $B : D \to \{0,1\}$ is defined by $B(x) = 1$ if $0 \leq \log_\alpha x \leq (p-1)/2$, and $B(x) = 0$ if $\log_\alpha x > (p-1)/2$. Assuming the intractability of the discrete logarithm problem in $\mathbb{Z}_p^*$ (§3.6; see also §3.9.1), the Blum-Micali generator was proven to satisfy the next-bit test. Long and Wigderson [772] improved the efficiency of the Blum-Micali generator by simultaneously extracting $O(\lg \lg p)$ bits (cf. §3.9.1) from each x_i. Kaliski [650, 651] modified the Blum-Micali generator so that the security depends on the discrete logarithm problem in the group of points on an elliptic curve defined over a finite field.

The RSA pseudorandom bit generator (Algorithm 5.35) and the improvement mentioned in Note 5.36 are due to Alexi et al. [23]. The Micali-Schnorr improvement of the RSA PRBG (Algorithm 5.37) is due to Micali and Schnorr [867], who also described a method that transforms any CSPRBG into one that can be accelerated by parallel evaluation. The method of parallelization is *perfect*: m parallel processors speed the generation of pseudorandom bits by a factor of m.

Algorithm 5.40 is due to Blum, Blum, and Shub [160], who showed that their pseudorandom bit generator is cryptographically secure assuming the intractability of the quadratic residuosity problem (§3.4). Vazirani and Vazirani [1218] established a stronger result regarding the security of this generator by proving it cryptographically secure under the weaker assumption that integer factorization is intractable. The improvement mentioned in

Note 5.41 is due to Vazirani and Vazirani. Alexi et al. [23] proved analogous results for the *modified-Rabin generator*, which differs as follows from the Blum-Blum-Shub generator: in step 3.1 of Algorithm 5.40, let $\overline{x} = x_{i-1}^2 \bmod n$; if $\overline{x} < n/2$, then $x_i = \overline{x}$; otherwise, $x_i = n - \overline{x}$.

Impagliazzo and Naor [569] devised efficient constructions for a CSPRBG and for a universal one-way hash function which are provably as secure as the subset sum problem. Fischer and Stern [411] presented a simple and efficient CSPRBG which is provably as secure as the *syndrome decoding problem*.

Yao [1258] showed how to obtain a CSPRBG using any one-way permutation. Levin [761] generalized this result and showed how to obtain a CSPRBG using any one-way function. For further refinements, see Goldreich, Krawczyk, and Luby [470], Impagliazzo, Levin, and Luby [568], and Håstad [545].

A *random function* $f: \{0,1\}^n \to \{0,1\}^n$ is a function which assigns independent and random values $f(x) \in \{0,1\}^n$ to all arguments $x \in \{0,1\}^n$. Goldreich, Goldwasser, and Micali [468] introduced a computational complexity measure of the randomness of functions. They defined a function to be *poly-random* if no polynomial-time algorithm can distinguish between values of the function and true random strings, even when the algorithm is permitted to select the arguments to the function. Goldreich, Goldwasser, and Micali presented an algorithm for constructing poly-random functions assuming the existence of one-way functions. This theory was applied by Goldreich, Goldwasser, and Micali [467] to develop provably secure protocols for the (essentially) storageless distribution of secret identification numbers, message authentication with timestamping, dynamic hashing, and identify friend or foe systems. Luby and Rackoff [776] showed how poly-random permutations can be efficiently constructed from poly-random functions. This result was used, together with some of the design principles of DES, to show how any CSPRBG can be used to construct a symmetric-key block cipher which is provably secure against chosen-plaintext attack. A simplified and generalized treatment of Luby and Rackoff's construction was given by Maurer [816].

Schnorr [1096] used Luby and Rackoff's poly-random permutation generator to construct a pseudorandom bit generator that was claimed to pass all statistical tests depending only on a small fraction of the output sequence, even when infinite computational resources are available. Rueppel [1079] showed that this claim is erroneous, and demonstrated that the generator can be distinguished from a truly random bit generator using only a small number of output bits. Maurer and Massey [821] extended Schnorr's work, and proved the existence of pseudorandom bit generators that pass all statistical tests depending only on a small fraction of the output sequence, even when infinite computational resources are available. The security of the generators does not rely on any unproved hypothesis, but rather on the assumption that the adversary can access only a limited number of bits of the generated sequence. This work is primarily of theoretical interest since no such polynomial-time generators are known.

Chapter 6

Stream Ciphers

Contents in Brief

6.1 Introduction

Stream ciphers are an important class of encryption algorithms. They encrypt individual characters (usually binary digits) of a plaintext message one at a time, using an encryption transformation which varies with time. By contrast, *block ciphers* (Chapter 7) tend to simultaneously encrypt groups of characters of a plaintext message using a fixed encryption transformation. Stream ciphers are generally faster than block ciphers in hardware, and have less complex hardware circuitry. They are also more appropriate, and in some cases mandatory (e.g., in some telecommunications applications), when buffering is limited or when characters must be individually processed as they are received. Because they have limited or no error propagation, stream ciphers may also be advantageous in situations where transmission errors are highly probable.

There is a vast body of theoretical knowledge on stream ciphers, and various design principles for stream ciphers have been proposed and extensively analyzed. However, there are relatively few fully-specified stream cipher algorithms in the open literature. This unfortunate state of affairs can partially be explained by the fact that most stream ciphers used in practice tend to be proprietary and confidential. By contrast, numerous concrete block cipher proposals have been published, some of which have been standardized or placed in the public domain. Nevertheless, because of their significant advantages, stream ciphers are widely used today, and one can expect increasingly more concrete proposals in the coming years.

Chapter outline

The remainder of §6.1 introduces basic concepts relevant to stream ciphers. Feedback shift registers, in particular linear feedback shift registers (LFSRs), are the basic building block in most stream ciphers that have been proposed; they are studied in §6.2. Three general techniques for utilizing LFSRs in the construction of stream ciphers are presented in §6.3: using a nonlinear combining function on the outputs of several LFSRs (§6.3.1), using a nonlinear filtering function on the contents of a single LFSR (§6.3.2), and using the output of one (or more) LFSRs to control the clock of one (or more) other LFSRs (§6.3.3). Two concrete proposals for clock-controlled generators, the alternating step generator and the shrinking generator are presented in §6.3.3. §6.4 presents a stream cipher not based on LFSRs, namely SEAL. §6.5 concludes with references and further chapter notes.

6.1.1 Classification

Stream ciphers can be either symmetric-key or public-key. The focus of this chapter is symmetric-key stream ciphers; the Blum-Goldwasser probabilistic public-key encryption scheme (§8.7.2) is an example of a public-key stream cipher.

6.1 Note (*block vs. stream ciphers*) Block ciphers process plaintext in relatively large blocks (e.g., $n \geq 64$ bits). The same function is used to encrypt successive blocks; thus (pure) block ciphers are *memoryless*. In contrast, stream ciphers process plaintext in blocks as small as a single bit, and the encryption function may vary as plaintext is processed; thus stream ciphers are said to have memory. They are sometimes called *state ciphers* since encryption depends on not only the key and plaintext, but also on the current state. This distinction between block and stream ciphers is not definitive (see Remark 7.25); adding a small amount of memory to a block cipher (as in the CBC mode) results in a stream cipher with large blocks.

(i) The one-time pad

Recall (Definition 1.39) that a *Vernam cipher* over the binary alphabet is defined by

$$c_i = m_i \oplus k_i \ \text{ for } i = 1, 2, 3 \ldots ,$$

where $m_1, m_2, m_3, \ldots$ are the plaintext digits, $k_1, k_2, k_3, \ldots$ (the *keystream*) are the key digits, $c_1, c_2, c_3, \ldots$ are the ciphertext digits, and $\oplus$ is the XOR function (bitwise addition modulo 2). Decryption is defined by $m_i = c_i \oplus k_i$. If the keystream digits are generated independently and randomly, the Vernam cipher is called a *one-time pad*, and is unconditionally secure (§1.13.3(i)) against a ciphertext-only attack. More precisely, if M, C, and K are random variables respectively denoting the plaintext, ciphertext, and secret key, and if $H()$ denotes the entropy function (Definition 2.39), then $H(M|C) = H(M)$. Equivalently, $I(M; C) = 0$ (see Definition 2.45): the ciphertext contributes no information about the plaintext.

Shannon proved that a necessary condition for a symmetric-key encryption scheme to be unconditionally secure is that $H(K) \geq H(M)$. That is, the uncertainty of the secret key must be at least as great as the uncertainty of the plaintext. If the key has bitlength k, and the key bits are chosen randomly and independently, then $H(K) = k$, and Shannon's necessary condition for unconditional security becomes $k \geq H(M)$. The one-time pad is unconditionally secure regardless of the statistical distribution of the plaintext, and is optimal in the sense that its key is the smallest possible among all symmetric-key encryption schemes having this property.

An obvious drawback of the one-time pad is that the key should be as long as the plaintext, which increases the difficulty of key distribution and key management. This motivates the design of stream ciphers where the keystream is *pseudorandomly* generated from a smaller secret key, with the intent that the keystream appears random to a computationally bounded adversary. Such stream ciphers do not offer unconditional security (since $H(K) << H(M)$), but the hope is that they are computationally secure (§1.13.3(iv)).

Stream ciphers are commonly classified as being *synchronous* or *self-synchronizing*.

(ii) Synchronous stream ciphers

6.2 Definition A *synchronous* stream cipher is one in which the keystream is generated independently of the plaintext message and of the ciphertext.

The encryption process of a synchronous stream cipher can be described by the equations

$$\sigma_{i+1} = f(\sigma_i, k),$$
$$z_i = g(\sigma_i, k),$$
$$c_i = h(z_i, m_i),$$

where σ_0 is the *initial state* and may be determined from the key k, f is the *next-state function*, g is the function which produces the *keystream* z_i, and h is the *output function* which combines the keystream and plaintext m_i to produce ciphertext c_i. The encryption and decryption processes are depicted in Figure 6.1. The OFB mode of a block cipher (see §7.2.2(iv)) is an example of a synchronous stream cipher.

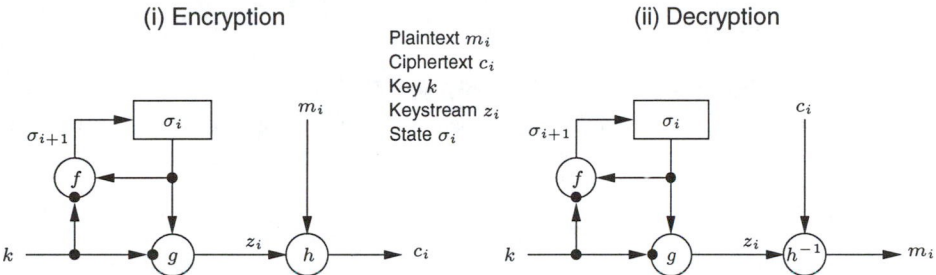

Figure 6.1: *General model of a synchronous stream cipher.*

6.3 Note (*properties of synchronous stream ciphers*)

(i) *synchronization requirements.* In a synchronous stream cipher, both the sender and receiver must be *synchronized* – using the same key and operating at the same position (state) within that key – to allow for proper decryption. If synchronization is lost due to ciphertext digits being inserted or deleted during transmission, then decryption fails and can only be restored through additional techniques for re-synchronization. Techniques for re-synchronization include re-initialization, placing special markers at regular intervals in the ciphertext, or, if the plaintext contains enough redundancy, trying all possible keystream offsets.

(ii) *no error propagation.* A ciphertext digit that is modified (but not deleted) during transmission does not affect the decryption of other ciphertext digits.

(iii) *active attacks.* As a consequence of property (i), the insertion, deletion, or replay of ciphertext digits by an active adversary causes immediate loss of synchronization, and hence might possibly be detected by the decryptor. As a consequence of property (ii), an active adversary might possibly be able to make changes to selected ciphertext digits, and know exactly what affect these changes have on the plaintext. This illustrates that additional mechanisms must be employed in order to provide data origin authentication and data integrity guarantees (see §9.5.4).

Most of the stream ciphers that have been proposed to date in the literature are additive stream ciphers, which are defined below.

6.4 Definition A *binary additive stream cipher* is a synchronous stream cipher in which the keystream, plaintext, and ciphertext digits are binary digits, and the output function h is the XOR function.

Binary additive stream ciphers are depicted in Figure 6.2. Referring to Figure 6.2, the *keystream generator* is composed of the next-state function f and the function g (see Figure 6.1), and is also known as the *running key generator*.

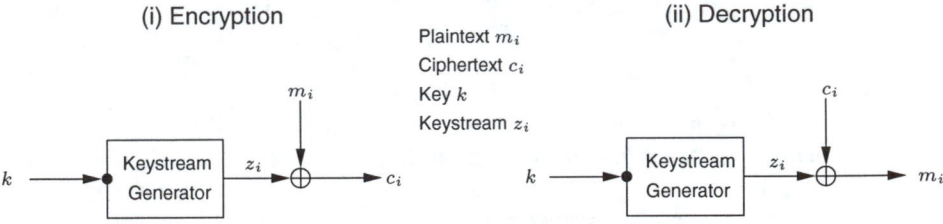

Figure 6.2: *General model of a binary additive stream cipher.*

(iii) Self-synchronizing stream ciphers

6.5 Definition A *self-synchronizing* or *asynchronous* stream cipher is one in which the keystream is generated as a function of the key and a fixed number of previous ciphertext digits.

The encryption function of a self-synchronizing stream cipher can be described by the equations

$$\begin{aligned} \sigma_i &= (c_{i-t}, c_{i-t+1}, \ldots, c_{i-1}), \\ z_i &= g(\sigma_i, k), \\ c_i &= h(z_i, m_i), \end{aligned}$$

where $\sigma_0 = (c_{-t}, c_{-t+1}, \ldots, c_{-1})$ is the (non-secret) *initial state*, k is the *key*, g is the function which produces the *keystream* z_i, and h is the *output function* which combines the keystream and plaintext m_i to produce ciphertext c_i. The encryption and decryption processes are depicted in Figure 6.3. The most common presently-used self-synchronizing stream ciphers are based on block ciphers in 1-bit cipher feedback mode (see §7.2.2(iii)).

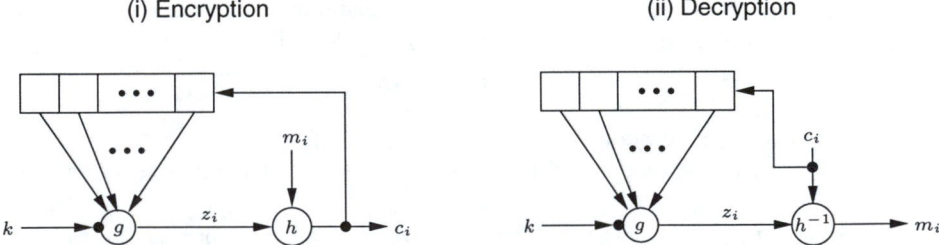

Figure 6.3: *General model of a self-synchronizing stream cipher.*

6.6 Note (*properties of self-synchronizing stream ciphers*)

    (i) *self-synchronization.* Self-synchronization is possible if ciphertext digits are deleted or inserted, because the decryption mapping depends only on a fixed number of preceding ciphertext characters. Such ciphers are capable of re-establishing proper decryption automatically after loss of synchronization, with only a fixed number of plaintext characters unrecoverable.

    (ii) *limited error propagation.* Suppose that the state of a self-synchronization stream cipher depends on t previous ciphertext digits. If a single ciphertext digit is modified (or even deleted or inserted) during transmission, then decryption of up to t subsequent ciphertext digits may be incorrect, after which correct decryption resumes.

    (iii) *active attacks.* Property (ii) implies that any modification of ciphertext digits by an active adversary causes several other ciphertext digits to be decrypted incorrectly, thereby improving (compared to synchronous stream ciphers) the likelihood of being detected by the decryptor. As a consequence of property (i), it is more difficult (than for synchronous stream ciphers) to detect insertion, deletion, or replay of ciphertext digits by an active adversary. This illustrates that additional mechanisms must be employed in order to provide data origin authentication and data integrity guarantees (see §9.5.4).

    (iv) *diffusion of plaintext statistics.* Since each plaintext digit influences the entire following ciphertext, the statistical properties of the plaintext are dispersed through the ciphertext. Hence, self-synchronizing stream ciphers may be more resistant than synchronous stream ciphers against attacks based on plaintext redundancy.

6.2 Feedback shift registers

Feedback shift registers, in particular linear feedback shift registers, are the basic components of many keystream generators. §6.2.1 introduces linear feedback shift registers. The linear complexity of binary sequences is studied in §6.2.2, while the Berlekamp-Massey algorithm for computing it is presented in §6.2.3. Finally, nonlinear feedback shift registers are discussed in §6.2.4.

6.2.1 Linear feedback shift registers

Linear feedback shift registers (LFSRs) are used in many of the keystream generators that have been proposed in the literature. There are several reasons for this:

    1. LFSRs are well-suited to hardware implementation;

    2. they can produce sequences of large period (Fact 6.12);

    3. they can produce sequences with good statistical properties (Fact 6.14); and

    4. because of their structure, they can be readily analyzed using algebraic techniques.

6.7 Definition A *linear feedback shift register* (LFSR) of length L consists of L *stages* (or *delay elements*) numbered $0, 1, \dots, L - 1$, each capable of storing one bit and having one input and one output; and a clock which controls the movement of data. During each unit of time the following operations are performed:

    (i) the content of stage 0 is output and forms part of the *output sequence*;

(ii) the content of stage i is moved to stage $i - 1$ for each i, $1 \leq i \leq L - 1$; and

(iii) the new content of stage $L - 1$ is the *feedback bit* s_j which is calculated by adding together modulo 2 the previous contents of a fixed subset of stages $0, 1, \ldots, L - 1$.

Figure 6.4 depicts an LFSR. Referring to the figure, each c_i is either 0 or 1; the closed semi-circles are AND gates; and the feedback bit s_j is the modulo 2 sum of the contents of those stages i, $0 \leq i \leq L - 1$, for which $c_{L-i} = 1$.

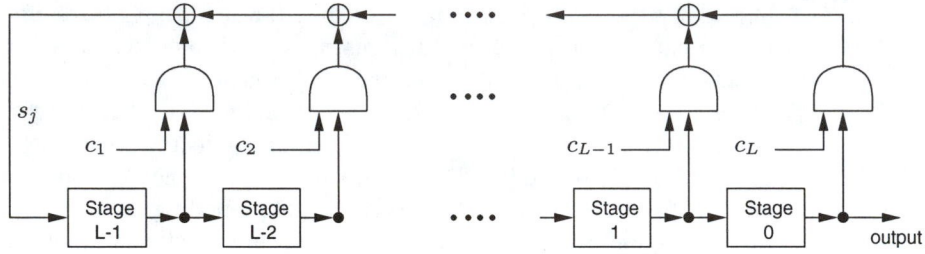

Figure 6.4: *A linear feedback shift register (LFSR) of length L.*

6.8 Definition The LFSR of Figure 6.4 is denoted $\langle L, C(D) \rangle$, where $C(D) = 1 + c_1 D + c_2 D^2 + \cdots + c_L D^L \in \mathbb{Z}_2[D]$ is the *connection polynomial*. The LFSR is said to be *non-singular* if the degree of $C(D)$ is L (that is, $c_L = 1$). If the initial content of stage i is $s_i \in \{0, 1\}$ for each i, $0 \leq i \leq L - 1$, then $[s_{L-1}, \ldots, s_1, s_0]$ is called the *initial state* of the LFSR.

6.9 Fact If the initial state of the LFSR in Figure 6.4 is $[s_{L-1}, \ldots, s_1, s_0]$, then the output sequence $s = s_0, s_1, s_2, \ldots$ is uniquely determined by the following recursion:

$$s_j = (c_1 s_{j-1} + c_2 s_{j-2} + \cdots + c_L s_{j-L}) \bmod 2 \quad \text{for } j \geq L.$$

6.10 Example (*output sequence of an LFSR*) Consider the LFSR $\langle 4, 1 + D + D^4 \rangle$ depicted in Figure 6.5. If the initial state of the LFSR is $[0, 0, 0, 0]$, the output sequence is the zero sequence. The following tables show the contents of the stages D_3, D_2, D_1, D_0 at the end of each unit of time t when the initial state is $[0, 1, 1, 0]$.

t	D_3	D_2	D_1	D_0
0	0	1	1	0
1	0	0	1	1
2	1	0	0	1
3	0	1	0	0
4	0	0	1	0
5	0	0	0	1
6	1	0	0	0
7	1	1	0	0

t	D_3	D_2	D_1	D_0
8	1	1	1	0
9	1	1	1	1
10	0	1	1	1
11	1	0	1	1
12	0	1	0	1
13	1	0	1	0
14	1	1	0	1
15	0	1	1	0

The output sequence is $s = 0, 1, 1, 0, 0, 1, 0, 0, 0, 1, 1, 1, 1, 0, 1, \ldots$, and is periodic with period 15 (see Definition 5.25). $\square$

The significance of an LFSR being non-singular is explained by Fact 6.11.

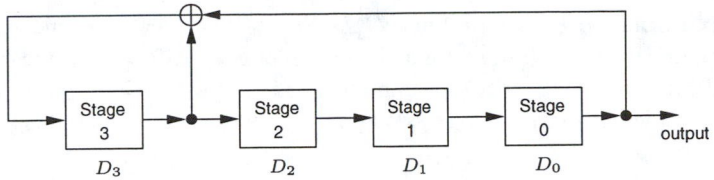

Figure 6.5: *The LFSR $\langle 4, 1 + D + D^4 \rangle$ of Example 6.10.*

6.11 Fact Every output sequence (i.e., for all possible initial states) of an LFSR $\langle L, C(D) \rangle$ is periodic if and only if the connection polynomial $C(D)$ has degree L.

If an LFSR $\langle L, C(D) \rangle$ is *singular* (i.e., $C(D)$ has degree less than L), then not all output sequences are periodic. However, the output sequences are *ultimately periodic*; that is, the sequences obtained by ignoring a certain finite number of terms at the beginning are periodic. For the remainder of this chapter, it will be assumed that all LFSRs are non-singular. Fact 6.12 determines the periods of the output sequences of some special types of non-singular LFSRs.

6.12 Fact (*periods of LFSR output sequences*) Let $C(D) \in \mathbb{Z}_2[D]$ be a connection polynomial of degree L.

(i) If $C(D)$ is irreducible over $\mathbb{Z}_2$ (see Definition 2.190), then each of the $2^L - 1$ non-zero initial states of the non-singular LFSR $\langle L, C(D) \rangle$ produces an output sequence with period equal to the least positive integer N such that $C(D)$ divides $1 + D^N$ in $\mathbb{Z}_2[D]$. (Note: it is always the case that this N is a divisor of $2^L - 1$.)

(ii) If $C(D)$ is a primitive polynomial (see Definition 2.228), then each of the $2^L - 1$ non-zero initial states of the non-singular LFSR $\langle L, C(D) \rangle$ produces an output sequence with maximum possible period $2^L - 1$.

A method for generating primitive polynomials over $\mathbb{Z}_2$ uniformly at random is given in Algorithm 4.78. Table 4.8 lists a primitive polynomial of degree m over $\mathbb{Z}_2$ for each m, $1 \leq m \leq 229$. Fact 6.12(ii) motivates the following definition.

6.13 Definition If $C(D) \in \mathbb{Z}_2[D]$ is a primitive polynomial of degree L, then $\langle L, C(D) \rangle$ is called a *maximum-length* LFSR. The output of a maximum-length LFSR with non-zero initial state is called an *m-sequence*.

Fact 6.14 demonstrates that the output sequences of maximum-length LFSRs have good statistical properties.

6.14 Fact (*statistical properties of m-sequences*) Let s be an m-sequence that is generated by a maximum-length LFSR of length L.

(i) Let k be an integer, $1 \leq k \leq L$, and let $\overline{s}$ be any subsequence of s of length $2^L + k - 2$. Then each non-zero sequence of length k appears exactly 2^{L-k} times as a subsequence of $\overline{s}$. Furthermore, the zero sequence of length k appears exactly $2^{L-k} - 1$ times as a subsequence of $\overline{s}$. In other words, the distribution of patterns having fixed length of at most L is almost uniform.

(ii) s satisfies Golomb's randomness postulates (§5.4.3). That is, every m-sequence is also a pn-sequence (see Definition 5.29).

6.15 Example (*m-sequence*) Since $C(D) = 1 + D + D^4$ is a primitive polynomial over $\mathbb{Z}_2$, the LFSR $\langle 4, 1 + D + D^4 \rangle$ is a maximum-length LFSR. Hence, the output sequence of this LFSR is an *m*-sequence of maximum possible period $N = 2^4 - 1 = 15$ (cf. Example 6.10). Example 5.30 verifies that this output sequence satisfies Golomb's randomness properties.

$\square$

6.2.2 Linear complexity

This subsection summarizes selected results about the linear complexity of sequences. All sequences are assumed to be binary sequences. Notation: s denotes an infinite sequence whose terms are $s_0, s_1, s_2, \ldots$; s^n denotes a finite sequence of length n whose terms are $s_0, s_1, \ldots, s_{n-1}$ (see Definition 5.24).

6.16 Definition An LFSR is said to *generate* a sequence s if there is some initial state for which the output sequence of the LFSR is s. Similarly, an LFSR is said to *generate* a finite sequence s^n if there is some initial state for which the output sequence of the LFSR has s^n as its first n terms.

6.17 Definition The *linear complexity* of an infinite binary sequence s, denoted $L(s)$, is defined as follows:

(i) if s is the zero sequence $s = 0, 0, 0, \ldots$, then $L(s) = 0$;
(ii) if no LFSR generates s, then $L(s) = \infty$;
(iii) otherwise, $L(s)$ is the length of the shortest LFSR that generates s.

6.18 Definition The *linear complexity* of a finite binary sequence s^n, denoted $L(s^n)$, is the length of the shortest LFSR that generates a sequence having s^n as its first n terms.

Facts 6.19 – 6.22 summarize some basic results about linear complexity.

6.19 Fact (*properties of linear complexity*) Let s and t be binary sequences.

(i) For any $n \geq 1$, the linear complexity of the subsequence s^n satisfies $0 \leq L(s^n) \leq n$.
(ii) $L(s^n) = 0$ if and only if s^n is the zero sequence of length n.
(iii) $L(s^n) = n$ if and only if $s^n = 0, 0, 0, \ldots, 0, 1$.
(iv) If s is periodic with period N, then $L(s) \leq N$.
(v) $L(s \oplus t) \leq L(s) + L(t)$, where $s \oplus t$ denotes the bitwise XOR of s and t.

6.20 Fact If the polynomial $C(D) \in \mathbb{Z}_2[D]$ is irreducible over $\mathbb{Z}_2$ and has degree L, then each of the $2^L - 1$ non-zero initial states of the non-singular LFSR $\langle L, C(D) \rangle$ produces an output sequence with linear complexity L.

6.21 Fact (*expectation and variance of the linear complexity of a random sequence*) Let s^n be chosen uniformly at random from the set of all binary sequences of length n, and let $L(s^n)$ be the linear complexity of s^n. Let $B(n)$ denote the parity function: $B(n) = 0$ if n is even; $B(n) = 1$ if n is odd.

(i) The expected linear complexity of s^n is

$$E(L(s^n)) = \frac{n}{2} + \frac{4 + B(n)}{18} - \frac{1}{2^n}\left(\frac{n}{3} + \frac{2}{9}\right).$$

Hence, for moderately large n, $E(L(s^n)) \approx \frac{n}{2} + \frac{2}{9}$ if n is even, and $E(L(s^n)) \approx \frac{n}{2} + \frac{5}{18}$ if n is odd.

(ii) The variance of the linear complexity of s^n is $\mathrm{Var}(L(s^n)) =$

$$\frac{86}{81} - \frac{1}{2^n}\left(\frac{14 - B(n)}{27}n + \frac{82 - 2B(n)}{81}\right) - \frac{1}{2^{2n}}\left(\frac{1}{9}n^2 + \frac{4}{27}n + \frac{4}{81}\right).$$

Hence, $\mathrm{Var}(L(s^n)) \approx \frac{86}{81}$ for moderately large n.

6.22 Fact (*expectation of the linear complexity of a random periodic sequence*) Let s^n be chosen uniformly at random from the set of all binary sequences of length n, where $n = 2^t$ for some fixed $t \geq 1$, and let s be the n-periodic infinite sequence obtained by repeating the sequence s^n. Then the expected linear complexity of s is $E(L(s^n)) = n - 1 + 2^{-n}$.

The linear complexity profile of a binary sequence is introduced next.

6.23 Definition Let $s = s_0, s_1, \ldots$ be a binary sequence, and let L_N denote the linear complexity of the subsequence $s^N = s_0, s_1, \ldots, s_{N-1}$, $N \geq 0$. The sequence $L_1, L_2, \ldots$ is called the *linear complexity profile of* s. Similarly, if $s^n = s_0, s_1, \ldots, s_{n-1}$ is a finite binary sequence, the sequence $L_1, L_2, \ldots, L_n$ is called the *linear complexity profile of* s^n.

The linear complexity profile of a sequence can be computed using the Berlekamp-Massey algorithm (Algorithm 6.30); see also Note 6.31. The following properties of the linear complexity profile can be deduced from Fact 6.29.

6.24 Fact (*properties of linear complexity profile*) Let $L_1, L_2, \ldots$ be the linear complexity profile of a sequence $s = s_0, s_1, \ldots$.

(i) If $j > i$, then $L_j \geq L_i$.

(ii) $L_{N+1} > L_N$ is possible only if $L_N \leq N/2$.

(iii) If $L_{N+1} > L_N$, then $L_{N+1} + L_N = N + 1$.

The linear complexity profile of a sequence s can be graphed by plotting the points (N, L_N), $N \geq 1$, in the $N \times L$ plane and joining successive points by a horizontal line followed by a vertical line, if necessary (see Figure 6.6). Fact 6.24 can then be interpreted as saying that the graph of a linear complexity profile is non-decreasing. Moreover, a (vertical) jump in the graph can only occur from below the line $L = N/2$; if a jump occurs, then it is symmetric about this line. Fact 6.25 shows that the expected linear complexity of a random sequence should closely follow the line $L = N/2$.

6.25 Fact (*expected linear complexity profile of a random sequence*) Let $s = s_0, s_1, \ldots$ be a random sequence, and let L_N be the linear complexity of the subsequence $s^N = s_0, s_1, \ldots,$ s_{N-1} for each $N \geq 1$. For any fixed index $N \geq 1$, the expected smallest j for which $L_{N+j} > L_N$ is 2 if $L_N \leq N/2$, or $2 + 2L_N - N$ if $L_N > N/2$. Moreover, the expected increase in linear complexity is 2 if $L_N \geq N/2$, or $N - 2L_N + 2$ if $L_N < N/2$.

6.26 Example (*linear complexity profile*) Consider the 20-periodic sequence s with cycle

$$s^{20} = 1, 0, 0, 1, 0, 0, 1, 1, 1, 1, 0, 0, 0, 1, 0, 0, 1, 1, 1, 0.$$

The linear complexity profile of s is $1, 1, 1, 3, 3, 3, 3, 5, 5, 5, 6, 6, 6, 8, 8, 8, 9, 9, 10, 10, 11,$ $11, 11, 11, 14, 14, 14, 14, 15, 15, 15, 17, 17, 17, 18, 18, 19, 19, 19, 19, \ldots$. Figure 6.6 shows the graph of the linear complexity profile of s. □

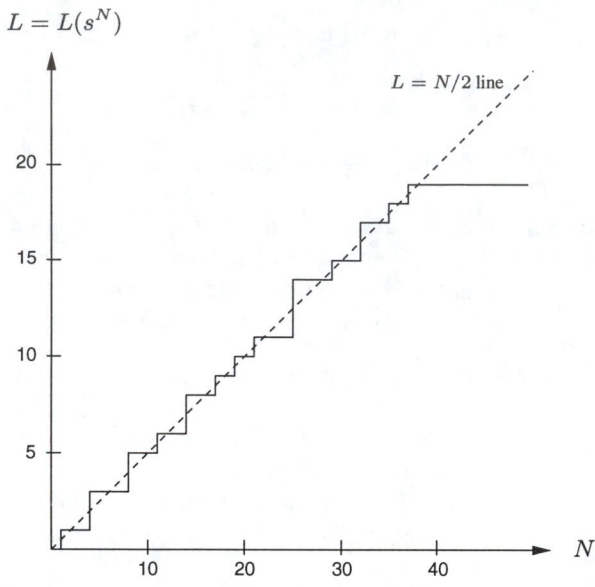

Figure 6.6: *Linear complexity profile of the 20-periodic sequence of Example 6.26.*

As is the case with all statistical tests for randomness (cf. §5.4), the condition that a sequence s have a linear complexity profile that closely resembles that of a random sequence is *necessary* but not *sufficient* for s to be considered random. This point is illustrated in the following example.

6.27 Example (*limitations of the linear complexity profile*) The linear complexity profile of the sequence s defined as

$$s_i = \begin{cases} 1, & \text{if } i = 2^j - 1 \text{ for some } j \geq 0, \\ 0, & \text{otherwise,} \end{cases}$$

follows the line $L = N/2$ as closely as possible. That is, $L(s^N) = \lfloor (N+1)/2 \rfloor$ for all $N \geq 1$. However, the sequence s is clearly non-random. $\square$

6.2.3 Berlekamp-Massey algorithm

The Berlekamp-Massey algorithm (Algorithm 6.30) is an efficient algorithm for determining the linear complexity of a finite binary sequence s^n of length n (see Definition 6.18). The algorithm takes n iterations, with the Nth iteration computing the linear complexity of the subsequence s^N consisting of the first N terms of s^n. The theoretical basis for the algorithm is Fact 6.29.

6.28 Definition Consider the finite binary sequence $s^{N+1} = s_0, s_1, \ldots, s_{N-1}, s_N$. For $C(D) = 1 + c_1 D + \cdots + c_L D^L$, let $\langle L, C(D) \rangle$ be an LFSR that generates the subsequence $s^N = s_0, s_1, \ldots, s_{N-1}$. The *next discrepancy* d_N is the difference between s_N and the $(N+1)^{\text{st}}$ term generated by the LFSR: $d_N = (s_N + \sum_{i=1}^{L} c_i s_{N-i}) \bmod 2$.

6.29 Fact Let $s^N = s_0, s_1, \ldots, s_{N-1}$ be a finite binary sequence of linear complexity $L = L(s^N)$, and let $\langle L, C(D) \rangle$ be an LFSR which generates s^N.

(i) The LFSR $\langle L, C(D) \rangle$ also generates $s^{N+1} = s_0, s_1, \ldots, s_{N-1}, s_N$ if and only if the next discrepancy d_N is equal to 0.

(ii) If $d_N = 0$, then $L(s^{N+1}) = L$.

(iii) Suppose $d_N = 1$. Let m the largest integer $< N$ such that $L(s^m) < L(s^N)$, and let $\langle L(s^m), B(D) \rangle$ be an LFSR of length $L(s^m)$ which generates s^m. Then $\langle L', C'(D) \rangle$ is an LFSR of smallest length which generates s^{N+1}, where

$$L' = \begin{cases} L, & \text{if } L > N/2, \\ N+1-L, & \text{if } L \leq N/2, \end{cases}$$

and $C'(D) = C(D) + B(D) \cdot D^{N-m}$.

6.30 Algorithm Berlekamp-Massey algorithm

INPUT: a binary sequence $s^n = s_0, s_1, s_2, \ldots, s_{n-1}$ of length n.

OUTPUT: the linear complexity $L(s^n)$ of s^n, $0 \leq L(s^n) \leq n$.

1. *Initialization.* $C(D) \leftarrow 1$, $L \leftarrow 0$, $m \leftarrow -1$, $B(D) \leftarrow 1$, $N \leftarrow 0$.
2. While $(N < n)$ do the following:

    2.1 *Compute the next discrepancy d.* $d \leftarrow (s_N + \sum_{i=1}^{L} c_i s_{N-i}) \bmod 2$.

    2.2 If $d = 1$ then do the following:
    $$T(D) \leftarrow C(D), \quad C(D) \leftarrow C(D) + B(D) \cdot D^{N-m}.$$
    If $L \leq N/2$ then $L \leftarrow N+1-L$, $m \leftarrow N$, $B(D) \leftarrow T(D)$.

    2.3 $N \leftarrow N+1$.
3. Return(L).

6.31 Note (*intermediate results in Berlekamp-Massey algorithm*) At the end of each iteration of step 2, $\langle L, C(D) \rangle$ is an LFSR of smallest length which generates s^N. Hence, Algorithm 6.30 can also be used to compute the linear complexity profile (Definition 6.23) of a finite sequence.

6.32 Fact The running time of the Berlekamp-Massey algorithm (Algorithm 6.30) for determining the linear complexity of a binary sequence of bitlength n is $O(n^2)$ bit operations.

6.33 Example (*Berlekamp-Massey algorithm*) Table 6.1 shows the steps of Algorithm 6.30 for computing the linear complexity of the binary sequence $s^n = 0, 0, 1, 1, 0, 1, 1, 1, 0$ of length $n = 9$. This sequence is found to have linear complexity 5, and an LFSR which generates it is $\langle 5, 1 + D^3 + D^5 \rangle$. $\square$

6.34 Fact Let s^n be a finite binary sequence of length n, and let the linear complexity of s^n be L. Then there is a unique LFSR of length L which generates s^n if and only if $L \leq \frac{n}{2}$.

An important consequence of Fact 6.34 and Fact 6.24(iii) is the following.

6.35 Fact Let s be an (infinite) binary sequence of linear complexity L, and let t be a (finite) subsequence of s of length at least $2L$. Then the Berlekamp-Massey algorithm with input t determines an LFSR of length L which generates s.

s_N	d	$T(D)$	$C(D)$	L	m	$B(D)$	N
—	—	—	1	0	-1	1	0
0	0	—	1	0	-1	1	1
0	0	—	1	0	-1	1	2
1	1	1	$1 + D^3$	3	2	1	3
1	1	$1 + D^3$	$1 + D + D^3$	3	2	1	4
0	1	$1 + D + D^3$	$1 + D + D^2 + D^3$	3	2	1	5
1	1	$1 + D + D^2 + D^3$	$1 + D + D^2$	3	2	1	6
1	0	$1 + D + D^2 + D^3$	$1 + D + D^2$	3	2	1	7
1	1	$1 + D + D^2$	$1 + D + D^2 + D^5$	5	7	$1 + D + D^2$	8
0	1	$1 + D + D^2 + D^5$	$1 + D^3 + D^5$	5	7	$1 + D + D^2$	9

Table 6.1: *Steps of the Berlekamp-Massey algorithm of Example 6.33.*

6.2.4 Nonlinear feedback shift registers

This subsection summarizes selected results about nonlinear feedback shift registers. A function with n binary inputs and one binary output is called a *Boolean function* of n variables; there are 2^{2^n} different Boolean functions of n variables.

6.36 Definition A (general) *feedback shift register* (FSR) of length L consists of L *stages* (or *delay elements*) numbered $0, 1, \dots, L - 1$, each capable of storing one bit and having one input and one output, and a clock which controls the movement of data. During each unit of time the following operations are performed:

(i) the content of stage 0 is output and forms part of the *output sequence*;

(ii) the content of stage i is moved to stage $i - 1$ for each i, $1 \le i \le L - 1$; and

(iii) the new content of stage $L - 1$ is the *feedback bit* $s_j = f(s_{j-1}, s_{j-2}, \dots, s_{j-L})$, where the *feedback function* f is a Boolean function and s_{j-i} is the previous content of stage $L - i$, $1 \le i \le L$.

If the initial content of stage i is $s_i \in \{0, 1\}$ for each $0 \le i \le L - 1$, then $[s_{L-1}, \dots, s_1, s_0]$ is called the *initial state* of the FSR.

Figure 6.7 depicts an FSR. Note that if the feedback function f is a linear function, then the FSR is an LFSR (Definition 6.7). Otherwise, the FSR is called a *nonlinear* FSR.

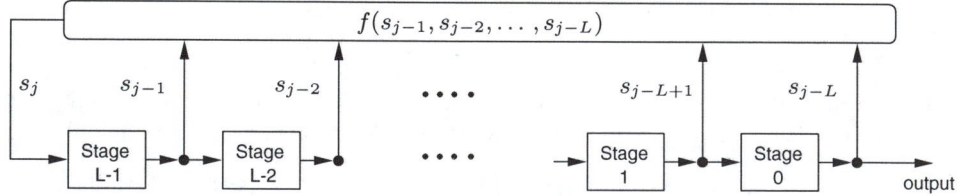

Figure 6.7: *A feedback shift register (FSR) of length L.*

6.37 Fact If the initial state of the FSR in Figure 6.7 is $[s_{L-1}, \dots, s_1, s_0]$, then the output sequence $s = s_0, s_1, s_2, \dots$ is uniquely determined by the following recursion:

$$s_j = f(s_{j-1}, s_{j-2}, \dots, s_{j-L}) \text{ for } j \ge L.$$

6.38 Definition An FSR is said to be *non-singular* if and only if every output sequence of the FSR (i.e., for all possible initial states) is periodic.

6.39 Fact An FSR with feedback function $f(s_{j-1}, s_{j-2}, \ldots, s_{j-L})$ is non-singular if and only if f is of the form $f = s_{j-L} \oplus g(s_{j-1}, s_{j-2}, \ldots, s_{j-L+1})$ for some Boolean function g.

The period of the output sequence of a non-singular FSR of length L is at most 2^L.

6.40 Definition If the period of the output sequence (for any initial state) of a non-singular FSR of length L is 2^L, then the FSR is called a *de Bruijn FSR*, and the output sequence is called a *de Bruijn sequence*.

6.41 Example (*de Bruijn sequence*) Consider the FSR of length 3 with nonlinear feedback function $f(x_1, x_2, x_3) = 1 \oplus x_2 \oplus x_3 \oplus x_1 x_2$. The following tables show the contents of the 3 stages of the FSR at the end of each unit of time t when the initial state is $[0, 0, 0]$.

t	Stage 2	Stage 1	Stage 0
0	0	0	0
1	1	0	0
2	1	1	0
3	1	1	1

t	Stage 2	Stage 1	Stage 0
4	0	1	1
5	1	0	1
6	0	1	0
7	0	0	1

The output sequence is the de Bruijn sequence with cycle $0, 0, 0, 1, 1, 1, 0, 1$. □

Fact 6.42 demonstrates that the output sequence of de Bruijn FSRs have good statistical properties (compare with Fact 6.14(i)).

6.42 Fact (*statistical properties of de Bruijn sequences*) Let s be a de Bruijn sequence that is generated by a de Bruijn FSR of length L. Let k be an integer, $1 \leq k \leq L$, and let $\bar{s}$ be any subsequence of s of length $2^L + k - 1$. Then each sequence of length k appears exactly 2^{L-k} times as a subsequence of $\bar{s}$. In other words, the distribution of patterns having fixed length of at most L is uniform.

6.43 Note (*converting a maximum-length LFSR to a de Bruijn FSR*) Let R_1 be a maximum-length LFSR of length L with (linear) feedback function $f(s_{j-1}, s_{j-2}, \ldots, s_{j-L})$. Then the FSR R_2 with feedback function $g(s_{j-1}, s_{j-2}, \ldots, s_{j-L}) = f \oplus \bar{s}_{j-1} \bar{s}_{j-2} \cdots \bar{s}_{j-L+1}$ is a de Bruijn FSR. Here, $\bar{s}_i$ denotes the complement of s_i. The output sequence of R_2 is obtained from that of R_1 by simply adding a 0 to the end of each subsequence of $L - 1$ 0's occurring in the output sequence of R_1.

6.3 Stream ciphers based on LFSRs

As mentioned in the beginning of §6.2.1, linear feedback shift registers are widely used in keystream generators because they are well-suited for hardware implementation, produce sequences having large periods and good statistical properties, and are readily analyzed using algebraic techniques. Unfortunately, the output sequences of LFSRs are also easily predictable, as the following argument shows. Suppose that the output sequence s of an LFSR has linear complexity L. The connection polynomial $C(D)$ of an LFSR of length L which generates s can be efficiently determined using the Berlekamp-Massey algorithm

(Algorithm 6.30) from any (short) subsequence t of s having length at least $n = 2L$ (cf. Fact 6.35). Having determined $C(D)$, the LFSR $\langle L, C(D) \rangle$ can then be initialized with any substring of t having length L, and used to generate the remainder of the sequence s. An adversary may obtain the required subsequence t of s by mounting a known or chosen-plaintext attack (§1.13.1) on the stream cipher: if the adversary knows the plaintext subsequence $m_1, m_2, \ldots, m_n$ corresponding to a ciphertext sequence $c_1, c_2, \ldots, c_n$, the corresponding keystream bits are obtained as $m_i \oplus c_i$, $1 \leq i \leq n$.

6.44 Note (*use of LFSRs in keystream generators*) Since a well-designed system should be secure against known-plaintext attacks, an LFSR should never be used by itself as a keystream generator. Nevertheless, LFSRs are desirable because of their very low implementation costs. Three general methodologies for destroying the linearity properties of LFSRs are discussed in this section:

  (i) using a nonlinear combining function on the outputs of several LFSRs (§6.3.1);
 (ii) using a nonlinear filtering function on the contents of a single LFSR (§6.3.2); and
(iii) using the output of one (or more) LFSRs to control the clock of one (or more) other LFSRs (§6.3.3).

Desirable properties of LFSR-based keystream generators

For essentially all possible secret keys, the output sequence of an LFSR-based keystream generator should have the following properties:

1. large period;
2. large linear complexity; and
3. good statistical properties (e.g., as described in Fact 6.14).

It is emphasized that these properties are only *necessary* conditions for a keystream generator to be considered cryptographically secure. Since mathematical proofs of security of such generators are not known, such generators can only be deemed *computationally secure* (§1.13.3(iv)) after having withstood sufficient public scrutiny.

6.45 Note (*connection polynomial*) Since a desirable property of a keystream generator is that its output sequences have large periods, component LFSRs should always be chosen to be maximum-length LFSRs, i.e., the LFSRs should be of the form $\langle L, C(D) \rangle$ where $C(D) \in \mathbb{Z}_2[D]$ is a primitive polynomial of degree L (see Definition 6.13 and Fact 6.12(ii)).

6.46 Note (*known vs. secret connection polynomial*) The LFSRs in an LFSR-based keystream generator may have *known* or *secret* connection polynomials. For known connections, the secret key generally consists of the initial contents of the component LFSRs. For secret connections, the secret key for the keystream generator generally consists of both the initial contents and the connections.

For LFSRs of length L with secret connections, the connection polynomials should be selected uniformly at random from the set of all primitive polynomials of degree L over $\mathbb{Z}_2$. Secret connections are generally recommended over known connections as the former are more resistant to certain attacks which use precomputation for analyzing the particular connection, and because the former are more amenable to statistical analysis. Secret connection LFSRs have the drawback of requiring extra circuitry to implement in hardware. However, because of the extra security possible with secret connections, this cost may sometimes be compensated for by choosing shorter LFSRs.

6.47 Note (*sparse vs. dense connection polynomial*) For implementation purposes, it is advantageous to choose an LFSR that is *sparse*; i.e., only a few of the coefficients of the connection polynomial are non-zero. Then only a small number of connections must be made between the stages of the LFSR in order to compute the feedback bit. For example, the connection polynomial might be chosen to be a primitive trinomial (cf. Table 4.8). However, in some LFSR-based keystream generators, special attacks can be mounted if sparse connection polynomials are used. Hence, it is generally recommended not to use sparse connection polynomials in LFSR-based keystream generators.

6.3.1 Nonlinear combination generators

One general technique for destroying the linearity inherent in LFSRs is to use several LFSRs in parallel. The keystream is generated as a nonlinear function f of the outputs of the component LFSRs; this construction is illustrated in Figure 6.8. Such keystream generators are called *nonlinear combination generators*, and f is called the *combining function*. The remainder of this subsection demonstrates that the function f must satisfy several criteria in order to withstand certain particular cryptographic attacks.

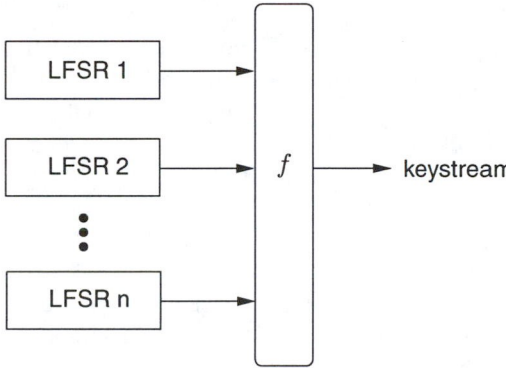

Figure 6.8: *A nonlinear combination generator. f is a nonlinear combining function.*

6.48 Definition A product of m distinct variables is called an mth *order product* of the variables. Every Boolean function $f(x_1, x_2, \ldots, x_n)$ can be written as a modulo 2 sum of distinct m^{th} order products of its variables, $0 \le m \le n$; this expression is called the *algebraic normal form* of f. The *nonlinear order* of f is the maximum of the order of the terms appearing in its algebraic normal form.

For example, the Boolean function $f(x_1, x_2, x_3, x_4, x_5) = 1 \oplus x_2 \oplus x_3 \oplus x_4 x_5 \oplus x_1 x_3 x_4 x_5$ has nonlinear order 4. Note that the maximum possible nonlinear order of a Boolean function in n variables is n. Fact 6.49 demonstrates that the output sequence of a nonlinear combination generator has high linear complexity, provided that a combining function f of high nonlinear order is employed.

6.49 Fact Suppose that n maximum-length LFSRs, whose lengths $L_1, L_2, \ldots, L_n$ are pairwise distinct and greater than 2, are combined by a nonlinear function $f(x_1, x_2, \ldots, x_n)$ (as in Figure 6.8) which is expressed in algebraic normal form. Then the linear complexity of the keystream is $f(L_1, L_2, \ldots, L_n)$. (The expression $f(L_1, L_2, \ldots, L_n)$ is evaluated over the integers rather than over $\mathbb{Z}_2$.)

6.50 Example (*Geffe generator*) The Geffe generator, as depicted in Figure 6.9, is defined by three maximum-length LFSRs whose lengths L_1, L_2, L_3 are pairwise relatively prime, with nonlinear combining function

$$f(x_1, x_2, x_3) \ = \ x_1 x_2 \oplus (1 + x_2) x_3 \ = \ x_1 x_2 \oplus x_2 x_3 \oplus x_3.$$

The keystream generated has period $(2^{L_1} - 1) \cdot (2^{L_2} - 1) \cdot (2^{L_3} - 1)$ and linear complexity $L = L_1 L_2 + L_2 L_3 + L_3$.

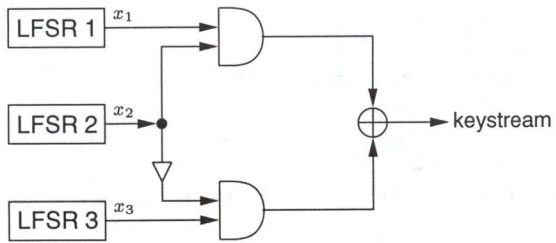

Figure 6.9: *The Geffe generator.*

The Geffe generator is cryptographically weak because information about the states of LFSR 1 and LFSR 3 leaks into the output sequence. To see this, let $x_1(t), x_2(t), x_3(t), z(t)$ denote the tth output bits of LFSRs 1, 2, 3 and the keystream, respectively. Then the *correlation probability* of the sequence $x_1(t)$ to the output sequence $z(t)$ is

$$
\begin{aligned}
P(z(t) = x_1(t)) \ &= \ P(x_2(t) = 1) + P(x_2(t) = 0) \cdot P(x_3(t) = x_1(t)) \\
&= \ \frac{1}{2} + \frac{1}{2} \cdot \frac{1}{2} = \frac{3}{4}.
\end{aligned}
$$

Similarly, $P(z(t) = x_3(t)) = \frac{3}{4}$. For this reason, despite having high period and moderately high linear complexity, the Geffe generator succumbs to correlation attacks, as described in Note 6.51.   □

6.51 Note (*correlation attacks*) Suppose that n maximum-length LFSRs $R_1, R_2, \ldots, R_n$ of lengths $L_1, L_2, \ldots, L_n$ are employed in a nonlinear combination generator. If the connection polynomials of the LFSRs and the combining function f are public knowledge, then the number of different keys of the generator is $\prod_{i=1}^{n}(2^{L_i} - 1)$. (A key consists of the initial states of the LFSRs.) Suppose that there is a correlation between the keystream and the output sequence of R_1, with correlation probability $p > \frac{1}{2}$. If a sufficiently long segment of the keystream is known (e.g., as is possible under a known-plaintext attack on a binary additive stream cipher), the initial state of R_1 can be deduced by counting the number of coincidences between the keystream and all possible shifts of the output sequence of R_1, until this number agrees with the correlation probability p. Under these conditions, finding the initial state of R_1 will take at most $2^{L_1} - 1$ trials. In the case where there is a correlation between the keystream and the output sequences of each of $R_1, R_2, \ldots, R_n$, the (secret) initial state of each LFSR can be determined independently in a total of about $\sum_{i=1}^{n}(2^{L_i} - 1)$ trials; this number is far smaller than the total number of different keys. In a similar manner, correlations between the output sequences of particular subsets of the LFSRs and the keystream can be exploited.

In view of Note 6.51, the combining function f should be carefully selected so that there is no statistical dependence between any small subset of the n LFSR sequences and the keystream. This condition can be satisfied if f is chosen to be m^{th}-order correlation immune.

6.52 Definition Let $X_1, X_2, \ldots, X_n$ be independent binary variables, each taking on the values 0 or 1 with probability $\frac{1}{2}$. A Boolean function $f(x_1, x_2, \ldots, x_n)$ is m^{th}-*order correlation immune* if for each subset of m random variables $X_{i_1}, X_{i_2}, \ldots, X_{i_m}$ with $1 \leq i_1 < i_2 < \cdots < i_m \leq n$, the random variable $Z = f(X_1, X_2, \ldots, X_n)$ is statistically independent of the random vector $(X_{i_1}, X_{i_2}, \ldots, X_{i_m})$; equivalently, $I(Z; X_{i_1}, X_{i_2}, \ldots, X_{i_m}) = 0$ (see Definition 2.45).

For example, the function $f(x_1, x_2, \ldots, x_n) = x_1 \oplus x_2 \oplus \cdots \oplus x_n$ is $(n-1)^{\text{th}}$-order correlation immune. In light of Fact 6.49, the following shows that there is a tradeoff between achieving high linear complexity and high correlation immunity with a combining function.

6.53 Fact If a Boolean function $f(x_1, x_2, \ldots, x_n)$ is m^{th}-order correlation immune, where $1 \leq m < n$, then the nonlinear order of f is at most $n - m$. Moreover, if f is *balanced* (i.e., exactly half of the output values of f are 0) then the nonlinear order of f is at most $n - m - 1$ for $1 \leq m \leq n - 2$.

The tradeoff between high linear complexity and high correlation immunity can be avoided by permitting *memory* in the nonlinear combination function f. This point is illustrated by the summation generator.

6.54 Example (*summation generator*) The combining function in the summation generator is based on the fact that integer addition, when viewed over $\mathbb{Z}_2$, is a nonlinear function with memory whose correlation immunity is maximum. To see this in the case $n = 2$, let $a = a_{m-1}2^{m-1} + \cdots + a_1 2 + a_0$ and $b = b_{m-1}2^{m-1} + \cdots + b_1 2 + b_0$ be the binary representations of integers a and b. Then the bits of $z = a + b$ are given by the recursive formula:

$$
\begin{aligned}
z_j &= f_1(a_j, b_j, c_{j-1}) = a_j \oplus b_j \oplus c_{j-1} \quad 0 \leq j \leq m, \\
c_j &= f_2(a_j, b_j, c_{j-1}) = a_j b_j \oplus (a_j \oplus b_j)c_{j-1}, \quad 0 \leq j \leq m-1,
\end{aligned}
$$

where c_j is the carry bit, and $c_{-1} = a_m = b_m = 0$. Note that f_1 is 2nd-order correlation immune, while f_2 is a *memoryless* nonlinear function. The carry bit c_{j-1} carries all the nonlinear influence of less significant bits of a and b (namely, $a_{j-1}, \ldots, a_1, a_0$ and $b_{j-1}, \ldots, b_1, b_0$).

The summation generator, as depicted in Figure 6.10, is defined by n maximum-length LFSRs whose lengths $L_1, L_2, \ldots, L_n$ are pairwise relatively prime. The secret key con-

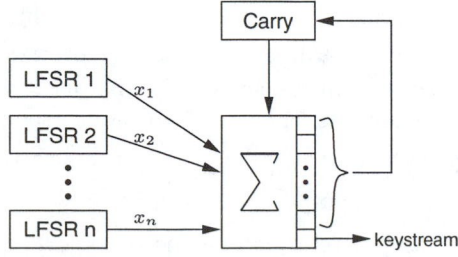

Figure 6.10: *The summation generator.*

sists of the initial states of the LFSRs, and an initial (integer) carry C_0. The keystream is generated as follows. At time j ($j \geq 1$), the LFSRs are stepped producing output bits $x_1, x_2, \ldots, x_n$, and the *integer* sum $S_j = \sum_{i=1}^{n} x_i + C_{j-1}$ is computed. The keystream bit is $S_j \bmod 2$ (the least significant bit of S_j), while the new carry is computed as $C_j = \lfloor S_j/2 \rfloor$ (the remaining bits of S_j). The period of the keystream is $\prod_{i=1}^{n}(2^{L_i} - 1)$, while its linear complexity is close to this number.

Even though the summation generator has high period, linear complexity, and correlation immunity, it is vulnerable to certain correlation attacks and a known-plaintext attack based on its 2-adic span (see page 218). $\square$

6.3.2 Nonlinear filter generators

Another general technique for destroying the linearity inherent in LFSRs is to generate the keystream as some nonlinear function of the stages of a single LFSR; this construction is illustrated in Figure 6.11. Such keystream generators are called *nonlinear filter generators*, and f is called the *filtering function*.

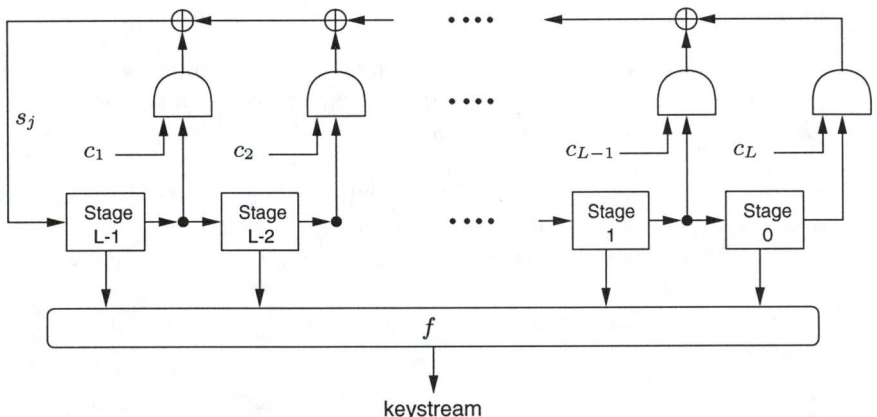

Figure 6.11: *A nonlinear filter generator. f is a nonlinear Boolean filtering function.*

Fact 6.55 describes the linear complexity of the output sequence of a nonlinear filter generator.

6.55 Fact Suppose that a nonlinear filter generator is constructed using a maximum-length LFSR of length L and a filtering function f of nonlinear order m (as in Figure 6.11).

(i) (*Key's bound*) The linear complexity of the keystream is at most $L_m = \sum_{i=1}^{m} \binom{L}{i}$.

(ii) For a fixed maximum-length LFSR of prime length L, the fraction of Boolean functions f of nonlinear order m which produce sequences of maximum linear complexity L_m is

$$P_m \approx \exp(-L_m/(L \cdot 2^L)) > e^{-1/L}.$$

Therefore, for large L, most of the generators produce sequences whose linear complexity meets the upper bound in (i).

The nonlinear function f selected for a filter generator should include many terms of each order up to the nonlinear order of f.

6.56 Example (*knapsack generator*) The knapsack keystream generator is defined by a maximum-length LFSR $\langle L, C(D) \rangle$ and a modulus $Q = 2^L$. The secret key consists of L knapsack integer weights $a_1, a_2, \ldots, a_L$ each of bitlength L, and the initial state of the LFSR. Recall that the subset sum problem (§3.10) is to determine a subset of the knapsack weights which add up to a given integer s, provided that such a subset exists; this problem is **NP**-hard (Fact 3.91). The keystream is generated as follows: at time j, the LFSR is stepped and the knapsack sum $S_j = \sum_{i=1}^{L} x_i a_i \bmod Q$ is computed, where $[x_L, \ldots, x_2, x_1]$ is the state of the LFSR at time j. Finally, selected bits of S_j (after S_j is converted to its binary representation) are extracted to form part of the keystream (the $\lceil \lg L \rceil$ least significant bits of S_j should be discarded). The linear complexity of the keystream is then virtually certain to be $L(2^L - 1)$.

Since the state of an LFSR is a binary vector, the function which maps the LFSR state to the knapsack sum S_j is indeed nonlinear. Explicitly, let the function f be defined by $f(x) = \sum_{i=1}^{L} x_i a_i \bmod Q$, where $x = [x_L, \ldots, x_2, x_1]$ is a state. If x and y are two states then, in general, $f(x \oplus y) \neq f(x) + f(y)$. □

6.3.3 Clock-controlled generators

In nonlinear combination generators and nonlinear filter generators, the component LFSRs are clocked regularly; i.e., the movement of data in all the LFSRs is controlled by the same clock. The main idea behind a *clock-controlled generator* is to introduce nonlinearity into LFSR-based keystream generators by having the output of one LFSR control the *clocking* (i.e., stepping) of a second LFSR. Since the second LFSR is clocked in an irregular manner, the hope is that attacks based on the regular motion of LFSRs can be foiled. Two clock-controlled generators are described in this subsection: (i) the alternating step generator and (ii) the shrinking generator.

(i) The alternating step generator

The alternating step generator uses an LFSR R_1 to control the stepping of two LFSRs, R_2 and R_3. The keystream produced is the XOR of the output sequences of R_2 and R_3.

6.57 Algorithm Alternating step generator

SUMMARY: a control LFSR R_1 is used to selectively step two other LFSRs, R_2 and R_3.
OUTPUT: a sequence which is the bitwise XOR of the output sequences of R_2 and R_3.
The following steps are repeated until a keystream of desired length is produced.

1. Register R_1 is clocked.
2. If the output of R_1 is 1 then:
   R_2 is clocked; R_3 is not clocked but its previous output bit is repeated.
   (For the first clock cycle, the "previous output bit" of R_3 is taken to be 0.)
3. If the output of R_1 is 0 then:
   R_3 is clocked; R_2 is not clocked but its previous output bit is repeated.
   (For the first clock cycle, the "previous output bit" of R_2 is taken to be 0.)
4. The output bits of R_2 and R_3 are XORed; the resulting bit is part of the keystream.

More formally, let the output sequences of LFSRs R_1, R_2, and R_3 be $a_0, a_1, a_2, \ldots$, $b_0, b_1, b_2, \ldots$, and $c_0, c_1, c_2 \ldots$, respectively. Define $b_{-1} = c_{-1} = 0$. Then the keystream produced by the alternating step generator is $x_0, x_1, x_2, \ldots$, where $x_j = b_{t(j)} \oplus c_{j-t(j)-1}$

and $t(j) = (\sum_{i=0}^{j} a_i) - 1$ for all $j \geq 0$. The alternating step generator is depicted in Figure 6.12.

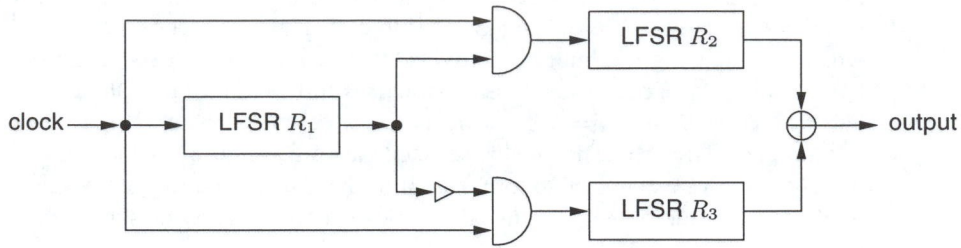

Figure 6.12: *The alternating step generator.*

6.58 Example (*alternating step generator with artificially small parameters*) Consider an alternating step generator with component LFSRs $R_1 = \langle 3, 1 + D^2 + D^3 \rangle$, $R_2 = \langle 4, 1 + D^3 + D^4 \rangle$, and $R_3 = \langle 5, 1 + D + D^3 + D^4 + D^5 \rangle$. Suppose that the initial states of R_1, R_2, and R_3 are $[0, 0, 1]$, $[1, 0, 1, 1]$, and $[0, 1, 0, 0, 1]$, respectively. The output sequence of R_1 is the 7-periodic sequence with cycle

$$a^7 = 1, 0, 0, 1, 0, 1, 1.$$

The output sequence of R_2 is the 15-periodic sequence with cycle

$$b^{15} = 1, 1, 0, 1, 0, 1, 1, 1, 1, 0, 0, 0, 1, 0, 0.$$

The output sequence of R_3 is the 31-periodic sequence with cycle

$$c^{31} = 1, 0, 0, 1, 0, 1, 0, 1, 1, 0, 0, 0, 0, 1, 1, 1, 0, 0, 1, 1, 0, 1, 1, 1, 1, 1, 1, 0, 1, 0, 0, 0.$$

The keystream generated is

$$x = 1, 0, 1, 1, 1, 0, 1, 0, 1, 0, 1, 0, 0, 0, 0, 0, 1, 0, 1, 1, 1, 1, 0, 1, 1, 0, 0, 0, 0, 1, 1, 1, 0, \ldots. \quad \square$$

Fact 6.59 establishes, under the assumption that R_1 produces a de Bruijn sequence (see Definition 6.40), that the output sequence of an alternating step generator satisfies the basic requirements of high period, high linear complexity, and good statistical properties.

6.59 Fact (*properties of the alternating step generator*) Suppose that R_1 produces a de Bruijn sequence of period 2^{L_1}. Furthermore, suppose that R_2 and R_3 are maximum-length LFSRs of lengths L_2 and L_3, respectively, such that $\gcd(L_2, L_3) = 1$. Let x be the output sequence of the alternating step generator formed by R_1, R_2, and R_3.

   (i) The sequence x has period $2^{L_1} \cdot (2^{L_2} - 1) \cdot (2^{L_3} - 1)$.

  (ii) The linear complexity $L(x)$ of x satisfies

$$(L_2 + L_3) \cdot 2^{L_1 - 1} < L(x) \leq (L_2 + L_3) \cdot 2^{L_1}.$$

 (iii) The distribution of patterns in x is almost uniform. More precisely, let P be any binary string of length t bits, where $t \leq \min(L_2, L_3)$. If $x(t)$ denotes any t consecutive bits in x, then the probability that $x(t) = P$ is $\left(\frac{1}{2}\right)^t + O(1/2^{L_2 - t}) + O(1/2^{L_3 - t})$.

Since a de Bruijn sequence can be obtained from the output sequence s of a maximum-length LFSR (of length L) by simply adding a 0 to the end of each subsequence of $L - 1$ 0's occurring in s (see Note 6.43), it is reasonable to expect that the assertions of high period, high linear complexity, and good statistical properties in Fact 6.59 also hold when R_1 is a maximum-length LFSR. Note, however, that this has not yet been proven.

6.60 Note (*security of the alternating step generator*) The LFSRs R_1, R_2, R_3 should be chosen to be maximum-length LFSRs whose lengths L_1, L_2, L_3 are pairwise relatively prime: $\gcd(L_1, L_2) = 1$, $\gcd(L_2, L_3) = 1$, $\gcd(L_1, L_3) = 1$. Moreover, the lengths should be about the same. If $L_1 \approx l$, $L_2 \approx l$, and $L_3 \approx l$, the best known attack on the alternating step generator is a divide-and-conquer attack on the control register R_1 which takes approximately 2^l steps. Thus, if $l \approx 128$, the generator is secure against all presently known attacks.

(ii) The shrinking generator

The shrinking generator is a relatively new keystream generator, having been proposed in 1993. Nevertheless, due to its simplicity and provable properties, it is a promising candidate for high-speed encryption applications. In the shrinking generator, a control LFSR R_1 is used to select a portion of the output sequence of a second LFSR R_2. The keystream produced is, therefore, a *shrunken* version (also known as an *irregularly decimated subsequence*) of the output sequence of R_2, as specified in Algorithm 6.61 and depicted in Figure 6.13.

6.61 Algorithm Shrinking generator

SUMMARY: a control LFSR R_1 is used to control the output of a second LFSR R_2. The following steps are repeated until a keystream of desired length is produced.

1. Registers R_1 and R_2 are clocked.
2. If the output of R_1 is 1, the output bit of R_2 forms part of the keystream.
3. If the output of R_1 is 0, the output bit of R_2 is discarded.

More formally, let the output sequences of LFSRs R_1 and R_2 be $a_0, a_1, a_2, \ldots$ and $b_0, b_1, b_2, \ldots$, respectively. Then the keystream produced by the shrinking generator is $x_0, x_1, x_2, \ldots$, where $x_j = b_{i_j}$, and, for each $j \geq 0$, i_j is the position of the jth 1 in the sequence $a_0, a_1, a_2, \ldots$.

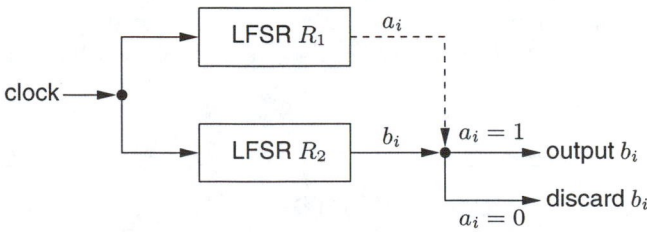

Figure 6.13: *The shrinking generator.*

6.62 Example (*shrinking generator with artificially small parameters*) Consider a shrinking generator with component LFSRs $R_1 = \langle 3, 1 + D + D^3 \rangle$ and $R_2 = \langle 5, 1 + D^3 + D^5 \rangle$. Suppose that the initial states of R_1 and R_2 are $[1, 0, 0]$ and $[0, 0, 1, 0, 1]$, respectively. The output sequence of R_1 is the 7-periodic sequence with cycle

$$a^7 = 0, 0, 1, 1, 1, 0, 1,$$

while the output sequence of R_2 is the 31-periodic sequence with cycle

$$b^{31} = 1,0,1,0,0,0,0,1,0,0,1,0,1,1,0,0,1,1,1,1,1,0,0,0,1,1,0,1,1,1,0.$$

The keystream generated is

$$x = 1,0,0,0,0,1,0,1,1,1,1,1,1,0,1,1,1,0,\ldots. \qquad \square$$

Fact 6.63 establishes that the output sequence of a shrinking generator satisfies the basic requirements of high period, high linear complexity, and good statistical properties.

6.63 Fact (*properties of the shrinking generator*) Let R_1 and R_2 be maximum-length LFSRs of lengths L_1 and L_2, respectively, and let x be an output sequence of the shrinking generator formed by R_1 and R_2.

   (i) If $\gcd(L_1, L_2) = 1$, then x has period $(2^{L_2} - 1) \cdot 2^{L_1-1}$.
   (ii) The linear complexity $L(x)$ of x satisfies

$$L_2 \cdot 2^{L_1-2} < L(x) \leq L_2 \cdot 2^{L_1-1}.$$

   (iii) Suppose that the connection polynomials for R_1 and R_2 are chosen uniformly at random from the set of all primitive polynomials of degrees L_1 and L_2 over $\mathbb{Z}_2$. Then the distribution of patterns in x is almost uniform. More precisely, if P is any binary string of length t bits and $x(t)$ denotes any t consecutive bits in x, then the probability that $x(t) = P$ is $(\frac{1}{2})^t + O(t/2^{L_2})$.

6.64 Note (*security of the shrinking generator*) Suppose that the component LFSRs R_1 and R_2 of the shrinking generator have lengths L_1 and L_2, respectively. If the connection polynomials for R_1 and R_2 are known (but not the initial contents of R_1 and R_2), the best attack known for recovering the secret key takes $O(2^{L_1} \cdot L_2^3)$ steps. On the other hand, if secret (and variable) connection polynomials are used, the best attack known takes $O(2^{2L_1} \cdot L_1 \cdot L_2)$ steps. There is also an attack through the linear complexity of the shrinking generator which takes $O(2^{L_1} \cdot L_2^2)$ steps (regardless of whether the connections are known or secret), but this attack requires $2^{L_1} \cdot L_2$ consecutive bits from the output sequence and is, therefore, infeasible for moderately large L_1 and L_2. For maximum security, R_1 and R_2 should be maximum-length LFSRs, and their lengths should satisfy $\gcd(L_1, L_2) = 1$. Moreover, secret connections should be used. Subject to these constraints, if $L_1 \approx l$ and $L_2 \approx l$, the shrinking generator has a security level approximately equal to 2^{2l}. Thus, if $L_1 \approx 64$ and $L_2 \approx 64$, the generator appears to be secure against all presently known attacks.

6.4 Other stream ciphers

While the LFSR-based stream ciphers discussed in §6.3 are well-suited to hardware implementation, they are not especially amenable to software implementation. This has led to several recent proposals for stream ciphers designed particularly for fast software implementation. Most of these proposals are either proprietary, or are relatively new and have not received sufficient scrutiny from the cryptographic community; for this reason, they are not presented in this section, and instead only mentioned in the chapter notes on page 222.

Two promising stream ciphers specifically designed for fast software implementation are SEAL and RC4. SEAL is presented in §6.4.1. RC4 is used in commercial products, and has a variable key-size, but it remains proprietary and is not presented here. Two other widely used stream ciphers not based on LFSRs are the Output Feedback (OFB; see §7.2.2(iv)) and Cipher Feedback (CFB; see §7.2.2(iii)) modes of block ciphers. Another class of keystream generators not based on LFSRs are those whose security relies on the intractability of an underlying number-theoretic problem; these generators are much slower than those based on LFSRs and are discussed in §5.5.

6.4.1 SEAL

SEAL (Software-optimized Encryption Algorithm) is a binary additive stream cipher (see Definition 6.4) that was proposed in 1993. Since it is relatively new, it has not yet received much scrutiny from the cryptographic community. However, it is presented here because it is one of the few stream ciphers that was specifically designed for efficient software implementation and, in particular, for 32-bit processors.

SEAL is a length-increasing pseudorandom function which maps a 32-bit *sequence number n* to an L-bit keystream under control of a 160-bit secret key a. In the preprocessing stage (step 1 of Algorithm 6.68), the key is stretched into larger tables using the table-generation function G_a specified in Algorithm 6.67; this function is based on the Secure Hash Algorithm SHA-1 (Algorithm 9.53). Subsequent to this preprocessing, keystream generation requires about 5 machine instructions per byte, and is an order of magnitude faster than DES (Algorithm 7.82).

The following notation is used in SEAL for 32-bit quantities A, B, C, D, X_i, and Y_j:

- $\overline{A}$: bitwise complement of A
- $A \wedge B$, $A \vee B$, $A \oplus B$: bitwise AND, inclusive-OR, exclusive-OR
- "$A \hookleftarrow s$": 32-bit result of rotating A left through s positions
- "$A \hookrightarrow s$": 32-bit result of rotating A right through s positions
- $A + B$: mod 2^{32} sum of the unsigned integers A and B
- $f(B,C,D) \stackrel{\text{def}}{=} (B \wedge C) \vee (\overline{B} \wedge D)$; $g(B,C,D) \stackrel{\text{def}}{=} (B \wedge C) \vee (B \wedge D) \vee (C \wedge D)$; $h(B,C,D) \stackrel{\text{def}}{=} B \oplus C \oplus D$
- $A \| B$: concatenation of A and B
- $(X_1, \ldots, X_j) \leftarrow (Y_1, \ldots, Y_j)$: simultaneous assignments $(X_i \leftarrow Y_i)$, where $(Y_1, \ldots, Y_j)$ is evaluated prior to any assignments.

6.65 Note (*SEAL 1.0 vs. SEAL 2.0*) The table-generation function (Algorithm 6.67) for the first version of SEAL (SEAL 1.0) was based on the Secure Hash Algorithm (SHA). SEAL 2.0 differs from SEAL 1.0 in that the table-generation function for the former is based on the modified Secure Hash Algorithm SHA-1 (Algorithm 9.53).

6.66 Note (*tables*) The table generation (step 1 of Algorithm 6.68) uses the compression function of SHA-1 to expand the secret key a into larger tables T, S, and R. These tables can be precomputed, but only after the secret key a has been established. Tables T and S are 2K bytes and 1K byte in size, respectively. The size of table R depends on the desired bitlength L of the keystream — each 1K byte of keystream requires 16 bytes of R.

6.67 Algorithm Table-generation function for SEAL 2.0

INPUT: a 160-bit string a and an unsigned integer i, $0 \le i < 2^{32}$.
OUTPUT: a 160-bit string, denoted $G_a(i)$.

1. *Definition of constants.* Define four 32-bit constants (in hex): $y_1 = 0x5a827999$, $y_2 = 0x6ed9eba1$, $y_3 = 0x8f1bbcdc$, $y_4 = 0xca62c1d6$.

2. *Table-generation function.*
   (initialize 80 32-bit words $X_0, X_1, \ldots, X_{79}$)
   Set $X_0 \leftarrow i$. For j from 1 to 15 do: $X_j \leftarrow 0x00000000$.
   For j from 16 to 79 do: $X_j \leftarrow ((X_{j-3} \oplus X_{j-8} \oplus X_{j-14} \oplus X_{j-16}) \hookleftarrow 1)$.
   (initialize working variables)
   Break up the 160-bit string a into five 32-bit words: $a = H_0 H_1 H_2 H_3 H_4$.
   $(A, B, C, D, E) \leftarrow (H_0, H_1, H_2, H_3, H_4)$.
   (execute four rounds of 20 steps, then update; t is a temporary variable)
   (Round 1) For j from 0 to 19 do the following:
   $t \leftarrow ((A \hookleftarrow 5) + f(B, C, D) + E + X_j + y_1)$,
   $(A, B, C, D, E) \leftarrow (t, A, B \hookleftarrow 30, C, D)$.
   (Round 2) For j from 20 to 39 do the following:
   $t \leftarrow ((A \hookleftarrow 5) + h(B, C, D) + E + X_j + y_2)$,
   $(A, B, C, D, E) \leftarrow (t, A, B \hookleftarrow 30, C, D)$.
   (Round 3) For j from 40 to 59 do the following:
   $t \leftarrow ((A \hookleftarrow 5) + g(B, C, D) + E + X_j + y_3)$,
   $(A, B, C, D, E) \leftarrow (t, A, B \hookleftarrow 30, C, D)$.
   (Round 4) For j from 60 to 79 do the following:
   $t \leftarrow ((A \hookleftarrow 5) + h(B, C, D) + E + X_j + y_4)$,
   $(A, B, C, D, E) \leftarrow (t, A, B \hookleftarrow 30, C, D)$.
   (update chaining values)
   $(H_0, H_1, H_2, H_3, H_4) \leftarrow (H_0 + A, H_1 + B, H_2 + C, H_3 + D, H_4 + E)$.
   (completion) The value of $G_a(i)$ is the 160-bit string $H_0 \| H_1 \| H_2 \| H_3 \| H_4$.

6.68 Algorithm Keystream generator for SEAL 2.0

SEAL(a,n)
INPUT: a 160-bit string a (the secret key), an unsigned (non-secret) integer n, $0 \le n < 2^{32}$ (the sequence number), and the desired bitlength L of the keystream.
OUTPUT: keystream y of bitlength L', where L' is the least multiple of 128 which is $\ge L$.

1. *Table generation.* Generate the tables T, S, and R, whose entries are 32-bit words. The function F used below is defined by $F_a(i) = H^i_{i \bmod 5}$, where $H^i_0 H^i_1 H^i_2 H^i_3 H^i_4 = G_a(\lfloor i/5 \rfloor)$, and where the function G_a is defined in Algorithm 6.67.

   1.1 For i from 0 to 511 do the following: $T[i] \leftarrow F_a(i)$.
   1.2 For j from 0 to 255 do the following: $S[j] \leftarrow F_a(0x00001000 + j)$.
   1.3 For k from 0 to $4 \cdot \lceil (L-1)/8192 \rceil - 1$ do: $R[k] \leftarrow F_a(0x00002000 + k)$.

2. *Initialization procedure.* The following is a description of the subroutine
   Initialize($n, l, A, B, C, D, n_1, n_2, n_3, n_4$) which takes as input a 32-bit word n and an integer l, and outputs eight 32-bit words A, B, C, D, n_1, n_2, n_3, and n_4. This subroutine is used in step 4.
   $A \leftarrow n \oplus R[4l]$, $B \leftarrow (n \hookrightarrow 8) \oplus R[4l+1]$, $C \leftarrow (n \hookrightarrow 16) \oplus R[4l+2]$,
   $D \leftarrow (n \hookrightarrow 24) \oplus R[4l+3]$.

For j from 1 to 2 do the following:

$P \leftarrow A \wedge \text{0x000007fc}$, $B \leftarrow B + T[P/4]$, $A \leftarrow (A \hookrightarrow 9)$,
$P \leftarrow B \wedge \text{0x000007fc}$, $C \leftarrow C + T[P/4]$, $B \leftarrow (B \hookrightarrow 9)$,
$P \leftarrow C \wedge \text{0x000007fc}$, $D \leftarrow D + T[P/4]$, $C \leftarrow (C \hookrightarrow 9)$,
$P \leftarrow D \wedge \text{0x000007fc}$, $A \leftarrow A + T[P/4]$, $D \leftarrow (D \hookrightarrow 9)$.

$(n_1, n_2, n_3, n_4) \leftarrow (D, B, A, C)$.
$P \leftarrow A \wedge \text{0x000007fc}$, $B \leftarrow B + T[P/4]$, $A \leftarrow (A \hookrightarrow 9)$.
$P \leftarrow B \wedge \text{0x000007fc}$, $C \leftarrow C + T[P/4]$, $B \leftarrow (B \hookrightarrow 9)$.
$P \leftarrow C \wedge \text{0x000007fc}$, $D \leftarrow D + T[P/4]$, $C \leftarrow (C \hookrightarrow 9)$.
$P \leftarrow D \wedge \text{0x000007fc}$, $A \leftarrow A + T[P/4]$, $D \leftarrow (D \hookrightarrow 9)$.

3. Initialize y to be the empty string, and $l \leftarrow 0$.

4. Repeat the following:

4.1 Execute the procedure $\texttt{Initialize}(n, l, A, B, C, D, n_1, n_2, n_3, n_4)$.

4.2 For i from 1 to 64 do the following:

$P \leftarrow A \wedge \text{0x000007fc}$, $B \leftarrow B + T[P/4]$, $A \leftarrow (A \hookrightarrow 9)$, $B \leftarrow B \oplus A$,
$Q \leftarrow B \wedge \text{0x000007fc}$, $C \leftarrow C \oplus T[Q/4]$, $B \leftarrow (B \hookrightarrow 9)$, $C \leftarrow C + B$,
$P \leftarrow (P + C) \wedge \text{0x000007fc}$, $D \leftarrow D + T[P/4]$, $C \leftarrow (C \hookrightarrow 9)$, $D \leftarrow D \oplus C$,
$Q \leftarrow (Q + D) \wedge \text{0x000007fc}$, $A \leftarrow A \oplus T[Q/4]$, $D \leftarrow (D \hookrightarrow 9)$, $A \leftarrow A + D$,
$P \leftarrow (P + A) \wedge \text{0x000007fc}$, $B \leftarrow B \oplus T[P/4]$, $A \leftarrow (A \hookrightarrow 9)$,
$Q \leftarrow (Q + B) \wedge \text{0x000007fc}$, $C \leftarrow C + T[Q/4]$, $B \leftarrow (B \hookrightarrow 9)$,
$P \leftarrow (P + C) \wedge \text{0x000007fc}$, $D \leftarrow D \oplus T[P/4]$, $C \leftarrow (C \hookrightarrow 9)$,
$Q \leftarrow (Q + D) \wedge \text{0x000007fc}$, $A \leftarrow A + T[Q/4]$, $D \leftarrow (D \hookrightarrow 9)$,
$y \leftarrow y \parallel (B + S[4i - 4]) \parallel (C \oplus S[4i - 3]) \parallel (D + S[4i - 2]) \parallel (A \oplus S[4i - 1])$.
If y is $\geq L$ bits in length then return(y) and stop.
If i is odd, set $(A, C) \leftarrow (A + n_1, C + n_2)$. Otherwise, $(A, C) \leftarrow (A + n_3, C + n_4)$.

4.3 Set $l \leftarrow l + 1$.

6.69 Note (*choice of parameter L*) In most applications of SEAL 2.0 it is expected that $L \leq 2^{19}$; larger values of L are permissible, but come at the expense of a larger table R. A preferred method for generating a longer keystream without requiring a larger table R is to compute the concatenation of the keystreams SEAL(a,0), SEAL(a,1), SEAL(a,2),.... Since the sequence number is $n < 2^{32}$, a keystream of length up to 2^{51} bits can be obtained in this manner with $L = 2^{19}$.

6.70 Example (*test vectors for SEAL 2.0*) Suppose the key a is the 160-bit (hexadecimal) string

```
67452301 efcdab89 98badcfe 10325476 c3d2e1f0,
```

$n = \text{0x013577af}$, and $L = 32768$ bits. Table R consists of words $R[0], R[1], \ldots, R[15]$:

```
5021758d ce577c11 fa5bd5dd 366d1b93 182cff72 ac06d7c6
2683ead8 fabe3573 82a10c96 48c483bd ca92285c 71fe84c0
bd76b700 6fdcc20c 8dada151 4506dd64
```

The table T consists of words $T[0], T[1], \ldots, R[511]$:

```
92b404e5 56588ced 6c1acd4e bf053f68 09f73a93 cd5f176a
b863f14e 2b014a2f 4407e646 38665610 222d2f91 4d941a21
........ ........ ........ ........ ........ ........
3af3a4bf 021e4080 2a677d95 405c7db0 338e4b1e 19ccf158
```

The table S consists of words $S[0], S[1], \ldots, S[255]$:

```
907c1e3d  ce71ef0a  48f559ef  2b7ab8bc  4557f4b8  033e9b05
4fde0efa  1a845f94  38512c3b  d4b44591  53765dce  469efa02
........  ........  ........  ........  ........  ........
bd7dea87  fd036d87  53aa3013  ec60e282  1eaef8f9  0b5a0949
```

The output y of Algorithm 6.68 consists of 1024 words $y[0], y[1], \ldots, y[1023]$:

```
37a00595  9b84c49c  a4be1e05  0673530f  0ac8389d  c5878ec8
da6666d0  6da71328  1419bdf2  d258bebb  b6a42a4d  8a311a72
........  ........  ........  ........  ........  ........
547dfde9  668d50b5  ba9e2567  413403c5  43120b5a  ecf9d062
```

The XOR of the 1024 words of y is 0x098045fc.   □

6.5 Notes and further references

§6.1

Although now dated, Rueppel [1075] provides a solid introduction to the analysis and design of stream ciphers. For an updated and more comprehensive survey, see Rueppel [1081]. Another recommended survey is that of Robshaw [1063].

The concept of unconditional security was introduced in the seminal paper by Shannon [1120]. Maurer [819] surveys the role of information theory in cryptography and, in particular, secrecy, authentication, and secret sharing schemes. Maurer [811] devised a *randomized stream cipher* that is unconditionally secure "with high probability". More precisely, an adversary is unable to obtain any information whatsoever about the plaintext with probability arbitrarily close to 1, unless the adversary can perform an infeasible computation. The cipher utilizes a publicly-accessible source of random bits whose length is much greater than that of all the plaintext to be encrypted, and can conceivably be made practical. Maurer's cipher is based on the impractical *Rip van Winkle cipher* of Massey and Ingermarsson [789], which is described by Rueppel [1081].

One technique for solving the re-synchronization problem with synchronous stream ciphers is to have the receiver send a resynchronization request to the sender, whereby a new internal state is computed as a (public) function of the original internal state (or key) and some public information (such as the time at the moment of the request). Daemen, Govaerts, and Vandewalle [291] showed that this approach can result in a total loss of security for some published stream cipher proposals. Proctor [1011] considered the trade-off between the security and error propagation problems that arise by varying the number of feedback ciphertext digits. Maurer [808] presented various design approaches for self-synchronizing stream ciphers that are potentially superior to designs based on block ciphers, both with respect to encryption speed and security.

§6.2

An excellent introduction to the theory of both linear and nonlinear shift registers is the book by Golomb [498]; see also Selmer [1107], Chapters 5 and 6 of Beker and Piper [84], and Chapter 8 of Lidl and Niederreiter [764]. A lucid treatment of m-sequences can be found in Chapter 10 of McEliece [830]. While the discussion in this chapter has been restricted to sequences and feedback shift registers over the binary field $\mathbb{Z}_2$, many of the results presented can be generalized to sequences and feedback shift registers over any finite field $\mathbb{F}_q$.

The results on the expected linear complexity and linear complexity profile of random sequences (Facts 6.21, 6.22, 6.24, and 6.25) are from Chapter 4 of Rueppel [1075]; they also appear in Rueppel [1077]. Dai and Yang [294] extended Fact 6.22 and obtained bounds for the expected linear complexity of an n-periodic sequence for each possible value of n. The bounds imply that the expected linear complexity of a random periodic sequence is close to the period of the sequence. The linear complexity profile of the sequence defined in Example 6.27 was established by Dai [293]. For further theoretical analysis of the linear complexity profile, consult the work of Niederreiter [927, 928, 929, 930].

Facts 6.29 and 6.34 are due to Massey [784]. The Berlekamp-Massey algorithm (Algorithm 6.30) is due to Massey [784], and is based on an earlier algorithm of Berlekamp [118] for decoding BCH codes. While the algorithm in §6.2.3 is only described for binary sequences, it can be generalized to find the linear complexity of sequences over any field. Further discussion and refinements of the Berlekamp-Massey algorithm are given by Blahut [144]. There are numerous other algorithms for computing the linear complexity of a sequence. For example, Games and Chan [439] and Robshaw [1062] present efficient algorithms for determining the linear complexity of binary sequences of period 2^n; these algorithms have limited practical use since they require an entire cycle of the sequence.

Jansen and Boekee [632] defined the *maximum order complexity* of a sequence to be the length of the shortest (not necessarily linear) feedback shift register (FSR) that can generate the sequence. The expected maximum order complexity of a random binary sequence of length n is approximately $2 \lg n$. An efficient linear-time algorithm for computing this complexity measure was also presented; see also Jansen and Boekee [631].

Another complexity measure, the *Ziv-Lempel complexity measure*, was proposed by Ziv and Lempel [1273]. This measure quantifies the rate at which new patterns appear in a sequence. Mund [912] used a heuristic argument to derive the expected Ziv-Lempel complexity of a random binary sequence of a given length. For a detailed study of the relative strengths and weaknesses of the linear, maximum order, and Ziv-Lempel complexity measures, see Erdmann [372].

Kolmogorov [704] and Chaitin [236] introduced the notion of so-called *Turing-Kolmogorov -Chaitin complexity*, which measures the minimum size of the input to a fixed universal Turing machine which can generate a given sequence; see also Martin-Löf [783]. While this complexity measure is of theoretical interest, there is no algorithm known for computing it and, hence, it has no apparent practical significance. Beth and Dai [124] have shown that the Turing-Kolmogorov-Chaitin complexity is approximately twice the linear complexity for most sequences of sufficient length.

Fact 6.39 is due to Golomb and Welch, and appears in the book of Golomb [498, p.115]. Lai [725] showed that Fact 6.39 is only true for the binary case, and established necessary and sufficient conditions for an FSR over a general finite field to be nonsingular.

Klapper and Goresky [677] introduced a new type of feedback register called a *feedback with carry shift register* (FCSR), which is equipped with auxiliary memory for storing the (integer) carry. An FCSR is similar to an LFSR (see Figure 6.4), except that the contents of the tapped stages of the shift register are added *as integers* to the current content of the memory to form a sum S. The least significant bit of S (i.e., $S \bmod 2$) is then fed back into the first (leftmost) stage of the shift register, while the remaining higher order bits (i.e., $\lfloor S/2 \rfloor$) are retained as the new value of the memory. If the FCSR has L stages, then the space required for the auxiliary memory is at most $\lg L$ bits. FCSRs can be conveniently analyzed using the algebra over the 2-adic numbers just as the algebra over finite fields is used to analyze LFSRs.

Any periodic binary sequence can be generated by a FCSR. The 2-*adic span* of a periodic sequence is the number of stages and memory bits in the smallest FCSR that generates the sequence. Let s be a periodic sequence having a 2-adic span of T; note that T is no more than the period of s. Klapper and Goresky [678] presented an efficient algorithm for finding an FCSR of length T which generates s, given $2T + 2\lceil \lg T \rceil + 4$ of the initial bits of s. A comprehensive treatment of FCSRs and the 2-adic span is given by Klapper and Goresky [676].

§6.3

Notes 6.46 and 6.47 on the selection of connection polynomials were essentially first pointed out by Meier and Staffelbach [834] and Chepyzhov and Smeets [256] in relation to fast correlation attacks on regularly clocked LFSRs. Similar observations were made by Coppersmith, Krawczyk, and Mansour [279] in connection with the shrinking generator. More generally, to withstand sophisticated correlation attacks (e.g., see Meier and Staffelbach [834]), the connection polynomials should not have low-weight polynomial multiples whose degrees are not sufficiently large.

Klapper [675] provides examples of binary sequences having high linear complexity, but whose linear complexity is low when considered as sequences (whose elements happen to be only 0 or 1) over a larger finite field. This demonstrates that high linear complexity (over $\mathbb{Z}_2$) by itself is inadequate for security. Fact 6.49 was proven by Rueppel and Staffelbach [1085].

The Geffe generator (Example 6.50) was proposed by Geffe [446]. The *Pless generator* (Arrangement D of [978]) was another early proposal for a nonlinear combination generator, and uses four J-K flip-flops to combine the output of eight LFSRs. This generator also succumbs to a divide-and-conquer attack, as was demonstrated by Rubin [1074].

The *linear syndrome attack* of Zeng, Yang, and Rao [1265] is a known-plaintext attack on keystream generators, and is based on earlier work of Zeng and Huang [1263]. It is effective when the known keystream B can be written in the form $B = A \oplus X$, where A is the output sequence of an LFSR with known connection polynomial, and the sequence X is unknown but sparse in the sense that it contains more 0's than 1's. If the connection polynomials of the Geffe generator are all known to an adversary, and are primitive trinomials of degrees not exceeding n, then the initial states of the three component LFSRs (i.e., the secret key) can be efficiently recovered from a known keystream segment of length $37n$ bits.

The correlation attack (Note 6.51) on nonlinear combination generators was first developed by Siegenthaler [1133], and estimates were given for the length of the observed keystream required for the attack to succeed with high probability. The importance of correlation immunity to nonlinear combining functions was pointed out by Siegenthaler [1132], who showed the tradeoff between high correlation immunity and high nonlinear order (Fact 6.53). Meier and Staffelbach [834] presented two new so-called *fast correlation attacks* which are more efficient than Siegenthaler's attack in the case where the component LFSRs have sparse feedback polynomials, or if they have low-weight polynomial multiples (e.g., each having fewer than 10 non-zero terms) of not too large a degree. Further extensions and refinements of correlation attacks can be found in the papers of Mihaljević and Golić [874], Chepyzhov and Smeets [256], Golić and Mihaljević [491], Mihaljević and J. Golić [875], Mihaljević [873], Clark, Golić, and Dawson [262], and Penzhorn and Kühn [967]. A comprehensive survey of correlation attacks on LFSR-based stream ciphers is the paper by Golić [486]; the cases where the combining function is memoryless or with memory, as well as when the LFSRs are clocked regularly or irregularly, are all considered.

The summation generator (Example 6.54) was proposed by Rueppel [1075, 1076]. Meier and Staffelbach [837] presented correlation attacks on combination generators having memory, cracked the summation generator having only two component LFSRs, and as a result recommended using several LFSRs of moderate lengths rather than just a few long LFSRs in the summation generator. As an example, if a summation generator employs two LFSRs each having length approximately 200, and if $50,000$ keystream bits are known, then Meier and Staffelbach's attack is expected to take less than 700 trials, where the dominant step in each trial involves solving a 400×400 system of binary linear equations. Dawson [312] presented another known-plaintext attack on summation generators having two component LFSRs, which requires fewer known keystream bits than Meier and Staffelbach's attack. Dawson's attack is only faster than that of Meier and Staffelbach in the case where both LFSRs are relatively short. Recently, Klapper and Goresky [678] showed that the summation generator has comparatively low 2-adic span (see page 218). More precisely, if a and b are two sequences of 2-adic span $\lambda_2(a)$ and $\lambda_2(b)$, respectively, and if s is the result of combining them with the summation generator, then the 2-adic span of s is at most $\lambda_2(a) + \lambda_2(b) + 2\lceil \lg(\lambda_2(a)) \rceil + 2\lceil \lg(\lambda_2(b)) \rceil + 6$. For example, if m-sequences of period $2^L - 1$ for $L = 7, 11, 13, 15, 16, 17$ are combined with the summation generator, then the resulting sequence has linear complexity nearly 2^{79}, but the 2-adic span is less than 2^{18}. Hence, the summation generator is vulnerable to a known-plaintext attack when the component LFSRs are all relatively short.

The probability distribution of the carry for addition of n random integers was analyzed by Staffelbach and Meier [1167]. It was proven that the carry is balanced for even n and biased for odd n. For $n = 3$ the carry is strongly biased, however, the bias converges to 0 as n tends to ∞. Golić [485] pointed out the importance of the correlation between linear functions of the output and input in general combiners with memory, and introduced the so-called *linear sequential circuit approximation method* for finding such functions that produce correlated sequences. Golić [488] used this as a basis for developing a *linear cryptanalysis* technique for stream ciphers, and in the same paper proposed a stream cipher called GOAL, incorporating principles of modified truncated linear congruential generators (see page 187), self-clock-control, and randomly generated combiners with memory.

Fact 6.55(i) is due to Key [670], while Fact 6.55(ii) was proven by Rueppel [1075]. Massey and Serconek [794] gave an alternate proof of Key's bound that is based on the Discrete Fourier Transform. Siegenthaler [1134] described a correlation attack on nonlinear filter generators. Forré [418] has applied fast correlation attacks to such generators. Anderson [29] demonstrated other correlations which may be useful in improving the success of correlation attacks. An attack called the *inversion attack*, proposed by Golić [490], may be more effective than Anderson's attack. Golić also provides a list of design criteria for nonlinear filter generators. Ding [349] introduced the notion of differential cryptanalysis for nonlinear filter generators where the LFSR is replaced by a simple counter having arbitrary period.

The *linear consistency attack* of Zeng, Yang, and Rao [1264] is a known-plaintext attack on keystream generators which can discover key redundancies in various generators. It is effective in situations where it is possible to single out a certain portion k_1 of the secret key k, and form a linear system equations $Ax = b$ where the matrix A is determined by k_1, and b is determined from the known keystream. The system of equations should have the property that it is consistent (and with high probability has a unique solution) if k_1 is the true value of the subkey, while it is inconsistent with high probability otherwise. In these circumstances, one can mount an exhaustive search for k_1, and subsequently mount a separate attack for the remaining bits of k. If the bitlengths of k_1 and k are l_1 and l, respectively, the attack demonstrates that the security level of the generator is $2^{l_1} + 2^{l-l_1}$, rather than 2^l.

The *multiplexer generator* was proposed by Jennings [637]. Two maximum-length LFSRs having lengths L_1, L_2 that are relatively prime are employed. Let h be a positive integer satisfying $h \leq \min(L_1, \lg L_2)$. After each clock cycle, the contents of a fixed subset of h stages of the first LFSR are selected, and converted to an integer t in the interval $[0, L_2 - 1]$ using a $1 - 1$ mapping θ. Finally, the content of stage t of the second LFSR is output as part of the keystream. Assuming that the connection polynomials of the LFSRs are known, the linear consistency attack provides a known-plaintext attack on the multiplexer generator requiring a known keystream sequence of length $N \geq L_1 + L_2 2^h$ and 2^{L_1+h} linear consistency tests. This demonstrates that the choice of the mapping θ and the second LFSR do not contribute significantly to the security of the generator.

The linear consistency attack has also been considered by Zeng, Yang, and Rao [1264] for the *multispeed inner-product generator* of Massey and Rueppel [793]. In this generator, two LFSRs of lengths L_1 and L_2 are clocked at different rates, and their contents combined at the lower clock rate by taking the inner-product of the $\min(L_1, L_2)$ stages of the two LFSRs. The paper by Zeng et al. [1266] is a readable survey describing the effectiveness of the linear consistency and linear syndrome attacks in cryptanalyzing stream ciphers.

The knapsack generator (Example 6.56) was proposed by Rueppel and Massey [1084] and extensively analyzed by Rueppel [1075], however, no concrete suggestions on selecting appropriate parameters (the length L of the LFSR and the knapsack weights) for the generator were given. No weaknesses of the knapsack generator have been reported in the literature.

The idea of using the output of a register to control the stepping of another register was used in several rotor machines during the second world war, for example, the German Lorenz SZ40 cipher. A description of this cipher, and also an extensive survey of clock-controlled shift registers, is provided by Gollmann and Chambers [496].

The alternating step generator (Algorithm 6.57) was proposed in 1987 by Günther [528], who also proved Fact 6.59 and described the divide-and-conquer attack mentioned in Note 6.60. The alternating step generator is based on the *stop-and-go* generator of Beth and Piper [126]. In the stop-and-go generator, a control register R_1 is used to control the stepping of another register R_2 as follows. If the output of R_1 is 1, then R_2 is clocked; if the output of R_1 is 0, then R_2 is not clocked, however, its previous output is repeated. The output of R_2 is then XORed with the output sequence of a third register R_3 which is clocked at the same rate as R_1. Beth and Piper showed how a judicious choice of registers R_1, R_2, and R_3 can guarantee that the output sequence has high linear complexity and period, and good statistical properties. Unfortunately, the generator succumbs to the linear syndrome attack of Zeng, Yang, and Rao [1265] (see also page 218): if the connection polynomials of R_1 and R_2 are primitive trinomials of degree not exceeding n, and known to the adversary, then the initial states of the three component LFSRs (i.e., the secret key) can be efficiently recovered from a known-plaintext segment of length $37n$ bits.

Another variant of the stop-and-go generator is the *step-1/step-2* generator due to Gollmann and Chambers [496]. This generator uses two maximum-length registers R_1 and R_2 of the same length. Register R_1 is used to control the stepping of R_2 as follows. If the output of R_1 is 0, then R_2 is clocked once; if the output of R_1 is 1, then R_2 is clocked twice before producing the next output bit. Živković [1274] proposed an *embedding correlation attack* on R_2 whose complexity of $O(2^{L_2})$, where L_2 is the length of R_2.

A *cyclic register* of length L is an LFSR with feedback polynomial $C(D) = 1 + D^L$. Gollmann [494] proposed *cascading* n cyclic registers of the same prime length p by arranging them serially in such a way that all except the first register are clock-controlled by their predecessors; the Gollmann *p-cycle cascade* can be viewed as an extension of the stop-and-go generator (page 220). The first register is clocked regularly, and its output bit is the input bit to the second register. In general, if the input bit to the ith register (for $i \geq 2$) at time t is a_t, then the ith register is clocked if $a_t = 1$; if $a_t = 0$, the register is not clocked but its previous output bit is repeated. The output bit of the ith register is then XORed with a_t, and the result becomes the input bit to the $(i+1)^{\text{st}}$ register. The output of the last register is the output of the p-cycle cascade. The initial (secret) stage of a component cyclic register should not be the all-0's vector or the all-1's vector. Gollmann proved that the period of the output sequence is p^n. Moreover, if p is a prime such that 2 is a generator of $\mathbb{Z}_p^*$, then the output sequence has linear complexity p^n. This suggests very strongly using long cascades (i.e., n large) of shorter registers rather than short cascades of longer registers. A variant of the Gollmann cascade, called an *m-sequence cascade*, has the cyclic registers replaced by maximum-length LFSRs of the same length L. Chambers [237] showed that the output sequence of such an m-sequence cascade has period $(2^L - 1)^n$ and linear complexity at least $L(2^L - 1)^{n-1}$. Park, Lee, and Goh [964] extended earlier work of Menicocci [845] and reported breaking 9-stage m-sequence cascades where each LFSR has length 100; they also suggested that 10-stage m-sequence cascades may be insecure. Chambers and Gollmann [239] studied an attack on p-cycle and m-sequence cascades called *lock-in*, which results in a reduction in the effective key space of the cascades.

The shrinking generator (Algorithm 6.61) was proposed in 1993 by Coppersmith, Krawczyk, and Mansour [279], who also proved Fact 6.63 and described the attacks mentioned in Note 6.64. The irregular output rate of the shrinking generator can be overcome by using a short buffer for the output; the influence of such a buffer is analyzed by Kessler and Krawczyk [669]. Krawczyk [716] mentions some techniques for improving software implementations. A throughput of 2.5 Mbits/sec is reported for a C language implementation on a 33MHz IBM workstation, when the two shift registers each have lengths in the range 61–64 bits and secret connections are employed. The security of the shrinking generator is studied further by Golić [487].

A key generator related to the shrinking generator is the *self-shrinking generator* (SSG) of Meier and Staffelbach [838]. The self-shrinking generator uses only one maximum-length LFSR R. The output sequence of R is partitioned into pairs of bits. The SSG outputs a 0 if a pair is 10, and outputs a 1 if a pair is 11; 01 and 00 pairs are discarded. Meier and Staffelbach proved that the self-shrinking generator can be implemented as a shrinking generator. Moreover, the shrinking generator can be implemented as a self-shrinking generator (whose component LFSR is not maximum-length). More precisely, if the component LFSRs of a shrinking generator have connection polynomials $C_1(D)$ and $C_2(D)$, its output sequence can be produced by a self-shrinking generator with connection polynomial $C(D) = C_1(D)^2 \cdot C_2(D)^2$. Meier and Staffelbach also proved that if the length of R is L, then the period and linear complexity of the output sequence of the SSG are at least $2^{\lfloor L/2 \rfloor}$ and $2^{\lfloor L/2 \rfloor - 1}$, respectively. Moreover, they provided strong evidence that this period and linear complexity is in fact about 2^{L-1}. Assuming a randomly chosen, but known, connection polynomial, the best attack presented by Meier and Staffelbach on the SSG takes $2^{0.79L}$ steps. More recently, Mihaljević [871] presented a significantly faster probabilistic attack on the SSG. For example, if $L = 100$, then the new attack takes 2^{57} steps and requires a portion of the output sequence of length 4.9×10^8. The attack does not have an impact on the security of the shrinking generator.

A recent survey of techniques for attacking clock-controlled generators is given by Gollmann [495]. For some newer attack techniques, see Mihaljević [872], Golić and O'Connor [492], and Golić [489]. Chambers [238] proposed a clock-controlled cascade composed of LFSRs each of length 32. Each 32-bit portion of the output sequence of a component LFSR is passed through an invertible scrambler box (*S-box*), and the resulting 32-bit sequence is used to control the clock of the next LFSR. Baum and Blackburn [77] generalized the notion of a clock-controlled shift register to that of a register based on a finite group.

§6.4

SEAL (Algorithm 6.68) was designed and patented by Rogaway and Coppersmith [281]. Rogaway and Coppersmith [1066] report an encryption speed of 7.2 Mbytes/sec for an assembly language implementation on a 50 MHz 486 processor with $L = 4096$ bits, assuming precomputed tables (cf. Note 6.66).

Although the stream cipher RC4 remains proprietary, alleged descriptions have been published which are output compatible with certified implementations of RC4; for example, see Schneier [1094]. Blöcher and Dichtl [156] proposed a fast software stream cipher called *FISH* (Fibonacci Shrinking generator), which is based on the shrinking generator principle applied to the lagged Fibonacci generator (also known as the additive generator) of Knuth [692, p.27]. Anderson [28] subsequently presented a known-plaintext attack on FISH which requires a few thousand 32-bit words of known plaintext and a work factor of about 2^{40} computations. Anderson also proposed a fast software stream cipher called *PIKE* based on the Fibonacci generator and the stream cipher A5; a description of A5 is given by Anderson [28].

Wolfram [1251, 1252] proposed a stream cipher based on one-dimensional cellular automata with nonlinear feedback. Meier and Staffelbach [835] presented a known-plaintext attack on this cipher which demonstrated that key lengths of 127 bits suggested by Wolfram [1252] are insecure; Meier and Staffelbach recommend key sizes of about 1000 bits.

Klapper and Goresky [679] presented constructions for FCSRs (see page 217) whose output sequences have nearly maximal period, are balanced, and are nearly de Bruijn sequences in the sense that for any fixed non-negative integer t, the number of occurrences of any two t-bit sequences as subsequences of a period differs by at most 2. Such FCSRs are good candidates for usage in the construction of secure stream ciphers, just as maximum-length LFSRs were used in §6.3. Goresky and Klapper [518] introduced a generalization of FCSRs called d-FCSRs, based on *ramified* extensions of the 2-adic numbers (d is the ramification).

7

Block Ciphers

Contents in Brief

7.1 Introduction and overview

Symmetric-key block ciphers are the most prominent and important elements in many cryptographic systems. Individually, they provide confidentiality. As a fundamental building block, their versatility allows construction of pseudorandom number generators, stream ciphers, MACs, and hash functions. They may furthermore serve as a central component in message authentication techniques, data integrity mechanisms, entity authentication protocols, and (symmetric-key) digital signature schemes. This chapter examines symmetric-key block ciphers, including both general concepts and details of specific algorithms. Public-key block ciphers are discussed in Chapter 8.

No block cipher is ideally suited for all applications, even one offering a high level of security. This is a result of inevitable tradeoffs required in practical applications, including those arising from, for example, speed requirements and memory limitations (e.g., code size, data size, cache memory), constraints imposed by implementation platforms (e.g., hardware, software, chipcards), and differing tolerances of applications to properties of various modes of operation. In addition, efficiency must typically be traded off against security. Thus it is beneficial to have a number of candidate ciphers from which to draw.

Of the many block ciphers currently available, focus in this chapter is given to a subset of high profile and/or well-studied algorithms. While not guaranteed to be more secure than other published candidate ciphers (indeed, this status changes as new attacks become known), emphasis is given to those of greatest practical interest. Among these, DES is paramount; FEAL has received both serious commercial backing and a large amount of independent cryptographic analysis; and IDEA (originally proposed as a DES replacement) is widely known and highly regarded. Other recently proposed ciphers of both high promise and high profile (in part due to the reputation of their designers) are SAFER and RC5. Additional ciphers are presented in less detail.

Chapter outline

Basic background on block ciphers and algorithm-independent concepts are presented in §7.2, including modes of operation, multiple encryption, and exhaustive search techniques. Classical ciphers and cryptanalysis thereof are addressed in §7.3, including historical details on cipher machines. Modern block ciphers covered in chronological order are DES (§7.4), FEAL (§7.5), and IDEA (§7.6), followed by SAFER, RC5, and other ciphers in §7.7, collectively illustrating a wide range of modern block cipher design approaches. Further notes, including details on additional ciphers (e.g., Lucifer) and references for the chapter, may be found in §7.8.

7.2 Background and general concepts

Introductory material on block ciphers is followed by subsections addressing modes of operation, and discussion of exhaustive key search attacks and multiple encryption.

7.2.1 Introduction to block ciphers

Block ciphers can be either symmetric-key or public-key. The main focus of this chapter is symmetric-key block ciphers; public-key encryption is addressed in Chapter 8.

(i) Block cipher definitions

A block cipher is a function (see §1.3.1) which maps n-bit plaintext blocks to n-bit ciphertext blocks; n is called the *blocklength*. It may be viewed as a simple substitution cipher with large character size. The function is parameterized by a k-bit key K,[1] taking values from a subset $\mathcal{K}$ (the *key space*) of the set of all k-bit vectors V_k. It is generally assumed that the key is chosen at random. Use of plaintext and ciphertext blocks of equal size avoids data expansion.

To allow unique decryption, the encryption function must be one-to-one (i.e., invertible). For n-bit plaintext and ciphertext blocks and a fixed key, the encryption function is a bijection, defining a permutation on n-bit vectors. Each key potentially defines a different bijection. The number of keys is $|\mathcal{K}|$, and the *effective key size* is $\lg |\mathcal{K}|$; this equals the key length if all k-bit vectors are valid keys ($\mathcal{K} = V_k$). If keys are equiprobable and each defines a different bijection, the *entropy* of the key space is also $\lg |\mathcal{K}|$.

7.1 Definition An n-bit *block cipher* is a function $E : V_n \times \mathcal{K} \to V_n$, such that for each key $K \in \mathcal{K}$, $E(P, K)$ is an invertible mapping (the *encryption function* for K) from V_n to V_n, written $E_K(P)$. The inverse mapping is the *decryption function*, denoted $D_K(C)$. $C = E_K(P)$ denotes that ciphertext C results from encrypting plaintext P under K.

Whereas block ciphers generally process plaintext in relatively large blocks (e.g., $n \geq 64$), stream ciphers typically process smaller units (see Note 6.1); the distinction, however, is not definitive (see Remark 7.25). For plaintext messages exceeding one block in length, various modes of operation for block ciphers are used (see §7.2.2).

The most general block cipher implements every possible substitution, as per Definition 7.2. To represent the key of such an n-bit (true) random block cipher would require

---

[1] This use of symbols k and K may differ from other chapters.

$\lg(2^n!) \approx (n - 1.44)2^n$ bits, or roughly 2^n times the number of bits in a message block. This excessive bitsize makes (true) random ciphers impractical. Nonetheless, it is an accepted design principle that the encryption function corresponding to a randomly selected key should *appear* to be a randomly chosen invertible function.

7.2 Definition A *(true) random cipher* is an n-bit block cipher implementing all $2^n!$ bijections on 2^n elements. Each of the $2^n!$ keys specifies one such permutation.

A block cipher whose block size n is too small may be vulnerable to attacks based on statistical analysis. One such attack involves simple frequency analysis of ciphertext blocks (see Note 7.74). This may be thwarted by appropriate use of modes of operation (e.g., Algorithm 7.13). Other such attacks are considered in Note 7.8. However, choosing too large a value for the blocksize n may create difficulties as the complexity of implementation of many ciphers grows rapidly with block size. In practice, consequently, for larger n, easily-implementable functions are necessary which *appear* to be random (without knowledge of the key).

An encryption function per Definition 7.1 is a deterministic mapping. Each pairing of plaintext block P and key K maps to a unique ciphertext block. In contrast, in a randomized encryption technique (Definition 7.3; see also Remark 8.22), each (P, K) pair is associated with a set $C_{(P,K)}$ of eligible ciphertext blocks; each time P is encrypted under K, an output R from a random source non-deterministically selects one of these eligible blocks. To ensure invertibility, for every fixed key K, the subsets $C_{(P,K)}$ over all plaintexts P must be disjoint. Since the encryption function is essentially one-to-many involving an additional parameter R (cf. homophonic substitution, §7.3.2), the requirement for invertibility implies data expansion, which is a disadvantage of randomized encryption and is often unacceptable.

7.3 Definition A *randomized encryption* mapping is a function E from a plaintext space V_n to a ciphertext space V_m, $m > n$, drawing elements from a space of random numbers $\mathcal{R} = V_t$. E is defined by $E : V_n \times \mathcal{K} \times \mathcal{R} \to V_m$, such that for each key $K \in \mathcal{K}$ and $R \in \mathcal{R}$, $E(P, K, R)$, also written $E_K^R(P)$, maps $P \in V_n$ to V_m; and an inverse (corresponding decryption) function exists, mapping $V_m \times \mathcal{K} \to V_n$.

(ii) Practical security and complexity of attacks

The objective of a block cipher is to provide confidentiality. The corresponding objective of an adversary is to recover plaintext from ciphertext. A block cipher is *totally broken* if a key can be found, and *partially broken* if an adversary is able to recover part of the plaintext (but not the key) from ciphertext.

7.4 Note *(standard assumptions)* To evaluate block cipher security, it is customary to always assume that an adversary (i) has access to all data transmitted over the ciphertext channel; and (ii) *(Kerckhoffs' assumption)* knows all details of the encryption function except the secret key (which security consequently rests entirely upon).

Under the assumptions of Note 7.4, attacks are classified based on what information a cryptanalyst has access to in addition to intercepted ciphertext (cf. §1.13.1). The most prominent classes of attack for symmetric-key ciphers are (for a fixed key):

1. *ciphertext-only* – no additional information is available.
2. *known-plaintext* – plaintext-ciphertext pairs are available.

3. *chosen-plaintext* – ciphertexts are available corresponding to plaintexts of the adversary's choice. A variation is an *adaptive chosen-plaintext* attack, where the choice of plaintexts may depend on previous plaintext-ciphertext pairs.

Additional classes of attacks are given in Note 7.6; while somewhat more hypothetical, these are nonetheless of interest for the purposes of analysis and comparison of ciphers.

7.5 Remark (*chosen-plaintext principle*) It is customary to use ciphers resistant to chosen-plaintext attack even when mounting such an attack is not feasible. A cipher secure against chosen-plaintext attack is secure against known-plaintext and ciphertext-only attacks.

7.6 Note (*chosen-ciphertext and related-key attacks*) A *chosen-ciphertext* attack operates under the following model: an adversary is allowed access to plaintext-ciphertext pairs for some number of ciphertexts of his choice, and thereafter attempts to use this information to recover the key (or plaintext corresponding to some new ciphertext). In a *related-key attack*, an adversary is assumed to have access to the encryption of plaintexts under both an unknown key and (unknown) keys chosen to have or known to have certain relationships with this key.

With few exceptions (e.g., the one-time pad), the best available measure of security for practical ciphers is the complexity of the best (currently) known attack. Various aspects of such complexity may be distinguished as follows:

1. *data complexity* – expected number of input data units required (e.g., ciphertext).
2. *storage complexity* – expected number of storage units required.
3. *processing complexity* – expected number of operations required to process input data and/or fill storage with data (at least one time unit per storage unit).

The *attack complexity* is the dominant of these (e.g., for linear cryptanalysis on DES, essentially the data complexity). When parallelization is possible, processing complexity may be divided across many processors (but not reduced), reducing attack time.

Given a data complexity of 2^n, an attack is always possible; this many different n-bit blocks completely characterize the encryption function for a fixed k-bit key. Similarly, given a processing complexity of 2^k, an attack is possible by exhaustive key search (§7.2.3). Thus as a minimum, the effective key size should be sufficiently large to preclude exhaustive key search, and the block size sufficiently large to preclude exhaustive data analysis. A block cipher is considered *computationally secure* if these conditions hold and no known attack has both data and processing complexity significantly less than, respectively, 2^n and 2^k. However, see Note 7.8 for additional concerns related to block size.

7.7 Remark (*passive vs. active complexity*) For symmetric-key block ciphers, data complexity is beyond the control of the adversary, and is *passive complexity* (plaintext-ciphertext pairs cannot be generated by the adversary itself). Processing complexity is *active complexity* which typically benefits from increased resources (e.g., parallelization).

7.8 Note (*attacks based on small block size*) Security concerns which arise if the block size n is too small include the feasibility of *text dictionary attacks* and *matching ciphertext attacks*. A text dictionary may be assembled if plaintext-ciphertext pairs become known for a fixed key. The more pairs available, the larger the dictionary and the greater the chance of locating a random ciphertext block therein. A complete dictionary results if 2^n plaintext-ciphertext pairs become known, and fewer suffice if plaintexts contain redundancy and a non-chaining mode of encryption (such as ECB) is used. Moreover, if about $2^{n/2}$ such pairs are known, and about $2^{n/2}$ ciphertexts are subsequently created, then by the birthday paradox one expects to locate a ciphertext in the dictionary. Relatedly, from ciphertext blocks alone, as the number of available blocks approaches $2^{n/2}$, one expects to find matching ciphertext blocks. These may reveal partial information about the corresponding plaintexts, depending on the mode of operation of the block cipher, and the amount of redundancy in the plaintext.

Computational and unconditional security are discussed in §1.13.3. Unconditional security is both unnecessary in many applications and impractical; for example, it requires as many bits of secret key as plaintext, and cannot be provided by a block cipher used to encrypt more than one block (due to Fact 7.9, since identical ciphertext implies matching plaintext). Nonetheless, results on unconditional security provide insight for the design of practical ciphers, and has motivated many of the principles of cryptographic practice currently in use (see Remark 7.10).

7.9 Fact A cipher provides *perfect secrecy* (unconditional security) if the ciphertext and plaintext blocks are statistically independent.

7.10 Remark (*theoretically-motivated principles*) The unconditional security of the one-time-pad motivates both additive stream ciphers (Chapter 6) and the frequent changing of cryptographic keys (§13.3.1). Theoretical results regarding the effect of redundancy on unicity distance (Fact 7.71) motivate the principle that for plaintext confidentiality, the plaintext data should be as random as possible, e.g., via data-compression prior to encryption, use of random-bit fields in message blocks, or randomized encryption (Definition 7.3). The latter two techniques may, however, increase the data length or allow covert channels.

(iii) Criteria for evaluating block ciphers and modes of operation

Many criteria may be used for evaluating block ciphers in practice, including:

1. *estimated security level*. Confidence in the (historical) security of a cipher grows if it has been subjected to and withstood expert cryptanalysis over a substantial time period, e.g., several years or more; such ciphers are certainly considered more secure than those which have not. This may include the performance of selected cipher components relative to various design criteria which have been proposed or gained favor in recent years. The amount of ciphertext required to mount practical attacks often vastly exceeds a cipher's unicity distance (Definition 7.69), which provides a theoretical estimate of the amount of ciphertext required to recover the unique encryption key.

2. *key size*. The effective bitlength of the key, or more specifically, the entropy of the key space, defines an upper bound on the security of a cipher (by considering exhaustive search). Longer keys typically impose additional costs (e.g., generation, transmission, storage, difficulty to remember passwords).

3. *throughput*. Throughput is related to the complexity of the cryptographic mapping (see below), and the degree to which the mapping is tailored to a particular implementation medium or platform.

4. *block size*. Block size impacts both security (larger is desirable) and complexity (larger is more costly to implement). Block size may also affect performance, for example, if padding is required.

5. *complexity of cryptographic mapping*. Algorithmic complexity affects the implementation costs both in terms of development and fixed resources (hardware gate

count or software code/data size), as well as real-time performance for fixed resources (throughput). Some ciphers specifically favor hardware or software implementations.

6. *data expansion*. It is generally desirable, and often mandatory, that encryption does not increase the size of plaintext data. Homophonic substitution and randomized encryption techniques result in data expansion.

7. *error propagation*. Decryption of ciphertext containing bit errors may result in various effects on the recovered plaintext, including propagation of errors to subsequent plaintext blocks. Different error characteristics are acceptable in various applications. Block size (above) typically affects error propagation.

7.2.2 Modes of operation

A block cipher encrypts plaintext in fixed-size n-bit blocks (often $n = 64$). For messages exceeding n bits, the simplest approach is to partition the message into n-bit blocks and encrypt each separately. This electronic-codebook (ECB) mode has disadvantages in most applications, motivating other methods of employing block ciphers (*modes of operation*) on larger messages. The four most common modes are ECB, CBC, CFB, and OFB. These are summarized in Figure 7.1 and discussed below.

In what follows, E_K denotes the encryption function of the block cipher E parameterized by key K, while E_K^{-1} denotes decryption (cf. Definition 7.1). A plaintext message $x = x_1 \ldots x_t$ is assumed to consist of n-bit blocks for ECB and CBC modes (see Algorithm 9.58 regarding padding), and r-bit blocks for CFB and OFB modes for appropriate fixed $r \leq n$.

(i) ECB mode

The *electronic codebook* (ECB) mode of operation is given in Algorithm 7.11 and illustrated in Figure 7.1(a).

7.11 Algorithm ECB mode of operation

INPUT: k-bit key K; n-bit plaintext blocks $x_1, \ldots, x_t$.
SUMMARY: produce ciphertext blocks $c_1, \ldots, c_t$; decrypt to recover plaintext.
1. Encryption: for $1 \leq j \leq t$, $c_j \leftarrow E_K(x_j)$.
2. Decryption: for $1 \leq j \leq t$, $x_j \leftarrow E_K^{-1}(c_j)$.

Properties of the ECB mode of operation:
1. Identical plaintext blocks (under the same key) result in identical ciphertext.
2. Chaining dependencies: blocks are enciphered independently of other blocks. Re-ordering ciphertext blocks results in correspondingly re-ordered plaintext blocks.
3. Error propagation: one or more bit errors in a single ciphertext block affect decipherment of that block only. For typical ciphers E, decryption of such a block is then random (with about 50% of the recovered plaintext bits in error). Regarding bits being deleted, see Remark 7.15.

7.12 Remark (*use of ECB mode*) Since ciphertext blocks are independent, malicious substitution of ECB blocks (e.g., insertion of a frequently occurring block) does not affect the decryption of adjacent blocks. Furthermore, block ciphers do not hide data patterns – identical ciphertext blocks imply identical plaintext blocks. For this reason, the ECB mode is not recommended for messages longer than one block, or if keys are reused for more than

a) Electronic Codebook (ECB)   b) Cipher-block Chaining (CBC)

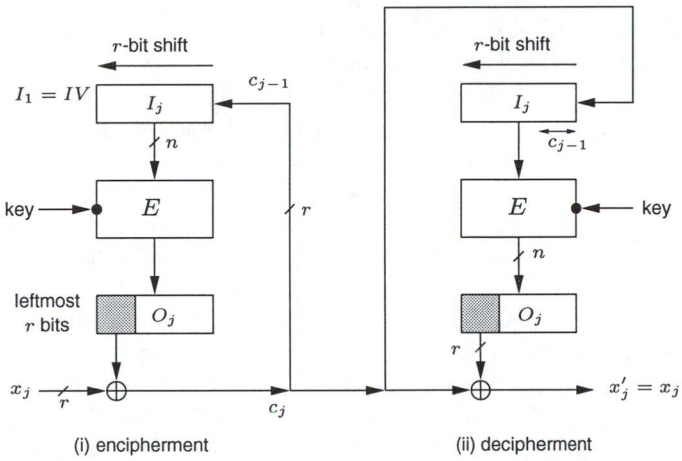

c) Cipher feedback (CFB), r-bit characters/r-bit feedback

d) Output feedback (OFB), r-bit characters/n-bit feedback

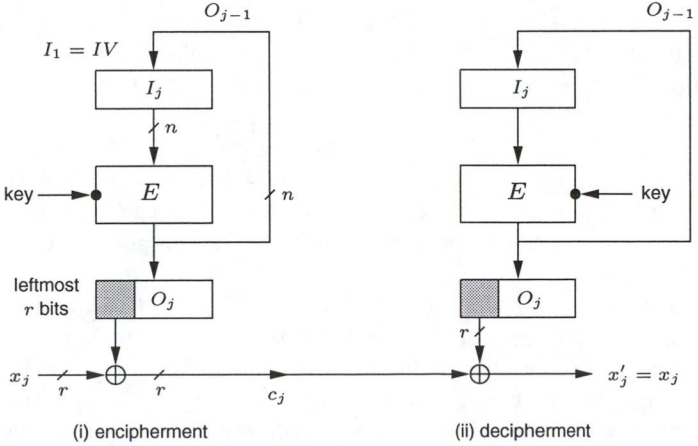

Figure 7.1: *Common modes of operation for an n-bit block cipher.*

a single one-block message. Security may be improved somewhat by inclusion of random padding bits in each block.

(ii) CBC mode

The *cipher-block chaining* (CBC) mode of operation, specified in Algorithm 7.13 and illustrated in Figure 7.1(b), involves use of an n-bit initialization vector, denoted IV.

7.13 Algorithm CBC mode of operation

INPUT: k-bit key K; n-bit IV; n-bit plaintext blocks $x_1, \ldots, x_t$.
SUMMARY: produce ciphertext blocks $c_1, \ldots, c_t$; decrypt to recover plaintext.
  1. Encryption: $c_0 \leftarrow IV$. For $1 \leq j \leq t$, $c_j \leftarrow E_K(c_{j-1} \oplus x_j)$.
  2. Decryption: $c_0 \leftarrow IV$. For $1 \leq j \leq t$, $x_j \leftarrow c_{j-1} \oplus E_K^{-1}(c_j)$.

Properties of the CBC mode of operation:
  1. Identical plaintexts: identical ciphertext blocks result when the same plaintext is enciphered under the same key and IV. Changing the IV, key, or first plaintext block (e.g., using a counter or random field) results in different ciphertext.
  2. Chaining dependencies: the chaining mechanism causes ciphertext c_j to depend on x_j and all preceding plaintext blocks (the entire dependency on preceding blocks is, however, contained in the value of the previous ciphertext block). Consequently, re-arranging the order of ciphertext blocks affects decryption. Proper decryption of a correct ciphertext block requires a correct preceding ciphertext block.
  3. Error propagation: a single bit error in ciphertext block c_j affects decipherment of blocks c_j and c_{j+1} (since x_j depends on c_j and c_{j-1}). Block x_j' recovered from c_j is typically totally random (50% in error), while the recovered plaintext x_{j+1}' has bit errors precisely where c_j did. Thus an adversary may cause predictable bit changes in x_{j+1} by altering corresponding bits of c_j. See also Remark 7.14.
  4. Error recovery: the CBC mode is *self-synchronizing* or *ciphertext autokey* (see Remark 7.15) in the sense that if an error (including loss of one or more entire blocks) occurs in block c_j but not c_{j+1}, c_{j+2} is correctly decrypted to x_{j+2}.

7.14 Remark (*error propagation in encryption*) Although CBC mode decryption recovers from errors in ciphertext blocks, modifications to a plaintext block x_j during encryption alter all subsequent ciphertext blocks. This impacts the usability of chaining modes for applications requiring random read/write access to encrypted data. The ECB mode is an alternative (but see Remark 7.12).

7.15 Remark (*self-synchronizing vs. framing errors*) Although self-synchronizing in the sense of recovery from bit errors, recovery from "lost" bits causing errors in block boundaries (*framing integrity errors*) is not possible in the CBC or other modes.

7.16 Remark (*integrity of IV in CBC*) While the IV in the CBC mode need not be secret, its integrity should be protected, since malicious modification thereof allows an adversary to make predictable bit changes to the first plaintext block recovered. Using a secret IV is one method for preventing this. However, if message integrity is required, an appropriate mechanism should be used (see §9.6.5); encryption mechanisms typically guarantee confidentiality only.

(iii) CFB mode

While the CBC mode processes plaintext n bits at a time (using an n-bit block cipher), some applications require that r-bit plaintext units be encrypted and transmitted without delay, for some fixed $r < n$ (often $r = 1$ or $r = 8$). In this case, the *cipher feedback* (CFB) mode may be used, as specified in Algorithm 7.17 and illustrated in Figure 7.1(c).

7.18 Algorithm CFB mode of operation (CFB-r)

7.17 Algorithm CFB mode of operation (CFB-r)

INPUT: k-bit key K; n-bit IV; r-bit plaintext blocks $x_1, \dots, x_u$ $(1 \le r \le n)$.
SUMMARY: produce r-bit ciphertext blocks $c_1, \dots, c_u$; decrypt to recover plaintext.
1. Encryption: $I_1 \leftarrow IV$. (I_j is the input value in a shift register.) For $1 \le j \le u$:
   (a) $O_j \leftarrow E_K(I_j)$. (Compute the block cipher output.)
   (b) $t_j \leftarrow$ the r leftmost bits of O_j. (Assume the leftmost is identified as bit 1.)
   (c) $c_j \leftarrow x_j \oplus t_j$. (Transmit the r-bit ciphertext block c_j.)
   (d) $I_{j+1} \leftarrow 2^r \cdot I_j + c_j \bmod 2^n$. (Shift c_j into right end of shift register.)
2. Decryption: $I_1 \leftarrow IV$. For $1 \le j \le u$, upon receiving c_j:
   $x_j \leftarrow c_j \oplus t_j$, where t_j, O_j and I_j are computed as above.

Properties of the CFB mode of operation:
1. Identical plaintexts: as per CBC encryption, changing the IV results in the same plaintext input being enciphered to a different output. The IV need not be secret (although an unpredictable IV may be desired in some applications).
2. Chaining dependencies: similar to CBC encryption, the chaining mechanism causes ciphertext block c_j to depend on both x_j and preceding plaintext blocks; consequently, re-ordering ciphertext blocks affects decryption. Proper decryption of a correct ciphertext block requires the preceding $\lceil n/k \rceil$ ciphertext blocks to be correct (so that the shift register contains the proper value).
3. Error propagation: one or more bit errors in any single r-bit ciphertext block c_j affects the decipherment of that and the next $\lceil n/k \rceil$ ciphertext blocks (i.e., until n bits of ciphertext are processed, after which the error block c_j has shifted entirely out of the shift register). The recovered plaintext x'_j will differ from x_j precisely in the bit positions c_j was in error; the other incorrectly recovered plaintext blocks will typically be random vectors, i.e., have 50% of bits in error. Thus an adversary may cause predictable bit changes in x_j by altering corresponding bits of c_j.
4. Error recovery: the CFB mode is self-synchronizing similar to CBC, but requires $\lceil n/k \rceil$ ciphertext blocks to recover.
5. Throughput: for $r < n$, throughput is decreased by a factor of n/r (vs. CBC) in that each execution of E yields only r bits of ciphertext output.

7.18 Remark (*CFB use of encryption only*) Since the encryption function E is used for both CFB encryption and decryption, the CFB mode must not be used if the block cipher E is a public-key algorithm; instead, the CBC mode should be used.

7.19 Example (*ISO variant of CFB*) The CFB mode of Algorithm 7.17 may be modified as follows, to allow processing of plaintext blocks (characters) whose bitsize s is less than the bitsize r of the feedback variable (e.g., 7-bit characters using 8-bit feedback; $s < r$). The leftmost s (rather than r) bits of O_j are assigned to t_j; the s-bit ciphertext character c_j is computed; the feedback variable is computed from c_j by pre-prepending (on the left) $r - s$ 1-bits; the resulting r-bit feedback variable is shifted into the least significant (LS) end of the shift register as before. $\square$

(iv) OFB mode

The *output feedback* (OFB) mode of operation may be used for applications in which all error propagation must be avoided. It is similar to CFB, and allows encryption of various block sizes (characters), but differs in that the output of the encryption block function E (rather than the ciphertext) serves as the feedback.

Two versions of OFB using an n-bit block cipher are common. The ISO version (Figure 7.1(d) and Algorithm 7.20) requires an n-bit feedback, and is more secure (Note 7.24). The earlier FIPS version (Algorithm 7.21) allows $r < n$ bits of feedback.

7.20 Algorithm OFB mode with full feedback (per ISO 10116)

INPUT: k-bit key K; n-bit IV; r-bit plaintext blocks $x_1, \ldots, x_u$ ($1 \le r \le n$).
SUMMARY: produce r-bit ciphertext blocks $c_1, \ldots, c_u$; decrypt to recover plaintext.

1. Encryption: $I_1 \leftarrow IV$. For $1 \le j \le u$, given plaintext block x_j:
   (a) $O_j \leftarrow E_K(I_j)$. (Compute the block cipher output.)
   (b) $t_j \leftarrow$ the r leftmost bits of O_j. (Assume the leftmost is identified as bit 1.)
   (c) $c_j \leftarrow x_j \oplus t_j$. (Transmit the r-bit ciphertext block c_j.)
   (d) $I_{j+1} \leftarrow O_j$. (Update the block cipher input for the next block.)
2. Decryption: $I_1 \leftarrow IV$. For $1 \le j \le u$, upon receiving c_j:
   $x_j \leftarrow c_j \oplus t_j$, where t_j, O_j, and I_j are computed as above.

7.21 Algorithm OFB mode with r-bit feedback (per FIPS 81)

INPUT: k-bit key K; n-bit IV; r-bit plaintext blocks $x_1, \ldots, x_u$ ($1 \le r \le n$).
SUMMARY: produce r-bit ciphertext blocks $c_1, \ldots, c_u$; decrypt to recover plaintext.
As per Algorithm 7.20, but with "$I_{j+1} \leftarrow O_j$" replaced by:
   $I_{j+1} \leftarrow 2^r \cdot I_j + t_j \bmod 2^n$. (Shift output t_j into right end of shift register.)

Properties of the OFB mode of operation:

1. Identical plaintexts: as per CBC and CFB modes, changing the IV results in the same plaintext being enciphered to a different output.
2. Chaining dependencies: the keystream is plaintext-independent (see Remark 7.22).
3. Error propagation: one or more bit errors in any ciphertext character c_j affects the decipherment of only that character, in the precise bit position(s) c_j is in error, causing the corresponding recovered plaintext bit(s) to be complemented.
4. Error recovery: the OFB mode recovers from ciphertext bit errors, but cannot self-synchronize after loss of ciphertext bits, which destroys alignment of the decrypting keystream (in which case explicit re-synchronization is required).
5. Throughput: for $r < n$, throughput is decreased as per the CFB mode. However, in all cases, since the keystream is independent of plaintext or ciphertext, it may be pre-computed (given the key and IV).

7.22 Remark (*changing IV in OFB*) The IV, which need not be secret, must be changed if an OFB key K is re-used. Otherwise an identical keystream results, and by XORing corresponding ciphertexts an adversary may reduce cryptanalysis to that of a running-key cipher with one plaintext as the running key (cf. Example 7.58 ff.).

Remark 7.18 on public-key block ciphers applies to the OFB mode as well as CFB.

7.23 Example (*counter mode*) A simplification of OFB involves updating the input block as a counter, $I_{j+1} = I_j + 1$, rather than using feedback. This both avoids the short-cycle problem of Note 7.24, and allows recovery from errors in computing E. Moreover, it provides a random-access property: ciphertext block i need not be decrypted in order to decrypt block $i + 1$. □

7.24 Note (*OFB feedback size*) In OFB with full n-bit feedback (Algorithm 7.20), the keystream is generated by the iterated function $O_j = E_K(O_{j-1})$. Since E_K is a permutation, and under the assumption that for random K, E_K is effectively a random choice among all $(2^n)!$ permutations on n elements, it can be shown that for a fixed (random) key and starting value, the expected cycle length before repeating any value O_j is about 2^{n-1}. On the other hand, if the number of feedback bits is $r < n$ as allowed in Algorithm 7.21, the keystream is generated by the iteration $O_j = f(O_{j-1})$ for some non-permutation f which, assuming it behaves as a random function, has an expected cycle length of about $2^{n/2}$. Consequently, it is strongly recommended to use the OFB mode with full n-bit feedback.

7.25 Remark (*modes as stream ciphers*) It is clear that both the OFB mode with full feedback (Algorithm 7.20) and the counter mode (Example 7.23) employ a block cipher as a keystream generator for a stream cipher. Similarly the CFB mode encrypts a character stream using the block cipher as a (plaintext-dependent) keystream generator. The CBC mode may also be considered a stream cipher with n-bit blocks playing the role of very large characters. Thus modes of operation allow one to define stream ciphers from block ciphers.

7.2.3 Exhaustive key search and multiple encryption

A fixed-size key defines an upper bound on the security of a block cipher, due to exhaustive key search (Fact 7.26). While this requires either known-plaintext or plaintext containing redundancy, it has widespread applicability since cipher operations (including decryption) are generally designed to be computationally efficient.

A design technique which complicates exhaustive key search is to make the task of changing cipher keys computationally expensive, while allowing encryption with a fixed key to remain relatively efficient. Examples of ciphers with this property include the block cipher Khufu and the stream cipher SEAL.

7.26 Fact (*exhaustive key search*) For an n-bit block cipher with k-bit key, given a small number (e.g., $\lceil (k+4)/n \rceil$) of plaintext-ciphertext pairs encrypted under key K, K can be recovered by exhaustive key search in an expected time on the order of 2^{k-1} operations.

Justification: Progress through the entire key space, decrypting a fixed ciphertext C with each trial key, and discarding those keys which do not yield the known plaintext P. The target key is among the undiscarded keys. The number of false alarms expected (non-target keys which map C to P) depends on the relative size of k and n, and follows from unicity distance arguments; additional (P', C') pairs suffice to discard false alarms. One expects to find the correct key after searching half the key space.

7.27 Example (*exhaustive DES key search*) For DES, $k = 56$, $n = 64$, and the expected requirement by Fact 7.26 is 2^{55} decryptions and a single plaintext-ciphertext pair. □

If the underlying plaintext is known to contain redundancy as in Example 7.28, then ciphertext-only exhaustive key search is possible with a relatively small number of ciphertexts.

7.28 Example (*ciphertext-only DES key search*) Suppose DES is used to encrypt 64-bit blocks of 8 ASCII characters each, with one bit per character serving as an even parity bit. Trial decryption with an incorrect key K yields all 8 parity bits correct with probability 2^{-8}, and correct parity for t different blocks (each encrypted by K) with probability 2^{-8t}. If this is used as a filter over all 2^{56} keys, the expected number of unfiltered incorrect keys is $2^{56}/2^{8t}$. For most practical purposes, $t = 10$ suffices.   □

(i) Cascades of ciphers and multiple encryption

If a block cipher is susceptible to exhaustive key search (due to inadequate keylength), encipherment of the same message block more than once may increase security. Various such techniques for multiple encryption of n-bit messages are considered here. Once defined, they may be extended to messages exceeding one block by using standard modes of operation (§7.2.2), with E denoting multiple rather than single encryption.

7.29 Definition A *cascade cipher* is the concatenation of $L \geq 2$ block ciphers (called *stages*), each with independent keys. Plaintext is input to first stage; the output of stage i is input to stage $i + 1$; and the output of stage L is the cascade's ciphertext output.

In the simplest case, all stages in a cascade cipher have k-bit keys, and the stage inputs and outputs are all n-bit quantities. The stage ciphers may differ (*general cascade of ciphers*), or all be identical (*cascade of identical ciphers*).

7.30 Definition *Multiple encryption* is similar to a cascade of L identical ciphers, but the stage keys need not be independent, and the stage ciphers may be either a block cipher E or its corresponding decryption function $D = E^{-1}$.

Two important cases of multiple encryption are double and triple encryption, as illustrated in Figure 7.2 and defined below.

(a) double encryption

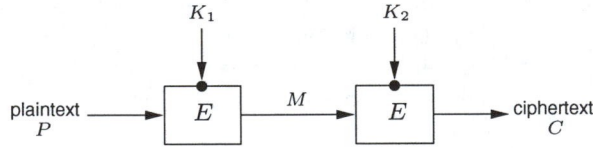

(b) triple encryption ($K_1 = K_3$ for two-key variant)

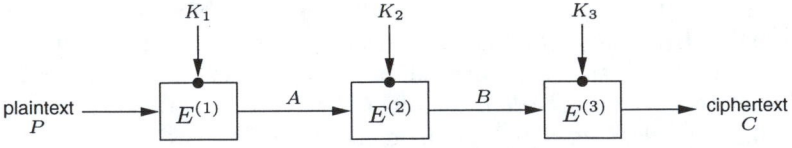

Figure 7.2: *Multiple encryption.*

7.31 Definition *Double encryption* is defined as $E(x) = E_{K_2}(E_{K_1}(x))$, where E_K denotes a block cipher E with key K.

7.32 Definition *Triple encryption* is defined as $E(x) = E_{K_3}^{(3)}(E_{K_2}^{(2)}(E_{K_1}^{(1)}(x)))$, where $E_K^{(j)}$ denotes either E_K or $D_K = E_K^{-1}$. The case $E(x) = E_{K_3}(D_{K_2}(E_{K_1}(x)))$ is called *E-D-E triple-encryption*; the subcase $K_1 = K_3$ is often called *two-key triple-encryption*.

Independent stage keys K_1 and K_2 are typically used in double encryption. In triple encryption (Definition 7.32), to save on key management and storage costs, dependent stage keys are often used. E-D-E triple-encryption with $K_1 = K_2 = K_3$ is backwards compatible with (i.e., equivalent to) single encryption.

(ii) Meet-in-the-middle attacks on multiple encryption

A naive exhaustive key search attack on double encryption tries all 2^{2k} key pairs. The attack of Fact 7.33 reduces time from 2^{2k}, at the cost of substantial space.

7.33 Fact For a block cipher with a k-bit key, a known-plaintext *meet-in-the-middle* attack defeats double encryption using on the order of 2^k operations and 2^k storage.

Justification (basic meet-in-the-middle): Noting Figure 7.2(a), given a (P, C) pair, compute $M_i = E_i(P)$ under all 2^k possible key values $K_1 = i$; store all pairs (M_i, i), sorted or indexed on M_i (e.g., using conventional hashing). Decipher C under all 2^k possible values $K_2 = j$, and for each pair (M_j, j) where $M_j = D_j(C)$, check for *hits* $M_j = M_i$ against entries M_i in the first table. (This can be done creating a second sorted table, or simply checking each M_j entry as generated.) Each hit identifies a candidate solution key pair (i, j), since $E_i(P) = M = D_j(C)$. Using a second known-plaintext pair (P', C') (cf. Fact 7.35), discard candidate key pairs which do not map P' to C'.

A concept analogous to unicity distance for ciphertext-only attack (Definition 7.69) can be defined for known-plaintext key search, based on the following strategy. Select a key; check if it is consistent with a given set (history) of plaintext-ciphertext pairs; if so, label the key a *hit*. A hit that is not the target key is a *false key hit*.

7.34 Definition The number of plaintext-ciphertext pairs required to uniquely determine a key under a known-plaintext key search is the *known-plaintext unicity distance*. This is the smallest integer t such that a history of length t makes false key hits improbable.

Using Fact 7.35, the (known-plaintext) unicity distance of a cascade of L random ciphers can be estimated. Less than one false hit is expected when $t > Lk/n$.

7.35 Fact For an L-stage cascade of random block ciphers with n-bit blocks and k-bit keys, the expected number of false key hits for a history of length t is about 2^{Lk-tn}.

Fact 7.35 holds with respect to random block ciphers defined as follows (cf. Definitions 7.2 and 7.70): given n and k, of the possible $(2^n)!$ permutations on 2^n elements, choose 2^k randomly and with equal probabilities, and associate these with the 2^k keys.

7.36 Example (*meet-in-the-middle – double-DES*) Applying Fact 7.33 to DES ($n = 64$, $k = 56$), the number of candidate key pairs expected for one (P, C) pair is $2^{48} = 2^k \cdot 2^k/2^n$, and the likelihood of a false key pair satisfying a second (P', C') sample is $2^{-16} = 2^{48}/2^n$. Thus with high probability, two (P, C) pairs suffice for key determination. This agrees with the unicity distance estimate of Fact 7.35: for $L = 2$, a history of length $t = 2$ yields 2^{-16} expected false key hits. □

A naive exhaustive attack on all key pairs in double-DES uses 2^{112} time and negligible space, while the meet-in-the-middle attack (Fact 7.33) requires 2^{56} time and 2^{56} space. Note 7.37 illustrates that the latter can be modified to yield a time-memory trade-off at any point between these two extremes, with the time-memory product essentially constant at 2^{112} (e.g., 2^{72} time, 2^{40} space).

7.37 Note (*time-memory tradeoff – double-encryption*) In the attack of Example 7.36, memory may be reduced (from tables of 2^{56} entries) by independently guessing s bits of each of K_1, K_2 (for any fixed s, $0 \leq s \leq k$). The tables then each have 2^{k-s} entries (fixing s key bits eliminates 2^s entries), but the attack must be run over $2^s \cdot 2^s$ pairs of such tables to allow all possible key pairs. The memory requirement is $2 \cdot 2^{k-s}$ entries (each $n+k-s$ bits, omitting s fixed key bits), while time is on the order of $2^{2s} \cdot 2^{k-s} = 2^{k+s}$. The time-memory product is 2^{2k+1}.

7.38 Note (*generalized meet-in-the-middle trade-off*) Variations of Note 7.37 allow time-space tradeoffs for meet-in-the-middle key search on any concatenation of $L \geq 2$ ciphers. For L even, meeting between the first and last $L/2$ stages results in requirements on the order of $2 \cdot 2^{(kL/2)-s}$ space and $2^{(kL/2)+s}$ time, $0 \leq s \leq kL/2$. For L odd, meeting after the first $(L-1)/2$ and before the last $(L+1)/2$ stages results in requirements on the order of $2 \cdot 2^{k(L-1)/2 - s}$ space and $2^{k(L+1)/2 + s}$ time, $1 \leq s \leq k(L-1)/2$.

For a block cipher with k-bit key, a naive attack on two-key triple encryption (Definition 7.32) involves trying all 2^{2k} key pairs. Fact 7.39 notes a chosen-plaintext alternative.

7.39 Fact For an n-bit block cipher with k-bit key, two-key triple encryption may be defeated by a chosen-plaintext attack requiring on the order of 2^k of each of the following: cipher operations, words of $(n+k)$-bit storage, and plaintext-ciphertext pairs with plaintexts chosen.

Justification (chosen-plaintext attack on two-key triple-encryption): Using 2^k chosen plaintexts, two-key triple encryption may be reduced to double-encryption as follows. Noting Figure 7.2(b), focus on the case where the result after the first encryption stage is the all-zero vector $A = 0$. For all 2^k values $K_1 = i$, compute $P_i = E_i^{-1}(A)$. Submit each resulting P_i as a chosen plaintext, obtaining the corresponding ciphertext C_i. For each, compute $B_i = E_i^{-1}(C_i)$, representing an intermediate result B after the second of three encryption stages. Note that the values P_i also represent candidate values B. Sort the values P_i and B_j in a table (using standard hashing for efficiency). Identify the keys corresponding to pairs $P_i = B_j$ as candidate solution key pairs $K_1 = i$, $K_2 = j$ to the given problem. Confirm these by testing each key pair on a small number of additional known plaintext-ciphertext pairs as required.

While generally impractical due to the storage requirement, the attack of Fact 7.39 is referred to as a *certificational attack* on two-key triple encryption, demonstrating it to be weaker than triple encryption. This motivates consideration of triple-encryption with three independent keys, although a penalty is a third key to manage.

Fact 7.40, stated specifically for DES ($n = 64$, $k = 56$), indicates that for the price of additional computation, the memory requirement in Fact 7.39 may be reduced and the chosen-plaintext condition relaxed to known-plaintext. The attack, however, appears impractical even with extreme parallelization; for example, for $\lg t = 40$, the number of operations is still 2^{80}.

7.40 Fact If t known plaintext-ciphertext pairs are available, an attack on two-key triple-DES requires $O(t)$ space and $2^{120-\lg t}$ operations.

(iii) Multiple-encryption modes of operation

In contrast to the *single modes* of operation in Figure 7.1, *multiple modes* are variants of multiple encryption constructed by concatenating selected single modes. For example, the combination of three single-mode CBC operations provides *triple-inner-CBC*; an alternative is *triple-outer-CBC*, the composite operation of triple encryption (per Definition 7.32) with one outer ciphertext feedback after the sequential application of three single-ECB operations. With replicated hardware, multiple modes such as triple-inner-CBC may be pipelined allowing performance comparable to single encryption, offering an advantage over triple-outer-CBC. Unfortunately (Note 7.41), they are often less secure.

7.41 Note (*security of triple-inner-CBC*) Many multiple modes of operation are weaker than the corresponding multiple-ECB mode (i.e., multiple encryption operating as a black box with only outer feedbacks), and in some cases multiple modes (e.g., ECB-CBC-CBC) are not significantly stronger than single encryption. In particular, under some attacks triple-inner-CBC is significantly weaker than triple-outer-CBC; against other attacks based on the block size (e.g., Note 7.8), it appears stronger.

(iv) Cascade ciphers

Counter-intuitively, it is possible to devise examples whereby cascading of ciphers (Definition 7.29) actually reduces security. However, Fact 7.42 holds under a wide variety of attack models and meaningful definitions of "breaking".

7.42 Fact A cascade of n (independently keyed) ciphers is at least as difficult to break as the first component cipher. Corollary: for stage ciphers which commute (e.g., additive stream ciphers), a cascade is at least as strong as the strongest component cipher.

Fact 7.42 does not apply to product ciphers consisting of component ciphers which may have dependent keys (e.g., two-key triple-encryption); indeed, keying dependencies across stages may compromise security entirely, as illustrated by a two-stage cascade wherein the components are two binary additive stream ciphers using an identical keystream – in this case, the cascade output is the original plaintext.

Fact 7.42 may suggest the following practical design strategy: cascade a set of keystream generators each of which relies on one or more different design principles. It is not clear, however, if this is preferable to one large keystream generator which relies on a single principle. The cascade may turn out to be less secure for a fixed set of parameters (number of key bits, block size), since ciphers built piecewise may often be attacked piecewise.

7.3 Classical ciphers and historical development

The term *classical ciphers* refers to encryption techniques which have become well-known over time, and generally created prior to the second half of the twentieth century (in some cases, many hundreds of years earlier). Many classical techniques are variations of simple substitution and simple transposition. Some techniques that are not technically block ciphers are also included here for convenience and context.

Classical ciphers and techniques are presented under §7.3 for historical and pedagogical reasons only. They illustrate important basic principles and common pitfalls. However, since these techniques are neither sophisticated nor secure against current cryptanalytic capabilities, *they are not generally suitable for practical use.*

7.3.1 Transposition ciphers (background)

For a *simple transposition* cipher with fixed period t, encryption involves grouping the plaintext into blocks of t characters, and applying to each block a single permutation e on the numbers 1 through t. More precisely, the ciphertext corresponding to plaintext block $m = m_1 \ldots m_t$ is $c = E_e(m) = m_{e(1)} \ldots m_{e(t)}$. The encryption key is e, which implicitly defines t; the key space $\mathcal{K}$ has cardinality $t!$ for a given value t. Decryption involves use of the permutation d which inverts e. The above corresponds to Definition 1.32.

The mathematical notation obscures the simplicity of the encryption procedure, as is evident from Example 7.43.

7.43 Example (*simple transposition*) Consider a simple transposition cipher with $t = 6$ and $e = (6\ 4\ 1\ 3\ 5\ 2)$. The message $m = $ CAESAR is encrypted to $c = $ RSCEAA. Decryption uses the inverse permutation $d = (3\ 6\ 4\ 2\ 5\ 1)$. The transposition may be represented by a two-row matrix with the second indicating the position to which the element indexed by the corresponding number of the first is mapped to: $\left(\begin{smallmatrix} 1 & 2 & 3 & 4 & 5 & 6 \\ 3 & 6 & 4 & 2 & 5 & 1 \end{smallmatrix} \right)$. Encryption may be done by writing a block of plaintext under headings "3 6 4 2 5 1", and then reading off the characters under the headings in numerical order.   □

7.44 Note (*terminology: transposition vs. permutation*) While the term "transposition" is traditionally used to describe a transposition cipher, the mapping of Example 7.43 may alternately be called a *permutation* on the set $\{1, 2, \ldots, 6\}$. The latter terminology is used, for example, in substitution-permutation networks, and in DES (§7.4).

A mnemonic keyword may be used in place of a key, although this may seriously decrease the key space entropy. For example, for $n = 6$, the keyword "CIPHER" could be used to specify the column ordering 1, 5, 4, 2, 3, 6 (by alphabetic priority).

7.45 Definition Sequential composition of two or more simple transpositions with respective periods $t_1, t_2, \ldots, t_i$ is called a *compound transposition.*

7.46 Fact The compound transposition of Definition 7.45 is equivalent to a simple transposition of period $t = \mathrm{lcm}(t_1, \ldots, t_i)$.

7.47 Note (*recognizing simple transposition*) Although simple transposition ciphers alter dependencies between consecutive characters, they are easily recognized because they preserve the frequency distribution of each character.

7.3.2 Substitution ciphers (background)

This section considers the following types of classical ciphers: simple (or mono-alphabetic) substitution, polygram substitution, and homophonic substitution. The difference between codes and ciphers is also noted. Polyalphabetic substitution ciphers are considered in §7.3.3.

(i) Mono-alphabetic substitution

Suppose the ciphertext and plaintext character sets are the same. Let $m = m_1 m_2 m_3 \ldots$ be a plaintext message consisting of juxtaposed characters $m_i \in \mathcal{A}$, where $\mathcal{A}$ is some fixed character alphabet such as $\mathcal{A} = \{A, B, \ldots, Z\}$. A *simple substitution cipher* or *mono-alphabetic substitution cipher* employs a permutation e over $\mathcal{A}$, with encryption mapping $E_e(m) = e(m_1)e(m_2)e(m_3) \ldots$. Here juxtaposition indicates concatenation (rather than multiplication), and $e(m_i)$ is the character to which m_i is mapped by e. This corresponds to Definition 1.27.

7.48 Example (*trivial shift cipher/Caesar cipher*) A *shift cipher* is a simple substitution cipher with the permutation e constrained to an alphabetic shift through k characters for some fixed k. More precisely, if $|\mathcal{A}| = s$, and m_i is associated with the integer value i, $0 \leq i \leq s - 1$, then $c_i = e(m_i) = m_i + k \bmod s$. The decryption mapping is defined by $d(c_i) = c_i - k \bmod s$. For English text, $s = 26$, and characters A through Z are associated with integers 0 through 25. For $k = 1$, the message $m = \text{HAL}$ is encrypted to $c = \text{IBM}$. According to folklore, Julius Caesar used the key $k = 3$. $\qquad\square$

The shift cipher can be trivially broken because there are only $s = |\mathcal{A}|$ keys (e.g., $s = 26$) to exhaustively search. A similar comment holds for affine ciphers (Example 7.49). More generally, see Fact 7.68.

7.49 Example (*affine cipher – historical*) The affine cipher on a 26-letter alphabet is defined by $e_K(x) = ax + b \bmod 26$, where $0 \leq a, b \leq 25$. The key is (a, b). Ciphertext $c = e_K(x)$ is decrypted using $d_K(c) = (c - b)a^{-1} \bmod 26$, with the necessary and sufficient condition for invertibility that $\gcd(a, 26) = 1$. Shift ciphers are a subclass defined by $a = 1$. $\qquad\square$

7.50 Note (*recognizing simple substitution*) Mono-alphabetic substitution alters the frequency of individual plaintext characters, but does not alter the frequency distribution of the overall character set. Thus, comparing ciphertext character frequencies to a table of expected letter frequencies (unigram statistics) in the plaintext language allows associations between ciphertext and plaintext characters. (E.g., if the most frequent plaintext character X occurred twelve times, then the ciphertext character that X maps to will occur twelve times).

(ii) Polygram substitution

A simple substitution cipher substitutes for single plaintext letters. In contrast, *polygram substitution ciphers* involve groups of characters being substituted by other groups of characters. For example, sequences of two plaintext characters (*digrams*) may be replaced by other digrams. The same may be done with sequences of three plaintext characters (*trigrams*), or more generally using n-grams.

In full digram substitution over an alphabet of 26 characters, the key may be any of the 26^2 digrams, arranged in a table with row and column indices corresponding to the first and second characters in the digram, and the table entries being the ciphertext digrams substituted for the plaintext pairs. There are then $(26^2)!$ keys.

7.51 Example (*Playfair cipher – historical*) A digram substitution may be defined by arranging the characters of a 25-letter alphabet (I and J are equated) in a 5×5 matrix M. Adjacent plaintext characters are paired. The pair (p_1, p_2) is replaced by the digram (c_3, c_4) as follows. If p_1 and p_2 are in distinct rows and columns, they define the corners of a submatrix (possibly M itself), with the remaining corners c_3 and c_4; c_3 is defined as the character in the same column as p_1. If p_1 and p_2 are in a common row, c_3 is defined as the character immediately to the right of p_1 and c_4 that immediately right of p_2 (the first column is viewed as being to the right of the last). If p_1 and p_2 are in the same column, the characters immediately (circularly) below them are c_3 and c_4. If $p_1 = p_2$, an infrequent plaintext character (e.g., X) is inserted between them and the plaintext is re-grouped. While cryptanalysis based on single character frequencies fails for the Playfair cipher (each letter may be replaced by any other), cryptanalysis employing digram frequencies succeeds.   □

The key for a Playfair cipher is the 5×5 square. A mnemonic aid may be used to more easily remember the square. An example is the use of a meaningful keyphrase, with repeated letters deleted and the remaining alphabet characters included alphabetically at the end. The keyphrase "PLAYFAIR IS A DIGRAM CIPHER" would define a square with rows PLAYF, IRSDG, MCHEB, KNOQT, VWXYZ. To avoid the trailing characters always being from the end of the alphabet, a further shift cipher (Example 7.48) could be applied to the resulting 25-character string.

Use of keyphrases may seriously reduce the key space entropy. This effect is reduced if the keyphrase is not directly written into the square. For example, the non-repeated keyphrase characters might be written into an 8-column rectangle (followed by the remaining alphabet letters), the trailing columns being incomplete. The 25-character string obtained by reading the columns vertically is then used to fill the 5×5 square row by row.

7.52 Example (*Hill cipher – historical*) An n-gram substitution may be defined using an invertible $n \times n$ matrix $A = a_{ij}$ as the key to map an n-character plaintext $m_1 \ldots m_n$ to a ciphertext n-gram $c_i = \sum_{j=1}^{n} a_{ij}m_j$, $i = 1, \ldots, n$. Decryption involves using A^{-1}. Here characters A–Z, for example, are associated with integers 0–25. This polygram substitution cipher is a linear transformation, and falls under known-plaintext attack.   □

(iii) Homophonic substitution

The idea of homophonic substitution, introduced in §1.5, is for each fixed key k to associate with each plaintext unit (e.g., character) m a set $S(k, m)$ of potential corresponding ciphertext units (generally all of common size). To encrypt m under k, randomly choose one element from this set as the ciphertext. To allow decryption, for each fixed key this one-to-many encryption function must be injective on ciphertext space. Homophonic substitution results in ciphertext data expansion.

In homophonic substitution, $|S(k, m)|$ should be proportional to the frequency of m in the message space. The motivation is to smooth out obvious irregularities in the frequency distribution of ciphertext characters, which result from irregularities in the plaintext frequency distribution when simple substitution is used.

While homophonic substitution complicates cryptanalysis based on simple frequency distribution statistics, sufficient ciphertext may nonetheless allow frequency analysis, in conjunction with additional statistical properties of plaintext manifested in the ciphertext. For example, in long ciphertexts each element of $S(k, m)$ will occur roughly the same number of times. Digram distributions may also provide information.

(iv) Codes vs. ciphers

A technical distinction is made between *ciphers* and *codes*. Ciphers are encryption techniques which are applied to plaintext units (bits, characters, or blocks) independent of their semantic or linguistic meaning; the result is called ciphertext. In contrast, cryptographic codes operate on linguistic units such as words, groups of words, or phrases, and substitute (replace) these by designated words, letter groups, or number groups called *codegroups*. The key is a dictionary-like *codebook* listing plaintext units and their corresponding codegroups, indexed by the former; a corresponding codebook for decoding is reverse-indexed.

When there is potential ambiguity, codes in this context (vs. ciphers) may be qualified as *cryptographic codebooks*, to avoid confusion with error-correcting codes (EC-codes) used to detect and/or correct non-malicious errors and authentication codes (A-codes, or MACs as per Definition 9.7) which provide data origin authentication.

Several factors suggest that codes may be more difficult to break than ciphers: the key (codebook) is vastly larger than typical cipher keys; codes may result in data compression (cf. Fact 7.71); and statistical analysis is complicated by the large plaintext unit block size (cf. Note 7.74). Opposing this are several major disadvantages: the coding operation not being easily automated (relative to an algorithmic mapping); and identical encryption of repeated occurrences of plaintext units implies susceptibility to known-plaintext attacks, and allows frequency analysis based on observed traffic. This implies a need for frequent rekeying (changing the codebook), which is both more costly and inconvenient. Consequently, codes are not commonly used to secure modern telecommunications.

7.3.3 Polyalphabetic substitutions and Vigenère ciphers (historical)

A simple substitution cipher involves a single mapping of the plaintext alphabet onto ciphertext characters. A more complex alternative is to use different substitution mappings (called *multiple alphabets*) on various portions of the plaintext. This results in so-called *polyalphabetic substitution* (also introduced in Definition 1.30). In the simplest case, the different alphabets are used sequentially and then repeated, so the position of each plaintext character in the source string determines which mapping is applied to it. Under different alphabets, the same plaintext character is thus encrypted to different ciphertext characters, precluding simple frequency analysis as per mono-alphabetic substitution (§7.3.5).

The simple Vigenère cipher is a polyalphabetic substitution cipher, introduced in Example 1.31. The definition is repeated here for convenience.

7.53 Definition A *simple Vigenère* cipher of period t, over an s-character alphabet, involves a t-character key $k_1 k_2 \ldots k_t$. The mapping of plaintext $m = m_1 m_2 m_3 \ldots$ to ciphertext $c = c_1 c_2 c_3 \ldots$ is defined on individual characters by $c_i = m_i + k_i \bmod s$, where subscript i in k_i is taken modulo t (the key is re-used).

The simple Vigenère uses t shift ciphers (see Example 7.48), defined by t shift values k_i, each specifying one of s (mono-alphabetic) substitutions; k_i is used on the characters in position i, $i + s$, $i + 2s$, $\ldots$. In general, each of the t substitutions is different; this is referred to as using t alphabets rather than a single substitution mapping. The shift cipher (Example 7.48) is a simple Vigenère with period $t = 1$.

7.54 Example (*Beaufort variants of Vigenère*) Compared to the simple Vigenère mapping $c_i = m_i + k_i \bmod s$, the *Beaufort cipher* has $c_i = k_i - m_i \bmod s$, and is its own inverse. The *variant Beaufort* has encryption mapping $c_i = m_i - k_i \bmod s$. □

7.55 Example (*compound Vigenère*) The compound Vigenère has encryption mapping $c_i = m_i + (k_i^1 + k_i^2 + \cdots + k_i^r) \bmod s$, where in general the keys k^j, $1 \leq j \leq r$, have distinct periods t_j, and the subscript i in k_i^j, indicating the ith character of k^j, is taken modulo t_j. This corresponds to the sequential application of r simple Vigenères, and is equivalent to a simple Vigenère of period $\operatorname{lcm}(t_1, \ldots, t_r)$. □

7.56 Example (*single mixed alphabet Vigenère*) A simple substitution mapping defined by a general permutation e (not restricted to an alphabetic shift), followed by a simple Vigenère, is defined by the mapping $c_i = e(m_i) + k_i \bmod s$, with inverse $m_i = e^{-1}(c_i - k_i) \bmod s$. An alternative is a simple Vigenère followed by a simple substitution: $c_i = e(m_i + k_i \bmod s)$, with inverse $m_i = e^{-1}(c_i) - k_i \bmod s$.   □

7.57 Example (*full Vigenère*) In a simple Vigenère of period t, replace the mapping defined by the shift value k_i (for shifting character m_i) by a general permutation e_i of the alphabet. The result is the substitution mapping $c_i = e_i(m_i)$, where the subscript i in e_i is taken modulo t. The key consists of t permutations $e_1, \ldots, e_t$.   □

7.58 Example (*running-key Vigenère*) If the keystream k_i of a simple Vigenère is as long as the plaintext, the cipher is called a *running-key cipher*. For example, the key may be meaningful text from a book.   □

While running-key ciphers prevent cryptanalysis by the Kasiski method (§7.3.5), if the key has redundancy, cryptanalysis exploiting statistical imbalances may nonetheless succeed. For example, when encrypting plaintext English characters using a meaningful text as a running key, cryptanalysis is possible based on the observation that a significant proportion of ciphertext characters results from the encryption of high-frequency running text characters with high-frequency plaintext characters.

7.59 Fact A running-key cipher can be strengthened by successively enciphering plaintext under two or more distinct running keys. For typical English plaintext and running keys, it can be shown that iterating four such encipherments appears unbreakable.

7.60 Definition An *auto-key cipher* is a cipher wherein the plaintext itself serves as the key (typically subsequent to the use of an initial priming key).

7.61 Example (*auto-key Vigenère*) In a running-key Vigenère (Example 7.58) with an s-character alphabet, define a *priming key* $k = k_1 k_2 \ldots k_t$. Plaintext characters m_i are encrypted as $c_i = m_i + k_i \bmod s$ for $1 \leq i \leq t$ (simplest case: $t = 1$). For $i > t$, $c_i = (m_i + m_{i-t}) \bmod s$. An alternative involving more keying material is to replace the simple shift by a full Vigenère with permutations e_i, $1 \leq i \leq s$, defined by the key k_i or character m_i: for $1 \leq i \leq t$, $c_i = e_{k_i}(m_i)$, and for $i > t$, $c_i = e_{m_{i-t}}(m_i)$.   □

An alternative to Example 7.61 is to auto-key a cipher using the resulting ciphertext as the key: for example, for $i > t$, $c_i = (m_i + c_{i-t}) \bmod s$. This, however, is far less desirable, as it provides an eavesdropping cryptanalyst the key itself.

7.62 Example (*Vernam viewed as a Vigenère*) Consider a simple Vigenère defined by $c_i = m_i + k_i \bmod s$. If the keystream is truly random and independent – as long as the plaintext and never repeated (cf. Example 7.58) – this yields the unconditionally secure Vernam cipher (Definition 1.39; §6.1.1), generalized from a binary to an arbitrary alphabet.   □

7.3.4 Polyalphabetic cipher machines and rotors (historical)

The *Jefferson cylinder* is a deceptively simple device which implements a polyalphabetic substitution cipher; conceived in the late 18th century, it had remarkable cryptographic strength for its time. Polyalphabetic substitution ciphers implemented by a class of rotor-based machines which were the dominant cryptographic tool in World War II. Such machines, including the Enigma machine and those of Hagelin, have an alphabet which changes continuously for a very long period before repeating; this provides protection against Kasiski analysis and methods based on the index of coincidence (§7.3.5).

(i) Jefferson cylinder

The *Jefferson cylinder* (Figure 7.3) implements a polyalphabetic substitution cipher while avoiding complex machinery, extensive user computations, and Vigenère tableaus. A solid cylinder 6 inches long is sliced into 36 disks. A rod inserted through the cylinder axis allows the disks to rotate. The periphery of each disk is divided into 26 parts. On each disk, the letters A–Z are inscribed in a (different) random ordering. Plaintext messages are encrypted in 36-character blocks. A reference bar is placed along the cylinder's length. Each of the 36 wheels is individually rotated to bring the appropriate character (matching the plaintext block) into position along the reference line. The 25 other parallel reference positions then each define a ciphertext, from which (in an early instance of randomized encryption) one is selected as the ciphertext to transmit.

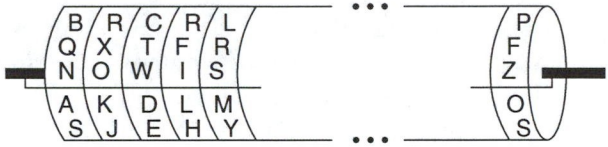

Figure 7.3: *The Jefferson cylinder.*

The second party possesses a cylinder with identically marked and ordered disks (1–36). The ciphertext is decrypted by rotating each of the 36 disks to obtain characters along a fixed reference line matching the ciphertext. The other 25 reference positions are examined for a recognizable plaintext. If the original message is not recognizable (e.g., random data), both parties agree beforehand on an index 1 through 25 specifying the offset between plaintext and ciphertext lines.

To accommodate plaintext digits 0–9 without extra disk sections, each digit is permanently assigned to one of 10 letters (a,e,i,o,u,y and f,l,r,s) which is encrypted as above but annotated with an overhead dot, identifying that the procedure must be reversed. Reordering disks (1 through 36) alters the polyalphabetic substitution key. The number of possible orderings is $36! \approx 3.72 \times 10^{41}$. Changing the ordering of letters on each disk affords $25!$ further mappings (per disk), but is more difficult in practice.

(ii) Rotor-based machines – technical overview

A simplified generic rotor machine (Figure 7.4) consists of a number of *rotors* (*wired codewheels*) each implementing a different fixed mono-alphabetic substitution, mapping a character at its input face to one on its output face. A plaintext character input to the first rotor generates an output which is input to the second rotor, and so on, until the final ciphertext character emerges from the last. For fixed rotor positions, the bank of rotors collectively implements a mono-alphabetic substitution which is the composition of the substitutions defined by the individual rotors.

To provide polyalphabetic substitution, the encipherment of each plaintext character causes various rotors to move. The simplest case is an odometer-like movement, with a single rotor stepped until it completes a full revolution, at which time it steps the adjacent

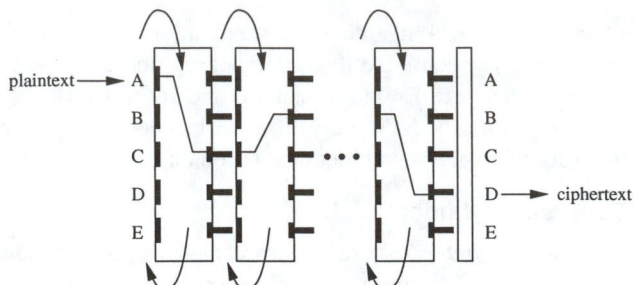

Figure 7.4: *A rotor-based machine.*

rotor one position, and so on. Stepping a rotor changes the mono-alphabetic substitution it defines (the *active* mapping). More precisely, each rotor R_i effects a mono-alphabetic substitution f_i. R_i can rotate into t_i positions (e.g., $t_i = 26$). When offset j places from a reference setting, R_i maps input a to $f_i(a - j) + j$, where both the input to f_i and the final output are reduced mod 26.

The cipher key is defined by the mono-alphabetic substitutions determined by the fixed wheel wirings and initial rotor positions. Re-arranging the order of rotors provides additional variability. Providing a machine with more rotors than necessary for operation at any one time allows further keying variation (by changing the active rotors).

7.63 Fact Two properties of rotor machines desirable for security-related reasons are: (1) long periods; and (2) state changes which are almost all "large".

The second property concerns the motion of rotors relative to each other, so that the sub-mappings between rotor faces change when the state changes. Rotor machines with odometer-like state changes fail to achieve this second property.

7.64 Note (*rotor machine output methods*) Rotor machines were categorized by their method of providing ciphertext output. In *indicating machines*, ciphertext output characters are indicated by means such as lighted lamps or displayed characters in output apertures. In *printing machines*, ciphertext is printed or typewritten onto an output medium such as paper. With *on-line machines*, output characters are produced in electronic form suitable for direct transmission over telecommunications media.

(iii) Rotor-based machines – historical notes

A number of individuals are responsible for the development of early machines based on rotor principles. In 1918, the American E.H. Hebern built the first rotor apparatus, based on an earlier typewriting machine modified with wired connections to generate a mono-alphabetic substitution. The output was originally by lighted indicators. The first rotor patent was filed in 1921, the year *Hebern Electric Code, Inc.* became the first U.S. cipher machine company (and first to bankrupt in 1926). The U.S. Navy (circa 1929-1930 and some years thereafter) used a number of Hebern's five-rotor machines.

In October 1919, H.A. Koch filed Netherlands patent no.10,700 ("Geheimschrijfmachine" – secret writing machine), demonstrating a deep understanding of rotor principles; no machine was built. In 1927, the patent rights were assigned to A. Scherbius.

The German inventor Scherbius built a rotor machine called the *Enigma*. Model A was replaced by Model B with typewriter output, and a portable Model C with indicator lamps.

The company set up in 1923 dissolved in 1934, but thereafter the Germans used the portable battery-powered Enigma, including for critical World War II operations.

In October 1919, three days after Koch, A.G. Damm filed Swedish patent no.52,279 describing a double-rotor device. His firm was joined by the Swede, B. Hagelin, whose 1925 modification yielded the B-21 rotor machine (with indicating lamps) used by the Swedish army. The B-21 had *keywheels* with varying number of teeth or gears, each of which was associated with a settable two-state pin. The period of the resulting polyalphabetic substitution was the product of the numbers of keywheel pins; the key was defined by the state of each pin and the initial keywheel positions. Hagelin later produced other models: B-211 (a printing machine); a more compact (phone-sized) model C-36 for the French in 1934; and based on alterations suggested by Friedman and others, model C-48 (of which over 140 000 were produced) which was called M-209 when used by the U.S. Army as a World War II field cipher. His 1948 Swiss factory later produced: model C-52, a strengthened version of M-209 (C-48) with period exceeding 2.75×10^9 (with keywheels of 47, 43, 41, 37, 31, 29 pins); CD-55, a pocket-size version of the C-52; and T-55, an on-line version of the same, modifiable to use a one-time tape. A further model was CD-57.

7.65 Note (*Enigma details*) The Enigma initially had three rotors R_i, each with 26 positions. R_1 stepped R_2 which stepped R_3 odometer-like, with R_2 also stepping itself; the period was $26 \cdot 25 \cdot 26 \approx 17\,000$. The key consisted of the initial positions of these rotors ($\approx 17\,000$ choices), their order ($3! = 6$ choices), and the state of a *plugboard*, which implemented a fixed but easily changed (e.g., manually, every hour) mono-alphabetic substitution ($26!$ choices), in addition to that carried out by rotor combinations.

7.66 Note (*Hagelin M-209 details*) The Hagelin M-209 rotor machine implements a polyalphabetic substitution using 6 keywheels – more specifically, a self-decrypting Beaufort cipher (Example 7.54), $E_{k_i}(m_i) = k_i - m_i \bmod 26$, of period $101\,405\,850 = 26 \cdot 25 \cdot 23 \cdot 21 \cdot 19 \cdot 17$ letters. Thus for a fixed ordered set of 6 keywheels, the cipher period exceeds 10^8. k_i may be viewed as the ith character in the key stream, as determined by a particular ordering of keywheels, their pin settings, and starting positions. All keywheels rotate one position forward after each character is enciphered. The wheels simultaneously return to their initial position only after a period equal to the least-common-multiple of their gear-counts, which (since these are co-prime) is their product. A ciphertext-only attack is possible with 1000-2000 characters, using knowledge of the machine's internal mechanical details, and assuming natural language redundancy in the plaintext; a known-plaintext attack is possible with 50-100 characters.

7.3.5 Cryptanalysis of classical ciphers (historical)

This section presents background material on redundancy and unicity distance, and techniques for cryptanalysis of classical ciphers,

(i) Redundancy

All natural languages are redundant. This redundancy results from linguistic structure. For example, in English the letter "E" appears far more frequently than "Z", "Q" is almost always followed by "U", and "TH" is a common digram.

An alphabet with 26 characters (e.g., Roman alphabet) can theoretically carry up to $\lg 26 = 4.7$ bits of information per character. Fact 7.67 indicates that, on average, far less information is actually conveyed by a natural language.

7.67 Fact The estimated average amount of information carried per character (per-character entropy) in meaningful English alphabetic text is 1.5 bits.

The per-character redundancy of English is thus about $4.7 - 1.5 = 3.2$ bits.

7.68 Fact Empirical evidence suggests that, for essentially any simple substitution cipher on a meaningful message (e.g., with redundancy comparable to English), as few as 25 ciphertext characters suffices to allow a skilled cryptanalyst to recover the plaintext.

(ii) Unicity distance and random cipher model

7.69 Definition The *unicity distance* of a cipher is the minimum amount of ciphertext (number of characters) required to allow a computationally unlimited adversary to recover the unique encryption key.

The unicity distance is primarily a theoretical measure, useful in relation to unconditional security. A small unicity distance does not necessarily imply that a block cipher is insecure in practice. For example, consider a 64-bit block cipher with a unicity distance of two ciphertext blocks. It may still be computationally infeasible for a cryptanalyst (of reasonable but bounded computing power) to recover the key, although theoretically there is sufficient information to allow this.

The random cipher model (Definition 7.70) is a simplified model of a block cipher providing a reasonable approximation for many purposes, facilitating results on block cipher properties not otherwise easily established (e.g., Fact 7.71).

7.70 Definition Let C and K be random variables, respectively, denoting the ciphertext block and the key, and let D denote the decryption function. Under the *random cipher model*, $D_K(C)$ is a random variable uniformly distributed over all possible pre-images of C (meaningful messages and otherwise, with and without redundancy).

In an intuitive sense, a random cipher as per the model of Definition 7.70 is a random mapping. (A more precise approximation would be as a random permutation.)

7.71 Fact Under the random cipher model, the expected unicity distance N_0 of a cipher is $N_0 = H(\mathcal{K})/D$, where $H(\mathcal{K})$ is the entropy of the key space (e.g., 64 bits for 2^{64} equiprobable keys), and D is the plaintext redundancy (in bits/character).

For a one-time pad, the unbounded entropy of the key space implies, by Fact 7.71, that the unicity distance is likewise unbounded. This is consistent with the one-time pad being theoretically unbreakable.

Data compression reduces redundancy. Fact 7.71 implies that data compression prior to encryption increases the unicity distance, thus increasing security. If the plaintext contains no redundancy whatsoever, then the unicity distance is infinite; that is, the system is theoretically unbreakable under a ciphertext-only attack.

7.72 Example (*unicity distance – transposition cipher*) The unicity distance of a simple transposition cipher of period t can be estimated under the random cipher model using Fact 7.71, and the assumption of plaintext redundancy of $D = 3.2$ bits/character. In this case, $H(\mathcal{K})/D = \lg(t!)/3.2$ and for $t = 12$ the estimated unicity distance is 9 characters, which is very crude, this being less than one 12-character block. For $t = 27$, the estimated unicity distance is a more plausible 29 characters; this can be computed using Stirling's approximation of Fact 2.57(iii) ($t! \approx \sqrt{2\pi t}(t/e)^t$, for large t and $e = 2.718$) as $H(\mathcal{K})/D = \lg(t!)/3.2 \approx (0.3t) \cdot \lg(t/e)$. $\qquad\square$

7.73 Example (*unicity distance – simple substitution*) The number of keys for a mono-alphabetic substitution cipher over alphabet $\mathcal{A}$ is $|\mathcal{K}| = s!$, where $s = |\mathcal{A}|$. For example, $s = 26$ (Roman alphabet) yields $26! \approx 4 \times 10^{26}$ keys. Assuming equiprobable keys, an estimate of the entropy of the key space is then (cf. Example 7.72) $H(\mathcal{K}) = \lg(26!) \approx 88.4$ bits. Assuming English text with $D = 3.2$ bits of redundancy per character (Fact 7.67), a theoretical estimate of the unicity distance of a simple substitution cipher is $H(\mathcal{K})/D = 88.4/3.2 \approx 28$ characters. This agrees closely with empirical evidence (Fact 7.68). □

(iii) Language statistics

Cryptanalysis of classical ciphers typically relies on redundancy in the source language (plaintext). In many cases a divide-and-conquer approach is possible, whereby the plaintext or key is recovered piece by piece, each facilitating further recovery.

Mono-alphabetic substitution on short plaintext blocks (e.g., Roman alphabet characters) is easily defeated by associating ciphertext characters with plaintext characters (Note 7.50). The frequency distribution of individual ciphertext characters can be compared to that of single characters in the source language, as given by Figure 7.5 (estimated from 1964 English text). This is facilitated by grouping plaintext letters by frequency into high, medium, low, and rare classes; focussing on the high-frequency class, evidence supporting trial letter assignments can be obtained by examining how closely hypothesized assignments match those of the plaintext language. Further evidence is available by examination of digram and trigram frequencies. Figure 7.6 gives the most common English digrams as a percentage of all digrams; note that of $26^2 = 676$ possible digrams, the top 15 account for 27% of all occurrences. Other examples of plaintext redundancy appearing in the ciphertext include associations of vowels with consonants, and repeated letters in *pattern words* (e.g., "that", "soon", "three").

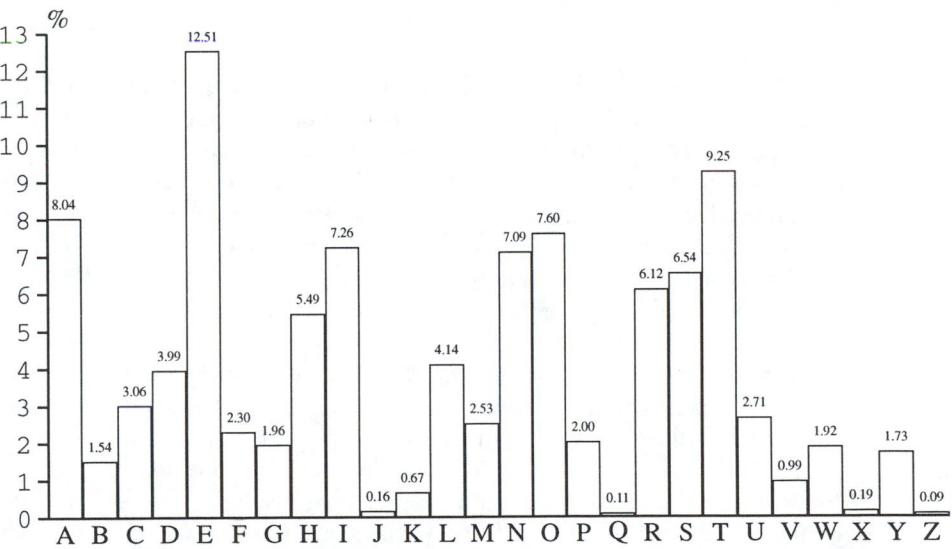

Figure 7.5: *Frequency of single characters in English text.*

7.74 Note (*large blocks preclude statistical analysis*) An n-bit block size implies 2^n plaintext units ("characters"). Compilation of frequency statistics on plaintext units thus becomes infeasible as the block size of the simple substitution increases; for example, this is clearly infeasible for DES (§7.4), where $n = 64$.

Cryptanalysis of simple transposition ciphers is similarly facilitated by source language statistics (see Note 7.47). Cryptanalyzing transposed blocks resembles solving an anagram. Attempts to reconstruct common digrams and trigrams are facilitated by frequency statistics. Solutions may be constructed piecewise, with the appearance of digrams and trigrams in trial decryptions confirming (partial) success.

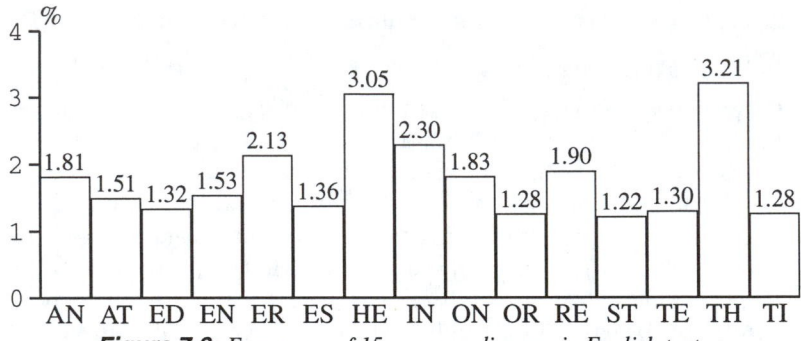

Figure 7.6: *Frequency of 15 common digrams in English text.*

Cryptanalysis of polyalphabetic ciphers is possible by various methods, including Kasiski's method and methods based on the index of coincidence, as discussed below.

(iv) Method of Kasiski (vs. polyalphabetic substitution)

Kasiski's method provides a general technique for cryptanalyzing polyalphabetic ciphers with repeated keywords, such as the simple Vigenère cipher (Definition 7.53), based on the following observation: repeated portions of plaintext encrypted with the same portion of the keyword result in identical ciphertext segments. Consequently one expects the number of characters between the beginning of repeated ciphertext segments to be a multiple of the keyword length. Ideally, it suffices to compute the greatest common divisor of the various distances between such repeated segments, but coincidental repeated ciphertext segments may also occur. Nonetheless, an analysis (*Kasiski examination*) of the common factors among all such distances is possible; the largest factor which occurs most commonly is the most likely keyword length. Repeated ciphertext segments of length 4 or longer are most useful, as coincidental repetitions are then less probable.

The number of letters in the keyword indicates the number of alphabets t in the polyalphabetic substitution. Ciphertext characters can then be partitioned into t sets, each of which is then the result of a mono-alphabetic substitution. Trial values for t are confirmed if the frequency distribution of the (candidate) mono-alphabetic groups matches the frequency distribution of the plaintext language. For example, the profile for plaintext English (Figure 7.5) exhibits a long trough characterizing $uvwxyz$, followed by a spike at a, and preceded by the triple-peak of rst. The resulting mono-alphabetic portions can be solved individually, with additional information available by combining their solution (based on digrams, probable words, etc.). If the source language is unknown, comparing the frequency distribution of ciphertext characters to that of candidate languages may allow determination of the source language itself.

(v) Index of coincidence (vs. polyalphabetic substitution)

The *index of coincidence* (*IC*) is a measure of the relative frequency of letters in a ciphertext sample, which facilitates cryptanalysis of polyalphabetic ciphers by allowing determination of the period t (as an alternative to Kasiski's method). For concreteness, consider a Vigenère cipher and assume natural language English plaintext.

Let the ciphertext alphabet be $\{a_0, a_1, \ldots, a_{n-1}\}$, and let p_i be the unknown probability that an arbitrarily chosen character in a random ciphertext is a_i. The *measure of roughness* measures the deviation of ciphertext characters from a flat frequency distribution as follows:

$$\text{MR} = \sum_{i=0}^{n-1} \left(p_i - \frac{1}{n} \right)^2 = \sum_{i=0}^{n-1} p_i{}^2 - \frac{1}{n} \tag{7.1}$$

The minimum value is $\text{MR}_{\min} = 0$, corresponding to a flat distribution (for equiprobable a_i, $p_i = 1/n$). The maximum value occurs when the frequency distribution of p_i has greatest variability, corresponding to a mono-alphabetic substitution (the plaintext frequency distribution is then manifested). Define this maximum value $\text{MR}_{\max} = \kappa_p - 1/n$, where κ_p corresponds to $\sum p_i{}^2$ when p_i are plaintext frequencies. For English as per Figure 7.5, the maximum value is $\text{MR} = \kappa_p - 1/n \approx 0.0658 - 0.0385 = 0.0273$. (This varies with letter frequency estimates; $\kappa_p = 0.0667$, yielding $\kappa_p - 1/n = 0.0282$ is commonly cited, and is used in Table 7.1.) While MR cannot be computed directly from a ciphertext sample (since the period t is unknown, the mono-alphabetic substitutions cannot be separated), it may be estimated from the frequency distribution of ciphertext characters as follows.

Let f_i denote the number of appearances of a_i in an L-character ciphertext sample (thus $\sum f_i = L$). The number of pairs of letters among these L is $L(L-1)/2$, of which $f_i(f_i - 1)/2$ are the pair (a_i, a_i) for any fixed character a_i. Define IC as the probability that two characters arbitrarily chosen from the *given* ciphertext sample are equal:

$$\text{IC} = \frac{\sum_{i=0}^{n-1} \binom{f_i}{2}}{\binom{L}{2}} = \frac{\sum_{i=0}^{n-1} f_i(f_i - 1)}{L(L-1)} \tag{7.2}$$

Independent of this given ciphertext sample, the probability that two randomly chosen ciphertext characters are equal is $\sum_{i=0}^{n-1} p_i{}^2$. Thus (comparing word definitions) IC is an estimate of $\sum p_i{}^2$, and by equation (7.1), thereby an estimate of $\text{MR} + 1/n$. Moreover, IC can be directly computed from a ciphertext sample, allowing estimation of MR itself. Since MR varies from 0 to $\kappa_p - 1/n$, one expects IC to range from $1/n$ (for polyalphabetic substitution with infinite period) to κ_p (for mono-alphabetic substitution). More precisely, the following result may be established.

7.75 Fact For a polyalphabetic cipher of period t, $E(\text{IC})$ as given below is the expected value of the index of coincidence for a ciphertext string of length L, where n is the number of alphabet characters, $\kappa_r = 1/n$, and κ_p is given in Table 7.1:

$$E(\text{IC}) = \frac{1}{t} \cdot \frac{L - t}{L - 1} \cdot \kappa_p + \frac{t - 1}{t} \cdot \frac{L}{L - 1} \cdot \kappa_r \tag{7.3}$$

(p in κ_p is intended to denote a plaintext frequency distribution, while the r in κ_r denotes a distribution for random characters.) For Roman-alphabet languages, $n = 26$ implies $\kappa_r = 0.03846$; for the Russian Cyrillic alphabet, $n = 30$.

7.76 Example (*estimating polyalphabetic period using IC*) Tabulating the expected values for IC for periods $t = 1, 2, \ldots$ using Equation (7.3) (which is essentially independent of L for large L and small t), and comparing this to that obtained from a particular ciphertext using Equation (7.2) allows a crude estimate of the period t of the cipher, e.g., whether it is mono-alphabetic or polyalphabetic with small period. Candidate values t in the range thus determined may be tested for correctness by partitioning ciphertext characters into groups of letters separated by t ciphertext positions, and in one or more such groups, comparing the character frequency distribution to that of plaintext. □

Language	κ_p
French	0.0778
Spanish	0.0775
German	0.0762
Italian	0.0738
English	0.0667
Russian	0.0529

Table 7.1: *Estimated roughness constant κ_p for various languages (see Fact 7.75).*

A polyalphabetic period t may be determined either by Example 7.76 or the alternative of Example 7.77, based on the same underlying ideas. Once t is determined, the situation is as per after successful completion of the Kasiski method.

7.77 Example (*determining period by ciphertext auto-correlation*) Given a sample of polyalphabetic ciphertext, the unknown period t may be determined by examining the number of coincidences when the ciphertext is auto-correlated. More specifically, given a ciphertext sample $c_1 c_2 \ldots c_L$, starting with $t = 1$, count the total number of occurrences $c_i = c_{i+t}$ for $1 \leq i \leq L - t$. Repeat for $t = 2, 3, \ldots$ and tabulate the counts (or plot a bar graph). The actual period t^* is revealed as follows: for values t that are a multiple of t^*, the counts will be noticeably higher (easily recognized as spikes on the bar graph). In fact, for L appropriately large, one expects approximately $L \cdot \kappa_p$ coincidences in this case, and significantly fewer in other cases.   □

In the auto-correlation method of coincidences of Example 7.77, the spikes on the bar graph reveal the period, independent of the source language. Once the period is determined, ciphertext characters from like alphabets can be grouped, and the profile of single-character letter frequencies among these, which differs for each language, may be used to determine the plaintext language.

7.4 DES

The Data Encryption Standard (DES) is the most well-known symmetric-key block cipher. Recognized world-wide, it set a precedent in the mid 1970s as the first commercial-grade modern algorithm with openly and fully specified implementation details. It is defined by the American standard FIPS 46–2.

7.4.1 Product ciphers and Feistel ciphers

The design of DES is related to two general concepts: product ciphers and Feistel ciphers. Each involves iterating a common sequence or round of operations.

The basic idea of a product cipher (see §1.5.3) is to build a complex encryption function by composing several simple operations which offer complementary, but individually insufficient, protection (note cascade ciphers per Definition 7.29 use independent keys). Basic operations include transpositions, translations (e.g., XOR) and linear transformations, arithmetic operations, modular multiplication, and simple substitutions.

7.78 Definition A *product cipher* combines two or more transformations in a manner intending that the resulting cipher is more secure than the individual components.

7.79 Definition A *substitution-permutation* (SP) *network* is a product cipher composed of a number of stages each involving substitutions and permutations (Figure 7.7).

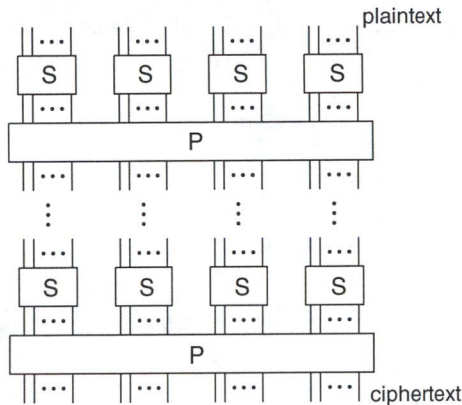

Figure 7.7: *Substitution-permutation (SP) network.*

Many SP networks are iterated ciphers as per Definition 7.80.

7.80 Definition An *iterated block cipher* is a block cipher involving the sequential repetition of an internal function called a *round function*. Parameters include the number of rounds r, the block bitsize n, and the bitsize k of the input key K from which r *subkeys* K_i (round keys) are derived. For invertibility (allowing unique decryption), for each value K_i the round function is a bijection on the round input.

7.81 Definition A *Feistel cipher* is an iterated cipher mapping a $2t$-bit plaintext (L_0, R_0), for t-bit blocks L_0 and R_0, to a ciphertext (R_r, L_r), through an r-round process where $r \geq 1$. For $1 \leq i \leq r$, round i maps $(L_{i-1}, R_{i-1}) \xrightarrow{K_i} (L_i, R_i)$ as follows: $L_i = R_{i-1}$, $R_i = L_{i-1} \oplus f(R_{i-1}, K_i)$, where each *subkey* K_i is derived from the cipher key K.

Typically in a Feistel cipher, $r \geq 3$ and often is even. The Feistel structure specifically orders the ciphertext output as (R_r, L_r) rather than (L_r, R_r); the blocks are exchanged from their usual order after the last round. Decryption is thereby achieved using the same r-round process but with subkeys used in reverse order, K_r through K_1; for example, the last round is undone by simply repeating it (see Note 7.84). The f function of the Feistel cipher may be a product cipher, though f itself need not be invertible to allow inversion of the Feistel cipher.

Figure 7.9(b) illustrates that successive rounds of a Feistel cipher operate on alternating halves of the ciphertext, while the other remains constant. Note the round function of Definition 7.81 may also be re-written to eliminate L_i: $R_i = R_{i-2} \oplus f(R_{i-1}, K_i)$. In this case, the final ciphertext output is (R_r, R_{r-1}), with input labeled (R_{-1}, R_0).

7.4.2 DES algorithm

DES is a Feistel cipher which processes plaintext blocks of $n = 64$ bits, producing 64-bit ciphertext blocks (Figure 7.8). The effective size of the secret key K is $k = 56$ bits; more precisely, the input key K is specified as a 64-bit key, 8 bits of which (bits $8, 16, \ldots, 64$) may be used as parity bits. The 2^{56} keys implement (at most) 2^{56} of the $2^{64}!$ possible bijections on 64-bit blocks. A widely held belief is that the parity bits were introduced to reduce the effective key size from 64 to 56 bits, to intentionally reduce the cost of exhaustive key search by a factor of 256.

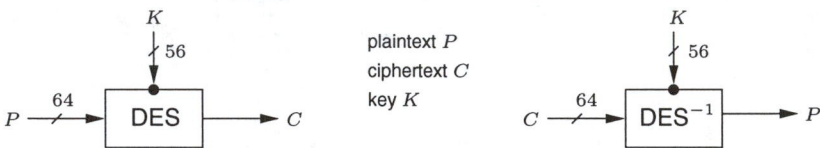

Figure 7.8: *DES input-output.*

Full details of DES are given in Algorithm 7.82 and Figures 7.9 and 7.10. An overview follows. Encryption proceeds in 16 stages or *rounds*. From the input key K, sixteen 48-bit subkeys K_i are generated, one for each round. Within each round, 8 fixed, carefully selected 6-to-4 bit substitution mappings (*S-boxes*) S_i, collectively denoted S, are used. The 64-bit plaintext is divided into 32-bit halves L_0 and R_0. Each round is functionally equivalent, taking 32-bit inputs L_{i-1} and R_{i-1} from the previous round and producing 32-bit outputs L_i and R_i for $1 \le i \le 16$, as follows:

$$L_i = R_{i-1}; \tag{7.4}$$
$$R_i = L_{i-1} \oplus f(R_{i-1},\ K_i), \quad \text{where } f(R_{i-1},\ K_i) = P(S(E(R_{i-1}) \oplus K_i)) \tag{7.5}$$

Here E is a fixed expansion permutation mapping R_{i-1} from 32 to 48 bits (all bits are used once; some are used twice). P is another fixed permutation on 32 bits. An initial bit permutation (IP) precedes the first round; following the last round, the left and right halves are exchanged and, finally, the resulting string is bit-permuted by the inverse of IP. Decryption involves the same key and algorithm, but with subkeys applied to the internal rounds in the reverse order (Note 7.84).

A simplified view is that the right half of each round (after expanding the 32-bit input to 8 characters of 6 bits each) carries out a key-dependent substitution on each of 8 characters, then uses a fixed bit transposition to redistribute the bits of the resulting characters to produce 32 output bits.

Algorithm 7.83 specifies how to compute the DES round keys K_i, each of which contains 48 bits of K. These operations make use of tables PC1 and PC2 of Table 7.4, which are called *permuted choice 1* and *permuted choice 2*. To begin, 8 bits ($k_8, k_{16}, \ldots, k_{64}$) of K are discarded (by PC1). The remaining 56 bits are permuted and assigned to two 28-bit variables C and D; and then for 16 iterations, both C and D are rotated either 1 or 2 bits, and 48 bits (K_i) are selected from the concatenated result.

7.82 Algorithm Data Encryption Standard (DES)

INPUT: plaintext $m_1 \ldots m_{64}$; 64-bit key $K = k_1 \ldots k_{64}$ (includes 8 parity bits).

OUTPUT: 64-bit ciphertext block $C = c_1 \ldots c_{64}$. (For decryption, see Note 7.84.)

1. (key schedule) Compute sixteen 48-bit round keys K_i from K using Algorithm 7.83.
2. $(L_0, R_0) \leftarrow \mathrm{IP}(m_1 m_2 \ldots m_{64})$. (Use IP from Table 7.2 to permute bits; split the result into left and right 32-bit halves $L_0 = m_{58} m_{50} \ldots m_8$, $R_0 = m_{57} m_{49} \ldots m_7$.)
3. (16 rounds) for i from 1 to 16, compute L_i and R_i using Equations (7.4) and (7.5) above, computing $f(R_{i-1}, K_i) = P(S(E(R_{i-1}) \oplus K_i))$ as follows:
    (a) Expand $R_{i-1} = r_1 r_2 \ldots r_{32}$ from 32 to 48 bits using E per Table 7.3: $T \leftarrow E(R_{i-1})$. (Thus $T = r_{32} r_1 r_2 \ldots r_{32} r_1$.)
    (b) $T' \leftarrow T \oplus K_i$. Represent T' as eight 6-bit character strings: $(B_1, \ldots, B_8) = T'$.
    (c) $T'' \leftarrow (S_1(B_1), S_2(B_2), \ldots S_8(B_8))$. (Here $S_i(B_i)$ maps $B_i = b_1 b_2 \ldots b_6$ to the 4-bit entry in row r and column c of S_i in Table 7.8, page 260 where $r = 2 \cdot b_1 + b_6$, and $b_2 b_3 b_4 b_5$ is the radix-2 representation of $0 \le c \le 15$. Thus $S_1(011011)$ yields $r = 1$, $c = 13$, and output 5, i.e., binary 0101.)
    (d) $T''' \leftarrow P(T'')$. (Use P per Table 7.3 to permute the 32 bits of $T'' = t_1 t_2 \ldots t_{32}$, yielding $t_{16} t_7 \ldots t_{25}$.)
4. $b_1 b_2 \ldots b_{64} \leftarrow (R_{16}, L_{16})$. (Exchange final blocks L_{16}, R_{16}.)
5. $C \leftarrow \mathrm{IP}^{-1}(b_1 b_2 \ldots b_{64})$. (Transpose using IP^{-1} from Table 7.2; $C = b_{40} b_8 \ldots b_{25}$.)

IP									IP^{-1}							
58	50	42	34	26	18	10	2		40	8	48	16	56	24	64	32
60	52	44	36	28	20	12	4		39	7	47	15	55	23	63	31
62	54	46	38	30	22	14	6		38	6	46	14	54	22	62	30
64	56	48	40	32	24	16	8		37	5	45	13	53	21	61	29
57	49	41	33	25	17	9	1		36	4	44	12	52	20	60	28
59	51	43	35	27	19	11	3		35	3	43	11	51	19	59	27
61	53	45	37	29	21	13	5		34	2	42	10	50	18	58	26
63	55	47	39	31	23	15	7		33	1	41	9	49	17	57	25

Table 7.2: *DES initial permutation and inverse (IP and IP^{-1}).*

E						P			
32	1	2	3	4	5	16	7	20	21
4	5	6	7	8	9	29	12	28	17
8	9	10	11	12	13	1	15	23	26
12	13	14	15	16	17	5	18	31	10
16	17	18	19	20	21	2	8	24	14
20	21	22	23	24	25	32	27	3	9
24	25	26	27	28	29	19	13	30	6
28	29	30	31	32	1	22	11	4	25

Table 7.3: *DES per-round functions: expansion E and permutation P.*

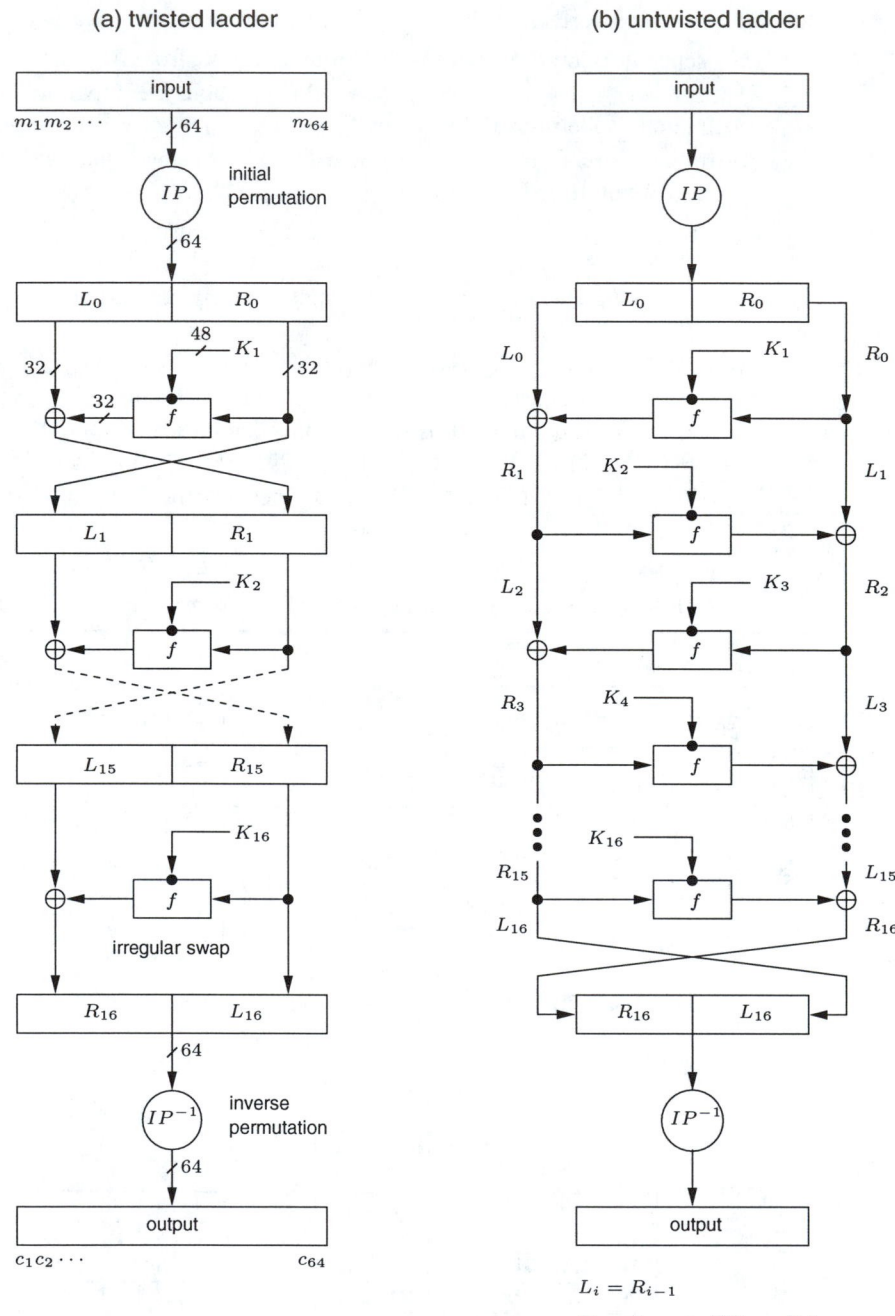

Figure 7.9: *DES computation path.*

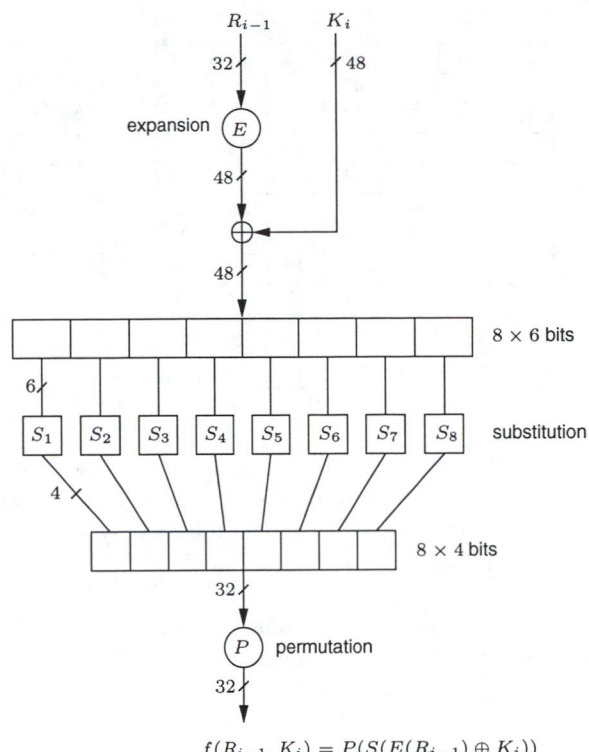

$$f(R_{i-1}, K_i) = P(S(E(R_{i-1}) \oplus K_i))$$

Figure 7.10: *DES inner function f.*

7.83 Algorithm DES key schedule

INPUT: 64-bit key $K = k_1 \ldots k_{64}$ (including 8 odd-parity bits).
OUTPUT: sixteen 48-bit keys K_i, $1 \leq i \leq 16$.

1. Define v_i, $1 \leq i \leq 16$ as follows: $v_i = 1$ for $i \in \{1, 2, 9, 16\}$; $v_i = 2$ otherwise. (These are left-shift values for 28-bit circular rotations below.)
2. $T \leftarrow \text{PC1}(K)$; represent T as 28-bit halves (C_0, D_0). (Use PC1 in Table 7.4 to select bits from K: $C_0 = k_{57}k_{49} \ldots k_{36}$, $D_0 = k_{63}k_{55} \ldots k_4$.)
3. For i from 1 to 16, compute K_i as follows: $C_i \leftarrow (C_{i-1} \hookleftarrow v_i)$, $D_i \leftarrow (D_{i-1} \hookleftarrow v_i)$, $K_i \leftarrow \text{PC2}(C_i, D_i)$. (Use PC2 in Table 7.4 to select 48 bits from the concatenation $b_1 b_2 \ldots b_{56}$ of C_i and D_i: $K_i = b_{14}b_{17} \ldots b_{32}$. '$\hookleftarrow$' denotes left circular shift.)

If decryption is designed as a simple variation of the encryption function, savings result in hardware or software code size. DES achieves this as outlined in Note 7.84.

7.84 Note (*DES decryption*) DES decryption consists of the encryption algorithm with the same key but reversed key schedule, using in order $K_{16}, K_{15}, \ldots, K_1$ (see Note 7.85). This works as follows (refer to Figure 7.9). The effect of IP^{-1} is cancelled by IP in decryption, leaving (R_{16}, L_{16}); consider applying round 1 to this input. The operation on the left half yields, rather than $L_0 \oplus f(R_0, K_1)$, now $R_{16} \oplus f(L_{16}, K_{16})$ which, since $L_{16} = R_{15}$ and $R_{16} = L_{15} \oplus f(R_{15}, K_{16})$, is equal to $L_{15} \oplus f(R_{15}, K_{16}) \oplus f(R_{15}, K_{16}) = L_{15}$. Thus round 1 decryption yields (R_{15}, L_{15}), i.e., inverting round 16. Note that the cancellation

PC1						
57	49	41	33	25	17	9
1	58	50	42	34	26	18
10	2	59	51	43	35	27
19	11	3	60	52	44	36
above for C_i; below for D_i						
63	55	47	39	31	23	15
7	62	54	46	38	30	22
14	6	61	53	45	37	29
21	13	5	28	20	12	4

PC2					
14	17	11	24	1	5
3	28	15	6	21	10
23	19	12	4	26	8
16	7	27	20	13	2
41	52	31	37	47	55
30	40	51	45	33	48
44	49	39	56	34	53
46	42	50	36	29	32

Table 7.4: *DES key schedule bit selections (PC1 and PC2).*

of each round is independent of the definition of f and the specific value of K_i; the swapping of halves combined with the XOR process is inverted by the second application. The remaining 15 rounds are likewise cancelled one by one in reverse order of application, due to the reversed key schedule.

7.85 Note (*DES decryption key schedule*) Subkeys $K_1, \ldots, K_{16}$ may be generated by Algorithm 7.83 and used in reverse order, or generated in reverse order directly as follows. Note that after K_{16} is generated, the original values of the 28-bit registers C and D are restored (each has rotated 28 bits). Consequently, and due to the choice of shift-values, modifying Algorithm 7.83 as follows generates subkeys in order $K_{16}, \ldots, K_1$: replace the left-shifts by right-shift rotates; change the shift value v_1 to 0.

7.86 Example (*DES test vectors*) The plaintext "Now is the time for all ", represented as a string of 8-bit hex characters (7-bit ASCII characters plus leading 0-bit), and encrypted using the DES key specified by the hex string $K = $ 0123456789ABCDEF results in the following plaintext/ciphertext:

$P = $ 4E6F772069732074 68652074696D6520 666F7220616C6C20

$C = $ 3FA40E8A984D4815 6A271787AB8883F9 893D51EC4B563B53. □

7.4.3 DES properties and strength

There are many desirable characteristics for block ciphers. These include: each bit of the ciphertext should depend on all bits of the key and all bits of the plaintext; there should be no statistical relationship evident between plaintext and ciphertext; altering any single plaintext or key bit should alter each ciphertext bit with probability $\frac{1}{2}$; and altering a ciphertext bit should result in an unpredictable change to the recovered plaintext block. Empirically, DES satisfies these basic objectives. Some known properties and anomalies of DES are given below.

(i) Complementation property

7.87 Fact Let E denote DES, and $\overline{x}$ the bitwise complement of x. Then $y = E_K(x)$ implies $\overline{y} = E_{\overline{K}}(\overline{x})$. That is, bitwise complementing both the key K and the plaintext x results in complemented DES ciphertext.

Justification: Compare the first round output (see Figure 7.10) to (L_0, R_0) for the uncomplemented case. The combined effect of the plaintext and key being complemented results in the inputs to the XOR preceding the S-boxes (the expanded R_{i-1} and subkey K_i) both being complemented; this double complementation cancels out in the XOR operation, resulting in S-box inputs, and thus an overall result $f(R_0, K_1)$, as before. This quantity is then XORed (Figure 7.9) to $\overline{L_0}$ (previously L_0), resulting in $\overline{L_1}$ (rather than L_1). The same effect follows in the remaining rounds.

The complementation property is normally of no help to a cryptanalyst in known-plaintext exhaustive key search. If an adversary has, for a fixed unknown key K, a chosen-plaintext set of (x, y) data (P_1, C_1), $(\overline{P_1}, C_2)$, then $C_2 = E_K(\overline{P_1})$ implies $\overline{C_2} = E_{\overline{K}}(P_1)$. Checking if the key K with plaintext P_1 yields either C_1 or $\overline{C_2}$ now rules out two keys with one encryption operation, thus reducing the expected number of keys required before success from 2^{55} to 2^{54}. This is not a practical concern.

(ii) Weak keys, semi-weak keys, and fixed points

If subkeys K_1 to K_{16} are equal, then the reversed and original schedules create identical subkeys: $K_1 = K_{16}$, $K_2 = K_{15}$, and so on. Consequently, the encryption and decryption functions coincide. These are called weak keys (and also: *palindromic keys*).

7.88 Definition A DES *weak key* is a key K such that $E_K(E_K(x)) = x$ for all x, i.e., defining an involution. A pair of DES *semi-weak* keys is a pair (K_1, K_2) with $E_{K_1}(E_{K_2}(x)) = x$.

Encryption with one key of a semi-weak pair operates as does decryption with the other.

7.89 Fact DES has four weak keys and six pairs of semi-weak keys.

The four DES weak keys are listed in Table 7.5, along with corresponding 28-bit variables C_0 and D_0 of Algorithm 7.83; here $\{0\}^j$ represents j repetitions of bit 0. Since C_0 and D_0 are all-zero or all-one bit vectors, and rotation of these has no effect, it follows that all subkeys K_i are equal and an involution results as noted above.

The six pairs of DES semi-weak keys are listed in Table 7.6. Note their defining property (Definition 7.88) occurs when subkeys K_1 through K_{16} of the first key, respectively, equal subkeys K_{16} through K_1 of the second. This requires that a 1-bit circular left-shift of each of C_0 and D_0 for the first 56-bit key results in the (C_0, D_0) pair for the second 56-bit key (see Note 7.84), and thereafter left-rotating C_i and D_i one or two bits for the first results in the same value as right-rotating those for the second the same number of positions. The values in Table 7.6 satisfy these conditions. Given any one 64-bit semi-weak key, its paired semi-weak key may be obtained by splitting it into two halves and rotating each half through 8 bits.

7.90 Fact Let E denote DES. For each of the four DES weak keys K, there exist 2^{32} *fixed points* of E_K, i.e., plaintexts x such that $E_K(x) = x$. Similarly, four of the twelve semi-weak keys K each have 2^{32} *anti-fixed points*, i.e., x such that $E_K(x) = \overline{x}$.

The four semi-weak keys of Fact 7.90 are in the upper portion of Table 7.6. These are called *anti-palindromic keys*, since for these $K_1 = \overline{K_{16}}$, $K_2 = \overline{K_{15}}$, and so on.

(iii) DES is not a group

For a fixed DES key K, DES defines a permutation from $\{0, 1\}^{64}$ to $\{0, 1\}^{64}$. The set of DES keys defines 2^{56} such (potentially different) permutations. If this set of permutations was closed under composition (i.e., given any two keys K_1, K_2, there exists a third key K_3 such that $E_{K_3}(x) = E_{K_2}(E_{K_1}(x))$ for all x) then multiple encryption would be equivalent to single encryption. Fact 7.91 states that this is not the case for DES.

weak key (hexadecimal)	C_0	D_0
0101 0101 0101 0101	$\{0\}^{28}$	$\{0\}^{28}$
FEFE FEFE FEFE FEFE	$\{1\}^{28}$	$\{1\}^{28}$
1F1F 1F1F 0E0E 0E0E	$\{0\}^{28}$	$\{1\}^{28}$
E0E0 E0E0 F1F1 F1F1	$\{1\}^{28}$	$\{0\}^{28}$

Table 7.5: *Four DES weak keys.*

C_0	D_0	semi-weak key pair (hexadecimal)	C_0	D_0
$\{01\}^{14}$	$\{01\}^{14}$	01FE 01FE 01FE 01FE, FE01 FE01 FE01 FE01	$\{10\}^{14}$	$\{10\}^{14}$
$\{01\}^{14}$	$\{10\}^{14}$	1FE0 1FE0 0EF1 0EF1, E01F E01F F10E F10E	$\{10\}^{14}$	$\{01\}^{14}$
$\{01\}^{14}$	$\{0\}^{28}$	01E0 01E0 01F1 01F1, E001 E001 F101 F101	$\{10\}^{14}$	$\{0\}^{28}$
$\{01\}^{14}$	$\{1\}^{28}$	1FFE 1FFE 0EFE 0EFE, FE1F FE1F FE0E FE0E	$\{10\}^{14}$	$\{1\}^{28}$
$\{0\}^{28}$	$\{01\}^{14}$	011F 011F 010E 010E, 1F01 1F01 0E01 0E01	$\{0\}^{28}$	$\{10\}^{14}$
$\{1\}^{28}$	$\{01\}^{14}$	E0FE E0FE F1FE F1FE, FEE0 FEE0 FEF1 FEF1	$\{1\}^{28}$	$\{10\}^{14}$

Table 7.6: *Six pairs of DES semi-weak keys (one pair per line).*

7.91 Fact The set of 2^{56} permutations defined by the 2^{56} DES keys is not closed under functional composition. Moreover, a lower bound on the size of the group generated by composing this set of permutations is 10^{2499}.

The lower bound in Fact 7.91 is important with respect to using DES for multiple encryption. If the group generated by functional composition was too small, then multiple encryption would be less secure than otherwise believed. .

(iv) Linear and differential cryptanalysis of DES

Assuming that obtaining enormous numbers of known-plaintext pairs is feasible, linear cryptanalysis provides the most powerful attack on DES to date; it is not, however, considered a threat to DES in practical environments. Linear cryptanalysis is also possible in a ciphertext-only environment if some underlying plaintext redundancy is known (e.g., parity bits or high-order 0-bits in ASCII characters).

Differential cryptanalysis is one of the most general cryptanalytic tools to date against modern iterated block ciphers, including DES, Lucifer, and FEAL among many others. It is, however, primarily a chosen-plaintext attack. Further information on linear and differential cryptanalysis is given in §7.8.

7.92 Note (*strength of DES*) The complexity (see §7.2.1) of the best attacks currently known against DES is given in Table 7.7; percentages indicate success rate for specified attack parameters. The 'processing complexity' column provides only an estimate of the expected cost (operation costs differ across the various attacks); for exhaustive search, the cost is in DES operations. Regarding storage complexity, both linear and differential cryptanalysis require only negligible storage in the sense that known or chosen texts can be processed individually and discarded, but in a practical attack, storage for accumulated texts would be required if ciphertext was acquired prior to commencing the attack.

attack method	data complexity		storage	processing
	known	chosen	complexity	complexity
exhaustive precomputation	—	1	2^{56}	1 (table lookup)
exhaustive search	1	—	negligible	2^{55}
linear cryptanalysis	2^{43} (85%)	—	for texts	2^{43}
	2^{38} (10%)	—	for texts	2^{50}
differential cryptanalysis	—	2^{47}	for texts	2^{47}
	2^{55}	—	for texts	2^{55}

Table 7.7: *DES strength against various attacks.*

7.93 Remark (*practicality of attack models*) To be meaningful, attack comparisons based on different models (e.g., Table 7.7) must appropriately weigh the feasibility of extracting (acquiring) enormous amounts of chosen (known) plaintexts, which is considerably more difficult to arrange than a comparable number of computing cycles on an adversary's own machine. Exhaustive search with one known plaintext-ciphertext pair (for ciphertext-only, see Example 7.28) and 2^{55} DES operations is significantly more feasible in practice (e.g., using highly parallelized custom hardware) than linear cryptanalysis (LC) requiring 2^{43} known pairs.

While exhaustive search, linear, and differential cryptanalysis allow recovery of a DES key and, therefore, the entire plaintext, the attacks of Note 7.8, which become feasible once about 2^{32} ciphertexts are available, may be more efficient if the goal is to recover only part of the text.

7.5 FEAL

The Fast Data Encipherment Algorithm (FEAL) is a family of algorithms which has played a critical role in the development and refinement of various advanced cryptanalytic techniques, including linear and differential cryptanalysis. FEAL-N maps 64-bit plaintext to 64-bit ciphertext blocks under a 64-bit secret key. It is an N-round Feistel cipher similar to DES (cf. Equations (7.4), (7.5)), but with a far simpler f-function, and augmented by initial and final stages which XOR the two data halves as well as XOR subkeys directly onto the data halves.

FEAL was designed for speed and simplicity, especially for software on 8-bit microprocessors (e.g., chipcards). It uses byte-oriented operations (8-bit addition mod 256, 2-bit left rotation, and XOR), avoids bit-permutations and table look-ups, and offers small code size. The initial commercially proposed version with 4 rounds (FEAL-4), positioned as a fast alternative to DES, was found to be considerably less secure than expected (see Table 7.10). FEAL-8 was similarly found to offer less security than planned. FEAL-16 or FEAL-32 may yet offer security comparable to DES, but throughput decreases as the number of rounds rises. Moreover, whereas the speed of DES implementations can be improved through very large lookup tables, this appears more difficult for FEAL.

Algorithm 7.94 specifies FEAL-8. The f-function $f(A, Y)$ maps an input pair of 32×16 bits to a 32-bit output. Within the f function, two byte-oriented data substitutions (S-boxes) S_0 and S_1 are each used twice; each maps a pair of 8-bit inputs to an 8-bit output

row	column number															
	[0]	[1]	[2]	[3]	[4]	[5]	[6]	[7]	[8]	[9]	[10]	[11]	[12]	[13]	[14]	[15]
S_1																
[0]	14	4	13	1	2	15	11	8	3	10	6	12	5	9	0	7
[1]	0	15	7	4	14	2	13	1	10	6	12	11	9	5	3	8
[2]	4	1	14	8	13	6	2	11	15	12	9	7	3	10	5	0
[3]	15	12	8	2	4	9	1	7	5	11	3	14	10	0	6	13
S_2																
[0]	15	1	8	14	6	11	3	4	9	7	2	13	12	0	5	10
[1]	3	13	4	7	15	2	8	14	12	0	1	10	6	9	11	5
[2]	0	14	7	11	10	4	13	1	5	8	12	6	9	3	2	15
[3]	13	8	10	1	3	15	4	2	11	6	7	12	0	5	14	9
S_3																
[0]	10	0	9	14	6	3	15	5	1	13	12	7	11	4	2	8
[1]	13	7	0	9	3	4	6	10	2	8	5	14	12	11	15	1
[2]	13	6	4	9	8	15	3	0	11	1	2	12	5	10	14	7
[3]	1	10	13	0	6	9	8	7	4	15	14	3	11	5	2	12
S_4																
[0]	7	13	14	3	0	6	9	10	1	2	8	5	11	12	4	15
[1]	13	8	11	5	6	15	0	3	4	7	2	12	1	10	14	9
[2]	10	6	9	0	12	11	7	13	15	1	3	14	5	2	8	4
[3]	3	15	0	6	10	1	13	8	9	4	5	11	12	7	2	14
S_5																
[0]	2	12	4	1	7	10	11	6	8	5	3	15	13	0	14	9
[1]	14	11	2	12	4	7	13	1	5	0	15	10	3	9	8	6
[2]	4	2	1	11	10	13	7	8	15	9	12	5	6	3	0	14
[3]	11	8	12	7	1	14	2	13	6	15	0	9	10	4	5	3
S_6																
[0]	12	1	10	15	9	2	6	8	0	13	3	4	14	7	5	11
[1]	10	15	4	2	7	12	9	5	6	1	13	14	0	11	3	8
[2]	9	14	15	5	2	8	12	3	7	0	4	10	1	13	11	6
[3]	4	3	2	12	9	5	15	10	11	14	1	7	6	0	8	13
S_7																
[0]	4	11	2	14	15	0	8	13	3	12	9	7	5	10	6	1
[1]	13	0	11	7	4	9	1	10	14	3	5	12	2	15	8	6
[2]	1	4	11	13	12	3	7	14	10	15	6	8	0	5	9	2
[3]	6	11	13	8	1	4	10	7	9	5	0	15	14	2	3	12
S_8																
[0]	13	2	8	4	6	15	11	1	10	9	3	14	5	0	12	7
[1]	1	15	13	8	10	3	7	4	12	5	6	11	0	14	9	2
[2]	7	11	4	1	9	12	14	2	0	6	10	13	15	3	5	8
[3]	2	1	14	7	4	10	8	13	15	12	9	0	3	5	6	11

Table 7.8: *DES S-boxes.*

(see Table 7.9). S_0 and S_1 add a single bit $d \in \{0,1\}$ to 8-bit arguments x and y, ignore the carry out of the top bit, and left rotate the result 2 bits (ROT2):

$$S_d(x,y) = ROT2(x + y + d \bmod 256) \tag{7.6}$$

The key schedule uses a function $f_K(A,B)$ similar to the f-function (see Table 7.9; A_i, B_i, Y_i, t_i, and U_i are 8-bit variables), mapping two 32-bit inputs to a 32-bit output.

	$U \leftarrow f(A,Y)$	$U \leftarrow f_K(A,B)$
$t_1 =$	$(A_0 \oplus A_1) \oplus Y_0$	$A_0 \oplus A_1$
$t_2 =$	$(A_2 \oplus A_3) \oplus Y_1$	$A_2 \oplus A_3$
$U_1 =$	$S_1(t_1, t_2)$	$S_1(t_1, t_2 \oplus B_0)$
$U_2 =$	$S_0(t_2, U_1)$	$S_0(t_2, U_1 \oplus B_1)$
$U_0 =$	$S_0(A_0, U_1)$	$S_0(A_0, U_1 \oplus B_2)$
$U_3 =$	$S_1(A_3, U_2)$	$S_1(A_3, U_2 \oplus B_3)$

Table 7.9: *Output $U = (U_0, U_1, U_2, U_3)$ for FEAL functions f, f_K (Algorithm 7.94).*

As the operations of 2-bit rotation and XOR are both linear, the only nonlinear elementary operation in FEAL is addition mod 256.

7.94 Algorithm Fast Data Encipherment Algorithm (FEAL-8)

INPUT: 64-bit plaintext $M = m_1 \ldots m_{64}$; 64-bit key $K = k_1 \ldots k_{64}$.
OUTPUT: 64-bit ciphertext block $C = c_1 \ldots c_{64}$. (For decryption, see Note 7.96.)

1. (key schedule) Compute sixteen 16-bit subkeys K_i from K using Algorithm 7.95.
2. Define $M_L = m_1 \cdots m_{32}$, $M_R = m_{33} \cdots m_{64}$.
3. $(L_0, R_0) \leftarrow (M_L, M_R) \oplus ((K_8, K_9), (K_{10}, K_{11}))$. (XOR initial subkeys.)
4. $R_0 \leftarrow R_0 \oplus L_0$.
5. For i from 1 to 8 do: $L_i \leftarrow R_{i-1}$, $R_i \leftarrow L_{i-1} \oplus f(R_{i-1}, K_{i-1})$. (Use Table 7.9 for $f(A,Y)$ with $A = R_{i-1} = (A_0, A_1, A_2, A_3)$ and $Y = K_{i-1} = (Y_0, Y_1)$.)
6. $L_8 \leftarrow L_8 \oplus R_8$.
7. $(R_8, L_8) \leftarrow (R_8, L_8) \oplus ((K_{12}, K_{13}), (K_{14}, K_{15}))$. (XOR final subkeys.)
8. $C \leftarrow (R_8, L_8)$. (Note the order of the final blocks is exchanged.)

7.95 Algorithm FEAL-8 key schedule

INPUT: 64-bit key $K = k_1 \ldots k_{64}$.
OUTPUT: 256-bit extended key (16-bit subkeys K_i, $0 \leq i \leq 15$).

1. (initialize) $U^{(-2)} \leftarrow 0$, $U^{(-1)} \leftarrow k_1 \ldots k_{32}$, $U^{(0)} \leftarrow k_{33} \ldots k_{64}$.
2. $U \stackrel{\text{def}}{=} (U_0, U_1, U_2, U_3)$ for 8-bit U_i. Compute $K_0, \ldots, K_{15}$ as i runs from 1 to 8:
   (a) $U \leftarrow f_K(U^{(i-2)}, U^{(i-1)} \oplus U^{(i-3)})$. ($f_K$ is defined in Table 7.9, where A and B denote 4-byte vectors (A_0, A_1, A_2, A_3), (B_0, B_1, B_2, B_3).)
   (b) $K_{2i-2} = (U_0, U_1)$, $K_{2i-1} = (U_2, U_3)$, $U^{(i)} \leftarrow U$.

7.96 Note (*FEAL decryption*) Decryption may be achieved using Algorithm 7.94 with the same key K and ciphertext $C = (R_8, L_8)$ as the plaintext input M, but with the key schedule reversed. More specifically, subkeys $((K_{12}, K_{13}), (K_{14}, K_{15}))$ are used for the initial XOR (step 3), $((K_8, K_9), (K_{10}, K_{11}))$ for the final XOR (step 7), and the round keys are used from K_7 back to K_0 (step 5). This is directly analogous to decryption for DES (Note 7.84).

7.97 Note (*FEAL-N*) FEAL with 64-bit key can be generalized to N-rounds, N even. $N = 2^x$ is recommended; $x = 3$ yields FEAL-8 (Algorithm 7.94). FEAL-N uses $N + 8$ sixteen-bit subkeys: $K_0, \ldots, K_{N-1}$, respectively, in round i; $K_N, \ldots, K_{N+3}$ for the initial XOR; and $K_{N+4}, \ldots K_{N+7}$ for the final XOR. The key schedule of Algorithm 7.95 is directly generalized to compute keys K_0 through K_{N+7} as i runs from 1 to $(N/2) + 4$.

7.98 Note (*FEAL-NX*) Extending FEAL-N to use a 128-bit key results in FEAL-NX, with altered key schedule as follows. The key is split into 64-bit halves (K_L, K_R). K_R is partitioned into 32-bit halves (K_{R1}, K_{R2}). For $1 \leq i \leq (N/2) + 4$, define $Q_i = K_{R1} \oplus K_{R2}$ for $i \equiv 1 \bmod 3$; $Q_i = K_{R1}$ for $i \equiv 2 \bmod 3$; and $Q_i = K_{R2}$ for $i \equiv 0 \bmod 3$. The second argument $(U^{(i-1)} \oplus U^{(i-3)})$ to f_K in step 2a of Algorithm 7.95 is replaced by $U^{(i-1)} \oplus U^{(i-3)} \oplus Q_i$. For $K_R = 0$, FEAL-NX matches FEAL-N with K_L as the 64-bit FEAL-N key K.

7.99 Example (*FEAL test vectors*) For hex plaintext $M = 00000000\ 00000000$ and hex key $K = 01234567\ 89ABCDEF$, Algorithm 7.95 generates subkeys $(K_0, \ldots, K_7) =$ DF3BCA36 F17C1AEC 45A5B9C7 26EBAD25, $(K_8, \ldots, K_{15})$ = 8B2AECB7 AC509D4C 22CD479B A8D50CB5. Algorithm 7.94 generates FEAL-8 ciphertext $C =$ CEEF2C86 F2490752. For FEAL-16, the corresponding ciphertext is $C' = $ 3ADE0D2A D84D0B6F; for FEAL-32, $C'' = $ 69B0FAE6 DDED6B0B. For 128-bit key (K_L, K_R) with $K_L = K_R = K$ as above, M has corresponding FEAL-8X ciphertext $C''' = $ 92BEB65D 0E9382FB.   □

7.100 Note (*strength of FEAL*) Table 7.10 gives various published attacks on FEAL; LC and DC denote linear and differential cryptanalysis, and times are on common personal computers or workstations.

attack	data complexity		storage	processing
method	known	chosen	complexity	complexity
FEAL-4 – LC	5	—	30K bytes	6 minutes
FEAL-6 – LC	100	—	100K bytes	40 minutes
FEAL-8 – LC	2^{24}			10 minutes
FEAL-8 – DC		2^7 pairs	280K bytes	2 minutes
FEAL-16 – DC	—	2^{29} pairs		2^{30} operations
FEAL-24 – DC	—	2^{45} pairs		2^{46} operations
FEAL-32 – DC	—	2^{66} pairs		2^{67} operations

Table 7.10: *FEAL strength against various attacks.*

7.6 IDEA

The cipher named IDEA (International Data Encryption Algorithm) encrypts 64-bit plaintext to 64-bit ciphertext blocks, using a 128-bit input key K. Based in part on a novel generalization of the Feistel structure, it consists of 8 computationally identical rounds followed by an output transformation (see Figure 7.11). Round r uses six 16-bit subkeys $K_i^{(r)}$, $1 \le i \le 6$, to transform a 64-bit input X into an output of four 16-bit blocks, which are input to the next round. The round 8 output enters the output transformation, employing four additional subkeys $K_i^{(9)}$, $1 \le i \le 4$ to produce the final ciphertext $Y = (Y_1, Y_2, Y_3, Y_4)$. All subkeys are derived from K.

A dominant design concept in IDEA is mixing operations from three different algebraic groups of 2^n elements. The corresponding group operations on sub-blocks a and b of bitlength $n = 16$ are bitwise XOR: $a{\oplus}b$; addition mod 2^n: $(a+b)$ AND $\texttt{0xFFFF}$, denoted $a{\boxplus}b$; and (modified) multiplication mod $2^n + 1$, with $0 \in \mathbb{Z}_{2^n}$ associated with $2^n \in \mathbb{Z}_{2^n+1}$: $a{\odot}b$ (see Note 7.104).

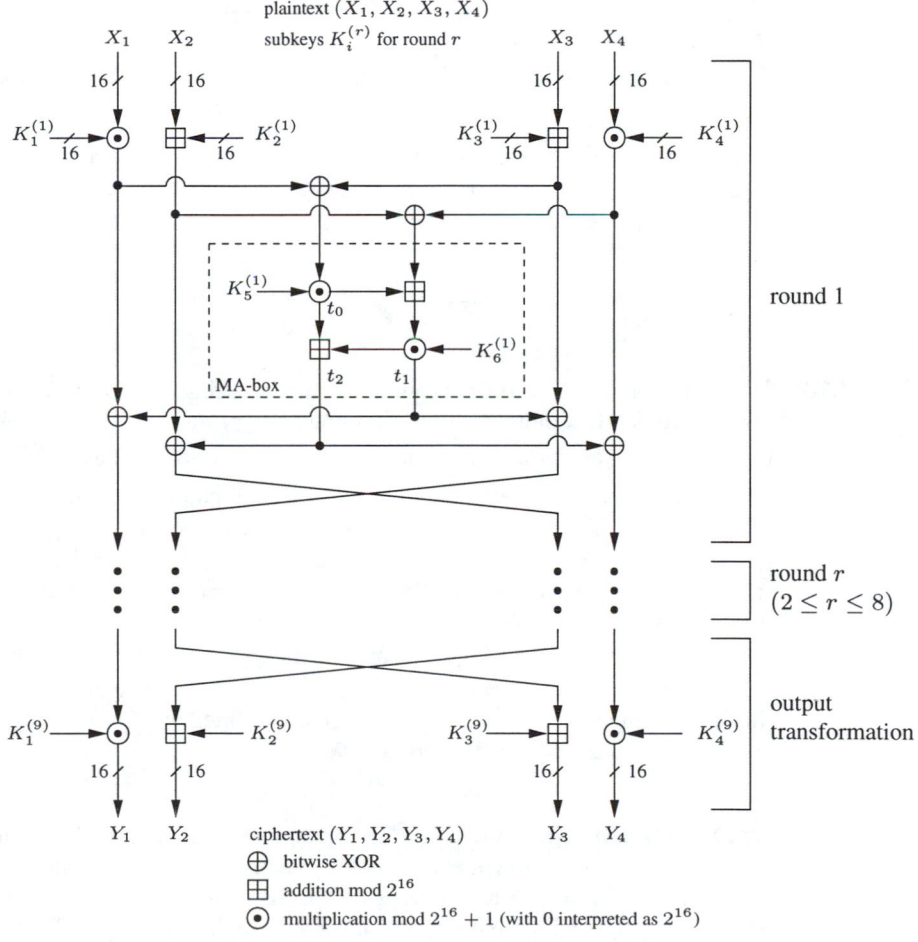

Figure 7.11: *IDEA computation path.*

7.101 Algorithm IDEA encryption

INPUT: 64-bit plaintext $M = m_1 \ldots m_{64}$; 128-bit key $K = k_1 \ldots k_{128}$.
OUTPUT: 64-bit ciphertext block $Y = (Y_1, Y_2, Y_3, Y_4)$. (For decryption, see Note 7.103.)

1. (key schedule) Compute 16-bit subkeys $K_1^{(r)}, \ldots, K_6^{(r)}$ for rounds $1 \leq r \leq 8$, and $K_1^{(9)}, \ldots, K_4^{(9)}$ for the output transformation, using Algorithm 7.102.
2. $(X_1, X_2, X_3, X_4) \leftarrow (m_1 \ldots m_{16}, m_{17} \ldots m_{32}, m_{33} \ldots m_{48}, m_{49} \ldots m_{64})$, where X_i is a 16-bit data store.
3. For round r from 1 to 8 do:
   (a) $X_1 \leftarrow X_1 \odot K_1^{(r)}$, $X_4 \leftarrow X_4 \odot K_4^{(r)}$, $X_2 \leftarrow X_2 \boxplus K_2^{(r)}$, $X_3 \leftarrow X_3 \boxplus K_3^{(r)}$.
   (b) $t_0 \leftarrow K_5^{(r)} \odot (X_1 \oplus X_3)$, $t_1 \leftarrow K_6^{(r)} \odot (t_0 \boxplus (X_2 \oplus X_4))$, $t_2 \leftarrow t_0 \boxplus t_1$.
   (c) $X_1 \leftarrow X_1 \oplus t_1$, $X_4 \leftarrow X_4 \oplus t_2$, $a \leftarrow X_2 \oplus t_2$, $X_2 \leftarrow X_3 \oplus t_1$, $X_3 \leftarrow a$.
4. (output transformation) $Y_1 \leftarrow X_1 \odot K_1^{(9)}$, $Y_4 \leftarrow X_4 \odot K_4^{(9)}$, $Y_2 \leftarrow X_3 \boxplus K_2^{(9)}$, $Y_3 \leftarrow X_2 \boxplus K_3^{(9)}$.

7.102 Algorithm IDEA key schedule (encryption)

INPUT: 128-bit key $K = k_1 \ldots k_{128}$.
OUTPUT: 52 16-bit key sub-blocks $K_i^{(r)}$ for 8 rounds r and the output transformation.

1. Order the subkeys $K_1^{(1)} \ldots K_6^{(1)}, K_1^{(2)} \ldots K_6^{(2)}, \ldots, K_1^{(8)} \ldots K_6^{(8)}, K_1^{(9)} \ldots K_4^{(9)}$.
2. Partition K into eight 16-bit blocks; assign these directly to the first 8 subkeys.
3. Do the following until all 52 subkeys are assigned: cyclic shift K left 25 bits; partition the result into 8 blocks; assign these blocks to the next 8 subkeys.

The key schedule of Algorithm 7.102 may be converted into a table which lists, for each of the 52 keys blocks, which 16 (consecutive) bits of the input key K form it.

7.103 Note (*IDEA decryption*) Decryption is achieved using Algorithm 7.101 with the ciphertext Y provided as input M, and the same encryption key K, but the following change to the key schedule. First use K to derive all encryption subkeys $K_i^{(r)}$; from these compute the decryption subkeys $K'_i^{(r)}$ per Table 7.11; then use $K'_i^{(r)}$ in place of $K_i^{(r)}$ in Algorithm 7.101. In Table 7.11, $-K_i$ denotes the additive inverse (mod 2^{16}) of K_i: the integer $u = (2^{16} - K_i)$ AND 0xFFFF, $0 \leq u \leq 2^{16} - 1$. K_i^{-1} denotes the multiplicative inverse (mod $2^{16} + 1$) of K_i, also in $\{0, 1, \ldots, 2^{16} - 1\}$, derivable by the Extended Euclidean algorithm (Algorithm 2.107), which on inputs $a \geq b \geq 0$ returns integers x and y such that $ax + by = \gcd(a, b)$. Using $a = 2^{16} + 1$ and $b = K_i$, the gcd is always 1 (except for $K_i = 0$, addressed separately) and thus $K_i^{-1} = y$, or $2^{16} + 1 + y$ if $y < 0$. When $K_i = 0$, this input is mapped to 2^{16} (since the inverse is defined by $K_i \odot K_i^{-1} = 1$; see Note 7.104) and $(2^{16})^{-1} = 2^{16}$ is then defined to give $K_i^{-1} = 0$.

7.104 Note (*definition of $\odot$*) In IDEA, $a \odot b$ corresponds to a (modified) multiplication, modulo $2^{16} + 1$, of unsigned 16-bit integers a and b, where $0 \in \mathbb{Z}_{2^{16}}$ is associated with $2^{16} \in \mathbb{Z}_{2^{16}+1}$ as follows:[2] if $a = 0$ or $b = 0$, replace it by 2^{16} (which is $\equiv -1 \bmod 2^{16} + 1$) prior to modular multiplication; and if the result is 2^{16}, replace this by 0. Thus, $\odot$ maps two 16-bit inputs to a 16-bit output. Pseudo-code for $\odot$ is as follows (cf. Note 7.105, for ordinary

---

[2]Thus the operands of $\odot$ are from a set of cardinality 2^{16} ($\mathbb{Z}_{2^{16}+1}^*$) as are those of $\oplus$ and $\boxplus$.

round r	$K'^{(r)}_1$	$K'^{(r)}_2$	$K'^{(r)}_3$	$K'^{(r)}_4$	$K'^{(r)}_5$	$K'^{(r)}_6$
$r = 1$	$(K_1^{(10-r)})^{-1}$	$-K_2^{(10-r)}$	$-K_3^{(10-r)}$	$(K_4^{(10-r)})^{-1}$	$K_5^{(9-r)}$	$K_6^{(9-r)}$
$2 \leq r \leq 8$	$(K_1^{(10-r)})^{-1}$	$-K_3^{(10-r)}$	$-K_2^{(10-r)}$	$(K_4^{(10-r)})^{-1}$	$K_5^{(9-r)}$	$K_6^{(9-r)}$
$r = 9$	$(K_1^{(10-r)})^{-1}$	$-K_2^{(10-r)}$	$-K_3^{(10-r)}$	$(K_4^{(10-r)})^{-1}$	—	—

Table 7.11: *IDEA decryption subkeys $K'^{(r)}_i$ derived from encryption subkeys $K^{(r)}_i$.*

multiplication mod $2^{16} + 1$), for c a 32-bit unsigned integer: if $(a = 0)$ $r \leftarrow$ (0x10001 $- b$) (since $2^{16}b \equiv -b$), elseif $(b = 0)$ $r \leftarrow$ (0x10001 $- a$) (by similar reasoning), else $\{c \leftarrow ab; r \leftarrow ((c \text{ AND } 0\text{xFFFF}) - (c >> 16))\}$; if $(r < 0)$ $r \leftarrow$ (0x10001 $+ r)\}$, with return value $(r \text{ AND } 0\text{xFFFF})$ in all 3 cases.

7.105 Note *(implementing $ab \bmod 2^n + 1$)* Multiplication mod $2^{16} + 1$ may be efficiently implemented as follows, for $0 \leq a, b \leq 2^{16}$ (cf. §14.3.4). Let $c = ab = c_0 \cdot 2^{32} + c_H \cdot 2^{16} + c_L$, where $c_0 \in \{0, 1\}$ and $0 \leq c_L, c_H < 2^{16}$. To compute $c' = c \bmod (2^{16} + 1)$, first obtain c_L and c_H by standard multiplication. For $a = b = 2^{16}$, note that $c_0 = 1$, $c_L = c_H = 0$, and $c' = (-1)(-1) = 1$, since $2^{16} \equiv -1 \bmod (2^{16} + 1)$; otherwise, $c_0 = 0$. Consequently, $c' = c_L - c_H + c_0$ if $c_L \geq c_H$, while $c' = c_L - c_H + (2^{16} + 1)$ if $c_L < c_H$ (since then $-2^{16} < c_L - c_H < 0$).

7.106 Example *(IDEA test vectors)* Sample data for IDEA encryption of 64-bit plaintext M using 128-bit key K is given in Table 7.12. All entries are 16-bit values displayed in hexadecimal. Table 7.13 details the corresponding decryption of the resulting 64-bit ciphertext C under the same key K. □

	128-bit key $K = (1, 2, 3, 4, 5, 6, 7, 8)$						64-bit plaintext $M = (0, 1, 2, 3)$			
r	$K^{(r)}_1$	$K^{(r)}_2$	$K^{(r)}_3$	$K^{(r)}_4$	$K^{(r)}_5$	$K^{(r)}_6$	X_1	X_2	X_3	X_4
1	0001	0002	0003	0004	0005	0006	00f0	00f5	010a	0105
2	0007	0008	0400	0600	0800	0a00	222f	21b5	f45e	e959
3	0c00	0e00	1000	0200	0010	0014	0f86	39be	8ee8	1173
4	0018	001c	0020	0004	0008	000c	57df	ac58	c65b	ba4d
5	2800	3000	3800	4000	0800	1000	8e81	ba9c	f77f	3a4a
6	1800	2000	0070	0080	0010	0020	6942	9409	e21b	1c64
7	0030	0040	0050	0060	0000	2000	99d0	c7f6	5331	620e
8	4000	6000	8000	a000	c000	e001	0a24	0098	ec6b	4925
9	0080	00c0	0100	0140	—	—	11fb	ed2b	0198	6de5

Table 7.12: *IDEA encryption sample: round subkeys and ciphertext (X_1, X_2, X_3, X_4).*

7.107 Note *(security of IDEA)* For the full 8-round IDEA, other than attacks on weak keys (see page 279), no published attack is better than exhaustive search on the 128-bit key space. The security of IDEA currently appears bounded only by the weaknesses arising from the relatively small (compared to its keylength) blocklength of 64 bits.

| | $K = (1, 2, 3, 4, 5, 6, 7, 8)$ | | | | | | $C = (11\text{fb}, \text{ed2b}, 0198, 6\text{de5})$ | | | |
r	$K'^{(r)}_1$	$K'^{(r)}_2$	$K'^{(r)}_3$	$K'^{(r)}_4$	$K'^{(r)}_5$	$K'^{(r)}_6$	X_1	X_2	X_3	X_4
1	fe01	ff40	ff00	659a	c000	e001	d98d	d331	27f6	82b8
2	fffd	8000	a000	cccc	0000	2000	bc4d	e26b	9449	a576
3	a556	ffb0	ffc0	52ab	0010	0020	0aa4	f7ef	da9c	24e3
4	554b	ff90	e000	fe01	0800	1000	ca46	fe5b	dc58	116d
5	332d	c800	d000	fffd	0008	000c	748f	8f08	39da	45cc
6	4aab	ffe0	ffe4	c001	0010	0014	3266	045e	2fb5	b02e
7	aa96	f000	f200	ff81	0800	0a00	0690	050a	00fd	1dfa
8	4925	fc00	fff8	552b	0005	0006	0000	0005	0003	000c
9	0001	fffe	fffd	c001	—	—	0000	0001	0002	0003

Table 7.13: *IDEA decryption sample: round subkeys and variables* (X_1, X_2, X_3, X_4).

7.7 SAFER, RC5, and other block ciphers

7.7.1 SAFER

SAFER K-64 (Secure And Fast Encryption Routine, with 64-bit key) is an iterated block cipher with 64-bit plaintext and ciphertext blocks. It consists of r identical rounds followed by an output transformation. The original recommendation of 6 rounds was followed by a recommendation to adopt a slightly modified key schedule (yielding SAFER SK-64, which should be used rather than SAFER K-64 – see Note 7.110) and to use 8 rounds (maximum $r = 10$). Both key schedules expand the 64-bit external key into $2r + 1$ subkeys each of 64-bits (two for each round plus one for the output transformation). SAFER consists entirely of simple byte operations, aside from byte-rotations in the key schedule; it is thus suitable for processors with small word size such as chipcards (cf. FEAL).

Details of SAFER K-64 are given in Algorithm 7.108 and Figure 7.12 (see also page 280 regarding SAFER K-128 and SAFER SK-128). The XOR-addition stage beginning each round (identical to the output transformation) XORs bytes 1, 4, 5, and 8 of the (first) round subkey with the respective round input bytes, and respectively adds (mod 256) the remaining 4 subkey bytes to the others. The XOR and addition (mod 256) operations are interchanged in the subsequent addition-XOR stage. The S-boxes are an invertible byte-to-byte substitution using one fixed 8-bit bijection (see Note 7.111). A linear transformation f (the *Pseudo-Hadamard Transform*) used in the 3-level linear layer was specially constructed for rapid diffusion. The introduction of additive key biases in the key schedule eliminates weak keys (cf. DES, IDEA). In contrast to Feistel-like and many other ciphers, in SAFER the operations used for encryption differ from those for decryption (see Note 7.113). SAFER may be viewed as an SP network (Definition 7.79).

Algorithm 7.108 uses the following definitions (L, R denote left, right 8-bit inputs):

1. $f(L, R) = (2L + R, \ L + R)$. Addition here is mod 256 (also denoted by $\boxplus$);
2. tables S and S_{inv}, and the constant table for *key biases* $B_i[j]$ as per Note 7.111.

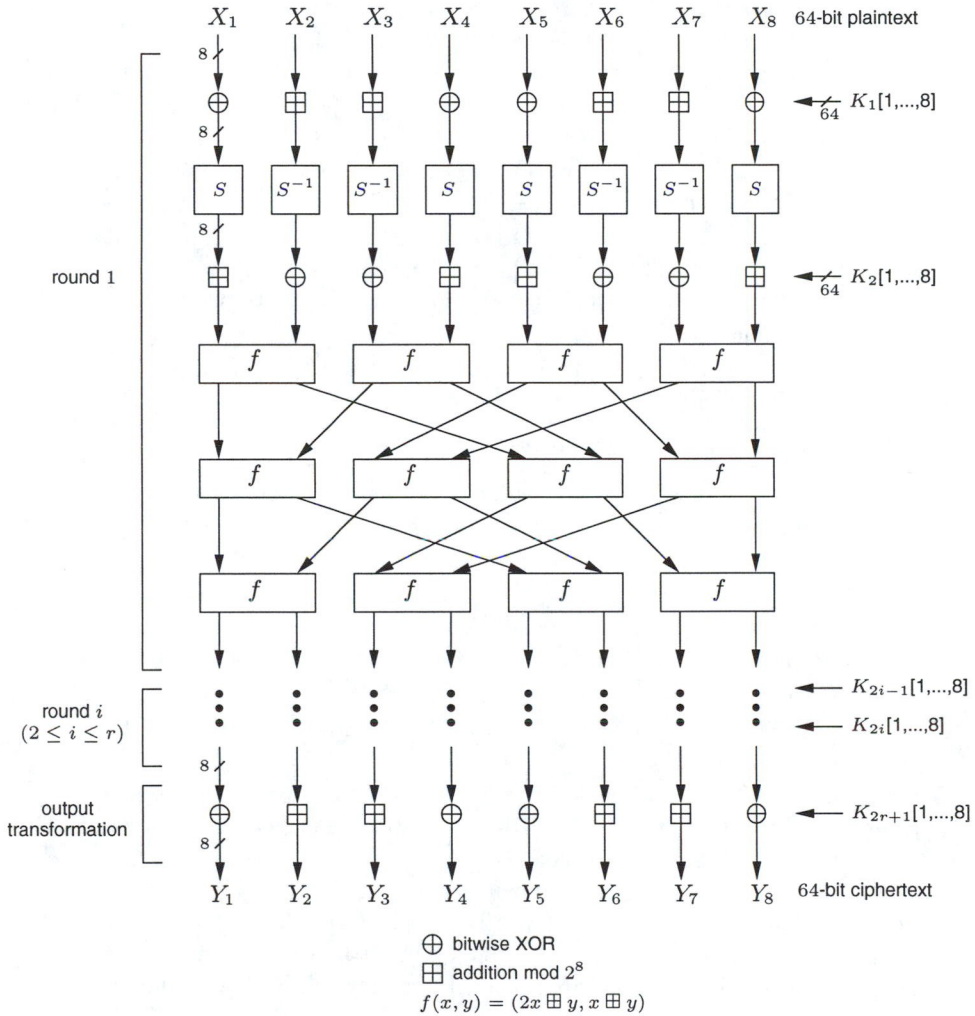

Figure 7.12: *SAFER K-64 computation path (r rounds).*

7.108 Algorithm SAFER K-64 encryption (r rounds)

INPUT: r, $6 \leq r \leq 10$; 64-bit plaintext $M = m_1 \cdots m_{64}$ and key $K = k_1 \cdots k_{64}$.
OUTPUT: 64-bit ciphertext block $Y = (Y_1, \ldots, Y_8)$. (For decryption, see Note 7.113.)

1. Compute 64-bit subkeys $K_1, \ldots, K_{2r+1}$ by Algorithm 7.109 with inputs K and r.
2. $(X_1, X_2, \ldots, X_8) \leftarrow (m_1 \cdots m_8, \ m_9 \cdots m_{16}, \ \ldots, \ m_{57} \cdots m_{64})$.
3. For i from 1 to r do: (XOR-addition, S-box, addition-XOR, and 3 linear layers)
    (a) For $j = 1, 4, 5, 8$: $X_j \leftarrow X_j \oplus K_{2i-1}[j]$.
        For $j = 2, 3, 6, 7$: $X_j \leftarrow X_j \boxplus K_{2i-1}[j]$.
    (b) For $j = 1, 4, 5, 8$: $X_j \leftarrow S[X_j]$. For $j = 2, 3, 6, 7$: $X_j \leftarrow S_{\text{inv}}[X_j]$.
    (c) For $j = 1, 4, 5, 8$: $X_j \leftarrow X_j \boxplus K_{2i}[j]$. For $j = 2, 3, 6, 7$: $X_j \leftarrow X_j \oplus K_{2i}[j]$.
    (d) For $j = 1, 3, 5, 7$: $(X_j, X_{j+1}) \leftarrow f(X_j, X_{j+1})$.
    (e) $(Y_1, Y_2) \leftarrow f(X_1, X_3)$, $(Y_3, Y_4) \leftarrow f(X_5, X_7)$,
        $(Y_5, Y_6) \leftarrow f(X_2, X_4)$, $(Y_7, Y_8) \leftarrow f(X_6, X_8)$.
        For j from 1 to 8 do: $X_j \leftarrow Y_j$.
    (f) $(Y_1, Y_2) \leftarrow f(X_1, X_3)$, $(Y_3, Y_4) \leftarrow f(X_5, X_7)$,
        $(Y_5, Y_6) \leftarrow f(X_2, X_4)$, $(Y_7, Y_8) \leftarrow f(X_6, X_8)$.
        For j from 1 to 8 do: $X_j \leftarrow Y_j$. (This mimics the previous step.)
4. (output transformation):
    For $j = 1, 4, 5, 8$: $Y_j \leftarrow X_j \oplus K_{2r+1}[j]$. For $j = 2, 3, 6, 7$: $Y_j \leftarrow X_j \boxplus K_{2r+1}[j]$.

7.109 Algorithm SAFER K-64 key schedule

INPUT: 64-bit key $K = k_1 \cdots k_{64}$; number of rounds r.
OUTPUT: 64-bit subkeys $K_1, \ldots, K_{2r+1}$. $K_i[j]$ is byte j of K_i (numbered left to right).

1. Let $R[i]$ denote an 8-bit data store and let $B_i[j]$ denote byte j of B_i (Note 7.111).
2. $(R[1], R[2], \ldots, R[8]) \leftarrow (k_1 \cdots k_8, \ k_9 \cdots k_{16}, \ \ldots, \ k_{57} \cdots k_{64})$.
3. $(K_1[1], K_1[2], \ldots, K_1[8]) \leftarrow (R[1], R[2], \ldots, R[8])$.
4. For i from 2 to $2r + 1$ do: (rotate key bytes left 3 bits, then add in the bias)
    (a) For j from 1 to 8 do: $R[j] \leftarrow (R[j] \hookleftarrow 3)$.
    (b) For j from 1 to 8 do: $K_i[j] \leftarrow R[j] \boxplus B_i[j]$. (See Note 7.110.)

7.110 Note (*SAFER SK-64 – strengthened key schedule*) An improved key schedule for Algorithm 7.108, resulting in SAFER SK-64, involves three changes as follows. (i) After initializing the $R[i]$ in step 1 of Algorithm 7.109, set $R[9] \leftarrow R[1] \oplus R[2] \oplus \cdots \oplus R[8]$. (ii) Change the upper bound on the loop index in step 4a from 8 to 9. (iii) Replace the iterated line in step 4b by: $K_i[j] \leftarrow R[((i + j - 2) \bmod 9) + 1] \boxplus B_i[j]$. Thus, key bytes $1, \ldots, 8$ of $R[\cdot]$ are used for K_1; bytes $2, \ldots, 9$ for K_2; bytes $3, \ldots 9, 1$ for K_3, etc. Here and originally, $\boxplus$ denotes addition mod 256. No attack against SAFER SK-64 better than exhaustive key search is known.

7.111 Note (*S-boxes and key biases in SAFER*) The S-box, inverse S-box, and key biases for Algorithm 7.108 are constant tables as follows. $g \leftarrow 45$. $S[0] \leftarrow 1$, $S_{\text{inv}}[1] \leftarrow 0$. for i from 1 to 255 do: $t \leftarrow g \cdot S[i - 1] \bmod 257$, $S[i] \leftarrow t$, $S_{\text{inv}}[t] \leftarrow i$. Finally, $S[128] \leftarrow 0$, $S_{\text{inv}}[0] \leftarrow 128$. (Since g generates $\mathbb{Z}_{257}^*$, $S[i]$ is a bijection on $\{0, 1, \ldots, 255\}$. (Note that $g^{128} \equiv 256 \pmod{257}$, and associating 256 with 0 makes S a mapping with 8-bit input and output.) The *additive key biases* are 8-bit constants used in the key schedule (Algorithm 7.109), intended to behave as random numbers, and defined $B_i[j] = S[S[9i + j]]$ for i from 2 to $2r + 1$ and j from 1 to 8. For example: $B_2 = (22, 115, 59, 30, 142, 112, 189, 134)$ and $B_{13} = (143, 41, 221, 4, 128, 222, 231, 49)$.

7.112 Remark (*S-box mapping*) The S-box of Note 7.111 is based on the function $S(x) = g^x$ mod 257 using a primitive element $g = 45 \in \mathbb{Z}_{257}$. This mapping is nonlinear with respect to both $\mathbb{Z}_{257}$ arithmetic and the vector space of 8-tuples over $\mathbb{F}_2$ under the XOR operation. The inverse S-box is based on the base-g logarithm function.

7.113 Note (*SAFER K-64 decryption*) For decryption of Algorithm 7.108, the same key K and subkeys K_i are used as for encryption. Each encryption step is undone in reverse order, from last to first. Begin with an input transformation (XOR-subtraction stage) with key K_{2r+1} to undo the output transformation, replacing modular addition with subtraction. Follow with r decryption rounds using keys K_{2r} through K_1 (two-per-round), inverting each round in turn. Each starts with a 3-stage inverse linear layer using $f_{\mathrm{inv}}(L, R) = (L - R, \; 2R - L)$, with subtraction here mod 256, in a 3-step sequence defined as follows (to invert the byte-permutations between encryption stages):
Level 1 (for $j = 1, 3, 5, 7$): $(X_j, X_{j+1}) \leftarrow f_{\mathrm{inv}}(X_j, X_{j+1})$.
Levels 2 and 3 (each): $(Y_1, Y_2) \leftarrow f_{\mathrm{inv}}(X_1, X_5)$, $(Y_3, Y_4) \leftarrow f_{\mathrm{inv}}(X_2, X_6)$,
$(Y_5, Y_6) \leftarrow f_{\mathrm{inv}}(X_3, X_7)$, $(Y_7, Y_8) \leftarrow f_{\mathrm{inv}}(X_4, X_8)$; for j from 1 to 8 do: $X_j \leftarrow Y_j$.
A subtraction-XOR stage follows (replace modular addition with subtraction), then an inverse substitution stage (exchange S and S^{-1}), and an XOR-subtraction stage.

7.114 Example (*SAFER test vectors*) Using 6-round SAFER K-64 (Algorithm 7.108) on the 64-bit plaintext $M = (1, 2, 3, 4, 5, 6, 7, 8)$ with the key $K = (8, 7, 6, 5, 4, 3, 2, 1)$ results in the ciphertext $C = (200, 242, 156, 221, 135, 120, 62, 217)$, written as 8 bytes in decimal. Using 6-round SAFER SK-64 (Note 7.110) on the plaintext M above with the key $K = (1, 2, 3, 4, 5, 6, 7, 8)$ results in the ciphertext $C = (95, 206, 155, 162, 5, 132, 56, 199)$. $\square$

7.7.2 RC5

The RC5 block cipher has a word-oriented architecture for variable word sizes $w = 16, 32$, or 64 bits. It has an extremely compact description, and is suitable for hardware or software. The number of rounds r and the key byte-length b are also variable. It is successively more completely identified as RC5–w, RC5–w/r, and RC5–$w/r/b$. RC5-32/12/16 is considered a common choice of parameters; $r = 12$ rounds are recommended for RC5–32, and $r = 16$ for RC5–64.

Algorithm 7.115 specifies RC5. Plaintext and ciphertext are blocks of bitlength $2w$. Each of r rounds updates both w-bit data halves, using 2 subkeys in an input transformation and 2 more for each round. The only operations used, all on w-bit words, are addition mod 2^w ($\boxplus$), XOR ($\oplus$), and rotations (left $\hookleftarrow$ and right $\hookrightarrow$). The XOR operation is linear, while the addition may be considered nonlinear depending on the metric for linearity. The data-dependent rotations featured in RC5 are the main nonlinear operation used: $x \hookleftarrow y$ denotes cyclically shifting a w-bit word left y bits; the rotation-count y may be reduced mod w (the low-order $\lg(w)$ bits of w suffice). The key schedule expands a key of b bytes into $2r + 2$ subkeys K_i of w bits each. Regarding packing/unpacking bytes into words, the byte-order is *little-endian*: for $w = 32$, the first plaintext byte goes in the low-order end of A, the fourth in A's high-order end, the fifth in B's low order end, and so on.

7.115 Algorithm RC5 encryption (w-bit wordsize, r rounds, b-byte key)

INPUT: $2w$-bit plaintext $M = (A, B)$; r; key $K = K[0]\ldots K[b-1]$.
OUTPUT: $2w$-bit plaintext C. (For decryption, see Note 7.117.)

1. Compute $2r + 2$ subkeys $K_0, \ldots, K_{2r+1}$ by Algorithm 7.116 from inputs K and r.
2. $A \leftarrow A \boxplus K_0$, $B \leftarrow B \boxplus K_1$. (Use addition modulo 2^w.)
3. For i from 1 to r do: $A \leftarrow ((A \oplus B) \hookleftarrow B) \boxplus K_{2i}$, $B \leftarrow ((B \oplus A) \hookleftarrow A) \boxplus K_{2i+1}$.
4. The output is $C \leftarrow (A, B)$.

7.116 Algorithm RC5 key schedule

INPUT: word bitsize w; number of rounds r; b-byte key $K[0]\ldots K[b-1]$.
OUTPUT: subkeys $K_0, \ldots, K_{2r+1}$ (where K_i is w bits).

1. Let $u = w/8$ (number of bytes per word) and $c = \lceil b/u \rceil$ (number of words K fills). Pad K on the right with zero-bytes if necessary to achieve a byte-count divisible by u (i.e., $K[j] \leftarrow 0$ for $b \le j \le c \cdot u - 1$). For i from 0 to $c - 1$ do: $L_i \leftarrow \sum_{j=0}^{u-1} 2^{8j} K[i \cdot u + j]$ (i.e., fill L_i low-order to high-order byte using each byte of $K[\cdot]$ once).
2. $K_0 \leftarrow P_w$; for i from 1 to $2r + 1$ do: $K_i \leftarrow K_{i-1} \boxplus Q_w$. (Use Table 7.14.)
3. $i \leftarrow 0, j \leftarrow 0, A \leftarrow 0, B \leftarrow 0, t \leftarrow \max(c, 2r + 2)$. For s from 1 to $3t$ do:
   (a) $K_i \leftarrow (K_i \boxplus A \boxplus B) \hookleftarrow 3$, $A \leftarrow K_i$, $i \leftarrow i + 1 \bmod (2r + 2)$.
   (b) $L_j \leftarrow (L_j \boxplus A \boxplus B) \hookleftarrow (A \boxplus B)$, $B \leftarrow L_j$, $j \leftarrow j + 1 \bmod c$.
4. The output is $K_0, K_1, \ldots, K_{2r+1}$. (The L_i are not used.)

7.117 Note (*RC5 decryption*) Decryption uses the Algorithm 7.115 subkeys, operating on ciphertext $C = (A, B)$ as follows (subtraction is mod 2^w, denoted $\boxminus$). For i from r down to 1 do: $B \leftarrow ((B \boxminus K_{2i+1}) \hookrightarrow A) \oplus A$, $A \leftarrow ((A \boxminus K_{2i}) \hookrightarrow B) \oplus B$. Finally $M \leftarrow (A \boxminus K_0, B \boxminus K_1)$.

w :	16	32	64	
P_w :	B7E1	B7E15163	B7E15162	8AED2A6B
Q_w :	9E37	9E3779B9	9E3779B9	7F4A7C15

Table 7.14: *RC5 magic constants (given as hex strings).*

7.118 Example (*RC5–32/12/16 test vectors*) For the hexadecimal plaintext $M = $ B278C165 CC97D184 and key $K = $ 5269F149 D41BA015 2497574D 7F153125, RC5 with $w = 32, r = 12$, and $b = 16$ generates ciphertext $C = $ 15E444EB 249831DA.   □

7.7.3 Other block ciphers

LOKI'91 (and earlier, LOKI'89) was proposed as a DES alternative with a larger 64-bit key, a matching 64-bit blocksize, and 16 rounds. It differs from DES mainly in key-scheduling and the f-function. The f-function of each round uses four identical 12-to-8 bit S-boxes,

4 input bits of which select one of 16 functions, each of which implements exponentiation with a fixed exponent in a different representation of $GF(2^8)$. While no significant exploitable weaknesses have been found in LOKI'91 when used for encryption, related-key attacks (see page 281) are viewed as a certificational weakness.

Khufu and Khafre are DES-like ciphers which were proposed as fast software-oriented alternatives to DES. They have 64-bit blocks, 8×32 bit S-boxes, and a variable number of rounds (typically 16, 24, or 32). Khufu keys may be up to 512 bits. Khafre keys have bitlength that is a multiple of 64 (64 and 128-bit keys are typical); 64 key bits are XORed onto the data block before the first and thereafter following every 8 rounds. Whereas a DES round involves eight 6-to-4 bit S-boxes, one round of Khufu involves a single 8-to-32 bit table look-up, with a different S-box for every 8 rounds. The S-boxes are generated pseudorandomly from the user key. Khafre uses fixed S-boxes generated pseudorandomly from an initial S-box constructed from random numbers published by the RAND corporation in 1955. Under the best currently known attacks, 16-round Khufu and 24-round Khafre are each more difficult to break than DES.

7.8 Notes and further references

§7.1

The extensive and particularly readable survey by Diffie and Hellman [347], providing a broad introduction to cryptography especially noteworthy for its treatment of Hagelin and rotor machines and the valuable annotated bibliography circa 1979, is a source for much of the material in §7.2, §7.3, and §7.4 herein. Aside from the appearance of DES [396] in the mid 1970s and FEAL [884] later in the 1980s, prior to 1990 few fully-specified serious symmetric block cipher proposals were widely available or discussed. (See Chapter 15 for Pohlig and Hellman's 1978 discrete exponentiation cipher.) With the increasing feasibility of exhaustive search on 56-bit DES keys, the period 1990-1995 resulted in a large number of proposals, beginning with PES [728], the preliminary version of IDEA [730]. The *Fast Software Encryption* workshops (Cambridge, U.K., Dec. 1993; Leuven, Belgium, Dec. 1994; and again Cambridge, Feb. 1996) were a major stimulus and forum for new proposals.

The most significant cryptanalytic advances over the 1990-1995 period were Matsui's linear cryptanalysis [796, 795], and the differential cryptanalysis of Biham and Shamir [138] (see also [134, 139]). Extensions of these included the differential-linear analysis by Langford and Hellman [741], and the truncated differential analysis of Knudsen [686]. For additional background on linear cryptanalysis, see Biham [132]; see also Matsui and Yamagishi [798] for a preliminary version of the method. Additional background on differential cryptanalysis is provided by many authors including Lai [726], Lai, Massey, and Murphy [730], and Coppersmith [271]; although more efficient 6-round attacks are known, Stinson [1178] provides detailed examples of attacks on 3-round and 6-round DES. Regarding both linear and differential cryptanalysis, see also Knudsen [684] and Kaliski and Yin [656].

§7.2

Lai [726, Chapter 2] provides an excellent concise introduction to block ciphers, including a lucid discussion of design principles (recommended for all block cipher designers). Regarding text dictionary and matching ciphertext attacks (Note 7.8), see Coppersmith, Johnson, and Matyas [278]. Rivest and Sherman [1061] provide a unified framework for randomized encryption (Definition 7.3); a common example is the use of random "salt" appended to passwords prior to password encryption in some operating systems (§10.2.3). Fact 7.9 is due to Shannon [1121], whose contributions are many (see below).

The four basic modes of operation (including k-bit OFB feedback) were originally defined specifically for DES in 1980 by FIPS 81 [398] and in 1983 by ANSI X3.106 [34], while ISO 8732 [578] and ISO/IEC 10116 [604], respectively, defined these modes for general 64-bit and general n-bit block ciphers, mandating n-bit OFB feedback (see also Chapter 15). Brassard [192] gives a concise summary of modes of operation; Davies and Price [308] provide a comprehensive discussion, including OFB cycling (Note 7.24; see also Jueneman [643] and Davies and Parkin [307]), and a method for encrypting incomplete CBC final blocks without data expansion, which is important if plaintext must be encrypted and returned into its original store. See Voydock and Kent [1225] for additional requirements on IVs. Recommending $r = s$ for maximum strength, ISO/IEC 10116 [604] specifies the CFB variation of Example 7.19, and provides extensive discussion of properties of the various modes. The counter mode (Example 7.23) was suggested by Diffie and Hellman [347].

The 1977 exhaustive DES key search machine (Example 7.27) proposed by Diffie and Hellman [346] contained 10^6 DES chips, with estimated cost US\$20 million (1977 technology) and 12-hour expected search time; Diffie later revised the estimate upwards one order of magnitude in a BNR Inc. report (US\$50 million machine, 2-day expected search time, 1980 technology). Diffie and Hellman noted the feasibility of a ciphertext-only attack (Example 7.28), and that attempting to preclude exhaustive search by changing DES keys more frequently, at best, doubles the expected search time before success.

Subsequently Wiener [1241] provided a gate-level design for a US\$1 million machine (1993 technology) using 57 600 DES chips with expected success in 3.5 hours. Each chip contains 16 pipelined stages, each stage completing in one clock tick at 50 MHz; a chip with full pipeline completes a key test every 20 nanoseconds, providing a machine $57\,600 \times 50$ times faster than the 1142 years noted in FIPS 74 [397] as the time required to check 2^{55} keys if one key can be tested each microsecond. Comparable key search machines of equivalent cost by Eberle [362] and Wayner [1231] are, respectively, 55 and 200 times slower, although the former does not require a chip design, and the latter uses a general-purpose machine. Wiener also noted adaptations of the ECB known-plaintext attack to other 64-bit modes (CBC, OFB, CFB) and 1-bit and 8-bit CFB.

Even and Goldreich [376] discuss the unicity distance of cascade ciphers under known-plaintext attack (Fact 7.35), present a generalized time-memory meet-in-the-middle trade-off (Note 7.38), and give several other concise results on cascades, including that under reasonable assumptions, the number of permutations realizable by a cascade of L random cipher stages is, with high probability, 2^{Lk}.

Diffie and Hellman [346] noted the meet-in-the-middle attack on double encryption (Fact 7.33), motivating their recommendation that multiple encipherment, if used, should be at least three-fold; Hoffman [558] credits them with suggesting E-E-E triple encryption with three independent keys. Merkle's June 1979 thesis [850] explains the attack on two-key triple-encryption of Fact 7.39 (see also Merkle and Hellman [858]), and after noting Tuchman's proposal of two-key E-D-E triple encryption in a June 1978 conference talk (*National Computer Conference*, Anaheim, CA; see also [1199]), recommended that E-D-E be used with three independent keys: $E_{K3}(E_{K2}^{-1}(E_{K1}(x)))$. The two-key E-D-E idea, adopted in ANSI X9.17 [37] and ISO 8732 [578], was reportedly conceived circa April 1977 by Tuchman's colleagues, Matyas and Meyer. The attack of Fact 7.40 is due to van Oorschot and Wiener [1206]. See Coppersmith, Johnson, and Matyas [278] for a proposed construction for a triple-DES algorithm. Other techniques intended to extend the strength of DES include the *DESX* proposal of Rivest as analyzed by Kilian and Rogaway [672], and the work of Biham and Biryukov [133].

Hellman [549] proposes a time-memory tradeoff for exhaustive key search on an n-bit cipher requiring a chosen-plaintext attack, $O(n^{2/3})$ time and $O(n^{2/3})$ space after an $O(n)$ precomputation; search time can be reduced somewhat by use of Rivest's suggestion of distinguished points (see Denning [326, p.100]). Kusuda and Matsumoto [722] recently extended this analysis. Fiat and Naor [393] pursue time-memory tradeoffs for more general functions. Amirazizi and Hellman [25] note that time-memory tradeoff with constant time-memory product offers no asymptotic cost advantage over exhaustive search; they examine tradeoffs between time, memory, and parallel processing, and using standard parallelization techniques, propose under a simplified model a search machine architecture for which doubling the machine budget (cost) increases the solution rate four-fold. This approach may be applied to exhaustive key search on double-encryption, as can the parallel collision search technique of van Oorschot and Wiener [1207, 1208]; see also Quisquater and Delescaille [1017, 1018].

Regarding Note 7.41, see Biham [131] (and earlier [130]) as well as Coppersmith, Johnson, and Matyas [278]. Biham's analysis on DES and FEAL shows that, in many cases, the use of intermediate data as feedback into an intermediate stage reduces security. 15 years earlier, reflecting on his chosen-plaintext attack on two-key triple-encryption, Merkle [850, p.149] noted "multiple encryption with any cryptographic system is liable to be much less secure than a system designed originally for the longer key".

Maurer and Massey [822] formalize Fact 7.42, where "break" means recovering plaintext from ciphertext (under a known-plaintext attack) or recovering the key; the results hold also for chosen-plaintext and chosen-ciphertext attack. They illustrate, however, that the earlier result and commonly-held belief proven by Even and Goldreich [376] – that a cascade is as strong as any of its component ciphers – requires the important qualifying (and non-practical) assumption that an adversary will not exploit statistics of the underlying plaintext; thus, the intuitive result is untrue for most practical ciphertext-only attacks.

§7.3

Kahn [648] is the definitive historical reference for classical ciphers and machines up to 1967, including much of §7.3 and the notes below. The selection of classical ciphers presented largely follows Shannon's lucid 1949 paper [1121]. Standard references for classical cryptanalysis include Friedman [423], Gaines [436], and Sinkov [1152]. More recent books providing expository material on classical ciphers, machines, and cryptanalytic examples include Beker and Piper [84], Meyer and Matyas [859], Denning [326], and Davies and Price [308].

Polyalphabetic ciphers were invented circa 1467 by the Florentine architect Alberti, who devised a cipher disk with a larger outer and smaller inner wheel, respectively indexed by plaintext and ciphertext characters. Letter alignments defined a simple substitution, modified by rotating the disk after enciphering a few words. The first printed book on cryptography, *Polygraphia*, written in 1508 by the German monk Trithemius and published in 1518, contains the first *tableau* – a square table on 24 characters listing all shift substitutions for a fixed ordering of plaintext alphabet characters. Tableau rows were used sequentially to substitute one plaintext character each for 24 letters, where-after the same tableau or one based on a different alphabet ordering was used. In 1553 Belaso (from Lombardy) suggested using an easily changed key (and key-phrases as memory aids) to define the fixed alphabetic (shift) substitutions in a polyalphabetic substitution. The 1563 book of Porta (from Naples) noted the ordering of tableau letters may define arbitrary substitutions (vs. simply shifted alphabets).

Various polyalphabetic auto-key ciphers, wherein the key changes with each message (the alteration depending on the message), were explored in the 16th century, most significantly by the Frenchman B. de Vigenère. His 1586 book *Traicté des Chiffres* proposed the combined use of a mixed tableau (mixed alphabet on both the tableau top and side) and an auto-keying technique (cf. Example 7.61). A single character served as a priming key to select the tableau row for the first character substitution, where-after the ith plaintext character determined the alphabet (tableau row) for substituting the next. The far less secure simple Vigenère cipher (Definition 7.53) is incorrectly attributed to Vigenère.

The Playfair cipher (Example 7.51), popularized by L. Playfair in England circa 1854 and invented by the British scientist C. Wheatstone, was used as a British field cipher [648, p.6]. J. Mauborgne (see also the Vernam and PURPLE ciphers below) is credited in 1914 with the first known solution of this digram cipher.

The Jefferson cylinder was designed by American statesman T. Jefferson, circa 1790-1800. In 1817, fellow American D. Wadsworth introduced the principle of plaintext and ciphertext alphabets of different lengths. His disk (cf. Alberti above) implemented a cipher similar to Trithemius' polyalphabetic substitution, but wherein the various alphabets were brought into play irregularly in a plaintext-dependent manner, foreshadowing both the polyalphabetic ciphers of later 20th century rotor machines, and the concept of chaining. The inner disk had 26 letters while the outer had an additional 7 digits; one full revolution of the larger caused the smaller to advance 7 characters into its second revolution. The driving disk was always turned in the same clockwise sense; when the character revealed through an aperture in the plaintext disk matched the next plaintext character, that visible through a corresponding ciphertext aperture indicated the resulting ciphertext. In 1867, Wheatstone displayed an independently devised similar device thereafter called the *Wheatstone disc*, receiving greater attention although less secure (having disks of respectively 26 and 27 characters, the extra character a plaintext space).

Vernam [1222] recorded his idea for telegraph encryption in 1917; a patent filed in September 1918 was issued July 1919. Vernam's device combined a stream of plaintext (5-bit Baudot coded) characters, via XOR, with a keystream of 5-bit (key) values, resulting in the *Vernam cipher* (a term often used for related techniques). This, the first polyalphabetic substitution automated using electrical impulses, had period equal to the length of the key stream; each 5-bit key value determined one of 32 fixed mono-alphabetic substitutions. Credit for the actual *one-time system* goes to J. Mauborgne (U.S. Army) who, after seeing Vernam's device with a repeated tape, realized that use of a random, non-repeated key improved security. While Vernam's device was a commercial failure, a related German system engineered by W. Kunze, R. Schauffler, and E. Langlotz was put into practice circa 1921-1923 for German diplomatic communications; their encryption system, which involved manually adding a key string to decimal-coded plaintext, was secured by using as the numerical key a random non-repeating decimal digit stream – the original *one-time pad*. Pads of 50 numbered sheets were used, each with 48 five-digit groups; no pads were repeated aside for one identical pad for a communicating partner, and no sheet was to be used twice; sheets were destroyed once used. The Vernam cipher proper, when used as a one-time system, involves only 32 alphabets, but provides more security than rotor machines with a far greater number of alphabets because the latter eventually repeat, whereas there is total randomness (for each plaintext character) in selecting among the 32 Vernam alphabets.

The matrix cipher of Example 7.52 was proposed in 1929 by Hill [557], providing a practical method for polygraphic substitution, albeit a linear transformation susceptible to known-plaintext attack. Hill also recognized that using an involution as the encryption mapping allowed the same function to provide decryption. Recent contributions on homophonic substitution include Günther [529] and Jendal, Kuhn, and Massey [636].

Among the unrivalled cryptanalytic contributions of the Russian-born American Friedman is his 1920 Riverbank Publication no.22 [426] on cryptanalysis using the index of coincidence. Friedman coined the term *cryptanalysis* in 1920, using it in his 1923 book *Elements of Cryptanalysis* [425], a 1944 expansion of which, *Military Cryptanalysis* [423], remains highly recommended. The method of Kasiski (from West Prussia) was originally published in 1863; see Kahn [648, pp.208-213] for a detailed example. The discussion on IC and MR follows that of Denning [326], itself based on Sinkov [1152]. Fact 7.75 follows from a standard expectation computation weighted by κ_p or κ_r depending on whether the second of a pair of randomly selected ciphertext characters is from the same ciphertext alphabet or one of the $t - 1$ remaining alphabets. The values in Table 7.1 are from Kahn [648], and vary somewhat over time as languages evolve.

Friedman teaches how to cryptanalyze running-key ciphers in his (circa 1918) Riverbank Publication no.16, *Methods for the Solution of Running-Key Ciphers*; the two basic techniques are outlined by Diffie and Hellman [347]. The first is a *probable word* attack wherein an attacker guesses an (e.g., 10 character) word hopefully present in underlying text, and subtracts that word (mod 26) from all possible starting locations in the ciphertext in hopes of finding a recognizable 10-character result, where-after the guessed word (as either partial running-key or plaintext) might be extended using context. Probable-word attacks also apply to polyalphabetic substitution. The second technique is based on the fact that each ciphertext letter c results from a pair of plaintext/running-key letters (m_i, m_i'), and is most likely to result from such pairs wherein both m_i and m_i' are high-frequency characters; one isolates the highest-probability pairs for each such ciphertext character value c, makes trial assumptions, and attempts to extend apparently successful guesses by similarly decrypting adjacent ciphertext characters; see Denning [326, p.83] for a partial example. Diffie and Hellman [347] note Fact 7.59 as an obvious method that is little-used (modern ciphers being more convenient); their suggestion that use of four iterative running keys is unbreakable follows from English being 75% redundant. They also briefly summarize various *scrambling* techniques (encryption via analog rather than digital methods), noting that analog scramblers are sometimes used in practice due to lower bandwidth and cost requirements, although such known techniques appear relatively insecure (possibly an inherent characteristic) and their use is waning as digital networks become prevalent.

Denning [326] tabulates digrams into high, medium, low, and rare classes. Konheim [705, p.24] provides transition probabilities $p(t|s)$, the probability that the next letter is t given that the current character is s in English text, in a table also presented by H. van Tilborg [1210]. Single-letter distributions in plaintext languages other than English are given by Davies and Price [308]. The letter frequencies in Figure 7.5, which should be interpreted only as an estimate, were derived by Meyer and Matyas [859] using excerpts totaling 4 million characters from the 1964 publication: W. Francis, *A Standard Sample of Present-Day Edited American English for Use with Digital Computers*, Linguistics Dept., Brown University, Providence, Rhode Island, USA. Figure 7.6 is based on data from Konheim [705, p.19] giving an estimated probability distribution of 2-grams in English, derived from a sample of size 67 320 digrams.

See Shannon [1122] and Cover and King [285] regarding redundancy and Fact 7.67. While not proven in any concrete manner, Fact 7.68 is noted by Friedman [424] and generally accepted. Unicity distance was defined by Shannon [1121]. Related issues are discussed in detail in various appendices of Meyer and Matyas [859]. Fact 7.71 and the random cipher model are due to Shannon [1121]; see also Hellman [548].

Diffie and Hellman [347] give an instructive overview of rotor machines (see also Denning [326]), and note their use in World War II by the Americans in their highest level system, the British, and the Germans (Enigma); they also give Fact 7.63 and the number of characters required under ciphertext-only and known-plaintext attacks (Note 7.66). Beker and Piper [84] provide technical details of the Hagelin M-209, as does Kahn [648, pp.427-431] who notes its remarkable compactness and weight: 3.25 x 5.5 x 7 inches and 6 lb. (including case); see also Barker [74], Morris [906], and Rivest [1053]. Davies and Price [308] briefly discuss the Enigma, noting it was cryptanalyzed during World War II in Poland, France, and then in the U.K. (Bletchley Park); see also Konheim [705].

The Japanese PURPLE cipher, used during World War II, was a polyalphabetic cipher cryptanalyzed August 1940 [648, p.18-23] by Friedman's team in the U.S. Signal Intelligence Service, under (Chief Signal Officer) Mauborgne. The earlier RED cipher used two rotor arrays; preceding it, the ORANGE system implemented a vowels-to-vowels, consonants-to-consonants cipher using sets of rotors.

§7.4

The concept of *fractionation*, related to product ciphers, is noted by Feistel [387], Shannon [1121], and Kahn [648, p.344] who identifies this idea in an early product cipher, the WWI German *ADFGVX* field cipher. As an example, an encryption function might operate on a block of $t = 8$ plaintext characters in three stages as follows: the first substitutes two symbols for each individual character; the second transposes (mixes) the substituted symbols among themselves; the third re-groups adjacent resulting symbols and maps them back to the plaintext alphabet. The action of the transposition on partial (rather than complete) characters contributes to the strength of the principle.

Shannon [1121, §5 and §23-26] explored the idea of the product of two ciphers, noted the principles of confusion and diffusion (Remark 1.36), and introduced the idea of a *mixing transformation F* (suggesting a preliminary transposition followed by a sequence of alternating substitution and simple linear operations), and combining ciphers in a product using an intervening transformation F. Transposition and substitution, respectively, rest on the principles of diffusion and confusion. Harpes, Kramer, and Massey [541] discuss a general model for iterated block ciphers (cf. Definition 7.80).

The name *Lucifer* is associated with two very different algorithms. The first is an SP network described by Feistel [387], which employs (bitwise nonlinear) 4×4 invertible S-boxes; the second, closely related to DES (albeit significantly weaker), is described by Smith [1160] (see also Sorkin [1165]). Principles related to both are discussed by Feistel, Notz, and Smith [388]; both are analyzed by Biham and Shamir [138], and the latter in greater detail by Ben-Aroya and Biham [108] whose extension of differential cryptanalysis allows, using 2^{36} chosen plaintexts and complexity, attack on 55% of the key space in Smith's Lucifer – still infeasible in practice, but illustrating inferiority to DES despite the longer 128-bit key.

Feistel's product cipher Lucifer [387], instantiated by a blocksize $n = 128$, consists of an unspecified number of alternating substitution and permutation (transposition) stages, using a fixed (unpublished) n-bit permutation P and 32 parallel identical S-boxes each effecting a mapping S_0 or S_1 (fixed but unpublished bijections on $\{0, 1\}^4$), depending on the value of one key bit; the unpublished key schedule requires 32-bits per S-box stage. Each stage operates on all n bits; decryption is by stage-wise inversion of P and S_i.

The structure of so-called Feistel ciphers (Definition 7.81) was first introduced in the Lucifer algorithm of Smith [1160], the direct predecessor of DES. This 16-round algorithm with 128-bit key operates on alternating half-blocks of a 128-bit message block with a simplified f function based on two published invertible 4×4 bit S-boxes S_0 and S_1 (cf. above). Feistel, Notz, and Smith [388] discuss both the abstract Feistel cipher structure (suggesting its use with non-invertible S-boxes) and SP networks based on invertible (distinct) S-boxes. Suggestions for SP networks include the use of single key bits to select one of two mappings (a fixed bijection or its inverse) from both S-boxes and permutation boxes; decryption then uses a reversed key schedule with complemented key. They also noted the multi-round *avalanche effect* of changing a single input bit, subsequently pursued by Kam and Davida [659] in relation to SP networks and S-boxes having a *completeness* property: for every pair of bit positions i, j, there must exist at least two input blocks x, y which differ only in bit i and whose outputs differ in at least bit j. More simply, a function is *complete* if each output bit depends on all input bits. Webster and Tavares [1233] proposed the more stringent *strict avalanche criterion*: whenever one input bit is changed, every output bit must change with probability 1/2.

DES resulted from IBM's submission to the 1974 U.S. National Bureau of Standards (NBS) solicitation for encryption algorithms for the protection of computer data. The original specification is the 1977 U.S. Federal Information Processing Standards Publication 46 [396], reprinted in its entirety as Appendix A in Meyer and Matyas [859]. DES is now specified in FIPS 46–2, which succeeded FIPS 46–1; the same cipher is defined in the American standard ANSI X3.92 [33] and referred to as the Data Encryption Algorithm (DEA). Differences between FIPS 46/46–1 and ANSI X3.92 included the following: these earlier FIPS required that DES be implemented in hardware and that the parity bits be used for parity; ANSI X3.92 specifies that the parity bits *may* be used for parity. Although no purpose was stated by the DES designers for the permutations IP and IP^{-1}, Preneel et al. [1008] provided some evidence of their cryptographic value in the CFB mode.

FIPS 81 [398] specifies the common modes of operation. Davies and Price [308] provide a comprehensive discussion of both DES and modes of operation; see also Diffie and Hellman [347], and the extensive treatment of Meyer and Matyas [859]. The survey of Smid and Branstad [1156] discusses DES, its history, and its use in the U.S. government. Test vectors for various modes of DES, including the ECB vectors of Example 7.86, may be found in ANSI X3.106 [34]. Regarding exhaustive cryptanalysis of DES and related issues, see also the notes under §7.2.

The 1981 publication FIPS 74 [397] notes that DES is not (generally) commutative under two keys, and summarizes weak and semi-weak keys using the term *dual keys* to include both (weak keys being self-dual); see also Davies [303] and Davies and Price [308]. Coppersmith [268] noted Fact 7.90; Moore and Simmons [900] pursue weak and semi-weak DES keys and related phenomena more rigorously.

The 56-bit keylength of DES was criticized from the outset as being too small (e.g., see Diffie and Hellman [346], and p.272 above). Claims which have repeatedly arisen and been denied (e.g., see Tuchman [1199]) over the past 20 years regarding built-in weaknesses of DES (e.g., trap-door S-boxes) remain un-substantiated. Fact 7.91 is significant in that if the permutation group were closed under composition, DES would fall to a known-plaintext attack requiring 2^{28} steps – see Kaliski, Rivest, and Sherman [654], whose cycling experiments provided strong evidence against this. Campbell and Wiener [229] prove the fact conclusively (and give the stated lower bound), through their own cycling experiments utilizing collision key search and an idea outlined earlier by Coppersmith [268] for establishing a lower bound on the group size; they attribute to Coppersmith the same result (in unpublished work), which may also be deduced from the cycle lengths published by Moore and Simmons [901].

Countless papers have analyzed various properties of DES; Davies and Price [308, pp.73-75] provide a partial summary to 1987. Subsequent to the discovery of differential cryptanalysis (DC) by Biham and Shamir, Coppersmith [271] explains how DES was specifically designed 15 years earlier to counter DC, citing national security concerns regarding the design team publishing neither the attack nor design criteria; then gives the (relevant) design criteria – some already noted by others, e.g., see Hellman et al. [552] – for DES S-boxes and the permutation P, explaining how these preclude DC. Coppersmith notes elements of DC were present in the work of den Boer [322], followed shortly by Murphy [913]. DES was not, however, specifically designed to preclude linear cryptanalysis (LC); Matsui [797] illustrates the order of the 8 DES S-boxes, while a strong (but not optimal) choice against DC, is relatively weak against LC, and that DES can be strengthened (vs. DC and LC) by carefully re-arranging these. Despite Remark 7.93, a DES key has actually been recovered by Matsui [795] using LC under experimental conditions (using 2^{43} known-plaintext pairs from randomly generated plaintexts, and 2^{43} complexity running twelve 99 MHz machines over 50 days); such a result remains to be published for exhaustive search or DC.

Ben-Aroya and Biham [108] note that often suggestions to redesign DES, some based on design criteria and attempts to specifically resist DC, have resulted in (sometimes far) weaker systems, including the RDES (randomized DES) proposal of Koyama and Terada [709], which fall to variant attacks. The lesson is that in isolation, individual design principles do not guarantee security.

DES alternatives are sought not only due to the desire for a keylength exceeding 56 bits, but also because its bit-oriented operations are inconvenient in conventional software implementations, often resulting in poor performance; this makes triple-DES less attractive. Regarding fast software implementations of DES, see Shepherd [1124], Pfitzmann and Aßmann [970], and Feldmeier and Karn [391].

§7.5

FEAL stimulated the development of a sequence of advanced cryptanalytic techniques of unparalleled richness and utility. While it appears to remain relatively secure when iterated a sufficient number of rounds (e.g., 24 or more), this defeats its original objective of speed. FEAL-4 as presented at Eurocrypt'87 (Abstracts of Eurocrypt'87, April 1987) was found to have certain vulnerabilities by den Boer (unpublished Eurocrypt'87 rump session talk), resulting in Shimizu and Miyaguchi [1126] (or see Miyaguchi, Shiraishi, and Shimizu [887]) increasing FEAL to 8 rounds in the final proceedings. In 1988 den Boer [322] showed FEAL-4 vulnerable to an adaptive chosen plaintext attack with 100 to 10 000 plaintexts. In 1990, Gilbert and Chassé [455] devised a chosen-plaintext attack (called a statistical meet-in-the-middle attack) on FEAL-8 requiring 10 000 pairs of plaintexts, the bitwise XOR of each pair being selected to be an appropriate constant (thus another early variant of differential cryptanalysis).

FEAL-N with N rounds, and its extension FEAL-NX with 128-bit key (Notes 7.97 and 7.98) were then published by Miyaguchi [884] (or see Miyaguchi et al. [885]), who nonetheless opined that chosen-plaintext attacks on FEAL-8 were not practical threats. However, improved chosen-plaintext attacks were subsequently devised, as well as known-plaintext attacks. Employing den Boer's G function expressing linearity in the FEAL f-function, Murphy [913] defeated FEAL-4 with 20 chosen plaintexts in under 4 hours (under 1 hour for most keys) on a Sun 3/60 workstation. A statistical method of Tardy-Corfdir and Gilbert [1187] then allowed a known-plaintext attack on FEAL-4 (1000 texts; or 200 in an announced improvement) and FEAL-6 ($2 \times 10\,000$ texts), involving linear approximation of FEAL S-boxes. Thereafter, the first version of linear cryptanalysis (LC) introduced by Matsui and Yamagishi [798] allowed known-plaintext attack of FEAL-4 (5 texts, 6 minutes on a 25MHz 68040 processor), FEAL-6 (100 texts, 40 minutes), and FEAL-8 (2^{28} texts, in time equivalent to exhaustive search on 50-bit keys); the latter betters the 2^{38} texts required for FEAL-8 by Biham and Shamir [136] in their known-plaintext conversion of differential cryptanalysis (DC). Biham and Shamir [138, p.101] later implemented a DC chosen-plaintext attack recovering FEAL-8 keys in two minutes on a PC using 128 chosen pairs, the program requiring 280K bytes of storage. Biham [132] subsequently used LC to defeat FEAL-8 with 2^{24} known-plaintexts in 10 minutes on a personal computer. Ohta and Aoki [943] suggest that FEAL-32 is as secure as DES against DC, while FEAL-16 is as secure as DES against certain restricted forms of LC.

Differential-linear cryptanalysis was introduced by Langford and Hellman [741], combining linear and differential cryptanalysis to allow a reduced 8-round version of DES to be attacked with fewer chosen-plaintexts than previous attacks. Aoki and Ohta [53] refined these ideas for FEAL-8 yielding a differential-linear attack requiring only 12 chosen texts and 35 days of computer time (cf. Table 7.10).

Test vectors for FEAL-N and FEAL-NX (Example 7.99) are given by Miyaguchi [884]. The DC attack of Biham and Shamir [137], which finds FEAL-N subkeys themselves, is equally as effective on FEAL-NX. Biham [132] notes that an LC attack on FEAL-N is possible with less than 2^{64} known plaintexts (and complexity) for up to $N = 20$. For additional discussion of properties of FEAL, see Biham and Shamir [138, §6.3].

§7.6

The primary reference for IDEA is Lai [726]. A preliminary version introduced by Lai and Massey [728] was named PES (Proposed Encryption Standard). Lai, Massey, and Murphy [730] showed that a generalization (see below) of differential cryptanalysis (DC) allowed recovery of PES keys, albeit requiring all 2^{64} possible ciphertexts (cf. exhaustive search of 2^{128} operations). Minor modifications resulted in IPES (Improved PES): in stage r, $1 \leq r \leq 9$, the group operations keyed by $K_2^{(r)}$ and $K_4^{(r)}$ ($\boxplus$ and $\odot$ in Figure 7.11) were reversed from PES; the permutation on 16-bit blocks after stage r, $1 \leq r \leq 9$, was altered; and necessary changes were made in the decryption (but not encryption) key schedule. IPES was commercialized under the name IDEA, and is patented (see Chapter 15).

The ingenious design of IDEA is supported by a careful analysis of the interaction and algebraic incompatibilities of operations across the groups ($\mathbb{F}_2{}^n$, $\oplus$), ($\mathbb{Z}_{2^n}$, $\boxplus$), and ($\mathbb{Z}_{2^n+1}^*$, $\odot$). The design of the MA structure (see Figure 7.11) results in IDEA being "complete" after a single round; for other security properties, see Lai [726]. Regarding mixing operations from different algebraic systems, see also the 1974 examination by Grossman [522] of transformations arising by alternating mod 2^n and mod 2 addition ($\oplus$), and the use of arithmetic modulo $2^{32} - 1$ and $2^{32} - 2$ in MAA (Algorithm 9.68).

Daemen [292, 289] identifies several classes of so-called *weak keys for IDEA*, and notes a small modification to the key schedule to eliminate them. The largest is a class of 2^{51} keys for which membership can be tested in two encryptions plus a small number of computations, whereafter the key itself can be recovered using 16 chosen plaintext-difference encryptions, on the order of 2^{16} group operations, plus 2^{17} key search encryptions. The probability of a randomly chosen key being in this class is $2^{51}/2^{128} = 2^{-77}$. A smaller number of weak key blocks were observed earlier by Lai [726], and dismissed as inconsequential. The analysis of Meier [832] revealed no attacks feasible against full 8-round IDEA, and supports the conclusion of Lai [726] that IDEA appears to be secure against DC after 4 of its 8 rounds (cf. Note 7.107). Daemen [289] also references attacks on reduced-round variants of IDEA. While linear cryptanalysis (LC) can be applied to any iterated block cipher,

Harpes, Kramer, and Massey [541] provide a generalization thereof; IDEA and SAFER K-64 are argued to be secure against this particular generalization.

Lai, Massey, and Murphy [730] (see also Lai [726]) generalized DC to apply to *Markov ciphers* (which they introduced for this purpose; DES, FEAL, and LOKI are all examples under the assumption of independent round keys) including IDEA; broadened the notion of a *difference* from that based on $\oplus$ to: $\Delta X = X \otimes (X^*)^{-1}$ where $\otimes$ is a specified group operation and $(X^*)^{-1}$ is the group inverse of an element X^*; and defined an *i-round differential* (as opposed to an *i*-round characteristic used by Biham and Shamir [138] on DES) to be a pair (α, β) such that two distinct plaintexts with difference $\Delta X = \alpha$ results in a pair of round *i* outputs with difference β.

Decimal values corresponding to Tables 7.12 and 7.13 may be found in Lai [726]. A table-based alternative for multiplication mod $2^{16} + 1$ (cf. Note 7.104) is to look up the anti-log of $\log_\alpha(a) + \log_\alpha(b) \bmod 2^{16}$, relative to a generator α of $\mathbb{Z}^*_{2^{16}+1}$; the required tables, however, are quite large.

§7.7

Massey [787] introduced SAFER K-64 with a 64-bit key and initially recommended 6 rounds, giving a reference implementation and test vectors (cf. Example 7.114). It is not patented. Massey [788] then published SAFER K-128 (with a reference implementation), differing only in its use of a non-proprietary (and backwards compatible) key schedule accommodating 128-bit keys, proposed by a Singapore group; 10 rounds were recommended (12 maximum). Massey [788] gave further justification for design components of SAFER K-64. Vaudenay [1215] showed SAFER K-64 is weakened if the S-box mapping (Remark 7.112) is replaced by a random permutation.

Knudsen [685] proposed the modified key schedule of Note 7.110 after finding a weakness in 6-round SAFER K-64 that, while not of practical concern for encryption (with 2^{45} chosen plaintexts, it finds 8 bits of the key), permitted collisions when using the cipher for hashing. This and a subsequent certificational attack on SAFER K-64 by S. Murphy (to be published) lead Massey ("Strengthened key schedule for the cipher SAFER", posted to the USENET newsgroup sci.crypt, September 9 1995) to advise adoption of the new key schedule, with the resulting algorithm distinguished as SAFER SK-64 with 8 rounds recommended (minimum 6, maximum 10); an analogous change to the 128-bit key schedule yields SAFER SK-128 for which 10 rounds remain recommended (maximum 12). A new variant of DC by Knudsen and Berson [687] using *truncated differentials* (building on Knudsen [686]) yields a certificational attack on 5-round SAFER K-64 with 2^{45} chosen plaintexts; the attack, which does not extend to 6 rounds, indicates that security is less than argued by Massey [788], who also notes that preliminary attempts at linear cryptanalysis of SAFER were unsuccessful.

RC5 was designed by Rivest [1056], and published along with a reference implementation. The magic constants of Table 7.14 are based on the golden ratio and the base of natural logarithms. The data-dependent rotations (which vary across rounds) distinguish RC5 from iterated ciphers which have identical operations each round; Madryga [779] proposed an earlier (less elegant) cipher involving data-dependent rotations. A preliminary examination by Kaliski and Yin [656] suggested that, while variations remain to be explored, standard linear and differential cryptanalysis appear impractical for RC5–32 (64-bit blocksize) for $r = 12$: their differential attacks on 9 and 12 round RC5 require, respectively, 2^{45}, 2^{62} chosen-plaintext pairs, while their linear attacks on 4, 5, and 6-round RC5–32 require, respectively, 2^{37}, 2^{47}, 2^{57} known plaintexts. Both attacks depend on the number of rounds and the blocksize, but not the byte-length of the input key (since subkeys are recovered directly). Knudsen and Meier [689] subsequently presented differential attacks on RC5 which improved on those of Kaliski and Yin by a factor up to 128, and showed that RC5 has so-called *weak keys* (independent of the key schedule) for which these differential attacks perform even better.

LOKI was introduced by Brown, Pieprzyk, and Seberry [215] and renamed LOKI'89 after the discovery of weaknesses lead to the introduction of LOKI'91 by Brown et al. [214]. Knudsen [682] noted each LOKI'89 key fell into a class of 16 equivalent keys, and the differential cryptanalysis of Biham and Shamir [137] was shown to be effective against reduced-round versions. LOKI'91 failed to succumb to differential analysis by Knudsen [683]; Tokita et al. [1193] later confirmed the optimality of Knudsen's characteristics, suggesting that LOKI'89 and LOKI'91 were resistant to both ordinary linear and differential cryptanalysis. However, neither should be used for hashing as originally proposed (see Knudsen [682]) or in other modes (see Preneel [1003]). Moreover, both are susceptible to *related-key attacks* (Note 7.6), popularized by Biham [128, 129]; but see also the earlier ideas of Knudsen [683]. Distinct from these are *key clustering attacks* (see Diffie and Hellman [347, p.410]), wherein a cryptanalyst first finds a key "close" to the correct key, and then searches a cluster of "nearby" keys to find the correct one.

8×32 bit S-boxes first appeared in the Snefru hash function of Merkle [854]; here such fixed S-boxes created from random numbers were used in its internal encryption mapping. Regarding large S-boxes, see also Gordon and Retkin [517], Adams and Tavares [7], and Biham [132]. Merkle [856] again used 8×32 S-boxes in Khufu and Khafre (see also §15.2.3(viii)). In this 1990 paper, Merkle gives a chosen-plaintext differential attack defeating 8 rounds of Khufu (with secret S-box). Regarding 16-round Khafre, a DC attack by Biham and Shamir [138, 137] requires somewhat over 1500 chosen plaintexts and one hour on a personal computer, and their known-plaintext differential attack requires $2^{37.5}$ plaintexts; for 24-round Khafre, they require 2^{53} chosen plaintexts or $2^{58.5}$ known plaintexts. Khufu with 16 rounds was examined by Gilbert and Chauvaud [456], who gave an attack using 2^{43} chosen plaintexts and about 2^{43} operations.

CAST is a design procedure for a family of DES-like ciphers, featuring fixed $m \times n$ bit S-boxes ($m < n$) based on bent functions. Adams and Tavares [7] examine the construction of large S-boxes resistant to differential cryptanalysis, and give a specific example (with 64-bit blocklength and 8×32 bit S-boxes) of a CAST cipher. CAST ciphers have variable keysize and numbers of rounds. Rijmen and Preneel [1049] presented a cryptanalytic technique applicable to Feistel ciphers with non-surjective round functions (e.g., LOKI'91 and an example CAST cipher), noting cases where 6 to 8 rounds is insufficient.

Blowfish is a 16-round DES-like cipher due to Schneier [1093], with 64-bit blocks and keys of length up to 448 bits. The computationally intensive key expansion phase creates eighteen 32-bit subkeys plus four 8×32 bit S-boxes derived from the input key (cf. Khafre above), for a total of 4168 bytes. See Vaudenay [1216] for a preliminary analysis of Blowfish.

3-WAY is a block cipher with 96-bit blocksize and keysize, due to Daemen [289] and introduced by Daemen, Govaerts, and Vandewalle [290] along with a reference C implementation and test vectors. It was designed for speed in both hardware and software, and to resist differential and linear attacks. Its core is a 3-bit nonlinear S-box and a linear mapping representable as polynomial multiplication in $\mathbb{Z}_2^{12}$.

SHARK is an SP-network block cipher due to Rijmen et al. [1048] (coordinates for a reference implementation are given) which may be viewed as a generalization of SAFER, employing highly nonlinear S-boxes and the idea of MDS codes (cf. Note 12.36) for diffusion to allow a small number of rounds to suffice. The block ciphers BEAR and LION of Anderson and Biham [30] are 3-round unbalanced Feistel networks, motivated by the earlier construction of Luby and Rackoff [776] (see also Maurer [816] and Lucks [777]) which provides a provably secure (under suitable assumptions) block cipher from pseudorandom functions using a 3-round Feistel structure. SHARK, BEAR, and LION all remain to be subjected to independent analysis in order to substantiate their conjectured security levels.

SKIPJACK is a classified block cipher whose specification is maintained by the U.S. National Security Agency (NSA). FIPS 185 [405] notes that its specification is available to organizations entering into a Memorandum of Agreement with the NSA, and includes interface details (e.g., it has an 80-bit secret key). A public report contains results of a preliminary security evaluation of this 64-bit block cipher ("SKIPJACK Review, Interim Report, The SKIPJACK Algorithm", 1993 July 28, by E.F. Brickell, D.E. Denning, S.T. Kent, D.P. Maher, and W. Tuchman). See also Roe [1064, p.312] regarding curious results on the cyclic closure tests on SKIPJACK, which give evidence related to the size of the cipher keyspace.

GOST 28147-89 is a Soviet government encryption algorithm with a 32-round Feistel structure and unspecified S-boxes; see Charnes et al. [241].

RC2 is a block cipher proprietary to RSA Data Security Inc. (as is the stream cipher RC4). WAKE is a block cipher due to Wheeler [1237] employing a key-dependent table, intended for fast encryption of bulk data on processors with 32-bit words. TEA (Tiny Encryption Algorithm) is a block cipher proposed by Wheeler and Needham [1238].

Chapter 8

Public-Key Encryption

Contents in Brief

8.1 Introduction

This chapter considers various techniques for public-key encryption, also referred to as *asymmetric encryption*. As introduced previously (§1.8.1), in public-key encryption systems each entity A has a *public key* e and a corresponding *private key* d. In secure systems, the task of computing d given e is computationally infeasible. The public key defines an *encryption transformation* E_e, while the private key defines the associated *decryption transformation* D_d. Any entity B wishing to send a message m to A obtains an authentic copy of A's public key e, uses the encryption transformation to obtain the ciphertext $c = E_e(m)$, and transmits c to A. To decrypt c, A applies the decryption transformation to obtain the original message $m = D_d(c)$.

The public key need not be kept secret, and, in fact, may be widely available – only its authenticity is required to guarantee that A is indeed the only party who knows the corresponding private key. A primary advantage of such systems is that providing authentic public keys is generally easier than distributing secret keys securely, as required in symmetric-key systems.

The main objective of public-key encryption is to provide *privacy* or *confidentiality*. Since A's encryption transformation is public knowledge, public-key encryption alone does not provide *data origin authentication* (Definition 9.76) or *data integrity* (Definition 9.75). Such assurances must be provided through use of additional techniques (see §9.6), including message authentication codes and digital signatures.

Public-key encryption schemes are typically substantially slower than symmetric-key encryption algorithms such as DES (§7.4). For this reason, public-key encryption is most commonly used in practice for the transport of keys subsequently used for bulk data encryption by symmetric algorithms and other applications including data integrity and authentication, and for encrypting small data items such as credit card numbers and PINs.

Public-key decryption may also provide authentication guarantees in entity authentication and authenticated key establishment protocols.

Chapter outline

The remainder of the chapter is organized as follows. The RSA public-key encryption scheme is presented in §8.2; related security and implementation issues are also discussed. Rabin's public-key encryption scheme, which is provably as secure as factoring, is the topic of §8.3. §8.4 considers the ElGamal encryption scheme; related security and implementation issues are also discussed. The McEliece public-key encryption scheme, based on error-correcting codes, is examined in §8.5. Although known to be insecure, the Merkle-Hellman knapsack public-key encryption scheme is presented in §8.6 for historical reasons – it was the first concrete realization of a public-key encryption scheme. Chor-Rivest encryption is also presented (§8.6.2) as an example of an as-yet unbroken public-key encryption scheme based on the subset sum (knapsack) problem. §8.7 introduces the notion of probabilistic public-key encryption, designed to meet especially stringent security requirements. §8.8 concludes with Chapter notes and references.

The number-theoretic computational problems which form the security basis for the public-key encryption schemes discussed in this chapter are listed in Table 8.1.

public-key encryption scheme	computational problem
RSA	integer factorization problem (§3.2)
	RSA problem (§3.3)
Rabin	integer factorization problem (§3.2)
	square roots modulo composite n (§3.5.2)
ElGamal	discrete logarithm problem (§3.6)
	Diffie-Hellman problem (§3.7)
generalized ElGamal	generalized discrete logarithm problem (§3.6)
	generalized Diffie-Hellman problem (§3.7)
McEliece	linear code decoding problem
Merkle-Hellman knapsack	subset sum problem (§3.10)
Chor-Rivest knapsack	subset sum problem (§3.10)
Goldwasser-Micali probabilistic	quadratic residuosity problem (§3.4)
Blum-Goldwasser probabilistic	integer factorization problem (§3.2)
	Rabin problem (3.9.3)

Table 8.1: *Public-key encryption schemes discussed in this chapter, and the related computational problems upon which their security is based.*

8.1.1 Basic principles

Objectives of adversary

The primary objective of an adversary who wishes to "attack" a public-key encryption scheme is to systematically recover plaintext from ciphertext intended for some other entity A. If this is achieved, the encryption scheme is informally said to have been *broken*. A more ambitious objective is *key recovery* – to recover A's private key. If this is achieved, the encryption scheme is informally said to have been *completely broken* since the adversary then has the ability to decrypt *all* ciphertext sent to A.

Types of attacks

Since the encryption transformations are public knowledge, a passive adversary can always mount a *chosen-plaintext attack* on a public-key encryption scheme (cf. §1.13.1). A stronger attack is a *chosen-ciphertext attack* where an adversary selects ciphertext of its choice, and then obtains by some means (from the victim A) the corresponding plaintext (cf. §1.13.1). Two kinds of these attacks are usually distinguished.

1. In an *indifferent* chosen-ciphertext attack, the adversary is provided with decryptions of any ciphertexts of its choice, but these ciphertexts must be chosen prior to receiving the (target) ciphertext c it actually wishes to decrypt.

2. In an *adaptive* chosen-ciphertext attack, the adversary may use (or have access to) A's decryption machine (but not the private key itself) even after seeing the target ciphertext c. The adversary may request decryptions of ciphertext which may be related to both the target ciphertext, and to the decryptions obtained from previous queries; a restriction is that it may not request the decryption of the target c itself.

Chosen-ciphertext attacks are of concern if the environment in which the public-key encryption scheme is to be used is subject to such an attack being mounted; if not, the existence of a chosen-ciphertext attack is typically viewed as a *certificational* weakness against a particular scheme, although apparently not directly exploitable.

Distributing public keys

The public-key encryption schemes described in this chapter assume that there is a means for the sender of a message to obtain an *authentic* copy of the intended receiver's public key. In the absence of such a means, the encryption scheme is susceptible to an *impersonation* attack, as outlined in §1.8.2. There are many techniques in practice by which authentic public keys can be distributed, including exchanging keys over a trusted channel, using a trusted public file, using an on-line trusted server, and using an off-line server and certificates. These and related methods are discussed in §13.4.

Message blocking

Some of the public-key encryption schemes described in this chapter assume that the message to be encrypted is, at most, some fixed size (bitlength). Plaintext messages longer than this maximum must be broken into *blocks*, each of the appropriate size. Specific techniques for breaking up a message into blocks are not discussed in this book. The component blocks can then be encrypted independently (cf. ECB mode in §7.2.2(i)). To prevent manipulation (e.g., re-ordering) of the blocks, the *Cipher Block Chaining* (CBC) mode may be used (cf. §7.2.2(ii)). Since the CFB and OFB modes (cf. §7.2.2(iii) and §7.2.2(iv)) employ only single-block encryption (and not decryption) for both message encryption and decryption, they cannot be used with public-key encryption schemes.

8.2 RSA public-key encryption

The RSA cryptosystem, named after its inventors R. Rivest, A. Shamir, and L. Adleman, is the most widely used public-key cryptosystem. It may be used to provide both secrecy and digital signatures and its security is based on the intractability of the integer factorization problem (§3.2). This section describes the RSA encryption scheme, its security, and some implementation issues; the RSA signature scheme is covered in §11.3.1.

8.2.1 Description

8.1 Algorithm Key generation for RSA public-key encryption

SUMMARY: each entity creates an RSA public key and a corresponding private key.
Each entity A should do the following:

1. Generate two large random (and distinct) primes p and q, each roughly the same size.
2. Compute $n = pq$ and $\phi = (p-1)(q-1)$. (See Note 8.5.)
3. Select a random integer e, $1 < e < \phi$, such that $\gcd(e, \phi) = 1$.
4. Use the extended Euclidean algorithm (Algorithm 2.107) to compute the unique integer d, $1 < d < \phi$, such that $ed \equiv 1 \pmod{\phi}$.
5. A's public key is (n, e); A's private key is d.

8.2 Definition The integers e and d in RSA key generation are called the *encryption exponent* and the *decryption exponent*, respectively, while n is called the *modulus*.

8.3 Algorithm RSA public-key encryption

SUMMARY: B encrypts a message m for A, which A decrypts.

1. *Encryption.* B should do the following:
   (a) Obtain A's authentic public key (n, e).
   (b) Represent the message as an integer m in the interval $[0, n-1]$.
   (c) Compute $c = m^e \bmod n$ (e.g., using Algorithm 2.143).
   (d) Send the ciphertext c to A.
2. *Decryption.* To recover plaintext m from c, A should do the following:
   (a) Use the private key d to recover $m = c^d \bmod n$.

Proof that decryption works. Since $ed \equiv 1 \pmod{\phi}$, there exists an integer k such that $ed = 1 + k\phi$. Now, if $\gcd(m, p) = 1$ then by Fermat's theorem (Fact 2.127),

$$m^{p-1} \equiv 1 \pmod{p}.$$

Raising both sides of this congruence to the power $k(q-1)$ and then multiplying both sides by m yields

$$m^{1+k(p-1)(q-1)} \equiv m \pmod{p}.$$

On the other hand, if $\gcd(m, p) = p$, then this last congruence is again valid since each side is congruent to 0 modulo p. Hence, in all cases

$$m^{ed} \equiv m \pmod{p}.$$

By the same argument,

$$m^{ed} \equiv m \pmod{q}.$$

Finally, since p and q are distinct primes, it follows that

$$m^{ed} \equiv m \pmod{n},$$

and, hence,

$$c^d \equiv (m^e)^d \equiv m \pmod{n}.$$

8.4 Example (*RSA encryption with artificially small parameters*)
Key generation. Entity A chooses the primes $p = 2357$, $q = 2551$, and computes $n = pq = 6012707$ and $\phi = (p-1)(q-1) = 6007800$. A chooses $e = 3674911$ and, using the extended Euclidean algorithm, finds $d = 422191$ such that $ed \equiv 1 \pmod{\phi}$. A's public key is the pair $(n = 6012707, e = 3674911)$, while A's private key is $d = 422191$.
Encryption. To encrypt a message $m = 5234673$, B uses an algorithm for modular exponentiation (e.g., Algorithm 2.143) to compute

$$c = m^e \bmod n = 5234673^{3674911} \bmod 6012707 = 3650502,$$

and sends this to A.
Decryption. To decrypt c, A computes

$$c^d \bmod n = 3650502^{422191} \bmod 6012707 = 5234673. \qquad \square$$

8.5 Note (*universal exponent*) The number $\lambda = \mathrm{lcm}(p-1, q-1)$, sometimes called the *universal exponent* of n, may be used instead of $\phi = (p-1)(q-1)$ in RSA key generation (Algorithm 8.1). Observe that λ is a proper divisor of ϕ. Using λ can result in a smaller decryption exponent d, which may result in faster decryption (cf. Note 8.9). However, if p and q are chosen at random, then $\gcd(p-1, q-1)$ is expected to be small, and consequently ϕ and λ will be roughly of the same size.

8.2.2 Security of RSA

This subsection discusses various security issues related to RSA encryption. Various attacks which have been studied in the literature are presented, as well as appropriate measures to counteract these threats.

(i) Relation to factoring

The task faced by a passive adversary is that of recovering plaintext m from the corresponding ciphertext c, given the public information (n, e) of the intended receiver A. This is called the *RSA problem* (RSAP), which was introduced in §3.3. There is no efficient algorithm known for this problem.

One possible approach which an adversary could employ to solving the RSA problem is to first factor n, and then compute ϕ and d just as A did in Algorithm 8.1. Once d is obtained, the adversary can decrypt any ciphertext intended for A.

On the other hand, if an adversary could somehow compute d, then it could subsequently factor n efficiently as follows. First note that since $ed \equiv 1 \pmod{\phi}$, there is an integer k such that $ed - 1 = k\phi$. Hence, by Fact 2.126(i), $a^{ed-1} \equiv 1 \pmod{n}$ for all $a \in \mathbb{Z}_n^*$. Let $ed - 1 = 2^s t$, where t is an odd integer. Then it can be shown that $a^{2^{s-1}t} \not\equiv \pm 1 \pmod{n}$ for at least half of all $a \in \mathbb{Z}_n^*$; if a is such an integer then $\gcd(a^{2^{s-1}t} - 1, n)$ is a non-trivial factor of n. Thus the adversary simply needs to repeatedly select random $a \in \mathbb{Z}_n^*$ and compute $\gcd(a^{2^{s-1}t} - 1, n)$; the expected number of trials before a non-trivial factor of n is obtained is 2. This discussion establishes the following.

8.6 Fact The problem of computing the RSA decryption exponent d from the public key (n, e), and the problem of factoring n, are computationally equivalent.

When generating RSA keys, it is imperative that the primes p and q be selected in such a way that factoring $n = pq$ is computationally infeasible; see Note 8.8 for more details.

(ii) Small encryption exponent e

In order to improve the efficiency of encryption, it is desirable to select a small encryption exponent e (see Note 8.9) such as $e = 3$. A group of entities may all have the same encryption exponent e, however, each entity in the group must have its own distinct modulus. If an entity A wishes to send the same message m to three entities whose public moduli are n_1, n_2, n_3, and whose encryption exponents are $e = 3$, then A would send $c_i = m^3 \bmod n_i$, for $i = 1, 2, 3$. Since these moduli are most likely pairwise relatively prime, an eavesdropper observing c_1, c_2, c_3 can use Gauss's algorithm (Algorithm 2.121) to find a solution x, $0 \le x < n_1 n_2 n_3$, to the three congruences

$$\begin{cases} x \equiv c_1 \pmod{n_1} \\ x \equiv c_2 \pmod{n_2} \\ x \equiv c_3 \pmod{n_3}. \end{cases}$$

Since $m^3 < n_1 n_2 n_3$, by the Chinese remainder theorem (Fact 2.120), it must be the case that $x = m^3$. Hence, by computing the integer cube root of x, the eavesdropper can recover the plaintext m.

Thus a small encryption exponent such as $e = 3$ should not be used if the same message, or even the same message with known variations, is sent to many entities. Alternatively, to prevent against such an attack, a pseudorandomly generated bitstring of appropriate length (for example, at least 64) should be appended to the plaintext message prior to encryption; the pseudorandom bitstring should be independently generated for each encryption. This process is sometimes referred to as *salting* the message.

Small encryption exponents are also a problem for small messages m, because if $m < n^{1/e}$, then m can be recovered from the ciphertext $c = m^e \bmod n$ simply by computing the integer e^{th} root of c; salting plaintext messages also circumvents this problem.

(iii) Forward search attack

If the message space is small or predictable, an adversary can decrypt a ciphertext c by simply encrypting all possible plaintext messages until c is obtained. Salting the message as described above is one simple method of preventing such an attack.

(iv) Small decryption exponent d

As was the case with the encryption exponent e, it may seem desirable to select a small decryption exponent d in order to improve the efficiency of decryption.[1] However, if $\gcd(p-1, q-1)$ is small, as is typically the case, and if d has up to approximately one-quarter as many bits as the modulus n, then there is an efficient algorithm (referenced on page 313) for computing d from the public information (n, e). This algorithm cannot be extended to the case where d is approximately the same size as n. Hence, to avoid this attack, the decryption exponent d should be roughly the same size as n.

(v) Multiplicative properties

Let m_1 and m_2 be two plaintext messages, and let c_1 and c_2 be their respective RSA encryptions. Observe that

$$(m_1 m_2)^e \equiv m_1^e m_2^e \equiv c_1 c_2 \pmod{n}.$$

---

[1] In this case, one would select d first and then compute e in Algorithm 8.1, rather than vice-versa.

In other words, the ciphertext corresponding to the plaintext $m = m_1 m_2 \bmod n$ is $c = c_1 c_2 \bmod n$; this is sometimes referred to as the *homomorphic property* of RSA. This observation leads to the following *adaptive chosen-ciphertext attack* on RSA encryption.

Suppose that an active adversary wishes to decrypt a particular ciphertext $c = m^e \bmod n$ intended for A. Suppose also that A will decrypt arbitrary ciphertext for the adversary, other than c itself. The adversary can conceal c by selecting a random integer $x \in \mathbb{Z}_n^*$ and computing $\bar{c} = cx^e \bmod n$. Upon presentation of $\bar{c}$, A will compute for the adversary $\bar{m} = (\bar{c})^d \bmod n$. Since

$$\bar{m} \equiv (\bar{c})^d \equiv c^d (x^e)^d \equiv mx \pmod{n},$$

the adversary can then compute $m = \bar{m} x^{-1} \bmod n$.

This adaptive chosen-ciphertext attack can be circumvented in practice by imposing some structural constraints on plaintext messages. If a ciphertext c is decrypted to a message not possessing this structure, then c is rejected by the decryptor as being fraudulent. Now, if a plaintext message m has this (carefully chosen) structure, then with high probability $mx \bmod n$ will not for $x \in \mathbb{Z}_n^*$. Thus the adaptive chosen-ciphertext attack described in the previous paragraph will fail because A will not decrypt $\bar{c}$ for the adversary. Note 8.63 provides a powerful technique for guarding against adaptive chosen-ciphertext and other kinds of attacks.

(vi) Common modulus attack

The following discussion demonstrates why it is imperative for each entity to choose its own RSA modulus n.

It is sometimes suggested that a central trusted authority should select a single RSA modulus n, and then distribute a distinct encryption/decryption exponent pair (e_i, d_i) to each entity in a network. However, as shown in (i) above, knowledge of any (e_i, d_i) pair allows for the factorization of the modulus n, and hence any entity could subsequently determine the decryption exponents of all other entities in the network. Also, if a single message were encrypted and sent to two or more entities in the network, then there is a technique by which an eavesdropper (any entity not in the network) could recover the message with high probability using only publicly available information.

(vii) Cycling attacks

Let $c = m^e \bmod n$ be a ciphertext. Let k be a positive integer such that $c^{e^k} \equiv c \pmod{n}$; since encryption is a permutation on the message space $\{0, 1, \dots, n-1\}$ such an integer k must exist. For the same reason it must be the case that $c^{e^{k-1}} \equiv m \pmod{n}$. This observation leads to the following *cycling attack* on RSA encryption. An adversary computes $c^e \bmod n, c^{e^2} \bmod n, c^{e^3} \bmod n, \dots$ until c is obtained for the first time. If $c^{e^k} \bmod n = c$, then the previous number in the cycle, namely $c^{e^{k-1}} \bmod n$, is equal to the plaintext m.

A *generalized cycling attack* is to find the smallest positive integer u such that $f = \gcd(c^{e^u} - c, n) > 1$. If

$$c^{e^u} \equiv c \pmod{p} \quad \text{and} \quad c^{e^u} \not\equiv c \pmod{q} \tag{8.1}$$

then $f = p$. Similarly, if

$$c^{e^u} \not\equiv c \pmod{p} \quad \text{and} \quad c^{e^u} \equiv c \pmod{q} \tag{8.2}$$

then $f = q$. In either case, n has been factored, and the adversary can recover d and then m. On the other hand, if both

$$c^{e^u} \equiv c \pmod{p} \quad \text{and} \quad c^{e^u} \equiv c \pmod{q}, \tag{8.3}$$

then $f = n$ and $c^{e^u} \equiv c \pmod{n}$. In fact, u must be the smallest positive integer k for which $c^{e^k} \equiv c \pmod{n}$. In this case, the basic cycling attack has succeeded and so $m = c^{e^{u-1}} \bmod n$ can be computed efficiently. Since (8.3) is expected to occur much less frequently than (8.1) or (8.2), the generalized cycling attack usually terminates before the cycling attack does. For this reason, the generalized cycling attack can be viewed as being essentially an algorithm for factoring n.

Since factoring n is assumed to be intractable, these cycling attacks do not pose a threat to the security of RSA encryption.

(viii) Message concealing

A plaintext message m, $0 \leq m \leq n-1$, in the RSA public-key encryption scheme is said to be *unconcealed* if it encrypts to itself; that is, $m^e \equiv m \pmod{n}$. There are always some messages which are unconcealed (for example $m = 0$, $m = 1$, and $m = n-1$). In fact, the number of unconcealed messages is exactly

$$[1 + \gcd(e-1, p-1)] \cdot [1 + \gcd(e-1, q-1)].$$

Since $e - 1$, $p - 1$ and $q - 1$ are all even, the number of unconcealed messages is always at least 9. If p and q are random primes, and if e is chosen at random (or if e is chosen to be a small number such as $e = 3$ or $e = 2^{16} + 1 = 65537$), then the proportion of messages which are unconcealed by RSA encryption will, in general, be negligibly small, and hence unconcealed messages do not pose a threat to the security of RSA encryption in practice.

8.2.3 RSA encryption in practice

There are numerous ways of speeding up RSA encryption and decryption in software and hardware implementations. Some of these techniques are covered in Chapter 14, including fast modular multiplication (§14.3), fast modular exponentiation (§14.6), and the use of the Chinese remainder theorem for faster decryption (Note 14.75). Even with these improvements, RSA encryption/decryption is substantially slower than the commonly used symmetric-key encryption algorithms such as DES (Chapter 7). In practice, RSA encryption is most commonly used for the transport of symmetric-key encryption algorithm keys and for the encryption of small data items.

The RSA cryptosystem has been patented in the U.S. and Canada. Several standards organizations have written, or are in the process of writing, standards that address the use of the RSA cryptosystem for encryption, digital signatures, and key establishment. For discussion of patent and standards issues related to RSA, see Chapter 15.

8.7 Note (*recommended size of modulus*) Given the latest progress in algorithms for factoring integers (§3.2), a 512-bit modulus n provides only marginal security from concerted attack. As of 1996, in order to foil the powerful quadratic sieve (§3.2.6) and number field sieve (§3.2.7) factoring algorithms, a modulus n of at least 768 bits is recommended. For long-term security, 1024-bit or larger moduli should be used.

8.8 Note (*selecting primes*)

(i) As mentioned in §8.2.2(i), the primes p and q should be selected so that factoring $n = pq$ is computationally infeasible. The major restriction on p and q in order to avoid the elliptic curve factoring algorithm (§3.2.4) is that p and q should be about the same bitlength, and sufficiently large. For example, if a 1024-bit modulus n is to be used, then each of p and q should be about 512 bits in length.

(ii) Another restriction on the primes p and q is that the difference $p - q$ should not be too small. If $p - q$ is small, then $p \approx q$ and hence $p \approx \sqrt{n}$. Thus, n could be factored efficiently simply by trial division by all odd integers close to $\sqrt{n}$. If p and q are chosen at random, then $p - q$ will be appropriately large with overwhelming probability.

(iii) In addition to these restrictions, many authors have recommended that p and q be strong primes. A prime p is said to be a *strong prime* (cf. Definition 4.52) if the following three conditions are satisfied:

(a) $p - 1$ has a large prime factor, denoted r;

(b) $p + 1$ has a large prime factor; and

(c) $r - 1$ has a large prime factor.

An algorithm for generating strong primes is presented in §4.4.2. The reason for condition (a) is to foil Pollard's $p - 1$ factoring algorithm (§3.2.3) which is efficient only if n has a prime factor p such that $p - 1$ is smooth. Condition (b) foils the $p + 1$ factoring algorithm mentioned on page 125 in §3.12, which is efficient only if n has a prime factor p such that $p + 1$ is smooth. Finally, condition (c) ensures that the cycling attacks described in §8.2.2(vii) will fail.

If the prime p is randomly chosen and is sufficiently large, then both $p - 1$ and $p + 1$ can be expected to have large prime factors. In any case, while strong primes protect against the $p - 1$ and $p + 1$ factoring algorithms, they do not protect against their generalization, the elliptic curve factoring algorithm (§3.2.4). The latter is successful in factoring n if a randomly chosen number of the same size as p (more precisely, this number is the order of a randomly selected elliptic curve defined over $\mathbb{Z}_p$) has only small prime factors. Additionally, it has been shown that the chances of a cycling attack succeeding are negligible if p and q are randomly chosen (cf. §8.2.2(vii)). Thus, strong primes offer little protection beyond that offered by random primes. Given the current state of knowledge of factoring algorithms, there is no compelling reason for requiring the use of strong primes in RSA key generation. On the other hand, they are no less secure than random primes, and require only minimal additional running time to compute; thus there is little real additional cost in using them.

8.9 Note (*small encryption exponents*)

(i) If the encryption exponent e is chosen at random, then RSA encryption using the repeated square-and-multiply algorithm (Algorithm 2.143) takes k modular squarings and an expected $k/2$ (less with optimizations) modular multiplications, where k is the bitlength of the modulus n. Encryption can be sped up by selecting e to be small and/or by selecting e with a small number of 1's in its binary representation.

(ii) The encryption exponent $e = 3$ is commonly used in practice; in this case, it is necessary that neither $p - 1$ nor $q - 1$ be divisible by 3. This results in a very fast encryption operation since encryption only requires 1 modular multiplication and 1 modular squaring. Another encryption exponent used in practice is $e = 2^{16} + 1 = 65537$. This number has only two 1's in its binary representation, and so encryption using the repeated square-and-multiply algorithm requires only 16 modular squarings and 1 modular multiplication. The encryption exponent $e = 2^{16} + 1$ has the advantage over $e = 3$ in that it resists the kind of attack discussed in §8.2.2(ii), since it is unlikely the same message will be sent to $2^{16} + 1$ recipients.

8.3 Rabin public-key encryption

A desirable property of any encryption scheme is a proof that breaking it is as difficult as solving a computational problem that is widely believed to be difficult, such as integer factorization or the discrete logarithm problem. While it is widely believed that breaking the RSA encryption scheme is as difficult as factoring the modulus n, no such equivalence has been proven. The Rabin public-key encryption scheme was the first example of a *provably secure* public-key encryption scheme – the problem faced by a passive adversary of recovering plaintext from some given ciphertext is computationally equivalent to factoring.

8.10 Algorithm Key generation for Rabin public-key encryption

SUMMARY: each entity creates a public key and a corresponding private key.
Each entity A should do the following:

1. Generate two large random (and distinct) primes p and q, each roughly the same size.
2. Compute $n = pq$.
3. A's public key is n; A's private key is (p, q).

8.11 Algorithm Rabin public-key encryption

SUMMARY: B encrypts a message m for A, which A decrypts.

1. *Encryption.* B should do the following:
    (a) Obtain A's authentic public key n.
    (b) Represent the message as an integer m in the range $\{0, 1, \dots, n-1\}$.
    (c) Compute $c = m^2 \bmod n$.
    (d) Send the ciphertext c to A.

2. *Decryption.* To recover plaintext m from c, A should do the following:
    (a) Use Algorithm 3.44 to find the four square roots m_1, m_2, m_3, and m_4 of c modulo n.[2] (See also Note 8.12.)
    (b) The message sent was either m_1, m_2, m_3, or m_4. A somehow (cf. Note 8.14) decides which of these is m.

8.12 Note (*finding square roots of c modulo n = pq when p ≡ q ≡ 3* (mod 4)) If p and q are both chosen to be $\equiv 3 \pmod 4$, then Algorithm 3.44 for computing the four square roots of c modulo n simplifies as follows:

1. Use the extended Euclidean algorithm (Algorithm 2.107) to find integers a and b satisfying $ap + bq = 1$. Note that a and b can be computed once and for all during the key generation stage (Algorithm 8.10).
2. Compute $r = c^{(p+1)/4} \bmod p$.
3. Compute $s = c^{(q+1)/4} \bmod q$.
4. Compute $x = (aps + bqr) \bmod n$.
5. Compute $y = (aps - bqr) \bmod n$.
6. The four square roots of c modulo n are x, $-x \bmod n$, y, and $-y \bmod n$.

---

[2]In the very unlikely case that $\gcd(m, n) \neq 1$, the ciphertext c does not have four distinct square roots modulo n, but rather only one or two.

8.13 Note (*security of Rabin public-key encryption*)

   (i) The task faced by a passive adversary is to recover plaintext m from the corresponding ciphertext c. This is precisely the SQROOT problem of §3.5.2. Recall (Fact 3.46) that the problems of factoring n and computing square roots modulo n are computationally equivalent. Hence, assuming that factoring n is computationally intractable, the Rabin public-key encryption scheme is *provably secure* against a passive adversary.

  (ii) While provably secure against a passive adversary, the Rabin public-key encryption scheme succumbs to a chosen-ciphertext attack (but see Note 8.14(ii)). Such an attack can be mounted as follows. The adversary selects a random integer $m \in \mathbb{Z}_n^*$ and computes $c = m^2 \bmod n$. The adversary then presents c to A's decryption machine, which decrypts c and returns some plaintext y. Since A does not know m, and m is randomly chosen, the plaintext y is not necessarily the same as m. With probability $\frac{1}{2}$, $y \not\equiv \pm m \bmod n$, in which case $\gcd(m - y, n)$ is one of the prime factors of n. If $y \equiv \pm m \bmod n$, then the attack is repeated with a new m.[3]

 (iii) The Rabin public-key encryption scheme is susceptible to attacks similar to those on RSA described in §8.2.2(ii), §8.2.2(iii), and §8.2.2(v). As is the case with RSA, attacks (ii) and (iii) can be circumvented by salting the plaintext message, while attack (v) can be avoided by adding appropriate redundancy prior to encryption.

8.14 Note (*use of redundancy*)

   (i) A drawback of Rabin's public-key scheme is that the receiver is faced with the task of selecting the correct plaintext from among four possibilities. This ambiguity in decryption can easily be overcome in practice by adding prespecified redundancy to the original plaintext prior to encryption. (For example, the last 64 bits of the message may be replicated.) Then, with high probability, exactly one of the four square roots m_1, m_2, m_3, m_4 of a legitimate ciphertext c will possess this redundancy, and the receiver will select this as the intended plaintext. If none of the square roots of c possesses this redundancy, then the receiver should reject c as fraudulent.

  (ii) If redundancy is used as above, Rabin's scheme is no longer susceptible to the chosen-ciphertext attack of Note 8.13(ii). If an adversary selects a message m having the required redundancy and gives $c = m^2 \bmod n$ to A's decryption machine, with very high probability the machine will return the plaintext m itself to the adversary (since the other three square roots of c will most likely not contain the required redundancy), providing no new information. On the other hand, if the adversary selects a message m which does not contain the required redundancy, then with high probability none of the four square roots of $c = m^2 \bmod n$ will possess the required redundancy. In this case, the decryption machine will fail to decrypt c and thus will not provide a response to the adversary. Note that the proof of equivalence of breaking the modified scheme by a passive adversary to factoring is no longer valid. However, if the natural assumption is made that Rabin decryption is composed of two processes, the first which finds the four square roots of $c \bmod n$, and the second which selects the distinguished square root as the plaintext, then the proof of equivalence holds. Hence, Rabin public-key encryption, suitably modified by adding redundancy, is of great practical interest.

---

[3]This chosen-ciphertext attack is an execution of the constructive proof of the equivalence of factoring n and the SQROOT problem (Fact 3.46), where A's decryption machine is used instead of the hypothetical polynomial-time algorithm for solving the SQROOT problem in the proof.

8.15 Example (*Rabin public-key encryption with artificially small parameters*)

Key generation. Entity A chooses the primes $p = 277$, $q = 331$, and computes $n = pq = 91687$. A's public key is $n = 91687$, while A's private key is $(p = 277, q = 331)$.

Encryption. Suppose that the last six bits of original messages are required to be replicated prior to encryption (cf. Note 8.14(i)). In order to encrypt the 10-bit message $\overline{m} = 1001111001$, B replicates the last six bits of $\overline{m}$ to obtain the 16-bit message $m = 1001111001111001$, which in decimal notation is $m = 40569$. B then computes

$$c = m^2 \bmod n = 40569^2 \bmod 91687 = 62111$$

and sends this to A.

Decryption. To decrypt c, A uses Algorithm 3.44 and her knowledge of the factors of n to compute the four square roots of $c \bmod n$:

$$m_1 = 69654, \quad m_2 = 22033, \quad m_3 = 40569, \quad m_4 = 51118,$$

which in binary are

$$m_1 = 10001000000010110, \qquad m_2 = 101011000010001,$$
$$m_3 = 1001111001111001, \qquad m_4 = 1100011110101110.$$

Since only m_3 has the required redundancy, A decrypts c to m_3 and recovers the original message $\overline{m} = 1001111001$. □

8.16 Note (*efficiency*) Rabin encryption is an extremely fast operation as it only involves a single modular squaring. By comparison, RSA encryption with $e = 3$ takes one modular multiplication and one modular squaring. Rabin decryption is slower than encryption, but comparable in speed to RSA decryption.

8.4 ElGamal public-key encryption

The security of the ElGamal public-key encryption scheme is based on the intractability of the discrete logarithm problem (see §3.6) and the Diffie-Hellman problem (§3.7). The basic ElGamal and generalized ElGamal encryption schemes are described in this section.

8.4.1 Basic ElGamal encryption

8.17 Algorithm Key generation for ElGamal public-key encryption

SUMMARY: each entity creates a public key and a corresponding private key.
Each entity A should do the following:

1. Generate a large random prime p and a generator α of the multiplicative group $\mathbb{Z}_p^*$ of the integers modulo p (using Algorithm 4.84).
2. Select a random integer a, $1 \leq a \leq p - 2$, and compute $\alpha^a \bmod p$ (using Algorithm 2.143).
3. A's public key is (p, α, α^a); A's private key is a.

8.18 Algorithm ElGamal public-key encryption

SUMMARY: B encrypts a message m for A, which A decrypts.

1. *Encryption.* B should do the following:
   (a) Obtain A's authentic public key (p, α, α^a).
   (b) Represent the message as an integer m in the range $\{0, 1, \ldots, p-1\}$.
   (c) Select a random integer k, $1 \leq k \leq p-2$.
   (d) Compute $\gamma = \alpha^k \bmod p$ and $\delta = m \cdot (\alpha^a)^k \bmod p$.
   (e) Send the ciphertext $c = (\gamma, \delta)$ to A.

2. *Decryption.* To recover plaintext m from c, A should do the following:
   (a) Use the private key a to compute $\gamma^{p-1-a} \bmod p$ (note: $\gamma^{p-1-a} = \gamma^{-a} = \alpha^{-ak}$).
   (b) Recover m by computing $(\gamma^{-a}) \cdot \delta \bmod p$.

Proof that decryption works. The decryption of Algorithm 8.18 allows recovery of original plaintext because

$$\gamma^{-a} \cdot \delta \equiv \alpha^{-ak} m \alpha^{ak} \equiv m \pmod{p}.$$

8.19 Example (*ElGamal encryption with artificially small parameters*)
Key generation. Entity A selects the prime $p = 2357$ and a generator $\alpha = 2$ of $\mathbb{Z}_{2357}^*$. A chooses the private key $a = 1751$ and computes

$$\alpha^a \bmod p = 2^{1751} \bmod 2357 = 1185.$$

A's public key is $(p = 2357, \alpha = 2, \alpha^a = 1185)$.
Encryption. To encrypt a message $m = 2035$, B selects a random integer $k = 1520$ and computes

$$\gamma = 2^{1520} \bmod 2357 = 1430$$

and

$$\delta = 2035 \cdot 1185^{1520} \bmod 2357 = 697.$$

B sends $\gamma = 1430$ and $\delta = 697$ to A.
Decryption. To decrypt, A computes

$$\gamma^{p-1-a} = 1430^{605} \bmod 2357 = 872,$$

and recovers m by computing

$$m = 872 \cdot 697 \bmod 2357 = 2035.$$ □

8.20 Note (*common system-wide parameters*) All entities may elect to use the same prime p and generator α, in which case p and α need not be published as part of the public key. This results in public keys of smaller sizes. An additional advantage of having a fixed base α is that exponentiation can then be expedited via precomputations using the techniques described in §14.6.2. A potential disadvantage of common system-wide parameters is that larger moduli p may be warranted (cf. Note 8.24).

8.21 Note (*efficiency*)

   (i) The encryption process requires two modular exponentiations, namely $\alpha^k \bmod p$ and $(\alpha^a)^k \bmod p$. These exponentiations can be sped up by selecting random exponents k having some additional structure, for example, having low Hamming weights. Care must be taken that the possible number of exponents is large enough to preclude a search via a baby-step giant-step algorithm (cf. Note 3.59).

  (ii) A disadvantage of ElGamal encryption is that there is *message expansion* by a factor of 2. That is, the ciphertext is twice as long as the corresponding plaintext.

8.22 Remark (*randomized encryption*) ElGamal encryption is one of many encryption schemes which utilizes randomization in the encryption process. Others include McEliece encryption (§8.5), and Goldwasser-Micali (§8.7.1), and Blum-Goldwasser (§8.7.2) probabilistic encryption. Deterministic encryption schemes such as RSA may also employ randomization in order to circumvent some attacks (e.g., see §8.2.2(ii) and §8.2.2(iii)). The fundamental idea behind randomized encryption techniques is to use randomization to increase the cryptographic security of an encryption process through one or more of the following methods:

   (i) increasing the effective size of the plaintext message space;

  (ii) precluding or decreasing the effectiveness of chosen-plaintext attacks by virtue of a one-to-many mapping of plaintext to ciphertext; and

 (iii) precluding or decreasing the effectiveness of statistical attacks by leveling the a priori probability distribution of inputs.

8.23 Note (*security of ElGamal encryption*)

   (i) The problem of breaking the ElGamal encryption scheme, i.e., recovering m given p, α, α^a, γ, and δ, is equivalent to solving the Diffie-Hellman problem (see §3.7). In fact, the ElGamal encryption scheme can be viewed as simply comprising a Diffie-Hellman key exchange to determine a session key α^{ak}, and then encrypting the message by multiplication with that session key. For this reason, the security of the ElGamal encryption scheme is said to be *based* on the discrete logarithm problem in $\mathbb{Z}_p^*$, although such an equivalence has not been proven.

  (ii) It is critical that different random integers k be used to encrypt different messages. Suppose the same k is used to encrypt two messages m_1 and m_2 and the resulting ciphertext pairs are (γ_1, δ_1) and (γ_2, δ_2). Then $\delta_1/\delta_2 = m_1/m_2$, and m_2 could be easily computed if m_1 were known.

8.24 Note (*recommended parameter sizes*) Given the latest progress on the discrete logarithm problem in $\mathbb{Z}_p^*$ (§3.6), a 512-bit modulus p provides only marginal security from concerted attack. As of 1996, a modulus p of at least 768 bits is recommended. For long-term security, 1024-bit or larger moduli should be used. For common system-wide parameters (cf. Note 8.20) even larger key sizes may be warranted. This is because the dominant stage in the index-calculus algorithm (§3.6.5) for discrete logarithms in $\mathbb{Z}_p^*$ is the precomputation of a database of factor base logarithms, following which individual logarithms can be computed relatively quickly. Thus computing the database of logarithms for one particular modulus p will compromise the secrecy of all private keys derived using p.

8.4.2 Generalized ElGamal encryption

The ElGamal encryption scheme is typically described in the setting of the multiplicative group $\mathbb{Z}_p^*$, but can be easily generalized to work in any finite cyclic group G.

As with ElGamal encryption, the security of the generalized ElGamal encryption scheme is *based* on the intractability of the discrete logarithm problem in the group G. The group G should be carefully chosen to satisfy the following two conditions:

1. for *efficiency*, the group operation in G should be relatively easy to apply; and
2. for *security*, the discrete logarithm problem in G should be computationally infeasible.

The following is a list of groups that appear to meet these two criteria, of which the first three have received the most attention.

1. The multiplicative group $\mathbb{Z}_p^*$ of the integers modulo a prime p.
2. The multiplicative group $\mathbb{F}_{2^m}^*$ of the finite field $\mathbb{F}_{2^m}$ of characteristic two.
3. The group of points on an elliptic curve over a finite field.
4. The multiplicative group $\mathbb{F}_q^*$ of the finite field $\mathbb{F}_q$, where $q = p^m$, p a prime.
5. The group of units $\mathbb{Z}_n^*$, where n is a composite integer.
6. The jacobian of a hyperelliptic curve defined over a finite field.
7. The class group of an imaginary quadratic number field.

8.25 Algorithm Key generation for generalized ElGamal public-key encryption

SUMMARY: each entity creates a public key and a corresponding private key.
Each entity A should do the following:

1. Select an appropriate cyclic group G of order n, with generator α. (It is assumed here that G is written multiplicatively.)
2. Select a random integer a, $1 \le a \le n - 1$, and compute the group element α^a.
3. A's public key is (α, α^a), together with a description of how to multiply elements in G; A's private key is a.

8.26 Algorithm Generalized ElGamal public-key encryption

SUMMARY: B encrypts a message m for A, which A decrypts.

1. *Encryption.* B should do the following:
   (a) Obtain A's authentic public key (α, α^a).
   (b) Represent the message as an element m of the group G.
   (c) Select a random integer k, $1 \le k \le n - 1$.
   (d) Compute $\gamma = \alpha^k$ and $\delta = m \cdot (\alpha^a)^k$.
   (e) Send the ciphertext $c = (\gamma, \delta)$ to A.
2. *Decryption.* To recover plaintext m from c, A should do the following:
   (a) Use the private key a to compute γ^a and then compute γ^{-a}.
   (b) Recover m by computing $(\gamma^{-a}) \cdot \delta$.

8.27 Note (*common system-wide parameters*) All entities may elect to use the same cyclic group G and generator α, in which case α and the description of multiplication in G do not have to be published as part of the public key (cf. Note 8.20).

8.28 Example (*ElGamal encryption using the multiplicative group of* $\mathbb{F}_{2^m}$, *with artificially small parameters*)

Key generation. Entity A selects the group G to be the multiplicative group of the finite field $\mathbb{F}_{2^4}$, whose elements are represented by the polynomials over $\mathbb{F}_2$ of degree less than 4, and where multiplication is performed modulo the irreducible polynomial $f(x) = x^4 + x + 1$ (cf. Example 2.231). For convenience, a field element $a_3 x^3 + a_2 x^2 + a_1 x + a_0$ is represented by the binary string $(a_3 a_2 a_1 a_0)$. The group G has order $n = 15$ and a generator is $\alpha = (0010)$.

A chooses the private key $a = 7$ and computes $\alpha^a = \alpha^7 = (1011)$. A's public key is $\alpha^a = (1011)$ (together with $\alpha = (0010)$ and the polynomial $f(x)$ which defines the multiplication in G, if these parameters are not common to all entities).

Encryption. To encrypt a message $m = (1100)$, B selects a random integer $k = 11$ and computes $\gamma = \alpha^{11} = (1110)$, $(\alpha^a)^{11} = (0100)$, and $\delta = m \cdot (\alpha^a)^{11} = (0101)$. B sends $\gamma = (1110)$ and $\delta = (0101)$ to A.

Decryption. To decrypt, A computes $\gamma^a = (0100)$, $(\gamma^a)^{-1} = (1101)$ and finally recovers m by computing $m = (\gamma^{-a}) \cdot \delta = (1100)$. $\square$

8.5 McEliece public-key encryption

The McEliece public-key encryption scheme is based on error-correcting codes. The idea behind this scheme is to first select a particular code for which an efficient decoding algorithm is known, and then to disguise the code as a general linear code. Since the problem of decoding an arbitrary linear code is **NP**-hard (Definition 2.73), a description of the original code can serve as the private key, while a description of the transformed code serves as the public key.

   The McEliece encryption scheme has resisted cryptanalysis to date. It is also notable as being the first public-key encryption scheme to use randomization in the encryption process. Although very efficient, the McEliece encryption scheme has received little attention in practice because of the very large public keys (see Remark 8.33).

8.29 Algorithm Key generation for McEliece public-key encryption

SUMMARY: each entity creates a public key and a corresponding private key.

1. Integers k, n, and t are fixed as common system parameters.
2. Each entity A should perform steps 3 – 7.
3. Choose a $k \times n$ generator matrix G for a binary (n, k)-linear code which can correct t errors, and for which an efficient decoding algorithm is known.
4. Select a random $k \times k$ binary non-singular matrix S.
5. Select a random $n \times n$ permutation matrix P.
6. Compute the $k \times n$ matrix $\widehat{G} = SGP$.
7. A's public key is $(\widehat{G}, t)$; A's private key is (S, G, P).

8.30 Algorithm McEliece public-key encryption

SUMMARY: B encrypts a message m for A, which A decrypts.

1. *Encryption.* B should do the following:
    (a) Obtain A's authentic public key $(\widehat{G}, t)$.
    (b) Represent the message as a binary string m of length k.
    (c) Choose a random binary error vector z of length n having at most t 1's.
    (d) Compute the binary vector $c = m\widehat{G} + z$.
    (e) Send the ciphertext c to A.

2. *Decryption.* To recover plaintext m from c, A should do the following:
    (a) Compute $\widehat{c} = cP^{-1}$, where P^{-1} is the inverse of the matrix P.
    (b) Use the decoding algorithm for the code generated by G to decode $\widehat{c}$ to $\widehat{m}$.
    (c) Compute $m = \widehat{m}S^{-1}$.

Proof that decryption works. Since

$$\widehat{c} = cP^{-1} = (m\widehat{G} + z)P^{-1} = (mSGP + z)P^{-1} = (mS)G + zP^{-1},$$

and zP^{-1} is a vector with at most t 1's, the decoding algorithm for the code generated by G corrects $\widehat{c}$ to $\widehat{m} = mS$. Finally, $\widehat{m}S^{-1} = m$, and, hence, decryption works.

A special type of error-correcting code, called a *Goppa code*, may be used in step 3 of the key generation. For each irreducible polynomial $g(x)$ of degree t over $\mathbb{F}_{2^m}$, there exists a binary Goppa code of length $n = 2^m$ and dimension $k \geq n - mt$ capable of correcting any pattern of t or fewer errors. Furthermore, efficient decoding algorithms are known for such codes.

8.31 Note (*security of McEliece encryption*) There are two basic kinds of attacks known.
   (i) From the public information, an adversary may try to compute the key G or a key G' for a Goppa code equivalent to the one with generator matrix G. There is no efficient method known for accomplishing this.
   (ii) An adversary may try to recover the plaintext m directly given some ciphertext c. The adversary picks k columns at random from $\widehat{G}$. If $\widehat{G}_k$, c_k and z_k denote the restriction of $\widehat{G}$, c and z, respectively, to these k columns, then $(c_k + z_k) = m\widehat{G}_k$. If $z_k = 0$ and if $\widehat{G}_k$ is non-singular, then m can be recovered by solving the system of equations $c_k = m\widehat{G}_k$. Since the probability that $z_k = 0$, i.e., the selected k bits were not in error, is only $\binom{n-t}{k}/\binom{n}{k}$, the probability of this attack succeeding is negligibly small.

8.32 Note (*recommended parameter sizes*) The original parameters suggested by McEliece were $n = 1024$, $t = 50$, and $k \geq 524$. Based on the security analysis (Note 8.31), an optimum choice of parameters for the Goppa code which maximizes the adversary's work factor appears to be $n = 1024$, $t = 38$, and $k \geq 644$.

8.33 Remark (*McEliece encryption in practice*) Although the encryption and decryption operations are relatively fast, the McEliece scheme suffers from the drawback that the public key is very large. A (less significant) drawback is that there is message expansion by a factor of n/k. For the recommended parameters $n = 1024$, $t = 38$, $k \geq 644$, the public key is about 2^{19} bits in size, while the message expansion factor is about 1.6. For these reasons, the scheme receives little attention in practice.

8.6 Knapsack public-key encryption

Knapsack public-key encryption schemes are based on the subset sum problem, which is **NP**-complete (see §2.3.3 and §3.10). The basic idea is to select an instance of the subset sum problem that is easy to solve, and then to disguise it as an instance of the general subset sum problem which is hopefully difficult to solve. The original knapsack set can serve as the private key, while the transformed knapsack set serves as the public key.

The Merkle-Hellman knapsack encryption scheme (§8.6.1) is important for historical reasons, as it was the first concrete realization of a public-key encryption scheme. Many variations have subsequently been proposed but most, including the original, have been demonstrated to be insecure (see Note 8.40), a notable exception being the Chor-Rivest knapsack scheme (§8.6.2).

8.6.1 Merkle-Hellman knapsack encryption

The Merkle-Hellman knapsack encryption scheme attempts to disguise an easily solved instance of the subset sum problem, called a *superincreasing subset sum problem*, by modular multiplication and a permutation. It is however not recommended for use (see Note 8.40).

8.34 Definition A *superincreasing sequence* is a sequence $(b_1, b_2, \ldots, b_n)$ of positive integers with the property that $b_i > \sum_{j=1}^{i-1} b_j$ for each i, $2 \leq i \leq n$.

Algorithm 8.35 efficiently solves the subset sum problem for superincreasing sequences.

8.35 Algorithm Solving a superincreasing subset sum problem

INPUT: a superincreasing sequence $(b_1, b_2, \ldots, b_n)$ and an integer s which is the sum of a subset of the b_i.
OUTPUT: $(x_1, x_2, \ldots, x_n)$ where $x_i \in \{0, 1\}$, such that $\sum_{i=1}^{n} x_i b_i = s$.
   1. $i \leftarrow n$.
   2. While $i \geq 1$ do the following:
      2.1 If $s \geq b_i$ then $x_i \leftarrow 1$ and $s \leftarrow s - b_i$. Otherwise $x_i \leftarrow 0$.
      2.2 $i \leftarrow i - 1$.
   3. Return$((x_1, x_2, \ldots, x_n))$.

8.36 Algorithm Key generation for basic Merkle-Hellman knapsack encryption

---

SUMMARY: each entity creates a public key and a corresponding private key.
   1. An integer n is fixed as a common system parameter.
   2. Each entity A should perform steps $3 - 7$.
   3. Choose a superincreasing sequence $(b_1, b_2, \ldots, b_n)$ and modulus M such that $M > b_1 + b_2 + \cdots + b_n$.
   4. Select a random integer W, $1 \leq W \leq M - 1$, such that $\gcd(W, M) = 1$.
   5. Select a random permutation π of the integers $\{1, 2, \ldots, n\}$.
   6. Compute $a_i = W b_{\pi(i)} \bmod M$ for $i = 1, 2, \ldots, n$.
   7. A's public key is $(a_1, a_2, \ldots, a_n)$; A's private key is $(\pi, M, W, (b_1, b_2, \ldots, b_n))$.

8.37 Algorithm Basic Merkle-Hellman knapsack public-key encryption

SUMMARY: B encrypts a message m for A, which A decrypts.

1. *Encryption.* B should do the following:

   (a) Obtain A's authentic public key $(a_1, a_2, \ldots, a_n)$.
   (b) Represent the message m as a binary string of length n, $m = m_1 m_2 \cdots m_n$.
   (c) Compute the integer $c = m_1 a_1 + m_2 a_2 + \cdots + m_n a_n$.
   (d) Send the ciphertext c to A.

2. *Decryption.* To recover plaintext m from c, A should do the following:

   (a) Compute $d = W^{-1}c \bmod M$.
   (b) By solving a superincreasing subset sum problem (Algorithm 8.35), find integers $r_1, r_2, \ldots, r_n, r_i \in \{0, 1\}$, such that $d = r_1 b_1 + r_2 b_2 + \cdots + r_n b_n$.
   (c) The message bits are $m_i = r_{\pi(i)}$, $i = 1, 2, \ldots, n$.

Proof that decryption works. The decryption of Algorithm 8.37 allows recovery of original plaintext because

$$d \equiv W^{-1}c \equiv W^{-1} \sum_{i=1}^{n} m_i a_i \equiv \sum_{i=1}^{n} m_i b_{\pi(i)} \pmod{M}.$$

Since $0 \le d < M$, $d = \sum_{i=1}^{n} m_i b_{\pi(i)} \bmod M$, and hence the solution of the superincreasing subset sum problem in step (b) of the decryption gives the message bits, after application of the permutation π.

8.38 Example (*basic Merkle-Hellman knapsack encryption with artificially small parameters*)
Key generation. Let $n = 6$. Entity A chooses the superincreasing sequence $(12, 17, 33, 74, 157, 316)$, $M = 737$, $W = 635$, and the permutation π of $\{1, 2, 3, 4, 5, 6\}$ defined by $\pi(1) = 3$, $\pi(2) = 6$, $\pi(3) = 1$, $\pi(4) = 2$, $\pi(5) = 5$, and $\pi(6) = 4$. A's public key is the knapsack set $(319, 196, 250, 477, 200, 559)$, while A's private key is $(\pi, M, W, (12, 17, 33, 74, 157, 316))$.
Encryption. To encrypt the message $m = 101101$, B computes

$$c = 319 + 250 + 477 + 559 = 1605$$

and sends this to A.
Decryption. To decrypt, A computes $d = W^{-1}c \bmod M = 136$, and solves the superincreasing subset sum problem

$$136 = 12r_1 + 17r_2 + 33r_3 + 74r_4 + 157r_5 + 316r_6$$

to get $136 = 12 + 17 + 33 + 74$. Hence, $r_1 = 1$, $r_2 = 1$, $r_3 = 1$, $r_4 = 1$, $r_5 = 0$, $r_6 = 0$, and application of the permutation π yields the message bits $m_1 = r_3 = 1$, $m_2 = r_6 = 0$, $m_3 = r_1 = 1$, $m_4 = r_2 = 1$, $m_5 = r_5 = 0$, $m_6 = r_4 = 1$. $\square$

Multiple-iterated Merkle-Hellman knapsack encryption

One variation of the basic Merkle-Hellman scheme involves disguising the easy superincreasing sequence by a series of modular multiplications. The key generation for this variation is as follows.

8.39 Algorithm Key generation for multiple-iterated Merkle-Hellman knapsack encryption

SUMMARY: each entity creates a public key and a corresponding private key.

1. Integers n and t are fixed as common system parameters.
2. Each entity A should perform steps 3 – 6.
3. Choose a superincreasing sequence $(a_1^{(0)}, a_2^{(0)}, \ldots, a_n^{(0)})$.
4. For j from 1 to t do the following:
   4.1 Choose a modulus M_j with $M_j > a_1^{(j-1)} + a_2^{(j-1)} + \cdots + a_n^{(j-1)}$.
   4.2 Select a random integer W_j, $1 \le W_j \le M_j - 1$, such that $\gcd(W_j, M_j) = 1$.
   4.3 Compute $a_i^{(j)} = a_i^{(j-1)} W_j \bmod M_j$ for $i = 1, 2, \ldots, n$.
5. Select a random permutation π of the integers $\{1, 2, \ldots, n\}$.
6. A's public key is $(a_1, a_2, \ldots, a_n)$, where $a_i = a_{\pi(i)}^{(t)}$ for $i = 1, 2, \ldots, n$; A's private key is $(\pi, M_1, \ldots, M_t, W_1, \ldots, W_t, a_1^{(0)}, a_2^{(0)}, \ldots, a_n^{(0)})$.

Encryption is performed in the same way as in the basic Merkle-Hellman scheme (Algorithm 8.37). Decryption is performed by successively computing $d_j = W_j^{-1} d_{j+1} \bmod M_j$ for $j = t, t-1, \ldots, 1$, where $d_{t+1} = c$. Finally, the superincreasing subset sum problem $d_1 = r_1 a_1^{(0)} + r_2 a_2^{(0)} + \cdots + r_n a_n^{(0)}$ is solved for r_i, and the message bits are recovered after application of the permutation π.

8.40 Note (*insecurity of Merkle-Hellman knapsack encryption*)

(i) A polynomial-time algorithm for breaking the basic Merkle-Hellman scheme is known. Given the public knapsack set, this algorithm finds a pair of integers U', M' such that U'/M' is close to U/M (where W and M are part of the private key, and $U = W^{-1} \bmod M$) and such that the integers $b_i' = U' a_i \bmod M$, $1 \le i \le n$, form a superincreasing sequence. This sequence can then be used by an adversary in place of $(b_1, b_2, \ldots, b_n)$ to decrypt messages.

(ii) The most powerful general attack known on knapsack encryption schemes is the technique discussed in §3.10.2 which reduces the subset sum problem to the problem of finding a short vector in a lattice. It is typically successful if the density (see Definition 3.104) of the knapsack set is less than 0.9408. This is significant because the density of a Merkle-Hellman knapsack set must be less than 1, since otherwise there will in general be many subsets of the knapsack set with the same sum, in which case some ciphertexts will not be uniquely decipherable. Moreover, since each iteration in the multiple-iterated scheme lowers the density, this attack will succeed if the knapsack set has been iterated a sufficient number of times.

Similar techniques have since been used to break most knapsacks schemes that have been proposed, including the multiple-iterated Merkle-Hellman scheme. The most prominent knapsack scheme that has resisted such attacks to date is the Chor-Rivest scheme (but see Note 8.44).

8.6.2 Chor-Rivest knapsack encryption

The Chor-Rivest scheme is the only known knapsack public-key encryption scheme that does not use some form of modular multiplication to disguise an easy subset sum problem.

8.41 Algorithm Key generation for Chor-Rivest public-key encryption

SUMMARY: each entity creates a public key and a corresponding private key.
Each entity A should do the following:

1. Select a finite field $\mathbb{F}_q$ of characteristic p, where $q = p^h$, $p \geq h$, and for which the discrete logarithm problem is feasible (see Note 8.45(ii)).
2. Select a random monic irreducible polynomial $f(x)$ of degree h over $\mathbb{Z}_p$ (using Algorithm 4.70). The elements of $\mathbb{F}_q$ will be represented as polynomials in $\mathbb{Z}_p[x]$ of degree less than h, with multiplication performed modulo $f(x)$.
3. Select a random primitive element $g(x)$ of the field $\mathbb{F}_q$ (using Algorithm 4.80).
4. For each ground field element $i \in \mathbb{Z}_p$, find the discrete logarithm $a_i = \log_{g(x)}(x+i)$ of the field element $(x + i)$ to the base $g(x)$.
5. Select a random permutation π on the set of integers $\{0, 1, 2, \ldots, p - 1\}$.
6. Select a random integer d, $0 \leq d \leq p^h - 2$.
7. Compute $c_i = (a_{\pi(i)} + d) \bmod (p^h - 1)$, $0 \leq i \leq p - 1$.
8. A's public key is $((c_0, c_1, \ldots, c_{p-1}), p, h)$; A's private key is $(f(x), g(x), \pi, d)$.

8.42 Algorithm Chor-Rivest public-key encryption

SUMMARY: B encrypts a message m for A, which A decrypts.

1. *Encryption.* B should do the following:
   (a) Obtain A's authentic public key $((c_0, c_1, \ldots, c_{p-1}), p, h)$.
   (b) Represent the message m as a binary string of length $\lfloor \lg \binom{p}{h} \rfloor$, where $\binom{p}{h}$ is a binomial coefficient (Definition 2.17).
   (c) Consider m as the binary representation of an integer. Transform this integer into a binary vector $M = (M_0, M_1, \ldots, M_{p-1})$ of length p having exactly h 1's as follows:
       i. Set $l \leftarrow h$.
       ii. For i from 1 to p do the following:
           If $m \geq \binom{p-i}{l}$ then set $M_{i-1} \leftarrow 1$, $m \leftarrow m - \binom{p-i}{l}$, $l \leftarrow l - 1$. Otherwise, set $M_{i-1} \leftarrow 0$. (Note: $\binom{n}{0} = 1$ for $n \geq 0$; $\binom{0}{l} = 0$ for $l \geq 1$.)
   (d) Compute $c = \sum_{i=0}^{p-1} M_i c_i \bmod (p^h - 1)$.
   (e) Send the ciphertext c to A.

2. *Decryption.* To recover plaintext m from c, A should do the following:
   (a) Compute $r = (c - hd) \bmod (p^h - 1)$.
   (b) Compute $u(x) = g(x)^r \bmod f(x)$ (using Algorithm 2.227).
   (c) Compute $s(x) = u(x) + f(x)$, a monic polynomial of degree h over $\mathbb{Z}_p$.
   (d) Factor $s(x)$ into linear factors over $\mathbb{Z}_p$: $s(x) = \prod_{j=1}^{h}(x + t_j)$, where $t_j \in \mathbb{Z}_p$ (cf. Note 8.45(iv)).
   (e) The components of the vector M that are 1 have indices $\pi^{-1}(t_j)$, $1 \leq j \leq h$. The remaining components are 0.
   (f) The message m is recovered from M as follows:
       i. Set $m \leftarrow 0$, $l \leftarrow h$.
       ii. For i from 1 to p do the following:
           If $M_{i-1} = 1$ then set $m \leftarrow m + \binom{p-i}{l}$ and $l \leftarrow l - 1$.

Proof that decryption works. Observe that

$$
\begin{aligned}
u(x) &= g(x)^r \bmod f(x) \\
&\equiv g(x)^{c-hd} \equiv g(x)^{(\sum_{i=0}^{p-1} M_i c_i)-hd} \pmod{f(x)} \\
&\equiv g(x)^{(\sum_{i=0}^{p-1} M_i(a_{\pi(i)}+d))-hd} \equiv g(x)^{\sum_{i=0}^{p-1} M_i a_{\pi(i)}} \pmod{f(x)} \\
&\equiv \prod_{i=0}^{p-1} [g(x)^{a_{\pi(i)}}]^{M_i} \equiv \prod_{i=0}^{p-1} (x+\pi(i))^{M_i} \pmod{f(x)}.
\end{aligned}
$$

Since $\prod_{i=0}^{p-1}(x+\pi(i))^{M_i}$ and $s(x)$ are monic polynomials of degree h and are congruent modulo $f(x)$, it must be the case that

$$
s(x) = u(x) + f(x) = \prod_{i=0}^{p-1}(x+\pi(i))^{M_i}.
$$

Hence, the h roots of $s(x)$ all lie in $\mathbb{Z}_p$, and applying π^{-1} to these roots gives the coordinates of M that are 1.

8.43 Example (*Chor-Rivest public-key encryption with artificially small parameters*)
Key generation. Entity A does the following:

1. Selects $p = 7$ and $h = 4$.
2. Selects the irreducible polynomial $f(x) = x^4 + 3x^3 + 5x^2 + 6x + 2$ of degree 4 over $\mathbb{Z}_7$. The elements of the finite field $\mathbb{F}_{7^4}$ are represented as polynomials in $\mathbb{Z}_7[x]$ of degree less than 4, with multiplication performed modulo $f(x)$.
3. Selects the random primitive element $g(x) = 3x^3 + 3x^2 + 6$.
4. Computes the following discrete logarithms:

$$
\begin{aligned}
a_0 &= \log_{g(x)}(x) &&= 1028 \\
a_1 &= \log_{g(x)}(x+1) &&= 1935 \\
a_2 &= \log_{g(x)}(x+2) &&= 2054 \\
a_3 &= \log_{g(x)}(x+3) &&= 1008 \\
a_4 &= \log_{g(x)}(x+4) &&= 379 \\
a_5 &= \log_{g(x)}(x+5) &&= 1780 \\
a_6 &= \log_{g(x)}(x+6) &&= 223.
\end{aligned}
$$

5. Selects the random permutation π on $\{0, 1, 2, 3, 4, 5, 6\}$ defined by $\pi(0) = 6, \pi(1) = 4, \pi(2) = 0, \pi(3) = 2, \pi(4) = 1, \pi(5) = 5, \pi(6) = 3$.
6. Selects the random integer $d = 1702$.
7. Computes

$$
\begin{aligned}
c_0 &= (a_6 + d) \bmod 2400 = 1925 \\
c_1 &= (a_4 + d) \bmod 2400 = 2081 \\
c_2 &= (a_0 + d) \bmod 2400 = 330 \\
c_3 &= (a_2 + d) \bmod 2400 = 1356 \\
c_4 &= (a_1 + d) \bmod 2400 = 1237 \\
c_5 &= (a_5 + d) \bmod 2400 = 1082 \\
c_6 &= (a_3 + d) \bmod 2400 = 310.
\end{aligned}
$$

8. A's public key is $((c_0, c_1, c_2, c_3, c_4, c_5, c_6), p = 7, h = 4)$, while A's private key is $(f(x), g(x), \pi, d)$.

Encryption. To encrypt a message $m = 22$ for A, B does the following:

(a) Obtains authentic A's public key.

(b) Represents m as a binary string of length 5: $m = 10110$. (Note that $\lfloor \lg \binom{7}{4} \rfloor = 5$.)

(c) Uses the method outlined in step 1(c) of Algorithm 8.42 to transform m to the binary vector $M = (1, 0, 1, 1, 0, 0, 1)$ of length 7.

(d) Computes $c = (c_0 + c_2 + c_3 + c_6) \bmod 2400 = 1521$.

(e) Sends $c = 1521$ to A.

Decryption. To decrypt the ciphertext $c = 1521$, A does the following:

(a) Computes $r = (c - hd) \bmod 2400 = 1913$.

(b) Computes $u(x) = g(x)^{1913} \bmod f(x) = x^3 + 3x^2 + 2x + 5$.

(c) Computes $s(x) = u(x) + f(x) = x^4 + 4x^3 + x^2 + x$.

(d) Factors $s(x) = x(x + 2)(x + 3)(x + 6)$ (so $t_1 = 0, t_2 = 2, t_3 = 3, t_4 = 6$).

(e) The components of M that are 1 have indices $\pi^{-1}(0) = 2$, $\pi^{-1}(2) = 3$, $\pi^{-1}(3) = 6$, and $\pi^{-1}(6) = 0$. Hence, $M = (1, 0, 1, 1, 0, 0, 1)$.

(f) Uses the method outlined in step 2(f) of Algorithm 8.42 to transform M to the integer $m = 22$, thus recovering the original plaintext. □

8.44 Note (*security of Chor-Rivest encryption*)

(i) When the parameters of the system are carefully chosen (see Note 8.45 and page 318), there is no feasible attack known on the Chor-Rivest encryption scheme. In particular, the density of the knapsack set $(c_0, c_1, \dots, c_{p-1})$ is $p/\lg(\max c_i)$, which is large enough to thwart the low-density attacks on the general subset sum problem (§3.10.2).

(ii) It is known that the system is insecure if portions of the private key are revealed, for example, if $g(x)$ and d in some representation of $\mathbb{F}_q$ are known, or if $f(x)$ is known, or if π is known.

8.45 Note (*implementation*)

(i) Although the Chor-Rivest scheme has been described only for the case p a prime, it extends to the case where the base field $\mathbb{Z}_p$ is replaced by a field of prime power order.

(ii) In order to make the discrete logarithm problem feasible in step 1 of Algorithm 8.41, the parameters p and h may be chosen so that $q = p^h - 1$ has only small factors. In this case, the Pohlig-Hellman algorithm (§3.6.4) can be used to efficiently compute discrete logarithms in the finite field $\mathbb{F}_q$.

(iii) In practice, the recommended size of the parameters are $p \approx 200$ and $h \approx 25$. One particular choice of parameters originally suggested is $p = 197$ and $h = 24$; in this case, the largest prime factor of $197^{24} - 1$ is 10316017, and the density of the knapsack set is about 1.077. Other parameter sets originally suggested are $\{p = 211, h = 24\}$, $\{p = 3^5, h = 24\}$ (base field $\mathbb{F}_{3^5}$), and $\{p = 2^8, h = 25\}$ (base field $\mathbb{F}_{2^8}$).

(iv) Encryption is a very fast operation. Decryption is much slower, the bottleneck being the computation of $u(x)$ in step 2b. The roots of $s(x)$ in step 2d can be found simply by trying all possibilities in $\mathbb{Z}_p$.

(v) A major drawback of the Chor-Rivest scheme is that the public key is fairly large, namely, about $(ph \cdot \lg p)$ bits. For the parameters $p = 197$ and $h = 24$, this is about $36,000$ bits.

(vi) There is message expansion by a factor of $\lg p^h / \lg \binom{p}{h}$. For $p = 197$ and $h = 24$, this is 1.797.

8.7 Probabilistic public-key encryption

A minimal security requirement of an encryption scheme is that it must be difficult, in essentially all cases, for a passive adversary to recover plaintext from the corresponding ciphertext. However, in some situations, it may be desirable to impose more stringent security requirements.

The RSA, Rabin, and knapsack encryption schemes are *deterministic* in the sense that under a fixed public key, a particular plaintext m is always encrypted to the same ciphertext c. A deterministic scheme has some or all of the following drawbacks.

1. The scheme is not secure for all probability distributions of the message space. For example, in RSA the messages 0 and 1 always get encrypted to themselves, and hence are easy to detect.

2. It is sometimes easy to compute partial information about the plaintext from the ciphertext. For example, in RSA if $c = m^e \bmod n$ is the ciphertext corresponding to a plaintext m, then

$$\left(\frac{c}{n}\right) = \left(\frac{m^e}{n}\right) = \left(\frac{m}{n}\right)^e = \left(\frac{m}{n}\right)$$

since e is odd, and hence an adversary can easily gain one bit of information about m, namely the Jacobi symbol $\left(\frac{m}{n}\right)$.

3. It is easy to detect when the same message is sent twice.

Of course, any deterministic encryption scheme can be converted into a randomized scheme by requiring that a portion of each plaintext consist of a randomly generated bitstring of a pre-specified length l. If the parameter l is chosen to be sufficiently large for the purpose at hand, then, in practice, the attacks listed above are thwarted. However, the resulting randomized encryption scheme is generally not provably secure against the different kinds of attacks that one could conceive.

Probabilistic encryption utilizes randomness to attain a *provable* and very strong level of security. There are two strong notions of security that one can strive to achieve.

8.46 Definition A public-key encryption scheme is said to be *polynomially secure* if no passive adversary can, in expected polynomial time, select two plaintext messages m_1 and m_2 and then correctly distinguish between encryptions of m_1 and m_2 with probability significantly greater than $\frac{1}{2}$.

8.47 Definition A public-key encryption scheme is said to be *semantically secure* if, for all probability distributions over the message space, whatever a passive adversary can compute in expected polynomial time about the plaintext given the ciphertext, it can also compute in expected polynomial time without the ciphertext.

Intuitively, a public-key encryption scheme is semantically secure if the ciphertext does not leak any partial information whatsoever about the plaintext that can be computed in expected polynomial time.

8.48 Remark (*perfect secrecy vs. semantic security*) In Shannon's theory (see §1.13.3(i)), an encryption scheme has *perfect secrecy* if a passive adversary, even with infinite computational resources, can learn nothing about the plaintext from the ciphertext, except possibly its length. The limitation of this notion is that perfect secrecy cannot be achieved unless the key is at least as long as the message. By contrast, the notion of semantic security can be viewed as a polynomially bounded version of perfect secrecy — a passive adversary with polynomially bounded computational resources can learn nothing about the plaintext from the ciphertext. It is then conceivable that there exist semantically secure encryption schemes where the keys are much shorter that the messages.

Although Definition 8.47 appears to be stronger than Definition 8.46, the next result asserts that they are, in fact, equivalent.

8.49 Fact A public-key encryption scheme is semantically secure if and only if it is polynomially secure.

8.7.1 Goldwasser-Micali probabilistic encryption

The Goldwasser-Micali scheme is a probabilistic public-key system which is semantically secure assuming the intractability of the quadratic residuosity problem (see §3.4).

8.50 Algorithm Key generation for Goldwasser-Micali probabilistic encryption

SUMMARY: each entity creates a public key and corresponding private key.
Each entity A should do the following:
1. Select two large random (and distinct) primes p and q, each roughly the same size.
2. Compute $n = pq$.
3. Select a $y \in \mathbb{Z}_n$ such that y is a quadratic non-residue modulo n and the Jacobi symbol $\left(\frac{y}{n}\right) = 1$ (y is a pseudosquare modulo n); see Remark 8.54.
4. A's public key is (n, y); A's private key is the pair (p, q).

8.51 Algorithm Goldwasser-Micali probabilistic public-key encryption

SUMMARY: B encrypts a message m for A, which A decrypts.
1. *Encryption.* B should do the following:
   (a) Obtain A's authentic public key (n, y).
   (b) Represent the message m as a binary string $m = m_1 m_2 \cdots m_t$ of length t.
   (c) For i from 1 to t do:
      i. Pick an $x \in \mathbb{Z}_n^*$ at random.
      ii. If $m_i = 1$ then set $c_i \leftarrow yx^2 \bmod n$; otherwise set $c_i \leftarrow x^2 \bmod n$.
   (d) Send the t-tuple $c = (c_1, c_2, \ldots, c_t)$ to A.
2. *Decryption.* To recover plaintext m from c, A should do the following:
   (a) For i from 1 to t do:
      i. Compute the Legendre symbol $e_i = \left(\frac{c_i}{p}\right)$ (using Algorithm 2.149).
      ii. If $e_i = 1$ then set $m_i \leftarrow 0$; otherwise set $m_i \leftarrow 1$.
   (b) The decrypted message is $m = m_1 m_2 \cdots m_t$.

Proof that decryption works. If a message bit m_i is 0, then $c_i = x^2 \bmod n$ is a quadratic residue modulo n. If a message bit m_i is 1, then since y is a pseudosquare modulo n, $c_i = yx^2 \bmod n$ is also a pseudosquare modulo n. By Fact 2.137, c_i is a quadratic residue modulo n if and only if c_i is a quadratic residue modulo p, or equivalently $\left(\frac{c_i}{p}\right) = 1$. Since A knows p, she can compute this Legendre symbol and hence recover the message bit m_i.

8.52 Note (*security of Goldwasser-Micali probabilistic encryption*) Since x is selected at random from $\mathbb{Z}_n^*$, $x^2 \bmod n$ is a random quadratic residue modulo n, and $yx^2 \bmod n$ is a random pseudosquare modulo n. Hence, an eavesdropper sees random quadratic residues and pseudosquares modulo n. Assuming that the quadratic residuosity problem is difficult, the eavesdropper can do no better that guess each message bit. More formally, if the quadratic residuosity problem is hard, then the Goldwasser-Micali probabilistic encryption scheme is semantically secure.

8.53 Note (*message expansion*) A major disadvantage of the Goldwasser-Micali scheme is the message expansion by a factor of $\lg n$ bits. Some message expansion is unavoidable in a probabilistic encryption scheme because there are many ciphertexts corresponding to each plaintext. Algorithm 8.56 is a major improvement of the Goldwasser-Micali scheme in that the plaintext is only expanded by a constant factor.

8.54 Remark (*finding pseudosquares*) A pseudosquare y modulo n can be found as follows. First find a quadratic non-residue a modulo p and a quadratic non-residue b modulo q (see Remark 2.151). Then use Gauss's algorithm (Algorithm 2.121) to compute the integer y, $0 \leq y \leq n-1$, satisfying the simultaneous congruences $y \equiv a \pmod{p}$, $y \equiv b \pmod{q}$. Since y ($\equiv a \pmod{p}$) is a quadratic non-residue modulo p, it is also a quadratic non-residue modulo n (Fact 2.137). Also, by the properties of the Legendre and Jacobi symbols (§2.4.5), $\left(\frac{y}{n}\right) = \left(\frac{y}{p}\right)\left(\frac{y}{q}\right) = (-1)(-1) = 1$. Hence, y is a pseudosquare modulo n.

8.7.2 Blum-Goldwasser probabilistic encryption

The Blum-Goldwasser probabilistic public-key encryption scheme is the most efficient probabilistic encryption scheme known and is comparable to the RSA encryption scheme, both in terms of speed and message expansion. It is semantically secure (Definition 8.47) assuming the intractability of the integer factorization problem. It is, however, vulnerable to a chosen-ciphertext attack. The scheme uses the Blum-Blum-Shub generator (§5.5.2) to generate a pseudorandom bit sequence which is then XORed with the plaintext. The resulting bit sequence, together with an encryption of the random seed used, is transmitted to the receiver who uses his trapdoor information to recover the seed and subsequently reconstruct the pseudorandom bit sequence and the plaintext.

8.55 Algorithm Key generation for Blum-Goldwasser probabilistic encryption

SUMMARY: each entity creates a public key and a corresponding private key.
Each entity A should do the following:
1. Select two large random (and distinct) primes p, q, each congruent to 3 modulo 4.
2. Compute $n = pq$.
3. Use the extended Euclidean algorithm (Algorithm 2.107) to compute integers a and b such that $ap + bq = 1$.
4. A's public key is n; A's private key is (p, q, a, b).

8.56 Algorithm Blum-Goldwasser probabilistic public-key encryption

SUMMARY: B encrypts a message m for A, which A decrypts.

1. *Encryption.* B should do the following:
   (a) Obtain A's authentic public key n.
   (b) Let $k = \lfloor \lg n \rfloor$ and $h = \lfloor \lg k \rfloor$. Represent the message m as a string $m = m_1 m_2 \cdots m_t$ of length t, where each m_i is a binary string of length h.
   (c) Select as a seed x_0, a random quadratic residue modulo n. (This can be done by selecting a random integer $r \in \mathbb{Z}_n^*$ and setting $x_0 \leftarrow r^2 \bmod n$.)
   (d) For i from 1 to t do the following:
       i. Compute $x_i = x_{i-1}^2 \bmod n$.
       ii. Let p_i be the h least significant bits of x_i.
       iii. Compute $c_i = p_i \oplus m_i$.
   (e) Compute $x_{t+1} = x_t^2 \bmod n$.
   (f) Send the ciphertext $c = (c_1, c_2, \ldots, c_t, x_{t+1})$ to A.

2. *Decryption.* To recover plaintext m from c, A should do the following:
   (a) Compute $d_1 = ((p+1)/4)^{t+1} \bmod (p-1)$.
   (b) Compute $d_2 = ((q+1)/4)^{t+1} \bmod (q-1)$.
   (c) Compute $u = x_{t+1}^{d_1} \bmod p$.
   (d) Compute $v = x_{t+1}^{d_2} \bmod q$.
   (e) Compute $x_0 = vap + ubq \bmod n$.
   (f) For i from 1 to t do the following:
       i. Compute $x_i = x_{i-1}^2 \bmod n$.
       ii. Let p_i be the h least significant bits of x_i.
       iii. Compute $m_i = p_i \oplus c_i$.

Proof that decryption works. Since x_t is a quadratic residue modulo n, it is also a quadratic residue modulo p; hence, $x_t^{(p-1)/2} \equiv 1 \pmod{p}$. Observe that

$$x_{t+1}^{(p+1)/4} \equiv (x_t^2)^{(p+1)/4} \equiv x_t^{(p+1)/2} \equiv x_t^{(p-1)/2} x_t \equiv x_t \pmod{p}.$$

Similarly, $x_t^{(p+1)/4} \equiv x_{t-1} \pmod{p}$ and so

$$x_{t+1}^{((p+1)/4)^2} \equiv x_{t-1} \pmod{p}.$$

Repeating this argument yields

$$u \equiv x_{t+1}^{d_1} \equiv x_{t+1}^{((p+1)/4)^{t+1}} \equiv x_0 \pmod{p}.$$

Analogously,

$$v \equiv x_{t+1}^{d_2} \equiv x_0 \pmod{q}.$$

Finally, since $ap + bq = 1$, $vap + ubq \equiv x_0 \pmod{p}$ and $vap + ubq \equiv x_0 \pmod{q}$. Hence, $x_0 = vap + ubq \bmod n$, and A recovers the same random seed that B used in the encryption, and consequently also recovers the original plaintext.

8.57 Example (*Blum-Goldwasser probabilistic encryption with artificially small parameters*)
Key generation. Entity A selects the primes $p = 499, q = 547$, each congruent to 3 modulo 4, and computes $n = pq = 272953$. Using the extended Euclidean algorithm, A computes

the integers $a = -57$, $b = 52$ satisfying $ap + bq = 1$. A's public key is $n = 272953$, while A's private key is (p, q, a, b).

Encryption. The parameters k and h have the values 18 and 4, respectively. B represents the message m as a string $m_1 m_2 m_3 m_4 m_5$ ($t = 5$) where $m_1 = 1001$, $m_2 = 1100$, $m_3 = 0001$, $m_4 = 0000$, $m_5 = 1100$. B then selects a random quadratic residue $x_0 = 159201$ ($= 399^2 \bmod n$), and computes:

i	$x_i = x_{i-1}^2 \bmod n$	p_i	$c_i = p_i \oplus m_i$
1	180539	1011	0010
2	193932	1100	0000
3	245613	1101	1100
4	130286	1110	1110
5	40632	1000	0100

and $x_6 = x_5^2 \bmod n = 139680$. B sends the ciphertext

$$c = (0010, 0000, 1100, 1110, 0100, 139680)$$

to A.

Decryption. To decrypt c, A computes

$$d_1 = ((p+1)/4)^6 \bmod (p-1) = 463$$
$$d_2 = ((q+1)/4)^6 \bmod (q-1) = 337$$
$$u = x_6^{463} \bmod p \qquad\qquad = 20$$
$$v = x_6^{337} \bmod q \qquad\qquad = 24$$
$$x_0 = vap + ubq \bmod n \qquad = 159201.$$

Finally, A uses x_0 to construct the x_i and p_i just as B did for encryption, and recovers the plaintext m_i by XORing the p_i with the ciphertext blocks c_i.   □

8.58 Note (*security of Blum-Goldwasser probabilistic encryption*)

  (i) Observe first that n is a Blum integer (Definition 2.156). An eavesdropper sees the quadratic residue x_{t+1}. Assuming that factoring n is difficult, the h least significant bits of the principal square root x_t of x_{t+1} modulo n are simultaneously secure (see Definition 3.82 and Fact 3.89). Thus the eavesdropper can do no better than to guess the pseudorandom bits p_t. More formally, if the integer factorization problem is hard, then the Blum-Goldwasser probabilistic encryption scheme is semantically secure. Note, however, that for a modulus n of a fixed bitlength (e.g., 1024 bits), this statement is no longer true, and the scheme should only be considered computationally secure.

  (ii) As of 1996, the modulus n should be at least 1024 bits in length if long-term security is desired (cf. Note 8.7). If n is a 1025-bit integer, then $k = 1024$ and $h = 10$.

  (iii) As with the Rabin encryption scheme (Algorithm 8.11), the Blum-Goldwasser scheme is also vulnerable to a chosen-ciphertext attack that recovers the private key from the public key. It is for this reason that the Blum-Goldwasser scheme has not received much attention in practice.

8.59 Note (*efficiency of Blum-Goldwasser probabilistic encryption*)

  (i) Unlike Goldwasser-Micali encryption, the ciphertext in Blum-Goldwasser encryption is only longer than the plaintext by a constant number of bits, namely $k + 1$ (the size in bits of the integer x_{t+1}).

(ii) The encryption process is quite efficient — it takes only 1 modular multiplication to encrypt h bits of plaintext. By comparison, the RSA encryption process (Algorithm 8.3) requires 1 modular exponentiation ($m^e \bmod n$) to encrypt k bits of plaintext. Assuming that the parameter e is randomly chosen and assuming that a modular exponentiation takes $3k/2$ modular multiplications, this translates to an encryption rate for RSA of $2/3$ bits per modular multiplication. If one chooses a special value for e, such as $e = 3$ (see Note 8.9), then RSA encryption is faster than Blum-Goldwasser encryption.

(iii) Blum-Goldwasser decryption (step 2 of Algorithm 8.56) is also quite efficient, requiring 1 exponentiation modulo $p-1$ (step 2a), 1 exponentiation modulo $q-1$ (step 2b), 1 exponentiation modulo p (step 2c), 1 exponentiation modulo q (step 2d), and t multiplications modulo n (step 2f) to decrypt ht ciphertext bits. (The time to perform step 2e is negligible.) By comparison, RSA decryption (step 2 of Algorithm 8.3) requires 1 exponentiation modulo n (which can be accomplished by doing 1 exponentiation modulo p and 1 exponentiation modulo q) to decrypt k ciphertext bits. Thus, for short messages ($< k$ bits), Blum-Goldwasser decryption is slightly slower than RSA decryption, while for longer messages, Blum-Goldwasser is faster.

8.7.3 Plaintext-aware encryption

While semantic security (Definition 8.47) is a strong security requirement for public-key encryption schemes, there are other measures of security.

8.60 Definition A public-key encryption scheme is said to be *non-malleable* if given a ciphertext, it is computationally infeasible to generate a different ciphertext such that the respective plaintexts are related in a known manner.

8.61 Fact If a public-key encryption scheme is non-malleable, it is also semantically secure.

An even stronger notion of security if that of being plaintext-aware. In Definition 8.62, *valid ciphertext* means those ciphertext which are the encryptions of legitimate plaintext messages (e.g. messages containing pre-specified forms of redundancy).

8.62 Definition A public-key encryption scheme is said to be *plaintext-aware* if it is computationally infeasible for an adversary to produce a valid ciphertext without knowledge of the corresponding plaintext.

The property of being plaintext-aware is indeed a strong one — it implies that the encryption scheme is non-malleable (and, hence, semantically secure) and also secure against adaptive chosen-ciphertext attacks. Note 8.63 gives one method of transforming any k-bit to k-bit trapdoor one-way permutation (such as RSA) into an encryption scheme that is plaintext-aware.

8.63 Note (*Bellare-Rogaway plaintext-aware encryption*) Let f be a k-bit to k-bit trapdoor one-way permutation (such as RSA). Let k_0 and k_1 be parameters such that 2^{k_0} and 2^{k_1} steps each represent infeasible amounts of work (e.g., $k_0 = k_1 = 128$). The length of the plaintext m is fixed to be $n = k - k_0 - k_1$ (e.g., for $k = 1024$, $n = 768$). Let $G : \{0,1\}^{k_0} \longrightarrow \{0,1\}^{n+k_1}$ and $H : \{0,1\}^{n+k_1} \longrightarrow \{0,1\}^{k_0}$ be random functions. Then the encryption function, as depicted in Figure 8.1, is

$$E(m) = f(\{m0^{k_1} \oplus G(r)\} \| \{r \oplus H(m0^{k_1} \oplus G(r))\}),$$

where $m0^{k_1}$ denotes m concatenated with a string of 0's of bitlength k_1, r is a random binary string of bitlength k_0, and $\|$ denotes concatenation.

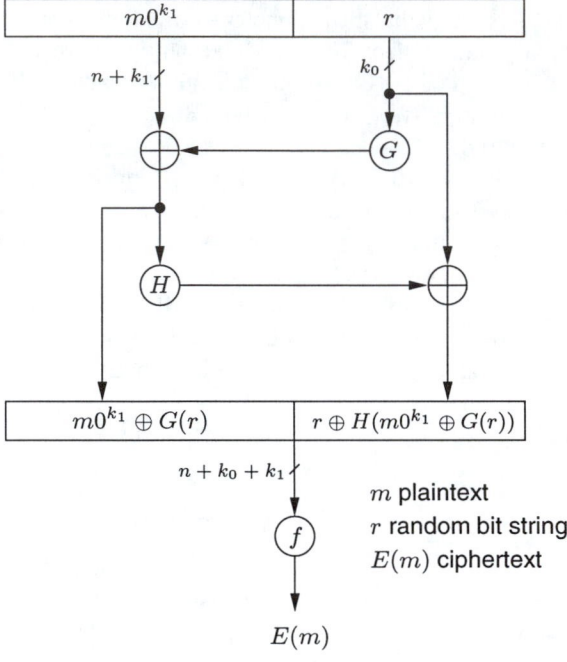

Figure 8.1: *Bellare-Rogaway plaintext-aware encryption scheme.*

Under the assumption that G and H are random functions, the encryption scheme E of Note 8.63 can be proven to be plaintext-aware. In practice, G and H can be derived from a standard cryptographic hash function such as the Secure Hash Algorithm (§9.4.2(iii)). In this case, the encryption scheme can no longer be proven to be plaintext-aware because the random function assumption is not true; however, such a scheme appears to provides greater security assurances than ones designed using ad hoc techniques.

8.8 Notes and further references

§8.1

For an introduction to public-key cryptography and public-key encryption in particular, see §1.8. A particularly readable introduction is the survey by Diffie [343]. Historical notes on public-key cryptography are given in the notes to §1.8 on page 47. A comparison of the features of public-key and symmetric-key encryption is given in §1.8.4; see also §13.2.5.

Other recent proposals for public-key encryption schemes include those based on finite automata (Renji [1032]); hidden field equations (Patarin [965]); and isomorphism of polynomials (Patarin [965]).

§8.2

The RSA cryptosystem was invented in 1977 by Rivest, Shamir, and Adleman [1060]. Kaliski and Robshaw [655] provide an overview of the major attacks on RSA encryption and signatures, and the practical methods of counteracting these threats.

The computational equivalence of computing the decryption exponent d and factoring n (§8.2.2(i)) was shown by DeLaurentis [320], based on earlier work by Miller [876].

The attack on RSA with small encryption exponent (§8.2.2(ii)) is discussed by Håstad [544], who showed more generally that sending the encryptions of more than $e(e+1)/2$ *linearly related messages* (messages of the form $(a_i m + b_i)$, where the a_i and b_i are known) enables an eavesdropper to recover the messages provided that the moduli n_i satisfy $n_i > 2^{(e+1)(e+2)/4}(e+1)^{(e+1)}$. Håstad also showed that sending three linearly related messages using the Rabin public-key encryption scheme (Algorithm 8.11) is insecure.

The attack on RSA with small decryption exponent d (§8.2.2(iv)) is due to Wiener [1240]. Wiener showed that his attack can be avoided if the encryption exponent e is chosen to be at least 50% longer than the modulus n. In this case, d should be at least 160 bits in length to avoid the square-root discrete logarithm algorithms such as Pollard's rho algorithm (Algorithm 3.60) and the parallelized variant of van Oorschot and Wiener [1207].

The adaptive chosen-ciphertext attack on RSA encryption (§8.2.2(v)) is due to Davida [302]. See also the related discussion in Denning [327]. Desmedt and Odlyzko [341] described an indifferent chosen-ciphertext attack in which the adversary has to obtain the plaintext corresponding to about $L_n[\frac{1}{2}, \frac{1}{2}]$ carefully chosen-ciphertext, subsequent to which it can decrypt all further ciphertext in $L_n[\frac{1}{2}, \frac{1}{2}]$ time without having to use the authorized user's decryption machine.

The common modulus attacks on RSA (§8.2.2(vi)) are due to DeLaurentis [320] and Simmons [1137].

The cycling attack (§8.2.2(vii)) was proposed by Simmons and Norris [1151]. Shortly after, Rivest [1052] showed that the cycling attack is extremely unlikely to succeed if the primes p and q are chosen so that: (i) $p - 1$ and $q - 1$ have large prime factors p' and q', respectively; and (ii) $p' - 1$ and $q' - 1$ have large prime factors p'' and q'', respectively. Maurer [818] showed that condition (ii) is unnecessary. Williams and Schmid [1249] proposed the generalized cycling attack and showed that this attack is really a factoring algorithm. Rivest [1051] provided heuristic evidence that if the primes p and q are selected at random, each having the same bitlength, then the expected time before the generalized cycling attack succeeds is at least $p^{1/3}$.

The note on message concealing (§8.2.2(viii)) is due to Blakley and Borosh [150], who also extended this work to all composite integers n and determined the number of *deranging* exponents for a fixed n, i.e., exponents e for which the number of unconcealed messages is the minimum possible. For further work see Smith and Palmer [1158].

Suppose that two or more plaintext messages which have a (known) polynomial relationship (e.g. m_1 and m_2 might be *linearly related*: $m_1 = a m_2 + b$) are encrypted with the same small encryption exponent (e.g. $e = 3$ or $e = 2^{16} + 1$). Coppersmith et al. [277] presented a new class of attacks on RSA which enable a passive adversary to recover such plaintext from the corresponding ciphertext. This attack is of practical significance because various cryptographic protocols have been proposed which require the encryption of polynomially related messages. Examples include the key distribution protocol of Tatebayashi, Matsuzaki, and Newman [1188], and the verifiable signature scheme of Franklin and Reiter [421]. Note that these attacks are different from those of §8.2.2(ii) and §8.2.2(vi) where the same plaintext is encrypted under different public keys.

Coppersmith [274] presented an efficient algorithm for finding a root of a polynomial of degree k over $\mathbb{Z}_n$, where n is an RSA-like modulus, provided that there there is a root smaller than $n^{1/k}$. The algorithm yielded the following two attacks on RSA with small encryption exponents. If $e = 3$ and if an adversary knows a ciphertext c and more than $2/3$ of the plaintext m corresponding to c, then the adversary can efficiently recover the rest of m. Suppose now that messages are padded with random bitstrings and encrypted with exponent $e = 3$. If an adversary knows two ciphertexts c_1 and c_2 which correspond to two encryptions of the same message m (with different padding), then the adversary can efficiently recovery m, provided that the padding is less than $1/9$ of the length of n. The latter attack suggests that caution must be exercised when using random padding in conjunction with a small encryption exponent.

Let $n = pq$ be a k-bit RSA modulus, where p and q are $k/2$-bit primes. Coppersmith [273] showed how n can be factored in polynomial time if the high order $k/4$ bits of p are known. This improves an algorithm of Rivest and Shamir [1058], which requires knowledge of the high order $k/3$ bits of p. For related theoretical work, see Maurer [814]. One implication of Coppersmith's result is that the method of Vanstone and Zuccherato [1214] for generating RSA moduli having a predetermined set of bits is insecure.

A trapdoor in the RSA cryptosystem was proposed by Anderson [26] whereby a hardware device generates the RSA modulus $n = pq$ in such a way that the hardware manufacturer can easily factor n, but factoring n remains difficult for all other parties. However, Kaliski [652] subsequently showed how to efficiently detect such trapdoors and, in some cases, to actually factor the modulus.

The arguments and recommendations about the use of strong primes in RSA key generation (Note 8.8) are taken from the detailed article by Rivest [1051].

Shamir [1117] proposed a variant of the RSA encryption scheme called *unbalanced RSA*, which makes it possible to enhance security by increasing the modulus size (e.g. from 500 bits to 5000 bits) without any deterioration in performance. In this variant, the public modulus n is the product of two primes p and q, where one prime (say p) is significantly larger in size than the other; plaintext messages m are in the interval $[0, p-1]$. For concreteness, consider the situation where p is a 500-bit prime, and q is a 4500-bit prime. Factoring such a 5000-bit modulus n is well beyond the reach of the special-purpose elliptic curve factoring algorithm of §3.2.4 (whose running time depends on the size of the smallest prime factor of n) and general-purpose factoring algorithms such as the number field sieve of §3.2.7. Shamir recommends that the encryption exponent e be in the range 20–100, which makes the encryption time with a 5000-bit modulus comparable to the decryption time with a 500-bit modulus. Decryption of the ciphertext $c \ (= m^d \bmod n)$ is accomplished by computing $m_1 = c^{d_1} \bmod p$, where $d_1 = d \bmod (p-1)$. Since $0 \le m < p$, m_1 is in fact equal to m. Decryption in unbalanced RSA thus only involves one exponentiation modulo a 500-bit prime, and takes the same time as decryption in ordinary RSA with a 500-bit modulus. This optimization does not apply to the RSA signature scheme (§11.3.1), since the verifier does not know the factor p of the public modulus n.

A *permutation polynomial* of $\mathbb{Z}_n$ is a polynomial $f(x) \in \mathbb{Z}_n[x]$ which induces a permutation of $\mathbb{Z}_n$ upon substitution of the elements of $\mathbb{Z}_n$; that is, $\{f(a) | a \in \mathbb{Z}_n\} = \mathbb{Z}_n$. In RSA encryption the permutation polynomial x^e of $\mathbb{Z}_n$ is used, where $\gcd(e, \phi) = 1$. Müller and Nöbauer [910] suggested replacing the polynomial x^e by the so-called *Dickson polynomials* to create a modified RSA encryption scheme called the *Dickson scheme*. The Dickson scheme was further studied by Müller and Nöbauer [909]. Other suitable classes of permutation polynomials were investigated by Lidl and Müller [763]. Smith and Lennon [1161] proposed an analogue of the RSA cryptosystem called LUC which is based on Lucas sequences. Due to the relationships between Dickson polynomials and the Lucas sequences, the LUC cryptosystem is closely related to the Dickson scheme. Bleichenbacher, Bosma, and Lenstra [154] presented a chosen-message attack on the LUC signature scheme, undermining the primary advantage claimed for LUC over RSA. Pinch [976, 977] extended the attacks on RSA with small encryption exponent (§8.2.2(ii)) and small decryption exponent (§8.2.2(iv)) to the LUC system.

An analogue of the RSA cryptosystem which uses special kinds of elliptic curves over $\mathbb{Z}_n$, where n is a composite integer, was proposed by Koyama et al. [708]. Demytko [321] presented an analogue where there is very little restriction on the types of elliptic curves that can be used. A new cryptosystem based on elliptic curves over $\mathbb{Z}_n$ in which the message is held in the exponent instead of the group element was proposed by Vanstone and Zuccherato [1213]. The security of all these schemes is based on the difficulty of factoring n. Kurosawa, Okada, and Tsujii [721] showed that the encryption schemes of Koyama et al. and Demytko are vulnerable to low exponent attacks (cf. §8.2.2(ii)); Pinch [977] demonstrated that the attack on RSA with small decryption exponent d (§8.2.2(iv)) also extends to these schemes. Kaliski [649] presented a chosen-ciphertext attack on the Demytko encryption scheme (and also a chosen-message attack on the corresponding signature scheme), and concluded that the present benefits of elliptic curve cryptosystems based on a composite modulus do not seem significant.

§8.3

The Rabin public-key encryption scheme (Algorithm 8.11) was proposed in 1979 by Rabin [1023]. In Rabin's paper, the encryption function was defined to be $E(m) = m(m + b) \bmod n$, where b and n comprise the public key. The security of this scheme is equivalent to the security of the scheme described in Algorithm 8.11 with encryption function $E(m) = m^2 \bmod n$. A related digital signature scheme is described in §11.3.4. Schwenk and Eisfeld [1104] consider public-key encryption and signature schemes whose security relies on the intractability of factoring polynomials over $\mathbb{Z}_n$.

Williams [1246] presented a public-key encryption scheme similar in spirit to Rabin's but using composite integers $n = pq$ with primes $p \equiv 3 \pmod 8$ and $q \equiv 7 \pmod 8$. Williams' scheme also has the property that breaking it (that is, recovering plaintext from some given ciphertext) is equivalent to factoring n, but has the advantage over Rabin's scheme that there is an easy procedure for identifying the intended message from the four roots of a quadratic polynomial. The restrictions on the forms of the primes p and q were removed later by Williams [1248]. A simpler and more efficient scheme also having the properties of provable security and unique decryption was presented by Kurosawa, Ito, and Takeuchi [720]. As with Rabin, all these schemes are vulnerable to a chosen-ciphertext attack (but see Note 8.14).

It is not the case that all public-key encryption schemes for which the decryption problem is provably as difficult as recovering the private key from the public key must succumb to a chosen-ciphertext attack. Goldwasser, Micali, and Rivest [484] were the first to observe this, and presented a digital signature scheme provably secure against an adaptive chosen-ciphertext attack (see §11.6.4). Naor and Yung [921] proposed the first concrete public-key encryption scheme that is semantically secure against *indifferent* chosen-ciphertext attack. The Naor-Yung scheme uses two independent keys of a probabilistic public-encryption scheme that is secure against a passive adversary (for example, the Goldwasser-Micali scheme of Algorithm 8.51) to encrypt the plaintext, and then both encryptions are sent along with a non-interactive zero-knowledge proof that the same message was encrypted with both keys. Following this work, Rackoff and Simon [1029] gave the first concrete construction for a public-key encryption scheme that is semantically secure against an *adaptive* chosen-ciphertext attack. Unfortunately, these schemes are all impractical because of the degree of message expansion.

Damgård [297] proposed simple and efficient methods for making public-key encryption schemes secure against indifferent chosen-ciphertext attacks. Zheng and Seberry [1269] noted that Damgård's schemes are insecure against an adaptive chosen-ciphertext attack, and proposed three practical schemes intended to resist such an attack. The Damgård and Zheng-Seberry schemes were not proven to achieve their claimed levels of security. Bellare and Rogaway [93] later proved that one of the Zheng-Seberry schemes is provably secure against adaptive chosen-ciphertext attacks for their *random oracle model*. Lim and Lee [766] proposed another method for making public-key schemes secure against adaptive chosen-ciphertext attacks; this scheme was broken by Frankel and Yung [419].

§8.4

The ElGamal cryptosystem was invented by ElGamal [368]. Haber and Lenstra (see Rueppel et al. [1083]) raised the possibility of a trapdoor in discrete logarithm cryptosystems whereby a modulus p is generated (e.g., by a hardware manufacturer) that is intentionally "weak"; cf. Note 4.58. Here, a "weak" prime p is one for which the discrete logarithm problem in $\mathbb{Z}_p^*$ is relatively easy. For example, $p - 1$ may contain only small prime factors, in which case the Pohlig-Hellman algorithm (§3.6.4) would be especially effective. Another example is a prime p for which the number field sieve for discrete logarithms (page 128) is especially well-suited. However, Gordon [509] subsequently showed how such trapdoors can be easily detected. Gordon also showed that the probability of a randomly chosen prime possessing such a trapdoor is negligibly small.

Rivest and Sherman [1061] gave an overview and unified framework for randomized encryption, including comments on chosen-plaintext and chosen-ciphertext attacks.

Elliptic curves were first proposed for use in public-key cryptography by Koblitz [695] and Miller [878]. Recent work on the security and implementation of elliptic curve systems is reported in Menezes [840]. Menezes, Okamoto, and Vanstone [843] showed that if the elliptic curve belongs to a special family called *supersingular curves*, then the discrete logarithm problem in the elliptic curve group can be reduced in expected polynomial time to the discrete logarithm problem in a small extension of the underlying finite field. Hence, if a supersingular elliptic curve is desired in practice, then it should be carefully chosen.

A modification of ElGamal encryption employing the group of units $\mathbb{Z}_n^*$, where n is a composite integer, was proposed by McCurley [825]; the scheme has the property that breaking it is *provably* at least as difficult as factoring the modulus n (see Fact 3.80). If a cryptanalyst somehow learns the factors of n, then in order to recover plaintext from ciphertext it is still left with the task of solving the Diffie-Hellman problem (§3.7) modulo the factors of n.

Hyperelliptic curve cryptosystems were proposed by Koblitz [696] but little research has since been done regarding their security and practicality.

The possibility of using the class group of an imaginary quadratic number field in public-key cryptography was suggested by Buchmann and Williams [218], however, the attractiveness of this choice was greatly diminished after the invention of a subexponential-time algorithm for computing discrete logarithms in these groups by McCurley [826].

Smith and Skinner [1162] proposed analogues of the Diffie-Hellman key exchange (called LUCDIF) and ElGamal encryption and digital signature schemes (called LUCELG) which use Lucas sequences modulo a prime p instead of modular exponentiation. Shortly thereafter, Laih, Tu, and Tai [733] and Bleichenbacher, Bosma, and Lenstra [154] showed that the analogue of the discrete logarithm problem for Lucas functions polytime reduces to the discrete logarithm problem in the multiplicative group of the finite field $\mathbb{F}_{p^2}$. Since there are subexponential-time algorithms known for the discrete logarithm problem in these fields (cf. §3.6), LUCDIF and LUCELG appear not to offer any advantages over the original schemes.

§8.5

The McEliece encryption scheme (Algorithm 8.30) was introduced in 1978 by McEliece [828]. For information on Goppa codes and their decoding algorithms, see MacWilliams and Sloane [778]. The problem of decoding an arbitrary linear code was shown to be **NP**-hard by Berlekamp, McEliece, and van Tilborg [120]. The security of the McEliece scheme has been studied by Adams and Meijer [6], Lee and Brickell [742], van Tilburg [1212], Gibson [451], and by Chabaud [235]. Gibson showed that there are, in fact, many trapdoors to a given McEliece encryption transformation, any of which may be used for decryption; this is contrary to the results of Adams and Meijer. However, Gibson notes that there are probably sufficiently few trapdoors that finding one by brute force is computationally infeasible. The cryptanalytic attack reported by Korzhik and Turkin [707] has not been published in its entirety, and is not believed to be an effective attack.

The strength of the McEliece encryption scheme can be severely weakened if the Goppa code is replaced with another type of error-correcting code. For example, Gabidulin, Paramonov, and Tretjakov [435] proposed a modification which uses maximum-rank-distance (MRD) codes in place of Goppa codes. This scheme, and a modification of it by Gabidulin [434], were subsequently shown to be insecure by Gibson [452, 453].

§8.6

The basic Merkle-Hellman knapsack scheme (Algorithm 8.37) was introduced by Merkle and Hellman [857]. An elementary overview of knapsack systems is given by Odlyzko [941].

The first polynomial-time attack on the basic Merkle-Hellman scheme (cf. Note 8.40(i)) was devised by Shamir [1114] in 1982. The attack makes use of H. Lenstra's algorithm for integer programming which runs in polynomial time when the number of variables is fixed, but is inefficient in practice. Lagarias [723] improved the practicality of the attack by reducing the main portion of the procedure to a problem of finding an unusually good simultaneous diophantine approximation; the latter can be solved by the more efficient L^3-lattice basis reduction algorithm (§3.10.1). The first attack on the multiple-iterated Merkle-Hellman scheme was by Brickell [200]. For surveys of the cryptanalysis of knapsack schemes, see Brickell [201] and Brickell and Odlyzko [209]. Orton [960] proposed a modification to the multiple-iterated Merkle-Hellman scheme that permits a knapsack density approaching 1, thus avoiding currently known attacks. The high density also allows for a fast digital signature scheme.

Shamir [1109] proposed a fast signature scheme based on the knapsack problem, later broken by Odlyzko [939] using the L^3-lattice basis reduction algorithm.

The Merkle-Hellman knapsack scheme illustrates the limitations of using an **NP**-complete problem to design a secure public-key encryption scheme. Firstly, Brassard [190] showed that under reasonable assumptions, the problem faced by the cryptanalyst cannot be **NP**-hard unless **NP=co-NP**, which would be a very surprising result in computational complexity theory. Secondly, complexity theory is concerned primarily with *asymptotic* complexity of a problem. By contrast, in practice one works with a problem instance of a *fixed* size. Thirdly, **NP**-completeness is a measure of the *worst-case* complexity of a problem. By contrast, cryptographic security should depend on the *average-case* complexity of the problem (or even better, the problem should be intractable for *essentially all* instances), since the cryptanalyst's task should be hard for virtually all instances and not merely in the worst case. There are many **NP**-complete problems that are known to have polynomial-time average-case algorithms, for example, the graph coloring problem; see Wilf [1243]. Another interesting example is provided by Even and Yacobi [379] who describe a symmetric-key encryption scheme based on the subset sum problem for which breaking the scheme (under a chosen-plaintext attack) is an **NP**-hard problem, yet an algorithm exists which solves most instances in polynomial time.

The Chor-Rivest knapsack scheme (Algorithm 8.42) was proposed by Chor and Rivest [261]. Recently, Schnorr and Hörner [1100] introduced new algorithms for lattice basis reduction that are improvements on the L^3-lattice basis reduction algorithm (Algorithm 3.101), and used these to break the Chor-Rivest scheme with parameters $\{p = 103, h = 12\}$. Since the density of such knapsack sets is 1.271, the attack demonstrated that subset sum problems with density greater than 1 can be solved via lattice basis reduction. Schnorr and Hörner also reported some success solving Chor-Rivest subset sum problems with parameters $\{p = 151, h = 16\}$. It remains to be seen whether the techniques of Schnorr and Hörner can be successfully applied to the recommended parameter case $\{p = 197, h = 24\}$.

Depending on the choice of parameters, the computation of discrete logarithms in the Chor-Rivest key generation stage (step 4 of Algorithm 8.41) may be a formidable task. A modified version of the scheme which does not require the computation of discrete logarithms in a field was proposed by H. Lenstra [758]. This modified scheme is called the *powerline system* and is not a knapsack system. It was proven to be at least as secure as the original Chor-Rivest scheme, and is comparable in terms of encryption and decryption speeds.

Qu and Vanstone [1013] showed how the Merkle-Hellman knapsack schemes can be viewed as special cases of certain knapsack-like encryption schemes arising from subset factorizations of finite groups. They also proposed an efficient public-key encryption scheme based on subset factorizations of the additive group $\mathbb{Z}_n$ of integers modulo n. Blackburn, Murphy, and Stern [143] showed that a simplified variant which uses subset factorizations of the n-dimensional vector space $\mathbb{Z}_2^n$ over $\mathbb{Z}_2$ is insecure.

§8.7

The notion of probabilistic public-key encryption was conceived by Goldwasser and Micali [479], who also introduced the notions of polynomial and semantic security. The equivalence of these two notions (Fact 8.49) was proven by Goldwasser and Micali [479] and Micali, Rackoff, and Sloan [865]. Polynomial security was also studied by Yao [1258], who referred to it as *polynomial-time indistinguishability*.

The Goldwasser-Micali scheme (Algorithm 8.51) can be described in a general setting by using the notion of a trapdoor predicate. Briefly, a *trapdoor predicate* is a Boolean function $B : \{0,1\}^* \longrightarrow \{0,1\}$ such that given a bit v it is easy to choose an x at random satisfying $B(x) = v$. Moreover, given a bitstring x, computing $B(x)$ correctly with probability significantly greater than $\frac{1}{2}$ is difficult; however, if certain trapdoor information is known, then it is easy to compute $B(x)$. If entity A's public key is a trapdoor predicate B, then any other entity encrypts a message bit m_i by randomly selecting an x_i such that $B(x_i) = m_i$, and then sends x_i to A. Since A knows the trapdoor information, she can compute $B(x_i)$ to recover m_i, but an adversary can do no better than guess the value of m_i. Goldwasser and Micali [479] proved that if trapdoor predicates exist, then this probabilistic encryption scheme is polynomially secure. Goldreich and Levin [471] simplified the work of Yao [1258], and showed how any trapdoor length-preserving permutation f can be used to obtain a trapdoor predicate, which in turn can be used to construct a probabilistic public-key encryption scheme.

The Blum-Goldwasser scheme (Algorithm 8.56) was proposed by Blum and Goldwasser [164]. The version given here follows the presentation of Brassard [192]. Two probabilistic public-key encryption schemes, one whose breaking is equivalent to solving the RSA problem (§3.3), and the other whose breaking is equivalent to factoring integers, were proposed by Alexi et al. [23]. The scheme based on RSA is as follows. Let $h = \lfloor \lg \lg n \rfloor$, where (n, e) is entity A's RSA public key. To encrypt an h-bit message m for A, choose a random $y \in \mathbb{Z}_n^*$ such that the h least significant bits of y equal m, and compute the ciphertext $c = y^e \bmod n$. A can recover m by computing $y = c^d \bmod n$, and extracting the h least significant bits of y. While both the schemes proposed by Alexi et al. are more efficient than the Goldwasser-Micali scheme, they suffer from large message expansion and are consequently not as efficient as the Blum-Goldwasser scheme.

The idea of non-malleable cryptography (Definition 8.60) was introduced by Dolev, Dwork, and Naor [357]. The paper gives the example of two contract bidders who encrypt their bids. It should not be possible for one bidder A to see the encrypted bid of the other bidder B and somehow be able to offer a bid that was slightly lower, even if A would not know what the resulting bid actually was. Bellare and Rogaway [95] introduced the notion of plaintext-aware encryption (Definition 8.62). They presented the scheme described in Note 8.63, improving upon earlier work of Johnson et al. [639]. Rigorous definitions and security proofs were provided, as well as a concrete instantiation of the plaintext-aware encryption scheme using RSA as the trapdoor permutation, and constructing the random functions G and H from the SHA-1 hash function (§9.4.2(iii)). Johnson and Matyas [640] presented some enhancements to the plaintext-aware encryption scheme. Bellare and Rogaway [93] presented various techniques for deriving appropriate random functions from standard cryptographic hash functions.

Chapter 9

Hash Functions and Data Integrity

Contents in Brief

9.1 Introduction

Cryptographic hash functions play a fundamental role in modern cryptography. While related to conventional hash functions commonly used in non-cryptographic computer applications – in both cases, larger domains are mapped to smaller ranges – they differ in several important aspects. Our focus is restricted to cryptographic hash functions (hereafter, simply hash functions), and in particular to their use for data integrity and message authentication.

Hash functions take a message as input and produce an output referred to as a *hash-code*, *hash-result*, *hash-value*, or simply *hash*. More precisely, a hash function h maps bit-strings of arbitrary finite length to strings of fixed length, say n bits. For a domain D and range R with $h : D \rightarrow R$ and $|D| > |R|$, the function is many-to-one, implying that the existence of *collisions* (pairs of inputs with identical output) is unavoidable. Indeed, restricting h to a domain of t-bit inputs ($t > n$), if h were "random" in the sense that all outputs were essentially equiprobable, then about 2^{t-n} inputs would map to each output, and two randomly chosen inputs would yield the same output with probability 2^{-n} (independent of t). The basic idea of cryptographic hash functions is that a hash-value serves as a compact representative image (sometimes called an *imprint*, *digital fingerprint*, or *message digest*) of an input string, and can be used as if it were uniquely identifiable with that string.

Hash functions are used for data integrity in conjunction with digital signature schemes, where for several reasons a message is typically hashed first, and then the hash-value, as a representative of the message, is signed in place of the original message (see Chapter 11). A distinct class of hash functions, called message authentication codes (MACs), allows message authentication by symmetric techniques. MAC algorithms may be viewed as hash functions which take two functionally distinct inputs, a message and a secret key, and produce a fixed-size (say n-bit) output, with the design intent that it be infeasible in practice to produce the same output without knowledge of the key. MACs can be used to provide data integrity and symmetric data origin authentication, as well as identification in symmetric-key schemes (see Chapter 10).

A typical usage of (unkeyed) hash functions for data integrity is as follows. The hash-value corresponding to a particular message x is computed at time T_1. The integrity of this hash-value (but not the message itself) is protected in some manner. At a subsequent time T_2, the following test is carried out to determine whether the message has been altered, i.e., whether a message x' is the same as the original message. The hash-value of x' is computed and compared to the protected hash-value; if they are equal, one accepts that the inputs are also equal, and thus that the message has not been altered. The problem of preserving the integrity of a potentially large message is thus reduced to that of a small fixed-size hash-value. Since the existence of collisions is guaranteed in many-to-one mappings, the unique association between inputs and hash-values can, at best, be in the computational sense. A hash-value should be uniquely identifiable with a single input *in practice*, and collisions should be *computationally* difficult to find (essentially never occurring in practice).

Chapter outline

The remainder of this chapter is organized as follows. §9.2 provides a framework including standard definitions, a discussion of the desirable properties of hash functions and MACs, and consideration of one-way functions. §9.3 presents a general model for iterated hash functions, some general construction techniques, and a discussion of security objectives and basic attacks (i.e., strategies an adversary may pursue to defeat the objectives of a hash function). §9.4 considers hash functions based on block ciphers, and a family of functions based on the MD4 algorithm. §9.5 considers MACs, including those based on block ciphers and customized MACs. §9.6 examines various methods of using hash functions to provide data integrity. §9.7 presents advanced attack methods. §9.8 provides chapter notes with references.

9.2 Classification and framework

9.2.1 General classification

At the highest level, hash functions may be split into two classes: *unkeyed hash functions*, whose specification dictates a single input parameter (a message); and *keyed hash functions*, whose specification dictates two distinct inputs, a message and a secret key. To facilitate discussion, a hash function is informally defined as follows.

9.1 Definition A *hash function* (in the unrestricted sense) is a function h which has, as a minimum, the following two properties:

1. *compression* — h maps an input x of arbitrary finite bitlength, to an output $h(x)$ of fixed bitlength n.
2. *ease of computation* — given h and an input x, $h(x)$ is easy to compute.

As defined here, *hash function* implies an unkeyed hash function. On occasion when discussion is at a generic level, this term is abused somewhat to mean both unkeyed and keyed hash functions; hopefully ambiguity is limited by context.

For actual use, a more goal-oriented classification of hash functions (beyond *keyed* vs. *unkeyed*) is necessary, based on further properties they provide and reflecting requirements of specific applications. Of the numerous categories in such a *functional classification*, two types of hash functions are considered in detail in this chapter:

1. *manipulation detection codes* (MDCs)
   Also known as *modification detection codes*, and less commonly as *message integrity codes* (MICs), the purpose of an MDC is (informally) to provide a representative image or *hash* of a message, satisfying additional properties as refined below. The end goal is to facilitate, in conjunction with additional mechanisms (see §9.6.4), data integrity assurances as required by specific applications. MDCs are a subclass of *unkeyed* hash functions, and themselves may be further classified; the specific classes of MDCs of primary focus in this chapter are (cf. Definitions 9.3 and 9.4):
   (i) *one-way hash functions* (OWHFs): for these, finding an input which hashes to a pre-specified hash-value is difficult;
   (ii) *collision resistant hash functions* (CRHFs): for these, finding any two inputs having the same hash-value is difficult.

2. *message authentication codes* (MACs)
   The purpose of a MAC is (informally) to facilitate, without the use of any additional mechanisms, assurances regarding both the source of a message and its integrity (see §9.6.3). MACs have two functionally distinct parameters, a message input and a secret key; they are a subclass of *keyed* hash functions (cf. Definition 9.7).

Figure 9.1 illustrates this simplified classification. Additional applications of unkeyed hash functions are noted in §9.2.6. Additional applications of keyed hash functions include use in challenge-response identification protocols for computing responses which are a function of both a secret key and a challenge message; and for key confirmation (Definition 12.7). Distinction should be made between a MAC algorithm, and the use of an MDC with a secret key included as part of its message input (see §9.5.2).

It is generally assumed that the algorithmic specification of a hash function is public knowledge. Thus in the case of MDCs, given a message as input, anyone may compute the hash-result; and in the case of MACs, given a message as input, anyone with knowledge of the key may compute the hash-result.

9.2.2 Basic properties and definitions

To facilitate further definitions, three potential properties are listed (in addition to *ease of computation* and *compression* as per Definition 9.1), for an unkeyed hash function h with inputs x, x' and outputs y, y'.

1. *preimage resistance* — for essentially all pre-specified outputs, it is computationally infeasible to find any input which hashes to that output, i.e., to find any preimage x' such that $h(x') = y$ when given any y for which a corresponding input is not known.[1]

2. *2nd-preimage resistance* — it is computationally infeasible to find any second input which has the same output as any specified input, i.e., given x, to find a 2nd-preimage $x' \neq x$ such that $h(x) = h(x')$.

---

[1] This acknowledges that an adversary may easily precompute outputs for any small set of inputs, and thereby invert the hash function trivially for such outputs (cf. Remark 9.35).

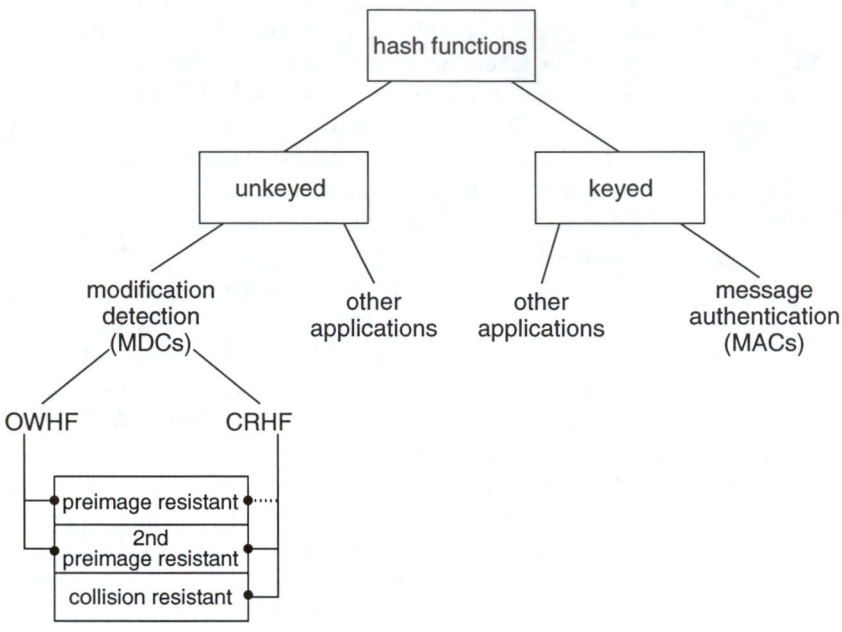

Figure 9.1: *Simplified classification of cryptographic hash functions and applications.*

3. *collision resistance* — it is computationally infeasible to find any two distinct inputs x, x' which hash to the same output, i.e., such that $h(x) = h(x')$. (Note that here there is free choice of both inputs.)

Here and elsewhere, the terms "easy" and "computationally infeasible" (or "hard") are intentionally left without formal definition; it is intended they be interpreted relative to an understood frame of reference. "Easy" might mean polynomial time and space; or more practically, within a certain number of machine operations or time units – perhaps seconds or milliseconds. A more specific definition of "computationally infeasible" might involve super-polynomial effort; require effort far exceeding understood resources; specify a lower bound on the number of operations or memory required in terms of a specified security parameter; or specify the probability that a property is violated be exponentially small. The properties as defined above, however, suffice to allow practical definitions such as Definitions 9.3 and 9.4 below.

9.2 Note (*alternate terminology*) Alternate terms used in the literature are as follows: preimage resistant ≡ *one-way* (cf. Definition 9.9); 2nd-preimage resistance ≡ *weak collision resistance*; collision resistance ≡ *strong collision resistance*.

For context, one motivation for each of the three major properties above is now given. Consider a digital signature scheme wherein the signature is applied to the hash-value $h(x)$ rather than the message x. Here h should be an MDC with 2nd-preimage resistance, otherwise, an adversary C may observe the signature of some party A on $h(x)$, then find an x' such that $h(x) = h(x')$, and claim that A has signed x'. If C is able to actually choose the message which A signs, then C need only find a collision pair (x, x') rather than the harder task of finding a second preimage of x; in this case, collision resistance is also necessary (cf. Remark 9.93). Less obvious is the requirement of preimage resistance for some public-key signature schemes; consider RSA (Chapter 11), where party A has public key

(e, n). C may choose a random value y, compute $z = y^e \bmod n$, and (depending on the particular RSA signature verification process used) claim that y is A's signature on z. This (existential) forgery may be of concern if C can find a preimage x such that $h(x) = z$, and for which x is of practical use.

9.3 Definition A *one-way hash function* (OWHF) is a hash function h as per Definition 9.1 (i.e., offering ease of computation and compression) with the following additional properties, as defined above: preimage resistance, 2nd-preimage resistance.

9.4 Definition A *collision resistant hash function* (CRHF) is a hash function h as per Definition 9.1 (i.e., offering ease of computation and compression) with the following additional properties, as defined above: 2nd-preimage resistance, collision resistance (cf. Fact 9.18).

Although in practice a CRHF almost always has the additional property of preimage resistance, for technical reasons (cf. Note 9.20) this property is not mandated in Definition 9.4.

9.5 Note (*alternate terminology for OWHF, CRHF*) Alternate terms used in the literature are as follows: OWHF $\equiv$ *weak one-way hash function* (but here preimage resistance is often not explicitly considered); CRHF $\equiv$ *strong one-way hash function*.

9.6 Example (*hash function properties*)

  (i) A simple modulo-32 checksum (32-bit sum of all 32-bit words of a data string) is an easily computed function which offers compression, but is not preimage resistant.
 (ii) The function $g(x)$ of Example 9.11 is preimage resistant but provides neither compression nor 2nd-preimage resistance.
(iii) Example 9.13 presents a function with preimage resistance and 2nd-preimage resistance (but not compression). □

9.7 Definition A *message authentication code (MAC)* algorithm is a family of functions h_k parameterized by a secret key k, with the following properties:

  1. *ease of computation* — for a known function h_k, given a value k and an input x, $h_k(x)$ is easy to compute. This result is called the *MAC-value* or *MAC*.
  2. *compression* — h_k maps an input x of arbitrary finite bitlength to an output $h_k(x)$ of fixed bitlength n.
     Furthermore, given a description of the function family h, for every fixed allowable value of k (unknown to an adversary), the following property holds:
  3. *computation-resistance* — given zero or more text-MAC pairs $(x_i, h_k(x_i))$, it is computationally infeasible to compute any text-MAC pair $(x, h_k(x))$ for any new input $x \neq x_i$ (including possibly for $h_k(x) = h_k(x_i)$ for some i).

If computation-resistance does not hold, a MAC algorithm is subject to *MAC forgery*. While computation-resistance implies the property of *key non-recovery* (it must be computationally infeasible to recover k, given one or more text-MAC pairs $(x_i, h_k(x_i))$ for that k), key non-recovery does not imply computation-resistance (a key need not always actually be recovered to forge new MACs).

9.8 Remark (*MAC resistance when key known*) Definition 9.7 does not dictate whether MACs need be preimage- and collision resistant for parties knowing the key k (as Fact 9.21 implies for parties without k).

(i) Objectives of adversaries vs. MDCs

The objective of an adversary who wishes to "attack" an MDC is as follows:

(a) to attack a OWHF: given a hash-value y, find a preimage x such that $y = h(x)$; or given one such pair $(x, h(x))$, find a second preimage x' such that $h(x') = h(x)$.

(b) to attack a CRHF: find any two inputs x, x', such that $h(x') = h(x)$.

A CRHF must be designed to withstand standard birthday attacks (see Fact 9.33).

(ii) Objectives of adversaries vs. MACs

The corresponding objective of an adversary for a MAC algorithm is as follows:

(c) to attack a MAC: without prior knowledge of a key k, compute a new text-MAC pair $(x, h_k(x))$ for some text $x \neq x_i$, given one or more pairs $(x_i, h_k(x_i))$.

Computation-resistance here should hold whether the texts x_i for which matching MACs are available are given to the adversary, or may be freely chosen by the adversary. Similar to the situation for signature schemes, the following attack scenarios thus exist for MACs, for adversaries with increasing advantages:

1. *known-text attack.* One or more text-MAC pairs $(x_i, h_k(x_i))$ are available.

2. *chosen-text attack.* One or more text-MAC pairs $(x_i, h_k(x_i))$ are available for x_i chosen by the adversary.

3. *adaptive chosen-text attack.* The x_i may be chosen by the adversary as above, now allowing successive choices to be based on the results of prior queries.

As a certificational checkpoint, MACs should withstand adaptive chosen-text attack regardless of whether such an attack may actually be mounted in a particular environment. Some practical applications may limit the number of interactions allowed over a fixed period of time, or may be designed so as to compute MACs only for inputs created within the application itself; others may allow access to an unlimited number of text-MAC pairs, or allow MAC verification of an unlimited number of messages and accept any with a correct MAC for further processing.

(iii) Types of forgery (selective, existential)

When MAC forgery is possible (implying the MAC algorithm has been technically defeated), the severity of the practical consequences may differ depending on the degree of control an adversary has over the value x for which a MAC may be forged. This degree is differentiated by the following classification of forgeries:

1. *selective forgery* – attacks whereby an adversary is able to produce a new text-MAC pair for a text of his choice (or perhaps partially under his control). Note that here the selected value is the text for which a MAC is forged, whereas in a chosen-text attack the chosen value is the text of a text-MAC pair used for analytical purposes (e.g., to forge a MAC on a distinct text).

2. *existential forgery* – attacks whereby an adversary is able to produce a new text-MAC pair, but with no control over the value of that text.

Key recovery of the MAC key itself is the most damaging attack, and trivially allows selective forgery. MAC forgery allows an adversary to have a forged text accepted as authentic. The consequences may be severe even in the existential case. A classic example is the replacement of a monetary amount known to be small by a number randomly distributed between 0 and $2^{32} - 1$. For this reason, messages whose integrity or authenticity is to be verified are often constrained to have pre-determined structure or a high degree of verifiable redundancy, in an attempt to preclude meaningful attacks.

Analogously to MACs, attacks on MDC schemes (primarily 2nd-preimage and collision attacks) may be classified as selective or existential. If the message can be partially controlled, then the attack may be classified as partially selective (e.g., see §9.7.1(iii)).

9.2.3 Hash properties required for specific applications

Because there may be costs associated with specific properties – e.g., CRHFs are in general harder to construct than OWHFs and have hash-values roughly twice the bitlength – it should be understood which properties are actually required for particular applications, and why. Selected techniques whereby hash functions are used for data integrity, and the corresponding properties required thereof by these applications, are summarized in Table 9.1.

In general, an MDC should be a CRHF if an untrusted party has control over the exact content of hash function inputs (see Remark 9.93); a OWHF suffices otherwise, including the case where there is only a single party involved (e.g., a store-and-retrieve application). Control over precise format of inputs may be eliminated by introducing into the message randomization that is uncontrollable by one or both parties. Note, however, that data integrity techniques based on a shared secret key typically involve mutual trust and do not address non-repudiation; in this case, collision resistance may or may not be a requirement.

Hash properties required → Integrity application ↓	Preimage resistant	2nd-preimage	Collision resistant	Details
MDC + asymmetric signature	yes	yes	yes†	page 324
MDC + authentic channel		yes	yes†	page 364
MDC + symmetric encryption				page 365
hash for one-way password file	yes			page 389
MAC (key unknown to attacker)	yes	yes	yes†	page 326
MAC (key known to attacker)		yes‡		page 325

Table 9.1: *Resistance properties required for specified data integrity applications.*
†Resistance required if attacker is able to mount a chosen message attack.
‡Resistance required in rare case of multi-cast authentication (see page 378).

9.2.4 One-way functions and compression functions

Related to Definition 9.3 of a OWHF is the following, which is unrestrictive with respect to a compression property.

9.9 Definition A *one-way function* (OWF) is a function f such that for each x in the domain of f, it is easy to compute $f(x)$; but for essentially all y in the range of f, it is computationally infeasible to find any x such that $y = f(x)$.

9.10 Remark (*OWF vs. domain-restricted OWHF*) A OWF as defined here differs from a OWHF with domain restricted to fixed-size inputs in that Definition 9.9 does not require 2nd-preimage resistance. Many one-way functions are, in fact, non-compressing, in which case most image elements have unique preimages, and for these 2nd-preimage resistance holds vacuously – making the difference minor (but see Example 9.11).

9.11 Example (*one-way functions and modular squaring*) The squaring of integers modulo a prime p, e.g., $f(x) = x^2 - 1 \bmod p$, behaves in many ways like a random mapping. However, $f(x)$ is not a OWF because finding square roots modulo primes is easy (§3.5.1). On the other hand, $g(x) = x^2 \bmod n$ is a OWF (Definition 9.9) for appropriate randomly chosen primes p and q where $n = pq$ and the factorization of n is unknown, as finding a preimage (i.e., computing a square root mod n) is computationally equivalent to factoring (Fact 3.46) and thus intractable. Nonetheless, finding a 2nd-preimage, and, therefore, collisions, is trivial (given x, $-x$ yields a collision), and thus g fits neither the definition of a OWHF nor a CRHF with domain restricted to fixed-size inputs.   □

9.12 Remark (*candidate one-way functions*) There are, in fact, no known instances of functions which are provably one-way (with no assumptions); indeed, despite known hash function constructions which are provably as secure as **NP**-complete problems, there is no assurance the latter are difficult. All instances of "one-way functions" to date should thus more properly be qualified as "conjectured" or "candidate" one-way functions. (It thus remains possible, although widely believed most unlikely, that one-way functions do not exist.) A proof of existence would establish $\mathbf{P} \neq \mathbf{NP}$, while non-existence would have devastating cryptographic consequences (see page 377), although not directly implying $\mathbf{P} = \mathbf{NP}$.

Hash functions are often used in applications (cf. §9.2.6) which require the one-way property, but not compression. It is, therefore, useful to distinguish three classes of functions (based on the relative size of inputs and outputs):

1. *(general) hash functions*. These are functions as per Definition 9.1, typically with additional one-way properties, which compress arbitrary-length inputs to n-bit outputs.
2. *compression functions* (fixed-size hash functions). These are functions as per Definition 9.1, typically with additional one-way properties, but with domain restricted to fixed-size inputs – i.e., compressing m-bit inputs to n-bit outputs, $m > n$.
3. *non-compressing one-way functions*. These are fixed-size hash functions as above, except that $n = m$. These include *one-way permutations*, and can be more explicitly described as computationally non-invertible functions.

9.13 Example (*DES-based OWF*) A one-way function can be constructed from DES or any block cipher E which behaves essentially as a random function (see Remark 9.14), as follows: $f(x) = E_k(x) \oplus x$, for any fixed known key k. The one-way nature of this construction can be proven under the assumption that E is a random permutation. An intuitive argument follows. For any choice of y, finding any x (and key k) such that $E_k(x) \oplus x = y$ is difficult because for any chosen x, $E_k(x)$ will be essentially random (for any key k) and thus so will $E_k(x) \oplus x$; hence, this will equal y with no better than random chance. By similar reasoning, if one attempts to use decryption and chooses an x, the probability that $E_k^{-1}(x \oplus y) = x$ is no better than random chance. Thus $f(x)$ appears to be a OWF. While $f(x)$ is not a OWHF (it handles only fixed-length inputs), it can be extended to yield one (see Algorithm 9.41).   □

9.14 Remark (*block ciphers and random functions*) Regarding random functions and their properties, see §2.1.6. If a block cipher behaved as a random function, then encryption and decryption would be equivalent to looking up values in a large table of random numbers; for a fixed input, the mapping from a key to an output would behave as a random mapping. However, block ciphers such as DES are bijections, and thus at best exhibit behavior more like random permutations than random functions.

9.15 Example (*one-wayness w.r.t. two inputs*) Consider $f(x, k) = E_k(x)$, where E represents DES. This is not a one-way function of the joint input (x, k), because given any function value $y = f(x, k)$, one can choose any key k' and compute $x' = E_{k'}^{-1}(y)$ yielding a preimage (x', k'). Similarly, $f(x, k)$ is not a one-way function of x if k is known, as given $y = f(x, k)$ and k, decryption of y using k yields x. (However, a "black-box" which computes $f(x, k)$ for fixed, externally-unknown k is a one-way function of x.) In contrast, $f(x, k)$ is a one-way function of k; given $y = f(x, k)$ and x, it is not known how to find a preimage k in less than about 2^{55} operations. (This latter concept is utilized in one-time digital signature schemes – see §11.6.2.) □

9.16 Example (*OWF - multiplication of large primes*) For appropriate choices of primes p and q, $f(p, q) = pq$ is a one-way function: given p and q, computing $n = pq$ is easy, but given n, finding p and q, i.e., *integer factorization*, is difficult. RSA and many other cryptographic systems rely on this property (see Chapter 3, Chapter 8). Note that contrary to many one-way functions, this function f does not have properties resembling a "random" function. □

9.17 Example (*OWF - exponentiation in finite fields*) For most choices of appropriately large primes p and any element $\alpha \in \mathbb{Z}_p^*$ of sufficiently large multiplicative order (e.g., a generator), $f(x) = \alpha^x \bmod p$ is a one-way function. (For example, p must not be such that all the prime divisors of $p - 1$ are small, otherwise the discrete log problem is feasible by the Pohlig-Hellman algorithm of §3.6.4.) $f(x)$ is easily computed given α, x, and p using the square-and-multiply technique (Algorithm 2.143), but for most choices p it is difficult, given (y, p, α), to find an x in the range $0 \leq x \leq p - 2$ such that $\alpha^x \bmod p = y$, due to the apparent intractability of the discrete logarithm problem (§3.6). Of course, for specific values of $f(x)$ the function can be inverted trivially. For example, the respective preimages of 1 and -1 are known to be 0 and $(p - 1)/2$, and by computing $f(x)$ for any small set of values for x (e.g., $x = 1, 2, \ldots, 10$), these are also known. However, for *essentially* all y in the range, the preimage of y is difficult to find. □

9.2.5 Relationships between properties

In this section several relationships between the hash function properties stated in the preceding section are examined.

9.18 Fact Collision resistance implies 2nd-preimage resistance of hash functions.

Justification. Suppose h has collision resistance. Fix an input x_j. If h does not have 2nd-preimage resistance, then it is feasible to find a distinct input x_i such that $h(x_i) = h(x_j)$, in which case (x_i, x_j) is a pair of distinct inputs hashing to the same output, contradicting collision resistance.

9.19 Remark (*one-way vs. preimage and 2nd-preimage resistant*) While the term "one-way" is generally taken to mean preimage resistant, in the hash function literature it is sometimes also used to imply that a function is 2nd-preimage resistant or computationally non-invertible. (*Computationally non-invertible* is a more explicit term for preimage resistance when preimages are unique, e.g., for one-way permutations. In the case that two or more preimages exist, a function fails to be computationally non-invertible if any one can be found.) This causes ambiguity as 2nd-preimage resistance does not guarantee preimage-resistance (Note 9.20), nor does preimage resistance guarantee 2nd-preimage resistance (Example 9.11); see also Remark 9.10. An attempt is thus made to avoid unqualified use of the term "one-way".

9.20 Note (*collision resistance does not guarantee preimage resistance*) Let g be a hash function which is collision resistant and maps arbitrary-length inputs to n-bit outputs. Consider the function h defined as (here and elsewhere, $\|$ denotes concatenation):

$$h(x) = \begin{cases} 1 & \| & x, & \text{if } x \text{ has bitlength } n \\ 0 & \| & g(x), & \text{otherwise.} \end{cases}$$

Then h is an $(n + 1)$-bit hash function which is collision resistant but not preimage resistant. As a simpler example, the identity function on fixed-length inputs is collision and 2nd-preimage resistant (preimages are unique) but not preimage resistant. While such pathological examples illustrate that collision resistance does not guarantee the difficulty of finding preimages of specific (or even most) hash outputs, for most CRHFs arising in practice it nonetheless appears reasonable to assume that collision resistance does indeed imply preimage resistance.

9.21 Fact (*implications of MAC properties*) Let h_k be a keyed hash function which is a MAC algorithm per Definition 9.7 (and thus has the property of computation-resistance). Then h_k is, against chosen-text attack by an adversary without knowledge of the key k, (i) both 2nd-preimage resistant and collision resistant; and (ii) preimage resistant (with respect to the hash-input).

Justification. For (i), note that computation-resistance implies hash-results should not even be computable by those without secret key k. For (ii), by way of contradiction, assume h were not preimage resistant. Then recovery of the preimage x for a randomly selected hash-output y violates computation-resistance.

9.2.6 Other hash function properties and applications

Most unkeyed hash functions commonly found in practice were originally designed for the purpose of providing data integrity (see §9.6), including digital fingerprinting of messages in conjunction with digital signatures (§9.6.4). The majority of these are, in fact, MDCs designed to have preimage, 2nd-preimage, or collision resistance properties. Because one-way functions are a fundamental cryptographic primitive, many of these MDCs, which typically exhibit behavior informally equated with one-wayness and randomness, have been proposed for use in various applications distinct from data integrity, including, as discussed below:

1. *confirmation of knowledge*
2. *key derivation*
3. *pseudorandom number generation*

Hash functions used for confirmation of knowledge facilitate commitment to data values, or demonstrate possession of data, without revealing such data itself (until possibly a later point in time); verification is possible by parties in possession of the data. This resembles the use of MACs where one also essentially demonstrates knowledge of a secret (but with the demonstration bound to a specific message). The property of hash functions required is preimage resistance (see also partial-preimage resistance below). Specific examples include use in password verification using unencrypted password-image files (Chapter 10); symmetric-key digital signatures (Chapter 11); key confirmation in authenticated key establishment protocols (Chapter 12); and document-dating or timestamping by hash-code registration (Chapter 13).

In general, use of hash functions for purposes other than which they were originally designed requires caution, as such applications may require additional properties (see below) these functions were not designed to provide; see Remark 9.22. Unkeyed hash functions having properties associated with one-way functions have nonetheless been proposed for a wide range of applications, including as noted above:

- *key derivation* – to compute sequences of new keys from prior keys (Chapter 13). A primary example is key derivation in point-of-sale (POS) terminals; here an important requirement is that the compromise of currently active keys must not compromise the security of previous transaction keys. A second example is in the generation of one-time password sequences based on one-way functions (Chapter 10).
- *pseudorandom number generation* – to generate sequences of numbers which have various properties of randomness. (A pseudorandom number generator can be used to construct a symmetric-key block cipher, among other things.) Due to the difficulty of producing cryptographically strong pseudorandom numbers (see Chapter 5), MDCs should not be used for this purpose unless the randomness requirements are clearly understood, and the MDC is verified to satisfy these.

For the applications immediately above, rather than hash functions, the cryptographic primitive which is needed may be a *pseudorandom function* (or keyed pseudorandom function).

9.22 Remark (*use of MDCs*) Many MDCs used in practice may appear to satisfy additional requirements beyond those for which they were originally designed. Nonetheless, the use of arbitrary hash functions cannot be recommended for any applications without careful analysis precisely identifying both the critical properties required by the application and those provided by the function in question (cf. §9.5.2).

Additional properties of one-way hash functions

Additional properties of one-way hash functions called for by the above-mentioned applications include the following.

1. *non-correlation*. Input bits and output bits should not be correlated. Related to this, an avalanche property similar to that of good block ciphers is desirable whereby every input bit affects every output bit. (This rules out hash functions for which preimage resistance fails to imply 2nd-preimage resistance simply due to the function effectively ignoring a subset of input bits.)
2. *near-collision resistance*. It should be hard to find any two inputs x, x' such that $h(x)$ and $h(x')$ differ in only a small number of bits.
3. *partial-preimage resistance* or *local one-wayness*. It should be as difficult to recover any substring as to recover the entire input. Moreover, even if part of the input is known, it should be difficult to find the remainder (e.g., if t input bits remain unknown, it should take on average 2^{t-1} hash operations to find these bits.)

Partial preimage resistance is an implicit requirement in some of the proposed applications of §9.5.2. One example where near-collision resistance is necessary is when only half of the output bits of a hash function are used.

Many of these properties can be summarized as requirements that there be neither local nor global statistical weaknesses; the hash function should not be weaker with respect to some parts of its input or output than others, and all bits should be equally hard. Some of these may be called *certificational properties* – properties which intuitively appear desirable, although they cannot be shown to be directly necessary.

9.3 Basic constructions and general results

9.3.1 General model for iterated hash functions

Most unkeyed hash functions h are designed as iterative processes which hash arbitrary-length inputs by processing successive fixed-size blocks of the input, as illustrated in Figure 9.2.

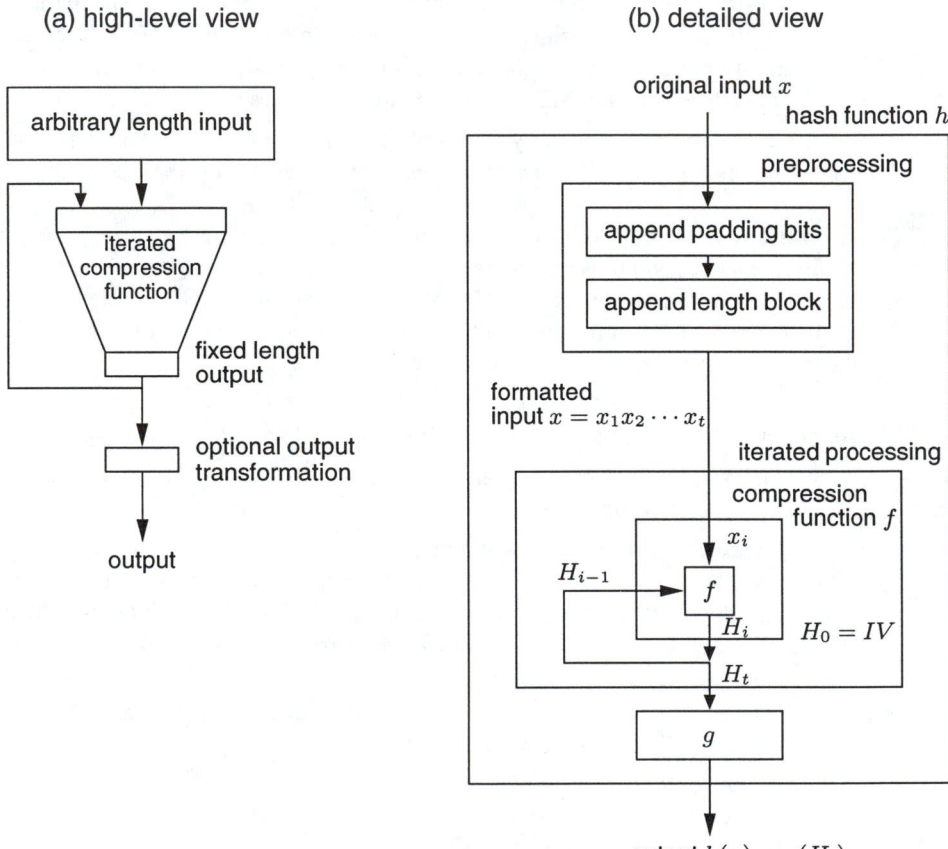

Figure 9.2: *General model for an iterated hash function.*

A hash input x of arbitrary finite length is divided into fixed-length r-bit blocks x_i. This preprocessing typically involves appending extra bits (*padding*) as necessary to attain an overall bitlength which is a multiple of the blocklength r, and often includes (for security reasons – e.g., see Algorithm 9.26) a block or partial block indicating the bitlength of the unpadded input. Each block x_i then serves as input to an internal fixed-size hash function f, the *compression function* of h, which computes a new intermediate result of bitlength n for some fixed n, as a function of the previous n-bit intermediate result and the next input block x_i. Letting H_i denote the partial result after stage i, the general process for an iterated hash function with input $x = x_1x_2 \ldots x_t$ can be modeled as follows:

$$H_0 = IV; \qquad H_i = f(H_{i-1}, x_i), \;\; 1 \leq i \leq t; \qquad h(x) = g(H_t). \qquad (9.1)$$

H_{i-1} serves as the n-bit *chaining variable* between stage $i - 1$ and stage i, and H_0 is a pre-defined starting value or *initializing value* (IV). An optional output transformation g (see Figure 9.2) is used in a final step to map the n-bit chaining variable to an m-bit result $g(H_t)$; g is often the identity mapping $g(H_t) = H_t$.

Particular hash functions are distinguished by the nature of the preprocessing, compression function, and output transformation.

9.3.2 General constructions and extensions

To begin, an example demonstrating an insecure construction is given. Several secure general constructions are then discussed.

9.23 Example (*insecure trivial extension of OWHF to CRHF*) In the case that an iterated OWHF h yielding n-bit hash-values is not collision resistant (e.g., when a $2^{n/2}$ birthday collision attack is feasible – see §9.7.1) one might propose constructing from h a CRHF using as output the concatenation of the last two n-bit chaining variables, so that a t-block message has hash-value $H_{t-1} \| H_t$ rather than H_t. This is insecure as the final message block x_t can be held fixed along with H_t, reducing the problem to finding a collision on H_{t-1} for h. □

Extending compression functions to hash functions

Fact 9.24 states an important relationship between collision resistant compression functions and collision resistant hash functions. Not only can the former be extended to the latter, but this can be done efficiently using Merkle's meta-method of Algorithm 9.25 (also called the Merkle-Damgård construction). This reduces the problem of finding such a hash function to that of finding such a compression function.

9.24 Fact (*extending compression functions*) Any compression function f which is collision resistant can be extended to a collision resistant hash function h (taking arbitrary length inputs).

9.25 Algorithm Merkle's meta-method for hashing

INPUT: compression function f which is collision resistant.
OUTPUT: unkeyed hash function h which is collision resistant.

1. Suppose f maps $(n + r)$-bit inputs to n-bit outputs (for concreteness, consider $n = 128$ and $r = 512$). Construct a hash function h from f, yielding n-bit hash-values, as follows.
2. Break an input x of bitlength b into blocks $x_1x_2 \ldots x_t$ each of bitlength r, padding out the last block x_t with 0-bits if necessary.
3. Define an extra final block x_{t+1}, the length-block, to hold the right-justified binary representation of b (presume that $b < 2^r$).
4. Letting 0^j represent the bitstring of j 0's, define the n-bit hash-value of x to be $h(x) = H_{t+1} = f(H_t \| x_{t+1})$ computed from:

$$H_0 = 0^n; \qquad H_i = f(H_{i-1} \| x_i), \;\; 1 \leq i \leq t + 1.$$

The proof that the resulting function h is collision resistant follows by a simple argument that a collision for h would imply a collision for f for some stage i. The inclusion of the length-block, which effectively encodes all messages such that no encoded input is the tail end of any other encoded input, is necessary for this reasoning. Adding such a length-block is sometimes called Merkle-Damgård strengthening (*MD-strengthening*), which is now stated separately for future reference.

9.26 Algorithm MD-strengthening

Before hashing a message $x = x_1 x_2 \ldots x_t$ (where x_i is a block of bitlength r appropriate for the relevant compression function) of bitlength b, append a final length-block, x_{t+1}, containing the (say) right-justified binary representation of b. (This presumes $b < 2^r$.)

Cascading hash functions

9.27 Fact (*cascading hash functions*) If *either* h_1 or h_2 is a collision resistant hash function, then $h(x) = h_1(x) \,||\, h_2(x)$ is a collision resistant hash function.

If both h_1 and h_2 in Fact 9.27 are n-bit hash functions, then h produces $2n$-bit outputs; mapping this back down to an n-bit output by an n-bit collision-resistant hash function (h_1 and h_2 are candidates) would leave the overall mapping collision-resistant. If h_1 and h_2 are independent, then finding a collision for h requires finding a collision for both simultaneously (i.e., on the same input), which one could hope would require the product of the efforts to attack them individually. This provides a simple yet powerful way to (almost surely) increase strength using only available components.

9.3.3 Formatting and initialization details

9.28 Note (*data representation*) As hash-values depend on exact bitstrings, different data representations (e.g., ASCII vs. EBCDIC) must be converted to a common format before computing hash-values.

(i) Padding and length-blocks

For block-by-block hashing methods, extra bits are usually appended to a hash input string before hashing, to pad it out to a number of bits which make it a multiple of the relevant block size. The padding bits need not be transmitted/stored themselves, provided the sender and recipient agree on a convention.

9.29 Algorithm Padding Method 1

INPUT: data x; bitlength n giving blocksize of data input to processing stage.
OUTPUT: padded data x', with bitlength a multiple of n.

1. Append to x as few (possibly zero) 0-bits as necessary to obtain a string x' whose bitlength is a multiple of n.

9.30 Algorithm Padding Method 2

INPUT: data x; bitlength n giving blocksize of data input to processing stage.
OUTPUT: padded data x', with bitlength a multiple of n.

1. Append to x a single 1-bit.

> 2. Then append as few (possibly zero) 0-bits as necessary to obtain a string x' whose bitlength is a multiple of n.

9.31 Remark (*ambiguous padding*) Padding Method 1 is *ambiguous* – trailing 0-bits of the original data cannot be distinguished from those added during padding. Such methods are acceptable if the length of the data (before padding) is known by the recipient by other means. Padding Method 2 is not ambiguous – each padded string x' corresponds to a unique unpadded string x. When the bitlength of the original data x is already a multiple of n, Padding Method 2 results in the creation of an extra block.

9.32 Remark (*appended length blocks*) Appending a logical length-block prior to hashing prevents collision and pseudo-collision attacks which find second messages of different length, including trivial collisions for random IVs (Example 9.96), long-message attacks (Fact 9.37), and fixed-point attacks (page 374). This further justifies the use of MD-strengthening (Algorithm 9.26).

Trailing length-blocks and padding are often combined. For Padding Method 2, a length field of pre-specified bitlength w may replace the final w 0-bits padded if padding would otherwise cause w or more redundant such bits. By pre-agreed convention, the length field typically specifies the bitlength of the original message. (If used instead to specify the number of padding bits appended, deletion of leading blocks cannot be detected.)

(ii) IVs

Whether the IV is fixed, is randomly chosen per hash function computation, or is a function of the data input, the same IV must be used to generate and verify a hash-value. If not known *a priori* by the verifier, it must be transferred along with the message. In the latter case, this generally should be done with guaranteed integrity (to cut down on the degree of freedom afforded to adversaries, in line with the principle that hash functions should be defined with a fixed or a small set of allowable IVs).

9.3.4 Security objectives and basic attacks

As a framework for evaluating the computational security of hash functions, the objectives of both the hash function designer and an adversary should be understood. Based on Definitions 9.3, 9.4, and 9.7, these are summarized in Table 9.2, and discussed below.

Hash type	Design goal	Ideal strength	Adversary's goal
OWHF	preimage resistance;	2^n	produce preimage;
	2nd-preimage resistance	2^n	find 2nd input, same image
CRHF	collision resistance	$2^{n/2}$	produce any collision
MAC	key non-recovery;	2^t	deduce MAC key;
	computation resistance	$P_f = \max(2^{-t}, 2^{-n})$	produce new (msg, MAC)

Table 9.2: *Design objectives for n-bit hash functions (t-bit MAC key). P_f denotes the probability of forgery by correctly guessing a MAC.*

Given a specific hash function, it is desirable to be able to prove a lower bound on the complexity of attacking it under specified scenarios, with as few or weak a set of assumptions as possible. However, such results are scarce. Typically the best guidance available regarding the security of a particular hash function is the complexity of the (most efficient) applicable known attack, which gives an *upper* bound on security. An attack of *complexity* 2^t is one which requires approximately 2^t operations, each being an appropriate unit of work (e.g., one execution of the compression function or one encryption of an underlying cipher). The storage complexity of an attack (i.e., storage required) should also be considered.

(i) Attacks on the bitsize of an MDC

Given a fixed message x with n-bit hash $h(x)$, a naive method for finding an input colliding with x is to pick a random bitstring x' (of bounded bitlength) and check if $h(x') = h(x)$. The cost may be as little as one compression function evaluation, and memory is negligible. Assuming the hash-code approximates a uniform random variable, the probability of a match is 2^{-n}. The implication of this is Fact 9.33, which also indicates the effort required to find collisions if x may itself be chosen freely. Definition 9.34 is motivated by the design goal that the best possible attack should require no less than such levels of effort, i.e., essentially brute force.

9.33 Fact (*basic hash attacks*) For an n-bit hash function h, one may expect a guessing attack to find a preimage or second preimage within 2^n hashing operations. For an adversary able to choose messages, a birthday attack (see §9.7.1) allows colliding pairs of messages x, x' with $h(x) = h(x')$ to be found in about $2^{n/2}$ operations, and negligible memory.

9.34 Definition An n-bit unkeyed hash function has *ideal security* if both: (1) given a hash output, producing each of a preimage and a 2nd-preimage requires approximately 2^n operations; and (2) producing a collision requires approximately $2^{n/2}$ operations.

(ii) Attacks on the MAC key space

An attempt may be made to determine a MAC key using exhaustive search. With a single known text-MAC pair, an attacker may compute the n-bit MAC on that text under all possible keys, and then check which of the computed MAC-values agrees with that of the known pair. For a t-bit key space this requires 2^t MAC operations, after which one expects $1 + 2^{t-n}$ candidate keys remain. Assuming the MAC behaves as a random mapping, it can be shown that one can expect to reduce this to a unique key by testing the candidate keys using just over t/n text-MAC pairs. Ideally, a MAC key (or information of cryptographically equivalent value) would not be recoverable in fewer than 2^t operations.

As a probabilistic attack on the MAC key space distinct from key recovery, note that for a t-bit key and a fixed input, a randomly guessed key will yield a correct (n-bit) MAC with probability $\approx 2^{-t}$ for $t > n$.

(iii) Attacks on the bitsize of a MAC

MAC forgery involves producing any input x and the corresponding correct MAC without having obtained the latter from anyone with knowledge of the key. For an n-bit MAC algorithm, either guessing a MAC for a given input, or guessing a preimage for a given MAC output, has probability of success about 2^{-n}, as for an MDC. A difference here, however, is that guessed MAC-values cannot be verified off-line without known text-MAC pairs – either knowledge of the key, or a "black-box" which provides MACs for given inputs (i.e., a chosen-text scenario) is required. Since recovering the MAC key trivially allows forgery, an attack on the t-bit key space (see above) must be also be considered here. Ideally, an adversary would be unable to produce new (correct) text-MAC pairs (x, y) with probability significantly better than $\max(2^{-t}, 2^{-n})$, i.e., the better of guessing a key or a MAC-value.

(iv) Attacks using precomputations, multiple targets, and long messages

9.35 Remark (*precomputation of hash values*) For both preimage and second preimage attacks, an opponent who precomputes a large number of hash function input-output pairs may trade off precomputation plus storage for subsequent attack time. For example, for a 64-bit hash value, if one randomly selects 2^{40} inputs, then computes their hash values and stores (hash value, input) pairs indexed by hash value, this precomputation of $O(2^{40})$ time and space allows an adversary to increase the probability of finding a preimage (per one subsequent hash function computation) from 2^{-64} to 2^{-24}. Similarly, the probability of finding a second preimage increases to r times its original value (when no stored pairs are known) if r input-output pairs of a OWHF are precomputed and tabulated.

9.36 Remark (*effect of parallel targets for OWHFs*) In a basic attack, an adversary seeks a second preimage for one fixed target (the image computed from a first preimage). If there are r targets and the goal is to find a second preimage for any one of these r, then the probability of success increases to r times the original probability. One implication is that when using hash functions in conjunction with keyed primitives such as digital signatures, repeated use of the keyed primitive may weaken the security of the combined mechanism in the following sense. If r signed messages are available, the probability of a hash collision increases r-fold (cf. Remark 9.35), and colliding messages yield equivalent signatures, which an opponent could not itself compute off-line.

Fact 9.37 reflects a related attack strategy of potential concern when using iterated hash functions on long messages.

9.37 Fact (*long-message attack for 2nd-preimage*) Let h be an iterated n-bit hash function with compression function f (as in equation (9.1), without MD-strengthening). Let x be a message consisting of t blocks. Then a 2nd-preimage for $h(x)$ can be found in time $(2^n/s) + s$ operations of f, and in space $n(s + \lg(s))$ bits, for any s in the range $1 \le s \le \min(t, 2^{n/2})$.

Justification. The idea is to use a birthday attack on the intermediate hash-results; a sketch for the choice $s = t$ follows. Compute $h(x)$, storing (H_i, i) for each of the t intermediate hash-results H_i corresponding to the t input blocks x_i in a table such that they may be later indexed by value. Compute $h(z)$ for random choices z, checking for a collision involving $h(z)$ in the table, until one is found; approximately $2^n/s$ values z will be required, by the birthday paradox. Identify the index j from the table responsible for the collision; the input $zx_{j+1}x_{j+2}\ldots x_t$ then collides with x.

9.38 Note (*implication of long messages*) Fact 9.37 implies that for "long" messages, a 2nd-preimage is generally easier to find than a preimage (the latter takes at most 2^n operations), becoming moreso with the length of x. For $t \ge 2^{n/2}$, computation is minimized by choosing $s = 2^{n/2}$ in which case a 2nd-preimage costs about $2^{n/2}$ executions of f (comparable to the difficulty of finding a collision).

9.3.5 Bitsizes required for practical security

Suppose that a hash function produces n-bit hash-values, and as a representative benchmark assume that 2^{80} (but not fewer) operations is acceptably beyond computational feasibility.[2] Then the following statements may be made regarding n.

---

[2] Circa 1996, 2^{40} simple operations is quite feasible, and 2^{56} is considered quite reachable by those with sufficient motivation (possibly using parallelization or customized machines).

1. For a OWHF, $n \geq 80$ is required. Exhaustive off-line attacks require at most 2^n operations; this may be reduced with precomputation (Remark 9.35).

2. For a CRHF, $n \geq 160$ is required. Birthday attacks are applicable (Fact 9.33).

3. For a MAC, $n \geq 64$ along with a MAC key of 64-80 bits is sufficient for most applications and environments (cf. Table 9.1). If a single MAC key remains in use, off-line attacks may be possible given one or more text-MAC pairs; but for a proper MAC algorithm, preimage and 2nd-preimage resistance (as well as collision resistance) should follow directly from lack of knowledge of the key, and thus security with respect to such attacks should depend on the keysize rather than n. For attacks requiring on-line queries, additional controls may be used to limit the number of such queries, constrain the format of MAC inputs, or prevent disclosure of MAC outputs for random (chosen-text) inputs. Given special controls, values as small as $n = 32$ or 40 may be acceptable; but caution is advised, since even with one-time MAC keys, the chance any randomly guessed MAC being correct is 2^{-n}, and the relevant factors are the total number of trials a system is subject to over its lifetime, and the consequences of a single successful forgery.

These guidelines may be relaxed somewhat if a lower threshold of computational infeasibility is assumed (e.g., 2^{64} instead of 2^{80}). However, an additional consideration to be taken into account is that for both a CRHF and a OWHF, not only can off-line attacks be carried out, but these can typically be parallelized. Key search attacks against MACs may also be parallelized.

9.4 Unkeyed hash functions (MDCs)

A move from general properties and constructions to specific hash functions is now made, and in this section the subclass of unkeyed hash functions known as manipulation detection codes (MDCs) is considered. From a structural viewpoint, these may be categorized based on the nature of the operations comprising their internal compression functions. From this viewpoint, the three broadest categories of iterated hash functions studied to date are hash functions *based on block ciphers*, *customized hash functions*, and hash functions *based on modular arithmetic*. Customized hash functions are those designed specifically for hashing, with speed in mind and independent of other system subcomponents (e.g., block cipher or modular multiplication subcomponents which may already be present for non-hashing purposes).

Table 9.3 summarizes the conjectured security of a subset of the MDCs subsequently discussed in this section. Similar to the case of block ciphers for encryption (e.g. 8- or 12-round DES vs. 16-round DES), security of MDCs often comes at the expense of speed, and tradeoffs are typically made. In the particular case of block-cipher-based MDCs, a provably secure scheme of Merkle (see page 378) with rate 0.276 (see Definition 9.40) is known but little-used, while MDC-2 is widely believed to be (but not provably) secure, has rate $= 0.5$, and receives much greater attention in practice.

9.4.1 Hash functions based on block ciphers

A practical motivation for constructing hash functions from block ciphers is that if an efficient implementation of a block cipher is already available within a system (either in hardware or software), then using it as the central component for a hash function may provide

↓Hash function	n	m	Preimage	Collision	Comments
Matyas-Meyer-Oseas[a]	n	n	2^n	$2^{n/2}$	for keylength $= n$
MDC-2 (with DES)[b]	64	128	$2 \cdot 2^{82}$	$2 \cdot 2^{54}$	rate 0.5
MDC-4 (with DES)	64	128	2^{109}	$4 \cdot 2^{54}$	rate 0.25
Merkle (with DES)	106	128	2^{112}	2^{56}	rate 0.276
MD4	512	128	2^{128}	2^{20}	Remark 9.50
MD5	512	128	2^{128}	2^{64}	Remark 9.52
RIPEMD-128	512	128	2^{128}	2^{64}	–
SHA-1, RIPEMD-160	512	160	2^{160}	2^{80}	–

[a]The same strength is conjectured for Davies-Meyer and Miyaguchi-Preneel hash functions.
[b]Strength could be increased using a cipher with keylength equal to cipher blocklength.

Table 9.3: *Upper bounds on strength of selected hash functions. n-bit message blocks are processed to produce m-bit hash-values. Number of cipher or compression function operations currently believed necessary to find preimages and collisions are specified, assuming no underlying weaknesses for block ciphers (figures for MDC-2 and MDC-4 account for DES complementation and weak key properties). Regarding rate, see Definition 9.40.*

the latter functionality at little additional cost. The (not always well-founded) hope is that a good block cipher may serve as a building block for the creation of a hash function with properties suitable for various applications.

Constructions for hash functions have been given which are "provably secure" assuming certain ideal properties of the underlying block cipher. However, block ciphers not do not possess the properties of random functions (for example, they are invertible – see Remark 9.14). Moreover, in practice block ciphers typically exhibit additional regularities or weaknesses (see §9.7.4). For example, for a block cipher E, double encryption using an encrypt-decrypt (E-D) cascade with keys K_1, K_2 results in the identity mapping when $K_1 = K_2$. In summary, while various necessary conditions are known, it is unclear exactly what requirements of a block cipher are sufficient to construct a secure hash function, and properties adequate for a block cipher (e.g., resistance to chosen-text attack) may not guarantee a good hash function.

In the constructions which follow, Definition 9.39 is used.

9.39 Definition An *(n,r) block cipher* is a block cipher defining an invertible function from n-bit plaintexts to n-bit ciphertexts using an r-bit key. If E is such a cipher, then $E_k(x)$ denotes the encryption of x under key k.

Discussion of hash functions constructed from n-bit block ciphers is divided between those producing *single-length* (n-bit) and *double-length* ($2n$-bit) hash-values, where single and double are relative to the size of the block cipher output. Under the assumption that computations of 2^{64} operations are infeasible,[3] the objective of single-length hash functions is to provide a OWHF for ciphers of blocklength near $n = 64$, or to provide CRHFs for cipher blocklengths near $n = 128$. The motivation for double-length hash functions is that many n-bit block ciphers exist of size approximately $n = 64$, and single-length hash-codes of this size are not collision resistant. For such ciphers, the goal is to obtain hash-codes of bitlength $2n$ which are CRHFs.

In the simplest case, the size of the key used in such hash functions is approximately the same as the blocklength of the cipher (i.e., n bits). In other cases, hash functions use

---

[3]The discussion here is easily altered for a more conservative bound, e.g., 2^{80} operations as used in §9.3.5. Here 2^{64} is more convenient for discussion, due to the omnipresence of 64-bit block ciphers.

larger (e.g., double-length) keys. Another characteristic to be noted in such hash functions is the number of block cipher operations required to produce a hash output of blocklength equal to that of the cipher, motivating the following definition.

9.40 Definition Let h be an iterated hash function constructed from a block cipher, with compression function f which performs s block encryptions to process each successive n-bit message block. Then the *rate* of h is $1/s$.

The hash functions discussed in this section are summarized in Table 9.4. The Matyas-Meyer-Oseas and MDC-2 algorithms are the basis, respectively, of the two generic hash functions in ISO standard 10118-2, each allowing use of any n-bit block cipher E and providing hash-codes of bitlength $m \leq n$ and $m \leq 2n$, respectively.

Hash function	(n, k, m)	Rate
Matyas-Meyer-Oseas	(n, k, n)	1
Davies-Meyer	(n, k, n)	k/n
Miyaguchi-Preneel	(n, k, n)	1
MDC-2 (with DES)	$(64, 56, 128)$	$1/2$
MDC-4 (with DES)	$(64, 56, 128)$	$1/4$

Table 9.4: *Summary of selected hash functions based on n-bit block ciphers. k = key bitsize (approximate); function yields m-bit hash-values.*

(i) Single-length MDCs of rate 1

The first three schemes described below, and illustrated in Figure 9.3, are closely related single-length hash functions based on block ciphers. These make use of the following predefined components:

1. a generic n-bit block cipher E_K parametrized by a symmetric key K;
2. a function g which maps n-bit inputs to keys K suitable for E (if keys for E are also of length n, g might be the identity function); and
3. a fixed (usually n-bit) initial value IV, suitable for use with E.

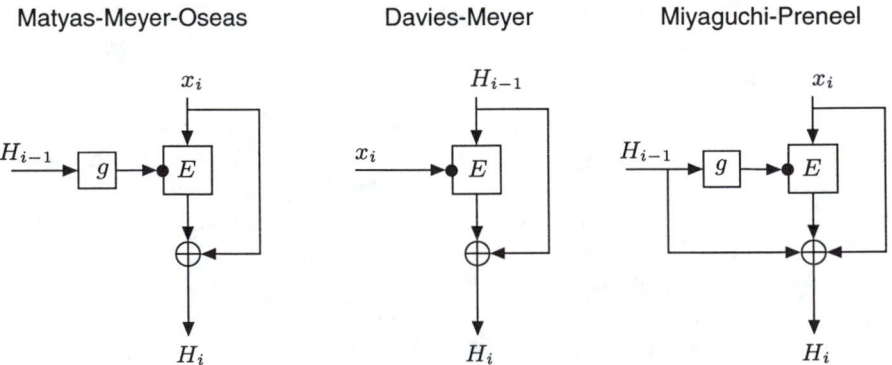

Figure 9.3: *Three single-length, rate-one MDCs based on block ciphers.*

9.41 Algorithm Matyas-Meyer-Oseas hash

INPUT: bitstring x.
OUTPUT: n-bit hash-code of x.

1. Input x is divided into n-bit blocks and padded, if necessary, to complete last block. Denote the padded message consisting of t n-bit blocks: $x_1 x_2 \ldots x_t$. A constant n-bit initial value IV must be pre-specified.
2. The output is H_t defined by: $H_0 = IV$; $H_i = E_{g(H_{i-1})}(x_i) \oplus x_i$, $1 \le i \le t$.

9.42 Algorithm Davies-Meyer hash

INPUT: bitstring x.
OUTPUT: n-bit hash-code of x.

1. Input x is divided into k-bit blocks where k is the keysize, and padded, if necessary, to complete last block. Denote the padded message consisting of t k-bit blocks: $x_1 x_2 \ldots x_t$. A constant n-bit initial value IV must be pre-specified.
2. The output is H_t defined by: $H_0 = IV$; $H_i = E_{x_i}(H_{i-1}) \oplus H_{i-1}$, $1 \le i \le t$.

9.43 Algorithm Miyaguchi-Preneel hash

This scheme is identical to that of Algorithm 9.41, except the output H_{i-1} from the previous stage is also XORed to that of the current stage. More precisely, H_i is redefined as: $H_0 = IV$; $H_i = E_{g(H_{i-1})}(x_i) \oplus x_i \oplus H_{i-1}$, $1 \le i \le t$.

9.44 Remark (*dual schemes*) The Davies-Meyer hash may be viewed as the 'dual' of the Matyas-Meyer-Oseas hash, in the sense that x_i and H_{i-1} play reversed roles. When DES is used as the block cipher in Davies-Meyer, the input is processed in 56-bit blocks (yielding rate $56/64 < 1$), whereas Matyas-Meyer-Oseas and Miyaguchi-Preneel process 64-bit blocks.

9.45 Remark (*black-box security*) Aside from heuristic arguments as given in Example 9.13, it appears that all three of Algorithms 9.41, 9.42, and 9.43 yield hash functions which are provably secure under an appropriate "black-box" model (e.g., assuming E has the required randomness properties, and that attacks may not make use of any special properties or internal details of E). "Secure" here means that finding preimages and collisions (in fact, pseudo-preimages and pseudo-collisions – see §9.7.2) require on the order of 2^n and $2^{n/2}$ n-bit block cipher operations, respectively. Due to their single-length nature, none of these three is collision resistant for underlying ciphers of relatively small blocklength (e.g., DES, which yields 64-bit hash-codes).

Several double-length hash functions based on block ciphers are considered next.

(ii) Double-length MDCs: MDC-2 and MDC-4

MDC-2 and MDC-4 are manipulation detection codes requiring 2 and 4, respectively, block cipher operations per block of hash input. They employ a combination of either 2 or 4 iterations of the Matyas-Meyer-Oseas (single-length) scheme to produce a double-length hash. When used as originally specified, using DES as the underlying block cipher, they produce 128-bit hash-codes. The general construction, however, can be used with other block ciphers. MDC-2 and MDC-4 make use of the following pre-specified components:

1. DES as the block cipher E_K of bitlength $n = 64$ parameterized by a 56-bit key K;
2. two functions g and $\tilde{g}$ which map 64-bit values U to suitable 56-bit DES keys as follows. For $U = u_1 u_2 \ldots u_{64}$, delete every eighth bit starting with u_8, and set the 2nd and 3rd bits to '10' for g, and '01' for $\tilde{g}$:

$$g(U) = u_1\, 1\, 0\, u_4 u_5 u_6 u_7 u_9 u_{10} \ldots u_{63}.$$
$$\tilde{g}(U) = u_1\, 0\, 1\, u_4 u_5 u_6 u_7 u_9 u_{10} \ldots u_{63}.$$

(The resulting values are guaranteed not to be weak or semi-weak DES keys, as all such keys have bit 2 = bit 3; see page 375. Also, this guarantees the security requirement that $g(IV) \neq \tilde{g}(\widetilde{IV})$.)

MDC-2 is specified in Algorithm 9.46 and illustrated in Figure 9.4.

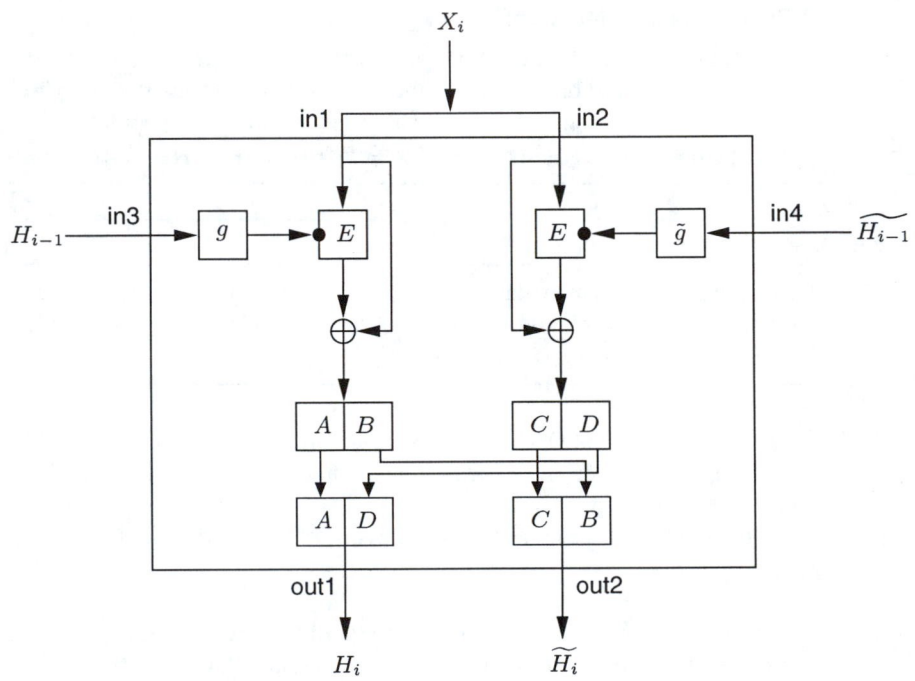

Figure 9.4: *Compression function of MDC-2 hash function. $E = DES$.*

9.46 Algorithm MDC-2 (DES-based)

INPUT: string x of bitlength $r = 64t$ for $t \geq 2$.
OUTPUT: 128-bit hash-code of x.

1. Partition x into 64-bit blocks x_i: $x = x_1 x_2 \ldots x_t$.
2. Choose the 64-bit non-secret constants IV, $\widetilde{IV}$ (the same constants must be used for MDC verification) from a set of recommended prescribed values. A default set of prescribed values is (in hexadecimal): $IV = 0x5252525252525252$, $\widetilde{IV} = 0x2525252525252525$.

3. Let $||$ denote concatenation, and C_i^L, C_i^R the left and right 32-bit halves of C_i. The output is $h(x) = H_t \parallel \widetilde{H_t}$ defined as follows (for $1 \le i \le t$):

$$H_0 = IV; \quad k_i = g(H_{i-1}); \quad C_i = E_{k_i}(x_i) \oplus x_i; \quad H_i = C_i^L \parallel \widetilde{C_i}^R$$

$$\widetilde{H_0} = \widetilde{IV}; \quad \widetilde{k_i} = \widetilde{g}(\widetilde{H_{i-1}}); \quad \widetilde{C_i} = E_{\widetilde{k_i}}(x_i) \oplus x_i; \quad \widetilde{H_i} = \widetilde{C_i}^L \parallel C_i^R .$$

In Algorithm 9.46, padding may be necessary to meet the bitlength constraint on the input x. In this case, an unambiguous padding method may be used (see Remark 9.31), possibly including MD-strengthening (see Remark 9.32).

MDC-4 (see Algorithm 9.47 and Figure 9.5) is constructed using the MDC-2 compression function. One iteration of the MDC-4 compression function consists of two sequential executions of the MDC-2 compression function, where:

1. the two 64-bit data inputs to the first MDC-2 compression are both the same next 64-bit message block;
2. the keys for the first MDC-2 compression are derived from the outputs (chaining variables) of the previous MDC-4 compression;
3. the keys for the second MDC-2 compression are derived from the outputs (chaining variables) of the first MDC-2 compression; and
4. the two 64-bit data inputs for the second MDC-2 compression are the outputs (chaining variables) from the opposite sides of the previous MDC-4 compression.

9.47 Algorithm MDC-4 (DES-based)

INPUT: string x of bitlength $r = 64t$ for $t \ge 2$. (See MDC-2 above regarding padding.)
OUTPUT: 128-bit hash-code of x.

1. As in step 1 of MDC-2 above.
2. As in step 2 of MDC-2 above.
3. With notation as in MDC-2, the output is $h(x) = G_t \parallel \widetilde{G_t}$ defined as follows (for $1 \le i \le t$):

$$G_0 = IV; \quad \widetilde{G_0} = \widetilde{IV};$$

$$k_i = g(G_{i-1}); \quad C_i = E_{k_i}(x_i) \oplus x_i; \quad H_i = C_i^L \parallel \widetilde{C_i}^R$$

$$\widetilde{k_i} = \widetilde{g}(\widetilde{G_{i-1}}); \quad \widetilde{C_i} = E_{\widetilde{k_i}}(x_i) \oplus x_i; \quad \widetilde{H_i} = \widetilde{C_i}^L \parallel C_i^R$$

$$j_i = g(H_i); \quad D_i = E_{j_i}(\widetilde{G_{i-1}}) \oplus \widetilde{G_{i-1}}; \quad G_i = D_i^L \parallel \widetilde{D_i}^R$$

$$\widetilde{j_i} = \widetilde{g}(\widetilde{H_i}); \quad \widetilde{D_i} = E_{\widetilde{j_i}}(G_{i-1}) \oplus G_{i-1}; \quad \widetilde{G_i} = \widetilde{D_i}^L \parallel D_i^R .$$

9.4.2 Customized hash functions based on MD4

Customized hash functions are those which are specifically designed "from scratch" for the explicit purpose of hashing, with optimized performance in mind, and without being constrained to reusing existing system components such as block ciphers or modular arithmetic. Those having received the greatest attention in practice are based on the MD4 hash function.

Number 4 in a series of hash functions (*Message Digest* algorithms), MD4 was designed specifically for software implementation on 32-bit machines. Security concerns motivated the design of MD5 shortly thereafter, as a more conservative variation of MD4.

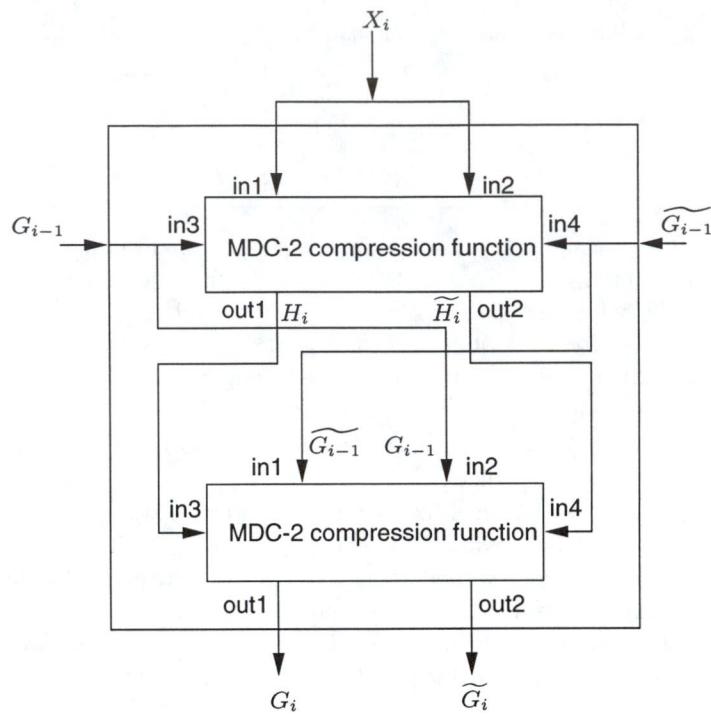

Figure 9.5: *Compression function of MDC-4 hash function*

Other important subsequent variants include the Secure Hash Algorithm (SHA-1), the hash function RIPEMD, and its strengthened variants RIPEMD-128 and RIPEMD-160. Parameters for these hash functions are summarized in Table 9.5. "Rounds × Steps per round" refers to operations performed on input blocks within the corresponding compression function. Table 9.6 specifies test vectors for a subset of these hash functions.

Notation for description of MD4-family algorithms

Table 9.7 defines the notation for the description of MD4-family algorithms described below. Note 9.48 addresses the implementation issue of converting strings of bytes to words in an unambiguous manner.

9.48 Note (*little-endian vs. big-endian*) For interoperable implementations involving byte-to-word conversions on different processors (e.g., converting between 32-bit words and groups of four 8-bit bytes), an unambiguous convention must be specified. Consider a stream of bytes B_i with increasing memory addresses i, to be interpreted as a 32-bit word with numerical value W. In *little-endian* architectures, the byte with the lowest memory address (B_1) is the least significant byte: $W = 2^{24}B_4 + 2^{16}B_3 + 2^8B_2 + B_1$. In *big-endian* architectures, the byte with the lowest address (B_1) is the most significant byte: $W = 2^{24}B_1 + 2^{16}B_2 + 2^8B_3 + B_4$.

(i) MD4

MD4 (Algorithm 9.49) is a 128-bit hash function. The original MD4 design goals were that breaking it should require roughly brute-force effort: finding distinct messages with the same hash-value should take about 2^{64} operations, and finding a message yielding a

Name	Bitlength	Rounds × Steps per round	Relative speed
MD4	128	3 × 16	1.00
MD5	128	4 × 16	0.68
RIPEMD-128	128	4 × 16 twice (in parallel)	0.39
SHA-1	160	4 × 20	0.28
RIPEMD-160	160	5 × 16 twice (in parallel)	0.24

Table 9.5: *Summary of selected hash functions based on MD4.*

Name	String	Hash value (as a hex byte string)
MD4	""	31d6cfe0d16ae931b73c59d7e0c089c0
	"a"	bde52cb31de33e46245e05fbdbd6fb24
	"abc"	a448017aaf21d8525fc10ae87aa6729d
	"abcdefghijklmnopqrstuvwxyz"	d79e1c308aa5bbcdeea8ed63df412da9
MD5	""	d41d8cd98f00b204e9800998ecf8427e
	"a"	0cc175b9c0f1b6a831c399e269772661
	"abc"	900150983cd24fb0d6963f7d28e17f72
	"abcdefghijklmnopqrstuvwxyz"	c3fcd3d76192e4007dfb496cca67e13b
SHA-1	""	da39a3ee5e6b4b0d3255bfef95601890afd80709
	"a"	86f7e437faa5a7fce15d1ddcb9eaeaea377667b8
	"abc"	a9993e364706816aba3e25717850c26c9cd0d89d
	"abcdefghijklmnopqrstuvwxyz"	32d10c7b8cf96570ca04ce37f2a19d84240d3a89
RIPEMD-160	""	9c1185a5c5e9fc54612808977ee8f548b2258d31
	"a"	0bdc9d2d256b3ee9daae347be6f4dc835a467ffe
	"abc"	8eb208f7e05d987a9b044a8e98c6b087f15a0bfc
	"abcdefghijklmnopqrstuvwxyz"	f71c27109c692c1b56bbdceb5b9d2865b3708dbc

Table 9.6: *Test vectors for selected hash functions.*

Notation	Meaning
u, v, w	variables representing 32-bit quantities
0x67452301	hexadecimal 32-bit integer (least significant byte: 01)
$+$	addition modulo 2^{32}
$\overline{u}$	bitwise complement
$u \hookleftarrow s$	result of rotating u left through s positions
uv	bitwise AND
$u \vee v$	bitwise inclusive-OR
$u \oplus v$	bitwise exclusive-OR
$f(u, v, w)$	$uv \vee \overline{u}w$
$g(u, v, w)$	$uv \vee uw \vee vw$
$h(u, v, w)$	$u \oplus v \oplus w$
$(X_1, \ldots, X_j) \leftarrow$ $(Y_1, \ldots, Y_j)$	simultaneous assignments $(X_i \leftarrow Y_i)$, where $(Y_1, \ldots, Y_j)$ is evaluated prior to any assignments

Table 9.7: *Notation for MD4-family algorithms.*

pre-specified hash-value about 2^{128} operations. It is now known that MD4 fails to meet this goal (Remark 9.50). Nonetheless, a full description of MD4 is included as Algorithm 9.49 for historical and cryptanalytic reference. It also serves as a convenient reference for describing, and allowing comparisons between, other hash functions in this family.

9.49 Algorithm MD4

INPUT: bitstring x of arbitrary bitlength $b \geq 0$. (For notation see Table 9.7.)
OUTPUT: 128-bit hash-code of x. (See Table 9.6 for test vectors.)

1. *Definition of constants.* Define four 32-bit initial chaining values (IVs):
   $h_1 = \text{0x67452301}, h_2 = \text{0xefcdab89}, h_3 = \text{0x98badcfe}, h_4 = \text{0x10325476}$.
   Define additive 32-bit constants:
   $y[j] = 0, 0 \leq j \leq 15$;
   $y[j] = \text{0x5a827999}, 16 \leq j \leq 31$; (constant = square-root of 2)
   $y[j] = \text{0x6ed9eba1}, 32 \leq j \leq 47$; (constant = square-root of 3)
   Define order for accessing source words (each list contains 0 through 15):
   $z[0..15] = [0, 1, 2, 3, 4, 5, 6, 7, 8, 9, 10, 11, 12, 13, 14, 15]$,
   $z[16..31] = [0, 4, 8, 12, 1, 5, 9, 13, 2, 6, 10, 14, 3, 7, 11, 15]$,
   $z[32..47] = [0, 8, 4, 12, 2, 10, 6, 14, 1, 9, 5, 13, 3, 11, 7, 15]$.
   Finally define the number of bit positions for left shifts (rotates):
   $s[0..15] = [3, 7, 11, 19, 3, 7, 11, 19, 3, 7, 11, 19, 3, 7, 11, 19]$,
   $s[16..31] = [3, 5, 9, 13, 3, 5, 9, 13, 3, 5, 9, 13, 3, 5, 9, 13]$,
   $s[32..47] = [3, 9, 11, 15, 3, 9, 11, 15, 3, 9, 11, 15, 3, 9, 11, 15]$.
2. *Preprocessing.* Pad x such that its bitlength is a multiple of 512, as follows. Append a single 1-bit, then append $r - 1$ (≥ 0) 0-bits for the smallest r resulting in a bitlength 64 less than a multiple of 512. Finally append the 64-bit representation of b mod 2^{64}, as two 32-bit words with least significant word first. (Regarding converting between streams of bytes and 32-bit words, the convention is little-endian; see Note 9.48.) Let m be the number of 512-bit blocks in the resulting string ($b + r + 64 = 512m = 32 \cdot 16m$). The formatted input consists of $16m$ 32-bit words: $x_0 x_1 \ldots x_{16m-1}$. Initialize: $(H_1, H_2, H_3, H_4) \leftarrow (h_1, h_2, h_3, h_4)$.
3. *Processing.* For each i from 0 to $m - 1$, copy the ith block of 16 32-bit words into temporary storage: $X[j] \leftarrow x_{16i+j}$, $0 \leq j \leq 15$, then process these as below in three 16-step rounds before updating the chaining variables:
   (*initialize working variables*) $(A, B, C, D) \leftarrow (H_1, H_2, H_3, H_4)$.
   (*Round 1*) For j from 0 to 15 do the following:
   $t \leftarrow (A + f(B, C, D) + X[z[j]] + y[j]), (A, B, C, D) \leftarrow (D, t \hookleftarrow s[j], B, C)$.
   (*Round 2*) For j from 16 to 31 do the following:
   $t \leftarrow (A + g(B, C, D) + X[z[j]] + y[j]), (A, B, C, D) \leftarrow (D, t \hookleftarrow s[j]), B, C)$.
   (*Round 3*) For j from 32 to 47 do the following:
   $t \leftarrow (A + h(B, C, D) + X[z[j]] + y[j]), (A, B, C, D) \leftarrow (D, t \hookleftarrow s[j]), B, C)$.
   (*update chaining values*) $(H_1, H_2, H_3, H_4) \leftarrow (H_1 + A, H_2 + B, H_3 + C, H_4 + D)$.
4. *Completion.* The final hash-value is the concatenation: $H_1 \| H_2 \| H_3 \| H_4$ (with first and last bytes the low- and high-order bytes of H_1, H_4, respectively).

9.50 Remark (*MD4 collisions*) Collisions have been found for MD4 in 2^{20} compression function computations (cf. Table 9.3). For this reason, MD4 is no longer recommended for use as a collision-resistant hash function. While its utility as a one-way function has not been studied in light of this result, it is prudent to expect a preimage attack on MD4 requiring fewer than 2^{128} operations will be found.

(ii) MD5

MD5 (Algorithm 9.51) was designed as a strengthened version of MD4, prior to actual MD4 collisions being found. It has enjoyed widespread use in practice. It has also now been found to have weaknesses (Remark 9.52).

The changes made to obtain MD5 from MD4 are as follows:

1. addition of a fourth round of 16 steps, and a Round 4 function
2. replacement of the Round 2 function by a new function
3. modification of the access order for message words in Rounds 2 and 3
4. modification of the shift amounts (such that shifts differ in distinct rounds)
5. use of unique additive constants in each of the 4×16 steps, based on the integer part of $2^{32} \cdot \sin(j)$ for step j (requiring overall, 256 bytes of storage)
6. addition of output from the previous step into each of the 64 steps.

9.51 Algorithm MD5

INPUT: bitstring x of arbitrary bitlength $b \geq 0$. (For notation, see Table 9.7.)
OUTPUT: 128-bit hash-code of x. (See Table 9.6 for test vectors.)
   MD5 is obtained from MD4 by making the following changes.

1. *Notation.* Replace the Round 2 function by: $g(u, v, w) \stackrel{\text{def}}{=} uw \vee v\overline{w}$.

   Define a Round 4 function: $k(u, v, w) \stackrel{\text{def}}{=} v \oplus (u \vee \overline{w})$.

2. *Definition of constants.* Redefine unique additive constants:
   $y[j] =$ first 32 bits of binary value abs$(\sin(j + 1))$, $0 \leq j \leq 63$, where j is in radians and "abs" denotes absolute value. Redefine access order for words in Rounds 2 and 3, and define for Round 4:
   $z[16..31] = [1, 6, 11, 0, 5, 10, 15, 4, 9, 14, 3, 8, 13, 2, 7, 12]$,
   $z[32..47] = [5, 8, 11, 14, 1, 4, 7, 10, 13, 0, 3, 6, 9, 12, 15, 2]$,
   $z[48..63] = [0, 7, 14, 5, 12, 3, 10, 1, 8, 15, 6, 13, 4, 11, 2, 9]$.
   Redefine number of bit positions for left shifts (rotates):
   $s[0..15] = [7, 12, 17, 22, 7, 12, 17, 22, 7, 12, 17, 22, 7, 12, 17, 22]$,
   $s[16..31] = [5, 9, 14, 20, 5, 9, 14, 20, 5, 9, 14, 20, 5, 9, 14, 20]$,
   $s[32..47] = [4, 11, 16, 23, 4, 11, 16, 23, 4, 11, 16, 23, 4, 11, 16, 23]$,
   $s[48..63] = [6, 10, 15, 21, 6, 10, 15, 21, 6, 10, 15, 21, 6, 10, 15, 21]$.

3. *Preprocessing.* As in MD4.

4. *Processing.* In each of Rounds 1, 2, and 3, replace "$B \leftarrow (t \hookleftarrow s[j])$" by "$B \leftarrow B + (t \hookleftarrow s[j])$". Also, immediately following Round 3 add:
   (*Round 4*) For j from 48 to 63 do the following:
   $t \leftarrow (A + k(B, C, D) + X[z[j]] + y[j]), (A, B, C, D) \leftarrow (D, B + (t \hookleftarrow s[j]), B, C)$.

5. *Completion.* As in MD4.

9.52 Remark (*MD5 compression function collisions*) While no collisions for MD5 have yet been found (cf. Table 9.3), collisions have been found for the MD5 compression function. More specifically, these are called collisions for random IV. (See §9.7.2, and in particular Definition 9.97 and Note 9.98.)

(iii) SHA-1

The Secure Hash Algorithm (SHA-1), based on MD4, was proposed by the U.S. National Institute for Standards and Technology (NIST) for certain U.S. federal government applications. The main differences of SHA-1 from MD4 are as follows:

1. The hash-value is 160 bits, and five (vs. four) 32-bit chaining variables are used.
2. The compression function has four rounds instead of three, using the MD4 step functions f, g, and h as follows: f in the first, g in the third, and h in both the second and fourth rounds. Each round has 20 steps instead of 16.
3. Within the compression function, each 16-word message block is expanded to an 80-word block, by a process whereby each of the last 64 of the 80 words is the XOR of 4 words from earlier positions in the expanded block. These 80 words are then input one-word-per-step to the 80 steps.
4. The core step is modified as follows: the only rotate used is a constant 5-bit rotate; the fifth working variable is added into each step result; message words from the expanded message block are accessed sequentially; and C is updated as B rotated left 30 bits, rather than simply B.
5. SHA-1 uses four non-zero additive constants, whereas MD4 used three constants only two of which were non-zero.

The byte ordering used for converting between streams of bytes and 32-bit words in the official SHA-1 specification is big-endian (see Note 9.48); this differs from MD4 which is little-endian.

9.53 Algorithm Secure Hash Algorithm – revised (SHA-1)

INPUT: bitstring x of bitlength $b \geq 0$. (For notation, see Table 9.7.)
OUTPUT: 160-bit hash-code of x. (See Table 9.6 for test vectors.)

SHA-1 is defined (with reference to MD4) by making the following changes.

1. *Notation.* As in MD4.
2. *Definition of constants.* Define a fifth IV to match those in MD4: $h_5 = $ 0xc3d2e1f0. Define per-round integer additive constants: $y_1 = $ 0x5a827999, $y_2 = $ 0x6ed9eba1, $y_3 = $ 0x8f1bbcdc, $y_4 = $ 0xca62c1d6. (No order for accessing source words, or specification of bit positions for left shifts is required.)
3. *Overall preprocessing.* Pad as in MD4, except the final two 32-bit words specifying the bitlength b is appended with most significant word preceding least significant. As in MD4, the formatted input is $16m$ 32-bit words: $x_0 x_1 \ldots x_{16m-1}$. Initialize chaining variables: $(H_1, H_2, H_3, H_4, H_5) \leftarrow (h_1, h_2, h_3, h_4, h_5)$.
4. *Processing.* For each i from 0 to $m - 1$, copy the ith block of sixteen 32-bit words into temporary storage: $X[j] \leftarrow x_{16i+j}$, $0 \leq j \leq 15$, and process these as below in four 20-step rounds before updating the chaining variables:
   (expand 16-word block into 80-word block; let X_j denote $X[j]$)
   for j from 16 to 79, $X_j \leftarrow ((X_{j-3} \oplus X_{j-8} \oplus X_{j-14} \oplus X_{j-16}) \hookleftarrow 1)$.
   (*initialize working variables*) $(A, B, C, D, E) \leftarrow (H_1, H_2, H_3, H_4, H_5)$.
   (*Round 1*) For j from 0 to 19 do the following:
   $t \leftarrow ((A \hookleftarrow 5) + f(B, C, D) + E + X_j + y_1)$,
   $(A, B, C, D, E) \leftarrow (t, A, B \hookleftarrow 30, C, D)$.
   (*Round 2*) For j from 20 to 39 do the following:
   $t \leftarrow ((A \hookleftarrow 5) + h(B, C, D) + E + X_j + y_2)$,
   $(A, B, C, D, E) \leftarrow (t, A, B \hookleftarrow 30, C, D)$.

(*Round 3*) For j from 40 to 59 do the following:
$t \leftarrow ((A \hookleftarrow 5) + g(B,C,D) + E + X_j + y_3)$,
$(A,B,C,D,E) \leftarrow (t, A, B \hookleftarrow 30, C, D)$.
(*Round 4*) For j from 60 to 79 do the following:
$t \leftarrow ((A \hookleftarrow 5) + h(B,C,D) + E + X_j + y_4)$,
$(A,B,C,D,E) \leftarrow (t, A, B \hookleftarrow 30, C, D)$.
(*update chaining values*)
$(H_1, H_2, H_3, H_4, H_5) \leftarrow (H_1 + A, H_2 + B, H_3 + C, H_4 + D, H_5 + E)$.

5. *Completion.* The hash-value is: $H_1 \| H_2 \| H_3 \| H_4 \| H_5$
(with first and last bytes the high- and low-order bytes of H_1, H_5, respectively).

9.54 Remark (*security of SHA-1*) Compared to 128-bit hash functions, the 160-bit hash-value of SHA-1 provides increased security against brute-force attacks. SHA-1 and RIPEMD-160 (see §9.4.2(iv)) presently appear to be of comparable strength; both are considered stronger than MD5 (Remark 9.52). In SHA-1, a significant effect of the expansion of 16-word message blocks to 80 words in the compression function is that any two distinct 16-word blocks yield 80-word values which differ in a larger number of bit positions, significantly expanding the number of bit differences among message words input to the compression function. The redundancy added by this preprocessing evidently adds strength.

(iv) RIPEMD-160

RIPEMD-160 (Algorithm 9.55) is a hash function based on MD4, taking into account knowledge gained in the analysis of MD4, MD5, and RIPEMD. The overall RIPEMD-160 compression function maps 21-word inputs (5-word chaining variable plus 16-word message block, with 32-bit words) to 5-word outputs. Each input block is processed in parallel by distinct versions (the *left line* and *right line*) of the compression function. The 160-bit outputs of the separate lines are combined to give a single 160-bit output.

Notation	Definition
$f(u,v,w)$	$u \oplus v \oplus w$
$g(u,v,w)$	$uv \vee \overline{u}w$
$h(u,v,w)$	$(u \vee \overline{v}) \oplus w$
$k(u,v,w)$	$uw \vee v\overline{w}$
$l(u,v,w)$	$u \oplus (v \vee \overline{w})$

Table 9.8: *RIPEMD-160 round function definitions.*

The RIPEMD-160 compression function differs from MD4 in the number of words of chaining variable, the number of rounds, the round functions themselves (Table 9.8), the order in which the input words are accessed, and the amounts by which results are rotated. The left and and right computation lines differ from each other in these last two items, in their additive constants, and in the order in which the round functions are applied. This design is intended to improve resistance against known attack strategies. Each of the parallel lines uses the same IV as SHA-1. When writing the IV as a bitstring, little-endian ordering is used for RIPEMD-160 as in MD4 (vs. big-endian in SHA-1; see Note 9.48).

9.55 Algorithm RIPEMD-160

INPUT: bitstring x of bitlength $b \geq 0$.
OUTPUT: 160-bit hash-code of x. (See Table 9.6 for test vectors.)

RIPEMD-160 is defined (with reference to MD4) by making the following changes.

1. *Notation.* See Table 9.7, with MD4 round functions f, g, h redefined per Table 9.8 (which also defines the new round functions k, l).

2. *Definition of constants.* Define a fifth IV: $h_5 = $ 0xc3d2e1f0. In addition:

   (a) Use the MD4 additive constants for the left line, renamed: $y_L[j] = 0$, $0 \leq j \leq 15$; $y_L[j] = $ 0x5a827999, $16 \leq j \leq 31$; $y_L[j] = $ 0x6ed9eba1, $32 \leq j \leq 47$. Define two further constants (square roots of 5,7): $y_L[j] = $ 0x8f1bbcdc, $48 \leq j \leq 63$; $y_L[j] = $ 0xa953fd4e, $64 \leq j \leq 79$.

   (b) Define five new additive constants for the right line (cube roots of 2,3,5,7):
   $y_R[j] = $ 0x50a28be6, $0 \leq j \leq 15$; $y_R[j] = $ 0x5c4dd124, $16 \leq j \leq 31$;
   $y_R[j] = $ 0x6d703ef3, $32 \leq j \leq 47$; $y_R[j] = $ 0x7a6d76e9, $48 \leq j \leq 63$;
   $y_R[j] = 0$, $64 \leq j \leq 79$.

   (c) See Table 9.9 for constants for step j of the compression function: $z_L[j]$, $z_R[j]$ specify the access order for source words in the left and right lines; $s_L[j]$, $s_R[j]$ the number of bit positions for rotates (see below).

3. *Preprocessing.* As in MD4, with addition of a fifth chaining variable: $H_5 \leftarrow h_5$.

4. *Processing.* For each i from 0 to $m - 1$, copy the ith block of sixteen 32-bit words into temporary storage: $X[j] \leftarrow x_{16i+j}$, $0 \leq j \leq 15$. Then:

   (a) Execute five 16-step rounds of the left line as follows:
   $(A_L, B_L, C_L, D_L, E_L) \leftarrow (H_1, H_2, H_3, H_4, H_5)$.
   (*left Round 1*) For j from 0 to 15 do the following:
   $t \leftarrow (A_L + f(B_L, C_L, D_L) + X[z_L[j]] + y_L[j])$,
   $(A_L, B_L, C_L, D_L, E_L) \leftarrow (E_L, A_L + (t \hookleftarrow s_L[j]), B_L, C_L \hookleftarrow 10, D_L)$.
   (*left Round 2*) For j from 16 to 31 do the following:
   $t \leftarrow (A_L + g(B_L, C_L, D_L) + X[z_L[j]] + y_L[j])$,
   $(A_L, B_L, C_L, D_L, E_L) \leftarrow (E_L, A_L + (t \hookleftarrow s_L[j]), B_L, C_L \hookleftarrow 10, D_L)$.
   (*left Round 3*) For j from 32 to 47 do the following:
   $t \leftarrow (A_L + h(B_L, C_L, D_L) + X[z_L[j]] + y_L[j])$,
   $(A_L, B_L, C_L, D_L, E_L) \leftarrow (E_L, A_L + (t \hookleftarrow s_L[j]), B_L, C_L \hookleftarrow 10, D_L)$.
   (*left Round 4*) For j from 48 to 63 do the following:
   $t \leftarrow (A_L + k(B_L, C_L, D_L) + X[z_L[j]] + y_L[j])$,
   $(A_L, B_L, C_L, D_L, E_L) \leftarrow (E_L, A_L + (t \hookleftarrow s_L[j]), B_L, C_L \hookleftarrow 10, D_L)$.
   (*left Round 5*) For j from 64 to 79 do the following:
   $t \leftarrow (A_L + l(B_L, C_L, D_L) + X[z_L[j]] + y_L[j])$,
   $(A_L, B_L, C_L, D_L, E_L) \leftarrow (E_L, E_L + (t \hookleftarrow s_L[j]), B_L, C_L \hookleftarrow 10, D_L)$.

   (b) Execute in parallel with the above five rounds an analogous right line with $(A_R, B_R, C_R, D_R, E_R)$, $y_R[j]$, $z_R[j]$, $s_R[j]$ replacing the corresponding quantities with subscript L; and the order of the round functions reversed so that their order is: l, k, h, g, and f. Start by initializing the right line working variables: $(A_R, B_R, C_R, D_R, E_R) \leftarrow (H_1, H_2, H_3, H_4, H_5)$.

   (c) After executing both the left and right lines above, update the chaining values as follows: $t \leftarrow H_1$, $H_1 \leftarrow H_2 + C_L + D_R$, $H_2 \leftarrow H_3 + D_L + E_R$, $H_3 \leftarrow H_4 + E_L + A_R$, $H_4 \leftarrow H_5 + A_L + B_R$, $H_5 \leftarrow t + B_L + C_R$.

5. *Completion.* The final hash-value is the concatenation: $H_1 \| H_2 \| H_3 \| H_4 \| H_5$ (with first and last bytes the low- and high-order bytes of H_1, H_5, respectively).

Variable	Value
$z_L[\ 0..15]$	[0, 1, 2, 3, 4, 5, 6, 7, 8, 9,10,11,12,13,14,15]
$z_L[16..31]$	[7, 4,13, 1,10, 6,15, 3,12, 0, 9, 5, 2,14,11, 8]
$z_L[32..47]$	[3,10,14, 4, 9,15, 8, 1, 2, 7, 0, 6,13,11, 5,12]
$z_L[48..63]$	[1, 9,11,10, 0, 8,12, 4,13, 3, 7,15,14, 5, 6, 2]
$z_L[64..79]$	[4, 0, 5, 9, 7,12, 2,10,14, 1, 3, 8,11, 6,15,13]
$z_R[\ 0..15]$	[5,14, 7, 0, 9, 2,11, 4,13, 6,15, 8, 1,10, 3,12]
$z_R[16..31]$	[6,11, 3, 7, 0,13, 5,10,14,15, 8,12, 4, 9, 1, 2]
$z_R[32..47]$	[15, 5, 1, 3, 7,14, 6, 9,11, 8,12, 2,10, 0, 4,13]
$z_R[48..63]$	[8, 6, 4, 1, 3,11,15, 0, 5,12, 2,13, 9, 7,10,14]
$z_R[64..79]$	[12,15,10, 4, 1, 5, 8, 7, 6, 2,13,14, 0, 3, 9,11]
$s_L[\ 0..15]$	[11,14,15,12, 5, 8, 7, 9,11,13,14,15, 6, 7, 9, 8]
$s_L[16..31]$	[7, 6, 8,13,11, 9, 7,15, 7,12,15, 9,11, 7,13,12]
$s_L[32..47]$	[11,13, 6, 7,14, 9,13,15,14, 8,13, 6, 5,12, 7, 5]
$s_L[48..63]$	[11,12,14,15,14,15, 9, 8, 9,14, 5, 6, 8, 6, 5,12]
$s_L[64..79]$	[9,15, 5,11, 6, 8,13,12, 5,12,13,14,11, 8, 5, 6]
$s_R[\ 0..15]$	[8, 9, 9,11,13,15,15, 5, 7, 7, 8,11,14,14,12, 6]
$s_R[16..31]$	[9,13,15, 7,12, 8, 9,11, 7, 7,12, 7, 6,15,13,11]
$s_R[32..47]$	[9, 7,15,11, 8, 6, 6,14,12,13, 5,14,13,13, 7, 5]
$s_R[48..63]$	[15, 5, 8,11,14,14, 6,14, 6, 9,12, 9,12, 5,15, 8]
$s_R[64..79]$	[8, 5,12, 9,12, 5,14, 6, 8,13, 6, 5,15,13,11,11]

Table 9.9: *RIPEMD-160 word-access orders and rotate counts (cf. Algorithm 9.55).*

9.4.3 Hash functions based on modular arithmetic

The basic idea of hash functions based on modular arithmetic is to construct an iterated hash function using mod M arithmetic as the basis of a compression function. Two motivating factors are re-use of existing software or hardware (in public-key systems) for modular arithmetic, and scalability to match required security levels. Significant disadvantages, however, include speed (e.g., relative to the customized hash functions of §9.4.2), and an embarrassing history of insecure proposals.

MASH

MASH-1 (*Modular Arithmetic Secure Hash, algorithm 1*) is a hash function based on modular arithmetic. It has been proposed for inclusion in a draft ISO/IEC standard. MASH-1 involves use of an RSA-like modulus M, whose bitlength affects the security. M should be difficult to factor, and for M of unknown factorization, the security is based in part on the difficulty of extracting modular roots (§3.5.2). The bitlength of M also determines the blocksize for processing messages, and the size of the hash-result (e.g., a 1025-bit modulus yields a 1024-bit hash-result). As a recent proposal, its security remains open to question (page 381). Techniques for reducing the size of the final hash-result have also been proposed, but their security is again undetermined as yet.

9.56 Algorithm MASH-1 (version of Nov. 1995)

INPUT: data x of bitlength $0 \leq b < 2^{n/2}$.
OUTPUT: n-bit hash of x (n is approximately the bitlength of the modulus M).

1. *System setup and constant definitions.* Fix an RSA-like modulus $M = pq$ of bitlength m, where p and q are randomly chosen secret primes such that the factorization of M is intractable. Define the bitlength n of the hash-result to be the largest multiple of 16 less than m (i.e., $n = 16n' < m$). $H_0 = 0$ is defined as an IV, and an n-bit integer constant $A = 0xf0\ldots0$. "$\vee$" denotes bitwise inclusive-OR; "$\oplus$" denotes bitwise exclusive-OR.

2. *Padding, blocking, and MD-strengthening.* Pad x with 0-bits, if necessary, to obtain a string of bitlength $t \cdot n/2$ for the smallest possible $t \geq 1$. Divide the padded text into $(n/2)$-bit blocks $x_1, \ldots, x_t$, and append a final block x_{t+1} containing the $(n/2)$-bit representation of b.

3. *Expansion.* Expand each x_i to an n-bit block y_i by partitioning it into (4-bit) nibbles and inserting four 1-bits preceding each, except for y_{t+1} wherein the inserted nibble is 1010 (not 1111).

4. *Compression function processing.* For $1 \leq i \leq t+1$, map two n-bit inputs (H_{i-1}, y_i) to one n-bit output as follows: $H_i \leftarrow ((((H_{i-1} \oplus y_i) \vee A)^2 \bmod M) \dashv n) \oplus H_{i-1}$. Here $\dashv n$ denotes keeping the rightmost n bits of the m-bit result to its left.

5. *Completion.* The hash is the n-bit block H_{t+1}.

MASH-2 is defined as per MASH-1 with the exponent $e = 2$ used for squaring in the compression function processing stage (step 4) replaced with $e = 2^8 + 1$.

9.5 Keyed hash functions (MACs)

Keyed hash functions whose specific purpose is message authentication are called message authentication code (MAC) algorithms. Compared to the large number of MDC algorithms, prior to 1995 relatively few MAC algorithms had been proposed, presumably because the original proposals, which were widely adopted in practice, were adequate. Many of these are for historical reasons block-cipher based. Those with relatively short MAC bitlengths (e.g., 32-bits for MAA) or short keys (e.g., 56 bits for MACs based on DES-CBC) may still offer adequate security, depending on the computational resources available to adversaries, and the particular environment of application.

Many iterated MACs can be described as iterated hash functions (see Figure 9.2, and equation (9.1) on page 333). In this case, the MAC key is generally part of the output transformation g; it may also be an input to the compression function in the first iteration, and be involved in the compression function f at every stage.

Fact 9.57 is a general result giving an upper bound on the security of MACs.

9.57 Fact (*birthday attack on MACs*) Let h be a MAC algorithm based on an iterated compression function, which has n bits of internal chaining variable, and is deterministic (i.e., the m-bit result is fully determined by the message). Then MAC forgery is possible using $O(2^{n/2})$ known text-MAC pairs plus a number v of chosen text-MAC pairs which (depending on h) is between 1 and about 2^{n-m}.

9.5.1 MACs based on block ciphers

CBC-based MACs

The most commonly used MAC algorithm based on a block cipher makes use of cipher-block-chaining (§7.2.2(ii)). When DES is used as the block cipher E, $n = 64$ in what follows, and the MAC key is a 56-bit DES key.

9.58 Algorithm CBC-MAC

INPUT: data x; specification of block cipher E; secret MAC key k for E.
OUTPUT: n-bit MAC on x (n is the blocklength of E).

1. *Padding and blocking.* Pad x if necessary (e.g., using Algorithm 9.30). Divide the padded text into n-bit blocks denoted $x_1, \ldots, x_t$.
2. *CBC processing.* Letting E_k denote encryption using E with key k, compute the block H_t as follows: $H_1 \leftarrow E_k(x_1)$; $H_i \leftarrow E_k(H_{i-1} \oplus x_i)$, $2 \leq i \leq t$. (This is standard cipher-block-chaining, $IV = 0$, discarding ciphertext blocks $C_i = H_i$.)
3. *Optional process to increase strength of MAC.* Using a second secret key $k' \neq k$, optionally compute: $H_t' \leftarrow E_{k'}^{-1}(H_t)$, $H_t \leftarrow E_k(H_t')$. (This amounts to using two-key triple-encryption on the last block; see Remark 9.59.)
4. *Completion.* The MAC is the n-bit block H_t.

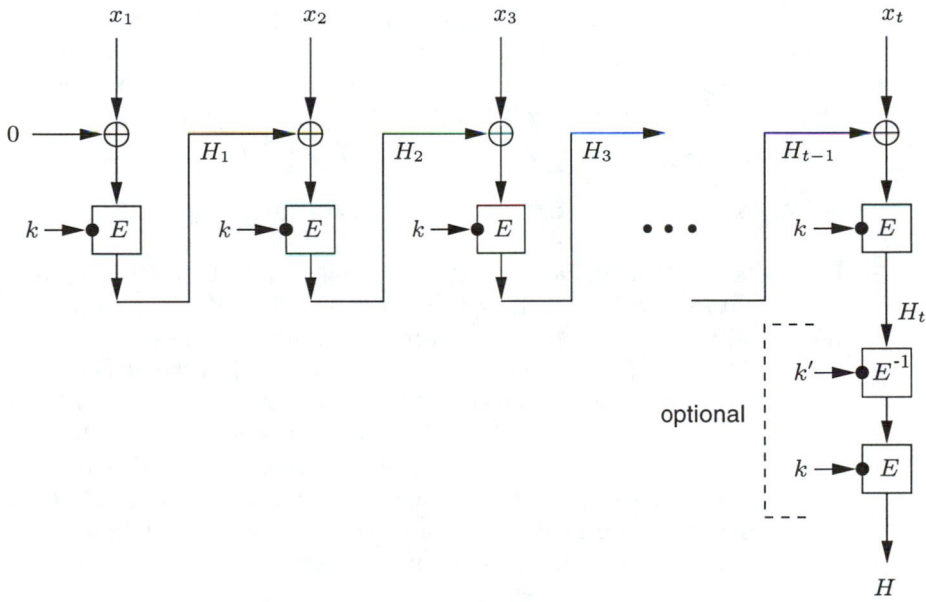

Figure 9.6: *CBC-based MAC algorithm.*

For CBC-MAC with $n = 64 = m$, Fact 9.57 applies with $v = 1$.

9.59 Remark (*CBC-MAC strengthening*) The optional process reduces the threat of exhaustive key search, and prevents chosen-text existential forgery (Example 9.62), without impacting the efficiency of the intermediate stages as would using two-key triple-encryption throughout. Alternatives to combat such forgery include prepending the input with a length block before the MAC computation; or using key K to encrypt the length m yielding $K' = E_K(m)$, before using K' as the key to MAC the message.

9.60 Remark (*truncated MAC outputs*) Exhaustive attack may, depending on the unicity distance of the MAC, be precluded (information-theoretically) by using less than n bits of the final output as the m-bit MAC. (This must be traded off against an increase in the probability of randomly guessing the MAC: 2^{-m}.) For $m = 32$ and $E = \text{DES}$, an exhaustive attack reduces the key space to about 2^{24} possibilities. However, even for $m < n$, a second text-MAC pair almost certainly determines a unique MAC key.

9.61 Remark (*CBC-MAC IV*) While a random IV in CBC encryption serves to prevent a codebook attack on the first ciphertext block, this is not a concern in a MAC algorithm.

9.62 Example (*existential forgery of CBC-MAC*) While CBC-MAC is secure for messages of a fixed number t of blocks, additional measures (beyond simply adding a trailing length-block) are required if variable length messages are allowed, otherwise (adaptive chosen-text) existential forgery is possible as follows. Assume x_i is an n-bit block, and let $\perp b$ denote the n-bit binary representation of b. Let (x_1, M_1) be a known text-MAC pair, and request the MAC M_2 for the one-block message $x_2 = M_1$; then $M_2 = E_k(E_k(x_1))$ is also the MAC for the 2-block message $(x_1 || \perp 0)$. As a less trivial example, given two known text-MAC pairs (x_1, H_1), (x_2, H_2) for one-block messages x_1, x_2, and requesting the MAC M on a chosen 2-block third message $(x_1 || z)$ for a third text-MAC pair $((x_1 || z), M)$, then $H_i = E_k(x_i)$, $M = E_k(H_1 \oplus z)$, and the MAC for the new 2-block message $X = x_2 || (H_1 \oplus z \oplus H_2)$ is known – it is M also. Moreover, MD-strengthening (Algorithm 9.26) does not address the problem: assume padding by Algorithm 9.29, replace the third message above by the 3-block message $(x_1 || \perp 64 || z)$, note

$$H_i' = E_k(E_k(x_i) \oplus \perp 64), \quad M_3 = E_k(E_k(E_k(E_k(x_1) \oplus \perp 64) \oplus z) \oplus \perp 192),$$

and M_3 is also the MAC for the new 3-block message $X = (x_2 || \perp 64 || H_1' \oplus H_2' \oplus z)$.   □

9.63 Example (*RIPE-MAC*) RIPE-MAC is a variant of CBC-MAC. Two versions RIPE-MAC1 and RIPE-MAC3, both producing 64-bit MACs, differ in their internal encryption function E being either single DES or two-key triple-DES, respectively, requiring a 56- or 112-bit key k (cf. Remark 9.59). Differences from Algorithm 9.58 are as follows: the compression function uses a non-invertible chaining best described as CBC with data feed-forward: $H_i \leftarrow E_k(H_{i-1} \oplus x_i) \oplus x_i$; after padding using Algorithm 9.30, a final 64-bit length-block (giving bitlength of original input) is appended; the optional process of Algorithm 9.58 is mandatory with final output block encrypted using key k' derived by complementing alternating nibbles of k: for $k = k_0 \ldots k_{63}$ a 56-bit DES key with parity bits $k_7 k_{15} \ldots k_{63}$, $k' = k \oplus \text{0xf0f0f0f0f0f0f0f0}$.   □

9.5.2 Constructing MACs from MDCs

A common suggestion is to construct a MAC algorithm from an MDC algorithm, by simply including a secret key k as part of the MDC input. A concern with this approach is that implicit but unverified assumptions are often made about the properties that MDCs have; in particular, while most MDCs are designed to provide one-wayness or collision resistance, the requirements of a MAC algorithm differ (Definition 9.7). Even in the case that a one-way hash function precludes recovery of a secret key used as a partial message input (cf. partial-preimage resistance, page 331), this does not guarantee the infeasibility of producing MACs for new inputs. The following examples suggest that construction of a MAC from a hash function requires careful analysis.

9.64 Example (*secret prefix method*) Consider a message $x = x_1 x_2 \ldots x_t$ and an iterated MDC h with compression function f, with definition: $H_0 = IV, H_i = f(H_{i-1}, x_i); h(x) = H_t$. (1) Suppose one attempts to use h as a MAC algorithm by prepending a secret key k, so that the proposed MAC on x is $M = h(k||x)$. Then, extending the message x by an arbitrary single block y, one may deduce $M' = h(k||x||y)$ as $f(M, y)$ without knowing the secret key k (the original MAC M serves as chaining variable). This is true even for hash functions whose preprocessing pads inputs with length indicators (e.g., MD5); in this case, the padding/length-block z for the original message x would appear as part of the extended message, $x||z||y$, but a forged MAC on the latter may nonetheless be deduced. (2) For similar reasons, it is insecure to use an MDC to construct a MAC algorithm by using the secret MAC key k as IV. If k comprises the entire first block, then for efficiency $f(IV, k)$ may be precomputed, illustrating that an adversary need only find a k' (not necessarily k) such that $f(IV, k) = f(IV, k')$; this is equivalent to using a secret IV. $\square$

9.65 Example (*secret suffix method*) An alternative proposal is to use a secret key as a suffix, i.e., the n-bit MAC on x is $M = h(x||k)$. In this case, a birthday attack applies (§9.7.1). An adversary free to choose the message x (or a prefix thereof) may, in $O(2^{n/2})$ operations, find a pair of messages x, x' for which $h(x) = h(x')$. (This can be done off-line, and does not require knowledge of k; the assumption here is that n is the size of both the chaining variable and the final output.) Obtaining a MAC M on x by legitimate means then allows an adversary to produce a correct text-MAC pair (x', M) for a new message x'. Note that this method essentially hashes and then encrypts the hash-value in the final iteration; in this weak form of MAC, the MAC-value depends only on the last chaining value, and the key is used in only one step. $\square$

The above examples suggest that a MAC key should be involved at both the start and the end of MAC computations, leading to Example 9.66.

9.66 Example (*envelope method with padding*) For a key k and MDC h, compute the MAC on a message x as: $h_k(x) = h(k \,||\, p \,||\, x \,||\, k)$. Here p is a string used to pad k to the length of one block, to ensure that the internal computation involves at least two iterations. For example, if h is MD5 and k is 128 bits, p is a 384-bit pad string. $\square$

Due to both a certificational attack against the MAC construction of Example 9.66 and theoretical support for that of Example 9.67 (see page 382), the latter construction is favored.

9.67 Example (*hash-based MAC*) For a key k and MDC h, compute the MAC on a message x as $\text{HMAC}(x) = h(k \,||\, p_1 \,||\, h(k \,||\, p_2 \,||\, x))$, where p_1, p_2 are distinct strings of sufficient length to pad k out to a full block for the compression function. The overall construction is quite efficient despite two calls to h, since the outer execution processes only (e.g., if h is MD5) a two-block input, independent of the length of x. $\square$

Additional suggestions for achieving MAC-like functionality by combining MDCs and encryption are discussed in §9.6.5.

9.5.3 Customized MACs

Two algorithms designed for the specific purpose of message authentication are discussed in this section: MAA and MD5-MAC.

Message Authenticator Algorithm (MAA)

The Message Authenticator Algorithm (MAA), dating from 1983, is a customized MAC algorithm for 32-bit machines, involving 32-bit operations throughout. It is specified as Algorithm 9.68 and illustrated in Figure 9.7. The main loop consists of two parallel inter-dependent streams of computation. Messages are processed in 4-byte blocks using 8 bytes of chaining variable. The execution time (excluding key expansion) is proportional to message length; as a rough guideline, MAA is twice as slow as MD4.

9.68 Algorithm Message Authenticator Algorithm (MAA)

INPUT: data x of bitlength $32j$, $1 \le j \le 10^6$; secret 64-bit MAC key $Z = Z[1]..Z[8]$.
OUTPUT: 32-bit MAC on x.

1. *Message-independent key expansion.* Expand key Z to six 32-bit quantities X, Y, V, W, S, T (X, Y are initial values; V, W are main loop variables; S, T are appended to the message) as follows.

   1.1 First replace any bytes 0x00 or 0xff in Z as follows. $P \leftarrow 0$; for i from 1 to 8 $(P \leftarrow 2P$; if $Z[i] = 0x00$ or 0xff then $(P \leftarrow P + 1; Z[i] \leftarrow Z[i]$ OR $P))$.

   1.2 Let J and K be the first 4 bytes and last 4 bytes of Z, and compute:[4]
   $X \leftarrow J^4 \pmod{2^{32} - 1} \oplus J^4 \pmod{2^{32} - 2}$
   $Y \leftarrow [K^5 \pmod{2^{32} - 1} \oplus K^5 \pmod{2^{32} - 2}](1 + P)^2 \pmod{2^{32} - 2}$
   $V \leftarrow J^6 \pmod{2^{32} - 1} \oplus J^6 \pmod{2^{32} - 2}$
   $W \leftarrow K^7 \pmod{2^{32} - 1} \oplus K^7 \pmod{2^{32} - 2}$
   $S \leftarrow J^8 \pmod{2^{32} - 1} \oplus J^8 \pmod{2^{32} - 2}$
   $T \leftarrow K^9 \pmod{2^{32} - 1} \oplus K^9 \pmod{2^{32} - 2}$

   1.3 Process the 3 resulting pairs (X, Y), (V, W), (S, T) to remove any bytes 0x00, 0xff as for Z earlier. Define the AND-OR constants: $A = 0x02040801$, $B = 0x00804021$, $C = 0xbfef7fdf$, $D = 0x7dfefbff$.

2. *Initialization and preprocessing.* Initialize the rotating vector: $v \leftarrow V$, and the chaining variables: $H_1 \leftarrow X$, $H_2 \leftarrow Y$. Append the key-derived blocks S, T to x, and let $x_1 \ldots x_t$ denote the resulting augmented segment of 32-bit blocks. (The final 2 blocks of the segment thus involve key-derived secrets.)

3. *Block processing.* Process each 32-bit block x_i (for i from 1 to t) as follows.
   $v \leftarrow (v \hookleftarrow 1)$, $\ U \leftarrow (v \oplus W)$
   $t_1 \leftarrow (H_1 \oplus x_i) \times_1 (((H_2 \oplus x_i) + U) \text{ OR } A) \text{ AND } C)$
   $t_2 \leftarrow (H_2 \oplus x_i) \times_2 (((H_1 \oplus x_i) + U) \text{ OR } B) \text{ AND } D)$
   $H_1 \leftarrow t_1, H_2 \leftarrow t_2$
   where $\times_i$ denotes special multiplication mod $2^{32} - i$ as noted above ($i = 1$ or 2); "+" is addition mod 2^{32}; and "$\hookleftarrow 1$" denotes rotation left one bit. (Each combined AND-OR operation on a 32-bit quantity sets 4 bits to 1, and 4 to 0, precluding 0-multipliers.)

4. *Completion.* The resulting MAC is: $H = H_1 \oplus H_2$.

---

[4]In ISO 8731-2, a well-defined but unconventional definition of multiplication mod $2^{32} - 2$ is specified, producing 32-bit results which in some cases are $2^{32} - 1$ or $2^{32} - 2$; for this reason, specifying e.g., J^6 here may be ambiguous; the standard should be consulted for exact details.

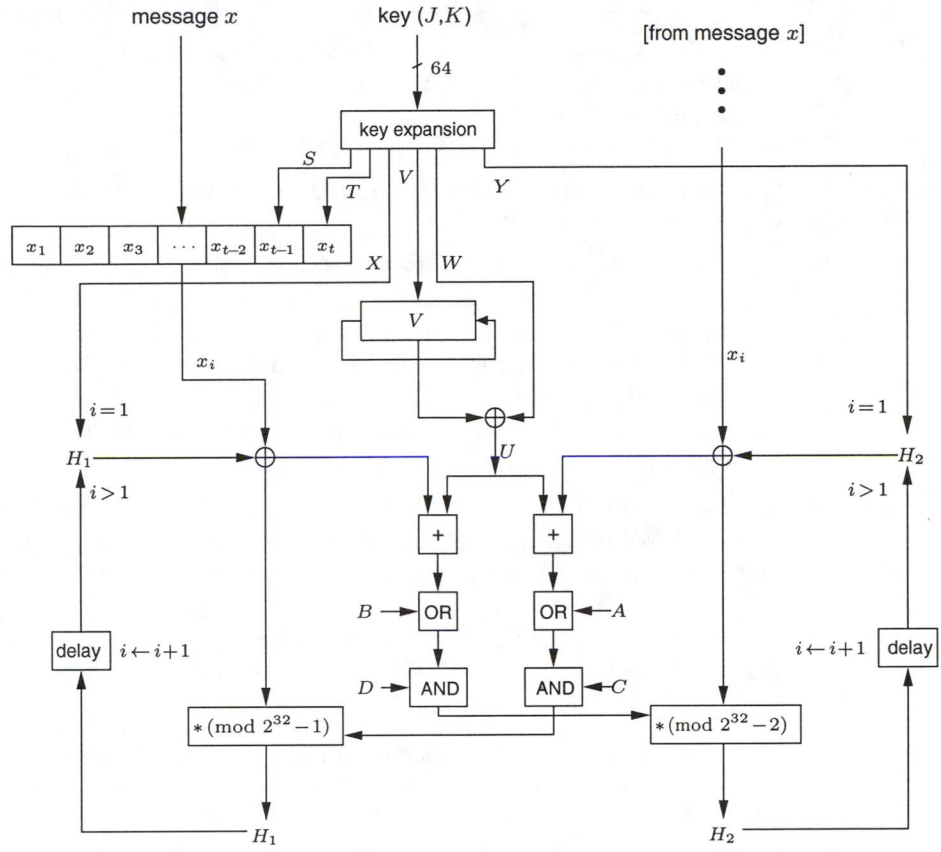

Figure 9.7: *The Message Authenticator Algorithm (MAA).*

Since the relatively complex key expansion stage is independent of the message, a one-time computation suffices for a fixed key. The mixing of various operations (arithmetic mod $2^{32} - i$, for $i = 0, 1$ and 2; XOR; and nonlinear AND-OR computations) is intended to strengthen the algorithm against arithmetic cryptanalytic attacks.

MD5-MAC

A more conservative approach (cf. Example 9.66) to building a MAC from an MDC is to arrange that the MAC compression function itself depend on k, implying the secret key be involved in all intervening iterations; this provides additional protection in the case that weaknesses of the underlying hash function become known. Algorithm 9.69 is such a technique, constructed using MD5. It provides performance close to that of MD5 (5-20% slower in software).

9.69 Algorithm MD5-MAC

INPUT: bitstring x of arbitrary bitlength $b \geq 0$; key k of bitlength ≤ 128.
OUTPUT: 64-bit MAC-value of x.

MD5-MAC is obtained from MD5 (Algorithm 9.51) by the following changes.

1. *Constants.* The constants U_i and T_i are as defined in Example 9.70.
2. *Key expansion.*
   (a) If k is shorter than 128 bits, concatenate k to itself a sufficient number of times, and redefine k to be the leftmost 128 bits.
   (b) Let $\overline{\text{MD5}}$ denote MD5 with both padding and appended length omitted. Expand k into three 16-byte subkeys K_0, K_1, and K_2 as follows: for i from 0 to 2, $K_i \leftarrow \overline{\text{MD5}}(k \,\|\, U_i \,\|\, k)$.
   (c) Partition each of K_0 and K_1 into four 32-bit substrings $K_j[i]$, $0 \leq i \leq 3$.
3. K_0 replaces the four 32-bit IV's of MD5 (i.e., $h_i = K_0[i]$).
4. $K_1[i]$ is added mod 2^{32} to each constant $y[j]$ used in Round i of MD5.
5. K_2 is used to construct the following 512-bit block, which is appended to the padded input x subsequent to the regular padding and length block as defined by MD5: $K_2 \,\|\, K_2 \oplus T_0 \,\|\, K_2 \oplus T_1 \,\|\, K_2 \oplus T_2$.
6. The MAC-value is the leftmost 64 bits of the 128-bit output from hashing this padded and extended input string using MD5 with the above modifications.

9.70 Example (*MD5-MAC constants/test vectors*) The 16-byte constants T_i and three test vectors $(x, \text{MD5-MAC}(x))$ for key $k = $ 00112233445566778899aabbccddeeff are given below. (The T_i themselves are derived using MD5 on pre-defined constants.) With subscripts in T_i taken mod 3, the 96-byte constants U_0, U_1, U_2 are defined:
$$U_i = T_i \,\|\, T_{i+1} \,\|\, T_{i+2} \,\|\, T_i \,\|\, T_{i+1} \,\|\, T_{i+2}.$$

```
T0:     97 ef 45 ac 29 0f 43 cd 45 7e 1b 55 1c 80 11 34
T1:     b1 77 ce 96 2e 72 8e 7c 5f 5a ab 0a 36 43 be 18
T2:     9d 21 b4 21 bc 87 b9 4d a2 9d 27 bd c7 5b d7 c3
("",                                  1f1ef2375cc0e0844f98e7e811a34da8)
("abc",                               e8013c11f7209d1328c0caa04fd012a6)
("abcdefghijklmnopqrstuvwxyz", 9172867eb60017884c6fa8cc88ebe7c9)
```
□

9.5.4 MACs for stream ciphers

Providing data origin authentication and data integrity guarantees for stream ciphers is particularly important due to the fact that bit manipulations in additive stream-ciphers may directly result in predictable modifications of the underlying plaintext (e.g., Example 9.83). While iterated hash functions process message data a block at a time (§9.3.1), MACs designed for use with stream ciphers process messages either one bit or one symbol (block) at a time, and those which may be implemented using linear feedback shift registers (LFSRs) are desirable for reasons of efficiency.

One such MAC technique, Algorithm 9.72 below, is based on cyclic redundancy codes (cf. Example 9.80). In this case, the polynomial division may be implemented using an LFSR. The following definition is of use in what follows.

9.71 Definition A (b, m) *hash-family* $\mathcal{H}$ is a collection of hash functions mapping b-bit messages to m-bit hash-values. A (b, m) hash-family is ε-*balanced* if for all messages $B \neq 0$ and all m-bit hash-values c, $prob_h(h(B) = c)) \leq \varepsilon$, where the probability is over all randomly selected functions $h \in \mathcal{H}$.

9.72 Algorithm CRC-based MAC

INPUT: b-bit message B; shared key (see below) between MAC source and verifier.
OUTPUT: m-bit MAC-value on B (e.g., $m = 64$).

1. *Notation.* Associate $B = B_{b-1} \ldots B_1 B_0$ with the polynomial $B(x) = \sum_{i=0}^{b-1} B_i x^i$.
2. *Selection of MAC key.*
   (a) Select a random binary irreducible polynomial $p(x)$ of degree m. (This represents randomly drawing a function h from a (b, m) hash-family.)
   (b) Select a random m-bit one-time key k (to be used as a one-time pad).

   The secret MAC key consists of $p(x)$ and k, both of which must be shared a priori between the MAC originator and verifier.
3. Compute $h(B) = coef(B(x) \cdot x^m \bmod p(x))$, the m-bit string of coefficients from the degree $m - 1$ remainder polynomial after dividing $B(x) \cdot x^m$ by $p(x)$.
4. The m-bit MAC-value for B is: $h(B) \oplus k$.

9.73 Fact (*security of CRC-based MAC*) For any values b and $m > 1$, the hash-family resulting from Algorithm 9.72 is ε-balanced for $\varepsilon = (b + m)/(2^{m-1})$, and the probability of MAC forgery is at most ε.

9.74 Remark (*polynomial reuse*) The hash function h in Algorithm 9.72 is determined by the irreducible polynomial $p(x)$. In practice, $p(x)$ may be re-used for different messages (e.g., within a session), but for each message a new random key k should be used.

9.6 Data integrity and message authentication

This section considers the use of hash functions for data integrity and message authentication. Following preliminary subsections, respectively, providing background definitions and distinguishing non-malicious from malicious threats to data integrity, three subsequent subsections consider three basic approaches to providing data integrity using hash functions, as summarized in Figure 9.8.

9.6.1 Background and definitions

This subsection discusses data integrity, data origin authentication (message authentication), and transaction authentication.

Assurances are typically required both that data actually came from its reputed source (data origin authentication), and that its state is unaltered (data integrity). These issues cannot be separated – data which has been altered effectively has a new source; and if a source cannot be determined, then the question of alteration cannot be settled (without reference to a source). Integrity mechanisms thus implicitly provide data origin authentication, and vice versa.

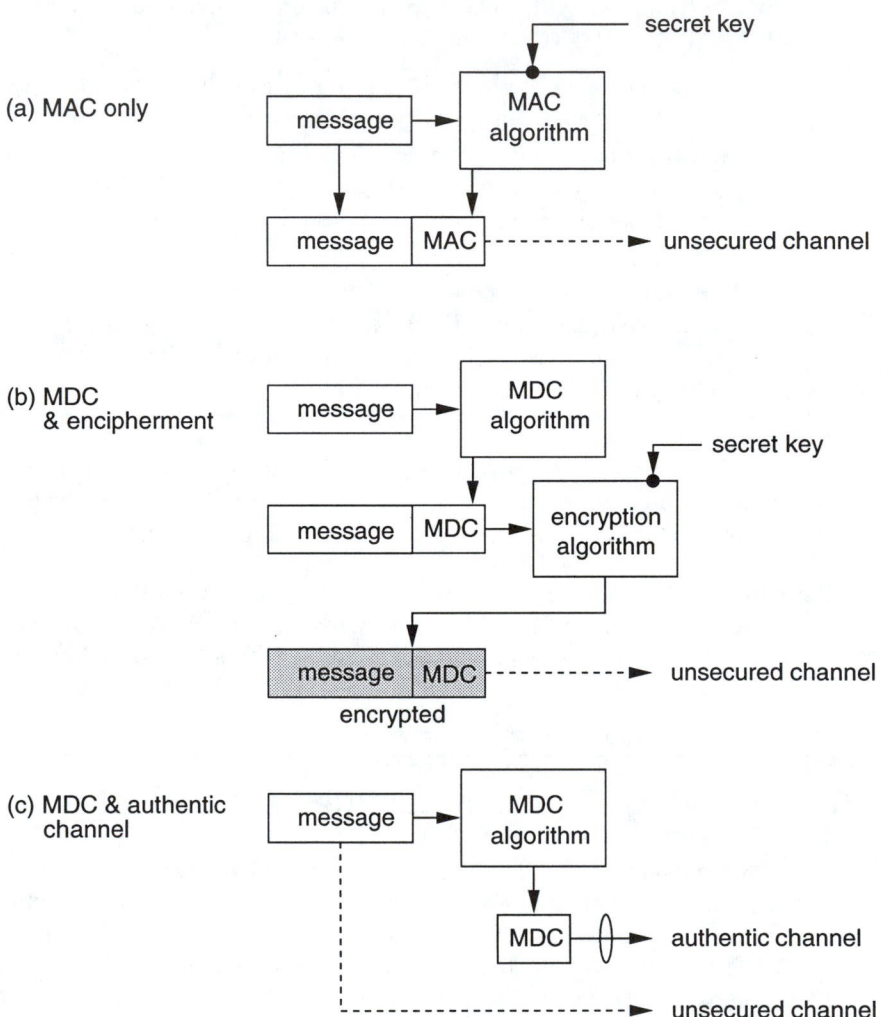

Figure 9.8: *Three methods for providing data integrity using hash functions. The second method provides encipherment simultaneously.*

(i) Data integrity

9.75 Definition *Data integrity* is the property whereby data has not been altered in an unauthorized manner since the time it was created, transmitted, or stored by an authorized source.

Verification of data integrity requires that only a subset of all candidate data items satisfies particular criteria distinguishing the acceptable from the unacceptable. Criteria allowing recognizability of data integrity include appropriate redundancy or expectation with respect to format. Cryptographic techniques for data integrity rely on either secret information or authentic channels (§9.6.4).

The specific focus of data integrity is on the bitwise composition of data (cf. transaction authentication below). Operations which invalidate integrity include: insertion of bits, including entirely new data items from fraudulent sources; deletion of bits (short of deleting entire data items); re-ordering of bits or groups of bits; inversion or substitution of bits; and any combination of these, such as message splicing (re-use of proper substrings to construct new or altered data items). Data integrity includes the notion that data items are complete. For items split into multiple blocks, the above alterations apply analogously with blocks envisioned as substrings of a contiguous data string.

(ii) Data origin authentication (message authentication)

9.76 Definition *Data origin authentication* is a type of authentication whereby a party is corroborated as the (original) source of specified data created at some (typically unspecified) time in the past.

By definition, data origin authentication includes data integrity.

9.77 Definition *Message authentication* is a term used analogously with data origin authentication. It provides data origin authentication with respect to the original message source (and data integrity, but no uniqueness and timeliness guarantees).

Methods for providing data origin authentication include the following:

1. message authentication codes (MACs)
2. digital signature schemes
3. appending (prior to encryption) a secret authenticator value to encrypted text.[5]

Data origin authentication mechanisms based on shared secret keys (e.g., MACs) do not allow a distinction to be made between the parties sharing the key, and thus (as opposed to digital signatures) do not provide non-repudiation of data origin – either party can equally originate a message using the shared key. If resolution of subsequent disputes is a potential requirement, either an on-line trusted third party in a notary role, or asymmetric techniques (see Chapter 11) may be used.

While MACs and digital signatures may be used to establish that data was generated by a specified party at some time in the past, they provide no inherent uniqueness or timeliness guarantees. These techniques alone thus cannot detect message re-use or replay, which is necessary in environments where messages may have renewed effect on second or subsequent use. Such message authentication techniques may, however, be augmented to provide these guarantees, as next discussed.

---

[5] Such a *sealed authenticator* (cf. a MAC, sometimes called an *appended authenticator*) is used along with an encryption method which provides error extension. While this resembles the technique of using encryption and an MDC (§9.6.5), whereas the MDC is a (known) function of the plaintext, a sealed authenticator is itself secret.

(iii) Transaction authentication

9.78 Definition *Transaction authentication* denotes message authentication augmented to additionally provide uniqueness and timeliness guarantees on data (thus preventing undetectable message replay).

The uniqueness and timeliness guarantees of Definition 9.78 are typically provided by appropriate use of time-variant parameters (TVPs). These include random numbers in challenge-response protocols, sequence numbers, and timestamps as discussed in §10.3.1. This may be viewed as a combination of message authentication and entity authentication (Definition 10.1). Loosely speaking,

$$\text{message authentication} + \text{TVP} = \text{transaction authentication.}$$

As a simple example, sequence numbers included within the data of messages authenticated by a MAC or digital signature algorithm allow replay detection (see Remark 9.79), and thus provide transaction authentication.

As a second example, for exchanges between two parties involving two or more messages, transaction authentication on each of the second and subsequent messages may be provided by including in the message data covered by a MAC a random number sent by the other party in the previous message. This chaining of messages through random numbers prevents message replay, since any MAC values in replayed messages would be incorrect (due to disagreement between the random number in the replayed message, and the most recent random number of the verifier).

Table 9.10 summarizes the properties of these and other types of authentication. Authentication in the broadest sense encompasses not only data integrity and data origin authentication, but also protection from all active attacks including fraudulent representation and message replay. In contrast, encryption provides protection only from passive attacks.

$\rightarrow$ Property $\downarrow$ Type of authentication	identification of source	data integrity	timeliness or uniqueness	defined in
message authentication	yes	yes	—	§9.6.1
transaction authentication	yes	yes	yes	§9.6.1
entity authentication	yes	—	yes	§10.1.1
key authentication	yes	yes	desirable	§12.2.1

Table 9.10: *Properties of various types of authentication.*

9.79 Remark (*sequence numbers and authentication*) Sequence numbers may provide uniqueness, but not (real-time) timeliness, and thus are more appropriate to detect message replay than for entity authentication. Sequence numbers may also be used to detect the deletion of entire messages; they thus allow data integrity to be checked over an ongoing sequence of messages, in addition to individual messages.

9.6.2 Non-malicious vs. malicious threats to data integrity

The techniques required to provide data integrity on noisy channels differ substantially from those required on channels subject to manipulation by adversaries.

Checksums provide protection against accidental or non-malicious errors on channels which are subject to transmission errors. The protection is non-cryptographic, in the sense that neither secret keys nor secured channels are used. Checksums generalize the idea of a parity bit by appending a (small) constant amount of message-specific redundancy. Both the data and the checksum are transmitted to a receiver, at which point the same redundancy computation is carried out on the received data and compared to the received checksum. Checksums can be used either for error detection or in association with higher-level error-recovery strategies (e.g., protocols involving acknowledgements and retransmission upon failure). Trivial examples include an arithmetic checksum (compute the running 32-bit sum of all 32-bit data words, discarding high-order carries), and a simple XOR (XOR all 32-bit words in a data string). *Error-correcting codes* go one step further than error-detecting codes, offering the capability to actually correct a limited number of errors without retransmission; this is sometimes called *forward error correction*.

9.80 Example (*CRCs*) *Cyclic redundancy codes* or CRCs are commonly used checksums. A k-bit CRC algorithm maps arbitrary length inputs into k-bit imprints, and provides significantly better error-detection capability than k-bit arithmetic checksums. The algorithm is based on a carefully chosen $(k + 1)$-bit vector represented as a binary polynomial; for $k = 16$, a commonly used polynomial (CRC-16) is $g(x) = 1 + x^2 + x^{15} + x^{16}$. A t-bit data input is represented as a binary polynomial $d(x)$ of degree $t - 1$, and the CRC-value corresponding to $d(x)$ is the 16-bit string represented by the polynomial remainder $c(x)$ when $x^{16} \cdot d(x)$ is divided by $g(x)$;[6] polynomial remaindering is analogous to computing integer remainders by long division. For all messages $d(x)$ with $t < 32\ 768$, CRC-16 can detect all errors that consist of only a single bit, two bits, three bits, or any odd number of bits, all burst errors of bitlength 16 or less, 99.997% $(1 - 2^{-15})$ of 17-bit burst errors, and 99.998% $(1 - 2^{-16})$ of all bursts 18 bits or longer. (A *burst error* of bitlength b is any bitstring of exactly b bits beginning and ending with a 1.) Analogous to the integer case, other data strings $d'(x)$ yielding the same remainder as $d(x)$ can be trivially found by adding multiples of the divisor $g(x)$ to $d(x)$, or inserting extra blocks representing a multiple of $g(x)$. CRCs thus do not provide one-wayness as required for MDCs; in fact, CRCs are a class of linear (error correcting) codes, with one-wayness comparable to an XOR-sum. $\square$

While of use for detection of random errors, k-bit checksums are not of cryptographic use, because typically a data string checksumming to any target value can be easily created. One method is to simply insert or append to any data string of choice a k-bit correcting-block c which has the effect of correcting the overall checksum to the desired value. For example, for the trivial XOR checksum, if the target checksum is c', insert as block c the XOR of c' and the XOR of all other blocks.

In contrast to checksums, data integrity mechanisms based on (cryptographic) hash functions are specifically designed to preclude undetectable intentional modification. The hash-values resulting are sometimes called *integrity check values (ICV)*, or *cryptographic check values* in the case of keyed hash functions. Semantically, it should not be possible for an adversary to take advantage of the willingness of users to associate a given hash output with a single, specific input, despite the fact that each such output typically corresponds to a large set of inputs. Hash functions should exhibit no predictable relationships or correlations between inputs and outputs, as these may allow adversaries to orchestrate unintended associations.

---

[6]A modification is typically used in practice (e.g., complementing $c(x)$) to address the combination of an input $d(x) = 0$ and a "stuck-at-zero" communications fault yielding a successful CRC check.

9.6.3 Data integrity using a MAC alone

Message Authentication Codes (MACs) as discussed earlier are designed specifically for applications where data integrity (but not necessarily privacy) is required. The originator of a message x computes a MAC $h_k(x)$ over the message using a secret MAC key k shared with the intended recipient, and sends both (effectively $x \parallel h_k(x)$). The recipient determines by some means (e.g., a plaintext identifier field) the claimed source identity, separates the received MAC from the received data, independently computes a MAC over this data using the shared MAC key, and compares the computed MAC to the received MAC. The recipient interprets the agreement of these values to mean the data is authentic and has integrity – that is, it originated from the other party which knows the shared key, and has not been altered in transit. This corresponds to Figure 9.8(a).

9.6.4 Data integrity using an MDC and an authentic channel

The use of a secret key is not essential in order to provide data integrity. It may be eliminated by hashing a message and protecting the authenticity of the hash via an authentic (but not necessarily private) channel. The originator computes a hash-code using an MDC over the message data, transmits the data to a recipient over an unsecured channel, and transmits the hash-code over an independent channel known to provide data origin authentication. Such authentic channels may include telephone (authenticity through voice recognition), any data medium (e.g., floppy disk, piece of paper) stored in a trusted place (e.g., locked safe), or publication over any difficult-to-forge public medium (e.g., daily newspaper). The recipient independently hashes the received data, and compares the hash-code to that received. If these values agree, the recipient accepts the data as having integrity. This corresponds to Figure 9.8(c).

Example applications include virus protection of software, and distribution of software or public keys via untrusted networks. For virus checking of computer source or object code, this technique is preferable to one resulting in encrypted text. A common example of combining an MDC with an authentic channel to provide data integrity is digital signature schemes such as RSA, which typically involve the use of MDCs, with the asymmetric signature providing the authentic channel.

9.6.5 Data integrity combined with encryption

Whereas digital signatures provide assurances regarding both integrity and authentication, in general, encryption alone provides neither. This issue is first examined, and then the question of how hash functions may be employed in conjunction with encryption to provide data integrity.

(i) Encryption alone does not guarantee data integrity

A common misconception is that encryption provides data origin authentication and data integrity, under the argument that if a message is decrypted with a key shared only with party A, and if the decrypted message is meaningful, then it must have originated from A. Here "meaningful" means the message contains sufficient redundancy or meets some other a priori expectation. While the intuition is that an attacker must know the secret key in order to manipulate messages, this is not always true. In some cases he may be able to choose the plaintext message, while in other cases he may be able to effectively manipulate plaintext despite not being able to control its specific content. The extent to which encrypted messages can be manipulated undetectably depends on many factors, as illustrated by the following examples.

9.81 Example (*re-ordering ECB blocks*) The ciphertext blocks of any block cipher used only in ECB mode are subject to re-ordering. □

9.82 Example (*encryption of random data*) If the plaintext corresponding to a given ciphertext contains no redundancy (e.g., a random key), then all attempted decryptions thereof are meaningful, and data integrity cannot be verified. Thus, some form of redundancy is always required to allow verification of integrity; moreover, to facilitate verification in practice, explicit redundancy verifiable by automated means is required. □

9.83 Example (*bit manipulations in additive stream ciphers*) Despite the fact that the one-time pad offers unconditional secrecy, an attacker can change any single bit of plaintext by modifying the corresponding bit of ciphertext. For known-plaintext attacks, this allows an attacker to substitute selected segments of plaintext by plaintext of his own choosing. An example target bit is the high-order bit in a numeric field known to represent a dollar value. Similar comments apply to any additive stream cipher, including the OFB mode of any block cipher. □

9.84 Example (*bit manipulation in DES ciphertext blocks*) Several standard modes of operation for any block cipher are subject to selective bit manipulation. Modifying the last ciphertext block in a CFB chain is undetectable. A ciphertext block in CFB mode which yields random noise upon decryption is an indication of possible selective bit-manipulation of the preceding ciphertext block. A ciphertext block in CBC mode which yields random noise upon decryption is an indication of possible selective bit-manipulation of the following ciphertext block. For further discussion regarding error extension in standard modes of operation, see §7.2.2. □

(ii) Data integrity using encryption and an MDC

If both confidentiality and integrity are required, then the following data integrity technique employing an m-bit MDC h may be used. The originator of a message x computes a hash value $H = h(x)$ over the message, appends it to the data, and encrypts the augmented message using a symmetric encryption algorithm E with shared key k, producing ciphertext

$$C = E_k(x \parallel h(x)) \qquad (9.2)$$

(Note that this differs subtly from enciphering the message and the hash separately as $(E_k(x), E_k(h(x)))$, which e.g. using CBC requires two IVs.) This is transmitted to a recipient, who determines (e.g., by a plaintext identifier field) which key to use for decryption, and separates the recovered data x' from the recovered hash H'. The recipient then independently computes the hash $h(x')$ of the received data x', and compares this to the recovered hash H'. If these agree, the recovered data is accepted as both being authentic and having integrity. This corresponds to Figure 9.8(b).

The intention is that the encryption protects the appended hash, and that it be infeasible for an attacker without the encryption key to alter the message without disrupting the correspondence between the decrypted plaintext and the recovered MDC. The properties required of the MDC here may be notably weaker, in general, than for an MDC used in conjunction with, say, digital signatures. Here the requirement, effectively a joint condition on the MDC and encryption algorithm, is that it not be feasible for an adversary to manipulate or create new ciphertext blocks so as to produce a new ciphertext C' which upon decryption will yield plaintext blocks having the same MDC as that recovered, with probability significantly greater than 1 in 2^m.

9.85 Remark (*separation of integrity and privacy*) While this approach appears to separate privacy and data integrity from a functional viewpoint, the two are not independent with respect to security. The security of the integrity mechanism is, at most, that of the encryption algorithm regardless of the strength of the MDC (consider exhaustive search of the encryption key). Thought should, therefore, be given to the relative strengths of the components.

9.86 Remark (*vulnerability to known-plaintext attack*) In environments where known-plaintext attacks are possible, the technique of equation (9.2) should not be used in conjunction with additive stream ciphers unless additional integrity techniques are used. In this scenario, an attacker can recover the key stream, then make plaintext changes, recompute a new MDC, and re-encrypt the modified message. Note this attack compromises the manner in which the MDC is used, rather than the MDC or encryption algorithm directly.

If confidentiality is not essential other than to support the requirement of integrity, an apparent option is to encrypt only either the message x or the MDC $h(x)$. Neither approach is common, for reasons including Remark 9.85, and the general undesirability to utilize encryption primitives in systems requiring only integrity or authentication services. The following further comments apply:

1. *encrypting the hash-code only*: $(x, \ E_k(h(x)))$
   Applying the key to the hash-value only (cf. Example 9.65) results in a property (typical for public-key signatures but) atypical for MACs: pairs of inputs x, x' with colliding outputs (MAC-values here) can be verifiably pre-determined without knowledge of k. Thus h must be collision-resistant. Other issues include: pairs of inputs having the same MAC-value under one key also do under other keys; if the blocklength of the cipher E_k is less than the bitlength m of the hash-value, splitting the latter across encryption blocks may weaken security; k must be reserved exclusively for this integrity function (otherwise chosen-text attacks on encryption allow selective MAC forgery); and E_k must not be an additive stream cipher (see Remark 9.86).
2. *encrypting the plaintext only*: $(E_k(x), \ h(x))$
   This offers little computational savings over encrypting both message and hash (except for very short messages) and, as above, $h(x)$ must be collision-resistant and thus twice the typical MAC bitlength. Correct guesses of the plaintext x may be confirmed (candidate values x' for x can be checked by comparing $h(x')$ to $h(x)$).

(iii) Data integrity using encryption and a MAC

It is sometimes suggested to use a MAC rather than the MDC in the mechanism of equation (9.2) on page 365. In this case, a MAC algorithm $h_{k'}$ replaces the MDC h, and rather than $C = E_k(x \ || \ h(x))$, the message sent is

$$C' = E_k(x \ || \ h_{k'}(x)) \tag{9.3}$$

The use of a MAC here offers the advantage (over an MDC) that should the encryption algorithm be defeated, the MAC still provides integrity. A drawback is the requirement of managing both an encryption key and a MAC key. Care must be exercised to ensure that dependencies between the MAC and encryption algorithms do not lead to security weaknesses, and as a general recommendation these algorithms should be independent (see Example 9.88).

9.87 Remark (*precluding exhaustive MAC search*) Encryption of the MAC-value in equation (9.3) precludes an exhaustive key search attack on the MAC key.

Two alternatives here include encrypting the plaintext first and then computing a MAC over the ciphertext, and encrypting the message and MAC separately. These are discussed in turn.

1. *computing a MAC over the ciphertext*: $(E_k(x), h_{k'}(E_k(x)))$.
   This allows message authentication without knowledge of the plaintext x (or ciphertext key). However, as the message authentication is on the ciphertext rather than the plaintext directly, there are no guarantees that the party creating the MAC knew the plaintext x. The recipient, therefore, must be careful about conclusions drawn – for example, if E_k is public-key encryption, the originator of x may be independent of the party sharing the key k' with the recipient.

2. *separate encryption and MAC*: $(E_k(x), h_{k'}(x))$.
   This alternative requires that neither the encryption nor the MAC algorithm compromises the objectives of the other. In particular, in this case an additional requirement on the algorithm is that the MAC on x must not compromise the confidentiality of x (cf. Definition 9.7). Keys (k, k') should also be independent here, e.g., to preclude exhaustive search on the weaker algorithm compromising the other (cf. Example 9.88). If k and k' are not independent, exhaustive key search is theoretically possible even without known plaintext.

(iv) Data integrity using encryption – examples

9.88 Example (*improper combination of CBC-MAC and CBC encryption*) Consider using the data integrity mechanism of equation (9.3) with E_k being CBC-encryption with key k and initialization vector IV, $h_{k'}(x)$ being CBC-MAC with k' and IV', and $k = k'$, $IV = IV'$. The data $x = x_1 x_2 \dots x_t$ can then be processed in a single CBC pass, since the CBC-MAC is equal to the last ciphertext block $c_t = E_k(c_{t-1} \oplus x_t)$, and the last data block is $x_{t+1} = c_t$, yielding final ciphertext block $c_{t+1} = E_k(c_t \oplus x_{t+1}) = E_k(0)$. The encrypted MAC is thus independent of both plaintext and ciphertext, rendering the integrity mechanism completely insecure. Care should thus be taken in combining a MAC with an encryption scheme. In general, it is recommended that distinct (and ideally, independent) keys be used. In some cases, one key may be derived from the other by a simple technique; a common suggestion for DES keys is complementation of every other nibble. However, arguments favoring independent keys include the danger of encryption algorithm weaknesses compromising authentication (or vice-versa), and differences between authentication and encryption keys with respect to key management life cycle. □

An efficiency drawback in using distinct keys for secrecy and integrity is the cost of two separate passes over the data. Example 9.89 illustrates a proposed data integrity mechanism (which appeared in a preliminary draft of U.S. Federal Standard 1026) which attempts this by using an essentially zero-cost linear checksum; it is, however, insecure.

9.89 Example (*CBC encryption with XOR checksum – CBCC*) Consider using the data integrity mechanism of equation (9.2) with E_k being CBC-encryption with key k, $x = x_1 x_2 \dots x_t$ a message of t blocks, and as MDC function the simple XOR of all plaintext blocks, $h(x) = \bigoplus_{i=1}^{i=t} x_i$. The quantity $M = h(x)$ which serves as MDC then becomes plaintext block x_{t+1}. The resulting ciphertext blocks using CBC encryption with $c_0 = IV$ are $c_i = E_k(x_i \oplus c_{i-1})$, $1 \le i \le t+1$. In the absence of manipulation, the recovered plaintext is $x_i = c_{i-1} \oplus D_k(c_i)$. To see that this scheme is insecure as an integrity mechanism, let c_i' denote the actual ciphertext blocks received by a recipient, resulting from possibly manipulated blocks c_i, and let x_i' denote the plaintext recovered by the recipient by CBC decryption with the proper IV. The MDC computed over the recovered plaintext blocks is then

$$M' = h(x') = \bigoplus_{i=1}^{i=t} x_i' = \bigoplus_{i=1}^{i=t} (c_{i-1}' \oplus D_k(c_i')) = IV \oplus \left(\bigoplus_{i=1}^{i=t-1} c_i' \right) \oplus \left(\bigoplus_{i=1}^{i=t} D_k(c_i') \right)$$

M' is compared for equality with $x_{t+1}'(= c_t' \oplus D_k(c_{t+1}'))$ as a check for data integrity, or equivalently, that $S = M' \oplus x_{t+1}' = 0$. By construction, $S = 0$ if there is no manipulation (i.e., if $c_i' = c_i$, which implies $x_i' = x_i$). Moreover, the sum S is invariant under any permutation of the values c_i', $1 \leq i \leq t$ (since $D_k(c_{t+1})$ appears as a term in S, but c_{t+1} does not, c_{t+1} must be excluded from the permutable set). Thus, any of the first t ciphertext blocks can be permuted without affecting the successful verification of the MDC. Furthermore, insertion into the ciphertext stream of any random block c_j^* twice, or any set of such pairs, will cancel itself out in the sum S, and thus also cannot be detected. □

9.90 Example (*CBC encryption with mod $2^n - 1$ checksum*) Consider as an alternative to Example 9.89 the simple MDC function $h(x) = \sum_{i=1}^{t} x_i$, the sum of plaintext blocks as n-bit integers with wrap-around carry (add overflow bits back into units bit), i.e., the sum modulo $2^n - 1$; consider $n = 64$ for ciphers of blocklength 64. The sum S from Example 9.89 in this case involves both XOR and addition modulo $2^n - 1$; both permutations of ciphertext blocks and insertions of pairs of identical random blocks are now detected. (This technique should not, however, be used in environments subject to chosen-plaintext attack.) □

9.91 Example (*PCBC encryption with mod 2^n checksum*) A non-standard, non-self-synchronizing mode of DES known as *plaintext-ciphertext block chaining* (PCBC) is defined as follows, for $i \geq 0$ and plaintext $x = x_1 x_2 \ldots x_t$: $c_{i+1} = E_k(x_{i+1} \oplus G_i)$ where $G_0 = IV$, $G_i = g(x_i, c_i)$ for $i \geq 1$, and g a simple function such as $g(x_i, c_i) = (x_i + c_i) \bmod 2^{64}$. A one-pass technique providing both encryption and integrity, which exploits the error-propagation property of this mode, is as follows. Append an additional plaintext block to provide redundancy, e.g., $x_{t+1} = IV$ (alternatively: a fixed constant or x_1). Encrypt all blocks of the augmented plaintext using PCBC encryption as defined above. The quantity $c_{t+1} = E_k(x_{t+1} \oplus g(x_t, c_t))$ serves as MAC. Upon decipherment of c_{t+1}, the receiver accepts the message as having integrity if the expected redundancy is evident in the recovered block x_{t+1}. (To avoid a known-plaintext attack, the function g in PCBC should not be a simple XOR for this integrity application.) □

9.7 Advanced attacks on hash functions

A deeper understanding of hash function security can be obtained through consideration of various general attack strategies. The resistance of a particular hash function to known general attacks provides a (partial) measure of security. A selection of prominent attack strategies is presented in this section, with the intention of providing an introduction sufficient to establish that designing (good) cryptographic hash functions is not an easily mastered art. Many other attack methods and variations exist; some are general methods, while others rely on peculiar properties of the internal workings of specific hash functions.

9.7.1 Birthday attacks

Algorithm-independent attacks are those which can be applied to any hash function, treating it as a black-box whose only significant characteristics are the output bitlength n (and MAC key bitlength for MACs), and the running time for one hash operation. It is typically assumed the hash output approximates a uniform random variable. Attacks falling under this category include those based on hash-result bitsize (page 336); exhaustive MAC key search (page 336); and birthday attacks on hash functions (including memoryless variations) as discussed below.

(i) Yuval's birthday attack on hash functions

Yuval's birthday attack was one of the first (and perhaps the most well-known) of many cryptographic applications of the birthday paradox arising from the classical occupancy distribution (§2.1.5): when drawing elements randomly, with replacement, from a set of N elements, with high probability a repeated element will be encountered after $O(\sqrt{N})$ selections. Such attacks are among those called *square-root attacks*.

The relevance to hash functions is that it is easier to find collisions for a one-way hash function than to find pre-images or second preimages of specific hash-values. As a result, signature schemes which employ one-way hash functions may be vulnerable to Yuval's attack outlined below. The attack is applicable to all unkeyed hash functions (cf. Fact 9.33), with running time $O(2^{m/2})$ varying with the bitlength m of the hash-value.

9.92 Algorithm Yuval's birthday attack

INPUT: legitimate message x_1; fraudulent message x_2; m-bit one-way hash function h.
OUTPUT: x_1', x_2' resulting from minor modifications of x_1, x_2 with $h(x_1') = h(x_2')$ (thus a signature on x_1' serves as a valid signature on x_2').

1. Generate $t = 2^{m/2}$ minor modifications x_1' of x_1.
2. Hash each such modified message, and store the hash-values (grouped with corresponding message) such that they can be subsequently searched on hash-value. (This can done in $O(t)$ total time using conventional hashing.)
3. Generate minor modifications x_2' of x_2, computing $h(x_2')$ for each and checking for matches with any x_1' above; continue until a match is found. (Each table lookup will require constant time; a match can be expected after about t candidates x_2'.)

9.93 Remark (*application of birthday attack*) The idea of this attack can be used by a dishonest signer who provides to an unsuspecting party his signature on x_1' and later repudiates signing that message, claiming instead that the message signed was x_2'; or by a dishonest verifier, who is able to convince an unsuspecting party to sign a prepared message x_1', and later claim that party's signature on x_2'. This remark generalizes to other schemes in which the hash of a message is taken to represent the message itself.

Regarding practicality, the collisions produced by the birthday attack are "real" (vs. pseudo-collisions or compression function collisions), and moreover of direct practical consequence when messages are constructed to be meaningful. The latter may often be done as follows: alter inputs via individual minor modifications which create semantically equivalent messages (e.g., substituting tab characters in text files for spaces, unprintable characters for each other, etc.). For 128-bit hash functions, 64 such potential modification points are required to allow 2^{64} variations. The attack then requires $O(2^{64})$ time (feasible with extreme parallelization); and while it requires space for $O(2^{64})$ messages (which is impractical), the memory requirement can be addressed as discussed below.

(ii) Memoryless variation of birthday attack

To remove the memory requirement of Algorithm 9.92, a deterministic mapping may be used which approximates a random walk through the hash-value space. By the birthday paradox, in a random walk through a space of 2^m points, one expects to encounter some point a second time (i.e., obtain a collision) after $O(2^{m/2})$ steps, after which the walk will repeat its previous path (and begin to cycle). General memoryless cycle-finding techniques may then be used to find this collision. (Here *memoryless* means requiring negligible memory, rather than in the stochastic sense.) These include Floyd's cycle-finding algorithm (§3.2.2) and improvements to it.

Following Algorithm 9.92, let g be a function such that $g(x_1, H) = x_1'$ is a minor modification, determined by the hash-value H, of message x_1 (each bit of H might define whether or not to modify x_1 at a pre-determined modification point). If x_1 is fixed, then g essentially maps a hash-result to a message and it is convenient to write $g_{x_1}(H) = x_1'$. Moreover, let g be injective so that distinct hashes H result in distinct x_1'. Then, with fixed messages x_1, x_2, and using some easily distinguishable property (e.g., parity) which splits the space of hash-values into two roughly equal-sized subsets, define a function r mapping hash-results to hash-results by:

$$r(H) = \begin{cases} h(g_{x_1}(H)) & \text{if } H \text{ is even} \\ h(g_{x_2}(H)) & \text{if } H \text{ is odd} \end{cases} \tag{9.4}$$

The memoryless collision search technique (see above) is then used to find two inputs to r which map to the same output (i.e., collide). If h behaves statistically as a random mapping then, with probability 0.5, the parity will differ in H and H' for the colliding inputs, in which case without loss of generality $h(g_{x_1}(H)) = h(g_{x_2}(H'))$. This yields a colliding pair of variations $x_1' = g_{x_1}(H)$, $x_2' = g_{x_2}(H')$ of distinct messages x_1, x_2, respectively, such that $h(x_1') = h(x_2')$, as per the output of Algorithm 9.92.

(iii) Illustrative application to MD5

Actual application of the above generic attack to a specific hash function raises additional technicalities. To illustrate how these may be addressed, such application is now examined, with assumptions and choices made for exposition only. Let h be an iterated hash function processing messages in 512-bit blocks and producing 128-bit hashes (e.g., MD5, RIPEMD-128). To minimize computational expense, restrict r (effectively g and h) in equation (9.4) to single 512-bit blocks of x_i, such that each iteration of r involves only the compression function f on inputs one message block and the current chaining variable.

Let the legitimate message input x_1 consist of s 512-bit blocks ($s \geq 1$, prior to MD-strengthening). Create a fraudulent message x_2 of equal bitlength. Allow x_2 to differ from x_1 up to and including the jth block, for any fixed $j \leq s-1$. Use the $(j+1)^{\text{st}}$ block of x_i, denoted B_i ($i = 1, 2$), as a matching/replacement block, to be replaced by the 512-bit blocks resulting from the collision search. Set all blocks in x_2 subsequent to B_i identically equal to those in x_1; x_i' will then differ from x_i only in the single block $(j+1)$. For maximum freedom in the construction of x_2, choose $j = s - 1$. Let c_1, c_2 be the respective 128-bit intermediate results (chaining variables) after the iterated hash operates on the first j blocks of x_1, x_2. Compression function f maps $(128 + 512 =)$ 640-bit inputs to 128-bit outputs. Since the chaining variables depend on x_i, $g_{x_i}(= g)$ may be defined independent of x_i here (cf. equation (9.4)); assume both entire blocks B_i may be replaced without practical implication. Let $g(H) = B$ denote an injective mapping from the space of 128-bit hash-values to the space of 512-bit potential replacement blocks, defined as follows: map each two-bit segment of H to one of four 8-bit values in the replacement block B. (A practical motivation for this is that if x_i is an ASCII message to be printed, and the four 8-bit values are selected to represent non-printable characters, then upon printing, the resulting blocks B are all indistinguishable, leaving no evidence of adversarial manipulation.)

The collision-finding function r for this specific example (corresponding to the generic equation (9.4)) is then:

$$r(H) = \begin{cases} f(c_1, \, g(H)) & \text{if } H \text{ is even} \\ f(c_2, \, g(H)) & \text{if } H \text{ is odd} \end{cases}$$

Collisions for MD5 (and similar hash functions) can thus be found in $O(2^{64})$ operations and without significant storage requirements.

9.7.2 Pseudo-collisions and compression function attacks

The exhaustive or brute force methods discussed in §9.3.4, producing preimages, 2nd-preimages, and collisions for hash functions, are always theoretically possible. They are not considered true "attacks" unless the number of operations required is significantly less than both the strength conjectured by the hash function designer and that of hash functions of similar parameters with ideal strength. An attack requiring such a reduced number of operations is informally said to *break* the hash function, whether or not this computational effort is feasible in practice. Any attack method which demonstrates that conjectured properties do not hold must be taken seriously; when this occurs, one must admit the possibility of additional weaknesses.

In addition to considering the complexity of finding (ordinary) preimages and collisions, it is common to examine the feasibility of attacks on slightly modified versions of the hash function in question, for reasons explained below. The most common case is examination of the difficulty of finding preimages or collisions if one allows free choice of IVs. Attacks on hash functions with unconstrained IVs dictate upper bounds on the security of the actual algorithms. Vulnerabilities found, while not direct weaknesses in the overall hash function, are nonetheless considered certificational weaknesses and cast suspicion on overall security. In some cases, restricted attacks can be extended to full attacks by standard techniques.

Table 9.11 lists the most commonly examined variations, including *pseudo-collisions* – collisions allowing different IVs for the different message inputs. In contrast to preimages and collisions, pseudo-preimages and pseudo-collisions are of limited direct practical significance.

9.94 Note (*alternate names for collision and preimage attacks*) Alternate names for those in Table 9.11 are as follows: preimage or 2nd-preimage $\equiv$ *target attack*; pseudo-preimage $\equiv$ *free-start target attack*; collision (fixed IV) $\equiv$ *collision attack*; collision (random IV) $\equiv$ *semi-free-start collision attack*; pseudo-collision $\equiv$ *free-start collision attack*.

9.95 Note (*relative difficulty of attacks*) Finding a collision can be no harder than finding a 2nd-preimage. Similarly, finding a pseudo-collision can be no harder than finding (two distinct) pseudo-preimages.

↓Type of attack	V	V'	x	x'	y	Find ...
preimage	V_0	—	*	—	y_0	x: $h(V_0, x) = y_0$
pseudo-preimage	*	—	*	—	y_0	x, V: $h(V, x) = y_0$
2nd-preimage	V_0	V_0	x_0	*	$h(V_0, x_0)$	x': $h(V_0, x_0) = h(V_0, x')$
collision (fixed IV)	V_0	V_0	*	*	—	x, x': $h(V_0, x) = h(V_0, x')$
collision (random IV)	*	V	*	*	—	x, x', V: $h(V, x) = h(V, x')$
pseudo-collision	*	*	*	*	—	x, x', V, V': $h(V, x) = h(V', x')$

Table 9.11: *Definition of preimage and collision attacks. V and V' denote (potentially different) IVs used for MDC h applied to inputs x and x', respectively; V_0 denotes the IV pre-specified in the definition of h, x_0 a pre-specified target input, and $y = y_0$ a pre-specified target output. * Denotes IVs or inputs which may be freely chosen by an attacker; $h(V_0, x_0)$ denotes the hash-code resulting from applying h with fixed IV $V = V_0$ to input $x = x_0$. — Means not applicable.*

9.96 Example (*trivial collisions for random IVs*) If free choice of IV is allowed, then trivial pseudo-collisions can be found by deleting leading blocks from a target message. For example, for an iterated hash (cf. equation (9.1) on page 333), $h(IV, x_1 x_2) = f(f(IV, x_1), x_2)$. Thus, for $IV' = f(IV, x_1)$, $h(IV', x_2) = h(IV, x_1 x_2)$ yields a pseudo-collision of h, independent of the strength of f. (MD-strengthening as per Algorithm 9.26 precludes this.) □

Another common analysis technique is to consider the strength of weakened variants of an algorithm, or attack specific subcomponents, akin to cryptanalyzing an 8-round version of DES in place of the full 16 rounds.

9.97 Definition An *attack on the compression function* of an iterated hash function is any attack as per Table 9.11 with $f(H_{i-1}, x_i)$ replacing $h(V_0, x)$ – the compression function f in place of hash function h, chaining variable H_{i-1} in place of initializing value V, and a single input block x_i in place of the arbitrary-length message x.

An attack on a compression function focuses on one fixed step i of the iterative function of equation (9.1). The entire message consists of a single block $x_i = x$ (without MD-strengthening), and the hash output is taken to be the compression function output so $h(x) = H_i$. The importance of such attacks arises from the following.

9.98 Note (*compression function vs. hash function attacks*) Any of the six attacks of Table 9.11 which is found for the compression function of an iterated hash can be extended to a similar attack of roughly equal complexity on the overall hash. An iterated hash function is thus in this regard at most as strong as its compression function. (However note, for example, an overall pseudo-collision is not always of practical concern, since most hash functions specify a fixed IV.)

For example, consider a message $x = x_1 x_2 \ldots x_t$. Suppose a successful 2nd-preimage attack on compression function f yields a 2nd-preimage $x_1' \neq x_1$ such that $f(IV, x_1') = f(IV, x_1)$. Then, $x' = x_1' x_2 \ldots x_t$ is a preimage of $h(x)$.

More positively, if MD-strengthening is used, the strength of an iterated hash with respect to the attacks of Table 9.11 is the same as that of its compression function (cf.

Fact 9.24). However, an iterated hash may certainly be weaker than its compression function (e.g., Example 9.96; Fact 9.37).

In summary, a compression function secure against preimage, 2nd-preimage, and collision (fixed IV) attacks is necessary and sometimes, but not always, sufficient for a secure iterated hash; and security against the other (i.e., free-start) attacks of Table 9.11 is desirable, but not always necessary for a secure hash function in practice. For this reason, compression functions are analyzed in isolation, and attacks on compression functions as per Definition 9.97 are considered. A further result motivating the study of pseudo-preimages is the following.

9.99 Fact (*pseudo-preimages yielding preimages*) If the compression function f of an n-bit iterated hash function h does not have ideal computational security (2^n) against pseudo-preimage attacks, then preimages for h can be found in fewer than 2^n operations (cf. §9.3.4, Table 9.2). This result is true even if h has MD-strengthening.

Justification. The attack requires messages of 3 or more blocks, with 2 or more unconstrained to allow a meet-in-the-middle attack (page 374). If pseudo-preimages can be found in 2^s operations, then $2^{(n+s)/2}$ forward points and $2^{(n-s)/2}$ backward points are employed (fewer backward points are used since they are more costly). Preimages can thus be found in $2 \cdot 2^{(n+s)/2}$ operations.

9.7.3 Chaining attacks

Chaining attacks are those which are based on the iterative nature of hash functions and, in particular, the use of chaining variables. These focus on the compression function f rather than the overall hash function h, and may be further classified as below. An example for context is first given.

9.100 Example (*chaining attack*) Consider a (candidate) collision resistant iterative hash function h producing a 128-bit hash-result, with a compression function f taking as inputs a 512-bit message block x_i and 128-bit chaining variable H_i ($H_0 = IV$) and producing output $H_{i+1} = f(H_i, x_i)$. For a fixed 10-block message x (640 bytes), consider $H = h(x)$. Suppose one picks any one of the 10 blocks, and wishes to replace it with another block without affecting the hash H. If h behaves like a random mapping, the number of such 512-bit blocks is approximately $2^{512}/2^{128} = 2^{384}$. Any efficient method for finding any one of these 2^{384} blocks distinct from the original constitutes an attack on h. The challenge is that such blocks are a sparse subset of all possible blocks, about 1 in 2^{128}. □

(i) Correcting-block chaining attacks

Using the example above for context, one could attempt to (totally) replace a message x with a new message x', such that $h(x) = h(x')$, by using a single unconstrained "correcting" block in x', designated ahead of time, to be determined later such that it produces a chaining value which results in the overall hash being equal to target value $h(x)$. Such a *correcting block attack* can be used to find both preimages and collisions. If the unconstrained block is the first (last) block in the message, it is called a *correcting first (last) block attack*. These attacks may be precluded by requiring per-block redundancy, but this results in an undesirable bandwidth penalty. Example 9.101 illustrates a correcting first block attack. The extension of Yuval's birthday attack (page 369), with message alterations restricted to the last block of candidate messages, resembles a correcting last block attack applied simultaneously to two messages, seeking a (birthday) collision rather than a fixed overall target hash-value.

9.101 Example (*correcting block attack on CBC cipher mode*) The CBC mode of encryption with non-secret key ($H_0 = IV$; $H_i = E_k(H_{i-1} \oplus x_i)$) is unsuitable as an MDC algorithm, because it fails to be one-way – the compression function is reversible when the encryption key is known. A message x', of unconstrained length (say t blocks) can be constructed to have any specified target hash-value H as follows. Let $x'_2, \ldots x'_t$ be $t - 1$ blocks chosen freely. Set $H'_t \leftarrow H$, then for i from t to 1 compute $H'_{i-1} \leftarrow D_k(H'_i) \oplus x'_i$. Finally, compute $x_1^* \leftarrow D_k(H'_1) \oplus IV$. Then, for $x' = x_1^* x'_2 \ldots x'_t$, $h(x') = H$ and all but block x_1^* (which will appear random) can be freely chosen by an adversary; even this minor drawback can be partially addressed by a meet-in-the-middle strategy (see below). Analogous remarks apply to the CFB mode. □

(ii) Meet-in-the-middle chaining attacks

These are birthday attacks similar to Yuval's (and which can be made essentially memory-less) but which seek collisions on intermediate results (i.e., chaining variables) rather than the overall hash-result. When applicable, they allow (unlike Yuval's attack) one to find a message with a pre-specified hash-result, for either a 2nd-preimage or a collision. An attack point is identified between blocks of a candidate (fraudulent) message. Variations of the blocks preceding and succeeding this point are generated. The variations are hashed forward from the algorithm-specified IV (computing $H_i = f(H_{i-1}, x_i)$ as usual) and backward from the target final hash-result (computing $H_i = f^{-1}(H_{i+1}, x_{i+1})$ for some H_{i+1}, x_{i+1}, ideally for x_{i+1} chosen by the adversary), seeking a collision in the chaining variable H_i at the attack point. For the attack to work, the attacker must be able to efficiently go backwards through the chain (certainly moreso than by brute force – e.g., see Example 9.102), i.e., invert the compression function in the following manner: given a value H_{i+1}, find a pair (H_i, x_{i+1}) such that $f(H_i, x_{i+1}) = H_{i+1}$.

9.102 Example (*meet-in-the-middle attack on invertible key chaining modes*) Chaining modes which allow easily derived stage keys result in reversible compression functions unsuitable for use in MDCs due to lack of one-wayness (cf. Example 9.101). An example of such *invertible key chaining* methods is Bitzer's scheme: $H_0 = IV$, $H_i = f(H_{i-1}, x_i) = E_{k_i}(H_{i-1})$ where $k_i = x_i \oplus s(H_{i-1})$ and $s(H_{i-1})$ is a function mapping chaining variables to the key space. For exposition, let s be the identity function. This compression function is unsuitable because it falls to a meet-in-the-middle attack as outlined above. The ability to move backwards through chaining variables, as required by such an attack, is possible here with the chaining variable H_i computed from H_{i+1} as follows. Choose a fixed value $k_{i+1} \leftarrow k$, compute $H_i \leftarrow D_k(H_{i+1})$, then choose as message block $x_{i+1} \leftarrow k \oplus H_i$. □

(iii) Fixed-point chaining attacks

A *fixed point* of a compression function is a pair (H_{i-1}, x_i) such that $f(H_{i-1}, x_i) = H_{i-1}$. For such a pair of message block and chaining value, the overall hash on a message is unchanged upon insertion of an arbitrary number of identical blocks x_i at the chain point at which that chaining value arises. Such attacks are thus of concern if it can be arranged that the chaining variable has a value for which a fixed point is known. This includes the following cases: if fixed points can be found and it can be easily arranged that the chaining variable take on a specific value; or if for arbitrary chaining values H_{i-1}, blocks x_i can be found which result in fixed-points. Fixed points allow 2nd-preimages and collisions to be produced; their effect can be countered by inclusion of a trailing length-block (Algorithm 9.26).

(iv) Differential chaining attacks

Differential cryptanalysis has proven to be a powerful tool for the cryptanalysis of not only block ciphers but also of hash functions (including MACs). For multi-round block ciphers this attack method examines input differences (XORs) to round functions and the corresponding output differences, searching for statistical anomalies. For hash functions, the examination is of input differences to compression functions and the corresponding output differences; a collision corresponds to an output difference of zero.

9.7.4 Attacks based on properties of underlying cipher

The implications of certain properties of block ciphers, which may be of no practical concern when used for encryption, must be carefully examined when such ciphers are used to construct iterated hash functions. The general danger is that such properties may facilitate adversarial manipulation of compression function inputs so as to allow prediction or greater control of outputs or relations between outputs of successive iterations. Included among block cipher properties of possible concern are the following (cf. Chapter 7):

1. *complementation property*: $y = E_k(x) \iff \overline{y} = E_{\overline{k}}(\overline{x})$, where $\overline{x}$ denotes bitwise complement. This makes it trivial to find key-message pairs of block cipher inputs whose outputs differ in a pre-determined manner. For example, for such a block cipher E, the compression function $f(H_{i-1}, x_i) = E_{H_{i-1} \oplus x_i}(x_i) \oplus x_i$ (a linear transformation of the Matyas-Meyer-Oseas function) produces the same output for x_i and its bitwise complement $\overline{x_i}$.

2. *weak keys*: $E_k(E_k(x)) = x$ (for all x). This property of involution of the block cipher may allow an adversary to easily create a two-step fixed point of the compression function f in the case that message blocks x_i have direct influence on the block cipher key input (e.g., if $f = E_{x_i}(H_{i-1})$, insert 2 blocks x_i containing a weak key). The threat is similar for *semi-weak keys*, where $E_{k'}(E_k(x)) = x$.

3. *fixed points*: $E_k(x) = x$. Block cipher fixed points may facilitate fixed-point attacks if an adversary can control the block cipher key input. For example, for the Davies-Meyer compression function $f(H_{i-1}, x_i) = E_{x_i}(H_{i-1}) \oplus H_{i-1}$, if H_{i-1} is a fixed point of the block cipher for key x_i (i.e., $E_{x_i}(H_{i-1}) = H_{i-1}$), then this yields a predictable compression function output $f(H_{i-1}, x_i) = 0$.

4. *key collisions*: $E_k(x) = E_{k'}(x)$. These may allow compression function collisions.

Although they may serve as distinguishing metrics, attacks which appear purely certificational in nature should be noted separately from others; for example, fixed point attacks appear to be of limited practical consequence.

9.103 Example (*DES-based hash functions*) Consider DES as the block cipher in question (see §7.4). DES has the complementation property; has 4 weak keys and 6 pairs of semi-weak keys (each with bit 2 equal to bit 3); each weak key has 2^{32} fixed points (thus a random plaintext is a fixed point of some weak key with probability 2^{-30}), as do 4 of the semi-weak keys; and key collisions can be found in 2^{32} operations. The security implications of these properties must be taken into account in the design of any DES-based hash function. Concerns regarding both weak keys and the complementation property can be eliminated by forcing key bits 2 and 3 to be 10 or 01 within the compression function. □

9.8 Notes and further references

§9.1

The definitive reference for cryptographic hash functions, and an invaluable source for the material in this chapter (including many otherwise unattributed results), is the comprehensive treatment of Preneel [1003, 1004]; see also the surveys of Preneel [1002] and Preneel, Govaerts, and Vandewalle [1006]. Davies and Price [308] also provide a solid treatment of message authentication and data integrity. An extensive treatment of conventional hashing, including historical discussion tracing origins back to IBM in 1953, is given by Knuth [693, p.506-549]. Independent of cryptographic application, *universal classes of hash functions* were introduced by Carter and Wegman [234] in the late 1970s, the idea being to find a class of hash functions such that for every pair of inputs, the probability was low that a randomly chosen function from the class resulted in that pair colliding. Shortly thereafter, Wegman and Carter [1234] noted the cryptographic utility of these hash functions, when combined with secret keys, for (unconditionally secure) *message authentication tag systems*; they formalized this concept, earlier considered by Gilbert, MacWilliams, and Sloane [454] (predating the concept of digital signatures) who attribute the problem to Simmons. Simmons ([1138],[1144]; see also Chapter 10 of Stinson [1178]) independently developed a general theory of unconditionally secure message authentication schemes and the subject of *authentication codes* (see also §9.5 below).

Rabin [1022, 1023] first suggested employing a one-way hash function (constructed by using successive message blocks to key an iterated block encryption) in conjunction with a one-time signature scheme and later in a public-key signature scheme; Rabin essentially noted the requirements of 2nd-preimage resistance and collision resistance. Merkle [850] explored further uses of one-way hash functions for authentication, including the idea of *tree authentication* [852] for both one-time signatures and authentication of public files.

§9.2

Merkle [850] (partially published as [853]) was the first to give a substantial (informal) definition of one-way hash functions in 1979, specifying the properties of preimage and 2nd-preimage resistance. Foreshadowing UOWHFs (see below), he suggested countering the effect of Remark 9.36 by using slightly different hash functions h over time; Merkle [850, p.16-18] also proposed a public key distribution method based on a one-way hash function (effectively used as a one-way pseudo-permutation) and the birthday paradox, in a precursor to his "puzzle system" (see page 537). The first formal definition of a CRHF was given in 1988 by Damgård [295] (an informal definition was later given by Merkle [855, 854]; see also [853]), who was first to explore collision resistant hash functions in a complexity-theoretic setting. Working from the idea of *claw-resistant pairs of trapdoor permutations* due to Goldwasser, Micali, and Rivest [484], Damgård defined *claw-resistant families of permutations* (without the trapdoor property). The term *claw-resistant* (originally: *claw-free*) originates from the pictorial representation of a functional mapping showing two distinct domain elements being mapped to the same range element under distinct functions $f^{(i)}$ and $f^{(j)}$ (colliding at $z = f^{(i)}(x) = f^{(j)}(y)$), thereby tracing out a claw.

Goldwasser et al. [484] established that the intractability of factoring suffices for the existence of claw-resistant pairs of permutations. Damgård showed that the intractability of the discrete logarithm problem likewise suffices. Using several reasonably efficient number-theoretic constructions for families of claw-resistant permutations, he gave the first provably collision resistant hash functions, under such intractability assumptions (for discrete logarithms, the assumption required is that taking *specific* discrete logarithms be difficult). Russell [1088] subsequently established that a collection of collision resistant hash functions exists if and only if there exists a collection of *claw-resistant pairs of pseudo-permutations*; a pseudo-permutation on a set is a function computationally indistinguishable from a permutation (pairs of elements demonstrating non-injectivity are hard to find). It remains open whether the existence of one-way functions suffices for the existence of collision resistant hash functions.

The definition of a one-way function (Definition 9.9) was given in the seminal paper of Diffie and Hellman [345], along with the use of the discrete exponential function modulo a prime as a candidate OWF, which they credit to Gill. The idea of providing the hash-value of some data, to indicate prior commitment to (or knowledge of) that data, was utilized in Lamport's one-time signature scheme (circa 1976); see page 485. The OWF of Example 9.13 was known to Matyas and Meyer circa 1979. As noted by Massey [786], the idea of one-wayness was published in 1873 by J.S. Jevons, who noted (preceding RSA by a century) that multiplying two primes is easy whereas factoring the result is not. Published work dated 1968 records the use of ciphers essentially as one-way functions (decryption was not required) in a technique to avoid storing cleartext computer account passwords in time-shared systems. These were referred to as *one-way ciphers* by Wilkes [1244] (p.91-93 in 1968 or 1972 editions; p.147 in 1975 edition), who credits Needham with the idea and an implementation thereof. The first proposal of a non-invertible function for the same purpose appears to be that of Evans, Kantrowitz, and Weiss [375], while Purdy [1012] proposed extremely high-degree, sparse polynomials over a prime field as a class of functions which were computationally difficult to invert. Foreshadowing later research into collision resistance, Purdy also defined the *degeneracy* of such a function to be the maximum number of preimages than any image could have, noting that "if the degeneracy is catastrophically large there may be no security at all".

Naor and Yung [920] introduced the cryptographic primitive known as a *universal one-way hash function (UOWHF)* family, and give a provably secure construction for a one-way hash function from a one-way hash function which compresses by a single bit ($t + 1$ to t bits); the main property of a UOWHF family is 2nd-preimage resistance as for a OWHF, but here an instance of the function is picked at random from a family of hash functions after fixing an input, as might be modeled in practice by using a random IV with a OWHF. Naor and Yung [920] also prove by construction that UOWHFs exist if and only if one-way permutations do, and show how to use UOWHFs to construct provably secure digital signature schemes assuming the existence of any one-way permutation. Building on this, Rompel [1068] showed how to construct a UOWHF family from any one-way function, and based signature schemes on such hash functions; combining this with the fact that a one-way function can be constructed from any secure signature scheme, the result is that the existence of one-way functions is necessary and sufficient for the existence of secure digital signature schemes. De Santis and Yung [318] proceed with more efficient reductions from one-way functions to UOWHFs, and show the equivalence of a number of complexity-theoretic definitions regarding collision resistance. Impagliazzo and Naor [569] give an efficient construction for a UOWHF and prove security equivalent to the subset-sum problem (an **NP**-hard problem whose corresponding decision problem is **NP**-complete); for parameters for which a random instance of subset-sum is hard, they argue that this UOWHF is secure (cf. Remark 9.12). Impagliazzo, Levin, and Luby [568] prove the existence of one-way functions is necessary and sufficient for that of secure pseudorandom generators.

Application-specific (often unprovable) hash function properties beyond collision resistance (but short of preimage resistance) may often be identified as necessary, e.g., for ordinary RSA signatures computed directly after hashing, the multiplicative RSA property dictates that for the hash function h used it be infeasible to find messages x, x_1, x_2 such that $h(x) = h(x_1) \cdot h(x_2)$. Anderson [27] discusses such additional requirements on hash functions. For a summary of requirements on a MAC in the special case of multi-cast authentication, see Preneel [1003]. Bellare and Rogaway [93] include discussion of issues related to the random nature of practical hash functions, and cryptographic uses thereof. Damgård [295] showed that the security of a digital signature scheme which is not existentially forgeable under an adaptive chosen-message attack will not be decreased if used in conjunction with a collision-resistant hash function.

Bellare, Goldreich, and Goldwasser [88] (see also [89]) introduce the idea of *incremental hashing*, involving computing a hash value over data and then updating the hash-value after changing the data; the objective is that the computation required for the update be proportional to the amount of change.

§9.3

Merkle's meta-method [854] (Algorithm 9.25) was based on ideas from his 1979 Ph.D. thesis [850]. An equivalent construction was given by Damgård [296], which Gibson [450] remarks on again yielding Merkle's method. Naor and Yung [920] give a related construction for a UOWHF. See Preneel [1003] for fundamental results (cf. Remarks 9.35 and 9.36, and Fact 9.27 on cascading hash functions which follow similar results on stream ciphers by Maurer and Massey [822]). The padding method of Algorithms 9.29 and 9.30 originated from ISO/IEC 10118-4 [608]. The basic idea of the long-message attack (Fact 9.37) is from Winternitz [1250].

§9.4

The hash function of Algorithm 9.42 and referred to as Davies-Meyer (as cited per Quisquater and Girault [1019]) has been attributed by Davies to Meyer; apparently known to Meyer and Matyas circa 1979, it was published along with Algorithm 9.41 by Matyas, Meyer, and Oseas [805]. The Miyaguchi-Preneel scheme (Algorithm 9.43) was proposed circa 1989 by Preneel [1003], and independently by Miyaguchi, Ohta, and Iwata [886]. The three single-length rate-one schemes discussed (Remark 9.44) are among 12 compression functions employing *non-invertible chaining* found through systematic analysis by Preneel et al. [1007] to be provably secure under black-box analysis, 8 being certificationally vulnerable to fixed-point attack nonetheless. These 12 are linear transformations on the message block and chaining variable (i.e., $[x', H'] = A[x, H]$ for any of the 6 invertible 2×2 binary matrices A) of the Matyas-Meyer-Oseas (Algorithm 9.41) and Miyaguchi-Preneel schemes; these latter two themselves are among the 4 recommended when the underlying cipher is resistant to differential cryptanalysis (e.g., DES), while Davies-Meyer is among the remaining 8 recommended otherwise (e.g., for FEAL). MDC-2 and MDC-4 are of IBM origin, proposed by Brachtl et al. [184], and reported by Meyer and Schilling [860]; details of MDC-2 are also reported by Matyas [803]. For a description of MDC-4, see Bosselaers and Preneel [178].

The DES-based hash function of Merkle [855] which is mentioned uses the meta-method and employs a compression function f mapping 119-bit input to 112-bit output in 2 DES operations, allowing 7-bit message blocks to be processed (with rate 0.055). An optimized version maps 234 bits to 128 bits in 6 DES operations, processing 106-bit message blocks (with rate 0.276); unfortunately, overheads related to "bit chopping" and the inconvenient block size are substantial in practice. This construction is provably as secure as the underlying block cipher assuming an unflawed cipher (cf. Table 9.3; Preneel [1003] shows that accounting for DES weak keys and complementation drops the rate slightly to 0.266). Winternitz [1250] considers the security of the Davies-Meyer hash under a black-box model (cf. Remark 9.45).

The search for secure double-length hash functions of rate 1 is ongoing, the goal being security better than single-length Matyas-Meyer-Oseas and approaching that of MDC-2. Quisquater and Girault [1019] proposed two functions, one (QG-original) appearing in the Abstracts of Eurocrypt'89 and a second (QG-revised) in the final proceedings altered to counter an attack of Coppersmith [276] on the first. The attack, restricted to the case of DES as underlying block cipher, uses fixed points resulting from weak keys to find collisions in 2^{36} DES operations. A general attack of Knudsen and Lai [688], which (unfortunately) applies to a large class of double-length (i.e., $2n$-bit) rate-one block-cipher-based hashes including QG-original, finds preimages in about 2^n operations plus 2^n storage. The systematic method used to establish this result was earlier used by Hohl et al. [560] to prove that pseudo-preimage and pseudo-collision attacks on a large class of double-length hash functions of rate 1/2 and 1, including MDC-2, are no more difficult than on the single-length rate-one Davies-Meyer hash; related results are summarized by Lai and Knudsen [727]. A second attack due to Coppersmith [276], not restricted to DES, employs 88 correcting blocks to find collisions for QG-revised in 2^{40} steps. Another modification of QG-original, the LOKI Double Hash Function (LOKI-DBH) of Brown, Pieprzyk, and Seberry [215], appears as a general construction to offer the same security as QG-revised (provided the underlying block cipher is not LOKI).

The speeds in Table 9.5 are normalized from the timings reported by Dobbertin, Bosselaers, and Preneel [355], relative to an assembly code MD4 implementation optimized for the Pentium processor, with a throughput (90 MHz clock) of 165.7 Mbit/s (optimized C code was roughly a factor of 2 slower). See Bosselaers, Govaerts, and Vandewalle [177] for a detailed MD5 implementation discussion.

MD4 and MD5 (Algorithms 9.49, 9.51) were designed by Rivest [1055, 1035]. An Australian extension of MD5 known as HAVAL has also been proposed by Zheng, Pieprzyk, and Seberry [1268]. The first published partial attack on MD4 was by den Boer and Bosselaers [324], who demonstrated collisions could be found when Round 1 (of the three) was omitted from the compression function, and confirmed unpublished work of Merkle showing that collisions could be found (for input pairs differing in only 3 bits) in under a millisecond on a personal computer if Round 3 was omitted. More devastating was the partial attack by Vaudenay [1215] on the full MD4, which provided only near-collisions, but allowed sets of inputs to be found for which, of the corresponding four 32-bit output words, three are constant while the remaining word takes on all possible 32-bit values. This revealed the word access-order in MD4 to be an unfortunate choice. Finally, late in 1995, using techniques related to those which earlier allowed a partial attack on RIPEMD (see below), Dobbertin [354] broke MD4 as a CRHF by finding not only collisions as stated in Remark 9.50 (taking a few seconds on a personal computer), but collisions for meaningful messages (in under one hour, requiring 20 free bytes at the start of the messages).

A first partial attack on MD5 was published by den Boer and Bosselaers [325], who found pseudo-collisions for its compression function f, which maps a 128-bit chaining variable and sixteen 32-bit words down to 128-bits; using 2^{16} operations, they found a 16-word message X and chaining variables $S_1 \neq S_2$ (these differing only in 4 bits, the most significant of each word), such that $f(S_1, X) = f(S_2, X)$. Because this specialized internal pseudo-collision could not be extended to an external collision due to the fixed initial chaining values (and due to the special relation between the inputs), this attack was considered by many to have little practical significance, although exhibiting a violation of the design goal to build a CRHF from a collision resistant compression function. But in May of 1996, using techniques related to his attack on MD4 above, Dobbertin (rump session, Eurocrypt'96) found MD5 compression function collisions (Remark 9.52) in 10 hours on a personal computer (about 2^{34} compress function computations).

Anticipating the feasibility of 2^{64} operations, Rivest [1055] proposed a method to extend MD4 to 256 bits by running two copies of MD4 in parallel over the input, with different initial chaining values and constants for the second, swapping the values of the variable A with the first after processing each 16-word block and, upon completion, concatenating the 128-bit hash-values from each copy. However, in October of 1995 Dobbertin [352] found collisions for the compression function of extended MD4 in 2^{26} compress function operations, and conjectured that a more sophisticated attack could find a collision for extended MD4 itself in $O(2^{40})$ operations.

MD2, an earlier and slower hash function, was designed in 1988 by Rivest; see Kaliski [1033] for a description. Rogier and Chauvaud [1067] demonstrated that collisions can be efficiently found for the compression function of MD2, and that the MD2 checksum block is necessary to preclude overall MD2 collisions.

RIPEMD [178] was designed in 1992 by den Boer and others under the European RACE Integrity Primitives Evaluation (RIPE) project. A version of MD4 strengthened to counter known attacks, its compression function has two parallel computation lines of three 16-step rounds. Nonetheless, early in 1995, Dobbertin [353] demonstrated that if the first or last (parallel) round of the 3-round RIPEMD compress function is omitted, collisions can be found in 2^{31} compress function computations (one day on a 66 MHz personal computer). This result coupled with concern about inherent limitations of 128-bit hash results motivated RIPEMD-160 (Algorithm 9.55) by Dobbertin, Bosselaers, and Preneel ([355]; but for corrections, see the directory `/pub/COSIC/bosselae/ripemd/` at ftp site `ftp.esat.kuleuven.ac.be`). Increased security is provided by five rounds (each with two lines) and greater independence between the parallel lines, at a performance penalty of a factor of 2. RIPEMD-128 (with 128-bit result and chaining variable) was simultaneously proposed as a drop-in upgrade for RIPEMD; it scales RIPEMD-160 back to four rounds (each with two lines).

SHA-1 (Algorithm 9.53) is a U.S. government standard [404]. It differs from the original standard SHA [403], which it supersedes, only in the inclusion of the 1-bit rotation in the block expansion from 16 to 80 words. For discussion of how this expansion in SHA is related to linear error correcting codes, see Preneel [1004].

Lai and Massey [729] proposed two hash functions of rate 1/2 with $2m$-bit hash values, *Tandem Davies-Meyer* and *Abreast Davies-Meyer*, based on an m-bit block cipher with $2m$-bit key (e.g., IDEA), and a third m-bit hash function using a similar block cipher. Merkle's public-domain hash function Snefru [854] and the FEAL-based N-Hash proposed by Miyaguchi, Ohta, and Iwata [886] are other hash functions which have attracted considerable attention. Snefru, one of the earliest proposals, is based on the idea of Algorithm 9.41, (typically) using as E the first 128 bits of output of a custom-designed symmetric 512-bit block cipher with fixed key $k = 0$. Differential cryptanalysis has been used by Biham and Shamir [137] to find collisions for Snefru with 2 passes, and is feasible for Snefru with 4 passes; Merkle currently recommends 8 passes (impacting performance). Cryptanalysis of the 128-bit hash N-Hash has been carried out by Biham and Shamir [136], with attacks on 3, 6, 9, and 12 rounds being of respective complexity 2^8, 2^{24}, 2^{40}, and 2^{56} for the more secure of the two proposed variations.

Despite many proposals, few hash functions based on modular arithmetic have withstood attack, and most that have (including those which are provably secure) tend to be relatively inefficient. MASH-1 (Algorithm 9.56), from Committee Draft ISO/IEC 10118-4 [608], evolved from a long line of related proposals successively broken and repaired, including contributions by Jueneman; Davies and Price; A. Jung; Girault [457] (which includes a summary); and members of ISO SC27/WG2 circa 1994-95 (e.g., in response to the cryptanalysis of the 1994 draft proposal, by Coppersmith and Preneel, in ISO/IEC JTC1/SC27 N1055, Attachment 12, "Comments on MASH-1 and MASH-2 (Feb.21 1995)"). Most prominent among prior proposals was the *sqmodn* algorithm (due to Jung) in informative Annex D of CCITT Recommendation X.509 (1988 version), which despite suffering ignominy at the hands of Coppersmith [275], was resurrected with modifications as the basis for MASH-1.

§9.5

Simmons [1146] notes that techniques for message authentication without secrecy (today called MACs) were known to Simmons, Stewart, and Stokes already in the early 1970s. In the open literature, the idea of using DES to provide a MAC was presented already in Feb. 1977 by Campbell [230], who wrote "... Each group of 64 message bits is passed through the algorithm after being combined with the output of the previous pass. The final DES output is thus a residue which is a cryptographic function of the entire message", and noted that to detect message replay or deletion each message could be made unique by using per-message keys or cryptographically protected sequence numbers. Page 121 of this same publication describes the use of encryption in conjunction with an appended redundancy check code for manipulation detection (cf. Figure 9.8(b)).

The term *MAC* itself evolved in the period 1979-1982 during development of ANSI X9.9 [36], where it is defined as "an eight-digit number in hexadecimal format which is the result of passing a financial message through the authentication algorithm using a specific key." FIPS 81 [398] standardizes MACs based on CBC and CFB modes (CFB-based MACs are little-used, having some disadvantages over CBC-MAC and apparently no advantages); see also FIPS 113 [400]. Algorithm 9.58 is generalized by ISO/IEC 9797 [597] to a CBC-based MAC for an n-bit block cipher providing an m-bit MAC, $m \leq n$, including an alternative to the optional strengthening process of Algorithm 9.58: a second key k' (possibly dependent on k) is used to encrypt the final output block. As discussed in Chapter 15, using ISO/IEC 9797 with DES to produce a 32-bit MAC and Algorithm 9.29 for padding is equivalent to the MAC specified in ISO 8731-1, ANSI X9.9 and required by ANSI X9.17. Regarding RIPE-MAC (Example 9.63) [178], other than the 2^{-64} probability of guessing a 64-bit MAC, and MAC forgery as applicable to all iterated MACs (see below), the best known attacks providing key recovery are linear cryptanalysis using 2^{42} known plaintexts for RIPE-MAC1, and a 2^{112} exhaustive search for RIPE-MAC3. Bellare, Kilian, and Rogaway [91] formally examine the security of CBC-based MACs and provide justification, establishing (via exact rather than asymptotic arguments) that pseudorandom functions are preserved under cipher block chaining; they also propose solutions to the problem of Example 9.62 (cf. Remark 9.59).

The MAA (Algorithm 9.68) was developed in response to a request by the Bankers Automated Clearing Services (U.K.), and first appeared as a U.K. National Physical Laboratory Report (NPL Report DITC 17/83 February 1983). It has been part of an ISO banking standard [577] since 1987, and is due to Davies and Clayden [306]; comments on its security (see also below) are offered by Preneel [1003], Davies [304], and Davies and Price [308], who note that its design follows the general principles of the Decimal Shift and Add (DSA) algorithm proposed by Sievi in 1980. As a consequence of the conjecture that MAA may show weaknesses in the case of very long messages, ISO 8731-2 specifies a special mode of operation for messages over 1024 bytes. For more recent results on MAA including exploration of a key recovery attack, see Preneel and van Oorschot [1010].

Methods for constructing a MAC algorithm from an MDC, including the secret prefix, suffix, and envelope methods, are discussed by Tsudik [1196]; Galvin, McCloghrie, and Davin [438] suggest addressing the message extension problem (Example 9.65) in the secret suffix method by using a prepended length field (this requires two passes over the message if the length is not known *a priori*). Preneel and van Oorschot [1009] compare the security of these methods; propose MD5-MAC (Algorithm 9.69) and similar constructions for customized MAC functions based on RIPEMD and SHA; and provide Fact 9.57, which applies to MAA ($n = 64 = 2m$) with $u = 2^{32.5}$ and $v = 2^{32.3}$, while for MD5-MAC ($n = 128 = 2m$) both u and v are on the order of 2^{64}. Remark 9.60 notwithstanding, the use of an n-bit internal chaining variable with a MAC-value of bitlength $m = n/2$ is supported by these results.

The envelope method with padding (Example 9.66) is discussed by Kaliski and Robshaw (*CryptoBytes* vol.1 no.1, Spring 1995). Preneel and van Oorschot [1010] proposed a key recovery attack on this method, which although clearly impractical by requiring over 2^{64} known text-MAC pairs (for MD5 with 128-bit key), reveals an architectural flaw. Bellare, Canetti, and Krawczyk [86] rigorously examined the security of a nested MAC construction (NMAC), and the practical variation HMAC thereof (Example 9.67), proving HMAC be secure provided the hash function used exhibits certain appropriate characteristics. Prior to this, the related construction $h(k_1 || h(k_2 || x))$ was considered in the note of Kaliski and Robshaw (see above).

Other recent proposals for practical MACs include the bucket hashing construction of Rogaway [1065], and the XOR MAC scheme of Bellare, Guérin, and Rogaway [90]. The latter is a provably secure construction for MACs under the assumption of the availability of a finite pseudorandom function, which in practice is instantiated by a block cipher or hash function; advantages include that it is parallelizable and incremental.

MACs intended to provide unconditional security are often called *authentication codes* (cf. §9.1 above), with an *authentication tag* (cf. MAC value) accompanying data to provide origin authentication (including data integrity). More formally, an authentication code involves finite sets $\mathcal{S}$ of source states (plaintext), $\mathcal{A}$ of authentication tags, and $\mathcal{K}$ of secret keys, and a set of rules such that each $k \in \mathcal{K}$ defines a mapping $e_K : \mathcal{S} \rightarrow \mathcal{A}$. An (authenticated) message, consisting of a source state and a tag, can be verified only by the intended recipient (as for MACs) possessing a pre-shared key. Wegman and Carter [1234] first combined one-time pads with hash functions for message authentication; this approach was pursued by Brassard [191] trading unconditional security for short keys.

This approach was further refined by Krawczyk [714] (see also [717]), whose CRC-based scheme (Algorithm 9.72) is a minor modification of a construction by Rabin [1026]. A second LFSR-based scheme proposed by Krawczyk for producing m-bit hashes (again combined with one-time pads as per Algorithm 9.72) improves on a technique of Wegman and Carter, and involves matrix-vector multiplication by an $m \times b$ binary *Toeplitz matrix A* (each left-to-right diagonal is fixed: $A_{i,j} = A_{k,l}$ for $k - i = l - j$), itself generated from a random binary irreducible polynomial of degree m (defining the LFSR), and m bits of initial state. Krawczyk proves that the probability of successful MAC forgery here for a b-bit message is at most $b/2^{m-1}$, e.g., less than 2^{-30} even for $m = 64$ and a 1 Gbyte message (cf. Fact 9.73). Earlier, Bierbrauer et al. [127] explored the relations between coding theory, universal hashing, and practical authentication codes with relatively short keys (see also Johansson, Kabatianskii, and Smeets [638]; and the survey of van Tilborg [1211]). These and other MAC constructions suitable for use with stream ciphers are very fast, scalable, and information-theoretically secure when the short keys they require are used as one-time pads; when used with key streams generated by pseudorandom generators, their security is dependent on the stream and (at best) computationally secure.

Desmedt [335] investigated authenticity in stream ciphers, and proposed both unconditionally secure authentication systems and stream ciphers providing authenticity. Lai, Rueppel, and Woollven [731] define an efficient MAC for use with stream ciphers (but see Preneel [1003] regarding a modification to address tampering with ends of messages). Part of an initial secret key is used to seed a key stream generator, each bit of which selectively routes message bits to one of two feedback shift registers (FSRs), the initial states of which are part of the secret key and the final states of which comprise the MAC. The number of pseudorandom bits required equals the number of message bits. Taylor [1189] proposes an alternate MAC technique for use with stream ciphers.

§9.6

Simmons [1144] notes the use of sealed authenticators by the U.S. military. An early presentation of MACs and authentication is given by Meyer and Matyas [859]; the third or later printings are recommended, and include the one-pass PCBC encryption-integrity method of Example 9.91. Example 9.89 was initially proposed by the U.S. National Bureau of Standards, and was subsequently found by Jueneman to have deficiencies; this is included in the extensive discussion by Jueneman, Matyas, and Meyer [645] of using MDCs for integrity, along with the idea of Example 9.90, which Davies and Price [308, p.124] also consider for $n = 16$. Later work by Jueneman [644] considers both MDCs and MACs; see also Meyer and Schilling [860]. Davies and Price also provide an excellent discussion of transaction authentication, noting additional techniques (cf. §9.6.1) addressing message replay including use of MAC values themselves from immediately preceding messages as chaining values in place of random number chaining. Subtle flaws in various fielded data integrity techniques are discussed by Stubblebine and Gligor [1179].

§9.7

The taxonomy of preimages and collisions is from Preneel [1003]. The alternate terminology of Note 9.94 is from Lai and Massey [729], who published the first systematic treatment of attacks on iterated hash functions, including relationships between fixed-start and free-start attacks, considered *ideal security*, and re-examined MD-strengthening. The idea of Algorithm 9.92 was published by Yuval [1262], but the implications of the birthday paradox were known to others at the time, e.g., see Merkle [850, p.12-13]. The details of the memoryless version are from van Oorschot and Wiener [1207], who also show the process can be perfectly parallelized (i.e., attaining a factor r speedup with r processors) using parallel collision search methods; related independent work (unpublished) has been reported by Quisquater.

Meet-in-the-middle chaining attacks can be extended to handle additional constraints and otherwise generalized. A "triple birthday" chaining attack, applicable when the compression function is invertible, is given by Coppersmith [267] and generalized by Girault, Cohen, Campana [460]; see also Jueneman [644]. For additional discussion of differential cryptanalysis of hash functions based on block ciphers, see Biham and Shamir [138], Preneel, Govaerts, and Vandewalle [1005], and Rijmen and Preneel [1050].

Chapter 10

Identification and Entity Authentication

Contents in Brief

10.1 Introduction

This chapter considers techniques designed to allow one party (the *verifier*) to gain assurances that the identity of another (the *claimant*) is as declared, thereby preventing impersonation. The most common technique is by the verifier checking the correctness of a message (possibly in response to an earlier message) which demonstrates that the claimant is in possession of a secret associated by design with the genuine party. Names for such techniques include *identification*, *entity authentication*, and (less frequently) *identity verification*. Related topics addressed elsewhere include message authentication (data origin authentication) by symmetric techniques (Chapter 9) and digital signatures (Chapter 11), and authenticated key establishment (Chapter 12).

A major difference between entity authentication and message authentication (as provided by digital signatures or MACs) is that message authentication itself provides no timeliness guarantees with respect to when a message was created, whereas entity authentication involves corroboration of a claimant's identity through actual communications with an associated verifier during execution of the protocol itself (i.e., in *real-time*, while the verifying entity awaits). Conversely, entity authentication typically involves no meaningful message other than the claim of being a particular entity, whereas message authentication does. Techniques which provide both entity authentication and key establishment are deferred to Chapter 12; in some cases, key establishment is essentially message authentication where the message is the key.

Chapter outline

The remainder of §10.1 provides introductory material. §10.2 discusses identification schemes involving fixed passwords including Personal Identification Numbers (PINs), and providing so-called weak authentication; one-time password schemes are also considered. §10.3 considers techniques providing so-called strong authentication, including challenge-response protocols based on both symmetric and public-key techniques. It includes discussion of time-variant parameters (TVPs), which may be used in entity authentication protocols and to provide uniqueness or timeliness guarantees in message authentication. §10.4 examines customized identification protocols based on or motivated by zero-knowledge techniques. §10.5 considers attacks on identification protocols. §10.6 provides references and further chapter notes.

10.1.1 Identification objectives and applications

The general setting for an identification protocol involves a *prover* or *claimant A* and a *verifier B*. The verifier is presented with, or presumes beforehand, the purported identity of the claimant. The goal is to corroborate that the identity of the claimant is indeed A, i.e., to provide entity authentication.

10.1 Definition *Entity authentication* is the process whereby one party is assured (through acquisition of corroborative evidence) of the identity of a second party involved in a protocol, and that the second has actually participated (i.e., is active at, or immediately prior to, the time the evidence is acquired).

10.2 Remark (*identification terminology*) The terms *identification* and *entity authentication* are used synonymously throughout this book. Distinction is made between weak, strong, and zero-knowledge based authentication. Elsewhere in the literature, sometimes identification implies only a claimed or stated identity whereas entity authentication suggests a corroborated identity.

(i) Objectives of identification protocols

From the point of view of the verifier, the outcome of an entity authentication protocol is either *acceptance* of the claimant's identity as authentic (completion with acceptance), or *termination without acceptance* (rejection). More specifically, the objectives of an identification protocol include the following.

1. In the case of honest parties A and B, A is able to successfully authenticate itself to B, i.e., B will complete the protocol having accepted A's identity.
2. (*transferability*) B cannot reuse an identification exchange with A so as to successfully impersonate A to a third party C.
3. (*impersonation*) The probability is negligible that any party C distinct from A, carrying out the protocol and playing the role of A, can cause B to complete and accept A's identity. Here *negligible* typically means "is so small that it is not of practical significance"; the precise definition depends on the application.
4. The previous points remain true even if: a (polynomially) large number of previous authentications between A and B have been observed; the adversary C has participated in previous protocol executions with either or both A and B; and multiple instances of the protocol, possibly initiated by C, may be run simultaneously.

The idea of zero-knowledge-based protocols is that protocol executions do not even reveal any partial information which makes C's task any easier whatsoever.

An identification (or entity authentication) protocol is a "real-time" process in the sense that it provides an assurance that the party being authenticated is operational at the time of protocol execution – that party is taking part, having carried out some action since the start of the protocol execution. Identification protocols provide assurances only at the particular instant in time of successful protocol completion. If ongoing assurances are required, additional measures may be necessary; see §10.5.

(ii) Basis of identification

Entity authentication techniques may be divided into three main categories, depending on which of the following the security is based:

1. *something known*. Examples include standard passwords (sometimes used to derive a symmetric key), Personal Identification Numbers (PINs), and the secret or private keys whose knowledge is demonstrated in challenge-response protocols.

2. *something possessed*. This is typically a physical accessory, resembling a passport in function. Examples include magnetic-striped cards, *chipcards* (plastic cards the size of credit cards, containing an embedded microprocessor or integrated circuit; also called *smart cards* or *IC cards*), and hand-held customized calculators (*password generators*) which provide time-variant passwords.

3. *something inherent* (to a human individual). This category includes methods which make use of human physical characteristics and involuntary actions (*biometrics*), such as handwritten signatures, fingerprints, voice, retinal patterns, hand geometries, and dynamic keyboarding characteristics. These techniques are typically non-cryptographic and are not discussed further here.

(iii) Applications of identification protocols

One of the primary purposes of identification is to facilitate access control to a resource, when an access privilege is linked to a particular identity (e.g., local or remote access to computer accounts; withdrawals from automated cash dispensers; communications permissions through a communications port; access to software applications; physical entry to restricted areas or border crossings). A password scheme used to allow access to a user's computer account may be viewed as the simplest instance of an *access control matrix*: each resource has a list of identities associated with it (e.g., a computer account which authorized entities may access), and successful corroboration of an identity allows access to the authorized resources as listed for that entity. In many applications (e.g., cellular telephony) the motivation for identification is to allow resource usage to be tracked to identified entities, to facilitate appropriate billing. Identification is also typically an inherent requirement in authenticated key establishment protocols (see Chapter 12).

10.1.2 Properties of identification protocols

Identification protocols may have many properties. Properties of interest to users include:

1. *reciprocity of identification*. Either one or both parties may corroborate their identities to the other, providing, respectively, *unilateral* or *mutual* identification. Some techniques, such as fixed-password schemes, may be susceptible to an entity posing as a verifier simply in order to capture a claimant's password.

2. *computational efficiency*. The number of operations required to execute a protocol.

3. *communication efficiency*. This includes the number of passes (message exchanges) and the bandwidth required (total number of bits transmitted).

    More subtle properties include:

4. *real-time involvement of a third party (if any)*. Examples of third parties include an on-line *trusted* third party to distribute common symmetric keys to communicating entities for authentication purposes; and an on-line (untrusted) directory service for distributing public-key certificates, supported by an off-line certification authority (see Chapter 13).

5. *nature of trust required in a third party (if any)*. Examples include trusting a third party to correctly authenticate and bind an entity's name to a public key; and trusting a third party with knowledge of an entity's private key.

6. *nature of security guarantees*. Examples include provable security and zero-knowledge properties (see §10.4.1).

7. *storage of secrets*. This includes the location and method used (e.g., software only, local disks, hardware tokens, etc.) to store critical keying material.

Relation between identification and signature schemes

Identification schemes are closely related to, but simpler than, digital signature schemes, which involve a variable message and typically provide a non-repudiation feature allowing disputes to be resolved by judges after the fact. For identification schemes, the semantics of the message are essentially fixed – a claimed identity at the current instant in time. The claim is either corroborated or rejected immediately, with associated privileges or access either granted or denied in real time. Identifications do not have "lifetimes" as signatures do[1] – disputes need not typically be resolved afterwards regarding a prior identification, and attacks which may become feasible in the future do not affect the validity of a prior identification. In some cases, identification schemes may also be converted to signature schemes using a standard technique (see Note 10.30).

10.2 Passwords (weak authentication)

Conventional password schemes involve time-invariant passwords, which provide so-called *weak authentication*. The basic idea is as follows. A *password*, associated with each user (entity), is typically a string of 6 to 10 or more characters the user is capable of committing to memory. The serves as a shared secret between the user and system. (Conventional password schemes thus fall under the category of symmetric-key techniques providing unilateral authentication.) To gain access to a system resource (e.g., computer account, printer, or software application), the user enters a (userid, password) pair, and explicitly or implicitly specifies a resource; here *userid* is a claim of identity, and *password* is the evidence supporting the claim. The system checks that the password matches corresponding data it holds for that userid, and that the stated identity is authorized to access the resource. Demonstration of knowledge of this secret (by revealing the password itself) is accepted by the system as corroboration of the entity's identity.

   Various password schemes are distinguished by the means by which information allowing password verification is stored within the system, and the method of verification. The collection of ideas presented in the following sections motivate the design decisions

---

[1] Some identification techniques involve, as a by-product, the granting of *tickets* which provide time-limited access to specified resources (see Chapter 13).

made in typical password schemes. A subsequent section summarizes the standard attacks these designs counteract. Threats which must be guarded against include: password disclosure (outside of the system) and line eavesdropping (within the system), both of which allow subsequent replay; and password guessing, including dictionary attacks.

10.2.1 Fixed password schemes: techniques

(i) Stored password files

The most obvious approach is for the system to store user passwords cleartext in a system password file, which is both read- and write-protected (e.g., via operating system access control privileges). Upon password entry by a user, the system compares the entered password to the password file entry for the corresponding userid; employing no secret keys or cryptographic primitives such as encryption, this is classified as a non-cryptographic technique. A drawback of this method is that it provides no protection against privileged insiders or *superusers* (special userids which have full access privileges to system files and resources). Storage of the password file on backup media is also a security concern, since the file contains cleartext passwords.

(ii) "Encrypted" password files

Rather than storing a cleartext user password in a (read- and write-protected) password file, a one-way function of each user password is stored in place of the password itself (see Figure 10.1). To verify a user-entered password, the system computes the one-way function of the entered password, and compares this to the stored entry for the stated userid. To preclude attacks suggested in the preceding paragraph, the password file need now only be write-protected.

10.3 Remark (*one-way function vs. encryption*) For the purpose of protecting password files, the use of a one-way function is generally preferable to reversible encryption; reasons include those related to export restrictions, and the need for keying material. However, in both cases, for historical reasons, the resulting values are typically referred to as "encrypted" passwords. Protecting passwords by either method before transmission over public communications lines addresses the threat of compromise of the password itself, but alone does not preclude disclosure or replay of the transmission (cf. Protocol 10.6).

(iii) Password rules

Since dictionary attacks (see §10.2.2(iii)) are successful against predictable passwords, some systems impose "password rules" to discourage or prevent users from using "weak" passwords. Typical password rules include a lower bound on the password length (e.g., 8 or 12 characters); a requirement for each password to contain at least one character from each of a set of categories (e.g., uppercase, numeric, non-alphanumeric); or checks that candidate passwords are not found in on-line or available dictionaries, and are not composed of account-related information such as userids or substrings thereof.

Knowing which rules are in effect, an adversary may use a modified dictionary attack strategy taking into account the rules, and targeting the weakest form of passwords which nonetheless satisfy the rules. The objective of password rules is to increase the entropy (rather than just the length) of user passwords beyond the reach of dictionary and exhaustive search attacks. *Entropy* here refers to the uncertainty in a password (cf. §2.2.1); if all passwords are equally probable, then the entropy is maximal and equals the base-2 logarithm of the number of possible passwords.

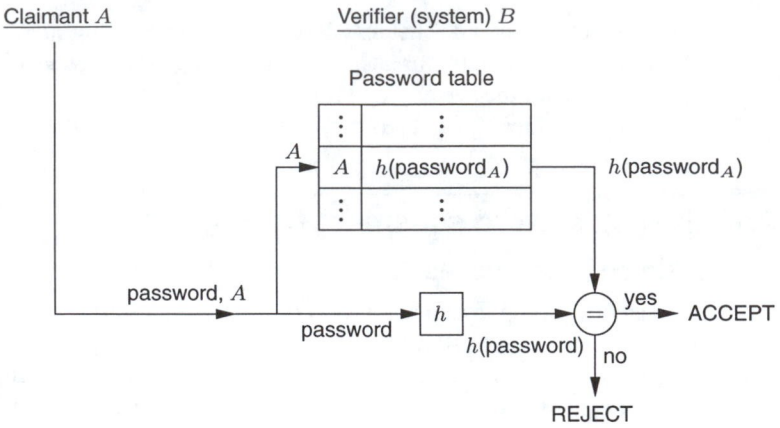

Figure 10.1: *Use of one-way function for password-checking.*

Another procedural technique intended to improve password security is *password aging*. A time period is defined limiting the lifetime of each particular password (e.g., 30 or 90 days). This requires that passwords be changed periodically.

(iv) Slowing down the password mapping

To slow down attacks which involve testing a large number of trial passwords (see §10.2.2), the password verification function (e.g., one-way function) may be made more computationally intensive, for example, by iterating a simpler function $t > 1$ times, with the output of iteration i used as the input for iteration $i + 1$. The total number of iterations must be restricted so as not to impose a noticeable or unreasonable delay for legitimate users. Also, the iterated function should be such that the iterated mapping does not result in a final range space whose entropy is significantly decimated.

(v) Salting passwords

To make dictionary attacks less effective, each password, upon initial entry, may be augmented with a t-bit random string called a *salt* (it alters the "flavor" of the password; cf. §10.2.3) before applying the one-way function. Both the hashed password and the salt are recorded in the password file. When the user subsequently enters a password, the system looks up the salt, and applies the one-way function to the entered password, as altered or augmented by the salt. The difficulty of exhaustive search on any particular user's password is unchanged by salting (since the salt is given in cleartext in the password file); however, salting increases the complexity of a dictionary attack against a large set of passwords simultaneously, by requiring the dictionary to contain 2^t variations of each trial password, implying a larger memory requirement for storing an encrypted dictionary, and correspondingly more time for its preparation. Note that with salting, two users who choose the same password have different entries in the system password file. In some systems, it may be appropriate to use an entity's userid itself as salt.

(vi) Passphrases

To allow greater entropy without stepping beyond the memory capacity of human users, passwords may be extended to *passphrases*; in this case, the user types in a phrase or sentence rather than a short "word". The passphrase is hashed down to a fixed-size value, which plays the same role as a password; here, it is important that the passphrase is not simply truncated by the system, as passwords are in some systems. The idea is that users can remember phrases easier than random character sequences. If passwords resemble English text, then since each character contains only about 1.5 bits of entropy (Fact 7.67), a passphrase provides greater security through increased entropy than a short password. One drawback is the additional typing requirement.

10.2.2 Fixed password schemes: attacks

(i) Replay of fixed passwords

A weakness of schemes using fixed, reusable passwords (i.e., the basic scheme of §10.2), is the possibility that an adversary learns a user's password by observing it as it is typed in (or from where it may be written down). A second security concern is that user-entered passwords (or one-way hashes thereof) are transmitted in cleartext over the communications line between the user and the system, and are also available in cleartext temporarily during system verification. An eavesdropping adversary may record this data, allowing subsequent impersonation.

Fixed password schemes are thus of use when the password is transmitted over trusted communications lines safe from monitoring, but are not suitable in the case that passwords are transmitted over open communications networks. For example, in Figure 10.1, the claimant A may be a user logging in from home over a telephone modem, to a remote office site B two (or two thousand) miles away; the cleartext password might then travel over an unsecured telephone network (including possibly a wireless link), subject to eavesdropping.

In the case that remote identity verification is used for access to a local resource, e.g., an automated cash dispenser with on-line identity verification, the system response (accept/reject) must be protected in addition to the submitted password, and must include variability to prevent trivial replay of a time-invariant accept response.

(ii) Exhaustive password search

A very naive attack involves an adversary simply (randomly or systematically) trying passwords, one at a time, on the actual verifier, in hope that the correct password is found. This may be countered by ensuring passwords are chosen from a sufficiently large space, limiting the number of invalid (on-line) attempts allowed within fixed time periods, and slowing down the password mapping or login-process itself as in §10.2.1(iv). *Off-line attacks*, involving a (typically large) computation which does not require interacting with the actual verifier until a final stage, are of greater concern; these are now considered.

Given a password file containing one-way hashes of user passwords, an adversary may attempt to defeat the system by testing passwords one at a time, and comparing the one-way hash of each to passwords in the encrypted password file (see §10.2.1(ii)). This is theoretically possible since both the one-way mapping and the (guessed) plaintext are known. (This could be precluded by keeping any or all of the details of the one-way mapping or the password file itself secret, but it is not considered prudent to base the security of the system on the assumption that such details remain secret forever.) The feasibility of the attack depends on the number of passwords that need be checked before a match is expected (which itself depends on the number of possible passwords), and the time required to test each (see Example 10.4, Table 10.1, and Table 10.2). The latter depends on the password mapping used, its implementation, the instruction execution time of the host processor, and the number of processors available (note exhaustive search is parallelizable). The time required to actually compare the image of each trial password to all passwords in a password file is typically negligible.

10.4 Example (*password entropy*) Suppose passwords consist of strings of 7-bit ASCII characters. Each has a numeric value in the range 0-127. (When 8-bit characters are used, values 128-255 compose the *extended character set*, generally inaccessible from standard keyboards.) ASCII codes 0-31 are reserved for control characters; 32 is a space character; 33-126 are keyboard-accessible printable characters; and 127 is a special character. Table 10.1 gives the number of distinct n-character passwords composed of typical combinations of characters, indicating an upper bound on the security of such password spaces.   □

$\to c$ $\downarrow n$	26 (lowercase)	36 (lowercase alphanumeric)	62 (mixed case alphanumeric)	95 (keyboard characters)
5	23.5	25.9	29.8	32.9
6	28.2	31.0	35.7	39.4
7	32.9	36.2	41.7	46.0
8	37.6	41.4	47.6	52.6
9	42.3	46.5	53.6	59.1
10	47.0	51.7	59.5	65.7

Table 10.1: *Bitsize of password space for various character combinations. The number of n-character passwords, given c choices per character, is c^n. The table gives the base-2 logarithm of this number of possible passwords.*

$\to c$ $\downarrow n$	26 (lowercase)	36 (lowercase alphanumeric)	62 (mixed case alphanumeric)	95 (keyboard characters)
5	0.67 hr	3.4 hr	51 hr	430 hr
6	17 hr	120 hr	130 dy	4.7 yr
7	19 dy	180 dy	22 yr	440 yr
8	1.3 yr	18 yr	1400 yr	42000 yr
9	34 yr	640 yr	86000 yr	4.0×10^6 yr
10	890 yr	23000 yr	5.3×10^6 yr	3.8×10^8 yr

Table 10.2: *Time required to search entire password space. The table gives the time T (in hours, days, or years) required to search or pre-compute over the entire specified spaces using a single processor (cf. Table 10.1). $T = c^n \cdot t \cdot y$, where t is the number of times the password mapping is iterated, and y the time per iteration, for $t = 25$, $y = 1/(125\ 000)$ sec. (This approximates the UNIX crypt command on a high-end PC performing DES at 1.0 Mbytes/s – see §10.2.3.)*

(iii) Password-guessing and dictionary attacks

To improve upon the expected probability of success of an exhaustive search, rather than searching through the space of all possible passwords, an adversary may search the space in order of decreasing (expected) probability. While ideally arbitrary strings of n characters would be equiprobable as user-selected passwords, most (unrestricted) users select passwords from a small subset of the full password space (e.g., short passwords; dictionary words; proper names; lowercase strings). Such weak passwords with low entropy are easily guessed; indeed, studies indicate that a large fraction of user-selected passwords are found in typical (intermediate) dictionaries of only 150 000 words, while even a large dictionary of 250 000 words represents only a tiny fraction of all possible n-character passwords (see Table 10.1).

Passwords found in any on-line or available list of words may be uncovered by an adversary who tries all words in this list, using a so-called *dictionary attack*. Aside from traditional dictionaries as noted above, on-line dictionaries of words from foreign languages, or on specialized topics such as music, film, etc. are available. For efficiency in repeated use by an adversary, an "encrypted" (hashed) list of dictionary or high-probability passwords may be created and stored on disk or tape; password images from system password files may then be collected, ordered (using a sorting algorithm or conventional hashing), and then compared to entries in the encrypted dictionary. Dictionary-style attacks are not generally successful at finding a particular user's password, but find many passwords in most systems.

10.2.3 Case study – UNIX passwords

The UNIX[2] operating system provides a widely known, historically important example of a fixed password system, implementing many of the ideas of §10.2.1. A UNIX password file contains a one-way function of user passwords computed as follows: each user password serves as the key to encrypt a known plaintext (64 zero-bits). This yields a one-way function of the key, since only the user (aside from the system, temporarily during password verification) knows the password. For the encryption algorithm, a minor modification of DES (§7.4) is used, as described below; variations may appear in products outside of the USA. The technique described relies on the conjectured property that DES is resistant to known-plaintext attacks – given cleartext and the corresponding ciphertext, it remains difficult to find the key.

The specific technique makes repeated use of DES, iterating the encipherment $t = 25$ times (see Figure 10.2). In detail, a user password is truncated to its first 8 ASCII characters. Each of these provides 7 bits for a 56-bit DES key (padded with 0-bits if less than 8 characters). The key is used to DES-encrypt the 64-bit constant 0, with the output fed back as input t times iteratively. The 64-bit result is repacked into 11 printable characters (a 64-bit output and 12 salt bits yields 76 bits; 11 ASCII characters allow 77). In addition, a non-standard method of password salting is used, intended to simultaneously complicate dictionary attacks and preclude use of off-the-shelf DES hardware for attacks:

1. *password salting*. UNIX password salting associates a 12-bit "random" salt (12 bits taken from the system clock at time of password creation) with each user-selected password. The 12 bits are used to alter the standard expansion function E of the DES mapping (see §7.4), providing one of 4096 variations. (The expansion E creates a 48-bit block; immediately thereafter, the salt bits collectively determine one of 4096 permutations. Each bit is associated with a pre-determined pair from the 48-bit block, e.g., bit 1 with block bits 1 and 25, bit 2 with block bits 2 and 26, etc. If the salt bit is 1, the block bits are swapped, and otherwise they are not.) Both the hashed password and salt are recorded in the system password file. Security of any particular user's password is unchanged by salting, but a dictionary attack now requires $2^{12} = 4096$ variations of each trial password.

2. *preventing use of off-the-shelf DES chips*. Because the DES expansion permutation E is dependent on the salt, standard DES chips can no longer be used to implement the UNIX password algorithm. An adversary wishing to use hardware to speed up an attack must build customized hardware rather than use commercially available chips. This may deter adversaries with modest resources.

The value stored for a given userid in the write-protected password file `/etc/passwd` is thus the iterated encryption of 0 under that user's password, using the salted modification of DES. The constant 0 here could be replaced by other values, but typically is not. The overall algorithm is called the UNIX *crypt* password algorithm.

---

[2] UNIX is a trademark of Bell Laboratories.

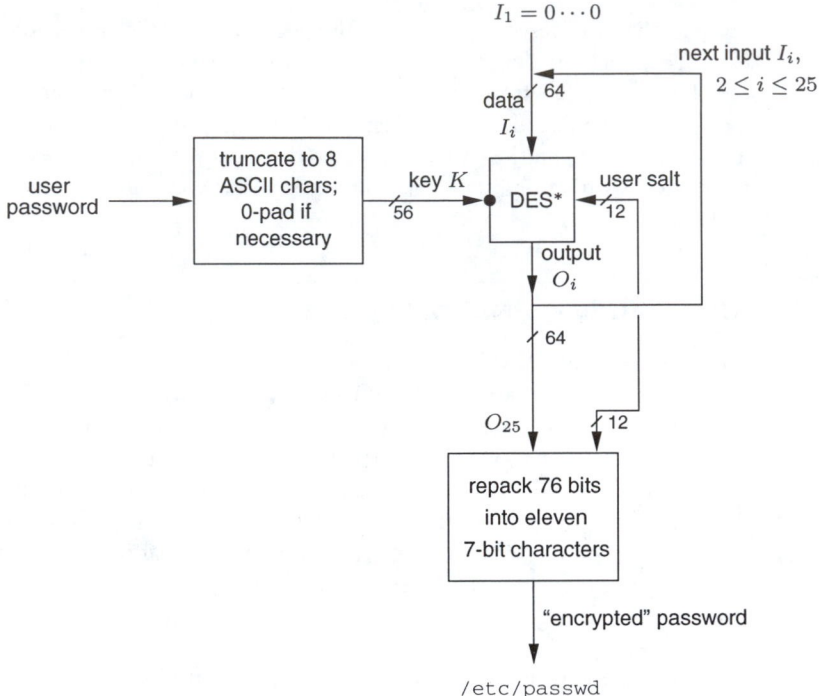

Figure 10.2: UNIX crypt *password mapping. DES* indicates DES with the expansion mapping E modified by a 12-bit salt.*

10.5 Remark (*performance advances*) While the UNIX *crypt* mapping with $t = 25$ iterations provided a reasonable measure of protection against exhaustive search when introduced in the 1970s, for equivalent security in a system designed today a more computationally intensive mapping would be provided, due to performance advances in both hardware and software.

10.2.4 PINs and passkeys

(i) PINs

Personal identification numbers (PINs) fall under the category of fixed (time-invariant) passwords. They are most often used in conjunction with "something possessed", typically a physical *token* such as a plastic banking card with a magnetic stripe, or a chipcard. To prove one's identity as the authorized user of the token, and gain access to the privileges associated therewith, entry of the correct PIN is required when the token is used. This provides a second level of security if the token is lost or stolen. PINs may also serve as the second level of security for entry to buildings which have an independent first level of security (e.g., a security guard or video camera).

For user convenience and historical reasons, PINs are typically short (relative to fixed password schemes) and numeric, e.g., 4 to 8 digits. To prevent exhaustive search through such a small key space (e.g., 10 000 values for a 4-digit numeric PIN), additional procedural constraints are necessary. For example, some automated cash dispenser machines accessed by banking cards confiscate a card if three incorrect PINs are entered successively; for others, incorrect entry of a number of successive PINs may cause the card to be "locked" or deactivated, thereafter requiring a longer PIN (e.g., 8 digits) for reactivation following such suspicious circumstances.

In an on-line system using PINs or reusable passwords, a claimed identity accompanied by a user-entered PIN may be verified by comparison to the PIN stored for that identity in a system database. An alternative is to use the PIN as a key for a MAC (see Chapter 9).

In an off-line system without access to a central database, information facilitating PIN verification must be stored on the token itself. If the PIN need not be user-selected, this may be done by defining the PIN to be a function of a secret key and the identity associated with the token; the PIN is then verifiable by any remote system knowing this master key.

In an off-line system, it may also be desirable to allow the PIN to be user-selectable, to facilitate PIN memorization by users. In this case, the PIN may be encrypted under a master key and stored on the token, with the master key known to all off-line terminals that need to be capable of verifying the token. A preferable design is to store a one-way function of the PIN, user identity, and master key on the token.

(ii) Two-stage authentication and password-derived keys

Human users have difficulty remembering secret keys which have sufficient entropy to provide adequate security. Two techniques which address this issue are now described.

When tokens are used with off-line PIN verification, a common technique is for the PIN to serve to verify the user to the token, while the token contains additional independent information allowing the token to authenticate itself to the system (as a valid token representing a legitimate user). The user is thereby indirectly authenticated to the system by a two-stage process. This requires the user have possession of the token but need remember only a short PIN, while a longer key (containing adequate entropy) provides cryptographic security for authentication over an unsecured link.

A second technique is for a user password to be mapped by a one-way hash function into a cryptographic key (e.g., a 56-bit DES key). Such password-derived keys are called *passkeys*. The passkey is then used to secure a communications link between the user and a system which also knows the user password. It should be ensured that the entropy of the user's password is sufficiently large that exhaustive search of the password space is not more efficient than exhaustive search of the passkey space (i.e., guessing passwords is not easier than guessing 56-bit DES keys); see Table 10.1 for guidance.

An alternative to having passkeys remain fixed until the password is changed is to keep a running sequence number on the system side along with each user's password, for use as a time-variant salt communicated to the user in the clear and incremented after each use. A fixed per-user salt could also be used in addition to a running sequence number.

Passkeys should be viewed as long-term keys, with use restricted to authentication and key management (e.g., rather than also for bulk encryption of user data). A disadvantage of using password-derived keys is that storing each user's password within the system requires some mechanism to protect the confidentiality of the stored passwords.

10.2.5 One-time passwords (towards strong authentication)

A natural progression from fixed password schemes to challenge-response identification protocols may be observed by considering one-time password schemes. As was noted in §10.2.2, a major security concern of fixed password schemes is eavesdropping and subsequent replay of the password. A partial solution is *one-time passwords*: each password is used only once. Such schemes are safe from passive adversaries who eavesdrop and later attempt impersonation. Variations include:

1. *shared lists of one-time passwords.* The user and the system use a sequence or set of t secret passwords, (each valid for a single authentication), distributed as a pre-shared list. A drawback is maintenance of the shared list. If the list is not used sequentially, the system may check the entered password against all remaining unused passwords. A variation involves use of a *challenge-response table*, whereby the user and the system share a table of matching challenge-response pairs, ideally with each pair valid at most once; this non-cryptographic technique differs from the cryptographic challenge-response of §10.3.

2. *sequentially updated one-time passwords.* Initially only a single secret password is shared. During authentication using password i, the user creates and transmits to the system a new password (password $i + 1$) encrypted under a key derived from password i. This method becomes difficult if communication failures occur.

3. *one-time password sequences based on a one-way function.* Lamport's one-time password scheme is described below. This method is more efficient (with respect to bandwidth) than sequentially updated one-time passwords, and may be viewed as a challenge-response protocol where the challenge is implicitly defined by the current position within the password sequence.

One-time passwords based on one-way functions (Lamport's scheme)

In Lamport's one-time password scheme, the user begins with a secret w. A one-way function (OWF) H is used to define the password sequence: $w, H(w), H(H(w)), \ldots, H^t(w)$. The password for the ith identification session, $1 \le i \le t$, is defined to be $w_i = H^{t-i}(w)$.

10.6 Protocol Lamport's OWF-based one-time passwords

SUMMARY: A identifies itself to B using one-time passwords from a sequence.

1. *One-time setup.*

    (a) User A begins with a secret w. Let H be a one-way function.

    (b) A constant t is fixed (e.g., $t = 100$ or 1000), defining the number of identifications to be allowed. (The system is thereafter restarted with a new w, to avoid replay attacks.)

    (c) A transfers (the *initial shared secret*) $w_0 = H^t(w)$, in a manner guaranteeing its authenticity, to the system B. B initializes its counter for A to $i_A = 1$.

2. *Protocol messages.* The ith identification, $1 \le i \le t$, proceeds as follows:

$$A \to B: \quad A, \ i, \ w_i \ (= H^{t-i}(w)) \quad (1)$$

Here $A \to B$: X denotes A sending the message X to B.

3. *Protocol actions.* To identify itself for session i, A does the following.

    (a) A's equipment computes $w_i = H^{t-i}(w)$ (easily done either from w itself, or from an appropriate intermediate value saved during the computation of $H^t(w)$ initially), and transmits (1) to B.

    (b) B checks that $i = i_A$, and that the received password w_i satisfies: $H(w_i) = w_{i-1}$. If both checks succeed, B accepts the password, sets $i_A \leftarrow i_A + 1$, and saves w_i for the next session verification.

10.7 Note (*pre-play attack*) Protocol 10.6 and similar one-time password schemes including that of Note 10.8 remain vulnerable to an active adversary who intercepts and traps (or impersonates the system in order to extract) an as-yet unused one-time password, for the purpose of subsequent impersonation. To prevent this, a password should be revealed only to a party which itself is known to be authentic. Challenge-response techniques (see §10.3) address this threat.

10.8 Note (*alternative one-time password scheme*) The following one-time-password alternative to Protocol 10.6 is suitable if storing actual passwords on the system side is acceptable (cf. Figure 10.1; compare also to §10.3.2(iii)). The claimant A has a shared password P with the system verifier B, to which it sends the data pair: $(r, H(r, P))$. The verifier computes the hash of the received value r and its local copy of P, and declares acceptance if this matches the received hash value. To avoid replay, r should be a sequence number, timestamp, or other parameter which can be easily guaranteed to be accepted only once.

10.3 Challenge-response identification (strong authentication)

The idea of cryptographic challenge-response protocols is that one entity (the claimant) "proves" its identity to another entity (the verifier) by demonstrating knowledge of a secret known to be associated with that entity, without revealing the secret itself to the verifier during the protocol.[3] This is done by providing a response to a time-variant challenge, where the response depends on both the entity's secret and the challenge. The *challenge* is typically a number chosen by one entity (randomly and secretly) at the outset of the protocol. If the communications line is monitored, the response from one execution of the identification protocol should not provide an adversary with useful information for a subsequent identification, as subsequent challenges will differ.

Before considering challenge-response identification protocols based on symmetric-key techniques (§10.3.2), public-key techniques (§10.3.3), and zero-knowledge concepts (§10.4), background on time-variant parameters is first provided.

10.3.1 Background on time-variant parameters

Time-variant parameters may be used in identification protocols to counteract replay and interleaving attacks (see §10.5), to provide uniqueness or timeliness guarantees, and to prevent certain chosen-text attacks. They may similarly be used in authenticated key establishment protocols (Chapter 12), and to provide uniqueness guarantees in conjunction with message authentication (Chapter 9).

Time-variant parameters which serve to distinguish one protocol instance from another are sometimes called *nonces*, *unique numbers*, or *non-repeating values*; definitions of these terms have traditionally been loose, as the specific properties required depend on the actual usage and protocol.

10.9 Definition A *nonce* is a value used no more than once for the same purpose. It typically serves to prevent (undetectable) replay.

---

[3] In some mechanisms, the secret is known to the verifier, and is used to verify the response; in others, the secret need not actually be known by the verifier.

The term *nonce* is most often used to refer to a "random" number in a challenge-response protocol, but the required randomness properties vary. Three main classes of time-variant parameters are discussed in turn below: random numbers, sequence numbers, and time-stamps. Often, to ensure protocol security, the integrity of such parameters must be guaranteed (e.g., by cryptographically binding them with other data in a challenge-response sequence). This is particularly true of protocols in which the only requirement of a time-variant parameter is uniqueness, e.g., as provided by a never-repeated sequential counter.[4]

Following are some miscellaneous points about time-variant parameters.

1. Verifiable timeliness may be provided through use of random numbers in challenge-response mechanisms, timestamps in conjunction with distributed timeclocks, or sequence numbers in conjunction with the maintenance of pairwise (claimant, verifier) state information.

2. To provide timeliness or uniqueness guarantees, the verifier in the protocol controls the time-variant parameter, either directly (through choice of a random number) or indirectly (through information maintained regarding a shared sequence, or logically through a common time clock).

3. To uniquely identify a message or sequence of messages (protocol instance), nonces drawn from a monotonically increasing sequence may be used (e.g., sequence or serial numbers, and timestamps, if guaranteed to be increasing and unique), or random numbers of sufficient size. Uniqueness is often required only within a given key lifetime or time window.

4. Combinations of time-variant parameters may be used, e.g., random numbers concatenated to timestamps or sequence numbers. This may guarantee that a pseudorandom number is not duplicated.

(i) Random numbers

Random numbers may be used in challenge-response mechanisms, to provide uniqueness and timeliness assurances, and to preclude certain replay and interleaving attacks (see §10.5, including Remark 10.42). Random numbers may also serve to provide unpredictability, for example, to preclude chosen-text attacks.

The term *random numbers*, when used in the context of identification and authentication protocols, includes pseudorandom numbers which are unpredictable to an adversary (see Remark 10.11); this differs from randomness in the traditional statistical sense. In protocol descriptions, "choose a random number" is usually intended to mean "pick a number with uniform distribution from a specified sample space" or "select from a uniform distribution".

Random numbers are used in challenge-response protocols as follows. One entity includes a (new) random number in an outgoing message. An incoming message subsequently received (e.g., the next protocol message of the same protocol instance), whose construction required knowledge of this nonce and to which this nonce is inseparably bound, is then deemed to be *fresh* (Remark 10.10) based on the reasoning that the random number links the two messages. The non-tamperable binding is required to prevent appending a nonce to an old message.

Random numbers used in this manner serve to fix a relative point in time for the parties involved, analogous to a shared timeclock. The maximum allowable time between protocol messages is typically constrained by a *timeout period*, enforced using local, independent countdown timers.

---

[4]Such predictable parameters differ from sequence numbers in that they might not be bound to any stored state. Without appropriate cryptographic binding, a potential concern then is a pre-play attack wherein an adversary obtains the response before the time-variant parameter is legitimately sent (see Note 10.7).

10.10 Remark (*freshness*) In the context of challenge-response protocols, *fresh* typically means recent, in the sense of having originated subsequent to the beginning of the current protocol instance. Note that such freshness alone does not rule out interleaving attacks using parallel sessions (see §10.5).

10.11 Remark (*birthday repetitions in random numbers*) In generating pseudorandom numbers for use as time-variant parameters, it suffices if the probability of a repeated number is acceptably low and if numbers are not intentionally reused. This may be achieved by selecting the random value from a sufficiently large sample space, taking into account coincidences arising from the birthday paradox. The latter may be addressed by either using a larger sample space, or by using a generation process guaranteed to avoid repetition (e.g., a bijection), such as using the counter or OFB mode of a block cipher (§7.2.2).

10.12 Remark (*disadvantages of random numbers*) Many protocols involving random numbers require the generation of cryptographically secure (i.e., unpredictable) random numbers. If pseudorandom number generators are used, an initial seed with sufficient entropy is required. When random numbers are used in challenge-response mechanisms in place of timestamps, typically the protocol involves one additional message, and the challenger must temporarily maintain state information, but only until the response is verified.

(ii) Sequence numbers

A sequence number (serial number, or counter value) serves as a unique number identifying a message, and is typically used to detect message replay. For stored files, sequence numbers may serve as *version numbers* for the file in question. Sequence numbers are specific to a particular pair of entities, and must explicitly or implicitly be associated with both the originator and recipient of a message; distinct sequences are customarily necessary for messages from A to B and from B to A.

Parties follow a pre-defined policy for message numbering. A message is accepted only if the sequence number therein has not been used previously (or not used previously within a specified time period), and satisfies the agreed policy. The simplest policy is that a sequence number starts at zero, is incremented sequentially, and each successive message has a number one greater than the previous one received. A less restrictive policy is that sequence numbers need (only) be monotonically increasing; this allows for lost messages due to non-malicious communications errors, but precludes detection of messages lost due to adversarial intervention.

10.13 Remark (*disadvantages of sequence numbers*) Use of sequence numbers requires an overhead as follows: each claimant must record and maintain long-term pairwise state information for each possible verifier, sufficient to determine previously used and/or still valid sequence numbers. Special procedures (e.g., for resetting sequence numbers) may be necessary following circumstances disrupting normal sequencing (e.g., system failures). Forced delays are not detectable in general. As a consequence of the overhead and synchronization necessary, sequence numbers are most appropriate for smaller, closed groups.

(iii) Timestamps

Timestamps may be used to provide timeliness and uniqueness guarantees, to detect message replay. They may also be used to implement time-limited access privileges, and to detect forced delays.

Timestamps function as follows. The party originating a message obtains a timestamp from its local (host) clock, and cryptographically binds it to a message. Upon receiving a time-stamped message, the second party obtains the current time from its own (host) clock, and subtracts the timestamp received. The received message is valid provided:

1. the timestamp difference is within the *acceptance window* (a fixed-size time interval, e.g., 10 milliseconds or 20 seconds, selected to account for the maximum message transit and processing time, plus clock skew); and

2. (optionally) no message with an identical timestamp has been previously received from the same originator. This check may be made by the verifier maintaining a list of all timestamps received from each source entity within the current acceptance window. Another method is to record the latest (valid) timestamp used by each source (in this case the verifier accepts only strictly increasing time values).

The security of timestamp-based verification relies on use of a common time reference. This requires that host clocks be available and both "loosely synchronized" and secured from modification. Synchronization is necessary to counter clock drift, and must be appropriate to accommodate the acceptance window used. The degree of clock skew allowed, and the acceptance window, must be appropriately small to preclude message replay if the above optional check is omitted. The timeclock must be secure to prevent adversarial resetting of a clock backwards so as to restore the validity of old messages, or setting a clock forward to prepare a message for some future point in time (cf. Note 10.7).

10.14 Remark (*disadvantages of timestamps*) Timestamp-based protocols require that time-clocks be both synchronized and secured. The preclusion of adversarial modification of local timeclocks is difficult to guarantee in many distributed environments; in this case, the security provided must be carefully re-evaluated. Maintaining lists of used timestamps within the current window has the drawback of a potentially large storage requirement, and corresponding verification overhead. While technical solutions exist for synchronizing distributed clocks, if synchronization is accomplished via network protocols, such protocols themselves must be secure, which typically requires authentication; this leads to a circular security argument if such authentication is itself timestamp-based.

10.15 Remark (*comparison of time-variant parameters*) Timestamps in protocols offer the advantage of fewer messages (typically by one), and no requirement to maintain pairwise long-term state information (cf. sequence numbers) or per-connection short-term state information (cf. random numbers). Minimizing state information is particularly important for servers in client-server applications. The main drawback of timestamps is the requirement of maintaining secure, synchronized distributed timeclocks. Timestamps in protocols may typically be replaced by a random number challenge plus a return message.

10.3.2 Challenge-response by symmetric-key techniques

Challenge-response mechanisms based on symmetric-key techniques require the claimant and the verifier to share a symmetric key. For closed systems with a small number of users, each pair of users may share a key a priori; in larger systems employing symmetric-key techniques, identification protocols often involve the use of a trusted on-line server with which each party shares a key. The on-line server effectively acts like the hub of a spoked wheel, providing a common session key to two parties each time one requests authentication with the other.

The apparent simplicity of the techniques presented below and in §10.3.3 is misleading. The design of such techniques is intricate and the security is brittle; those presented have been carefully selected.

(i) Challenge-response based on symmetric-key encryption

Both the Kerberos protocol (Protocol 12.24) and the Needham-Schroeder shared-key protocol (Protocol 12.26) provide entity authentication based on symmetric encryption and involve use of an on-line trusted third party. These are discussed in Chapter 12, as they additionally provide key establishment.

Below, three simple techniques based on ISO/IEC 9798-2 are described. They assume the prior existence of a shared secret key (and no further requirement for an on-line server). In this case, two parties may carry out unilateral entity authentication in one pass using timestamps or sequence numbers, or two passes using random numbers; mutual authentication requires, respectively, two and three passes. The claimant corroborates its identity by demonstrating knowledge of the shared key by encrypting a challenge (and possibly additional data) using the key. These techniques are similar to those given in §12.3.1.

10.16 Remark (*data integrity*) When encipherment is used in entity authentication protocols, data integrity must typically also be guaranteed to ensure security. For example, for messages spanning more than one block, the rearrangement of ciphertext blocks cannot be detected in the ECB mode of block encryption, and even CBC encryption may provide only a partial solution. Such data integrity should be provided through use of an accepted data integrity mechanism (see §9.6; cf. Remark 12.19).

9798-2 mechanisms: Regarding notation: r_A and t_A, respectively, denote a random number and a timestamp, generated by A. (In these mechanisms, the timestamp t_A may be replaced by a sequence number n_A, providing slightly different guarantees.) E_K denotes a symmetric encryption algorithm, with a key K shared by A and B; alternatively, distinct keys K_{AB} and K_{BA} may be used for unidirectional communication. It is assumed that both parties are aware of the claimed identity of the other, either by context or by additional (unsecured) cleartext data fields. Optional message fields are denoted by an asterisk (*), while a comma (,) within the scope of E_K denotes concatenation.

1. *unilateral authentication, timestamp-based*:

$$A \to B : E_K(t_A, B^*) \quad (1)$$

Upon reception and decryption, B verifies that the timestamp is acceptable, and optionally verifies the received identifier as its own. The identifier B here prevents an adversary from re-using the message immediately on A, in the case that a single bi-directional key K is used.

2. *unilateral authentication, using random numbers*:
To avoid reliance on timestamps, the timestamp may be replaced by a random number, at the cost of an additional message:

$$A \leftarrow B : r_B \qquad (1)$$
$$A \to B : E_K(r_B, B^*) \quad (2)$$

B decrypts the received message and checks that the random number matches that sent in (1). Optionally, B checks that the identifier in (2) is its own; this prevents a reflection attack in the case of a bi-directional key K. To prevent chosen-text attacks on the encryption scheme E_K, A may (as below) embed an additional random number in (2) or, alternately, the form of the challenges can be restricted; the critical requirement is that they be non-repeating.

3. *mutual authentication, using random numbers*:

$$A \leftarrow B : r_B \qquad (1)$$
$$A \rightarrow B : E_K(r_A, r_B, B^*) \qquad (2)$$
$$A \leftarrow B : E_K(r_B, r_A) \qquad (3)$$

Upon reception of (2), B carries out the checks as above and, in addition, recovers the decrypted r_A for inclusion in (3). Upon decrypting (3), A checks that both random numbers match those used earlier. The second random number r_A in (2) serves both as a challenge and to prevent chosen-text attacks.

10.17 Remark (*doubling unilateral authentication*) While mutual authentication may be obtained by running any of the above unilateral authentication mechanisms twice (once in each direction), such an ad-hoc combination suffers the drawback that the two unilateral authentications, not being linked, cannot logically be associated with a single protocol run.

(ii) Challenge-response based on (keyed) one-way functions

The encryption algorithm in the above mechanisms may be replaced by a one-way or non-reversible function of the shared key and challenge, e.g., having properties similar to a MAC (Definition 9.7). This may be preferable in situations where encryption algorithms are otherwise unavailable or undesirable (e.g., due to export restrictions or computational costs). The modifications required to the 9798-2 mechanisms above (yielding the analogous mechanisms of ISO/IEC 9798-4) are the following:

1. the encryption function E_K is replaced by a MAC algorithm h_K;
2. rather than decrypting and verifying that fields match, the recipient now independently computes the MAC value from known quantities, and accepts if the computed MAC matches the received MAC value; and
3. to enable independent MAC computation by the recipient, the additional cleartext field t_A must be sent in message (1) of the one-pass mechanism. r_A must be sent as an additional cleartext field in message (2) of the three-pass mechanism.

The revised three-pass challenge-response mechanism based on a MAC h_K, with actions as noted above, provides mutual identification. Essentially the same protocol, called *SKID3*, has messages as follows:

$$A \leftarrow B : \quad r_B \qquad (1)$$
$$A \rightarrow B : \quad r_A, \ h_K(r_A, r_B, B) \qquad (2)$$
$$A \leftarrow B : \quad h_K(r_B, r_A, A) \qquad (3)$$

Note that the additional field A is included in message (3). The protocol *SKID2*, obtained by omitting the third message, provides unilateral entity authentication.

(iii) Implementation using hand-held passcode generators

Answering a challenge in challenge-response protocols requires some type of computing device and secure storage for long-term keying material (e.g., a file on a trusted local disk, perhaps secured under a local password-derived key). For additional security, a device such as a chipcard (and corresponding card reader) may be used for both the key storage and response computation. In some cases, a less expensive option is a passcode generator.

Passcode generators are hand-held devices, resembling thin calculators in both size and display, and which provide time-variant passwords or *passcodes* (see Figure 10.3). The generator contains a device-specific secret key. When a user is presented with a challenge (e.g., by a system displaying it on a computer terminal), the challenge is keyed into the generator. The generator displays a passcode, computed as a function of the secret key and the challenge; this may be either an asymmetric function, or a symmetric function (e.g., encryption or MAC as discussed above). The user returns the response (e.g., keys the passcode in at his terminal), which the system verifies by comparison to an independently computed response, using the same information stored on the system side.

For further protection against misplaced generators, the response may also depend on a user-entered PIN. Simpler passcode generators omit the user keypad, and use as an implicit challenge a time value (with a typical granularity of one minute) defined by a timeclock loosely synchronized automatically between the system and the passcode generator. A more sophisticated device combines implicit synchronization with explicit challenges, presenting an explicit challenge only when synchronization is lost.

A drawback of systems using passcode generators is, as per §10.2.1(i), the requirement to provide confidentiality for user passwords stored on the system side.

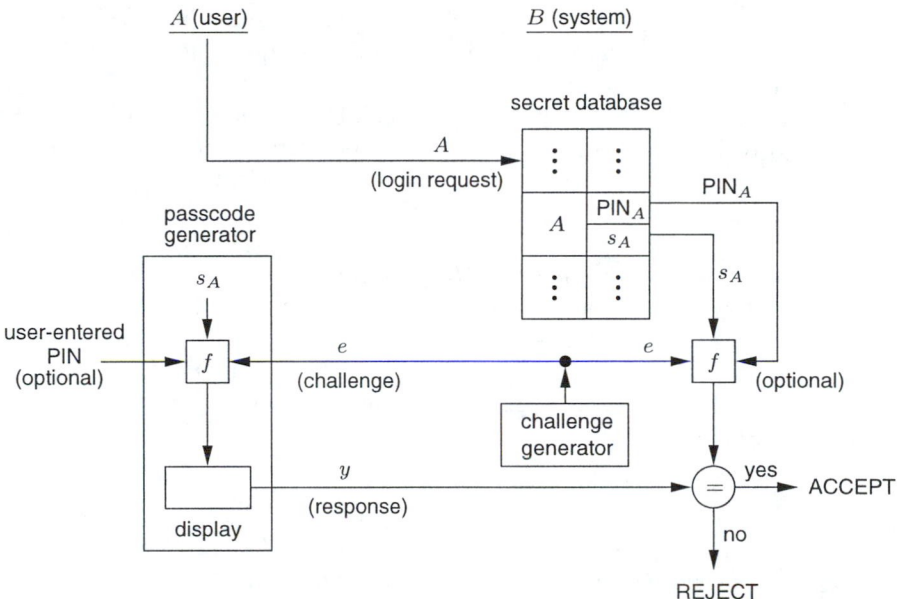

Figure 10.3: *Functional diagram of a hand-held passcode generator. s_A is A's user-specific secret. f is a one-way function. The (optional) PIN could alternatively be locally verified in the passcode generator only, making y independent of it.*

10.3.3 Challenge-response by public-key techniques

Public-key techniques may be used for challenge-response based identification, with a claimant demonstrating knowledge of its private key in one of two ways (cf. §12.5):

1. the claimant decrypts a challenge encrypted under its public key;
2. the claimant digitally signs a challenge.

Ideally, the public-key pair used in such mechanisms should not be used for other purposes, since combined usage may compromise security (Remark 10.40). A second caution is that the public-key system used should not be susceptible to chosen-ciphertext attacks,[5]

---

[5] Both chosen-ciphertext and chosen-plaintext attacks are of concern for challenge-response techniques based on symmetric-key encryption.

as an adversary may attempt to extract information by impersonating a verifier and choosing strategic rather than random challenges. (See Notes 8.13 and 8.58 regarding the Rabin/Williams and Blum-Goldwasser schemes.)

Incorporating a self-generated random number or *confounder* (§10.5) into the data over which the response is computed may address both of these concerns. Such data may be made available to the verifier in cleartext to allow verification.

(i) Challenge-response based on public-key decryption

Identification based on PK decryption and witness. Consider the following protocol:

$$A \leftarrow B : \quad h(r), B, P_A(r, B) \quad (1)$$
$$A \rightarrow B : \quad r \quad\quad\quad\quad\quad\quad (2)$$

B chooses a random r, computes the *witness* $x = h(r)$ (x demonstrates knowledge of r without disclosing it – cf. §10.4.1), and computes the challenge $e = P_A(r, B)$. Here P_A denotes the public-key encryption (e.g., RSA) algorithm of A, and h denotes a one-way hash function. B sends (1) to A. A decrypts e to recover r' and B', computes $x' = h(r')$, and quits if $x' \neq x$ (implying $r' \neq r$) or if B' is not equal to its own identifier B. Otherwise, A sends $r = r'$ to B. B succeeds with (unilateral) entity authentication of A upon verifying the received r agrees with that sent earlier. The use of the witness precludes chosen-text attacks.

Modified Needham-Schroeder PK protocol for identification. The modified Needham-Schroeder public-key protocol of Note 12.39 provides key transport of distinct keys k_1, k_2 from A to B and B to A, respectively, as well as mutual authentication. If the key establishment feature is not required, k_1 and k_2 may be omitted. With P_B denoting the public-key encryption algorithm for B (e.g., RSA), the messages in the modified protocol for identification are then as follows:

$$A \rightarrow B : \quad P_B(r_1, A) \quad (1)$$
$$A \leftarrow B : \quad P_A(r_1, r_2) \quad (2)$$
$$A \rightarrow B : \quad r_2 \quad\quad\quad\quad (3)$$

Verification actions are analogous to those of Note 12.39.

(ii) Challenge-response based on digital signatures

X.509 mechanisms based on digital signatures. The ITU-T (formerly CCITT) X.509 two- and three-way strong authentication protocols specify identification techniques based on digital signatures and, respectively, timestamps and random number challenges. These are described in §12.5.2, and optionally provide key establishment in addition to entity authentication.

9798-3 mechanisms. Three challenge-response identification mechanisms based on signatures are given below, analogous to those in §10.3.2(i) based on symmetric-key encryption, but, in this case, corresponding to techniques in ISO/IEC 9798-3. Regarding notation (cf. 9798-2 above): r_A and t_A, respectively, denote a random number and timestamp generated by A. S_A denotes A's signature mechanism; if this mechanism provides message recovery, some of the cleartext fields listed below are redundant and may be omitted. $cert_A$ denotes the public-key certificate containing A's signature public key. (In these mechanisms, if the verifier has the authentic public key of the claimant a priori, certificates may be omitted; otherwise, it is assumed that the verifier has appropriate information to verify the validity of the public key contained in a received certificate – see Chapter 13.) Remark 10.17 also applies here.

1. *unilateral authentication with timestamps*:

$$A \to B : cert_A, t_A, B, S_A(t_A, B) \quad (1)$$

Upon reception, B verifies that the timestamp is acceptable, the received identifier B is its own, and (using A's public key extracted from $cert_A$ after verifying the latter) checks that the signature over these two fields is correct.

2. *unilateral authentication with random numbers*: Reliance on timestamps may be replaced by a random number, at the cost of an additional message:

$$A \leftarrow B : r_B \qquad\qquad\qquad\qquad (1)$$
$$A \to B : cert_A, r_A, B, S_A(r_A, r_B, B) \quad (2)$$

B verifies that the cleartext identifier is its own, and using a valid signature public key for A (e.g., from $cert_A$), verifies that A's signature is valid over the cleartext random number r_A, the same number r_B as sent in (1), and this identifier. The signed r_A explicitly prevents chosen-text attacks.

3. *mutual authentication with random numbers*:

$$A \leftarrow B : r_B \qquad\qquad\qquad\qquad (1)$$
$$A \to B : cert_A, r_A, B, S_A(r_A, r_B, B) \quad (2)$$
$$A \leftarrow B : cert_B, A, S_B(r_B, r_A, A) \quad (3)$$

Processing of (1) and (2) is as above; (3) is processed analogously to (2).

10.4 Customized and zero-knowledge identification protocols

This section considers protocols specifically designed to achieve identification, which use asymmetric techniques but do not rely on digital signatures or public-key encryption, and which avoid use of block ciphers, sequence numbers, and timestamps. They are similar in some regards to the challenge-response protocols of §10.3, but are based on the ideas of interactive proof systems and zero-knowledge proofs (see §10.4.1), employing random numbers not only as challenges, but also as *commitments* to prevent cheating.

10.4.1 Overview of zero-knowledge concepts

A disadvantage of simple password protocols is that when a claimant A (called a *prover* in the context of zero-knowledge protocols) gives the verifier B her password, B can thereafter impersonate A. Challenge-response protocols improve on this: A responds to B's challenge to demonstrate knowledge of A's secret in a time-variant manner, providing information not directly reusable by B. This might nonetheless reveal some partial information about the claimant's secret; an adversarial verifier might also be able to strategically select challenges to obtain responses providing such information (see chosen-text attacks, §10.5).

Zero-knowledge (ZK) protocols are designed to address these concerns, by allowing a prover to demonstrate knowledge of a secret while revealing no information whatsoever (beyond what the verifier was able to deduce prior to the protocol run) of use to the verifier in conveying this demonstration of knowledge to others. The point is that only a single bit of information need be conveyed – namely, that the prover actually does know the secret.

More generally, a zero-knowledge protocol allows a proof of the truth of an assertion, while conveying no information whatsoever (this notion can be quantified in a rigorous sense) about the assertion itself other than its actual truth. In this sense, a zero-knowledge proof is similar to an answer obtained from a (trusted) *oracle*.

(i) Interactive proof systems and zero-knowledge protocols

The ZK protocols to be discussed are instances of *interactive proof systems*, wherein a prover and verifier exchange multiple messages (challenges and responses), typically dependent on random numbers (ideally: the outcomes of fair coin tosses) which they may keep secret. The prover's objective is to convince (*prove* to) the verifier the truth of an assertion, e.g., claimed knowledge of a secret. The verifier either accepts or rejects the *proof*. The traditional mathematical notion of a proof, however, is altered to an interactive game wherein proofs are *probabilistic* rather than absolute; a proof in this context need be correct only with bounded probability, albeit possibly arbitrarily close to 1. For this reason, an interactive proof is sometimes called a *proof by protocol*.

Interactive proofs used for identification may be formulated as proofs of knowledge. A possesses some secret s, and attempts to convince B it has *knowledge* of s by correctly responding to queries (involving publicly known inputs and agreed upon functions) which require knowledge of s to answer. Note that proving knowledge of s differs from proving that such s exists – for example, proving knowledge of the prime factors of n differs from proving that n is composite.

An interactive proof is said to be a *proof of knowledge* if it has both the properties of completeness and soundness. Completeness may be viewed as the customary requirement that a protocol functions properly given honest participants.

10.18 Definition (*completeness property*) An interactive proof (protocol) is *complete* if, given an honest prover and an honest verifier, the protocol succeeds with overwhelming probability (i.e., the verifier accepts the prover's claim). The definition of *overwhelming* depends on the application, but generally implies that the probability of failure is not of practical significance.

10.19 Definition (*soundness property*) An interactive proof (protocol) is *sound* if there exists an expected polynomial-time algorithm M with the following property: if a dishonest prover (impersonating A) can with non-negligible probability successfully execute the protocol with B, then M can be used to extract from this prover knowledge (essentially equivalent to A's secret) which with overwhelming probability allows successful subsequent protocol executions.

An alternate explanation of the condition in Definition 10.19 is as follows: the prover's secret s together with public data satisfies some polynomial-time predicate, and another solution of this predicate (possibly the same) can be extracted, allowing successful execution of subsequent protocol instances.

Since any party capable of impersonating A must know the equivalent of A's secret knowledge (M can be used to extract it from this party in polynomial time), soundness guarantees that the protocol does indeed provide a proof of knowledge – knowledge equivalent to that being queried is required to succeed. Soundness thus prevents a dishonest prover from convincing an honest verifier (but does does not by itself guarantee that acquiring the prover's secret is difficult; see Remark 10.23). A standard method to establish the soundness of a particular protocol is to assume the existence of a dishonest prover capable of successfully executing the protocol, and show how this allows one to compute the real prover's secret.

While an interactive proof of knowledge (or protocol based thereon) must be sound to be of cryptographic use, the main property of zero-knowledge protocols is the zero-knowledge aspect itself. For what follows, define a *transcript* (or view) to be the collection of messages resulting from protocol execution.

10.20 Definition (*zero-knowledge property*) A protocol which is a proof of knowledge has the *zero-knowledge property* if it is simulatable in the following sense: there exists an expected polynomial-time algorithm (*simulator*) which can produce, upon input of the assertion(s) to be proven but without interacting with the real prover, transcripts indistinguishable from those resulting from interaction with the real prover.

The zero-knowledge property implies that a prover executing the protocol (even when interacting with a malicious verifier) does not release any information (about its secret knowledge, other than that the particular assertion itself is true) not otherwise computable in polynomial time from public information alone. Thus, participation does not increase the chances of subsequent impersonation.

10.21 Remark (*simulated ZK protocols and protocol observers*) Consider an observer C who witnesses a zero-knowledge interactive proof (ZKIP) involving a prover A convincing a verifier B ($B \neq C$) of some knowledge A has. The "proof" to B does not provide any guarantees to C. (Indeed, A and B might have a prior agreement, conspiring against C, on the challenges to be issued.) Similarly, a recorded ZKIP conveys no guarantees upon playback. This is fundamental to the idea of the zero-knowledge property and the condition that proofs be simulatable by a verifier alone. Interactive proofs convey knowledge only to (interactive) verifiers able to select their own random challenges.

10.22 Definition (*computational vs. perfect zero-knowledge*) A protocol is *computationally* zero-knowledge if an observer restricted to probabilistic polynomial-time tests cannot distinguish real from simulated transcripts. For *perfect* zero-knowledge, the probability distributions of the transcripts must be identical. By convention, when not further qualified, *zero-knowledge* means computational zero-knowledge.

In the case of computational zero-knowledge, real and simulated transcripts are said to be *polynomially indistinguishable* (indistinguishable using polynomial-time algorithms). Any information extracted by a verifier through interaction with a prover provides no advantage to the verifier within polynomial time.

10.23 Remark (*ZK property and soundness vs. security*) The zero-knowledge property (Definition 10.20) does not guarantee that a protocol is secure (i.e., that the probability of it being easily defeated is negligible). Similarly, the soundness property (Definition 10.19) does not guarantee that a protocol is secure. Neither property has much value unless the underlying problem faced by an adversary is computationally hard.

(ii) Comments on zero-knowledge vs. other asymmetric protocols

The following observations may be made regarding zero-knowledge (ZK) techniques, as compared with other public-key (PK) techniques.

1. *no degradation with usage*: protocols proven to have the ZK property do not suffer degradation of security with repeated use, and resist chosen-text attacks. This is perhaps the most appealing practical feature of ZK techniques.
   A ZK technique which is not provably secure may or may not be viewed as more desirable than a PK technique which is provably secure (e.g., as difficult as factoring).

2. *encryption avoided*: many ZK techniques avoid use of explicit encryption algorithms. This may offer political advantages (e.g., with respect to export controls).

3. *efficiency*: while some ZK-based techniques are extremely efficient (see §10.4.5), protocols which formally have the zero-knowledge property typically have higher communications and/or computational overheads than PK protocols which do not. The computational efficiency of the more practical ZK-based schemes arises from their nature as interactive proofs, rather than their zero-knowledge aspect.

4. *unproven assumptions*: many ZK protocols ("proofs of knowledge") themselves rely on the same unproven assumptions as PK techniques (e.g., the intractability of factoring or quadratic residuosity).

5. *ZK-based vs. ZK*: although supported by prudent underlying principles, many techniques based on zero-knowledge concepts fall short of formally being zero-knowledge and/or formally sound in practice, due to parameter selection for reasons of efficiency, or for other technical reasons (cf. Notes 10.33 and 10.38). In fact, many such concepts are asymptotic, and do not apply directly to practical protocols (Remark 10.34).

(iii) Example of zero-knowledge proof: Fiat-Shamir identification protocol

The general idea of a zero-knowledge (ZK) proof is illustrated by the basic version of the Fiat-Shamir protocol. The basic version is presented here for historical and illustrative purposes (Protocol 10.24). In practice, one would use a more efficient variation, such as Protocol 10.26, with multiple "questions" per iteration rather than as here, where B poses only a single one-bit challenge per iteration.

The objective is for A to identify itself by proving knowledge of a secret s (associated with A through authentic public data) to any verifier B, without revealing any information about s not known or computable by B prior to execution of the protocol (see Note 10.25). The security relies on the difficulty of extracting square roots modulo large composite integers n of unknown factorization, which is equivalent to that of factoring n (Fact 3.46).

10.24 Protocol Fiat-Shamir identification protocol (basic version)

SUMMARY: A proves knowledge of s to B in t executions of a 3-pass protocol.

1. *One-time setup*.
   (a) A trusted center T selects and publishes an RSA-like modulus $n = pq$ but keeps primes p and q secret.
   (b) Each claimant A selects a secret s coprime to n, $1 \leq s \leq n - 1$, computes $v = s^2 \bmod n$, and registers v with T as its public key.[6]

2. *Protocol messages*. Each of t rounds has three messages with form as follows.

$$A \rightarrow B : \quad x = r^2 \bmod n \qquad (1)$$
$$A \leftarrow B : \quad e \in \{0, 1\} \qquad (2)$$
$$A \rightarrow B : \quad y = r \cdot s^e \bmod n \qquad (3)$$

---

[6]Technically, T should verify the condition $\gcd(s, n) = 1$ or equivalently $\gcd(v, n) = 1$, for this to be a sound proof of knowledge; and B should stop with failure if $\gcd(y, n) \neq 1$, where y is A's response in the third message. But either condition failing would allow the factorization of n, violating the assumption that n cannot be factored.

3. *Protocol actions.* The following steps are iterated t times (sequentially and independently). B accepts the proof if all t rounds succeed.

   (a) A chooses a random (*commitment*) r, $1 \leq r \leq n - 1$, and sends (the *witness*) $x = r^2 \bmod n$ to B.

   (b) B randomly selects a (*challenge*) bit $e = 0$ or $e = 1$, and sends e to A.

   (c) A computes and sends to B (the *response*) y, either $y = r$ (if $e = 0$) or $y = rs \bmod n$ (if $e = 1$).

   (d) B rejects the proof if $y = 0$, and otherwise accepts upon verifying $y^2 \equiv x \cdot v^e$ (mod n). (Depending on e, $y^2 = x$ or $y^2 = xv \bmod n$, since $v = s^2 \bmod n$. Note that checking for $y = 0$ precludes the case $r = 0$.)

Protocol 10.24 may be explained and informally justified as follows. The challenge (or *exam*) e requires that A be capable of answering two questions, one of which demonstrates her knowledge of the secret s, and the other an easy question (for honest provers) to prevent cheating. An adversary impersonating A might try to cheat by selecting any r and setting $x = r^2/v$, then answering the challenge $e = 1$ with a "correct" answer $y = r$; but would be unable to answer the exam $e = 0$ which requires knowing a square root of $x \bmod n$. A prover A knowing s can answer both questions, but otherwise can at best answer one of the two questions, and so has probability only $1/2$ of escaping detection. To decrease the probability of cheating arbitrarily to an acceptably small value of 2^{-t} (e.g., $t = 20$ or $t = 40$), the protocol is iterated t times, with B accepting A's identity only if all t questions (over t rounds) are successfully answered.

10.25 Note (*secret information revealed by A*) The response $y = r$ is independent of A's secret s, while the response $y = rs \bmod n$ also provides no information about s because the random r is unknown to B. Information pairs (x, y) extracted from A could equally well be simulated by a verifier B alone by choosing y randomly, then defining $x = y^2$ or $y^2/v \bmod n$. While this is not the method by which A would construct such pairs, such pairs (x, y) have a probability distribution which is indistinguishable from those A would produce; this establishes the zero-knowledge property. Despite the ability to simulate proofs, B is unable to impersonate A because B cannot predict the real-time challenges.

As a minor technical point, however, the protocol does reveal a bit of information: the answer $y = rs$ provides supporting evidence that v is indeed a square modulo n, and the soundness of the protocol allows one to conclude, after t successful iterations, that this is indeed the case.

(iv) General structure of zero-knowledge protocols

Protocol 10.24 illustrates the general structure of a large class of three-move zero-knowledge protocols:

$$
\begin{aligned}
A \rightarrow B : \quad & witness \\
A \leftarrow B : \quad & challenge \\
A \rightarrow B : \quad & response
\end{aligned}
$$

The prover claiming to be A selects a random element from a pre-defined set as its secret commitment (providing hidden randomization or "private coin tosses"), and from this computes an associated (public) witness. This provides initial randomness for variation from other protocol runs, and essentially defines a set of questions all of which the prover claims to be able to answer, thereby a priori constraining her forthcoming response. By protocol design, only the legitimate party A, with knowledge of A's secret, is truly capable of answering all the questions, and the answer to any one of these provides no information about

A's long-term secret. B's subsequent challenge selects one of these questions. A provides its response, which B checks for correctness. The protocol is iterated, if necessary, to improve the bound limiting the probability of successful cheating.

Zero-knowledge interactive protocols thus combine the ideas of *cut-and-choose* protocols (this terminology results from the standard method by which two children share a piece of cake: one cuts, the other chooses) and challenge-response protocols. A responds to at most one challenge (question) for a given witness, and should not reuse any witness; in many protocols, security (possibly of long-term keying material) may be compromised if either of these conditions is violated.

10.4.2 Feige-Fiat-Shamir identification protocol

The basic version of the Fiat-Shamir protocol is presented as Protocol 10.24. This can be generalized, and the Feige-Fiat-Shamir (FSS) identification protocol (Protocol 10.26) is a minor variation of such a generalization. The FFS protocol involves an entity identifying itself by proving knowledge of a secret using a zero-knowledge proof; the protocol reveals no partial information whatsoever regarding the secret identification value(s) of A (cf. Definition 10.20). It requires limited computation (a small fraction of that required by RSA – see §10.4.5), and is thus well-suited for applications with low-power processors (e.g., 8-bit chipcard microprocessors).

10.26 Protocol Feige-Fiat-Shamir identification protocol

SUMMARY: A proves its identity to B in t executions of a 3-pass protocol.

1. *Selection of system parameters.* A trusted center T publishes the common modulus $n = pq$ for all users, after selecting two secret primes p and q each congruent to 3 mod 4, and such that n is computationally infeasible to factor. (Consequently, n is a Blum integer per §2.4.6, and -1 is a quadratic non-residue mod n with Jacobi symbol $+1$.) Integers k and t are defined as security parameters (see Note 10.28).

2. *Selection of per-entity secrets.* Each entity A does the following.
   (a) Select k random integers $s_1, s_2, \dots, s_k$ in the range $1 \leq s_i \leq n - 1$, and k random bits $b_1, \dots, b_k$. (For technical reasons, $\gcd(s_i, n) = 1$ is required, but is almost surely guaranteed as its failure allows factorization of n.)
   (b) Compute $v_i = (-1)^{b_i} \cdot (s_i^2)^{-1} \bmod n$ for $1 \leq i \leq k$. (This allows v_i to range over all integers coprime to n with Jacobi symbol $+1$, a technical condition required to prove that no secret information is "leaked"; by choice of n, precisely one signed choice for v_i has a square root.)
   (c) A identifies itself by non-cryptographic means (e.g., photo id) to T, which thereafter registers A's public key $(v_1, \dots, v_k; n)$, while only A knows its private key $(s_1, \dots, s_k)$ and n. (To guarantee the bounded probability of attack specified per Note 10.28, T may confirm that each v_i indeed does have Jacobi symbol $+1$ relative to n.) This completes the one-time set-up phase.

3. *Protocol messages.* Each of t rounds has three messages with form as follows.

$$
\begin{aligned}
A \rightarrow B : & \quad x \, (= \pm r^2 \bmod n) & (1) \\
A \leftarrow B : & \quad (e_1, \dots, e_k), \ e_i \in \{0, 1\} & (2) \\
A \rightarrow B : & \quad y \, (= r \cdot \textstyle\prod_{e_j=1} s_j \bmod n) & (3)
\end{aligned}
$$

4. *Protocol actions.* The following steps are executed t times; B accepts A's identity if all t rounds succeed. Assume B has A's authentic public key $(v_1, \dots, v_k; n)$; otherwise, a certificate may be sent in message (1), and used as in Protocol 10.36.

(a) A chooses a random integer r, $1 \leq r \leq n - 1$, and a random bit b; computes $x = (-1)^b \cdot r^2 \bmod n$; and sends x (the *witness*) to B.

(b) B sends to A (the *challenge*,) a random k-bit vector $(e_1, \ldots, e_k)$.

(c) A computes and sends to B (the *response*): $y = r \cdot \prod_{j=1}^{k} s_j^{e_j} \bmod n$ (the product of r and those s_j specified by the challenge).

(d) B computes $z = y^2 \cdot \prod_{j=1}^{k} v_j^{e_j} \bmod n$, and verifies that $z = \pm x$ and $z \neq 0$. (The latter precludes an adversary succeeding by choosing $r = 0$.)

10.27 Example (*Feige-Fiat-Shamir protocol with artificially small parameters*)

1. The trusted center T selects the primes $p = 683$, $q = 811$, and publishes $n = pq = 553913$. Integers $k = 3$ and $t = 1$ are defined as security parameters.

2. Entity A does the following.

   (a) Selects 3 random integers $s_1 = 157$, $s_2 = 43215$, $s_3 = 4646$, and 3 bits $b_1 = 1$, $b_2 = 0$, $b_3 = 1$.

   (b) Computes $v_1 = 441845$, $v_2 = 338402$, and $v_3 = 124423$.

   (c) A's public key is $(441845, 338402, 124423; 553913)$ and private key is $(157, 43215, 4646)$.

3. See Protocol 10.26 for a summary of the messages exchanged.

4. (a) A chooses $r = 1279$, $b = 1$, computes $x = 25898$, and sends this to B.

   (b) B sends to A the 3-bit vector $(0, 0, 1)$.

   (c) A computes and sends to B $y = r \cdot s_3 \bmod n = 403104$.

   (d) B computes $z = y^2 \cdot v_3 \bmod n = 25898$ and accepts A's identity since $z = +x$ and $z \neq 0$. □

10.28 Note (*security of Feige-Fiat-Shamir identification protocol*)

(i) *probability of forgery.* Protocol 10.26 is provably secure against chosen message attack in the following sense: provided that factoring n is difficult, the best attack has a probability 2^{-kt} of successful impersonation.

(ii) *security assumption required.* The security relies on the difficulty of extracting square roots modulo large composite integers n of unknown factorization. This is equivalent to that of factoring n (see Fact 3.46).

(iii) *zero-knowledge and soundness.* The protocol is, relative to a trusted server, a (sound) zero-knowledge proof of knowledge provided $k = O(\log \log n)$ and $t = \Theta(\log n)$. See Remark 10.34 regarding the practical significance of such constraints. A simplistic view for fixed k is that the verifier, interested in soundness, favors larger t (more iterations) for a decreased probability of fraud; while the prover, interested in zero-knowledge, favors smaller t.

(iv) *parameter selection.* Choosing k and t such that $kt = 20$ allows a 1 in a million chance of impersonation, which suffices in the case that an identification attempt requires a personal appearance by a would-be impersonator (see §10.5). Computation, memory, and communication can be traded off; $1 \leq k \leq 18$ was originally suggested as appropriate. Specific parameter choices might be, for security 2^{-20}: $k = 5, t = 4$; for 2^{-30}: $k = 6, t = 5$.

(v) *security trade-off.* Both computation and communication may be reduced by trading off security parameters to yield a single iteration ($t = 1$), holding the product kt constant and increasing k while decreasing t; however, in this case the protocol is no longer a zero-knowledge proof of knowledge.

10.29 Note (*modifications to Feige-Fiat-Shamir*)

    (i) As an alternative to step 1 of Protocol 10.26, each user may pick its own such modulus n. T is still needed to associate each user with its modulus.

    (ii) The communication complexity can be reduced if A sends B (e.g., 128 bits of) a hash value $h(x)$ instead of x in message (1), with B's verification modified accordingly.

    (iii) The scheme can be made *identity-based* as follows (cf. §13.4.3). T assigns a distinguishing identifying string I_A to each party A (e.g., A's name, address, or other information which a verifier may wish to corroborate). A's public values v_i, $1 \le i \le k$ are then derived by both T and other parties B as $v_i = f(I_A, i)$ using an appropriate function f. Then the trusted center, knowing the factorization of n, computes a square root s_i of each v_i and gives these to A.

       As an example of f, consider, for a randomly chosen but known value c, $f(I_A, i) = I_A + i + c \bmod n$. Since a square root of $f_i = f(I_A, i)$ is required, any f_i with Jacobi symbol $-1 \bmod n$ may be multiplied by a fixed number with Jacobi symbol -1. A non-residue f_i with Jacobi $+1$ may be either discarded (A must then indicate to B, e.g., in message (3), which values i allow computation of the v_j); or mapped to a residue via multiplication by -1, again with an indication to B of this to allow computation of v_j. Note that both cases for dealing with a non-residue f_i with Jacobi $+1$ reveal some (non-useful) information.

    (iv) The *parallel version* of the protocol, in which each of three messages contains the respective data for all t rounds simultaneously, can be shown to be secure (it releases no "transferable information"), but for technical reasons loses the zero-knowledge property. Such parallel execution (as opposed to *sequential iteration*) in interactive proofs allows the probability of error (forgery) to be decreased without increasing the number of rounds.

10.30 Note (*converting identification to signature scheme*) The following general technique may be used to convert an identification scheme involving a witness-challenge-response sequence to a signature scheme: replace the random challenge e of the verifier by the one-way hash $e = h(x\|m)$, of the concatenation of the witness x and the message m to be signed (h essentially plays the role of verifier). As this converts an interactive identification scheme to a non-interactive signature scheme, the bitsize of the challenge e must typically be increased to preclude off-line attacks on the hash function.

10.4.3 GQ identification protocol

The Guillou-Quisquater (GQ) identification scheme (Protocol 10.31) is an extension of the Fiat-Shamir protocol. It allows a reduction in both the number of messages exchanged and memory requirements for user secrets and, like Fiat-Shamir, is suitable for applications in which the claimant has limited power and memory. It involves three messages between a claimant A whose identity is to be corroborated, and a verifier B.

10.31 Protocol GQ identification protocol

SUMMARY: A proves its identity (via knowledge of s_A) to B in a 3-pass protocol.

    1. *Selection of system parameters.*

        (a) An authority T, trusted by all parties with respect to binding identities to public keys, selects secret random RSA-like primes p and q yielding a modulus $n = pq$. (As for RSA, it must be computationally infeasible to factor n.)

(b) T defines a public exponent $v \geq 3$ with $\gcd(v, \phi) = 1$ where $\phi = (p-1)(q-1)$, and computes its private exponent $s = v^{-1} \bmod \phi$. (See Note 10.33.)

(c) System parameters (v, n) are made available (with guaranteed authenticity) for all users.

2. *Selection of per-user parameters.*

(a) Each entity A is given a unique identity I_A, from which (the *redundant identity*) $J_A = f(I_A)$, satisfying $1 < J_A < n$, is derived using a known redundancy function f. (See Note 10.35. Assuming that factoring n is difficult implies $\gcd(J_A, \phi) = 1$.)

(b) T gives to A the secret (*accreditation data*) $s_A = (J_A)^{-s} \bmod n$.

3. *Protocol messages.* Each of t rounds has three messages as follows (often $t = 1$).

$$
\begin{array}{lll}
A \rightarrow B: & I_A, \; x = r^v \bmod n & (1) \\
A \leftarrow B: & e \; (\text{where } 1 \leq e \leq v) & (2) \\
A \rightarrow B: & y = r \cdot s_A{}^e \bmod n & (3)
\end{array}
$$

4. *Protocol actions.* A proves its identity to B by t executions of the following; B accepts the identity only if all t executions are successful.

(a) A selects a random secret integer r (the *commitment*), $1 \leq r \leq n - 1$, and computes (the *witness*) $x = r^v \bmod n$.

(b) A sends to B the pair of integers (I_A, x).

(c) B selects and sends to A a random integer e (the *challenge*), $1 \leq e \leq v$.

(d) A computes and sends to B (the *response*) $y = r \cdot s_A{}^e \bmod n$.

(e) B receives y, constructs J_A from I_A using f (see above), computes $z = J_A{}^e \cdot y^v \bmod n$, and accepts A's proof of identity if both $z = x$ and $z \neq 0$. (The latter precludes an adversary succeeding by choosing $r = 0$.)

10.32 Example (*GQ identification protocol with artificially small parameters and $t = 1$*)

1. (a) The authority T selects primes $p = 569$, $q = 739$, and computes $n = pq = 420491$.

(b) T computes $\phi = (p - 1)(q - 1) = 419184$, selects $v = 54955$, and computes $s = v^{-1} \bmod \phi = 233875$.

(c) System parameters $(54955, 420491)$ are made available for all users.

2. (a) Suppose that A's redundant identity is $J_A = 34579$.

(b) T gives to A the accreditation data $s_A = (J_A)^{-s} \bmod n = 403154$.

3. See Protocol 10.31 for a summary of the messages exchanged.

4. (a) A selects $r = 65446$ and computes $x = r^v \bmod n = 89525$.

(b) A sends to B the pair $(I_A, 89525)$.

(c) B sends to A the random challenge $e = 38980$.

(d) A sends $y = r \cdot s_A{}^e \bmod n = 83551$ to B.

(e) B computes $z = J_A{}^e \cdot y^v \bmod n = 89525$ and accepts A's identity since $z = x$ and $z \neq 0$. $\square$

10.33 Note (*security of GQ identification protocol*)

(i) *probability of forgery.* In Protocol 10.31, v determines the security level (cf. Fiat-Shamir where $v = 2$ but there are many rounds); some values such as $v = 2^{16}+1$ may offer computational advantages. A fraudulent claimant can defeat the protocol with a 1 in v chance by guessing e correctly a priori (and then forming $x = J_A{}^e \cdot y^v$ as the verifier would). The recommended bitlength of v thus depends on the environment under which attacks could be mounted (see §10.5).

(ii) *security assumption required.* Extracting v^{th} roots modulo the composite integer n (i.e., solving the RSA problem – §3.3) appears necessary to defeat the protocol; this is no harder than factoring n (Fact 3.30), and appears computationally intractable without knowing the factors of n.

(iii) *soundness.* In practice, GQ with $t = 1$ and a k-bit prime v is often suggested. For generalized parameters (n, v, t), the probability of forgery is v^{-t}. If v is constant, then technically for soundness, t must grow asymptotically faster than $\log \log n$. (For soundness, $v^{-t} = O(e^{-kt})$ must be smaller than inverse-polynomial in $\log n$; only polynomial security is provided if for a constant c, $v^t = O((\log n)^c)$. See also Remark 10.34.)

(iv) *zero-knowledge property.* In opposition to the soundness requirement, for GQ to be zero-knowledge apparently requires $tv = O((\log n)^c)$ for constant c, imposing an upper bound on t asymptotically: for v constant, t must be no larger than polynomial in $\log n$.

10.34 Remark (*asymptotic concepts vs. practical protocols*) The asymptotic conditions for soundness specified in Note 10.33 have little meaning in practice, e.g., because big-O notation is not applicable once fixed values are assigned to parameters. Indeed, zero-knowledge is a theoretical concept; while complexity-theoretic definitions offer guidance in selecting practical security parameters, their significance diminishes when parameters are fixed. Regarding Note 10.33, if $t = 1$ is viewed as the instantiation of a non-constant parameter (e.g., the iterated logarithm of n), then $t = 1$ will suffice for all practical purposes; consider $n = 1024$, $t = \lceil \lg^4 n \rceil = 1$.

10.35 Note (*redundancy function for identity-based GQ*)

(i) The protocol as given is an identity-based version (cf. Note 10.29), where A's public key is reconstructed from identifier I_A sent in message (1). Alternatively, a certified public key may be used, distributed in a certificate as per Protocol 10.36.

(ii) One example of the redundancy function f is the redundancy mapping of the preprocessing stage of ISO/IEC 9796 (see §11.3.5). A second example is a single function value of f as in Note 10.29, for an appropriate value i.

(iii) The purpose of the redundancy is to preclude an adversary computing false accreditation data corresponding to a plausible identity; this would be equivalent to forging a certificate in certificate-based schemes.

10.4.4 Schnorr identification protocol

The Schnorr identification protocol is an alternative to the Fiat-Shamir and GQ protocols. Its security is based on the intractability of the discrete logarithm problem. The design allows pre-computation, reducing the real-time computation for the claimant to one multiplication modulo a prime q; it is thus particularly suitable for claimants of limited computational ability. A further important computational efficiency results from the use of a subgroup of order q of the multiplicative group of integers modulo p, where $q|(p-1)$; this also reduces the required number of transmitted bits. Finally, the protocol was designed to require only three passes, and a low communications bandwidth (e.g., compared to Fiat-Shamir).

The basic idea is that A proves knowledge of a secret a (without revealing it) in a time-variant manner (depending on a challenge e), identifying A through the association of a with the public key v via A's authenticated certificate.

10.36 Protocol Schnorr identification protocol

SUMMARY: A proves its identity to B in a 3-pass protocol.

1. *Selection of system parameters.*

   (a) A suitable prime p is selected such that $p - 1$ is divisible by another prime q. (Discrete logarithms modulo p must be computationally infeasible – see §3.6; e.g., $p \approx 2^{1024}$, $q \geq 2^{160}$.)

   (b) An element β is chosen, $1 \leq \beta \leq p - 1$, having multiplicative order q. (For example, for α a generator mod p, $\beta = \alpha^{(p-1)/q} \bmod p$; see Note 4.81.)

   (c) Each party obtains an authentic copy of the system parameters (p, q, β) and the verification function (public key) of the trusted party T, allowing verification of T's signatures $S_T(m)$ on messages m. (S_T involves a suitable known hash function prior to signing, and may be any signature mechanism.)

   (d) A parameter t (e.g., $t \geq 40$), $2^t < q$, is chosen (defining a security level 2^t).

2. *Selection of per-user parameters.*

   (a) Each claimant A is given a unique identity I_A.

   (b) A chooses a private key a, $0 \leq a \leq q - 1$, and computes $v = \beta^{-a} \bmod p$.

   (c) A identifies itself by conventional means (e.g., passport) to T, transfers v to T with integrity, and obtains a certificate $cert_A = (I_A, \ v, \ S_T(I_A, v))$ from T binding I_A with v.

3. *Protocol messages.* The protocol involves three messages.

$$\begin{aligned}
A \to B &: \quad cert_A, \ x = \beta^r \bmod p & (1) \\
A \leftarrow B &: \quad e \ (\text{where } 1 \leq e \leq 2^t < q) & (2) \\
A \to B &: \quad y = ae + r \bmod q & (3)
\end{aligned}$$

4. *Protocol actions.* A identifies itself to verifier B as follows.

   (a) A chooses a random r (the *commitment*), $1 \leq r \leq q-1$, computes (the *witness*) $x = \beta^r \bmod p$, and sends (1) to B.

   (b) B authenticates A's public key v by verifying T's signature on $cert_A$, then sends to A a (never previously used) random e (the *challenge*), $1 \leq e \leq 2^t$.

   (c) A checks $1 \leq e \leq 2^t$ and sends B (the *response*) $y = ae + r \bmod q$.

   (d) B computes $z = \beta^y v^e \bmod p$, and accepts A's identity provided $z = x$.

10.37 Example (*Schnorr identification protocol with artificially small parameters*)

1. (a) The prime $p = 48731$ is selected, where $p-1$ is divisible by the prime $q = 443$.

   (b) A generator mod 48731 is $\alpha = 6$; β is computed as $\alpha^{(p-1)/q} \bmod p = 11444$.

   (c) The system parameters are $(48731, 443, 11444)$.

   (d) The parameter $t = 8$ is chosen.

2. (b) A chooses a private key $a = 357$ and computes $v = \beta^{-a} \bmod p = 7355$.

3. See Protocol 10.36 for a summary of the messages exchanged.

4. (a) A chooses $r = 274$ and sends $x = \beta^r \bmod p = 37123$ to B.

   (b) B sends to A the random challenge $e = 129$.

   (c) A sends B the number $y = ae + r \bmod q = 255$.

   (d) B computes $z = \beta^y v^e \bmod p = 37123$ and accept's A's identity since $z = x$. $\qquad\square$

10.38 Note (*security of Schnorr identification protocol*)

(i) *probability of forgery*. In Protocol 10.36, t must be sufficiently large to make the probability 2^{-t} of correctly guessing the challenge e negligible. $t = 40$, $q \geq 2^{2t} = 2^{80}$ was originally suggested in the case that a response is required within seconds (see §10.5); larger q may be necessary to preclude time-memory trade-offs, and $q \geq 2^{160}$ is recommended to preclude other off-line discrete log attacks. Correctly guessing e allows an adversary to impersonate A by choosing any y, sending $x = \beta^y v^e \bmod p$ to B in (1), then sending y in (3).

(ii) *soundness*. It can be shown that the protocol is a proof of knowledge of a, i.e., any party completing the protocol as A must be capable of computing a. Informally, the protocol reveals "no useful information" about a because x is a random number, and y is perturbed by the random number r. (However, this does not prove that adversarial discovery of a is difficult.)

(iii) *zero-knowledge property*. The protocol is not zero-knowledge for large e, because through interaction, B obtains the solution (x, y, e) to the equation $x = \beta^y v^e \bmod p$, which B itself might not be able to compute (e.g., if e were chosen to depend on x).

10.39 Note (*reducing transmission bandwidth*) The number of bits transmitted in the protocol can be reduced by replacing x in message (1) by t pre-specified bits of x (e.g., the least significant t bits), and having B compare this to t corresponding bits of z.

10.4.5 Comparison: Fiat-Shamir, GQ, and Schnorr

The protocols of Feige-Fiat-Shamir, Guillou-Quisquater, and Schnorr all provide solutions to the identification problem. Each has relative advantages and disadvantages with respect to various performance criteria and for specific applications. To compare the protocols, a typical set of selected parameters must be chosen for each providing comparable estimated security levels. The protocols may then be compared based on the following criteria:

1. *communications*: number of messages exchanged, and total bits transferred;
2. *computations*: number of modular multiplications for each of prover and verifier (noting on-line and off-line computations);
3. *memory*: storage requirements for secret keys (and signature size, in the case of signature schemes);
4. *security guarantees*: comparisons should consider security against forgery by guessing (soundness), possible disclosure of secret information (zero-knowledge property), and status regarding provable security; and
5. *trust required in third party*: variations of the protocols may require different trust assumptions in the trusted party involved.

The number of criteria and potential parameter choices precludes a comparison which is both definitive and concise. The following general comments may, however, be made.

1. *computational efficiency*. Fiat-Shamir requires between one and two orders of magnitude fewer full modular multiplications (steps) by the prover than an RSA private-key operation (cf. §10.3.3). When $kt = 20$ and n is 512 bits, Fiat-Shamir uses from about 11 to about 30 steps ($k = 20$, $t = 1$; and $k = 1$, $t = 20$); GQ requires about 60 steps (for $t = 1$, $m = 20 = \log_2(v)$), or somewhat fewer if v has low Hamming weight; and full exponentiation in unoptimized RSA takes 768 steps.

2. *off-line computations.* Schnorr identification has the advantage of requiring only a single on-line modular multiplication by the claimant, provided exponentiation may be done as a precomputation. (Such a trade-off of on-line for off-line computation is possible in some applications; in others, the total computation must be considered.) However, significant computation is required by the verifier compared to Fiat-Shamir and GQ.

3. *bandwidth and memory for secrets.* GQ allows the simultaneous reduction of both memory (parameter k) and transmission bandwidth (parameter t) with $k = t = 1$, by introducing the public exponent $v > 2$ with the intention that the probability of successful cheating becomes v^{-kt}; this simultaneous reduction is not possible in Fiat-Shamir, which requires k user secrets and t iterations for an estimated security (probability of cheating) of 2^{-kt}. Regarding other tradeoffs, see Note 10.28.

4. *security assumptions.* The protocols require the assumptions that the following underlying problems are intractable, for a composite (RSA) integer n: Fiat-Shamir – extracting square roots mod n; GQ – extracting v^{th} roots mod n (i.e., the RSA problem); Schnorr identification – computing discrete logs modulo a prime p.

10.5 Attacks on identification protocols

The methods an adversary may employ in an attempt to defeat identification protocols are a subset of those discussed in Chapter 12 for authenticated key establishment, and the types of adversaries may be similarly classified (e.g., passive vs. active, insider vs. outsider); for a discussion of attacks on simple password schemes, see §10.2.2. Identification is, however, less complex than authenticated key establishment, as there is no issue of an adversary learning a previous session key, or forcing an old key to be reused. For conciseness, the following definitions are made:

1. *impersonation:* a deception whereby one entity purports to be another.

2. *replay attack:* an impersonation or other deception involving use of information from a single previous protocol execution, on the same or a different verifier. For stored files, the analogue of a replay attack is a *restore* attack, whereby a file is replaced by an earlier version.

3. *interleaving attack:* an impersonation or other deception involving selective combination of information from one or more previous or simultaneously ongoing protocol executions (*parallel sessions*), including possible origination of one or more protocol executions by an adversary itself.

4. *reflection attack:* an interleaving attack involving sending information from an ongoing protocol execution back to the originator of such information.

5. *forced delay:* a forced delay occurs when an adversary intercepts a message (typically containing a sequence number), and relays it at some later point in time. Note the delayed message is not a replay.

6. *chosen-text attack:* an attack on a challenge-response protocol wherein an adversary strategically chooses challenges in an attempt to extract information about the claimant's long-term key.
   Chosen-text attacks are sometimes referred to as using the claimant as an *oracle*, i.e., to obtain information not computable from knowledge of a claimant's public key alone. The attack may involve chosen-plaintext if the claimant is required to sign,

encrypt, or MAC the challenge, or chosen-ciphertext if the requirement is to decrypt a challenge.

Potential threats to identification protocols include impersonation by any of the following attacks: replay, interleaving, reflection, or forced delay. Impersonation is also trivial if an adversary is able to discover an entity's long-term (secret or private) keying material, for example, using a chosen-text attack. This may be possible in protocols which are not zero-knowledge, because the claimant uses its private key to compute its response, and thus a response may reveal partial information. In the case of an active adversary, attacks may involve the adversary itself initiating one or more new protocol runs, and creating, injecting, or otherwise altering new or previous messages. Table 10.3 summarizes counter-measures for these attacks.

Type of attack	Principles to avoid attack
replay	use of challenge-response techniques; use of nonces; embed target identity in response
interleaving	linking together all messages from a protocol run (e.g., using chained nonces)
reflection	embed identifier of target party in challenge responses; construct protocols with each message of different form (avoid message symmetries); use of uni-directional keys
chosen-text	use of zero-knowledge techniques; embed in each challenge response a self-chosen random number (*confounder*)
forced delay	combined use of random numbers with short response time-outs; timestamps plus appropriate additional techniques

Table 10.3: *Identification protocol attacks and counter-measures.*

10.40 Remark (*use of keys for multiple purposes*) Caution is advised if any cryptographic key is used for more than one purpose. For example, using an RSA key for both entity authentication and signatures may compromise security by allowing a chosen-text attack. Suppose authentication here consists of B challenging A with a random number r_B RSA-encrypted under A's public key, and A is required to respond with the decrypted random number. If B challenges A with $r_B = h(x)$, A's response to this authentication request may (unwittingly) provide to B its RSA signature on the hash value of the (unknown to A) message x. See also Example 9.88, where a DES key used for both CBC encryption and CBC-MAC leads to a security flaw; and Remark 13.32.

10.41 Remark (*adversary acting "as a wire"*) In any identification protocol between A and B, an adversary C may step into the communications path and simply relay (without changing) the messages between legitimates parties A and B, itself acting as a part of the communications link. Typically in practice, this is not considered a true "attack", in the sense that it does not alter the aliveness assurance delivered by the protocol; however, in some special applications, this may be a concern (see Remark 10.42).

10.42 Remark (*grandmaster postal-chess problem*) Identification protocols do not provide assurances about the physical location of the authenticated party. Therefore, Remark 10.41 notwithstanding, a concern may arise in the special case that the following is possible: an adversary C attempts to impersonate B, is challenged (to prove it is B) by A, and is able to relay (in real time, without detection or noticeable delay, and pretending to be A) the challenge on to the real B, get a proper response from B, and pass this response along back to A. In this case, additional measures are necessary to prevent a challenged entity from eliciting aid in computing responses. This is related to the so-called *grandmaster postal-chess problem*, whereby an amateur's chess rating may unfairly be improved by engaging in two simultaneous chess games with distinct grandmasters, playing black in one game and white in the second, and using the grandmaster's moves from each game in the other. Either two draws, or a win and a loss, are guaranteed, both of which will improve the amateur's rating.

For further discussion of protocol attacks including specific examples of flawed entity authentication protocols, see §12.9.

(i) Maintaining authenticity

Identification protocols provide assurances corroborating the identity of an entity only at a given instant in time. If the continuity of such an assurance is required, additional techniques are necessary to counteract active adversaries. For example, if identification is carried out at the beginning of a communications session to grant communications permissions, a potential threat is an adversary who "cuts in" on the communications line immediately after the successful identification of the legitimate party. Approaches to prevent this include:

1. performing re-authentication periodically, or for each discrete resource requested (e.g., each file access). A remaining threat here is an adversary who "steps out" every time re-authentication is performed, allowing the legitimate party to perform this task, before re-entering.
2. tying the identification process to an ongoing integrity service. In this case, the identification process should be integrated with a key establishment mechanism, such that a by-product of successful identification is a session key appropriate for use in a subsequent ongoing integrity mechanism.

(ii) Security level required for on-line vs. off-line attacks

The security level required for identification protocols depends on the environment and the specific application at hand. The probability of success of "guessing attacks" should be considered, and distinguished from the amount of computation required to mount on-line or off-line attacks (using the best techniques known). Some illustrative notes follow (see also Note 10.28).

1. *Local attacks.* Selecting security parameters which limit the probability of successful impersonation of a guessing attack (an adversary simply guesses a legitimate party's secret) to a 1 in 2^{20} chance (20 bits of security) may suffice if, for each attempted impersonation, a local appearance is required by the would-be impersonator and there is a penalty for failed attempts. Depending on the potential loss resulting relative to the penalty, 10 to 30 bits or more of security may be required.
2. *Remote attacks.* A higher level of security is required in environments where unlimited identification attempts, each involving minimal computational effort, are possible by remote electronic communications, by an anonymous claimant interacting with an on-line system, with no penalties for failed attempts. 20 to 40 bits of security or more may be called for here, unless the number of interactions may be somehow limited.
3. *Off-line or non-interactive attacks.* Selecting security parameters such that an attack requires 2^{40} computations in real-time (during a protocol execution) may be acceptable, but a bound of 2^{60} to 2^{80} computations (the latter should be adequate in all

cases) may be called for if the computations can be carried out off-line, and the attack is *verifiable* (i.e., the adversary can confirm, before interacting with the on-line system, that his probability of successful impersonation is near 1; or can recover a long-term secret by off-line computations subsequent to an interaction).

10.6 Notes and further references

§10.1

Davies and Price [308] and Ford [414] provide extensive discussion of authentication and identification; see also the former for biometric techniques, as well as Everett [380]. The comprehensive survey on login protocols by de Waleffe and Quisquater [319] is highly recommended. Crépeau and Goutier provide a lucid concise summary of user identification techniques with Brassard [192]. For standardized entity authentication mechanisms, see ISO/IEC 9798 [598, 599, 600, 601, 602].

§10.2

See the §9.2 notes on page 377 for historical discussion of using a one-way function (*one-way cipher*) for "encrypted" password files. Morris and Thompson [907] introduce the notion of password salting in their 1979 report on UNIX passwords; in one study of 3289 user passwords unconstrained by password rules, 86% fell within an easily-searched subset of passwords. Feldmeier and Karn [391] give an update 10 years later, indicating 30% of passwords they encountered fell to their attack using a precomputed encrypted dictionary, sorted on tapes by salt values. See also Klein [680] and Lomas et al. [771]. Password salting is related to randomized encryption; the idea of padding plaintext with random bits before encryption may also be used to prevent *forward search* attacks on public-key encryption with small plaintext spaces. Password rules and procedures have been published by the U.S. Departments of Commerce [399] and Defense [334].

Methods for computing password-derived keys (§10.2.4) are specified in the Kerberos Authentication Service [1041] and PKCS #5 [1072]. A concern related to password-derived keys is that known plaintext allows password-guessing attacks; protocols specifically designed to prevent such attacks are mentioned in Chapter 12 notes on §12.6. The idea of chaining one-time passwords by a one-way function (Protocol 10.6) is due to Lamport [739]; for related practical applications, see RFC 1938 [1047]. Davies and Price [308, p.176] note a questionnaire-based identification technique related to fixed challenge-response tables, wherein the user is challenged by a random subset of previously answered questions.

§10.3

Needham and Schroeder [923] stimulated much early work in the area of authentication protocols in the late 1970s, and Needham was again involved with Burrows and Abadi [227] in the BAN logic work which stimulated considerable interest in protocol analysis beginning in the late 1980s; see Chapter 12 notes for further discussion.

Gong [501] provides an overview of both time variant parameters and message replay; see also Neuman and Stubblebine [925], and the annexes of parts of ISO/IEC 9798 (e.g., [600]). For security arguments against the use of timestamps and a discussion of implementation difficulties, see Bellovin and Merritt [103]; Gaarder and Snekkenes [433]; Diffie, van Oorschot, and Wiener [348]; and Gong [500], who considers postdated timestamps. See also §12.3 notes. Lam and Beth [734] note that timestamp-based protocols are appropriate for connectionless interactions whereas challenge-response suits connection-oriented communications, and suggest challenge-response techniques be used to securely synchronize timeclocks with applications themselves using timestamp-based authentication.

ISO/IEC 9798 [598] parts 2 through 5 specify entity authentication protocols respectively based on symmetric encryption [599], digital signatures [600], keyed one-way functions [601], and zero-knowledge techniques [602]; a subset of these are presented in this chapter. FIPS JJJ [407] is a subset of 9798-3 containing the unilateral and mutual authentication protocols involving challenge-response with random numbers.

Several parts of 9798 were influenced by the SKID2 and SKID3 (*Secret Key IDentification*) protocols from the RACE/RIPE project [178], which leave the keyed hash function unspecified but recommend RIPE-MAC with 64-bit random-number challenges. Diffie [342, 345] notes that two-pass challenge-response identification based on encryption and random challenges has been used since the 1950s in military *Identification Friend or Foe* (IFF) systems to distinguish friendly from hostile aircraft. Mao and Boyd [781] discuss the danger of improperly using encryption in authentication protocols, specifically the CBC mode without an integrity mechanism (cf. Remark 10.16). Stubblebine and Gligor [1179] discuss attacks involving this same mode; see also the much earlier paper by Akl [20].

Davies and Price [308] give a concise discussion of password generators. The identification technique in §10.3.3(i) based on public-key decryption and witness is derived from a Danish contribution to the 4th Working Draft of ISO/IEC 9798-5, specifying a protocol called COMSET and motivated in part by Brandt et al. [188], and related to ideas noted earlier by Blum et al. [163].

§10.4

A refreshingly non-mathematical introduction to zero-knowledge proofs is provided by Quisquater, Guillou, and Berson [1020], who document the secret of Ali Baba's legendary cave, and its rediscovery by Mick Ali. Mitropoulos and Meijer [883] give an exceptionally readable and comprehensive survey (circa 1990) of interactive proofs and zero knowledge, with a focus on identification. Other overviews include Johnson [641]; Stinson [1178, Ch.13]; and Brassard, Chaum, and Crépeau [193] (or [192]) for a discussion of *minimum disclosure* proofs, based on *bit commitment* and the primitive of a *blob*. Brassard and Crépeau [195] provide a user-friendly discussion of various definitions of zero-knowledge, while Goldreich and Oren [475] examine properties and relationships between various definitions of ZK proof systems.

Rabin [1022] employed the idea of *cut-and-choose protocols* for cryptographic applications as early as 1978. While Babai (with Moran) [60, 61] independently developed a theory of randomized interactive proofs known as *Arthur-Merlin games* in an attempt to "formalize the notion of efficient provability by overwhelming statistical evidence", interactive proof systems and the notion of zero-knowledge (ZK) proofs were formalized in 1985 by Goldwasser, Micali, and Rackoff [481] in the context of an interactive *proof of membership* of a string x in a language $\mathcal{L}$; they showed that the languages of quadratic-residues and of quadratic non-residues each have ZK interactive proof (ZKIP) systems revealing only a single bit of knowledge, namely, that $x \in \mathcal{L}$. Goldreich, Micali, and Wigderson [473, 474] prove likewise for graph non-isomorphism (known not to be in **NP**) and graph isomorphism, and that assuming the existence of secure encryption schemes, every language in **NP** has a ZKIP; see also Chaum [244], and Brassard and Crépeau [194].

Motivated by cryptographic applications and identification in particular, Feige, Fiat, and Shamir [383] adapted the concepts of interactive pro(:s of membership to interactive *proofs of knowledge*, including reformulated definitions for completeness, soundness, and zero-knowledge; while proofs of membership reveal one bit of set membership information, proofs of knowledge reveal only one bit about the prover's *state* of knowledge. The definitions given in §10.4.1 are based on these. These authors refine the original scheme of Fiat and Shamir [395] to yield that of Protocol 10.26; both may be converted to identity-based schemes (Note 10.29) in the sense of Shamir [1115]. The Fiat-Shamir scheme is related to (but more efficient than) an earlier protocol for proving quadratic residuosity (presented at Eurocrypt'84, but unpublished) by Fischer, Micali, and Rackoff [412]. The Fiat-Shamir protocol as per Protocol 10.24 includes an improvement noted by Desmedt et al. [340] to avoid inverses in the derivation of user secrets; this optimization may also be made to Protocol 10.26.

Related to definitions in §10.4.1, Bellare and Goldreich [87] noted that Goldwasser, Micali, and Rackoff [481] did not formally propose a definition for a proof of knowledge, and suggested that the formal definitions of Feige, Fiat, and Shamir [383] and Tompa and Woll [1194] were unsatisfactory for some applications. To address these issues they proposed a new definition, having some common aspects with that of Feige and Shamir [384], but offering additional advantages.

Micali and Shamir [868] provide preliminary notes on reducing computation in the Fiat-Shamir protocol by choosing the public keys v_i, $1 \leq i \leq k$ to be the first k prime numbers; each user then has an independent modulus n. A modification of Fiat-Shamir identification by Ong and Schnorr [957] decreases computational complexity, signature size, and the number of communications required, condensing t Fiat-Shamir iterations into one iteration while leaving each user with k private keys (cf. the $k = 1$ extension below); for computational efficiency, they suggest using as secret keys (not too) small integers.

The idea of generalizing Fiat-Shamir identification in other ways, including "replacing square roots by cubic or higher roots", was suggested in the original paper; using higher roots allows users to reduce their number of private keys k, including to the limiting case $k = 1$. Guillou and Quisquater [524] proposed a specific formulation of this idea of "using deep coin tosses" as the GQ scheme (Protocol 10.31); apparently independently, Ohta and Okamoto [945, 944] proposed a similar formulation, including security analysis.

The Ohta-Okamoto (OO) version of this *extended Fiat-Shamir* scheme differs from the GQ version (Protocol 10.31) as follows: (1) in OO, rather than T computing s_A from identity I_A, A chooses its own secret $s_A \in \mathbb{Z}_n$ and publishes $I_A = s_A{}^v \bmod n$; and (2) the verification relation $x \equiv J_A{}^e \cdot y^v \pmod{n}$ becomes $y^v \equiv x \cdot I_A{}^e$. OO is more general in that, as originally proposed, it avoids the GQ (RSA) constraint that $\gcd(v, \phi(n)) = 1$. Subsequent analysis by Burmester and Desmedt [221] suggests that additional care may be required when v is not prime. While the OO version precludes an identity-based variation, a further subsequent version of extended Fiat-Shamir (GQ variation) by Okamoto [949] ("Scheme 3" of 5 protocols therein) is provably as secure as factoring, only slightly less efficient, and is amenable to an identity-based variation.

The zero-knowledge interactive protocols of Chaum et al. [248, 249] for proving possession of discrete logarithms, provided a basis for Protocol 10.36 which is due to Schnorr [1097, 1098]. Schnorr also proposed a preprocessing scheme to reduce real-time computation, but see de Rooij [314] regarding its security. The Schnorr identification and signature schemes must not both be used with the same parameters β, p [1098] (cf. Remark 10.40). Schnorr's protocol is related to the log-based identification scheme of Beth [123] also proven to be zero-knowledge. Burmester et al. [223] analyze (cf. Note 10.33) a generalized identification protocol encompassing all the well-known variations related to Fiat-Shamir and including those of both Chaum et al. and Beth noted above. Van de Graaf and Peralta [1200] give a ZK interactive protocol for proving that a Blum integer is a Blum integer.

Brickell and McCurley [207] propose a modification of Schnorr's identification scheme, in which q is kept secret and exponent computations are reduced modulo $p - 1$ rather than q; it has provable security if factoring $p - 1$ is difficult, and moreover security equivalent to that of Schnorr's scheme otherwise; a drawback is that almost 4 times as much computation is required by the claimant. Another variant of Schnorr's scheme by Girault [458, 461] was the first identity-based identification scheme based on discrete logs; it uses a composite modulus, and features the user choosing its own secret key, which remains unknown to the trusted party (cf. implicitly-certified public keys, §12.6.2). A further variation of Schnorr's identification protocol by Okamoto [949] ("Scheme 1") uses two elements β_1 and β_2, of order q, and is provably secure, assuming the computational infeasibility of computing the $\mathbb{Z}_p$ discrete logarithm $\log_{\beta_1} \beta_2$ of β_2 relative to β_1; it does, however, involve some additional computation.

Aside from the above protocols based on the computational intractability of the standard number-theoretic problems (factoring and discrete logarithms), a number of very efficient identification protocols have more recently been proposed based on **NP**-hard problems. Shamir [1116] proposed a zero-knowledge identification protocol based on the **NP**-hard *permuted kernel problem*: given an $m \times n$ matrix A over $\mathbb{Z}_p$, p prime (and relatively small, e.g., $p = 251$), and an n-vector V, find a permutation π on $\{1, \dots, n\}$ such that $V_\pi \in ker(A)$, where $ker(A)$ is the kernel of A consisting of all n-vectors W such that $AW = [0 \dots 0] \bmod p$. Patarin and Chauvaud [966] discuss attacks on the permuted kernel problem which are feasible for the smallest of parameter choices originally suggested, while earlier less efficient attacks are presented by Baritaud et al. [73] and Georgiades [447]. Stern [1176] proposed a practical zero-knowledge identification scheme based on the **NP**-hard *syndrome decoding* problem, following an earlier less practical scheme of Stern [1174] based on intractable problems in coding theory. Stern [1175] proposed another practical identification scheme based on an **NP**-hard combinatorial *constrained linear equations* problem, offering a very short key length, which is of particular interest in specific applications. Pointcheval [983] proposed another such scheme based on the **NP**-hard *perceptrons problem*: given an $m \times n$ matrix M with entries ± 1, find an n-vector y with entries ± 1 such that $My \geq 0$.

Goldreich and Krawczyk [469] pursue the fact that the original definition of ZK of Goldwasser, Micali, and Rackoff is not closed under sequential composition (this was noted earlier by D. Simon), establishing the importance of the stronger definitions of ZK formulated subsequently (e.g., *auxiliary-input* zero-knowledge – see Goldreich and Oren [475]), for which closure under sequential composition has been proven. They prove that even these strong formulations of ZK are not, however, closed under parallel composition (thus motivating the definition of weaker notions of zero-knowledge), and that 3-pass interactive ZK proofs of membership that are *black-box simulation ZK* exist only for languages in **BPP** (Definition 2.77); while the definition of "black-box simulation ZK" is more restrictive than the original definition of ZK, all known ZK protocols are ZK by this definition also. Consequently, protocols that are (formally) ZK are less practical than their corresponding 3-pass parallel versions.

As a replacement for the security requirement of zero knowledge in many protocols, Feige and Shamir [384] proposed *witness indistinguishability* and the related notion of *witness hiding protocols*. Unlike zero knowledge, witness indistinguishability is preserved under arbitrary composition of protocols.

Methods have been proposed to reduce the communication complexity of essentially all customized identification protocols, including the use of hash values in the first message (cf. Note 10.29; Note 10.39). Girault and Stern [462] examine the security implications of the length of such hash values, note that collision-resistance of the hash function suffices for the typically claimed security levels, and examine further optimizations of the communication complexity of such protocols, including use of *r-collision resistant* hash functions.

Blum, Feldman, and Micali [163] introduced the idea of non-interactive (or more clearly: *mono-directional*) ZK proofs, separating the notions of interactive proof systems and zero-knowledge protocols; here the prover and verifier share a random string, and communication is restricted to one-way (or the prover may simply publish a proof, for verification at some future time). De Santis, Micali, and Persiano [317] improve these results employing a weaker complexity assumption; Blum et al. [162] provide a summary and further improvements. While the technique of Remark 10.30, due to Fiat and Shamir [395], allows a zero-knowledge identification scheme to be converted to a signature scheme, the latter cannot be a sound zero-knowledge signature scheme because the very simulatability of the identification which establishes the ZK property would allow signature forgery (e.g., see Okamoto [949]).

A further flavor of zero-knowledge (cf. Definition 10.22) is *statistical* (or *almost perfect*) zero-knowledge; here the probability distributions of the transcripts must be *statistically indistinguishable* (indistinguishable by an examiner with unlimited computing power but given only polynomially many samples). Pursuing other characterizations, interactive protocols in which the assurance a verifier obtains is based on some unproven assumption may be distinguished as *arguments* (see Brassard and Crépeau [195]), with *proofs* then required to be free of any unproven assumptions, although possibly probabilistic.

For performance comparisons and tradeoffs for the Fiat-Shamir, Guillou-Quisquater, and Schnorr schemes, see Fiat and Shamir [395], Schnorr [1098], Okamoto [949], and Lim and Lee [768], among others. For an overview of chipcard technology and the use thereof for identification, see Guillou, Ugon, and Quisquater [527]; an earlier paper on chipcards is by Guillou and Ugon [526]. Knobloch [681] describes a preliminary chipcard implementation of the Fiat-Shamir protocol.

§10.5

Bauspiess and Knobloch [78] discuss issues related to Remark 10.41, including taking over a communications line after entity authentication has completed. Bengio et al. [113] discuss implementation issues related to identification schemes such as the Fiat-Shamir protocol, including Remark 10.42. Classes of replay attacks are discussed in several papers, e.g., see Syverson [1182] and the ISO/IEC 10181-2 authentication framework [610]. For further references on the analysis of entity authentication protocols and attacks, see the §12.9 notes.

11

Digital Signatures

Contents in Brief

11.1 Introduction

This chapter considers techniques designed to provide the digital counterpart to a handwritten signature. A *digital signature* of a message is a number dependent on some secret known only to the signer, and, additionally, on the content of the message being signed. Signatures must be verifiable; if a dispute arises as to whether a party signed a document (caused by either a lying signer trying to *repudiate* a signature it did create, or a fraudulent claimant), an unbiased third party should be able to resolve the matter equitably, without requiring access to the signer's secret information (private key).

Digital signatures have many applications in information security, including authentication, data integrity, and non-repudiation. One of the most significant applications of digital signatures is the certification of public keys in large networks. Certification is a means for a trusted third party (TTP) to bind the identity of a user to a public key, so that at some later time, other entities can authenticate a public key without assistance from a trusted third party.

The concept and utility of a digital signature was recognized several years before any practical realization was available. The first method discovered was the RSA signature scheme, which remains today one of the most practical and versatile techniques available. Subsequent research has resulted in many alternative digital signature techniques. Some offer significant advantages in terms of functionality and implementation. This chapter is an account of many of the results obtained to date, with emphasis placed on those developments which are practical.

Chapter outline

§11.2 provides terminology used throughout the chapter, and describes a framework for digital signatures that permits a useful classification of the various schemes. It is more abstract than succeeding sections. §11.3 provides an indepth discussion of the RSA signature scheme, as well as closely related techniques. Standards which have been adopted to implement RSA and related signature schemes are also considered here. §11.4 looks at methods which arise from identification protocols described in Chapter 10. Techniques based on the intractability of the discrete logarithm problem, such as the Digital Signature Algorithm (DSA) and ElGamal schemes, are the topic of §11.5. One-time signature schemes, many of which arise from symmetric-key cryptography, are considered in §11.6. §11.7 describes arbitrated digital signatures and the ESIGN signature scheme. Variations on the basic concept of digital signatures, including blind, undeniable, and fail-stop signatures, are discussed in §11.8. Further notes, including subtle points on schemes documented in the chapter and variants (e.g., designated confirmer signatures, convertible undeniable signatures, group signatures, and electronic cash) may be found in §11.9.

11.2 A framework for digital signature mechanisms

§1.6 provides a brief introduction to the basic ideas behind digital signatures, and §1.8.3 shows how these signatures can be realized through reversible public-key encryption techniques. This section describes two general models for digital signature schemes. A complete understanding of the material in this section is not necessary in order to follow subsequent sections; the reader unfamiliar with some of the more concrete methods such as RSA (§11.3) and ElGamal (§11.5) is well advised not to spend an undue amount of time. The idea of a redundancy function is necessary in order to understand the algorithms which give digital signatures with message recovery. The notation provided in Table 11.1 will be used throughout the chapter.

11.2.1 Basic definitions

1. A *digital signature* is a data string which associates a message (in digital form) with some originating entity.

2. A *digital signature generation algorithm* (or *signature generation algorithm*) is a method for producing a digital signature.

3. A *digital signature verification algorithm* (or *verification algorithm*) is a method for verifying that a digital signature is authentic (i.e., was indeed created by the specified entity).

4. A *digital signature scheme* (or *mechanism*) consists of a signature generation algorithm and an associated verification algorithm.

5. A *digital signature signing process* (or *procedure*) consists of a (mathematical) digital signature generation algorithm, along with a method for formatting data into messages which can be signed.

6. A *digital signature verification process* (or *procedure*) consists of a verification algorithm, along with a method for recovering data from the message.[1]

---

[1] Often little distinction is made between the terms scheme and process, and they are used interchangeably.

This chapter is, for the most part, concerned simply with digital signature schemes. In order to use a digital signature scheme in practice, it is necessary to have a digital signature process. Several processes related to various schemes have emerged as commercially relevant standards; two such processes, namely ISO/IEC 9796 and PKCS #1, are described in §11.3.5 and §11.3.6, respectively. Notation used in the remainder of this chapter is provided in Table 11.1. The sets and functions listed in Table 11.1 are all publicly known.

Notation	Meaning
$\mathcal{M}$	a set of elements called the *message space*.
$\mathcal{M}_S$	a set of elements called the *signing space*.
$\mathcal{S}$	a set of elements called the *signature space*.
R	a $1-1$ mapping from $\mathcal{M}$ to $\mathcal{M}_S$ called the *redundancy function*.
$\mathcal{M}_R$	the image of R (i.e., $\mathcal{M}_R = \mathrm{Im}(R)$).
R^{-1}	the inverse of R (i.e., $R^{-1}\colon \mathcal{M}_R \longrightarrow \mathcal{M}$).
$\mathcal{R}$	a set of elements called the *indexing set for signing*.
h	a one-way function with domain $\mathcal{M}$.
$\mathcal{M}_h$	the image of h (i.e., $h\colon \mathcal{M} \longrightarrow \mathcal{M}_h$); $\mathcal{M}_h \subseteq \mathcal{M}_S$ called the *hash value space*.

Table 11.1: *Notation for digital signature mechanisms.*

11.1 Note (*comments on Table 11.1*)

(i) (*messages*) $\mathcal{M}$ is the set of elements to which a signer can affix a digital signature.

(ii) (*signing space*) $\mathcal{M}_S$ is the set of elements to which the signature transformations (to be described in §11.2.2 and §11.2.3) are applied. The signature transformations are not applied directly to the set $\mathcal{M}$.

(iii) (*signature space*) $\mathcal{S}$ is the set of elements associated to messages in $\mathcal{M}$. These elements are used to bind the signer to the message.

(iv) (*indexing set*) $\mathcal{R}$ is used to identify specific signing transformations.

A classification of digital signature schemes

§11.2.2 and §11.2.3 describe two general classes of digital signature schemes, which can be briefly summarized as follows:

1. Digital signature schemes with appendix require the original message as input to the verification algorithm. (See Definition 11.3.)

2. Digital signature schemes with message recovery do not require the original message as input to the verification algorithm. In this case, the original message is recovered from the signature itself. (See Definition 11.7.)

These classes can be further subdivided according to whether or not $|\mathcal{R}| = 1$, as noted in Definition 11.2.

11.2 Definition A digital signature scheme (with either message recovery or appendix) is said to be a *randomized digital signature scheme* if $|\mathcal{R}| > 1$; otherwise, the digital signature scheme is said to be *deterministic*.

Figure 11.1 illustrates this classification. Deterministic digital signature mechanisms can be further subdivided into *one-time signature schemes* (§11.6) and *multiple-use schemes*.

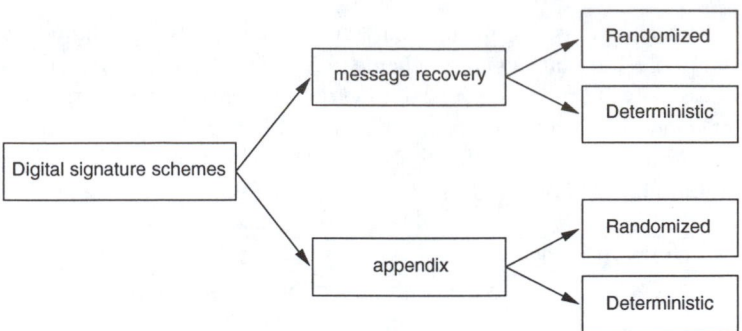

Figure 11.1: *A taxonomy of digital signature schemes.*

11.2.2 Digital signature schemes with appendix

Digital signature schemes with appendix, as discussed in this section, are the most commonly used in practice. They rely on cryptographic hash functions rather than customized redundancy functions, and are less prone to existential forgery attacks (§11.2.4).

11.3 Definition Digital signature schemes which require the message as input to the verification algorithm are called *digital signature schemes with appendix*.

Examples of mechanisms providing digital signatures with appendix are the DSA (§11.5.1), ElGamal (§11.5.2), and Schnorr (§11.5.3) signature schemes. Notation for the following discussion is given in Table 11.1.

11.4 Algorithm Key generation for digital signature schemes with appendix

---

SUMMARY: each entity creates a private key for signing messages, and a corresponding public key to be used by other entities for verifying signatures.

1. Each entity A should select a private key which defines a set $\mathcal{S}_A = \{S_{A,k} : k \in \mathcal{R}\}$ of transformations. Each $S_{A,k}$ is a 1-1 mapping from $\mathcal{M}_h$ to $\mathcal{S}$ and is called a *signing transformation*.

2. $\mathcal{S}_A$ defines a corresponding mapping V_A from $\mathcal{M}_h \times \mathcal{S}$ to {true, false} such that

$$V_A(\widetilde{m}, s^*) = \begin{cases} \text{true,} & \text{if } S_{A,k}(\widetilde{m}) = s^*, \\ \text{false,} & \text{otherwise,} \end{cases}$$

for all $\widetilde{m} \in \mathcal{M}_h$, $s^* \in \mathcal{S}$; here, $\widetilde{m} = h(m)$ for $m \in \mathcal{M}$. V_A is called a *verification transformation* and is constructed such that it may be computed without knowledge of the signer's private key.

3. A's public key is V_A; A's private key is the set $\mathcal{S}_A$.

11.5 Algorithm Signature generation and verification (digital signature schemes with appendix)

SUMMARY: entity A produces a signature $s \in \mathcal{S}$ for a message $m \in \mathcal{M}$, which can later be verified by any entity B.

1. *Signature generation.* Entity A should do the following:
   (a) Select an element $k \in \mathcal{R}$.
   (b) Compute $\widetilde{m} = h(m)$ and $s^* = S_{A,k}(\widetilde{m})$.
   (c) A's signature for m is s^*. Both m and s^* are made available to entities which may wish to verify the signature.

2. *Verification.* Entity B should do the following:
   (a) Obtain A's authentic public key V_A.
   (b) Compute $\widetilde{m} = h(m)$ and $u = V_A(\widetilde{m}, s^*)$.
   (c) Accept the signature if and only if $u = $ true.

Figure 11.2 provides a schematic overview of a digital signature scheme with appendix. The following properties are required of the signing and verification transformations:

(i) for each $k \in \mathcal{R}$, $S_{A,k}$ should be efficient to compute;
(ii) V_A should be efficient to compute; and
(iii) it should be computationally infeasible for an entity other than A to find an $m \in \mathcal{M}$ and an $s^* \in \mathcal{S}$ such that $V_A(\widetilde{m}, s^*) = $ true, where $\widetilde{m} = h(m)$.

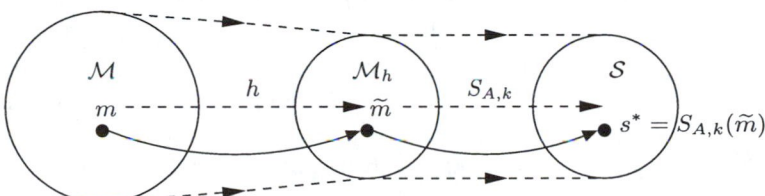

(a) The signing process

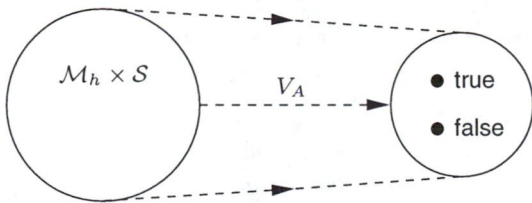

(b) The verification process

Figure 11.2: *Overview of a digital signature scheme with appendix.*

11.6 Note (*use of hash functions*) Most digital signature schemes with message recovery (§11.2.3) are applied to messages of a fixed length, while digital signatures with appendix are applied to messages of arbitrary length. The one-way function h in Algorithm 11.5 is typically selected to be a collision-free hash function (see Definition 9.3). An alternative to hashing is to break the message into blocks of a fixed length which can be individually signed using a signature scheme with message recovery. Since signature generation is relatively slow for many schemes, and since reordering of multiple signed blocks presents a security risk, the preferred method is to hash.

11.2.3 Digital signature schemes with message recovery

The digital signature schemes described in this section have the feature that the message signed can be recovered from the signature itself. In practice, this feature is of use for short messages (see §11.3.3(viii)).

11.7 Definition A *digital signature scheme with message recovery* is a digital signature scheme for which a priori knowledge of the message is not required for the verification algorithm.

Examples of mechanisms providing digital signatures with message recovery are RSA (§11.3.1), Rabin (§11.3.4), and Nyberg-Rueppel (§11.5.4) public-key signature schemes.

11.8 Algorithm Key generation for digital signature schemes with message recovery

SUMMARY: each entity creates a private key to be used for signing messages, and a corresponding public key to be used by other entities for verifying signatures.

1. Each entity A should select a set $\mathcal{S}_A = \{S_{A,k} : k \in \mathcal{R}\}$ of transformations. Each $S_{A,k}$ is a 1-1 mapping from $\mathcal{M}_S$ to $\mathcal{S}$ and is called a *signing transformation*.
2. $\mathcal{S}_A$ defines a corresponding mapping V_A with the property that $V_A \circ S_{A,k}$ is the identity map on $\mathcal{M}_S$ for all $k \in \mathcal{R}$. V_A is called a *verification transformation* and is constructed such that it may be computed without knowledge of the signer's private key.
3. A's public key is V_A; A's private key is the set $\mathcal{S}_A$.

11.9 Algorithm Signature generation and verification for schemes with message recovery

SUMMARY: entity A produces a signature $s \in \mathcal{S}$ for a message $m \in \mathcal{M}$, which can later be verified by any entity B. The message m is recovered from s.

1. *Signature generation.* Entity A should do the following:
    (a) Select an element $k \in \mathcal{R}$.
    (b) Compute $\widetilde{m} = R(m)$ and $s^* = S_{A,k}(\widetilde{m})$. ($R$ is a redundancy function; see Table 11.1 and Note 11.10.)
    (c) A's signature is s^*; this is made available to entities which may wish to verify the signature and recover m from it.
2. *Verification.* Entity B should do the following:
    (a) Obtain A's authentic public key V_A.
    (b) Compute $\widetilde{m} = V_A(s^*)$.
    (c) Verify that $\widetilde{m} \in \mathcal{M}_R$. (If $\widetilde{m} \notin \mathcal{M}_R$, then reject the signature.)
    (d) Recover m from $\widetilde{m}$ by computing $R^{-1}(\widetilde{m})$.

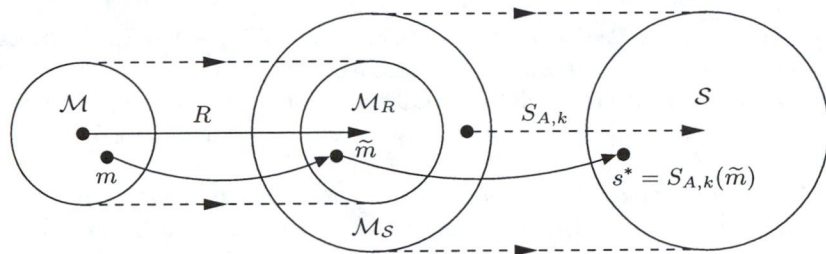

Figure 11.3: *Overview of a digital signature scheme with message recovery.*

Figure 11.3 provides a schematic overview of a digital signature scheme with message recovery. The following properties are required of the signing and verification transformations:

   (i) for each $k \in \mathcal{R}$, $S_{A,k}$ should be efficient to compute;

  (ii) V_A should be efficient to compute; and

 (iii) it should be computationally infeasible for an entity other than A to find any $s^* \in \mathcal{S}$ such that $V_A(s^*) \in \mathcal{M}_R$.

11.10 Note (*redundancy function*) The redundancy function R and its inverse R^{-1} are publicly known. Selecting an appropriate R is critical to the security of the system. To illustrate this point, suppose that $\mathcal{M}_R = \mathcal{M}_S$. Suppose R and $S_{A,k}$ are bijections from $\mathcal{M}$ to $\mathcal{M}_R$ and $\mathcal{M}_S$ to $\mathcal{S}$, respectively. This implies that $\mathcal{M}$ and $\mathcal{S}$ have the same number of elements. Then for any $s^* \in \mathcal{S}$, $V_A(s^*) \in \mathcal{M}_R$, and it is trivial to find messages m and corresponding signatures s^* which will be accepted by the verification algorithm (step 2 of Algorithm 11.9) as follows.

   1. Select random $k \in \mathcal{R}$ and random $s^* \in \mathcal{S}$.

   2. Compute $\widetilde{m} = V_A(s^*)$.

   3. Compute $m = R^{-1}(\widetilde{m})$.

The element s^* is a valid signature for the message m and was created without knowledge of the set of signing transformations $\mathcal{S}_A$.

11.11 Example (*redundancy function*) Suppose $\mathcal{M} = \{m \colon m \in \{0,1\}^n\}$ for some fixed positive integer n and $\mathcal{M}_S = \{t \colon t \in \{0,1\}^{2n}\}$. Define $R \colon \mathcal{M} \longrightarrow \mathcal{M}_S$ by $R(m) = m \| m$, where $\|$ denotes concatenation; that is, $\mathcal{M}_R = \{m \| m \colon m \in \mathcal{M}\} \subseteq \mathcal{M}_S$. For large values of n, the quantity $|\mathcal{M}_R|/|\mathcal{M}_S| = (\frac{1}{2})^n$ is a negligibly small fraction. This redundancy function is suitable provided that no judicious choice of s^* on the part of an adversary will have a non-negligible probability of yielding $V_A(s^*) \in \mathcal{M}_R$.   □

11.12 Remark (*selecting a redundancy function*) Even though the redundancy function R is public knowledge and R^{-1} is easy to compute, selection of R is critical and should not be made independently of the choice of the signing transformations in $\mathcal{S}_A$. Example 11.21 provides a specific example of a redundancy function which compromises the security of the signature scheme. An example of a redundancy function which has been accepted as an international standard is given in §11.3.5. This redundancy function is not appropriate for all digital signature schemes with message recovery, but does apply to the RSA (§11.3.1) and Rabin (§11.3.4) digital signature schemes.

11.13 Remark (*a particular class of message recovery schemes*) §1.8.3 describes a class of digital signature schemes with message recovery which arise from reversible public-key encryption methods. Examples include the RSA (§8.2) and Rabin (§8.3) encryption schemes. The corresponding signature mechanisms are discussed in §11.3.1 and §11.3.4, respectively.

11.14 Note (*signatures with appendix from schemes providing message recovery*) Any digital signature scheme with message recovery can be turned into a digital signature scheme with appendix by simply hashing the message and then signing the hash value. The message is now required as input to the verification algorithm. A schematic for this situation can be derived from Figure 11.3 and is illustrated in Figure 11.4. The redundancy function R is no longer critical to the security of the signature scheme, and can be any $1-1$ function from $\mathcal{M}_h$ to $\mathcal{M}_S$.

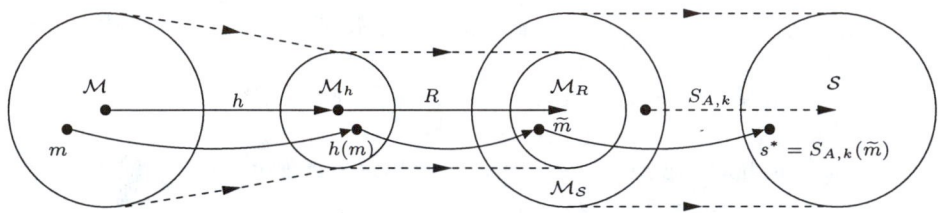

Figure 11.4: *Signature scheme with appendix obtained from one providing message recovery.*

11.2.4 Types of attacks on signature schemes

The goal of an adversary is to *forge* signatures; that is, produce signatures which will be accepted as those of some other entity. The following provides a set of criteria for what it means to break a signature scheme.

1. *total break.* An adversary is either able to compute the private key information of the signer, or finds an efficient signing algorithm functionally equivalent to the valid signing algorithm. (For example, see §11.3.2(i).)
2. *selective forgery.* An adversary is able to create a valid signature for a particular message or class of messages chosen a priori. Creating the signature does not directly involve the legitimate signer. (See Example 11.21.)
3. *existential forgery.* An adversary is able to forge a signature for at least one message. The adversary has little or no control over the message whose signature is obtained, and the legitimate signer may be involved in the deception (for example, see Note 11.66(iii)).

There are two basic attacks against public-key digital signature schemes.

1. *key-only attacks.* In these attacks, an adversary knows only the signer's public key.
2. *message attacks.* Here an adversary is able to examine signatures corresponding either to known or chosen messages. Message attacks can be further subdivided into three classes:
   (a) *known-message attack.* An adversary has signatures for a set of messages which are known to the adversary but not chosen by him.

(b) *chosen-message attack*. An adversary obtains valid signatures from a chosen list of messages before attempting to break the signature scheme. This attack is *non-adaptive* in the sense that messages are chosen before any signatures are seen. Chosen-message attacks against signature schemes are analogous to chosen-ciphertext attacks against public-key encryption schemes (see §1.13.1).

(c) *adaptive chosen-message attack*. An adversary is allowed to use the signer as an oracle; the adversary may request signatures of messages which depend on the signer's public key and he may request signatures of messages which depend on previously obtained signatures or messages.

11.15 Note (*adaptive chosen-message attack*) In principle, an adaptive chosen-message attack is the most difficult type of attack to prevent. It is conceivable that given enough messages and corresponding signatures, an adversary could deduce a pattern and then forge a signature of its choice. While an adaptive chosen-message attack may be infeasible to mount in practice, a well-designed signature scheme should nonetheless be designed to protect against the possibility.

11.16 Note (*security considerations*) The level of security required in a digital signature scheme may vary according to the application. For example, in situations where an adversary is only capable of mounting a key-only attack, it may suffice to design the scheme to prevent the adversary from being successful at selective forgery. In situations where the adversary is capable of a message attack, it is likely necessary to guard against the possibility of existential forgery.

11.17 Note (*hash functions and digital signature processes*) When a hash function h is used in a digital signature scheme (as is often the case), h should be a fixed part of the signature process so that an adversary is unable to take a valid signature, replace h with a weak hash function, and then mount a selective forgery attack.

11.3 RSA and related signature schemes

This section describes the RSA signature scheme and other closely related methods. The security of the schemes presented here relies to a large degree on the intractability of the integer factorization problem (see §3.2). The schemes presented include both digital signatures with message recovery and appendix (see Note 11.14).

11.3.1 The RSA signature scheme

The message space and ciphertext space for the RSA public-key encryption scheme (§8.2) are both $\mathbb{Z}_n = \{0, 1, 2, \ldots, n - 1\}$ where $n = pq$ is the product of two randomly chosen distinct prime numbers. Since the encryption transformation is a bijection, digital signatures can be created by reversing the roles of encryption and decryption. The RSA signature scheme is a deterministic digital signature scheme which provides message recovery (see Definition 11.7). The signing space $\mathcal{M}_S$ and signature space $\mathcal{S}$ are both $\mathbb{Z}_n$ (see Table 11.1 for notation). A redundancy function $R\colon \mathcal{M} \longrightarrow \mathbb{Z}_n$ is chosen and is public knowledge.

11.18 Algorithm Key generation for the RSA signature scheme

SUMMARY: each entity creates an RSA public key and a corresponding private key.
Each entity A should do the following:

1. Generate two large distinct random primes p and q, each roughly the same size (see §11.3.2).
2. Compute $n = pq$ and $\phi = (p-1)(q-1)$.
3. Select a random integer e, $1 < e < \phi$, such that $\gcd(e, \phi) = 1$.
4. Use the extended Euclidean algorithm (Algorithm 2.107) to compute the unique integer d, $1 < d < \phi$, such that $ed \equiv 1 \pmod{\phi}$.
5. A's public key is (n, e); A's private key is d.

11.19 Algorithm RSA signature generation and verification

SUMMARY: entity A signs a message $m \in \mathcal{M}$. Any entity B can verify A's signature and recover the message m from the signature.

1. *Signature generation.* Entity A should do the following:
   (a) Compute $\widetilde{m} = R(m)$, an integer in the range $[0, n-1]$.
   (b) Compute $s = \widetilde{m}^d \bmod n$.
   (c) A's signature for m is s.
2. *Verification.* To verify A's signature s and recover the message m, B should:
   (a) Obtain A's authentic public key (n, e).
   (b) Compute $\widetilde{m} = s^e \bmod n$.
   (c) Verify that $\widetilde{m} \in \mathcal{M}_R$; if not, reject the signature.
   (d) Recover $m = R^{-1}(\widetilde{m})$.

Proof that signature verification works. If s is a signature for a message m, then $s \equiv \widetilde{m}^d \bmod n$ where $\widetilde{m} = R(m)$. Since $ed \equiv 1 \pmod{\phi}$, $s^e \equiv \widetilde{m}^{ed} \equiv \widetilde{m} \pmod n$. Finally, $R^{-1}(\widetilde{m}) = R^{-1}(R(m)) = m$.

11.20 Example (*RSA signature generation with artificially small parameters*)
Key generation. Entity A selects primes $p = 7927$, $q = 6997$, and computes $n = pq = 55465219$ and $\phi = 7926 \times 6996 = 55450296$. A chooses $e = 5$ and solves $ed = 5d \equiv 1 \pmod{55450296}$, yielding $d = 44360237$. A's public key is $(n = 55465219, e = 5)$; A's private key is $d = 44360237$.
Signature generation. For the sake of simplicity (but see §11.3.3(ii)), assume that $\mathcal{M} = \mathbb{Z}_n$ and that the redundancy function $R \colon \mathcal{M} \longrightarrow \mathbb{Z}_n$ is the identity map $R(m) = m$ for all $m \in \mathcal{M}$. To sign a message $m = 31229978$, A computes $\widetilde{m} = R(m) = 31229978$, and computes the signature $s = \widetilde{m}^d \bmod n = 31229978^{44360237} \bmod 55465219 = 30729435$.
Signature verification. B computes $\widetilde{m} = s^e \bmod n = 30729435^5 \bmod 55465219 = 31229978$. Finally, B accepts the signature since $\widetilde{m}$ has the required redundancy (i.e., $\widetilde{m} \in \mathcal{M}_R$), and recovers $m = R^{-1}(\widetilde{m}) = 31229978$. $\qquad\square$

11.3.2 Possible attacks on RSA signatures

(i) Integer factorization

If an adversary is able to factor the public modulus n of some entity A, then the adversary can compute ϕ and then, using the extended Euclidean algorithm (Algorithm 2.107), deduce the private key d from ϕ and the public exponent e by solving $ed \equiv 1 \pmod{\phi}$. This constitutes a total break of the system. To guard against this, A must select p and q so that factoring n is a computationally infeasible task. For further information, see §8.2.2(i) and Note 8.8.

(ii) Multiplicative property of RSA

The RSA signature scheme (as well as the encryption method, cf. §8.2.2(v)) has the following multiplicative property, sometimes referred to as the *homomorphic property*. If $s_1 = m_1^d \bmod n$ and $s_2 = m_2^d \bmod n$ are signatures on messages m_1 and m_2, respectively (or more properly on messages with redundancy added), then $s = s_1 s_2 \bmod n$ has the property that $s = (m_1 m_2)^d \bmod n$. If $m = m_1 m_2$ has the proper redundancy (i.e., $m \in \mathcal{M}_R$), then s will be a valid signature for it. Hence, it is important that the redundancy function R is not multiplicative, i.e., for essentially all pairs $a, b \in \mathcal{M}$, $R(a \cdot b) \neq R(a)R(b)$. As Example 11.21 shows, this condition on R is necessary but not sufficient for security.

11.21 Example (*insecure redundancy function*) Let n be an RSA modulus and d the private key. Let $k = \lceil \lg n \rceil$ be the bitlength of n, and let t be a fixed positive integer such that $t < k/2$. Let $w = 2^t$ and let messages be integers m in the interval $[1, n2^{-t} - 1]$. The redundancy function R is taken to be $R(m) = m2^t$ (the least significant t bits of the binary representation of $R(m)$ are 0's). For most choices of n, R will not have the multiplicative property. The general existential forgery attack described in Note 11.10 would have a probability of success of $(\frac{1}{2})^t$. But for this redundancy function, a selective forgery attack (which is more serious) is possible, as is now explained.

Suppose that an adversary wishes to forge a signature on a message m. The adversary knows n but not d. The adversary can mount the following chosen-message attack to obtain the signature on m. Apply the extended Euclidean algorithm (Algorithm 2.107) to n and $\tilde{m} = R(m) = m2^t = mw$. At each stage of the extended Euclidean algorithm, integers x, y, and r are computed such that $xn + y\tilde{m} = r$. It can be shown that at some stage there exists a y and r such that $|y| < n/w$ and $r < n/w$, provided $w \leq \sqrt{n}$. If $y > 0$, form integers $m_2 = rw$ and $m_3 = yw$. If $y < 0$, form integers $m_2 = rw$ and $m_3 = -yw$. In either case, m_2 and m_3 have the required redundancy. If signatures $s_2 = m_2^d \bmod n$ and $s_3 = m_3^d \bmod n$ are obtained from the legitimate signer, then the adversary can compute a signature for m as follows:

- if $y > 0$, compute $\frac{s_2}{s_3} = \frac{m_2^d}{m_3^d} = (\frac{rw}{yw})^d = (\frac{r}{y})^d = \tilde{m}^d \bmod n$;

- if $y < 0$, compute $\frac{s_2}{-s_3} = \frac{m_2^d}{(-m_3)^d} = (\frac{rw}{yw})^d = (\frac{r}{y})^d = \tilde{m}^d \bmod n$.

In either case, the adversary has a signed message of its choice with the required redundancy. This attack is an example of a chosen-message attack providing selective forgery. It emphasizes the requirement for judicious choice of the redundancy function R.   □

11.3.3 RSA signatures in practice

(i) Reblocking problem

One suggested use of RSA is to sign a message and then encrypt the resulting signature. One must be concerned about the relative sizes of the moduli involved when implementing this procedure. Suppose that A wishes to sign and then encrypt a message for B. Suppose that (n_A, e_A) and (n_B, e_B) are A's and B's public keys, respectively. If $n_A > n_B$, then there is a chance that the message cannot be recovered by B, as illustrated in Example 11.22.

11.22 Example (*reblocking problem*) Let $n_A = 8387 \times 7499 = 62894113$, $e_A = 5$, and $d_A = 37726937$; and $n_B = 55465219$, $e_B = 5$, $d_B = 44360237$. Notice that $n_A > n_B$. Suppose $m = 1368797$ is a message with redundancy to be signed under A's private key and then encrypted using B's public key. A computes the following:

1. $s = m^{d_A} \bmod n_A = 1368797^{37726937} \bmod 62894113 = 59847900$.
2. $c = s^{e_B} \bmod n_B = 59847900^5 \bmod 55465219 = 38842235$.

To recover the message and verify the signature, B computes the following:

1. $\widehat{s} = c^{d_B} \bmod n_B = 38842235^{44360237} \bmod 55465219 = 4382681$.
2. $\widehat{m} = \widehat{s}^{e_A} \bmod n_A = 4382681^5 \bmod 62894113 = 54383568$.

Observe that $m \neq \widehat{m}$. The reason for this is that s is larger than the modulus n_B. Here, the probability of this problem occurring is $(n_A - n_B)/n_A \approx 0.12$.   □

There are various ways to overcome the reblocking problem.

1. *reordering*. The problem of incorrect decryption will never occur if the operation using the smaller modulus is performed first. That is, if $n_A > n_B$, then entity A should first encrypt the message using B's public key, and then sign the resulting ciphertext using A's private key. The preferred order of operations, however, is always to sign the message first and then encrypt the signature; for if A encrypts first and then signs, an adversary could remove the signature and replace it with its own signature. Even though the adversary will not know what is being signed, there may be situations where this is advantageous to the adversary. Thus, reordering is not a prudent solution.

2. *two moduli per entity*. Have each entity generate separate moduli for encrypting and for signing. If each user's signing modulus is smaller than all of the possible encrypting moduli, then incorrect decryption never occurs. This can be guaranteed by requiring encrypting moduli to be $(t + 1)$-bit numbers and signing moduli t-bit numbers.

3. *prescribing the form of the modulus*. In this method, one selects the primes p and q so that the modulus n has a special form: the highest-order bit is a 1 and the k following bits are all 0's. A t-bit modulus n of this form can be found as follows. For n to have the required form, $2^{t-1} \leq n < 2^{t-1} + 2^{t-k-1}$. Select a random $\lceil t/2 \rceil$-bit prime p, and search for a prime q in the interval between $\lceil 2^{t-1}/p \rceil$ and $\lfloor (2^{t-1} + 2^{t-k-1})/p \rfloor$; then $n = pq$ is a modulus of the required type (see Example 11.23). This choice for the modulus n does not completely prevent the incorrect decryption problem, but it can reduce the probability of its occurrence to a negligibly small number. Suppose that n_A is such a modulus and $s = m^{d_A} \bmod n_A$ is a signature on m. Suppose further that s has a 1 in one of the high-order $k + 1$ bit positions, other than the highest. Then s, since it is smaller than n_A, must have a 0 in the highest-order bit position and so is necessarily smaller than any other modulus of a similar form. The probability that s does not have any 1's in the high-order $k + 1$ bit positions, other than the highest, is less than $(\frac{1}{2})^k$, which is negligibly small if k is selected to be around 100.

11.23 Example (*prescribing the form of the modulus*) Suppose one wants to construct a 12-bit modulus n such that the high order bit is a 1 and the next $k = 3$ bits are 0's. Begin by selecting a 6-bit prime $p = 37$. Select a prime q in the interval between $\lceil 2^{11}/p \rceil = 56$ and $\lfloor (2^{11} + 2^8)/p \rfloor = 62$. The possibilities for q are 59 and 61. If $q = 59$ is selected, then $n = 37 \times 59 = 2183$, having binary representation 100010000111. If $q = 61$ is selected, then $n = 37 \times 61 = 2257$, having binary representation 100011010001.   □

(ii) Redundancy functions

In order to avoid an existential forgery attack (see §11.2.4) on the RSA signature scheme, a suitable redundancy function R is required. §11.3.5 describes one such function which has been accepted as an international standard. Judicious choice of a redundancy function is crucial to the security of the system (see §11.3.2(ii)).

(iii) The RSA digital signature scheme with appendix

Note 11.14 describes how any digital signature scheme with message recovery can be modified to give a digital signature scheme with appendix. For example, if MD5 (Algorithm 9.51) is used to hash messages of arbitrary bitlengths to bitstrings of length 128, then Algorithm 11.9 could be used to sign these hash values. If n is a k-bit RSA modulus, then a suitable redundancy function R is required to assign 128-bit integers to k-bit integers. §11.3.6 describes a method for doing this which is often used in practice.

(iv) Performance characteristics of signature generation and verification

Let $n = pq$ be a $2k$-bit RSA modulus where p and q are each k-bit primes. Computing a signature $s = m^d \bmod n$ for a message m requires $O(k^3)$ bit operations (regarding modular multiplication, see §14.3; and for modular exponentiation, §14.6). Since the signer typically knows p and q, she can compute $s_1 = m^d \bmod p$, $s_2 = m^d \bmod q$, and determine s by using the Chinese remainder theorem (see Note 14.75). Although the complexity of this procedure remains $O(k^3)$, it is considerably more efficient in some situations.

Verification of signatures is significantly faster than signing if the public exponent is chosen to be a small number. If this is done, verification requires $O(k^2)$ bit operations. Suggested values for e in practice are 3 or $2^{16} + 1$;[2] of course, p and q must be chosen so that $\gcd(e, (p-1)(q-1)) = 1$.

The RSA signature scheme is thus ideally suited to situations where signature verification is the predominant operation being performed. For example, when a trusted third party creates a public-key certificate for an entity A, this requires only one signature generation, and this signature may be verified many times by various other entities (see §13.4.2).

(v) Parameter selection

As of 1996, a minimum of 768 bits is recommended for RSA signature moduli. A modulus of at least 1024 bits is recommended for signatures which require much longer lifetimes or which are critical to the overall security of a large network. It is prudent to remain aware of progress in integer factorization, and to be prepared to adjust parameters accordingly.

No weaknesses in the RSA signature scheme have been reported when the public exponent e is chosen to be a small number such as 3 or $2^{16} + 1$. It is not recommended to restrict the size of the private exponent d in order to improve the efficiency of signature generation (cf. §8.2.2(iv)).

(vi) Bandwidth efficiency

Bandwidth efficiency for digital signatures with message recovery refers to the ratio of the logarithm (base 2) of the size of the signing space $\mathcal{M}_S$ to the logarithm (base 2) of the size of $\mathcal{M}_R$, the image space of the redundancy function. Hence, the bandwidth efficiency is determined by the redundancy R. For RSA (and the Rabin digital signature scheme, §11.3.4), the redundancy function specified by ISO/IEC 9796 (§11.3.5) takes k-bit messages and encodes them to $2k$-bit elements in $\mathcal{M}_S$ from which a $2k$-bit signature is formed. The bandwidth

---

[2]The choice of $e = 2^{16} + 1$ is based on the fact that e is a prime number, and $\tilde{m}^e \bmod n$ can be computed with only 16 modular squarings and one modular multiplication (see §14.6.1).

efficiency in this case is $\frac{1}{2}$. For example, with a modulus of size 1024 bits, the maximum size of a message which can be signed is 512 bits.

(vii) System-wide parameters

Each entity must have a distinct RSA modulus; it is insecure to use a system-wide modulus (see §8.2.2(vi)). The public exponent e can be a system-wide parameter, and is in many applications (see Note 8.9(ii)).

(viii) Short vs. long messages

Suppose n is a $2k$-bit RSA modulus which is used in Algorithm 11.19 to sign k-bit messages (i.e., the bandwidth efficiency is $\frac{1}{2}$). Suppose entity A wishes to sign a kt-bit message m. One approach is to partition m into k-bit blocks such that $m = m_1 \| m_2 \| \cdots \| m_t$ and sign each block individually (but see Note 11.6 regarding why this is not recommended). The bandwidth requirement for this is $2kt$ bits. Alternatively, A could hash message m to a bitstring of length $l \leq k$ and sign the hash value. The bandwidth requirement for this signature is $kt + 2k$, where the term kt comes from sending the message m. Since $kt + 2k \leq 2kt$ whenever $t \geq 2$, it follows that the most bandwidth efficient method is to use RSA digital signatures with appendix. For a message of size at most k-bits, RSA with message recovery is preferred.

11.3.4 The Rabin public-key signature scheme

The Rabin public-key signature scheme is similar to RSA (Algorithm 11.19), but it uses an even public exponent e. [3] For the sake of simplicity, it will be assumed that $e = 2$. The signing space $\mathcal{M}_S$ is Q_n (the set of quadratic residues modulo n — see Definition 2.134) and signatures are square roots of these. A redundancy function R from the message space $\mathcal{M}$ to $\mathcal{M}_S$ is selected and is public knowledge.

Algorithm 11.25 describes the basic version of the Rabin public-key signature scheme. A more detailed version (and one more useful in practice) is presented in Algorithm 11.30.

11.24 Algorithm Key generation for the Rabin public-key signature scheme

---

SUMMARY: each entity creates a public key and corresponding private key.
Each entity A should do the following:

1. Generate two large distinct random primes p and q, each roughly the same size.
2. Compute $n = pq$.
3. A's public key is n; A's private key is (p, q).

11.25 Algorithm Rabin signature generation and verification

SUMMARY: entity A signs a message $m \in \mathcal{M}$. Any entity B can verify A's signature and recover the message m from the signature.

1. *Signature generation.* Entity A should do the following:
   (a) Compute $\widetilde{m} = R(m)$.
   (b) Compute a square root s of $\widetilde{m}$ mod n (using Algorithm 3.44).
   (c) A's signature for m is s.

---

[3] Since p and q are distinct primes in an RSA modulus, $\phi = (p-1)(q-1)$ is even. In RSA, the public exponent e must satisfy $\gcd(e, \phi) = 1$ and so must be odd.

2. *Verification.* To verify A's signature s and recover the message m, B should:

    (a) Obtain A's authentic public key n.

    (b) Compute $\widetilde{m} = s^2 \bmod n$.

    (c) Verify that $\widetilde{m} \in \mathcal{M}_R$; if not, reject the signature.

    (d) Recover $m = R^{-1}(\widetilde{m})$.

11.26 Example (*Rabin signature generation with artificially small parameters*)

Key generation. Entity A selects primes $p = 7$, $q = 11$, and computes $n = 77$. A's public key is $n = 77$; A's private key is $(p = 7, q = 11)$. The signing space is $\mathcal{M}_S = Q_{77} = \{1, 4, 9, 15, 16, 23, 25, 36, 37, 53, 58, 60, 64, 67, 71\}$. For the sake of simplicity (but see Note 11.27), take $\mathcal{M} = \mathcal{M}_S$ and the redundancy function R to be the identity map (i.e., $\widetilde{m} = R(m) = m$).

Signature generation. To sign a message $m = 23$, A computes $R(m) = \widetilde{m} = 23$, and then finds a square root of $\widetilde{m}$ modulo 77. If s denotes such a square root, then $s \equiv \pm 3 \pmod{7}$ and $s \equiv \pm 1 \pmod{11}$, implying $s = 10, 32, 45$, or 67. The signature for m is chosen to be $s = 45$. (The signature could be any one of the four square roots.)

Signature verification. B computes $\widetilde{m} = s^2 \bmod 77 = 23$. Since $\widetilde{m} = 23 \in \mathcal{M}_R$, B accepts the signature and recovers $m = R^{-1}(\widetilde{m}) = 23$. $\square$

11.27 Note (*redundancy*)

    (i) As with the RSA signature scheme (Example 11.21), an appropriate choice of a redundancy function R is crucial to the security of the Rabin signature scheme. For example, suppose that $\mathcal{M} = \mathcal{M}_S = Q_n$ and $R(m) = m$ for all $m \in \mathcal{M}$. If an adversary selects any integer $s \in \mathbb{Z}_n^*$ and squares it to get $\widetilde{m} = s^2 \bmod n$, then s is a valid signature for $\widetilde{m}$ and is obtained without knowledge of the private key. (Here, the adversary has little control over what the message will be.) In this situation, existential forgery is trivial.

    (ii) In most practical applications of digital signature schemes with message recovery, the message space $\mathcal{M}$ consists of bitstrings of some fixed length. For the Rabin scheme, determining a redundancy function R is a challenging task. For example, if a message m is a bitstring, R might assign it to the integer whose binary representation is the message. There is, however, no guarantee that the resulting integer is a quadratic residue modulo n, and so computing a square root might be impossible. One might try to append a small number of random bits to m and apply R again in the hope that $R(m) \in Q_n$. On average, two such attempts would suffice, but a deterministic method would be preferable.

Modified-Rabin signature scheme

To overcome the problem discussed in Note 11.27 (ii), a modified version of the basic Rabin signature scheme is provided. The technique presented is similar to that used in the ISO/IEC 9796 digital signature standard (§11.3.5). It provides a deterministic method for associating messages with elements in the signing space $\mathcal{M}_S$, such that computing a square root (or something close to it) is always possible. An understanding of this method will facilitate the reading of §11.3.5.

11.28 Fact Let p and q be distinct primes each congruent to 3 modulo 4, and let $n = pq$.

    (i) If $\gcd(x, n) = 1$, then $x^{(p-1)(q-1)/2} \equiv 1 \pmod{n}$.

    (ii) If $x \in Q_n$, then $x^{(n-p-q+5)/8} \bmod n$ is a square root of x modulo n.

(iii) Let x be an integer having Jacobi symbol $\left(\frac{x}{n}\right) = 1$, and let $d = (n - p - q + 5)/8$. Then

$$x^{2d} \bmod n = \left\{ \begin{array}{ll} x, & \text{if } x \in Q_n, \\ n - x, & \text{if } x \notin Q_n. \end{array} \right.$$

(iv) If $p \not\equiv q \pmod 8$, then $\left(\frac{2}{n}\right) = -1$. Hence, multiplication of any integer x by 2 or $2^{-1} \bmod n$ reverses the Jacobi symbol of x. (Integers of the form $n = pq$ where $p \equiv q \equiv 3 \pmod 4$ and $p \not\equiv q \pmod 8$ are sometimes called *Williams integers*.)

Algorithm 11.30 is a modified version of the Rabin digital signature scheme. Messages to be signed are from $\mathcal{M}_S = \{m \in \mathbb{Z}_n : m \equiv 6 \pmod{16}\}$. Notation is given in Table 11.2. In practice, the redundancy function R should be more complex to prevent existential forgery (see §11.3.5 for an example).

Symbol	Term	Description
$\mathcal{M}$	message space	$\{m \in \mathbb{Z}_n : m \le \lfloor (n-6)/16 \rfloor\}$
$\mathcal{M}_S$	signing space	$\{m \in \mathbb{Z}_n : m \equiv 6 \pmod{16}\}$
$\mathcal{S}$	signature space	$\{s \in \mathbb{Z}_n : (s^2 \bmod n) \in \mathcal{M}_S\}$
R	redundancy function	$R(m) = 16m + 6$ for all $m \in \mathcal{M}$
$\mathcal{M}_R$	image of R	$\{m \in \mathbb{Z}_n : m \equiv 6 \pmod{16}\}$

Table 11.2: *Definition of sets and functions for Algorithm 11.30.*

11.29 Algorithm Key generation for the modified-Rabin signature scheme

SUMMARY: each entity creates a public key and corresponding private key.
Each entity A should do the following:

1. Select random primes $p \equiv 3 \pmod 8$, $q \equiv 7 \pmod 8$ and compute $n = pq$.
2. A's public key is n; A's private key is $d = (n - p - q + 5)/8$.

11.30 Algorithm Modified-Rabin public-key signature generation and verification

SUMMARY: entity A signs a message $m \in \mathcal{M}$. Any entity B can verify A's signature and recover the message m from the signature.

1. *Signature generation.* Entity A should do the following:
   (a) Compute $\widetilde{m} = R(m) = 16m + 6$.
   (b) Compute the Jacobi symbol $J = \left(\frac{\widetilde{m}}{n}\right)$ (using Algorithm 2.149).
   (c) If $J = 1$ then compute $s = \widetilde{m}^d \bmod n$.
   (d) If $J = -1$ then compute $s = (\widetilde{m}/2)^d \bmod n$. [4]
   (e) A's signature for m is s.

2. *Verification.* To verify A's signature s and recover the message m, B should:
   (a) Obtain A's authentic public key n.
   (b) Compute $m' = s^2 \bmod n$. (Note the original message m itself is not required.)
   (c) If $m' \equiv 6 \pmod 8$, take $\widetilde{m} = m'$.
   (d) If $m' \equiv 3 \pmod 8$, take $\widetilde{m} = 2m'$.

---

[4]If $J \ne 1$ or -1 then $J = 0$, implying $\gcd(\widetilde{m}, n) \ne 1$. This leads to a factorization of n. In practice, the probability that this will ever occur is negligible.

(e) If $m' \equiv 7 \pmod 8$, take $\tilde{m} = n - m'$.

(f) If $m' \equiv 2 \pmod 8$, take $\tilde{m} = 2(n - m')$.

(g) Verify that $\tilde{m} \in \mathcal{M}_R$ (see Table 11.2); if not, reject the signature.

(h) Recover $m = R^{-1}(\tilde{m}) = (\tilde{m} - 6)/16$.

Proof that signature verification works. The signature generation phase signs either $v = \tilde{m}$ or $v = \tilde{m}/2$ depending upon which has Jacobi symbol 1. By Fact 11.28(iv), exactly one of $\tilde{m}$, $\tilde{m}/2$ has Jacobi symbol 1. The value v that is signed is such that $v \equiv 3$ or $6 \pmod 8$. By Fact 11.28(iii), $s^2 \bmod n = v$ or $n - v$ depending on whether or not $v \in Q_n$. Since $n \equiv 5 \pmod 8$, these cases can be uniquely distinguished.

11.31 Example (*modified-Rabin signature scheme with artificially small parameters*)

Key generation. A chooses $p = 19$, $q = 31$, and computes $n = pq = 589$ and $d = (n - p - q + 5)/8 = 68$. A's public key is $n = 589$, while A's private key is $d = 68$. The signing space $\mathcal{M}_S$ is given in the following table, along with the Jacobi symbol of each element.

m	6	22	54	70	86	102	118	134	150	166
$\left(\frac{m}{589}\right)$	-1	1	-1	-1	1	1	1	1	-1	1
m	182	198	214	230	246	262	278	294	326	358
$\left(\frac{m}{589}\right)$	-1	1	1	1	1	-1	1	-1	-1	-1
m	374	390	406	422	438	454	470	486	502	518
$\left(\frac{m}{589}\right)$	-1	-1	-1	1	1	1	-1	-1	1	-1
m	534	550	566	582						
$\left(\frac{m}{589}\right)$	-1	1	-1	1						

Signature generation. To sign a message $m = 12$, A computes $\tilde{m} = R(12) = 198$, $\left(\frac{\tilde{m}}{n}\right) = \left(\frac{198}{589}\right) = 1$, and $s = 198^{68} \bmod 589 = 102$. A's signature for $m = 12$ is $s = 102$.

Signature verification. B computes $m' = s^2 \bmod n = 102^2 \bmod 589 = 391$. Since $m' \equiv 7 \pmod 8$, B takes $\tilde{m} = n - m' = 589 - 391 = 198$. Finally, B computes $m = R^{-1}(\tilde{m}) = (198 - 6)/16 = 12$, and accepts the signature. $\square$

11.32 Note (*security of modified-Rabin signature scheme*)

(i) When using Algorithm 11.30, one should never sign a value v having Jacobi symbol -1, since this leads to a factorization of n. To see this, observe that $y = v^{2d} = s^2$ must have Jacobi symbol 1; but $y^2 \equiv (v^2)^{2d} \equiv v^2 \pmod n$ by Fact 11.28(iii). Therefore, $(v-y)(v+y) \equiv 0 \pmod n$. Since v and y have opposite Jacobi symbols, $v \not\equiv y \pmod n$ and thus $\gcd(v - y, n) = p$ or q.

(ii) Existential forgery is easily accomplished for the modified-Rabin scheme as it was for the original Rabin scheme (see Note 11.27(i)). One only needs to find an s, $1 \leq s \leq n - 1$, such that either s^2 or $n - s^2$ or $2s^2$ or $2(n - s^2) \bmod n$ is congruent to 6 modulo 16. In any of these cases, s is a valid signature for $m' = s^2 \bmod n$.

11.33 Note (*performance characteristics of the Rabin signature scheme*) Algorithm 11.25 requires a redundancy function from $\mathcal{M}$ to $\mathcal{M}_S = Q_n$ which typically involves computing a Jacobi symbol (Algorithm 2.149). Signature generation then involves computing at least one Jacobi symbol (see Note 11.27) and a square root modulo n. The square root computation is comparable to an exponentiation modulo n (see Algorithm 3.44). Since computing the Jacobi symbol is equivalent to a small number of modular multiplications, Rabin signature generation is not significantly more computationally intensive than an RSA signature generation with the same modulus size. Signature verification is very fast if $e = 2$; it requires only one modular multiplication. Squaring can be performed slightly more efficiently than a general modular multiplication (see Note 14.18). This, too, compares favorably with RSA signature verification even when the RSA public exponent is $e = 3$. The modified Rabin scheme (Algorithm 11.30) specifies the message space and redundancy function. Signature generation requires the evaluation of a Jacobi symbol and one modular exponentiation.

11.34 Note (*bandwidth efficiency*) The Rabin digital signature scheme is similar to the RSA scheme with respect to bandwidth efficiency (see §11.3.3(vi)).

11.3.5 ISO/IEC 9796 formatting

ISO/IEC 9796 was published in 1991 by the International Standards Organization as the first international standard for digital signatures. It specifies a digital signature process which uses a digital signature mechanism providing message recovery.

The main features of ISO/IEC 9796 are: (i) it is based on public-key cryptography; (ii) the particular signature algorithm is not specified but it must map k bits to k bits; (iii) it is used to sign messages of limited length and does not require a cryptographic hash function; (iv) it provides message recovery (see Note 11.14); and (v) it specifies the message padding, where required. Examples of mechanisms suitable for the standard are RSA (Algorithm 11.19) and modified-Rabin (Algorithm 11.30). The specific methods used for padding, redundancy, and truncation in ISO/IEC 9796 prevent various means to forge signatures. Table 11.3 provides notation for this subsection.

Symbol	Meaning
k	the bitlength of the signature.
d	the bitlength of the message m to be signed; it is required that $d \leq 8 \lfloor (k+3)/16 \rfloor$.
z	the number of bytes in the padded message; $z = \lceil d/8 \rceil$.
r	one more than the number of padding bits; $r = 8z - d + 1$.
t	the least integer such that a string of $2t$ bytes includes at least $k - 1$ bits; $t = \lceil (k-1)/16 \rceil$.

Table 11.3: *ISO/IEC 9796 notation.*

11.35 Example (*sample parameter values for ISO/IEC 9796*) The following table lists sample values of parameters in the signing process for a 150-bit message and a 1024-bit signature.

Parameter	k (bits)	d (bits)	z (bytes)	r (bits)	t (bytes)
Value	1024	150	19	3	64

□

(i) Signature process for ISO/IEC 9796

The signature process consists of 5 steps as per Figure 11.5(a).

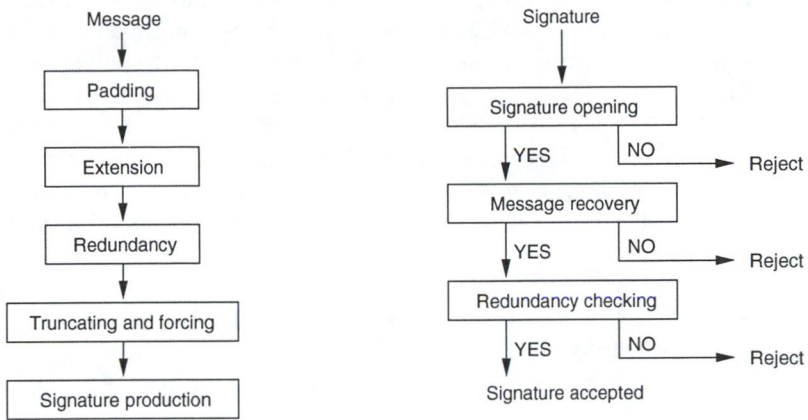

Figure 11.5: *Signature and verification processes for ISO/IEC 9796.*

1. *padding.* If m is the message, form the padded message $MP = 0^{r-1}\|m$ where $1 \leq r \leq 8$, such that the number of bits in MP is a multiple of 8. The number of bytes in MP is z: $MP = m_z\|m_{z-1}\| \cdots \|m_2\|m_1$ where each m_i is a byte.

2. *message extension.* The extended message, denoted ME, is obtained from MP by repeated concatenation on the left of MP with itself until t bytes are in the string: $ME = ME_t\|ME_{t-1}\| \cdots \|ME_2\|ME_1$ (each ME_i is a byte). If t is not a multiple of z, then the last bytes to be concatenated are a partial set of bytes from MP, where these bytes are consecutive bytes of MP from the right. More precisely, $ME_{i+1} = m_{(i \bmod z)+1}$ for $0 \leq i \leq t - 1$.

3. *message redundancy.* Redundancy is added to ME to get the byte string $MR = MR_{2t}\|MR_{2t-1}\| \cdots \|MR_2\|MR_1$ as follows. MR is obtained by interleaving the t bytes of ME with t redundant bytes and then adjusting byte MR_{2z} of the resulting string. More precisely, $MR_{2i-1} = ME_i$ and $MR_{2i} = S(ME_i)$ for $1 \leq i \leq t$, where $S(u)$ is called the *shadow function* of the byte u, and is defined as follows. If $u = u_2\|u_1$ where u_1 and u_2 are nibbles (strings of bitlength 4), then $S(u) = \pi(u_2)\|\pi(u_1)$ where π is the permutation

$$\pi = \begin{pmatrix} 0 & 1 & 2 & 3 & 4 & 5 & 6 & 7 & 8 & 9 & A & B & C & D & E & F \\ E & 3 & 5 & 8 & 9 & 4 & 2 & F & 0 & D & B & 6 & 7 & A & C & 1 \end{pmatrix}.$$

(For brevity, π is written with nibbles represented by hexadecimal characters.) Finally, MR is obtained by replacing MR_{2z} with $r \oplus MR_{2z}$.[5]

4. *truncation and forcing.* Form the k-bit intermediate integer IR from MR as follows:

   (a) to the least significant $k - 1$ bits of MR, append on the left a single bit 1;

   (b) modify the least significant byte $u_2\|u_1$ of the result, replacing it by $u_1\|0110$. (This is done to ensure that $IR \equiv 6 \pmod{16}$.)

---

[5]The purpose of MR_{2z} is to permit the verifier of a signature to recover the length d of the message. Since $d = 8z - r + 1$, it suffices to know z and r. These values can be deduced from MR.

5. *signature production.* A signature mechanism is used which maps k-bit integers to k-bit integers (and allows message recovery). IR is signed using this mechanism; let s denote the resulting signature.

11.36 Note *(RSA, Rabin)* ISO/IEC 9796 was intended for use with the RSA (Algorithm 11.19)[6] and Rabin (Algorithm 11.25)[7] digital signature mechanisms. For these particular schemes, signature production is stated more explicitly. Let e be the public exponent for the RSA or Rabin algorithms, n the modulus, and d the private exponent. First form the representative element RR which is: (i) IR if e is odd, or if e is even and the Jacobi symbol of IR (treated as an integer) with respect to the modulus n is 1; (ii) $IR/2$ if e is even and the Jacobi symbol of IR with respect to n is -1. The signature for m is $s = (RR)^d \bmod n$. ISO/IEC 9796 specifies that the signature s should be the lesser of $(RR)^d \bmod n$ and $n-((RR)^d \bmod n)$.

(ii) Verification process for ISO/IEC 9796

The verification process for an ISO/IEC 9796 digital signature can be separated into three stages, as per Figure 11.5(b).

1. *signature opening.* Let s be the signature. Then the following steps are performed.

    (a) Apply the public verification transformation to s to recover an integer IR'.

    (b) Reject the signature if IR' is not a string of k bits with the most significant bit being a 1, or if the least significant nibble does not have value 0110.

2. *message recovery.* A string MR' of $2t$ bytes is constructed from IR' by performing the following steps.

    (a) Let X be the least significant $k - 1$ bits of IR'.

    (b) If $u_4 \| u_3 \| u_2 \| 0110$ are the four least significant nibbles of X, replace the least significant byte of X by $\pi^{-1}(u_4) \| u_2$.

    (c) MR' is obtained by padding X with between 0 and 15 zero bits so that the resulting string has $2t$ bytes.

    The values z and r are computed as follows.

    (a) From the $2t$ bytes of MR', compute the t sums $MR'_{2i} \oplus S(MR'_{2i-1})$, $1 \leq i \leq t$. If all sums are 0, reject the signature.

    (b) Let z be the smallest value of i for which $MR'_{2i} \oplus S(MR'_{2i-1}) \neq 0$.

    (c) Let r be the least significant nibble of the sum found in step (b). Reject the signature if the hexadecimal value of r is not between 1 and 8.

    From MR', the z-byte string MP' is constructed as follows.

    (a) $MP'_i = MR'_{2i-1}$ for $1 \leq i \leq z$.

    (b) Reject the signature if the $r - 1$ most significant bits of MP' are not all 0's.

    (c) Let M' be the $8z - r + 1$ least significant bits of MP'.

3. *redundancy checking.* The signature s is verified as follows.

    (a) From M' construct a string MR'' by applying the message padding, message extension, and message redundancy steps of the signing process.

    (b) Accept the signature if and only if the $k - 1$ least significant bits of MR'' are equal to the $k - 1$ least significant bits of MR'.

---

[6]Since steps 1 through 4 of the signature process describe the redundancy function R, $\tilde{m}$ in step 1a of Algorithm 11.19 is taken to be IR.

[7]$\tilde{m}$ is taken to be IR in step 1 of Algorithm 11.25.

11.3.6 PKCS #1 formatting

Public-key cryptography standards (PKCS) are a suite of specifications which include techniques for RSA encryption and signatures (see §15.3.6). This subsection describes the digital signature process specified in PKCS #1 ("RSA Encryption Standard").

The digital signature mechanism in PKCS #1 does not use the message recovery feature of the RSA signature scheme. It requires a hashing function (either MD2, or MD5 — see Algorithm 9.51) and, therefore, is a digital signature scheme with appendix. Table 11.4 lists notation used in this subsection. Capital letters refer to octet strings. If X is an octet string, then X_i is octet i counting from the left.

Symbol	Meaning	Symbol	Meaning
k	the length of n in octets ($k \geq 11$)	EB	encryption block
n	the modulus, $2^{8(k-1)} \leq n < 2^{8k}$	ED	encrypted data
p, q	the prime factors of n	octet	a bitstring of length 8
e	the public exponent	ab	hexadecimal octet value
d	the private exponent	BT	block type
M	message	PS	padding string
MD	message digest	S	signature
MD$'$	comparative message digest	$\|X\|$	length of X in octets

Table 11.4: PKCS #1 notation.

(i) PKCS #1 data formatting

The data is an octet string D, where $\|D\| \leq k - 11$. BT is a single octet whose hexadecimal representation is either 00 or 01. PS is an octet string with $\|PS\| = k - 3 - \|D\|$. If BT $= 00$, then all octets in PS are 00; if BT $= 01$, then all octets in PS are ff. The formatted data block (called the *encryption block*) is EB $= 00\|BT\|PS\|00\|D$.

11.37 Note (*data formatting rationale*)

(i) The leading 00 block ensures that the octet string EB, when interpreted as an integer, is less than the modulus n.

(ii) If the block type is BT $= 00$, then either D must begin with a non-zero octet or its length must be known, in order to permit unambiguous parsing of EB.

(iii) If BT $= 01$, then unambiguous parsing is always possible.

(iv) For the reason given in (iii), and to thwart certain potential attacks on the signature mechanism, BT $= 01$ is recommended.

11.38 Example (*PKCS #1 data formatting for particular values*) Suppose that n is a 1024-bit modulus (so $k = 128$). If $\|D\| = 20$ octets, then $\|PS\| = 105$ octets, and $\|EB\| = 128$ octets. $\square$

(ii) Signature process for PKCS #1

The signature process involves the steps as per Figure 11.6(a).

The input to the signature process is the message M, and the signer's private exponent d and modulus n.

1. *message hashing.* Hash the message M using the selected message-digest algorithm to get the octet string MD.

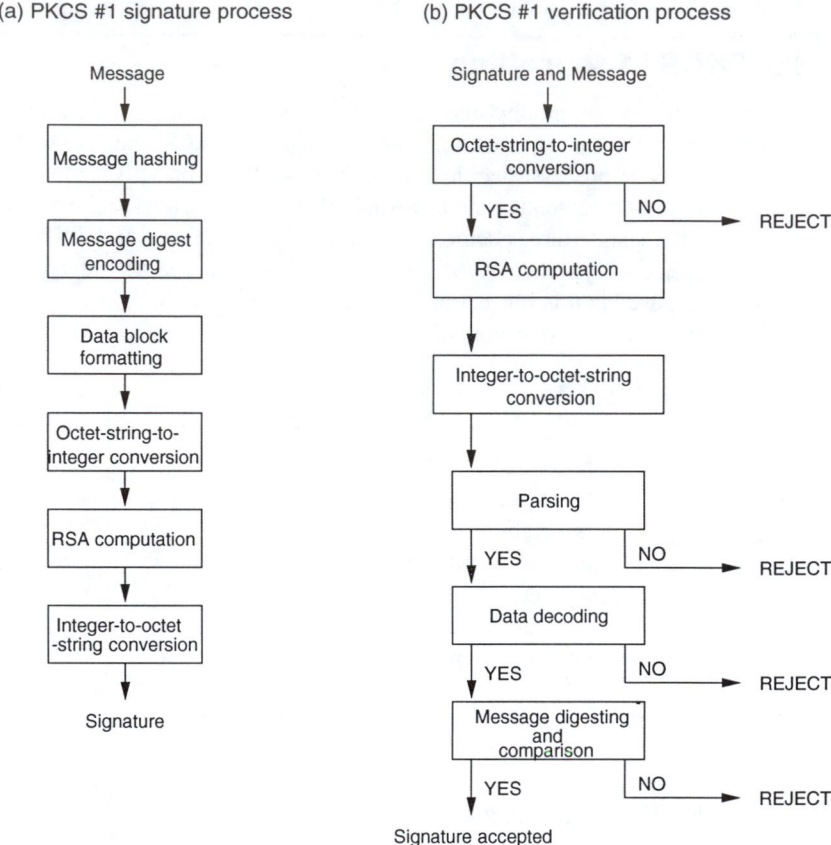

Figure 11.6: *Signature and verification processes for PKCS #1.*

2. *message digest encoding.* MD and the hash algorithm identifier are combined into an ASN.1 (*abstract syntax notation*) value and then BER-encoded (*basic encoding rules*) to give an octet data string D.

3. *data block formatting.* With data string input D, use the data formatting from §11.3.6(i) to form octet string EB.

4. *octet-string-to-integer conversion.* Let the octets of EB be $EB_1 \| EB_2 \| \cdots \| EB_k$. Define $\widetilde{EB}_i$ to be the integer whose binary representation is the octet EB_i (least significant bit is on the right). The integer representing EB is $m = \sum_{i=1}^{k} 2^{8(k-i)} \widetilde{EB}_i$. [8]

5. *RSA computation.* Compute $s = m^d \bmod n$.

6. *integer-to-octet-string conversion.* Convert s to an octet string $ED = ED_1 \| ED_2 \| \cdots \| ED_k$, where the octets ED_i satisfy $s = \sum_{i=1}^{k} 2^{8(k-i)} \widetilde{ED}_i$. The signature is S = ED.

(iii) Verification process for PKCS #1

The verification process involves the steps as per Figure 11.6(b). The input to the verification process is the message M, the signature S, the public exponent e, and modulus n.

1. *octet-string-to-integer conversion.*

    (a) Reject S if the bitlength of S is not a multiple of 8.

---

[8]Since $EB_1 = 00$ and $n \geq 2^{8(k-1)}$, then $0 \leq m < n$.

    (b) Convert S to an integer s as in step 4 of the signature process.

    (c) Reject the signature if $s > n$.

2. *RSA computation.* Compute $m = s^e \bmod n$.

3. *integer-to-octet-string conversion.* Convert m to an octet string EB of length k octets as in step 6 of the signature process.

4. *parsing.* Parse EB into a block type BT, a padding string PS, and the data D.

    (a) Reject the signature if EB cannot be parsed unambiguously.

    (b) Reject the signature if BT is not one of 00 or 01.

    (c) Reject the signature if PS consists of < 8 octets or is inconsistent with BT.

5. *data decoding.*

    (a) BER-decode D to get a message digest MD and a hash algorithm identifier.

    (b) Reject the signature if the hashing algorithm identifier does not identify one of MD2 or MD5.

6. *message digesting and comparison.*

    (a) Hash the message M with the selected message-digest algorithm to get MD'.

    (b) Accept the signature S on M if and only if $MD' = MD$.

11.4 Fiat-Shamir signature schemes

As described in Note 10.30, any identification scheme involving a witness-challenge response sequence can be converted to a signature scheme by replacing the random challenge of the verifier with a one-way hash function. This section describes two signature mechanisms which arise in this way. The basis for this methodology is the Fiat-Shamir identification protocol (Protocol 10.24).

11.4.1 Feige-Fiat-Shamir signature scheme

The Feige-Fiat-Shamir signature scheme is a modification of an earlier signature scheme of Fiat and Shamir, and requires a one-way hash function $h\colon \{0,1\}^* \longrightarrow \{0,1\}^k$ for some fixed positive integer k. Here $\{0,1\}^k$ denotes the set of bitstrings of bitlength k, and $\{0,1\}^*$ denotes the set of all bitstrings (of arbitrary bitlengths). The method provides a digital signature with appendix, and is a randomized mechanism.

11.39 Algorithm Key generation for the Feige-Fiat-Shamir signature scheme

---

SUMMARY: each entity creates a public key and corresponding private key.

Each entity A should do the following:

1. Generate random distinct secret primes p, q and form $n = pq$.

2. Select a positive integer k and distinct random integers $s_1, s_2, \ldots, s_k \in \mathbb{Z}_n^*$.

3. Compute $v_j = s_j^{-2} \bmod n$, $1 \le j \le k$.

4. A's public key is the k-tuple $(v_1, v_2, \ldots, v_k)$ and the modulus n; A's private key is the k-tuple $(s_1, s_2, \ldots, s_k)$.

11.40 Algorithm Feige-Fiat-Shamir signature generation and verification

SUMMARY: entity A signs a binary message m of arbitrary length. Any entity B can verify this signature by using A's public key.

1. *Signature generation.* Entity A should do the following:
   (a) Select a random integer r, $1 \le r \le n-1$.
   (b) Compute $u = r^2 \bmod n$.
   (c) Compute $e = (e_1, e_2, \dots, e_k) = h(m\|u)$; each $e_i \in \{0,1\}$.
   (d) Compute $s = r \cdot \prod_{j=1}^{k} s_j^{e_j} \bmod n$.
   (e) A's signature for m is (e, s).

2. *Verification.* To verify A's signature (e, s) on m, B should do the following:
   (a) Obtain A's authentic public key $(v_1, v_2, \dots, v_k)$ and n.
   (b) Compute $w = s^2 \cdot \prod_{j=1}^{k} v_j^{e_j} \bmod n$.
   (c) Compute $e' = h(m\|w)$.
   (d) Accept the signature if and only if $e = e'$.

Proof that signature verification works.

$$w \equiv s^2 \cdot \prod_{j=1}^{k} v_j^{e_j} \equiv r^2 \cdot \prod_{j=1}^{k} s_j^{2e_j} \prod_{j=1}^{k} v_j^{e_j} \equiv r^2 \cdot \prod_{j=1}^{k} (s_j^2 v_j)^{e_j} \equiv r^2 \equiv u \pmod{n}.$$

Hence, $w = u$ and therefore $e = e'$.

11.41 Example (*Feige-Fiat-Shamir signature generation with artificially small parameters*)
Key generation. Entity A generates primes $p = 3571$, $q = 4523$, and computes $n = pq = 16151633$. The following table displays the selection of s_j (A's private key) and integers v_j (A's public key) along with intermediate values s_j^{-1}.

j	1	2	3	4	5
s_j	42	73	85	101	150
$s_j^{-1} \bmod n$	4999315	885021	6270634	13113207	11090788
$v_j = s_j^{-2} \bmod n$	503594	4879739	7104483	1409171	6965302

Signature generation. Suppose $h\colon \{0,1\}^* \longrightarrow \{0,1\}^5$ is a hash function. A selects a random integer $r = 23181$ and computes $u = r^2 \bmod n = 4354872$. To sign message m, A evaluates $e = h(m\|u) = 10110$ (the hash value has been contrived for this example). A forms $s = rs_1s_3s_4 \bmod n = (23181)(42)(85)(101) \bmod n = 7978909$; the signature for m is $(e = 10110, s = 7978909)$.
Signature verification. B computes $s^2 \bmod n = 2926875$ and $v_1v_3v_4 \bmod n = (503594)(7104483)(1409171) \bmod n = 15668174$. B then computes $w = s^2v_1v_3v_4 \bmod n = 4354872$. Since $w = u$, it follows that $e' = h(m\|w) = h(m\|u) = e$ and, hence, B accepts the signature. □

11.42 Note (*security of Feige-Fiat-Shamir signature scheme*)
   (i) Unlike the RSA signature scheme (Algorithm 11.19), all entities may use the same modulus n (cf. §8.2.2(vi)). In this scenario, a trusted third party (TTP) would need to generate the primes p and q and also public and private keys for each entity.

(ii) The security of the Feige-Fiat-Shamir scheme is based on the intractability of computing square roots modulo n (see §3.5.2). It has been proven to be secure against an adaptive chosen-message attack, provided that factoring is intractable, h is a random function, and the s_i's are distinct.

11.43 Note (*parameter selection and key storage requirements*) If n is a t-bit integer, the private key constructed in Algorithm 11.39 is kt bits in size. This may be reduced by selecting the random values s_j, $1 \leq j \leq k$, as numbers of bitlength $t' < t$; t', however, should not be chosen so small that guessing the s_j is feasible. The public key is $(k+1)t$ bits in size. For example, if $t = 768$ and $k = 128$, then the private key requires 98304 bits and the public key requires 99072 bits.

11.44 Note (*identity-based Feige-Fiat-Shamir signatures*) Suppose a TTP constructs primes p and q and modulus n; the modulus is common to all entities in the system. Algorithm 11.39 can be modified so that the scheme is identity-based. Entity A's bitstring I_A contains information which identifies A. The TTP computes $v_j = f(I_A \| j)$, $1 \leq j \leq k$, where f is a one-way hash function from $\{0, 1\}^*$ to Q_n and j is represented in binary, and computes a square root s_j of v_j^{-1} modulo n, $1 \leq j \leq k$. A's public key is simply the identity information I_A, while A's private key (transported securely and secretly by the TTP to A) is the k-tuple $(s_1, s_2, \ldots, s_k)$. The functions h, f, and the modulus n are system-wide quantities.

This procedure has the advantage that the public key generated in Algorithm 11.39 might be generated from a smaller quantity I_A, potentially reducing the storage and transmission cost. It has the disadvantages that the private keys of entities are known to the TTP, and the modulus n is system-wide, making it a more attractive target.

11.45 Note (*small prime variation of Feige-Fiat-Shamir signatures*) This improvement aims to reduce the size of the public key and increase the efficiency of signature verification. Unlike the modification described in Note 11.44, each entity A generates its own modulus n_A and a set of k small primes $v_1, v_2, \ldots, v_k \in Q_n$ (each prime will require around 2 bytes to represent). Entity A selects one of the square roots s_j of v_j^{-1} modulo n for each j, $1 \leq j \leq k$; these form the private key. The public key consists of n_A and the values $v_1, v_2, \ldots, v_k$. Verification of signatures proceeds more efficiently since computations are done with much smaller numbers.

11.46 Note (*performance characteristics of Feige-Fiat-Shamir signatures*) With the RSA scheme and a modulus of length $t = 768$, signature generation using naive techniques requires, on average, 1152 modular multiplications (more precisely, 768 squarings and 384 multiplications). Signature generation for the Feige-Fiat-Shamir scheme (Algorithm 11.40) requires, on average, $k/2$ modular multiplications. To sign a message with this scheme, a modulus of length $t = 768$ and $k = 128$ requires, on average, 64 modular multiplications, or less than 6% of the work required by a naive implementation of RSA. Signature verification requires only one modular multiplication for RSA if the public exponent is $e = 3$, and 64 modular multiplications, on average, for Feige-Fiat-Shamir. For applications where signature generation must be performed quickly and key space storage is not limited, the Feige-Fiat-Shamir scheme (or DSA-like schemes — see §11.5) may be preferable to RSA.

11.4.2 GQ signature scheme

The Guillou-Quisquater (GQ) identification protocol (§10.4.3) can be turned into a digital signature mechanism (Algorithm 11.48) if the challenge is replaced with a one-way hash function. Let $h\colon \{0,1\}^* \longrightarrow \mathbb{Z}_n$ be a hash function where n is a positive integer.

11.47 Algorithm Key generation for the GQ signature scheme

SUMMARY: each entity creates a public key (n, e, J_A) and corresponding private key a.
Entity A should do the following:

1. Select random distinct secret primes p, q and form $n = pq$.
2. Select an integer $e \in \{1, 2, \dots, n-1\}$ such that $\gcd(e, (p-1)(q-1)) = 1$. (See Note 11.50 for guidance on selecting e.)
3. Select an integer J_A, $1 < J_A < n$, which serves as an identifier for A and such that $\gcd(J_A, n) = 1$. (The binary representation of J_A could be used to convey information about A such as name, address, driver's license number, etc.)
4. Determine an integer $a \in \mathbb{Z}_n$ such that $J_A a^e \equiv 1 \pmod{n}$ as follows:
   4.1 Compute $J_A^{-1} \bmod n$.
   4.2 Compute $d_1 = e^{-1} \bmod (p-1)$ and $d_2 = e^{-1} \bmod (q-1)$.
   4.3 Compute $a_1 = (J_A^{-1})^{d_1} \bmod p$ and $a_2 = (J_A^{-1})^{d_2} \bmod q$.
   4.4 Find a solution a to the simultaneous congruences $a \equiv a_1 \pmod{p}$, $a \equiv a_2 \pmod{q}$.
5. A's public key is (n, e, J_A); A's private key is a.

11.48 Algorithm GQ signature generation and verification

SUMMARY: entity A signs a binary message m of arbitrary length. Any entity B can verify this signature by using A's public key.

1. *Signature generation.* Entity A should do the following:
   (a) Select a random integer k and compute $r = k^e \bmod n$.
   (b) Compute $l = h(m\|r)$.
   (c) Compute $s = ka^l \bmod n$.
   (d) A's signature for m is the pair (s, l).
2. *Verification.* To verify A's signature (s, l) on m, B should do the following:
   (a) Obtain A's authentic public key (n, e, J_A).
   (b) Compute $u = s^e J_A{}^l \bmod n$ and $l' = h(m\|u)$.
   (c) Accept the signature if and only if $l = l'$.

Proof that signature verification works. Note that $u \equiv s^e J_A{}^l \equiv (ka^l)^e J_A{}^l \equiv k^e (a^e J_A)^l \equiv k^e \equiv r \pmod{n}$. Hence, $u = r$ and therefore $l = l'$.

11.49 Example (*GQ signature generation with artificially small parameters*)
Key generation. Entity A chooses primes $p = 20849$, $q = 27457$, and computes $n = pq = 572450993$. A selects an integer $e = 47$, an identifier $J_A = 1091522$, and solves the congruence $J_A a^e \equiv 1 \pmod{n}$ to get $a = 214611724$. A's public key is ($n = 572450993$, $e = 47$, $J_A = 1091522$), while A's private key is $a = 214611724$.
Signature generation. To sign the message $m = 1101110001$, A selects a random integer

$k = 42134$ and computes $r = k^e \bmod n = 297543350$. A then computes $l = h(m\|r) = 2713833$ (the hash value has been contrived for this example) and $s = ka^l \bmod n = (42134)214611724^{2713833} \bmod n = 252000854$. A's signature for m is the pair ($s = 252000854, l = 2713833$).

Signature verification. B computes $s^e \bmod n = 252000854^{47} \bmod n = 398641962$, $J_A{}^l \bmod n = 1091522^{2713833} \bmod n = 110523867$, and finally $u = s^e J_A{}^l \bmod n = 297543350$. Since $u = r$, $l' = h(m\|u) = h(m\|r) = l$, and so B accepts the signature. $\square$

11.50 Note (*security of GQ signature scheme*) In Algorithm 11.47, e must be sufficiently large to exclude the possibility of forgery based on the birthday paradox (see §2.1.5). The potential attack proceeds along the following lines. The adversary selects a message m and computes $l = h(m\|J_A{}^t)$ for sufficiently many values of t until $l \equiv t \pmod{e}$; this is expected to occur within $O(\sqrt{e})$ trials. Having determined such a pair (l, t), the adversary determines an integer x such that $t = xe + l$ and computes $s = J_A{}^x \bmod n$. Observe that $s^e J_A{}^l \equiv (J_A{}^x)^e J_A{}^l \equiv J_A{}^{xe+l} \equiv J_A{}^t \pmod{n}$, and, hence, $h(m\|J_A{}^t) = l$. Thus, (s, l) is a valid (forged) signature for message m.

11.51 Note (*parameter selection*) Current methods (as of 1996) for integer factorization suggest that a modulus n of size at least 768 bits is prudent. Note 11.50 suggests that e should be at least 128 bits in size. Typical values for the outputs of secure hash functions are 128 or 160 bits. With a 768-bit modulus and a 128-bit e, the public key for the GQ scheme is $896 + u$ bits in size, where u is the number of bits needed to represent J_A. The private key a is 768 bits in size.

11.52 Note (*performance characteristics of GQ signatures*) Signature generation for GQ (Algorithm 11.48) requires two modular exponentiations and one modular multiplication. Using a 768-bit modulus n, a 128-bit value e, and a hash function with a 128-bit output l, signature generation (using naive techniques for exponentiation) requires on average 384 modular multiplications (128 squarings and 64 multiplications for each of e and l). Signature verification requires a similar amount of work. Compare this with RSA (naively 1152 modular multiplications) and Feige-Fiat-Shamir (64 modular multiplications) for signature generation (see Note 11.46). GQ is computationally more intensive than Feige-Fiat-Shamir but requires significantly smaller key storage space (see Note 11.51).

11.53 Note (*message recovery variant of GQ signatures*) Algorithm 11.48 can be modified as follows to provide message recovery. Let the signing space be $\mathcal{M}_\mathcal{S} = \mathbb{Z}_n$, and let $m \in \mathcal{M}_\mathcal{S}$. In signature generation, select a random k such that $\gcd(k, n) = 1$ and compute $r = k^e \bmod n$ and $l = mr \bmod n$. The signature is $s = ka^l \bmod n$. Verification gives $s^e J_A{}^l \equiv k^e a^{el} J_A{}^l \equiv k^e \equiv r \pmod{n}$. Message m is recovered from $lr^{-1} \bmod n$. As for all digital signature schemes with message recovery, a suitable redundancy function R is required to guard against existential forgery.

11.5 The DSA and related signature schemes

This section presents the Digital Signature Algorithm (DSA) and several related signature schemes. Most of these are presented over $\mathbb{Z}_p^*$ for some large prime p, but all of these mechanisms can be generalized to any finite cyclic group; this is illustrated explicitly for the El-

Gamal signature scheme in §11.5.2. All of the methods discussed in this section are randomized digital signature schemes (see Definition 11.2). All give digital signatures with appendix and can be modified to provide digital signatures with message recovery (see Note 11.14). A necessary condition for the security of all of the signature schemes described in this section is that computing logarithms in $\mathbb{Z}_p^*$ be computationally infeasible. This condition, however, is not necessarily sufficient for the security of these schemes; analogously, it remains unproven that RSA signatures are secure even if factoring integers is hard.

11.5.1 The Digital Signature Algorithm (DSA)

In August of 1991, the U.S. National Institute of Standards and Technology (NIST) proposed a digital signature algorithm (DSA). The DSA has become a U.S. Federal Information Processing Standard (FIPS 186) called the *Digital Signature Standard* (DSS), and is the first digital signature scheme recognized by any government. The algorithm is a variant of the ElGamal scheme (§11.5.2), and is a digital signature scheme with appendix.

The signature mechanism requires a hash function $h\colon \{0,1\}^* \longrightarrow \mathbb{Z}_q$ for some integer q. The DSS explicitly requires use of the Secure Hash Algorithm (SHA-1), given by Algorithm 9.53.

11.54 Algorithm Key generation for the DSA

SUMMARY: each entity creates a public key and corresponding private key.
Each entity A should do the following:

1. Select a prime number q such that $2^{159} < q < 2^{160}$.
2. Choose t so that $0 \le t \le 8$, and select a prime number p where $2^{511+64t} < p < 2^{512+64t}$, with the property that q divides $(p-1)$.
3. (Select a generator α of the unique cyclic group of order q in $\mathbb{Z}_p^*$.)

   3.1 Select an element $g \in \mathbb{Z}_p^*$ and compute $\alpha = g^{(p-1)/q} \bmod p$.
   3.2 If $\alpha = 1$ then go to step 3.1.

4. Select a random integer a such that $1 \le a \le q-1$.
5. Compute $y = \alpha^a \bmod p$.
6. A's public key is (p, q, α, y); A's private key is a.

11.55 Note (*generation of DSA primes p and q*) In Algorithm 11.54 one must select the prime q first and then try to find a prime p such that q divides $(p-1)$. The algorithm recommended by the DSS for accomplishing this is Algorithm 4.56.

11.56 Algorithm DSA signature generation and verification

SUMMARY: entity A signs a binary message m of arbitrary length. Any entity B can verify this signature by using A's public key.

1. *Signature generation.* Entity A should do the following:

   (a) Select a random secret integer k, $0 < k < q$.
   (b) Compute $r = (\alpha^k \bmod p) \bmod q$ (e.g., using Algorithm 2.143).
   (c) Compute $k^{-1} \bmod q$ (e.g., using Algorithm 2.142).
   (d) Compute $s = k^{-1}\{h(m) + ar\} \bmod q$.
   (e) A's signature for m is the pair (r, s).

2. *Verification.* To verify A's signature (r, s) on m, B should do the following:

    (a) Obtain A's authentic public key (p, q, α, y).

    (b) Verify that $0 < r < q$ and $0 < s < q$; if not, then reject the signature.

    (c) Compute $w = s^{-1} \bmod q$ and $h(m)$.

    (d) Compute $u_1 = w \cdot h(m) \bmod q$ and $u_2 = rw \bmod q$.

    (e) Compute $v = (\alpha^{u_1} y^{u_2} \bmod p) \bmod q$.

    (f) Accept the signature if and only if $v = r$.

Proof that signature verification works. If (r, s) is a legitimate signature of entity A on message m, then $h(m) \equiv -ar + ks \pmod{q}$ must hold. Multiplying both sides of this congruence by w and rearranging gives $w \cdot h(m) + arw \equiv k \pmod{q}$. But this is simply $u_1 + au_2 \equiv k \pmod{q}$. Raising α to both sides of this equation yields $(\alpha^{u_1} y^{u_2} \bmod p) \bmod q = (\alpha^k \bmod p) \bmod q$. Hence, $v = r$, as required.

11.57 Example (*DSA signature generation with artificially small parameters*)
Key generation. A selects primes $p = 124540019$ and $q = 17389$ such that q divides $(p - 1)$; here, $(p - 1)/q = 7162$. A selects a random element $g = 110217528 \in \mathbb{Z}_p^*$ and computes $\alpha = g^{7162} \bmod p = 10083255$. Since $\alpha \neq 1$, α is a generator for the unique cyclic subgroup of order q in $\mathbb{Z}_p^*$. A next selects a random integer $a = 12496$ satisfying $1 \leq a \leq q - 1$, and computes $y = \alpha^a \bmod p = 10083255^{12496} \bmod 124540019 = 119946265$. A's public key is $(p = 124540019, q = 17389, \alpha = 10083255, y = 119946265)$, while A's private key is $a = 12496$.
Signature generation. To sign m, A selects a random integer $k = 9557$, and computes $r = (\alpha^k \bmod p) \bmod q = (10083255^{9557} \bmod 124540019) \bmod 17389 = 27039929 \bmod 17389 = 34$. A then computes $k^{-1} \bmod q = 7631$, $h(m) = 5246$ (the hash value has been contrived for this example), and finally $s = (7631)\{5246 + (12496)(34)\} \bmod q = 13049$. The signature for m is the pair $(r = 34, s = 13049)$.
Signature verification. B computes $w = s^{-1} \bmod q = 1799$, $u_1 = w \cdot h(m) \bmod q = (5246)(1799) \bmod 17389 = 12716$, and $u_2 = rw \bmod q = (34)(1799) \bmod 17389 = 8999$. B then computes $v = (\alpha^{u_1} y^{u_2} \bmod p) \bmod q = (10083255^{12716} \cdot 119946265^{8999} \bmod 124540019) \bmod 17389 = 27039929 \bmod 17389 = 34$. Since $v = r$, B accepts the signature. $\square$

11.58 Note (*security of DSA*) The security of the DSA relies on two distinct but related discrete logarithm problems. One is the logarithm problem in $\mathbb{Z}_p^*$ where the powerful index-calculus methods apply; the other is the logarithm problem in the cyclic subgroup of order q, where the best current methods run in "square-root" time. For further discussion, see §3.6.6. Since the DSA is a special case of ElGamal signatures (§11.5.2) with respect to the equation for s, security considerations for the latter are pertinent here (see Note 11.66).

11.59 Note (*recommended parameter sizes*) The size of q is fixed by Algorithm 11.54 (as per FIPS 186) at 160 bits, while the size of p can be any multiple of 64 between 512 and 1024 bits inclusive. A 512-bit prime p provides marginal security against a concerted attack. As of 1996, a modulus of at least 768 bits is recommended. FIPS 186 does not permit primes p larger than 1024 bits.

11.60 Note (*performance characteristics of the DSA*) For concreteness, suppose p is a 768-bit integer. Signature generation requires one modular exponentiation, taking on average (using naive techniques for exponentiation) 240 modular multiplications, one modular inverse with a 160-bit modulus, two 160-bit modular multiplications, and one addition. The 160-bit operations are relatively minor compared to the exponentiation. The DSA has the advantage that the exponentiation can be precomputed and need not be done at the time of signature generation. By comparison, no precomputation is possible with the RSA signature scheme. The major portion of the work for signature verification is two exponentiations modulo p, each to 160-bit exponents. On average, these each require 240 modular multiplications or 480 in total. Some savings can be realized by doing the two exponentiations simultaneously (cf. Note 14.91); the cost, on average, is then 280 modular multiplications.

11.61 Note (*system-wide parameters*) It is not necessary for each entity to select its own primes p and q. The DSS permits p, q, and α to be system-wide parameters. This does, however, present a more attractive target for an adversary.

11.62 Note (*probability of failure*) Verification requires the computation of $s^{-1} \bmod q$. If $s = 0$, then s^{-1} does not exist. To avoid this situation, the signer may check that $s \neq 0$; but if s is assumed to be a random element in $\mathbb{Z}_q$, then the probability that $s = 0$ is $(\frac{1}{2})^{160}$. In practice, this is extremely unlikely ever to occur. The signer may also check that $r \neq 0$. If the signer detects that either $r = 0$ or $s = 0$, a new value of k should be generated.

11.5.2 The ElGamal signature scheme

The ElGamal signature scheme is a randomized signature mechanism. It generates digital signatures with appendix on binary messages of arbitrary length, and requires a hash function $h\colon \{0,1\}^* \longrightarrow \mathbb{Z}_p$ where p is a large prime number. The DSA (§11.5.1) is a variant of the ElGamal signature mechanism.

11.63 Algorithm Key generation for the ElGamal signature scheme

SUMMARY: each entity creates a public key and corresponding private key.
Each entity A should do the following:

1. Generate a large random prime p and a generator α of the multiplicative group $\mathbb{Z}_p^*$ (using Algorithm 4.84).
2. Select a random integer a, $1 \leq a \leq p - 2$.
3. Compute $y = \alpha^a \bmod p$ (e.g., using Algorithm 2.143).
4. A's public key is (p, α, y); A's private key is a.

11.64 Algorithm ElGamal signature generation and verification

SUMMARY: entity A signs a binary message m of arbitrary length. Any entity B can verify this signature by using A's public key.

1. *Signature generation.* Entity A should do the following:
    (a) Select a random secret integer k, $1 \leq k \leq p - 2$, with $\gcd(k, p - 1) = 1$.
    (b) Compute $r = \alpha^k \bmod p$ (e.g., using Algorithm 2.143).
    (c) Compute $k^{-1} \bmod (p - 1)$ (e.g., using Algorithm 2.142).
    (d) Compute $s = k^{-1}\{h(m) - ar\} \bmod (p - 1)$.
    (e) A's signature for m is the pair (r, s).
2. *Verification.* To verify A's signature (r, s) on m, B should do the following:

(a) Obtain A's authentic public key (p, α, y).

(b) Verify that $1 \leq r \leq p - 1$; if not, then reject the signature.

(c) Compute $v_1 = y^r r^s \bmod p$.

(d) Compute $h(m)$ and $v_2 = \alpha^{h(m)} \bmod p$.

(e) Accept the signature if and only if $v_1 = v_2$.

Proof that signature verification works. If the signature was generated by A, then $s \equiv k^{-1}\{h(m) - ar\} \pmod{p-1}$. Multiplying both sides by k gives $ks \equiv h(m) - ar \pmod{p-1}$, and rearranging yields $h(m) \equiv ar + ks \pmod{p - 1}$. This implies $\alpha^{h(m)} \equiv \alpha^{ar+ks} \equiv (\alpha^a)^r r^s \pmod{p}$. Thus, $v_1 = v_2$, as required.

11.65 Example (*ElGamal signature generation with artificially small parameters*)

Key generation. A selects the prime $p = 2357$ and a generator $\alpha = 2$ of $\mathbb{Z}_{2357}^*$. A chooses the private key $a = 1751$ and computes $y = \alpha^a \bmod p = 2^{1751} \bmod 2357 = 1185$. A's public key is $(p = 2357, \alpha = 2, y = 1185)$.

Signature generation. For simplicity, messages will be integers from $\mathbb{Z}_p$ and $h(m) = m$ (i.e., for this example only, take h to be the identity function). To sign the message $m = 1463$, A selects a random integer $k = 1529$, computes $r = \alpha^k \bmod p = 2^{1529} \bmod 2357 = 1490$, and $k^{-1} \bmod (p-1) = 245$. Finally, A computes $s = 245\{1463 - 1751(1490)\} \bmod 2356 = 1777$. A's signature for $m = 1463$ is the pair $(r = 1490, s = 1777)$.

Signature verification. B computes $v_1 = 1185^{1490} \cdot 1490^{1777} \bmod 2357 = 1072$, $h(m) = 1463$, and $v_2 = 2^{1463} \bmod 2357 = 1072$. B accepts the signature since $v_1 = v_2$.   □

11.66 Note (*security of ElGamal signatures*)

(i) An adversary might attempt to forge A's signature (per Algorithm 11.64) on m by selecting a random integer k and computing $r = \alpha^k \bmod p$. The adversary must then determine $s = k^{-1}\{h(m) - ar\} \bmod (p - 1)$. If the discrete logarithm problem is computationally infeasible, the adversary can do no better than to choose an s at random; the success probability is only $\frac{1}{p}$, which is negligible for large p.

(ii) A different k must be selected for each message signed; otherwise, the private key can be determined with high probability as follows. Suppose $s_1 = k^{-1}\{h(m_1) - ar\} \bmod (p - 1)$ and $s_2 = k^{-1}\{h(m_2) - ar\} \bmod (p - 1)$. Then $(s_1 - s_2)k \equiv (h(m_1) - h(m_2)) \pmod{p - 1}$. If $s_1 - s_2 \not\equiv 0 \pmod{p - 1}$, then $k = (s_1 - s_2)^{-1}(h(m_1) - h(m_2)) \bmod (p - 1)$. Once k is known, a is easily found.

(iii) If no hash function h is used, the signing equation is $s = k^{-1}\{m - ar\} \bmod (p - 1)$. It is then easy for an adversary to mount an existential forgery attack as follows. Select any pair of integers (u, v) with $\gcd(v, p - 1) = 1$. Compute $r = \alpha^u y^v \bmod p = \alpha^{u+av} \bmod p$ and $s = -rv^{-1} \bmod (p - 1)$. The pair (r, s) is a valid signature for the message $m = su \bmod (p - 1)$, since $(\alpha^m \alpha^{-ar})^{s^{-1}} = \alpha^u y^v = r$.

(iv) Step 2b in Algorithm 11.64 requires the verifier to check that $0 < r < p$. If this check is not done, then an adversary can sign messages of its choice provided it has one valid signature created by entity A, as follows. Suppose that (r, s) is a signature for message m produced by A. The adversary selects a message m' of its choice and computes $h(m')$ and $u = h(m') \cdot [h(m)]^{-1} \bmod (p-1)$ (assuming $[h(m)]^{-1} \bmod (p-1)$ exists). It then computes $s' = su \bmod (p-1)$ and r' such that $r' \equiv ru \pmod{p-1}$ and $r' \equiv r \pmod{p}$. The latter is always possible by the Chinese Remainder Theorem (Fact 2.120). The pair (r', s') is a signature for message m' which would be accepted by the verification algorithm (Algorithm 11.64) if step 2b were ignored.

11.67 Note (*security based on parameter selection*)

(i) (*index-calculus attack*) The prime p should be sufficiently large to prevent efficient use of the index-calculus methods (§3.6.5).

(ii) (*Pohlig-Hellman attack*) $p - 1$ should be divisible by a prime number q sufficiently large to prevent a Pohlig-Hellman discrete logarithm attack (§3.6.4).

(iii) (*weak generators*) Suppose the generator α satisfies the following conditions:

(a) α divides $(p - 1)$; and

(b) computing logarithms in the subgroup S of order α in $\mathbb{Z}_p^*$ can be efficiently done (for example, if a Pohlig-Hellman attack (§3.6.4) can be mounted in S).

It is then possible for an adversary to construct signatures (without knowledge of A's private key) which will be accepted by the verification algorithm (step 2 of Algorithm 11.64). To see this, suppose that $p - 1 = \alpha q$. To sign a message m the adversary does the following:

(a) Compute $t = (p - 3)/2$ and set $r = q$.

(b) Determine z such that $\alpha^{qz} \equiv y^q \pmod{p}$ where y is A's public key. (This is possible since α^q and y^q are elements of S and α^q is a generator of S.)

(c) Compute $s = t \cdot \{h(m) - qz\} \bmod (p - 1)$.

(d) (r, s) is a signature on m which will be accepted by step 2 of Algorithm 11.64.

This attack works because the verification equation $r^s y^r \equiv \alpha^{h(m)} \pmod{p}$ is satisfied. To see this, first observe that $\alpha q \equiv -1 \pmod{p}$, $\alpha \equiv -q^{-1} \pmod{p}$, and that $q^{(p-1)/2} \equiv -1 \pmod{p}$. (The latter congruence follows from the fact that α is a generator of $\mathbb{Z}_p^*$ and $q \equiv -\alpha^{-1} \pmod{p}$.) From these, one deduces that $q^t = q^{(p-1)/2} q^{-1} \equiv -q^{-1} \equiv \alpha \pmod{p}$. Now $r^s y^r = (q^t)^{[h(m)-qz]} y^q \equiv \alpha^{h(m)} \alpha^{-qz} y^q \equiv \alpha^{h(m)} y^{-q} y^q = \alpha^{h(m)} \pmod{p}$. Notice in the case where $\alpha = 2$ is a generator that the conditions specified in (iii) above are trivially satisfied.

The attack of Note 11.67(iii) can be avoided if α is selected as a generator for a subgroup of $\mathbb{Z}_p^*$ of prime order rather than a generator for $\mathbb{Z}_p^*$ itself.

11.68 Note (*performance characteristics of ElGamal signatures*)

(i) Signature generation by Algorithm 11.64 is relatively fast, requiring one modular exponentiation ($\alpha^k \bmod p$), the extended Euclidean algorithm (for computing $k^{-1} \bmod (p - 1)$), and two modular multiplications. (Modular subtraction is negligible when compared with modular multiplication.) The exponentiation and application of the extended Euclidean algorithm can be done off-line, in which case signature generation (in instances where precomputation is possible) requires only two (on-line) modular multiplications.

(ii) Signature verification is more costly, requiring three exponentiations. Each exponentiation (using naive techniques) requires $\frac{3}{2}\lceil \lg p \rceil$ modular multiplications, on average, for a total cost of $\frac{9}{2}\lceil \lg p \rceil$ multiplications. The computing costs can be reduced by modifying the verification slightly. Compute $v_1 = \alpha^{-h(m)} y^r r^s \bmod p$, and accept the signature as valid if and only if $v_1 = 1$. Now, v_1 can be computed more efficiently by doing the three exponentiations simultaneously (see Note 14.91); the total cost is now about $\frac{15}{8}\lceil \lg p \rceil$ modular multiplications, almost 2.5 times as cost efficient as before.

(iii) Signature verification calculations are all performed modulo p, while signature generation calculations are done modulo p and modulo $(p - 1)$.

11.69 Note (*recommended parameter sizes*) Given the latest progress on the discrete logarithm problem in $\mathbb{Z}_p^*$ (§3.6), a 512-bit modulus p provides only marginal security from concerted attack. As of 1996, a modulus p of at least 768 bits is recommended. For long-term security, 1024-bit or larger moduli should be used.

11.70 Note (*system-wide parameters*) All entities may elect to use the same prime number p and generator α, in which case p and α are not required to be part of the public key (cf. Note 11.61).

(i) Variations of the ElGamal scheme

Many variations of the basic ElGamal signature scheme (Algorithm 11.64) have been proposed. Most of these alter what is commonly referred to as the *signing equation* (given in step 1d of Algorithm 11.64). After suitable rearrangement, this signing equation can be written as $u = av + kw \bmod (p-1)$ where $u = h(m)$, $v = r$, and $w = s$ (i.e., $h(m) = ar + ks \bmod (p-1)$). Other signing equations can be obtained by permitting u, v, and w to take on the values s, r, and $h(m)$ in different orders. Table 11.5 lists the 6 possibilities.

	u	v	w	Signing equation	Verification
1	$h(m)$	r	s	$h(m) = ar + ks$	$\alpha^{h(m)} = (\alpha^a)^r r^s$
2	$h(m)$	s	r	$h(m) = as + kr$	$\alpha^{h(m)} = (\alpha^a)^s r^r$
3	s	r	$h(m)$	$s = ar + kh(m)$	$\alpha^s = (\alpha^a)^r r^{h(m)}$
4	s	$h(m)$	r	$s = ah(m) + kr$	$\alpha^s = (\alpha^a)^{h(m)} r^r$
5	r	s	$h(m)$	$r = as + kh(m)$	$\alpha^r = (\alpha^a)^s r^{h(m)}$
6	r	$h(m)$	s	$r = ah(m) + ks$	$\alpha^r = (\alpha^a)^{h(m)} r^s$

Table 11.5: *Variations of the ElGamal signing equation. Signing equations are computed modulo* $(p-1)$; *verification is done modulo p.*

11.71 Note (*comparing variants of the ElGamal signature scheme*)

(i) Some of the signing equations listed in Table 11.5 are more efficient to compute than the original ElGamal equation in Algorithm 11.64. For example, equations (3) and (4) of Table 11.5 do not require the computation of an inverse to determine the signature s. Equations (2) and (5) require the signer to compute $a^{-1} \bmod (p-1)$, but this fixed quantity need only be computed once.

(ii) Verification equations (2) and (4) involve the expression r^r. Part of the security of signature schemes based on these signing equations is the intractability of finding solutions to an expression of the form $x^x \equiv c \pmod{p}$ for fixed c. This problem appears to be intractable for large values of p, but has not received the same attention as the discrete logarithm problem.

(ii) The generalized ElGamal signature scheme

The ElGamal digital signature scheme, originally described in the setting of the multiplicative group $\mathbb{Z}_p^*$, can be generalized in a straightforward manner to work in any finite abelian group G. The introductory remarks for §8.4.2 are pertinent to the algorithm presented in this section. Algorithm 11.73 requires a cryptographic hash function $h: \{0,1\}^* \longrightarrow \mathbb{Z}_n$

where n is the number of elements in G. It is assumed that each element $r \in G$ can be represented in binary so that $h(r)$ is defined.[9]

11.72 Algorithm Key generation for the generalized ElGamal signature scheme

SUMMARY: each entity selects a finite group G; generator of G; public and private keys. Each entity A should do the following:

1. Select an appropriate cyclic group G of order n, with generator α. (Assume that G is written multiplicatively.)
2. Select a random secret integer a, $1 \leq a \leq n-1$. Compute the group element $y = \alpha^a$.
3. A's public key is (α, y), together with a description of how to multiply elements in G; A's private key is a.

11.73 Algorithm Generalized ElGamal signature generation and verification

SUMMARY: entity A signs a binary message m of arbitrary length. Any entity B can verify this signature by using A's public key.

1. *Signature generation.* Entity A should do the following:
   (a) Select a random secret integer k, $1 \leq k \leq n - 1$, with $\gcd(k, n) = 1$.
   (b) Compute the group element $r = \alpha^k$.
   (c) Compute $k^{-1} \bmod n$.
   (d) Compute $h(m)$ and $h(r)$.
   (e) Compute $s = k^{-1}\{h(m) - ah(r)\} \bmod n$.
   (f) A's signature for m is the pair (r, s).

2. *Verification.* To verify A's signature (r, s) on m, B should do the following:
   (a) Obtain A's authentic public key (α, y).
   (b) Compute $h(m)$ and $h(r)$.
   (c) Compute $v_1 = y^{h(r)} \cdot r^s$.
   (d) Compute $v_2 = \alpha^{h(m)}$.
   (e) Accept the signature if and only if $v_1 = v_2$.

11.74 Example (*generalized ElGamal signatures with artificially small parameters*)

Key generation. Consider the finite field $\mathbb{F}_{2^5}$ constructed from the irreducible polynomial $f(x) = x^5 + x^2 + 1$ over $\mathbb{F}_2$. (See Example 2.231 for examples of arithmetic in the field $\mathbb{F}_{2^4}$.) The elements of this field are the 31 binary 5-tuples displayed in Table 11.6, along with 00000. The element $\alpha = (00010)$ is a generator for $G = \mathbb{F}_{2^5}^*$, the multiplicative cyclic group of the field. The order of this group G is $n = 31$. Let $h\colon \{0,1\}^* \longrightarrow \mathbb{Z}_{31}$ be a hash function. Entity A selects the private key $a = 19$ and computes $y = \alpha^a = (00010)^{19} = (00110)$. A's public key is $(\alpha = (00010), y = (00110))$.

Signature generation. To sign the message $m = 10110101$, A selects a random integer $k = 24$, and computes $r = \alpha^{24} = (11110)$ and $k^{-1} \bmod 31 = 22$. A then computes $h(m) = 16$ and $h(r) = 7$ (the hash values have been contrived for this example) and $s = 22 \cdot \{16 - (19)(7)\} \bmod 31 = 30$. A's signature for message m is $(r = (11110), s = 30)$.

Signature verification. B computes $h(m) = 16$, $h(r) = 7$, $v_1 = y^{h(r)}r^s = (00110)^7 \cdot (11110)^{30} = (11011)$, and $v_2 = \alpha^{h(m)} = \alpha^{16} = (11011)$. B accepts the signature since $v_1 = v_2$. $\square$

---

[9]More precisely, one would define a function $f\colon G \longrightarrow \{0,1\}^*$ and write $h(f(r))$ instead of $h(r)$.

i	α^i
0	00001
1	00010
2	00100
3	01000
4	10000
5	00101
6	01010
7	10100

i	α^i
8	01101
9	11010
10	10001
11	00111
12	01110
13	11100
14	11101
15	11111

i	α^i
16	11011
17	10011
18	00011
19	00110
20	01100
21	11000
22	10101
23	01111

i	α^i
24	11110
25	11001
26	10111
27	01011
28	10110
29	01001
30	10010

Table 11.6: *The elements of* $\mathbb{F}_{2^5}$ *as powers of a generator* α.

11.75 Note (*security of generalized ElGamal*) Much of the security of Algorithm 11.73 relies on the intractability of the discrete logarithm problem in the group G (see §3.6). Most of the security comments in Note 11.66 apply to the generalized ElGamal scheme.

11.76 Note (*signing and verification operations*) Signature generation requires computations in the group G (i.e., $r = \alpha^k$) and computations in $\mathbb{Z}_n$. Signature verification only requires computations in the group G.

11.77 Note (*generalized ElGamal using elliptic curves*) One of the most promising implementations of Algorithm 11.73 is the case where the finite abelian group G is constructed from the set of points on an elliptic curve over a finite field $\mathbb{F}_q$. The discrete logarithm problem in groups of this type appears to be more difficult than the discrete logarithm problem in the multiplicative group of a finite field $\mathbb{F}_q$. This implies that q can be chosen smaller than for corresponding implementations in groups such as $G = \mathbb{F}_q^*$.

11.5.3 The Schnorr signature scheme

Another well-known variant of the ElGamal scheme (Algorithm 11.64) is the Schnorr signature scheme. As with the DSA (Algorithm 11.56), this technique employs a subgroup of order q in $\mathbb{Z}_p^*$, where p is some large prime number. The method also requires a hash function $h\colon \{0,1\}^* \longrightarrow \mathbb{Z}_q$. Key generation for the Schnorr signature scheme is the same as DSA key generation (Algorithm 11.54), except that there are no constraints on the sizes of p and q.

11.78 Algorithm Schnorr signature generation and verification

SUMMARY: entity A signs a binary message m of arbitrary length. Any entity B can verify this signature by using A's public key.

1. *Signature generation.* Entity A should do the following:
   (a) Select a random secret integer k, $1 \le k \le q - 1$.
   (b) Compute $r = \alpha^k \bmod p$, $e = h(m\|r)$, and $s = ae + k \bmod q$.
   (c) A's signature for m is the pair (s, e).

2. *Verification.* To verify A's signature (s, e) on m, B should do the following:

   (a) Obtain A's authentic public key (p, q, α, y).

   (b) Compute $v = \alpha^s y^{-e} \bmod p$ and $e' = h(m\|v)$.

   (c) Accept the signature if and only if $e' = e$.

Proof that signature verification works. If the signature was created by A, then $v \equiv \alpha^s y^{-e} \equiv \alpha^s \alpha^{-ae} \equiv \alpha^k \equiv r \pmod{p}$. Hence, $h(m\|v) = h(m\|r)$ and $e' = e$.

11.79 Example (*Schnorr's signature scheme with artificially small parameters*)

Key generation. A selects primes $p = 129841$ and $q = 541$; here, $(p-1)/q = 240$. A then selects a random integer $g = 26346 \in \mathbb{Z}_p^*$ and computes $\alpha = 26346^{240} \bmod p = 26$. Since $\alpha \neq 1$, α generates the unique cyclic subgroup of order 541 in $\mathbb{Z}_p^*$. A then selects the private key $a = 423$ and computes $y = 26^{423} \bmod p = 115917$. A's public key is $(p = 129841, q = 541, \alpha = 26, y = 115917)$.

Signature generation. To sign the message $m = 11101101$, A selects a random number $k = 327$ such that $1 \leq k \leq 540$, and computes $r = 26^{327} \bmod p = 49375$ and $e = h(m\|r) = 155$ (the hash value has been contrived for this example). Finally, A computes $s = 423 \cdot 155 + 327 \bmod 541 = 431$. The signature for m is $(s = 431, e = 155)$.

Signature verification. B computes $v = 26^{431} \cdot 115917^{-155} \bmod p = 49375$ and $e' = h(m\|v) = 155$. B accepts the signature since $e = e'$. $\qquad\square$

11.80 Note (*performance characteristics of the Schnorr scheme*) Signature generation in Algorithm 11.78 requires one exponentiation modulo p and one multiplication modulo q. The exponentiation modulo p could be done off-line. Depending on the hash algorithm used, the time to compute $h(m\|r)$ should be relatively small. Verification requires two exponentiations modulo p. These two exponentiations can be computed by Algorithm 14.88 at a cost of about 1.17 exponentiations. Using the subgroup of order q does not significantly enhance computational efficiency over the ElGamal scheme of Algorithm 11.64, but does provide smaller signatures (for the same level of security) than those generated by the ElGamal method.

11.5.4 The ElGamal signature scheme with message recovery

The ElGamal scheme and its variants (§11.5.2) discussed so far are all randomized digital signature schemes with appendix (i.e., the message is required as input to the verification algorithm). In contrast, the signature mechanism of Algorithm 11.81 has the feature that the message can be recovered from the signature itself. Hence, this ElGamal variant provides a randomized digital signature with message recovery.

For this scheme, the signing space is $\mathcal{M}_S = \mathbb{Z}_p^*$, p a prime, and the signature space is $\mathcal{S} = \mathbb{Z}_p \times \mathbb{Z}_q$, q a prime, where q divides $(p-1)$. Let R be a redundancy function from the set of messages $\mathcal{M}$ to $\mathcal{M}_S$ (see Table 11.1). Key generation for Algorithm 11.81 is the same as DSA key generation (Algorithm 11.54), except that there are no constraints on the sizes of p and q.

11.81 Algorithm Nyberg-Rueppel signature generation and verification

---

SUMMARY: entity A signs a message $m \in \mathcal{M}$. Any entity B can verify A's signature and recover the message m from the signature.

1. *Signature generation.* Entity A should do the following:
   (a) Compute $\widetilde{m} = R(m)$.
   (b) Select a random secret integer k, $1 \le k \le q-1$, and compute $r = \alpha^{-k} \bmod p$.
   (c) Compute $e = \widetilde{m}r \bmod p$.
   (d) Compute $s = ae + k \bmod q$.
   (e) A's signature for m is the pair (e, s).

2. *Verification.* To verify A's signature (e, s) on m, B should do the following:
   (a) Obtain A's authentic public key (p, q, α, y).
   (b) Verify that $0 < e < p$; if not, reject the signature.
   (c) Verify that $0 \le s < q$; if not, reject the signature.
   (d) Compute $v = \alpha^s y^{-e} \bmod p$ and $\widetilde{m} = ve \bmod p$.
   (e) Verify that $\widetilde{m} \in \mathcal{M}_R$; if $\widetilde{m} \notin \mathcal{M}_R$ then reject the signature.
   (f) Recover $m = R^{-1}(\widetilde{m})$.

Proof that signature verification works. If A created the signature, then $v \equiv \alpha^s y^{-e} \equiv \alpha^{s-ae} \equiv \alpha^k \pmod{p}$. Thus $ve \equiv \alpha^k \widetilde{m} \alpha^{-k} \equiv \widetilde{m} \pmod{p}$, as required.

11.82 Example (*Nyberg-Rueppel signature generation with artificially small parameters*)
Key generation. Entity A selects primes $p = 1256993$ and $q = 3571$, where q divides $(p-1)$; here, $(p-1)/q = 352$. A then selects a random number $g = 42077 \in \mathbb{Z}_p^*$ and computes $\alpha = 42077^{352} \bmod p = 441238$. Since $\alpha \ne 1$, α generates the unique cyclic subgroup of order 3571 in $\mathbb{Z}_p^*$. Finally, A selects a random integer $a = 2774$ and computes $y = \alpha^a \bmod p = 1013657$. A's public key is $(p = 1256993, q = 3571, \alpha = 441238, y = 1013657)$, while A's private key is $a = 2774$.
Signature generation. To sign a message m, A computes $\widetilde{m} = R(m) = 1147892$ (the value $R(m)$ has been contrived for this example). A then randomly selects $k = 1001$, computes $r = \alpha^{-k} \bmod p = 441238^{-1001} \bmod p = 1188935$, $e = \widetilde{m}r \bmod p = 138207$, and $s = (2774)(138207) + 1001 \bmod q = 1088$. The signature for m is $(e = 138207, s = 1088)$.
Signature verification. B computes $v = 441238^{1088} \cdot 1013657^{-138207} \bmod 1256993 = 504308$, and $\widetilde{m} = v \cdot 138207 \bmod 1256993 = 1147892$. B verifies that $\widetilde{m} \in \mathcal{M}_R$ and recovers $m = R^{-1}(\widetilde{m})$. $\square$

11.83 Note (*security of the Nyberg-Rueppel signature scheme*)
   (i) Since Algorithm 11.81 is a variant of the basic ElGamal scheme (Algorithm 11.64), the security considerations of Note 11.66 apply. Like DSA (Algorithm 11.56), this ElGamal mechanism with message recovery relies on the difficulty of two related but distinct discrete logarithm problems (see Note 11.58).
   (ii) Since Algorithm 11.81 provides message recovery, a suitable redundancy function R is required (see Note 11.10) to guard against existential forgery. As is the case with RSA, the multiplicative nature of this signature scheme must be carefully considered when choosing a redundancy function R. The following possible attack should be kept in mind. Suppose $m \in \mathcal{M}$, $\widetilde{m} = R(m)$, and (e, s) is a signature for m. Then $e = \widetilde{m}\alpha^{-k} \bmod p$ for some integer k and $s = ae + k \bmod q$. Let $\widetilde{m}^* = \widetilde{m}\alpha^l \bmod p$ for some integer l. If $s^* = s + l \bmod q$ and $\widetilde{m}^* \in \mathcal{M}_R$, then (e, s^*)

is a valid signature for $m^* = R^{-1}(\widetilde{m}^*)$. To see this, consider the verification algorithm (step 2 of Algorithm 11.81). $v \equiv \alpha^{s^*} y^{-e} \equiv \alpha^{s+l} \alpha^{-ae} \equiv \alpha^{k+l} \pmod{p}$, and $ve \equiv \alpha^{k+l} \widetilde{m} \alpha^{-k} \equiv \widetilde{m} \alpha^l \equiv \widetilde{m}^* \pmod{p}$. Since $\widetilde{m}^* \in \mathcal{M}_R$, the forged signature (e, s^*) will be accepted as a valid signature for m^*.

(iii) The verification that $0 < e < p$ given in step 2b of Algorithm 11.81 is crucial. Suppose (e, s) is A's signature for the message m. Then $e = \widetilde{m}r \bmod p$ and $s = ae + k \bmod q$. An adversary can use this signature to compute a signature on a message m^* of its choice. It determines an e^* such that $e^* \equiv \widetilde{m}^* r \pmod{p}$ and $e^* \equiv e \pmod{q}$. (This is possible by the Chinese Remainder Theorem (Fact 2.120).) The pair (e^*, s) will pass the verification algorithm provided that $0 < e^* < p$ is not checked.

11.84 Note (*a generalization of ElGamal signatures with message recovery*) The expression $e = \widetilde{m}r \bmod p$ in step 1c of Algorithm 11.81 provides a relatively simple way to encrypt $\widetilde{m}$ with key r and could be generalized to any symmetric-key algorithm. Let $E = \{E_r : r \in \mathbb{Z}_p\}$ be a set of encryption transformations where each E_r is indexed by an element $r \in \mathbb{Z}_p^*$ and is a bijection from $\mathcal{M}_S = \mathbb{Z}_p^*$ to $\mathbb{Z}_p^*$. For any $m \in \mathcal{M}$, select a random integer k, $1 \le k \le q - 1$, compute $r = \alpha^k \bmod p$, $e = E_r(\widetilde{m})$, and $s = ae + k \bmod q$. The pair (e, s) is a signature for m. The fundamental signature equation $s = ae + k \bmod q$ is a means to bind entity A's private key and the message m to a symmetric key which can then be used to recover the message by any other entity at some later time.

11.6 One-time digital signatures

One-time digital signature schemes are digital signature mechanisms which can be used to sign, at most, one message; otherwise, signatures can be forged. A new public key is required for each message that is signed. The public information necessary to verify one-time signatures is often referred to as *validation parameters*. When one-time signatures are combined with techniques for authenticating validation parameters, multiple signatures are possible (see §11.6.3 for a description of authentication trees).

Most, but not all, one-time digital signature schemes have the advantage that signature generation and verification are very efficient. One-time digital signature schemes are useful in applications such as chipcards, where low computational complexity is required.

11.6.1 The Rabin one-time signature scheme

Rabin's one-time signature scheme was one of the first proposals for a digital signature of any kind. It permits the signing of a single message. The verification of a signature requires interaction between the signer and verifier. Unlike other digital signature schemes, verification can be done only once. While not practical, it is presented here for historical reasons. Notation used in this section is given in Table 11.7.

Symbol	Meaning
M_0	0^l = the all 0's string of bitlength l.
$M_0(i)$	$0^{l-e}\|b_{e-1}\cdots b_1 b_0$ where $b_{e-1}\cdots b_1 b_0$ is the binary representation of i.
$\mathcal{K}$	a set of l-bit strings.
E	a set of encryption transformations indexed by a key space $\mathcal{K}$.
E_t	an encryption transformation belonging to E with $t \in \mathcal{K}$. Each E_t maps l-bit strings to l-bit strings.
h	a publicly-known one-way hash function from $\{0,1\}^*$ to $\{0,1\}^l$.
n	a fixed positive integer which serves as a security parameter.

Table 11.7: *Notation for the Rabin one-time signature scheme.*

11.85 Algorithm Key generation for the Rabin one-time signature scheme

SUMMARY: each entity A selects a symmetric-key encryption scheme E, generates $2n$ random bitstrings, and creates a set of validation parameters.
Each entity A should do the following:

1. Select a symmetric-key encryption scheme E (e.g., DES).
2. Generate $2n$ random secret strings $k_1, k_2, \ldots, k_{2n} \in \mathcal{K}$, each of bitlength l.
3. Compute $y_i = E_{k_i}(M_0(i))$, $1 \le i \le 2n$.
4. A's public key is $(y_1, y_2, \ldots, y_{2n})$; A's private key is $(k_1, k_2, \ldots, k_{2n})$.

11.86 Algorithm Rabin one-time signature generation and verification

SUMMARY: entity A signs a binary message m of arbitrary length. Signature verification is interactive with A.

1. *Signature generation.* Entity A should do the following:
   (a) Compute $h(m)$.
   (b) Compute $s_i = E_{k_i}(h(m))$, $1 \le i \le 2n$.
   (c) A's signature for m is $(s_1, s_2, \ldots, s_{2n})$.
2. *Verification.* To verify A's signature $(s_1, s_2, \ldots, s_{2n})$ on m, B should:
   (a) Obtain A's authentic public key $(y_1, y_2, \ldots, y_{2n})$.
   (b) Compute $h(m)$.
   (c) Select n distinct random numbers r_j, $1 \le r_j \le 2n$, for $1 \le j \le n$.
   (d) Request from A the keys k_{r_j}, $1 \le j \le n$.
   (e) Verify the authenticity of the received keys by computing $z_j = E_{k_{r_j}}(M_0(r_j))$ and checking that $z_j = y_{r_j}$, for each $1 \le j \le n$.
   (f) Verify that $s_{r_j} = E_{k_{r_j}}(h(m))$, $1 \le j \le n$.

11.87 Note (*key sizes for Rabin's one-time signatures*) Since E_t outputs l bits (see Table 11.7), the public and private keys in Algorithm 11.86 each consist of $2nl$ bits. For $n = 80$ and $l = 64$, the keys are each 1280 bytes long.

11.88 Note (*resolution of disputes*) To resolve potential disputes between the signer A and the verifier B using Algorithm 11.86, the following procedure is followed:

1. B provides a trusted third party (TTP) with m and the signature $(s_1, s_2, \ldots, s_{2n})$.

2. The TTP obtains $k_1, k_2, \ldots, k_{2n}$ from A.

3. The TTP verifies the authenticity of the private key by computing $z_i = E_{k_i}(M_0(i))$ and checking that $y_i = z_i$, $1 \leq i \leq 2n$. If this fails, the TTP rules in favor of B (i.e., the signature is deemed to be valid).

4. The TTP computes $u_i = E_{k_i}(h(m))$, $1 \leq i \leq 2n$. If $u_i = s_i$ for at most n values of i, $1 \leq i \leq 2n$, the signature is declared a forgery and the TTP rules in favor of A (who denies having created the signature). If $n + 1$ or more values of i give $u_i = s_i$, the signature is deemed valid and the TTP rules in favor of B.

11.89 Note (*rationale for dispute resolution protocol*) The rationale for adjudicating disputes in Rabin's one-time signature scheme, as outlined in Note 11.88, is as follows. If B has attempted to forge A's signature on a new message m', B either needs to determine at least one more key k' so that at least $n + 1$ values of i give $u_i = s_i$, or determine m' such that $h(m) = h(m')$. This should be infeasible if the symmetric-key algorithm and hash function are chosen appropriately. If A attempts to create a signature which it can later disavow, A must ensure that $u_i = s_i$ for precisely n values of i and hope that B chooses these n values in step 2c of the verification procedure, the probability of which is only $1/\binom{2n}{n}$.

11.90 Note (*one-timeness of Algorithm 11.86*) A can sign at most one message with a given private key in Rabin's one-time scheme; otherwise, A will (with high probability) reveal $n + 1$ or more of the private key values and enable B (and perhaps collaborators) to forge signatures on new messages (see Note 11.89). A signature can only be verified once without revealing (with high probability) more than n of the $2n$ private values.

11.6.2 The Merkle one-time signature scheme

Merkle's one-time digital signature scheme (Algorithm 11.92) differs substantially from that of Rabin (Algorithm 11.86) in that signature verification is not interactive with the signer. A TTP or some other trusted means is required to authenticate the validation parameters constructed in Algorithm 11.91.

11.91 Algorithm Key generation for the Merkle one-time signature scheme

---

SUMMARY: to sign n-bit messages, A generates $t = n + \lfloor \lg n \rfloor + 1$ validation parameters. Each entity A should do the following:

1. Select $t = n + \lfloor \lg n \rfloor + 1$ random secret strings $k_1, k_2, \ldots, k_t$ each of bitlength l.

2. Compute $v_i = h(k_i)$, $1 \leq i \leq t$. Here, h is a preimage-resistant hash function $h\colon \{0,1\}^* \longrightarrow \{0,1\}^l$ (see §9.2.2).

3. A's public key is $(v_1, v_2, \ldots, v_t)$; A's private key is $(k_1, k_2, \ldots, k_t)$.

To sign an n-bit message m, a bitstring $w = m \| c$ is formed where c is the binary representation for the number of 0's in m. c is assumed to be a bitstring of bitlength $\lfloor \lg n \rfloor + 1$ with high-order bits padded with 0's, if necessary. Hence, w is a bitstring of bitlength $t = n + \lfloor \lg n \rfloor + 1$.

11.92 Algorithm Merkle one-time signature generation and verification

---

SUMMARY: entity A signs a binary message m of bitlength n. Any entity B can verify this signature by using A's public key.

1. *Signature generation.* Entity A should do the following:
   (a) Compute c, the binary representation for the number of 0's in m.
   (b) Form $w = m\|c = (a_1 a_2 \cdots a_t)$.
   (c) Determine the coordinate positions $i_1 < i_2 < \cdots < i_u$ in w such that $a_{i_j} = 1$, $1 \leq j \leq u$.
   (d) Let $s_j = k_{i_j}, 1 \leq j \leq u$.
   (e) A's signature for m is $(s_1, s_2, \ldots, s_u)$.

2. *Verification.* To verify A's signature $(s_1, s_2, \ldots, s_u)$ on m, B should:
   (a) Obtain A's authentic public key $(v_1, v_2, \ldots, v_t)$.
   (b) Compute c, the binary representation for the number of 0's in m.
   (c) Form $w = m\|c = (a_1 a_2 \cdots a_t)$.
   (d) Determine the coordinate positions $i_1 < i_2 < \cdots < i_u$ in w such that $a_{i_j} = 1$, $1 \leq j \leq u$.
   (e) Accept the signature if and only if $v_{i_j} = h(s_j)$ for all $1 \leq j \leq u$.

11.93 Note (*security of Merkle's one-time signature scheme*) Let m be a message, $w = m\|c$ the bitstring formed in step 1b of Algorithm 11.92, and $(s_1, s_2, \ldots, s_u)$ a signature for m. If h is a preimage-resistant hash function, the following argument shows that no signature for a message $m' \neq m$ can be forged. Let $w' = m'\|c'$ where c' is the $(\lfloor \lg n \rfloor + 1)$-bit string which is the binary representation for the number of 0's in m'. Since an adversary has access to only that portion of the signer's private key which consists of $(s_1, s_2, \ldots, s_u)$, the set of coordinate positions in m' having a 1 must be a subset of the coordinate positions in m having a 1 (otherwise, m' will have a 1 in some position where m has a 0 and the adversary will require an element of the private key not revealed by the signer). But this means that m' has more 0's than m and that $c' > c$ (when considered as integers). In this case, c' will have a 1 in some position where c has a 0. The adversary would require a private key element, corresponding to this position, which was not revealed by the signer.

11.94 Note (*storage and computational requirements of Algorithm 11.92*)
   (i) To sign an n-bit message m which has k ones requires $l \cdot (n + \lfloor \lg n \rfloor + 1)$ bits of storage for the validation parameters (public key), and $l \cdot (n + \lfloor \lg n \rfloor + 1)$ bits for the private key. The signature requires $l \cdot (k + k')$ bits of storage, where k' is the number of 1's in the binary representation of $n - k$. For example, if $n = 128$, $l = 64$, and $k = 72$, then the public and private keys each require 8704 bits (1088 bytes). The signature requires 4800 bits (600 bytes).
   (ii) The private key can be made smaller by forming the k_i's from a single *seed* value. For example, if k^* is a bitstring of bitlength at least l, then form $k_i = h(k^*\|i), 1 \leq i \leq t$. Since only the seed k^* need be stored, the size of the private key is drastically reduced.
   (iii) Signature generation is very fast, requiring no computation. Signature verification requires the evaluation of the hash function for fewer than $n + \lfloor \lg n \rfloor + 1$ values.

11.95 Note (*improving efficiency of Merkle's one-time scheme*) Algorithm 11.92 requires $l \cdot (n + \lfloor \lg n \rfloor + 1)$ bits for each of the public and private keys. The public key must necessarily be this large because the signing algorithm considers individual bits of the message. The scheme can be made more efficient if the signing algorithm considers more than one bit at a time. Suppose entity A wishes to sign a kt-bit message m. Write $m = m_1 \| m_2 \| \cdots \| m_t$ where each m_i has bitlength k and each represents an integer between 0 and $2^k - 1$ inclusive. Define $U = \sum_{i=1}^{t} (2^k - m_i) \leq t2^k$. U can be represented by $\lg U \leq \lfloor \lg t \rfloor + 1 + k$ bits. If $r = \lceil (\lfloor \lg t \rfloor + 1 + k)/k \rceil$, then U can be written in binary as $U = u_1 \| u_2 \| \cdots \| u_r$, where each u_i has bitlength k. Form the bitstring $w = m_1 \| m_2 \| \cdots m_t \| u_1 \| u_2 \| \cdots \| u_r$. Generate $t+r$ random bitstrings $k_1, k_2, \ldots, k_{t+r}$ and compute $v_i = h(k_i)$, $1 \leq i \leq t+r$. The private key for the modified scheme is $(k_1, k_2, \ldots, k_{t+r})$ and the public key is $(v_1, v_2, \ldots, v_{t+r})$. The signature for m is $(s_1, s_2, \ldots, s_{t+r})$ where $s_i = h^{m_i}(k_i)$, $1 \leq i \leq t$, and $s_i = h^{u_i}(k_{t+i})$, $1 \leq i \leq r$. Here, h^c denotes the c-fold composition of h with itself. As with the original scheme (Algorithm 11.92), the bits appended to the message act as a check-sum (see Note 11.93) as follows. Given an element $s_i = h^a(k_j)$, an adversary can easily compute $h^{a+\delta}(k_j)$ for $0 \leq \delta \leq 2^k - a$, but is unable to compute $h^{a-\delta}$ for any $\delta > 0$ if h is a one-way hash function. To forge a signature on a new message, an adversary can only reduce the value of the check-sum, which will make it impossible for him to compute the required hash values on the appended kr bits.

11.96 Example (*signing more than one bit at a time*) This example illustrates the modification of the Merkle scheme described in Note 11.95. Let $m = m_1 \| m_2 \| m_3 \| m_4$ where $m_1 = 1011$, $m_2 = 0111$, $m_3 = 1010$, and $m_4 = 1101$. m_1, m_2, m_3, and m_4 are the binary representations of 11, 7, 10, and 13, respectively. $U = (16 - m_1) + (16 - m_2) + (16 - m_3) + (16 - m_4) = 5 + 9 + 6 + 3 = 23$. In binary, $U = 10111$. Form $w = m \| 0001\,0111$. The signature is $(s_1, s_2, s_3, s_4, s_5, s_6)$ where $s_1 = h^{11}(k_1)$, $s_2 = h^7(k_2)$, $s_3 = h^{10}(k_3)$, $s_4 = h^{13}(k_4)$, $s_5 = h^1(x_5)$, and $s_6 = h^7(x_6)$. If an adversary tries to alter the message, he can only apply the function h to some s_i. This causes the sum of the exponents used (i.e., $\sum m_i$) to increase and, hence, $t2^d - \sum m_i$ to decrease. An adversary would be unable to modify the last two blocks since h^{-1} is required to decrease the sum. But, since h is preimage-resistant, h^{-1} cannot be computed by the adversary. $\square$

11.6.3 Authentication trees and one-time signatures

§13.4.1 describes the basic structure of an authentication tree and relates how such a tree could be used, among other things, to authenticate a large number of public validation parameters for a one-time signature scheme. This section describes how an authentication tree can be used in conjunction with a one-time signature scheme to provide a scheme which allows multiple signatures. A small example will serve to illustrate how this is done.

11.97 Example (*an authentication tree for Merkle's one-time scheme*) Consider the one-time signature scheme of Algorithm 11.92 for signing n-bit messages. Let $h: \{0,1\}^* \longrightarrow \{0,1\}^l$ be a preimage-resistant hash function and $t = n + \lfloor \lg n \rfloor + 1$. Figure 11.7 illustrates a 5-vertex binary tree created by an entity A in the course of signing five messages m_0, m_1, m_2, m_3, m_4. Each vertex in the tree is associated with one of the five messages. For the vertex associated with message m_i, A has selected $X_i = (x_{1i}, x_{2i}, \ldots, x_{ti})$, $U_i = (u_{1i}, u_{2i}, \ldots, u_{ti})$ and $W_i = (w_{1i}, w_{2i}, \ldots, w_{ti})$, $0 \leq i \leq 4$, the elements of which are random bitstrings. From these lists, A has computed $Y_i = (h(x_{ji}): 1 \leq j \leq t)$, $V_i = (h(u_{ji}): 1 \leq j \leq t)$, and $Z_i = (h(w_{ji}): 1 \leq j \leq t)$. Define $h(Y_i) =$

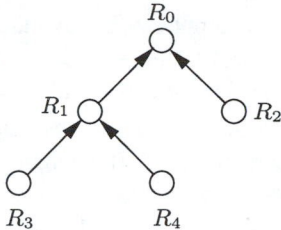

Figure 11.7: *An authentication tree for the Merkle one-time signature scheme (cf. Example 11.97).*

$h(h(x_{1i})\|h(x_{2i})\|\cdots\|h(x_{ti}))$ for $0 \leq i \leq 4$, and define $h(V_i)$ and $h(Z_i)$ analogously. Denote the Merkle one-time signature of m_i using private key X_i by $S_A(m_i, X_i)$, $0 \leq i \leq 4$. Y_i is the set of validation parameters for the signature $S_A(m_i, X_i)$. Finally, let $R_i = h(h(Y_i)\|h(V_i)\|h(Z_i))$, $0 \leq i \leq 4$. Table 11.8 summarizes the parameters associated with the vertex R_i. The sets U_i and W_i are used to sign the labels of the children

message	m_i
private parameters	X_i, U_i, W_i
public parameters	Y_i, V_i, Z_i
hash values	$h(Y_i), h(V_i), h(Z_i)$
R_i	$h(h(Y_i)\|h(V_i)\|h(Z_i))$
signature	$S_A(m_i, X_i)$
validation parameters	Y_i

Table 11.8: *Parameters and signature associated with vertex R_i, $0 \leq i \leq 4$ (cf. Figure 11.7).*

of vertex R_i. The signature on vertex R_0 is that of a trusted third party (TTP). Table 11.9 summarizes the parameters and signatures associated with each vertex label of the binary tree. To describe how the tree is used to verify signatures, consider message m_4 and signa-

Message	Vertex Label	Signature on Vertex Label	Authentication Parameters
m_0	R_0	Signature of TTP	—
m_1	R_1	$S_A(R_1, U_0)$	$V_0, h(Y_0), h(Z_0)$
m_2	R_2	$S_A(R_2, W_0)$	$Z_0, h(Y_0), h(V_0)$
m_3	R_3	$S_A(R_3, U_1)$	$V_1, h(Y_1), h(Z_1)$
m_4	R_4	$S_A(R_4, W_1)$	$Z_1, h(Y_1), h(V_1)$

Table 11.9: *Parameters and signatures associated with vertices of the binary tree (cf. Figure 11.7).*

ture $S_A(m_4, X_4)$. The signer A first provides the verifier B with the validation parameters Y_4. The verifier checks the Merkle one-time signature using step 2 of Algorithm 11.92. B must then be convinced that Y_4 is an authentic set of validation parameters created by A. To accomplish this, A provides B with a sequence of values enumerated in the steps below:

1. $h(V_4)$, $h(Z_4)$; B computes $h(Y_4)$ and then $R_4 = h(h(Y_4)\|h(V_4)\|h(Z_4))$.
2. $S_A(R_4, W_1)$ and Z_1; B verifies the signature on R_4 using Algorithm 11.92.
3. $h(Y_1)$, $h(V_1)$; B computes $h(Z_1)$ and then $R_1 = h(h(Y_1)\|h(V_1)\|h(Z_1))$.
4. $S_A(R_1, U_0)$ and V_0; B verifies the signature using Algorithm 11.92.

5. $h(Y_0)$, $h(Z_0)$; B computes $h(V_0)$ and then $R_0 = h(h(Y_0)\|h(V_0)\|h(Z_0))$.
6. the signature of the TTP for R_0; B verifies the TTP's signature using an algorithm appropriate to the signature mechanism for the TTP.

The binary tree on 5 vertices (Figure 11.7) could be extended indefinitely from any leaf as more signatures are created by A. The length of a longest authentication path (or equivalently, the depth of the tree) determines the maximum amount of information which A must provide B in order for B to verify the signature of a message associated with a vertex. $\square$

11.6.4 The GMR one-time signature scheme

The Goldwasser, Micali, and Rivest (GMR) scheme (Algorithm 11.102) is a one-time signature scheme which requires a pair of claw-free permutations (see Definition 11.98). When combined with a tree authentication procedure, it provides a mechanism for signing more than one message. The GMR scheme is noteworthy as it was the first digital signature mechanism proven to be secure against an adaptive chosen-message attack. Although the GMR scheme is not practical, variations of it have been proposed which suggest that the concept is not purely of theoretical importance.

11.98 Definition Let $g_i\colon X \longrightarrow X$, $i = 0, 1$, be two permutations defined on a finite set X. g_0 and g_1 are said to be a *claw-free pair* of permutations if it is computationally infeasible to find $x, y \in X$ such that $g_0(x) = g_1(y)$. A triple (x, y, z) of elements from X with $g_0(x) = g_1(y) = z$ is called a *claw*. If both g_i, $i = 0$, 1, have the property that given additional information it is computationally feasible to determine g_0^{-1}, g_1^{-1}, respectively, the permutations are called a *trapdoor claw-free pair* of permutations.

In order for g_0, g_1 to be a claw-free pair, computing $g_i^{-1}(x)$, for both $i = 0$ and 1, must be computationally infeasible for essentially all $x \in X$. For, if g_1^{-1} (and similarly for g_0^{-1}) could be efficiently computed, one could select an $x \in X$, compute $g_0(x) = z$ and $g_1^{-1}(z) = y$, to obtain a claw (x, y, z).

11.99 Example (*trapdoor claw-free permutation pair*) Let $n = pq$ where $p \equiv 3 \pmod 4$ and $q \equiv 7 \pmod 8$. For this choice of p and q, $\left(\frac{-1}{n}\right) = 1$ but $-1 \notin Q_n$, and $\left(\frac{2}{n}\right) = -1$. Here, $\left(\frac{\cdot}{n}\right)$ denotes the Jacobi symbol (Definition 2.147). Define $D_n = \{x\colon \left(\frac{x}{n}\right) = 1$ and $0 < x < \frac{n}{2}\}$. Define $g_0\colon D_n \longrightarrow D_n$ and $g_1\colon D_n \longrightarrow D_n$ by

$$g_0(x) = \begin{cases} x^2 \bmod n, & \text{if } x^2 \bmod n < \frac{n}{2}, \\ -x^2 \bmod n, & \text{if } x^2 \bmod n > \frac{n}{2}, \end{cases}$$

$$g_1(x) = \begin{cases} 4x^2 \bmod n, & \text{if } 4x^2 \bmod n < \frac{n}{2}, \\ -4x^2 \bmod n, & \text{if } 4x^2 \bmod n > \frac{n}{2}. \end{cases}$$

If factoring n is intractable, then g_0, g_1 form a trapdoor claw-free pair of permutations; this can be seen as follows.

(i) (*g_0 and g_1 are permutations on D_n*) If $g_0(x) = g_0(y)$, then $x^2 \equiv y^2 \pmod n$ ($x^2 \equiv -y^2 \pmod n$ is not possible since $-1 \notin Q_n$), whence $x \equiv \pm y \pmod n$. Since $0 < x, y < n/2$, then $x = y$, and hence g_0 is a permutation on D_n. A similar argument shows that g_1 is a permutation on D_n.

(ii) (*g_0 and g_1 are claw-free*) Suppose that there is an efficient method for finding $x, y \in D_n$ such that $g_0(x) = g_1(y)$. Then $x^2 \equiv 4y^2 \pmod n$ ($x^2 \equiv -4y^2 \pmod n$ is impossible since $-1 \notin Q_n$), whence $(x-2y)(x+2y) \equiv 0 \pmod{n}$. Since $\left(\frac{x}{n}\right) = 1$ and $\left(\frac{\pm 2y}{n}\right) = -1$, $x \not\equiv \pm 2y \pmod{n}$ and, hence, $\gcd(x-2y, n)$ yields a non-trivial factor of n. This contradicts the assumption that factoring n is intractable.

(iii) (g_0, g_1 *is a trapdoor claw-free pair*) Knowing the factorization of n permits one to compute g_0^{-1} and g_1^{-1}. Hence, g_0, g_1 is a trapdoor claw-free permutation pair. □

The following example illustrates the general construction given in Example 11.99.

11.100 Example (*pair of claw-free permutations for artificially small parameters*) Let $p = 11$, $q = 7$, and $n = pq = 77$. $D_{77} = \{x \colon \left(\frac{x}{n}\right) = 1 \text{ and } 0 < x < 38\} = \{1, 4, 6, 9, 10, 13, 15, 16, 17, 19, 23, 24, 25, 36, 37\}$. The following table describes g_0 and g_1.

x	1	4	6	9	10	13	15	16	17	19	23	24	25	36	37
$g_0(x)$	1	16	36	4	23	15	6	25	19	24	10	37	9	13	17
$g_1(x)$	4	13	10	16	15	17	24	23	1	19	37	6	36	25	9

Notice that g_0 and g_1 are permutations on D_{77}. □

11.101 Algorithm Key generation for the GMR one-time signature scheme

SUMMARY: each entity selects a pair of trapdoor claw-free permutations and a validation parameter.
Each entity A should do the following:
1. Select a pair g_0, g_1 of trapdoor claw-free permutations on some set X. (It is "trapdoor" in that A itself can compute g_0^{-1} and g_1^{-1}.)
2. Select a random element $r \in X$. (r is called a *validation parameter*.)
3. A's public key is (g_0, g_1, r); A's private key is (g_0^{-1}, g_1^{-1}).

In the following, the notation for the composition of functions g_0, g_1 usually denoted $g_0 \circ g_1$ (see Definition 1.33) is simplified to g_0g_1. Also, $(g_0g_1)(r)$ will be written as $g_0g_1(r)$. The signing space $\mathcal{M}_S$ consists of binary strings which are prefix-free (see Note 11.103).

11.102 Algorithm GMR one-time signature generation and verification

SUMMARY: A signs a binary string $m = m_1m_2 \cdots m_t$. B verifies using A's public key.
1. *Signature generation.* Entity A should do the following:
   (a) Compute $S_r(m) = \prod_{i=0}^{t-1} g_{m_{t-i}}^{-1}(r)$.
   (b) A's signature for m is $S_r(m)$.
2. *Verification.* To verify A's signature $S_r(m)$ on m, B should do the following:
   (a) Obtain A's authentic public key (g_0, g_1, r).
   (b) Compute $r' = \prod_{i=1}^{t} g_{m_i}(S_r(m))$.
   (c) Accept the signature if and only if $r' = r$.

Proof that signature verification works.

$$r' = \prod_{i=1}^{t} g_{m_i}(S_r(m)) = \prod_{i=1}^{t} g_{m_i} \prod_{j=0}^{t-1} g_{m_{t-j}}^{-1}(r)$$
$$= g_{m_1} \circ g_{m_2} \circ \cdots \circ g_{m_t} \circ g_{m_t}^{-1} \circ g_{m_{t-1}}^{-1} \circ \cdots \circ g_{m_1}^{-1}(r) = r.$$

Thus $r' = r$, as required.

11.103 Note (*message encoding and security*) The set of messages which can be signed using Algorithm 11.102 must come from a set of binary strings which are *prefix-free*. (For example, 101 and 10111 cannot be in the same space since 101 is a prefix of 10111.) One method to accomplish this is to encode a binary string $b_1 b_2 \cdots b_l$ as $b_1 b_1 b_2 b_2 \cdots b_l b_l 01$. To see why the prefix-free requirement is necessary, suppose $m = m_1 m_2 \cdots m_t$ is a message whose signature is $S_r(m) = \prod_{i=0}^{t-1} g_{m_{t-i}}^{-1}(r)$. If $m' = m_1 m_2 \cdots m_u$, $u < t$, then an adversary can easily find a valid signature for m' from $S_r(m)$ by computing

$$S_r(m') = \prod_{j=u+1}^{t} g_{m_j}(S_r(m)) = \prod_{i=0}^{u-1} g_{m_{u-i}}^{-1}(r).$$

11.104 Note (*one-timeness of Algorithm 11.102*) To see that the GMR signature scheme is a one-time scheme, suppose that two prefix-free messages $m = m_1 m_2 \cdots m_t$ and $m' = n_1 n_2 \cdots n_u$ are both signed with the same validation parameter r. Then $S_r(m) = \prod_{i=0}^{t-1} g_{m_{t-i}}^{-1}(r)$ and $S_r(m') = \prod_{i=0}^{u-1} g_{n_{u-i}}^{-1}(r)$. Therefore, $\prod_{i=1}^{t} g_{m_i}(S_r(m)) = r = \prod_{i=1}^{u} g_{n_i}(S_r(m'))$. Since the message space is prefix-free, there is a smallest index $h \geq 1$ for which $m_h \neq n_h$. Since each g_j is a bijection, it follows that

$$\prod_{i=h}^{t} g_{m_i}(S_r(m)) = \prod_{i=h}^{u} g_{n_i}(S_r(m'))$$

or

$$g_{m_h} \prod_{i=h+1}^{t} g_{m_i}(S_r(m)) = g_{n_h} \prod_{i=h+1}^{u} g_{n_i}(S_r(m')).$$

Taking $x = \prod_{i=h+1}^{t} g_{m_i}(S_r(m))$, and $y = \prod_{i=h+1}^{u} g_{n_i}(S_r(m'))$, the adversary has a claw $(x, y, g_{m_h}(x))$. This violates the basic premise that it is computationally infeasible to find a claw. It should be noted that this does not necessarily mean that a signature for a new message can be forged. In the particular case given in Example 11.99, finding a claw factors the modulus n and permits anyone to sign an unlimited number of new messages (i.e., a total break of the system is possible).

11.105 Example (*GMR with artificially small parameters.*)
Key generation. Let n, p, q, g_0, g_1 be those given in Example 11.100. A selects the validation parameter $r = 15 \in D_{77}$.
Signature generation. Let $m = 1011000011$ be the message to be signed. Then

$$S_r(m) = g_1^{-1} \circ g_1^{-1} \circ g_0^{-1} \circ g_0^{-1} \circ g_0^{-1} \circ g_0^{-1} \circ g_1^{-1} \circ g_1^{-1} \circ g_0^{-1} \circ g_1^{-1}(15) = 23.$$

A's signature for message m is 23.
Signature verification. To verify the signature, B computes

$$r' = g_1 \circ g_0 \circ g_1 \circ g_1 \circ g_0 \circ g_0 \circ g_0 \circ g_0 \circ g_1 \circ g_1(23) = 15.$$

Since $r = r'$, B accepts the signature. $\square$

GMR scheme with authentication trees

In order to sign multiple messages using the GMR one-time signature scheme (Algorithm 11.102), authentication trees (see §13.4.1) are required. Although conceptually similar to the method described in §11.6.3, only the leaves are used to produce the signature. Before giving details, an overview and some additional notation are necessary.

11.106 Definition A *full binary tree* with k *levels* is a binary tree which has $2^{k+1} - 1$ vertices and 2^k leaves. The leaves are said to be at level k of the tree.

Let T be a full binary tree with k levels. Select public parameters $Y_1, Y_2, \ldots, Y_n$ where $n = 2^k$. Form an authentication tree T^* from T with root label R (see below). R is certified by a TTP and placed in a publicly available file. T^* can now be used to authenticate any of the Y_i by providing the authentication path values associated with the authentication path for Y_i. Each Y_i can now be used as the public parameter r for the GMR scheme. The details for constructing the authentication tree T^* now follow.

The tree T^* is constructed recursively. For the root vertex, select a value r and two t-bit binary strings r_L and r_R. Sign the string $r_L \| r_R$ with the GMR scheme using the public value r. The label for the root consists of the values r, r_L, r_R, and $S_r(r_L \| r_R)$. To authenticate the children of the root vertex, select t-bit binary strings b_{0L}, b_{1L}, b_{0R}, and b_{1R}. The label for the left child of the root is the set of values r_L, b_{0L}, b_{1L}, $S_{r_L}(b_{0L} \| b_{1L})$ and the label for the right child is r_R, b_{0R}, b_{1R}, $S_{r_R}(b_{0R} \| b_{1R})$. Using the strings b_{0L}, b_{1L}, b_{0R}, and b_{1R} as public values for the signing mechanism, one can construct labels for the children of the children of the root. Continuing in this manner, each vertex of T^* can be labeled. The method is illustrated in Figure 11.8.

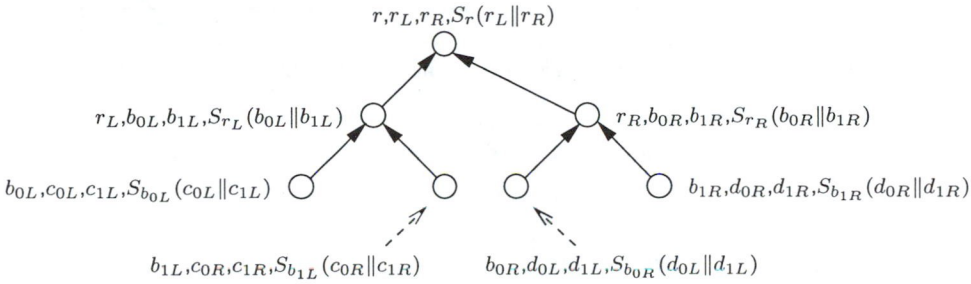

Figure 11.8: *A full binary authentication tree of level 2 for the GMR scheme.*

Each leaf of the authentication tree T^* can be used to sign a different binary message m. The signing procedure uses a pair of claw-free permutations g_0, g_1. If m is the binary message to be signed, and x is the public parameter in the label of a leaf which has not been used to sign any other message, then the signature for m consists of both $S_x(m)$ and the authentication path labels.

11.7 Other signature schemes

The signature schemes described in this section do not fall naturally into the general settings of §11.3 (RSA and related signature schemes), §11.4 (Fiat-Shamir signature schemes), §11.5 (DSA and related signature schemes), or §11.6 (one-time digital signatures).

11.7.1 Arbitrated digital signatures

11.107 Definition An *arbitrated digital signature scheme* is a digital signature mechanism requiring an unconditionally trusted third party (TTP) as part of the signature generation and verification.

Algorithm 11.109 requires a symmetric-key encryption algorithm $E = \{E_k \colon k \in \mathcal{K}\}$ where $\mathcal{K}$ is the key space. Assume that the inputs and outputs of each E_k are l-bit strings, and let $h \colon \{0,1\}^* \longrightarrow \{0,1\}^l$ be a one-way hash function. The TTP selects a key $k_T \in \mathcal{K}$ which it keeps secret. In order to verify a signature, an entity must share a symmetric key with the TTP.

11.108 Algorithm Key generation for arbitrated signatures

SUMMARY: each entity selects a key and transports it secretly with authenticity to the TTP.
Each entity A should do the following:

1. Select a random secret key $k_A \in \mathcal{K}$.
2. Secretly and by some authentic means, make k_A available to the TTP.

11.109 Algorithm Signature generation and verification for arbitrated signatures

SUMMARY: entity A generates signatures using E_{k_A}. Any entity B can verify A's signature with the cooperation of the TTP.

1. *Signature generation.* To sign a message m, entity A should do the following:
    (a) A computes $H = h(m)$.
    (b) A encrypts H with E to get $u = E_{k_A}(H)$.
    (c) A sends u along with some identification string I_A to the TTP.
    (d) The TTP computes $E_{k_A}^{-1}(u)$ to get H.
    (e) The TTP computes $s = E_{k_T}(H\|I_A)$ and sends s to A.
    (f) A's signature for m is s.
2. *Verification.* Any entity B can verify A's signature s on m by doing the following:
    (a) B computes $v = E_{k_B}(s)$.
    (b) B sends v and some identification string I_B to the TTP.
    (c) The TTP computes $E_{k_B}^{-1}(v)$ to get s.
    (d) The TTP computes $E_{k_T}^{-1}(s)$ to get $H\|I_A$.
    (e) The TTP computes $w = E_{k_B}(H\|I_A)$ and sends w to B.
    (f) B computes $E_{k_B}^{-1}(w)$ to get $H\|I_A$.
    (g) B computes $H' = h(m)$ from m.
    (h) B accepts the signature if and only if $H' = H$.

11.110 Note (*security of arbitrated signature scheme*) The security of Algorithm 11.109 is based on the symmetric-key encryption scheme chosen and the ability to distribute keys to participants in an authentic manner. §13.3 discusses techniques for distributing confidential keys.

11.111 Note (*performance characteristics of arbitrated signatures*) Since symmetric-key algorithms are typically much faster than public-key techniques, signature generation and verification by Algorithm 11.109 are (relatively) very efficient. A drawback is that interaction with the TTP is required, which places a much higher burden on the TTP and requires additional message exchanges between entities and the TTP.

11.7.2 ESIGN

ESIGN (an abbreviation for Efficient digital SIGNature) is another digital signature scheme whose security relies on the difficulty of factoring integers. It is a signature scheme with appendix and requires a one-way hash function $h: \{0,1\}^* \longrightarrow \mathbb{Z}_n$.

11.112 Algorithm Key generation for ESIGN

SUMMARY: each entity creates a public key and corresponding private key.
Each entity A should do the following:

1. Select random primes p and q such that $p \geq q$ and p, q are roughly of the same bitlength.
2. Compute $n = p^2 q$.
3. Select a positive integer $k \geq 4$.
4. A's public key is (n, k); A's private key is (p, q).

11.113 Algorithm ESIGN signature generation and verification

SUMMARY: the signing algorithm computes an integer s such that $s^k \bmod n$ lies in a certain interval determined by the message. Verification demonstrates that $s^k \bmod n$ does indeed lie in the specified interval.

1. *Signature generation.* To sign a message m which is a bitstring of arbitrary length, entity A should do the following:
   (a) Compute $v = h(m)$.
   (b) Select a random secret integer x, $0 \leq x < p$.
   (c) Compute $w = \lceil ((v - x^k) \bmod n)/(pq) \rceil$ and $y = w \cdot (kx^{k-1})^{-1} \bmod p$.
   (d) Compute $s = x + ypq \bmod n$.
   (e) A's signature for m is s.
2. *Verification.* To verify A's signature s on m, B should do the following:
   (a) Obtain A's authentic public key (n, k).
   (b) Compute $u = s^k \bmod n$ and $z = h(m)$.
   (c) If $z \leq u \leq z + 2^{\lceil \frac{2}{3} \lg n \rceil}$, accept the signature; else reject it.

Proof that signature verification works. Note that $s^k \equiv (x+ypq)^k \equiv \sum_{i=0}^{k} \binom{k}{i} x^{k-i} (ypq)^i$
$\equiv x^k + kypqx^{k-1} \pmod{n}$. But $kx^{k-1}y \equiv w \pmod{p}$ and, thus, $kx^{k-1}y = w + lp$ for some $l \in \mathbb{Z}$. Hence, $s^k \equiv x^k + pq(w + lp) \equiv x^k + pqw \equiv x^k + pq \left\lceil \frac{(h(m)-x^k)\bmod n}{pq} \right\rceil \equiv$
$x^k + pq \left(\frac{h(m)-x^k+jn+\epsilon}{pq} \right) \pmod{n}$, where $\epsilon = (x^k - h(m)) \bmod pq$. Therefore, $s^k \equiv$
$x^k + h(m) - x^k + \epsilon \equiv h(m) + \epsilon \pmod{n}$. Since $0 \leq \epsilon < pq$, it follows that $h(m) \leq$
$s^k \bmod n \leq h(m) + pq \leq h(m) + 2^{\lceil \frac{2}{3} \lg n \rceil}$, as required.

11.114 Example (*ESIGN for artificially small parameters*) In Algorithm 11.113, take messages to be integers m, $0 \le m < n$, and the hash function h to be $h(m) = m$.

Key generation. A selects primes $p = 17389$ and $q = 15401$, $k = 4$, and computes $n = p^2 q = 4656913120721$. A's public key is $(n = 4656913120721, k = 4)$; A's private key is $(p = 17389, q = 15401)$.

Signature generation. To sign the message $m = 3111527988477$, A computes $v = h(m) = 3111527988477$, and selects $x = 14222$ such that $0 \le x < p = 17389$. A then computes $w = \lceil ((v - x^k) \bmod n)/(pq) \rceil = \lceil 2848181921806/267807989 \rceil = \lceil 10635.16414 \rceil = 10636$ and $y = w(kx^{k-1})^{-1} \bmod p = 10636(4 \times 14222^3)^{-1} \bmod 17389 = 9567$. Finally, A computes the signature $s = x + ypq \bmod n = 2562119044985$.

Signature verification. B obtains A's public key $(n = 4656913120721, k = 4)$, and computes $u = s^k \bmod n = 3111751837675$. Since $3111527988477 \le 3111751837675 \le 3111527988477 + 2^{29}$, B accepts the signature (here, $\lceil \frac{2}{3} \lg n \rceil = 29$). $\square$

11.115 Note (*security of ESIGN*)

    (i) The modulus $n = p^2 q$ in Algorithm 11.113 differs from an RSA modulus by having a repeated factor of p. It is unknown whether or not moduli of this form are easier to factor than integers which are simply the product of two distinct primes.

    (ii) Given a valid signature s for a message m, an adversary could forge a signature for a message m' if $h(m')$ is such that $h(m') \le u \le h(m') + 2^{\lceil \frac{2}{3} \lg n \rceil}$ (where $u = s^k \bmod n$). If an m' with this property is found, then s will be a signature for it. This will occur if $h(m)$ and $h(m')$ agree in the high-order $(\lg n)/3$ bits. Assuming that h behaves like a random function, one would expect to try $2^{(\lg n)/3}$ different values of m' before observing this.

    (iii) Another possible approach to forging is to find a pair of messages m and m' such that $h(m)$ and $h(m')$ agree in the high-order $(\lg n)/3$ bits. By the birthday paradox (Fact 2.27(ii)), one can expect to find such a pair in $O(2^{(\lg n)/6})$ trials. If an adversary is able to get the legitimate signer to sign m, the same signature will be a signature for m'.

    (iv) For the size of the integer n necessary to make the factorization of n infeasible, (ii) and (iii) above are extremely unlikely possibilities.

11.116 Note (*performance characteristics of ESIGN signatures*) Signature generation in Algorithm 11.113 is very efficient. For small values of k (e.g., $k = 4$), the most computationally intensive part is the modular inverse required in step 1c. Depending on the implementation, this corresponds to a small number of modular multiplications with modulus p. For $k = 4$ and a 768-bit modulus n, ESIGN signature generation may be between one and two orders of magnitude (10 to 100 times) faster than RSA signature generation with an equivalent modulus size. Signature verification is also very efficient and is comparable to RSA with a small public exponent.

11.8 Signatures with additional functionality

The mechanisms described in this section provide functionality beyond authentication and non-repudiation. In most instances, they combine a basic digital signature scheme (e.g., RSA) with a specific protocol to achieve additional features which the basic method does not provide.

11.8.1 Blind signature schemes

Rather than signature schemes as described in §11.2, *blind signature schemes* are two-party protocols between a *sender* A and a *signer* B. The basic idea is the following. A sends a piece of information to B which B signs and returns to A. From this signature, A can compute B's signature on an a priori message m of A's choice. At the completion of the protocol, B knows neither the message m nor the signature associated with it.

The purpose of a blind signature is to prevent the signer B from observing the message it signs and the signature; hence, it is later unable to associate the signed message with the sender A.

11.117 Example (*applications of blind signatures*) Blind signature schemes have applications where the sender A (the customer) does not want the signer B (the bank) to be capable of associating a postiori a message m and a signature $S_B(m)$ to a specific instance of the protocol. This may be important in electronic cash applications where a message m might represent a monetary value which A can spend. When m and $S_B(m)$ are presented to B for payment, B is unable to deduce which party was originally given the signed value. This allows A to remain anonymous so that spending patterns cannot be monitored. □

A blind signature protocol requires the following components:

1. A digital signature mechanism for signer B. $S_B(x)$ denotes the signature of B on x.
2. Functions f and g (known only to the sender) such that $g(S_B(f(m))) = S_B(m)$. f is called a *blinding function*, g an *unblinding function*, and $f(m)$ a *blinded message*.

Property 2 places many restrictions on the choice of S_B and g.

11.118 Example (*blinding function based on RSA*) Let $n = pq$ be the product of two large random primes. The signing algorithm S_B for entity B is the RSA signature scheme (Algorithm 11.19) with public key (n, e) and private key d. Let k be some fixed integer with $\gcd(n, k) = 1$. The blinding function $f \colon \mathbb{Z}_n \longrightarrow \mathbb{Z}_n$ is defined by $f(m) = m \cdot k^e \bmod n$ and the unblinding function $g \colon \mathbb{Z}_n \longrightarrow \mathbb{Z}_n$ by $g(m) = k^{-1}m \bmod n$. For this choice of f, g, and S_B, $g(S_B(f(m))) = g(S_B(mk^e \bmod n)) = g(m^d k \bmod n) = m^d \bmod n = S_B(m)$, as required by property 2. □

Protocol 11.119 presents a blind signature scheme which uses the digital signature mechanism and functions f and g described in Example 11.118.

11.119 Protocol Chaum's blind signature protocol

SUMMARY: sender A receives a signature of B on a blinded message. From this, A computes B's signature on a message m chosen a priori by A, $0 \le m \le n - 1$. B has no knowledge of m nor the signature associated with m.

1. *Notation.* B's RSA public and private keys are (n, e) and d, respectively. k is a random secret integer chosen by A satisfying $0 \le k \le n - 1$ and $\gcd(n, k) = 1$.
2. *Protocol actions.*
   (a) (*blinding*) A computes $m^* = mk^e \bmod n$ and sends this to B.
   (b) (*signing*) B computes $s^* = (m^*)^d \bmod n$ which it sends to A.
   (c) (*unblinding*) A computes $s = k^{-1}s^* \bmod n$, which is B's signature on m.

11.8.2 Undeniable signature schemes

Undeniable signature schemes are distinct from digital signatures in the sense of §11.2 in that the signature verification protocol requires the cooperation of the signer. The following example describes two scenarios where an undeniable signature could be applied.

11.120 Example (*scenarios for undeniable signatures*)

(i) Entity A (the customer) wishes to gain access to a secured area controlled by entity B (the bank). The secured area might, for example, be a safety-deposit box room. B requires A to sign a time and date document before access is granted. If A uses an undeniable signature, then B is unable to prove (at some later date) to anyone that A used the facility without A's direct involvement in the signature verification process.

(ii) Suppose some large corporation A creates a software package. A signs the package and sells it to entity B, who decides to make copies of this package and resell it to a third party C. C is unable to verify the authenticity of the software without the cooperation of A. Of course, this scenario does not prevent B from re-signing the package with its own signature but the marketing advantage associated with corporation A's name is lost to B. It will also be easier to trace the fraudulent activity of B.   □

11.121 Algorithm Key generation for Algorithm 11.122

SUMMARY: each entity selects a private key and corresponding public key.
Each entity A should do the following:

1. Select a random prime $p = 2q + 1$ where q is also a prime.
2. (Select a generator α for the subgroup of order q in $\mathbb{Z}_p^*$.)

   2.1 Select a random element $\beta \in \mathbb{Z}_p^*$ and compute $\alpha = \beta^{(p-1)/q} \bmod p$.
   2.2 If $\alpha = 1$ then go to step 2.1.

3. Select a random integer $a \in \{1, 2, \ldots, q-1\}$ and compute $y = \alpha^a \bmod p$.
4. A's public key is (p, α, y); A's private key is a.

11.122 Algorithm Chaum-van Antwerpen undeniable signature scheme

SUMMARY: A signs a message m belonging to the subgroup of order q in $\mathbb{Z}_p^*$. Any entity B can verify this signature with the cooperation of A.

1. *Signature generation.* Entity A should do the following:

   (a) Compute $s = m^a \bmod p$.
   (b) A's signature on message m is s.

2. *Verification.* The protocol for B to verify A's signature s on m is the following:

   (a) B obtains A's authentic public key (p, α, y).
   (b) B selects random secret integers $x_1, x_2 \in \{1, 2, \ldots, q-1\}$.
   (c) B computes $z = s^{x_1} y^{x_2} \bmod p$ and sends z to A.
   (d) A computes $w = (z)^{a^{-1}} \bmod p$ (where $aa^{-1} \equiv 1 \pmod{q}$) and sends w to B. ·
   (e) B computes $w' = m^{x_1} \alpha^{x_2} \bmod p$ and accepts the signature if and only if $w = w'$.

Proof that signature verification works.

$$w \equiv (z)^{a^{-1}} \equiv (s^{x_1}y^{x_2})^{a^{-1}} \equiv (m^{ax_1}\alpha^{ax_2})^{a^{-1}} \equiv m^{x_1}\alpha^{x_2} \equiv w' \bmod p,$$

as required.

Fact 11.123 states that, with high probability, an adversary is unable to cause B to accept a fraudulent signature.

11.123 Fact (*detecting forgeries of undeniable signatures*) Suppose that s is a forgery of A's signature for a message m, i.e., $s \neq m^a \bmod p$. Then the probability of B accepting the signature in Algorithm 11.122 is only $1/q$; this probability is independent of the adversary's computational resources.

11.124 Note (*disavowing signatures*) The signer A could attempt to disavow a (valid) signature constructed by Algorithm 11.122 in one of three ways:

(i) refuse to participate in the verification protocol of Algorithm 11.122;
(ii) perform the verification protocol incorrectly; or
(iii) claim a signature a forgery even though the verification protocol is successful.

Disavowing a signature by following (i) would be considered as an obvious attempt at (wrongful) repudiation. (ii) and (iii) are more difficult to guard against, and require a disavowal protocol (Protocol 11.125).

Protocol 11.125 essentially applies the verification protocol of Algorithm 11.122 twice and then performs a check to verify that A has performed the protocol correctly.

11.125 Protocol Disavowal protocol for Chaum-van Antwerpen undeniable signature scheme

---

SUMMARY: this protocol determines whether the signer A is attempting to disavow a valid signature s using Algorithm 11.122, or whether the signature is a forgery.

1. B obtains A's authentic public key (p, α, y).
2. B selects random secret integers $x_1, x_2 \in \{1, 2, \ldots, q-1\}$, and computes $z = s^{x_1}y^{x_2} \bmod p$, and sends z to A.
3. A computes $w = (z)^{a^{-1}} \bmod p$ (where $aa^{-1} \equiv 1 \pmod{q}$) and sends w to B.
4. If $w = m^{x_1}\alpha^{x_2} \bmod p$, B accepts the signature s and the protocol halts.
5. B selects random secret integers $x_1', x_2' \in \{1, 2, \ldots, q-1\}$, and computes $z' = s^{x_1'}y^{x_2'} \bmod p$, and sends z' to A.
6. A computes $w' = (z')^{a^{-1}} \bmod p$ and sends w' to B.
7. If $w' = m^{x_1'}\alpha^{x_2'} \bmod p$, B accepts the signature s and the protocol halts.
8. B computes $c = (w\alpha^{-x_2})^{x_1'} \bmod p$ and $c' = (w'\alpha^{-x_2'})^{x_1} \bmod p$. If $c = c'$, then B concludes that s is a forgery; otherwise, B concludes that the signature is valid and A is attempting to disavow the signature s.

Fact 11.126 states that Protocol 11.125 achieves its desired objectives.

11.126 Fact Let m be a message and suppose that s is A's (purported) signature on m.

(i) If s is a forgery, i.e., $s \neq m^a \bmod p$, and if A and B follow Protocol 11.125 correctly, then $w = w'$ (and hence, B's conclusion that s is a forgery is correct).
(ii) Suppose that s is indeed A's signature for m, i.e., $s = m^a \bmod p$. Suppose that B follows Protocol 11.125 correctly, but that A does not. Then the probability that $w = w'$ (and hence A succeeds in disavowing the signature) is only $1/q$.

11.127 Note (*security of undeniable signatures*)

    (i) The security of Algorithm 11.122 is dependent on the intractability of the discrete logarithm problem in the cyclic subgroup of order q in $\mathbb{Z}_p^*$ (see §3.6.6).

    (ii) Suppose verifier B records the messages exchanged in step 2 of Algorithm 11.122, and also the random values x_1, x_2 used in the protocol. A third party C should never accept this transcript from B as a verification of signature s. To see why this is the case, it suffices to show how B could contrive a successful transcript of step 2 of Algorithm 11.122 without the signer A's participation. B chooses a message m, integers x_1, x_2 and l in the interval $[1, q-1]$, and computes $s = ((m^{x_1}\alpha^{x_2})^{l^{-1}}y^{-x_2})^{x_1^{-1}}$ mod p. The protocol message from B to A would be $z = s^{x_1}y^{x_2}$ mod p, and from A to B would be $w = z^l$ mod p. Algorithm 11.122 will accept s as a valid signature of A for message m. This argument demonstrates that signatures can only be verified by interacting directly with the signer.

11.8.3 Fail-stop signature schemes

Fail-stop digital signatures are digital signatures which permit an entity A to prove that a signature purportedly (but not actually) signed by A is a forgery. This is done by showing that the underlying assumption on which the signature mechanism is based has been compromised. The ability to prove a forgery does not rely on any cryptographic assumption, but may fail with some small probability; this failure probability is independent of the computing power of the forger. Fail-stop signature schemes have the advantage that even if a very powerful adversary can forge a single signature, the forgery can be detected and the signing mechanism no longer used. Hence, the term *fail-then-stop* is also appropriate. A fail-stop signature scheme should have the following properties:

1. If a signer signs a message according to the mechanism, then a verifier upon checking the signature should accept it.
2. A forger cannot construct signatures that pass the verification algorithm without doing an exponential amount of work.
3. If a forger succeeds in constructing a signature which passes the verification test then, with high probability, the true signer can produce a proof of forgery.
4. A signer cannot construct signatures which are at some later time claimed to be forgeries.

Algorithm 11.130 is an example of a fail-stop mechanism. As described, it is a one-time signature scheme, but there are ways to generalize it to allow multiple signings; using authentication trees is one possibility (see §11.6.3). The proof-of-forgery algorithm is presented in Algorithm 11.134.

11.128 Algorithm Key generation for Algorithm 11.130

SUMMARY: key generation is divided between entity A and a trusted third party (TTP).

1. The TTP should do the following:
   - (a) Select primes p and q such that q divides $(p-1)$ and the discrete logarithm problem in $\mathbb{Z}_q^*$ is intractable.
   - (b) (Select a generator α for the cyclic subgroup G of $\mathbb{Z}_p^*$ having order q.)
     - (i) Select a random element $g \in \mathbb{Z}_p^*$ and compute $\alpha = g^{(p-1)/q}$ mod p.
     - (ii) If $\alpha = 1$ then go to step (i).

(c) Select a random integer a, $1 \leq a \leq q - 1$, and compute $\beta = \alpha^a \bmod p$. The integer a is kept secret by the TTP.

(d) Send (p, q, α, β) in the clear to entity A.

2. Entity A should do the following:

(a) Select random secret integers x_1, x_2, y_1, y_2 in the interval $[0, q - 1]$.

(b) Compute $\beta_1 = \alpha^{x_1} \beta^{x_2}$ and $\beta_2 = \alpha^{y_1} \beta^{y_2} \bmod p$.

(c) A's public key is $(\beta_1, \beta_2, p, q, \alpha, \beta)$; A's private key is the quadruple $\overline{x} = (x_1, x_2, y_1, y_2)$.

11.129 Note (*TTP's secret information*) Assuming that the discrete logarithm problem in the subgroup of order q in $\mathbb{Z}_p^*$ is intractable in Algorithm 11.128, the only entity which knows a, the discrete logarithm of β to the base α, is the TTP.

11.130 Algorithm Fail-stop signature scheme (van Heyst-Pedersen)

SUMMARY: this is a one-time digital signature scheme whose security is based on the discrete logarithm problem in the subgroup of order q in $\mathbb{Z}_p^*$.

1. *Signature generation.* To sign a message $m \in [0, q - 1]$, A should do the following:

(a) Compute $s_{1,m} = x_1 + my_1 \bmod q$ and $s_{2,m} = x_2 + my_2 \bmod q$.

(b) A's signature for m is $(s_{1,m}, s_{2,m})$.

2. *Verification.* To verify A's signature $(s_{1,m}, s_{2,m})$ on m, B should do the following:

(a) Obtain A's authentic public key $(\beta_1, \beta_2, p, q, \alpha, \beta)$.

(b) Compute $v_1 = \beta_1 \beta_2^m \bmod p$ and $v_2 = \alpha^{s_{1,m}} \beta^{s_{2,m}} \bmod p$.

(c) Accept the signature if and only if $v_1 = v_2$.

Proof that signature verification works.

$$v_1 \equiv \beta_1 \beta_2^m \equiv (\alpha^{x_1} \beta^{x_2})(\alpha^{y_1} \beta^{y_2})^m \equiv \alpha^{x_1 + my_1} \beta^{x_2 + my_2}$$
$$\equiv \alpha^{s_{1,m}} \beta^{s_{2,m}} \equiv v_2 \pmod{p}.$$

Algorithm 11.130 is a one-time signature scheme since A's private key $\overline{x}$ can be computed if two messages are signed using $\overline{x}$. Before describing the algorithm for proof of forgery (Algorithm 11.134), a number of facts are needed. These are given in Fact 11.131 and illustrated in Example 11.132.

11.131 Fact (*number of distinct quadruples representing a public key and a signature*) Suppose that A's public key in Algorithm 11.130 is $(\beta_1, \beta_2, p, q, \alpha, \beta)$ and private key is the quadruple $\overline{x} = (x_1, x_2, y_1, y_2)$.

(i) There are exactly q^2 quadruples $\overline{x}' = (x_1', x_2', y_1', y_2')$ with $x_1', x_2', y_1', y_2' \in \mathbb{Z}_q$ which yield the same portion (β_1, β_2) of the public key.

(ii) Let T be the set of q^2 quadruples which yield the same portion of the public key (β_1, β_2). For each $m \in \mathbb{Z}_q$, there are exactly q quadruples in T which give the same signature $(s_{1,m}, s_{2,m})$ for m (where a signature is as described in Algorithm 11.130). Hence, the q^2 quadruples in T give exactly q different signatures for m.

(iii) Let $m' \in \mathbb{Z}_q$ be a message different from m. Then the q quadruples in T which yield A's signature $(s_{1,m}, s_{2,m})$ for m, yield q different signatures for m'.

11.132 Example *(illustration of Fact 11.131)* Let $p = 29$ and $q = 7$. $\alpha = 16$ is a generator of the subgroup of order q in $\mathbb{Z}_p^*$. Take $\beta = \alpha^5 \bmod 29 = 23$. Suppose A's private key is $\overline{x} = (2, 3, 5, 2)$; A's public key is $\beta_1 = \alpha^2 \beta^3 \bmod 29 = 7$, $\beta_2 = \alpha^5 \beta^2 \bmod 29 = 16$. The following table lists the $q^2 = 49$ quadruples which give the same public key.

1603	2303	3003	4403	5103	6503	0203
1610	2310	3010	4410	5110	6510	0210
1624	2324	3024	4424	5124	6524	0224
1631	2331	3031	4431	5131	6531	0231
1645	2345	3045	4445	5145	6545	0245
1652	2352	3052	4452	5152	6552	0252
1666	2366	3066	4466	5166	6566	0266

If the 49 quadruples of this table are used to sign the message $m = 1$, exactly $q = 7$ signature pairs $(s_{1,m}, s_{2,m})$ arise. The next table lists the possibilities and those quadruples which generate each signature.

signature pair	$(2, 6)$	$(3, 3)$	$(4, 0)$	$(5, 4)$	$(6, 1)$	$(0, 5)$	$(1, 2)$
quadruples	1610	1624	1631	1645	1652	1666	1603
	2303	2310	2324	2331	2345	2352	2366
	3066	3003	3010	3024	3031	3045	3052
	4452	4466	4403	4410	4424	4431	4445
	5145	5152	5166	5103	5110	5124	5131
	6531	6545	6552	6566	6503	6510	6524
	0224	0231	0245	0252	0266	0203	0210

The next table lists, for each message $m' \in \mathbb{Z}_7$, all signature pairs for the 7 quadruples which yield A's signature $(0, 5)$ for $m = 1$.

quadruple	m'						
	0	1	2	3	4	5	6
1666	16	05	64	53	42	31	20
2352	23	05	50	32	14	66	41
3045	30	05	43	11	56	24	62
4431	44	05	36	60	21	52	13
5124	51	05	22	46	63	10	34
6510	65	05	15	25	35	45	55
0203	02	05	01	04	00	03	06

□

11.133 Note *(probability of successful forgery in Algorithm 11.130)* Suppose that an adversary (the forger) wishes to derive A's signature on some message m'. There are two possibilities to consider.

    (i) The forger has access only to the signer's public key (i.e., the forger is not in possession of a message and valid signature). By Fact 11.131(ii), the probability that the signature created by the adversary is the same as A's signature for m' is only $q/q^2 = 1/q$; this probability is independent of the adversary's computational resources.

    (ii) The forger has access to a message m and a signature $(s_{1,m}, s_{2,m})$ created by the signer. By Fact 11.131(iii), the probability that the signature created by the adversary is the same as A's signature for m' is only $1/q$; again, this probability is independent of the adversary's computational resources.

Suppose now that an adversary has forged A's signature on a message, and the signature passed the verification stage in Algorithm 11.130. The objective is that A should be able to prove that this signature is a forgery. The following algorithm shows how A can, with high probability, use the forged signature to derive the secret a. Since a was supposed to have been known only to the TTP (Note 11.129), it serves as proof of forgery.

11.134 Algorithm Proof-of-forgery algorithm for Algorithm 11.130

SUMMARY: to prove that a signature $s' = (s'_{1,m}, s'_{2,m})$ on a message m is a forgery, the signer derives the integer $a = \log_\alpha \beta$ which serves as proof of forgery.
The signer (entity A) should do the following:

1. Compute a signature pair $s = (s_{1,m}, s_{2,m})$ for message m using its private key $\overline{x}$ (see Algorithm 11.128).
2. If $s = s'$ return to step 1.
3. Compute $a = (s_{1,m} - s'_{1,m}) \cdot (s_{2,m} - s'_{2,m})^{-1} \bmod q$.

Proof that Algorithm 11.134 works. By Fact 11.131(iii), the probability that $s = s'$ in step 1 of Algorithm 11.134 is $1/q$. From the verification algorithm (Algorithm 11.130), $\alpha^{s_{1,m}} \beta^{s_{2,m}} \equiv \alpha^{s'_{1,m}} \beta^{s'_{2,m}} \pmod{p}$ or $\alpha^{s_{1,m} - s'_{1,m}} \equiv \alpha^{a(s'_{2,m} - s_{2,m})} \pmod{p}$ or $s_{1,m} - s'_{1,m} \equiv a(s'_{2,m} - s_{2,m}) \pmod{q}$. Hence, $a = (s_{1,m} - s'_{1,m}) \cdot (s_{2,m} - s'_{2,m})^{-1} \bmod q$. $\square$

11.135 Remark (*disavowing signatures*) In order for a signer to disavow a signature that it created with Algorithm 11.134, an efficient method for computing logarithms is required.

11.9 Notes and further references

§11.1

The concept of a digital signature was introduced in 1976 by Diffie and Hellman [344, 345]. Although the idea of a digital signature was clearly articulated, no practical realization emerged until the 1978 paper by Rivest, Shamir, and Adleman [1060]. Digital signatures appear to have been independently discovered by Merkle [849, 850] but not published until 1978. One of Merkle's contributions is discussed in §11.6.2. Other early research was due to Lamport [738], Rabin [1022, 1023], and Matyas [801].

A detailed survey on digital signatures is given by Mitchell, Piper, and Wild [882]. A thorough discussion of a selected subset of topics in the area is provided by Stinson [1178]. Other sources which provide a good overview are Meyer and Matyas [859], Goldwasser, Micali, and Rivest [484], Rivest [1054], and Schneier [1094].

§11.2

The original proposal for a digital signature scheme by Diffie and Hellman [344] considered only digital signatures with message recovery. The first discussion of digital signature schemes with appendix (although the term was not used per se) appears to be in the patent by Merkle and Hellman [553]. Davies and Price [308] and Denning [326] give brief introductions to digital signatures but restrict the discussion to digital signature schemes with message recovery and one-time digital signature schemes. Mitchell, Piper, and Wild [882] and Stinson [1178] give abstract definitions of digital signature schemes somewhat less general than those given in §11.2.

Excellent discussions on attacks against signature schemes are provided by Goldwasser, Micali, and Rivest [484] and Rivest [1054]. The former refers to the discovery of a functionally equivalent signing algorithm as *universal forgery*, and separates chosen-message attacks into *generic chosen-message attacks* and *directed chosen-message attacks*.

Many proposed digital signature schemes have been shown to be insecure. Among the most prominent of these are the Merkle-Hellman knapsack scheme proposed by Merkle and Hellman [857], shown to be totally breakable by Shamir [1114]; the Shamir fast signature scheme [1109], shown to be totally breakable by Odlyzko [939]; and the Ong-Schnorr-Shamir (OSS) scheme [958], shown to be totally breakable by Pollard (see Pollard and Schnorr [988]). Naccache [914] proposed a modification of the Ong-Schnorr-Shamir scheme to avoid the earlier attacks.

§11.3

The RSA signature scheme (Algorithm 11.19), discovered by Rivest, Shamir, and Adleman [1060], was the first practical signature scheme based on public-key techniques.

The multiplicative property of RSA (§11.3.2(ii)) was first exploited by Davida [302]. Denning [327] reports and expands on Davida's attack and credits Moore with a simplification. Gordon [515] uses the multiplicative property of RSA to show how to create public-key parameters and associated (forged) certificates if the signing authority does not take adequate precautions. The existential attack on RSA signatures having certain types of redundancy (Example 11.21) is due to de Jonge and Chaum [313]. Evertse and van Heyst [381] consider other types of attacks on RSA signatures which also rely on the multiplicative property.

The reblocking problem (§11.3.3(i)) is discussed by Davies and Price [308], who attribute the method of prescribing the form of the modulus to Guillou. An alternate way of constructing an (even) t-bit modulus $n = pq$ having a 1 in the high-order position followed by k 0's is the following. Construct an integer $u = 2^t + w2^{t/2}$ for some randomly selected $(t/2 - k)$-bit integer w. Select a $(t/2)$-bit prime p, and divide p into u to get a quotient q and a remainder r (i.e., $u = pq + r$). If q is a prime number, then $n = pq$ is an RSA modulus of the required type. For example, if $t = 14$ and $k = 3$, let $u = 2^{14} + w2^7$ where $w = 11$. If $p = 89$, then $q = 199$ and $n = pq = 17711$. The binary representation of n is 100010100101111.

The Rabin public-key signature scheme (Algorithm 11.25) is due to Rabin [1023]. Verification of signatures using the Rabin scheme is efficient since only one modular multiplication is required (cf. Note 11.33). Beller and Yacobi [101] take advantage of this aspect in their authenticated key transport protocol (see §12.5.3).

The modified-Rabin signature scheme (Algorithm 11.30) is derived from the RSA variant proposed by Williams [1246] (see also page 315). The purpose of the modification is to provide a deterministic procedure for signing. A similar methodology is incorporated in ISO/IEC 9796 (§11.3.5). The modified scheme can be generalized to other even public exponents besides $e = 2$. If $\gcd(e, (p-1)(q-1)/4) = 1$, then exponentiation by e is a permutation of Q_n.

ISO/IEC 9796 [596] became an international standard in October of 1991. This standard provides examples based on both the RSA and Rabin digital signature mechanisms. Although the standard permits the use of any digital signature scheme with message recovery which provides a t-bit signature for a $\lfloor \frac{t}{2} \rfloor$-bit message, the design was specifically tailored for the RSA and Rabin mechanisms. For design motivation, see Guillou et al. [525]. At the time of publication of ISO/IEC 9796, no other digital signature schemes providing message recovery were known, but since then several have been found; see Koyama et al. [708].

ISO/IEC 9796 is effective for signing messages which do not exceed a length determined by the signature process. Quisquater [1015] proposed a method for extending the utility of ISO/IEC 9796 to longer messages. Briefly, the modified scheme is as follows. Select a one-way hash function h which maps bitstrings of arbitrary length to k-bitstrings. If the signing capability of ISO/IEC 9796 is t bits and m is an n-bit message where $n > t$, then m is partitioned into two bitstrings m_c and m_s, where m_c is $(n-t+k)$ bits long. Compute $d = h(m)$ and form $m' = m_s\|d$; m' is a string of bitlength t. Sign m' using ISO/IEC 9796 to get J. The signature on message m is $m_c\|J$. This provides a randomized digital signature mechanism with message recovery, where the hash function provides the randomization.

§11.3.6 is from PKCS #1 [1072]. This document describes formatting for both encryption and digital signatures but only those details pertinent to digital signatures are mentioned here. The specification does not include message recovery as ISO/IEC 9796 does. It also does not specify the size of the primes, how they should be generated, nor the size of public and private keys. It is suggested that $e = 3$ or $e = 2^{16} + 1$ are widely used. The only attacks mentioned in PKCS #1 (which the formatting attempts to prevent) are those by den Boer and Bosselaers [324], and Desmedt and Odlyzko [341].

§11.4

The Feige-Fiat-Shamir digital signature scheme (Algorithm 11.40), proposed by Feige, Fiat, and Shamir [383], is a minor improvement of the Fiat-Shamir signature scheme [395], requiring less computation and providing a smaller signature. Fiat and Shamir [395] prove that their scheme is secure against existential forgery provided that factoring is intractable and that h is a truly random function. Feige, Fiat, and Shamir [383] prove that their modification has the same property.

Note 11.44 was suggested by Fiat and Shamir [395]. Note 11.45 is due to Micali and Shamir [868], who suggest that only the modulus n_A of entity A needs to be public if $v_1, v_2, \ldots, v_k$ are system-wide parameters. Since all entities have distinct moduli, it is unlikely that $v_j \in Q_n, 1 \le j \le k$, for many different values of n. To overcome this problem, Micali and Shamir claim that some perturbation of k public values is possible to ensure that the resulting values are quadratic residues with respect to a particular modulus, but do not specify any method which provides the necessary perturbation.

The GQ signature scheme (Algorithm 11.48) is due to Guillou and Quisquater [524].

§11.5

The DSA (Algorithm 11.56) is due to Kravitz [711] and was proposed as a Federal Information Processing Standard in August of 1991 by the U.S. National Institute for Science and Technology. It became the Digital Signature Standard (DSS) in May 1994, as specified in FIPS 186 [406]. Smid and Branstad [1157] comment that the DSA was selected based on a number of important factors: the level of security provided, the applicability of patents, the ease of export from the U.S., the impact on national security and law enforcement, and the efficiency in a number of government and commercial applications. They provide a comparison of the computational efficiencies of DSA and RSA and address a number of negative responses received during the FIPS public comment period.

Naccache et al. [916] describe a number of techniques for improving the efficiency of the DSA. For example, the computation of $k^{-1} \bmod q$ in step 1c of Algorithm 11.56 can be replaced by the random generation of an integer b, the computation of $u = bk \bmod q$ and $s = b \cdot \{h(m) + ar\} \bmod q$. The signature is (r, s, u). The verifier computes $u^{-1} \bmod q$ and $u^{-1}s \bmod q = \tilde{s}$. Verification of the signature $(r, \tilde{s})$ now proceeds as in Algorithm 11.56. This variant might be useful for signature generation in chipcard applications where computing power is limited. Naccache et al. also propose the idea of *use and throw coupons* which eliminate the need to compute $r = (\alpha^k \bmod p) \bmod q$. Since this exponentiation is the most computationally intensive portion of DSA signature generation, use and throw coupons greatly improve efficiency. Coupons require storage, and only one signature can be created for each coupon. If storage is limited (as is often the case), only a fixed number of DSA signatures can be created with this method.

Béguin and Quisquater [82] show how to use an insecure server to aid in computations associated with DSA signature generation and verification. The method accelerates the computation of modular multiplication and exponentiation by using an untrusted auxiliary device to provide the majority of the computing. As such, it also applies to schemes other than DSA. Arazi [54] shows how to integrate a Diffie-Hellman key exchange into the DSA.

The ElGamal digital signature scheme (Algorithm 11.64) was proposed in 1984 by ElGamal [368]. ElGamal [368], Mitchell, Piper, and Wild [882], and Stinson [1178] comment further on its security.

Note 11.66 (iv) is due to Bleichenbacher [153], as is Note 11.67 (iii), which is a special case of the following more general result. Suppose p is a prime, α is a generator of $\mathbb{Z}_p^*$, and y is the public key of entity A for an ElGamal signature scheme (Algorithm 11.64). Suppose $p - 1 = bq$ and logarithms in the subgroup of order b in $\mathbb{Z}_p^*$ can be efficiently computed. Finally, suppose that a generator $\beta = cq$ for some $c, 0 < c < b$, and an integer t are known such that $\beta^t \equiv \alpha \pmod{p}$. For message m, the pair (r, s) with $r = \beta$ and $s = t \cdot \{h(m) - cqz\} \bmod (p - 1)$ where z satisfies $\alpha^{qz} \equiv y^q \pmod{p}$ is a signature for message m which will be accepted by Algorithm 11.64. Bleichenbacher also describes how a trapdoor could be constructed for the ElGamal signature scheme (§11.5.2) when system-wide parameters p and α are selected by a fraudulent trusted third party.

Variations of the ElGamal signing equation described in §11.5.2 were proposed by ElGamal [366], Agnew, Mullin, and Vanstone [19], Kravitz [711], Schnorr [1098], and Yen and Laih [1259]. Nyberg and Rueppel [938] and, independently, Horster and Petersen [564], placed these variations in a much more general framework and compared their various properties.

ElGamal signatures based on elliptic curves over finite fields were first proposed by Koblitz [695] and independently by Miller [878] in 1985. A variation of the DSA based on elliptic curves and referred to as the ECDSA is currently being drafted for an IEEE standard.

The Schnorr signature scheme (Algorithm 11.78), due to Schnorr [1098], is derived from an identification protocol given in the same paper (see §10.4.4). Schnorr proposed a preprocessing method to improve the efficiency of the signature generation in Algorithm 11.78. Instead of generating a random integer k and computing $\alpha^k \bmod p$ for each signature, a small number of integers k_i and $\alpha^{k_i} \bmod p$, $1 \le i \le t$, are precomputed and stored, and subsequently combined and refreshed for each signature. De Rooij [315] showed that this preprocessing is insecure if t is small.

Brickell and McCurley [207] proposed a variant of the Schnorr scheme. Their method uses a prime p such that $p-1$ is hard to factor, a prime divisor q of $p-1$, and an element α of order q in $\mathbb{Z}_p^*$. The signing equation is $s = ae+k \bmod (p - 1)$ as opposed to the Schnorr equation $s = ae+k \bmod q$. While computationally less efficient than Schnorr's, this variant has the advantage that its security is based on the difficulty of two hard problems: (i) computing logarithms in the cyclic subgroup of order q in $\mathbb{Z}_p^*$; and (ii) factoring $p - 1$. If either of these problems is hard, then the problem of computing logarithms in $\mathbb{Z}_p^*$ is also hard.

Okamoto [949] describes a variant of the Schnorr scheme which he proves to be secure, provided that the discrete logarithm problem in $\mathbb{Z}_p^*$ is intractable and that correlation-free hash functions exist (no instance of a correlation-free hash function is yet known). Signature generation and verification are not significantly more computationally intensive than in the Schnorr scheme; however, the public key is larger.

The Nyberg-Rueppel scheme (Algorithm 11.81) is due to Nyberg and Rueppel [936]. For an extensive treatment including variants, see Nyberg and Rueppel [938]. They note that unlike RSA, this signature scheme cannot be used for encryption since the signing transformation S has a left inverse, namely, the verification transformation V, but S is not the left inverse of V; in other words, $V(S(m)) = m$ for all $m \in \mathbb{Z}_p$, but $S(V(m)) \neq m$ for most $m \in \mathbb{Z}_p$. The second paper also defines the notion of strong equivalence between signature schemes (two signature schemes are called *strongly equivalent* if the signature on a message m in one scheme can be transformed into the corresponding signature in the other scheme, without knowledge of the private key), and discusses how to modify DSA to provide message recovery.

Some digital signature schemes make it easy to conceal information in the signature which can only be recovered by entities privy to the concealment method. Information communicated this way is said to be *subliminal* and the conveying mechanism is called a *subliminal channel*. Among the papers on this subject are those of Simmons [1139, 1140, 1147, 1149]. Simmons [1139] shows that if a signature requires l_1 bits to convey and provides l_2 bits of security, then $l_1 - l_2$ bits are available for the subliminal channel. This does not imply that all $l_1 - l_2$ bits can, in fact, be used by the channel; this depends on the signature mechanism. If a large proportion of these bits are available, the subliminal channel is said to be *broadband*; otherwise, it is *narrowband*. Simmons [1149] points out that ElGamal-like signature schemes provide a broadband subliminal channel. For example, if the signing equation is $s = k^{-1} \cdot \{h(m) - ar\} \bmod (p - 1)$ where a is the private key known to both the signer and the recipient of the signature, then k can be used to carry the subliminal message. This has the disadvantage that the signer must provide the recipient with the private key, allowing the recipient to sign messages that will be accepted as having originated with the signer. Simmons [1147] describes narrowband channels for the DSA.

§11.6

Rabin [1022] proposed the first one-time signature scheme (Algorithm 11.86) in 1978. Lamport [738] proposed a similar mechanism, popularized by Diffie and Hellman [347], which does not require interaction with the signer for verification. Diffie suggested the use of a one-way hash function to improve the efficiency of the method. For this reason, the mechanism is often referred to as the *Diffie-Lamport scheme*. Lamport [738] also describes a more efficient method for one-time digital signatures, which was rediscovered by Bos and Chaum [172]. Bos and Chaum provide more substantial modifications which lead to a scheme that can be proven to be existentially unforgeable under adaptive chosen-message attack, provided RSA is secure.

Merkle's one-time signature scheme (Algorithm 11.92) is due to Merkle [853]. The modification described in Note 11.95 is attributed by Merkle [853] to Winternitz. Bleichenbacher and Maurer [155] generalize the methods of Lamport, Merkle, and Winternitz through directed acyclic graphs and one-way functions.

Authentication trees were introduced by Merkle [850, 852, 853] at the time when public-key cryptography was in its infancy. Since public-key cryptography and, in particular, digital signatures had not yet been carefully scrutinized, it seemed prudent to devise alternate methods for providing authentication over insecure channels. Merkle [853] suggests that authentication trees provide as much versatility as public-key techniques and can be quite practical. An authentication tree, constructed by a single user to authenticate a large number of public values, requires the user to either regenerate the authentication path values at the time of use or to store all authentication paths and values in advance. Merkle [853] describes a method to minimize the storage requirements if public values are used in a prescribed order.

The GMR scheme (Algorithm 11.102) is due to Goldwasser, Micali, and Rivest [484], who introduced the notion of a claw-free pair of permutations, and described the construction of a claw-free pair of permutations (Example 11.99) based on the integer factorization problem. Combining the one-time signature scheme with tree authentication gives a digital signature mechanism which Goldwasser, Micali and Rivest prove existentially unforgeable under an adaptive chosen-message attack. In order to make their scheme more practical, the tree authentication structure is constructed in such a way that the system must retain some information about preceding signatures (i.e., *memory history* is required). Goldreich [465] suggested modifications to both the general scheme and the example based on integer factorization (Example 11.99), removing the memory constraint and, in the latter, improving the efficiency of the signing procedure. Bellare and Micali [92] generalized the GMR scheme by replacing the claw-free pair of permutations by any trapdoor one-way permutation (the latter requiring a weaker cryptographic assumption). Naor and Yung [920] further generalized the scheme by requiring only the existence of a one-way permutation. The most general result is due to Rompel [1068], who proved that digital signature schemes which are secure against an adaptive chosen-message attack exist if and only if one-way functions exist. Although attractive in theory (due to the fact that secure digital signatures can be reduced to the study of a single structure), none of these methods seem to provide techniques as efficient as RSA and other methods which, although their security has yet to be proven rigorously, have withstood all attacks to date.

On-line/off-line digital signatures were introduced by Even, Goldreich, and Micali [377, 378] as a means to speed up the signing process in applications where computing resources are limited and time to sign is critical (e.g., chipcard applications). The method uses both one-time digital signatures and digital signatures arising from public-key techniques (e.g., RSA, Rabin, DSA). The off-line portion of the signature generation is to create a set of validation parameters for a one-time signature scheme such as the Merkle scheme (Algorithm 11.92), and to hash this set and sign the resulting hash value using a public-key signature scheme. Since the public-key signature scheme is computationally more intensive, it is done off-line. The off-line computations are independent of the message to be signed. The on-line portion is to sign the message using the one-time signature scheme and the validation parameters which were constructed off-line; this part of the signature process is very efficient. Signatures are much longer than would be the case if only the public-key signature mechanism were used to sign the message directly and, consequently, bandwidth requirements are a disadvantage of this procedure.

§11.7

The arbitrated digital signature scheme of Algorithm 11.109 is from Davies and Price [308], based on work by Needham and Schroeder [923].

ESIGN (Algorithm 11.113) was proposed by Okamoto and Shiraishi [953] and was motivated by the digital signature mechanism OSS devised by Ong, Schnorr, and Shamir [958]. The OSS scheme was shown to be insecure by Pollard in a private communication. Ong, Schnorr, and Shamir [958] modified their original scheme but this too was shown insecure by Estes et al. [374]. ESIGN bases its security on the integer factorization problem and the problem of solving polynomial inequalities. The original version of ESIGN [953] proposed $k = 2$ as the appropriate value for the public key. Brickell and DeLaurentis [202] demonstrated that this choice was insecure. Their attack also extends to the case $k = 3$; see Brickell and Odlyzko [209, p.516]. Okamoto [948] revised the method by requiring

$k \geq 4$. No weaknesses for these values of k have been reported in the literature. Fujioka, Okamoto, and Miyaguchi [428] describe an implementation of ESIGN which suggests that it is twenty times faster than RSA signatures with comparable key and signature lengths.

§11.8

Blind signatures (§11.8.1) were introduced by Chaum [242], who described the concept, desired properties, and a protocol for untraceable payments. The first concrete realization of the protocol (Protocol 11.119) was by Chaum [243]. Chaum and Pedersen [251] provide a digital signature scheme which is a variant of the ElGamal signature mechanism (§11.5.2), using a signing equation similar to the Schnorr scheme (§11.5.3), but computationally more intensive for both signing and verification. This signature technique is then used to provide a blind signature scheme.

The concept of a blind signature was extended by Chaum [245] to blinding for unanticipated signatures. Camenisch, Piveteau, and Stadler [228] describe a blind signature protocol based on the DSA (Algorithm 11.56) and one based on the Nyberg-Rueppel scheme (Algorithm 11.81). Horster, Petersen, and Michels [563] consider a number of variants of these protocols. Stadler, Piveteau, and Camenisch [1166] extend the idea of a blind signature to a *fair blind signature* where the signer in cooperation with a trusted third party can link the message and signature, and trace the sender.

Chaum, Fiat, and Naor [250] propose a scheme for *untraceable electronic cash*, which allows a participant A to receive an electronic cash token from a bank. A can subsequently spend the token at a shop B, which need not be on-line with the bank to accept and verify the authenticity of the token. When the token is cashed at the bank by B, the bank is unable to associate it with A. If, however, A attempts to spend the token twice (*double-spending*), A's identity is revealed. Okamoto [951] proposes a divisible electronic cash scheme. A *divisible electronic coin* is an element which has some monetary value associated with it, and which can be used to make electronic purchases many times, provided the total value of all transactions does not exceed the value of the coin.

Undeniable signatures (§11.8.2) were first introduced by Chaum and van Antwerpen [252], along with a disavowal protocol (Protocol 11.125). Chaum [246] shows how to modify the verification protocol for undeniable signatures (step 2 of Algorithm 11.122) to obtain a zero-knowledge verification.

One shortcoming of undeniable signature schemes is the possibility that the signer is unavailable or refuses to co-operate so that the signature cannot be verified by a recipient. Chaum [247] proposed the idea of a *designated confirmer signature* where the signer designates some entity as a confirmer of its signature. If the signer is unavailable or refuses to co-operate, the confirmer has the ability to interact with a recipient of a signature in order to verify it. The confirmer is unable to create signatures for the signer. Chaum [247] describes an example of designated confirmer signatures based on RSA encryption. Okamoto [950] provides a more indepth analysis of this technique and gives other realizations.

A *convertible undeniable digital signature*, introduced by Boyar et al. [181], is an undeniable signature (§11.8.2) with the property that the signer A can reveal a secret piece of information, causing all undeniable signatures signed by A to become ordinary digital signatures. These ordinary digital signatures can be verified by anyone using only the public key of A and requiring no interaction with A in the verification process; i.e., the signatures become *self-authenticating*. This secret information which is made available should not permit anyone to create new signatures which will be accepted as originating from A. As an application of this type of signature, consider the following scenario. Entity A signs all documents during her lifetime with convertible undeniable signatures. The secret piece of information needed to convert these signatures to self-authenticating signatures is placed in trust with her lawyer B. After the death of A, the lawyer can make the secret information public knowledge and all signatures can be verified. B does not have the ability to alter or create new signatures on behalf of A. Boyar et al. [181] give a realization of the concept of convertible undeniable signatures using ElGamal signatures (§11.5.2) and describe how one can reveal information selectively to convert some, but not all, previously created signatures to self-authenticating ones.

Chaum, van Heijst, and Pfitzmann [253] provide a method for constructing undeniable signatures which are unconditionally secure for the signer.

Fail-stop signatures were introduced by Waidner and Pfitzmann [1227] and formally defined by Pfitzmann and Waidner [971]. The first constructions for fail-stop signatures used claw-free pairs of permutations (Definition 11.98) and one-time signature methods (see Pfitzmann and Waidner [972]). More efficient techniques were provided by van Heyst and Pedersen [1202], whose construction is the basis for Algorithm 11.130; they describe three methods for extending the one-time nature of the scheme to multiple signings. van Heyst, Pedersen, and Pfitzmann [1201] extended the idea of van Heyst and Pedersen to fail-stop signatures based on the integer factorization problem.

Damgård [298] proposed a signature scheme in which the signer can gradually and verifiably release the signature to a verifier.

Chaum and van Heyst [254] introduced the concept of a *group signature*. A group signature has the following properties: (i) only members of a predefined group can sign messages; (ii) anyone can verify the validity of a signature but no one is able to identify which member of the group signed; and (iii) in case of disputes, the signature can be opened (with or without the help of group members) to reveal the identity of the group member who signed it. Chen and Pedersen [255] extended this idea to provide group signatures with additional functionality.

Key Establishment Protocols

Contents in Brief

12.1 Introduction

This chapter considers key establishment protocols and related cryptographic techniques which provide shared secrets between two or more parties, typically for subsequent use as symmetric keys for a variety of cryptographic purposes including encryption, message authentication, and entity authentication. The main focus is two-party key establishment, with the aid of a trusted third party in some cases. While many concepts extend naturally to multi-party key establishment including conference keying protocols, such protocols rapidly become more complex, and are considered here only briefly, as is the related area of secret sharing. Broader aspects of key management, including distribution of public keys, certificates, and key life cycle issues, are deferred to Chapter 13.

Relationships to other cryptographic techniques. Key establishment techniques known as key transport mechanisms directly employ symmetric encryption (Chapter 7) or public-key encryption (Chapter 8). Authenticated key transport may be considered a special case of message authentication (Chapter 9) with privacy, where the message includes a cryptographic key. Many key establishment protocols based on public-key techniques employ digital signatures (Chapter 11) for authentication. Others are closely related to techniques for identification (Chapter 10).

Chapter outline

The remainder of this chapter is organized as follows. §12.2 provides background material including a general classification, basic definitions and concepts, and a discussion of objectives. §12.3 and §12.4 discuss key transport and agreement protocols, respectively, based on symmetric techniques; the former includes several protocols involving an on-line trusted third party. §12.5 and §12.6 discuss key transport and agreement protocols, respectively, based on asymmetric techniques; the former includes protocols based on public-key encryption, some of which also employ digital signatures, while the latter includes selected variations of Diffie-Hellman key agreement. §12.7 and §12.8 consider secret sharing and conference keying, respectively. §12.9 addresses the analysis of key establishment protocols and standard attacks which must be countered. §12.10 contains chapter notes with references.

The particular protocols discussed provide a representative subset of the large number of practical key establishment protocols proposed to date, selected according to a number of criteria including historical significance, distinguishing merits, and practical utility, with particular emphasis on the latter.

12.2 Classification and framework

12.2.1 General classification and fundamental concepts

12.1 Definition A *protocol* is a multi-party algorithm, defined by a sequence of steps precisely specifying the actions required of two or more parties in order to achieve a specified objective.

12.2 Definition *Key establishment* is a process or protocol whereby a shared secret becomes available to two or more parties, for subsequent cryptographic use.

Key establishment may be broadly subdivided into *key transport* and *key agreement*, as defined below and illustrated in Figure 12.1.

12.3 Definition A *key transport* protocol or mechanism is a key establishment technique where one party creates or otherwise obtains a secret value, and securely transfers it to the other(s).

12.4 Definition A *key agreement* protocol or mechanism is a key establishment technique in which a shared secret is derived by two (or more) parties as a function of information contributed by, or associated with, each of these, (ideally) such that no party can predetermine the resulting value.

Additional variations beyond key transport and key agreement exist, including various forms of *key update*, such as *key derivation* in §12.3.1.

Key establishment protocols involving authentication typically require a set-up phase whereby authentic and possibly secret initial keying material is distributed. Most protocols have as an objective the creation of distinct keys on each protocol execution. In some cases, the initial keying material pre-defines a fixed key which will result every time the protocol is executed by a given pair or group of users. Systems involving such static keys are insecure under known-key attacks (Definition 12.17).

12.5 Definition *Key pre-distribution* schemes are key establishment protocols whereby the resulting established keys are completely determined *a priori* by initial keying material. In contrast, *dynamic key establishment* schemes are those whereby the key established by a fixed pair (or group) of users varies on subsequent executions.

Dynamic key establishment is also referred to as *session key establishment*. In this case the session keys are dynamic, and it is usually intended that the protocols are immune to known-key attacks.

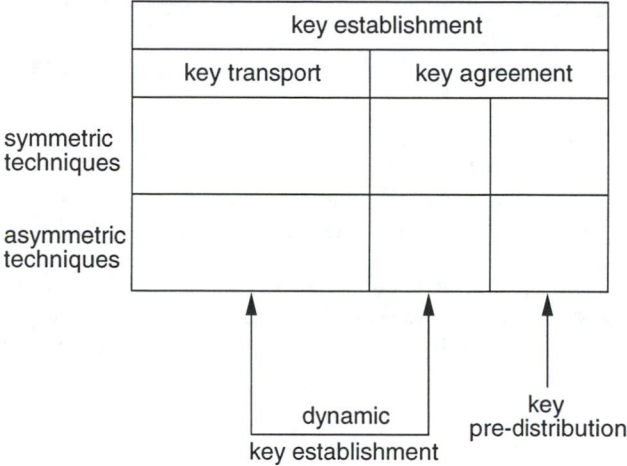

Figure 12.1: *Simplified classification of key establishment techniques.*

Use of trusted servers

Many key establishment protocols involve a centralized or trusted party, for either or both initial system setup and on-line actions (i.e., involving real-time participation). This party is referred to by a variety of names depending on the role played, including: *trusted third party*, *trusted server*, *authentication server*, *key distribution center* (KDC), *key translation center* (KTC), and *certification authority (CA)*. The various roles and functions of such trusted parties are discussed in greater detail in Chapter 13. In the present chapter, discussion is limited to the actions required of such parties in specific key establishment protocols.

Entity authentication, key authentication, and key confirmation

It is generally desired that each party in a key establishment protocol be able to determine the true identity of the other(s) which could possibly gain access to the resulting key, implying preclusion of any unauthorized additional parties from deducing the same key. In this case, the technique is said (informally) to provide *secure key establishment*. This requires both secrecy of the key, and identification of those parties with access to it. Furthermore, the identification requirement differs subtly, but in a very important manner, from that of entity authentication – here the requirement is knowledge of the identity of parties which may gain access to the key, rather than corroboration that actual communication has been established with such parties. Table 12.1 distinguishes various such related concepts, which are highlighted by the definitions which follow.

While *authentication* may be informally defined as the process of verifying that an identity is as claimed, there are many aspects to consider, including who, what, and when. *Entity authentication* is defined in Chapter 10 (Definition 10.1), which presents protocols providing entity authentication alone. *Data origin authentication* is defined in Chapter 9 (Definition 9.76), and is quite distinct.

Authentication term	Central focus
authentication	depends on context of usage
entity authentication	identity of a party, and aliveness at a given instant
data origin authentication	identity of the source of data
(implicit) key authentication	identity of party which may possibly share a key
key confirmation	evidence that a key is possessed by some party
explicit key authentication	evidence an identified party possesses a given key

Table 12.1: *Authentication summary – various terms and related concepts.*

12.6 Definition *Key authentication* is the property whereby one party is assured that no other party aside from a specifically identified second party (and possibly additional identified trusted parties) may gain access to a particular secret key.

Key authentication is independent of the actual possession of such key by the second party, or knowledge of such actual possession by the first party; in fact, it need not involve any action whatsoever by the second party. For this reason, it is sometimes referred to more precisely as *(implicit) key authentication.*

12.7 Definition *Key confirmation* is the property whereby one party is assured that a second (possibly unidentified) party actually has possession of a particular secret key.

12.8 Definition *Explicit key authentication* is the property obtained when both (implicit) key authentication and key confirmation hold.

In the case of explicit key authentication, an identified party is known to actually possess a specified key, a conclusion which cannot otherwise be drawn. Encryption applications utilizing key establishment protocols which offer only implicit key authentication often begin encryption with an initial known data unit serving as an integrity check-word, thus moving the burden of key confirmation from the establishment mechanism to the application.

The focus in key authentication is the identity of the second party rather than the value of the key, whereas in key confirmation the opposite is true. Key confirmation typically involves one party receiving a message from a second containing evidence demonstrating the latter's possession of the key. In practice, possession of a key may be demonstrated by various means, including producing a one-way hash of the key itself, use of the key in a (keyed) hash function, and encryption of a known quantity using the key. These techniques may reveal some information (albeit possibly of no practical consequence) about the value of the key itself; in contrast, methods using zero-knowledge techniques (cf. §10.4.1) allow demonstration of possession of a key while providing no additional information (beyond that previously known) regarding its value.

Entity authentication is not a requirement in all protocols. Some key establishment protocols (such as unauthenticated Diffie-Hellman key agreement) provide *none* of entity authentication, key authentication, and key confirmation. Unilateral key confirmation may always be added e.g., by including a one-way hash of the derived key in a final message.

12.9 Definition An *authenticated key establishment* protocol is a key establishment protocol (Definition 12.2) which provides key authentication (Definition 12.6).

12.10 Remark (*combining entity authentication and key establishment*) In a key establishment protocol which involves entity authentication, it is critical that the protocol be constructed to guarantee that the party whose identity is thereby corroborated is the same party with which the key is established. When this is not so, an adversary may enlist the aid of an unsuspecting authorized party to carry out the authentication aspect, and then impersonate that party in key establishment (and subsequent communications).

Identity-based and non-interactive protocols

Motivation for identity-based systems is provided in §13.4.3.

12.11 Definition A key establishment protocol is said to be *identity-based* if identity information (e.g., name and address, or an identifying index) of the party involved is used as the party's public key. A related idea (see §13.4.4) involves use of identity information as an input to the function which determines the established key.

Identity-based authentication protocols may be defined similarly.

12.12 Definition A two-party key establishment protocol is said to be *message-independent* if the messages sent by each party are independent of any per-session time-variant data (dynamic data) received from other parties.

Message-independent protocols which furthermore involve no dynamic data in the key computation are simply key pre-distribution schemes (Definition 12.5). In general, dynamic data (e.g., that received from another party) is involved in the key computation, even in message-independent protocols.

12.13 Remark (*message-independent vs. non-interactive*) Message-independent protocols include non-interactive protocols (zero-pass and one-pass protocols, i.e., those involving zero or one message but no reply), as well as some two-pass protocols. Regarding inter-party communications, some specification (explicit or otherwise) of the parties involved in key establishment is necessary even in zero-pass protocols. More subtly, in protocols involving t users identified by a vector $(i_1, \ldots, i_t)$, the ordering of indices may determine distinct keys. In other protocols (e.g., basic Diffie-Hellman key agreement or Protocol 12.53), the cryptographic data in one party's message is independent of both dynamic data in other parties' messages and of all party-specific data including public keys and identity information.

12.2.2 Objectives and properties

Cryptographic protocols involving message exchanges require precise definition of both the messages to be exchanged and the actions to be taken by each party. The following types of protocols may be distinguished, based on objectives as indicated:

1. *authentication protocol* – to provide to one party some degree of assurance regarding the identity of another with which it is purportedly communicating;
2. *key establishment protocol* – to establish a shared secret;
3. *authenticated key establishment protocol* – to establish a shared secret with a party whose identity has been (or can be) corroborated.

Motivation for use of session keys

Key establishment protocols result in shared secrets which are typically called, or used to derive, *session keys*. Ideally, a session key is an *ephemeral* secret, i.e., one whose use is restricted to a short time period such as a single telecommunications connection (or session), after which all trace of it is eliminated. Motivation for ephemeral keys includes the following:

1. to limit available ciphertext (under a fixed key) for cryptanalytic attack;
2. to limit exposure, with respect to both time period and quantity of data, in the event of (session) key compromise;
3. to avoid long-term storage of a large number of distinct secret keys (in the case where one terminal communicates with a large number of others), by creating keys only when actually required;
4. to create independence across communications sessions or applications.

It is also desirable in practice to avoid the requirement of maintaining state information across sessions.

Types of assurances and distinguishing protocol characteristics

When designing or selecting a key establishment technique for use, it is important to consider what assurances and properties an intended application requires. Distinction should be made between functionality provided to a user, and technical characteristics which distinguish mechanisms at the implementation level. (The latter are typically of little interest to the user, aside from cost and performance implications.) Characteristics which differentiate key establishment techniques include:

1. *nature* of the authentication. Any combination of the following may be provided: entity authentication, key authentication, and key confirmation.
2. *reciprocity* of authentication. When provided, each of entity authentication, key authentication, and key confirmation may be *unilateral* or *mutual* (provided to one or both parties, respectively).
3. *key freshness*. A key is *fresh* (from the viewpoint of one party) if it can be guaranteed to be new, as opposed to possibly an old key being reused through actions of either an adversary or authorized party. This is related to key control (below).
4. *key control*. In some protocols (key transport), one party chooses a key value. In others (key agreement), the key is derived from joint information, and it may be desirable that neither party be able to control or predict the value of the key.
5. *efficiency*. Considerations include:
   (a) number of message exchanges (*passes*) required between parties;
   (b) bandwidth required by messages (total number of bits transmitted);
   (c) complexity of computations by each party (as it affects execution time); and
   (d) possibility of precomputation to reduce on-line computational complexity.
6. *third party requirements*. Considerations include (see §13.2.4):
   (a) requirement of an on-line (real-time), off-line, or no third party;
   (b) degree of trust required in a third party (e.g., trusted to certify public keys vs. trusted not to disclose long-term secret keys).
7. *type of certificate used, if any*. More generally, one may consider the manner by which initial keying material is distributed, which may be related to third party requirements. (This is often not of direct concern to a user, being an implementation detail typically providing no additional functionality.)

8. *non-repudiation.* A protocol may provide some type of receipt that keying material has been exchanged.

12.14 Remark (*efficiency vs. security*) The efficiency and security of cryptographic techniques are often related. For example, in some protocols a basic step is executed repeatedly, and security increases with the number of repetitions; in this case, the level of security attainable given a fixed amount of time depends on the efficiency of the basic step.

In the description of protocol messages, it is assumed that when the claimed source identity or source network address of a message is not explicitly included as a message field, these are known by context or otherwise available to the recipient, possibly by (unspecified) additional cleartext fields.

12.2.3 Assumptions and adversaries in key establishment protocols

To clarify the threats protocols may be subject to, and to motivate the need for specific protocol characteristics, one requires (as a minimum) an informal model for key establishment protocols, including an understanding of underlying assumptions. Attention here is restricted to two-party protocols, although the definitions and models may be generalized.

Adversaries in key establishment protocols

Communicating parties or entities in key establishment protocols are formally called *principals*, and assumed to have unique names. In addition to legitimate parties, the presence of an unauthorized "third" party is hypothesized, which is given many names under various circumstances, including: *adversary, intruder, opponent, enemy, attacker, eavesdropper,* and *impersonator.*

When examining the security of protocols, it is assumed that the underlying cryptographic mechanisms used, such as encryption algorithms and digital signatures schemes, are secure. If otherwise, then there is no hope of a secure protocol. An adversary is hypothesized to be not a *cryptanalyst* attacking the underlying mechanisms directly, but rather one attempting to subvert the protocol objectives by defeating the manner in which such mechanisms are combined, i.e., attacking the protocol itself.

12.15 Definition A *passive attack* involves an adversary who attempts to defeat a cryptographic technique by simply recording data and thereafter analyzing it (e.g., in key establishment, to determine the session key). An *active attack* involves an adversary who modifies or injects messages.

It is typically assumed that protocol messages are transmitted over *unprotected* (*open*) networks, modeled by an adversary able to completely control the data therein, with the ability to record, alter, delete, insert, redirect, reorder, and reuse past or current messages, and inject new messages. To emphasize this, legitimate parties are modeled as receiving messages exclusively via intervening adversaries (on every communication path, or on some subset of t of n paths), which have the option of either relaying messages unaltered to the intended recipients, or carrying out (with no noticeable delay) any of the above actions. An adversary may also be assumed capable of engaging unsuspecting authorized parties by initiating new protocol executions.

An adversary in a key establishment protocol may pursue many strategies, including attempting to:

1. deduce a session key using information gained by eavesdropping;
2. participate covertly in a protocol initiated by one party with another, and influence it, e.g., by altering messages so as to be able to deduce the key;
3. initiate one or more protocol executions (possibly simultaneously), and combine (*interleave*) messages from one with another, so as to masquerade as some party or carry out one of the above attacks;
4. without being able to deduce the session key itself, deceive a legitimate party regarding the identity of the party with which it shares a key. A protocol susceptible to such an attack is not resilient (see Definition 12.82).

In unauthenticated key establishment, impersonation is (by definition) possible. In entity authentication, where there is no session key to attack, an adversary's objective is to arrange that one party receives messages which satisfy that party that the protocol has been run successfully with a party other than the adversary.

Distinction is sometimes made between adversaries based on the type of information available to them. An *outsider* is an adversary with no special knowledge beyond that generally available, e.g., by eavesdropping on protocol messages over open channels. An *insider* is an adversary with access to additional information (e.g., session keys or secret partial information), obtained by some privileged means (e.g., physical access to private computer resources, conspiracy, etc.). A *one-time insider* obtains such information at one point in time for use at a subsequent time; a *permanent insider* has continual access to privileged information.

Perfect forward secrecy and known-key attacks

In analyzing key establishment protocols, the potential impact of compromise of various types of keying material should be considered, even if such compromise is not normally expected. In particular, the effect of the following is often considered:

1. compromise of long-term secret (symmetric or asymmetric) keys, if any;
2. compromise of past session keys.

12.16 Definition A protocol is said to have *perfect forward secrecy* if compromise of long-term keys does not compromise past session keys.

The idea of perfect forward secrecy (sometimes called *break-backward protection*) is that previous traffic is locked securely in the past. It may be provided by generating sessions keys by Diffie-Hellman key agreement (e.g., Protocol 12.57), wherein the Diffie-Hellman exponentials are based on short-term keys. If long-term secret keys are compromised, future sessions are nonetheless subject to impersonation by an active adversary.

12.17 Definition A protocol is said to be vulnerable to a *known-key attack* if compromise of past session keys allows either a passive adversary to compromise future session keys, or impersonation by an active adversary in the future.

Known-key attacks on key establishment protocols are analogous to known-plaintext attacks on encryption algorithms. One motivation for their consideration is that in some environments (e.g., due to implementation and engineering decisions), the probability of compromise of session keys may be greater than that of long-term keys. A second motivation is that when using cryptographic techniques of only moderate strength, the possibility exists that over time extensive cryptanalytic effort may uncover past session keys. Finally, in some systems, past session keys may be deliberately uncovered for various reasons (e.g., after authentication, to possibly detect use of the authentication channel as a covert or hidden channel).

12.3 Key transport based on symmetric encryption

This section presents a selection of key establishment protocols based on key transport (i.e., transfer of a specific key chosen *a priori* by one party) using symmetric encryption. Related techniques involving non-reversible functions are also presented. Discussion is subdivided into protocols with and without the use of a trusted server, as summarized in Table 12.2. Some of these use time-variant parameters (timestamps, sequence numbers, or random numbers) or nonces as discussed in §10.3.1.

$\rightarrow$ Properties $\downarrow$ Protocol	server type	use of timestamps	number of messages
point-to-point key update	none	optional	1-3
Shamir's no-key protocol	none	no	3
Kerberos	KDC	yes	4
Needham-Schroeder shared-key	KDC	no	5
Otway-Rees	KDC	no	4
Protocol 13.12	KTC	no	3

Table 12.2: *Key transport protocols based on symmetric encryption.*

12.3.1 Symmetric key transport and derivation without a server

Server-less key transport based on symmetric techniques may either require that the two parties in the protocol initially share a long-term pairwise secret or not, respectively illustrated below by point-to-point key update techniques and Shamir's no-key algorithm. Other illustrative techniques are also given.

(i) Point-to-point key update using symmetric encryption

Point-to-point key update techniques based on symmetric encryption make use of a long-term symmetric key K shared *a priori* by two parties A and B. This key, initially distributed over a secure channel or resulting from a key pre-distribution scheme (e.g., see Note 12.48), is used repeatedly to establish new session keys W. Representative examples of point-to-point key transport techniques follow.

Notation: r_A, t_A, and n_A, respectively, denote a random number, timestamp, and sequence number generated by A (see §10.3.1). E denotes a symmetric encryption algorithm (see Remark 12.19). Optional message fields are denoted by an asterisk (*).

1. *key transport with one pass*:

$$A \rightarrow B : E_K(r_A) \quad (1)$$

The session key used is $W = r_A$, and both A and B obtain implicit key authentication. Additional optional fields which might be transferred in the encrypted portion include: a timestamp or sequence number to provide a freshness guarantee to B (see Remark 12.18); a field containing redundancy, to provide explicit key authentication to B or facilitate message modification detection (see Remark 12.19); and a target identifier to prevent undetectable message replay back on A immediately. Thus:

$$A \to B : E_K(r_A, t_A^*, B^*) \quad (1')$$

If it is desired that both parties contribute to the session key, B may send A an analogous message, with the session key computed as $f(r_A, r_B)$. Choosing f to be a one-way function precludes control of the final key value by either party, or an adversary who acquires one of r_A, r_B.

2. *key transport with challenge-response*:

$$A \leftarrow B : n_B \qquad (1)$$
$$A \to B : E_K(r_A, n_B, B^*) \quad (2)$$

If a freshness guarantee is desired but reliance on timestamps is not, a random number or sequence number, denoted n_B here, may be used to replace the timestamp in the one-pass technique; the cost is an additional message. The session key is again $W = r_A$.

If it is required that the session key W be a function of inputs from both parties, A may insert a nonce n_A preceding n_B in (2), and a third message may be added as below. (Here r_A, r_B are random numbers serving as keying material, while n_A, n_B are nonces for freshness.)

$$A \leftarrow B : n_B \qquad\qquad\qquad (1)$$
$$A \to B : E_K(r_A, n_A, n_B, B^*) \quad (2)$$
$$A \leftarrow B : E_K(r_B, n_B, n_A, A^*) \quad (3)$$

12.18 Remark (*key update vulnerabilities*) The key update techniques above do not offer perfect forward secrecy, and fail completely if the long-term key K is compromised. For this reason they may be inappropriate for many applications. The one-pass protocol is also subject to replay unless a timestamp is used.

12.19 Remark (*integrity guarantees within encryption*) Many authentication protocols which employ encryption, including the above key update protocols and Protocols 12.24, 12.26, and 12.29, require for security reasons that the encryption function has a built-in data integrity mechanism (see Figure 9.8(b) for an example, and Definition §9.75) to detect message modification.

(ii) Point-to-point key update by key derivation and non-reversible functions

Key update may be achieved by key transport as above, or by *key derivation* wherein the derived session key is based on per-session random input provided by one party. In this case, there is also a single message:

$$A \to B : r_A \quad (1)$$

The session key is computed as $W = E_K(r_A)$. The technique provides to both A and B implicit key authentication. It is, however, susceptible to known-key attacks; Remark 12.18 similarly applies. The random number r_A here may be replaced by other time-variant parameters; for example, a timestamp t_A validated by the recipient by comparison to its local clock provides an implicit key freshness property, provided the long-term key is not compromised.

Here A could control the value of W, forcing it to be x by choosing $r_A = D_K(x)$. Since the technique itself does not require decryption, E may be replaced by an appropriate keyed pseudorandom function h_K, in which case the session key may be computed as $W = h_K(r_A)$, with r_A a time-variant parameter as noted above.

In the other techniques of §12.3.1(i) employing an encryption function E, the confidentiality itself of the encrypted fields other than the session key W is not critical. A key derivation protocol which entirely avoids the use of an encryption function may offer potential advantages with respect to export restrictions. Protocol 12.20 is such a technique, which also provides authentication guarantees as stated. It uses two distinct functions h and h' (generating outputs of different bitlengths), respectively, for message authentication and key derivation.

12.20 Protocol Authenticated Key Exchange Protocol 2 (AKEP2)

SUMMARY: A and B exchange 3 messages to derive a session key W.

RESULT: mutual entity authentication, and implicit key authentication of W.

1. *Setup*: A and B share long-term symmetric keys K, K' (these should differ but need not be independent). h_K is a MAC (keyed hash function) used for entity authentication. $h'_{K'}$ is a pseudorandom permutation or keyed one-way function used for key derivation.

2. *Protocol messages.* Define $T = (B, A, r_A, r_B)$.

$$
\begin{aligned}
A \to B : &\quad r_A & (1) \\
A \leftarrow B : &\quad T,\ h_K(T) & (2) \\
A \to B : &\quad (A, r_B),\ h_K(A, r_B) & (3) \\
W = h'_{K'}(r_B)
\end{aligned}
$$

3. *Protocol actions.* Perform the following steps for each shared key required.

   (a) A selects and sends to B a random number r_A.

   (b) B selects a random number r_B and sends to A the values (B, A, r_A, r_B), along with a MAC over these quantities generated using h with key K.

   (c) Upon receiving message (2), A checks the identities are proper, that the r_A received matches that in (1), and verifies the MAC.

   (d) A then sends to B the values (A, r_B), along with a MAC thereon.

   (e) Upon receiving (3), B verifies that the MAC is correct, and that the received value r_B matches that sent earlier.

   (f) Both A and B compute the session key as $W = h'_{K'}(r_B)$.

12.21 Note (*AKEP1 variant of Protocol 12.20*) The following modification of AKEP2 results in AKEP1 (Authenticated Key Exchange Protocol 1). B explicitly generates a random session key W and probabilistically encrypts it using h' under K' and random number r. The quantity $(r, W \oplus h'_{K'}(r))$ is now included as a final extra field within T and $h_K(T)$ in (2), and from which A may recover W. As an optimization, $r = r_B$.

(iii) Key transport without a priori shared keys

Shamir's no-key algorithm (Protocol 12.22) is a key transport protocol which, using only symmetric techniques (although involving modular exponentiation), allows key establishment over an open channel without requiring either shared or public keys. Each party has only its own local symmetric key. The protocol provides protection from passive adversaries only; it does not provide authentication. It thus solves the same problem as basic

Diffie-Hellman (Protocol 12.47) – two parties sharing no *a priori* keying material end up with a shared secret key, secure against passive adversaries – although differences include that it uses three messages rather than two, and provides key transport.

12.22 Protocol Shamir's no-key protocol

SUMMARY: users A and B exchange 3 messages over a public channel.
RESULT: secret K is transferred with privacy (but no authentication) from A to B.

1. *One-time setup (definition and publication of system parameters).*

    (a) Select and publish for common use a prime p chosen such that computation of discrete logarithms modulo p is infeasible (see Chapter 3).

    (b) A and B choose respective secret random numbers a, b, with $1 \leq a, b \leq p - 2$, each coprime to $p - 1$. They respectively compute a^{-1} and $b^{-1} \bmod p - 1$.

2. *Protocol messages.*

$$A \rightarrow B : K^a \bmod p \qquad (1)$$
$$A \leftarrow B : (K^a)^b \bmod p \qquad (2)$$
$$A \rightarrow B : (K^{ab})^{a^{-1}} \bmod p \qquad (3)$$

3. *Protocol actions.* Perform the following steps for each shared key required.

    (a) A chooses a random key K for transport to B, $1 \leq K \leq p - 1$. A computes $K^a \bmod p$ and sends B message (1).

    (b) B exponentiates $(\bmod\ p)$ the received value by b, and sends A message (2).

    (c) A exponentiates $(\bmod\ p)$ the received value by $a^{-1} \bmod p - 1$, effectively "undoing" its previous exponentiation and yielding $K^b \bmod p$. A sends the result to B as message (3).

    (d) B exponentiates $(\bmod\ p)$ the received value by $b^{-1} \bmod p - 1$, yielding the newly shared key $K \bmod p$.

Use of ElGamal encryption for key transport (as per §12.5.1) with an uncertified public key sent in a first message (which would by definition be safe from passive attack) achieves in two passes the same goals as the above three-pass algorithm. In this case, the key is transported *from* the recipient of the first message *to* the originator.

12.23 Remark (*choice of cipher in Protocol 12.22*) While it might appear that any commutative cipher (i.e., cipher wherein the order of encryption and decryption is interchangeable) would suffice in place of modular exponentiation in Protocol 12.22, caution is advised. For example, use of the Vernam cipher (§1.5.4) would be totally insecure here, as the XOR of the three exchanged messages would equal the key itself.

12.3.2 Kerberos and related server-based protocols

The key transport protocols discussed in this section are based on symmetric encryption, and involve two communicating parties, A and B, and a trusted server with which they share long-term pairwise secret keys *a priori*. In such protocols, the server either plays the role of a *key distribution center* (KDC) and itself supplies the session key, or serves as a *key translation center* (KTC), and makes a key chosen by one party available to the other, by re-encrypting (translating) it under a key shared with the latter. KDCs and KTCs are discussed further in §13.2.3.

(i) Kerberos authentication protocol

Kerberos is the name given to all of the following: the distributed authentication service originating from MIT's Project Athena, which includes specifications for data integrity and encryption; the software which implements it, and the processes executing such software; and the specific authentication protocol used therein. Focus here, and use of the term "Kerberos", is restricted to the protocol itself, which supports both entity authentication and key establishment using symmetric techniques and a third party.

The basic Kerberos protocol involves A (the *client*), B (the *server* and *verifier*), and a trusted server T (the Kerberos *authentication server*). At the outset A and B share no secret, while T shares a secret with each (e.g., a user password, transformed into a cryptographic key by an appropriate function). The primary objective is for B to verify A's identity; the establishment of a shared key is a side effect. Options include a final message providing mutual entity authentication and establishment of an additional secret shared by A and B (a *subsession key* not chosen by T).

The protocol proceeds as follows. A requests from T appropriate *credentials* (data items) to allow it to authenticate itself to B. T plays the role of a KDC, returning to A a session key encrypted for A and a *ticket* encrypted for B. The ticket, which A forwards on to B, contains the session key and A's identity; this allows authentication of A to B when accompanied by an appropriate message (the *authenticator*) created by A containing a timestamp recently encrypted under that session key.

12.24 Protocol Basic Kerberos authentication protocol (simplified)[1]

SUMMARY: A interacts with trusted server T and party B.
RESULT: entity authentication of A to B (optionally mutual), with key establishment.

1. *Notation.* Optional items are denoted by an asterisk (*).
   E is a symmetric encryption algorithm (see Remark 12.19).
   N_A is a nonce chosen by A; T_A is a timestamp from A's local clock.
   k is the session-key chosen by T, to be shared by A and B.
   L indicates a validity period (called the "lifetime").

2. *One-time setup.* A and T share a key K_{AT}; similarly, B and T share K_{BT}. Define
   $$\text{ticket}_B \overset{\text{def}}{=} E_{K_{BT}}(k, A, L); \quad \text{authenticator} \overset{\text{def}}{=} E_k(A, T_A, A^*_{\text{subkey}}).$$

3. *Protocol messages.*
$$
\begin{aligned}
A \rightarrow T : &\quad A, B, N_A & (1)\\
A \leftarrow T : &\quad \text{ticket}_B, E_{K_{AT}}(k, N_A, L, B) & (2)\\
A \rightarrow B : &\quad \text{ticket}_B, \text{authenticator} & (3)\\
A \leftarrow B : &\quad E_k(T_A, B^*_{\text{subkey}}) & (4)
\end{aligned}
$$

4. *Protocol actions.* Algorithm E includes a built-in integrity mechanism, and protocol failure results if any decryption yields an integrity check failure.

   (a) A generates a nonce N_A and sends to T message (1).
   (b) T generates a new session key k, and defines a validity period (lifetime L) for the ticket, consisting of an ending time and optional starting time. T encrypts k, the received nonce, lifetime, and received identifier (B) using A's key. T also creates a ticket secured using B's key containing k, received identifier (A), and lifetime. T sends to A message (2).

---

[1]The basic Kerberos (version 5) protocol between client and authentication server is given, with messages simplified (some non-cryptographic fields omitted) to allow focus on cryptographic aspects.

(c) A decrypts the non-ticket part of message (2) using K_{AT} to recover: k, N_A, lifetime L, and the identifier of the party for which the ticket was actually created. A verifies that this identifier and N_A match those sent in message (1), and saves L for reference. A takes its own identifier and fresh timestamp T_A, optionally generates a secret A_{subkey}, and encrypts these using k to form the authenticator. A sends to B message (3).

(d) B receives message (3), decrypts the ticket using K_{BT} yielding k to allow decryption of the authenticator. B checks that:

    i. the identifier fields (A) in the ticket and authenticator match;

    ii. the timestamp T_A in the authenticator is valid (see §10.3.1); and

    iii. B's local time is within the lifetime L specified in the ticket.

If all checks pass, B declares authentication of A successful, and saves A_{subkey} (if present) as required.

(e) (Optionally for mutual entity authentication:) B constructs and sends to A message (4) containing A's timestamp from the authenticator (specifically excluding the identifier A, to distinguish it from the authenticator), encrypted using k. B optionally includes a subkey to allow negotiation of a subsession key.

(f) (Optionally for mutual entity authentication:) A decrypts message (4). If the timestamp within matches that sent in message (3), A declares authentication of B successful and saves B_{subkey} (if present) as required.

12.25 Note (*security and options in Kerberos protocol*)

   (i) Since timestamps are used, the hosts on which this protocol runs must provide both secure and synchronized clocks (see §10.3.1).

  (ii) If, as is the case in actual implementations, the initial shared keys are password-derived, then the protocol is no more secure than the secrecy of such passwords or their resistance to password-guessing attacks.

 (iii) Optional parameters A_{subkey} and B_{subkey} allow transfer of a key (other than k) from A to B or vice-versa, or the computation of a combined key using some function $f(A_{\text{subkey}}, B_{\text{subkey}})$.

 (iv) The lifetime within the ticket is intended to allow A to re-use the ticket over a limited time period for multiple authentications to B without additional interaction with T, thus eliminating messages (1) and (2). For each such re-use, A creates a new authenticator with a fresh timestamp and the same session key k; the optional subkey field is of greater use in this case.

(ii) Needham-Schroeder shared-key protocol

The Needham-Schroeder shared-key protocol is important primarily for historical reasons. It is the basis for many of the server-based authentication and key distribution protocols proposed since 1978, including Kerberos and Otway-Rees. It is an example of a protocol independent of timestamps, providing both entity authentication assurances and key establishment with key confirmation. However, it is no longer recommended (see Remark 12.28).

12.26 Protocol Needham-Schroeder shared-key protocol

SUMMARY: A interacts with trusted server T and party B.
RESULT: entity authentication (A with B); key establishment with key confirmation.

1. *Notation.* E is a symmetric encryption algorithm (see Remark 12.19).
   N_A and N_B are nonces chosen by A and B, respectively.
   k is a session key chosen by the trusted server T for A and B to share.
2. *One-time setup.* A and T share a symmetric key K_{AT}; B and T share K_{BT}.
3. *Protocol messages.*

$$
\begin{aligned}
A \to T : &\quad A, B, N_A &\quad (1)\\
A \leftarrow T : &\quad E_{K_{AT}}(N_A, B, k, E_{K_{BT}}(k, A)) &\quad (2)\\
A \to B : &\quad E_{K_{BT}}(k, A) &\quad (3)\\
A \leftarrow B : &\quad E_k(N_B) &\quad (4)\\
A \to B : &\quad E_k(N_B - 1) &\quad (5)
\end{aligned}
$$

4. *Protocol actions.* Aside from verification of nonces, actions are essentially analogous to those in Kerberos (Protocol 12.24), and are not detailed here.

12.27 Note (*functionality and options in Needham-Schroeder shared-key protocol*)

(i) The protocol provides A and B with a shared key k with key authentication (due to the trusted server).

(ii) Messages (4) and (5) provide entity authentication of A to B; entity authentication of B to A can be obtained provided A can carry out some redundancy check on N_B upon decrypting message (4).

(iii) If it is acceptable for A to re-use a key k with B, A may securely cache the data sent in message (3) along with k. Upon subsequent re-use, messages (1) and (2) may then be omitted, but now to prevent replay of old messages (4), an encrypted nonce $E_k(N_A')$ should be appended to message (3), and message (4) should be replaced by $E_k(N_A' - 1, N_B)$ allowing A to verify B's current knowledge of k (thereby providing entity authentication).

12.28 Remark (*Needham-Schroeder weakness vs. Kerberos*) The essential differences between Protocol 12.26 and Kerberos (Protocol 12.24) are as follows: the Kerberos lifetime parameter is not present; the data of message (3), which corresponds to the Kerberos ticket, is unnecessarily double-encrypted in message (2) here; and authentication here employs nonces rather than timestamps. A weakness of the Needham-Schroeder protocol is that since B has no way of knowing if the key k is fresh, should a session key k ever be compromised, any party knowing it may both resend message (3) and compute a correct message (5) to impersonate A to B. This situation is ameliorated in Kerberos by the lifetime parameter which limits exposure to a fixed time interval.

(iii) Otway-Rees protocol

The Otway-Rees protocol is a server-based protocol providing authenticated key transport (with key authentication and key freshness assurances) in only 4 messages – the same as Kerberos, but here without the requirement of timestamps. It does not, however, provide entity authentication or key confirmation.

12.29 Protocol Otway-Rees protocol

SUMMARY: B interacts with trusted server T and party A.
RESULT: establishment of fresh shared secret K between A and B.

1. *Notation.* E is a symmetric encryption algorithm (see Remark 12.19). k is a session key T generates for A and B to share. N_A and N_B are nonces chosen by A and B, respectively, to allow verification of key freshness (thereby detecting replay). M is a second nonce chosen by A which serves as a transaction identifier.

2. *One-time setup.* T shares symmetric keys K_{AT} and K_{BT} with A, B, respectively.

3. *Protocol messages.*

$$
\begin{aligned}
A \to B: &\quad M, A, B, E_{K_{AT}}(N_A, M, A, B) &\quad (1)\\
B \to T: &\quad M, A, B, E_{K_{AT}}(N_A, M, A, B), E_{K_{BT}}(N_B, M, A, B) &\quad (2)\\
B \leftarrow T: &\quad E_{K_{AT}}(N_A, k), E_{K_{BT}}(N_B, k) &\quad (3)\\
A \leftarrow B: &\quad E_{K_{AT}}(N_A, k) &\quad (4)
\end{aligned}
$$

4. *Protocol actions.* Perform the following steps each time a shared key is required.

   (a) A encrypts data for the server containing two nonces, N_A and M, and the identities of itself and the party B to whom it wishes the server to distribute a key. A sends this and some plaintext to B in message (1).

   (b) B creates its own nonce N_B and an analogous encrypted message (with the same M), and sends this along with A's message to T in message (2).

   (c) T uses the cleartext identifiers in message (2) to retrieve K_{AT} and K_{BT}, then verifies the cleartext (M A, B) matches that recovered upon decrypting both parts of message (2). (Verifying M in particular confirms the encrypted parts are linked.) If so, T inserts a new key k and the respective nonces into distinct messages encrypted for A and B, and sends both to B in message (3).

   (d) B decrypts the second part of message (3), checks N_B matches that sent in message (2), and if so passes the first part on to A in message (4).

   (e) A decrypts message (4) and checks N_A matches that sent in message (1).

If all checks pass, each of A and B are assured that k is fresh (due to their respective nonces), and trust that the other party T shared k with is the party bound to their nonce in message (2). A knows that B is active as verification of message (4) implies B sent message (2) recently; B however has no assurance that A is active until subsequent use of k by A, since B cannot determine if message (1) is fresh.

12.30 Remark (*nonces in Otway-Rees protocol*) The use of two nonces generated by A is redundant (N_A could be eliminated in messages (1) and (2), and replaced by M in (3) and (4)), but nonetheless allows M to serve solely as an administrative transaction identifier, while keeping the format of the encrypted messages of each party identical. (The latter is generally considered desirable from an implementation viewpoint, but dubious from a security viewpoint.)

12.31 Remark (*extension of Otway-Rees protocol*) Protocol 12.29 may be extended to provide both key confirmation and entity authentication in 5 messages. Message (4) could be augmented to both demonstrate B's timely knowledge of k and transfer a nonce to A (e.g., appending $E_k(N_A, N_B)$), with a new fifth message ($A \to B : E_k(N_B)$) providing B reciprocal assurances.

12.4 Key agreement based on symmetric techniques

This section presents ideas related to key agreement based on symmetric techniques. It also presents a key pre-distribution system which is in some ways a symmetric-key analogue to Diffie-Hellman key agreement with fixed exponentials (Note 12.48).

12.32 Definition A *key distribution system* (KDS) is a method whereby, during an initialization stage, a trusted server generates and distributes secret data values (*pieces*) to users, such that any pair of users may subsequently compute a shared key unknown to all others (aside from the server).

For fixed pairwise keys, a KDS is a key pre-distribution scheme. A trivial KDS is as follows: the trusted server chooses distinct keys for each pair among the n users, and by some secure means initially distributes to each user its $n - 1$ keys appropriately labeled. This provides unconditional security (perfect security in the information-theoretic sense); an outside adversary can do no better than guess the key. However, due to the large amount of storage required, alternate methods are sought, at the price of losing unconditional security against arbitrarily large groups of colluding users.

12.33 Definition A KDS is said to be *j-secure* if, given a specified pair of users, any coalition of j or fewer users (disjoint from the two), pooling their pieces, can do no better at computing the key shared by the two than a party which guesses the key without any pieces whatsoever.

A j-secure KDS is thus unconditionally secure against coalitions of size j or smaller.

12.34 Fact (*Blom's KDS bound*) In any j-secure KDS providing m-bit pairwise session keys, the secret data stored by each user must be at least $m \cdot (j + 1)$ bits.

The trivial KDS described above is optimal with respect to the number of secret key bits stored, assuming collusion by all parties other than the two directly involved. This corresponds to meeting the lower bound of Fact 12.34 for $j = n - 2$.

Blom's symmetric key pre-distribution system

Blom's scheme (Mechanism 12.35) is a KDS which can be used to meet the bound of Fact 12.34 for values $j < n - 2$. It is non-interactive; each party requires only an index i, $1 \le i \le n$, which uniquely identifies the party with which it is to form a joint key (the scheme is identity-based in this regard). Each user is assigned a secret vector of initial keying material (*base key*) from which it is then able to compute a pairwise secret (*derived key*) with each other user.

As outlined in Remark 12.37, the scheme may be engineered to provide unconditional security against coalitions of a specified maximum size. The initial keying material assigned to each user (a row of S, corresponding to k keys) allows computation of a larger number of derived keys (a row of K, providing n keys), one per each other user. Storage savings results from choosing k less than n. The derived keys of different user pairs, however, are not statistically independent.

12.35 Mechanism Blom's symmetric key pre-distribution system

SUMMARY: each of n users is given initial secret keying material and public data.
RESULT: each pair of users U_i, U_j may compute an m-bit pairwise secret key $K_{i,j}$.

1. A $k \times n$ generator matrix G of an (n, k) MDS code over a finite field $\mathbb{F}_q$ of order q is made known to all n system users (see Note 12.36).
2. A trusted party T creates a random secret $k \times k$ symmetric matrix D over $\mathbb{F}_q$.
3. T gives to each user U_i the secret key S_i, defined as row i of the $n \times k$ matrix $S = (DG)^T$. (S_i is a k-tuple over $\mathbb{F}_q$ of $k \cdot \lg(q)$ bits, allowing U_i to compute any entry in row i of $(DG)^T G$.)
4. Users U_i and U_j compute the common secret $K_{i,j} = K_{j,i}$ of bitlength $m = \lg(q)$ as follows. Using S_i and column j of G, U_i computes the (i, j) entry of the $n \times n$ symmetric matrix $K = (DG)^T G$. Using S_j and column i of G, U_j similarly computes the (j, i) entry (which is equal to the (i, j) entry since K is symmetric).

12.36 Note (*background on MDS codes*) The motivation for Mechanism 12.35 arises from well-known concepts in linear error-correcting codes, summarized here. Let $G = [I_k A]$ be a $k \times n$ matrix where each row is an n-tuple over $\mathbb{F}_q$ (for q a prime or prime power). I_k is the $k \times k$ identity matrix. The set of n-tuples obtained by taking all linear combinations (over $\mathbb{F}_q$) of rows of G is the *linear code* C. Each of these q^k n-tuples is a *codeword*, and $C = \{c : c = mG, m = (m_1 m_2 \ldots m_k), m_i \in \mathbb{F}_q\}$. G is a *generator matrix* for the linear (n, k) code C. The *distance* between two codewords c, c' is the number of components they differ in; the distance d of the code is the minimum such distance over all pairs of distinct codewords. A code of distance d can correct $e = \lfloor (d-1)/2 \rfloor$ component errors in a codeword, and for linear codes $d \le n - k + 1$ (the *Singleton bound*). Codes meeting this bound with equality ($d = n - k + 1$) have the largest possible distance for fixed n and k, and are called *maximum distance separable* (MDS) codes.

12.37 Remark (*choice of k in Blom's scheme*) The condition $d = n - k + 1$ defining MDS codes can be shown equivalent to the condition that every set of k columns of G is linearly independent. From this, two facts follow about codewords of MDS codes: (i) any k components uniquely define a codeword; and (ii) any $j \le k - 1$ components provide no information about other components. For Mechanism 12.35, the choice of k is governed by the fact that if k or more users conspire, they are able to recover the secret keys of all other users. (k conspirators may compute k rows of K, or equivalently k columns, corresponding to k components in each row. Each row is a codeword in the MDS code generated by G, and corresponds to the key of another user, and by the above remark k components thus define all remaining components of that row.) However, if fewer than k users conspire, they obtain no information whatsoever about the keys of any other user (by similar reasoning). Thus Blom's scheme is j-secure for $j \le k - 1$, and relative to Fact 12.34, is optimal with respect to the amount of initial keying material required.

12.5 Key transport based on public-key encryption

Key transport based on public-key encryption involves one party choosing a symmetric key, and transferring it to a second, using that party's encryption public key. This provides key authentication to the originator (only the intended recipient has the private key allowing decryption), but the originator itself obtains neither entity authentication nor key confirmation. The second party receives no source authentication. Such additional assurances may be obtained through use of further techniques including: additional messages (§12.5.1); digital signatures (§12.5.2); and symmetric encryption in addition to signatures (§12.5.3).

Authentication assurances can be provided with or without the use of digital signatures, as follows:

1. *entity authentication via public-key decryption* (§12.5.1). The intended recipient authenticates itself by returning some time-variant value which it alone may produce or recover. This may allow authentication of both the entity and a transferred key.

2. *data origin authentication via digital signatures* (§12.5.2). Public-key encryption is combined with a digital signature, providing key transport with source identity assurances.

The distinction between entity authentication and data origin authentication is that the former provides a timeliness assurance, whereas the latter need not. Table 12.3 summarizes the protocols presented.

→ Properties ↓ Protocol	signatures required‡	entity authentication	number of messages
basic PK encryption (1-pass)	no	no	1
Needham-Schroeder PK	no	mutual	3
encrypting signed keys	yes	data origin only†	1
separate signing, encrypting	yes	data origin only†	1
signing encrypted keys	yes	data origin only†	1
X.509 (2-pass) – timestamps	yes	mutual	2
X.509 (3-pass) – random #'s	yes	mutual	3
Beller-Yacobi (4-pass)	yes	mutual	4
Beller-Yacobi (2-pass)	yes	unilateral	2

Table 12.3: *Selected key transport protocols based on public-key encryption.*
†Unilateral entity authentication may be achieved if timestamps are included.
‡Schemes using public keys transported by certificates require signatures for verification thereof, but signatures are not required within protocol messages.

12.5.1 Key transport using PK encryption without signatures

One-pass key transport by public-key encryption

One-pass protocols are appropriate for one-way communications and store-and-forward applications such as electronic mail and fax. Basic key transport using public-key encryption can be achieved in a one-pass protocol, assuming the originator A possesses *a priori* an authentic copy of the encryption public key of the intended recipient B. Using B's public encryption key, A encrypts a randomly generated key k, and sends the result $P_B(k)$ to B. Public-key encryption schemes P_B of practical interest here include RSA encryption, Rabin encryption, and ElGamal encryption (see Chapter 8).

The originator A obtains no entity authentication of the intended recipient B (and indeed, does not know if B even receives the message), but is assured of implicit key authentication – no one aside from B could possibly recover the key. On the other hand, B has no assurances regarding the source of the key, which remains true even in the case

$A \rightarrow B : P_B(k, A)$. A timeliness guarantee may be provided using timestamps, for example, $A \rightarrow B : P_B(k, T_A)$. This is necessary if security against known-key attacks is required, as this technique is otherwise vulnerable to message replay (cf. Remark 12.18).

Maintaining the restriction of using public-key encryption alone (i.e., without signatures), assurances in addition to unilateral key authentication, namely, mutual entity authentication, and mutual key authentication, may be obtained through additional messages as illustrated by Protocol 12.38 below.

Needham-Schroeder public-key protocol

The Needham-Schroeder public-key protocol provides mutual entity authentication and mutual key transport (A and B each transfer a symmetric key to the other). The transported keys may serve both as nonces for entity authentication and secret keys for further use. Combination of the resulting shared keys allows computation of a joint key to which both parties contribute.

12.38 Protocol Needham-Schroeder public-key protocol

SUMMARY: A and B exchange 3 messages.
RESULT: entity authentication, key authentication, and key transport (all mutual).

1. *Notation.* $P_X(Y)$ denotes public-key encryption (e.g., RSA) of data Y using party X's public key; $P_X(Y_1, Y_2)$ denotes the encryption of the concatenation of Y_1 and Y_2. k_1, k_2 are secret symmetric session keys chosen by A, B, respectively.

2. *One-time setup.* Assume A, B possess each other's authentic public-key. (If this is not the case, but each party has a certificate carrying its own public key, then one additional message is required for certificate transport.)

3. *Protocol messages.*

$$\begin{array}{rll} A \rightarrow B : & P_B(k_1, A) & (1) \\ A \leftarrow B : & P_A(k_1, k_2) & (2) \\ A \rightarrow B : & P_B(k_2) & (3) \end{array}$$

4. *Protocol actions.*

   (a) A sends B message (1).

   (b) B recovers k_1 upon receiving message (1), and returns to A message (2).

   (c) Upon decrypting message (2), A checks the key k_1 recovered agrees with that sent in message (1). (Provided k_1 has never been previously used, this gives A both entity authentication of B and assurance that B knows this key.) A sends B message (3).

   (d) Upon decrypting message (3), B checks the key k_2 recovered agrees with that sent in message (2). The session key may be computed as $f(k_1, k_2)$ using an appropriate publicly known non-reversible function f.

12.39 Note (*modification of Needham-Schroeder protocol*) Protocol 12.38 may be modified to eliminate encryption in the third message. Let r_1 and r_2 be random numbers generated respectively by A and B. Then, with checks analogous to those in the basic protocol, the messages in the modified protocol are:

$$\begin{array}{rll} A \rightarrow B : & P_B(k_1, A, r_1) & (1') \\ A \leftarrow B : & P_A(k_2, r_1, r_2) & (2') \\ A \rightarrow B : & r_2 & (3') \end{array}$$

12.5.2 Protocols combining PK encryption and signatures

While privacy of keying material is a requirement in key transport protocols, source authentication is also typically needed. Encryption and signature primitives may respectively be used to provide these properties. Key transport protocols involving both public-key encryption and signatures include:

1. those which sign the key, then public-key encrypt the signed key;
2. those which sign the key, and separately public-key encrypt the (unsigned) key;
3. those which public-key encrypt the key, then sign the encrypted key; and
4. those using symmetric encryption in addition to public-key encryption and signatures.

The first three types are discussed in this subsection (as noted in §12.5.2(ii), the second is secure only in certain circumstances); the fourth is discussed in §12.5.3. The signature schemes S_A of greatest practical interest are RSA, Rabin signatures, and ElGamal-family signatures (see Chapter 11). The public-key encryption schemes P_B of greatest practical interest are RSA, Rabin encryption, and ElGamal encryption (see Chapter 8).

Notation. For data input y, in what follows, $S_A(y)$ and $P_B(y)$ denote the data values resulting, respectively, from the signature operation on y using A's signature private key, and the encryption operation on y using B's encryption public key. As a default, it is assumed that the signature scheme does not provide message recovery, i.e., the input y cannot be recovered from the signature $S_A(y)$, and y must be sent explicitly in addition to $S_A(y)$ to allow signature verification. (This is the case for DSA, or RSA following input hashing; see Chapter 11. However, in the case of encrypting and signing separately, any secret data y must remain confidential.) If y consists of multiple data values $y = (y_1, \dots, y_n)$, then the input is taken to be the bitwise concatenation of these multiple values.

(i) Encrypting signed keys

One option for combining signatures and public-key encryption is to encrypt signed blocks:

$$A \to B : \quad P_B(k, \ t_A{}^*, \ S_A(B, k, t_A{}^*))$$

The asterisk denotes that the timestamp t_A of A is optional; inclusion facilitates entity authentication of A to B and provides a freshness property. The identifier B within the scope of the signature prevents B from sending the signed key on to another party and impersonating A. A disadvantage of this method over the "signing encrypted keys" alternative (§12.5.2(iii)) is that here the data to be public-key encrypted is larger, implying the possible requirement of adjusting the block size of the public-key encryption scheme, or the use of techniques such as cipher-block-chaining. In the case of signature schemes with message recovery (e.g., ordinary RSA), the above can be simplified to:

$$A \to B : \quad P_B(S_A(B, k, t_A{}^*))$$

(ii) Encrypting and signing separately

For signature schemes without message recovery, a variation of the above option is to sign the key and encrypt the key, but not to encrypt the signature itself. This is acceptable only if the signature scheme is such that no information regarding plaintext data can be deduced from the signature itself on that data (e.g., when the signature operation involves preliminary one-way hashing). This is critical because, in general, data may be recovered from a signature on it (e.g., RSA without hashing). A summary of this case is then as follows:

$$A \to B : \quad P_B(k, \ t_A{}^*), \ S_A(B, k, t_A{}^*)$$

If the key k is used solely to encrypt a data file y, then the signature S_A may be over y instead of k. This is suitable in *store-and-forward* environments. The encrypted file may then be transferred along with the key establishment information, in which case y is first recovered by using k to decrypt the file, and then the signature on y is verified.

(iii) Signing encrypted keys

In contrast to encrypting signed keys, one may sign encrypted keys:

$$A \to B: \quad t_A{}^*, \; P_B(A, k), \; S_A(B, t_A{}^*, P_B(A, k))$$

The asterisk denotes that the timestamp t_A of A is optional; inclusion facilitates entity authentication of A to B. The parameter A within the scope of the public-key encryption prevents *signature stripping* – simply signing a publicly-encrypted key, e.g., $S_A(P_B(k))$ is vulnerable to a third party C extracting the encrypted quantity $P_B(k)$ and then oversigning with its own key, thus defeating authentication (cf. Note 12.42). Furthermore, the encryption mechanism must ensure that an adversary C without access to k, cannot change $P_B(A, k)$ to $P_B(C, k)$; see Remark 12.19. It is desirable and assumed that the combined length of the parameters A and k not exceed the blocklength of the public-key encryption scheme, to limit computation to a single block encryption.

Mutual entity authentication using timestamps. The message format given above can be used for key establishment in a one-pass protocol, although this provides no entity authentication of the recipient to the originator. For mutual entity authentication, two messages of this form may be used, yielding essentially X.509 strong two-way authentication (Protocol 12.40).

Mutual entity authentication using challenge-response. The 2-pass key transport protocol discussed in the previous paragraph requires the use of timestamps, in which case security relies on the assumption of secure, synchronized clocks. This requirement can be eliminated by using a 3-pass protocol with random numbers for challenge-response (essentially the X.509 strong three-way authentication protocol; cf. Protocol 12.43):

$$
\begin{aligned}
A \to B: &\quad r_A \\
A \leftarrow B: &\quad r_B, \; P_A(B, k_1), \; S_B(r_B, r_A, A, P_A(B, k_1)) \\
A \to B: &\quad P_B(A, k_2), \; S_A(r_A, r_B, B, P_B(A, k_2))
\end{aligned}
$$

A and B may compute a joint key k as some function of k_1 and k_2; alternately, one of $P_A(B, k_1)$ and $P_B(A, k_2)$ may be omitted from the second or third message. The identifiers within the scope of the encryption blocks remain necessary as above; the identifiers within the scope of (only) the signature are, however, redundant, both here and in the case of signing encrypted keys above – it may be assumed they must match those corresponding to the public-key encryption.

(iv) X.509 strong authentication protocols

This subsection considers in greater detail a fully-specified protocol involving public-key transport using the general technique of §12.5.2(iii), namely, signing encrypted keys.

The X.509 recommendation defines both "strong two-way" and "strong three-way" authentication protocols, providing mutual entity authentication with optional key transport. Here *strong* distinguishes these from simpler password-based methods, and *two-* and *three-way* refers to protocols with two and three passes (message exchanges), using timestamps and challenge-response based on random numbers, respectively.

Both protocols were designed to provide the assurances listed below to the responder B (and reciprocal assurances intended for the originator A); here *token* refers to cryptographically protected data:

1. the identity of A, and that the token received by B was constructed by A (and not thereafter altered);
2. that the token received by B was specifically intended for B;
3. that the token received by B has "freshness" (has not been used previously, and originated within an acceptably recent timeframe);
4. the mutual secrecy of the transferred key.

12.40 Protocol X.509 strong two-way authentication (two-pass)

SUMMARY: A sends B one message, and B responds with one message.
RESULT: mutual entity authentication and key transport with key authentication.

1. *Notation.*
   $P_X(y)$ denotes the result of applying X's encryption public key to data y.
   $S_X(y)$ denotes the result of applying X's signature private key to y.
   r_A, r_B are never re-used numbers (to detect replay and impersonation).
   $cert_X$ is a certificate binding party X to a public key suitable for both encryption and signature verification (see Remark 12.41).

2. *System setup.*
   (a) Each party has its public key pair for signatures and encryption.
   (b) A must acquire (and authenticate) the encryption public key of B *a priori*. (This may require additional messages and computation.)

3. *Protocol messages.* (An asterisk denotes items are optional.)
   Let $D_A = (t_A, r_A, B, \text{data}_1{}^*, P_B(k_1)^*)$, $D_B = (t_B, r_B, A, r_A, \text{data}_2{}^*, P_A(k_2)^*)$.

$$A \rightarrow B : \quad cert_A, D_A, S_A(D_A) \quad (1)$$
$$A \leftarrow B : \quad cert_B, D_B, S_B(D_B) \quad (2)$$

4. *Protocol actions.*
   (a) A obtains a timestamp t_A indicating an expiry time, generates r_A, optionally obtains a symmetric key k_1 and sends to B message (1). (data_1 is optional data for which data origin authentication is desired.)
   (b) B verifies the authenticity of $cert_A$ (checking the signature thereon, expiry date, etc.), extracts A's signature public key, and verifies A's signature on the data block D_A. B then checks that the identifier in message (1) specifies itself as intended recipient, that the timestamp is valid, and checks that r_A has not been replayed. (r_A includes a sequential component which B checks, against locally maintained state information, for uniqueness within the validity period defined by t_A.)
   (c) If all checks succeed, B declares the authentication of A successful, decrypts k_1 using its private decryption key, and saves this now-shared key. (This terminates the protocol if only unilateral authentication is desired.) B then obtains timestamp t_B, generates r_B, and sends A message (2). (data_2 is optional data, and k_2 is an optional symmetric key provided for A.)
   (d) A carries out actions analogous to those carried out by B. If all checks succeed, A declares the authentication of B successful, and saves key k_2 for subsequent use. A and B share mutual secrets k_1 and k_2.

12.41 Remark (*separate keys in X.509*) The X.509 standard assumes a public-key scheme such as RSA, whereby the same key pair may be used for both encryption and signature functionality. The protocol, however, is easily adapted for separate signature and encryption keys, and, indeed, it is prudent to use separate keys.

12.42 Note (*criticism of X.509 protocol*) Since Protocol 12.40 does not specify inclusion of an identifier (e.g., A) within the scope of the encryption P_B within D_A, one cannot guarantee that the signing party actually knows (or was the source of) the plaintext key.

12.43 Protocol X.509 strong three-way authentication (three-pass)

SUMMARY: A and B exchange 3 messages.
RESULT: as in Protocol 12.40, without requiring timestamps.
The protocol differs from Protocol 12.40 as follows:

1. Timestamps t_A and t_B may be set to zero, and need not be checked.
2. Upon receiving (2), A checks the received r_A matches that in message (1).
3. A third message is sent from A to B:

$$A \to B: \quad (r_B, B), \ S_A(r_B, B) \quad (3)$$

4. Upon receiving (3), B verifies the signature matches the received plaintext, that plaintext identifier B is correct, and that plaintext r_B received matches that in (2).

12.5.3 Hybrid key transport protocols using PK encryption

In contrast to the preceding key transport protocols, the Beller-Yacobi protocol uses symmetric encryption in addition to both PK encryption and digital signatures. Such protocols using both asymmetric and symmetric techniques are called *hybrid protocols*.

Beller-Yacobi protocol (4-pass)

The key transport protocol of Beller and Yacobi, which provides mutual entity authentication and explicit key authentication, was designed specifically for applications where there is an imbalance in processing power between two parties; the goal is to minimize the computational requirements of the weaker party. (Candidate applications include transactions involving chipcards, and wireless communications involving a low-power telephone handset.) Another feature of the protocol is that the identity of one of the parties (the weaker, here A) remains concealed from eavesdroppers.

Essentially, A authenticates itself to B by signing a random challenge m, while B authenticates itself to A by demonstrating knowledge of a key K only B itself could recover. For simplicity of exposition, the protocol is described using RSA with public exponent 3, although Rabin's scheme is more efficient and recommended in practice (but see Note 8.13 regarding chosen-ciphertext attack).

12.44 Protocol Beller-Yacobi key transport (4-pass)

SUMMARY: A transfers key K to B in a 4-pass protocol.
RESULT: mutual entity authentication and mutual explicit key authentication.

1. *Notation.*

   $E_K(y)$ denotes symmetric encryption of y using key K and algorithm E.
   $P_X(y)$ denotes the result of applying X's public-key function to y.
   $S_X(y)$ denotes the result of applying X's private-key function to y.
   I_X denotes an identifying string for party X.
   $h(y)$ denotes the hash of y, used in association with the signature scheme.
   If $y = (y_1, \ldots, y_n)$, the input is the concatenation of these multiple values.

2. *System setup.*

   (a) *Selection of system parameters.* An appropriate prime n_S and generator α for the multiplicative group of integers modulo n_S are fixed as ElGamal system parameters. A trusted server T chooses appropriate primes p and q yielding public modulus $n_T = pq$ for RSA signatures, then for public exponent $e_T = 3$ computes a private key d_T satisfying: $e_T d_T \equiv 1 \bmod (p-1)(q-1)$.

   (b) *Distribution of system parameters.* Each party (A and B) is given an authentic copy of T's public key and the system parameters: n_T, (n_S, α). T assigns to each party X a unique *distinguished name* or identifying string I_X (e.g., X's name and address).

   (c) *Initialization of terminal.* Each party playing the role of A (*terminal*) selects a random integer a, $1 < a \le n_S - 2$, and computes its ElGamal signature public key $u_A = \alpha^a \bmod n_S$. A keeps its corresponding private key a secret, and transfers an authentic copy of u_A to T, identifying itself to T by out-of-band means (e.g., in person). T constructs and returns to A the public-key certificate: $cert_A = (I_A, u_A, G_A)$. (The certificate contains A's identity and ElGamal signature public key, plus T's RSA signature G_A over these: $G_A = S_T(I_A, u_A) = (h(I_A, u_A))^{d_T} \bmod n_T$.)

   (d) *Initialization of server.* Each party playing the role of B (*server*) creates an encryption private key and corresponding public key based on RSA with public exponent $e_B = 3$. B chooses a public-key modulus n_B as the product of two appropriate secret primes, and itself computes the corresponding RSA private key d_B. B transfers n_B to T, identifying itself to T by out-of-band means. T then constructs and returns to B the public-key certificate: $cert_B = (I_B, n_B, G_B)$. (The certificate contains B's identity and RSA encryption public key n_B, plus T's RSA signature over these: $G_B = S_T(I_B, n_B) = (h(I_B, n_B))^{d_T} \bmod n_T$.)

3. *Protocol messages.*

$$
\begin{aligned}
A \leftarrow B : \quad & cert_B = (I_B, n_B, G_B) \quad &(1) \\
A \rightarrow B : \quad & P_B(K) = K^3 \bmod n_B \quad &(2) \\
A \leftarrow B : \quad & E_K(m, \{0\}^t) \quad &(3) \\
A \rightarrow B : \quad & E_K((v, w), cert_A) \quad &(4)
\end{aligned}
$$

4. *Protocol actions.* The following steps are performed each time a shared key is required. The protocol is aborted (with result of failure) if any check fails.

   (a) *Precomputation by terminal.* A selects a random x, $1 \le x \le n_S - 2$, and computes three values: $v = \alpha^x \bmod n_S$; $x^{-1} \bmod (n_S - 1)$; and $av \bmod (n_S - 1)$. (For the security of ElGamal signatures, x must be new for each signature, and be co-prime to $n_S - 1$ to ensure x^{-1} exists.)

(b) B sends to A message (1).

(c) A checks the authenticity of n_B by confirming: $h(I_B, n_B) = G_B{}^3 \bmod n_T$. A chooses a random key $1 < K < n_B - 1$ and sends B message (2), where $Y = P_B(K)$.

(d) B recovers $K = S_B(Y) = Y^{d_B} \bmod n_B$. (The final two messages will be encrypted using K.) B chooses a random integer m as a challenge, extends it with t (say $t \approx 50$) least significant zeros, symmetrically encrypts this using key K, and sends A message (3).

(e) A decrypts the received message, and checks it has t trailing zeros; if so, A accepts that it originated from B and that B knows key K. A takes the decrypted challenge m, concatenates it to the identity I_B of the party whose public key it used to share K in message (2), forming the concatenated quantity $M = (m, I_B)$, then computes w satisfying: $w \equiv (M - av) \cdot x^{-1} \bmod (n_S - 1)$, and sends B message (4). (Here (v, w) is A's ElGamal signature on M, and $cert_A = (I_A, u_A, G_A)$. The identity I_B in M is essential to preclude an intruder-in-the-middle attack – see §12.9.)

(f) B decrypts the received message, and verifies the authenticity of u_A by checking that: $h(I_A, u_A) = G_A{}^3 \bmod n_T$. Finally, B constructs the concatenated quantity $M = (m, I_B)$ from the challenge m remembered from message (3) and its own identity, then verifies A's signature on the challenge by checking that: $\alpha^M \equiv u_A{}^v \cdot v^w \bmod n_S$. If all checks succeed, B accepts the party A associated with identity I_A as the source of key K.

12.45 Note (*on Beller-Yacobi key transport protocol*)

(i) To achieve mutual authentication here requires that each party carry out at least one private-key operation (showing knowledge of its private key), and one or two public-key operations (related to verifying the other's identity, and its public key if not known *a priori*).

(ii) The novelty here is careful selection of two separate public-key schemes, each requiring only an inexpensive computation by the computationally limited party, in this case A. Choosing RSA with exponent 3 or Rabin with exponent 2 results in an inexpensive public-key operation (2 or 1 modular multiplications, respectively), for encryption and signature verification. Choosing ElGamal-family signatures, the private-key operation is inexpensive (a single modular multiplication, assuming precomputation).

(iii) DSA signatures (Chapter 11) or others with similar properties could be used in place of ElGamal signatures.

12.46 Remark (*signature scheme used to certify public keys*) Protocol 12.44 requires an ElGamal public key be certified using an RSA signature. This is done for reasons of efficiency, and highlights an advantage of allowing signature public keys from one system to be certified using signatures of a different type.

Beller-Yacobi protocol (2-pass)

Protocol 12.44 can be modified to yield a 2-pass protocol as illustrated in Figure 12.2. The modified protocol is obtained by essentially combining the pair of messages each party sends into a single message, as now described using notation as in Protocol 12.44.

B generates a random challenge m and sends to A: $m, cert_B$. A computes its ElGamal signature (v, w) on the concatenation $M = (m, I_B)$, and using part v of the signature as the session key $K = v,$[2] sends to B: $P_B(v)$, $E_v(cert_A, w)$. B recovers v ($= K$) via public-key decryption, uses it to recover $cert_A$ and w, then verifies $cert_A$ and A's signature (v, w) on $M = (m, I_B)$.

The 2-pass protocol has slightly weaker authentication assurances: B obtains entity authentication of A and obtains a key K that A alone knows, while A has key authentication with respect to B. For A to obtain explicit key authentication of B (implying entity authentication also), a third message may be added whereby B exhibits knowledge through use of K on a challenge or standard message (e.g., $\{0\}^t$). All three of A's asymmetric operations remain inexpensive.

terminal A		server B
precompute $x, v = \alpha^x \bmod n_S$		select random challenge m
verify $cert_B$ via $P_T(G_B)$	$\longleftarrow$	send $m, cert_B$
compute $(v, w) = S_A(m, I_B)$		$cert_B = (I_B, n_B, G_B)$
send $P_B(v), E_v(cert_A, w)$	$\longrightarrow$	recover v, set $K = v$
$cert_A = (I_A, u_A, G_A)$		verify $cert_A$, signature (v, w)

Figure 12.2: *Summary of Beller-Yacobi protocol (2-pass).*

In Figure 12.2, an alternative to using $K = v$ as the session key is to set $K = w$. This results in the property that both parties influence the value of K (as w is a function of both m and x).

12.6 Key agreement based on asymmetric techniques

Diffie-Hellman key agreement (also called *exponential key exchange*) is a fundamental technique providing unauthenticated key agreement. This section discusses key establishment protocols based on exponential key agreement, as well as the concept of implicitly-certified public keys and their use in Diffie-Hellman protocols.

12.6.1 Diffie-Hellman and related key agreement protocols

This section considers the basic Diffie-Hellman protocol and related protocols providing various authentication assurances (see Table 12.4).

(i) Diffie-Hellman key agreement

Diffie-Hellman key agreement provided the first practical solution to the key distribution problem, allowing two parties, never having met in advance or shared keying material, to establish a shared secret by exchanging messages over an open channel. The security rests on the intractability of the Diffie-Hellman problem and the related problem of computing discrete logarithms (§3.6). The basic version (Protocol 12.47) provides protection in the form of secrecy of the resulting key from passive adversaries (eavesdroppers), but not from

---

[2]A side effect of using $K = v$ is that A no longer directly controls the key value, transforming the key transport protocol into a key agreement. Alternately, a random x could be chosen by A and used as key $K = x$, and x could be sent encrypted alongside w.

→ Properties ↓ Protocol	key authentication	entity authentication	number of messages
Diffie-Hellman	none	none	2
ElGamal key agreement	unilateral	none	1
MTI/A0	mutual – implicit	none	2
Günther (see Remark 12.63)	mutual – implicit	none	2
STS	mutual – explicit	mutual	3

Table 12.4: *Selected key agreement protocols.*

active adversaries capable of intercepting, modifying, or injecting messages. Neither party has assurances of the source identity of the incoming message or the identity of the party which may know the resulting key, i.e., entity authentication or key authentication.

12.47 Protocol Diffie-Hellman key agreement (basic version)

SUMMARY: A and B each send the other one message over an open channel.
RESULT: shared secret K known to both parties A and B.

1. *One-time setup.* An appropriate prime p and generator α of $\mathbb{Z}_p^*$ ($2 \leq \alpha \leq p - 2$) are selected and published.

2. *Protocol messages.*

$$A \rightarrow B : \alpha^x \bmod p \quad (1)$$
$$A \leftarrow B : \alpha^y \bmod p \quad (2)$$

3. *Protocol actions.* Perform the following steps each time a shared key is required.

   (a) A chooses a random secret x, $1 \leq x \leq p - 2$, and sends B message (1).
   (b) B chooses a random secret y, $1 \leq y \leq p - 2$, and sends A message (2).
   (c) B receives α^x and computes the shared key as $K = (\alpha^x)^y \bmod p$.
   (d) A receives α^y and computes the shared key as $K = (\alpha^y)^x \bmod p$.

12.48 Note (*Diffie-Hellman with fixed exponentials*) A variation of Protocol 12.47 provides mutual key authentication. Fix α^x and $\alpha^y \bmod p$ as long-term public keys of the respective parties, and distribute these using signed certificates, thus fixing the long-term shared key for this user pair to $K = \alpha^{xy}$. If such certificates are available *a priori*, this becomes a zero-pass key agreement (no cryptographic messages need be exchanged). The time-invariant nature of this key K, however, is a drawback; Protocol 12.53 provides one resolution. A second solution involves use of key update techniques as in §12.3.1(ii).

12.49 Remark (*Diffie-Hellman in other groups*) The Diffie-Hellman protocol, and those based on it, can be carried out in any group in which both the discrete logarithm problem is hard and exponentiation is efficient. The most common examples of such groups used in practice are the multiplicative group $\mathbb{Z}_p^*$ of $\mathbb{Z}_p$, the analogous multiplicative group of $\mathbb{F}_{2^m}$, and the group of points defined by an elliptic curve over a finite field.

12.50 Note (*control over Diffie-Hellman key*) While it may appear as though Diffie-Hellman key agreement allows each party to guarantee key freshness and preclude key control, use of an exponential with small multiplicative order restricts the order (and thereby value) of the overall key. The most degenerate case for $\mathbb{Z}_p$ would be selection of 0 as private exponent, yielding an exponential with order 1 and the multiplicative identity itself as the resulting key. Thus, either participant may force the resulting key into a subset of the original (naively assumed) range set. Relatedly, some variants of Diffie-Hellman involving unauthenticated exponentials are vulnerable to the following active attack. Assume α generates $\mathbb{Z}_p^*$ where $p = Rq + 1$ (consider $R = 2$ and q prime). Then $\beta = \alpha^q = \alpha^{(p-1)/R}$ has order R ($\beta = -1$ for $R = 2$). If A and B exchange unauthenticated short-term exponentials α^x and α^y, an adversary may replace these by $(\alpha^x)^q$ and $(\alpha^y)^q$, forcing the shared key to be $K = \alpha^{xyq} = \beta^{xy}$, which takes one of only R values ($+1$ or -1 for $R = 2$). K may thus be found by exhaustive trial of R values. A more direct attack involves simply replacing the exchanged exponentials by $+1$ or $p - 1 = -1$. This general class of attacks may be prevented by authenticating the exchanged exponentials, e.g., by a digital signature.

(ii) ElGamal key agreement in one-pass

ElGamal key agreement is a Diffie-Hellman variant providing a one-pass protocol with unilateral key authentication (of the intended recipient to the originator), provided the public key of the recipient is known to the originator *a priori*. While related to ElGamal encryption (§8.4), the protocol is more simply Diffie-Hellman key agreement wherein the public exponential of the recipient is fixed and has verifiable authenticity (e.g., is embedded in a certificate).

12.51 Protocol ElGamal key agreement (half-certified Diffie-Hellman)

SUMMARY: A sends to B a single message allowing one-pass key agreement.
RESULT: shared secret K known to both parties A and B.

1. *One-time setup (key generation and publication).* Each user B does the following:
   Pick an appropriate prime p and generator α of $\mathbb{Z}_p^*$.
   Select a random integer b, $1 \le b \le p - 2$, and compute $\alpha^b \bmod p$.
   B publishes its public key (p, α, α^b), keeping private key b secret.
2. *Protocol messages.*

$$A \to B : \quad \alpha^x \bmod p \quad (1)$$

3. *Protocol actions.* Perform the following steps each time a shared key is required.
   (a) A obtains an authentic copy of B's public key (p, α, α^b).
      A chooses a random integer x, $1 \le x \le p - 2$, and sends B message (1).
      A computes the key as $K = (\alpha^b)^x \bmod p$.
   (b) B computes the same key on receipt of message (1) as $K = (\alpha^x)^b \bmod p$.

12.52 Remark (*assurances in one-pass ElGamal*) The recipient in Protocol 12.51 has no corroboration of whom it shares the secret key with, nor any key freshness assurances. Neither party obtains entity authentication or key confirmation.

(iii) MTI two-pass key agreement protocols

The MTI/A0 variant (Protocol 12.53) of Diffie-Hellman key agreement yields, in two messages (neither requiring signatures), time-variant session keys with mutual (implicit) key authentication against passive attacks. As in ElGamal key agreement (Protocol 12.51), A sends to B a single message, resulting in the shared key K. B independently initiates an analogous protocol with A, resulting in the shared key K'. Each of A and B then computes $k = KK' \bmod p$ (p and α are global parameters now). Neither entity authentication nor key confirmation is provided. Although appropriate for applications where only passive attacks are possible, this protocol is vulnerable to certain active attacks (see Note 12.54).

12.53 Protocol MTI/A0 key agreement

SUMMARY: two-pass Diffie-Hellman key agreement secure against passive attacks.
RESULT: shared secret K known to both parties A and B.

1. *One-time setup.* Select and publish (in a manner guaranteeing authenticity) an appropriate system prime p and generator α of $\mathbb{Z}_p^*$, $2 \leq \alpha \leq p - 2$. A selects as a long-term private key a random integer a, $1 \leq a \leq p - 2$, and computes a long-term public key $z_A = \alpha^a \bmod p$. (B has analogous keys b, z_B.) A and B have access to authenticated copies of each other's long-term public key.

2. *Protocol messages.*

$$A \rightarrow B : \quad \alpha^x \bmod p \quad (1)$$
$$A \leftarrow B : \quad \alpha^y \bmod p \quad (2)$$

3. *Protocol actions.* Perform the following steps each time a shared key is required.

   (a) A chooses a random secret x, $1 \leq x \leq p - 2$, and sends B message (1).
   (b) B chooses a random secret y, $1 \leq y \leq p - 2$, and sends A message (2).
   (c) A computes the key $k = (\alpha^y)^a z_B^x \bmod p$.
   (d) B computes the key $k = (\alpha^x)^b z_A^y \bmod p$. (Both parties now share the key $k = \alpha^{bx+ay} \bmod p$.)

Table 12.5 summarizes Protocol 12.53 and three related two-pass protocols. All four of these MTI protocols provide mutual key authentication without key confirmation or entity authentication, and are role-symmetric: each party executes directly analogous operations. The protocols are also message-independent per Definition 12.12 (neither party requires receipt of the other's message before sending its own), although three of the four require *a priori* access to the other party's authentic public key. The remaining protocol – MTI/A0 – does not, and requires no additional passes (or communications delays) if this is not true; public keys may be exchanged e.g., via certificates included with the existing protocol messages. Thus in MTI/A0, the content of both messages sent is also independent (e.g., of the identity and public key) of the intended recipient.

↓Protocol	m_{AB}	m_{BA}	K_A	K_B	key K
MTI/A0	α^x	α^y	$m_{BA}{}^a z_B{}^x$	$m_{AB}{}^b z_A{}^y$	α^{bx+ay}
MTI/B0	$z_B{}^x$	$z_A{}^y$	$m_{BA}{}^{a^{-1}} \alpha^x$	$m_{AB}{}^{b^{-1}} \alpha^y$	α^{x+y}
MTI/C0	$z_B{}^x$	$z_A{}^y$	$m_{BA}{}^{a^{-1}x}$	$m_{AB}{}^{b^{-1}y}$	α^{xy}
MTI/C1	$z_B{}^{xa}$	$z_A{}^{yb}$	$m_{BA}{}^x$	$m_{AB}{}^y$	α^{abxy}

Table 12.5: *Selected MTI key agreement protocols. A and B have long-term secrets a and b, respectively, verifiably authentic corresponding long-term public keys $z_A = \alpha^a$, $z_B = \alpha^b \bmod p$, and random per-session secrets x and y, respectively. m_{AB} denotes the message A sends to B; m_{BA} is analogous. K_A and K_B are the final key K as computed by A and B.*

12.54 Note (*source-substitution attack on MTI/A0*) As a general rule in all public-key protocols (including Table 12.5), prior to accepting the authenticated public key of a party A, a party B should have assurance (either direct or through a trusted third party) that A actually knows the corresponding private key. Otherwise, an adversary C may claim A's public key as its own, allowing possible attacks, such as that on MTI/A0 as follows. Assume that in a particular implementation, A sends to B its certified public key in a certificate appended to message (1). C registers A's public key as its own (legitimately proving its own identity to the certificate-creating party). When A sends B message (1), C replaces A's certificate with its own, effectively changing the source indication (but leaving the exponential α^x sent by A to B unchanged). C forwards B's response α^y to A. B concludes that subsequently received messages encrypted by the key $k = \alpha^{bx+ay}$ originated from C, whereas, in fact, it is only A who knows k and can originate such messages.

A more complicated attack achieves the same, with C's public key differing from A's public key z_A. C selects an integer e, computes $(z_A)^e = \alpha^{ae}$, and registers the public key α^{ae}. C then modifies α^y sent by B in message (2) to $(\alpha^y)^e$. A and B each compute the key $k = \alpha^{aey}\alpha^{xb}$, which A believes is shared with B (and is), while B believes it is shared with C.

In both variations, C is not actually able to compute k itself, but rather causes B to have false beliefs. Such attacks may be prevented by modifying the protocol such that the exponentials are authenticated (cf. Note 12.50), and binding key confirmation evidence to an authenticated source indication, e.g., through a digital signature (cf. Remark 12.58). The MTI protocols are, however, also subject to certain theoretical known-key attacks (see p.538).

12.55 Remark (*implications of message independence*) Protocols such as MTI/A0 "leak" no information about long-term secrets, since the exchanged messages are independent thereof. However, such protocols in which each party's message is independent of the other's, and yet the session key depends on fresh input from each, cannot provide mutual explicit key authentication.

12.56 Remark (*computational complexity of MTI protocols*) The A0 and B0 protocols require 3 exponentiations by each party, whereas the C0 and C1 protocols require only 2. C1 has the additional advantage over B0 and C0 that no inverses are needed; however, these fixed long-term values may be precomputed.

(iv) Station-to-Station protocol (STS)

The following three-pass variation of the basic Diffie-Hellman protocol allows the establishment of a shared secret key between two parties with mutual entity authentication and mutual explicit key authentication. The protocol also facilitates anonymity – the identities of A and B may be protected from eavesdroppers. The method employs digital signatures; the description below is for the specific case of RSA signatures.

12.57 Protocol Station-to-Station protocol (STS)

SUMMARY: parties A and B exchange 3 messages.
RESULT: key agreement, mutual entity authentication, explicit key authentication.

1. *Notation.* E is a symmetric encryption algorithm.
   $S_A(m)$ denotes A's signature on m, defined as: $S_A(m) = (H(m))^{d_A} \bmod n_A$ (i.e., RSA preceded by an appropriate one-way hash function H, $H(m) < n_A$).
2. *One-time setup (definition and publication of system parameters).*
   (a) Select and publish an appropriate system prime p and generator α of $\mathbb{Z}_p^*$, $2 \le \alpha \le p - 2$. (For additional security, each party may have its own unique such parameters as part of its public key.)
   (b) Each user A selects RSA public and private signature keys (e_A, n_A) and d_A, respectively (B has analogous keys). Assume each party has access to authentic

copies of the other's public key (if not, certificates can be included in existing messages (2) and (3)).

3. *Protocol messages.*

$$
\begin{aligned}
A \rightarrow B : \quad & \alpha^x \bmod p & (1) \\
A \leftarrow B : \quad & \alpha^y \bmod p, \ E_k(S_B(\alpha^y, \ \alpha^x)) & (2) \\
A \rightarrow B : \quad & E_k(S_A(\alpha^x, \ \alpha^y)) & (3)
\end{aligned}
$$

4. *Protocol actions.* Perform the following steps each time a shared key is required. The protocol is aborted (with failure) immediately upon any signature failure.

   (a) A generates a secret random x, $1 \leq x \leq p - 2$, and sends B message (1).

   (b) B generates a secret random y, $1 \leq y \leq p - 2$, and computes the shared key $k = (\alpha^x)^y \bmod p$. B signs the concatenation of both exponentials ordered as in (2), encrypts this using the computed key, and sends A message (2).

   (c) A computes the shared key $k = (\alpha^y)^x \bmod p$, decrypts the encrypted data, and uses B's public key to verify the received value as the signature on the hash of the cleartext exponential received and the exponential sent in message (1). Upon successful verification, A accepts that k is actually shared with B, and sends B an analogous message (3).

   (d) B similarly decrypts the received message (3) and verifies A's signature therein. If successful, B accepts that k is actually shared with A.

The attack of Note 12.50 is precluded in the STS protocol due to the signatures over the exchanged exponentials.

12.58 Remark (*key confirmation in STS protocol*) Encryption under key k provides mutual key confirmation plus allows the conclusion that the party knowing the key is that which signed the exponentials. The optimal use of this protocol occurs when all subsequent messages are also to be encrypted under key k; if this is not the case, alternate means of key confirmation avoiding encryption may be preferable. One alternative is to use a MAC in messages (2) and (3), e.g., for $s = S_A(\alpha^x, \alpha^y)$, $A \rightarrow B : (s, \mathrm{MAC}_k(s))$. A second alternative is inclusion of a one-way hash of k within the signed messages, e.g., $A \rightarrow B : S_A(\alpha^x, \ \alpha^y, \ h(k))$ where here $h(k)$ may be replaced by k alone if the signature process itself employs an appropriate one-way hash.

12.6.2 Implicitly-certified public keys

In contrast both to systems which use public-key certificates (§13.4.2) and to identity-based systems (§13.4.3), an alternate approach to distributing public keys involves *implicitly-certified public keys*, for which a framework is provided in §13.4.4. Use of the word *implicit* here is consistent with that in the term (implicit) key authentication. The current section presents several specific techniques involving implicitly-certified public keys.

(i) Implicitly-certified public keys (of Günther)

Mechanism 12.59 provides a method by which a trusted party may create a Diffie-Hellman public key $r^s \bmod p$ for an entity, with the key being implicitly-certified. Such public keys, which may be reconstructed from public data, may be used in key agreement protocols requiring certified Diffie-Hellman public keys (e.g., z_A in Protocol 12.53) as an alternative to transporting these keys by public-key certificates, or in customized protocols such as Protocol 12.62.

12.59 Mechanism Günther's implicitly-certified (identity-based) public keys

SUMMARY: a trusted party T creates an implicitly-certified, publicly-recoverable Diffie-Hellman public key for A, and transfers to A the corresponding private key.

1. A trusted server T selects an appropriate fixed public prime p and generator α of $\mathbb{Z}_p^*$. T selects a random integer t, with $1 \leq t \leq p - 2$ and $\gcd(t, p - 1) = 1$ as its private key, and publishes its public key $u = \alpha^t \bmod p$, along with α, p.

2. T assigns to each party A a unique *distinguished name* or identifying string I_A (e.g., name and address), and a random integer k_A with $\gcd(k_A, p - 1) = 1$. T then computes $P_A = \alpha^{k_A} \bmod p$. ($P_A$ is A's *reconstruction public data*, allowing other parties to compute $(P_A)^a$ below. The gcd condition ensures that P_A itself is a generator.)

3. Using a suitable hash function h, T solves the following equation for a (restarting with a new k_A if $a = 0$):

$$h(I_A) \equiv t \cdot P_A + k_A \cdot a \pmod{p - 1}. \tag{12.1}$$

4. T securely transmits to A the pair $(r,\ s) = (P_A,\ a)$, which is T's ElGamal signature (see Chapter 11) on I_A. (a is A's private key for Diffie-Hellman key-agreement.)

5. Any other party can then reconstruct A's (Diffie-Hellman) public key $P_A{}^a\ (= \alpha^{k_A a})$ entirely from publicly available information (α, I_A, u, P_A, p) by computing (since $\alpha^{h(I_A)} \equiv u^{P_A} \cdot P_A{}^a$):

$$P_A{}^a \equiv \alpha^{h(I_A)} \cdot u^{-P_A} \bmod p. \tag{12.2}$$

The above mechanism can be generalized to be independent of ElGamal signatures, by using any suitable alternate method to generate a pair $(r,\ s)$ where r is used as the reconstruction public data, the secret s is used as a (key-agreement) private key, and whereby the reconstructed public key $r^s \bmod p$ can be computed from public information alone.

12.60 Remark (*optimization of ElGamal signatures*) Equation (12.1) can be replaced by using the following optimization of the ElGamal signature scheme, where $\gcd(t, p - 1) = 1$:

$$h(I_A) \equiv t \cdot a + k_A \cdot P_A \pmod{p - 1}.$$

To solve for a then requires a one-time inverse computation $(t^{-1} \bmod p - 1)$ rather than the per-signature inverse computation $((k_A)^{-1} \bmod p - 1)$ required by the original signature scheme. With this modification, A's key-agreement public key is $u^a\ (= \alpha^{ta})$ rather than $P_A{}^a\ (= \alpha^{k_A a})$, correspondingly recovered by computing

$$\alpha^{h(I_A)} \cdot P_A^{-P_A} \bmod p\ (= \alpha^{ta} \bmod p). \tag{12.3}$$

(ii) Self-certified public keys (of Girault)

Mechanism 12.61, which is employed in several protocols in §12.6.3, presents a technique for creating implicitly-certified public keys. It differs from that of Mechanism 12.59 in that it allows users to "self-certify" the keys, in the sense that the user itself is the only party knowing the private key (as opposed to the trusted party having access to each party's private key).

12.61 Mechanism Girault's self-certified public keys

SUMMARY: a trusted party T creates an implicitly-certified, publicly-recoverable Diffie-Hellman public key for party A, without learning the corresponding private key.

1. A trusted server T selects secret primes p and q for an RSA integer $n = pq$, an element α of maximal order in $\mathbb{Z}_n^*$ (see Algorithm 4.83), and appropriate integers e and d as a (public, private) RSA key pair for n.

2. T assigns to each party A a unique *distinguished name* or identifying string I_A (e.g., name and address).

3. Party A itself chooses a private key a, and provides the public key $\alpha^a \bmod n$ to T in an authenticatable manner. (α^a is A's key-agreement public key.) Moreover, A provides proof to T that it knows the corresponding secret a. (This is necessary to prevent a certain forgery attack by A in some ways analogous to that of Note 12.54, and might be done by A producing for T a Diffie-Hellman key based on α^a and an exponential chosen by T.)

4. T computes A's reconstruction public data (essentially replacing a certificate) as $P_A = (\alpha^a - I_A)^d \bmod n$. (Thus $(P_A{}^e + I_A) \bmod n = \alpha^a \bmod n$, and from public information alone, any party can compute A's public key, $\alpha^a \bmod p$.)

12.6.3 Diffie-Hellman protocols using implicitly-certified keys

The authenticity of Diffie-Hellman exponentials used as public keys in authenticated key agreement protocols can be established by distributing them via public-key certificates, or by reconstructing them as implicitly-certified public keys (e.g., using Mechanisms of §12.6.2) from publicly available parameters. Protocol 12.62 is one example of the latter. The idea may be adopted to other Diffie-Hellman based protocols as further illustrated by Examples 12.64, 12.65, and 12.66 respectively corresponding to the fixed-key Diffie-Hellman, ElGamal, and MTI/A0 key agreement protocols of §12.6.1.

12.62 Protocol Günther's key agreement protocol

SUMMARY: Diffie-Hellman based key agreement protocol between A and B.
RESULT: A and B establish shared secret K with key authentication.

1. *One-time setup (definition of global parameters).* Using Mechanism 12.59, a trusted party T constructs ElGamal signatures (P_A, a) and (P_B, b) on the identities I_A and I_B of A and B, respectively, and gives these signatures respectively to A and B as secrets, along with the following authentic public system parameters as per Mechanism 12.59: a prime p, generator α of $\mathbb{Z}_p^*$, and T's public key u.

2. *Protocol messages.*

$$
\begin{aligned}
A \rightarrow B: \quad & I_A, P_A & (1) \\
A \leftarrow B: \quad & I_B, P_B, (P_A)^y \bmod p & (2) \\
A \rightarrow B: \quad & (P_B)^x \bmod p & (3)
\end{aligned}
$$

3. *Protocol actions.* Perform the following steps each time a shared key is required.

   (a) A sends B message (1).
   (b) B generates a random integer y, $1 \leq y \leq p - 2$, and sends A message (2).
   (c) A generates a random integer x, $1 \leq x \leq p - 2$, and sends B message (3).

> (d) *Key computation.* As per Mechanism 12.59, A and B respectively construct the other's identity-based public key (equivalent to $(P_B)^b$ and $(P_A)^a \bmod p$, respectively). The common key-agreement key K $(= \alpha^{k_A y a + k_B b x})$ is established as A and B respectively compute $K = (P_A{}^y)^a \cdot (P_B{}^b)^x$, $K = (P_A{}^a)^y \cdot (P_B{}^x)^b \bmod p$.

Protocol 12.62 is subject to theoretical known-key attacks similar to those which apply to the MTI protocols (Note 12.54).

12.63 Remark (*two-pass Günther protocol*) In Protocol 12.62, a party's identity information and long-term public key (respectively, I_A and P_A) are long-term parameters. If these are known to parties *a priori*, then this three-pass protocol reduces to two passes. The reduced protocol provides the same assurances, namely, key agreement with key authentication, as Protocol 12.62 and the two-pass MTI schemes of Table 12.5, and closely resembles MTI/A0 with respect to the logarithm of the final key.

12.64 Example (*Protocol G0*) Fixed-key Diffie-Hellman key-agreement (Note 12.48) may be modified to use implicitly-certified keys as follows. Using the setup and notation as in Girault's self-certified public keys (Mechanism 12.61), A and B establish the time-invariant joint key K by respectively computing $(P_B)^e + I_B \bmod n$ $(= \alpha^b)$ and $(P_A)^e + I_A \bmod n$ $(= \alpha^a)$, from which they effectively compute

$$K = (\alpha^b)^a \quad \text{and} \quad K = (\alpha^a)^b \bmod n. \qquad (12.4)$$

Alternatively, the same protocol may be modified to use Günther's ID-based public keys assuming the setup and notation as in Mechanism 12.59 with modified ElGamal signatures as per Remark 12.60. In this case, A and B respectively compute the other's key-agreement public keys α^{tb} and α^{ta} by (12.3), in place of α^b and α^a in (12.4). □

12.65 Example (*Protocol G1*) The one-pass ElGamal key agreement of Protocol 12.51 may be modified to use implicitly-certified keys as follows. Using the setup and notation as in Girault's self-certified public keys (Mechanism 12.61), A chooses a random integer x and sends to B: $\alpha^x \bmod n$. A computes $P_B{}^e + I_B \bmod n$ $(= \alpha^b)$. A and B establish the time-variant joint key $K = \alpha^{bx} \bmod n$, by respectively computing, effectively,

$$K = (\alpha^b)^x \quad \text{and} \quad K = (\alpha^x)^b \bmod n. \qquad (12.5)$$

The protocol may be modified to use Günther's ID-based public keys as follows: rather than sending $\alpha^x \bmod n$ to B, A sends $P_B{}^x \bmod p$, with P_B (and p, b, u, etc.) defined as in Mechanism 12.59. B then computes $K = (P_B{}^x)^b \bmod p$, while A effectively computes $K = (P_B{}^b)^x \bmod p$, having reconstructed $P_B{}^b$ via equation (12.2) on page 521. The resulting protocol is essentially one-half of the Günther key agreement of Protocol 12.62. A related modification utilizing Remark 12.60 involves A sending to B $u^x \bmod p$ in place of $P_B{}^x$, the joint key now being $K = u^{bx} \bmod p$, computed by A as $K = (u^b)^x$ with u^b computed per (12.3), and B computing $K = (u^x)^b \bmod p$. This final protocol then resembles (one-half of) Protocol MTI/A0 in that, since the message A sends is independent of the recipient B, it may be computed ahead of time before the recipient is determined. □

12.66 Example (*Protocol G2*) The two-pass MTI/A0 key agreement (Protocol 12.53) may be modified to use implicitly-certified keys as follows. Using the setup and notation as in Girault's self-certified public keys (Mechanism 12.61), A chooses a random integer x and sends to B: $\alpha^x \bmod n$. Analogously, B chooses a random integer y and sends to A: α^y

mod n. A computes $P_B{}^e + I_B$ mod n $(= \alpha^b)$; B computes $P_A{}^e + I_A$ mod n $(= \alpha^a)$. A and B then establish the time-variant common key $K = \alpha^{ay+bx}$ (mod n) by respectively computing $K = (\alpha^y)^a (P_B{}^e + I_B)^x$ and $K = (\alpha^x)^b (P_A{}^e + I_A)^y$ mod n. Alternatively, this protocol may be modified to use Günther's ID-based public keys in a manner directly analogous to that of Example 12.64. $\qquad\square$

12.67 Example (*self-certified version of Günther's ID-based keys*) The following modification of Mechanism 12.59 transforms it into a "self-certified" public-key scheme (i.e., one in which the third party does not learn users' private keys). A chooses a secret random v, $1 \leq v \leq p-1$ with $\gcd(v, p-1) = 1$, computes $w = \alpha^v$ mod p, and gives w to T. While v is not given to T, A should demonstrate knowledge of v to T (cf. Note 12.54). T chooses k_A as before but computes $P_A = w^{k_A}$ mod p (instead of: $P_A = \alpha^{k_A}$). T solves equation (12.1) for a as before (using the new P_A) and again gives A the pair $(r, s) = (P_A, a)$. A then calculates $a' = a \cdot v^{-1}$ mod $(p-1)$; it follows that (P_A, a') is now T's ElGamal signature on I_A (it is easily verified that $u^{P_A} \cdot P_A{}^{a'} \equiv \alpha^{h(I_A)}$), and T does not know a'. $\square$

12.7 Secret sharing

Secret sharing schemes are multi-party protocols related to key establishment. The original motivation for secret sharing was the following. To safeguard cryptographic keys from loss, it is desirable to create backup copies. The greater the number of copies made, the greater the risk of security exposure; the smaller the number, the greater the risk that all are lost. Secret sharing schemes address this issue by allowing enhanced reliability without increased risk. They also facilitate distributed trust or shared control for critical activities (e.g., signing corporate cheques; opening bank vaults), by gating the critical action on cooperation by t of n users.

The idea of *secret sharing* is to start with a secret, and divide it into pieces called *shares* which are distributed amongst users such that the pooled shares of specific subsets of users allow reconstruction of the original secret. This may be viewed as a key pre-distribution technique, facilitating one-time key establishment, wherein the recovered key is pre-determined (static), and, in the basic case, the same for all groups.

A secret sharing scheme may serve as a *shared control scheme* if inputs (shares) from two or more users are required to enable a critical action (perhaps the recovered key allows this action to trigger, or the recovery itself is the critical action). In what follows, simple shared-control schemes introduced in §12.7.1 are a subset of threshold schemes discussed in §12.7.2, which are themselves a subclass of generalized secret sharing schemes as described in §12.7.3.

12.7.1 Simple shared control schemes

(i) Dual control by modular addition

If a secret number S, $0 \leq S \leq m-1$ for some integer m, must be entered into a device (e.g., a seed key), but for operational reasons, it is undesirable that any single individual (other than a trusted party) know this number, the following scheme may be used. A trusted party T generates a random number $1 \leq S_1 \leq m-1$, and gives the values S_1 and $S - S_1$ mod m to two parties A and B, respectively. A and B then separately enter their values into the device, which sums them modulo m to recover S. If A and B are trusted not to collude, then neither one has any information about S, since the value each possesses is a random number between 0 and $m-1$. This is an example of a *split-knowledge* scheme – knowledge of the secret S is split among two people. Any action requiring S is said to be under *dual control* – two people are required to trigger it.

(ii) Unanimous consent control by modular addition

The dual control scheme above is easily generalized so that the secret S may be divided among t users, all of whom are required in order to recover S, as follows: T generates $t-1$ independent random numbers S_i, $0 \leq S_i \leq m-1$, $1 \leq i \leq t-1$. Parties P_1 through P_{t-1} are given S_i, while P_t is given $S_t = S - \sum_{i=1}^{t-1} S_i \bmod m$. The secret is recovered as $S = \sum_{i=1}^{t} S_i \bmod m$. Both here and in the dual control scheme above, modulo m operations may be replaced by exclusive-OR, using data values S and S_i of fixed bit-length $\lg(m)$.

12.68 Remark (*technique for splitting keys*) The individual key components in a split control scheme should be full-length. This provides greater security than partitioning an r-bit key into t pieces of r/t bits each. For example, for $r = 56$ and $t = 2$, if two parties are each given 28 bits of the key, exhaustive search by one party requires only 2^{28} trials, while if each party is given a 56-bit piece, 2^{56} trials are necessary.

12.7.2 Threshold schemes

12.69 Definition A (t, n) *threshold scheme* $(t \leq n)$ is a method by which a trusted party computes secret *shares* S_i, $1 \leq i \leq n$ from an initial secret S, and securely distributes S_i to user P_i, such that the following is true: any t or more users who pool their shares may easily recover S, but any group knowing only $t-1$ or fewer shares may not. A *perfect* threshold scheme is a threshold scheme in which knowing only $t-1$ or fewer shares provide no advantage (no information about S whatsoever, in the information-theoretic sense) to an opponent over knowing no pieces.

The split-knowledge scheme of §12.7.1(i) is an example of a $(2, 2)$ threshold scheme, while the unanimous consent control of §12.7.1(ii) is a (t, t) threshold scheme.

12.70 Remark (*use of threshold schemes*) If a threshold scheme is to be reused without decreased security, controls are necessary to prevent participants from deducing the shares of other users. One method is to prevent group members themselves from accessing the value of the recovered secret, as may be done by using a trusted combining device. This is appropriate for systems where the objective is shared control, and participants need only see that an action is triggered, rather than have access to the key itself. For example, each share might be stored on a chipcard, and each user might swipe its card through a trusted card reader which computes the secret, thereby enabling the critical action of opening an access door.

Shamir's threshold scheme

Shamir's threshold scheme is based on polynomial interpolation, and the fact that a univariate polynomial $y = f(x)$ of degree $t-1$ is uniquely defined by t points (x_i, y_i) with distinct x_i (since these define t linearly independent equations in t unknowns).

12.71 Mechanism Shamir's (t, n) threshold scheme

SUMMARY: a trusted party distributes shares of a secret S to n users.
RESULT: any group of t users which pool their shares can recover S.

1. *Setup*. The trusted party T begins with a secret integer $S \geq 0$ it wishes to distribute among n users.

   (a) T chooses a prime $p > \max(S, n)$, and defines $a_0 = S$.
   (b) T selects $t - 1$ random, independent coefficients $a_1, \ldots, a_{t-1}, 0 \leq a_j \leq p - 1$, defining the random polynomial over $\mathbb{Z}_p$, $f(x) = \sum_{j=0}^{t-1} a_j x^j$.
   (c) T computes $S_i = f(i) \bmod p, 1 \leq i \leq n$ (or for any n distinct points $i, 1 \leq i \leq p - 1$), and securely transfers the share S_i to user P_i, along with public index i.

2. *Pooling of shares*. Any group of t or more users pool their shares (see Remark 12.70). Their shares provide t distinct points $(x, y) = (i, S_i)$ allowing computation of the coefficients $a_j, 1 \leq j \leq t - 1$ of $f(x)$ by Lagrange interpolation (see below). The secret is recovered by noting $f(0) = a_0 = S$.

The coefficients of an unknown polynomial $f(x)$ of degree at most t, defined by points $(x_i, y_i), 1 \leq i \leq t$, are given by the Lagrange interpolation formula:

$$f(x) = \sum_{i=1}^{t} y_i \prod_{1 \leq j \leq t, j \neq i} \frac{x - x_j}{x_i - x_j}.$$

Since $f(0) = a_0 = S$, the shared secret may be expressed as:

$$S = \sum_{i=1}^{t} c_i y_i, \quad \text{where } c_j = \prod_{1 \leq j \leq t, j \neq i} \frac{x_j}{x_j - x_i}.$$

Thus each group member may compute S as a linear combination of t shares y_i, since the c_i are non-secret constants (which for a fixed group of t users may be pre-computed).

12.72 Note (*properties of Shamir's threshold scheme*) Properties of Mechanism 12.71 include:

1. *perfect*. Given knowledge of any $t - 1$ or fewer shares, all values $0 \leq S \leq p - 1$ of the shared secret remain equally probable (see Definition 12.69).
2. *ideal*. The size of one share is the size of the secret (see Definition 12.76).
3. *extendable for new users*. New shares (for new users) may be computed and distributed without affecting shares of existing users.
4. *varying levels of control possible*. Providing a single user with multiple shares bestows more control upon that individual. (In the terminology of §12.7.3, this corresponds to changing the access structure.)
5. *no unproven assumptions*. Unlike many cryptographic schemes, its security does not rely on any unproven assumptions (e.g., about the difficulty of number-theoretic problems).

12.7.3 Generalized secret sharing

The idea of a threshold scheme may be broadened to a *generalized secret sharing scheme* as follows. Given a set P of users, define $\mathcal{A}$ (the *access structure*) to be a set of subsets, called the *authorized subsets* of P. Shares are computed and distributed such that the pooling of shares corresponding to any authorized subset $A \in \mathcal{A}$ allows recovery of the secret S, but the pooling of shares corresponding to any unauthorized subset $B \subseteq P, B \notin \mathcal{A}$ does not.

Threshold schemes are a special class of generalized secret sharing schemes, in which the access structure consists of precisely all t-subsets of users. An access structure is called *monotone* if, whenever a particular subset A of users is an authorized subset, then any subset of P containing A is also authorized. Monotone access structures are a requirement in many applications, and most natural schemes are monotone. Perfect secret sharing schemes have a monotone access structure as a consequence of the entropy formulation in Definition 12.73.

12.73 Definition A secret sharing scheme is *perfect* if the shares corresponding to each unauthorized subset provide absolutely no information, in the information-theoretic sense, about the shared secret (cf. Definition 12.69). More formally, where H denotes entropy (see §2.2.1), and A, B are sets of users using the above notation: $H(S|A) = 0$ for any $A \in \mathcal{A}$, while $H(S|B) = H(S)$ for any $B \notin \mathcal{A}$.

The efficiency of a secret sharing scheme is measured by its information rate.

12.74 Definition For secret sharing schemes, the *information rate* for a particular user is the bit-size ratio (size of the shared secret)/(size of that user's share). The *information rate* for a secret sharing scheme itself is the minimum such rate over all users.

12.75 Fact (*perfect share bound*) In any perfect secret sharing scheme the following holds for all user shares: (size of a user share) $\geq$ (size of the shared secret). Consequently, all perfect secret sharing schemes must have information rate ≤ 1.

Justification. If any user P_i had a share of bit-size less than that of the secret, knowledge of the shares (excepting that of P_i) corresponding to any authorized set to which P_i belonged, would reduce the uncertainty in the secret to at most that in P_i's share. Thus by definition, the scheme would not be perfect.

12.76 Definition Secret sharing schemes of rate 1 (see Definition 12.74) are called *ideal*.

As per Note 12.72, Shamir's threshold scheme is an example of an ideal secret sharing scheme. Examples of access structures are known for which it has been proven that ideal schemes do not exist.

Secret sharing schemes with extended capabilities

Secret sharing schemes with a variety of extended capabilities exist, including:

1. *pre-positioned secret sharing schemes*. All necessary secret information is put in place excepting a single (constant) share which must later be communicated, e.g., by broadcast, to activate the scheme.
2. *dynamic secret sharing schemes*. These are pre-positioned schemes wherein the secrets reconstructed by various authorized subsets vary with the value of communicated activating shares.
3. *multi-secret threshold schemes*. In these secret sharing schemes different secrets are associated with different authorized subsets.
4. *detection of cheaters*, and *verifiable secret sharing*. These schemes respectively address *cheating* by one or more group members, and the distributor of the shares.

    5. *secret sharing with disenrollment.* These schemes address the issue that when a secret share of a (t, n) threshold scheme is made public, it becomes a $(t - 1, n)$ scheme.

12.8 Conference keying

12.77 Definition A *conference keying protocol* is a generalization of two-party key establishment to provide three or more parties with a shared secret key.

    Despite superficial resemblance, conference keying protocols differ from dynamic secret sharing schemes in fundamental aspects. General requirements for conference keying include that distinct groups recover distinct keys (session keys); that session keys are dynamic (excepting key pre-distribution schemes); that the information exchanged between parties is non-secret and transferred over open channels; and that each party individually computes the session key (vs. pooling shares in a black box). A typical application is telephone conference calls. The group able to compute a session key is called the *privileged subset.* When a central point enables members of a (typically large) privileged subset to share a key by broadcasting one or more messages, the process resembles pre-positioned secret sharing somewhat and is called *broadcast encryption.*

    An obvious method to establish a conference key K for a set of $t \geq 3$ parties is to arrange that each party share a unique symmetric key with a common trusted party. Thereafter the trusted party may choose a new random key and distribute it by symmetric key transport individually to each member of the conference group. Disadvantages of this approach include the requirement of an on-line trusted third party, and the communication and computational burden on this party.

    A related approach not requiring a trusted party involves a designated group member (the *chair*) choosing a key K, computing pairwise Diffie-Hellman keys with each other group member, and using such keys to securely send K individually to each. A drawback of this approach is the communication and computational burden on the chair, and the lack of protocol symmetry (balance). Protocol 12.78 offers an efficient alternative, albeit more complex in design.

Burmester-Desmedt conference keying protocol

The following background is of use in understanding Protocol 12.78. t users U_0 through U_{t-1} with individual Diffie-Hellman exponentials $z_i = \alpha^{r_i}$ will form a conference key $K = \alpha^{r_0 r_1 + r_1 r_2 + r_2 r_3 + \cdots + r_{t-1} r_0}$. Define $A_j = \alpha^{r_j r_{j+1}} = z_j^{r_{j+1}}$ and $X_j = \alpha^{r_{j+1} r_j - r_j r_{j-1}}$. Noting $A_j = A_{j-1} X_j$, K may equivalently be written as (with subscripts taken modulo t)

$$\begin{aligned} K_i &= A_0 A_1 \cdots A_{t-1} = A_{i-1} A_i A_{i+1} \cdots A_{i+(t-2)} \\ &= A_{i-1} \cdot (A_{i-1} X_i) \cdot (A_{i-1} X_i X_{i+1}) \cdots (A_{i-1} X_i X_{i+1} \cdots X_{i+(t-2)}). \end{aligned}$$

Noting $A_{i-1}{}^t = (z_{i-1})^{t r_i}$, this is seen to be equivalent to K_i as in equation (12.6) of Protocol 12.78.

12.78 Protocol Burmester-Desmedt conference keying

SUMMARY: $t \geq 2$ users derive a common conference key K.
RESULT: K is secure from attack by passive adversaries.

    1. *One-time setup.* An appropriate prime p and generator α of $\mathbb{Z}_p^*$ are selected, and authentic copies of these are provided to each of n system users.

2. *Conference key generation.* Any group of $t \leq n$ users (typically $t \ll n$), derive a common conference key K as follows. (Without loss of generality, the users are labeled U_0 through U_{t-1}, and all indices j indicating users are taken modulo t.)

    (a) Each U_i selects a random integer r_i, $1 \leq r_i \leq p-2$, computes $z_i = \alpha^{r_i} \bmod p$, and sends z_i to each of the other $t-1$ group members. (Assume that U_i has been notified *a priori*, of the indices j identifying other conference members.)

    (b) Each U_i, after receiving z_{i-1} and z_{i+1}, computes $X_i = (z_{i+1}/z_{i-1})^{r_i} \bmod p$ (note $X_i = \alpha^{r_{i+1}r_i - r_i r_{i-1}}$), and sends X_i to each of the other $t-1$ group members.

    (c) After receiving X_j, $1 \leq j \leq t$ excluding $j = i$, U_i computes $K = K_i$ as

$$K_i = (z_{i-1})^{tr_i} \cdot X_i^{t-1} \cdot X_{i+1}^{t-2} \cdots X_{i+(t-3)}^{2} \cdot X_{i+(t-2)}^{1} \bmod p \qquad (12.6)$$

For small conferences (small t), the computation required by each party is small, since all but one exponentiation in equation (12.6) involves an exponent between 1 and t. The protocol requires an order be established among users in the privileged subset (for indexing). For $t = 2$, the resulting key is $K = (\alpha^{r_1 r_2})^2$, the square of the standard Diffie-Hellman key. It is provably as difficult for a passive adversary to deduce the conference key K in Protocol 12.78 as to solve the Diffie-Hellman problem.

Attention above has been restricted to unauthenticated conference keying; additional measures are required to provide authentication in the presence of active adversaries. Protocol 12.78 as presented assumes a broadcast model (each user exchanges messages with all others); it may also be adapted for a bi-directional ring (wherein each user transmits only to two neighbors).

Unconditionally secure conference keying

While conference keying schemes such as Protocol 12.78 provide computational security, protocols with the goal of unconditional security are also of theoretical interest. Related to this, a generalization of Fact 12.34 is given below, for conferences of fixed size (t participants from among n users) which are information-theoretically secure against conspiracies of up to j non-participants. The model for this result is a non-interactive protocol, and more specifically a key pre-distribution scheme: each conference member computes the conference key solely from its own secret data (pre-distributed by a server) and an identity vector specifying (an ordered sequence of) indices of the other conference members.

12.79 Fact (*Blundo's conference KDS bound*) In any j-secure conference KDS providing m-bit conference keys to privileged subsets of fixed size t, the secret data stored by each user must be at least $m \cdot \binom{j+t-1}{t-1}$ bits.

Fact 12.79 with $t = 2$ and $j = n - 2$ corresponds to the trivial scheme (see p.505) where each user has $n - 1$ shared keys each of m bits, one for each other user. A non-trivial scheme meeting the bound of Fact 12.79 can be constructed as a generalization of Mechanism 12.35 (see p.540).

12.80 Remark (*refinement of Fact 12.79*) A more precise statement of Fact 12.79 requires consideration of entropy; the statement holds if the conference keys in question have m bits of entropy.

12.9 Analysis of key establishment protocols

The main objective of this section is to highlight the delicate nature of authenticated key establishment protocols, and the subtlety of design flaws. Examples of flawed protocols are included to illustrate typical attack strategies, and to discourage protocol design by the novice.

12.9.1 Attack strategies and classic protocol flaws

The study of successful attacks which have uncovered flaws in protocols allows one to learn from previous design errors, understand general attack methods and strategies, and formulate design principles. This both motivates and allows an understanding of various design features of protocols. General attack strategies are discussed in §12.2.3. In the specific examples below, A and B are the legitimate parties, and E is an adversary (enemy). Two of the protocols discussed are, in fact, authentication-only protocols (i.e., do not involve key establishment), but are included in this discussion because common principles apply.

Attack 1: Intruder-in-the-middle

The classic "intruder-in-the-middle" attack on unauthenticated Diffie-Hellman key agreement is as follows.

$$
\begin{array}{ccccccc}
A & & E & & B \\
\rightarrow & \alpha^x & \rightarrow & \alpha^{x'} & \rightarrow \\
\leftarrow & \alpha^{y'} & \leftarrow & \alpha^y & \leftarrow
\end{array}
$$

A and B have private keys x and y, respectively. E creates keys x' and y'. E intercepts A's exponential and replaces it by $\alpha^{x'}$; and intercepts B's exponential, replacing it with $\alpha^{y'}$. A forms session key $K_A = \alpha^{xy'}$, while B forms session key $K_B = \alpha^{x'y}$. E is able to compute both these keys. When A subsequently sends a message to B encrypted under K_A, E deciphers it, re-enciphers under K_B, and forwards it to B. Similarly E deciphers messages encrypted by B (for A) under K_B, and re-enciphers them under K_A. A and B believe they communicate securely, while E reads all traffic.

Attack 2: Reflection attack

Suppose A and B share a symmetric key K, and authenticate one another on the basis of demonstrating knowledge of this key by encrypting or decrypting a challenge as follows.

$$
\begin{array}{cccc}
A & & B & \\
\rightarrow & r_A & & (1) \\
& E_K(r_A, r_B) & \leftarrow & (2) \\
\rightarrow & r_B & & (3)
\end{array}
$$

An adversary E can impersonate B as follows. Upon A sending (1), E intercepts it, and initiates a new protocol, sending the identical message r_A back to A as message (1) purportedly from B. In this second protocol, A responds with message (2'): $E_K(r_A, r_A')$, which E again intercepts and simply replays back on A as the answer (2) in response to the challenge r_A in the original protocol. A then completes the first protocol, and believes it has successfully authenticated B, while in fact B has not been involved in any communications.

$$
\begin{array}{llll}
A & & E & \\
\rightarrow & r_A & & (1) \\
& r_A & \leftarrow & (1') \\
\rightarrow & E_K(r_A, r_A') & & (2') \\
& E_K(r_A, r_B = r_A') & \leftarrow & (2) \\
\rightarrow & r_B & & (3) \\
\end{array}
$$

The attack can be prevented by using distinct keys K and K' for encryptions from A to B and B to A, respectively. An alternate solution is to avoid message symmetry, e.g., by including the identifier of the originating party within the encrypted portion of (2).

Attack 3: Interleaving attack

Consider the following (flawed) authentication protocol, where s_A denotes the signature operation of party A, and it is assumed that all parties have authentic copies of all others' public keys.

$$
\begin{array}{llll}
A & & B & \\
\rightarrow & r_A & & (1) \\
& r_B, s_B(r_B, r_A, A) & \leftarrow & (2) \\
\rightarrow & r_A', s_A(r_A', r_B, B) & & (3) \\
\end{array}
$$

The intention is that the random numbers chosen by A and B, respectively, together with the signatures, provide a guarantee of freshness and entity authentication. However, an enemy E can initiate one protocol with B (pretending to be A), and another with A (pretending to be B), as shown below, and use a message from the latter protocol to successfully complete the former, thereby deceiving B into believing E is A (and that A initiated the protocol).

$$
\begin{array}{llllll}
A & & E & & B & \\
& & \rightarrow & r_A & & (1) \\
& & & r_B, s_B(r_B, r_A, A) & \leftarrow & (2) \\
& r_B & \leftarrow & & & (1') \\
\rightarrow & r_A', s_A(r_A', r_B, B) & & & & (2') \\
& & \rightarrow & r_A', s_A(r_A', r_B, B) & & (3) \\
\end{array}
$$

This attack is possible due to the message symmetry of (2) and (3), and may be prevented by making their structures differ, securely binding an identifier to each message indicating a message number, or simply requiring the original r_A take the place of r_A' in (3).

The implications of this attack depend on the specific objectives the protocol was assumed to provide. Such specific objectives are, however, (unfortunately) often not explicitly stated.

Attack 4: Misplaced trust in server

The Otway-Rees protocol (Protocol 12.29) has messages as follows:

$$
\begin{array}{llll}
A \rightarrow B: & M, A, B, E_{K_{AT}}(N_A, M, A, B) & & (1) \\
B \rightarrow T: & M, A, B, E_{K_{AT}}(N_A, M, A, B), E_{K_{BT}}(N_B, M, A, B) & & (2) \\
B \leftarrow T: & E_{K_{AT}}(N_A, k), E_{K_{BT}}(N_B, k) & & (3) \\
A \leftarrow B: & E_{K_{AT}}(N_A, k) & & (4) \\
\end{array}
$$

Upon receiving message (2), the server must verify that the encrypted fields (M, A, B) in both parts of (2) match, and in addition that these fields match the cleartext (M, A, B). If the latter check is not carried out, the protocol is open to attack by an enemy E (who is another authorized system user) impersonating B as follows. E modifies (2), replacing cleartext B

by E (but leaving both enciphered versions of both identifiers A and B intact), replacing nonce N_B by its own nonce N_E, and using key K_{ET} (which E shares *a priori* with T) in place of K_{BT}. Based on the cleartext identifier E, T then encrypts part of message (3) under K_{ET} allowing E to recover k; but A believes, as in the original protocol, that k is shared with B. The attack is summarized as follows.

$$
\begin{aligned}
A \rightarrow B: &\quad M, A, B, E_{K_{AT}}(N_A, M, A, B) & (1)\\
B \rightarrow E: &\quad M, A, B, E_{K_{AT}}(N_A, M, A, B), E_{K_{BT}}(N_B, M, A, B) & (2)\\
E \rightarrow T: &\quad M, A, E, E_{K_{AT}}(N_A, M, A, B), E_{K_{ET}}(N_E, M, A, B) & (2')\\
E \leftarrow T: &\quad E_{K_{AT}}(N_A, k), E_{K_{ET}}(N_E, k) & (3)\\
A \leftarrow E: &\quad E_{K_{AT}}(N_A, k) & (4)
\end{aligned}
$$

The attack is possible due to the subtle manner by which A infers the identity of the other party to which k is made available: in (4), A has no direct indication of the other party to which T has made k available, but relies on the nonce N_A in (4) and its association with the pair (N_A, B) within the protected part of (1). Thus, A relies on (or delegates trust to) the server to make k available only to the party requested by A, and this can be assured only by T making use of the protected fields (M, A, B).

12.9.2 Analysis objectives and methods

The primary aim of protocol analysis is to establish confidence in the cryptographic security of a protocol. The following definitions aid discussion of protocol analysis.

12.81 Definition A key establishment protocol is *operational* (or *compliant*) if, in the absence of active adversaries and communications errors, honest participants who comply with its specification always complete the protocol having computed a common key and knowledge of the identities of the parties with whom the key is shared.

The most obvious objectives and properties of key establishment protocols, namely authenticity and secrecy of keys, are discussed in §12.2.2.

12.82 Definition A key establishment protocol is *resilient* if it is impossible for an active adversary to mislead honest participants as to the final outcome.

Protocol analysis should confirm that a protocol meets all claimed objectives. As a minimum, for a key establishment protocol this should include being operational (note this implies no security guarantees), providing both secrecy and authenticity of the key, and being resilient. Key authenticity implies the identities of the parties sharing the key are understood and corroborated, thus addressing impersonation and substitution. Resilience differs subtlely from authentication, and is a somewhat broader requirement (e.g., see the attack of Note 12.54). Additional objectives beyond authenticated key establishment may include key confirmation, perfect forward secrecy, detection of key re-use, and resistance to known-key attacks (see §12.2.3).

In addition to verifying objectives are met, additional benefits of analysis include:

1. explicit identification of assumptions on which the security of a protocol is based;
2. identification of protocol properties, and precise statement of its objectives (this facilitates comparison with other protocols, and determining appropriateness);
3. examination of protocol efficiency (with respect to bandwidth and computation).

Essentially all protocol analysis methods require the following (implicitly or explicitly):

1. *protocol specification* – an unambiguous specification of protocol messages, when they are sent, and the actions to be taken upon reception thereof;

2. *goals* – an unambiguous statement of claimed assurances upon completion;

3. *assumptions and initial state* – a statement of assumptions and initial conditions;

4. *proof* – some form of argument that, given the assumptions and initial state, the specified protocol steps lead to a final state meeting the claimed goals.

Analysis methods

Common approaches for analyzing cryptographic protocols include the following:

1. *ad hoc and practical analysis.* This approach consists of any variety of convincing arguments that any successful protocol attack requires a resource level (e.g., time or space) greater than the resources of the perceived adversary. Protocols which survive such analysis are said to have *heuristic security*, with security here typically in the computational sense and adversaries assumed to have fixed resources. Arguments often presuppose secure building blocks. Protocols are typically designed to counter standard attacks, and shown to follow accepted principles. Practical arguments (paralleling complexity-theoretic arguments) involving constructions which assemble basic building blocks may justify security claims.

    While perhaps the most commonly used and practical approach, it is in some ways the least satisfying. This approach may uncover protocol flaws thereby establishing that a protocol is bad. However, claims of security may remain questionable, as subtle flaws in cryptographic protocols typically escape ad hoc analysis; unforeseen attacks remain a threat.

2. *reducibility from hard problems.* This technique consists of proving that any successful protocol attack leads directly to the ability to solve a well-studied reference problem (Chapter 3), itself considered computationally infeasible given current knowledge and an adversary with bounded resources. Such analysis yields so-called *provably secure protocols*, although the security is conditional on the reference problem being truly (rather than presumably) difficult.

    A challenge in this approach is to establish that all possible attacks have been taken into account, and can in fact be equated to solving the identified reference problems. This approach is considered by some to be as good a practical analysis technique as exists. Such provably secure protocols belong to the larger class of techniques which are *computationally secure*.

3. *complexity-theoretic analysis.* An appropriate model of computation is defined, and adversaries are modeled as having polynomial computational power (they may mount attacks involving time and space polynomial in the size of appropriate security parameters). A security proof relative to the model is then constructed. The existence of underlying cryptographic primitives with specified properties is typically assumed. An objective is to design cryptographic protocols which require the fewest cryptographic primitives, or the weakest assumptions.

    As the analysis is asymptotic, care is required to determine when proofs have practical significance. Polynomial attacks which are feasible under such a model may nonetheless in practice be computationally infeasible. Asymptotic analysis may be of limited relevance to concrete problems in practice, which have finite size. Despite these issues, complexity-theoretic analysis is invaluable for formulating fundamental principles and confirming intuition.

4. *information-theoretic analysis.* This approach uses mathematical proofs involving entropy relationships to prove protocols are *unconditionally secure*. In some cases,

this includes the case where partial secrets are disclosed (e.g., for unconditional security against coalitions of fixed size). Adversaries are modeled to have unbounded computing resources.

While unconditional security is ultimately desirable, this approach is not applicable to most practical schemes for several reasons. These include: many schemes, such as those based on public-key techniques, can at best be computationally secure; and information-theoretic schemes typically either involve keys of impractically large size, or can only be used once. This approach cannot be combined with computational complexity arguments because it allows unlimited computation.

5. *formal methods*. So-called *formal* analysis and verification methods include logics of authentication (cryptographic protocol logics), term re-writing systems, expert systems, and various other methods which combine algebraic and state-transition techniques. The most popular protocol logic is the Burrows-Abadi-Needham (BAN) logic. Logic-based methods attempt to reason that a protocol is correct by evolving a set of beliefs held by each party, to eventually derive a belief that the protocol goals have been obtained.

This category of analysis is somewhat disjoint from the first four. Formal methods have proven to be of utility in finding flaws and redundancies in protocols, and some are automatable to varying degrees. On the other hand, the "proofs" provided are proofs within the specified formal system, and cannot be interpreted as absolute proofs of security. A one-sidedness remains: the absence of discovered flaws does not imply the absence of flaws. Some of these techniques are also unwieldy, or applicable only to a subset of protocols or classes of attack. Many require (manually) converting a concrete protocol into a formal specification, a critical process which itself may be subject to subtle flaws.

12.10 Notes and further references

§12.1

While the literature is rife with proposals for key establishment protocols, few comprehensive treatments exist and many proposed protocols are supported only by ad hoc analysis.

§12.2

Much of §12.2 builds on the survey of Rueppel and van Oorschot [1086]. Fumy and Munzert [431] discuss properties and principles for key establishment. While encompassing the majority of key establishment as currently used in practice, Definition 12.2 gives a somewhat restricted view which excludes a rich body of research. More generally, key establishment may be defined as a process or mechanism which provides a shared capability (rather than simply a shared secret) between specified sets of participants, facilitating some operation for which the intention is that other sets of participants cannot execute. This broader definition includes many protocols in the area of *threshold cryptography*, introduced independently by Desmedt [336], Boyd [182], and Croft and Harris [288]; see the comprehensive survey of Desmedt [337].

The term *perfect forward secrecy* (Definition 12.16) was coined by Günther [530]; see also Diffie, van Oorschot, and Wiener [348]. Here "perfect" does not imply any properties of information-theoretic security (cf. Definition 12.73). The concept of known-key attacks (Definition 12.17), developed by Yacobi and Shmuely [1256] (see also Yacobi [1255]), is related to that of Denning and Sacco [330] on the use of timestamps to prevent message replay (see page 535).

Among items not discussed in detail in this chapter is *quantum cryptography*, based on the uncertainty principle of quantum physics, and advanced by Bennett et al. [114] building on the idea of quantum coding first described by Wiesner [1242] *circa* 1970. Although not providing digital signatures or non-repudiation, quantum cryptography allows key distribution (between two parties who share no *a priori* secret keying material), which is provably secure against adversaries with unlimited computing power, provided the parties have access to (aside from the quantum channel) a conventional channel subject to only passive adversaries. For background on the basic quantum channel for key distribution (*quantum key distribution*), see Brassard [192]; Phoenix and Townsend [973] survey developments in this area including experimental implementations.

Mitchell [879] presented a key agreement system based on use of a public broadcast channel transmitting data at a rate so high that an eavesdropper cannot store all data sent over a specified time interval. This is closely related to work of Maurer [815] regarding secret key agreement using only publicly available information, in turn motivated by Wyner's *wire-tap channel* [1254], which addresses the rate at which secret information can be conveyed to a communicating partner with security against a passive eavesdropper whose channel is subject to additional noise.

§12.3

Regarding point-to-point techniques presented, those based on symmetric encryption are essentially from ISO/IEC 11770-2 [617], while AKEP1 and AKEP2 (Note 12.21; Protocol 12.20) are derived from Bellare and Rogaway [94] (see also §12.9 below). The idea of key derivation allowing key establishment by symmetric techniques based on a one-way function (without encryption), was noted briefly by Matsumoto, Takashima and Imai [800]; see also the proposals of Gong [499], and related techniques in the *KryptoKnight* suite [891, 141, 142].

Shamir's no-key protocol (Protocol 12.22); also called *Shamir's three-pass protocol*), including exponentiation-based implementation, is attributed to Shamir by Konheim [705, p.345]. Massey [786, p.35] notes that Omura [792], aware of Shamir's generic protocol, later independently proposed implementing it with an exponentiation-based cipher as per Protocol 12.22. See also Massey and Omura [956] (discussed in Chapter 15).

Version 5 of Kerberos (V5), the development of which began in 1989, was specified by Kohl and Neuman [1041]; for a high-level overview, see Neuman and Ts'o [926] who also note that a typical timestamp window is 5 minutes (centered around the verifier's time). The original design of Kerberos V4 was by Miller and Neuman, with contributions by Saltzer and Schiller [877]; an overview is given by Steiner, Neuman, and Schiller [1171], while V4 issues are noted by Kohl [701] and the critique of Bellovin and Merritt [103]. The basic protocol originates from the shared-key protocol of Needham and Schroeder [923], with timestamps (which Needham and Schroeder explicitly avoided) later proposed by Denning and Sacco [330], reducing the number of messages at the expense of secure and synchronized clocks. Bauer, Berson, and Feiertag [76] addressed symmetric assurances of freshness, recovery from single-key compromise, and reduction of messages through per-participant use of a local counter called an *event marker*; they also extended the Needham-Schroeder setting to multiple security domains (each with a separate KDC) and connectionless environments. Bellare and Rogaway [96] presented an efficient 4-pass server-based key transfer protocol with implicit key authentication, and key freshness properties secure against known-key attacks; significantly, their treatment (the first of its kind) shows the protocol to be provably secure (assuming a pseudorandom function). Advantages and disadvantages of using timestamps are discussed in §10.3.1.

Protocol 12.29 is due to Otway and Rees [961]. Kehne, Schönwälder, and Langendörfer [663] discuss a 5-message nonce-based protocol with the same features as Kerberos (Protocol 12.24), without requiring distributed timeclocks. Excluding the optional re-authentication capability (as per Kerberos), it is essentially that of Mechanism 9 in ISO/IEC DIS 11770-2 [617], and similar to the 5-message Otway-Rees protocol as augmented per Remark 12.30 (with one fewer encryption by each of A and B); but see also the analysis of Neuman and Stubblebine [925]. A 5-message authentication protocol included in ISO/IEC 9798-2 [599] provides key transport using a trusted server, with mutual entity authentication and mutual key confirmation, without timestamps; Needham and Schroeder [924] propose a 7-message protocol with similar properties.

§12.4

Mechanism 12.35 and Fact 12.34 are due to Blom [158]; a simpler polynomial formulation is noted under §12.8 below. For background in coding theory, see MacWilliams and Sloane [778]. Mitchell and Piper [881] consider the use of combinatorial block designs and finite incidence structures called *key distribution patterns* to construct a class of non-interactive KDS. Each user is given a set of secret subkeys (with no algebraic structure as per Blom's scheme), from which each pair of users may compute a common key by combining appropriate subkeys via a public function. The question of reducing key storage was considered earlier by Blom [157], including security against coalitions of fixed size and the use of commutative functions (later generalized to symmetric functions by Blundo et al. [169]; see also §12.8 below). For related work, see Quinn [1014], Gong and Wheeler [506], and §12.7 below.

§12.5

Protocol 12.38, the public-key protocol of Needham and Schroeder [923], was originally specified to include 4 additional messages whereby signed public keys were requested from an on-line certification authority. Asymmetric key transport protocols involving various combinations of encryption and signatures are given in ISO/IEC CD 11770-3 [618]. The three-pass encrypt-then-sign protocol of §12.5.2(iii) originates from ISO/IEC 9798-3 [600]; it is closely related to the STS protocol (Protocol 12.57) which transfers Diffie-Hellman exponentials in place of random numbers. I'Anson and Mitchell [567] critique (e.g., see Note 12.42) the X.509 protocols [595]; see also the formal analysis of Gaarder and Snekkenes [433]. Protocol 12.44 and the related 2-pass key agreement of Figure 12.2 are due to Beller and Yacobi [101, 100], building on work of Beller, Chang, and Yacobi [99, 98, 97].

A two-pass key transport protocol called COMSET, based on public-key encryption, was adopted by the European community RACE Integrity Primitives Evaluation (RIPE) project [178]. Arising from zero-knowledge considerations studied by Brandt et al. [188], it employs Williams' variant of the Rabin public-key encryption (§8.3), and is similar in some aspects to the Needham-Schroeder public-key and Beller-Yacobi protocols. The protocol specified in Note 12.39 combines concepts of COMSET and the Needham-Schroeder protocol.

§12.6

The landmark 1976 paper of Whitfield Diffie and Martin Hellman [345] is the standard reference for both the seminal idea of public-key cryptography and the fundamental technique of exponential key agreement. An earlier conference paper of Diffie and Hellman [344], written in December 1975 and presented in June 1976, conceived the concept of public key agreement and the use of public-key techniques for identification and digital signatures.

Diffie [342] reports that amidst joint work on the problem for some time, Hellman distilled exponential key agreement in May 1976, and this was added to their June 1976 conference presentation (but not the written paper). Preceding this, in the fall of 1974, Merkle independently conceived a particular method for key agreement using the same abstract concepts. Merkle's *puzzle system* [849], submitted for publication in 1975 and appearing in April 1978, is as follows. Alice constructs m *puzzles*, each of which is a cryptogram Bob can solve in n steps (exhaustively trying n keys until a recognizable plaintext is found). Alice sends all m puzzles to Bob over an unsecured channel. Bob picks one of these, solves it (cost: n steps), and treats the plaintext therein as the agreed key, which he then uses to encrypt and send to Alice a known message. The encrypted message, now a puzzle which Alice must solve, takes Alice n steps (by exhaustively trying n keys). For $m \approx n$, each of Alice and Bob require $O(n)$ steps for key agreement, while an opponent requires $O(n^2)$ steps to deduce the key. An appropriate value n is chosen such that n steps is computationally feasible, but n^2 is not.

Rueppel [1078] explores the use of function composition to generalize Diffie-Hellman key agreement. Shmuely [1127] and McCurley [825] consider *composite Diffie-Hellman*, i.e., Diffie-Hellman key agreement with a composite modulus. McCurley presents a variation thereof, with an RSA-like modulus m of specific form and particular base α of high order in $\mathbb{Z}_m^*$, which is provably as secure (under passive attack) as the more difficult of factoring m and solving the discrete logarithm problem modulo the factors of m.

Regarding Diffie-Hellman key agreement, van Oorschot and Wiener [1209] note that use of "short" private exponents in conjunction with a random prime modulus p (e.g., 256-bit exponents with 1024-bit p) makes computation of discrete logarithms easy. They also document the attack of Note 12.50, which is related to issues explored by Simmons [1150] concerning a party's ability to control the resulting Diffie-Hellman key, and more general issues of unfairness in protocols. Waldvogel and Massey [1228] carefully examine the probability distribution and entropy of Diffie-Hellman keys under various assumptions. When private exponents are chosen independently and uniformly at random from the invertible elements of $\mathbb{Z}_{p-1}$, the $\phi(p-1)$ keys which may result are equiprobable. When private exponents are chosen independently and uniformly at random from $\{0, \dots, p-2\}$ (as is customary in practice), in the best case (when p is a *safe prime*, $p = 2q + 1$, q prime) the most probable Diffie-Hellman key is only 6 times more likely than the least probable, and the key entropy is less than 2 bits shy of the maximum, $\lg(p-1)$; while in the worst case (governed by a particular factorization pattern of $p-1$) the distribution is still sufficiently good to preclude significant cryptanalytic advantage, for p of industrial size or larger.

The one-pass key agreement of Protocol 12.51 was motivated by the work of ElGamal [368]. The MTI protocols of Table 12.5 were published in 1986 by Matsumoto, Takashima, and Imai [800]. MTI/A0 is closely related to a scheme later patented by Goss [519]; in the latter, exclusive-OR is used in place of modular multiplication to combine partial keys. Matsumoto et al. equate the computational complexity of passive attacks (excluding known-key attacks) on selected key agreement protocols to that of one or two Diffie-Hellman problems. Active attacks related to Note 12.54 are considered by Diffie, van Oorschot, and Wiener [348], and Menezes, Qu, and Vanstone [844]. Yacobi and Shmuely [1256] note two time-variant versions of Diffie-Hellman key agreement which are insecure against known-key attack. A similar protocol which falls to known-key attack was discussed by Yacobi [1255], subsequently rediscovered by Alexandris et al. [21], and reexamined by Nyberg and Rueppel [937]. Yacobi [1255] proves that the MTI/A0 protocol with composite-modulus is provably secure (security equivalent to composite Diffie-Hellman) under known-key attack by a passive adversary; Desmedt and Burmester [339], however, note the security is only heuristic under known-key attack by an active adversary. A formal-logic security comparison of the protocols of Goss (essentially Protocol 12.53), Günther (Protocol 12.62), and STS (Protocol 12.57) is given by van Oorschot [1204]. Burmester [220] identifies known-key *triangle attacks* which may be mounted on the former two and related protocols which provide only implicit key authentication (including MTI protocols, cf. Note 12.54). Known-key attacks were also one motivation for Denning and Sacco [330] to modify the Needham-Schroeder protocol as discussed above (cf. p.534).

Protocol 12.57 (STS) evolved from earlier work on ISDN telephone security as outlined by Diffie [342, p.568], who also reports on STU-III telephones. Variations of STS and an informal model for authentication and authenticated key establishment are discussed by Diffie, van Oorschot, and Wiener [348]. Bellovin and Merritt [104, 105] (see also the patent [102]) propose another hybrid protocol (*Encrypted Key Exchange* – EKE), involving exponential key agreement with authentication based on a shared password, designed specifically to protect against password-guessing attacks by precluding easy verification of guessed passwords; Steiner, Tsudik, and Waidner [1172] provide further analysis and extensions. A hybrid protocol with similar goals is given Gong et al. [504], including discussion of its relationship to EKE, and expanding the earlier work of Lomas et al. [771].

Blom [157] was apparently the first to propose an identity-based (or more accurately, index-based) key establishment protocol. Shamir [1115] proposed the more general idea of *identity-based systems* wherein a user's public key may be a commonly known name and address. For further discussion of ID-based schemes, see the chapter notes on §13.4. Self-certified public keys (Mechanism 12.61) are discussed by Girault [459], who credits earlier work by others, and provides the self-certified version of Günther's ID-based keys (Example 12.67). The parenthetical forgery attack mentioned in Mechanism 12.61 is outlined by Stinson [1178]. Key agreement protocols as in Examples 12.64 and 12.65, using both ID-based public keys of Günther [530] (Mechanism 12.59) and modified ElGamal signatures, are given by Horster and Knobloch [562]. The optimization of ElGamal signatures noted in Remark 12.60 is by Agnew, Mullin, and Vanstone [19]. Rabin's signature scheme (Chapter 11) may be used in place of RSA to reduce the computations required in schemes based on Girault's implicitly-certified public keys. Maurer and Yacobi [824] (modifying their earlier proposal [823]) propose an identity-based one-pass key pre-distribution scheme using composite modulus Diffie-Hellman, featuring implicitly-certified public key-agreement keys essentially consisting of a user's identity (or email address); the corresponding private key is the discrete logarithm of this, computed by a trusted authority which, knowing the factorization of an appropriately chosen modulus n, can thereby compute logarithms.

Nyberg and Rueppel [936] note their signature scheme (Chapter 11) may be used to create implicitly certified, identity-based public keys with properties similar to those of Girault (Mechanism 12.61), as well as key agreement protocols; Nyberg [935] presents an improved one-pass key agreement based on these ideas. Okamoto and Tanaka [946] propose identity-based key agreement protocols combining exponential key agreement and RSA, including one using timestamps and providing entity authentication, and a simpler protocol providing (implicit) key authentication.

§12.7

The idea of split control has long been known (e.g., see Sykes [1180]). Shamir [1110] and Blakley [148] independently proposed the idea of threshold schemes, the latter based on vector subspaces. The simplest example of the Blakley's idea is a $(2, n)$ threshold scheme where the shares (here called *shadows*) distributed to parties are non-collinear lines in a common plane; the shared secret of any two parties is the intersection of their lines. For a $(3, n)$ scheme, the shadows consist of non-parallel planes, any two of which intersect in a line, and any three of which intersect in a point. While Shamir's threshold scheme is *perfect*, Blakley's vector scheme is not (the set of possible values of the shared secret narrows as subsequent shares are added). Karnin, Greene, and Hellman [662] discuss the unanimous consent control scheme of §12.7.1; see also Diffie and Hellman [344, p.110].

Generalized secret sharing schemes and the idea of access structures were first studied by Ito, Saito, and Nishizeki [625], who provided a construction illustrating that any monotone access structure can be realized by a perfect secret sharing scheme. Benaloh and Leichter [112] provided more elegant constructions. A comprehensive discussion of secret sharing including adaptations providing shared control capabilities of arbitrary complexity, and many of the extended capabilities including pre-positioned schemes, is given by Simmons [1145, 1141, 1142], mainly with geometric illustration. An exposition by Stinson [1177] addresses information rate in particular. Ingemarsson and Simmons [570] consider secret sharing schemes which do not require a trusted party.

Laih et al. [732] consider dynamic secret sharing schemes. Blundo et al. [168] consider pre-positioned schemes, dynamic secret sharing, and bounds on share sizes and broadcast messages therein; Jackson, Martin, and O'Keefe [629] examine related multi-secret threshold schemes. Blakley et al. [147] consider threshold schemes with disenrollment.

Tompa and Woll [1195] note that an untrustworthy participant U may cheat in Shamir's threshold scheme by submitting a share different than its own, but carefully computed such that pooling of shares provides other participants with no information about the secret S, while allowing U to recover S. They propose modifications which (with high probability) allow detection of cheating, and which prevent a cheater U from actually obtaining the secret.

The related problem of *verifiable secret sharing*, which is of broader interest in secure distributed computation, was introduced by Chor et al. [259]; see also Benaloh [110] and Feldman [390], as well as Rabin and Ben-Or [1028]. Here the trusted party distributing shares might also cheat, and the goal is to verify that all distributed shares are consistent in the sense that appropriate subsets of shares define the same secret. For applications of verifiable secret sharing to key escrow, see Micali [863].

Fact 12.75 is based on the definition of perfect secret sharing and information-theoretic security, as is the majority of research in secret sharing. *Ramp schemes* with shares shorter than the secret were examined by Blakley and Meadows [151]; while trading off perfect security for shorter shares, their examination is nonetheless information-theoretic. In practice, a more appropriate goal may be computationally secure secret sharing; here the objective is that if one or more shares is missing, an opponent has insufficient information to (computationally) recover the shared secret. This idea was elegantly addressed by Krawczyk [715] as follows. To share a large s-bit secret $S = P$ (e.g., a plaintext file) among n users, first encrypt it under a k-bit symmetric key K as $C = E_K(P)$; using a perfect secret sharing scheme such as Shamir's (t, n) scheme, split K into n k-bit shares $K_1, \ldots, K_n$; then using Rabin's *information dispersal algorithm* (IDA) [1027] split C into n pieces $C_1, \ldots, C_n$ each of (s/t) bits; finally, distribute to user U_i the secret share $S_i = (K_i, C_i)$. Any t participants who pool their shares can then recover K by secret sharing, C by IDA, and $P = S$ by decrypting C using K. By the remarkable property of IDA, the sum of the sizes of the t pieces C_i used is exactly the size of the recovered secret S itself (which cannot be bettered); globally, the only space overhead is that for the short keys K_i, whose size k is independent of the large secret S.

The clever idea of *visual cryptography* to facilitate sharing (or encryption) of pictures is due to Naor and Shamir [919]. The pixels of a (secret) picture are treated as individual secrets to be shared. The picture is split into two or more images each of which contains one share for each original pixel. Each original pixel is split into shares by subdivision into subpixels of appropriate size, with selection of appropriate combinations of subpixel shadings (black and white) such that stacking the images on transparencies reveals the original, while each individual image appears random. Picture recovery requires no computation (it is visual); anyone with all but one of the images still has (provably) no information.

§12.8

An early investigation of conference keying schemes based on Diffie-Hellman key agreement was undertaken by Ingemarsson, Tang and Wong [571]. The protocol of Burmester and Desmedt [222] (Protocol 12.78) is the most efficient of those which have been proposed and are provably secure; their work includes a review of alternate proposals and a thorough bibliography. Research in this area with particular emphasis on digital telephony includes that of Brickell, Lee, and Yacobi [205]; Steer et al. [1169]; and Heiman [547].

Matsumoto and Imai [799] systematically define (symmetric-key) key pre-distribution schemes, based on symmetric functions, for conferences of two or more parties. Their proposals are non-interactive and ID-based, following the original idea of two-party non-interactive ID-based schemes by Blom [157, 158], including consideration of information-theoretic security against coalitions of fixed size. Tsujii and Chao [1197], among many others, propose schemes in a similar setting. Blundo et al. [169] both specialize the work of Matsumoto and Imai, and generalize Blom's symmetric key distribution (Mechanism 12.35) and bounds from two-party key pre-distribution to non-interactive j-secure conference keying schemes of fixed size; prove Fact 12.79; and provide a scheme meeting this bound. Their generalization uses symmetric polynomials in t variables for privileged subsets of size t, yielding in the two-party case ($t = 2$) an equivalent but simpler formulation of Blom's scheme: the trusted party selects an appropriate secret symmetric polynomial $f(x, y)$ and gives party i the secret univariate polynomial $f(i, y)$, allowing parties i and j to share the pairwise key $f(i, j) = f(j, i)$. They also consider an interactive model. Further examination of interactive vs. non-interactive conferencing is undertaken by Beimel and Chor [83]. Fiat and Naor [394] consider j-secure broadcast encryption schemes, and practical schemes requiring less storage; for the former, Blundo and Cresti [167] establish lower bounds on the number of keys held and the size of user secrets.

Berkovits [116] gives constructions for creating *secret broadcasting schemes* (conference keying schemes where all messages are broadcast) from (t, n) threshold schemes. Essentially, for conferences with t members, a new $(t + 1, 2t + 1)$ threshold scheme with secret K is created from the old, and t new shares are publicly broadcast such that each of the t pre-assigned secret shares of the intended conference members serves as share $t + 1$, allowing recovery of the conference key K in the new scheme. For related work involving use of polynomial interpolation, key distribution involving a trusted party, and broadcasting keys, see Gong [502] and Just et al. [647].

§12.9

The intruder-in-the-middle attack (Attack 1) is discussed by Rivest and Shamir [1057], who propose an "interlock protocol" to allow its detection; but see also Bellovin and Merritt [106]. The reflection attack (Attack 2) is discussed by Mitchell [880]. Attack 4 on the Otway-Rees protocol is discussed by Boyd and Mao [183] and van Oorschot [1205]. The interleaving attack (Attack 3) is due to Wiener circa June 1991 (document ISO/IEC JTC1/SC27 N313, 2 October 1991), and discussed by Diffie, van Oorschot, and Wiener [348] along with attacks on sundry variations of Diffie-Hellman key agreement. Bird et al. [140] systematically examine interleaving attacks on symmetric-key protocols, consider exhaustive analysis to detect such attacks, and propose a protocol resistant thereto (namely 2PP, included in the IBM prototype *KryptoKnight* [891]; see also [141, 142]).

Bellare and Rogaway [94], building on the work of earlier informal models, present a complexity-theoretic communications model and formal definitions for secure symmetric-key two-party mutual authentication and authenticated key establishment, taking known-key attacks into account. They prove AKEP1 (Note 12.21) and AKEP2 (Protocol 12.20) secure relative to this model, for parameters of appropriate size and assuming h and h' are pseudorandom functions or pseudorandom permutations; they also suggest practical constructions for pseudorandom functions based on DES and MD5. Gong [503] examines the efficiency of various authentication protocols and proposes lower bounds (e.g., on the number of message-passes required).

The examples illustrating attacks on flawed protocols are only a few of countless documented in the literature. Moore [898] provides an excellent survey on protocol failure; see also Anderson and Needham [31] and Abadi and Needham [1] for sound engineering principles. A large number of authenticated key establishment protocols with weaknesses are analyzed using the BAN logic in the highly recommended report of Burrows, Abadi, and Needham [227] (and by the same title: [224, 226, 225]). Gligor et al. [463] discuss the limitations of authentication logics. Syverson [1181] examines the goals of formal logics for protocol analysis and the utility of formal semantics as a reasoning tool. Among the authentication logics evolving from BAN are those of Abadi and Tuttle [2], Gong, Needham, and Yahalom [505], and Syverson and van Oorschot [1183]. The work of Abadi and Tuttle is notable for its model of computation and formal semantics relative to this model. Lampson et al. [740] both provide a theory of authentication in distributed systems (including delegation and revocation) and discuss a practical system based on this theory.

One of the first contributions to formal protocol analysis was that of Dolev and Yao [359], whose formal model, which focuses on two-party protocols for transmitting secret plaintexts, facilitates precise discussion of security issues. This approach was augmented with respect to message authentication and information leakage by Book and Otto [170]. Three general approaches to protocol analysis are discussed by Kemmerer, Meadows, and Millen [664] (see also Simmons [1148]): an algebraic approach, a state transition approach, and a logical approach (which can be given a state-transition semantics). They illustrate several methods on a protocol with known flaws (the infamous TMN protocol of Tatebayashi, Matsuzaki, and Newman [1188]). Other recent surveys on formal methods include that of Meadows [831], and the comprehensive survey of Rubin and Honeyman [1073]. An extensive bibliographic tour of authentication literature is provided by Liebl [765].

13

Key Management Techniques

Contents in Brief

13.1 Introduction

This chapter considers key management techniques for controlling the distribution, use, and update of cryptographic keys. Whereas Chapter 12 focuses on details of specific key establishment protocols which provide shared secret keys, here the focus is on communications models for key establishment and use, classification and control of keys based on their intended use, techniques for the distribution of public keys, architectures supporting automated key updates in distributed systems, and the roles of trusted third parties. Systems providing cryptographic services require techniques for initialization and key distribution as well as protocols to support on-line update of keying material, key backup/recovery, revocation, and for managing certificates in certificate-based systems. This chapter examines techniques related to these issues.

Chapter outline

The remainder of this chapter is organized as follows. §13.2 provides context including background definitions, classification of cryptographic keys, simple models for key establishment, and a discussion of third party roles. §13.3 considers techniques for distributing confidential keys, including key layering, key translation centers, and symmetric-key certificates. §13.4 summarizes techniques for distributing and authenticating public keys including authentication trees, public-key certificates, the use of identity-based systems, and implicitly-certified keys. §13.5 presents techniques for controlling the use of keying material, including key notarization and control vectors. §13.6 considers methods for establishing trust in systems involving multiple domains, certification authority trust models, and certification chains. The key management life cycle is summarized in §13.7, while §13.8 discusses selected specialized third party services, including trusted timestamping and notary services supporting non-repudiation of digital signatures, and key escrow. Notes and sources for further information are provided in §13.9.

13.2 Background and basic concepts

A *keying relationship* is the state wherein communicating entities share common data (*keying material*) to facilitate cryptographic techniques. This data may include public or secret keys, initialization values, and additional non-secret parameters.

13.1 Definition *Key management* is the set of techniques and procedures supporting the establishment and maintenance of keying relationships between authorized parties.

Key management encompasses techniques and procedures supporting:

1. initialization of system users within a domain;
2. generation, distribution, and installation of keying material;
3. controlling the use of keying material;
4. update, revocation, and destruction of keying material; and
5. storage, backup/recovery, and archival of keying material.

13.2.1 Classifying keys by algorithm type and intended use

The terminology of Table 13.1 is used in reference to keying material. A *symmetric cryptographic system* is a system involving two transformations – one for the originator and one for the recipient – both of which make use of either the same secret key (symmetric key) or two keys easily computed from each other. An *asymmetric cryptographic system* is a system involving two related transformations – one defined by a public key (the public transformation), and another defined by a private key (the private transformation) – with the property that it is computationally infeasible to determine the private transformation from the public transformation.

Term	Meaning
private key, public key	paired keys in an asymmetric cryptographic system
symmetric key	key in a symmetric (single-key) cryptographic system
secret	adjective used to describe private or symmetric key

Table 13.1: *Private, public, symmetric, and secret keys.*

Table 13.2 indicates various types of algorithms commonly used to achieve the specified cryptographic objectives. Keys associated with these algorithms may be correspondingly classified, for the purpose of controlling key usage (§13.5). The classification given requires specification of both the type of algorithm (e.g., encryption vs. signature) and the intended use (e.g., confidentiality vs. entity authentication).

↓ Cryptographic objective (usage)	Algorithm type	
	public-key	symmetric-key
confidentiality†	encryption	encryption
data origin authentication‡	signature	MAC
key agreement	Diffie-Hellman	various methods
entity authentication (by challenge-response protocols)	1. signature 2. decryption 3. customized	1. MAC 2. encryption

Table 13.2: *Types of algorithms commonly used to meet specified objectives.*
†May include data integrity, and includes key transport; see also §13.3.1.
‡Includes data integrity; and in the public-key case, non-repudiation.

13.2.2 Key management objectives, threats, and policy

Key management plays a fundamental role in cryptography as the basis for securing cryptographic techniques providing confidentiality, entity authentication, data origin authentication, data integrity, and digital signatures. The goal of a good cryptographic design is to reduce more complex problems to the proper management and safe-keeping of a small number of cryptographic keys, ultimately secured through trust in hardware or software by physical isolation or procedural controls. Reliance on physical and procedural security (e.g., secured rooms with isolated equipment), tamper-resistant hardware, and trust in a large number of individuals is minimized by concentrating trust in a small number of easily monitored, controlled, and trustworthy elements.

Keying relationships in a communications environment involve at least two parties (a sender and a receiver) in real-time. In a storage environment, there may be only a single party, which stores and retrieves data at distinct points in time.

The objective of key management is to maintain keying relationships and keying material in a manner which counters relevant threats, such as:

1. compromise of confidentiality of secret keys.
2. compromise of authenticity of secret or public keys. Authenticity requirements include knowledge or verifiability of the true identity of the party a key is shared or associated with.
3. unauthorized use of secret or public keys. Examples include using a key which is no longer valid, or for other than an intended purpose (see Remark 13.32).

In practice, an additional objective is conformance to a relevant security policy.

Security policy and key management

Key management is usually provided within the context of a specific *security policy*. A security policy explicitly or implicitly defines the threats a system is intended to address. The policy may affect the stringency of cryptographic requirements, depending on the susceptibility of the environment in question to various types of attack. Security policies typically also specify:

1. practices and procedures to be followed in carrying out technical and administrative aspects of key management, both automated and manual;
2. the responsibilities and accountability of each party involved; and
3. the types of records (*audit trail information*) to be kept, to support subsequent reports or reviews of security-related events.

13.2.3 Simple key establishment models

The following key distribution problem motivates more efficient key establishment models.

The n^2 key distribution problem

In a system with n users involving symmetric-key techniques, if each pair of users may potentially need to communicate securely, then each pair must share a distinct secret key. In this case, each party must have $n - 1$ secret keys; the overall number of keys in the system, which may need to be centrally backed up, is then $n(n - 1)/2$, or approximately n^2. As the size of a system increases, this number becomes unacceptably large.

In systems based on symmetric-key techniques, the solution is to use centralized key servers: a star-like or spoked-wheel network is set up, with a trusted third party at the center or hub of communications (see Remark 13.3). This addresses the n^2 key distribution problem, at the cost of the requirement of an on-line trusted server, and additional communications with it. Public-key techniques offer an alternate solution.

Point-to-point and centralized key management

Point-to-point communications and *centralized key management*, using key distribution centers or key translation centers, are examples of simple key distribution (communications) models relevant to symmetric-key systems. Here "simple" implies involving at most one third party. These are illustrated in Figure 13.1 and described below, where K_{XY} denotes a symmetric key shared by X and Y.

(a) Point-to-point key distribution

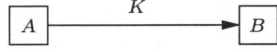

(b) Key distribution center (KDC)

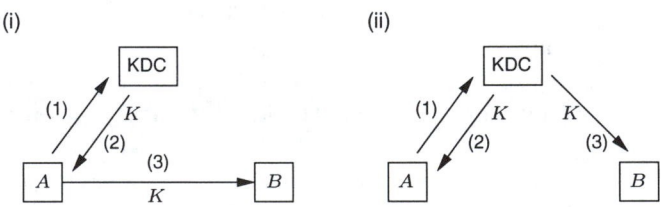

(c) Key translation center (KTC)

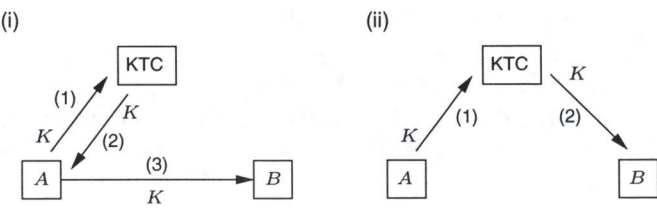

Figure 13.1: *Simple key distribution models (symmetric-key).*

1. *point-to-point* mechanisms. These involve two parties communicating directly (see §12.3.1).
2. *key distribution centers* (KDCs). KDCs are used to distribute keys between users which share distinct keys with the KDC, but not with each other.

   A basic KDC protocol proceeds as follows.[1] Upon request from A to share a key with B, the KDC T generates or otherwise acquires a key K, then sends it encrypted under K_{AT} to A, along with a copy of K (for B) encrypted under K_{BT}. Alternatively, T may communicate K (secured under K_{BT}) to B directly.
3. *key translation centers* (KTCs). The assumptions and objectives of KTCs are as for KDCs above, but here one of the parties (e.g., A) supplies the session key rather than the trusted center.

   A basic KTC protocol proceeds as follows.[2] A sends a key K to the KTC T encrypted under K_{AT}. The KTC deciphers and re-enciphers K under K_{BT}, then returns this to A (to relay to B) or sends it to B directly.

KDCs provide centralized key generation, while KTCs allow distributed key generation. Both are centralized techniques in that they involve an on-line trusted server.

13.2 Note (*initial keying requirements*) Point-to-point mechanisms require that A and B share a secret key *a priori*. Centralized key management involving a trusted party T requires that A and B each share a secret key with T. These shared long-term keys are initially established by non-cryptographic, out-of-band techniques providing confidentiality and authenticity (e.g., in person, or by trusted courier). By comparison, with public keys confidentiality is not required; initial distribution of these need only guarantee authenticity.

13.3 Remark (*centralized key management – pros and cons*) Centralized key management involving third parties (KDCs or KTCs) offers the advantage of key-storage efficiency: each party need maintain only one long-term secret key with the trusted third party (rather than one for each potential communications partner). Potential disadvantages include: vulnerability to loss of overall system security if the central node is compromised (providing an attractive target to adversaries); a performance bottleneck if the central node becomes overloaded; loss of service if the central node fails (a critical reliability point); and the requirement of an on-line trusted server.

13.2.4 Roles of third parties

Below, trusted third parties (TTPs) are first classified based on their real-time interactions with other entities. Key management functions provided by third parties are then discussed.

(i) In-line, on-line, and off-line third parties

From a communications viewpoint, three categories of third parties T can be distinguished based on relative location to and interaction with the communicating parties A and B (see Figure 13.2):

1. *in-line*: T is an intermediary, serving as the real-time means of communication between A and B.
2. *on-line*: T is involved in real-time during each protocol instance (communicating with A or B or both), but A and B communicate directly rather than through T.

---

[1] For specific examples of such protocols including Kerberos (Protocol 12.24), see §12.3.2.
[2] A specific example is the message-translation protocol, Protocol 13.12, with $M = K$.

3. *off-line*: T is not involved in the protocol in real-time, but prepares information *a priori*, which is available to A or B or both and used during protocol execution.

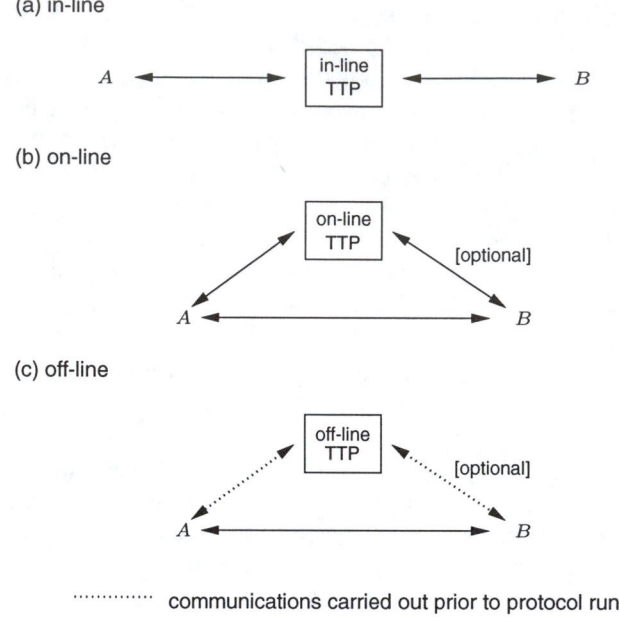

Figure 13.2: *In-line, on-line, and off-line third parties.*

In-line third parties are of particular interest when A and B belong to different security domains or cannot otherwise interact directly due to non-interoperable security mechanisms. Examples of an in-line third party include a KDC or KTC which provides the communications path between A and B, as in Figure 13.1(b)(ii) or (c)(ii). Parts (b)(i) and (c)(i) illustrate examples of on-line third parties which are not in-line. An example of an off-line third party is a certification authority producing public-key certificates and placing them in a public directory; here, the directory may be an on-line third party, but the certification authority is not.

13.4 Remark (*pros and cons: in-line, on-line, off-line*) Protocols with off-line third parties usually involve fewer real-time message exchanges, and do not require real-time availability of third parties. Revocation of privileges (e.g., if a secret key is compromised) is more easily handled by in-line or on-line third parties.

(ii) Third party functions related to public-key certificates

Potential roles played by third parties within a key management system involving public-key certificates (§13.4.2) are listed below. Their relationship is illustrated in Figure 13.3.

1. *certification authority* (CA) – responsible for establishing and vouching for the authenticity of public keys. In certificate-based systems (§13.4.2), this includes binding public keys to distinguished names through signed certificates, managing certificate serial numbers, and certificate revocation.[3]

---

[3]Certificate creation requires verification of the authenticity of the entity to be associated with the public key. This authentication may be delegated to a registration authority. The CA may carry out the combined functions of a registration authority, name server, and key generation facility; such a combined facility is called either a CA or a *key management facility*.

2. *name server* – responsible for managing a name space of unique user names (e.g., unique relative to a CA).

3. *registration authority* – responsible for authorizing entities, distinguished by unique names, as members of a security domain. User registration usually involves associating keying material with the entity.

4. *key generator* – creates public/private key pairs (and symmetric keys or passwords). This may be part of the user entity, part of the CA, or an independent trusted system component.

5. *certificate directory* – a certificate database or server accessible for read-access by users. The CA may supply certificates to (and maintain) the database, or users may manage their own database entries (under appropriate access control).

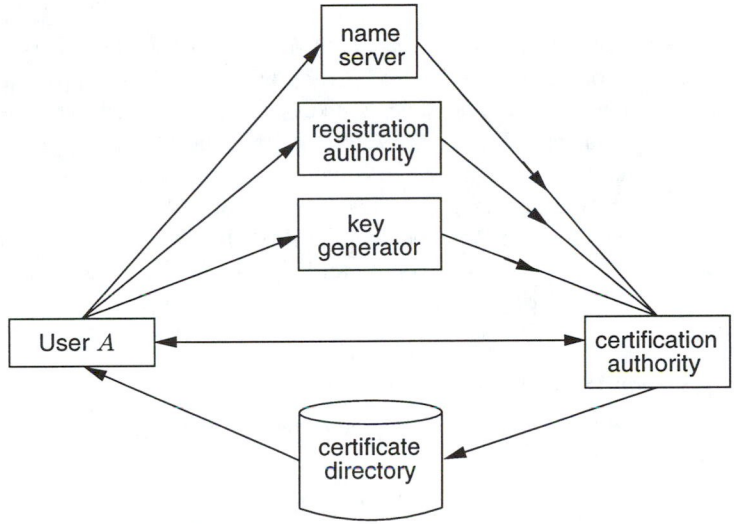

Figure 13.3: *Third party services related to public-key certification.*

(iii) Other basic third party functions

Additional basic functions a trusted third party may provide include:

1. *key server (authentication server)* – facilitates key establishment between other parties, including for entity authentication. Examples include *KDCs* and *KTCs* (§13.2.3).

2. *key management facility* – provides a number of services including storage and archival of keys, audit collection and reporting tools, and (in conjunction with a certification authority or CA) enforcement of life cycle requirements including updating and revoking keys. The associated key server or certification authority may provide a record (audit trail) of all events related to key generation and update, certificate generation and revocation, etc.

13.5 Note *(key access server)* A key server may be generalized to a *key access server*, providing shared keys under controlled access to individual members of groups of two or more parties, as follows. A key K is securely deposited with the server by party A along with an access control list specifying entities authorized to access it. The server stores the key and the associated list. Subsequently, entities contact the server and request the key by referencing a key identifier supplied by A. Upon entity authentication, the server grants access to the keying material (using KTC-like functionality) if the entity is authorized.

13.6 Note *(digital enveloping of files)* A key access server may be employed to store a key K used to symmetrically encrypt a file. The source party A may make the (encrypted) file available by attaching it to the encrypted key, posting it to a public site, or communicating it independently over a distinct (unsecured) channel. Retrieval of the key from the server by an authorized party then allows that party access to the (decrypted) file. The same end goal can be attained by public-key techniques directly, without key access servers, as follows: A encrypts the file under K as above; asymmetrically encrypts K using the intended recipient's public encryption key (or recipients' keys); and includes the encrypted key(s) in a header field preceding the encrypted file.

13.7 Remark *(levels of trust vs. competency)* Various third party services require different types of trust and competency in the third party. For example, a third party possessing secret decryption keys (or entity authentication keys) must be trusted not to disclose encrypted information (or impersonate users). A third party required (only) to bind an encryption public key to an identity must still be trusted not to create false associations and thereafter impersonate an entity. In general, three levels of trust in a third party T responsible for certifying credentials for users may be distinguished. Level 1: T knows each user's secret key. Level 2: T does not know users' secret keys, but can create false credentials without detection. Level 3: T does not know users' secret keys, and generation of false credentials is detectable.

(iv) Advanced third party functions

Advanced service roles which may be provided by trusted third parties, discussed further in §13.8, include:

1. *timestamp agent* – used to assert the existence of a specified document at a certain point in time, or affix a trusted date to a transaction or digital message.
2. *notary agent* – used to verify digital signatures at a given point in time to support non-repudiation, or more generally establish the truth of any statement (which it is trusted on or granted jurisdiction over) at a given point in time.
3. *key escrow agent* – used to provide third-party access to users' secret keys under special circumstances. Here distinction is usually made between key types; for example, encryption private keys may need to be escrowed but not signature private keys (cf. Remark 13.32).

13.2.5 Tradeoffs among key establishment protocols

A vast number of key establishment protocols are available (Chapter 12). To choose from among these for a particular application, many factors aside from cryptographic security may be relevant. §12.2.2 discusses different types of assurances provided, and characteristics useful in comparing protocols.

In selected key management applications, hybrid protocols involving both symmetric and asymmetric techniques offer the best alternative (e.g., Protocol 12.44; see also Note 13.6). More generally, the optimal use of available techniques generally involves combining symmetric techniques for bulk encryption and data integrity with public-key techniques for signatures and key management.

Public-key vs. symmetric-key techniques (in key management)

Primary advantages offered by public-key (vs. symmetric-key) techniques for applications related to key management include:

1. *simplified key management*. To encrypt data for another party, only the encryption public key of that party need be obtained. This simplifies key management as only authenticity of public keys is required, not their secrecy. Table 13.3 illustrates the case for encryption keys. The situation is analogous for other types of public-key pairs, e.g., signature key pairs.

2. *on-line trusted server not required*. Public-key techniques allow a trusted on-line server to be replaced by a trusted off-line server plus any means for delivering authentic public keys (e.g., public-key certificates and a public database provided by an untrusted on-line server). For applications where an on-line trusted server is not mandatory, this may make the system more amenable to scaling, to support very large numbers of users.

3. *enhanced functionality*. Public-key cryptography offers functionality which typically cannot be provided cost-effectively by symmetric techniques (without additional on-line trusted third parties or customized secure hardware). The most notable such features are non-repudiation of digital signatures, and true (single-source) data origin authentication.

	Symmetric keys		Asymmetric keys	
	secrecy	authenticity	secrecy	authenticity
encryption key	yes	yes	no	yes
decryption key	yes	yes	yes	yes

Table 13.3: *Key protection requirements: symmetric-key vs. public-key systems.*

Figure 13.4 compares key management for symmetric-key and public-key encryption. The pairwise secure channel in Figure 13.4(a) is often a trusted server with which each party communicates. The pairwise authentic channel in Figure 13.4(b) may be replaced by a public directory through which public keys are available via certificates; the public key in this case is typically used to encrypt a symmetric data key (cf. Note 13.6).

13.3 Techniques for distributing confidential keys

Various techniques and protocols are available to distribute cryptographic keys whose confidentiality must be preserved (both private keys and symmetric keys). These include the use of key layering (§13.3.1) and symmetric-key certificates (§13.3.2).

13.3.1 Key layering and cryptoperiods

Table 13.2 (page 545) may be used to classify keys based on usage. The class "confidentiality" may be sub-classified on the nature of the information being protected: user data vs. keying material. This suggests a natural *key layering* as follows:

1. *master keys* – keys at the highest level in the hierarchy, in that they themselves are not cryptographically protected. They are distributed manually or initially installed and protected by procedural controls and physical or electronic isolation.

(a) Symmetric-key encryption

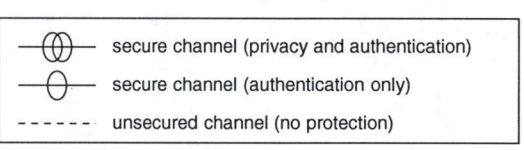

(b) Public-key encryption

Figure 13.4: *Key management: symmetric-key vs. public-key encryption.*

2. *key-encrypting keys* – symmetric keys or encryption public keys used for key transport or storage of other keys, e.g., in the key transport protocols of Chapter 12. These may also be called *key-transport keys*, and may themselves be secured under other keys.

3. *data keys* – used to provide cryptographic operations on user data (e.g., encryption, authentication). These are generally short-term symmetric keys; however, asymmetric signature private keys may also be considered data keys, and these are usually longer-term keys.

The keys at one layer are used to protect items at a lower level. This constraint is intended to make attacks more difficult, and to limit exposure resulting from compromise of a specific key, as discussed below.

13.8 Note (*protection of key-encrypting keys*) Compromise of a key-encrypting key (and moreover, a master key as a special case thereof) affects all keys protected thereunder. Consequently, special measures are used to protect master keys, including severely limiting access and use, hardware protection, and providing access to the key only under shared control (§12.7.1).

13.9 Example (*key layering with master and terminal keys*) Assume each terminal X from a predefined set shares a key-encrypting key (*terminal key*) K_X with a trusted central node C, and that C stores an encrypted list of all terminal keys under a master key K_M. C may then provide a session key to terminals X and Y as follows. C obtains a random value R (possibly from an external source) and defines the session key to be $S = D_{K_M}(R)$, the decryption of R under K_M. Using K_M, C decrypts the key list to obtain K_X, computes S

from R, then encrypts S under K_X and transmits it to X. S is analogously transmitted to Y, and can be recovered by both X and Y. □

Cryptoperiods, long-term keys, and short-term keys

13.10 Definition The *cryptoperiod* of a key is the time period over which it is valid for use by legitimate parties.

Cryptoperiods may serve to:

1. limit the information (related to a specific key) available for cryptanalysis;
2. limit exposure in the case of compromise of a single key;
3. limit the use of a particular technology to its estimated effective lifetime; and
4. limit the time available for computationally intensive cryptanalytic attacks (in applications where long-term key protection is not required).

In addition to the key layering hierarchy above, keys may be classified based on temporal considerations as follows.

1. *long-term keys*. These include master keys, often key-encrypting keys, and keys used to facilitate key agreement.
2. *short-term* keys. These include keys established by key transport or key agreement, and often used as data keys or *session keys* for a single communications session. See Remark 13.11.

In general, communications applications involve short-term keys, while data storage applications require longer-term keys. Long-term keys typically protect short-term keys. Diffie-Hellman keys are an exception in some cases (see §12.6.1). Cryptoperiods limit the use of keys to fixed periods, after which they must be replaced.

13.11 Remark (*short-term use vs. protection*) The term *short* as used in short-term keys refers to the intended time of the key usage by legitimate parties, rather than the *protection lifetime* (cf. §13.7.1). For example, an encryption key used for only a single session might nonetheless be required to provide protection sufficient to withstand long-term attack (perhaps 20 years), whereas if signatures are verified immediately and never checked again, a signature key may need to provide protection only for a relatively short period of time. The more severe the consequences of a secret key being disclosed, the greater the reward to an adversary for obtaining access to it, and the greater the time or level of effort an adversary will invest to do so. (See also §12.2.2, and §12.2.3 on perfect forward secrecy.)

13.3.2 Key translation centers and symmetric-key certificates

Further to centralized key management discussed in §13.2.3, this section considers techniques involving key translation centers, including use of symmetric-key certificates.

(i) Key translation centers

A key translation center (KTC) T is a trusted server which allows two parties A and B, which do not directly share keying material, to establish secure communications through use of long-term keys K_{AT} and K_{BT} they respectively share with T. A may send a confidential message M to B using Protocol 13.12. If M is a key K, this provides a key transfer protocol (cf. §13.2.3); thus, KTCs provide translation of keys or messages.

13.12 Protocol Message translation protocol using a KTC

SUMMARY: A interacts with a trusted server (KTC) T and party B.
RESULT: A transfers a secret message M (or session key) to B. See Note 13.13.

1. *Notation.* E is a symmetric encryption algorithm. M may be a session key K.
2. *One-time setup.* A and T share key K_{AT}. Similarly B and T share K_{BT}.
3. *Protocol messages.*

$$
\begin{aligned}
A \to T : \quad & A,\ E_{K_{AT}}(B, M) \quad &(1) \\
A \leftarrow T : \quad & E_{K_{BT}}(M, A) \quad &(2) \\
A \to B : \quad & E_{K_{BT}}(M, A) \quad &(3)
\end{aligned}
$$

4. *Protocol actions.*

   (a) A encrypts M (along with the identifier of the intended recipient) under K_{AT}, and sends this to T with its own identifier (to allow T to look up K_{AT}).
   (b) Upon decrypting the message, T determines it is intended for B, looks up the key (K_{BT}) of the indicated recipient, and re-encrypts M for B.
   (c) T returns the translated message for A to send to (or post in a public site for) B; alternatively, T may send the response to B directly.

Only one of A and B need communicate with T. As an alternative to the protocol as given, A may send the first message to B directly, which B would then relay to T for translation, with T responding directly to B.

13.13 Note (*security of Protocol 13.12*)

   (i) The identifier A, corresponding to the key under which message (1) was encrypted, is included in message (2) as a secure indication (to B) of the original source. Key notarization (§13.5.2) offers a more robust method of preventing key substitution.
  (ii) A recognizable distinction (e.g., re-ordering the message and identifier fields) between the format of messages (1) and (2) is required to prevent an adversary from reflecting (1) back to A as a message (3) purportedly originating from B.
 (iii) Message replay is possible; attacks may be detected through the use of timestamps or sequence numbers within M. The protocol as given provides no entity authentication.
  (iv) An integrity check mechanism on the encrypted text should be used to allow T to detect tampering of the cleartext identifier A in (1), as well as in (2) and (3).
   (v) A chosen-text attack on key K_{BT} in (2) may be prevented by an encryption mode such as CBC, and inserting an initial field containing a random number.

(ii) Symmetric-key certificates

Symmetric-key certificates provide a means for a KTC to avoid the requirement of either maintaining a secure database of user secrets (or duplicating such a database for multiple servers), or retrieving such keys from a database upon translation requests.

As before, associated with each party B is a key K_{BT} shared with T, which is now embedded in a *symmetric-key certificate* $E_{K_T}(K_{BT}, B)$ encrypted under a symmetric master key K_T known only to T. (A lifetime parameter L could also be included in the certificate as a validity period.) The certificate serves as a memo from T to itself (who alone can open it), and is given to B so that B may subsequently present it back to T precisely when required to access B's symmetric key K_{BT} for message translation. Rather than storing all user keys, T now need securely store only K_T.

Symmetric-key certificates may be used in Protocol 13.12 by changing only the first message as below, where $SCert_A = E_{K_T}(K_{AT}, A)$, $SCert_B = E_{K_T}(K_{BT}, B)$:

$$A \rightarrow T: \quad SCert_A, \ E_{K_{AT}}(B, M), \ SCert_B \quad (1)$$

A public database may be established with an entry specifying the name of each user and its corresponding symmetric-key certificate. To construct message (1), A retrieves B's symmetric-key certificate and includes this along with its own. T carries out the translation as before, retrieving K_{AT} and K_{BT} from these certificates, but now also verifies that A's intended recipient B as specified in $E_{K_{AT}}(B, M)$ matches the identifier in the supplied certificate $SCert_B$.

13.14 Remark (*public-key functionality via symmetric techniques*) The trusted third party functionality required when using symmetric-key certificates may be provided by per-user tamper-resistant hardware units keyed with a common (user-inaccessible) master key K_T. The trusted hardware unit H_A of each user A generates a symmetric-key certificate $SCert_A = E_{K_T}(K_{AT}, A)$, which is made available to B when required. H_B decrypts the certificate to recover K_{AT} (inaccessible to B) and the identity A (accessible to B). By design, H_B is constrained to use other users' keys $K_{AT} = K_A$ solely for verification functions (e.g., MAC verification, message decryption). K_A then functions as A's public key (cf. Example 13.36), allowing data origin authentication with non-repudiation; an adjudicator may resolve disputes given a hardware unit containing K_T, a disputed (message, signature) pair, and the authentic value $SCert_A$ from H_A.

13.15 Remark (*symmetric-key vs. public-key certificates*) Symmetric-key certificates differ from public-key certificates as follows: they are symmetric-key encrypted under T's master key (vs. signed using T's private key); the symmetric key within may be extracted only by T (vs. many parties being able to verify a public-key certificate); and T is required to be on-line for key translation (vs. an off-line certification authority). In both cases, certificates may be stored in a public directory.

13.4 Techniques for distributing public keys

Protocols involving public-key cryptography are typically described assuming *a priori* possession of (authentic) public keys of appropriate parties. This allows full generality among various options for acquiring such keys. Alternatives for distributing explicit public keys with guaranteed or verifiable authenticity, including public exponentials for Diffie-Hellman key agreement (or more generally, public parameters), include the following.

1. *Point-to-point delivery over a trusted channel.* Authentic public keys of other users are obtained directly from the associated user by personal exchange, or over a direct channel, originating at that user, and which (procedurally) guarantees integrity and authenticity (e.g., a trusted courier or registered mail). This method is suitable if used infrequently (e.g., one-time user registration), or in small closed systems. A related method is to exchange public keys and associated information over an untrusted electronic channel, and provide authentication of this information by communicating a hash thereof (using a collision-resistant hash function) via an independent, lower-bandwidth authentic channel, such as a registered mail.

Drawbacks of this method include: inconvenience (elapsed time); the requirement of non-automated key acquisition prior to secured communications with each new party (chronological timing); and the cost of the trusted channel.

2. *Direct access to a trusted public file (public-key registry).* A public database, the integrity of which is trusted, may be set up to contain the name and authentic public key of each system user. This may be implemented as a public-key registry operated by a trusted party. Users acquire keys directly from this registry.

   While remote access to the registry over unsecured channels is acceptable against passive adversaries, a secure channel is required for remote access in the presence of active adversaries. One method of authenticating a public file is by tree authentication of public keys (§13.4.1).

3. *Use of an on-line trusted server.* An on-line trusted server provides access to the equivalent of a public file storing authentic public keys, returning requested (individual) public keys in signed transmissions; confidentiality is not required. The requesting party possesses a copy of the server's signature verification public key, allowing verification of the authenticity of such transmissions.

   Disadvantages of this approach include: the trusted server must be on-line; the trusted server may become a bottleneck; and communications links must be established with both the intended communicant and the trusted server.

4. *Use of an off-line server and certificates.* In a one-time process, each party A contacts an off-line trusted party referred to as a *certification authority* (CA), to register its public key and obtain the CA's signature verification public key (allowing verification of other users' certificates). The CA certifies A's public key by binding it to a string identifying A, thereby creating a certificate (§13.4.2). Parties obtain authentic public keys by exchanging certificates or extracting them from a public directory.

5. *Use of systems implicitly guaranteeing authenticity of public parameters.* In such systems, including identity-based systems (§13.4.3) and those using implicitly certified keys (§13.4.4), by algorithmic design, modification of public parameters results in detectable, non-compromising failure of cryptographic techniques (see Remark 13.26).

The following subsections discuss the above techniques in greater detail. Figure 13.7 (page 564) provides a comparison of the certificate-based approach, identity-based systems, and the use of implicitly-certified public keys.

13.4.1 Authentication trees

Authentication trees provide a method for making public data available with verifiable authenticity, by using a tree structure in conjunction with a suitable hash function, and authenticating the root value. Applications include:

1. *authentication of public keys* (as an alternative to public-key certificates). An authentication tree created by a trusted third party, containing users' public keys, allows authentication of a large number of such keys.

2. *trusted timestamping service.* Creation of an authentication tree by a trusted third party, in a similar way, facilitates a trusted timestamping service (see §13.8.1).

3. *authentication of user validation parameters.* Creation of a tree by a single user allows that user to publish, with verifiable authenticity, a large number of its own public validation parameters, such as required in one-time signature schemes (see §11.6.3).

To facilitate discussion of authentication trees, binary trees are first introduced.

Binary trees

A *binary tree* is a structure consisting of vertices and directed edges. The vertices are divided into three types:

1. a *root vertex*. The root has two edges directed towards it, a left and a right edge.
2. *internal vertices*. Each internal vertex has three edges incident to it – an upper edge directed away from it, and left and right edges directed towards it.
3. *leaves*. Each leaf vertex has one edge incident to it, and directed away from it.

The vertices incident with the left and right edges of an internal vertex (or the root) are called the *children* of the internal vertex. The internal (or root) vertex is called the *parent* of the associated children. Figure 13.5 illustrates a binary tree with 7 vertices and 6 edges.

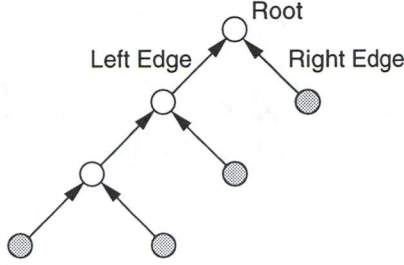

Figure 13.5: *A binary tree (with 4 shaded leaves and 3 internal vertices).*

13.16 Fact There is a unique directed path from any non-root vertex in a binary tree to the root vertex.

Constructing and using authentication trees

Consider a binary tree T which has t leaves. Let h be a collision-resistant hash function. T can be used to authenticate t public values, $Y_1, Y_2, \ldots, Y_t$, by constructing an *authentication tree T^** as follows.

1. Label each of the t leaves by a unique public value Y_i.
2. On the edge directed away from the leaf labeled Y_i, put the label $h(Y_i)$.
3. If the left and right edge of an internal vertex are labeled h_1 and h_2, respectively, label the upper edge of the vertex $h(h_1 \| h_2)$.
4. If the edges directed toward the root vertex are labeled u_1 and u_2, label the root vertex $h(u_1 \| u_2)$.

Once the public values are assigned to leaves of the binary tree, such a labeling is well-defined. Figure 13.6 illustrates an authentication tree with 4 leaves. Assuming some means to authenticate the label on the root vertex, an authentication tree provides a means to authenticate any of the t public leaf values Y_i, as follows. For each public value Y_i, there is a unique path (the *authentication path*) from Y_i to the root. Each edge on the path is a left or right edge of an internal vertex or the root. If e is such an edge directed towards vertex x, record the label on the other edge (not e) directed toward x. This sequence of labels (the *authentication path values*) used in the correct order provides the authentication of Y_i, as illustrated by Example 13.17. Note that if a single leaf value (e.g., Y_1) is altered, maliciously or otherwise, then authentication of that value will fail.

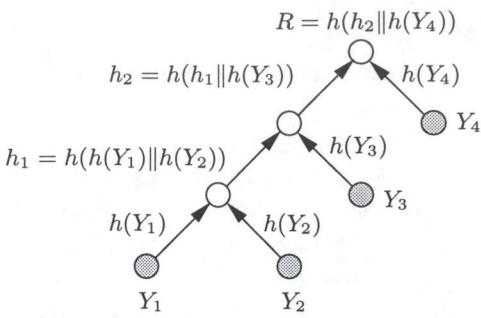

Figure 13.6: *An authentication tree.*

13.17 Example *(key verification using authentication trees)* Refer to Figure 13.6. The public value Y_1 can be authenticated by providing the sequence of labels $h(Y_2), h(Y_3), h(Y_4)$. The authentication proceeds as follows: compute $h(Y_1)$; next compute $h_1 = h(h(Y_1)\|h(Y_2))$; then compute $h_2 = h(h_1\|h(Y_3))$; finally, accept Y_1 as authentic if $h(h_2\|h(Y_4)) = R$, where the root value R is known to be authentic.   □

The advantage of authentication trees is evident by considering the storage required to allow authentication of t public values using the following (very simple) alternate approach: an entity A authenticates t public values $Y_1, Y_2, \ldots, Y_t$ by registering each with a trusted third party. This approach requires registration of t public values, which may raise storage issues at the third party when t is large. In contrast, an authentication tree requires only a single value be registered with the third party.

If a public key Y_i of an entity A is the value corresponding to a leaf in an authentication tree, and A wishes to provide B with information allowing B to verify the authenticity of Y_i, then A must (store and) provide to B both Y_i and all hash values associated with the authentication path from Y_i to the root; in addition, B must have prior knowledge and trust in the authenticity of the root value R. These values collectively guarantee authenticity, analogous to the signature on a public-key certificate. The number of values each party must store (and provide to others to allow verification of its public key) is $\lg(t)$, as per Fact 13.19.

13.18 Fact *(depth of a binary tree)* Consider the length of (or number of edges in) the path from each leaf to the root in a binary tree. The length of the longest such path is minimized when the tree is *balanced*, i.e., when the tree is constructed such that all such paths differ in length by at most one. The length of the path from a leaf to the root in a balanced binary tree containing t leaves is about $\lg(t)$.

13.19 Fact *(length of authentication paths)* Using a balanced binary tree (Fact 13.18) as an authentication tree with t public values as leaves, authenticating a public value therein may be achieved by hashing $\lg(t)$ values along the path to the root.

13.20 Remark *(time-space tradeoff)* Authentication trees require only a single value (the root value) in a tree be registered as authentic, but verification of the authenticity of any particular leaf value requires access to and hashing of all values along the authentication path from leaf to root.

13.21 Remark *(changing leaf values)* To change a public (leaf) value or add more values to an authentication tree requires recomputation of the label on the root vertex. For large balanced trees, this may involve a substantial computation. In all cases, re-establishing trust of all users in this new root value (i.e., its authenticity) is necessary.

The computational cost involved in adding more values to a tree (Remark 13.21) may motivate constructing the new tree as an unbalanced tree with the new leaf value (or a sub-tree of such values) being the right child of the root, and the old tree, the left. Another motivation for allowing unbalanced trees arises when some leaf values are referenced far more frequently than others.

13.4.2 Public-key certificates

Public-key certificates are a vehicle by which public keys may be stored, distributed or forwarded over unsecured media without danger of undetectable manipulation. The objective is to make one entity's public key available to others such that its authenticity (i.e., its status as the true public key of that entity) and validity are verifiable. In practice, X.509 certificates are commonly used (see page 587). Further details regarding public-key certificates follow.

13.22 Definition A *public-key certificate* is a data structure consisting of a *data part* and a *signature part*. The data part contains cleartext data including, as a minimum, a public key and a string identifying the party (*subject entity*) to be associated therewith. The signature part consists of the digital signature of a certification authority over the data part, thereby binding the subject entity's identity to the specified public key.

The *Certification Authority* (CA) is a trusted third party whose signature on the certificate vouches for the authenticity of the public key bound to the subject entity. The significance of this binding (e.g., what the key may be used for) must be provided by additional means, such as an attribute certificate or policy statement. Within the certificate, the string which identifies the subject entity must be a unique name within the system (*distinguished name*), which the CA typically associates with a real-world entity. The CA requires its own signature key pair, the authentic public key of which is made available to each party upon registering as an authorized system user. This CA public key allows any system user, through certificate acquisition and verification, to transitively acquire trust in the authenticity of the public key in any certificate signed by that CA.

Certificates are a means for transferring trust, as opposed to establishing trust originally. The authenticity of the CA's public key may be originally provided by non-cryptographic means including personal acquisition, or through trusted couriers; authenticity is required, but not secrecy.

Examples of additional information which the certificate data part might contain include:

1. a validity period of the public key;
2. a serial number or key identifier identifying the certificate or key;
3. additional information about the subject entity (e.g., street or network address);
4. additional information about the key (e.g., algorithm and intended use);
5. quality measures related to the identification of the subject entity, the generation of the key pair, or other policy issues;
6. information facilitating verification of the signature (e.g., a signature algorithm identifier, and issuing CA's name);
7. the status of the public key (cf. revocation certificates, §13.6.3).

(i) Creation of public-key certificates

Before creating a public-key certificate for a subject entity A, the certification authority should take appropriate measures (relative to the security level required, and customary business practices), typically non-cryptographic in nature, to verify the claimed identity of A and the fact that the public key to be certified is actually that of A. Two cases may be distinguished.

Case 1: trusted party creates key pair. The trusted party creates a public-key pair, assigns it to a specific entity, and includes the public key and the identity of that entity in the certificate. The entity obtains a copy of the corresponding private key over a secure (authentic and private) channel after proving its identity (e.g., by showing a passport or trusted photo-id, in person). All parties subsequently using this certificate essentially delegate trust to this prior verification of identity by the trusted party.

Case 2: entity creates own key pair. The entity creates its own public-key pair, and securely transfers the public key to the trusted party in a manner which preserves authenticity (e.g., over a trusted channel, or in person). Upon verification of the authenticity (source) of the public key, the trusted party creates the public-key certificate as above.

13.23 Remark (*proof of knowledge of private key*) In Case 2 above, the certification authority should require proof of knowledge of the corresponding private key, to preclude (among other possible attacks) an otherwise legitimate party from obtaining, for malicious purposes, a public-key certificate binding its name to the public key of another party. For the case of signature public keys, this might be done by the party providing its own signature on a subset of the data part of the certificate; or by responding to a challenge r_1 randomized by the party itself e.g., signing $h(r_1 \| r_2)$ for an appropriate hash function h and a random number r_2 chosen by the signer.

(ii) Use and verification of public-key certificates

The overall process whereby a party B uses a public-key certificate to obtain the authentic public key of a party A may be summarized as follows:

1. (One-time) acquire the authentic public key of the certification authority.
2. Obtain an identifying string which uniquely identifies the intended party A.
3. Acquire over some unsecured channel (e.g. from a central public database of certificates, or from A directly), a public-key certificate corresponding to subject entity A and agreeing with the previous identifying string.
4.   (a) Verify the current date and time against the validity period (if any) in the certificate, relying on a local trusted time/day-clock;
      (b) Verify the current validity of the CA's public key itself;
      (c) Verify the signature on A's certificate, using the CA's public key;
      (d) Verify that the certificate has not been revoked (§13.6.3).
5. If all checks succeed, accept the public key in the certificate as A's authentic key.

13.24 Remark (*life cycle reasons for single-key certificates*) Due to differing life cycle requirements for different types of keys (e.g., differing cryptoperiods, backup, archival, and other lifetime protection requirements – see §13.7), separate certificates are recommended for separate keys, as opposed to including several keys in a single certificate. See also Remark 13.32.

(iii) Attribute certificates

Public-key certificates bind a public key and an identity, and include additional data fields necessary to clarify this binding, but are not intended for certifying additional information. *Attribute certificates* are similar to public-key certificates, but specifically intended to allow specification of information (*attributes*) other than public keys (but related to a CA, entity, or public key), such that it may also be conveyed in a trusted (verifiable) manner. Attribute certificates may be associated with a specific public key by binding the attribute information to the key by the method by which the key is identified, e.g., by the serial number of a corresponding public-key certificate, or to a hash-value of the public key or certificate.

Attribute certificates may be signed by an *attribute certification authority*, created in conjunction with an *attribute registration authority*, and distributed in conjunction with an *attribute directory service* (cf. Figure 13.3). More generally, any party with a signature key and appropriate recognizable authority may create an attribute certificate. One application is to certify authorization information related to a public key. More specifically, this may be used, for example, to limit liability resulting from a digital signature, or to constrain the use of a public key (e.g., to transactions of limited values, certain types, or during certain hours).

13.4.3 Identity-based systems

Identity-based systems resemble ordinary public-key systems, involving a private transformation and a public transformation, but users do not have explicit public keys as before. Instead, the public key is effectively replaced by (or constructed from) a user's publicly available identity information (e.g., name and network or street address). Any publicly available information which uniquely identifies a user and can be undeniably associated with the user, may serve as the identity information.

13.25 Definition An identity-based cryptographic system (*ID-based system*) is an asymmetric system wherein an entity's public identification information (unique name) plays the role of its public key, and is used as input by a trusted authority T (along with T's private key) to compute the entity's corresponding private key.

After computing it, T transfers the entity's private key to the entity over a secure (authentic and private) channel. This private key is computed from not only the entity's identity information, but must also be a function of some privileged information known only to T (T's private key). This is necessary to prevent forgery and impersonation – it is essential that only T be able to create valid private keys corresponding to given identification information. Corresponding (authentic) publicly available system data must be incorporated in the cryptographic transformations of the ID-based system, analogous to the certification authority's public key in certificate-based systems. Figure 13.7(b) on page 564 illustrates the design of an identity-based system. In some cases, additional system-defined public data D_A must be associated with each user A in addition to its *a priori* identity ID_A (see Remark 13.27); such systems are no longer "purely" identity-based, although neither the authenticity of D_A nor ID_A need be explicitly verified.

13.26 Remark (*authenticity in ID-based systems*) Id-based systems differ from public-key systems in that the authenticity of user-specific public data is not (and need not be) explicitly verified, as is necessary for user public keys in certificate-based systems. The inherent redundancy of user public data in ID-based systems (derived through the dependence of the corresponding private key thereon), together with the use of authentic public system data, implicitly protects against forgery; if incorrect user public data is used, the cryptographic transformations simply fail. More specifically: signature verification fails, entity authentication fails, public-key encryption results in undecipherable text, and key-agreement results in parties establishing different keys, respectively, for (properly constructed) identity-based signature, authentication, encryption, and key establishment mechanisms.

The motivation behind ID-based systems is to create a cryptographic system modeling an ideal mail system wherein knowledge of a person's name alone suffices to allow mail to be sent which that person alone can read, and to allow verification of signatures that person alone could have produced. In such an ideal cryptographic system:

1. users need exchange neither symmetric keys nor public keys;
2. public directories (files of public keys or certificates) need not be kept; and
3. the services of a trusted authority are needed solely during a set-up phase (during which users acquire authentic public system parameters, to be maintained).

13.27 Remark (*ideal vs. actual ID-based systems*) A drawback in many concrete proposals of ID-based systems is that the required user-specific identity data includes additional data (an integer or public data value), denoted D_A in Figure 13.7(b), beyond an *a priori* identity ID_A. For example, see Note 10.29(ii) on Feige-Fiat-Shamir identification. Ideally, D_A is not required, as a primary motivation for identity-based schemes is to eliminate the need to transmit public keys, to allow truly non-interactive protocols with identity information itself sufficing as an authentic public key. The issue is less significant in signature and identification schemes where the public key of a claimant is not required until receiving a message from that claimant (in this case D_A is easily provided); but in this case, the advantage of identity-based schemes diminishes. It is more critical in key agreement and public-key encryption applications where another party's public key is needed at the outset. See also Remark 13.31.

13.28 Example (*ID-based system implemented using chipcards*) A simplified ID-based system based on chipcards may be run as follows. A third party T, acting as a trusted key generation system, is responsible solely for providing each user a chipcard during a set-up phase, containing that party's ID-based private key, after carrying out a thorough identity check. If no further users need be added, T may publish the public system data and cease to exist. Users are responsible for not disclosing their private keys or losing their cards. □

13.4.4 Implicitly-certified public keys

Another variation of public-key systems is asymmetric systems with *implicitly-certified public keys*. Here explicit user public keys exist (see Figure 13.7(c)), but they must be reconstructed rather than transported by public-key certificates as per certificate-based systems. For other advantages, see Remark 13.30. Examples of specific such mechanisms are given in §12.6.2. Systems with implicitly-certified public keys are designed such that:

1. Entities' public keys may be reconstructed (by other parties) from public data (which essentially replace a certificate).
2. The public data from which a public key is reconstructed includes:
   (a) public (i.e., system) data associated with a trusted party T;
   (b) the user entity's identity (or identifying information, e.g., name and address);
   (c) additional per-user public data (*reconstruction public data*).

3. The integrity of a reconstructed public key is not directly verifiable, but a "correct" public key can be recovered only from authentic user public data.

Regarding authenticity of reconstructed public keys, the system design must guarantee:

1. Alteration of either a user's identity or reconstruction public data results in recovery of a corrupted public key, which causes denial of service but not cryptographic exposure (as per Remark 13.26).

2. It is computationally infeasible for an adversary (without knowledge of T's private data) to compute a private key corresponding to any party's public key, or to construct a matching user identity and reconstruction public data for which a corresponding private key may also be computed. Reconstructed public keys are thus implicitly authenticated by construction.

13.29 Remark (*applications of implicitly-certified keys*) Implicitly-certified public keys may be used as an alternate means for distributing public keys (e.g., Diffie-Hellman keys – see §12.6.3) in various key agreement protocols, or in conjunction with identification protocols, digital signature schemes, and public-key encryption schemes.

Classes of implicitly-certified public keys

Two classes of implicitly-certified public keys may be distinguished:

1. *identity-based public keys (Class 1)*. The private key of each entity A is computed by a trusted party T, based on A's identifying information and T's private key; it is also a function of A's user-specific reconstruction public data, which is fixed *a priori* by T. A's private key is then securely transferred by T to A. An example is Mechanism 12.59.

2. *self-certified public keys (Class 2)*. Each entity A itself computes its private key and corresponding public key. A's reconstruction public data (rather than A's private key, as in Class 1) is computed by T as a function of the public key (transferred to T by A), A's identifying information, and T's private key. An example is Mechanism 12.61.

Class 1 requires more trust in the third party, which therein has access to users' private keys. This differs from Class 2, as emphasized by the term "self" in "self-certified", which refers to the knowledge of this key being restricted to the entity itself.

13.4.5 Comparison of techniques for distributing public keys

§13.4 began with an overview of techniques for addressing authenticity in public key distribution. The basic approaches of §13.4.2, §13.4.3, and §13.4.4 are discussed further here. Figure 13.7 illustrates corresponding classes of asymmetric signature systems, contrasting public-key systems (with explicit public keys), identity-based systems (the public key is a user's identity information), and systems with implicitly-certified public keys (an explicit public key is reconstructed from user public data).[4] The main differences are as follows:

1. Certificate-based public-key systems have explicit public keys, while ID-based systems do not; in implicitly-certified systems explicit public keys are reconstructed. The explicit public key in public-key systems (Figure 13.7(a)) is replaced by:

   (a) the triplet (D_A, ID_A, P_T) for identity-based systems (Figure 13.7(b)). ID_A is an identifying string for A, D_A is additional public data (defined by T and related to ID_A and A's private key), and P_T consists of the trusted public key (or system parameters) of a trusted authority T.

---

[4]While the figure focuses (for concreteness) on signature systems, concepts carry over analogously for asymmetric entity authentication, key establishment, and encryption systems.

(a) Public key system (explicit public keys)

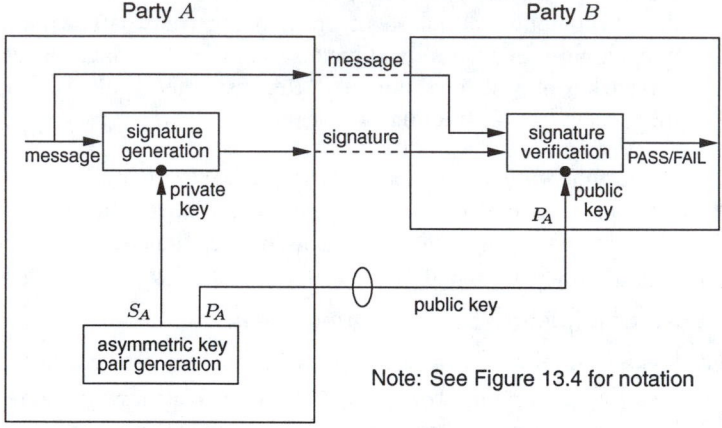

(b) Identity-based system
S_T, P_T are T's private, public keys; D_A is A's public data

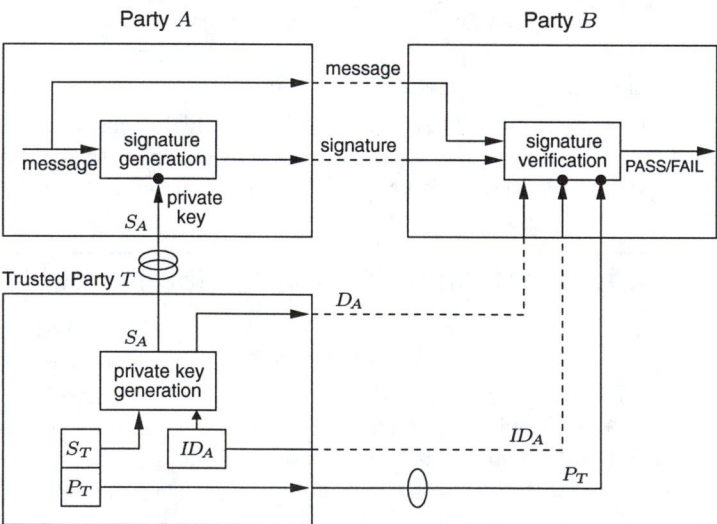

Figure 13.7: *Key management in different classes of asymmetric signature systems.*

(c) System with implicitly-certified public keys

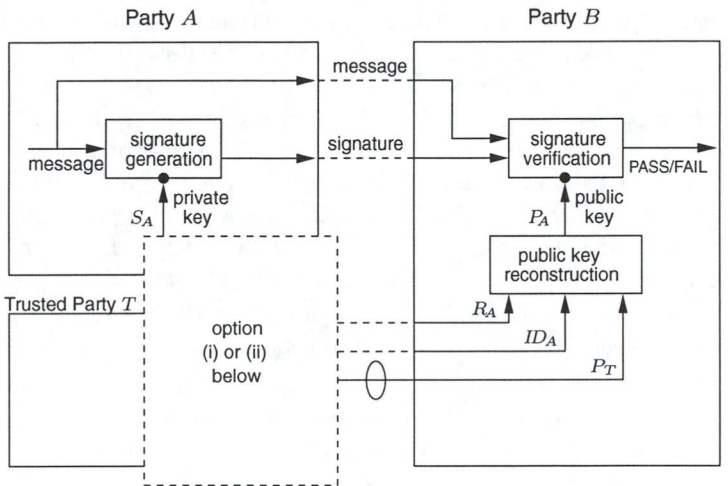

(i) identity-based public keys

(ii) self-certified public keys

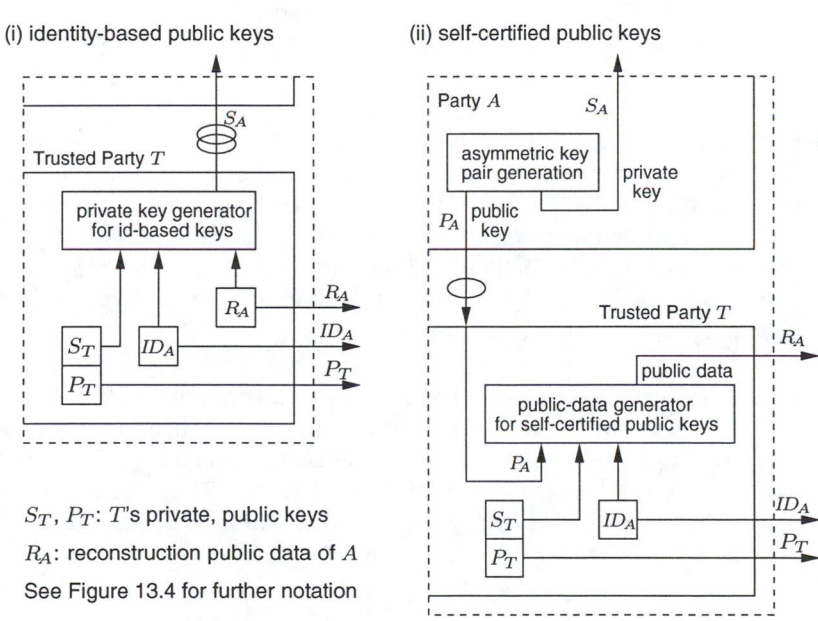

S_T, P_T: T's private, public keys

R_A: reconstruction public data of A

See Figure 13.4 for further notation

Figure 13.7: *(cont'd) Key management in different classes of asymmetric signature systems.*

(b) the triplet (R_A, ID_A, P_T) for systems with implicitly-certified public keys (Figure 13.7(c)). In this case, an explicit public key P_A is reconstructed from these parameters. The reconstruction public data R_A plays a role analogous to the public data D_A in Figure 13.7(b).

2. The authenticity of public keys can (and must) be explicitly verified in certificate-based systems, but not (and need not) in ID-based or implicitly-certified systems.

3. The trusted authority need not know users' private keys in certificate-based public-key systems or implicitly-certified systems with self-certified public keys; but does in ID-based systems, and in implicitly-certified systems with ID-based keys.

4. Similar to identity-based systems (§13.4.3), implicitly-certified public keys (of both classes) depend on an entity's identifying information, and in this sense are also "identity-based". However, ID-based systems avoid explicit public keys entirely (a user's identity data is essentially its public key), while implicitly-certified public keys are not restricted to user identities and may be explicitly computed (and thus more easily used in conjunction with ordinary public-key schemes).

5. The two classes of implicitly-certified public keys (Figure 13.7(c)) differ in their relationship between users' reconstruction public data and private keys as follows.

   (a) Class 1: a user's private key is computed as a function of the reconstruction data, and this private key is computed by the trusted authority;

   (b) Class 2: the reconstruction data is computed as a function of the user's public key, and the corresponding private key is computed by the party itself.

6. In all three approaches, at some stage a third party which is trusted to some level (cf. Note 13.7) is required to provide a link transferring trust between users who may have never met each other and may share nothing in common other than authentic system parameters (and possibly knowledge of other users' identities).

13.30 Remark (*implicitly-certified public keys vs. public-key certificates*) Advantages of implicitly-certified public keys over public-key certificates include: possibly reduced space requirements (signed certificates require storage for signatures); possible computational savings (signature verification, as required for certificates, is avoided); and possible communications savings (e.g. if identity-based and the identity is known *a priori*). Countering these points, computation is actually required to reconstruct a public key; and additional reconstruction public data is typically required.

13.31 Remark (*key revocation in ID-based systems*) Revocation of public keys may be addressed in ID-based schemes and systems using implicitly-certified public keys by incorporating information such as a key validity period or serial number into the identification string used to compute an entity's public key (cf. Remark 13.27). The revocation issue is then analogous to that for public-key certificates. Additional information, e.g., pertaining to key usage or an associated security policy, may similarly be incorporated.

13.5 Techniques for controlling key usage

This section considers techniques for restricting keys to pre-authorized uses.

13.5.1 Key separation and constraints on key usage

Information that may be associated with cryptographic keys includes both attributes which restrict their use, and other information of operational use. These include:

1. owner of key
2. validity period (intended cryptoperiod)
3. key identifier (allowing non-cryptographic reference to the key)
4. intended use (see Table 13.2 for a coarse selection)
5. specific algorithm
6. system or environment of intended use, or authorized users of key
7. names of entities associated with key generation, registration, and certification
8. integrity checksum on key (usually part of authenticity requirement)

Key separation and the threat of key misuse

In simple key management systems, information associated with keys, including authorized uses, are inferred by context. For additional clarity or control, information explicitly specifying allowed uses may accompany distributed keys and be enforced by verification, at the time of use, that the attempted uses are authorized. If control information is subject to manipulation, it should be bound to the key by a method which guarantees integrity and authenticity, e.g., through signatures (cf. public-key certificates) or an encryption technique providing data integrity.

The principle of *key separation* is that keys for different purposes should be cryptographically separated (see Remark 13.32). The threat of key misuse may be addressed by techniques which ensure that keys are used only for those purposes pre-authorized at the time of key creation. Restrictions on key usage may be enforced by procedural techniques, physical protection (tamper-resistant hardware), or cryptographic techniques as discussed below.

Discussion of other methods in §13.5.2 includes key tags, which allow key separation with explicitly-defined uses; key variants, which separate keys without explicitly defining authorized uses; and key notarization and control vectors, which bind control information into the process by which keys are derived.

13.32 Remark (*cryptographic reasons for key separation*) A principle of sound cryptographic design is to avoid use of the same cryptographic key for multiple purposes. A key-encrypting key should not be used interchangeably as a data encryption key, since decrypted keys are not generally made available to application programs, whereas decrypted data is. Distinct asymmetric encryption and signature keys are also generally used, due to both differing life cycle requirements and cryptographic prudence. Flaws also potentially arise if: asymmetric keys are used for both signatures and challenge-response entity authentication (Remark 10.40); keys are used for both encryption and challenge-response entity authentication (chosen-text attacks); symmetric keys are used for both encryption and message authentication (Example 9.88). See also Remark 13.24.

13.5.2 Techniques for controlling use of symmetric keys

The main technique discussed below is the use of control vectors. For historical context, key tags/key variants and key notarization are also discussed.

(i) Key tags and key variants

Key tags provide a simplified method for specifying allowed uses of keys (e.g., data-encrypting vs. key-encrypting keys). A key tag is a bit-vector or structured field which accompanies and remains associated with a key over its lifetime. The tag bits are encrypted jointly with the key and thereby bound to it, appearing in plaintext form only when the key is decrypted. If the combination of tag bits and key are sufficiently short to allow encryption in a single block operation (e.g., a 56-bit key with an 8-bit tag for a 64-bit block cipher), then the inherent integrity provided by encryption precludes meaningful manipulation of the tag.

A naive method for providing key separation is to derive separate keys from a single *base key* (or *derivation key*) using additional non-secret parameters and a non-secret function. The resulting keys are called *key variants* or *derived keys*.

One technique for varying keys is *key offsetting*, whereby a key-encrypting key K is modified on a per-use basis by a counter N incremented after each use. This may prevent replay of encrypted keys. The modified key $K \oplus N$ is used to encrypt another (e.g., session) key. The recipient likewise modifies K to decrypt the session key. A second technique, complementing alternate 4-bit blocks of K commencing with the first 4 bits, is a special case of fixed-mask offsetting (Example 13.33).

13.33 Example (*key variants using fixed-mask offsets*) Suppose exactly three classes of keys are desired. Construct keys by using variations K_1 and K_2 of a master key K, with $K_1 = K \oplus v_1$, $K_2 = K \oplus v_2$, and v_1, v_2 nonsecret mask values. Using K, K_1, and K_2 to encrypt other keys then allows key separation of the latter into three classes. □

If the derivation process is invertible, the base key can be recovered from the derived key. Ideally, the derivation technique is non-reversible (one-way), implying that compromise of one derived key would not compromise the base key or other derived keys (cf. §13.7.1 on security impacts of related keys). Yet another example of key derivation (see §12.3.1) has this property: compute $K_i = E_K(r_i)$ where r_i is a random number, or replace the encryption function E by a MAC, or simply hash K and r_i using a hash function h with suitable properties.

(ii) Key notarization

Key notarization is a technique intended to prevent key substitution by requiring explicit specification of the identities of parties involved in a keying relationship. A key is authenticated with respect to these identities (preventing impersonation) by modifying a key-encrypting key such that the correct identities must be specified to properly recover the protected key. The key is said to be *sealed* with these identities. Preventing key substitution is a requirement in all (authenticated) key establishment protocols. Notarization requires proper control information for accurate recovery of encrypted keys, providing implicit protection analogous to implicitly-certified public keys (§13.4.4).

The basic technique (*simple key notarization*) involves a trusted server (notary), or one of the parties sharing the key, using a key-encrypting key K to encrypt a session key S, intended for use with the originating party i and the recipient j, as: $E_{K \oplus (i||j)}(S)$. Here i and j are assumed to identify unique entities in the given system. The party intending to recover

S from this must share K and explicitly specify i and j in the correct order, otherwise a random key will be recovered. The analogy to a notary originated from the assumption that the third party properly authenticates the identities of the intended parties, and then provides a session key which may only be recovered by these parties. A more involved process, key notarization with offsetting, is given in Example 13.34

13.34 Example (*key notarization with offsetting*) Let E be a block cipher operating on 64-bit blocks with 64-bit key, $K = K_L \| K_R$ be a 128-bit key-encrypting key, N a 64-bit counter, and $i = i_L \| i_R, j = j_L \| j_R$ 128-bit source and destination identifiers. For key notarization with offsetting, compute: $K_1 = E_{K_R \oplus i_L}(j_R) \oplus K_L \oplus N$, $K_2 = E_{K_L \oplus j_L}(i_R) \oplus K_R \oplus N$. The resulting 128-bit *notarized key* (K_1, K_2) then serves as a key-encrypting key in two-key triple-encryption. The leftmost terms $f_1(K_R, i, j)$ and $f_2(K_L, i, j)$ in the computation of K_1, K_2 above are called *notary seals*, which, when combined with K_L and K_R, respectively, result in quantities analogous to those used in simple key notarization (i.e., functions of K, i, j). For K a 64-bit (single-length) key, the process is modified as follows: using $K_L = K_R = K$, compute the notary seals $f_1(K_R, i, j)$, $f_2(K_L, i, j)$ as above, concatenate the leftmost 32 bits of f_1 with the rightmost of f_2 to obtain f, then compute $f \oplus K \oplus N$ as the notarized key. □

(iii) Control vectors

While key notarization may be viewed as a mechanism for establishing authenticated keys, *control vectors* provide a method for controlling the use of keys, by combining the idea of key tags with the mechanism of simple key notarization. Associated with each key S is a control vector C, which is a data field (similar to a key tag) defining the authorized uses of the key (effectively *typing* the key). It is bound to S by varying a key-encrypting key K before encryption: $E_{K \oplus C}(S)$.

Key decryption thus requires the control vector be properly specified, as well as the correct key-encrypting key; if the combined quantity $K \oplus C$ is incorrect, a spurious key of no advantage to an adversary is recovered. Cryptographically binding the control vector C to S at the time of key generation prevents unauthorized manipulation of C, assuming only authorized parties have access to the key-encrypting key K.

Control vectors may encompass key notarization by using one or more fields in C to specify identities. In relation to standard models for access control (Note 13.35), a control vector may be used to specify a subject's identity (S_i) and privileges ($A_{i,j}$) regarding the use of a key (K_j).

At time of use for a specific cryptographic operation, the control vector is input as well as the protected key. At this time, a check is made that the requested operation complies with the control vector; if so, the key is decrypted using the control vector. If the control vector does not match that bound to the protected key (or if K is incorrect), the recovered key $S' \neq S$ will be spurious. Security here is dependent on the assumption that checking is inseparable from use, and done within a trusted subsystem.

If the bitsize of the control vector C differs from that of the key K, a collision-resistant hash function may be used prior to coupling. This allows arbitrary length control vectors. Thus a 128-bit key K and a hash function h with 128-bit output may be used to encrypt S as: $E_{K \oplus h(C)}(S)$.

13.35 Note (*models for access control*) Several methods are available to control access to resources. The *access matrix model* uses a 2-dimensional matrix $A_{i \times j}$ with a row for each subject (S_i) and a column for each object (O_j), and relies on proper identification of subjects S_i. Each access record $A_{i,j}$ specifies the privileges entity S_i has on object O_j (e.g., an application program may have read, write, modify, or execute privileges on a file). Column j may alternately serve as an *access list* for object O_j, having entries (S_i, P_{ij}) where $P_{ij} = A_{i,j}$ specifies privileges. Another method of resource protection uses the idea of capabilities: a *capability* (O, P) specifies an object O and privilege set P related to O, and functions as a *ticket* – possession of capability (O, P) grants the holder the specified privileges, without further validation or ticket-holder identification.

13.36 Example (*sample uses of control vectors*) Control vectors may be used to provide a public-key like functionality as follows (cf. Remark 13.14). Two copies of a symmetric key are distributed, one typed to allow encryption only (or MAC generation), and a second allowing decryption only (or MAC verification). Other sample uses of control fields include: allowing random number generation; allowing ciphertext translation (e.g., in KTCs); distinguishing data encryption and key encryption keys; or incorporation of any field within a public-key certificate. □

13.37 Remark (*key verification and preventing replay*) Replay of keys distributed by key transport protocols may be countered by the same techniques used to provide uniqueness/timeliness and prevent replay of messages – sequence numbers, timestamps, and challenge-response techniques (§10.3.1). Before a key resulting from a key derivation, notarization, or control vector technique is actually used, verification of its integrity may be desirable (cf. key confirmation, §12.2). This can be achieved using standard techniques for data integrity (Figure 9.8). A simple method involves the originator sending the encryption (under the key in question) of a data item which the recipient can recognize.

13.6 Key management involving multiple domains

This section considers key management models for systems involving multiple domains or authorities, as opposed to the simpler single-domain models of §13.2.3.

13.38 Definition A *security domain* (domain) is defined as a (sub)system under the control of a single authority which the entities therein trust. The security policy in place over a domain is defined either implicitly or explicitly by its authority.

The trust that each entity in a domain has in its authority originates from, and is maintained through, an entity-specific shared secret key or password (in the symmetric case), or possession of the authority's authentic public key (in the asymmetric case). This allows secure communications channels (with guaranteed authenticity and/or confidentiality) to be established between the entity and authority, or between two entities in the same domain. Security domains may be organized (e.g., hierarchically) to form larger domains.

13.6.1 Trust between two domains

Two parties A and B, belonging to distinct security domains D_A and D_B with respective trusted authorities T_A and T_B, may wish to communicate securely (or A may wish to access resources from a distinct domain D_B). This can be reduced to the requirement that A and B either:

1. (*share a symmetric key*) establish a shared secret key K_{AB} which both trust as being known only to the other (and possibly trusted authorities); or

2. (*share trusted public keys*) acquire trust in one or more common public keys which may be used to bridge trust between the domains, e.g., allowing verification of the authenticity of messages purportedly from the other, or ensure the confidentiality of messages sent to the other.

Either of these is possible provided T_A and T_B have an existing trust relationship, based on either trusted public keys or shared secret keys.

If T_A and T_B do have an existing trust relationship, either requirement may be met by using this and other initial pairwise trust relationships, which allow secure communications channels between the pairs (A, T_A), (T_A, T_B), and (T_B, B), to be successively used to establish the objective trust relationship (A, B). This may be done by A and B essentially delegating to their respective authorities the task of acquiring trust in an entity under the other authority (as detailed below).

If T_A and T_B do not share an existing trust relationship directly, a third authority T_C, in which they both do trust, may be used as an intermediary to achieve the same end result. This is analogous to a *chain of trust* in the public-key case (§13.6.2). The two numbered options beginning this subsection are now discussed in further detail.

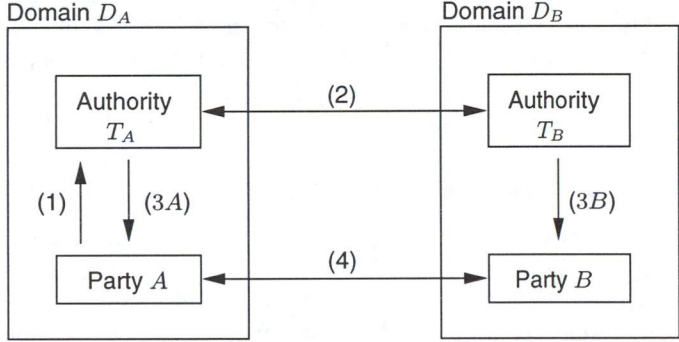

Figure 13.8: *Establishing trust between users in distinct domains.*

1. *trusted symmetric key*: Trust in a shared secret key may be acquired through a variety of authenticated key establishment techniques (see §12.3 for detailed protocols). An outline of steps by which parties A and B above may do so follows, with reference to Figure 13.8.

   (a) A makes a request to T_A to obtain a key to share with B (1).

   (b) T_A and T_B establish a short-term secret key K_{AB} (2).

   (c) T_A and T_B, respectively, distribute K_{AB} to A and B, guaranteeing secrecy and authenticity (3A, 3B).

   (d) A uses K_{AB} for secure direct communications with B (4). Message (3B) may be eliminated if its contents are relayed by T_B to A via T_A as part of the existing messages (2), (3A).

In this case, from A's viewpoint the composition of T_A, T_B and the trust relationship (T_A, T_B) may be seen as a single (composite) authority, which A communicates with through T_A, and which plays the role of the (simple) authority in the standard case of a KDC or KTC (see §13.2.3).

2. *trusted public key*: Trust in a public key may be acquired, based on existing trust relationships, through data origin authentication by standard techniques such as digital signatures or message authentication codes. A may acquire the trusted public key of party B described above as follows (cf. Figure 13.8).

    (a) A requests from T_A the trusted public key of user B (1).

    (b) T_A acquires this from T_B, with guaranteed authenticity (2).

    (c) T_A transfers this public key to A, with guaranteed authenticity (3A).

    (d) A uses this public key to secure direct communications with B (4).

13.39 Definition A *cross-certificate* (or *CA-certificate*) is a certificate created by one certification authority (CA), certifying the public key of another CA.

13.40 Remark (*user-specific vs. domain cross-trust*) Method 2 above transfers to A trust specifically in the public key of B; this may be called a user-specific transfer of trust. Alternatively, a general transfer of trust between domains is possible as follows, assuming T_B has created a certificate C_B containing the identity and public key of B. In this case, T_A creates a cross-certificate containing the identity and public key of T_B. A, possessing the trusted signature verification key of T_A, may verify the signature on this latter certificate, thereby acquiring trust in T_B's signature verification key, and allowing A to verify and thereby trust B's public key within C_B (or the public key in any other certificate signed by T_B). Thus, user A from domain D_A (with authority T_A) acquires trust in public keys certified in D_B by T_B.

13.6.2 Trust models involving multiple certification authorities

Many alternatives exist for organizing trust relationships between certification authorities (CAs) in public-key systems involving multiple CAs. These are called *trust models* or *certification topologies*, and are logically distinct from (although possibly coincident with) communications models. (In particular, a communications link does not imply a trust relationship.) Trust relationships between CAs determine how certificates issued by one CA may be utilized or verified by entities certified by distinct CAs (in other domains). Before discussing various trust models, certificate chains are first introduced.

(i) Certificate chains and certification paths

Public-key certificates provide a means for obtaining authenticated public keys, provided the verifier has a trusted verification public key of the CA which signed the certificate. In the case of multiple certification authorities, a verifier may wish to obtain an authentic public key by verifying a certificate signed by a CA other than one for which it (originally) possesses a trusted public key. In this case, the verifier may still do so provided a *chain of certificates* can be constructed which corresponds to an unbroken chain of trust from the CA public key which the verifier does trust, to the public key it wishes to obtain trust in.

    Certificate chains correspond to directed paths in the graphical representation of a CA trust model (see Figure 13.9). The goal is to find a sequence of certificates corresponding to a directed path (*certification path*) starting at the node corresponding to the CA whose public key a verifier trusts *a priori*, and ending at the CA which has signed the certificate of the public key to be verified.

13.41 Example (*illustration of certificate chain*) Consider Figure 13.9(e) on page 574. Suppose an entity A in possession of the public key P_5 of CA_5 wishes to verify the certificate of an entity B signed by CA_3, and thereby obtain trust in P_B. A directed path (CA_5, CA_4, CA_3) exists. Let $CA_5\{CA_4\}$ denote a certificate signed by CA_5 binding the name CA_4 to the public key P_4. Then the certificate chain $(CA_5\{CA_4\}, CA_4\{CA_3\})$, along with initial trust in P_5, allows A to verify the signature on $CA_5\{CA_4\}$ to extract a trusted copy of P_4, use P_4 to verify the signature on $CA_4\{CA_3\}$ to extract a trusted copy of P_3, and then use P_3 to verify the authenticity of (the certificate containing) P_B. □

Given an initial trusted public key and a certificate to be verified, if a certificate chain is not provided to the verifier, a method is required to find (build) the appropriate chain from publicly available data, prior to actual cryptographic chain verification. This non-cryptographic task resembles that of routing in standard communications networks.

13.42 Example (*building certificate chains using cross-certificate pairs*) One search technique for finding the certification path given in Example 13.41 involves cross-certificate pairs. In a public directory, in the directory entry for each CA X, for every CA Y that either cross-certifies X or that X cross-certifies, store the certificate pair (forward, reverse) = $(CA_Y\{CA_X\}, CA_X\{CA_Y\})$, called a *cross-certificate pair*. Here notation is as in Example 13.41, the pair consists of the forward and reverse certificates of CA_X (see page 575), and at least one of the two certificates is present. In the absence of more advanced techniques or routing tables, any existent certification path could be found by depth-first or breadth-first search of the reverse certificates in cross-certificate pairs starting at the CA whose public key the verifier possesses initially. □

As part of signature verification with certificate chains, verification of cross-certificates requires checking they themselves have not been revoked (see §13.6.3).

(ii) Trust with separate domains

Figure 13.9 illustrates a number of possible trust models for certification, which are discussed below, beginning with the case of separated domains.

Simple public-key systems involve a single certification authority (CA). Larger systems involve two or more CAs. In this case, a trust relationship between CAs must be specified in order for users under different CAs to interoperate cryptographically. By default, two distinct CAs define separate security domains as in Figure 13.9(a), with no trust relationship between domains. Users in one domain are unable to verify the authenticity of certificates originating in a separate domain.

(iii) Strict hierarchical trust model

The first solution to the lack of cryptographic interoperability between separate domains is the idea of a strict hierarchy, illustrated by Figure 13.9(b). Each entity starts with the public key of the root node – e.g., entity $E_1^{(1)}$ is now given CA_5's public key at registration, rather than that of CA_1 as in figure (a). This model is called the *rooted chain* model, as all trust chains begin at the root. It is a *centralized trust* model.

Several such rooted trees, each being a strict hierarchy, may be combined in a trust model supporting *multiple rooted trees* as in Figure 13.9(c). In this case, a cross-certificate is allowed between the roots of the trees, illustrated by a bi-directional arrow between roots. The arrow directed from CA_X to CA_Y denotes a certificate for the public key of CA_Y created by CA_X. This allows users in the tree under CA_X to obtain trust in certificates under CA_Y through certificate chains which start at CA_X and cross over to CA_Y.

In the strict hierarchical model, all entities are effectively in a single domain (defined by the root). Despite the fact that, for example, CA_1 signs the public-key certificate of $E_1^{(1)}$,

(a) Separate domains

(b) Strict hierarchy

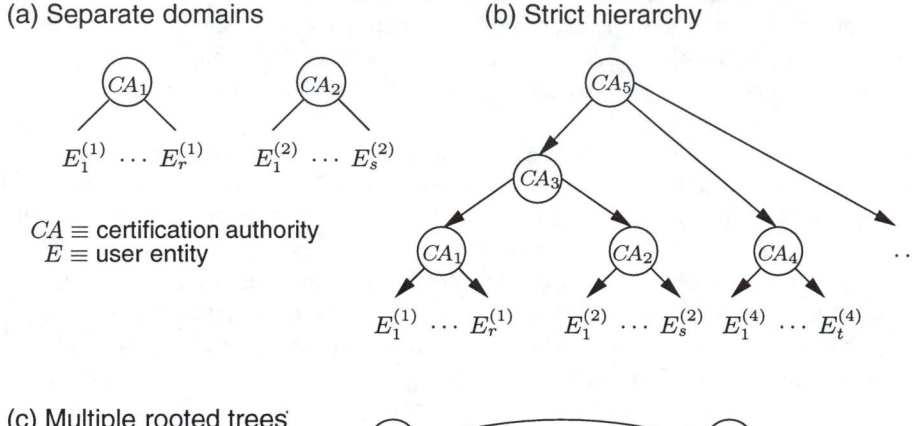

(c) Multiple rooted trees

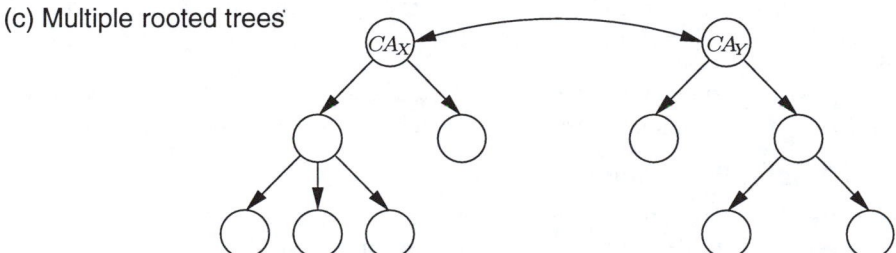

(d) Hierarchy with reverse certificates   (e) Directed graph (digraph) trust model

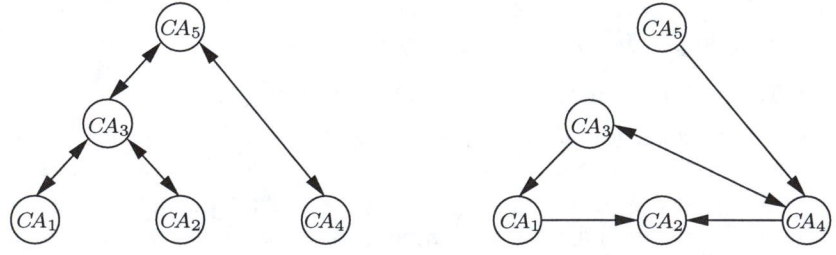

Figure 13.9: *Trust models for certification.*

$E_1^{(1)}$ trusts the root (CA_5) directly but not CA_1. $E_1^{(1)}$ trusts CA_1 only indirectly through the root. Potential drawbacks of this model include:

1. all trust in the system is dependent on the root key
2. certificate chains are required even for two entities under the same CA
3. certificate chains become long in deep hierarchies
4. a more natural model in some organizations is for trust to begin at a local node (the parent CA) rather than a distant node (the root).

(iv) Reverse certificates and the general digraph trust model

A more general hierarchical model, a *hierarchy with reverse certificates*, is illustrated in Figure 13.9(d). This resembles the strict hierarchy of Figure 13.9(b), but now each CA lower in the hierarchy also creates certificates certifying the public keys of its directly superior (*parent*) CA. Two types of certificates may then be distinguished in a hierarchy:

1. *forward certificate.* A forward certificate (relative to CA_X) is created by the CA directly above CA_X signing the public key of CA_X, and illustrated in the hierarchy by a downward arrow towards CA_X.
2. *reverse certificate.* A reverse certificate (relative to CA_X) is created by CA_X signing the public key of its immediately superior CA, and illustrated in the hierarchy by an upward arrow originating from CA_X.

In this model, each entity starts not with the public key of the root, but rather with the public key of the CA which created its own certificate, i.e., its local CA (parent). All trust chains now begin at an entity's local CA. The shortest trust chain from any entity A to any other entity B is now the path in the tree which travels upwards from A to the *least-common-ancestor* of A and B, and downwards from that node on to B.

A drawback of the hierarchical model with reverse certificates is that long certificate chains may arise between entities which are under distinct CAs even if these entities communicate frequently (e.g., consider entities under CA_1 and CA_4 in Figure 13.9(d). This situation can be ameliorated by allowing CA_1 to cross-certify CA_4 directly, even though this edge is not in the hierarchy. This is the most general model, the *directed graph (digraph) trust* model as illustrated in Figure 13.9(e). The analogy to graph theory is as follows: CAs are represented by nodes or vertices in a graph, and trust relationships by directed edges. (The *complete graph* on n vertices, with a directed edge from each vertex to every other, corresponds to complete trust, with each CA cross-certifying every other directly.)

The digraph model is a *distributed trust* model. There is no central node or root, any CA may cross-certify any other, and each user-entity begins with the trusted public key of its local CA. The concept of a hierarchy remains useful as a reference for organizing trust relationships. This model may be used to implement the other trust models discussed above, including strict hierarchies if variation is permitted in the trusted public key(s) end-user entities are provided with initially.

13.43 Remark (*assigning end-users to CAs*) In hierarchical models, one option is to specify that only CAs at the lowest level certify end-users, while internal CAs serve (only) to cross-certify other CAs. In the general digraph model, where all CAs are considered equal, it is more natural to allow every CA to certify end-users.

(v) Constraints in trust models

Trust obtained through certificate chains requires successful verification of each certificate forming a link in the chain. Once a CA (CA_X) cross-certifies the public key of another CA (CA_Y), in the absence of additional constraints, this trust extended by CA_X is transitively granted to all authorities which may be reached by certificate chains originating from

CA_Y. To limit the scope of trust extended by a single cross-certificate, a CA may impose constraints on cross-certificates it signs. Such constraints would be enforced during verification of certificate chains, and might be recorded explicitly through additional certificate fields indicating specific policies, or through attribute certificates (§13.4.2). Examples of simple constraints on cross-certificates include:

1. *limiting chain length.* A constraint may be imposed on the length of the certificate chain which may follow the cross-certificate in question. For example, a CA may limit the extent of trust granted to CAs which it directly cross-certifies by specifying, in all cross-certificates it signs, that that certificate must be the last CA-certificate in any trust chain.

2. *limiting the set of valid domains.* A set of CAs (or domain names) may be specified as valid with respect to a given cross-certificate. All CAs in a certificate chain following the cross-certificate in question may be required to belong to this set.

Certification may also be carried out relative to a *certification policy* specifying the conditions under which certification took place, including e.g., the type of authentication carried out on the certificate subject before certifying a key, and the method used to guarantee unique subject names in certificates.

13.6.3 Certificate distribution and revocation

A certificate directory (cf. §13.2.4) is a database which implements a *pull* model – users extract (pull) certificates from the database as necessary. A different model of certificate distribution, the *push* model, involves certificates being sent out (pushed) to all users upon certificate creation or periodically; this may be suitable for closed systems. Alternatively, individual users may provide their certificates to others when specifically needed, e.g., for signature verification. In certificate-based systems with certificate revocation lists (CRLs – see below), a method for distribution of CRLs as well as certificates is required.

A certificate directory is usually viewed as an *unsecured* third party. While access control to the directory in the form of write and delete protection is necessary to allow maintenance and update without denial of service, certificates themselves are individually secured by the signatures thereon, and need not be transferred over secured channels. An exception is *on-line certificates*, which are created by a certification authority in real-time on request and have no on-going lifetime, or are distributed by a trusted party which guarantees they have not been revoked.

Certificate or CRL *caching* may be used, whereby frequently referenced items are saved in short-term local storage to avoid the cost of repeated retrievals. Cached CRLs must be refreshed sufficiently often to ensure recent revocations are known.

Certificate revocation and CRLs

Upon compromise of a secret key, damage may be minimized by preventing subsequent use of or trust in the associated keying material. (Note the implications differ between signature and encryption keys.) Here *compromise* includes any situation whereby an adversary gains knowledge of secret data. If public keys must be obtained in real-time from a trusted on-line server, the keys in question may be immediately removed or replaced. The situation involving certificates is more difficult, as all distributed copies must be effectively retracted. While (suspected or actual) key compromise may be rare, there may be other reasons a CA will prematurely dissolve its binding of a public key to a user name (i.e., *revoke* the certificate). Reasons for early termination of keying material include the associated entity leaving or changing its role within an organization, or ceasing to require authorization as a user. Techniques for addressing the problem of revoked keys include:

1. *expiration dates within certificates.* Expiration dates limit exposure following compromise. The extreme case of short validity periods resembles on-line certificates which expire essentially immediately. Short-term certificates without CRLs may be compared to long-term certificates with frequently updated CRLs.

2. *manual notification.* All system users are informed of the revoked key by out-of-band means or special channels. This may be feasible in small or closed systems.

3. *public file of revoked keys.* A public file is maintained identifying revoked keys, to be checked by all users before key use. (The authenticity of data extracted from the file may be provided by similar techniques as for public keys – see §13.4.)

4. *certificate revocation lists* (CRLs). A CRL is one method of managing a public file of revoked keys (see below).

5. *revocation certificates.* An alternative to CRLs, these may be viewed as public-key certificates containing a revocation flag and a time of revocation, serving to cancel the corresponding certificate. The original certificate may be removed from the certificate directory and replaced by the revocation certificate.

A CRL is a signed list of entries corresponding to revoked public keys, with each entry indicating the serial number of the associated certificate, the time the revocation was first made, and possibly other information such as the revocation reason. The list signature, guaranteeing its authenticity, is generated by the CA which originally issued the certificates; the CRL typically includes this name also. Inclusion of a date on the overall CRL provides an indication of its freshness. If CRLs are distributed using a pull model (e.g., via a public database), they should be issued at regular intervals (or intervals as advertised within the CRL itself) even if there are no changes, to prevent new CRLs being maliciously replaced by old CRLs.

Revoked cross-certificates may be specified on separate *authority revocation lists* (ARLs), analogous to CRLs (which are then restricted to revoked end-user certificates).

13.44 Note (*CRL segmenting*) For reasons of operational efficiency when large CRLs may arise, an option is to distribute CRLs in pieces. One technique is to use *delta-CRLs*: upon each CRL update, only new entries which have been revoked since the last issued CRL are included. This requires end-users maintain (and update) secured, local images of the current CRL. A second technique is to partition a CRL into segments based on revocation reason. A third is to segment a CRL by pre-assigning each certificate (upon creation) to a specified sub-list, with a limit $n_{\max}$ on the number of certificates pre-assigned to any segment and new segments created as required. In all cases, for each certificate, available information must indicate which CRL segment must be consulted.

13.7 Key life cycle issues

Key management is simplest when all cryptographic keys are fixed for all time. Cryptoperiods necessitate the update of keys. This imposes additional requirements, e.g., on certification authorities which maintain and update user keys. The set of stages through which a key progresses during its existence, referred to as the *life cycle* of keys, is discussed in this section.

13.7.1 Lifetime protection requirements

Controls are necessary to protect keys both during usage (cf. §13.5.2) and storage. Regarding long-term storage of keys, the duration of protection required depends on the cryptographic function (e.g., encryption, signature, data origin authentication/integrity) and the time-sensitivity of the data in question.

Security impact of dependencies in key updates

Keying material should be updated prior to cryptoperiod expiry (see Definition 13.10). Update involves use of existing keying material to establish new keying material, through appropriate key establishment protocols (Chapter 12) and key layering (§13.3.1).

To limit exposure in case of compromise of either long term secret keys or past session keys, dependencies among keying material should be avoided. For example, securing a new session key by encrypting it under the old session key is not recommended (since compromise of the old key compromises the new). See §12.2.3 regarding perfect forward secrecy and known-key attacks.

Lifetime storage requirements for various types of keys

Stored secret keys must be secured so as to provide both confidentiality and authenticity. Stored public keys must be secured such that their authenticity is verifiable. Confidentiality and authenticity guarantees, respectively countering the threats of disclosure and modification, may be provided by cryptographic techniques, procedural (trust-based) techniques, or physical protection (tamper-resistant hardware).

Signature verification public keys may require archival to allow signature verification at future points in time, including possibly after the private key ceases to be used. Some applications may require that signature private keys neither be backed up nor archived: such keys revealed to any party other than the owner potentially invalidates the property of non-repudiation. Note here that loss (without compromise) of a signature private key may be addressed by creation of a new key, and is non-critical as such a private key is not needed for access to past transactions; similarly, public encryption keys need not be archived. On the other hand, decryption private keys may require archival, since past information encrypted thereunder might otherwise be lost.

Keys used for entity authentication need not be backed up or archived. All secret keys used for encryption or data origin authentication should remain secret for as long as the data secured thereunder requires continued protection (the *protection lifetime*), and backup or archival is required to prevent loss of this data or verifiability should the key be lost.

13.7.2 Key management life cycle

Except in simple systems where secret keys remain fixed for all time, cryptoperiods associated with keys require that keys be updated periodically. Key update necessitates additional procedures and protocols, often including communications with third parties in public-key systems. The sequence of states which keying material progresses through over its lifetime is called the *key management life cycle*. Life cycle stages, as illustrated in Figure 13.10, may include:

1. *user registration* – an entity becomes an authorized member of a security domain. This involves acquisition, or creation and exchange, of initial keying material such as shared passwords or PINs by a secure, one-time technique (e.g., personal exchange, registered mail, trusted courier).

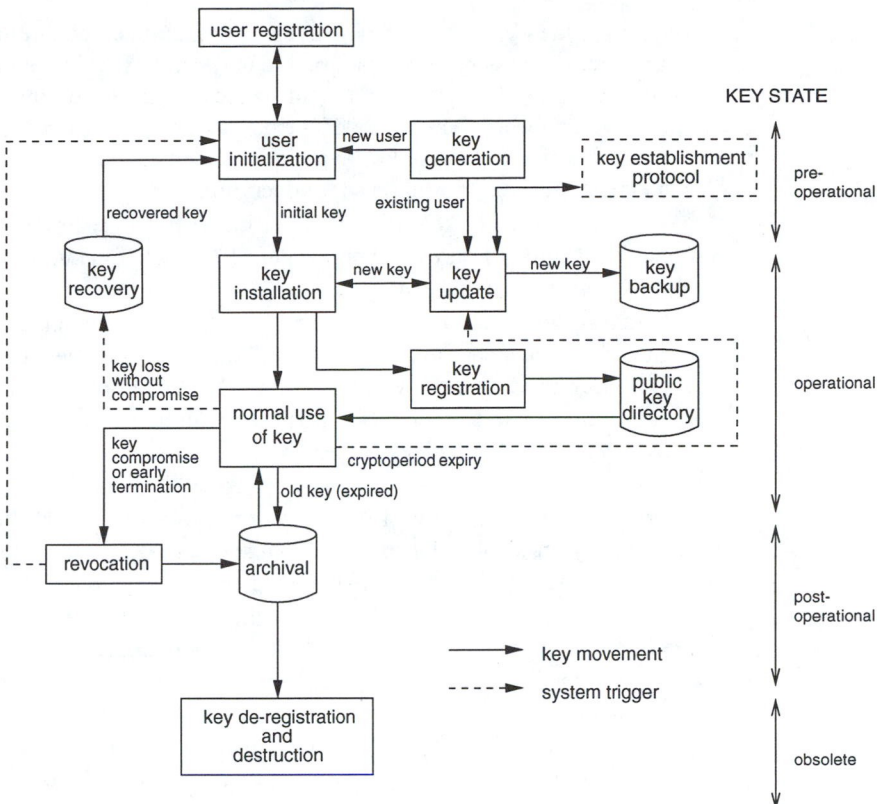

Figure 13.10: *Key management life cycle.*

2. *user initialization* – an entity initializes its cryptographic application (e.g., installs and initializes software or hardware), involving use or installation (see below) of initial keying material obtained during user registration.

3. *key generation* – generation of cryptographic keys should include measures to ensure appropriate properties for the intended application or algorithm and randomness in the sense of being predictable (to adversaries) with negligible probability (see Chapter 5). An entity may generate its own keys, or acquire keys from a trusted system component.

4. *key installation* – keying material is installed for operational use within an entity's software or hardware, by a variety of techniques including one or more of the following: manual entry of a password or PIN, transfer of a disk, read-only-memory device, chipcard or other hardware token or device (e.g., key-loader). The initial keying material may serve to establish a secure on-line session through which working keys are established. During subsequent updates, new keying material is installed to replace that in use, ideally through a secure on-line update technique.

5. *key registration* – in association with key installation, keying material may be officially recorded (by a registration authority) as associated with a unique name which distinguishes an entity. For public keys, public-key certificates may be created by a certification authority (which serves as guarantor of this association), and made available to others through a public directory or other means (see §13.4).

6. *normal use* – the objective of the life cycle is to facilitate operational availability of keying material for standard cryptographic purposes (cf. §13.5 regarding control of keys during usage). Under normal circumstances, this state continues until cryptoperiod expiry; it may also be subdivided – e.g., for encryption public-key pairs, a point may exist at which the public key is no longer deemed valid for encryption, but the private key remains in (normal) use for decryption.

7. *key backup* – backup of keying material in independent, secure storage media provides a data source for key recovery (point 11 below). Backup refers to short-term storage during operational use.

8. *key update* – prior to cryptoperiod expiry, operational keying material is replaced by new material. This may involve some combination of key generation, key derivation (§13.5.2), execution of two-party key establishment protocols (Chapter 12), or communications with a trusted third party. For public keys, update and registration of new keys typically involves secure communications protocols with certification authorities.

9. *archival* – keying material no longer in normal use may be archived to provide a source for key retrieval under special circumstances (e.g., settling disputes involving repudiation). Archival refers to off-line long-term storage of post-operational keys.

10. *key de-registration and destruction* – once there are no further requirements for the value of a key or maintaining its association with an entity, the key is de-registered (removed from all official records of existing keys), and all copies of the key are destroyed. In the case of secret keys, all traces are securely erased.

11. *key recovery* – if keying material is lost in a manner free of compromise (e.g., due to equipment failure or forgotten passwords), it may be possible to restore the material from a secure backup copy.

12. *key revocation* – it may be necessary to remove keys from operational use prior to their originally scheduled expiry, for reasons including key compromise. For public keys distributed by certificates, this involves revoking certificates (see §13.6.3).

Of the above stages, all are regularly scheduled, except key recovery and key revocation which arise under special situations.

13.45 Remark (*public-key vs. symmetric-key life cycle*) The life cycle depicted in Figure 13.10 applies mainly to public-key pairs, and involves keying material of only a single party. The life cycle of symmetric keys (including key-encrypting and session keys) is generally less complex; for example, session keys are typically not registered, backed up, revoked, or archived.

Key states within life cycle

The typical events involving keying material over the lifetime of the key define stages of the life cycle. These may be grouped to define a smaller set of *states* for cryptographic keys, related to their availability for use. One classification of key states is as follows (cf. Figure 13.10):

1. *pre-operational*. The key is not yet available for normal cryptographic operations.

2. *operational*. The key is available, and in normal use.

3. *post-operational*. The key is no longer in normal use, but off-line access to it is possible for special purposes.

4. *obsolete*. The key is no longer available. All records of the key value are deleted.

System initialization and key installation

Key management systems require an initial keying relationship to provide an initial secure channel and optionally support the establishment of subsequent working keys (long-term and short-term) by automated techniques. The initialization process typically involves non-cryptographic one-time procedures such as transfer of keying material in person, by trusted courier, or over other trusted channels.

The security of a properly architected system is reduced to the security of keying material, and ultimately to the security of initial key installation. For this reason, initial key installation may involve dual or split control, requiring co-operation of two or more independent trustworthy parties (cf. Note §13.8).

13.8 Advanced trusted third party services

This section provides further details on trusted third party services of a more advanced nature, introduced briefly in §13.2.4.

13.8.1 Trusted timestamping service

A trusted timestamping service provides a user with a dated receipt (upon presentation of a document), which thereafter can be verified by others to confirm the presentation or existence of the document at the (earlier) date of receipt. Specific applications include establishing the time of existence of documents such as signed contracts or lab notes related to patent claims, or to support non-repudiation of digital signatures (§13.8.2).

The basic idea is as follows. A trusted third party T (the *timestamp agent*) appends a timestamp t_1 to a submitted digital document or data file D, signs the composite document (thereby vouching for the time of its existence), and returns the signed document including t_1 to the submitter. Subsequent verification of T's signature then establishes, based on trust in T, the existence of the document at the time t_1.

If the data submitted for timestamping is the hash of a document, then the document content itself need not be disclosed at the time of timestamping. This also provides privacy protection from eavesdroppers in the case of submissions over an unsecured channel, and reduces bandwidth and storage costs for large documents.

13.46 Remark (*non-cryptographic timestamp service*) A similar service may be provided by non-cryptographic techniques as follows. T stores D along with a timestamp t_1, and is trusted to maintain the integrity of this record by procedural techniques. Later some party A submits the document again (now D'), and T compares D' to D on file. If these match, T declares that D' existed at the time t_1 of the retrieved timestamp.

The timestamp agent T is trusted not to disclose its signing key, and also to competently create proper signatures. An additional desirable feature is *prevention of collusion*: T should be unable to successfully collude (with any party) to undetectably back-date a document. This may be ensured using Mechanism 13.47, which combines digital signatures with tree authentication based on hashing.

13.47 Mechanism Trusted timestamping service based on tree authentication

---

SUMMARY: party A interacts with a trusted timestamping agent T.
RESULT: A obtains a timestamp on a digital document D.

1. A submits the hash value $h(D)$ to T. (h is a collision-resistant hash function.)
2. T notes the date and time t_1 of receipt, digitally signs the concatenation of $h(D)$ and t_1, and returns t_1 and the signature to A. (The signature is called the *certified time-stamp*.) A may verify the signature to confirm T's competence.
3. At the end of each fixed period (e.g., one day), or more frequently if there is a large number n of certified timestamps, T:
   (i) computes from these an authentication tree T^* with root label R (see §13.4.1);
   (ii) returns to A the authentication path values to its certified timestamp; and
   (iii) makes the root value R widely available through a means allowing both verifiable authenticity and establishment of the time of creation t_c of T^* (e.g., publishing in a trusted dated medium such as a newspaper).
4. To allow any other party B to verify (with T's verification public key) that D was submitted at time t_1, A produces the certified timestamp. If trust in T itself is challenged (with respect to backdating t_1), A provides the authentication path values from its certified timestamp to the root R, which B may verify (see §13.4.1) against an independently obtained authentic root value R for the period t_c.

To guarantee verifiability, A should itself verify the authentication path upon receiving the path values in step 3.

13.8.2 Non-repudiation and notarization of digital signatures

The timestamping service of §13.8.1 is a document certification or document notarization service. A *notary service* is a more general service capable not only of ascertaining the existence of a document at a certain time, but of vouching for the truth of more general statements at specified points in time. The terminology originates from the dictionary definition of a *notary public* – a public official (usually a solicitor) legally authorized to administer oaths, and attest and certify certain documents. No specific legal connotation is intended in the cryptographic use of this term.

The non-repudiation aspect of digital signatures is a primary advantage of public-key cryptography. By this property, a signer is prevented from signing a document and subsequently being able to successfully deny having done so. A non-repudiation service requires specification of precise details including an adjudication process and adjudicator (judge), what evidence would be submitted to the adjudicator, and what precise process the adjudicator is to follow to render judgement on disputes. The role of an adjudicator is distinct from that of a timestamp agent or notary which generates evidence.

13.48 Remark (*origin authentication vs. non-repudiable signature*) A fundamental distinction exists between a party A being able to convince itself of the validity of a digital signature s at a point in time t_0, and that party being able to convince others at some time $t_1 \geq t_0$ that s was valid at time t_0. The former resembles data origin authentication as typically provided by symmetric-key origin authentication mechanisms, and may be accepted by a verifier as a form of authorization in an environment of mutual trust. This differs from digital signatures which are non-repudiable in the future.

Data origin authentication as provided by a digital signature is valid only while the secrecy of the signer's private key is maintained. A threat which must be addressed is a signer who intentionally discloses his private key, and thereafter claims that a previously valid signature was forged. (A similar problem exists with credit cards and other methods of authorization.) This threat may be addressed by:

1. *preventing direct access to private keys.* Preventing users from obtaining direct access to their own private keys precludes intentional disclosure. As an example, the private keys may be stored in tamper-resistant hardware, and by system design never available outside thereof.

2. *use of a trusted timestamp agent.* The party obtaining a signature on a critical document submits the signature to a timestamp agent, which affixes a timestamp to signature and then signs the concatenation of these. This establishes a time t_1 at which the critical signature may be ascertained to have existed. If the private signature key corresponding to this signature is subsequently compromised, and the compromise occurred after t_1, then the critical signature may still be considered valid relative to t_1. For reasons as given in Remark 13.49, use of a notary agent (below) may be preferable.

3. *use of a trusted notary agent.* The party obtaining a signature on a critical document (or hash thereof) submits the signature (and document or hash thereof) to an agent for *signature notarization.* The agent verifies the signature and notarizes the result by appending a statement (confirming successful signature verification) to the signature, as well as a timestamp, and signing the concatenation of the three. A reasonable period of time (clearance period) may be allowed for declarations of lost private keys, after which the notary's record of verification must be accepted (by all parties who trust the notary and verify its signature) as the truth regarding the validity of the critical signature at that point in time,[5] even should the private key corresponding to the critical signature subsequently be compromised.

For signed messages having short lifetimes (i.e., whose significance does not extend far into the future), non-repudiation is less important, and notarization may be unnecessary. For other messages, the requirement for a party to be able to re-verify signatures at a later point in time (including during or after signature keys have been updated or revoked), as well as the adjudication process related to non-repudiation of signatures, places additional demands on practical key management systems. These may include the storage or archival of keying material (e.g., keys, certificates, CRLs) possibly required as evidence at a future point in time.

A related support service is that of maintaining a record (*audit trail*) of security-related events including registration, certificate generation, key update, and revocation. Audit trails may provide sufficient information to allow resolution of disputed signatures by non-automated procedures.

13.49 Remark (*reconstructing past trust*) Both signature re-verification (relative to a past point in time) and resolution of disputes may require reconstruction of chains of trust from a past point in time. This requires access to keying material and related information for (re)constructing past chains of trust. Direct reconstruction of such past chains is unnecessary if a notarizing agent was used. The original verification of the notary establishes existence of a trust chain at that point in time, and subsequently its record thereof serves as proof of prior validity. It may be of interest (for audit purposes) to record the details of the original trust chain.

---

[5]More generally, the truth of the appended statement must be accepted, relative to the timestamp.

13.8.3 Key escrow

The objective of a key escrow encryption system is to provide encryption of user traffic (e.g., voice or data) such that the session keys used for traffic encryption are available to properly authorized third parties under special circumstances ("emergency access"). This grants third parties which have monitored user traffic the capability to decrypt such traffic. Wide-scale public interest in such systems arose when law enforcement agencies promoted their use to facilitate legal wiretapping of telephone calls to combat criminal activities. However, other uses in industry include recovery of encrypted data following loss of keying material by a legitimate party, or destruction of keying material due to equipment failure or malicious activities. One example of a key escrow system is given below, followed by more general issues.

(i) The Clipper key escrow system

The Clipper key escrow system involves use of the Clipper chip (or a similar tamper-resistant hardware device – generically referred to below as an escrow chip) in conjunction with certain administrative procedures and controls. The basic idea is to deposit two key components, which jointly determine an encryption key, with two trusted third parties (escrow agents), which subsequently allow (upon proper authorization) recovery of encrypted user data.

More specifically, encryption of telecommunications between two users proceeds as follows. Each party has a telephone combined with a key escrow chip. The users negotiate or otherwise establish a session key K_S which is input to the escrow chip of the party encrypting data (near end). As a function of K_S and an initialization vector (IV), the chip creates by an undisclosed method a data block called a *law enforcement access field* (LEAF). The LEAF and IV are transmitted to the far end during call set-up of a communications session. The near end escrow chip then encrypts the user data D under K_S producing $E_{K_S}(D)$, by a U.S. government classified symmetric algorithm named SKIPJACK. The far end escrow chip decrypts the traffic only if the transmitted LEAF validates properly. Such verification requires that this far end chip has access to a common family key K_F (see below) with the near end chip.

The LEAF (see Figure 13.11) contains a copy of the session key encrypted under a device-specific key K_U. K_U is generated and data-filled into the chip at the time of chip manufacture, but prior to the chip being embedded in a security product. The system meets its objective by providing third party access under proper authorization (as defined by the Key Escrow System) to the device key K_U of targeted individuals.

To derive the key K_U embedded in an escrow chip with identifier UID, two key components (K_{C1}, K_{C2}) are created whose XOR is K_U. Each component is encrypted under a key $K_{CK} = K_{N1} \oplus K_{N2}$, where K_{Ni} is input to the chip programming facility by the first and second trusted *key escrow agent*, respectively. (Used to program a number of chips, K_{Ni} is stored by the escrow agent for subsequent recovery of K_{CK}.) One encrypted key component is then given to each escrow agent, which stores it along with UID to service later requests. Stored data from both agents must subsequently be obtained by an authorized official to allow recovery of K_U (by recovering first K_{CK}, and then K_{C1}, K_{C2}, and $K_U = K_{C1} \oplus K_{C2}$).

Disclosed details of the LEAF are given in Figure 13.11. Each escrow chip contains a 32-bit device unique identifier (UID), an 80-bit device unique key (K_U), and an 80-bit family key (K_F) common to a larger collection of devices. The LEAF contains a copy of the 80-bit session key K_S encrypted under K_U, the UID, and a 16-bit encryption authenticator (EA) created by an undisclosed method; these are then encrypted under K_F. Recovery of K_S from the LEAF thus requires both K_F and K_U. The encryption authenticator is a checksum designed to allow detection of LEAF tampering (e.g., by an adversary attempting to prevent authorized recovery of K_S and thereby D).

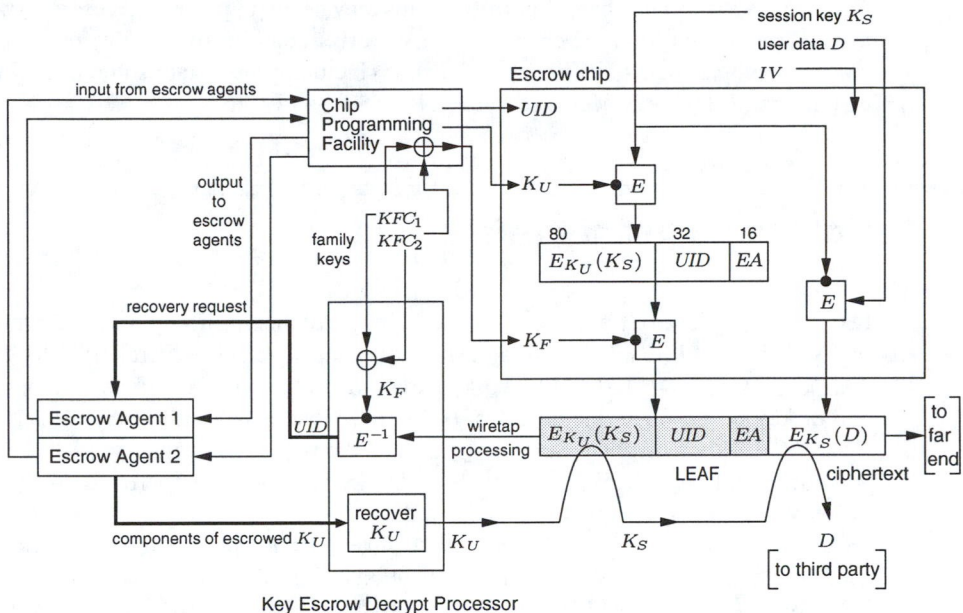

Key Escrow Decrypt Processor

Schematic representation:

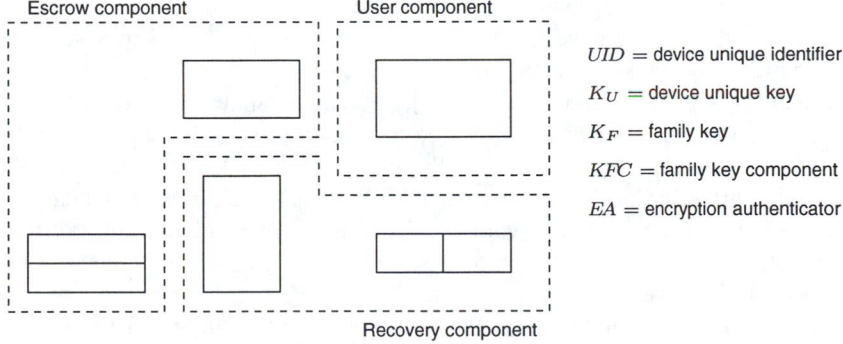

UID = device unique identifier

K_U = device unique key

K_F = family key

KFC = family key component

EA = encryption authenticator

Figure 13.11: *Creation and use of LEAF for key escrow data recovery.*

(ii) Issues related to key escrow

Key escrow encryption systems may serve a wide variety of applications, and a corresponding range of features exists. Distinguishing properties of escrow systems include:

1. applicability to store-and-forward vs. real-time user communications
2. capability of real-time decryption of user traffic
3. requirement of tamper-resistant hardware or hardware with trusted clock
4. capability of user selection of escrow agents

5. user input into value of escrowed key
6. varying trust requirements in escrow agents
7. extent of user data uncovered by one escrow access (e.g., limited to one session or fixed time period) and implications thereof (e.g., hardware replacement necessary).

Threshold systems and shared control systems may be put in place to access escrowed keying information, to limit the chances of unauthorized data recovery. Key escrow systems may be combined with other life cycle functions including key establishment, and key backup and archival (cf. key access servers – Notes 13.5 and 13.6).

13.9 Notes and further references

§13.1

Davies and Price [308] provide a comprehensive treatment of key management, including overviews of ISO 8732 [578] and techniques introduced in several 1978 *IBM Systems Journal* papers [364, 804]. Early work addressing protection in communications networks and/or key management includes that of Feistel, Notz, and Smith [388], Branstad [189], Kent [665], Needham and Schroeder [923], and the surveys of Popek and Kline [998] and Voydock and Kent [1225]. Security issues in electronic funds transfer (EFT) systems for point-of-sale (POS) terminals differ from those for remote banking machines due to the weaker physical security of the former; special key management techniques such as unique (derived) transaction keys reduce the implications of terminal key compromise – see Beker and Walker [85], Davies [305], and Davies and Price [308, Ch.10]. See Meyer and Matyas [859] for general symmetric-key techniques, EFT applications, and PIN management; and Ford [414] for directory services and standards, including the X.500 Directory and X.509 Authentication Framework [626].

For an overview of key management concepts and life cycles aspects, see Fumy and Landrock [429]. Fumy and Leclerc [430] consider placement of key distribution protocols within the ISO Open Systems Interconnection (OSI) architecture. Regarding key management principles, see Abadi and Needham [1], and Anderson and Needham [31]. See Vedder [1220] for security issues and architectures relevant to wireless communications, including European digital cellular (Global System for Mobile Communications – GSM) and the Digital European Cordless Telephone (DECT) system. Regarding key management for security (authentication and encryption) in North American digital cellular systems, see IS-54 Rev B [365]. ISO 11166-1 [586] (see also comments by Rueppel [1082]) specifies key management techniques and life cycle principles for use in banking systems, and is used by the Society for Worldwide Interbank Financial Telecommunications (SWIFT).

§13.2

Various parts of ISO/IEC 11770 [616, 617, 618] contain background material on key management; Figure 13.3 is derived from an early draft of 11770-3. KDCs and KTCs were popularized by ANSI X9.17 [37]. Related to tradeoffs, Needham and Schroeder [923] compare symmetric and public-key techniques; the formalization proposed by Rueppel [1080] allows analysis of security architectures to distinguish complexity-increasing from complexity-reducing techniques.

The Kerberos authentication service (§12.3.2) includes a *ticket-granting service* whereby a client may re-authenticate itself multiple times using its long-term secret only once. The client A first acquires a *ticket-granting-ticket* through a protocol with an Authentication Server (AS). Thereafter, using a variation of Protocol 12.24, A may obtain authentication credentials for a server B from a Ticket-Granting-Server (TGS), extracting a TGS session key from the time-limited ticket to secure protocol messages with the TGS. A's long-term secret (password) need neither be cached for an extended period in memory nor re-entered, reducing the threat of its compromise; compromise of a TGS session key has time-restricted impact. See RFC 1510 [1041] for details.

Ford and Wiener [417] describe key access servers (Note 13.6), effectively an access control mechanism where the resource is a *key package*. Girault [459] mentions the three levels of trust of Remark 13.7. Digital envelopes (Note 13.6) are discussed in PKCS #7 [1072].

§13.3

Example 13.9 is from Tuchman [1198]. Davis and Swick [310] discuss symmetric-key certificates as defined herein under the name *private-key certificates* (crediting Abadi, Burrows, and Lampson) and propose protocols for their use with trusted third parties, including a password-based initial registration protocol. Predating this, Davies and Price [308, p.259] note that tamper-resistant hardware may replace the trusted third party requirement of symmetric-key certificates (Note 13.14). A generalization of Protocol 13.12 appears as Mechanism 11 of ISO/IEC 11770-2 [617], along with related KTC protocols offering additional authenticity guarantees (cf. Note 13.13(iii)); these provide KTC variations of the KDC protocols of §12.3.2).

§13.4

Diffie and Hellman [345] suggested using a trusted public file, maintained by a trusted authority with which each communicant registers once (in person), and from which authentic public keys of other users can be obtained. To secure requests by one party for the public key of another, Rivest, Shamir, and Adleman [1060] and Needham and Schroeder [923] note the trusted authority may respond via signed messages (essentially providing on-line certificates).

Authentication trees were first discussed in Merkle's thesis [851, p.126-131] (see also [852, 853]). For security requirements on hash functions used for tree authentication, see Preneel [1003, p.38]. Public-key certificates were first proposed in the 1978 B.Sc. thesis of Kohnfelder [703]; the overall thesis considers implementation and systems issues related to using RSA in practice. Kohnfelder's original certificate was an ordered triple containing a party's name, public-key information, and an authenticator, with the authenticator a signature over the value resulting from encrypting the name with the public key/algorithm in question.

X.509 certificates [626] were defined in 1988 and modified in 1993 (yielding Version 2 certificates); an *extensions* field was added by a technical corrigendum [627] in 1995 (yielding Version 3 certificates). Standard extensions for Version 3 certificates appear in an amendment to X.509 [628]; these accommodate information related to key identifiers, key usage, certificate policy, alternate names (vs. X.500 names) and name attributes, certification path constraints, and enhancements for certificate revocation including revocation reasons and CRL partitioning. For details, see Ford [416]. ANSI X9.45 [49] addresses attribute certificates. The alternative of including hard-coded attribute fields within public-key certificates is proposed in PKCS #6 [1072]; suggested attributes are listed in PKCS #9 [1072].

In 1984 Shamir [1115] formulated the general idea of asymmetric systems employing user's identities in place of public keys (*identity-based systems*), giving a concrete proposal for an ID-based signature system, and the model for an ID-based encryption scheme. Fiat and Shamir [395] combined this idea with that of zero-knowledge interactive proofs, yielding interactive identification and signature protocols. T. Okamoto [947] (based on a January 1984 paper in Japanese by Okamoto, Shiraishi, and Kawaoka [954]) independently proposed a specific entity-authentication scheme wherein a trusted center T distributes to a claimant A a secret accreditation value computed as a function of T's private key and A's identity (or unique index value). The identity-based key-agreement scheme of Maurer and Yacobi [824] (cf. §12.6 notes on page 538) is an exception to Remark 13.27: extra public data D_A is avoided, as ideally desired.

Günther [530] proposed a protocol for key agreement (Protocol 12.62) wherein users' private keys are constructed by a trusted authority T based on their identities, with corresponding Diffie-Hellman public keys reconstructed from public data provided by T (herein called implicitly-certified public keys, identity-based subclass). The protocol introduced by Girault [459], based on the key agreement protocol of Paillès and Girault [962] (itself updated by Girault and Paillès [461] and Girault [458]) similar to a protocol of Tanaka and E. Okamoto [1184], involved what he christened *self-certified public keys* (herein called implicitly-certified public keys, self-certified subclass); see Mechanism 12.61.

Related to self-certified public keys, Brands [185] has proposed *secret-key certificates* for use in so-called restrictive blind signature schemes. These involve a data triple consisting of a private key, matching public key, and an explicit (secret-key) certificate created by a trusted third party to certify the public key. Users can themselves create pairs of public keys and matching (secret-key) certificates, but cannot create valid triples. As with self-certified keys, performance of a cryptographic action relative to the public key (e.g., signing) implicitly demonstrates that the performing party knows the private key and hence that the corresponding public key was indeed issued by the trusted third party.

§13.5

Key tags are due to Jones [642]. Key separation as in Example 13.33 is based on Ehrsam et al. [364], which outlines the use of master keys, key variants, key- and data-encrypting keys. Smid [1153] introduced key notarization in the Key Notarization System (KNS), a key management system designed by the U.S. National Bureau of Standards (now NIST), and based on a Key Notarization Facility (KNF) – a KTC-like system component trusted to handle master keys, and to generate and notarize symmetric keys. Key notarization with key offsetting (Example 13.34) is from ISO 8732 [578], which is derived from ANSI X9.17 [37].

The generalization of key notarization to control vectors is due to Matyas, Meyer, and Brachtl [806], and described by Matyas [803] (also [802]), including an efficient method for allowing arbitrary length control vectors that does not penalize short vectors. The IBM proposal specifies E as two-key triple-DES, as per ANSI X9.17. Matyas notes that a second approach to implement control vectors, using a MAC computed on the control vector and the key (albeit requiring additional processing), has the property that both the control vector and the recovered key may be authenticated before the key is used. The notion of a capability (Note 13.35) was introduced in 1966 by Dennis and Van Horn [332], who also considered the access matrix model.

§13.6

Key distribution between domains is discussed in ISO/IEC 11770-1 [616]; see also Kohl and Neuman [1041] with respect to Kerberos V5, and Davis and Swick [310]. A Kerberos domain is called a *realm*; authentication of clients in one realm to servers in others is supported in V5 by *inter-realm keys*, with a concept of authentication paths analogous to public-key certification paths.

Kent [666] overviews the design and implementation of *Privacy Enhanced Mail* (PEM) (see RFC 1421-1424 [1036, 1037, 1038, 1039]), a prototyped method for adding security to Internet mail. Encryption and signature capabilities are provided. The PEM infrastructure of

RFC 1422 is based on a strict hierarchy of certification authorities, and includes specification of *Policy Certification Authorities* (PCAs) which define policies with respect to which certificates are issued. Regarding certification paths, see Tarah and Huitema [1185].

The 1988 version of X.509 [626] defines forward and reverse certificates, certificate chains, and cross-certificate pairs, allowing support for the general digraph trust model. The formal analysis of Gaarder and Snekkenes [433] highlights a practical difficulty in verifying the validity of certificates – the requirement of trusted timeclocks. For reports on implementations based on X.509 certificates, see Tardo and Alagappan [1186] and others [660, 72, 839]. Techniques for segmenting CRLs (Note 13.44) are included in the above-cited work on Version 3 certificate extensions [628]. Kohnfelder [703] noted many of the issues regarding certificate revocation in 1978 when use of certificates was first proposed, and suggested techniques to address revocation including manual notification, maintaining a public file identifying revoked keys, and use of certificate expiration dates (cf. Denning [326, p.170]).

§13.7

Matyas and Meyer [804] consider several life cycle aspects. ISO 11770-1 [616] provides a general overview of key management issues including key life cycle. ANSI X9.57 [52] provides broad discussion on certificate management, including trust models, registration, certificate chains, and life cycle aspects. ISO 10202-7 [584] specifies a key management life cycle for chipcards.

§13.8

Davies and Price [308] discuss practical issues related to registries of public keys, non-repudiation, and revocation, including the use of timestamps and notarization; see also the original works of Kohnfelder [703] and Merkle [851], which include discussion of notaries. Haber and Stornetta [535] propose two additional techniques for timestamping digital data (one enhanced by Bayer, Haber, and Stornetta [79]), although tree authentication, due to Merkle [852], appears to be preferable in practice. Benaloh and de Mare [111] introduce one-way accumulators to address the same problem.

Although key backup/archive functionality existed in earlier commercial products, the widespread study of key escrow systems began circa 1992, and combines issues related to secret sharing, key establishment, and key life cycle. For practical aspects including commercial key recovery and backup, see Walker et al. [1229] and Maher [780]. Denning and Branstad [329] provide an excellent overview of the numerous proposals to date, including a taxonomy. Among such proposals and results are those of Micali [863] (see also [862]), Leighton and Micali [745], Beth et al. [125], Desmedt [338] (but see also Knudsen and Pederson [690]), Jefferies, Mitchell, and Walker [635], Lenstra, Winkler, and Yacobi [755], Kilian and Leighton [671], Frankel and Yung [420], and Micali and Sidney [869]. In some systems, it is required that escrow agents be able to verify that (partial) keys received are authentic, raising issues of verifiable secret sharing (see Chor et al. [259]).

The Clipper chip is a tamper-resistant hardware encryption device compliant with FIPS 185 [405], a voluntary U.S. government standard intended for sensitive but unclassified phone (voice and data) communications. FIPS 185 specifies use of the SKIPJACK encryption algorithm (80-bit key, 64-bit blocks) and LEAF creation method, the details of both of which remain classified. The two initial key escrow agents named by the U.S. Government are the National Institute of Standards and Technology (NIST) and the Department of the Treasury, Automated Systems Division. Denning and Smid [331] describe the operation of an initial key escrow system employing a chip in accordance with FIPS 185. The *Capstone* chip, a more advanced device than Clipper, implements in addition a public key agreement algorithm, DSA, SHA, high-speed general-purpose exponentiation, and a (pure noise source) random number generator; it is used in the U.S. government Multilevel Information Security System Initiative (MISSI) for secure electronic mail and other applications. Blaze [152] demonstrated that a protocol attack is possible on Clipper, requiring at most 2^{16} trial LEAF values to construct a bogus LEAF with a valid EA; Denning and Smid note this is not a threat in practical systems. For a debate on issues related to U.S. digital telephony legislation passed in October 1994 as the Communications Assistance for Law Enforcement Act (CALEA), requiring telephone companies to provide technical assistance facilitating authorized wiretapping, see Denning [328].

Chapter 14

Efficient Implementation

Contents in Brief

14.1 Introduction

Many public-key encryption and digital signature schemes, and some hash functions (see §9.4.3), require computations in $\mathbb{Z}_m$, the integers modulo m (m is a large positive integer which may or may not be a prime). For example, the RSA, Rabin, and ElGamal schemes require efficient methods for performing multiplication and exponentiation in $\mathbb{Z}_m$. Although $\mathbb{Z}_m$ is prominent in many aspects of modern applied cryptography, other algebraic structures are also important. These include, but are not limited to, polynomial rings, finite fields, and finite cyclic groups. For example, the group formed by the points on an elliptic curve over a finite field has considerable appeal for various cryptographic applications. The efficiency of a particular cryptographic scheme based on any one of these algebraic structures will depend on a number of factors, such as parameter size, time-memory tradeoffs, processing power available, software and/or hardware optimization, and mathematical algorithms.

This chapter is concerned primarily with mathematical algorithms for efficiently carrying out computations in the underlying algebraic structure. Since many of the most widely implemented techniques rely on $\mathbb{Z}_m$, emphasis is placed on efficient algorithms for performing the basic arithmetic operations in this structure (addition, subtraction, multiplication, division, and exponentiation).

In some cases, several algorithms will be presented which perform the same operation. For example, a number of techniques for doing modular multiplication and exponentiation are discussed in §14.3 and §14.6, respectively. Efficiency can be measured in numerous ways; thus, it is difficult to definitively state which algorithm is the best. An algorithm may be efficient in the time it takes to perform a certain algebraic operation, but quite inefficient in the amount of storage it requires. One algorithm may require more code space than another. Depending on the environment in which computations are to be performed, one algorithm may be preferable over another. For example, current chipcard technology provides very limited storage for both precomputed values and program code. For such applications, an algorithm which is less efficient in time but very efficient in memory requirements may be preferred.

The algorithms described in this chapter are those which, for the most part, have received considerable attention in the literature. Although some attempt is made to point out their relative merits, no detailed comparisons are given.

Chapter outline

§14.2 deals with the basic arithmetic operations of addition, subtraction, multiplication, squaring, and division for multiple-precision integers. §14.3 describes the basic arithmetic operations of addition, subtraction, and multiplication in $\mathbb{Z}_m$. Techniques described for performing modular reduction for an arbitrary modulus m are the classical method (§14.3.1), Montgomery's method (§14.3.2), and Barrett's method (§14.3.3). §14.3.4 describes a reduction procedure ideally suited to moduli of a special form. Greatest common divisor (gcd) algorithms are the topic of §14.4, including the binary gcd algorithm (§14.4.1) and Lehmer's gcd algorithm (§14.4.2). Efficient algorithms for performing extended gcd computations are given in §14.4.3. Modular inverses are also considered in §14.4.3. Garner's algorithm for implementing the Chinese remainder theorem can be found in §14.5. §14.6 is a treatment of several of the most practical exponentiation algorithms. §14.6.1 deals with exponentiation in general, without consideration of any special conditions. §14.6.2 looks at exponentiation when the base is variable and the exponent is fixed. §14.6.3 considers algorithms which take advantage of a fixed-base element and variable exponent. Techniques involving representing the exponent in non-binary form are given in §14.7; recoding the exponent may allow significant performance enhancements. §14.8 contains further notes and references.

14.2 Multiple-precision integer arithmetic

This section deals with the basic operations performed on multiple-precision integers: addition, subtraction, multiplication, squaring, and division. The algorithms presented in this section are commonly referred to as the *classical methods*.

14.2.1 Radix representation

Positive integers can be represented in various ways, the most common being *base* 10. For example, $a = 123$ base 10 means $a = 1 \cdot 10^2 + 2 \cdot 10^1 + 3 \cdot 10^0$. For machine computations, *base* 2 (*binary representation*) is preferable. If $a = 1111011$ base 2, then $a = 2^6 + 2^5 + 2^4 + 2^3 + 0 \cdot 2^2 + 2^1 + 2^0$.

14.1 Fact If $b \geq 2$ is an integer, then any positive integer a can be expressed uniquely as $a = a_n b^n + a_{n-1} b^{n-1} + \cdots + a_1 b + a_0$, where a_i is an integer with $0 \leq a_i < b$ for $0 \leq i \leq n$, and $a_n \neq 0$.

14.2 Definition The representation of a positive integer a as a sum of multiples of powers of b, as given in Fact 14.1, is called the *base b* or *radix b* representation of a.

14.3 Note (*notation and terminology*)

(i) The base b representation of a positive integer a given in Fact 14.1 is usually written as $a = (a_n a_{n-1} \cdots a_1 a_0)_b$. The integers a_i, $0 \le i \le n$, are called *digits*. a_n is called the *most significant digit* or *high-order digit*; a_0 the *least significant digit* or *low-order digit*. If $b = 10$, the standard notation is $a = a_n a_{n-1} \cdots a_1 a_0$.

(ii) It is sometimes convenient to pad high-order digits of a base b representation with 0's; such a padded number will also be referred to as the base b representation.

(iii) If $(a_n a_{n-1} \cdots a_1 a_0)_b$ is the base b representation of a and $a_n \ne 0$, then the *precision* or *length* of a is $n+1$. If $n = 0$, then a is called a *single-precision integer*; otherwise, a is a *multiple-precision integer*. $a = 0$ is also a single-precision integer.

The division algorithm for integers (see Definition 2.82) provides an efficient method for determining the base b representation of a non-negative integer, for a given base b. This provides the basis for Algorithm 14.4.

14.4 Algorithm Radix b representation

INPUT: integers a and b, $a \ge 0$, $b \ge 2$.
OUTPUT: the base b representation $a = (a_n \cdots a_1 a_0)_b$, where $n \ge 0$ and $a_n \ne 0$ if $n \ge 1$.

1. $i \leftarrow 0$, $x \leftarrow a$, $q \leftarrow \lfloor \frac{x}{b} \rfloor$, $a_i \leftarrow x - qb$. ($\lfloor \cdot \rfloor$ is the floor function; see page 49.)
2. While $q > 0$, do the following:
    2.1 $i \leftarrow i + 1$, $x \leftarrow q$, $q \leftarrow \lfloor \frac{x}{b} \rfloor$, $a_i \leftarrow x - qb$.
3. Return($(a_i a_{i-1} \cdots a_1 a_0)$).

14.5 Fact If $(a_n a_{n-1} \cdots a_1 a_0)_b$ is the base b representation of a and k is a positive integer, then $(u_l u_{l-1} \cdots u_1 u_0)_{b^k}$ is the base b^k representation of a, where $l = \lceil (n+1)/k \rceil - 1$, $u_i = \sum_{j=0}^{k-1} a_{ik+j} b^j$ for $0 \le i \le l - 1$, and $u_l = \sum_{j=0}^{n-lk} a_{lk+j} b^j$.

14.6 Example (*radix b representation*) The base 2 representation of $a = 123$ is $(1111011)_2$. The base 4 representation of a is easily obtained from its base 2 representation by grouping digits in pairs from the right: $a = ((1)_2 (11)_2 (10)_2 (11)_2)_4 = (1323)_4$. □

Representing negative numbers

Negative integers can be represented in several ways. Two commonly used methods are:

1. *signed-magnitude representation*
2. *complement representation*.

These methods are described below. The algorithms provided in this chapter all assume a signed-magnitude representation for integers, with the sign digit being implicit.

(i) Signed-magnitude representation

The *sign* of an integer (i.e., either positive or negative) and its *magnitude* (i.e., absolute value) are represented separately in a *signed-magnitude representation*. Typically, a positive integer is assigned a sign digit 0, while a negative integer is assigned a sign digit $b - 1$. For n-digit radix b representations, only $2b^{n-1}$ sequences out of the b^n possible sequences are utilized: precisely $b^{n-1} - 1$ positive integers and $b^{n-1} - 1$ negative integers can be represented, and 0 has two representations. Table 14.1 illustrates the binary signed-magnitude representation of the integers in the range $[7, -7]$.

Signed-magnitude representation has the drawback that when certain operations (such as addition and subtraction) are performed, the sign digit must be checked to determine the appropriate manner to perform the computation. Conditional branching of this type can be costly when many operations are performed.

(ii) Complement representation

Addition and subtraction using *complement representation* do not require the checking of the sign digit. Non-negative integers in the range $[0, b^{n-1} - 1]$ are represented by base b sequences of length n with the high-order digit being 0. Suppose x is a positive integer in this range represented by the sequence $(x_n x_{n-1} \cdots x_1 x_0)_b$ where $x_n = 0$. Then $-x$ is represented by the sequence $\overline{x} = (\overline{x}_n \overline{x}_{n-1} \cdots \overline{x}_1 \overline{x}_0) + 1$ where $\overline{x}_i = b - 1 - x_i$ and $+$ is the standard addition with carry. Table 14.1 illustrates the binary complement representation of the integers in the range $[-7, 7]$. In the binary case, complement representation is referred to as *two's complement representation*.

Sequence	Signed-magnitude	Two's complement	Sequence	Signed-magnitude	Two's complement
0111	7	7	1111	-7	-1
0110	6	6	1110	-6	-2
0101	5	5	1101	-5	-3
0100	4	4	1100	-4	-4
0011	3	3	1011	-3	-5
0010	2	2	1010	-2	-6
0001	1	1	1001	-1	-7
0000	0	0	1000	-0	-8

Table 14.1: *Signed-magnitude and two's complement representations of integers in $[-7, 7]$.*

14.2.2 Addition and subtraction

Addition and subtraction are performed on two integers having the same number of base b digits. To add or subtract two integers of different lengths, the smaller of the two integers is first padded with 0's on the left (i.e., in the high-order positions).

14.7 Algorithm Multiple-precision addition

INPUT: positive integers x and y, each having $n + 1$ base b digits.
OUTPUT: the sum $x + y = (w_{n+1} w_n \cdots w_1 w_0)_b$ in radix b representation.

1. $c \leftarrow 0$ (c is the *carry* digit).
2. For i from 0 to n do the following:
   2.1 $w_i \leftarrow (x_i + y_i + c) \bmod b$.
   2.2 If $(x_i + y_i + c) < b$ then $c \leftarrow 0$; otherwise $c \leftarrow 1$.
3. $w_{n+1} \leftarrow c$.
4. Return($(w_{n+1} w_n \cdots w_1 w_0)$).

14.8 Note (*computational efficiency*) The base b should be chosen so that $(x_i + y_i + c) \bmod b$ can be computed by the hardware on the computing device. Some processors have instruction sets which provide an add-with-carry to facilitate multiple-precision addition.

14.9 Algorithm Multiple-precision subtraction

INPUT: positive integers x and y, each having $n + 1$ base b digits, with $x \geq y$.
OUTPUT: the difference $x - y = (w_n w_{n-1} \cdots w_1 w_0)_b$ in radix b representation.

1. $c \leftarrow 0$.
2. For i from 0 to n do the following:

    2.1 $w_i \leftarrow (x_i - y_i + c) \bmod b$.
    2.2 If $(x_i - y_i + c) \geq 0$ then $c \leftarrow 0$; otherwise $c \leftarrow -1$.
3. Return$((w_n w_{n-1} \cdots w_1 w_0))$.

14.10 Note (*eliminating the requirement $x \geq y$*) If the relative magnitudes of the integers x and y are unknown, then Algorithm 14.9 can be modified as follows. On termination of the algorithm, if $c = -1$, then repeat Algorithm 14.9 with $x = (00 \cdots 00)_b$ and $y = (w_n w_{n-1} \cdots w_1 w_0)_b$. Conditional checking on the relative magnitudes of x and y can also be avoided by using a complement representation (§14.2.1(ii)).

14.11 Example (*modified subtraction*) Let $x = 3996879$ and $y = 4637923$ in base 10, so that $x < y$. Table 14.2 shows the steps of the modified subtraction algorithm (cf. Note 14.10). □

First execution of Algorithm 14.9								Second execution of Algorithm 14.9						
i	6	5	4	3	2	1	0	i	6	5	4	3	2	1	0
x_i	3	9	9	6	8	7	9	x_i	0	0	0	0	0	0	0
y_i	4	6	3	7	9	2	3	y_i	9	3	5	8	9	5	6
w_i	9	3	5	8	9	5	6	w_i	0	6	4	1	0	4	4
c	−1	0	0	−1	−1	0	0	c	−1	−1	−1	−1	−1	−1	−1

Table 14.2: *Modified subtraction (see Example 14.11).*

14.2.3 Multiplication

Let x and y be integers expressed in radix b representation: $x = (x_n x_{n-1} \cdots x_1 x_0)_b$ and $y = (y_t y_{t-1} \cdots y_1 y_0)_b$. The product $x \cdot y$ will have at most $(n + t + 2)$ base b digits. Algorithm 14.12 is a reorganization of the standard pencil-and-paper method taught in grade school. A *single-precision* multiplication means the multiplication of two base b digits. If x_j and y_i are two base b digits, then $x_j \cdot y_i$ can be written as $x_j \cdot y_i = (uv)_b$, where u and v are base b digits, and u may be 0.

14.12 Algorithm Multiple-precision multiplication

INPUT: positive integers x and y having $n + 1$ and $t + 1$ base b digits, respectively.
OUTPUT: the product $x \cdot y = (w_{n+t+1} \cdots w_1 w_0)_b$ in radix b representation.

1. For i from 0 to $(n + t + 1)$ do: $w_i \leftarrow 0$.
2. For i from 0 to t do the following:

    2.1 $c \leftarrow 0$.
    2.2 For j from 0 to n do the following:
       Compute $(uv)_b = w_{i+j} + x_j \cdot y_i + c$, and set $w_{i+j} \leftarrow v$, $c \leftarrow u$.
    2.3 $w_{i+n+1} \leftarrow u$.
3. Return$((w_{n+t+1} \cdots w_1 w_0))$.

14.13 Example (*multiple-precision multiplication*) Take $x = x_3 x_2 x_1 x_0 = 9274$ and $y = y_2 y_1 y_0 = 847$ (base 10 representations), so that $n = 3$ and $t = 2$. Table 14.3 shows the steps performed by Algorithm 14.12 to compute $x \cdot y = 7855078$.   □

i	j	c	$w_{i+j} + x_j y_i + c$	u	v	w_6	w_5	w_4	w_3	w_2	w_1	w_0
0	0	0	$0 + 28 + 0$	2	8	0	0	0	0	0	0	8
	1	2	$0 + 49 + 2$	5	1	0	0	0	0	0	1	8
	2	5	$0 + 14 + 5$	1	9	0	0	0	0	9	1	8
	3	1	$0 + 63 + 1$	6	4	0	0	6	4	9	1	8
1	0	0	$1 + 16 + 0$	1	7	0	0	6	4	9	7	8
	1	1	$9 + 28 + 1$	3	8	0	0	6	4	8	7	8
	2	3	$4 + 8 + 3$	1	5	0	0	6	5	8	7	8
	3	1	$6 + 36 + 1$	4	3	0	4	3	5	8	7	8
2	0	0	$8 + 32 + 0$	4	0	0	4	3	5	0	7	8
	1	4	$5 + 56 + 4$	6	5	0	4	3	5	0	7	8
	2	6	$3 + 16 + 6$	2	5	0	4	5	5	0	7	8
	3	2	$4 + 72 + 2$	7	8	7	8	5	5	0	7	8

Table 14.3: *Multiple-precision multiplication (see Example 14.13).*

14.14 Remark (*pencil-and-paper method*) The pencil-and-paper method for multiplying $x = 9274$ and $y = 847$ would appear as

$$
\begin{array}{rrrrrrrl}
 &   & 9 & 2 & 7 & 4 &   & \\
 & \times & 8 & 4 & 7 &   &   & \\
\hline
 & 6 & 4 & 9 & 1 & 8 &   & \text{(row 1)} \\
 3 & 7 & 0 & 9 & 6 &   &   & \text{(row 2)} \\
7 & 4 & 1 & 9 & 2 &   &   & \text{(row 3)} \\
\hline
7 & 8 & 5 & 5 & 0 & 7 & 8 & \\
\end{array}
$$

The shaded entries in Table 14.3 correspond to row 1, row 1 + row 2, and row 1 + row 2 + row 3, respectively.

14.15 Note (*computational efficiency of Algorithm 14.12*)

    (i) The computationally intensive portion of Algorithm 14.12 is step 2.2. Computing $w_{i+j} + x_j \cdot y_i + c$ is called the *inner-product operation*. Since w_{i+j}, x_j, y_i and c are all base b digits, the result of an inner-product operation is at most $(b-1) + (b-1)^2 + (b-1) = b^2 - 1$ and, hence, can be represented by two base b digits.

    (ii) Algorithm 14.12 requires $(n+1)(t+1)$ single-precision multiplications.

    (iii) It is assumed in Algorithm 14.12 that single-precision multiplications are part of the instruction set on a processor. The quality of the implementation of this instruction is crucial to an efficient implementation of Algorithm 14.12.

14.2.4 Squaring

In the preceding algorithms, $(uv)_b$ has both u and v as single-precision integers. This notation is abused in this subsection by permitting u to be a double-precision integer, such that $0 \le u \le 2(b-1)$. The value v will always be single-precision.

14.16 Algorithm Multiple-precision squaring

INPUT: positive integer $x = (x_{t-1}x_{t-2}\cdots x_1x_0)_b$.
OUTPUT: $x \cdot x = x^2$ in radix b representation.

1. For i from 0 to $(2t-1)$ do: $w_i \leftarrow 0$.
2. For i from 0 to $(t-1)$ do the following:
   2.1 $(uv)_b \leftarrow w_{2i} + x_i \cdot x_i$, $w_{2i} \leftarrow v$, $c \leftarrow u$.
   2.2 For j from $(i+1)$ to $(t-1)$ do the following:
       $(uv)_b \leftarrow w_{i+j} + 2x_j \cdot x_i + c$, $w_{i+j} \leftarrow v$, $c \leftarrow u$.
   2.3 $w_{i+t} \leftarrow u$.
3. $(uv)_b \leftarrow w_{2t-2} + x_{t-1} \cdot x_{t-1}$, $w_{2t-2} \leftarrow v$, $w_{2t-1} \leftarrow u$.
4. Return($(w_{2t-1}w_{2t-2}\ldots w_1w_0)_b$).

14.17 Note (*computational efficiency of Algorithm 14.16*)

(i) (*overflow*) In step 2.2, u can be larger than a single-precision integer. Since w_{i+j} is always set to v, $w_{i+j} \leq b-1$. If $c \leq 2(b-1)$, then $w_{i+j} + 2x_jx_i + c \leq (b-1) + 2(b-1)^2 + 2(b-1) = (b-1)(2b+1)$, implying $0 \leq u \leq 2(b-1)$. This value of u may exceed single-precision, and must be accommodated.

(ii) (*number of operations*) The computationally intensive part of the algorithm is step 2. The number of single-precision multiplications is about $(t^2 + t)/2$, discounting the multiplication by 2. This is approximately one half of the single-precision multiplications required by Algorithm 14.12 (cf. Note 14.15(ii)).

14.18 Note (*squaring vs. multiplication in general*) Squaring a positive integer x (i.e., computing x^2) can at best be no more than twice as fast as multiplying distinct integers x and y. To see this, consider the identity $xy = ((x+y)^2 - (x-y)^2)/4$. Hence, $x \cdot y$ can be computed with two squarings (i.e., $(x+y)^2$ and $(x-y)^2$). Of course, a speed-up by a factor of 2 can be significant in many applications.

14.19 Example (*squaring*) Table 14.4 shows the steps performed by Algorithm 14.16 in squaring $x = 989$. Here, $t = 3$ and $b = 10$. □

i	j	$w_{2i} + x_i^2$	$w_{i+j} + 2x_jx_i + c$	u	v	w_5	w_4	w_3	w_2	w_1	w_0
0	–	$0 + 81$	–	8	1	0	0	0	0	0	1
	1	–	$0 + 2 \cdot 8 \cdot 9 + 8$	15	2	0	0	0	0	2	1
	2	–	$0 + 2 \cdot 9 \cdot 9 + 15$	17	7	0	0	0	7	2	1
				17	7	0	0	17	7	2	1
1	–	$7 + 64$	–	7	1	0	0	17	1	2	1
	2	–	$17 + 2 \cdot 9 \cdot 8 + 7$	16	8	0	0	8	1	2	1
				16	8	0	16	8	1	2	1
2	–	$16 + 81$	–	9	7	0	7	8	1	2	1
				9	7	9	7	8	1	2	1

Table 14.4: *Multiple-precision squaring (see Example 14.19).*

14.2.5 Division

Division is the most complicated and costly of the basic multiple-precision operations. Algorithm 14.20 computes the quotient q and remainder r in radix b representation when x is divided by y.

14.20 Algorithm Multiple-precision division

INPUT: positive integers $x = (x_n \cdots x_1 x_0)_b$, $y = (y_t \cdots y_1 y_0)_b$ with $n \geq t \geq 1$, $y_t \neq 0$.
OUTPUT: the quotient $q = (q_{n-t} \cdots q_1 q_0)_b$ and remainder $r = (r_t \cdots r_1 r_0)_b$ such that $x = qy + r$, $0 \leq r < y$.

1. For j from 0 to $(n - t)$ do: $q_j \leftarrow 0$.
2. While $(x \geq yb^{n-t})$ do the following: $q_{n-t} \leftarrow q_{n-t} + 1$, $x \leftarrow x - yb^{n-t}$.
3. For i from n down to $(t + 1)$ do the following:

    3.1 If $x_i = y_t$ then set $q_{i-t-1} \leftarrow b - 1$; otherwise set $q_{i-t-1} \leftarrow \lfloor (x_i b + x_{i-1})/y_t \rfloor$.
    3.2 While $(q_{i-t-1}(y_t b + y_{t-1}) > x_i b^2 + x_{i-1} b + x_{i-2})$ do: $q_{i-t-1} \leftarrow q_{i-t-1} - 1$.
    3.3 $x \leftarrow x - q_{i-t-1} y b^{i-t-1}$.
    3.4 If $x < 0$ then set $x \leftarrow x + yb^{i-t-1}$ and $q_{i-t-1} \leftarrow q_{i-t-1} - 1$.

4. $r \leftarrow x$.
5. Return(q,r).

14.21 Example (*multiple-precision division*) Let $x = 721948327$, $y = 84461$, so that $n = 8$ and $t = 4$. Table 14.5 illustrates the steps in Algorithm 14.20. The last row gives the quotient $q = 8547$ and the remainder $r = 60160$. ☐

i	q_4	q_3	q_2	q_1	q_0	x_8	x_7	x_6	x_5	x_4	x_3	x_2	x_1	x_0
−	0	0	0	0	0	7	2	1	9	4	8	3	2	7
8	0	9	0	0	0	7	2	1	9	4	8	3	2	7
		8	0	0	0			4	6	2	6	0	3	2	7
7		8	5	0	0				4	0	2	9	8	2	7
6		8	5	5	0				4	0	2	9	8	2	7
		8	5	4	0					6	5	1	3	8	7
5		8	5	4	8					6	5	1	3	8	7
		8	5	4	7						6	0	1	6	0

Table 14.5: *Multiple-precision division (see Example 14.21).*

14.22 Note (*comments on Algorithm 14.20*)

  (i) Step 2 of Algorithm 14.20 is performed at most once if $y_t \geq \lfloor \frac{b}{2} \rfloor$ and b is even.
  (ii) The condition $n \geq t \geq 1$ can be replaced by $n \geq t \geq 0$, provided one takes $x_j = y_j = 0$ whenever a subscript $j < 0$ in encountered in the algorithm.

14.23 Note (*normalization*) The estimate for the quotient digit q_{i-t-1} in step 3.1 of Algorithm 14.20 is never less than the true value of the quotient digit. Furthermore, if $y_t \geq \lfloor \frac{b}{2} \rfloor$, then step 3.2 is repeated no more than twice. If step 3.1 is modified so that $q_{i-t-1} \leftarrow \lfloor (x_i b^2 + x_{i-1} b + x_{i-2})/(y_t b + y_{t-1}) \rfloor$, then the estimate is almost always correct and step 3.2 is never repeated more than once. One can always guarantee that $y_t \geq \lfloor \frac{b}{2} \rfloor$ by replacing the integers x, y by λx, λy for some suitable choice of λ. The quotient of λx divided by λy is the same as that of x by y; the remainder is λ times the remainder of x divided by y. If the base b is a power of 2 (as in many applications), then the choice of λ should be a power of 2; multiplication by λ is achieved by simply left-shifting the binary representations of x and y. Multiplying by a suitable choice of λ to ensure that $y_t \geq \lfloor \frac{b}{2} \rfloor$ is called *normalization*. Example 14.24 illustrates the procedure.

14.24 Example (*normalized division*) Take $x = 73418$ and $y = 267$. Normalize x and y by multiplying each by $\lambda = 3$: $x' = 3x = 220254$ and $y' = 3y = 801$. Table 14.6 shows the steps of Algorithm 14.20 as applied to x' and y'. When x' is divided by y', the quotient is 274, and the remainder is 780. When x is divided by y, the quotient is also 274 and the remainder is $780/3 = 260$. □

i	q_3	q_2	q_1	q_0	x_5	x_4	x_3	x_2	x_1	x_0
−	0	0	0	0	2	2	0	2	5	4
5	0	2	0	0		6	0	0	5	4
4		2	7	0			3	9	8	4
3			2	7	4				7	8	0

Table 14.6: *Multiple-precision division after normalization (see Example 14.24).*

14.25 Note (*computational efficiency of Algorithm 14.20 with normalization*)

(i) (*multiplication count*) Assuming that normalization extends the number of digits in x by 1, each iteration of step 3 requires $1 + (t + 2) = t + 3$ single-precision multiplications. Hence, Algorithm 14.20 with normalization requires about $(n - t)(t + 3)$ single-precision multiplications.

(ii) (*division count*) Since step 3.1 of Algorithm 14.20 is executed $n - t$ times, at most $n - t$ single-precision divisions are required when normalization is used.

14.3 Multiple-precision modular arithmetic

§14.2 provided methods for carrying out the basic operations (addition, subtraction, multiplication, squaring, and division) with multiple-precision integers. This section deals with these operations in $\mathbb{Z}_m$, the integers modulo m, where m is a multiple-precision positive integer. (See §2.4.3 for definitions of $\mathbb{Z}_m$ and related operations.)

Let $m = (m_n m_{n-1} \cdots m_1 m_0)_b$ be a positive integer in radix b representation. Let $x = (x_n x_{n-1} \cdots x_1 x_0)_b$ and $y = (y_n y_{n-1} \cdots y_1 y_0)_b$ be non-negative integers in base b representation such that $x < m$ and $y < m$. Methods described in this section are for computing $x + y \bmod m$ (*modular addition*), $x - y \bmod m$ (*modular subtraction*), and $x \cdot y \bmod m$ (*modular multiplication*). Computing $x^{-1} \bmod m$ (*modular inversion*) is addressed in §14.4.3.

14.26 Definition If z is any integer, then $z \bmod m$ (the integer remainder in the range $[0, m-1]$ after z is divided by m) is called the *modular reduction* of z with respect to modulus m.

Modular addition and subtraction

As is the case for ordinary multiple-precision operations, addition and subtraction are the simplest to compute of the modular operations.

14.27 Fact Let x and y be non-negative integers with $x, y < m$. Then:

(i) $x + y < 2m$;

(ii) if $x \geq y$, then $0 \leq x - y < m$; and

(iii) if $x < y$, then $0 \leq x + m - y < m$.

If $x, y \in \mathbb{Z}_m$, then modular addition can be performed by using Algorithm 14.7 to add x and y as multiple-precision integers, with the additional step of subtracting m if (and only if) $x + y \geq m$. Modular subtraction is precisely Algorithm 14.9, provided $x \geq y$.

14.3.1 Classical modular multiplication

Modular multiplication is more involved than multiple-precision multiplication (§14.2.3), requiring both multiple-precision multiplication and some method for performing modular reduction (Definition 14.26). The most straightforward method for performing modular reduction is to compute the remainder on division by m, using a multiple-precision division algorithm such as Algorithm 14.20; this is commonly referred to as the *classical algorithm* for performing modular multiplication.

14.28 Algorithm Classical modular multiplication

INPUT: two positive integers x, y and a modulus m, all in radix b representation.

OUTPUT: $x \cdot y \bmod m$.

1. Compute $x \cdot y$ (using Algorithm 14.12).
2. Compute the remainder r when $x \cdot y$ is divided by m (using Algorithm 14.20).
3. Return(r).

14.3.2 Montgomery reduction

Montgomery reduction is a technique which allows efficient implementation of modular multiplication without explicitly carrying out the classical modular reduction step.

Let m be a positive integer, and let R and T be integers such that $R > m$, $\gcd(m, R) = 1$, and $0 \leq T < mR$. A method is described for computing $TR^{-1} \bmod m$ without using the classical method of Algorithm 14.28. $TR^{-1} \bmod m$ is called a *Montgomery reduction* of T modulo m *with respect to* R. With a suitable choice of R, a Montgomery reduction can be efficiently computed.

Suppose x and y are integers such that $0 \leq x, y < m$. Let $\tilde{x} = xR \bmod m$ and $\tilde{y} = yR \bmod m$. The Montgomery reduction of $\tilde{x}\tilde{y}$ is $\tilde{x}\tilde{y}R^{-1} \bmod m = xyR \bmod m$. This observation is used in Algorithm 14.94 to provide an efficient method for modular exponentiation.

To briefly illustrate, consider computing $x^5 \bmod m$ for some integer $x, 1 \leq x < m$. First compute $\tilde{x} = xR \bmod m$. Then compute the Montgomery reduction of $\tilde{x}\tilde{x}$, which is $A = \tilde{x}^2R^{-1} \bmod m$. The Montgomery reduction of A^2 is $A^2R^{-1} \bmod m = \tilde{x}^4R^{-3} \bmod m$. Finally, the Montgomery reduction of $(A^2R^{-1} \bmod m)\tilde{x}$ is $(A^2R^{-1})\tilde{x}R^{-1} \bmod m = \tilde{x}^5R^{-4} \bmod m = x^5R \bmod m$. Multiplying this value by $R^{-1} \bmod m$ and reducing modulo m gives $x^5 \bmod m$. Provided that Montgomery reductions are more efficient to compute than classical modular reductions, this method may be more efficient than computing $x^5 \bmod m$ by repeated application of Algorithm 14.28.

If m is represented as a base b integer of length n, then a typical choice for R is b^n. The condition $R > m$ is clearly satisfied, but $\gcd(R, m) = 1$ will hold only if $\gcd(b, m) = 1$. Thus, this choice of R is not possible for all moduli. For those moduli of practical interest (such as RSA moduli), m will be odd; then b can be a power of 2 and $R = b^n$ will suffice.

Fact 14.29 is basic to the Montgomery reduction method. Note 14.30 then implies that $R = b^n$ is sufficient (but not necessary) for efficient implementation.

14.29 Fact (*Montgomery reduction*) Given integers m and R where $\gcd(m, R) = 1$, let $m' = -m^{-1} \bmod R$, and let T be any integer such that $0 \leq T < mR$. If $U = Tm' \bmod R$, then $(T + Um)/R$ is an integer and $(T + Um)/R \equiv TR^{-1} \pmod{m}$.

Justification. $T + Um \equiv T \pmod{m}$ and, hence, $(T + Um)R^{-1} \equiv TR^{-1} \pmod{m}$. To see that $(T + Um)R^{-1}$ is an integer, observe that $U = Tm' + kR$ and $m'm = -1 + lR$ for some integers k and l. It follows that $(T + Um)/R = (T + (Tm' + kR)m)/R = (T + T(-1 + lR) + kRm)/R = lT + km$.

14.30 Note (*implications of Fact 14.29*)

(i) $(T + Um)/R$ is an estimate for $TR^{-1} \bmod m$. Since $T < mR$ and $U < R$, then $(T + Um)/R < (mR + mR)/R = 2m$. Thus either $(T + Um)/R = TR^{-1} \bmod m$ or $(T + Um)/R = (TR^{-1} \bmod m) + m$ (i.e., the estimate is within m of the residue). Example 14.31 illustrates that both possibilities can occur.

(ii) If all integers are represented in radix b and $R = b^n$, then $TR^{-1} \bmod m$ can be computed with two multiple-precision multiplications (i.e., $U = T \cdot m'$ and $U \cdot m$) and simple right-shifts of $T + Um$ in order to divide by R.

14.31 Example (*Montgomery reduction*) Let $m = 187, R = 190$. Then $R^{-1} \bmod m = 125$, $m^{-1} \bmod R = 63$, and $m' = 127$. If $T = 563$, then $U = Tm' \bmod R = 61$ and $(T + Um)/R = 63 = TR^{-1} \bmod m$. If $T = 1125$ then $U = Tm' \bmod R = 185$ and $(T + Um)/R = 188 = (TR^{-1} \bmod m) + m$. □

Algorithm 14.32 computes the Montgomery reduction of $T = (t_{2n-1} \cdots t_1 t_0)_b$ when $R = b^n$ and $m = (m_{n-1} \cdots m_1 m_0)_b$. The algorithm makes implicit use of Fact 14.29 by computing quantities which have similar properties to $U = Tm' \bmod R$ and $T + Um$, although the latter two expressions are not computed explicitly.

14.32 Algorithm Montgomery reduction

INPUT: integers $m = (m_{n-1} \cdots m_1 m_0)_b$ with $\gcd(m, b) = 1$, $R = b^n$, $m' = -m^{-1} \bmod b$, and $T = (t_{2n-1} \cdots t_1 t_0)_b < mR$.
OUTPUT: $TR^{-1} \bmod m$.
   1. $A \leftarrow T$. (Notation: $A = (a_{2n-1} \cdots a_1 a_0)_b$.)
   2. For i from 0 to $(n-1)$ do the following:
      2.1 $u_i \leftarrow a_i m' \bmod b$.
      2.2 $A \leftarrow A + u_i m b^i$.
   3. $A \leftarrow A/b^n$.
   4. If $A \geq m$ then $A \leftarrow A - m$.
   5. Return(A).

14.33 Note (*comments on Montgomery reduction*)

    (i) Algorithm 14.32 does not require $m' = -m^{-1} \bmod R$, as Fact 14.29 does, but rather $m' = -m^{-1} \bmod b$. This is due to the choice of $R = b^n$.

    (ii) At step 2.1 of the algorithm with $i = l$, A has the property that $a_j = 0$, $0 \le j \le l-1$. Step 2.2 does not modify these values, but does replace a_l by 0. It follows that in step 3, A is divisible by b^n.

    (iii) Going into step 3, the value of A equals T plus some multiple of m (see step 2.2); here $A = (T + km)/b^n$ is an integer (see (ii) above) and $A \equiv TR^{-1} \pmod{m}$. It remains to show that A is less than $2m$, so that at step 4, a subtraction (rather than a division) will suffice. Going into step 3, $A = T + \sum_{i=0}^{n-1} u_i b^i m$. But $\sum_{i=0}^{n-1} u_i b^i m < b^n m = Rm$ and $T < Rm$; hence, $A < 2Rm$. Going into step 4 (after division of A by R), $A < 2m$ as required.

14.34 Note (*computational efficiency of Montgomery reduction*) Step 2.1 and step 2.2 of Algorithm 14.32 require a total of $n + 1$ single-precision multiplications. Since these steps are executed n times, the total number of single-precision multiplications is $n(n + 1)$. Algorithm 14.32 does not require any single-precision divisions.

14.35 Example (*Montgomery reduction*) Let $m = 72639$, $b = 10$, $R = 10^5$, and $T = 7118368$. Here $n = 5$, $m' = -m^{-1} \bmod 10 = 1$, $T \bmod m = 72385$, and $TR^{-1} \bmod m = 39796$. Table 14.7 displays the iterations of step 2 in Algorithm 14.32.   □

i	$u_i = a_i m' \bmod 10$	$u_i m b^i$	A
–	–	–	7118368
0	8	581112	7699480
1	8	5811120	13510600
2	6	43583400	57094000
3	4	290556000	347650000
4	5	3631950000	3979600000

Table 14.7: *Montgomery reduction algorithm (see Example 14.35).*

Montgomery multiplication

Algorithm 14.36 combines Montgomery reduction (Algorithm 14.32) and multiple-precision multiplication (Algorithm 14.12) to compute the Montgomery reduction of the product of two integers.

14.36 Algorithm Montgomery multiplication

INPUT: integers $m = (m_{n-1} \cdots m_1 m_0)_b$, $x = (x_{n-1} \cdots x_1 x_0)_b$, $y = (y_{n-1} \cdots y_1 y_0)_b$ with $0 \le x, y < m$, $R = b^n$ with $\gcd(m, b) = 1$, and $m' = -m^{-1} \bmod b$.
OUTPUT: $xyR^{-1} \bmod m$.

    1. $A \leftarrow 0$. (Notation: $A = (a_n a_{n-1} \cdots a_1 a_0)_b$.)
    2. For i from 0 to $(n - 1)$ do the following:
        2.1 $u_i \leftarrow (a_0 + x_i y_0) m' \bmod b$.
        2.2 $A \leftarrow (A + x_i y + u_i m)/b$.
    3. If $A \ge m$ then $A \leftarrow A - m$.
    4. Return(A).

14.37 Note (*partial justification of Algorithm 14.36*) Suppose at the ith iteration of step 2 that $0 \le A < 2m - 1$. Step 2.2 replaces A with $(A + x_i y + u_i m)/b$; but $(A + x_i y + u_i m)/b \le (2m - 2 + (b-1)(m-1) + (b-1)m)/b = 2m - 1 - (1/b)$. Hence, $A < 2m - 1$, justifying step 3.

14.38 Note (*computational efficiency of Algorithm 14.36*) Since $A + x_i y + u_i m$ is a multiple of b, only a right-shift is required to perform a division by b in step 2.2. Step 2.1 requires two single-precision multiplications and step 2.2 requires $2n$. Since step 2 is executed n times, the total number of single-precision multiplications is $n(2 + 2n) = 2n(n + 1)$.

14.39 Note (*computing $xy \bmod m$ with Montgomery multiplication*) Suppose x, y, and m are n-digit base b integers with $0 \le x, y < m$. Neglecting the cost of the precomputation in the input, Algorithm 14.36 computes $xyR^{-1} \bmod m$ with $2n(n+1)$ single-precision multiplications. Neglecting the cost to compute $R^2 \bmod m$ and applying Algorithm 14.36 to $xyR^{-1} \bmod m$ and $R^2 \bmod m$, $xy \bmod m$ is computed in $4n(n+1)$ single-precision operations. Using classical modular multiplication (Algorithm 14.28) would require $2n(n+1)$ single-precision operations and no precomputation. Hence, the classical algorithm is superior for doing a single modular multiplication; however, Montgomery multiplication is very effective for performing modular exponentiation (Algorithm 14.94).

14.40 Remark (*Montgomery reduction vs. Montgomery multiplication*) Algorithm 14.36 (Montgomery multiplication) takes as input two n-digit numbers and then proceeds to interleave the multiplication and reduction steps. Because of this, Algorithm 14.36 is not able to take advantage of the special case where the input integers are equal (i.e., squaring). On the other hand, Algorithm 14.32 (Montgomery reduction) assumes as input the product of two integers, each of which has at most n digits. Since Algorithm 14.32 is independent of multiple-precision multiplication, a faster squaring algorithm such as Algorithm 14.16 may be used prior to the reduction step.

14.41 Example (*Montgomery multiplication*) In Algorithm 14.36, let $m = 72639$, $R = 10^5$, $x = 5792$, $y = 1229$. Here $n = 5$, $m' = -m^{-1} \bmod 10 = 1$, and $xyR^{-1} \bmod m = 39796$. Notice that m and R are the same values as in Example 14.35, as is $xy = 7118368$. Table 14.8 displays the steps in Algorithm 14.36. □

i	x_i	$x_i y_0$	u_i	$x_i y$	$u_i m$	A
0	2	18	8	2458	581112	58357
1	9	81	8	11061	581112	65053
2	7	63	6	8603	435834	50949
3	5	45	4	6145	290556	34765
4	0	0	5	0	363195	39796

Table 14.8: *Montgomery multiplication (see Example 14.41).*

14.3.3 Barrett reduction

Barrett reduction (Algorithm 14.42) computes $r = x \bmod m$ given x and m. The algorithm requires the precomputation of the quantity $\mu = \lfloor b^{2k}/m \rfloor$; it is advantageous if many reductions are performed with a single modulus. For example, each RSA encryption for one entity requires reduction modulo that entity's public key modulus. The precomputation takes a fixed amount of work, which is negligible in comparison to modular exponentiation cost. Typically, the radix b is chosen to be close to the word-size of the processor. Hence, assume $b > 3$ in Algorithm 14.42 (see Note 14.44 (ii)).

14.42 Algorithm Barrett modular reduction

INPUT: positive integers $x = (x_{2k-1} \cdots x_1 x_0)_b$, $m = (m_{k-1} \cdots m_1 m_0)_b$ (with $m_{k-1} \neq 0$), and $\mu = \lfloor b^{2k}/m \rfloor$.
OUTPUT: $r = x \bmod m$.
1. $q_1 \leftarrow \lfloor x/b^{k-1} \rfloor$, $q_2 \leftarrow q_1 \cdot \mu$, $q_3 \leftarrow \lfloor q_2/b^{k+1} \rfloor$.
2. $r_1 \leftarrow x \bmod b^{k+1}$, $r_2 \leftarrow q_3 \cdot m \bmod b^{k+1}$, $r \leftarrow r_1 - r_2$.
3. If $r < 0$ then $r \leftarrow r + b^{k+1}$.
4. While $r \geq m$ do: $r \leftarrow r - m$.
5. Return(r).

14.43 Fact By the division algorithm (Definition 2.82), there exist integers Q and R such that $x = Qm + R$ and $0 \leq R < m$. In step 1 of Algorithm 14.42, the following inequality is satisfied: $Q - 2 \leq q_3 \leq Q$.

14.44 Note (*partial justification of correctness of Barrett reduction*)
(i) Algorithm 14.42 is based on the observation that $\lfloor x/m \rfloor$ can be written as $Q = \lfloor (x/b^{k-1})(b^{2k}/m)(1/b^{k+1}) \rfloor$. Moreover, Q can be approximated by the quantity $q_3 = \lfloor \lfloor x/b^{k-1} \rfloor \mu/b^{k+1} \rfloor$. Fact 14.43 guarantees that q_3 is never larger than the true quotient Q, and is at most 2 smaller.
(ii) In step 2, observe that $-b^{k+1} < r_1 - r_2 < b^{k+1}$, $r_1 - r_2 \equiv (Q - q_3)m + R$ (mod b^{k+1}), and $0 \leq (Q - q_3)m + R < 3m < b^{k+1}$ since $m < b^k$ and $3 < b$. If $r_1 - r_2 \geq 0$, then $r_1 - r_2 = (Q - q_3)m + R$. If $r_1 - r_2 < 0$, then $r_1 - r_2 + b^{k+1} = (Q - q_3)m + R$. In either case, step 4 is repeated at most twice since $0 \leq r < 3m$.

14.45 Note (*computational efficiency of Barrett reduction*)
(i) All divisions performed in Algorithm 14.42 are simple right-shifts of the base b representation.
(ii) q_2 is only used to compute q_3. Since the $k + 1$ least significant digits of q_2 are not needed to determine q_3, only a partial multiple-precision multiplication (i.e., $q_1 \cdot \mu$) is necessary. The only influence of the $k + 1$ least significant digits on the higher order digits is the carry from position $k + 1$ to position $k + 2$. Provided the base b is sufficiently large with respect to k, this carry can be accurately computed by only calculating the digits at positions k and $k+1$. [1] Hence, the $k-1$ least significant digits of q_2 need not be computed. Since μ and q_1 have at most $k + 1$ digits, determining q_3 requires at most $(k + 1)^2 - \binom{k}{2} = (k^2 + 5k + 2)/2$ single-precision multiplications.
(iii) In step 2 of Algorithm 14.42, r_2 can also be computed by a partial multiple-precision multiplication which evaluates only the least significant $k + 1$ digits of $q_3 \cdot m$. This can be done in at most $\binom{k+1}{2} + k$ single-precision multiplications.

14.46 Example (*Barrett reduction*) Let $b = 4$, $k = 3$, $x = (313221)_b$, and $m = (233)_b$ (i.e., $x = 3561$ and $m = 47$). Then $\mu = \lfloor 4^6/m \rfloor = 87 = (1113)_b$, $q_1 = \lfloor (313221)_b/4^2 \rfloor = (3132)_b$, $q_2 = (3132)_b \cdot (1113)_b = (10231302)_b$, $q_3 = (1023)_b$, $r_1 = (3221)_b$, $r_2 = (1023)_b \cdot (233)_b \bmod b^4 = (3011)_b$, and $r = r_1 - r_2 = (210)_b$. Thus $x \bmod m = 36$. $\square$

---

[1] If $b > k$, then the carry computed by simply considering the digits at position $k - 1$ (and ignoring the carry from position $k - 2$) will be in error by at most 1.

14.3.4 Reduction methods for moduli of special form

When the modulus has a special (customized) form, reduction techniques can be employed to allow more efficient computation. Suppose that the modulus m is a t-digit base b positive integer of the form $m = b^t - c$, where c is an l-digit base b positive integer (for some $l < t$). Algorithm 14.47 computes $x \bmod m$ for any positive integer x by using only shifts, additions, and single-precision multiplications of base b numbers.

14.47 Algorithm Reduction modulo $m = b^t - c$

INPUT: a base b positive integer x and a modulus $m = b^t - c$, where c is an-l digit base b integer for some $l < t$.
OUTPUT: $r = x \bmod m$.

1. $q_0 \leftarrow \lfloor x/b^t \rfloor$, $r_0 \leftarrow x - q_0 b^t$, $r \leftarrow r_0$, $i \leftarrow 0$.
2. While $q_i > 0$ do the following:

    2.1 $q_{i+1} \leftarrow \lfloor q_i c/b^t \rfloor$, $r_{i+1} \leftarrow q_i c - q_{i+1} b^t$.
    2.2 $i \leftarrow i + 1$, $r \leftarrow r + r_i$.

3. While $r \geq m$ do: $r \leftarrow r - m$.
4. Return(r).

14.48 Example (*reduction modulo $b^t - c$*) Let $b = 4$, $m = 935 = (32213)_4$, and $x = 31085 = (13211231)_4$. Since $m = 4^5 - (1121)_4$, take $c = (1121)_4$. Here $t = 5$ and $l = 4$. Table 14.9 displays the quotients and remainders produced by Algorithm 14.47. At the beginning of step 3, $r = (102031)_4$. Since $r > m$, step 3 computes $r - m = (3212)_4$.   □

i	$q_{i-1}c$	q_i	r_i	r
0	$-$	$(132)_4$	$(11231)_4$	$(11231)_4$
1	$(221232)_4$	$(2)_4$	$(21232)_4$	$(33123)_4$
2	$(2302)_4$	$(0)_4$	$(2302)_4$	$(102031)_4$

Table 14.9: *Reduction modulo $m = b^t - c$ (see Example 14.48).*

14.49 Fact (*termination*) For some integer $s \geq 0$, $q_s = 0$; hence, Algorithm 14.47 terminates.

Justification. $q_i c = q_{i+1} b^t + r_{i+1}$, $i \geq 0$. Since $c < b^t$, $q_i = (q_{i+1} b^t/c) + (r_{i+1}/c) > q_{i+1}$. Since the q_i's are non-negative integers which strictly decrease as i increases, there is some integer $s \geq 0$ such that $q_s = 0$.

14.50 Fact (*correctness*) Algorithm 14.47 terminates with the correct residue modulo m.

Justification. Suppose that s is the smallest index i for which $q_i = 0$ (i.e., $q_s = 0$). Now, $x = q_0 b^t + r_0$ and $q_i c = q_{i+1} b^t + r_{i+1}$, $0 \leq i \leq s - 1$. Adding these equations gives $x + \left(\sum_{i=0}^{s-1} q_i \right) c = \left(\sum_{i=0}^{s-1} q_i \right) b^t + \sum_{i=0}^{s} r_i$. Since $b^t \equiv c \pmod{m}$, it follows that $x \equiv \sum_{i=0}^{s} r_i \pmod{m}$. Hence, repeated subtraction of m from $r = \sum_{i=0}^{s} r_i$ gives the correct residue.

14.51 Note (*computational efficiency of reduction modulo* $b^t - c$)

(i) Suppose that x has $2t$ base b digits. If $l \leq t/2$, then Algorithm 14.47 executes step 2 at most $s = 3$ times, requiring 2 multiplications by c. In general, if l is approximately $(s-2)t/(s-1)$, then Algorithm 14.47 executes step 2 about s times. Thus, Algorithm 14.47 requires about sl single-precision multiplications.

(ii) If c has few non-zero digits, then multiplication by c will be relatively inexpensive. If c is large but has few non-zero digits, the number of iterations of Algorithm 14.47 will be greater, but each iteration requires a very simple multiplication.

14.52 Note (*modifications*) Algorithm 14.47 can be modified if $m = b^t + c$ for some positive integer $c < b^t$: in step 2.2, replace $r \leftarrow r + r_i$ with $r \leftarrow r + (-1)^i r_i$.

14.53 Remark (*using moduli of a special form*) Selecting RSA moduli of the form $b^t \pm c$ for small values of c limits the choices of primes p and q. Care must also be exercised when selecting moduli of a special form, so that factoring is not made substantially easier; this is because numbers of this form are more susceptible to factoring by the special number field sieve (see §3.2.7). A similar statement can be made regarding the selection of primes of a special form for cryptographic schemes based on the discrete logarithm problem.

14.4 Greatest common divisor algorithms

Many situations in cryptography require the computation of the greatest common divisor (gcd) of two positive integers (see Definition 2.86). Algorithm 2.104 describes the classical Euclidean algorithm for this computation. For multiple-precision integers, Algorithm 2.104 requires a multiple-precision division at step 1.1 which is a relatively expensive operation. This section describes three methods for computing the gcd which are more efficient than the classical approach using multiple-precision numbers. The first is non-Euclidean and is referred to as the *binary gcd algorithm* (§14.4.1). Although it requires more steps than the classical algorithm, the binary gcd algorithm eliminates the computationally expensive division and replaces it with elementary shifts and additions. Lehmer's gcd algorithm (§14.4.2) is a variant of the classical algorithm more suited to multiple-precision computations. A binary version of the extended Euclidean algorithm is given in §14.4.3.

14.4.1 Binary gcd algorithm

14.54 Algorithm Binary gcd algorithm

INPUT: two positive integers x and y with $x \geq y$.
OUTPUT: $\gcd(x, y)$.

1. $g \leftarrow 1$.
2. While both x and y are even do the following: $x \leftarrow x/2$, $y \leftarrow y/2$, $g \leftarrow 2g$.
3. While $x \neq 0$ do the following:
   3.1 While x is even do: $x \leftarrow x/2$.
   3.2 While y is even do: $y \leftarrow y/2$.
   3.3 $t \leftarrow |x - y|/2$.
   3.4 If $x \geq y$ then $x \leftarrow t$; otherwise, $y \leftarrow t$.
4. Return($g \cdot y$).

14.55 Example (*binary gcd algorithm*) The following table displays the steps performed by Algorithm 14.54 for computing $\gcd(1764, 868) = 28$. □

x	1764	441	112	7	7	7	7	7	0
y	868	217	217	217	105	49	21	7	7
g	1	4	4	4	4	4	4	4	4

14.56 Note (*computational efficiency of Algorithm 14.54*)

(i) If x and y are in radix 2 representation, then the divisions by 2 are simply right-shifts.

(ii) Step 3.3 for multiple-precision integers can be computed using Algorithm 14.9.

14.4.2 Lehmer's gcd algorithm

Algorithm 14.57 is a variant of the classical Euclidean algorithm (Algorithm 2.104) and is suited to computations involving multiple-precision integers. It replaces many of the multiple-precision divisions by simpler single-precision operations.

Let x and y be positive integers in radix b representation, with $x \geq y$. Without loss of generality, assume that x and y have the same number of base b digits throughout Algorithm 14.57; this may necessitate padding the high-order digits of y with 0's.

14.57 Algorithm Lehmer's gcd algorithm

INPUT: two positive integers x and y in radix b representation, with $x \geq y$.
OUTPUT: $\gcd(x, y)$.

1. While $y \geq b$ do the following:
   1.1 Set $\widetilde{x}, \widetilde{y}$ to be the high-order digit of x, y, respectively ($\widetilde{y}$ could be 0).
   1.2 $A \leftarrow 1$, $B \leftarrow 0$, $C \leftarrow 0$, $D \leftarrow 1$.
   1.3 While $(\widetilde{y} + C) \neq 0$ and $(\widetilde{y} + D) \neq 0$ do the following:
   $q \leftarrow \lfloor (\widetilde{x} + A)/(\widetilde{y} + C) \rfloor$, $q' \leftarrow \lfloor (\widetilde{x} + B)/(\widetilde{y} + D) \rfloor$.
   If $q \neq q'$ then go to step 1.4.
   $t \leftarrow A - qC$, $A \leftarrow C$, $C \leftarrow t$, $t \leftarrow B - qD$, $B \leftarrow D$, $D \leftarrow t$.
   $t \leftarrow \widetilde{x} - q\widetilde{y}$, $\widetilde{x} \leftarrow \widetilde{y}$, $\widetilde{y} \leftarrow t$.
   1.4 If $B = 0$, then $T \leftarrow x \bmod y$, $x \leftarrow y$, $y \leftarrow T$;
   otherwise, $T \leftarrow Ax + By$, $u \leftarrow Cx + Dy$, $x \leftarrow T$, $y \leftarrow u$.

2. Compute $v = \gcd(x, y)$ using Algorithm 2.104.

3. Return(v).

14.58 Note (*implementation notes for Algorithm 14.57*)

(i) T is a multiple-precision variable. A, B, C, D, and t are signed single-precision variables; hence, one bit of each of these variables must be reserved for the sign.

(ii) The first operation of step 1.3 may result in overflow since $0 \leq \widetilde{x} + A, \widetilde{y} + D \leq b$. This possibility needs to be accommodated. One solution is to reserve two bits more than the number of bits in a digit for each of $\widetilde{x}$ and $\widetilde{y}$ to accommodate both the sign and the possible overflow.

(iii) The multiple-precision additions of step 1.4 are actually subtractions, since $AB \leq 0$ and $CD \leq 0$.

14.59 Note (*computational efficiency of Algorithm 14.57*)

(i) Step 1.3 attempts to simulate multiple-precision divisions by much simpler single-precision operations. In each iteration of step 1.3, all computations are single precision. The number of iterations of step 1.3 depends on b.

(ii) The modular reduction in step 1.4 is a multiple-precision operation. The other operations are multiple-precision, but require only linear time since the multipliers are single precision.

14.60 Example (*Lehmer's gcd algorithm*) Let $b = 10^3$, $x = 768\,454\,923$, and $y = 542\,167\,814$. Since $b = 10^3$, the high-order digits of x and y are $\widetilde{x} = 768$ and $\widetilde{y} = 542$, respectively. Table 14.10 displays the values of the variables at various stages of Algorithm 14.57. The single-precision computations (Step 1.3) when $q = q'$ are shown in Table 14.11. Hence $\gcd(x, y) = 1$.   □

14.4.3 Binary extended gcd algorithm

Given integers x and y, Algorithm 2.107 computes integers a and b such that $ax + by = v$, where $v = \gcd(x, y)$. It has the drawback of requiring relatively costly multiple-precision divisions when x and y are multiple-precision integers. Algorithm 14.61 eliminates this requirement at the expense of more iterations.

14.61 Algorithm Binary extended gcd algorithm

INPUT: two positive integers x and y.
OUTPUT: integers a, b, and v such that $ax + by = v$, where $v = \gcd(x, y)$.

1. $g \leftarrow 1$.
2. While x and y are both even, do the following: $x \leftarrow x/2$, $y \leftarrow y/2$, $g \leftarrow 2g$.
3. $u \leftarrow x$, $v \leftarrow y$, $A \leftarrow 1$, $B \leftarrow 0$, $C \leftarrow 0$, $D \leftarrow 1$.
4. While u is even do the following:

    4.1 $u \leftarrow u/2$.
    4.2 If $A \equiv B \equiv 0 \pmod 2$ then $A \leftarrow A/2$, $B \leftarrow B/2$; otherwise, $A \leftarrow (A + y)/2$, $B \leftarrow (B - x)/2$.

5. While v is even do the following:

    5.1 $v \leftarrow v/2$.
    5.2 If $C \equiv D \equiv 0 \pmod 2$ then $C \leftarrow C/2$, $D \leftarrow D/2$; otherwise, $C \leftarrow (C + y)/2$, $D \leftarrow (D - x)/2$.

6. If $u \geq v$ then $u \leftarrow u - v$, $A \leftarrow A - C$, $B \leftarrow B - D$;
   otherwise, $v \leftarrow v - u$, $C \leftarrow C - A$, $D \leftarrow D - B$.
7. If $u = 0$, then $a \leftarrow C$, $b \leftarrow D$, and return$(a, b, g \cdot v)$; otherwise, go to step 4.

14.62 Example (*binary extended gcd algorithm*) Let $x = 693$ and $y = 609$. Table 14.12 displays the steps in Algorithm 14.61 for computing integers a, b, v such that $693a + 609b = v$, where $v = \gcd(693, 609)$. The algorithm returns $v = 21$, $a = -181$, and $b = 206$.   □

x	y	q	q'	precision	reference
768 454 923	542 167 814	1	1	single	Table 14.11(i)
89 593 596	47 099 917	1	1	single	Table 14.11(ii)
42 493 679	4 606 238	10	8	multiple	
4 606 238	1 037 537	5	2	multiple	
1 037 537	456 090	–	–	multiple	
456 090	125 357	3	3	single	Table 14.11(iii)
34 681	10 657	3	3	single	Table 14.11(iv)
10 657	2 710	5	3	multiple	
2 710	2 527	1	0	multiple	
2 527	183				Algorithm 2.104
183	148				Algorithm 2.104
148	35				Algorithm 2.104
35	8				Algorithm 2.104
8	3				Algorithm 2.104
3	2				Algorithm 2.104
2	1				Algorithm 2.104
1	0				Algorithm 2.104

Table 14.10: *Lehmer's gcd algorithm (see Example 14.60).*

	$\widetilde{x}$	$\widetilde{y}$	A	B	C	D	q	q'
(i)	768	542	1	0	0	1	1	1
	542	226	0	1	1	−1	2	2
	226	90	1	−1	−2	3	2	2
	90	46	−2	3	5	−7	1	2
(ii)	89	47	1	0	0	1	1	1
	47	42	0	1	1	−1	1	1
	42	5	1	−1	−1	2	10	5
(iii)	456	125	1	0	0	1	3	3
	125	81	0	1	1	−3	1	1
	81	44	1	−3	−1	4	1	1
	44	37	−1	4	2	−7	1	1
	37	7	2	−7	−3	11	9	1
(iv)	34	10	1	0	0	1	3	3
	10	4	0	1	1	−3	2	11

Table 14.11: *Single-precision computations (see Example 14.60 and Table 14.10).*

u	v	A	B	C	D
693	609	1	0	0	1
84	609	1	-1	0	1
42	609	305	-347	0	1
21	609	457	-520	0	1
21	588	457	-520	-457	521
21	294	457	-520	76	-86
21	147	457	-520	38	-43
21	126	457	-520	-419	477
21	63	457	-520	95	-108
21	42	457	-520	-362	412
21	21	457	-520	-181	206
0	21	638	-726	-181	206

Table 14.12: *The binary extended gcd algorithm with $x = 693$, $y = 609$ (see Example 14.62).*

14.63 Note (*computational efficiency of Algorithm 14.61*)

   (i) The only multiple-precision operations needed for Algorithm 14.61 are addition and subtraction. Division by 2 is simply a right-shift of the binary representation.

   (ii) The number of bits needed to represent either u or v decreases by (at least) 1, after at most two iterations of steps $4 - 7$; thus, the algorithm takes at most $2(\lfloor \lg x \rfloor + \lfloor \lg y \rfloor + 2)$ such iterations.

14.64 Note (*multiplicative inverses*) Given positive integers m and a, it is often necessary to find an integer $z \in \mathbb{Z}_m$ such that $az \equiv 1 \pmod{m}$, if such an integer exists. z is called the multiplicative inverse of a modulo m (see Definition 2.115). For example, constructing the private key for RSA requires the computation of an integer d such that $ed \equiv 1 \pmod{(p-1)(q-1)}$ (see Algorithm 8.1). Algorithm 14.61 provides a computationally efficient method for determining z given a and m, by setting $x = m$ and $y = a$. If $\gcd(x, y) = 1$, then, at termination, $z = D$ if $D > 0$, or $z = m - D$ if $D < 0$; if $\gcd(x, y) \neq 1$, then a is not invertible modulo m. Notice that if m is odd, it is not necessary to compute the values of A and C. It would appear that step 4 of Algorithm 14.61 requires both A and B in order to decide which case in step 4.2 is executed. But if m is odd and B is even, then A must be even; hence, the decision can be made using the parities of B and m.

    Example 14.65 illustrates Algorithm 14.61 for computing a multiplicative inverse.

14.65 Example (*multiplicative inverse*) Let $m = 383$ and $a = 271$. Table 14.13 illustrates the steps of Algorithm 14.61 for computing $271^{-1} \bmod 383 = 106$. Notice that values for the variables A and C need not be computed.   □

14.5 Chinese remainder theorem for integers

Fact 2.120 introduced the Chinese remainder theorem (CRT) and Fact 2.121 outlined an algorithm for solving the associated system of linear congruences. Although the method described there is the one found in most textbooks on elementary number theory, it is not the

iteration:	1	2	3	4	5	6	7	8	9	10
u	383	112	56	28	14	7	7	7	7	7
v	271	271	271	271	271	271	264	132	66	33
B	0	-1	-192	-96	-48	-24	-24	-24	-24	-24
D	1	1	1	1	1	1	25	-179	-281	-332
iteration:	11	12	13	14	15	16	17	18	19	
u	7	7	7	7	4	2	1	1	1	
v	26	13	6	3	3	3	3	2	1	
B	-24	-24	-24	-24	41	-171	-277	-277	-277	
D	-308	-154	-130	-65	-65	-65	-65	212	106	

Table 14.13: *Inverse computation using the binary extended gcd algorithm (see Example 14.65).*

method of choice for large integers. Garner's algorithm (Algorithm 14.71) has some computational advantages. §14.5.1 describes an alternate (non-radix) representation for non-negative integers, called a *modular representation*, that allows some computational advantages compared to standard radix representations. Algorithm 14.71 provides a technique for converting numbers from modular to base b representation.

14.5.1 Residue number systems

In previous sections, non-negative integers have been represented in radix b notation. An alternate means is to use a mixed-radix representation.

14.66 Fact Let B be a fixed positive integer. Let $m_1, m_2, \dots, m_t$ be positive integers such that $\gcd(m_i, m_j) = 1$ for all $i \neq j$, and $M = \prod_{i=1}^{t} m_i \geq B$. Then each integer x, $0 \leq x < B$, can be uniquely represented by the sequence of integers $v(x) = (v_1, v_2, \dots, v_t)$, where $v_i = x \bmod m_i$, $1 \leq i \leq t$.

14.67 Definition Referring to Fact 14.66, $v(x)$ is called the *modular representation* or *mixed-radix representation* of x for the moduli $m_1, m_2, \dots, m_t$. The set of modular representations for all integers x in the range $0 \leq x < B$ is called a *residue number system*.

If $v(x) = (v_1, v_2, \dots, v_t)$ and $v(y) = (u_1, u_2, \dots, u_t)$, define $v(x) + v(y) = (w_1, w_2, \dots, w_t)$ where $w_i = v_i + u_i \bmod m_i$, and $v(x) \cdot v(y) = (z_1, z_2, \dots, z_t)$ where $z_i = v_i \cdot u_i \bmod m_i$.

14.68 Fact If $0 \leq x, y < M$, then $v((x + y) \bmod M) = v(x) + v(y)$ and $v((x \cdot y) \bmod M) = v(x) \cdot v(y)$.

14.69 Example (*modular representation*) Let $M = 30 = 2 \times 3 \times 5$; here, $t = 3$, $m_1 = 2$, $m_1 = 3$, and $m_3 = 5$. Table 14.14 displays each residue modulo 30 along with its associated modular representation. As an example of Fact 14.68, note that $21 + 27 \equiv 18 \pmod{30}$ and $(101) + (102) = (003)$. Also $22 \cdot 17 \equiv 14 \pmod{30}$ and $(012) \cdot (122) = (024)$. $\square$

14.70 Note (*computational efficiency of modular representation for RSA decryption*) Suppose that $n = pq$, where p and q are distinct primes. Fact 14.68 implies that $x^d \bmod n$ can be computed in a modular representation as $v^d(x)$; that is, if $v(x) = (v_1, v_2)$ with respect to moduli $m_1 = p$, $m_2 = q$, then $v^d(x) = (v_1^d \bmod p, v_2^d \bmod q)$. In general, computing

x	$v(x)$	x	$v(x)$	x	$v(x)$	x	$v(x)$	x	$v(x)$
0	(000)	6	(001)	12	(002)	18	(003)	24	(004)
1	(111)	7	(112)	13	(113)	19	(114)	25	(110)
2	(022)	8	(023)	14	(024)	20	(020)	26	(021)
3	(103)	9	(104)	15	(100)	21	(101)	27	(102)
4	(014)	10	(010)	16	(011)	22	(012)	28	(013)
5	(120)	11	(121)	17	(122)	23	(123)	29	(124)

Table 14.14: *Modular representations (see Example 14.69).*

$v_1^d \bmod p$ and $v_2^d \bmod q$ is faster than computing $x^d \bmod n$. For RSA, if p and q are part of the private key, modular representation can be used to improve the performance of both decryption and signature generation (see Note 14.75).

Converting an integer x from a base b representation to a modular representation is easily done by applying a modular reduction algorithm to compute $v_i = x \bmod m_i$, $1 \le i \le t$. Modular representations of integers in $\mathbb{Z}_M$ may facilitate some computational efficiencies, provided conversion from a standard radix to modular representation and back are relatively efficient operations. Algorithm 14.71 describes one way of converting from modular representation back to a standard radix representation.

14.5.2 Garner's algorithm

Garner's algorithm is an efficient method for determining x, $0 \le x < M$, given $v(x) = (v_1, v_2, \dots, v_t)$, the residues of x modulo the pairwise co-prime moduli $m_1, m_2, \dots, m_t$.

14.71 Algorithm Garner's algorithm for CRT

INPUT: a positive integer $M = \prod_{i=1}^{t} m_i > 1$, with $\gcd(m_i, m_j) = 1$ for all $i \ne j$, and a modular representation $v(x) = (v_1, v_2, \dots, v_t)$ of x for the m_i.
OUTPUT: the integer x in radix b representation.

1. For i from 2 to t do the following:

    1.1 $C_i \leftarrow 1$.

    1.2 For j from 1 to $(i-1)$ do the following:

        $u \leftarrow m_j^{-1} \bmod m_i$ (use Algorithm 14.61).

        $C_i \leftarrow u \cdot C_i \bmod m_i$.

2. $u \leftarrow v_1$, $x \leftarrow u$.

3. For i from 2 to t do the following: $u \leftarrow (v_i - x) C_i \bmod m_i$, $x \leftarrow x + u \cdot \prod_{j=1}^{i-1} m_j$.

4. Return(x).

14.72 Fact x returned by Algorithm 14.71 satisfies $0 \le x < M$, $x \equiv v_i \pmod{m_i}$, $1 \le i \le t$.

14.73 Example (*Garner's algorithm*) Let $m_1 = 5$, $m_2 = 7$, $m_3 = 11$, $m_4 = 13$, $M = \prod_{i=1}^{4} m_i = 5005$, and $v(x) = (2, 1, 3, 8)$. The constants C_i computed are $C_2 = 3$, $C_3 = 6$, and $C_4 = 5$. The values of (i, u, x) computed in step 3 of Algorithm 14.71 are $(1, 2, 2)$, $(2, 4, 22)$, $(3, 7, 267)$, and $(4, 5, 2192)$. Hence, the modular representation $v(x) = (2, 1, 3, 8)$ corresponds to the integer $x = 2192$. $\square$

14.74 Note (*computational efficiency of Algorithm 14.71*)

    (i) If Garner's algorithm is used repeatedly with the same modulus M and the same factors of M, then step 1 can be considered as a precomputation, requiring the storage of $t - 1$ numbers.

   (ii) The classical algorithm for the CRT (Algorithm 2.121) typically requires a modular reduction with modulus M, whereas Algorithm 14.71 does not. Suppose M is a kt-bit integer and each m_i is a k-bit integer. A modular reduction by M takes $O((kt)^2)$ bit operations, whereas a modular reduction by m_i takes $O(k^2)$ bit operations. Since Algorithm 14.71 only does modular reduction with m_i, $2 \leq i \leq t$, it takes $O(tk^2)$ bit operations in total for the reduction phase, and is thus more efficient.

14.75 Note (*RSA decryption and signature generation*)

    (i) (*special case of two moduli*) Algorithm 14.71 is particularly efficient for RSA moduli $n = pq$, where $m_1 = p$ and $m_2 = q$ are distinct primes. Step 1 computes a single value $C_2 = p^{-1} \bmod q$. Step 3 is executed once: $u = (v_2 - v_1)C_2 \bmod q$ and $x = v_1 + up$.

   (ii) (*RSA exponentiation*) Suppose p and q are t-bit primes, and let $n = pq$. Let d be a $2t$-bit RSA private key. RSA decryption and signature generation compute $x^d \bmod n$ for some $x \in \mathbb{Z}_n$. Suppose that modular multiplication and squaring require k^2 bit operations for k-bit inputs, and that exponentiation with a k-bit exponent requires about $\frac{3}{2}k$ multiplications and squarings (see Note 14.78). Then computing $x^d \bmod n$ requires about $\frac{3}{2}(2t)^3 = 12t^3$ bit operations. A more efficient approach is to compute $x^{d_p} \bmod p$ and $x^{d_q} \bmod q$ (where $d_p = d \bmod (p-1)$ and $d_q = d \bmod (q-1)$), and then use Garner's algorithm to construct $x^d \bmod pq$. Although this procedure takes two exponentiations, each is considerably more efficient because the moduli are smaller. Assuming that the cost of Algorithm 14.71 is negligible with respect to the exponentiations, computing $x^d \bmod n$ is about $\frac{3}{2}(2t)^3/2(\frac{3}{2}t^3) = 4$ times faster.

14.6 Exponentiation

One of the most important arithmetic operations for public-key cryptography is exponentiation. The RSA scheme (§8.2) requires exponentiation in $\mathbb{Z}_m$ for some positive integer m, whereas Diffie-Hellman key agreement (§12.6.1) and the ElGamal encryption scheme (§8.4) use exponentiation in $\mathbb{Z}_p$ for some large prime p. As pointed out in §8.4.2, ElGamal encryption can be generalized to any finite cyclic group. This section discusses methods for computing the *exponential* g^e, where the *base* g is an element of a finite group G (§2.5.1) and the *exponent* e is a non-negative integer. A reader uncomfortable with the setting of a general group may consider G to be $\mathbb{Z}_m^*$; that is, read g^e as $g^e \bmod m$.

    An efficient method for multiplying two elements in the group G is essential to performing efficient exponentiation. The most naive way to compute g^e is to do $e - 1$ multiplications in the group G. For cryptographic applications, the order of the group G typically exceeds 2^{160} elements, and may exceed 2^{1024}. Most choices of e are large enough that it would be infeasible to compute g^e using $e - 1$ successive multiplications by g.

    There are two ways to reduce the time required to do exponentiation. One way is to decrease the time to multiply two elements in the group; the other is to reduce the number of multiplications used to compute g^e. Ideally, one would do both.

    This section considers three types of exponentiation algorithms.

1. *basic techniques for exponentiation.* Arbitrary choices of the base g and exponent e are allowed.

2. *fixed-exponent exponentiation algorithms.* The exponent e is fixed and arbitrary choices of the base g are allowed. RSA encryption and decryption schemes benefit from such algorithms.

3. *fixed-base exponentiation algorithms.* The base g is fixed and arbitrary choices of the exponent e are allowed. ElGamal encryption and signatures schemes and Diffie-Hellman key agreement protocols benefit from such algorithms.

14.6.1 Techniques for general exponentiation

This section includes general-purpose exponentiation algorithms referred to as *repeated square-and-multiply* algorithms.

(i) Basic binary and k-ary exponentiation

Algorithm 14.76 is simply Algorithm 2.143 restated in terms of an arbitrary finite abelian group G with identity element 1.

14.76 Algorithm Right-to-left binary exponentiation

INPUT: an element $g \in G$ and integer $e \geq 1$.
OUTPUT: g^e.

1. $A \leftarrow 1$, $S \leftarrow g$.
2. While $e \neq 0$ do the following:
   2.1 If e is odd then $A \leftarrow A \cdot S$.
   2.2 $e \leftarrow \lfloor e/2 \rfloor$.
   2.3 If $e \neq 0$ then $S \leftarrow S \cdot S$.
3. Return(A).

14.77 Example (*right-to-left binary exponentiation*) The following table displays the values of A, e, and S during each iteration of Algorithm 14.76 for computing g^{283}. □

A	1	g	g^3	g^3	g^{11}	g^{27}	g^{27}	g^{27}	g^{27}	g^{283}
e	283	141	70	35	17	8	4	2	1	0
S	g	g^2	g^4	g^8	g^{16}	g^{32}	g^{64}	g^{128}	g^{256}	–

14.78 Note (*computational efficiency of Algorithm 14.76*) Let $t + 1$ be the bitlength of the binary representation of e, and let $\text{wt}(e)$ be the number of 1's in this representation. Algorithm 14.76 performs t squarings and $\text{wt}(e) - 1$ multiplications. If e is randomly selected in the range $0 \leq e < |G| = n$, then about $\lfloor \lg n \rfloor$ squarings and $\frac{1}{2}(\lfloor \lg n \rfloor + 1)$ multiplications can be expected. (The assignment $1 \cdot x$ is not counted as a multiplication, nor is the operation $1 \cdot 1$ counted as a squaring.) If squaring is approximately as costly as an arbitrary multiplication (cf. Note 14.18), then the expected amount of work is roughly $\frac{3}{2} \lfloor \lg n \rfloor$ multiplications.

Algorithm 14.76 computes $A \cdot S$ whenever e is odd. For some choices of g, $A \cdot g$ can be computed more efficiently than $A \cdot S$ for arbitrary S. Algorithm 14.79 is a left-to-right binary exponentiation which replaces the operation $A \cdot S$ (for arbitrary S) by the operation $A \cdot g$ (for fixed g).

14.79 Algorithm Left-to-right binary exponentiation

INPUT: $g \in G$ and a positive integer $e = (e_t e_{t-1} \cdots e_1 e_0)_2$.
OUTPUT: g^e.

1. $A \leftarrow 1$.
2. For i from t down to 0 do the following:
   2.1 $A \leftarrow A \cdot A$.
   2.2 If $e_i = 1$, then $A \leftarrow A \cdot g$.
3. Return(A).

14.80 Example (*left-to-right binary exponentiation*) The following table displays the values of A during each iteration of Algorithm 14.79 for computing g^{283}. Note that $t = 8$ and $283 = (100011011)_2$. □

i	8	7	6	5	4	3	2	1	0
e_i	1	0	0	0	1	1	0	1	1
A	g	g^2	g^4	g^8	g^{17}	g^{35}	g^{70}	g^{141}	g^{283}

14.81 Note (*computational efficiency of Algorithm 14.79*) Let $t + 1$ be the bitlength of the binary representation of e, and let $\mathrm{wt}(e)$ be the number of 1's in this representation. Algorithm 14.79 performs $t + 1$ squarings and $\mathrm{wt}(e) - 1$ multiplications by g. The number of squarings and multiplications is the same as in Algorithm 14.76 but, in this algorithm, multiplication is always with the fixed value g. If g has a special structure, this multiplication may be substantially easier than multiplying two arbitrary elements. For example, a frequent operation in ElGamal public-key schemes is the computation of $g^k \bmod p$, where g is a generator of $\mathbb{Z}_p^*$ and p is a large prime number. The multiple-precision computation $A \cdot g$ can be done in linear time if g is chosen so that it can be represented by a single-precision integer (e.g., $g = 2$). If the radix b is sufficiently large, there is a high probability that such a generator exists.

Algorithm 14.82, sometimes referred to as the *window method* for exponentiation, is a generalization of Algorithm 14.79 which processes more than one bit of the exponent per iteration.

14.82 Algorithm Left-to-right k-ary exponentiation

INPUT: g and $e = (e_t e_{t-1} \cdots e_1 e_0)_b$, where $b = 2^k$ for some $k \geq 1$.
OUTPUT: g^e.

1. *Precomputation.*
   1.1 $g_0 \leftarrow 1$.
   1.2 For i from 1 to $(2^k - 1)$ do: $g_i \leftarrow g_{i-1} \cdot g$. (Thus, $g_i = g^i$.)
2. $A \leftarrow 1$.
3. For i from t down to 0 do the following:
   3.1 $A \leftarrow A^{2^k}$.
   3.2 $A \leftarrow A \cdot g_{e_i}$.
4. Return(A).

In Algorithm 14.83, Algorithm 14.82 is modified slightly to reduce the amount of pre-computation. The following notation is used: for each i, $0 \leq i \leq t$, if $e_i \neq 0$, then write $e_i = 2^{h_i} u_i$ where u_i is odd; if $e_i = 0$, then let $h_i = 0$ and $u_i = 0$.

14.83 Algorithm Modified left-to-right k-ary exponentiation

INPUT: g and $e = (e_t e_{t-1} \cdots e_1 e_0)_b$, where $b = 2^k$ for some $k \geq 1$.
OUTPUT: g^e.

1. *Precomputation.*

     1.1 $g_0 \leftarrow 1$, $g_1 \leftarrow g$, $g_2 \leftarrow g^2$.
     1.2 For i from 1 to $(2^{k-1} - 1)$ do: $g_{2i+1} \leftarrow g_{2i-1} \cdot g_2$.

2. $A \leftarrow 1$.
3. For i from t down to 0 do: $A \leftarrow (A^{2^{k-h_i}} \cdot g_{u_i})^{2^{h_i}}$.
4. Return(A).

14.84 Remark (*right-to-left k-ary exponentiation*) Algorithm 14.82 is a generalization of Algorithm 14.79. In a similar manner, Algorithm 14.76 can be generalized to the k-ary case. However, the optimization given in Algorithm 14.83 is not possible for the generalized right-to-left k-ary exponentiation method.

(ii) Sliding-window exponentiation

Algorithm 14.85 also reduces the amount of precomputation compared to Algorithm 14.82 and, moreover, reduces the average number of multiplications performed (excluding squarings). k is called the *window size*.

14.85 Algorithm Sliding-window exponentiation

INPUT: g, $e = (e_t e_{t-1} \cdots e_1 e_0)_2$ with $e_t = 1$, and an integer $k \geq 1$.
OUTPUT: g^e.

1. *Precomputation.*

     1.1 $g_1 \leftarrow g$, $g_2 \leftarrow g^2$.
     1.2 For i from 1 to $(2^{k-1} - 1)$ do: $g_{2i+1} \leftarrow g_{2i-1} \cdot g_2$.

2. $A \leftarrow 1$, $i \leftarrow t$.
3. While $i \geq 0$ do the following:

     3.1 If $e_i = 0$ then do: $A \leftarrow A^2$, $i \leftarrow i - 1$.
     3.2 Otherwise ($e_i \neq 0$), find the longest bitstring $e_i e_{i-1} \cdots e_l$ such that $i - l + 1 \leq k$ and $e_l = 1$, and do the following:

$$A \leftarrow A^{2^{i-l+1}} \cdot g_{(e_i e_{i-1} \dots e_l)_2}, \quad i \leftarrow l - 1.$$

4. Return(A).

14.86 Example (*sliding-window exponentiation*) Take $e = 11749 = (10110111100101)_2$ and $k = 3$. Table 14.15 illustrates the steps of Algorithm 14.85. Notice that the sliding-window method for this exponent requires three multiplications, corresponding to $i = 7$, 4, and 0. Algorithm 14.79 would have required four multiplications for the same values of k and e. $\square$

i	A	Longest bitstring
13	1	101
10	g^5	101
7	$(g^5)^8 g^5 = g^{45}$	111
4	$(g^{45})^8 g^7 = g^{367}$	–
3	$(g^{367})^2 = g^{734}$	–
2	$(g^{734})^2 = g^{1468}$	101
0	$(g^{1468})^8 g^5 = g^{11749}$	–

Table 14.15: *Sliding-window exponentiation with $k = 3$ and exponent $e = (10110111100101)_2$.*

14.87 Note (*comparison of exponentiation algorithms*) Let $t + 1$ be the bitlength of e, and let $l + 1$ be the number of k-bit words formed from e; that is, $l = \lceil (t+1)/k \rceil - 1 = \lfloor t/k \rfloor$. Table 14.16 summarizes the number of squarings and multiplications required by Algorithms 14.76, 14.79, 14.82, and 14.83. Analysis of the number of squarings and multiplications for Algorithm 14.85 is more difficult, although it is the recommended method.

(i) (*squarings for Algorithm 14.82*) The number of squarings for Algorithm 14.82 is lk. Observe that $lk = \lfloor t/k \rfloor k = t - (t \bmod k)$. It follows that $t - (k-1) \leq lk \leq t$ and that Algorithm 14.82 can save up to $k - 1$ squarings over Algorithms 14.76 and 14.79. An optimal value for k in Algorithm 14.82 will depend on t.

(ii) (*squarings for Algorithm 14.83*) The number of squarings for Algorithm 14.83 is $lk + h_l$ where $0 \leq h_l \leq t \bmod k$. Since $t - (k-1) \leq lk \leq lk + h_l \leq lk + (t \bmod k) = t$ or $t - (k-1) \leq lk + h_l \leq t$, the number of squarings for this algorithm has the same bounds as Algorithm 14.82.

Algorithm	Precomputation		squarings	Multiplications	
	sq	mult		worst case	average case
14.76	0	0	t	t	$t/2$
14.79	0	0	t	t	$t/2$
14.82	1	$2^k - 3$	$t - (k-1) \leq lk \leq t$	$l - 1$	$l(2^k - 1)/2^k$
14.83	1	$2^{k-1} - 1$	$t - (k-1) \leq lk + h_l \leq t$	$l - 1$	$l(2^k - 1)/2^k$

Table 14.16: *Number of squarings (sq) and multiplications (mult) for exponentiation algorithms.*

(iii) Simultaneous multiple exponentiation

There are a number of situations which require computation of the product of several exponentials with distinct bases and distinct exponents (for example, verification of ElGamal signatures; see Note 14.91). Rather than computing each exponential separately, Algorithm 14.88 presents a method to do them simultaneously.

Let $e_0, e_1, \ldots, e_{k-1}$ be positive integers each of bitlength t; some of the high-order bits of some of the exponents might be 0, but there is at least one e_i whose high-order bit is 1. Form a $k \times t$ array EA (called the *exponent array*) whose rows are the binary representations of the exponents e_i, $0 \leq i \leq k - 1$. Let I_j be the non-negative integer whose binary representation is the jth column, $1 \leq j \leq t$, of EA, where low-order bits are at the top of the column.

14.88 Algorithm Simultaneous multiple exponentiation

INPUT: group elements $g_0, g_1, \ldots, g_{k-1}$ and non-negative t-bit integers $e_0, e_1, \ldots e_{k-1}$.
OUTPUT: $g_0^{e_0} g_1^{e_1} \cdots g_{k-1}^{e_{k-1}}$.

1. *Precomputation.* For i from 0 to $(2^k - 1)$: $G_i \leftarrow \prod_{j=0}^{k-1} g_j^{i_j}$ where $i = (i_{k-1} \cdots i_0)_2$.
2. $A \leftarrow 1$.
3. For i from 1 to t do the following: $A \leftarrow A \cdot A$, $A \leftarrow A \cdot G_{I_i}$.
4. Return(A).

14.89 Example (*simultaneous multiple exponentiation*) In this example, $g_0^{30} g_1^{10} g_2^{24}$ is computed using Algorithm 14.88. Let $e_0 = 30 = (11110)_2$, $e_1 = 10 = (01010)_2$, and $e_2 = 24 = (11000)_2$. The 3×5 array EA is:

$$\begin{array}{|ccccc|}
\hline
1 & 1 & 1 & 1 & 0 \\
0 & 1 & 0 & 1 & 0 \\
1 & 1 & 0 & 0 & 0 \\
\hline
\end{array}$$

The next table displays precomputed values from step 1 of Algorithm 14.88.

i	0	1	2	3	4	5	6	7
G_i	1	g_0	g_1	$g_0 g_1$	g_2	$g_0 g_2$	$g_1 g_2$	$g_0 g_1 g_2$

Finally, the value of A at the end of each iteration of step 3 is shown in the following table. Here, $I_1 = 5$, $I_2 = 7$, $I_3 = 1$, $I_4 = 3$, and $I_5 = 0$.

i	1	2	3	4	5
A	$g_0 g_2$	$g_0^3 g_1 g_2^3$	$g_0^7 g_1^2 g_2^6$	$g_0^{15} g_1^5 g_2^{12}$	$g_0^{30} g_1^{10} g_2^{24}$

$\square$

14.90 Note (*computational efficiency of Algorithm 14.88*)

(i) Algorithm 14.88 computes $g_0^{e_0} g_1^{e_1} \cdots g_{k-1}^{e_{k-1}}$ (where each e_i is represented by t bits) by performing $t - 1$ squarings and at most $(2^k - 2) + t - 1$ multiplications. The multiplication is trivial for any column consisting of all 0's.

(ii) Not all of the G_i, $0 \leq i \leq 2^k - 1$, need to be precomputed, but only for those i whose binary representation is a column of EA.

14.91 Note (*ElGamal signature verification*) The signature verification equation for the ElGamal signature scheme (Algorithm 11.64) is $\alpha^{h(m)} (\alpha^{-a})^r \equiv r^s \pmod{p}$ where p is a large prime, α a generator of $\mathbb{Z}_p^*$, α^a is the public key, and (r, s) is a signature for message m. It would appear that three exponentiations and one multiplication are required to verify the equation. If $t = \lceil \lg p \rceil$ and Algorithm 11.64 is applied, the number of squarings is $3(t - 1)$ and the number of multiplications is, on average, $3t/2$. Hence, one would expect to perform about $(9t - 4)/2$ multiplications and squarings modulo p. Algorithm 14.88 can reduce the number of computations substantially if the verification equation is rewritten as $\alpha^{h(m)} (\alpha^{-a})^r r^{-s} \equiv 1 \pmod{p}$. Taking $g_0 = \alpha$, $g_1 = \alpha^{-a}$, $g_2 = r$, and $e_0 = h(m) \bmod (p-1)$, $e_1 = r \bmod (p-1)$, $e_2 = -s \bmod (p-1)$ in Algorithm 14.88, the expected number of multiplications and squarings is $(t-1) + (6 + (7t/8)) = (15t + 40)/8$. (For random exponents, one would expect that, on average, $\frac{7}{8}$ of the columns of EA will be non-zero and necessitate a non-trivial multiplication.) This is only about 25% more costly than a single exponentiation computed by Algorithm 14.79.

(iv) Additive notation

Algorithms 14.76 and 14.79 have been described in the setting of a multiplicative group. Algorithm 14.92 uses the methodology of Algorithm 14.79 to perform efficient multiplication in an additive group G. (For example, the group formed by the points on an elliptic curve over a finite field uses additive notation.) Multiplication in an additive group corresponds to exponentiation in a multiplicative group.

14.92 Algorithm Left-to-right binary multiplication in an additive group

---

INPUT: $g \in G$, where G is an additive group, and a positive integer $e = (e_t e_{t-1} \cdots e_1 e_0)_2$.
OUTPUT: $e \cdot g$.

1. $A \leftarrow 0$.
2. For i from t down to 0 do the following:
    2.1 $A \leftarrow A + A$.
    2.2 If $e_i = 1$ then $A \leftarrow A + g$.
3. Return(A).

14.93 Note (*the additive group* $\mathbb{Z}_m$)

(i) If G is the additive group $\mathbb{Z}_m$, then Algorithm 14.92 provides a method for doing modular multiplication. For example, if $a, b \in \mathbb{Z}_m$, then $a \cdot b \bmod m$ can be computed using Algorithm 14.92 by taking $g = a$ and $e = b$, provided b is written in binary.

(ii) If $a, b \in \mathbb{Z}_m$, then $a < m$ and $b < m$. The accumulator A in Algorithm 14.92 never contains an integer as large as $2m$; hence, modular reduction of the value in the accumulator can be performed by a simple subtraction when $A \geq m$; thus no divisions are required.

(iii) Algorithms 14.82 and 14.83 can also be used for modular multiplication. In the case of the additive group $\mathbb{Z}_m$, the time required to do modular multiplication can be improved at the expense of precomputing a table of residues modulo m. For a left-to-right k-ary exponentiation scheme, the table will contain $2^k - 1$ residues modulo m.

(v) Montgomery exponentiation

The introductory remarks to §14.3.2 outline an application of the Montgomery reduction method for exponentiation. Algorithm 14.94 below combines Algorithm 14.79 and Algorithm 14.36 to give a *Montgomery exponentiation algorithm* for computing $x^e \bmod m$. Note the definition of m' requires that $\gcd(m, R) = 1$. For integers u and v where $0 \leq u, v < m$, define $\text{Mont}(u, v)$ to be $uvR^{-1} \bmod m$ as computed by Algorithm 14.36.

14.94 Algorithm Montgomery exponentiation

INPUT: $m = (m_{l-1} \cdots m_0)_b$, $R = b^l$, $m' = -m^{-1} \bmod b$, $e = (e_t \cdots e_0)_2$ with $e_t = 1$, and an integer x, $1 \le x < m$.
OUTPUT: $x^e \bmod m$.

1. $\tilde{x} \leftarrow \text{Mont}(x, R^2 \bmod m)$, $A \leftarrow R \bmod m$. ($R \bmod m$ and $R^2 \bmod m$ may be provided as inputs.)
2. For i from t down to 0 do the following:
   2.1 $A \leftarrow \text{Mont}(A, A)$.
   2.2 If $e_i = 1$ then $A \leftarrow \text{Mont}(A, \tilde{x})$.
3. $A \leftarrow \text{Mont}(A, 1)$.
4. Return(A).

14.95 Example (*Montgomery exponentiation*) Let x, m, and R be integers suitable as inputs to Algorithm 14.94. Let $e = 11 = (1011)_2$; here, $t = 3$. The following table displays the values of $A \bmod m$ at the end of each iteration of step 2, and after step 3.   □

i	3	2	1	0	Step 3
$A \bmod m$	$\tilde{x}$	$\tilde{x}^2 R^{-1}$	$\tilde{x}^5 R^{-4}$	$\tilde{x}^{11} R^{-10}$	$\text{Mont}(A, 1) = \tilde{x}^{11} R^{-11} = x^{11}$

14.96 Note (*computational efficiency of Montgomery exponentiation*)
   (i) Table 14.17 displays the average number of single-precision multiplications required for each step of Algorithm 14.94. The expected number of single-precision multiplications to compute $x^e \bmod m$ by Algorithm 14.94 is $3l(l+1)(t+1)$.
  (ii) Each iteration of step 2 in Algorithm 14.94 applies Algorithm 14.36 at a cost of $2l(l+1)$ single-precision multiplications but no single-precision divisions. A similar algorithm for modular exponentiation based on classical modular multiplication (Algorithm 14.28) would similarly use $2l(l+1)$ single-precision multiplications per iteration but also l single-precision divisions.
 (iii) Any of the other exponentiation algorithms discussed in §14.6.1 can be combined with Montgomery reduction to give other Montgomery exponentiation algorithms.

Step	1	2	3
Number of Montgomery multiplications	1	$\frac{3}{2}t$	1
Number of single-precision multiplications	$2l(l+1)$	$3tl(l+1)$	$l(l+1)$

Table 14.17: *Average number of single-precision multiplications per step of Algorithm 14.94.*

14.6.2 Fixed-exponent exponentiation algorithms

There are numerous situations in which a number of exponentiations by a fixed exponent must be performed. Examples include RSA encryption and decryption, and ElGamal decryption. This subsection describes selected algorithms which improve the repeated square-and-multiply algorithms of §14.6.1 by reducing the number of multiplications.

(i) Addition chains

The purpose of an addition chain is to minimize the number of multiplications required for an exponentiation.

14.97 Definition An *addition chain* V of length s for a positive integer e is a sequence $u_0, u_1, \ldots, u_s$ of positive integers, and an associated sequence $w_1, \ldots, w_s$ of pairs $w_i = (i_1, i_2)$, $0 \le i_1, i_2 < i$, having the following properties:

(i) $u_0 = 1$ and $u_s = e$; and

(ii) for each u_i, $1 \le i \le s$, $u_i = u_{i_1} + u_{i_2}$.

14.98 Algorithm Addition chain exponentiation

INPUT: a group element g, an addition chain $V = (u_0, u_1, \ldots, u_s)$ of length s for a positive integer e, and the associated sequence $w_1, \ldots, w_s$, where $w_i = (i_1, i_2)$.
OUTPUT: g^e.

1. $g_0 \leftarrow g$.
2. For i from 1 to s do: $g_i \leftarrow g_{i_1} \cdot g_{i_2}$.
3. Return(g_s).

14.99 Example (*addition chain exponentiation*) An addition chain of length 5 for $e = 15$ is $u_0 = 1$, $u_1 = 2$, $u_2 = 3$, $u_3 = 6$, $u_4 = 12$, $u_5 = 15$. The following table displays the values of w_i and g_i during each iteration of Algorithm 14.98 for computing g^{15}. □

i	0	1	2	3	4	5
w_i	$-$	$(0,0)$	$(0,1)$	$(2,2)$	$(3,3)$	$(2,4)$
g_i	g	g^2	g^3	g^6	g^{12}	g^{15}

14.100 Remark (*addition chains and binary representations*) Given the binary representation of an exponent e, it is a relatively simple task to construct an addition chain directly from this representation. Chains constructed in this way generally do not provide the shortest addition chain possible for the given exponent. The methods for exponentiation described in §14.6.1 could be phrased in terms of addition chains, but this is typically not done.

14.101 Note (*computational efficiency of addition chain exponentiation*) Given an addition chain of length s for the positive integer e, Algorithm 14.98 computes g^e for any $g \in G$, $g \ne 1$, using exactly s multiplications.

14.102 Fact If l is the length of a shortest addition chain for a positive integer e, then $l \ge (\lg e + \lg \text{wt}(e) - 2.13)$, where $\text{wt}(e)$ is the number of 1's in the binary representation of e. An upper bound of $(\lfloor \lg e \rfloor + \text{wt}(e) - 1)$ is obtained by constructing an addition chain for e from its binary representation. Determining a shortest addition chain for e is known to be an **NP**-hard problem.

(ii) Vector-addition chains

Algorithms 14.88 and 14.104 are useful for computing $g_0^{e_0} g_1^{e_1} \cdots g_{k-1}^{e_{k-1}}$ where $g_0, g_1, \ldots, g_{k-1}$ are arbitrary elements in a group G and $e_0, e_1, \ldots, e_{k-1}$ are fixed positive integers. These algorithms can also be used to advantage when the exponents are not necessarily fixed values (see Note 14.91). Algorithm 14.104 makes use of vector-addition chains.

14.103 Definition Let s and k be positive integers and let v_i denote a k-dimensional vector of non-negative integers. An ordered set $V = \{v_i : -k+1 \leq i \leq s\}$ is called a *vector-addition chain* of length s and dimension k if V satisfies the following:

(i) Each v_i, $-k+1 \leq i \leq 0$, has a 0 in each coordinate position, except for coordinate position $i + k - 1$, which is a 1. (Coordinate positions are labeled 0 through $k - 1$.)

(ii) For each v_i, $1 \leq i \leq s$, there exists an associated pair of integers $w_i = (i_1, i_2)$ such that $-k+1 \leq i_1, i_2 < i$ and $v_i = v_{i_1} + v_{i_2}$ ($i_1 = i_2$ is allowed).

Example 14.105 illustrates a sample vector-addition chain. Let $V = \{v_i : -k+1 \leq i \leq s\}$ be a vector-addition chain of length s and dimension k with associated sequence $w_1, \ldots, w_s$. Algorithm 14.104 computes $g_0^{e_0} g_1^{e_1} \cdots g_{k-1}^{e_{k-1}}$ where $v_s = (e_0, e_1, \ldots, e_{k-1})$.

14.104 Algorithm Vector-addition chain exponentiation

INPUT: group elements $g_0, g_1, \ldots, g_{k-1}$ and a vector-addition chain V of length s and dimension k with associated sequence $w_1, \ldots, w_s$, where $w_i = (i_1, i_2)$.
OUTPUT: $g_0^{e_0} g_1^{e_1} \cdots g_{k-1}^{e_{k-1}}$ where $v_s = (e_0, e_1, \ldots, e_{k-1})$.

1. For i from $(-k+1)$ to 0 do: $a_i \leftarrow g_{i+k-1}$.
2. For i from 1 to s do: $a_i \leftarrow a_{i_1} \cdot a_{i_2}$.
3. Return(a_s).

14.105 Example (*vector-addition chain exponentiation*) A vector-addition chain V of length $s = 9$ and dimension $k = 3$ is displayed in the following table.

v_{-2}	v_{-1}	v_0	v_1	v_2	v_3	v_4	v_5	v_6	v_7	v_8	v_9
1	0	0	1	2	2	3	5	6	12	15	30
0	1	0	0	0	1	1	2	2	4	5	10
0	0	1	1	2	2	2	4	5	10	12	24

The following table displays the values of w_i and a_i during each iteration of step 2 in Algorithm 14.104 for computing $g_0^{30} g_1^{10} g_2^{24}$. Nine multiplications are required. □

i	1	2	3	4	5	6	7	8	9
w_i	$(-2,0)$	$(1,1)$	$(-1,2)$	$(-2,3)$	$(3,4)$	$(1,5)$	$(6,6)$	$(4,7)$	$(8,8)$
a_i	$g_0 g_2$	$g_0^2 g_2^2$	$g_0^2 g_1 g_2^2$	$g_0^3 g_1 g_2^2$	$g_0^5 g_1^2 g_2^4$	$g_0^6 g_1^2 g_2^5$	$g_0^{12} g_1^4 g_2^{10}$	$g_0^{15} g_1^5 g_2^{12}$	$g_0^{30} g_1^{10} g_2^{24}$

14.106 Note (*computational efficiency of vector-addition chain exponentiation*)

(i) (*multiplications*) Algorithm 14.104 performs exactly s multiplications for a vector-addition chain of length s. To compute $g_0^{e_0} g_1^{e_1} \cdots g_{k-1}^{e_{k-1}}$ using Algorithm 14.104, one would like to find a vector-addition chain of length s and dimension k with $v_s = (e_0, e_1, \ldots, e_{k-1})$, where s is as small as possible (see Fact 14.107).

(ii) (*storage*) Algorithm 14.104 requires intermediate storage for the elements a_i, $-k + 1 \leq i < t$, at the tth iteration of step 2. If not all of these are required for succeeding iterations, then they need not be stored. Algorithm 14.88 provides a special case of Algorithm 14.104 where the intermediate storage is no larger than $2^k - 1$ vectors of dimension k.

14.107 Fact The minimum value of s in Note 14.106(i) satisfies the following bound, where $M = \max\{e_i : 0 \leq i \leq k - 1\}$ and c is a constant:

$$s \leq k - 1 + \lg M + ck \cdot \lg M / \lg \lg(M + 2).$$

14.108 Example (*vector-addition chains from binary representations*) The vector-addition chain implicit in Algorithm 14.88 is not necessarily of minimum length. The vector-addition chain associated with Example 14.89 is displayed in Table 14.18. This chain is longer than the one used in Example 14.105. The advantage of Algorithm 14.88 is that the vector-addition chain does not have to be explicitly provided to the algorithm. In view of this, Algorithm 14.88 can be applied more generally to situations where the exponents are not necessarily fixed. □

v_{-2}	v_{-1}	v_0	v_1	v_2	v_3	v_4	v_5	v_6	v_7	v_8	v_9	v_{10}
1	0	0	1	1	1	2	3	6	7	14	15	30
0	1	0	1	1	0	0	1	2	2	4	5	10
0	0	1	0	1	1	2	3	6	6	12	12	24

Table 14.18: *Binary vector-addition chain exponentiation (see Example 14.108).*

14.6.3 Fixed-base exponentiation algorithms

Three methods are presented for exponentiation when the base g is fixed and the exponent e varies. With a fixed base, precomputation can be done once and used for many exponentiations. For example, Diffie-Hellman key agreement (Protocol 12.47) requires the computation of α^x, where α is a fixed element in $\mathbb{Z}_p^*$.

For each of the algorithms described in this section, $\{b_0, b_1, \ldots, b_t\}$ is a set of integers for some $t \geq 0$, such that any exponent $e \geq 1$ (suitably bounded) can be written as $e = \sum_{i=0}^{t} e_i b_i$, where $0 \leq e_i < h$ for some fixed positive integer h. For example, if e is any $(t + 1)$-digit base b integer with $b \geq 2$, then $b_i = b^i$ and $h = b$ are possible choices.

Algorithms 14.109 and 14.113 are two fixed-base exponentiation methods. Both require precomputation of the exponentials $g^{b_0}, g^{b_1}, \ldots, g^{b_t}$, e.g., using one of the algorithms from §14.6.1. The precomputation needed for Algorithm 14.117 is more involved and is explicitly described in Algorithm 14.116.

(i) Fixed-base windowing method

Algorithm 14.109 takes as input the precomputed exponentials $g_i = g^{b_i}$, $0 \leq i \leq t$, and positive integers h and $e = \sum_{i=0}^{t} e_i b_i$ where $0 \leq e_i < h$, $0 \leq i \leq t$. The basis for the algorithm is the observation that $g^e = \prod_{i=0}^{t} g_i^{e_i} = \prod_{j=1}^{h-1} (\prod_{e_i=j} g_i)^j$.

14.109 Algorithm Fixed-base windowing method for exponentiation

INPUT: $\{g^{b_0}, g^{b_1}, \ldots, g^{b_t}\}$, $e = \sum_{i=0}^{t} e_i b_i$, and h.
OUTPUT: g^e.

1. $A \leftarrow 1$, $B \leftarrow 1$.
2. For j from $(h-1)$ down to 1 do the following:
   2.1 For each i for which $e_i = j$ do: $B \leftarrow B \cdot g^{b_i}$.
   2.2 $A \leftarrow A \cdot B$.
3. Return(A).

14.110 Example (*fixed-base windowing exponentiation*) Precompute the group elements g^1, g^4, g^{16}, g^{64}, g^{256}. To compute g^e for $e = 862 = (31132)_4$, take $t = 4$, $h = 4$, and $b_i = 4^i$ for $0 \le i \le 4$, in Algorithm 14.109. The following table displays the values of A and B at the end of each iteration of step 2. $\square$

j	$-$	3	2	1
B	1	$g^4 g^{256} = g^{260}$	$g^{260} g = g^{261}$	$g^{261} g^{16} g^{64} = g^{341}$
A	1	g^{260}	$g^{260} g^{261} = g^{521}$	$g^{521} g^{341} = g^{862}$

14.111 Note (*computational efficiency of fixed-base windowing exponentiation*)
   (i) (*number of multiplications*) Suppose $t + h \ge 2$. Only multiplications where both operands are distinct from 1 are counted. Step 2.2 is executed $h - 1$ times, but at least one of these multiplications involves an operand with value 1 (A is initialized to 1). Since B is also initially 1, at most t multiplications are done in step 2.1. Thus, Algorithm 14.109 computes g^e with at most $t + h - 2$ multiplications (cf. Note 14.112).
   (ii) (*storage*) Storage is required for the $t + 1$ group elements g_i, $0 \le i \le t$.

14.112 Note (*a particular case*) The most obvious application of Algorithm 14.109 is the case where the exponent e is represented in radix b. If $e = \sum_{i=0}^{t} e_i b^i$, then $g_i = g^{b^i}$, $0 \le i \le t$, are precomputed. If e is randomly selected from $\{0, 1, \ldots, m-1\}$, then $t + 1 \le \lceil \log_b m \rceil$ and, on average, $\frac{1}{b}$ of the base b digits in e will be 0. In this case, the expected number of multiplications is $\frac{b-1}{b} \lceil \log_b m \rceil + b - 3$. If m is a 512-bit integer and $b = 32$, then 128.8 multiplications are needed on average, 132 in the worst case; 103 values must be stored.

(ii) Fixed-base Euclidean method

Let $\{x_0, x_1, \ldots, x_t\}$ be a set of integers with $t \ge 2$. Define M to be an integer in the interval $[0, t]$ such that $x_M \ge x_i$ for all $0 \le i \le t$. Define N to be an integer in the interval $[0, t]$, $N \ne M$, such that $e_N \ge e_i$ for all $0 \le i \le t$, $i \ne M$.

14.113 Algorithm Fixed-base Euclidean method for exponentiation

INPUT: $\{g^{b_0}, g^{b_1}, \ldots, g^{b_t}\}$ and $e = \sum_{i=0}^{t} e_i b_i$.
OUTPUT: g^e.

1. For i from 0 to t do the following: $g_i \leftarrow g^{b_i}$, $x_i \leftarrow e_i$.
2. Determine the indices M and N for $\{x_0, x_1, \ldots, x_t\}$.
3. While $x_N \ne 0$ do the following:
   3.1 $q \leftarrow \lfloor x_M / x_N \rfloor$, $g_N \leftarrow (g_M)^q \cdot g_N$, $x_M \leftarrow x_M \bmod x_N$.

3.2 Determine the indices M and N for $\{x_0, x_1, \ldots, x_t\}$.

4. Return($g_M^{x_M}$).

14.114 Example (*fixed-base Euclidean method*) This example repeats the computation of g^e, $e = 862$ done in Example 14.110, but now uses Algorithm 14.113. Take $b_0 = 1$, $b_1 = 16$, $b_2 = 256$. Then $e = (3, 5, 14)_{16}$. Precompute g^1, g^{16}, g^{256}. Table 14.19 illustrates the steps performed by Algorithm 14.113. Notice that for this example, Algorithm 14.113 does 8

x_0	x_1	x_2	M	N	q	g_0	g_1	g_2
14	5	3	0	1	2	g	g^{18}	g^{256}
4	5	3	1	0	1	g^{19}	g^{18}	g^{256}
4	1	3	0	2	1	g^{19}	g^{18}	g^{275}
1	1	3	2	1	3	g^{19}	g^{843}	g^{275}
1	1	0	0	1	1	g^{19}	g^{862}	g^{275}
0	1	0	1	0	–	g^{19}	g^{862}	g^{275}

Table 14.19: *Fixed-base Euclidean method to compute g^{862} (see Example 14.114).*

multiplications, whereas Algorithm 14.109 needs only 6 to do the same computation. Storage requirements for Algorithm 14.113 are, however, smaller. The vector-addition chain (Definition 14.103) corresponding to this example is displayed in the following table. □

v_{-2}	v_{-1}	v_0	v_1	v_2	v_3	v_4	v_5	v_6	v_7	v_8
1	0	0	2	2	3	3	6	9	11	14
0	1	0	0	1	1	1	2	3	4	5
0	0	1	0	0	0	1	2	3	3	3

14.115 Note (*fixed-base Euclidean vs. fixed-base windowing methods*)

(i) In most cases, the quotient q computed in step 3.1 of Algorithm 14.113 is 1. For a given base b, the computational requirements of this algorithm are not significantly greater than those of Algorithm 14.109.

(ii) Since the division algorithm is logarithmic in the size of the inputs, Algorithm 14.113 can take advantage of a larger value of h than Algorithm 14.109. This results in less storage for precomputed values.

(iii) Fixed-base comb method

Algorithm 14.117 computes g^e where $e = (e_t e_{t-1} \cdots e_1 e_0)_2$, $t \geq 1$. Select an integer h, $1 \leq h \leq t+1$ and compute $a = \lceil (t+1)/h \rceil$. Select an integer v, $1 \leq v \leq a$, and compute $b = \lceil a/v \rceil$. Clearly, $ah \geq t + 1$. Let $X = R_{h-1}||R_{h-2}|| \cdots ||R_0$ be a bitstring formed from e by padding (if necessary) e on the left with 0's, so that X has bitlength ah and each R_i, $0 \leq i \leq h - 1$, is a bitstring of length a. Form an $h \times a$ array EA (called the *exponent array*) where row i of EA is the bitstring R_i, $0 \leq i \leq h - 1$. Algorithm 14.116 is the precomputation required for Algorithm 14.117.

14.116 Algorithm Precomputation for Algorithm 14.117

INPUT: group element g and parameters h, v, a, and b (defined above).
OUTPUT: $\{G[j][i] : 1 \le i < 2^h, 0 \le j < v\}$.

1. For i from 0 to $(h-1)$ do: $g_i \leftarrow g^{2^{ia}}$.
2. For i from 1 to $(2^h - 1)$ (where $i = (i_{h-1} \cdots i_0)_2$), do the following:
   2.1 $G[0][i] \leftarrow \prod_{j=0}^{h-1} g_j^{i_j}$.
   2.2 For j from 1 to $(v-1)$ do: $G[j][i] \leftarrow (G[0][i])^{2^{jb}}$.
3. Return($\{G[j][i] : 1 \le i < 2^h, 0 \le j < v\}$).

Let $I_{j,k}$, $0 \le k < b$, $0 \le j < v$, be the integer whose binary representation is column $(jb+k)$ of EA, where column 0 is on the right and the least significant bits of a column are at the top.

14.117 Algorithm Fixed-base comb method for exponentiation

INPUT: g, e and $\{G[j][i] : 1 \le i < 2^h, 0 \le j < v\}$ (precomputed in Algorithm 14.116).
OUTPUT: g^e.

1. $A \leftarrow 1$.
2. For k from $(b-1)$ down to 0 do the following:
   2.1 $A \leftarrow A \cdot A$.
   2.2 For j from $(v-1)$ down to 0 do: $A \leftarrow G[j][I_{j,k}] \cdot A$.
3. Return(A).

14.118 Example (*fixed-base comb method for exponentiation*) Let $t = 9$ and $h = 3$; then $a = \lceil 10/3 \rceil = 4$. Let $v = 2$; then $b = \lceil a/v \rceil = 2$. Suppose the exponent input to Algorithm 14.117 is $e = (e_9 e_8 \cdots e_1 e_0)_2$. Form the bitstring $X = x_{11} x_{10} \cdots x_1 x_0$ where $x_i = e_i$, $0 \le i \le 9$, and $x_{11} = x_{10} = 0$. The following table displays the exponent array EA.

$I_{1,1}$	$I_{1,0}$	$I_{0,1}$	$I_{0,0}$
x_3	x_2	x_1	x_0
x_7	x_6	x_5	x_4
x_{11}	x_{10}	x_9	x_8

The precomputed values from Algorithm 14.116 are displayed below. Recall that $g_i = g^{2^{ia}}$, $0 \le i < 3$.

i	1	2	3	4	5	6	7
$G[0][i]$	g_0	g_1	$g_1 g_0$	g_2	$g_2 g_0$	$g_2 g_1$	$g_2 g_1 g_0$
$G[1][i]$	g_0^4	g_1^4	$g_1^4 g_0^4$	g_2^4	$g_2^4 g_0^4$	$g_2^4 g_1^4$	$g_2^4 g_1^4 g_0^4$

Finally, the following table displays the steps in Algorithm 14.117 for EA.

k	j	l_0	l_1	l_2
			$A = g_0^{l_0} g_1^{l_1} g_2^{l_2}$	
1	–	0	0	0
1	1	$4x_3$	$4x_7$	$4x_{11}$
1	0	$4x_3 + x_1$	$4x_7 + x_5$	$4x_{11} + x_9$
0	–	$8x_3 + 2x_1$	$8x_7 + 2x_5$	$8x_{11} + 2x_9$
0	1	$8x_3 + 2x_1 + 4x_2$	$8x_7 + 2x_5 + 4x_6$	$8x_{11} + 2x_9 + 4x_{10}$
0	0	$8x_3 + 2x_1 + 4x_2 + x_0$	$8x_7 + 2x_5 + 4x_6 + x_4$	$8x_{11} + 2x_9 + 4x_{10} + x_8$

The last row of the table corresponds to $g^{\sum_{i=0}^{11} x_i 2^i} = g^e$. $\qquad\square$

14.119 Note (*computational efficiency of fixed-base comb method*)

    (i) (*number of multiplications*) Algorithm 14.117 requires at most one multiplication for each column of EA. The right-most column of EA requires a multiplication with the initial value 1 of the accumulator A. The algorithm also requires a squaring of the accumulator A for each k, $0 \leq k < b$, except for $k = b - 1$ when A has value 1. Discounting multiplications by 1, the total number of non-trivial multiplications (including squarings) is, at most, $a + b - 2$.

    (ii) (*storage*) Algorithm 14.117 requires storage for the $v(2^h - 1)$ precomputed group elements (Algorithm 14.116). If squaring is a relatively simple operation compared to multiplication in the group, then some space-saving can be achieved by storing only $2^h - 1$ group elements (i.e., only those elements computed in step 2.1 of Algorithm 14.116).

    (iii) (*trade-offs*) Since h and v are independent of the number of bits in the exponent, selection of these parameters can be made based on the amount of storage available vs. the amount of time (determined by multiplication) to do the computation.

14.7 Exponent recoding

Another approach to reducing the number of multiplications in the basic repeated square-and-multiply algorithms (§14.6.1) is to replace the binary representation of the exponent e with a representation which has fewer non-zero terms. Since the binary representation is unique (Fact 14.1), finding a representation with fewer non-zero components necessitates the use of digits besides 0 and 1. Transforming an exponent from one representation to another is called *exponent recoding*. Many techniques for exponent recoding have been proposed in the literature. This section describes two possibilities: signed-digit representation (§14.7.1) and string-replacement representation (§14.7.2).

14.7.1 Signed-digit representation

14.120 Definition If $e = \sum_{i=0}^{t} d_i 2^i$ where $d_i \in \{0, 1, -1\}$, $0 \leq i \leq t$, then $(d_t \cdots d_1 d_0)_{SD}$ is called a *signed-digit representation with radix* 2 for the integer e.

    Unlike the binary representation, the signed-digit representation of an integer is not unique. The binary representation is an example of a signed-digit representation. Let e be a positive integer whose binary representation is $(e_{t+1} e_t e_{t-1} \cdots e_1 e_0)_2$, with $e_{t+1} = e_t = 0$. Algorithm 14.121 constructs a signed-digit representation for e having at most $t + 1$ digits and the smallest possible number of non-zero terms.

14.121 Algorithm Signed-digit exponent recoding

INPUT: a positive integer $e = (e_{t+1}e_t e_{t-1} \cdots e_1 e_0)_2$ with $e_{t+1} = e_t = 0$.
OUTPUT: a signed-digit representation $(d_t \cdots d_1 d_0)_{SD}$ for e. (See Definition 14.120.)

1. $c_0 \leftarrow 0$.
2. For i from 0 to t do the following:
   2.1 $c_{i+1} \leftarrow \lfloor (e_i + e_{i+1} + c_i)/2 \rfloor$, $d_i \leftarrow e_i + c_i - 2c_{i+1}$.
3. Return$((d_t \cdots d_1 d_0)_{SD})$.

14.122 Example (*signed-digit exponent recoding*) Table 14.20 lists all possible inputs to the ith iteration of step 2, and the corresponding outputs. If $e = (1101110111)_2$, then Algorithm 14.121 produces the signed-digit representation $e = (100\bar{1}000\bar{1}00\bar{1})_{SD}$ where $\bar{1} = -1$. Note that $e = 2^9 + 2^8 + 2^6 + 2^5 + 2^4 + 2^2 + 2 + 1 = 2^{10} - 2^7 - 2^3 - 1$. $\square$

inputs	e_i	0	0	0	0	1	1	1	1
	c_i	0	0	1	1	0	0	1	1
	e_{i+1}	0	1	0	1	0	1	0	1
outputs	c_{i+1}	0	0	0	1	0	1	1	1
	d_i	0	0	1	-1	1	-1	0	0

Table 14.20: *Signed-digit exponent recoding (see Example 14.122).*

14.123 Definition A signed-digit representation of an integer e is said to be *sparse* if no two non-zero entries are adjacent in the representation.

14.124 Fact (*sparse signed-digit representation*)
   (i) Every integer e has a unique sparse signed-digit representation.
   (ii) A sparse signed-digit representation for e has the smallest number of non-zero entries among all signed-digit representations for e.
   (iii) The signed-digit representation produced by Algorithm 14.121 is sparse.

14.125 Note (*computational efficiency of signed-digit exponent recoding*)
   (i) Signed-digit exponent recoding as per Algorithm 14.121 is very efficient, and can be done by table look-up (using Table 14.20).
   (ii) When e is given in a signed-digit representation, computing g^e requires both g and g^{-1}. If g is a fixed base, then g^{-1} can be precomputed. For a variable base g, unless g^{-1} can be computed very quickly, recoding an exponent to signed-digit representation may not be worthwhile.

14.7.2 String-replacement representation

14.126 Definition Let $k \geq 1$ be a positive integer. A non-negative integer e is said to have a *k-ary string-replacement representation* $(f_{t-1}f_{t-2} \cdots f_1 f_0)_{SR(k)}$, denoted $SR(k)$, if $e = \sum_{i=0}^{t-1} f_i 2^i$ and $f_i \in \{2^j - 1 : 0 \leq j \leq k\}$ for $0 \leq i \leq t - 1$.

14.127 Example (*non-uniqueness of string-replacement representations*) A string-replacement representation for a non-negative integer is generally not unique. The binary representation is a 1-ary string-replacement representation. If $k = 3$ and $e = 987 = (1111011011)_2$, then some other string-replacements of e are $(303003003)_{SR(3)}$, $(1007003003)_{SR(3)}$, and $(71003003)_{SR(3)}$. □

14.128 Algorithm k-ary string-replacement representation

INPUT: $e = (e_{t-1}e_{t-2} \cdots e_1 e_0)_2$ and positive integer $k \geq 2$.
OUTPUT: $e = (f_{t-1}f_{t-2} \cdots f_1 f_0)_{SR(k)}$.

1. For i from k down to 2 do the following: starting with the most significant digit of $e = (e_{t-1}e_{t-2} \cdots e_1 e_0)_2$, replace each consecutive string of i ones with a string of length i consisting of $i - 1$ zeros in the high-order string positions and the integer $2^i - 1$ in the low-order position.
2. Return($(f_{t-1}f_{t-2} \cdots f_1 f_0)_{SR(k)}$).

14.129 Example (*k-ary string-replacement*) Suppose $e = (110111110011101)_2$ and $k = 3$. The $SR(3)$ representations of e at the end of each of the two iterations of Algorithm 14.128 are $(110007110000701)_{SR(3)}$ and $(030007030000701)_{SR(3)}$. □

14.130 Algorithm Exponentiation using an $SR(k)$ representation

INPUT: an integer $k \geq 2$, an element $g \in G$, and $e = (f_{t-1}f_{t-2} \cdots f_1 f_0)_{SR(k)}$.
OUTPUT: g^e.

1. *Precomputation.* Set $g_1 \leftarrow g$. For i from 2 to k do: $g_{2^i-1} \leftarrow (g_{2^{i-1}-1})^2 \cdot g$.
2. $A \leftarrow 1$.
3. For i from $(t - 1)$ down to 0 do the following:
   3.1 $A \leftarrow A \cdot A$.
   3.2 If $f_i \neq 0$ then $A \leftarrow A \cdot g_{f_i}$.
4. Return(A).

14.131 Example (*$SR(k)$ vs. left-to-right binary exponentiation*) Let $e = 987 = (1111011011)_2$ and consider the 3-ary string-replacement representation $(0071003003)_{SR(3)}$. Computing g^e using Algorithm 14.79 requires 9 squarings and 7 multiplications. Algorithm 14.130 requires 2 squarings and 2 multiplications for computing g^3 and g^7, and then 7 squarings and 3 multiplications for the main part of the algorithm. In total, the $SR(3)$ for e computes g^e with 9 squarings and 5 multiplications. □

14.132 Note (*computational efficiency of Algorithm 14.130*) The precomputation requires $k - 1$ squarings and $k - 1$ multiplications. Algorithm 14.128 is not guaranteed to produce an $SR(k)$ representation with a minimum number of non-zero entries, but in practice it seems to give representations which are close to minimal. Heuristic arguments indicate that a randomly selected t-bit exponent will be encoded with a suitably chosen value of k to an $SR(k)$ representation having about $t/4$ non-zero entries; hence, one expects to perform $t - 1$ squarings in step 3, and about $t/4$ multiplications.

14.8 Notes and further references

§14.1

This chapter deals almost exclusively with methods to perform operations in the integers and the integers modulo some positive integer. When p is a prime number, $\mathbb{Z}_p$ is called a finite field (Fact 2.184). There are other finite fields which have significance in cryptography. Of particular importance are those of characteristic two, $\mathbb{F}_{2^m}$. Perhaps the most useful property of these structures is that squaring is a *linear operator* (i.e., if $\alpha, \beta \in \mathbb{F}_{2^m}$, then $(\alpha + \beta)^2 = \alpha^2 + \beta^2$). This property leads to efficient methods for exponentiation and for inversion. Characteristic two finite fields have been used extensively in connection with error-correcting codes; for example, see Berlekamp [118] and Lin and Costello [769]. For error-correcting codes, m is typically quite small (e.g., $1 \leq m \leq 16$); for cryptographic applications, m is usually much larger (e.g., $m \geq 100$).

The majority of the algorithms presented in this chapter are best suited to software implementations. There is a vast literature on methods to perform modular multiplication and other operations in hardware. The basis for most hardware implementations for modular multiplication is efficient methods for integer addition. In particular, *carry-save adders* and *delayed-carry adders* are at the heart of the best methods to perform modular multiplication. The concept of a delayed-carry adder was proposed by Norris and Simmons [933] to produce a hardware modular multiplier which computes the product of two t-bit operands modulo a t-bit modulus in $2t$ clock cycles. Brickell [199] improved the idea to produce a modular multiplier requiring only $t + 7$ clock cycles. Enhancements of Brickell's method were given by Walter [1230]. Koç [699] gives a comprehensive survey of hardware methods for modular multiplication.

§14.2

For a treatment of radix representations including *mixed-radix representations*, see Knuth [692]. Knuth describes efficient methods for performing radix conversions. Representing and manipulating negative numbers is an important topic; for an introduction, consult the book by Koren [706].

The techniques described in §14.2 are commonly referred to as the *classical algorithms* for multiple-precision addition, subtraction, multiplication, and division. These algorithms are the most useful for integers of the size used for cryptographic purposes. For much larger integers (on the order of thousands of decimal digits), more efficient methods exist. Although not of current practical interest, some of these may become more useful as security requirements force practitioners to increase parameter sizes. A brief description of one of these methods follows.

The classical algorithm for multiplication (Algorithm 14.12) takes $O(n^2)$ bit operations for multiplying two n-bit integers. A recursive algorithm due to Karatsuba and Ofman [661] reduces the complexity of multiplying two n-bit integers to $O(n^{1.58})$. This *divide-and-conquer* method is based on the following simple observation. Suppose that x and y are n-bit integers and $n = 2t$. Then $x = 2^t x_1 + x_0$ and $y = 2^t y_1 + y_0$, where x_1, y_1 are the t high-order bits of x and y, respectively, and x_0, y_0 are the t low-order bits. Furthermore, $x \cdot y = u_2 2^{2t} + u_1 2^t + u_0$ where $u_0 = x_0 \cdot y_0$, $u_2 = x_1 \cdot y_1$ and $u_1 = (x_0 + x_1) \cdot (y_0 + y_1) - u_0 - u_2$. It follows that $x \cdot y$ can be computed by performing three multiplications of t-bit integers (as opposed to one multiplication with $2t$-bit integers) along with two additions and two subtractions. For large values of t, the cost of the additions and subtractions is insignificant relative to the cost of the multiplications. With appropriate modifications, u_0, u_1 and

$(x_0 + x_1) \cdot (y_0 + y_1)$ can each be computed similarly. This procedure is continued on the intermediate values until the size of the integers reaches the word size of the computing device, and multiplication can be efficiently accomplished. Due to the recursive nature of the algorithm, a number of intermediate results must be stored which can add significant overhead, and detract from the algorithm's efficiency for relatively small integers. Combining the Karatsuba-Ofman method with classical multiplication may have some practical significance. For a more detailed treatment of the Karatsuba-Ofman algorithm, see Knuth [692], Koç [698], and Geddes, Czapor, and Labahn [445].

Another commonly used method for multiple-precision integer multiplication is the *discrete Fourier transform* (DFT). Although mathematically elegant and asymptotically better than the classical algorithm, it does not appear to be superior for the size of integers of practical importance to cryptography. Lipson [770] provides a well-motivated and easily readable treatment of this method.

The identity given in Note 14.18 was known to Karatsuba and Ofman [661].

§14.3

There is an extensive literature on methods for multiple-precision modular arithmetic. A detailed treatment of methods for performing modular multiplication can be found in Knuth [692]. Koç [698] and Bosselaers, Govaerts, and Vandewalle [176] provide comprehensive but brief descriptions of the classical method for modular multiplication.

Montgomery reduction (Algorithm 14.32) is due to Montgomery [893], and is one of the most widely used methods in practice for performing modular exponentiation (Algorithm 14.94). Dussé and Kaliski [361] discuss variants of Montgomery's method. Montgomery reduction is a generalization of a much older technique due to Hensel (see Shand and Vuillemin [1119] and Bosselaers, Govaerts, and Vandewalle [176]). Hensel's observation is the following. If m is an odd positive integer less than 2^k (k a positive integer) and T is some integer such that $2^k \leq T < 2^{2k}$, then $R_0 = (T + q_0 m)/2$, where $q_0 = T \bmod 2$ is an integer and $R_0 \equiv T2^{-1} \bmod m$. More generally, $R_i = (R_{i-1} + q_i m)/2$, where $q_i = R_{i-1} \bmod 2$ is an integer and $R_i \equiv N2^{-i+1} \bmod m$. Since $T < 2^{2k}$, it follows that $R_{k-1} < 2m$.

Barrett reduction (Algorithm 14.42) is due to Barrett [75]. Bosselaers, Govaerts, and Vandewalle [176] provide a clear and concise description of the algorithm along with motivation and justification for various choices of parameters and steps, and compare three alternative methods: classical (§14.3.1), Montgomery reduction (§14.3.2), and Barrett reduction (§14.3.3). This comparison indicates that there is not a significant difference in performance between the three methods, provided the precomputation necessary for Montgomery and Barrett reduction is ignored. Montgomery exponentiation is shown to be somewhat better than the other two methods. The conclusions are based on both theoretical analysis and machine implementation for various sized moduli. Koç, Acar, and Kaliski [700] provide a more detailed comparison of various Montgomery multiplication algorithms; see also Naccache, M'Raïhi, and Raphaeli [915]. Naccache and M'silti [917] provide proofs for the correctness of Barrett reduction along with a possible optimization.

Mohan and Adiga [890] describe a special case of Algorithm 14.47 where $b = 2$.

Hong, Oh, and Yoon [561] proposed new methods for modular multiplication and modular squaring. They report improvements of 50% and 30%, respectively, on execution times over Montgomery's method for multiplication and squaring. Their approach to modular multiplication interleaves multiplication and modular reduction and uses precomputed tables such that one operand is always single-precision. Squaring uses recursion and precomputed tables and, unlike Montgomery's method, also integrates the multiplication and reduction steps.

§14.4

The binary gcd algorithm (Algorithm 14.54) is due to Stein [1170]. An analysis of the algorithm is given by Knuth [692]. Harris [542] proposed an algorithm for computing gcd's which combines the classical Euclidean algorithm (Algorithm 2.104) and binary operations; the method is called the *binary Euclidean algorithm*.

Lehmer's gcd algorithm (Algorithm 14.57), due to Lehmer [743], determines the gcd of two positive multiple-precision integers using mostly single-precision operations. This has the advantage of using the hardware divide in the machine and only periodically resorting to an algorithm such as Algorithm 14.20 for a multiple-precision divide. Knuth [692] gives a comprehensive description of the algorithm along with motivation of its correctness. Cohen [263] provides a similar discussion, but without motivation. Lehmer's gcd algorithm is readily adapted to the extended Euclidean algorithm (Algorithm 2.107).

According to Sorenson [1164], the binary gcd algorithm is the most efficient method for computing the greatest common divisor. Jebelean [633] suggests that Lehmer's gcd algorithm is more efficient. Sorenson [1164] also describes a k-ary version of the binary gcd algorithm, and proves a worst-case running time of $O(n^2/\lg n)$ bit operations for computing the gcd of two n-bit integers.

The binary extended gcd algorithm was first described by Knuth [692], who attributes it to Penk. Algorithm 14.61 is due to Bach and Shallit [70], who also give a comprehensive and clear analysis of several gcd and extended gcd algorithms. Norton [934] described a version of the binary extended gcd algorithm which is somewhat more complicated than Algorithm 14.61. Gordon [516] proposed a method for computing modular inverses, derived from the classical extended Euclidean algorithm (Algorithm 2.107) with multiple-precision division replaced by an approximation to the quotient by an appropriate power of 2; no analysis of the expected running time is given, but observed results on moduli of specific sizes are described.

The *Montgomery inverse* of $a \bmod m$ is defined to be $a^{-1}2^t \bmod m$ where t is the bitlength of m. Kaliski [653] extended ideas of Guyot [534] on the right-shift binary extended Euclidean algorithm, and presented an algorithm for computing the Montgomery inverse.

§14.5

Let m_i, $1 \le i \le t$, be a set of pairwise relatively prime positive integers which define a residue number system (RNS). If $n = \prod_{i=1}^{t} m_i$ then this RNS provides an effective method for computing the product of integers modulo n where the integers and the product are represented in the RNS. If n is a positive integer where the m_i do not necessarily divide n, then a method for performing arithmetic modulo n entirely within the RNS is not obvious. Couveignes [284] and Montgomery and Silverman [895] propose an interesting method for accomplishing this. Further research in the area is required to determine if this approach is competitive with or better than the modular multiplication methods described in §14.3.

Algorithm 14.71 is due to Garner [443]. A detailed discussion of this algorithm and variants of it are given by Knuth [692]; see also Cohen [263]. Algorithm 2.121 for applying the Chinese remainder theorem is due to Gauss; see Bach and Shallit [70]. Gauss's algorithm is a special case of the following result due to Nagasaka, Shiue, and Ho [918]. The solution to the system of linear congruences $x \equiv a_i \pmod{m_i}$, $1 \le i \le t$, for pairwise relative prime moduli m_i, is equivalent to the solution to the single linear congruence $(\sum_{i=1}^{t} b_i M_i)x \equiv \sum_{i=1}^{t} a_i b_i M_i \pmod{M}$ where $M = \prod_{i=1}^{t} m_i$, $M_i = M/m_i$

for $1 \leq i \leq t$, for any choice of integers b_i where gcd $(b_i, M_i) = 1$. Notice that if $\sum_{i=1}^{t} b_i M_i \equiv 1 \pmod{M}$, then $b_i \equiv M_i^{-1} \pmod{m_i}$, giving the special case discussed in Algorithm 2.121. Quisquater and Couvreur [1016] were the first to apply the Chinese remainder theorem to RSA decryption and signature generation.

§14.6

Knuth [692] and Bach and Shallit [70] describe the right-to-left binary exponentiation method (Algorithm 14.76). Cohen [263] provides a more comprehensive treatment of the right-to-left and left-to-right (Algorithm 14.79) binary methods along with their generalizations to the k-ary method. Koç [698] discusses these algorithms in the context of the RSA public-key cryptosystem. Algorithm 14.92 is the basis for Blakley's modular multiplication algorithm (see Blakley [149] and Koç [698]). The generalization of Blakley's method to process more than one bit per iteration (Note 14.93(iii)) is due to Quisquater and Couvreur [1016]. Quisquater and Couvreur describe an algorithm for modular exponentiation which makes use of the generalization and precomputed tables to accelerate multiplication in $\mathbb{Z}_m$.

For a comprehensive and detailed discussion of addition chains, see Knuth [692], where various methods for constructing addition chains (such as the *power tree* and *factor* methods) are described. Computing the shortest addition chain for a positive integer was shown to be an **NP**-hard problem by Downey, Leong, and Sethi [360]. The lower bound on the length of a shortest addition chain (Fact 14.102) was proven by Schönhage [1101].

An *addition sequence* for positive integers $a_1 < a_2 < \cdots < a_k$ is an addition chain for a_k in which $a_1, a_2, \ldots, a_{k-1}$ appear. Yao [1257] proved that there exists an addition sequence for $a_1 < a_2 < \cdots < a_k$ of length less than $\lg a_k + ck \cdot \lg a_k / \lg \lg(a_k + 2)$ for some constant c. Olivos [955] established a 1-1 correspondence between addition sequences of length l for $a_1 < a_2 < \cdots < a_k$ and vector-addition chains of length $l + k - 1$ where $v_{l+k-1} = (a_1, a_2, \ldots, a_k)$. These results are the basis for the inequality given in Fact 14.107. Bos and Coster [173] described a heuristic method for computing vector-addition chains. The special case of Algorithm 14.104 (Algorithm 14.88) is attributed by ElGamal [368] to Shamir.

The fixed-base windowing method (Algorithm 14.109) for exponentiation is due to Brickell et al. [204], who describe a number of variants of the basic algorithm. For b a positive integer, let S be a set of integers with the property that any integer can be expressed in base b using only coefficients from S. S is called a *basic digit set* for the base b. Brickell et al. show how basic digit sets can be used to reduce the amount of work in Algorithm 14.109 without large increases in storage requirements. De Rooij [316] proposed the fixed-base Euclidean method (Algorithm 14.113) for exponentiation; compares this algorithm to Algorithm 14.109; and provides a table of values for numbers of practical importance. The fixed-base comb method (Algorithm 14.117) for exponentiation is due to Lim and Lee [767]. For a given exponent size, they discuss various possibilities for the choice of parameters h and v, along with a comparison of their method to fixed-base windowing.

§14.7

The signed-digit exponent recoding algorithm (Algorithm 14.121) is due to Reitwiesner [1031]. A simpler description of the algorithm was given by Hwang [566]. Booth [171] described another algorithm for producing a signed-digit representation, but not necessarily one with the minimum possible non-zero components. It was originally given in terms of the additive group of integers where exponentiation is referred to as multiplication. In this case, $-g$ is easily computed from g. The additive abelian group formed from the points on an elliptic curve over a finite field is another example where signed-digit representation is very useful (see Morain and Olivos [904]). Zhang [1267] described a modified signed-digit representation which requires on average $t/3$ multiplications for a square-and-multiply algorithm for t-bit exponents. A slightly more general version of Algorithm 14.121, given by Jedwab and Mitchell [634], does not require as input a binary representation of the exponent e but simply a signed-digit representation. For binary inputs, the algorithms of Reitwiesner and Jedwab-Mitchell are the same. Fact 14.124 is due to Jedwab and Mitchell [634].

String-replacement representations were introduced by Gollmann, Han, and Mitchell [497], who describe Algorithms 14.128 and 14.130. They also provide an analysis of the expected number of non-zero entries in an $SR(k)$ representation for a randomly selected t-bit exponent (see Note 14.132), as well as a complexity analysis of Algorithm 14.130 for various values of t and k. Lam and Hui [735] proposed an alternate string-replacement algorithm. The idea is to precompute all odd powers $g, g^3, g^5, \ldots, g^{2^k-1}$ for some fixed positive integer k. Given a t-bit exponent e, start at the most significant bit, and look for the longest bitstring of bitlength at most k whose last digit is a 1 (i.e., this substring represents an odd positive integer between 1 and $2^k - 1$). Applying a left-to-right square-and-multiply exponentiation algorithm based on this scanning process results in an algorithm which requires, at most, $\lceil t/k \rceil$ multiplications. Lam and Hui proved that as t increases, the average number of multiplications approaches $\lceil t/(k+1) \rceil$.

Chapter 15

Patents and Standards

Contents in Brief

15.1 Introduction

This chapter discusses two topics which have significant impact on the use of cryptography in practice: patents and standards. At their best, cryptographic patents make details of significant new processes and efficient techniques publicly available, thereby increasing awareness and promoting use; at their worst, they limit or stifle the use of such techniques due to licensing requirements. Cryptographic standards serve two important goals: facilitating widespread use of cryptographically sound and well-accepted techniques; and promoting interoperability between components involving security mechanisms in various systems.

An overview of patents is given in §15.2. Standards are pursued in §15.3. Notes and further references follow in §15.4.

15.2 Patents on cryptographic techniques

A vast number of cryptographic patents have been issued, of widely varying significance and use. Here attention is focused on a subset of these with primary emphasis on unexpired patents of industrial interest, involving fundamental techniques and specific algorithms and protocols. In addition, some patents of historical interest are noted.

Where appropriate, a brief description of major claims or disclosed techniques is given. Inclusion herein is intended to provide reference information to practitioners on the existence and content of well-known patents, and to illustrate the nature of cryptographic patents in general. There is no intention to convey any judgement on the validity of any claims.

Because most patents are eventually filed in the United States, U.S. patent numbers and associated details are given. Additional information including related filings in other countries may be found in patent databases. For further technical details, the original patents should be consulted (see §15.2.4). Where details of patented techniques and algorithms appear elsewhere in this book, cross-references are given.

Expiry of patents

U.S. patents are valid for 17 years from the date of issue, or 20 years from the date a patent application was filed. For applications filed before June 8 1995 (and unexpired at that point), the longer period applies; the 20-year rule applies for applications filed after this date.

Priority data

Many countries require that a patent be filed before any public disclosure of the invention; in the USA, the filing must be within one year of disclosure. A large number of countries are parties to a patent agreement which recognizes *priority dates*. A patent filed in such a country, and filed in another such country within one year thereof, may claim the date of the first filing as a priority date for the later filing.

Outline of patents section

The discussion of patents is broken into three main subsections. §15.2.1 notes five fundamental patents, including DES and basic patents on public-key cryptography. §15.2.2 addresses ten prominent patents including those on well-known block ciphers, hash functions, identification and signature schemes. §15.2.3 includes ten additional patents addressing various techniques, of historical or practical interest. Finally, §15.2.4 provides information on ordering patents.

15.2.1 Five fundamental patents

Table 15.1 lists five basic cryptographic patents which are fundamental to current cryptographic practice, three involving basic ideas of public-key cryptography. These patents are discussed in chronological order.

Inventors	Patent #	Issue date	Ref.	Major claim or area
Ehrsam et al.	3,962,539	Jun. 08 1976	[363]	DES
Hellman-Diffie-Merkle	4,200,770	Apr. 29 1980	[551]	Diffie-Hellman agreement
Hellman-Merkle	4,218,582	Aug. 19 1980	[553]	public-key systems
Merkle	4,309,569	Jan. 05 1982	[848]	tree authentication
Rivest-Shamir-Adleman	4,405,829	Sep. 20 1983	[1059]	RSA system

Table 15.1: *Five fundamental U.S. cryptographic patents.*

(i) DES block cipher

The patent of Ehrsam et al. (3,962,539) covers the algorithm which later became well-known as DES (§7.4). Filed on February 24 1975 and now expired, the patent was assigned to the International Business Machines Corporation (IBM). Its background section comments briefly on 1974 product cipher patents of Feistel (3,798,359) and Smith (3,796,830), respectively filed June 30 1971 and November 2 1971. It notes that while the Feistel patent discloses a product cipher which combines key-dependent linear and nonlinear transformations, it fails to disclose specific details including precisely how key bits are used, regarding the nonlinear transformation within S-boxes, and regarding a particular permutation. In addition, the effect of key bits is limited by the particular grouping used. The background section comments further on the cipher of Smith's patent, noting its inherently serial nature as a performance drawback, and that both it and that of Feistel have only two types of substitution boxes, which are selected as a function of a single key bit. Thus, apparently, the need for a new cipher. The patent contains ten (10) claims.

(ii) Diffie-Hellman key agreement

The first public-key patent issued, on April 29 1980, was the Hellman-Diffie-Merkle patent (4,200,770). Filed on September 6 1977, it was assigned to Stanford University (Stanford, California). It is generally referred to as the *Diffie-Hellman patent*, as it covers Diffie-Hellman key agreement (§12.6.1). There are two major objects of the patent. The first is a method for communicating securely over an insecure channel without *a priori* shared keys; this can be done by Diffie-Hellman key agreement. The second is a method allowing authentication of an identity over insecure channels; this can be done using authentic, long-term Diffie-Hellman public keys secured in a public directory, with derivation and use of the resulting Diffie-Hellman secret keys providing the authentication. The patent contains eight (8) claims including the idea of establishing a session key by public-key distribution, e.g., using message exchanges as in two-pass Diffie-Hellman key agreement. Claim 8 is the most specific, specifying Diffie-Hellman using a prime modulus q and exponents x_i and x_j in $[1, q-1]$.

(iii) Merkle-Hellman knapsacks and public-key systems

The Hellman-Merkle patent (4,218,582) was filed October 6 1977 and assigned to the Board of Trustees of the Leland Stanford Junior University (Stanford, California). It covers public-key cryptosystems based on the subset-sum problem, i.e., Merkle-Hellman trapdoor knapsacks (now known to be insecure – see §8.6.1), in addition to various claims on public-key encryption and public-key signatures. The objects of the invention are to allow private conversations over channels subject to interception by eavesdroppers; to allow authentication of a receiver's identity (through its ability to use a key only it would be able to compute); and to allow data origin authentication without the threat of dispute (i.e., via public-key techniques, rather than a shared secret key). There are seventeen (17) claims, with Claims 1–6 broadly applying to public-key systems, and Claims 7–17 more narrowly focused on knapsack systems. The broad claims address aspects of general methods using public-private key pairs for public-key encryption, public-key signatures, and the use of public-key encryption to provide authentication of a receiver via the receiver transmitting back to the sender a representation of the enciphered message.

(iv) Tree authentication method of validating parameters

Merkle's 1982 patent (4,309,569) covers tree authentication (§13.4.1). It was filed September 5 1979, and assigned to the Board of Trustees of the Leland Stanford Junior University (Stanford, California). The main motivation cited was to eliminate the large storage requirement inherent in prior one-time signature schemes, although the idea has wider application. The main ideas are to use a binary tree and a one-way hash function to allow authentication of leaf values Y_i associated with each user i. Modifications cited include: use of a ternary or k-ary tree in place of a binary tree; use of the tree for not only public values of one-time signatures, but for authenticating arbitrary public values for alternate purposes; and use of a distinct authentication tree for each user i, the root R_i of which replaces Y_i above, thereby allowing authentication of all values in i's tree, rather than just a single Y_i. The epitome of conciseness, this patent contains a single figure and just over two pages of text including four (4) claims.

(v) RSA public-key encryption and signature system

The Rivest-Shamir-Adleman patent (4,405,829) was filed December 14 1977, and assigned to the Massachusetts Institute of Technology. It covers the RSA public-key encryption (§8.2.1) and digital signature method (§11.3.1). Also mentioned are generalizations, including: use of a modulus n which is a product of three or more primes (not necessarily distinct); and using an encryption public key e to encrypt a message M to a ciphertext C by evaluating a polynomial $\sum_{i=0}^{t} a_i M^e \bmod n$ where e and a_i, $0 \leq i \leq t$, are integers, and recovering the plaintext M by "utilizing conventional root-finding techniques, choosing which of any roots is the proper decoded version, for example, by the internal redundancy of the message". Other variations mentioned include using RSA encipherment in CFB mode, or as a pseudorandom number generator to generate key pads; signing a compressed version of the message rather than the message itself; and using RSA encryption for key transfer, the key thereby transferred to be used in another encryption method. This patent has the distinction of a claims section, with forty (40) claims, which is longer than the remainder of the patent.

15.2.2 Ten prominent patents

Ten prominent patents are discussed in this section, in order as per Table 15.2.

Inventors	Patent #	Issue date	Ref.	Major claim or area
Okamoto et al.	4,625,076	Nov. 25 1986	[952]	ESIGN signatures
Shamir-Fiat	4,748,668	May 31 1988	[1118]	Fiat-Shamir identification
Matyas et al.	4,850,017	Jul. 18 1989	[806]	control vectors
Shimizu-Miyaguchi	4,850,019	Jul. 18 1989	[1125]	FEAL cipher
Brachtl et al.	4,908,861	Mar. 13 1990	[184]	MDC-2, MDC-4 hashing
Schnorr	4,995,082	Feb. 19 1991	[1095]	Schnorr signatures
Guillou-Quisquater	5,140,634	Aug. 18 1992	[523]	GQ identification
Massey-Lai	5,214,703	May 25 1993	[791]	IDEA cipher
Kravitz	5,231,668	Jul. 27 1993	[711]	DSA signatures
Micali	5,276,737	Jan. 04 1994	[861, 862]	'fair' key escrow

Table 15.2: *Ten prominent U.S. cryptographic patents.*

(i) ESIGN signatures

The Okamoto-Miyaguchi-Shiraishi-Kawaoka patent (4,625,076) covers the original ES-IGN signature scheme (see §11.7.2). The patent was filed March 11 1985 and assigned to the Nippon Telegraph and Telephone Corporation (Tokyo), with priority data listed as March 19 1984 (Japanese patent office). The objective is to provide a signature scheme faster than RSA. The patent contains twenty-five (25) claims.

(ii) Fiat-Shamir identification and signatures

The Shamir-Fiat patent (4,748,668) covers Fiat-Shamir identification (§10.4.2) and signatures (§11.4.1). It was filed July 9 1986, and assigned to Yeda Research and Development Co. Ltd. (Israel). For identification, the inventors suggest a typical number of rounds t as 1 to 4, and parameter selections including $k = 5$ (secrets), $t = 4$ for a 2^{-20} probability of forgery, and $k = 6$, $t = 5$ for 2^{-30}. A range of parameters k, t for $kt = 72$ is tabulated for the corresponding signature scheme, showing tradeoffs between key storage, signature size, and real-time operations required. Noted features relative to prior art include being able to pipeline computations, and being able to change the security level after the key is selected (e.g., by changing t). Generalizations noted include replacing square roots by cubic or higher roots. There are forty-two (42) claims.

(iii) Control vectors for key management

The Matyas-Meyer-Brachtl patent (4,850,017) is one of several in the area of control vectors for key management, in this case allowing a sending node to constrain the use of keys at a receiving node. It was filed May 29 1987 and assigned to the IBM Corporation. Control vectors reduce the probability of key misuse. Two general methods are distinguished. In the first method, the key and a control value are authenticated before use through verification of a special authentication code, the key for which is part of the data being authenticated. In the second method (see §13.5.2), the key and control value are cryptographically bound at the time of key generation, such that recovery of the key requires specification of the correct control vector. In each method, additional techniques may be employed to control which users may use the key in question. The patent contains twenty-two (22) claims.

(iv) FEAL block cipher

The Shimizu-Miyaguchi patent (4,850,019) gives the originally proposed ideas of the FEAL block cipher (see §7.5). It was filed November 3 1986 and assigned to the Nippon Telegraph and Telephone Corporation (Tokyo), with priority data listed as November 8 1985 (Japanese patent office). Embodiments of FEAL with various numbers of rounds are described, with figures including four- and six-round FEAL (now known to be insecure – see Note 7.100), and discussion of key lengths including 128 bits. The patent makes twenty-six (26) claims.

(v) MDC-2/MDC-4 hash functions

The patent of Brachtl et al. (4,908,861) covers the MDC-2 and MDC-4 hash functions (§9.4.1). It was filed August 28 1987 and assigned to the IBM Corporation. The patent notes that interchanging internal key halves, as is done at a particular stage in both algorithms, is actually required for security in MDC-2 but not MDC-4; however, the common design was nonetheless used, to allow MDC-4 to be implemented using MDC-2 twice. A preliminary section of the patent discusses alternatives for providing message authentication (see §9.6), as well as estimates of the security of the new hash functions, and justification for fixing certain bits within the specification to avoid effects of weak DES keys. There are twenty-one (21) claims, mainly on building $2N$-bit hash functions from N-bit block ciphers.

(vi) Schnorr identification and signatures

The Schnorr patent (4,995,082) covers Schnorr's identification (§10.4.4) and signature (§11.5.3) schemes, and optimizations thereof involving specific pre-processing. It was filed February 23 1990, with no assignee listed, and priority data given as February 24 1989 (European patent office). There are eleven (11) claims. Part of Claim 6 covers a specific variation of the Fiat-Shamir identification method using a prime modulus p, such that $p - 1$ is divisible by a prime q, and using a base β of order q.

(vii) GQ identification and signatures

The Guillou-Quisquater patent (5,140,634) addresses GQ identification (Protocol 10.31) and signatures (Algorithm 11.48). It was filed October 9 1991, as a continuation-in-part of two abandoned applications, the first filed September 7 1988. The original assignee was the U.S. Philips Corporation (New York). The disclosed techniques allow for authentication of so-called *accreditation information*, authentication of messages, and the signing of messages. The central authentication protocol involves a commitment-challenge-response method and is closely related to the zero-knowledge-based identification technique of Fiat and Shamir (Protocol 10.24). However, it requires only a single protocol execution and single accreditation value, rather than a repetition of executions and a plurality of accreditation values. The cited advantages over previous methods include smaller memory requirements, and shorter overall duration due to fewer total message exchanges. The main applications cited are those involving chipcards in banking applications. There are twenty-three (23) claims, including specific claims involving the use of chipcards.

(viii) IDEA block cipher

The Massey-Lai patent (5,214,703) covers the IDEA block cipher (§7.6), proposed as a European or international alternative to DES offering greater key bitlength (and thereby, hopefully greater security). It was filed May 16 1991, and assigned to Ascom Tech AG (Bern), with priority data given as May 18 1990 from the original Swiss patent. A key concept in the cipher is the use of at least two different types of arithmetic and logical operations, with emphasis on different operations in successive stages. Three such types of operation are proposed: addition mod 2^m, multiplication mod $2^m + 1$, and bitwise exclusive-or (XOR). Symbols denoting these operations, hand-annotated in the European version of the patent (WO 91/18459, dated 28 November 1991, in German), appear absent in the text of the U.S. patent, making the latter difficult to read. There are fourteen (14) figures and ten (10) multipart claims.

(ix) DSA signature scheme

The patent of Kravitz (5,231,668), titled "Digital Signature Algorithm", has become widely known and adopted as the DSA (§11.5.1). It was filed July 26 1991, and assigned to "The United States of America as represented by the Secretary of Commerce, Washington, D.C." The background section includes a detailed discussion of ElGamal signatures and Schnorr signatures, including their advantage relative to RSA – allowing more efficient on-line signatures by using off-line precomputation. Schnorr signatures are noted as more efficient than ElGamal for communication and signature verification, although missing some "desirable features of ElGamal" and having the drawback that cryptanalytic experience and confidence associated with the ElGamal system do not carry over. DSA is positioned as having all the efficiencies of the Schnorr model, while remaining compatible with the ElGamal model from an analysis perspective. In the exemplary specification of DSA, the hash function used was MD4. The patent makes forty-four (44) claims.

(x) Fair cryptosystems and key escrow

Micali's patent (5,276,737) and its continuation-in-part (5,315,658), respectively filed April 20 1992 and April 19 1993 (with no assignees listed), cover key escrow systems called "fair cryptosystems" (cf. §13.8.3). The subject of the first is a method involving a public-key cryptosystem, for allowing third-party monitoring of communications (e.g., government wiretapping). A number of shares (see secret-sharing – §12.7) created from a user-selected private key are given to a set of trustees. By some method of verifiable secret sharing, the trustees independently verify the authenticity of the shares and communicate this to an authority, which approves a user's public key upon receiving all such trustee approvals. Upon proper authorization (e.g., a court order), the trustees may then subsequently provide their shares to the authority to allow reconstruction of a user private key. Exemplary systems include transforming Diffie-Hellman (see paragraph below) and RSA public-key systems into fair cryptosystems. Modifications require only k out of n trustees to contribute shares to recover a user secret and prevent trustees from learning the identity of a user whose share is requested. The patent contains eighteen (18) claims, the first 14 being restricted to public-key systems.

A fair cryptosystem for Diffie-Hellman key agreement modulo p, with a generator g and n trustees, may be constructed as follows. Each user A selects n integers $s_1, \ldots, s_n$ in the interval $[1, p-1]$, and computes $s = \sum_{i=1}^{n} s_i \bmod p$, public shares $y_i = g^{s_i} \bmod p$, and a public key $y = g^s \bmod p$. Trustee T_i, $1 \leq i \leq n$, is given y, public shares $y_1, \ldots, y_n$, and the secret share s_i to be associated with A. Upon verifying $y_i = g^{s_i}$, T_i stores (A, y, s_i), and sends the authority a signature on $(i, y, y_1, \ldots, y_n)$. Upon receiving such valid signatures from all n trustees, verifying the y_i in the signed messages are identical, and that $y = \prod y_i \bmod p$, the authority authorizes y as A's Diffie-Hellman public key.

The continuation-in-part pursues time-bounded monitoring in greater detail, including use of tamper-proof chips with internal clocks. Methods are also specified allowing an authority (hereafter, the government) access to session keys, including users employing a master key to allow such access. A further method allows verification, without monitoring content, that transmitted messages originated from government-approved devices. This may involve tamper-proof chips in each communicating device, containing and employing a government master key K_M. Such devices allow verification by transmitting a redundant data string dependent on this key. The continuation-in-part has thirteen (13) claims, with the first two (2) restricted to public-key systems. Claims 11 and 12 pursue methods for verifying that messages originate from a tamper-proof device using an authorized encryption algorithm.

15.2.3 Ten selected patents

Ten additional patents are discussed in this section, as listed in Table 15.3. These provide a selective sample of the wide array of existing cryptographic patents.

Inventors	Patent #	Issue date	Ref.	Major claim or area
Feistel	3,798,359	Mar.19 1974	[385]	Lucifer cipher
Smid-Branstad	4,386,233	May 31 1983	[1154]	key notarization
Hellman-Pohlig	4,424,414	Jan. 03 1984	[554]	Pohlig-Hellman cipher
Massey, Omura	4,567,600	Jan 28 1986	[792, 956]	normal basis arithmetic
Hellman-Bach	4,633,036	Dec. 30 1986	[550]	generating strong primes
Merkle	4,881,264	Nov. 14 1989	[846]	one-time signatures
Goss	4,956,863	Sep. 11 1990	[519]	Diffie-Hellman variation
Merkle	5,003,597	Mar. 26 1991	[847]	Khufu, Khafre ciphers
Micali et al.	5,016,274	May 14 1991	[864]	on-line/off-line signing
Brickell et al.	5,299,262	Mar. 29 1994	[203]	exponentiation method

Table 15.3: *Ten selected U.S. cryptographic patents.*

(i) Lucifer cipher

Feistel's patent (3,798,359) is of historical interest. Filed June 30 1971 and assigned to the IBM Corporation, it has now expired. The background section cites a number of earlier cipher patents including ciphering wheel devices and key stream generators. The patent discloses a block cipher, more specifically a product cipher noted as being under the control of subscriber keys, and designed to resist cryptanalysis "not withstanding ... knowledge of the structure of the system" (see Chapter 7 notes on §7.4). It is positioned as distinct from prior art systems, none of which "utilized the advantages of a digital processor and its inherent speed." The patent has 31 figures supporting (only) six pages of text plus one page of thirteen (13) claims.

(ii) Key notarization

The Smid-Branstad patent (4,386,233) addresses key notarization (§13.5.2). It was filed September 29 1980, with no assignee listed. A primary objective of key notarization is to prevent key substitution attacks. The patent contains twenty-one (21) claims.

(iii) Pohlig-Hellman exponentiation cipher

The Hellman-Pohlig patent (4,424,414) was filed May 1 1978 (four and one-half months after the RSA patent), and assigned to the Board of Trustees of the Leland Stanford Junior University (Stanford, California). It covers the Pohlig-Hellman symmetric-key exponentiation cipher, wherein a prime q is chosen, along with a secret key K, $1 \leq K \leq q-2$, from which a second key D, $1 \leq D \leq q-2$, is computed such that $KD \equiv 1 \bmod (q-1)$. A message M is enciphered as $C = M^K \bmod q$, and the plaintext is recovered by computing $C^D \bmod q = M$. Two parties make use of this by arranging, *a priori*, to share the symmetric-keys K and D. The patent contains two (2) claims, specifying a method and an apparatus for implementing this block cipher. Although of limited practical significance, this patent is often confused with the three well-known public-key patents of Table 15.1.

(iv) Arithmetic in $\mathbb{F}_{2^m}$ using normal bases

Two patents of Massey and Omura are discussed here. The Omura-Massey patent (4,587,627) teaches a method for efficient multiplication of elements of a finite field $\mathbb{F}_{2^m}$ by exploiting normal bases representations. It was filed September 14 1982, with priority data November 30 1981 (European patent office), and was issued May 6 1986 with the assignee being OMNET Associates (Sunnyvale, California). The customary method for representing a field element $\beta \in \mathbb{F}_{2^m}$ involves a polynomial basis $1, x, x^2, x^3, \ldots, x^{m-1}$, with $\beta = \sum_{i=0}^{m-1} a_i x^i$, $a_i \in \{0,1\}$ (see §2.6.3). Alternatively, using a normal basis $x, x^2, x^4, \ldots, x^{2^{m-1}}$ (with x selected such that these are linearly independent) allows one to represent β as $\beta = \sum_{i=0}^{m-1} b_i x^{2^i}$, $b_i \in \{0,1\}$. The inventors note that this representation "is unconventional, but results in much simpler logic circuitry". For example, squaring in this representation is particularly efficient (noted already by Magleby in 1963) – it requires simply a rotation of the coordinate representation from $[b_{m-1} \ldots b_1 b_0]$ to $[b_{m-2} \ldots b_1 b_0 b_{m-1}]$. This follows since $x^{2^m} \equiv 1$ and squaring in $\mathbb{F}_{2^m}$ is a linear operation in the sense that $(B+C)^2 = B^2 + C^2$; furthermore, $D = B \times C$ implies $D^2 = B^2 \times C^2$. From this, the main object of the patent follows directly: to multiply two elements B and C to yield $D = B \times C = [d_{m-1} \ldots d_1 d_0]$, the same method used for computing d_{m-1} can be used to sequentially produce d_i, $m-2 \leq i \leq 0$, by applying it to one-bit rotations of the representations of B and C. Alternatively, m such identical processes can be used to compute the m components d_i in parallel. The patent makes twenty-four (24) claims.

The closely related Massey-Omura patent (4,567,600) includes claims on exponentiation in $\mathbb{F}_{2^m}$ using normal bases. It was likewise filed September 14 1982 and assigned to OMNET Associates (Sunnyvale, California), with priority date February 2 1982 (European patent office). Its foundation is the observation that using a normal basis representation allows efficient exponentiation in $\mathbb{F}_{2^m}$ (Claim 16), since the cost of squaring (see above) in the customary square-and-multiply exponentiation technique is eliminated. A second subject is the implementation of Shamir's three-pass protocol (Protocol 12.22) using modular exponentiation in $\mathbb{F}_{2^m}$ as the ciphering operation along with a normal basis representation for elements; and subsequently employing a shared key, established by this method, as the key in an $\mathbb{F}_{2^m}$ exponentiation cipher (cf. Hellman-Pohlig patent) again using normal bases. A further object is a method for computing pairs of integers e, d such that $ed \equiv 1 \bmod 2^m - 1$. Whereas customarily e is selected and, from it, d is computed via the extended Euclidean algorithm (which involves division), the new technique selects a group element H of high order, then chooses a random integer R in $[1, 2^m - 2]$, and computes $e = H^R, d = H^{-R}$. The patent includes twenty-six (26) claims in total.

(v) Generation of strong primes

The Hellman-Bach patent (4,633,036) covers a method for generating RSA primes p and q and an RSA modulus $n = pq$ satisfying certain conditions such that factoring n is believed to be computationally infeasible. The patent was filed May 31 1984 and assigned to Martin E. Hellman. The standard strong prime conditions (Definition 4.52) are embedded: $p - 1$ requiring a large prime factor r; $p + 1$ requiring a large prime factor s; and $r - 1$ requiring a large prime factor r'. A new requirement according to the invention was that $s - 1$ have a large prime factor s', with cited justification that the (then) best known factoring methods exploiting small s' required s' operations. The patent includes twenty-four (24) claims, but is now apparently of historical interest only, as the best-known factoring techniques no longer depend on the cited properties (cf. §4.4.2).

(vi) Efficient one-time signatures using expanding trees

Merkle's 1989 patent (4,881,264), filed July 30 1987 with no assignee listed on the issued patent, teaches how to construct authentication trees which may be expanded arbitrarily, without requiring a large computation when a new tree is constructed (or expanded). The primary cited use of such a tree is for making available public values y (corresponding to secret values x) of a user A in a one-time signature scheme (several of which are summarized). In such schemes, additional public values are continually needed over time. The key idea is to associate with each node in the tree three vectors of public information, each of which contains sufficient public values to allow one one-time signature; call these the LEFT, RIGHT, and MESSAGE vectors. The combined hash value H_i of all three of these vectors serves as the hash value of the node i. The root hash value H_1 is made widely available, as per the root value of ordinary authentication trees (§13.4.1). A new message M may be signed by selecting a previously unused node of the tree (e.g., H_1), using the associated MESSAGE vector for a one-time signature thereon. The tree may be expanded downward from node i (e.g., $i = 1$), to provide additional (verifiably authentic) public values in a new left sub-node $2i$ or a right sub-node $2i + 1$, by respectively using the LEFT and RIGHT vectors at node i to (one-time) sign the hashes H_{2i} and H_{2i+1} of the newly created public values in the respective new nodes. Full details are given in the patent; there are nine (9) claims.

The one-time signatures themselves are based on a symmetric cipher such as DES; the associated one-way function F of a private value x may be created by computing $y = F(x) = DES_x(0)$, i.e., encrypting a constant value using x as key; and a hash function for the authentication tree may also be constructed using DES. Storage requirements on user A for its own tree are further reduced by noting that only x values need be stored; and that these may be pseudorandomly generated, for example, letting $J = 0, 1, 2$ denote the LEFT, RIGHT, and MESSAGE vectors, and assuming that K public values are needed per one-time signature, the Kth value x in a vector of public values at node I may be defined as $x[I, J, K] = DES_{K_A}(I||J||K)$, where K_A is A's secret key and "$||$" denotes concatenation.

(vii) Goss variation of Diffie-Hellman

The patent of Goss (4,956,863) covers a variation of Diffie-Hellman key agreement essentially the same as Protocol 12.53. It was filed April 17 1989 and assigned to TRW Inc. (Redondo Beach, California). The primary application cited is an authenticated key establishment technique, completely transparent to end-users, for facsimile (FAX) machines on existing telephone networks. At the time of manufacture, a unique device identifier and a signed certificate binding this to a long-term Diffie-Hellman public key (public exponential) is embedded in each device. The identity in the certificate, upon verification, may be used as the basis on which to accept or terminate communications channels. Such a protocol allows new session keys for each FAX call, while basing authentication on long-term certified keys (cf. Remark 12.48; but regarding security, see also Note 12.54). The patent makes sixteen (16) claims.

(viii) Khufu and Khafre block ciphers

Merkle's 1991 patent (5,003,597) covers two symmetric-key block ciphers named Khufu and Khafre (see §7.7.3). These were designed specifically as fast software-oriented alternatives to DES, which itself was designed with hardware performance in mind. The patent was filed December 21 1989 and assigned to the Xerox Corporation. Khufu and Khafre have block size 64 bits and a user-selectable number of rounds. Khufu has key bitlength up to 512 bits, and S-boxes derived from the input key; it encrypts 64-bit blocks faster than Khafre. Khafre has fixed S-boxes, and a key of selectable size (with no upper bound), though larger keys impact throughput. The majority of the patent consists of C-code listings specifying the ciphers. The patent contains twenty-seven (27) claims.

(ix) On-line/off-line digital signatures

The Micali-Goldreich-Even patent (5,016,274) teaches on-line/off-line digital signature schemes. The patent was filed November 8 1988, with no assignee listed. The basic idea is to carry out a precomputation to reduce real-time requirements for signing a particular message m. The pre-computation, executed during idle time and independent of m, involves generation of matching one-time public and private keying material for a fast (one-time) first signature scheme, and using a second underlying signature scheme to create a signature s_2 over the one-time public key. This key from the first scheme is then used to create a signature s_1 on m. The overall signature on m is (s_1, s_2). Appropriate hash functions can be used as usual to allow signing of a hash value $h(m)$ rather than m. In the exemplary method, Rabin's scheme is the underlying signature scheme, and DES is used both to build a one-time signature scheme and for hashing. Regarding security of the overall scheme, a one-time scheme, if secure, is presumed secure against chosen-text attack (since it is used only once); the underlying scheme is secure against chosen-text attack because it signs only strings independent of a message m. The method thus may convert any signature scheme into one secure against chosen-text attacks (should this be a concern), or convert any underlying signature scheme to one with smaller real-time requirements. The patent contains thirty-three (33) claims.

(x) Efficient exponentiation for fixed base

The Brickell-Gordon-McCurley patent (5,299,262) teaches a method for fast exponentiation for the case where a fixed base is re-used. This has application in systems such as the ElGamal, Schnorr, and DSA signature schemes. The patent was filed August 13 1992, issued March 29 1994, and assigned to "The United States of America as represented by the United States Department of Energy, Washington, D.C." The method is presented in Algorithm 14.109. The patent contains nine (9) claims.

15.2.4 Ordering and acquiring patents

Any American patent may be ordered by patent number from the U.S. Patent and Trademark Office (PTO). Written requests should be posted to: PTO, Washington, D.C., 20231, USA. Telephone requests may also be made at +703-305-4350, with payment by credit card. A nominal fee applies (e.g., US$3 for patents returned by postal mail; or US$6 for returns by fax, usually the same day). For on-line information on recent patents, consult URL `http://www.micropatent.com` (e.g., specifying patent class code 380 for cryptography).

15.3 Cryptographic standards

This section summarizes cryptographic and security standards of practical interest. These facilitate widespread use of cryptographically sound techniques, and interoperability of systems and system components. Tables 15.4–15.11 present an overview allowing relevant standards to be located and identified, and access to formal title information allowing acquisition of particular standards. These tables may also be used to locate standards addressing particular areas (e.g., key management). For specific details of techniques and algorithms, the original standards should be consulted. Where relevant technical details appear elsewhere in the book, cross-references are given.

Outline of standards section

§15.3.1 presents international (ISO and ISO/IEC) application-independent standards on cryptographic techniques. §15.3.2 summarizes banking security standards, subdivided into ANSI and ISO standards. §15.3.3 considers international security architectures and frameworks (ISO and X.509). §15.3.4 summarizes security-related standards for use by U.S. federal government departments. §15.3.5 addresses selected Internet specifications, while §15.3.6 notes selected de facto industry standards. §15.3.7 provides information allowing acquisition of standards.

15.3.1 International standards – cryptographic techniques

The International Organization for Standardization (ISO) and the International Electrotechnical Commission (IEC) develop standards individually and jointly. Joint standards are developed under the joint technical committee ISO/IEC JTC 1. ISO and ISO/IEC standards progress through the following draft stages before maturing to the International Standard status: Working Draft (WD); Committee Draft (CD); and Draft International Standard (DIS). Each ISO and ISO/IEC standard is reviewed every five years, at which time it is either reaffirmed, revised, or retracted. The ISO/IEC subcommittee responsible for standardizing generic cryptographic techniques is SC 27 (ISO/IEC JTC 1 SC 27). Table 15.4 lists selected ISO and ISO/IEC standards on cryptographic techniques.

ISO 8372: This standard specifies the four well-known modes of operation of a block cipher – electronic codebook (ECB), cipher block chaining (CBC), cipher feedback (CFB), and output feedback (OFB). These modes were originally standardized for DES in FIPS 81 (1980) and ANSI X3.106 (1983). ISO 8372 (first published in 1987) specifies these modes for general 64-bit block ciphers (cf. ISO/IEC 10116).

ISO #	Subject	Ref.
8372	modes of operation for a 64-bit cipher	[574]
9796	signatures with message recovery (e.g., RSA)	[596]
9797	data integrity mechanism (MAC)	[597]
9798–1	entity authentication – introduction	[598]
9798–2	— using symmetric encipherment	[599]
9798–3	— using public-key techniques	[600]
9798–4	— using keyed one-way functions	[601]
9798–5	— using zero-knowledge techniques	[602]
9979	register of cryptographic algorithms	[603]
10116	modes of operation for an n-bit cipher	[604]
10118–1	hash functions – introduction	[605]
10118–2	— using block ciphers	[606]
10118–3	— customized algorithms	[607]
10118–4	— using modular arithmetic	[608]
11770–1	key management – introduction	[616]
11770–2	— symmetric techniques	[617]
11770–3	— asymmetric techniques	[618]
13888–1	non-repudiation – introduction	[619]
13888–2	— symmetric techniques	[620]
13888–3	— asymmetric techniques	[621]
14888–1	signatures with appendix – introduction	[622]
14888–2	— identity-based mechanisms	[623]
14888–3	— certificate-based mechanisms	[624]

Table 15.4: *ISO and ISO/IEC standards for generic cryptographic techniques.*

ISO/IEC 9796: This standard specifies a generic mechanism for digital signature schemes giving message recovery (see §11.3.5 and ANSI X9.31–1; cf. ISO/IEC 14888). Examples are given in its Annex B corresponding to RSA and Rabin's variant thereof (with encryption exponent 2). The main part of the standard is a redundancy scheme, intended to be generically applicable to a large class of signature schemes, although specifically designed to preclude attacks on schemes such as RSA and Rabin which have a multiplicative property.

ISO/IEC 9797: This standard defines a message authentication code (MAC) based on the CBC mode of operation of a block cipher, similar to the MAC algorithms of ISO 8731–1, ISO 9807, ANSI X9.9, and ANSI X9.19 (see Algorithm 9.58).[1] Relative to these, in 9797 the m-bit MAC result is constrained only by $m \leq n$ (the leftmost or most significant bits are retained), the block cipher is unspecified but has n-bit blocks, and a second padding method is specified. These other MAC algorithms may be viewed as special cases of 9797; for example, the specific values $n = 64$ and $m = 32$ along with use of the first padding method (see below) and DES as the block cipher yields the MAC of X9.9.

In 9797, one of two specified padding methods must be selected (Algorithms 9.29, 9.30). The first pads the data input by appending zero or more 0-bits, as few as necessary, to obtain a string whose bitlength is a multiple of n. The second method always appends to the data input a single 1-bit, and then zero or more 0-bits, as few as necessary, to obtain

---

[1]Specific technical details are provided for MAC standards in this chapter moreso than for other standards, in an attempt to clarify the differences between the large number of CBC-MAC standards which differ only in fine details.

a string whose bitlength is a multiple of n. Annex A specifies two optional processes; Annex B provides examples. The first optional process is the optional process as described under ANSI X9.19 in §15.3.2; this reduces the threat of exhaustive key search and chosen-plaintext attacks, and is recommended when $m = n$ (see Remark 9.59). The alternative second optional process, providing protection against chosen-plaintext attacks, employs a second key K' (possibly derived from K) to encrypt the (previously final) output block, before extracting the m-bit MAC result.

ISO/IEC 9798: Parts subsequent to the introduction (9798–1) of this standard specify entity authentication mechanisms based on: symmetric encryption algorithms (9798–2); public-key signature algorithms (9798–3); a cryptographic check function or MAC (9798–4); and other customized techniques (9798–5), historically referred to by academics as zero-knowledge techniques. The mechanisms use timestamps, sequence numbers, and random numbers as time-variant parameters (§10.3.1). The 9798-3 mechanisms are functionally analogous to those of X.509, and the 9798-3 two-pass and three-pass techniques based on random number challenge-response are the source for those in (draft) FIPS JJJ.

9798-2 specifies four entity authentication mechanisms (as given in §10.3.2) involving two parties A and B and requiring that they share a symmetric key *a priori*, for use in a symmetric encryption algorithm. When timestamps or sequence numbers are used, these mechanisms require one and two messages, respectively, for unilateral and mutual entity authentication; using challenge-response based on random numbers, one additional message is required in each case. 9798-3 includes four analogous mechanisms (see §10.3.3) wherein the role of the symmetric encryption algorithm is replaced by a digital signature algorithm, and the requirement of shared symmetric keys is replaced by that of possession of authentic (or the capability to authenticate) public keys. 9798-4 specifies four analogous mechanisms (again see §10.3.2) where symmetric encryption as used in 9798-2 is replaced by a cryptographic check function or MAC. 9798-2 specifies two additional mutual authentication mechanisms for the case that A and B do not share a key *a priori*, but each does share a key with a trusted third party T; these require two further messages (for communication with T) beyond those for the respective mutual entity authentication mechanisms above. 9798-5 (draft) includes an identity-based identification protocol of which Fiat-Shamir (cf. Protocol 10.24) and GQ identification (Protocol 10.31) are special cases, and a protocol based on public-key decryption with witness (see §10.3.3).

ISO/IEC 9979: This standard specifies procedures allowing certain entities (e.g., ISO member bodies and liaison organizations) to register encryption algorithms in an official ISO register of such algorithms. Registration involves no security evaluation or assessment (the policy of ISO/IEC is to not standardize encryption algorithms themselves). The standard specifies the formats required for such register entries, and registration results in the assignment of a unique identifier to each algorithm, e.g., to allow interoperability. For further information, see page 660.

ISO/IEC 10116: This standard specifies the same four modes of block-cipher operation as ISO 8372, but subsumes that standard by allowing general n-bit block ciphers. ISO/IEC 10116 also provides greater detail regarding various properties of the modes, and sample calculations based on DES.

ISO/IEC 10118: This is a multi-part standard on cryptographic hashing algorithms. 10118–1 specifies common definitions and general requirements. 10118–2 specifies two generic constructions based on n-bit block ciphers: the Matyas-Meyer-Oseas hash function (Algorithm 9.41) and a block-cipher independent MDC-2 (cf. Algorithm 9.46). The draft standard 10118–3 includes SHA–1 (Algorithm 9.53), RIPEMD-128 and RIPEMD-160 (Algorithm 9.55). The draft 10118–4 includes MASH-1 and MASH-2 (see Algorithm 9.56).

ISO/IEC 11770: This multi-part standard addresses generic key management and specifies key establishment mechanisms. 11770–1 is a key management framework and overview including discussion of the key life cycle, protection requirements for keying material, and roles of third parties in key establishment. 11770-2 specifies key establishment mechanisms based on symmetric techniques, including those wherein two parties communicate point-to-point (as in §12.3.1), those similar to the Kerberos and Otway-Rees protocols involving a trusted server or key distribution center (§12.3.2), and those involving a key translation center (e.g., Protocol 13.12). 11770-3 specifies key establishment mechanisms based on asymmetric techniques. These are divided into key agreement protocols, practical instantiations of which are based on Diffie-Hellman and similar techniques (§12.6.1); and key transfer protocols, which typically involve both public-key encryption and digital signatures (§12.5.2) including adaptations of the random number based ISO/IEC 9798-3 mechanisms involving transfer of an embedded encrypted key.

ISO/IEC 13888: This multi-part (draft) standard addresses non-repudiation services (protection against false denials) related to the transfer of a message from an originator to a recipient. Mechanisms are specified for non-repudiation of origin (denial of being the originator of a message), non-repudiation of delivery (denial of having received a message), and non-repudiation associated with the actions of a third party acting as a transfer agent on behalf of others. 13888–1 (draft) provides a non-repudiation model and overview. 13888-2 (draft) specifies mechanisms involving symmetric techniques (encipherment and keyed one-way functions). 13888-3 (draft) specifies mechanisms involving asymmetric techniques and the use of digital signatures.

ISO/IEC 14888: This multi-part (draft) standard addresses schemes for signature with appendix (see §11.2.2 and ANSI X9.30–1; cf. ISO/IEC 9796). 14888–1 (draft) provides common definitions and a general overview including models outlining the steps required for signature generation and various classes of verification processes. 14888–2 (draft) addresses identity-based signature mechanisms, wherein the signature verification key is a public function of the signer's identity. 14888–3 (draft) addresses certificate-based mechanisms, wherein this public key is explicitly specified and, for example, distributed by means of a certificate. These may include DSA and similar signature mechanisms such as ElGamal, Schnorr signatures, and RSA.

15.3.2 Banking security standards (ANSI, ISO)

This section considers banking security standards developed by ANSI and by ISO. Banking security standards are typically divided into wholesale and retail banking (see Table 15.5). *Wholesale banking* involves transactions between financial institutions. *Retail banking* involves transactions between institutions and private individuals, including automated teller machine (ATM) and point-of-sale (POS) transactions, and credit authorizations.

category	transaction volume	average transaction value
retail	high (millions per day)	$50
wholesale	low (thousands per day)	$3 million

Table 15.5: *Retail vs. wholesale banking characteristics.*

(i) ANSI encryption standards

The American National Standards Institute (ANSI) develops standards through various Accredited Standards Committees (ASCs). Accreditation implies that standards developed under a particular committee become ANSI standards. Accredited committees include ASC X3 – Information Processing Systems; ASC X9 – Financial Services; and ASC X12 – Electronic Business Data Interchange. Table 15.6 lists selected ANSI encryption and banking security standards developed under X3 and X9.

ANSI X3.92: This standard specifies the DES algorithm, which ANSI standards refer to as the Data Encryption Algorithm (DEA). X3.92 is technically the same as FIPS 46.

ANSI X3.106: This standard specifies DES modes of operation, or DEA modes of operation as referred to in ANSI standards. X3.106 is technically the same as FIPS 81 (cf. ISO 8372). An appendix in FIPS 81 contains additional background information on the various modes.

(ii) ANSI banking security standards

ASC X9 subcommittee X9F develops information security standards for the financial services industry. Banking security standards include cryptographic and operational requirements, with a heavy emphasis on controls, audit, sound business practices, and interoperability. Among the working groups under X9F, most of the cryptographic work is in X9F1 (public key cryptography and cryptographic tools) and X9F3 (security in wholesale financial telecommunications).

ANSI #	Subject	Ref.
X3.92	data encryption algorithm (DEA)	[33]
X3.106	data encryption algorithm (DEA) modes	[34]
X9.8	PIN management and security	[35]
X9.9	message authentication (wholesale)	[36]
X9.17	key management (wholesale; symmetric)	[37]
X9.19	message authentication (retail)	[38]
X9.23	encryption of messages (wholesale)	[39]
X9.24	key management (retail)	[40]
X9.26	sign-on authentication (wholesale)	[41]
X9.28	multi-center key management (wholesale)	[42]
X9.30–1	digital signature algorithm (DSA)	[43]
X9.30–2	secure hash algorithm (SHA) for DSA	[44]
X9.31–1	RSA signature algorithm	[45]
X9.31–2	hashing algorithms for RSA	[46]
X9.42	key management using Diffie-Hellman	[47]
X9.45	attribute certificates and other controls	[49]
X9.52	triple DES and modes of operation	[50]
X9.55	certificate extensions (v3) and CRLs	[51]
X9.57	certificate management	[52]

Table 15.6: *ANSI encryption and banking security standards.*

ANSI X9.8: This standard addresses PIN management and security. It consists of ISO 9564 reproduced in its entirety, with clearly marked "X9 Notes" added where required to adapt the text for use as an ANSI X9 standard. A standard means for interchanging PIN data is specified. Annex A of 9564 (procedures for the approval of an encipherment algorithm) is included; the only currently specified approved algorithm is DES. Annex B (general principles for key management) is also retained from 9564, but noted as superseded by X9.24 (retail key management).

ANSI X9.9: This standard specifies a DES-based message authentication code (MAC) algorithm for wholesale banking as summarized below (cf. X9.19 for retail banking). If data is protected by both authentication and encryption mechanisms, a different key is required for each purpose. Message replay is precluded by use of date and message identifier fields. Appendix B includes sample MAC computations. X9.9 requires key management in accordance with ANSI X9.17, and also addresses implementation issues including coded character sets and representations, field delimiters, and message normalization (e.g., replacing carriage returns or line feeds by space characters, and multiple spaces by single spaces), and notes other practical concerns such as escape sequences beyond the scope of a MAC causing over-writing of authenticated data fields on display devices.

The X9.9 MAC algorithm may be implemented using either the cipher-block chaining (CBC) or 64-bit cipher feedback (CFB-64) mode, initialized to produce the same result (see Note 15.1). Final data blocks with fewer than 64 bits are left-justified and zero-bits are appended to complete the block before processing. The MAC result is specified to be the leftmost 32 bits of the final DES output. X9.9 states that the capability to generate 48-bit and 64-bit MAC values should also exist.

15.1 Note (*CBC-MAC and equivalent CFB-64 MAC*) For data blocks $D_1, \ldots, D_t$ and a fixed MAC key K, equivalent MACs may be generated using either the CBC or 64-bit cipher feedback (CFB-64) modes. In the CBC case, the MAC C_t is defined by $C_i = E_K(D_i \oplus C_{i-1})$ for $1 \leq i \leq t$ and $C_0 = IV = 0$. For the CFB-64 case, let $O_i = E_K(I_i)$ be the output from the block encryption at stage i for $1 \leq i \leq t$, where $I_i = D_i \oplus O_{i-1}$ for $2 \leq i \leq t$ and $I_1 = D_1$ (the first 8 data bytes serve as IV). Note $O_t = C_t$ from above. (A block $D_{t+1} = 0$ may be introduced if the CFB implementation interface requires the final output O_t be XORed to a data block before release.)

ANSI X9.17: This standard, which was the basis for ISO 8732, specifies manual and automated methods (symmetric-based) for wholesale banking key management, including key establishment techniques and protection of keys in key management facilities. A key management hierarchy is defined consisting of manually-distributed key-encrypting keys, electronically-distributed key-encrypting keys, and electronically-distributed data or transaction keys for authentication or encryption. Key management techniques include the use of key counters, key offsetting, and key notarization. Key establishment settings include direct exchange between two nodes (point-to-point), and both key distribution centers (KDCs) and key translation centers (KTCs).

ANSI X9.19: This standard specifies a DES-based message authentication code (MAC) algorithm for retail banking (cf. X9.9 for wholesale banking). Implementation and other issues are addressed as per X9.9, and the MAC algorithm itself is essentially the same as X9.9, differing in that the MAC result is the leftmost m bits of the final 64-bit output, where m is to be specified by the application. An optional X9.19 procedure using a second key K' is specified for increased protection against exhaustive key determination: the (previously) final output is decrypted using K' and then re-encrypted under the original key. The resulting algorithm is widely referred to as the *retail MAC*; see Figure 9.6.

ANSI X9.23: This standard addresses message formatting and representation issues related to the use of DES encryption in wholesale banking transactions. These include field delimiting and padding, as well as filtering methods required to prevent ciphertext bit sequences from interfering with communications protocols when inadvertently interpreted as control characters (e.g., end-of-transmission).

ANSI X9.24: This standard, which motivated ISO 11568, specifies manual and automated methods for retail key management, addressing authentication and (DES-based) encryption of PINs, keys, and other data. Guidelines include protection requirements at various stages in the key management life cycle. Appendices provide additional information, including (Appendix D) methods providing unique per-transaction keys, updated after each transaction as a one-way function of the current key and transaction-specific details; and (Appendix E) how to derive a large number of different terminal keys (for distinct terminals) from a common base key, simplifying key management for servers which must communicate with all terminals. Such derived keys may be combined with the unique per-transaction key methods.

ANSI X9.26: This standard specifies two main classes of entity authentication mechanisms of use for access control. The first involves user passwords. The second involves cryptographic keys used in DES-based challenge-response protocols (e.g., a time-variant parameter challenge must be ECB-encrypted). The latter class is subdivided, on the basis of granularity, into user-unique and node-unique keys.

ANSI X9.28: This standard extends X9.17 to allow the distribution of keying material (using X9.17 protocols) between entities (subscriber nodes) which neither share a common key, nor share a key with a common central server (KDC or KTC). Two or more key centers form a *multiple-center group* to provide a more general key distribution service allowing the establishment of keying material between any two subscribers sharing a key with at least one center in the group. As there are no known or proposed implementations of this standard, it appears destined to be withdrawn from the ANSI suite.

ANSI X9.30: The first in a suite of ANSI public-key standards, X9.30–1 and X9.30–2 specify DSA and SHA for the financial services industry, as per FIPS 186 and FIPS 180, respectively.

ANSI X9.31: The (draft) standard X9.31–1 parallels X9.30–1, and specifies a signature mechanism based on an RSA signature algorithm, more specifically the ISO/IEC 9796 variant combined with a hashing algorithm. The (draft) standard X9.31–2 defines hash functions for use with Part 1, including MDC-2.

ANSI X9.42: This (draft) standard specifies several variations of unauthenticated Diffie-Hellman key agreement, providing shared symmetric keys for subsequent cryptographic use.

ANSI X9.45: This (draft) standard employs a particular type of attribute certificate (§13.4.2) called an *authorization certificate*, and other techniques from ANSI X9.57, to allow a party to determine whether a received message or signed document is authorized with respect to relevant rules or limits, e.g., as specified in the authorization certificate.

ANSI X9.52: This (draft) standard for encryption offers improvements over DES security by specifying a number of modes of operation for triple-DES encryption, including the four basic modes of ISO 8372, enhanced modes intended to provide additional protection against advanced cryptanalytic attacks, and message-interleaved and pipelined modes intended to allow increased throughput in multi-processor systems.

ANSI X9.55: This (draft) standard specifies extensions to the certificate definitions of ANSI X9.57 corresponding to, and aligned with, ISO certificate extensions for ITU-T X.509 Version 3 certificates (see page 660).

ANSI X9.57: This (draft) certificate management standard includes both technical specifications defining public-key certificates (based on ITU-T X.509) for electronic commerce, and business controls necessary to employ this technology. The initial version is defined for use with DSA certificates, in conjunction with ANSI X9.30–1.

(iii) ISO banking security standards

ISO banking security standards are developed under the ISO technical committee TC68 – Banking and Related Financial Services. TC68 subcommittees include TC68/SC2 (wholesale banking security) and TC68/SC6 (retail banking security and smart card security). Table 15.7 lists selected ISO banking security standards.

ISO #	Subject	Ref.
8730	message authentication – requirements (W)	[575]
8731–1	message authentication – CBC-MAC	[576]
8731–2	message authentication – MAA	[577]
8732	key management/symmetric (W)	[578]
9564	PIN management and security	[579]
9807	message authentication – requirements (R)	[581]
10126	message encipherment (W)	[582]
10202–7	key management for smart cards	[584]
11131	sign-on authentication	[585]
11166–1	key management/asymmetric – overview	[586]
11166–2	key management using RSA	[587]
11568	key management (R), in 6 parts	[588]

Table 15.7: *ISO banking security standards (W–wholesale; R–retail).*

ISO 8730: Together with ISO 8731, this wholesale banking standard for message authentication code (MAC) algorithms forms the international equivalent of ANSI X9.9. ISO 8730 is algorithm-independent, and specifies methods and requirements for the use of MACs including data formatting and representation issues, and a method by which specific algorithms are to be approved.

ISO 8731: ISO 8731–1 and 8731–2 specify particular MAC algorithms complementary to the companion standard ISO 8730. 8731–1 specifies a DES-based CBC-MAC with $m = 32$ (cf. ISO/IEC 9797). 8731–2 specifies the Message Authenticator Algorithm, MAA (Algorithm 9.68).

ISO 8732: This standard for key management in wholesale banking was derived from ANSI X9.17, and is its international equivalent.

ISO 9564: This standard, used as the basis for ANSI X9.8, specifies minimum measures for the management and security of Personal Identification Numbers (PINs). Part 1 specifies principles and techniques to protect against disclosure of PINs to unauthorized parties during the PIN life cycle. Part 2 specifies encipherment algorithms approved to protect PINs.

ISO 9807: This standard for message authentication in retail banking is analogous to ANSI X9.19 (cf. ISO 8730/8731–1 vs. ANSI X9.9), but does not address data representation issues, and names two approved algorithms in Annex A – the CBC-MAC of 8731–1 (allowing optional final processing as per X9.19), and the MAA of 8731-2.

ISO 10126: This multi-part standard is the international equivalent of X9.23 addressing confidentiality protection of (parts of) financial messages. ISO 10126–1 provides general principles; 10126–2 defines a specific algorithm – DES.

ISO 10202: This eight-part standard addresses security architecture issues for integrated circuit cards (chipcards) used for financial transactions. In particular, ISO 10202-7 specifies key management aspects.

ISO 11131: This standard for sign-on authentication is the international (non-DES specific) analogue of ANSI X9.26.

ISO 11166: This multi-part standard for banking key management specifies asymmetric techniques for distributing keys for symmetric algorithms. It was developed from ISO

8732, which uses symmetric techniques only. Part 1 specifies general principles, procedures, and formats, including background regarding key protection during its life cycle, certification of keying material, key distribution by either key exchange (e.g., Diffie-Hellman) or key transport, and cryptographic service messages. Further parts are intended to define approved algorithms for use with the procedures of Part 1. Part 2 specifies the RSA algorithm for both encipherment and digital signatures; RSA formatting differs from both ISO/IEC 9796 and PKCS #1.

ISO 11568: This multi-part standard addresses retail key management and life cycle issues. It originated from X9.24, but is generalized for international use (e.g., it is no longer DES-specific), and addresses both symmetric and public-key techniques.

15.3.3 International security architectures and frameworks

Table 15.8 lists selected ISO standards on security frameworks and architectures. Some of these are developed by SC21 (ISO/IEC JTC 1 SC21), which includes activities on Open Systems Interconnection (OSI) projects. The International Telecommunication Union (ITU) develops common-text specifications with JTC 1 for some standards in this area.

ISO #	Subject	Ref.
7498-2	OSI security architecture	[573]
9594-8	authentication framework (X.509)	[595]
10181	OSI security frameworks	[609]

Table 15.8: *ISO and ISO/IEC security architectures and frameworks.*

ISO 7498-2 (X.800): The OSI basic reference model of ISO 7498 defines a communications protocol stack with seven layers: application (layer 7), presentation (6), session (5), transport (4), network (3), data-link (2), and physical layers (1). ISO 7498-2 specifies the security architecture for the basic reference model, including the placement of security services and mechanisms within these layers. It also provides a general description of the basic OSI security services: authentication (peer-entity and data-origin); access control; data confidentiality; data integrity; and non-repudiation (with proof of origin, or with proof of delivery). Specific mechanisms are used to implement these services; for example, encipherment is a mechanism for providing confidentiality.

ISO/IEC 9594-8 (X.509): This standard is the same as ITU-T (formerly CCITT) Recommendation X.509. It defines both simple authentication techniques (based on passwords) and so-called strong authentication techniques (wherein secret values themselves are not revealed to the verifier). The strong techniques included are the two-pass and three-pass X.509 exchanges (see §12.5.2) based on digital signatures and the use of time-variant parameters. An implicit assumption is the use of an algorithm such as RSA which may serve as both an encryption and a signature mechanism; the specification may, however, be modified (e.g., to use DSA). The standard also specifies techniques, including X.509 certificates, for acquiring or distributing authentic public keys; and addresses cross-certificates, and the use of certificate chains (§13.6.2(i)).

ISO/IEC 10181 (X.810 through X.816): This specification is a series of security frameworks intended to provide context and background, consisting of the following parts: security frameworks overview (1); authentication framework (2); access control framework (3); non-repudiation framework (4); integrity framework (5); confidentiality framework (6); and security audit framework (7).

15.3.4 U.S. government standards (FIPS)

Table 15.9 lists selected security-related Federal Information Processing Standards (FIPS) publications. These are developed under the National Institute of Standards and Technology (NIST), for use by U.S. federal government departments.

FIPS #	Subject	Ref.
FIPS 46–2	DES	[396]
FIPS 74	guidelines for using DES	[397]
FIPS 81	DES modes of operation	[398]
FIPS 112	password usage	[399]
FIPS 113	data authentication (CBC-MAC)	[400]
FIPS 140–1	cryptomodule security requirements	[401]
FIPS 171	key management using X9.17	[402]
FIPS 180–1	secure hash standard (SHA–1)	[404]
FIPS 185	key escrow (Clipper & SKIPJACK)	[405]
FIPS 186	digital signature standard (DSA)	[406]
FIPS JJJ	entity authentication (asymmetric)	[407]

Table 15.9: *Selected security-related U.S. FIPS Publications.*

FIPS 46: This standard specifies the DES algorithm (cf. ANSI X3.92).

FIPS 74: This standard provides guidelines for implementing and using DES.

FIPS 81: This standard specifies 4 basic DES modes of operation (cf. ANSI X3.106).

FIPS 112: This standard provides guidelines on password management and usage.

FIPS 113: This standard specifies the customary DES-based CBC-MAC algorithm (see ISO/IEC 9797), referring to it as the Data Authentication Algorithm (DAA). The MAC result is called a Data Authentication Code (DAC). The last data bock, if incomplete, is left-justified and zero-padded before processing; the result is the leftmost m output bits, where m is a multiple of 8, and $16 \leq m \leq 64$. Implementation may be either by the CBC mode with $IV = 0$, or CFB-64 mode with $IV = D_1$, the first data block (see Note 15.1). 7-bit ASCII-coded data to be authenticated by the DAA is preprocessed into 8-bit characters with leading bit 0.

FIPS 140–1: This standard specifies security requirements for the design and implementation of cryptographic modules for protecting (U.S. government) unclassified information, including hardware, firmware, software modules, and combinations thereof. Four grades of increasing security are specified as Levels 1 through 4, covering a wide range of security applications and environments. A FIPS 140–1 validation program is run by NIST to determine if cryptomodules meet the stated requirements.

FIPS 171: FIPS 171 specifies, for use by (U.S.) federal government departments, a subset of the key distribution techniques of ANSI X9.17. The objective of specifying a subset is to increase interoperability and decrease system costs.

FIPS 180 and 180–1: The hash algorithm specified in the original standard FIPS 180 is the Secure Hash Algorithm, SHA. A revised version was specified shortly thereafter in FIPS 180–1 (Algorithm 9.53), and denoted SHA–1. SHA–1 differs from SHA as noted in §9.8.

FIPS 185: This Escrowed Encryption Standard (EES) specifies the parameters and use of the SKIPJACK symmetric-key block cipher, and a method of creating Law Enforcement Access Fields (LEAFs) for use with the Clipper key escrow system (§13.8.3). The purpose is to allow wiretapping under lawful authorization. Internal details of the SKIPJACK algorithm are not publicly available, although its interface specification is (§13.8.3(i)).

FIPS 186: This standard is the Digital Signature Standard (DSS), which specifies the Digital Signature Algorithm (DSA). The hash function originally mandated for use with DSA is defined in FIPS 180 (SHA), which was superseded by FIPS 180–1 (SHA–1).

FIPS JJJ: This (draft) standard on entity authentication using asymmetric techniques was derived from the two-pass and three-pass random-number based mechanisms of ISO/IEC 9798-3. It includes additional expository and implementation details.

15.3.5 Internet standards and RFCs

Documents called *Requests for Comments* (RFCs) are official working notes of the Internet research and development community. A subset of these are specifications which are candidates for standardization within the community as Internet Standards.

The Internet Engineering Steering Group (IESG) of the Internet Engineering Task Force (IETF) is responsible for making recommendations regarding progression of "standards-track" specifications from Proposed Standard (PS) to Draft Standard (DS) to Standard (STD). RFCs may also correspond to the following types of documents: Experimental (E) protocols which may be part of early research efforts; Informational (I) protocols published for convenience of the community; and Historical (H) protocols which have been superseded, expired, or abandoned.

The E, I, and H categories are not on the standards track, and the IESG does not make recommendations on these. Less mature, less stable, or less widely circulated documents are typically available as an Internet-Draft (I-D); these are considered to be "work in progress", and should be cited as such.

RFC	Status	Subject	Ref.
1319	I	MD2 hash function	[1033]
1320	I	MD4 hash function	[1034]
1321	I	MD5 hash function	[1035]
1421	PS	PEM – encryption, authentication	[1036]
1422	PS	PEM – certificates, key management	[1037]
1423	PS	PEM – algorithms, modes, identifiers	[1038]
1424	PS	PEM – key certification and services	[1039]
1508	PS	Generic Security Service API	[1040]
1510	PS	Kerberos V5 network authentication	[1041]
1828	PS	keyed MD5 (as a MAC)	[1044]
1847	PS	security multiparts for MIME	[1045]
1848	PS	MIME Object Security Services (MOSS)	[1046]
1938	PS	one-time password system	[1047]

Table 15.10: *Selected Internet RFCs (May 1996 status).*

Table 15.10 lists selected security-related Internet RFCs. The hashing algorithms MD2, MD4, and MD5 are specified in RFCs 1319-1321, respectively. The Internet Privacy-Enhanced Mail (PEM) specifications are given in RFCs 1421-1424.

The Generic Security Service Application Program Interface (GSS-API) of RFC 1508 is a high-level security API which isolates application code from implementation details; for example, the interface provides functions such as *sign* and *seal* (e.g., as opposed to "seal using a 32-bit DES CBC-MAC and this particular key"). Specific implementation mechanisms must be provided beneath GSS-API; options include Kerberos V5 as per RFC 1510 for symmetric-based techniques, and SPKM for public-key based techniques (see page 661).

RFC 1828 specifies a method for using keyed MD5 as a MAC (cf. §9.5.2). RFC 1848 defines MIME Object Security Services (MOSS), where MIME denotes Multipurpose Internet Mail Extensions. MOSS makes use of the RFC 1847 framework of multipart/signed and multipart/encrypted MIME messages, and facilitates encryption and signature services for MIME including key management based on asymmetric techniques. RFC 1938 specifies an authentication technique based on Lamport's one-time password scheme (Protocol 10.6).

15.3.6 De facto standards

Various security specifications arising through informal processes become de facto standards. This section mentions one such class of specifications: the PKCS suite.

PKCS specifications

A suite of specifications called *The Public-Key Cryptography Standards* (PKCS) has parts as listed in Table 15.11. The original PKCS #2 and PKCS #4 have been incorporated into PKCS #1. PKCS #11 is referred to as *CRYPTOKI*.

No.	PKCS title
1	RSA encryption standard
3	Diffie-Hellman key-agreement standard
5	Password-based encryption standard
6	Extended-certificate syntax standard
7	Cryptographic message syntax standard
8	Private-key information syntax standard
9	Selected attribute types
10	Certification request syntax standard
11	Cryptographic token interface standard

Table 15.11: *PKCS specifications.*

15.3.7 Ordering and acquiring standards

ISO and ISO/IEC standards may be obtained from (member body) national standards organizations such as ANSI, the British Standards Institution (BSI), and the Standards Council of Canada (SCC). To purchase standards directly from ISO, contact ISO Central Secretariat, Case postale 56, CH-1211 Geneva 20, Switzerland; telephone +41.22.749.01.11.

ANSI X9 standards are published by EDI Support Services Incorporated; to purchase standards, telephone 1-800-334-4912 (from within the USA) or +216-974-7650 (from outside the USA).

FIPS PUBS may be purchased from the National Technical Information Service, U.S. Department of Commerce, 5285 Port Royal Road, Springfield, Virginia 22161 (USA); telephone +703-487-4650, fax +703-321-8547. To obtain copies of specifications of proposed (draft) FIPS, contact the Standards Processing Coordinator, National Institute of Standards and Technology, Technology Building, Room B–64, Gaithersburg, Maryland 20899 (USA); telephone +301-975-2816. Alternatively, consult URL http://csrc.ncsl.nist.gov/.

Internet RFCs and Internet-Drafts are available on-line via anonymous FTP from numerous ftp sites (e.g., ds.internic.net); further information can be obtained by sending an email message to rfc-info@isi.edu with the message body "help: ways_to_get_rfcs". RFCs are typically under the directory rfc/ as rfcXXXX.txt (e.g. rfc1321.txt), and an RFC index is available as rfc-index.txt. RFCs can also be obtained via electronic mail by sending an email message to rfc-info@isi.edu whose body includes "Retrieve: RFC" and "Doc-ID: RFCnnnn" on separate lines.

The PKCS suite is published by RSA Laboratories, 100 Marine Parkway, Suite 500, Redwood City, California 94065-1031 (telephone +415-595-7703), and is available by anonymous FTP from rsa.com under the directory pub/pkcs/.

15.4 Notes and further references

§15.1

Levine [762] compiled a comprehensive list of American cryptographic patents issued between 1861 and 1981, citing patent number, name of principal inventor, date granted, and patent title; this provides an insightful perspective of the history of cryptography over this period. Kahn [648] discusses many patents in his historical tour, including many related to rotor machines (cf. Chapter 7). Contact information regarding the current assignees of some cryptographic patents may be found throughout the book of Schneier [1094].

Davies and Price [308] provide both general discussion of standards, and detailed technical discussion of selected standards. Preneel [1001] gives background on worldwide, European, and North American standardization organizations, and an overview of activities therein. Ford [414] provides a comprehensive overview of information security standards including extensive background information on various standardization processes and organizations, including technical committees ISO TC 68 and ISO/IEC JTC 1 and their subcommittees; ITU; ANSI; and national, regional, and international standardization bodies. For a more recent overview of security standards for open systems, see Fumy and Rietenspiess [432]. A status update of selected standards is also provided by Ford [415].

§15.2

One of the earliest and most important cryptographic patents was U.S. Patent No. 1,310,719 [1221] issued to Vernam on July 22 1919 for the *Vernam cipher* (cf. the one-time pad, Chapter 7; see also Kahn [648, p.401]). Two other patents by Vernam, titled "Ciphering device", were granted May 23 1922 (1,416,765) and January 8 1924 (1,479,846).

In consideration of ANSI making DES a standard, IBM made the DES patent of Ehrsam et al. (3,962,539) [363] available free of license fees in the U.S. when used to implement ANSI standards.

The first widespread published disclosure of public-key cryptography was through the conference paper of Diffie and Hellman [344], presented June 8 1976, fifteen months prior to the filing of the Hellman-Diffie-Merkle patent [551]. Merkle independently conceived the idea of deriving a secret key over a public channel in 1974 (see §12.10); his paper [849], first submitted to *Communications of the ACM* in 1975, was rejected several times before final publication in 1978. Meanwhile, the 1976 Diffie-Hellman conference paper introduced the concept of a digital signature as well as public-key cryptography and public-key authentication. Although Diffie and Hellman noted: "At present we have neither a proof that public key systems exist, nor a demonstration system", the existence of public-key systems was postulated, and three suggestions were offered supporting the general idea. The first involved matrix inversion, which is more difficult than multiplication by a factor $O(n)$ for $n \times n$ matrices; this offers a degree of security for very large n. The second involved compiling a function described in a high-level language into machine code; this makes it difficult to recover the original function. The third suggestion involved obscuring the input-output relationships between, e.g., 100 input and 100 output bits (wires) in an invertible hardware circuit originally implementing the identity mapping, by, e.g., inserting 4-by-4 bit invertible S-boxes into randomly selected sets of 4 wires; re-arranging the particular mappings of input lines into S-boxes then makes inverting the resulting circuit difficult.

The Hellman-Merkle patent [553] was filed sixteen months after the above Diffie-Hellman conference paper was presented. A major reason why the RSA patent [1059] took almost 6 years from application filing to issue date was so-called interference proceedings between it and some of the Stanford patents. The subject of the authentication trees patent of Merkle [848] is discussed in his thesis [851, p.126-131] and in the open literature [852, 853].

The signature technique of the ESIGN patent [952] is discussed in the literature by Okamoto [948]; see also Fujioka, Okamoto, and Miyaguchi [428]. The identification and signature technique of the Shamir-Fiat patent [1118] is described by Fiat and Shamir [395]. Regarding the Guillou-Quisquater patent [523], see Guillou and Quisquater [524]. The identification and signature schemes patented by Schnorr [1095] are discussed in the literature by Schnorr [1097, 1098]; the preprocessing scheme proposed therein, however, was shown to be insecure by de Rooij [314, 315].

In its announcement of the proposed FIPS for DSS (*Federal Register* vol.56 no.169, August 30 1991, 42980-42982), NIST noted its intent to make the DSA patent of Kravitz [711] available world-wide on a royalty-free basis. In a letter to the Director of the Computer System Laboratories at NIST dated October 30 1991, Schnorr stated that DSA infringed on Claim 6 of his patent (4,995,082). FIPS 186 itself (1994) states that "The Department of Commerce is not aware of any patents that would be infringed by this standard".

MDC-2 and MDC-4 [184] (see also Bosselaers and Preneel [178]) are discussed in §9.4.1. For further discussion of FEAL [1125], see §7.5. A patent on IDEA was originally filed in Switzerland and subsequently as a European patent [790], prior to being filed as a U.S. patent [791]; for literature references, see Chapter 7.

Related to the Matyas-Meyer-Brachtl patent [806] on control vectors, the October 7 1980 patent of Ehrsam et al. (4,227,253), "Cryptographic communication security for multiple domain networks", describes use of a master key and two variants obtained by inverting designated bits of the master key, equivalent to an XOR of the master with fixed mask values. Also related is the key notarization method of the patent by Smid and Branstad [1154], which controls which parties use a key, but not the uses. The key notarization technique is essentially identical – involving concatenation of various quantities (user identities), which are then XOR'd with a key-encryption key – but control vectors have broader functionality.

Fair cryptosystems [861, 862] are discussed in the literature by Micali [863]; but see also Kilian and Leighton [671], who remark on a critical weakness.

Interest in product cipher systems was stimulated by the product ciphers described in Shannon's 1949 paper [1121]. Meyer and Matyas [859] note that Lucifer was the name of the cryptographic system in which the product cipher of Feistel's patent (3,798,359) [385] was implemented, and from which the IBM team lead by Tuchman derived DES. The 1974 patent of Smith [1159] is also related to Lucifer. A second 1974 patent of Feistel [386] on a "step code ciphering system" was filed and issued with dates matching the Lucifer algorithm patent. Sorkin [1165] states that Lucifer is the subject of all three of these patents, plus a fourth: "Centralized verification system" (3,798,605) granted March 19 1974 to H. Feistel. Feistel gives a high-level background discussion on a first variation of Lucifer in his 1973 *Scientific American* article [387], which appeared prior to his 1974 patents being issued. A description of the second variation of Lucifer (which lead to the design of DES) is given by Sorkin [1165]; see also Biham and Shamir [138].

Related to the Massey-Omura [792] and Omura-Massey [956] patents is that of Onyszchuk, Mullin, and Vanstone [959]. It was filed May 30 1985 and issued May 17 1988 with no assignee listed. The patent teaches the construction of a multiplier for elements in $\mathbb{F}_{2^m}$, stated to be a significant improvement over the method of Omura-Massey. The patent also tabulates those values m, $2 \leq m \leq 2493$, for which so-called optimal normal bases exist; in these fields, the disclosed normal-basis multipliers for $\mathbb{F}_{2^m}$ are more efficient than in others. Shamir's three-pass protocol was first proposed by Shamir, as indicated by Konheim [705]. Massey [786] notes that Shamir also specifically proposed implementing the three-pass protocol using exponentiation as the ciphering operation, an idea later independently proposed by Omura (cf. §12.3 notes on page 535).

In contrast to the prime generation methods of Shawe-Taylor and Maurer (§4.4.4) which result in guaranteed primes, the prime generation method of the Hellman-Bach patent [550] uses probabilistic primality tests, and is related to that presented by Gordon at Eurocrypt in April of 1984 [514], and which appeared (dated April 26 1984) in the June 7 1984 issue (vol.20 no.12) of *Electronics Letters* [513].

The protocol patented by Goss [519], filed April 17 1989, combines exponentials by an XOR operation. An essentially identical protocol published in 1986 by Matsumoto, Takashima, and Imai [800] uses modular multiplication (cf. Protocol 12.53).

The exponentiation cipher of the Hellman-Pohlig patent [554] is discussed in the literature by Pohlig and Hellman [982]. The ciphers Khufu and Khafre [847] are similarly discussed by Merkle [856]; on-line/off-line digital signatures [864] by Even, Goldreich, and Micali [377, 378]; and the techniques of the patent on efficient exponentiation [203] are presented by Brickell et al. [204] (for more recent work, see Hong, Oh, and Yoon [561]).

A patent by Crandall (5,159,632) [286] includes twelve (12) claims on specific implementations of elliptic curves using primes p of special form (e.g., $p = 2^q - C$ for C small) allowing fast multiplication using shifts and adds alone (cf. Mohan and Adiga, 1985), and specific use of Fast Fourier Transforms (FFT) for optimized modular multiplication in this case. The patent, filed September 17 1991, was issued October 27 1992 and assigned to NeXT Computer, Inc. (Redwood City, California); see also its continuation-in-part, (5,271,061) [287]. Another patent in this area is the Miyaji-Tatebayashi patent (5,272,755) [888] filed June 26 1992, with priority data June 28 1991 (Japanese patent office). Issued December 21 1993, and assigned to the Matsushita Electric Industrial Co. (Osaka), it contains six (6) claims in the area of selecting elliptic curves over $\mathbb{F}_p$ whose order is precisely p. This covers a small subset of possible curves of this order over $\mathbb{F}_p$, and one particular method for selecting from among these; see also its continuation-in-part, (5,351,297) [889].

Regarding other block ciphers discussed in this book, a patent application has been filed for the RC5 cipher (§7.7.2). Adams [3] is the inventor for a patent on the CAST block cipher design procedure (see p.281); the assignee, Northern Telecom Limited (Montreal), will, however, make a CAST cipher available free of license fees.

The SEAL stream cipher (§6.4.1) of Coppersmith and Rogaway is also patented [281].

§15.3

A draft standard in development under the IEEE Microprocessor Standards Committee group is *IEEE P1363: Standard for RSA, Diffie-Hellman and related public-key cryptography*, which includes specifications for elliptic curve systems.

Theoretical justification for the redundancy scheme used in ISO/IEC 9796 is given by Guillou et al. [525]. The customary 5-year review of this standard in 1996 resulted in a title change and the creation of a second part. The original standard (with content unchanged) will be re-titled *Digital signature schemes giving message recovery – Part 1: Mechanisms using redundancy*. The second part, a working draft (WD) as of April 1996 titled *Part 2: Mechanisms using a hash function*, specifies mechanisms utilizing the idea that when a signature algorithm such as RSA is used with a hash function, and the RSA modulus (say 1024 bits) is much larger than a hash value (say 160 bits), the remaining bits may be used to carry message text which can be recovered upon signature verification. This *partial message recovery* mode of the signature algorithm decreases the amount of accompanying cleartext required, which is of interest in bandwidth or memory-limited applications, and those wherein the text being signed is relatively small.

The Registration Authority designated by ISO/IEC to maintain the register of cryptographic algorithms of ISO/IEC 9979 is the National Computer Centre, Oxford Road, Manchester, M1 7ED, United Kingdom (telephone +44-161-228-6333, fax +44-161-228-1636). Twelve algorithms were registered as of October 1995: BARAS, B-Crypt, CDMF, DES, FEAL, IDEA, LUC, MULTI2, RC2, RC4, SXAL/MBAL, and SKIPJACK. An alternative for obtaining unique algorithm identifiers is the *object identifier* (OID) and registration scheme of the Abstract Syntax Notation One (ASN.1) standard ISO/IEC 8824; for more information, see Ford [414, pp.478-480].

For a history of DES-related standards from an American perspective, including ANSI standards, see Smid and Branstad [1156]. ANSI X9.24, Annex C contains a convenient six-page summary of ANSI X9.17. A revision of X9.30–2:1993 is to specify FIPS 180–1 (SHA–1) in place of SHA. An ANSI standard in development, but currently "on hold" pending resolution of patent issues, is (draft) X9.44 [48], which specifies a key transport technique based on RSA. An enhanced mode of triple-DES encryption included in the draft ANSI X9.52 [50] is *cipher block chaining with output feedback masking*. The draft ANSI X9.57 [52] is intended for use with X9.30 and (draft) X9.31, although the initial version addresses X9.30 (DSA) certificates. ITU-T X.509 v3 certificates and certificate extensions to which ANSI X9.55 is aligned are discussed below. Both (draft) X9.45 and (draft) X9.55 may eventually be incorporated into X9.57. Related to attribute certificates, see Fischer [410] regarding electronic document authorization and related patents [408, 409].

The ISO 11568 retail key management project includes six parts [588, 589, 590, 591, 592, 593]. Among these, 11568-3 specifies the key life cycle for symmetric encryption algorithms; 11568–4 addresses key management techniques for public-key cryptosystems, including certificate management and (in Annex C) attribute certificates; and 11568–5 addresses key life cycle for public-key cryptosystems.

ISO/IEC 9594-8 (X.509) is one part of a series of specifications outlining directory services for Open Systems Interconnection (OSI) and other systems. The *Directory* is a logical database of information with directory entries arranged in a tree structure, the *Directory Information Tree* (DIT), as introduced in ISO/IEC 9594–1 (ITU-T Recommendation X.500) [594], which also provides an overview of directory services. For extension discussion, see Chapter 14 of Ford [414]. The 1988 version of X.509 (equivalent to ISO/IEC 9594-8:1990) was updated in 1993 [626] (equivalent to ISO/IEC 9594-8:1995). A 1995 technical corrigendum [627] added a certificate *extensions* field, yielding Version 3 (v3) certificates. Standard extensions for v3 certificates are defined in a further amendment [628] (see §13.9). The OSI security frameworks project is specified in seven parts of ISO 10181 [609, 610, 611, 612, 613, 614, 615].

FIPS 140–1 [401] supersedes FIPS 140, *General Security Requirements for Equipment Using the Data Encryption Standard* (formerly Federal Standard 1027, April 1982). Information on FS 1027 is provided by Davies and Price [308]. In May 1994, NIST announced a weakness in SHA [403], resulting from unpublished analysis carried out by the U.S. National Security Agency; the formal revision was published as FIPS 180–1 [404].

The PKCS standards, developed by industrial collaboration lead by RSA Laboratories (a Division of RSA Data Security Inc.), are widely used in practice, and periodically updated. PKCS #1,3,5,6,7,8,9,10 [1072] and PKCS #11 [1071] are currently available (e.g., from URL http://www.rsa.com/).

For an overview of Internet security standards, see Kent [667]. Linn's GSS-API (RFC 1508) [1040] is an API suitable for session-oriented applications. An analogous specification for store-and-forward applications is the IDUP-GSS-API (Independent Data Unit Protection GSS-API) interface. Implementation mechanisms which have been specified to plug in beneath GSS-API include a symmetric-key mechanism based on Kerberos (the Kerberos Version 5 GSS-API mechanism), and a public-key based mechanism SPKM (*Simple Public-Key Mechanism*). For an overview of these work-in-progress items under development in the Common Authentication Technologies (CAT) group of the IETF, see Adams [4].

Work-in-progress in the IP Security (IPSEC) working group of the IETF includes two items using Diffie-Hellman key exchange for session key establishment over the Internet – the Photuris protocol of Karn and Simpson, and the SKIP protocol of Aziz. Krawczyk [718] notes these and presents an alternative (SKEME).

MIME, specified in RFC 1521 [1042], is designed to facilitate multipart textual and non-textual mail, i.e., mail messages whose bodies may contain multiple objects of a variety of content types including non-ASCII text, multi-font text, and audio and image fragments. An alternative to the MOSS proposal of RFC 1848 [1046] is S/MIME [1191], which adds signature and/or encryption services to MIME messages, using PKCS specifications.

Many other standards, both formal and informal, have been developed or are undergoing development. A collection of cryptographic algorithms and protocols recommended for use in Europe is that resulting from the European RACE Integrity Primitives Evaluation (RIPE) project; see Bosselaers and Preneel [178]. Pretty Good Privacy (PGP) is a popular, widely available software package originally developed by Zimmermann [1272] (see Garfinkel [442] for additional perspective), currently employing RSA signatures, MD5 hashing, and IDEA encipherment.

Examples of pseudorandom number generators (PRNGs) which appear in U.S. standards include a DES-based PRNG in ANSI X9.17 (Appendix C), and two further methods in FIPS 186 (Appendix 3) based on both the Secure Hash Algorithm (SHA) and DES.

Bibliography of Papers from Selected Cryptographic Forums

Contents in Brief

A.1 Asiacrypt/Auscrypt Proceedings

Advances in Cryptology – **AUSCRYPT '90**. Springer-Verlag LNCS 453 (1990).
Editors: J. Seberry and J. Pieprzyk.

V.S. Alagar, *Range equations and range matrices: A study in statistical database security,* 360–385.

M. Ames, *Secure cryptographic initialization,* 451–462.

M.H.G. Anthony, K.M. Martin, J. Seberry, P. Wild, *Some remarks on authentication systems,* 122–139.

L. Brown, J. Pieprzyk, J. Seberry, *LOKI – a cryptographic primitive for authentication and secrecy applications,* 229–236.

L. Brown, J. Seberry, *Key scheduling in DES type cryptosystems,* 221–228.

J.M. Carroll, *The three faces of information security,* 433–450.

D. Chaum, *Showing credentials without identification: Transferring signatures between unconditionally unlinkable pseudonyms,* 246–264.

R.H. Cooper, W. Patterson, *RSA as a benchmark for multiprocessor machines,* 356–359.

Z.-D. Dai, K. Zeng, *Continued fractions and Berlekamp-Massey algorithm,* 24–31.

E. Dawson, B. Goldburg, *Universal logic sequences,* 426–432.

C. Ding, *Lower bounds on the weight complexities of cascaded binary sequences,* 39–43.

R. Ferreira, *The practical application of state of the art security in real environments,* 334–355.

K. Gaarder, E. Snekkenes, *On the formal analysis of PKCS authentication protocols,* 106–121.

W. Geiselmann, D. Gollmann, *VLSI design for exponentiation in $GF(2^n)$,* 398–405.

M. Girault, *A (non-practical) three-pass identification protocol using coding theory,* 265–272.

G. Guang, *Nonlinear generators of binary sequences with controllable complexity and double key,* 32–36.

H. Gustafson, E. Dawson, B. Caelli, *Comparison of block ciphers,* 208–220.

T. Hardjono, *Record encryption in distributed databases,* 386–395.

B. Hayes, *Anonymous one-time signatures and flexible untraceable electronic cash,* 294–305.

C.J.A. Jansen, D.E. Boekee, *A binary sequence generator based on Ziv-Lempel source coding*, 156–164.

C.J.A. Jansen, D.E. Boekee, *On the significance of the directed acyclic word graph in cryptology*, 318–326.

S.J. Knapskog, *Formal specification and verification of secure communication protocols*, 58–73.

K. Koyama, *Direct demonstration of the power to break public-key cryptosystems*, 14–21.

P.J. Lee, *Secure user access control for public networks*, 46–57.

R. Lidl, W.B. Müller, *A note on strong Fibonacci pseudoprimes*, 311–317.

A. Menezes, S. Vanstone, *The implementation of elliptic curve cryptosystems*, 2–13.

M.J. Mihaljević, J.D. Golić, *A fast iterative algorithm for a shift register initial state reconstruction given the noisy output sequence*, 165–175.

H. Morita, *A fast modular-mulitplication module for smart cards*, 406–409.

M. Newberry, *Minòs: Extended user authentication*, 410–423.

K. Ohta, K. Koyama, *Meet-in-the-middle attack on digital signature schemes*, 140–154.

J. Pieprzyk, X.-M. Zhang, *Permutation generators of alternating groups*, 237–244.

R. Safavi-Naini, *Parallel generation of pseudo-random sequences*, 176–193.

H. Shizuya, K. Koyama, T. Itoh, *Demonstrating possession without revealing factors and its application*, 273–293.

J.C.A. van der Lubbe, D.E. Boekee, *KEYMEX: An expert system for the design of key management schemes*, 96–103.

V. Varadharajan, *Network security policy models*, 74–95.

Y.Y. Xian, *Dyadic matrices and their potential significance in cryptography*, 308–310.

Y.Y. Xian, *K-M sequence is forwardly predictable*, 37–38.

K. Zeng, M. Huang, *Solving equations in sequences*, 327–332.

K. Zeng, C.H. Yang, T.R.N. Rao, *Large primes in stream cipher cryptography*, 194–205.

Advances in Cryptology – ASIACRYPT '91. Springer-Verlag LNCS 739 (1993).
Editors: H. Imai, R.L. Rivest, and T. Matsumoto.

J. Brandt, I. Damgård, P. Landrock, *Speeding up prime number generation*, 440–449.

L. Brown, M. Kwan, J. Pieprzyk, J. Seberry, *Improving resistance to differential cryptanalysis and the redesign of LOKI*, 36–50.

J. Daemen, *Limitations of the Even-Mansour construction*, 495–498.

J. Daemen, A. Bosselaers, R. Govaerts, J. Vandewalle, *Collisions for Schnorr's hash function FFT-Hash presented at Crypto'91*, 477–480.

J. Daemen, R. Govaerts, J. Vandewalle, *A framework for the design of one-way hash functions including cryptanalysis of Damgård's one-way function based on a cellular automaton*, 82–96.

D.W. Davies, *The transition from mechanisms to electronic computers, 1940 to 1950*, 1–21.

Y. Desmedt, M. Burmester, *An efficient zero-knowledge scheme for the discrete logarithm based on smooth numbers*, 360–367.

S. Even, Y. Mansour, *A construction of a cipher from a single pseudorandom permutation*, 210–224.

J. Feigenbaum, R. Ostrovsky, *A note on one-prover, instance-hiding zero-knowledge proof systems*, 352–359.

L. Fortnow, M. Szegedy, *On the power of two-local random reductions*, 346–351.

B. Goldburg, E. Dawson, S. Sridharan, *A secure analog speech scrambler using the discrete cosine transform*, 299–311.

L. Harn, H.-Y. Lin, *An oblivious transfer protocol and its application for the exchange of secrets*, 312–320.

T. Itoh, K. Sakurai, *On the complexity of constant round ZKIP of possession of knowledge*, 331–345.

T. Itoh, K. Sakurai, H. Shizuya, *Any language in IP has a divertible ZKIP*, 382–396.

A. Joux, J. Stern, *Cryptanalysis of another knapsack cryptosystem*, 470–476.

T. Kaneko, *A known-plaintext attack of FEAL-4 based on the system of linear equations on difference*, 485–488.

K. Kim, *Construction of DES-like S-boxes based on Boolean functions satisfying the SAC*, 59–72.

A. Klapper, M. Goresky, *Revealing information with partial period correlations*, 277–287.

L.R. Knudsen, *Cryptanalysis of LOKI*, 22–35.

M. Kwan, *Simultaneous attacks in differential cryptanalysis (getting more pairs per encryption)*, 489–492.

M. Kwan, J. Pieprzyk, *A general purpose technique for locating key scheduling weaknesses in DES-like cryptosystems,* 237–246.

C.-S. Laih, L. Harn, *Generalized threshold cryptosystems,* 159–166.

C.-S. Laih, S.-M. Yen, L. Harn, *Two efficient server-aided secret computation protocols based on the addition sequence,* 450–459.

H.-Y. Lin, L. Harn, *A generalized secret sharing scheme with cheater detection,* 149–158.

J. Meijers, J. van Tilburg, *Extended majority voting and private-key algebraic-code encryptions,* 288–298.

A. Miyaji, *On ordinary elliptic curve cryptosystems,* 460–469.

H. Miyano, *A method to estimate the number of ciphertext pairs for differential cryptanalysis,* 51–58.

J.-I. Mizusawa, *IC-cards and telecommunication services,* 253–264.

S. Mjølsnes, *Privacy, cryptographic pseudonyms, and the state of health,* 493–494.

H. Morita, K. Ohta, S. Miyaguchi, *Results of switching-closure-test on FEAL,* 247–252.

W. Ogata, K. Kurosawa, *On claw free families,* 111–123.

K. Ohta, T. Okamoto, *A digital multisignature scheme based on the Fiat-Shamir scheme,* 139–148.

T. Okamoto, *An extension of zero-knowledge proofs and its applications,* 368–381.

J. Pieprzyk, B. Sadeghiyan, *Optimal perfect randomizers,* 225–236.

M.Y. Rhee, *Research activities on cryptology in Korea,* 179–193.

R.L. Rivest, *Cryptography and machine learning,* 427–439.

R.L. Rivest, *On NIST's proposed digital signature standard,* 481–484.

B. Sadeghiyan, J. Pieprzyk, *On necessary and sufficient conditions for the construction of super pseudorandom permutations,* 194–209.

B. Sadeghiyan, Y. Zheng, J. Pieprzyk, *How to construct a family of strong one-way permutations,* 97–110.

R. Safavi-Naini, *Feistel type authentication codes,* 167–178.

T. Saito, K. Kurosawa, K. Sakurai, *4 move perfect ZKIP of knowledge with no assumption,* 321–330.

A. Shimbo, S.-I. Kawamura, *Cryptanalysis of several conference key distribution schemes,* 265–276.

C. Shu, T. Matsumoto, H. Imai, *A multi-purpose proof system – for identity and membership proofs,* 397–411.

M.-J. Toussaint, *Formal verification of probabilistic properties in cryptographic protocols,* 412–426.

J.-H. Yang, Z.-D. Dai, K.-C. Zeng, *The data base of selected permutations,* 73–81.

Y. Zheng, T. Hardjono, J. Pieprzyk, *Sibling intractable function families and their applications,* 124–138.

Advances in Cryptology – **AUSCRYPT '92**. Springer-Verlag LNCS 718 (1993).
Editors: J. Seberry and Y. Zheng.

M. Bertilsson, I. Ingemarsson, *A construction of practical secret sharing schemes using linear block codes,* 67–79.

M. Cerecedo, T. Matsumoto, H. Imai, *Non-interactive generation of shared pseudorandom sequences,* 385–396.

C.-C. Chang, T.-C. Wu, C.-P. Chen, *The design of a conference key distribution system,* 459–466.

C. Charnes, J. Pieprzyk, *Linear nonequivalence versus nonlinearity,* 156–164.

L. Condie, *Prime generation with the Demytko-Miller-Trbovich algorithm,* 413–421.

E. Dawson, *Cryptanalysis of summation generator,* 209–215.

Y. Desmedt, *Threshold cryptosystems,* 3–14.

Y. Desmedt, J. Seberry, *Practical proven secure authentication with arbitration,* 27–32.

J. Detombe, S.E. Tavares, *Constructing large cryptographically strong S-boxes,* 165–181.

A. Fujioka, T. Okamoto, K. Ohta, *A practical secret voting scheme for large scale elections,* 244–251.

T. Hardjono, Y. Zheng, *A practical digital multisignature scheme based on discrete logarithms,* 122–132.

L. Harn, S. Yang, *Group-oriented undeniable signature schemes without the assistance of a mutually trusted party,* 133–142.

L. Harn, S. Yang, *Public-key cryptosystem based on the discrete logarithm problem,* 469–476.

A.P.L. Hiltgen, *Construction of feebly-one-way families of permutations,* 422–434.

W.-A. Jackson, K.M. Martin, *Cumulative arrays and geometric secret sharing schemes,* 48–55.

A. Klapper, *The vulnerability of geometric sequences based on fields of odd characteristic,* 327–338.

L.R. Knudsen, *Cryptanalysis of LOKI91,* 196–208.

V. Korzhik, V. Yakovlev, *Nonasymptotic estimates of information protection efficiency for the wire-tap channel concept*, 185–195.

X. Lai, R.A. Rueppel, J. Woollven, *A fast cryptographic checksum algorithm based on stream ciphers*, 339–348.

C.-S. Laih, S.-M. Yen, *Secure addition sequence and its applications on the server-aided secret computation protocols*, 219–230.

R. Lidl, W.B. Müller, *Primality testing with Lucas functions*, 539–542.

C.H. Lim, P.J. Lee, *Modified Maurer-Yacobi's scheme and its applications*, 308–323.

T. Matsumoto, H. Imai, C.-S. Laih, S.-M. Yen, *On verifiable implicit asking protocols for RSA computation*, 296–307.

M. Mihaljević, *An approach to the initial state reconstruction of a clock-controlled shift register based on a novel distance measure*, 349–356.

A. Miyaji, *Elliptic curves over F_p suitable for cryptosystems*, 479–491.

B.B. Nieh, S.E. Tavares, *Modelling and analyzing cryptographic protocols using Petri nets*, 275–295.

W. Ogata, K. Kurosawa, S. Tsujii, *Nonperfect secret sharing schemes*, 56–66.

C.M. O'Keefe, *A comparison of key distribution patterns constructed from circle geometries*, 517–527.

J.C. Paillès, *New protocols for electronic money*, 263–274.

M. Portz, *A generalized description of DES-based and Benes-based permutation generators*, 397–409.

B. Preneel, R. Govaerts, J. Vandewalle, *An attack on two hash functions by Zheng-Matsumoto-Imai*, 535–538.

B. Preneel, R. Govaerts, J. Vandewalle, *On the power of memory in the design of collision resistant hash functions*, 105–121.

M. Rezny, E. Trimarchi, *A block cipher method using combinations of different methods under the control of the user key*, 531–534.

R. Safavi-Naini, L. Tombak, *Authentication codes under impersonation attack*, 35–47.

K. Sakurai, T. Itoh, *On bit correlations among preimages of "many to one" one-way functions – a new approach to study on randomness and hardness of one-way functions*, 435–446.

K. Sakurai, T. Itoh, *Subliminal channels for signature transfer and their application to signature distribution schemes*, 231–243.

T. Satoh, K. Kurosawa, S. Tsujii, *Privacy for multi-party protocols*, 252–260.

J. Sauerbrey, *A modular exponentiation unit based on systolic arrays*, 505–516.

J. Seberry, X.-M. Zhang, *Highly nonlinear 0-1 balanced Boolean functions satisfying strict avalanche criterion*, 145–155.

J. Snare, *Information technology security standards – an Australian perspective*, 367–384.

L. Tombak, R. Safavi-Naini, *Authentication codes with perfect protection*, 15–26.

C.P. Waldvogel, J.L. Massey, *The probability distribution of the Diffie-Hellman key*, 492–504.

J.-H. Yang, Z.-D. Dai, *Construction of m-ary de Bruijn sequences*, 357–363.

S.-M. Yen, C.-S. Laih, *The fast cascade exponentiation algorithm and its applications on cryptography*, 447–456.

Y. Zheng, J. Pieprzyk, J. Seberry, *HAVAL – a one-way hashing algorithm with variable length of output*, 83–104.

E. Zuk, *Remarks on "The design of a conference key distribution system"*, 467–468.

Advances in Cryptology – **ASIACRYPT '94**. Springer-Verlag LNCS 917 (1995).
Editors: J. Pieprzyk and R. Safavi-Naini.

M. Abe, H. Morita, *Higher radix nonrestoring modular multiplication algorithm and public-key LSI architecture with limited hardware resources*, 365–375.

M. Alabbadi, S.B. Wicker, *A digital signature scheme based on linear error-correcting block codes*, 238–248.

D. Atkins, M. Graff, A.K. Lenstra, P.C. Leyland, *The magic words are SQUEAMISH OSSIFRAGE*, 263–277.

D. Beaver, *Factoring: The DNA solution*, 419–423.

P. Béguin, J.-J. Quisquater, *Secure acceleration of DSS signatures using insecure server*, 249–259.

T. Beth, *Multifeature security through homomorphic encryption*, 1–17.

E. Biham, *Cryptanalysis of multiple modes of operation*, 278–292.

E. Biham, A. Biryukov, *How to strengthen DES using existing hardware*, 398–412.

C. Boyd, W. Mao, *Design and analysis of key exchange protocols via secure channel identification*, 171–181.

G. Carter, A. Clark, L. Nielsen, *DESV-1: A variation of the data encryption standard (DES)*, 427–430.

X. Chang, Z.-D. Dai, G. Gong, *Some cryptographic properties of exponential functions*, 415–418.

C. Charnes, J. Pieprzyk, *Attacking the SL_2 hashing scheme*, 322–330.

S. Chee, S. Lee, K. Kim, *Semi-bent functions*, 107–118.

A. De Santis, T. Okamoto, G. Persiano, *Zero-knowledge proofs of computational power in the shared string model*, 182–192.

Y. Desmedt, G. Di Crescenzo, M. Burmester, *Multiplicative non-abelian sharing schemes and their application to threshold cryptography*, 21–32.

A. Fúster-Sabater, P. Caballero-Gil, *On the linear complexity of nonlinearly filtered PN-sequences*, 80–90.

J.D. Golić, *Intrinsic statistical weakness of keystream generators*, 91–103.

P. Horster, M. Michels, H. Petersen, *Meta-message recovery and meta-blind signature schemes based on the discrete logarithm problem and their applications*, 224–237.

H. Imai, *Information security aspects of spread spectrum systems*, 193–208.

W.-A. Jackson, K.M. Martin, C.M. O'Keefe, *On sharing many secrets*, 42–54.

K. Kurosawa, K. Okada, *Combinatorial interpretation of secret sharing schemes*, 55–64.

K. Kurosawa, K. Okada, K. Sakano, *Security of the center in key distribution schemes*, 333–341.

K. Kurosawa, K. Okada, S. Tsujii, *Low exponent attack against elliptic curve RSA*, 376–383.

T. Matsumoto, *Incidence structures for key sharing*, 342–353.

C.A. Meadows, *Formal verification of cryptographic protocols: a survey*, 133–150.

M. Mihaljević, *A correlation attack on the binary sequence generators with time-varying output function*, 67–79.

V. Niemi, A. Renvall, *How to prevent buying of votes in computer elections*, 164–170.

L. O'Connor, J.D. Golić, *A unified Markov approach to differential and linear cryptanalysis*, 387–397.

K. Okada, K. Kurosawa, *Lower bound on the size of shares of nonperfect secret sharing schemes*, 33–41.

J. Patarin, *Collisions and inversions for Damgård's whole hash function*, 307–321.

R. Safavi-Naini, L. Tombak, *Combinatorial structure of A-codes with r-fold security*, 211–223.

J. Seberry, X.-M. Zhang, Y. Zheng, *Structures of cryptographic functions with strong avalanche characteristics*, 119–132.

P. Smith, C. Skinner, *A public-key cryptosystem and a digital signature system based on the Lucas function analogue to discrete logarithms*, 357–364.

J. Stern, *Can one design a signature scheme based on error-correcting codes?*, 424–426.

T. Tokita, T. Sorimachi, M. Matsui, *Linear cryptanalysis of LOKI and $s^2 DES$*, 293–303.

Y. Yacobi, *Efficient electronic money*, 153–163.

A.2 Crypto Proceedings

ADVANCES IN CRYPTOGRAPHY – A Report on **CRYPTO 81**. ECE Rept No 82-04, Dept. of Electrical & Computer Engineering, University of California, Santa Barbara, CA, U.S.A., 1982. Editor: A. Gersho.

L.M. Adleman, *Primality testing* (abstract only), 10.

H.R. Amirazizi, M.E. Hellman, *Time-memory-processor tradeoffs* (abstract only), 7–9.

H.R. Amirazizi, E.D. Karnin, J.M. Reyneri, *Compact knapsacks are polynomially solvable* (abstract only), 17–19.

H.J. Beker, *Stream ciphers: Applications and techniques*, 121–123.

T.A. Berson, R.K. Bauer, *Local network cryptosystem architecture*, 73–78.

G.R. Blakley, *Key management from a security viewpoint* (abstract only), 82.

M. Blum, *Coin flipping by telephone: A protocol for solving impossible problems*, 11–15.

G. Brassard, *An optimally secure relativized cryptosystem*, 54–58.

D.L. Chaum, *Silo watching*, 138–139.

D.W. Davies, *Some regular properties of the DES* (abstract only), 41.

R.A. DeMillo, N.A. Lynch, M.J. Merritt, *The design and analysis of cryptographic protocols* (abstract only), 71.

W. Diffie, *Cryptographic technology: Fifteen year forecast*, 84–108.

S. Even, *A protocol for signing contracts*, 148–153.

M. Gasser, *Limitations of encryption to enforce mandatory security*, 130–134.

J.A. Gordon, *Towards a design procedure for cryptosecure substitution boxes* (abstract only), 53.

M.E. Hellman, E.D. Karnin, J. Reyneri, *On the necessity of cryptanalytic exhaustive search*, 2–6.

P.S. Henry, R.D. Nash, *Fast decryption algorithm for the knapsack cipher* (abstract only), 16.

E. Henze, *The solution of the general equation for public key distribution systems*, 140–141.

T. Herlestam, *On the feasibility of computing discrete logarithms using Adleman's subexponential algorithm*, 142–147.

I. Ingemarsson, *Are all injective knapsacks partly solvable after multiplication modulo q?*, 20–24.

J.P. Jordan, *A variant of a public key cryptosystem based on Goppa codes*, 25–30.

S.C. Kak, *Scrambling and randomization*, 59–63.

S.T. Kent, *Cryptographic techniques for protecting storage* (abstract only), 80.

A.G. Konheim, *A one-way sequence for transaction verification* (abstract only), 38.

A.L. Lang Jr., J. Vasak, *A methodology for evaluating the relative security of commercial COMSEC devices*, 124–129.

Y.A. Lau, T.R. McPherson, *Implementation of a hybrid RSA/DES key management system* (abstract only), 83.

L.-S. Lee, G.-C. Chou, *New results on sampling-based scrambling techniques for secure speech communications*, 115–119.

H. Meijer, S. Akl, *Digital signature schemes*, 65–70.

D.R. Morrison, *Subtractive encryptors – alternatives to the DES*, 42–52.

J.M. Nye, *Current market: Products, costs, trends*, 110–114.

J.M. Nye, *The import/export dilemma* (abstract only), 135–137.

S. Porter, *A password extension for improved human factors* (abstract only), 81.

G. Purdy, G. Simmons, J. Studier, *Software protection using "communal-key-cryptosystems"* (abstract only), 79.

B.P. Schanning, *MEMO: A hybrid approach to encrypted electronic mail* (abstract only), 64.

A. Shamir, *The generation of cryptographically strong pseudo-random sequences* (abstract only), 1.

G.J. Simmons, *A system for point-of-sale or access user authentication and identification*, 31–37.

M.E. Smid, *DES 81: An update*, 39–40.

S.B. Weinstein, *Security mechanism in electronic cards* (abstract only), 109.

A.D. Wyner, *Some thoughts on speech encryption* (abstract only), 120.

Advances in Cryptology – Proceedings of **CRYPTO 82**. Plenum Press (1983).
Editors: D. Chaum, R.L. Rivest, and A.T. Sherman.

L.M. Adleman, *Implementing an electronic notary public*, 259–265.

L.M. Adleman, *On breaking the iterated Merkle-Hellman public-key cryptosystem*, 303–308.

S.G. Akl, P.D. Taylor, *Cryptographic solution to a multilevel security problem*, 237–249.

G.M. Avis, S.E. Tavares, *Using data uncertainty to increase the crypto-complexity of simple private key enciphering schemes*, 139–143.

C.H. Bennett, G. Brassard, S. Breidbart, S. Wiesner, *Quantum cryptography, or unforgeable subway tokens*, 267–275.

T.A. Berson, *Local network cryptosystem architecture: Access control*, 251–258.

T.A. Berson, *Long key variants of DES*, 311–313.

G.R. Blakley, L. Swanson, *Infinite structures in information theory*, 39–50.

R. Blom, *Non-public key distribution*, 231–236.

L. Blum, M. Blum, M. Shub, *Comparison of two pseudo-random number generators*, 61–78.

G. Brassard, *On computationally secure authentication tags requiring short secret shared keys*, 79–86.

E.F. Brickell, *A fast modular multiplication algorithm with applications to two key cryptography*, 51–60.

E.F. Brickell, J.A. Davis, G.J. Simmons, *A preliminary report on the cryptanalysis of Merkle-Hellman knapsack cryptosystems*, 289–301.

E.F. Brickell, J.H. Moore, *Some remarks on the Herlestam-Johannesson algorithm for computing logarithms over* $GF(2^p)$, 15–19.

D. Chaum, *Blind signatures for untraceable payments*, 199–203.

D.W. Davies, *Some regular properties of the 'Data Encryption Standard' algorithm*, 89–96.

D.W. Davies, G.I.P. Parkin, *The average cycle size of the key stream in output feedback encipherment*, 97–98.

D. Dolev, S. Even, R.M. Karp, *On the security of ping-pong protocols*, 177–186.

D. Dolev, A. Wigderson, *On the security of multi-party protocols in distributed systems*, 167–175.

S. Even, O. Goldreich, *On the security of multi-party ping-pong protocols*, 315.

S. Even, O. Goldreich, A. Lempel, *A randomized protocol for signing contracts*, 205–210.

S. Goldwasser, S. Micali, A. Yao, *On signatures and authentication*, 211–215.

M.E. Hellman, J.M. Reyneri, *Drainage and the DES*, 129–131.

M.E. Hellman, J.M. Reyneri, *Fast computation of discrete logarithms in* $GF(q)$, 3–13.

R. Janardan, K.B. Lakshmanan, *A public-key cryptosystem based on the matrix cover NP-complete problem*, 21–37.

R.R. Jueneman, *Analysis of certain aspects of output feedback mode*, 99–127.

L. Longpré, *The use of public-key cryptography for signing checks*, 187–197.

M. Merritt, *Key reconstruction*, 321–322.

C. Mueller-Schloer, N.R. Wagner, *Cryptographic protection of personal data cards*, 219–229.

C. Nicolai, *Nondeterministic cryptography*, 323–326.

J.B. Plumstead, *Inferring a sequence produced by a linear congruence*, 317–319.

R.L. Rivest, *A short report on the RSA chip*, 327.

R.L. Rivest, A.T. Sherman, *Randomized encryption techniques*, 145–163.

A. Shamir, *A polynomial time algorithm for breaking the basic Merkle-Hellman cryptosystem*, 279–288.

R.S. Winternitz, *Security of a keystrem cipher with secret initial value*, 133–137.

Advances in Cryptology – Proceedings of **CRYPTO 83**. Plenum Press (1984).
Editor: D. Chaum.

S.G. Akl, *On the security of compressed encodings*, 209–230.

M. Blum, U.V. Vazirani, V.V. Vazirani, *Reducibility among protocols*, 137–146.

E.F. Brickell, *Solving low density knapsacks*, 25–37.

E.F. Brickell, J.C. Lagarias, A.M. Odlyzko, *Evaluation of the Adleman attack on multiply iterated knapsack cryptosystems*, 39–42.

D. Chaum, *Blind signature system*, 153.

D. Chaum, *Design concepts for tamper responding systems*, 387–392.

D.W. Davies, *Use of the 'signature token' to create a negotiable document*, 377–382.

M. Davio, Y. Desmedt, M. Fosséprez, R. Govaerts, J. Hulsbosch, P. Neutjens, P. Piret, J.-J. Quisquater, J. Vandewalle, P. Wouters, *Analytical characteristics of the DES*, 171–202.

J.A. Davis, D.B. Holdridge, *Factorization using the quadratic sieve algorithm*, 103–113.

D.E. Denning, *Field encryption and authentication*, 231–247.

T. ElGamal, *A subexponential-time algorithm for computing discrete logarithms over* $GF(p^2)$, 275–292.

S. Even, O. Goldreich, *Electronic wallet*, 383–386.

S. Even, O. Goldreich, *On the power of cascade ciphers*, 43–50.

B.W. Fam, *Improving the security of exponential key exchange*, 359–368.

O. Goldreich, *A simple protocol for signing contracts*, 133–136.

H. Jürgensen, D.E. Matthews, *Some results on the information theoretic analysis of cryptosystems*, 303–356.

J.C. Lagarias, *Knapsack public key cryptosystems and diophantine approximation*, 3–23.

R. Lidl, W.B. Müller, *Permutation polynomials in RSA-cryptosystems*, 293–301.

H. Ong, C.P. Schnorr, *Signatures through approximate respresentations by quadratic forms*, 117–131.

C. Pomerance, J.W. Smith, S.S. Wagstaff Jr., *New ideas for factoring large integers*, 81–85.

J.A. Reeds, N.J.A. Sloane, *Shift-register synthesis (modulo m)*, 249.

J.E. Sachs, S. Berkovits, *Probabilistic analysis and performance modelling of the 'Swedish' algorithm and modifications*, 253–273.

G.J. Simmons, *The prisoners' problem and the subliminal channel*, 51–67.

M.E. Spencer, S.E. Tavares, *A layered broadcaset cryptographic system*, 157–170.

T. Tedrick, *How to exchange half a bit*, 147–151.

U.V. Vazirani, V.V. Vazirani, *RSA bits are .732 + ϵ secure*, 369–375.

H.C. Williams, *An overview of factoring*, 71–80.

R.S. Winternitz, *Producing a one-way hash function from DES*, 203–207.

M.C. Wunderlich, *Factoring numbers on the massively parallel computer*, 87–102.

Advances in Cryptology – Proceedings of **CRYPTO 84**. Springer-Verlag LNCS 196 (1985). Editors: G.R. Blakley and D. Chaum.

S.G. Akl, H. Meijer, *A fast pseudo random permutation generator with applications to cryptology*, 269–275.

H. Beker, M. Walker, *Key management for secure electronic funds transfer in a retail environment*, 401–410.

C.H. Bennett, G. Brassard, *An update on quantum cryptography*, 475–480.

I.F. Blake, R.C. Mullin, S.A. Vanstone, *Computing logarithms in $GF(2^n)$*, 73–82.

G.R. Blakley, *Information theory without the finiteness assumption, I: Cryptosystems as group-theoretic objects*, 314–338.

G.R. Blakley, C. Meadows, *Security of ramp schemes*, 242–268.

M. Blum, S. Goldwasser, *An efficient probabilistic public-key encryption scheme which hides all partial information*, 289–299.

E.F. Brickell, *Breaking iterated knapsacks*, 342–358.

D. Chaum, *How to keep a secret alive: Extensible partial key, key safeguarding, and threshold systems*, 481–485.

D. Chaum, *New secret codes can prevent a computerized big brother*, 432–433.

S.-S. Chen, *On rotation group and encryption of analog signals*, 95–100.

B. Chor, O. Goldreich, *RSA/Rabin least significant bits are $1/2 + 1/poly(\log n)$ secure*, 303–313.

B. Chor, R.L. Rivest, *A knapsack type public key cryptosystem based on arithmetic in finite fields*, 54–65.

D.W. Davies, *A message authenticator algorithm suitable for a mainframe computer*, 393–400.

M. Davio, Y. Desmedt, J. Goubert, F. Hoornaert, J.-J. Quisquater, *Efficient hardware and software implementations for the DES*, 144–146.

J.A. Davis, D.B. Holdridge, *An update on factorization at Sandia National Laboratories*, 114.

Y. Desmedt, J.-J. Quisquater, M. Davio, *Dependence of output on input in DES: Small avalanche characteristics*, 359–376.

T. ElGamal, *A public key cryptosystem and a signature scheme based on discrete logarithms*, 10–18.

R.C. Fairfield, A. Matusevich, J. Plany, *An LSI digital encryption processor (DEP)*, 115–143.

R.C. Fairfield, R.L. Mortenson, K.B. Coulthart, *An LSI random number generator (RNG)*, 203–230.

S. Fortune, M. Merritt, *Poker protocols*, 454–464.

O. Goldreich, S. Goldwasser, S. Micali, *On the cryptographic applications of random functions*, 276–288.

S. Goldwasser, S. Micali, R.L. Rivest, *A "paradoxical" solution to the signature problem*, 467.

F. Hoornaert, J. Goubert, Y. Desmedt, *Efficient hardware implementation of the DES*, 147–173.

B.S. Kaliski, *Wyner's analog encryption scheme: Results of a simulation*, 83–94.

A.G. Konheim, *Cryptanalysis of ADFGVX enciperment systems*, 339–341.

S.C. Kothari, *Generalized linear threshold scheme*, 231–241.

A.C. Leighton, S.M. Matyas, *The history of book ciphers*, 101–113.

A.K. Leung, S.E. Tavares, *Sequence complexity as a test for cryptographic systems*, 468–474.

H. Ong, C.P. Schnorr, A. Shamir, *Efficient signature schemes based on polynomial equations*, 37–46.

N. Proctor, *A self-synchronizing cascaded cipher system with dynamic control of error propagation*, 174–190.

J.A. Reeds, J.L. Manferdelli, *DES has no per round linear factors*, 377–389.

S.C. Serpell, C.B. Brookson, B.L. Clark, *A prototype encryption system using public key*, 3–9.

A. Shamir, *Identity-based cryptosystems and signature schemes*, 47–53.

G.J. Simmons, *Authentication theory/coding theory*, 411–431.

T. Tedrick, *Fair exchange of secrets*, 434–438.

U.V. Vazirani, V.V. Vazirani, *Efficient and secure pseudo-random number generation*, 193–202.

N.R. Wagner, M.R. Magyarik, *A public key cryptosystem based on the word problem*, 19–36.

H.C. Williams, *Some public key crypto-functions as intractable as factorization*, 66–70.

M. Yung, *Cryptoprotocols: Subscription to a public key, the secret blocking and the multi-player mental poker game*, 439–453.

Advances in Cryptology – **CRYPTO '85**. Springer-Verlag LNCS 218 (1986).
Editor: H.C. Williams.

C.H. Bennett, G. Brassard, J.-M. Robert, *How to reduce your enemy's information*, 468–476.

R. Berger, S. Kannan, R. Peralta, *A framework for the study of cryptographic protocols*, 87–103.

G.R. Blakley, *Information theory without the finiteness assumption, II. Unfolding the DES*, 282–337.

G.R. Blakley, C. Meadows, G.B. Purdy, *Fingerprinting long forgiving messages*, 180–189.

E.F. Brickell, J.M. DeLaurentis, *An attack on a signature scheme proposed by Okamoto and Shiraishi*, 28–32.

D. Chaum, J.-H. Evertse, *Cryptanalysis of DES with a reduced number of rounds – sequences of linear factors in block ciphers*, 192–211.

B. Chor, O. Goldreich, S. Goldwasser, *The bit security of modular squaring given partial factorization of the modulos*, 448–457.

D. Coppersmith, *Another birthday attack*, 14–17.

D. Coppersmith, *Cheating at mental poker*, 104–107.

D. Coppersmith, *The real reason for Rivest's phenomenon*, 535–536.

C. Crépeau, *A secure poker protocol that minimizes the effect of player coalitions*, 73–86.

W. de Jonge, D. Chaum, *Attacks on some RSA signatures*, 18–27.

Y. Desmedt, *Unconditionally secure authentication schemes and practical and theoretical consequences*, 42–55.

Y. Desmedt, A.M. Odlyzko, *A chosen text attack on the RSA cryptosystem and some discrete logarithm schemes*, 516–522.

W. Diffie, *Security for the DoD transmission control protocol*, 108–127.

T. ElGamal, *On computing logarithms over finite fields*, 396–402.

D. Estes, L.M. Adleman, K. Kompella, K.S. McCurley, G.L. Miller, *Breaking the Ong-Schnorr-Shamir signature scheme for quadratic number fields*, 3–13.

S. Even, O. Goldreich, A. Shamir, *On the security of ping-pong protocols when implemented using the RSA*, 58–72.

J. Feigenbaum, *Encrypting problem instances: Or . . . , can you take advantage of someone without having to trust him?*, 477–488.

H. Fell, W. Diffie, *Analysis of a public key approach based on polynomial substitution*, 340–349.

Z. Galil, S. Haber, M. Yung, *Symmetric public-key encryption*, 128–137.

P. Godlewski, G.D. Cohen, *Some cryptographic aspects of Womcodes*, 458–467.

J.R. Gosler, *Software protection: Myth or reality?*, 140–157.

J. Håstad, *On using RSA with low exponent in a public key network*, 403–408.

W. Haemers, *Access control at the Netherlands Postal and Telecommunications Services*, 543–544.

A. Herzberg, S. Pinter, *Public protection of software*, 158–179.

B.S. Kaliski Jr., R.L. Rivest, A.T. Sherman, *Is DES a pure cipher? (Results of more cycling experiments on DES)*, 212–226.

M. Kochanski, *Developing an RSA chip*, 350–357.

M. Luby, C. Rackoff, *How to construct pseudo-random permutations from pseudo-random functions*, 447.

V.S. Miller, *Use of elliptic curves in cryptography*, 417–426.

T.E. Moore, S.E. Tavares, *A layered approach to the design of private key cryptosystems*, 227–245.

E. Okamoto, K. Nakamura, *Lifetimes of keys in cryptographic key management systems*, 246–259.

J.-J. Quisquater, Y. Desmedt, M. Davio, *The importance of "good" key scheduling schemes (how to make a secure DES scheme with $\leq$ 48 bit keys?),* 537–542.

J.H. Reif, J.D. Tygar, *Efficient parallel pseudo-random number generation,* 433–446.

R.A. Rueppel, *Correlation immunity and the summation generator,* 260–272.

A. Shamir, *On the security of DES,* 280–281.

T. Siegenthaler, *Design of combiners to prevent divide and conquer attacks,* 273–279.

G.J. Simmons, *A secure subliminal channel (?),* 33–41.

N.M. Stephens, *Lenstra's factorisation method based on elliptic curves,* 409–416.

J. van Tilburg, D.E. Boekee, *Divergence bounds on key equivocation and error probability in cryptanalysis,* 489–513.

V. Varadharajan, *Trapdoor rings and their use in cryptography,* 369–395.

A.F. Webster, S.E. Tavares, *On the design of S-boxes,* 523–534.

H.C. Williams, *An M^3 public-key encryption scheme,* 358–368.

S. Wolfram, *Cryptography with cellular automata,* 429–432.

Advances in Cryptology – **CRYPTO '86**. Springer-Verlag LNCS 263 (1987).
Editor: A.M. Odlyzko.

P. Barrett, *Implementing the Rivest Shamir and Adleman public key encryption algorithm on a standard digital signal processor,* 311–323.

P. Beauchemin, G. Brassard, C. Crépeau, C. Goutier, *Two observations on probabilistic primality testing,* 443–450.

J.C. Benaloh, *Cryptographic capsules: A disjunctive primitive for interactive protocols,* 213–222.

J.C. Benaloh, *Secret sharing homomorphisms: Keeping shares of a secret secret,* 251–260.

T. Beth, B.M. Cook, D. Gollmann, *Architectures for exponentiation in $GF(2^n)$,* 302–310.

G.R. Blakley, R.D. Dixon, *Smallest possible message expansion in threshold schemes,* 266–274.

G. Brassard, C. Crépeau, *Zero-knowledge simulation of Boolean circuits,* 223–233.

G. Brassard, C. Crépeau, J.-M. Robert, *All-or-nothing disclosure of secrets,* 234–238.

E.F. Brickell, J.H. Moore, M.R. Purtill, *Structure in the S-boxes of the DES,* 3–8.

J.J. Cade, *A modification of a broken public-key cipher,* 64–83.

A.H. Chan, R.A. Games, *On the linear span of binary sequences obtained from finite geometries,* 405–417.

D. Chaum, *Demonstrating that a public predicate can be satisfied without revealing any information about how,* 195–199.

D. Chaum, J.-H. Evertse, *A secure and privacy-protecting protocol for transmitting personal information between organizations,* 118–167.

D. Chaum, J.-H. Evertse, J. van de Graaf, R. Peralta, *Demonstrating possession of a discrete logarithm without revealing it,* 200–212.

C. Crépeau, *A zero-knowledge poker protocol that achieves confidentiality of the players' strategy or how to achieve an electronic poker face,* 239–247.

W. de Jonge, D. Chaum, *Some variations on RSA signatures and their security,* 49–59.

Y. Desmedt, *Is there an ultimate use of cryptography?,* 459–463.

Y. Desmedt, J.-J. Quisquater, *Public-key systems based on the difficulty of tampering (Is there a difference between DES and RSA?),* 111–117.

A. Fiat, A. Shamir, *How to prove yourself: Practical solutions to identification and signature problems,* 186–194.

O. Goldreich, *Towards a theory of software protection,* 426–439.

O. Goldreich, *Two remarks concerning the Goldwasser-Micali-Rivest signature scheme,* 104–110.

O. Goldreich, S. Micali, A. Wigderson, *How to prove all NP statements in zero-knowledge, and a methodology of cryptographic protocol design,* 171–185.

L.C. Guillou, M. Ugon, *Smart card – a highly reliable and portable security device,* 464–479.

R. Gyoery, J. Seberry, *Electronic funds transfer point of sale in Australia,* 347–377.

N.S. James, R. Lidl, H. Niederreiter, *Breaking the Cade cipher,* 60–63.

R.R. Jueneman, *A high speed manipulation detection code,* 327–346.

B.S. Kaliski Jr., *A pseudo-random bit generator based on elliptic logarithms,* 84–103.

S.M. Matyas, *Public-key registration,* 451–458.

S. Micali, C. Rackoff, B. Sloan, *The notion of security for probabilistic cryptosystems*, 381–392.

J.H. Moore, G.J. Simmons, *Cycle structure of the DES with weak and semi-weak keys*, 9–32.

G.A. Orton, M.P. Roy, P.A. Scott, L.E. Peppard, S.E. Tavares, *VLSI implementation of public-key encryption algorithms*, 277–301.

G. Rankine, *THOMAS - a complete single chip RSA device*, 480–487.

T.R.N. Rao, K.-H. Nam, *Private-key algebraic-coded cryptosystems*, 35–48.

D.R. Stinson, *Some constructions and bounds for authentication codes*, 418–425.

M. Tompa, H. Woll, *How to share a secret with cheaters*, 261–265.

N.R. Wagner, P.S. Putter, M.R. Cain, *Large-scale randomization techniques*, 393–404.

Advances in Cryptology – **CRYPTO '87**. Springer-Verlag LNCS 293 (1988).
Editor: C. Pomerance.

C.M. Adams, H. Meijer, *Security-related comments regarding McEliece's public-key cryptosystem*, 224–228.

P. Beauchemin, G. Brassard, *A generalization of Hellman's extension of Shannon's approach to cryptography*, 461.

G.R. Blakley, W. Rundell, *Cryptosystems based on an analog of heat flow*, 306–329.

E.F. Brickell, D. Chaum, I.B. Damgård, J. van de Graaf, *Gradual and verifiable release of a secret*, 156–166.

E.F. Brickell, P.J. Lee, Y. Yacobi, *Secure audio teleconference*, 418–426.

D. Chaum, C. Crépeau, I. Damgård, *Multiparty unconditionally secure protocols*, 462.

D. Chaum, I.B. Damgård, J. van de Graaf, *Multiparty computations ensuring privacy of each party's input and correctness of the result*, 87–119.

C. Crépeau, *Equivalence between two flavours of oblivious transfers*, 350–354.

G.I. Davida, F.B. Dancs, *A crypto-engine*, 257–268.

G.I. Davida, B.J. Matt, *Arbitration in tamper proof systems (If DES $\approx$ RSA then what's the difference between true signature and arbitrated signature schemes?)*, 216–222.

A. De Santis, S. Micali, G. Persiano, *Non-interactive zero-knowledge proof systems*, 52–72.

J.M. DeLaurentis, *Components and cycles of a random function*, 231–242.

Y. Desmedt, *Society and group oriented cryptography: A new concept*, 120–127.

Y. Desmedt, C. Goutier, S. Bengio, *Special uses and abuses of the Fiat-Shamir passport protocol*, 21–39.

F.A. Feldman, *Fast spectral tests for measuring nonrandomness and the DES*, 243–254.

W. Fumy, *On the F-function of FEAL*, 434–437.

Z. Galil, S. Haber, M. Yung, *Cryptographic computation: Secure fault-tolerant protocols and the public-key model*, 135–155.

O. Goldreich, R. Vainish, *How to solve any protocol problem - an efficient improvement*, 73–86.

L. Guillou, J.-J. Quisquater, *Efficient digital public-key signatures with shadow*, 223.

M.P. Herlihy, J.D. Tygar, *How to make replicated data secure*, 379–391.

R. Impagliazzo, M. Yung, *Direct minimum-knowledge computations*, 40–51.

R.A. Kemmerer, *Analyzing encryption protocols using formal verification techniques*, 289–305.

K. Koyama, K. Ohta, *Identity-based conference key distribution systems*, 175–184.

M. Luby, C. Rackoff, *A study of password security*, 392–397.

Y. Matias, A. Shamir, *A video scrambling technique based on space filling curves*, 398–417.

T. Matsumoto, H. Imai, *On the key predistribution system: A practical solution to the key distribution problem*, 185–193.

R.C. Merkle, *A digital signature based on a conventional encryption function*, 369–378.

J.H. Moore, *Strong practical protocols*, 167–172.

E. Okamoto, *Key distribution systems based on identification information*, 194–202.

K. Presttun, *Integrating cryptography in ISDN*, 9–18.

W.L. Price, *Standards for data security – a change of direction*, 3–8.

J.-J. Quisquater, *Secret distribution of keys for public-key systems*, 203–208.

J.-J. Quisquater, J.-P. Delescaille, *Other cycling tests for DES*, 255–256.

T.R.N. Rao, *On Struik-Tilburg cryptanalysis of Rao-Nam scheme*, 458–460.

G.J. Simmons, *An impersonation-proof identity verification scheme,* 211–215.

G.J. Simmons, *A natural taxonomy for digital information authentication schemes,* 269–288.

D.R. Stinson, *A construction for authentication/secrecy codes from certain combinatorial designs,* 355–366.

D.R. Stinson, S.A. Vanstone, *A combinatorial approach to threshold schemes,* 330–339.

R. Struik, J. van Tilburg, *The Rao-Nam scheme is insecure against a chosen-plaintext attack,* 445–457.

H. Tanaka, *A realization scheme for the identity-based cryptosystem,* 340–349.

J. van de Graaf, R. Peralta, *A simple and secure way to show the validity of your public key,* 128–134.

Y. Yacobi, *Attack on the Koyama-Ohta identity based key distribution scheme,* 429–433.

K.C. Zeng, J.H. Yang, Z.T. Dai, *Patterns of entropy drop of the key in an S-box of the DES,* 438–444.

Advances in Cryptology – **CRYPTO '88**. Springer-Verlag LNCS 403 (1990).
Editor: S. Goldwasser.

M. Abadi, E. Allender, A. Broder, J. Feigenbaum, L.A Hemachandra, *On generating solved instances of computational problems,* 297–310.

L.M. Adleman, *An abstract theory of computer viruses,* 354–374.

E. Bach, *Intractable problems in number theory,* 77–93.

M. Bellare, S. Micali, *How to sign given any trapdoor function,* 200–215.

M. Ben-Or, O. Goldreich, S. Goldwasser, J. Håstad, J. Kilian, S. Micali, P. Rogaway, *Everything provable is provable in zero-knowledge,* 37–56.

J. Benaloh, J. Leichter, *Generalized secret sharing and monotone functions,* 27–35.

M. Blum, P. Feldman, S. Micali, *Proving security against chosen ciphertext attacks,* 256–268.

J. Brandt, I.B. Damgård, P. Landrock, T. Pedersen, *Zero-knowledge authentication scheme with secret key exchange,* 583–588.

G. Brassard, I.B. Damgård, *"Practical IP" ⊆ MA,* 580–582.

E.F. Brickell, D.R. Stinson, *The detection of cheaters in threshold schemes,* 564–577.

D. Chaum, A. Fiat, M. Naor, *Untraceable electronic cash,* 319–327.

C. Crépeau, J. Kilian, *Weakening security assumptions and oblivious transfer,* 2–7.

I.B. Damgård, *On the randomness of Legendre and Jacobi sequences,* 163–172.

I.B. Damgård, *Payment systems and credential mechanisms with provable security against abuse by individuals,* 328–335.

A. De Santis, S. Micali, G. Persiano, *Non-interactive zero-knowledge with preprocessing,* 269–282.

M. De Soete, *Bounds and constructions for authentication-secrecy codes with splitting,* 311–317.

B. den Boer, *Diffie-Hellman is as strong as discrete log for certain primes,* 530–539.

Y. Desmedt, *Abuses in cryptography and how to fight them,* 375–389.

C. Dwork, L. Stockmeyer, *Zero-knowledge with finite state verifiers,* 71–75.

U. Feige, A. Shamir, M. Tennenholtz, *The noisy oracle problem,* 284–296.

R. Forré, *The strict avalanche criterion: Spectral properties of Boolean functions and an extended definition,* 450–468.

M. Girault, P. Toffin, B. Vallée, *Computation of approximate L-th roots modulo n and application to cryptography,* 100–117.

O. Goldreich, H. Krawczyk, M. Luby, *On the existence of pseudorandom generators,* 146–162.

O. Goldreich, E. Kushilevitz, *A perfect zero-knowledge proof for a problem equivalent to discrete logarithm,* 57–70.

L.C. Guillou, J.-J. Quisquater, *A "paradoxical" identity-based signature scheme resulting from zero-knowledge,* 216–231.

B.J. Herbison, *Developing Ethernet enhanced-security system,* 507–519.

M.-D.A. Huang, S.-H. Teng, *A universal problem in secure and verifiable distributed computation,* 336–352.

T. Hwang, T.R.N. Rao, *Secret error-correcting codes (SECC),* 540–563.

R. Impagliazzo, S. Rudich, *Limits on the provable consequences of one-way permutations,* 8–26.

N. Koblitz, *A family of Jacobians suitable for discrete log cryptosystems,* 94–99.

S.A. Kurtz, S.R. Mahaney, J.S. Royer, *On the power of 1-way functions,* 578–579.

R.T.C. Kwok, M. Beale, *Aperiodic linear complexities of de Bruijn sequences,* 479–482.

M. Lucks, *A constraint satisfaction algorithm for the automated decryption of simple substitution ciphers*, 132–144.

T. Matsumoto, K. Kato, H. Imai, *Speeding up secret computations with insecure auxiliary devices*, 497–506.

S. Micali, C.P. Schnorr, *Efficient, perfect random number generators*, 173–198.

S. Micali, A. Shamir, *An improvement of the Fiat-Shamir identification and signature scheme*, 244–247.

K. Ohta, T. Okamoto, *A modification of the Fiat-Shamir scheme*, 232–243.

C. Rackoff, *A basic theory of public and private cryptosystems*, 249–255.

J.R. Sherwood, V.A. Gallo, *The application of smart cards for RSA digital signatures in a network comprising both interactive and store-and-forwarded facilities*, 484–496.

G.J. Simmons, *How to (really) share a secret*, 390–448.

D.G. Steer, L. Strawczynski, W. Diffie, M. Wiener, *A secure audio teleconference system*, 520–528.

J. van Tilburg, *On the McEliece public-key cryptosystem*, 119–131.

K. Zeng, M. Huang, *On the linear syndrome method in cryptanalysis*, 469–478.

Advances in Cryptology – **CRYPTO '89**. Springer-Verlag LNCS 435 (1990).
Editor: G. Brassard.

C. Adams, S. Tavares, *Good S-boxes are easy to find*, 612–615.

P. Barrett, R. Eisele, *The smart diskette – a universal user token and personal crypto-engine*, 74–79.

D. Beaver, *Multiparty protocols tolerating half faulty processors*, 560–572.

D. Beaver, S. Goldwasser, *Multiparty computation with faulty majority*, 589–590.

M. Bellare, L. Cowen, S. Goldwasser, *On the structure of secret key exchange protocols*, 604–605.

M. Bellare, S. Goldwasser, *New paradigms for digital signatures and message authentication based on non-interactive zero knowledge proofs*, 194–211.

M. Bellare, S. Micali, *Non-interactive oblivious transfer and applications*, 547–557.

M. Ben-Or, S. Goldwasser, J. Kilian, A. Wigderson, *Efficient identification schemes using two prover interactive proofs*, 498–506.

A. Bender, G. Castagnoli, *On the implementation of elliptic curve cryptosystems*, 186–192.

J. Bos, M. Coster, *Additon chain heuristics*, 400–407.

J. Boyar, R. Peralta, *On the concrete complexity of zero-knowledge proofs*, 507–525.

R.L. Brand, *Problems with the normal use of cryptography for providing security on unclassified networks*, 30–34.

E.F. Brickell, *A survey of hardware implementations of RSA*, 368–370.

E.F. Brickell, D.M. Davenport, *On the classification of ideal secret sharing schemes*, 278–285.

J.A. Buchmann, H.C. Williams, *A key exchange system based on real quadratic fields*, 335–343.

A.H. Chan, R.A. Games, *On the quadratic spans of periodic sequences*, 82–89.

D. Chaum, *The Spymasters double-agent problem: Multiparty computations secure unconditionally from minorities and cryptographically from majorities*, 591–602.

D. Chaum, H. van Antwerpen, *Undeniable signatures*, 212–216.

G.C. Chick, S.E. Tavares, *Flexible access control with master keys*, 316–322.

B. Chor, E. Kushilevitz, *Secret sharing over infinite domains*, 299–306.

R. Cleve, *Controlled gradual disclosure schemes for random bits and their applications*, 573–588.

I.B. Damgård, *A design principle for hash functions*, 416–427.

I.B. Damgård, *On the existence of bit commitment schemes and zero-knowledge proofs*, 17–27.

M. De Soete, J.-J. Quisquater, K. Vedder, *A signature with shared verification scheme*, 253–262.

Y.G. Desmedt, *Making conditionally secure cryptosystems unconditionally abuse-free in a general context*, 6–16.

Y.G. Desmedt, Y. Frankel, *Threshold cryptosystems*, 307–315.

S. Even, O. Goldreich, S. Micali, *On-line/off-line digital signatures*, 263–275.

U. Feige, A. Shamir, *Zero knowledge proofs of knowledge in two rounds*, 526–544.

D.C. Feldmeier, P.R. Karn, *UNIX password security – ten years later*, 44–63.

A. Fiat, *Batch RSA*, 175–185.

P.A. Findlay, B.A. Johnson, *Modular exponentiation using recursive sums of residues*, 371–386.

O. Goldreich, H. Krawczyk, *Sparse pseudorandom distributions*, 113–127.

C.J.A. Jansen, D.E. Boekee, *The shortest feedback shift register that can generate a given sequence*, 90–99.

D. Kahn, *Keying the German navy's Enigma*, 2–5.

J. Kilian, S. Micali, R. Ostrovsky, *Minimum resource zero-knowledge proofs*, 545–546.

J.T. Kohl, *The use of encryption in Kerberos for network authentication*, 35–43.

H. Krawczyk, *How to predict congruential generators*, 138–153.

C.-S. Laih, L. Harn, J.-Y. Lee, T. Hwang, *Dynamic threshold scheme based on the definition of cross-product in an n-dimensional linear space*, 286–298.

S.S. Magliveras, N.D. Memon, *Properties of cryptosystem PGM*, 447–460.

U.M. Maurer, J.L. Massey, *Perfect local randomness in pseudo-random sequences*, 100–112.

R.C. Merkle, *A certified digital signature*, 218–238.

R.C. Merkle, *One way hash functions and DES*, 428–446.

S. Miyaguchi, *The FEAL - 8 cryptosystem and a call for attack*, 624–627.

H. Morita, *A fast modular-multiplication algorithm based on a higher radix*, 387–399.

M. Naor, *Bit commitment using pseudo-randomness*, 128–136.

R. Nelson, J. Heimann, *SDNS architecture and end-to-end encryption*, 356–366.

T. Okamoto, K. Ohta, *Disposable zero-knowledge authentications and their applications to untraceable electronic cash*, 481–496.

R. Ostrovsky, *An efficient software protection scheme*, 610–611.

B. Preneel, A. Bosselaers, R. Govaerts, J. Vandewalle, *A chosen text attack on the modified cryptographic checksum algorithm of Cohen and Huang*, 154–163.

W.L. Price, *Progress in data security standardisation*, 620–623.

J.-J. Quisquater, J.-P. Delescaille, *How easy is collision search. New results and applications to DES*, 408–413.

J.-J. Quisquater, L. Guillou, T. Berson, *How to explain zero-knowledge protocols to your children*, 628–631.

C.P. Schnorr, *Efficient identification and signatures for smart cards*, 239–252.

A. Shamir, *An efficient identification scheme based on permuted kernels*, 606–609.

J.M. Smith, *Practical problems with a cryptographic protection scheme*, 64–73.

M. Tatebayashi, N. Matsuzaki, D.B. Newman Jr., *Key distribution protocol for digital mobile communication systems*, 324–334.

S.R. White, *Covert distributed processing with computer viruses*, 616–619.

Y. Yacobi, Z. Shmuely, *On key distribution systems*, 344–355.

K. Zeng, C.H. Yang, T.R.N. Rao, *On the linear consistency test (LCT) in cryptanalysis with applications*, 164–174.

Y. Zheng, T. Matsumoto, H. Imai, *On the construction of block ciphers provably secure and not relying on any unproved hypotheses*, 461–480.

Advances in Cryptology – **CRYPTO '90**. Springer-Verlag LNCS 537 (1991).
Editors: A.J. Menezes and S.A. Vanstone.

D. Beaver, J. Feigenbaum, J. Kilian, P. Rogaway, *Security with low communication overhead*, 62–76.

D. Beaver, J. Feigenbaum, V. Shoup, *Hiding instances in zero-knowledge proof systems*, 326–338.

T. Beth, Y. Desmedt, *Identification tokens – or: Solving the chess grandmaster problem*, 169–176.

E. Biham, A. Shamir, *Differential cryptanalysis of DES-like cryptosystems*, 2–21.

J. Boyar, D. Chaum, I.B. Damgård, T. Pedersen, *Convertible undeniable signatures*, 189–205.

G. Brassard, C. Crépeau, *Quantum bit commitment and coin tossing protocols*, 49–61.

G. Brassard, M. Yung, *One-way group actions*, 94–107.

E.F. Brickell, D.R. Stinson, *Some improved bounds on the information rate of perfect secret sharing schemes*, 242–252.

J. Buchmann, S. Düllmann, *On the computation of discrete logarithms in class groups*, 134–139.

D. Chaum, S. Roijakkers, *Unconditionally-secure digital signatures*, 206–214.

C.-C. Chuang, J.G. Dunham, *Matrix extensions of the RSA algorithm*, 140–155.

R. Cleve, *Complexity theoretic issues concerning block ciphers related to D.E.S.*, 530–544.

T.W. Cusick, M.C. Wood, *The REDOC II cryptosystem*, 545–563.

A. De Santis, M. Yung, *Cryptographic applications of the non-interactive metaproof and many-prover systems*, 366–377.

D. de Waleffe, J.-J. Quisquater, *CORSAIR: A smart card for public key cryptosystems*, 502–513.

Y. Desmedt, M. Yung, *Arbitrated unconditionally secure authentication can be unconditionally protected against arbiter's attacks*, 177–188.

S. Even, *Systolic modular multiplication*, 619–624.

W. Fumy, M. Munzert, *A modular approach to key distribution*, 274–283.

H. Gilbert, G. Chassé, *A statistical attack of the Feal-8 cryptosystem*, 22–33.

S. Goldwasser, L. Levin, *Fair computation of general functions in presence of immoral majority*, 77–93.

S. Haber, W.S. Stornetta, *How to time-stamp a digital document*, 437–455.

J. Kilian, *Achieving zero-knowledge robustly*, 313–325.

J. Kilian, *Interactive proofs with provable security against honest verifiers*, 378–392.

K. Kim, T. Matsumoto, H. Imai, *A recursive construction method of S-boxes satisfying strict avalanche criterion*, 564–574.

N. Koblitz, *Constructing elliptic curve cryptosystems in characteristic 2*, 156–167.

K. Kompella, L. Adleman, *Fast checkers for cryptography*, 515–529.

K. Koyama, R. Terada, *Nonlinear parity circuits and their cryptographic applications*, 582–600.

K. Kurosawa, S. Tsujii, *Multi-language zero knowledge interactive proof systems*, 339–352.

B.A. LaMacchia, A.M. Odlyzko, *Computation of discrete logarithms in prime fields*, 616–618.

B.A. LaMacchia, A.M. Odlyzko, *Solving large sparse linear systems over finite fields*, 109–133.

D. Lapidot, A. Shamir, *Publicly verifiable non-interactive zero-knowledge proofs*, 353–365.

U.M. Maurer, *A universal statistical test for random bit generators*, 409–420.

J.L. McInnes, B. Pinkas, *On the impossibility of private key cryptography with weakly random keys*, 421–435.

R.C. Merkle, *Fast software encryption functions*, 476–501.

S. Micali, T. Rabin, *Collective coin tossing without assumptions nor broadcasting*, 253–266.

S. Miyaguchi, *The FEAL cipher family*, 627–638.

T. Okamoto, K. Ohta, *How to utilize the randomness of zero-knowledge proofs*, 456–475.

R.L. Rivest, *Finding four million large random primes*, 625–626.

R.L. Rivest, *The MD4 message digest algorithm*, 303–311.

A.W. Schrift, A. Shamir, *On the universality of the next bit test*, 394–408.

G.J. Simmons, *Geometric shared secret and/or shared control schemes*, 216–241.

O. Staffelbach, W. Meier, *Cryptographic significance of the carry for ciphers based on integer addition*, 601–614.

P. van Oorschot, *A comparison of practical public-key cryptosystems based on integer factorization and discrete logarithms*, 576–581.

Y. Yacobi, *Discrete-log with compressible exponents*, 639–643.

Y. Yacobi, *A key distribution "paradox"*, 268–273.

K. Zeng, C.H. Yang, T.R.N. Rao, *An improved linear syndrome algorithm in cryptanalysis with applications*, 34–47.

Y. Zheng, T. Matsumoto, H. Imai, *Structural properties of one-way hash functions*, 285–302.

Advances in Cryptology – **CRYPTO '91**. Springer-Verlag LNCS 576 (1992).
Editor: J. Feigenbaum.

M. Abadi, M. Burrows, B. Lampson, G. Plotkin, *A calculus for access control in distributed systems*, 1–23.

D. Beaver, *Efficient multiparty protocols using circuit randomization*, 420–432.

D. Beaver, *Foundations of secure interactive computing*, 377–391.

C.H. Bennett, G. Brassard, C. Crépeau, M.-H. Skubiszewska, *Practical quantum oblivious transfer*, 351–366.

E. Biham, A. Shamir, *Differential cryptanalysis of Snefru, Khafre, REDOC-II, LOKI, and Lucifer*, 156–171.

R. Bird, I. Gopal, A. Herzberg, P. Janson, S. Kutten, R. Molva, M. Yung, *Systematic design of two-party authentication protocols*, 44–61.

A.G. Broscius, J.M. Smith, *Exploiting parallelism in hardware implementation of the DES*, 367–376.

P. Camion, C. Carlet, P. Charpin, N. Sendrier, *On correlation-immune functions*, 86–100.

R.M. Capocelli, A. De Santis, L. Gargano, U. Vaccaro, *On the size of shares for secret sharing schemes*, 101–113.

D. Chaum, E. van Heijst, B. Pfitzmann, *Cryptographically strong undeniable signatures, unconditionally secure for the signer*, 470–484.

Y.M. Chee, A. Joux, J. Stern, *The cryptanalysis of a new public-key cryptosystem based on modular knapsacks*, 204–212.

I.B. Damgård, *Towards practical public key systems secure against chosen ciphertext attacks*, 445–456.

B. den Boer, A. Bosselaers, *An attack on the last two rounds of MD4*, 194–203.

Y. Desmedt, Y. Frankel, *Shared generation of authenticators and signatures*, 457–469.

C. Dwork, *On verification in secret sharing*, 114–128.

M.J. Fischer, R.N. Wright, *Multiparty secret key exchange using a random deal of cards*, 141–155.

K.R. Iversen, *A cryptographic scheme for computerized general elections*, 405–419.

J. Kilian, R. Rubinfeld, *Interactive proofs with space bounded provers*, 225–231.

N. Koblitz, *CM-Curves with good cryptographic properties*, 279–287.

K. Koyama, U.M. Maurer, T. Okamoto, S.A. Vanstone, *New public-key schemes based on elliptic curves over the ring Z_n*, 252–266.

D. Lapidot, A. Shamir, *A one-round, two-prover, zero-knowledge protocol for NP*, 213–224.

M. Luby, *Pseudo-random generators from one-way functions*, 300.

S. Micali, P. Rogaway, *Secure computation*, 392–404.

H. Morita, K. Ohta, S. Miyaguchi, *A switching closure test to analyze cryptosystems*, 183–193.

T. Okamoto, K. Ohta, *Universal electronic cash*, 324–337.

T. Okamoto, K. Sakurai, *Efficient algorithms for the construction of hyperelliptic cryptosystems*, 267–278.

J. Patarin, *New results on pseudorandom permutation generators based on the DES scheme*, 301–312.

T.P. Pedersen, *Non-interactive and information-theoretic secure verifiable secret sharing*, 129–140.

B. Pfitzmann, M. Waidner, *How to break and repair a "provably secure" untraceable payment system*, 338–350.

C. Rackoff, D.R. Simon, *Non-interactive zero-knowledge proof of knowledge and chosen ciphertext attack*, 433–444.

S. Rudich, *The use of interaction in public cryptosystems*, 242–251.

D.R. Stinson, *Combinatorial characterizations of authentication codes*, 62–73.

D.R. Stinson, *Universal hashing and authentication codes*, 74–85.

A. Tardy-Corfdir, H. Gilbert, *A known plaintext attack of FEAL-4 and FEAL-6*, 172–182.

S.-H. Teng, *Functional inversion and communication complexity*, 232–241.

M.-J. Toussaint, *Deriving the complete knowledge of participants in cryptographic protocols*, 24–43.

S. Tsujii, J. Chao, *A new ID-based key sharing system*, 288–299.

C.D. Walter, *Faster modular multiplication by operand scaling*, 313–323.

Advances in Cryptology – **CRYPTO '92**. Springer-Verlag LNCS 740 (1993).
Editor: E.F. Brickell.

T. Baritaud, M. Campana, P. Chauvaud, H. Gilbert, *On the security of the permuted kernel identification scheme*, 305–311.

A. Beimel, B. Chor, *Universally ideal secret sharing schemes*, 183–195.

M. Bellare, O. Goldreich, *On defining proofs of knowledge*, 390–420.

M. Bellare, M. Yung, *Certifying cryptographic tools: The case of trapdoor permutations*, 442–460.

E. Biham, A. Shamir, *Differential cryptanalysis of the full 16-round DES*, 487–496.

B. Blakley, G.R. Blakley, A.H. Chan, J.L. Massey, *Threshold schemes with disenrollment*, 540–548.

C. Blundo, A. De Santis, L. Gargano, U. Vaccaro, *On the information rate of secret sharing schemes*, 148–167.

C. Blundo, A. De Santis, A. Herzberg, S. Kutten, U. Vaccaro, M. Yung, *Perfectly-secure key distribution for dynamic conferences*, 471–486.

J.N.E. Bos, D. Chaum, *Provably unforgeable signatures*, 1–14.

J. Brandt, I. Damgård, *On generation of probable primes by incremental search*, 358–370.

K.W. Campbell, M.J. Wiener, *DES is not a group*, 512–520.

C. Carlet, *Partially-bent functions*, 280–291.

D. Chaum, T.P. Pedersen, *Wallet databases with observers*, 89–105.

C. Dwork, U. Feige, J. Kilian, M. Naor, M. Safra, *Low communication 2-prover zero-knowledge proofs for NP*, 215–227.

C. Dwork, M. Naor, *Pricing via processing or combatting junk mail*, 139–147.

H. Eberle, *A high-speed DES implementation for network applications*, 521–539.

M. Fellows, N. Koblitz, *Kid krypto*, 371–389.

Y. Frankel, Y. Desmedt, M. Burmester, *Non-existence of homomorphic general sharing schemes for some key spaces*, 549–557.

S. Goldwasser, R. Ostrovsky, *Invariant signatures and non-interactive zero-knowledge proofs are equivalent*, 228–245.

D.M. Gordon, *Designing and detecting trapdoors for discrete log cryptosystems*, 66–75.

D.M. Gordon, K.S. McCurley, *Massively parallel computations of discrete logarithms*, 312–323.

L. Harn, H.-Y. Lin, *An l-span generalized secret sharing scheme*, 558–565.

A. Herzberg, M. Luby, *Public randomness in cryptography*, 421–432.

R. Hirschfeld, *Making electronic refunds safer*, 106–112.

L.R. Knudsen, *Iterative characteristics of DES and s^2-DES*, 497–511.

K. Koyama, Y. Tsuruoka, *Speeding up elliptic cryptosystems by using a signed binary window method*, 345–357.

U.M. Maurer, *Protocols for secret key agreement by public discussion based on common information*, 461–470.

W. Meier, O. Staffelbach, *Efficient multiplication on certain nonsupersingular elliptic curves*, 333–344.

S. Micali, *Fair public-key cryptosystems*, 113–138.

M. Naor, R. Ostrovsky, R. Venkatesan, M. Yung, *Perfect zero-knowledge arguments for NP can be based on general complexity assumptions*, 196–214.

K. Nyberg, L.R. Knudsen, *Provable security against differential cryptanalysis*, 566–574.

T. Okamoto, *Provably secure and practical identification schemes and corresponding signature schemes*, 31–53.

T. Okamoto, A. Fujioka, E. Fujisaki, *An efficient digital signature scheme based on an elliptic curve over the ring Z_n*, 54–65.

R. Peralta, *A quadratic sieve on the n-dimensional cube*, 324–332.

A. Russell, *Necessary and sufficient conditions for collision-free hashing*, 433–441.

K. Sakurai, T. Itoh, *On the discrepancy between serial and parallel of zero-knowledge protocols*, 246–259.

M. Sivabalan, S. Tavares, L.E. Peppard, *On the design of SP networks from an information theoretic point of view*, 260–279.

M.E. Smid, D.K. Branstad, *Response to comments on the NIST proposed digital signature standard*, 76–88.

D.R. Stinson, *New general lower bounds on the information rate of secret sharing schemes*, 168–182.

E. van Heijst, T.P. Pedersen, B. Pfitzmann, *New constructions of fail-stop signatures and lower bounds*, 15–30.

S. Vaudenay, *FFT-Hash-II is not yet collision-free*, 587–593.

P.C. Wayner, *Content-addressable search engines and DES-like systems*, 575–586.

Y. Zheng, J. Seberry, *Practical approaches to attaining security against adaptively chosen ciphertext attacks*, 292–304.

Advances in Cryptology – CRYPTO '93. Springer-Verlag LNCS 773 (1994).
Editor: D.R. Stinson.

L.M. Adleman, J. Demarrais, *A subexponential algorithm for discrete logarithms over all finite fields*, 147–158.

Y. Aumann, U. Feige, *One message proof systems with known space verifiers*, 85–99.

A. Beimel, B. Chor, *Interaction in key distribution schemes*, 444–455.

M. Bellare, P. Rogaway, *Entity authentication and key distribution*, 232–249.

I. Ben-Aroya, E. Biham, *Differential cyptanalysis of Lucifer*, 187–199.

J. Bierbrauer, T. Johansson, G. Kabatianskii, B. Smeets, *On families of hash functions via geometric codes and concatenation*, 331–342.

A. Blum, M. Furst, M. Kearns, R.J. Lipton, *Cryptographic primitives based on hard learning problems*, 278–291.

C. Blundo, A. Cresti, A. De Santis, U. Vaccaro, *Fully dynamic secret sharing schemes*, 110–125.

A. Bosselaers, R. Govaerts, J. Vandewalle, *Comparison of three modular reduction functions*, 175–186.

S. Brands, *Untraceable off-line cash in wallets with observers*, 302–318.

J. Buchmann, J. Loho, J. Zayer, *An implementation of the general number field sieve*, 159–165.

D. Coppersmith, H. Krawczyk, Y. Mansour, *The shrinking generator*, 22–39.

D. Coppersmith, J. Stern, S. Vaudenay, *Attacks on the birational permutation signature schemes*, 435–443.

C. Crépeau, J. Kilian, *Discreet solitary games*, 319–330.

J. Daemen, R. Govaerts, J. Vandewalle, *Weak keys for IDEA*, 224–231.

I.B. Damgård, *Interactive hashing can simplify zero-knowledge protocol design without computational assumptions*, 100–109.

I.B. Damgård, T.P. Pedersen, B. Pfitzmann, *On the existence of statistically hiding bit commitment schemes and fail-stop signatures*, 250–265.

A. De Santis, G. Di Crescenzo, G. Persiano, *Secret sharing and perfect zero knowledge*, 73–84.

T. Denny, B. Dodson, A.K. Lenstra, M.S. Manasse, *On the factorization of RSA-120*, 166–174.

N. Ferguson, *Extensions of single-term coins*, 292–301.

A. Fiat, M. Naor, *Broadcast encryption*, 480–491.

M. Franklin, S. Haber, *Joint encryption and message-efficient secure computation*, 266–277.

P. Gemmell, M. Naor, *Codes for interactive authentication*, 355–367.

W. Hohl, X. Lai, T. Meier, C. Waldvogel, *Security of iterated hash functions based on block ciphers*, 379–390.

T. Itoh, M. Hoshi, S. Tsujii, *A low communication competitive interactive proof system for promised quadratic residuosity*, 61–72.

W.-A. Jackson, K.M. Martin, C.M. O'Keefe, *Multisecret threshold schemes*, 126–135.

T. Johansson, *On the construction of perfect authentication codes that permit arbitration*, 343–354.

H. Krawczyk, *Secret sharing made short*, 136–146.

T. Leighton, S. Micali, *Secret-key agreement without public-key cryptography*, 456–479.

C.-M. Li, T. Hwang, N.-Y. Lee, *Remark on the threshold RSA signature scheme*, 413–419.

C.H. Lim, P.J. Lee, *Another method for attaining security against adaptively chosen ciphertext attacks*, 420–434.

L. O'Connor, *On the distribution of characteristics in composite permutations*, 403–412.

K. Ohta, M. Matsui, *Differential attack on message authentication codes*, 200–211.

J. Patarin, P. Chauvaud, *Improved algorithms for the permuted kernel problem*, 391–402.

B. Preneel, R. Govaerts, J. Vandewalle, *Hash functions based on block ciphers: A synthetic approach*, 368–378.

B. Preneel, M. Nuttin, V. Rijmen, J. Buelens, *Cryptanalysis of the CFB mode of the DES with a reduced number of rounds*, 212–223.

J. Seberry, X.-M. Zhang, Y. Zheng, *Nonlinearly balanced Boolean functions and their propagation characteristics*, 49–60.

A. Shamir, *Efficient signature schemes based on birational permutations*, 1–12.

J. Stern, *A new identification scheme based on syndrome decoding*, 13–21.

R. Taylor, *An integrity check value algorithm for stream ciphers*, 40–48.

Advances in Cryptology – **CRYPTO '94**. Springer-Verlag LNCS 839 (1994).
Editor: Y.G. Desmedt.

M. Bellare, O. Goldreich, S. Goldwasser, *Incremental cryptography: The case of hashing and signing*, 216–233.

M. Bellare, J. Kilian, P. Rogaway, *The security of cipher block chaining*, 341–358.

T. Beth, D.E. Lazic, A. Mathias, *Cryptanalysis of cryptosystems based on remote chaos replication*, 318–331.

I. Biehl, J. Buchmann, C. Thiel, *Cryptographic protocols based on discrete logarithms in real-quadratic orders*, 56–60.

J. Bierbrauer, K. Gopalakrishnan, D.R. Stinson, *Bounds for resilient functions and orthogonal arrays*, 247–256.

D. Bleichenbacher, U.M. Maurer, *Directed acyclic graphs, one-way functions and digital signatures*, 75–82.

C. Blundo, A. De Santis, G. Di Crescenzo, A.G. Gaggia, U. Vaccaro, *Multi-secret sharing schemes*, 150–163.

M. Burmester, *On the risk of opening distributed keys*, 308–317.

R. Canetti, A. Herzberg, *Maintaining security in the presence of transient faults*, 425–438.

J. Chao, K. Tanada, S. Tsujii, *Design of elliptic curves with controllable lower boundary of extension degree for reduction attacks*, 50–55.

B. Chor, A. Fiat, M. Naor, *Tracing traitors*, 257–270.

D. Coppersmith, *Attack on the cryptographic scheme NIKS-TAS*, 294–307.

R. Cramer, I. Damgård, B. Schoenmakers, *Proofs of partial knowledge and simplified design of witness hiding protocols*, 174–187.

D. Davis, R. Ihaka, P. Fenstermacher, *Cryptographic randomness from air turbulence in disk drives*, 114–120.

O. Delos, J.-J. Quisquater, *An identity-based signature scheme with bounded life-span*, 83–94.

C. Dwork, M. Naor, *An efficient existentially unforgeable signature scheme and its applications*, 234–246.

C. Gehrmann, *Cryptanalysis of the Gemmell and Naor multiround authentication protocol*, 121–128.

H. Gilbert, P. Chauvaud, *A chosen plaintext attack of the 16-round Khufu cryptosystem*, 359–368.

M. Girault, J. Stern, *On the length of cryptographic hash-values used in identification schemes*, 202–215.

T. Horváth, S.S. Magliveras, T. van Trung, *A parallel permutation multiplier for a PGM crypto-chip*, 108–113.

T. Itoh, Y. Ohta, H. Shizuya, *Language dependent secure bit commitment*, 188–201.

B.S. Kaliski Jr., M.J.B. Robshaw, *Linear cryptanalysis using multiple approximations*, 26–39.

H. Krawczyk, *LFSR-based hashing and authentication*, 129–139.

K. Kurosawa, *New bound on authentication code with arbitration*, 140–149.

E. Kushilevitz, A. Rosén, *A randomness-rounds tradeoff in private computation*, 397–410.

C.H. Lim, P.J. Lee, *More flexible exponentiation with precomputation*, 95–107.

J.L. Massey, S. Serconek, *A Fourier transform approach to the linear complexity of nonlinearly filtered sequences*, 332–340.

M. Matsui, *The first experimental cryptanalysis of the Data Encryption Standard*, 1–11.

U.M. Maurer, *Towards the equivalence of breaking the Diffie-Hellman protocol and computing discrete logarithms*, 271–281.

P. Mihailescu, *Fast generation of provable primes using search in arithmetic progressions*, 282–293.

K. Ohta, K. Aoki, *Linear cryptanalysis of the Fast Data Encipherment Algorithm*, 12–16.

T. Okamoto, *Designated confirmer signatures and public-key encryption are equivalent*, 61–74.

K. Sako, J. Kilian, *Secure voting using partially compatible homomorphisms*, 411–424.

J. Seberry, X.-M. Zhang, Y. Zheng, *Pitfalls in designing substitution boxes*, 383–396.

J. Stern, *Designing identification schemes with keys of short size*, 164–173.

J.-P. Tillich, G. Zémor, *Hashing with SL_2*, 40–49.

Y. Tsunoo, E. Okamoto, T. Uyematsu, *Ciphertext only attack for one-way function of the MAP using one ciphertext*, 369–382.

Advances in Cryptology – **CRYPTO '95**. Springer-Verlag LNCS 963 (1995).
Editor: D. Coppersmith.

R. Anderson, R. Needham, *Robustness principles for public key protocols*, 236–247.

D. Beaver, *Precomputing oblivious transfer*, 97–109.

P. Béguin, J.-J. Quisquater, *Fast server-aided RSA signatures secure against active attacks*, 57–69.

A. Beimel, B. Chor, *Secret sharing with public reconstruction*, 353–366.

M. Bellare, R. Guérin, P. Rogaway, *XOR MACs: New methods for message authentication using finite pseudorandom functions*, 15–28.

G.R. Blakley, G.A. Kabatianskii, *On general perfect secret sharing schemes*, 367–371.

D. Bleichenbacher, W. Bosma, A.K. Lenstra, *Some remarks on Lucas-based cryptosystems*, 386–396.

D. Boneh, R.J. Lipton, *Quantum cryptanalysis of hidden linear functions*, 424–437.

D. Boneh, J. Shaw, *Collusion-secure fingerprinting for digital data*, 452–465.

R. Cramer, I. Damgård, *Secure signature schemes based on interactive protocols*, 297–310.

C. Crépeau, J. van de Graaf, A. Tapp, *Committed oblivious transfer and private multi-party computation*, 110–123.

I. Damgård, O. Goldreich, T. Okamoto, A. Wigderson, *Honest verifier vs. dishonest verifier in public coin zero-knowledge proofs*, 325–338.

B. Dodson, A.K. Lenstra, *NFS with four large primes: An explosive experiment*, 372–385.

Y. Frankel, M. Yung, *Cryptanalysis of the immunized LL public key systems*, 287–296.

Y. Frankel, M. Yung, *Escrow encryption systems visited: Attacks, analysis and designs*, 222–235.

S. Halevi, *Efficient commitment schemes with bounded sender and unbounded receiver*, 84–96.

A. Herzberg, S. Jarecki, H. Krawczyk, M. Yung, *Proactive secret sharing or: How to cope with perpetual leakage*, 339–352.

B.S. Kaliski Jr., Y.L. Yin, *On differential and linear cryptanalysis of the RC5 encryption algorithm*, 171–184.

J. Kilian, *Improved efficient arguments*, 311–324.

J. Kilian, T. Leighton, *Fair cryptosystems, revisited: A rigorous approach to key-escrow*, 208–221.

A. Klapper, M. Goresky, *Cryptanalysis based on 2-adic rational approximation*, 262–273.

L.R. Knudsen, *A key-schedule weakness in SAFER K-64*, 274–286.

K. Kurosawa, S. Obana, W. Ogata, *t-cheater identifiable (k, n) threshold secret sharing schemes*, 410–423.

S.K. Langford, *Threshold DSS signatures without a trusted party*, 397–409.

A.K. Lenstra, P. Winkler, Y. Yacobi, *A key escrow system with warrant bounds*, 197–207.

C.H. Lim, P.J. Lee, *Security and performance of server-aided RSA computation protocols*, 70–83.

D. Mayers, *On the security of the quantum oblivious transfer and key distribution protocols*, 124–135.

S. Micali, R. Sidney, *A simple method for generating and sharing pseudo-random functions, with applications to Clipper-like key escrow systems*, 185–196.

K. Ohta, S. Moriai, K. Aoki, *Improving the search algorithm for the best linear expression*, 157–170.

T. Okamoto, *An efficient divisible electronic cash scheme*, 438–451.

S.-J. Park, S.-J. Lee, S.-C. Goh, *On the security of the Gollmann cascades*, 148–156.

J. Patarin, *Cryptanalysis of the Matsumoto and Imai public key scheme of Eurocrypt '88*, 248–261.

B. Preneel, P. van Oorschot, *MDx-MAC and building fast MACs from hash functions*, 1–14.

P. Rogaway, *Bucket hashing and its application to fast message authentication*, 29–42.

R. Schroeppel, H. Orman, S. O'Malley, O. Spatscheck, *Fast key exchange with elliptic curve systems*, 43–56.

T. Theobald, *How to break Shamir's asymmetric basis*, 136–147.

Advances in Cryptology – CRYPTO '96. Springer-Verlag LNCS 1109 (1996).
Editor: N. Koblitz.

M. Atici, D. Stinson, *Universal hashing and multiple authentication,* 16–30.

M. Bellare, R. Canetti, H. Krawczyk, *Keying hash functions for message authenticaion,* 1–15.

C. Blundo, L. Mattos, D. Stinson, *Trade-offs between communication and storage in unconditionally secure schemes for broadcast encryption and interactive key distribution,* 388–401.

D. Boneh, R. Lipton, *Algorithms for black-box fields and their application to cryptography,* 283–297.

D. Boneh, R. Venkatesan, *Hardness of computing the most significant bits of secret keys in Diffie-Hellman and related schemes,* 129–142.

A. Bosselaers, R. Govaerts, J. Vandewalle, *Fast hashing on the Pentium,* 298–312.

P. Camion, A. Canteaut, *Generalization of Siegenthaler inequality and Schnorr–Vaudenay multipermutations,* 373–387.

R. Cramer, I. Damgård, *New generation of secure and practical RSA-based signatures,* 173–185.

S. Droste, *New results on visual cryptography,* 402–416.

R. Gennaro, S. Jarecki, H. Krawczyk, T. Rabin, *Robust and efficient sharing of RSA functions,* 157–172.

S. Halevi, S. Micali, *Practical and provably-secure commitment schemes from collision-free hashing,* 201–215.

T. Helleseth, T. Johansson, *Universal hash functions from exponential sums over finite fields and Galois rings,* 31–44.

R. Hughes, G. Luther, G. Morgan, C. Peterson, C. Simmons, *Quantum cryptography over underground optical fibers,* 329–343.

M. Jakobsson, M. Yung, *Proving without knowing: On oblivious, agnostic and blindfolded provers,* 186–200.

J. Kelsey, B. Schneier, D. Wagner, *Key-schedule cryptanalysis of IDEA, G-DES, GOST, SAFER, and Triple-DES,* 237–251.

J. Kilian, P. Rogaway, *How to protect DES against exhaustive key search,* 252–267.

L. Knudsen, W. Meier, *Improved differential attacks on RC5,* 216–228.

P. Kocher, *Timing attacks on implementations of Diffie-Hellman, RSA, DSS, and other systems,* 104–113.

S. Langford, *Weaknesses in some threshold cryptosystems,* 74–82.

J. Massey, S. Serconek, *Linear complexity of periodic sequences: A general theory,* 359–372.

U. Maurer, S. Wolf, *Diffie-Hellman oracles,* 268–282.

D. Mayers, *Quantum key distribution and string oblivious transfer in noisy channels,* 344–358.

M. Näslund, *All bits in $ax + b \bmod p$ are hard,* 114–128.

J. Patarin, *Asymmetric cryptography with a hidden monomial,* 45–60.

C. Schnorr, *Security of 2^t-root identification and signatures,* 143–156.

V. Shoup, *On fast and provably secure message authentication based on universal hashing,* 313–328.

D. Simon, *Anonymous communication and anonymous cash,* 61–73.

P. van Oorschot, M. Wiener, *Improving implementable meet-in-the-middle attacks by orders of magnitude,* 229–236.

S. Vaudenay, *Hidden collisions on DSS,* 83–88.

A. Young, M. Yung, *The dark side of 'black-box' cryptography, or: Why should we trust Capstone?,* 89–103.

A.3 Eurocrypt Proceedings

Cryptography – Proceedings of the Workshop on Cryptography, Burg Feuerstein, Germany, 1982.
Springer-Verlag LNCS 149 (1983).
Editor: T. Beth.

No Author, *Introduction*, 1–28.

No Author, *Mechanical cryptographic devices*, 47–48.

F.L. Bauer, *Cryptology-methods and maxims*, 31–46.

H.J. Beker, *Analogue speech security systems*, 130–146.

D.W. Davies, G.I.P. Parkin, *The average cycle size of the key stream in output feedback encipherment*, 263–279.

M. Davio, J.-M. Goethals, J.-J. Quisquater, *Authentication procedures*, 283–288.

A. Ecker, *Finite semigroups and the RSA-cryptosystem*, 353–369.

R. Eier, H. Lagger, *Trapdoors in knapsack cryptosystems*, 316–322.

J.A. Gordon, H. Retkin, *Are big S-boxes best?*, 257–262.

L. Győrfi, I. Kerekes, *Analysis of multiple access channel using multiple level FSK*, 165–172.

T. Herlestam, *On using prime polynomials in crypto generators*, 207–216.

P. Hess, K. Wirl, *A voice scrambling system for testing and demonstration*, 147–156.

L. Horbach, *Privacy and data protection in medicine*, 228–232.

I. Ingemarsson, *A new algorithm for the solution of the knapsack problem*, 309–315.

S.M. Jennings, *Multiplexed sequences: Some properties of the minimum polynomial*, 189–206.

A.G. Konheim, *Cryptanalysis of a Kryha machine*, 49–64.

M. Mignotte, *How to share a secret*, 371–375.

M.R. Oberman, *Communication security in remote controlled computer systems*, 219–227.

F. Pichler, *Analog scrambling by the general fast Fourier transform*, 173–178.

F.C. Piper, *Stream ciphers*, 181–188.

J. Sattler, C.P. Schnorr, *Ein effizienzvergleich der faktorisierungsverfahren von Morrison-Brillhart und Schroeppel*, 331–351.

I. Schaumüller-Bichl, *Cryptanalysis of the Data Encryption Standard by the method of formal coding*, 235–255.

C.P. Schnorr, *Is the RSA-scheme safe?*, 325–329.

P. Schöbi, J.L. Massey, *Fast authentication in a trapdoor-knapsack public key cryptosystem*, 289–306.

H.-R. Schuchmann, *Enigma variations*, 65–68.

N.J.A. Sloane, *Encrypting by random rotations*, 71–128.

K.-P. Timmann, *The rating of understanding in secure voice communications systems*, 157–163.

Advances in Cryptology – Proceedings of **EUROCRYPT 84**, Paris, France.
Springer-Verlag LNCS 209 (1985).
Editors: T. Beth, N. Cot, and I. Ingemarsson.

G.B. Agnew, *Secrecy and privacy in a local area network environment*, 349–363.

R. Berger, R. Peralta, T. Tedrick, *A provably secure oblivious transfer protocol*, 379–386.

T. Beth, F.C. Piper, *The stop-and-go generator*, 88–92.

R. Blom, *An optimal class of symmetric key generation systems*, 335–338.

A. Bouckaert, *Security of transportable computerized files*, 416–425.

O. Brugia, S. Improta, W. Wolfowicz, *An encryption and authentification procedure for tele-surveillance systems*, 437–445.

M. Davio, Y. Desmedt, J.-J. Quisquater, *Propogation characteristics of the DES*, 62–73.

J.A. Davis, D.B. Holdridge, G.J. Simmons, *Status report on factoring (at the Sandia National Labs)*, 183–215.

P. Delsarte, Y. Desmedt, A. Odlyzko, P. Piret, *Fast cryptanalysis of the Matsumoto-Imai public key scheme*, 142–149.

A. Ecker, *Time-division multiplexing scramblers: Selecting permutations and testing the systems*, 399–415.

Y. Girardot, *Bull CP8 smart card uses in cryptology*, 464–469.

O. Goldreich, *On concurrent identification protocols*, 387–396.

O. Goldreich, *On the number of close-and-equal pairs of bits in a string (with implications on the security of RSA's L.S.B)*, 127–141.

D. Gollmann, *Pseudo random properties of cascade connections of clock controlled shift registers*, 93–98.

R.M.F. Goodman, A.J. McAuley, *A new trapdoor knapsack public-key cryptosystem*, 150–158.

J. Gordon, *Strong primes are easy to find*, 216–223.

J. Goutay, *Smart card applications in security and data protection*, 459–463.

H. Groscot, *Estimation of some encryption functions implemented into smart cards*, 470–479.

L.C. Guillou, *Smart cards and conditional access*, 480–489.

S. Harari, *Non-linear, non-commutative functions for data integrity*, 25–32.

R.W. Jones, *User functions for the generation and distribution of encipherment keys*, 317–334.

R. Lidl, *On cryptosystems based on polynomials and finite fields*, 10–15.

J.L. Massey, R.A. Rueppel, *Linear ciphers and random sequence generators with multiple clocks*, 74–87.

A.M. Odlyzko, *Discrete logarithms in finite fields and their cryptographic significance*, 224–314.

L.H. Ozarow, A.D. Wyner, *Wire-tap channel II*, 33–50.

J.P. Pieprzyk, *Algebraical structures of cryptographic transformations*, 16–24.

C. Pomerance, *The quadratic sieve factoring algorithm*, 169–182.

R. Rivest, *RSA chips (past/present/future)*, 159–165.

G. Ruggiu, *Cryptology and complexity theories*, 3–9.

I. Schaumueller-Bichl, E. Piller, *A method of software protection based on the use of smart cards and cryptographic techniques*, 446–454.

C.P. Schnorr, W. Alexi, *RSA-bits are* $0.5 + \epsilon$ *secure*, 113–126.

S.C. Serpell, C.B. Brookson, *Encryption and key management for the ECS satellite service*, 426–436.

A. Sgarro, *Equivocations for homophonic ciphers*, 51–61.

G.J. Simmons, *The subliminal channel and digital signatures*, 364–378.

B.J.M. Smeets, *On the use of the binary multiplying channel in a private communication system*, 339–348.

A. Turbat, *Session on smart cards – introductory remarks*, 457–458.

R. Vogel, *On the linear complexity of cascaded sequences*, 99–109.

Advances in Cryptology – **EUROCRYPT '85**, Linz, Austria. Springer-Verlag LNCS 219 (1986).
Editor: F. Pichler.

G.B. Agnew, *Modeling of encryption techniques for secrecy and privacy in multi-user networks*, 221–230.

J. Bernasconi, C.G. Günther, *Analysis of a nonlinear feedforward logic for binary sequence generators*, 161–166.

R.V. Book, F. Otto, *The verifiability of two-party protocols*, 254–260.

R.L. Bradey, I.G. Graham, *Full encryption in a personal computer system*, 231–240.

L. Brynielsson, *On the linear complexity of combined shift register sequences*, 156–160.

D. Chaum, *Showing credentials without identification signatures transferred between unconditionally unlinkable pseudonyms*, 241–244.

D.-S. Chen, Z.-D. Dai, *On feedforward transforms and p-fold periodic p-arrays*, 130–134.

D.W. Davies, W.L. Price, *Engineering secure information systems*, 191–199.

P. Godlewski, G.D. Cohen, *Authorized writing for "write-once" memories*, 111–115.

T. Herlestam, *On functions of linear shift register sequences*, 119–129.

O.J. Horak, *The contribution of E.B. Fleissner and A. Figl for today's cryptography*, 3–17.

R.W. Jones, M.S.J. Baxter, *The role of encipherment services in distributed systems*, 214–220.

B.S. Kaliski Jr., R.L. Rivest, A.T. Sherman, *Is the Data Encryption Standard a group?*, 81–95.

M. Kowatsch, B.O. Eichinger, F.J. Seifert, *Message protection by spread spectrum modulation in a packet voice radio link*, 273–277.

T. Krivachy, *The chipcard – an identification card with cryptographic protection*, 200–207.

M.-L. Liu, Z.-X. Wan, *Generalized multiplexed sequences*, 135–141.

H. Meijer, S. Akl, *Two new secret key cryptosystems*, 96–102.

W.B. Müller, R. Nöbauer, *Cryptanalysis of the Dickson-scheme*, 50–61.

H. Niederreiter, *A public-key cryptosystem based on shift register sequences*, 35–39.

R. Peralta, *Simultaneous security of bits in the discrete log*, 62–72.

A. Pfitzmann, M. Waidner, *Networks without user observability – design options*, 245–253.

J.P. Pieprzyk, *On public-key cryptosystems built using polynomial rings*, 73–78.

U. Rimensberger, *Encryption: Needs, requirements and solutions in banking networks*, 208–213.

R.L. Rivest, A. Shamir, *Efficient factoring based on partial information*, 31–34.

R.A. Rueppel, *Linear complexity and random sequences*, 167–188.

T. Siegenthaler, *Cryptanalysts representation of nonlinearly filtered ML-sequences*, 103–110.

G.J. Simmons, *The practice of authentication*, 261–272.

B. Smeets, *A comment on Niederreiter's public key cryptosystem*, 40–42.

B. Smeets, *A note on sequences generated by clock controlled shift registers*, 142–148.

T. Tedrick, *On the history of cryptography during WW2, and possible new directions for cryptographic research*, 18–28.

J. Vandewalle, R. Govaerts, W. De Becker, M. Decroos, G. Speybrouck, *Implementation study of public key cryptographic protection in an existing electronic mail and document handling system*, 43–49.

N.R. Wagner, P.S. Putter, M.R. Cain, *Using algorithms as keys in stream ciphers*, 149–155.

EUROCRYPT 86, Linköping, Sweden.
Abstracts of papers (no conference proceedings were published).
Program Chair: J.L. Massey.

G. Agnew, *Another look at redundancy in cryptographic systems.*

A. Bauval, *Crypanalysis of pseudo-random number sequences generated by a linear congruential recurrence of given order.*

M. Beale, *Properties of de Bruijn sequences generated by a cross-join technique.*

A. Beutelspacher, *Geometric structures as threshold schemes.*

E.F. Brickell, *Cryptanalysis of the Yagisawa public key cryptosystem.*

D.D. Buckley, M. Beale, *Public key encryption of stream ciphers.*

H. Cloetens, Y. Desmedt, L. Bierens, J. Vandewalle, R. Govaerts, *Additional properties in the S-boxes of the DES.*

G.I. Davida, Y.-S. Yeh, *Multilevel cryptosecure relational databases.*

Y. Desmedt, F. Hoornaert, J.-J Quisquater, *Several exhaustive key search machines and DES.*

G. Dial, F. Pessoa, *Sharma-Mittal entropy and Shannon's random cipher result.*

A. Ecker, *Tactical configurations and threshold schemes.*

V. Fåk, *Activities of IFIP working group 11:4 on crypto management.*

O. Frank, P. Weidenman, *Controlling individual information in statistics by coding.*

A.S. Glass, *Could the smart card be dumb?*

D. Gollmann, *Linear complexity of sequences with period p^n.*

C.G. Günther, *On some properties of the sum of two pseudorandom generators.*

F.-P. Heider, D. Kraus, M. Welschenbach, *Some preliminary remarks on the decimal, shift and add-algorithm (DSA).*

T. Herlestam, *On linear shift registers with permuted feedback.*

N.S. James, R. Lidl, H. Niederreiter, *A cryptanalytic attack on the CADE cryptosystem.*

C.J.A. Jansen, *Protection against active eavesdropping.*

R.A. Kemmerer, *Analyzing encryption protocols using formal verification techniques.*

D.S.P. Khoo, G.J. Bird, J. Seberry, *Encryption exponent 3 and the security of RSA.*

J.H. Moore, *Cycle structure of the weak and semi-weak DES keys.*

W.B. Müller, R. Nöbauer, *On commutative semigroups of polynomials and their applications in cryptography.*

Q.A. Nguyen, *Elementary proof of Rueppel's linear complexity conjecture.*

R. Peralta, *A simple and fast probabilistic algorithm for computing square roots modulo a prime number.*

F. Pichler, *On the Walsh-Fourier analysis of correlation-immune switching functions.*

D. Pinkas, B. Transac, *The need for a standardized compression algorithm for digital signatures.*

W.L. Price, *The NPL intelligent token and its application.*

R.A. Rueppel, O.J. Staffelbach, *Products of linear recurring sequence with maximum complexity.*

P. Schöbi, *Perfect authentication systems for data sources with arbitrary statistics.*

T. Siegenthaler, *Correlation-immune polynomials over finite fields.*

B. Smeets, *Some properties of sequences generated by a windmill machine.*

M.Z. Wang, J.L. Massey, *The characterization of all binary sequences with perfect linear complexity profiles.*

Advances in Cryptology – **EUROCRYPT '87**, Amsterdam, The Netherlands.
Springer-Verlag LNCS 304 (1988).
Editors: D. Chaum and W.L. Price.

G.B. Agnew, *Random sources for cryptographic systems,* 77–81.

D.P. Anderson, P.V. Rangan, *High-performance interface architectures for cryptographic hardware,* 301–309.

H.J. Beker, G.M. Cole, *Message authentication and dynamic passwords,* 171–175.

A. Beutelspacher, *Perfect and essentially perfect authentication schemes,* 167–170.

E.F. Brickell, Y. Yacobi, *On privacy homomorphisms,* 117–125.

D. Chaum, *Blinding for unanticipated signatures,* 227–233.

D. Chaum, J.-H. Evertse, J. van de Graaf, *An improved protocol for demonstrating possession of discrete logarithms and some generalizations,* 127–141.

A.J. Clark, *Physical protection of cryptographic devices,* 83–93.

I.B. Damgård, *Collision free hash functions and public key signature schemes,* 203–216.

G.I. Davida, G.G. Walter, *A public key analog cryptosystem,* 143–147.

J.-H. Evertse, *Linear structures in blockciphers,* 249–266.

M. Girault, *Hash-functions using modulo-n operations,* 217–226.

C.G. Günther, *Alternating step generators controlled by de Bruijn sequences,* 5–14.

C.J.A. Jansen, D.E. Boekee, *Modes of blockcipher algorithms and their protection against active eavesdropping,* 281–286.

F. Jorissen, J. Vandewalle, R. Govaerts, *Extension of Brickell's algorithm for breaking high density knapsacks,* 109–115.

J.L. Massey, U. Maurer, M. Wang, *Non-expanding, key-minimal, robustly-perfect, linear and bilinear ciphers,* 237–247.

S. Mund, D. Gollmann, T. Beth, *Some remarks on the cross correlation analysis of pseudo random generators,* 25–35.

H. Niederreiter, *Sequences with almost perfect linear complexity profile,* 37–51.

F. Pichler, *Finite state machine modelling of cryptographic systems in loops,* 65–73.

R.A. Rueppel, *When shift registers clock themselves,* 53–64.

I. Schaumüller-Bichl, *IC-Cards in high-security applications,* 177–199.

H. Sedlak, *The RSA cryptography processor,* 95–105.

A. Shimizu, S. Miyaguchi, *Fast data encipherment algorithm FEAL,* 267–278.

T. Siegenthaler, A.W. Kleiner, R. Forré, *Generation of binary sequences with controllable complexity and ideal r-tupel distribution,* 15–23.

G.J. Simmons, *Message authentication with arbitration of transmitter/receiver disputes,* 151–165.

I. Verbauwhede, F. Hoornaert, J. Vandewalle, H. De Man, *Security considerations in the design and implementation of a new DES chip,* 287–300.

Advances in Cryptology – **EUROCRYPT '88**, Davos, Switzerland. Springer-Verlag LNCS 330 (1988). Editor: C. Günther.

G.B. Agnew, R.C. Mullin, S.A. Vanstone, *Fast exponentiation in* $GF(2^n)$, 251–255.

G.B. Agnew, R.C. Mullin, S.A. Vanstone, *An interactive data exchange protocol based on discrete exponentiation,* 159–166.

T. Beth, *Efficient zero-knowledge identification scheme for smart cards,* 77–84.

C. Boyd, *Some applications of multiple key ciphers,* 455–467.

J. Brandt, I.B. Damgård, P. Landrock, *Anonymous and verifiable registration in databases,* 167–176.

E.F. Brickell, D.R. Stinson, *Authentication codes with multiple arbiters,* 51–55.

W.G. Chambers, D. Gollmann, *Lock-in effect in cascades of clock-controlled shift-registers,* 331–343.

D. Chaum, *Elections with unconditionally-secret ballots and disruption equivalent to breaking RSA,* 177–182.

G.I. Davida, Y.G. Desmedt, *Passports and visas versus ID's,* 183–188.

J.A. Davis, D.B. Holdridge, *Factorization of large integers on a massively parallel computer,* 235–243.

M. De Soete, *Some constructions for authentication-secrecy codes,* 57–75.

M. De Soete, K. Vedder, *Some new classes of geometric threshold schemes,* 389–401.

B. den Boer, *Cryptanalysis of F.E.A.L.,* 293–299.

Y. Desmedt, *Subliminal-free authentication and signature,* 23–33.

A. Di Porto, P. Filipponi, *A probabilistic primality test based on the properties of certain generalized Lucas numbers,* 211–223.

C. Ding, *Proof of Massey's conjectured algorithm,* 345–349.

M. Girault, R. Cohen, M. Campana, *A generalized birthday attack,* 129–156.

P. Godlewski, P. Camion, *Manipulations and errors, detection and localization,* 97–106.

R.N. Gorgui-Naguib, S.S. Dlay, *Properties of the Euler totient function modulo 24 and some of its cryptographic implications,* 267–274.

L.C. Guillou, J.-J. Quisquater, *A practical zero-knowledge protocol fitted to security microprocessor minimizing both transmission and memory,* 123–128.

C.G. Günther, *A universal algorithm for homophonic coding,* 405–414.

F. Hoornaert, M. Decroos, J. Vandewalle, R. Govaerts, *Fast RSA-hardware: Dream or reality?,* 257–264.

H. Jingmin, L. Kaicheng, *A new probabilistic encryption scheme,* 415–418.

S. Kawamura, K. Hirano, *A fast modular arithmetic algorithm using a residue table,* 245–250.

S.J. Knapskog, *Privacy protected payments - realization of a protocol that guarantees payer anonymity,* 107–122.

H.-J. Knobloch, *A smart card implementation of the Fiat-Shamir identification scheme,* 87–95.

K. Koyama, K. Ohta, *Security of improved identity-based conference key distribution systems,* 11–19.

P.J. Lee, E.F. Brickell, *An observation on the security of McEliece's public-key cryptosystem,* 275–280.

D. Lin, M. Liu, *Linear recurring* m-*arrays,* 351–357.

T. Matsumoto, H. Imai, *Public quadratic polynomial-tuples for efficient signature-verification and message-encryption,* 419–453.

W. Meier, O. Staffelbach, *Fast correlation attacks on stream ciphers,* 301–314.

H. Niederreiter, *The probabilistic theory of linear complexity,* 191–209.

E. Okamoto, *Substantial number of cryptographic keys and its application to encryption designs,* 361–373.

R.A. Rueppel, *Key agreements based on function composition,* 3–10.

C.P. Schnorr, *On the construction of random number generators and random function generators,* 225–232.

A. Sgarro, *A measure of semiequivocation,* 375–387.

G.J. Simmons, G.B. Purdy, *Zero-knowledge proofs of identity and veracity of transaction receipts,* 35–49.

B.J.M. Smeets, W.G. Chambers, *Windmill generators: A generalization and an observation of how many there are,* 325–330.

S. Tezuka, *A new class of nonlinear functions for running-key generators,* 317–324.

B. Vallée, M. Girault, P. Toffin, *How to break Okamoto's cryptosystem by reducing lattice bases,* 281–291.

Advances in Cryptology – EUROCRYPT '89, Houthalen, Belgium. Springer-Verlag LNCS 434 (1990). Editors: J.-J. Quisquater and J. Vandewalle.

G.B. Agnew, R.C. Mullin, S.A. Vanstone, *A fast elliptic curve cryptosystem*, 706–708.

M. Antoine, J.-F Brakeland, M. Eloy, Y. Poullet, *Legal requirements facing new signature technologies*, 273–287.

F. Bauspieß, H.-J. Knobloch, *How to keep authenticity alive in a computer network*, 38–46.

M. Bertilsson, E.F. Brickell, I. Ingemarsson, *Cryptanalysis of video encryption based on space-filling curves*, 403–411.

T. Beth, Z.-D. Dai, *On the complexity of pseudo-random sequences – or: If you can describe a sequence it can't be random*, 533–543.

A. Beutelspacher, *How to say "no"*, 491–496.

J. Bos, B. den Boer, *Detection of disrupters in the DC protocol*, 320–327.

W. Bosma, M.-P van der Hulst, *Faster primality testing*, 652–656.

J. Boyar, K. Friedl, C. Lund, *Practical zero-knowledge proofs: Giving hints and using deficiencies*, 155–172.

C. Boyd, *A new multiple key cipher and an improved voting scheme*, 617–625.

G. Brassard, *How to improve signature schemes*, 16–22.

G. Brassard, C. Crépeau, *Sorting out zero-knowledge*, 181–191.

G. Brassard, C. Crépeau, M. Yung, *Everything in NP can be argued in perfect zero-knowledge in a bounded number of rounds*, 192–195.

E.F. Brickell, *Some ideal secret sharing schemes*, 468–475.

L. Brown, J. Seberry, *On the design of permutation P in DES type cryptosystems*, 696–705.

J.A. Buchmann, S. Düllmann, H.C. Williams, *On the complexity and efficiency of a new key exchange system*, 597–616.

M.V.D. Burmester, Y. Desmedt, F. Piper, M. Walker, *A general zero-knowledge scheme*, 122–133.

G. Carter, *Some conditions on the linear complexity profiles of certain binary sequences*, 691–695.

A.H. Chan, M. Goresky, A. Klapper, *On the linear complexity of feedback registers*, 563–570.

D. Chaum, *Online cash checks*, 288–293.

D. Chaum, B. den Boer, E. van Heyst, S. Mjølsnes, A. Steenbeek, *Efficient offline electronic checks*, 294–301.

H. Cnudde, *CRYPTEL – the practical protection of an existing electronic mail system*, 237–242.

C. Crépeau, *Verifiable disclosure of secrets and applications*, 150–154.

Z.-D. Dai, K.C. Zeng, *Feedforward functions defined by de Bruijn sequences*, 544–548.

G. Davida, Y. Desmedt, R. Peralta, *A key distribution system based on any one-way function*, 75–79.

M. De Soete, K. Vedder, M. Walker, *Cartesian authentication schemes*, 476–490.

B. den Boer, *More efficient match-making and satisfiability. The five card trick*, 208–217.

W. Diffie, *The adolescence of public-key cryptography*, 2.

J. Domingo i Ferrer, L. Huguet i Rotger, *Full secure key exchange and authentication with no previously shared secrets*, 665–669.

Y. Duhoux, *Deciphering bronze age scripts of Crete. The case of linear A*, 649–650.

P. Flajolet, A. Odlyzko, *Random mapping statistics*, 329–354.

R. Forré, *A fast correlation attack on nonlinearly feedforward filtered shift-register sequences*, 586–595.

Y. Frankel, *A practical protocol for large group oriented networks*, 56–61.

Z. Galil, S. Haber, M. Yung, *A secure public-key authentication scheme*, 3–15.

P. Godlewski, C. Mitchell, *Key minimal authentication systems for unconditional secrecy*, 497–501.

D. Gollmann, W.G. Chambers, *A cryptanalysis of $\text{step}_{k,m}$-cascades*, 680–687.

C.G. Günther, *An identity-based key-exchange protocol*, 29–37.

C.G. Günther, *Parallel generation of recurring sequences*, 503–522.

T. Hwang, T.R.N. Rao, *Private-key algebraic-code cryptosystems with high information rates*, 657–661.

H. Isselhorst, *The use of fractions in public-key cryptosystems*, 47–55.

W.J. Jaburek, *A generalization of El Gamal's public-key cryptosystem*, 23–28.

H.N. Jendal, Y.J.B. Kuhn, J.L. Massey, *An information-theoretic treatment of homophonic substitution*, 382–394.

A.K. Lenstra, M.S. Manasse, *Factoring by electronic mail*, 355–371.

S. Lloyd, *Counting functions satisfying a higher order strict avalanche criterion*, 63–74.

U.M. Maurer, *Fast generation of secure RSA-moduli with almost maximal diversity*, 636–647.

W. Meier, O. Staffelbach, *Nonlinearity criteria for cryptographic functions*, 549–562.

S.F. Mjølsnes, *A simple technique for diffusing cryptoperiods*, 110–120.

F. Morain, *Atkin's test: News from the front*, 626–635.

H. Niederreiter, *Keystream sequences with a good linear complexity profile for every starting point*, 523–532.

T. Okamoto, K. Ohta, *Divertible zero-knowledge interactive proofs and commutative random self-reducibility*, 134–149.

B. Pfitzmann, A. Pfitzmann, *How to break the direct RSA-implementation of MIXes*, 373–381.

J.P. Pieprzyk, *Non-linearity of exponent permutations*, 80–92.

J.-J. Quisquater, A. Bouckaert, *Zero-knowledge procedures for confidential access to medical records*, 662–664.

J.-J. Quisquater, J.-P. Delescaille, *How easy is collision search? Application to DES*, 429–434.

J.-J. Quisquater, M. Girault, $2n$-*bit hash-functions using* n-*bit symmetric block cipher algorithms*, 102–109.

Y. Roggeman, *Varying feedback shift registers*, 670–679.

R.A. Rueppel, *On the security of Schnorr's pseudo random generator*, 423–428.

C.P. Schnorr, *Efficient identification and signatures for smart cards*, 688–689.

A. Sgarro, *Informational divergence bounds for authentication codes*, 93–101.

G.J. Simmons, *Prepositioned shared secret and/or shared control schemes*, 436–467.

C. Siuda, *Security in open distributed processing*, 249–266.

J. Stern, *An alternative to the Fiat-Shamir protocol*, 173–180.

J. Van Auseloos, *Technical security: The starting point*, 243–248.

A. Vandemeulebroecke, E. Vanzieleghem, T. Denayer, P.G.A. Jespers, *A single chip 1024 bits RSA processor*, 219–236.

J. Vandewalle, D. Chaum, W. Fumy, C. Jansen, P. Landrock, G. Roelofsen, *A European call for cryptographic Algorithms: RIPE; RACE Integrity Primitives Evaluation*, 267–271.

M. Waidner, *Unconditional sender and recipient untraceability in spite of active attacks*, 302–319.

M. Waidner, B. Pfitzmann, *The dining cryptographers in the disco: Unconditional sender and recipient untraceability with computationally secure serviceability*, 690.

M. Wang, *Linear complexity profiles and continued fractions*, 571–585.

P. Wichmann, *Cryptanalysis of a modified rotor machine*, 395–402.

M.J. Wiener, *Cryptanalysis of short RSA secret exponents*, 372.

M. Yung, *Zero-knowledge proofs of computational power*, 196–207.

Y. Zheng, T. Matsumoto, H. Imai, *Impossibility and optimality results on constructing pseudorandom permutations*, 412–422.

Advances in Cryptology – **EUROCRYPT '90**, Aarhus, Denmark. Springer-Verlag LNCS 473 (1991).
Editor: I.B. Damgård.

F. Bauspieß, H.-J. Knobloch, P. Wichmann, *Inverting the pseudo exponentiation*, 344–351.

C.H. Bennett, F. Bessette, G. Brassard, L. Salvail, J. Smolin, *Experimental quantum cryptography*, 253–265.

A. Beutelspacher, U. Rosenbaum, *Essentially l-fold secure authentication systems*, 294–305.

G. Bleumer, B. Pfitzmann, M. Waidner, *A remark on a signature scheme where forgery can be proved*, 441–445.

E.F. Brickell, K.S. McCurley, *An interactive identification scheme based on discrete logarithms and factoring*, 63–71.

M.V.D. Burmester, *A remark on the efficiency of identification schemes*, 493–495.

M.V.D. Burmester, Y. Desmedt, *All languages in NP have divertible zero-knowledge proofs and arguments under cryptographic assumptions*, 1–10.

A.H. Chan, M. Goresky, A. Klapper, *Correlation functions of geometric sequences*, 214–221.

D. Chaum, *Zero-knowledge undeniable signatures*, 458–464.

Z.-D. Dai, T. Beth, D. Gollmann, *Lower bounds for the linear complexity of sequences over residue rings*, 189–195.

G. Davida, Y. Desmedt, R. Peralta, *On the importance of memory resources in the security of key exchange protocols*, 11–15.

A. De Santis, G. Persiano, *Public-randomness in public-key cryptography*, 46–62.

A. De Santis, M. Yung, *On the design of provably secure cryptographic hash functions*, 412–431.

B. den Boer, *Oblivious transfer protecting secrecy – an implementation for oblivious transfer protecting secrecy almost unconditionally and a bitcommitment based on factoring protecting secrecy unconditionally*, 31–45.

J. Domingo-Ferrer, *Software run-time protection: A cryptographic issue*, 474–480.

S.R. Dussé, B.S. Kaliski Jr., *A cryptographic library for the Motorola DSP 56000*, 230–244.

J.-H. Evertse, E. van Heyst, *Which new RSA signatures can be computed from some given RSA signatures?*, 83–97.

M. Girault, *An identity-based identification scheme based on discrete logarithms modulo a composite number*, 481–486.

J.D. Golić, M.J. Mihaljević, *A noisy clock-controlled shift register cryptanalysis concept based on sequence comparison approach*, 487–491.

L.C. Guillou, J.-J. Quisquater, M. Walker, P. Landrock, C. Shaer, *Precautions taken against various potential attacks in ISO/IEC DIS 9796*, 465–473.

T. Hwang, *Cryptosystems for group oriented cryptography*, 352–360.

I. Ingemarsson, G.J. Simmons, *A protocol to set up shared secret schemes without the assistance of a mutually trusted party*, 266–282.

C.J.A. Jansen, *On the construction of run permuted sequences*, 196–203.

B.S. Kaliski Jr., *The MD4 message digest algorithm*, 492.

K. Kurosawa, Y. Katayama, W. Ogata, S. Tsujii, *General public key residue cryptosystems and mental poker protocols*, 374–388.

X. Lai, J.L. Massey, *A proposal for a new block encryption standard*, 389–404.

A.K. Lenstra, M.S. Manasse, *Factoring with two large primes*, 72–82.

S. Lloyd, *Properties of binary functions*, 124–139.

U. Maurer, *A provably-secure strongly-randomized cipher*, 361–373.

W. Meier, O. Staffelbach, *Correlation properties of combiners with memory in stream ciphers*, 204–213.

G. Meister, *On an implementation of the Mohan-Adiga algorithm*, 496–500.

S. Miyaguchi, K. Ohta, M. Iwata, *Confirmation that some hash functions are not collision free*, 326–343.

F. Morain, *Distributed primality proving and the primality of $(2^{3539} + 1)/3$*, 110–123.

H. Niederreiter, *The linear complexity profile and the jump complexity of keystream sequences*, 174–188.

V. Niemi, *A new trapdoor in knapsacks*, 405–411.

K. Nyberg, *Constructions of bent functions and difference sets*, 151–160.

K. Ohta, T. Okamoto, K. Koyama, *Membership authentication for hierarchical multigroups using the extended Fiat-Shamir scheme*, 446–457.

H. Ong, C.P. Schnorr, *Fast signature generation with a Fiat Shamir-like scheme*, 432–440.

H. Orup, E. Svendsen, E. Andreasen, *VICTOR - an efficient RSA hardware implementation*, 245–252.

J. Pieprzyk, *How to construct pseudorandom permutations from single pseudorandom functions*, 140–150.

B. Preneel, W. Van Leekwijck, L. Van Linden, R. Govaerts, J. Vandewalle, *Propagation characteristics of Boolean functions*, 161–173.

R. Scheidler, J.A. Buchmann, H.C. Williams, *Implementation of a key exchange protocol using real quadratic fields*, 98–109.

A. Sgarro, *Lower bounds for authentication codes with splitting*, 283–293.

S. Shinozaki, T. Itoh, A. Fujioka, S. Tsujii, *Provably secure key-updating schemes in identity-based systems*, 16–30.

B. Smeets, P. Vanrose, Z.-X. Wan, *On the construction of authentication codes with secrecy and codes withstanding spoofing attacks of order $L \geq 2$*, 306–312.

J. Stern, P. Toffin, *Cryptanalysis of a public-key cryptosystem based on approximations by rational numbers*, 313–317.

P.C. van Oorschot, M.J. Wiener, *A known-plaintext attack on two-key triple encryption*, 318–325.

Y. Yacobi, *Exponentiating faster with addition chains*, 222–229.

Advances in Cryptology – **EUROCRYPT '91**, Brighton, UK. Springer-Verlag LNCS 547 (1991).
Editor: D.W. Davies.

S. Berkovits, *How to broadcast a secret*, 535–541.

T. Beth, F. Schaefer, *Non supersingular elliptic curves for public key cryptosystems*, 316–327.

E. Biham, *Cryptanalysis of the chaotic-map cryptosystem suggested at EUROCRYPT '91*, 532–534.

E. Biham, A. Shamir, *Differential cryptanalysis of Feal and N-Hash*, 1–16.

C. Boyd, *Enhancing secrecy by data compression: Theoretical and practical aspects*, 266–280.

L. Brynielsson, *The information leakage through a randomly generated function*, 552–553.

M. Burmester, Y. Desmedt, *Broadcast interactive proofs*, 81–95.

P. Camion, J. Patarin, *The knapsack hash function proposed at Crypto '89 can be broken*, 39–53.

W.G. Chambers, Z.-D. Dai, *On binary sequences from recursions "modulo 2^e" made non-linear by the bit-by-bit "XOR" function*, 200–204.

D. Chaum, *Some weaknesses of "Weaknesses of undeniable signatures"*, 554–556.

D. Chaum, E. van Heyst, *Group signatures*, 257–265.

V. Chepyzhov, B. Smeets, *On a fast correlation attack on certain stream ciphers*, 176–185.

M.J. Coster, B.A. LaMacchia, A.M. Odlyzko, C.P. Schnorr, *An improved low-density subset sum algorithm*, 54–67.

C. Crépeau, M. Sántha, *On the reversibility of oblivious transfer*, 106–113.

Z.-D. Dai, J.-H. Yang, *Linear complexity of periodically repeated random sequences*, 168–175.

M.H. Dawson, S.E. Tavares, *An expanded set of S-box design criteria based on information theory and its relation to differential-like attacks*, 352–367.

P. de Rooij, *On the security of the Schnorr scheme using preprocessing*, 71–80.

Y. Desmedt, M. Yung, *Weaknesses of undeniable signature schemes*, 205–220.

A. Fujioka, T. Okamoto, S. Miyaguchi, *ESIGN: An efficient digital signature implementation for smart cards*, 446–457.

A. Fujioka, T. Okamoto, K. Ohta, *Interactive bi-proof systems and undeniable signature schemes*, 243–256.

E.M. Gabidulin, A.V. Paramonov, O.V. Tretjakov, *Ideals over a non-commutative ring and their application in cryptology*, 482–489.

J.K. Gibson, *Equivalent Goppa codes and trapdoors to McEliece's public key cryptosystem*, 517–521.

M. Girault, *Self-certified public keys*, 490–497.

B. Goldburg, E. Dawson, S. Sridharan, *The automated cryptanalysis of analog speech scramblers*, 422–430.

J.D. Golić, *The number of output sequences of a binary sequence generator*, 160–167.

T. Habutsu, Y. Nishio, I. Sasase, S. Mori, *A secret key cryptosystem by iterating a chaotic map*, 127–140.

P. Horster, H.-J. Knobloch, *Discrete logarithm based protocols*, 399–408.

K. Huber, *Some considerations concerning the selection of RSA moduli*, 294–301.

C.J.A. Jansen, *The maximum order complexity of sequence ensembles*, 153–159.

V.I. Korzhik, A.I. Turkin, *Cryptanalysis of McEliece's public-key cryptosystem*, 68–70.

X. Lai, J.L. Massey, S. Murphy, *Markov ciphers and differential cryptanalysis*, 17–38.

T. Matsumoto, H. Imai, *Human identification through insecure channel*, 409–421.

U.M. Maurer, *New approaches to the design of self-synchronizing stream ciphers*, 458–471.

U.M. Maurer, Y. Yacobi, *Non-interactive public-key cryptography*, 498–507.

W. Meier, O. Staffelbach, *Analysis of pseudo random sequences generated by cellular automata*, 186–199.

M.J. Mihaljević, J.D. Golić, *A comparison of cryptanalytic principles based on iterative error-correction*, 527–531.

F. Morain, *Building cyclic elliptic curves modulo large primes*, 328–336.

W.B. Müller, A. Oswald, *Dickson pseudoprimes and primality testing*, 512–516.

S. Mund, *Ziv-Lempel complexity for periodic sequences and its cryptographic application*, 114–126.

K. Nyberg, *Perfect nonlinear S-boxes*, 378–386.

L. O'Connor, *Enumerating nondegenerate permutations*, 368–377.

T. Okamoto, D. Chaum, K. Ohta, *Direct zero knowledge proofs of computational power in five rounds*, 96–105.

T.P. Pedersen, *Distributed provers with applications to undeniable signatures*, 221–242.

T.P. Pedersen, *A threshold cryptosystem without a trusted party*, 522–526.

J. Pieprzyk, *Probabilistic analysis of elementary randomizers*, 542–546.

J. Pieprzyk, R. Safavi-Naini, *Randomized authentication systems*, 472–481.

M. Portz, *On the use of interconnection networks in cryptography*, 302–315.

B. Preneel, D. Chaum, W. Fumy, C.J.A. Jansen, P. Landrock, G. Roelofsen, *Race Integrity Primitives Evaluation (RIPE): A status report*, 547–551.

B. Preneel, R. Govaerts, J. Vandewalle, *Boolean functions satisfying higher order propagation criteria*, 141–152.

R.A. Rueppel, *A formal approach to security architectures*, 387–398.

B. Sadeghiyan, J. Pieprzyk, *A construction for one way hash functions and pseudorandom bit generators*, 431–445.

C.P. Schnorr, *Factoring integers and computing discrete logarithms via diophantine approximation*, 281–293.

H. Shizuya, T. Itoh, K. Sakurai, *On the complexity of hyperelliptic discrete logarithm problem*, 337–351.

G. Zémor, *Hash functions and graphs with large girths*, 508–511.

Advances in Cryptology – **EUROCRYPT '92**, Balantonfüred, Hungary.
Springer-Verlag LNCS 658 (1993).
Editor: R.A. Rueppel.

G.B. Agnew, R.C. Mullin, S.A. Vanstone, *On the development of a fast elliptic curve cryptosystem*, 482–487.

P. Barbaroux, *Uniform results in polynomial-time security*, 297–306.

T. Baritaud, H. Gilbert, M. Girault, *FFT hashing is not collision-free*, 35–44.

D. Beaver, *How to break a "secure" oblivious transfer protocol*, 285–296.

D. Beaver, S. Haber, *Cryptographic protocols provably secure against dynamic adversaries*, 307–323.

M.J. Beller, Y. Yacobi, *Batch Diffie-Hellman key agreement systems and their application to portable communications*, 208–220.

T.A. Berson, *Differential cryptanalysis* mod 2^{32} *with applications to MD5*, 71–80.

I. Biehl, J. Buchmann, B. Meyer, C. Thiel, C. Thiel, *Tools for proving zero knowledge*, 356–365.

C. Blundo, A. De Santis, D.R. Stinson, U. Vaccaro, *Graph decompositions and secret sharing schemes*, 1–24.

E.F. Brickell, D.M. Gordon, K.S. McCurley, D.B. Wilson, *Fast exponentiation with precomputation*, 200–207.

D. Chaum, T.P. Pedersen, *Transferred cash grows in size*, 390–407.

L. Chen, I. Damgård, *Security bounds for parallel versions of identification protocols*, 461–466.

I. Damgård, *Non-interactive circuit based proofs and non-interactive perfect zero-knowledge with preprocessing*, 341–355.

B. Dixon, A.K. Lenstra, *Massively parallel elliptic curve factoring*, 183–193.

J.-H. Evertse, E. van Heyst, *Which new RSA signatures can be computed from RSA signatures, obtained in a specific interactive protocol?*, 378–389.

Y. Frankel, Y. Desmedt, *Classification of ideal homomorphic threshold schemes over finite abelian groups*, 25–34.

J.D. Golić, *Correlation via linear sequential circuit approximation of combiners with memory*, 113–123.

J.D. Golić, S.V. Petrović, *A generalized correlation attack with a probabilistic constrained edit distance*, 472–476.

G. Harper, A. Menezes, S. Vanstone, *Public-key cryptosystems with very small key lengths*, 163–173.

R. Heiman, *A note on discrete logarithms with special structure*, 454–457.

R. Heiman, *Secure audio teleconferencing: A practical solution*, 437–448.

K. Iwamura, T. Matsumoto, H. Imai, *High-speed implementation methods for RSA scheme*, 221–238.

K. Iwamura, T. Matsumoto, H. Imai, *Systolic arrays for modular exponentiation using Montgomery method*, 477–481.

K. Koyama, *Secure conference key distribution schemes for conspiracy attacks*, 449–453.

X. Lai, J.L. Massey, *Hash functions based on block ciphers*, 55–70.

M. Matsui, A. Yamagishi, *A new method for known plaintext attack of FEAL cipher*, 81–91.

U.M. Maurer, *Factoring with an oracle,* 429–436.

U.M. Maurer, *A simplified and generalized treatment of Luby-Rackoff pseudorandom permutation generators,* 239–255.

U.M. Maurer, Y. Yacobi, *A remark on a non-interactive public-key distribution system,* 458–460.

M. Mihaljević, J.D. Golić, *Convergence of a Bayesian iterative error-correction procedure on a noisy shift register sequence,* 124–137.

D. Naccache, *A Montgomery-suitable Fiat-Shamir-like authentication scheme,* 488–491.

H. Niederreiter, C.P. Schnorr, *Local randomness in candidate one-way functions,* 408–419.

K. Nyberg, *On the construction of highly nonlinear permutations,* 92–98.

L. O'Connor, T. Snider, *Suffix trees and string complexity,* 138–152.

K. Ohta, T. Okamoto, A. Fujioka, *Secure bit commitment function against divertibility,* 324–340.

T. Okamoto, K. Sakurai, H. Shizuya, *How intractable is the discrete logarithm for a general finite group,* 420–428.

J. Patarin, *How to construct pseudorandom and super pseudorandom permutations from one single pseudorandom function,* 256–266.

B. Pfitzmann, M. Waidner, *Attacks on protocols for server-aided RSA computation,* 153–162.

R. Rueppel, A. Lenstra, M. Smid, K. McCurley, Y. Desmedt, A. Odlyzko, P. Landrock, *The Eurocrypt '92 controversial issue: trapdoor primes and moduli,* 194–199.

B. Sadeghiyan, J. Pieprzyk, *A construction for super pseudorandom permutations from a single pseudorandom function,* 267–284.

J. Sauerbrey, A. Dietel, *Resource requirements for the application of addition chains in modulo exponentiation,* 174–182.

C.P. Schnorr, *FFT-Hash II, efficient cryptographic hashing,* 45–54.

A. Sgarro, *Information-theoretic bounds for authentication frauds,* 467–471.

E. van Heyst, T.P. Pedersen, *How to make efficient fail-stop signatures,* 366–377.

R. Wernsdorf, *The one-round functions of the DES generate the alternating group,* 99–112.

Advances in Cryptology – **EUROCRYPT '93**, Lofthus, Norway. Springer-Verlag LNCS 765 (1994). Editor: T. Helleseth.

D. Beaver, N. So, *Global, unpredictable bit generation without broadcast,* 424–434.

J. Benaloh, M. de Mare, *One-way accumulators: A decentralized alternative to digital signatures,* 274–285.

T. Beth, C. Ding, *On almost perfect nonlinear permutations,* 65–76.

E. Biham, *New types of cryptanalytic attacks using related keys,* 398–409.

S. Blackburn, S. Murphy, J. Stern, *Weaknesses of a public-key cryptosystem based on factorizations of finite groups,* 50–54.

C. Boyd, W. Mao, *On a limitation of BAN logic,* 240–247.

S. Brands, D. Chaum, *Distance-bounding protocols,* 344–359.

G. Brassard, L. Salvail, *Secret key reconciliation by public discussion,* 410–423.

M. Burmester, *Cryptanalysis of the Chang-Wu-Chen key distribution system,* 440–442.

C. Carlet, *Two new classes of bent functions,* 77–101.

M. Carpentieri, A. De Santis, U. Vaccaro, *Size of shares and probability of cheating in threshold schemes,* 118–125.

R.J.F. Cramer, T.P. Pedersen, *Improved privacy in wallets with observers,* 329–343.

T.W. Cusick, *Boolean functions satisfying a higher order strict avalanche criterion,* 102–117.

J. Daemen, R. Govaerts, J. Vandewalle, *Resynchronization weaknesses in synchronous stream ciphers,* 159–167.

I.B. Damgård, *Practical and provably secure release of a secret and exchange of signatures,* 200–217.

I.B. Damgård, L.R. Knudsen, *The breaking of the AR hash function,* 286–292.

P. de Rooij, *On Schnorr's preprocessing for digital signature schemes,* 435–439.

N. Demytko, *A new elliptic curve based analogue of RSA,* 40–49.

B. den Boer, A. Bosselaers, *Collisions for the compression function of MD5,* 293–304.

B. Dixon, A.K. Lenstra, *Factoring integers using SIMD sieves,* 28–39.

J. Domingo-Ferrer, *Untransferable rights in a client-independent server environment,* 260–266.

N. Ferguson, *Single term off-line coins*, 318–328.

R.A. Games, J.J. Rushanan, *Blind synchronization of m-sequences with even span*, 168–180.

R. Göttfert, H. Niederreiter, *On the linear complexity of products of shift-register sequences*, 151–158.

G. Hornauer, W. Stephan, R. Wernsdorf, *Markov ciphers and alternating groups*, 453–460.

T. Johansson, G. Kabatianskii, B. Smeets, *On the relation between A-codes and codes correcting independent errors*, 1–11.

K. Kurosawa, K. Okada, K. Sakano, W. Ogata, S. Tsujii, *Nonperfect secret sharing schemes and matroids*, 126–141.

M. Matsui, *Linear cryptanalysis method for DES cipher*, 386–397.

W. Meier, *On the security of the IDEA block cipher*, 371–385.

D. Naccache, *Can O.S.S. be repaired? – proposal for a new practical signature scheme*, 233–239.

K. Nyberg, *Differentially uniform mappings for cryptography*, 55–64.

L. O'Connor, *On the distribution of characteristics in bijective mappings*, 360–370.

R. Ostrovsky, R. Venkatesan, M. Yung, *Interactive hashing simplifies zero-knowledge protocol design*, 267–273.

C. Park, K. Itoh, K. Kurosawa, *Efficient anonymous channel and all/nothing election scheme*, 248–259.

C. Park, K. Kurosawa, T. Okamoto, S. Tsujii, *On key distribution and authentication in mobile radio networks*, 461–465.

J. Patarin, *How to find and avoid collisions for the knapsack hash function*, 305–317.

R. Safavi-Naini, L. Tombak, *Optimal authentication systems*, 12–27.

J. Seberry, X.-M. Zhang, Y. Zheng, *On constructions and nonlinearity of correlation immune functions*, 181–199.

E.S. Selmer, *From the memoirs of a Norwegian cryptologist*, 142–150.

G.J. Simmons, *The consequences of trust in shared secret schemes*, 448–452.

G.J. Simmons, *Subliminal communication is easy using the DSA*, 218–232.

P.C. van Oorschot, *An alternate explanation of two BAN-logic "failures"*, 443–447.

Advances in Cryptology – **EUROCRYPT '94**, Perugia, Italy. Springer-Verlag LNCS 950 (1995).
Editor: A. De Santis

M. Bellare, P. Rogaway, *Optimal asymmetric encryption*, 92–111.

E. Biham, *On Matsui's linear cryptanalysis*, 341–355.

E. Biham, A. Biryukov, *An improvement of Davies' attack on DES*, 461–467.

C. Blundo, A. Cresti, *Space requirements for broadcast encryption*, 287–298.

C. Blundo, A. Giorgio Gaggia, D.R. Stinson, *On the dealer's randomness required in secret sharing schemes*, 35–46.

M. Burmester, Y. Desmedt, *A secure and efficient conference key distribution system*, 275–286.

C. Cachin, U.M. Maurer, *Linking information reconciliation and privacy amplification*, 266–274.

J.L. Camenisch, J.-M. Piveteau, M.A. Stadler, *Blind signatures based on the discrete logarithm problem*, 428–432.

F. Chabaud, *On the security of some cryptosystems based on error-correcting codes*, 131–139.

F. Chabaud, S. Vaudenay, *Links between differential and linear cryptanalysis*, 356–365.

C. Charnes, L. O'Connor, J. Pieprzyk, R. Safavi-Naini, Y. Zheng, *Comments on Soviet encryption algorithm*, 433–438.

D. Chaum, *Designated confirmer signatures*, 86–91.

L. Chen, I.B. Damgård, T.P. Pedersen, *Parallel divertibility of proofs of knowledge*, 140–155.

L. Chen, T.P. Pedersen, *New group signature schemes*, 171–181.

L. Csirmaz, *The size of a share must be large*, 13–22.

S. D'Amiano, G. Di Crescenzo, *Methodology for digital money based on general cryptographic tools*, 156–170.

F. Damm, F.-P. Heider, G. Wambach, *MIMD-factorisation on hypercubes*, 400–409.

P. de Rooij, *Efficient exponentiation using precomputation and vector addition chains*, 389–399.

T. Eng, T. Okamoto, *Single-term divisible electronic coins*, 306–319.

M. Franklin, M. Yung, *The blinding of weak signatures*, 67–76.

J.D. Golić, L. O'Connor, *Embedding and probabilistic correlation attacks on clock-controlled shift registers*, 230–243.

M. Goresky, A. Klapper, *Feedback registers based on ramified extensions of the 2-adic numbers*, 215–222.

R. Göttfert, H. Niederreiter, *A general lower bound for the linear complexity of the product of shift-register sequences*, 223–229.

J. Hruby, *Q-deformed quantum cryptography*, 468–472.

M. Jakobsson, *Blackmailing using undeniable signatures*, 425–427.

T. Johansson, B. Smeets, *On A^2-codes including arbiter's attacks*, 456–460.

A. Joux, L. Granboulan, *A practical attack against knapsack based hash functions*, 58–66.

L.R. Knudsen, *New potentially 'weak' keys for DES and LOKI*, 419–424.

L.R. Knudsen, X. Lai, *New attacks on all double block length hash functions of hash rate 1, including the parallel-DM*, 410–418.

C.-M. Li, T. Hwang, N.-Y. Lee, *Threshold-multisignature schemes where suspected forgery implies traceability of adversarial shareholders*, 194–204.

M. Matsui, *On correlation between the order of S-boxes and the strength of DES*, 366–375.

W. Meier, O. Staffelbach, *The self-shrinking generator*, 205–214.

R. Menicocci, *A systematic attack on clock controlled cascades*, 450–455.

D. Naccache, D. M'Raïhi, S. Vaudenay, D. Raphaeli, *Can D.S.A. be improved? Complexity trade-offs with the digital signature standard*, 77–85.

M. Naor, A. Shamir, *Visual cryptography*, 1–12.

K. Nyberg, *Linear approximation of block ciphers*, 439–444.

K. Nyberg, R.A. Rueppel, *Message recovery for signature schemes based on the discrete logarithm problem*, 182–193.

G. Orton, *A multiple-iterated trapdoor for dense compact knapsacks*, 112–130.

B. Pfitzmann, *Breaking an efficient anonymous channel*, 332–340.

R. Safavi-Naini, L. Tombak, *Authentication codes in plaintext and chosen-content attacks*, 254–265.

C.P. Schnorr, S. Vaudenay, *Black box cryptanalysis of hash networks based on multipermutations*, 47–57.

J. Seberry, X.-M. Zhang, Y. Zheng, *Relationships among nonlinearity criteria*, 376–388.

A. Shamir, *Memory efficient variants of public-key schemes for smart card applications*, 445–449.

P. Syverson, C. Meadows, *Formal requirements for key distribution protocols*, 320–331.

R. Taylor, *Near optimal unconditionally secure authentication*, 244–253.

M. van Dijk, *A linear construction of perfect secret sharing schemes*, 23–34.

Y. Zheng, *How to break and repair Leighton and Micali's key agreement protocol*, 299–305.

Advances in Cryptology – **EUROCRYPT '95**, Saint-Malo, France. Springer-Verlag LNCS 921 (1995). Editors: L.C. Guillou and J.-J. Quisquater

P. Béguin, A. Cresti, *General short computational secret sharing schemes*, 194–208.

J. Bierbrauer, *A^2-codes from universal hash classes*, 311–318.

S. Brands, *Restrictive blinding of secret-key certificates*, 231–247.

L. Chen, T.P. Pedersen, *On the efficiency of group signatures providing information-theoretic anonymity*, 39–49.

C. Crépeau, L. Salvail, *Quantum oblivious mutual identification*, 133–146.

S. D'Amiano, G. Di Crescenzo, *Anonymous NIZK proofs of knowledge with preprocessing*, 413–416.

Y. Desmedt, *Securing traceability of ciphertexts – Towards a secure software key escrow system*, 147–157.

G. Di Crescenzo, *Recycling random bits in composed perfect zero-knowledge*, 367–381.

M.K. Franklin, M.K. Reiter, *Verifiable signature sharing*, 50–63.

C. Gehrmann, *Secure multiround authentication protocols*, 158–167.

R. Gennaro, S. Micali, *Verifiable secret sharing as secure computation*, 168–182.

J.D. Golić, *Towards fast correlation attacks on irregularly clocked shift registers*, 248–262.

C. Harpes, G.G. Kramer, J.L. Massey, *A generalization of linear cryptanalysis and the applicability of Matsui's piling-up lemma*, 24–38.

W.-A. Jackson, K.M. Martin, C.M. O'Keefe, *Efficient secret sharing without a mutually trusted authority*, 183–193.

M. Jakobsson, *Ripping coins for a fair exchange*, 220–230.

A. Klapper, M. Goresky, *Large period nearly de Bruijn FCSR sequences*, 263–273.

K. Koyama, *Fast RSA-type schemes based on singular cubic curves $y^2 + axy \equiv x^3 \pmod{n}$*, 329–340.

H. Krawczyk, *New hash functions for message authentication*, 301–310.

K. Kurosawa, S. Obana, *Combinatorial bounds for authentication codes with arbitration*, 289–300.

R. Lercier, F. Morain, *Counting the number of points on elliptic curves over finite fields: strategies and performances*, 79–94.

C.H. Lim, P.J. Lee, *Server (prover/signer)-aided verification of identity proofs and signatures*, 64–78.

P.L. Montgomery, *A block Lanczos algorithm for finding dependencies over $GF(2)$*, 106–120.

D. Naccache, D. M'raïhi, W. Wolfowicz, A. di Porto, *Are crypto-accelerators really inevitable? 20 bit zero-knowledge in less than a second on simple 8-bit microcontrollers*, 404–409.

M. Näslund, *Universal hash functions & hard core bits*, 356–366.

L. O'Connor, *Convergence in differential distributions*, 13–23.

B. Pfitzmann, M. Schunter, M. Waidner, *How to break another "provably secure" payment system*, 121–132.

D. Pointcheval, *A new identification scheme based on the perceptrons problem*, 319–328.

K. Sako, J. Kilian, *Receipt-free mix-type voting scheme – A practical solution to the implementation of a voting booth*, 393–403.

K. Sakurai, H. Shizuya, *Relationships among the computational powers of breaking discrete log cryptosystems*, 341–355.

C.P. Schnorr, H.H. Hörner, *Attacking the Chor-Rivest cryptosystem by improved lattice reduction*, 1–12.

M. Stadler, J.-M. Piveteau, J. Camenisch, *Fair blind signatures*, 209–219.

C.-H. Wang, T. Hwang, J.-J. Tsai, *On the Matsumoto and Imai's human identification scheme*, 382–392.

D. Weber, *An implementation of the general number field sieve to compute discrete logarithms mod p*, 95–105.

X.-M. Zhang, Y. Zheng, *On nonlinear resilient functions*, 274–288.

Advances in Cryptology – **EUROCRYPT '96**, Zaragoza, Spain. Springer-Verlag LNCS 1070 (1996).
Editor: U.M. Maurer

W. Aiello, R. Venkatesan, *Foiling birthday attacks in length-doubling transformations*, 307–320.

D. Beaver, *Equivocable oblivious transfer*, 119–130.

M. Bellare, P. Rogaway, *The exact security of digital signatures – how to sign with RSA and Rabin*, 399–416.

S. Blackburn, M. Burmester, Y. Desmedt, P. Wild, *Efficient multiplicative sharing schemes*, 107–118.

D. Bleichenbacher, *Generating ElGamal signatures without knowing the secret key*, 10–18.

J. Boyar, R. Peralta, *Short discreet proofs*, 131–142.

M. Burmester, *Homomorphisms of secret sharing schemes: A tool for verifiable signature sharing*, 96–106.

P. Camion, A. Canteaut, *Constructions of t-resilient functions over a finite alphabet*, 283–293.

D. Coppersmith, *Finding a small root of a bivariate integer equation; factoring with high bits known*, 178–189.

D. Coppersmith, *Finding a small root of a univariate modular equation*, 155–165.

D. Coppersmith, M. Franklin, J. Patarin, M. Reiter, *Low-exponent RSA with related messages*, 1–9.

R. Cramer, M. Franklin, B. Schoenmakers, M. Yung, *Multi-authority secret-ballot elections with linear work*, 72–83.

I.B. Damgård, T.P. Pedersen, *New convertible undeniable signature schemes*, 372–386.

J.-B. Fischer, J. Stern, *An efficient pseudo-random generator provably as secure as syndrome decoding*, 245–255.

R. Gennaro, S. Jarecki, H. Krawczyk, T. Rabin, *Robust threshold DSS signatures*, 354–371.

K. Gibson, *The security of the Gabidulin public key cryptosystem*, 212–223.

J. Golić, *Fast low order approximation of cryptographic functions*, 268–282.

S.-M. Hong, S.-Y. Oh, H. Yoon, *New modular multiplication algorithms for fast modular exponentiation*, 166–177.

M. Jakobsson, K. Sako, R. Impagliazzo, *Designated verifier proofs and their applications*, 143–154.

A. Klapper, *On the existence of secure feedback registers*, 256–267.

L.R. Knudsen, T.P. Pedersen, *On the difficulty of software key escrow*, 237–244.

L.R. Knudsen, M.J.B. Robshaw, *Non-linear approximations in linear cryptanalysis*, 224–236.

B. Meyer, V. Müller, *A public key cryptosystem based on elliptic curves over $\mathbb{Z}/n\mathbb{Z}$ equivalent to factoring*, 49–59.

W. Ogata, K. Kurosawa, *Optimum secret sharing scheme secure against cheating*, 200–211.

J. Patarin, *Hidden fields equations (HFE) and isomorphisms of polynomials (IP): Two new families of asymmetric algorithms*, 33–48.

B. Pfitzmann, M. Schunter, *Asymmetric fingerprinting*, 84–95.

D. Pointcheval, J. Stern, *Security proofs for signature schemes*, 387–398.

B. Preneel, P.C. van Oorschot, *On the security of two MAC algorithms*, 19–32.

F. Schwenk, J. Eisfeld, *Public key encryption and signature schemes based on polynomials over $\mathbb{Z}_n$*, 60–71.

V. Shoup, *On the security of a practical identification scheme*, 344–353.

V. Shoup, A. Rubin, *Session key distribution using smart cards*, 321–331.

M. Stadler, *Publicly verifiable secret sharing*, 190–199.

P.C. van Oorschot, M.J. Wiener, *On Diffie-Hellman key agreement with short exponents*, 332–343.

X.-M. Zhang, Y. Zheng, *Auto-correlations and new bounds on the nonlinearity of Boolean functions*, 294–306.

A.4 Fast Software Encryption Proceedings

Fast Software Encryption: Cambridge Security Workshop, Cambridge, UK., December 1993. Springer-Verlag LNCS 809 (1994).
Editor: R. Anderson

R. Anderson, *A modern rotor machine*, 47–50.

E. Biham, *On modes of operation*, 116–120.

U. Blöcher, M. Dichtl, *Fish: A fast software stream cipher*, 41–44.

W.G. Chambers, *Two stream ciphers*, 51–55.

A. Chan, R. Games, J. Rushanan, *On quadratic m-sequences*, 166–173.

J. Daemen, R. Govaerts, J. Vandewalle, *A new approach to block cipher design*, 18–32.

A. Di Porto, W. Wolfowicz, *VINO: A block cipher including variable permutations*, 205–210.

C. Ding, *The differential cryptanalysis and design of natural stream ciphers*, 101–115.

J. Golić, *On the security of shift register based keystream generators*, 90–100.

D. Gollmann, *Cryptanalysis of clock controlled shift registers*, 121–126.

B.S. Kaliski Jr., M.J.B. Robshaw, *Fast block cipher proposal*, 33–40.

A. Klapper, M. Goresky, *2-Adic shift registers*, 174–178.

L.R. Knudsen, *Practically secure Feistel ciphers*, 211–221.

H. Krawczyk, *The shrinking generator: Some practical considerations*, 45–46.

X. Lai, L.R. Knudsen, *Attacks on double block length hash functions*, 157–165.

M. Lomas, *Encrypting network traffic*, 64–70.

N. Maclaren, *Cryptographic pseudo-random numbers in simulation*, 185–190.

J. Massey, *SAFER K-64: A byte-oriented block-ciphering algorithm*, 1–17.

K. Nyberg, *New bent mappings suitable for fast implementation*, 179–184.

B. Preneel, *Design principles for dedicated hash functions*, 71–82.

T. Renji, *On finite automaton one-key cryptosystems*, 135–148.

M. Roe, *Performance of symmetric ciphers and one-way hash functions*, 83–89.

P. Rogaway, D. Coppersmith, *A software-optimized encryption algorithm*, 56–63.

B. Schneier, *Description of a new variable-length key, 64-bit block cipher (Blowfish)*, 191–204.

C. Schnorr, S. Vaudenay, *Parallel FFT-hashing*, 149–156.

D. Wheeler, *A bulk data encryption algorithm*, 127–134.

Fast Software Encryption: Second International Workshop, Leuven, Belgium, December 1994. Springer-Verlag LNCS 1008 (1995).
Editor: B. Preneel

R. Anderson, *On Fibonacci keystream generators*, 346–352.

R. Anderson, *Searching for the optimum correlation attack*, 137–143.

U. Baum, S. Blackburn, *Clock-controlled pseudorandom generators on finite groups*, 6–21.

E. Biham, P.C. Kocher, *A known plaintext attack on the PKZIP stream cipher*, 144–153.

M. Blaze, B. Schneier, *The MacGuffin block cipher algorithm*, 97–110.

U. Blöcher, M. Dichtl, *Problems with the linear cryptanalysis of DES using more than one active S-box per round*, 265–274.

W.G. Chambers, *On random mappings and random permutations*, 22–28.

J. Daemen, R. Govaerts, J. Vandewalle, *Correlation matrices*, 275–285.

C. Ding, *Binary cyclotomic generators*, 29–60.

H. Dobbertin, *Construction of bent functions and balanced Boolean functions with high nonlinearity*, 61–74.

J.D. Golić, *Linear cryptanalysis of stream ciphers*, 154–169.

B.S. Kaliski Jr., M.J.B. Robshaw, *Linear cryptanalysis using multiple approximations and FEAL*, 249–264.

A. Klapper, *Feedback with carry shift registers over finite fields*, 170–178.

L.R. Knudsen, *Truncated and higher order differentials*, 196–211.

X. Lai, *Additive and linear structures of cryptographic functions*, 75–85.

S. Lucks, *How to exploit the intractability of exact TSP for cryptography*, 298–304.

D.J.C. MacKay, *A free energy minimization framework for inference problems in modulo 2 arithmetic*, 179–195.

J.L. Massey, *SAFER K-64: One year later*, 212–241.

K. Nyberg, *S-boxes and round functions with controllable linearity and differential uniformity*, 111–130.

L. O'Connor, *Properties of linear approximation tables*, 131–136.

W.T. Penzhorn, *A fast homophonic coding algorithm based on arithmetic coding*, 329–345.

B. Preneel, *Introduction*, 1–5.

V. Rijmen, B. Preneel, *Cryptanalysis of McGuffin*, 353–358.

V. Rijmen, B. Preneel, *Improved characteristics for differential cryptanalysis of hash functions based on block ciphers*, 242–248.

R.L. Rivest, *The RC5 encryption algorithm*, 86–96.

M. Roe, *How to reverse engineer an EES device*, 305–328.

M. Roe, *Performance of block ciphers and hash functions – one year later*, 359–362.

S. Vaudenay, *On the need for multipermutations: Cryptanalysis of MD4 and SAFER*, 286–297.

D.J. Wheeler, R.M. Needham, *TEA, a tiny encryption algorithm*, 363–366.

Fast Software Encryption: Third International Workshop, Cambridge, UK., February 1996. Springer-Verlag LNCS 1039 (1996).
Editor: D. Gollmann

R. Anderson, E. Biham, *Tiger: a fast new hash function*, 89–97.

R. Anderson, E. Biham, *Two practical and provably secure block ciphers: BEAR and LION*, 113–120.

M. Blaze, *High-bandwidth encryption with low-bandwidth smartcards*, 33-40.

A. Clark, J.D. Golić, E. Dawson, *A comparison of fast correlation attacks*, 145–157.

H. Dobbertin, *Cryptanalysis of MD4*, 53–69.

H. Dobbertin, A. Bosselaers, B. Preneel, *RIPEMD-160: a strengthened version of RIPEMD*, 71–82.

W. Geiselmann, *A note on the hash function of Tillich and Zémor*, 51-52.

J.D. Golić, *On the security of nonlinear filter generators*, 173–188.

R. Jenkins Jr., *ISAAC*, 41-49.

L.R. Knudsen, T.A. Berson, *Truncated differentials of SAFER*, 15-26.

X. Lai, R.A. Rueppel, *Attacks on the HKM/HFX cryptosystem*, 1–14.

S. Lucks, *Faster Luby-Rackoff ciphers*, 189–203.

M. Matsui, *New structure of block ciphers with provable security against differential and linear cryptanalysis*, 205–218.

K. Nyberg, *Fast accumulated hashing*, 83–87.

W.T. Penzhorn, *Correlation attacks on stream ciphers: computing low-weight parity checks based on error-correcting codes*, 159–172.

V. Rijmen, J. Daemen, B. Preneel, A. Bosselaers, E. De Win, *The cipher SHARK*, 99-111.

B. Schneier, J. Kelsey, *Unbalanced Feistel networks and block cipher design*, 121–144.

S. Vaudenay, *On the weak keys of Blowfish*, 27-32.

A.5 Journal of Cryptology papers

Journal of Cryptology papers (Volume 1 No.1 – Volume 9 No.3, 1988-1996)

M. Abadi, J. Feigenbaum, *Secure circuit evaluation*, 2 (1990), 1–12.

C. Adams, S. Tavares, *The structured design of cryptographically good S-Boxes*, 3 (1990), 27–41.

G.B. Agnew, T. Beth, R.C. Mullin, S.A. Vanstone, *Arithmetic operations in $GF(2^m)$*, 6 (1993), 3–13.

G.B. Agnew, R.C. Mullin, I.M. Onyszchuk, S.A. Vanstone, *An implementation for a fast public-key cryptosystem*, 3 (1991), 63–79.

P. Beauchemin, G. Brassard, *A generalization of Hellman's extension to Shannon's approach to cryptography*, 1 (1988), 129–131.

P. Beauchemin, G. Brassard, C. Crépeau, C. Goutier, C. Pomerance, *The generation of random numbers that are probably prime*, 1 (1988), 53–64.

D. Beaver, *Secure multiparty protocols and zero-knowledge proof systems tolerating a faulty minority*, 4 (1991), 75–122.

M. Bellare, M. Yung, *Certifying permutations: Noninteractive zero-knowledge based on any trapdoor permutation*, 9 (1996), 149–166.

I. Ben-Aroya, E. Biham, *Differential cryptanalysis of Lucifer*, 9 (1996), 21–34.

S. Bengio, G. Brassard, Y.G. Desmedt, C. Goutier, J.-J. Quisquater, *Secure implementation of identification systems*, 4 (1991), 175–183.

C.H. Bennett, F. Bessette, G. Brassard, L. Salvail, J. Smolin, *Experimental quantum cryptography*, 5 (1992), 3–28.

E. Biham, *New types of cryptanalytic attacks using related keys*, 7 (1994), 229–246.

E. Biham, A. Shamir, *Differential cryptanalysis of DES-like cryptosystems*, 4 (1991), 3–72.

S. Blackburn, S. Murphy, J. Stern, *The cryptanalysis of a public-key implementation of finite group mappings*, 8 (1995), 157–166.

C. Blundo, A. De Santis, D.R. Stinson, U. Vaccaro, *Graph decompositions and secret sharing schemes*, 8 (1995), 39–64.

J. Boyar, *Inferring sequences produced by a linear congruential generator missing low-order bits*, 1 (1989), 177–184.

J. Boyar, K. Friedl, C. Lund, *Practical zero-knowledge proofs: Giving hints and using deficiencies*, 4 (1991), 185–206.

J. Boyar, C. Lund, R. Peralta, *On the communication complexity of zero-knowledge proofs*, 6 (1993), 65–85.

J.F. Boyar, S.A. Kurtz, M.W. Krentel, *A discrete logarithm implementation of perfect zero-knowledge blobs*, 2 (1990), 63–76.

E.F. Brickell, D.M. Davenport, *On the classification of ideal secret sharing schemes*, 4 (1991), 123–134.

E.F. Brickell, K.S. McCurley, *An interactive identification scheme based on discrete logarithms and factoring*, 5 (1992), 29–39.

E.F. Brickell, D.R. Stinson, *Some improved bounds on the information rate of perfect secret sharing schemes*, 5 (1992), 153–166.

J. Buchmann, H.C. Williams, *A key-exchange system based on imaginary quadratic fields*, 1 (1988), 107–118.

R.M. Capocelli, A. De Santis, L. Gargano, U. Vaccaro, *On the size of shares for secret sharing schemes,* 6 (1993), 157–167.

D. Chaum, *The dining cryptographers problem: Unconditional sender and recipient untraceability,* 1 (1988), 65–75.

B. Chor, M. Geréb-Graus, E. Kushilevitz, *On the structure of the privacy hierarchy,* 7 (1994), 53–60.

B. Chor, E. Kushilevitz, *Secret sharing over infinite domains,* 6 (1993), 87–95.

D. Coppersmith, *Modifications to the number field sieve,* 6 (1993), 169–180.

Z.-D. Dai, *Binary sequences derived from ML-Sequences over rings, I: Periods and minimal polynomials,* 5 (1992), 193–207.

D.W. Davies, S. Murphy, *Pairs and triplets of DES S-boxes,* 8 (1995), 1–25.

A. De Santis, G. Persiano, *The power of preprocessing in zero-knowledge proofs of knowledge,* 9 (1996), 129–148.

M. De Soete, *New bounds and constructions for authentication/secrecy codes with splitting,* 3 (1991), 173–186.

M. Dyer, T. Fenner, A. Frieze, A. Thomason, *On key storage in secure networks,* 8 (1995), 189–200.

S. Even, O. Goldreich, S. Micali, *On-line/off-line digital signatures,* 9 (1996), 35–67.

J.-H. Evertse, E. van Heyst, *Which new RSA-signatures can be computed from certain given RSA-signatures?,* 5 (1992), 41–52.

U. Feige, A. Fiat, A. Shamir, *Zero-knowledge proofs of identity,* 1 (1988), 77–94.

M. Fischer, R. Wright, *Bounds on secret key exchange using a random deal of cards,* 9 (1996), 71–99.

M.J. Fischer, S. Micali, C. Rackoff, *A secure protocol for the oblivious transfer,* 9 (1996), 191–195.

R. Forré, *Methods and instruments for designing S-Boxes,* 2 (1990), 115–130.

K. Gaarder, E. Snekkenes, *Applying a formal analysis technique to the CCITT X.509 strong two-way authentication protocol,* 3 (1991), 81–98.

J. Georgiades, *Some remarks on the security of the identification scheme based on permuted kernels,* 5 (1992), 133–137.

P. Godlewski, C. Mitchell, *Key-minimal cryptosystems for unconditional secrecy,* 3 (1990), 1–25.

O. Goldreich, *A uniform-complexity treatment of encryption and zero-knowledge,* 6 (1993), 21–53.

O. Goldreich, A. Kahan, *How to construct constant-round zero-knowledge proof systems for NP,* 9 (1996), 167–189.

O. Goldreich, E. Kushilevitz, *A perfect zero-knowledge proof system for a problem equivalent to the discrete logarithm,* 6 (1993), 97–116.

O. Goldreich, Y. Oren, *Definitions and properties of zero-knowledge proof systems,* 7 (1994), 1–32.

J. Golić, *Correlation properties of a general binary combiner with memory,* 9 (1996), 111–126.

J. Golić, M. Mihaljević, *A generalized correlation attack on a class of stream ciphers based on the Levenshtein distance,* 3 (1991), 201–212.

L. Gong, D.J. Wheeler, *A matrix key-distribution scheme,* 2 (1990), 51–59.

S. Haber, W.S. Stornetta, *How to time-stamp a digital document,* 3 (1991), 99–111.

H. Heys, S. Tavares, *Substitution-permutation networks resistant to differential and linear cryptanalysis,* 9 (1996), 1–19.

M. Ito, A. Saito, T. Nishizeki, *Multiple assignment scheme for sharing secret,* 6 (1993), 15–20.

T. Itoh, M. Hoshi, S. Tsujii, *A low communication competitive interactive proof system for promised quadratic residuosity,* 9 (1996), 101–109.

B.S. Kaliski Jr., *One-way permutations on elliptic curves,* 3 (1991), 187–199.

B.S. Kaliski Jr., R.L. Rivest, A.T. Sherman, *Is the Data Encryption Standard a group? (Results of cycling experiments on DES),* 1 (1988), 3–36.

R. Kemmerer, C. Meadows, J. Millen, *Three systems for cryptographic protocol analysis,* 7 (1994), 79–130.

A. Klapper, *The vulnerability of geometric sequences based on fields of odd characteristic,* 7 (1994), 33–51.

N. Koblitz, *Hyperelliptic cryptosystems,* 1 (1989), 139–150.

N. Koblitz, *Elliptic curve implementation of zero-knowledge blobs,* 4 (1991), 207–213.

A.K. Lenstra, Y. Yacobi, *User impersonation in key certification schemes,* 6 (1993), 225–232.

H.W. Lenstra Jr., *On the Chor-Rivest knapsack cryptosystem,* 3 (1991), 149–155.

S. Lloyd, *Counting binary functions with certain cryptographic properties,* 5 (1992), 107–131.

J.H. Loxton, D.S.P. Khoo, G.J. Bird, J. Seberry, *A cubic RSA code equivalent to factorization,* 5 (1992), 139–150.

M. Luby, C. Rackoff, *A study of password security,* 1 (1989), 151–158.

S.S. Magliveras, N.D. Memon, *Algebraic properties of cryptosystem PGM,* 5 (1992), 167–183.

S.M. Matyas, *Key processing with control vectors*, 3 (1991), 113–136.

U. Maurer, *Conditionally-perfect secrecy and a provably-secure randomized cipher*, 5 (1992), 53–66.

U. Maurer, *A universal statistical test for random bit generators*, 5 (1992), 89–105.

U. Maurer, *Fast generation of prime numbers and secure public-key cryptographic parameters*, 8 (1995), 123–155.

U. Maurer, J.L. Massey, *Local randomness in pseudorandom sequences*, 4 (1991), 135–149.

U. Maurer, J.L. Massey, *Cascade ciphers: The importance of being first*, 6 (1993), 55–61.

K.S. McCurley, *A key distribution system equivalent to factoring*, 1 (1988), 95–105.

W. Meier, O. Staffelbach, *Fast correlation attacks on certain stream ciphers*, 1 (1989), 159–176.

W. Meier, O. Staffelbach, *Correlation properties of combiners with memory in stream ciphers*, 5 (1992), 67–86.

A. Menezes, S. Vanstone, *Elliptic curve cryptosystems and their implementation*, 6 (1993), 209–224.

R.C. Merkle, *A fast software one-way hash function*, 3 (1990), 43–58.

S. Micali, C.P. Schnorr, *Efficient, perfect polynomial random number generators*, 3 (1991), 157–172.

C. Mitchell, *Enumerating Boolean functions of cryptographic significance*, 2 (1990), 155–170.

S. Murphy, *The cryptanalysis of FEAL-4 with 20 chosen plaintexts*, 2 (1990), 145–154.

S. Murphy, K. Paterson, P. Wild, *A weak cipher that generates the symmetric group*, 7 (1994), 61–65.

M. Naor, *Bit commitment using pseudorandomness*, 4 (1991), 151–158.

H. Niederreiter, *A combinatorial approach to probabilistic results on the linear-complexity profile of random sequences*, 2 (1990), 105–112.

K. Nishimura, M. Sibuya, *Probability to meet in the middle*, 2 (1990), 13–22.

K. Nyberg, L.R. Knudsen, *Provable security against a differential attack*, 8 (1995), 27–37.

L. O'Connor, *An analysis of a class of algorithms for S-box construction*, 7 (1994), 133–151.

L. O'Connor, *On the distribution of characteristics in bijective mappings*, 8 (1995), 67–86.

L. O'Connor, A. Klapper, *Algebraic nonlinearity and its applications to cryptography*, 7 (1994), 213–227.

G. Orton, L. Peppard, S. Tavares, *A design of a fast pipelined modular multiplier based on a diminished-radix algorithm*, 6 (1993), 183–208.

J. Pastor, CRYPTOPOST^{TM}–*a cryptographic application to mail processing*, 3 (1991), 137–146.

D. Pei, *Information-theoretic bounds for authentication codes and block designs*, 8 (1995), 177–188.

S.J. Phillips, N.C. Phillips, *Strongly ideal secret sharing schemes*, 5 (1992), 185–191.

F. Piper, M. Walker, *Linear ciphers and spreads*, 1 (1989), 185–188.

M. Qu, S.A. Vanstone, *Factorizations in the elementary abelian p-group and their cryptographic significance*, 7 (1994), 201–212.

U. Rosenbaum, *A lower bound on authentication after having observed a sequence of messages*, 6 (1993), 135–156.

A. Russell, *Necessary and sufficient conditions for collision-free hashing*, 8 (1995), 87–99.

R. Scheidler, J.A. Buchmann, H.C. Williams, *A key-exchange protocol using real quadratic fields*, 7 (1994), 171–199.

C.P. Schnorr, *Efficient signature generation by smart cards*, 4 (1991), 161–174.

A.W. Schrift, A. Shamir, *Universal tests for nonuniform distributions*, 6 (1993), 119–133.

G.J. Simmons, *A cartesian product construction for unconditionally secure authentication codes that permit arbitration*, 2 (1990), 77–104.

G.J. Simmons, *Proof of soundness (integrity) of cryptographic protocols*, 7 (1994), 69–77.

D.R. Stinson, *A construction for authentication/secrecy codes from certain combinatorial designs*, 1 (1988), 119–127.

D.R. Stinson, *Some constructions and bounds for authentication codes*, 1 (1988), 37–51.

D.R. Stinson, *The combinatorics of authentication and secrecy codes*, 2 (1990), 23–49.

D.R. Stinson, J.L. Massey, *An infinite class of counterexamples to a conjecture concerning nonlinear resilient functions*, 8 (1995), 167–173.

P. Syverson, *Knowledge, belief and semantics in the analysis of cryptographic protocols*, 1 (1992), 317–334.

S.-H. Teng, *Functional inversion and communication complexity*, 7 (1994), 153–170.

M. Tompa, H. Woll, *How to share a secret with cheaters*, 1 (1988), 133–138.

S.A. Vanstone, R.J. Zuccherato, *Short RSA keys and their generation*, 8 (1995), 101–114.

M. Walker, *Information-theoretic bounds for authentication schemes*, 2 (1990), 131–143.

Y.-X. Yang, B. Guo, *Further enumerating boolean functions of cryptographic significance*, 8 (1995), 115–122.

References

[1] M. ABADI AND R. NEEDHAM, "Prudent engineering practice for cryptographic protocols", DEC SRC report #125, Digital Equipment Corporation, Palo Alto, CA, 1994.

[2] M. ABADI AND M.R. TUTTLE, "A semantics for a logic of authentication", *Proceedings of the Tenth Annual ACM Symposium on Principles of Distributed Computing*, 201–216, 1991.

[3] C. ADAMS, "Symmetric cryptographic system for data encryption", U.S. Patent # 5,511,123, 23 Apr 1996.

[4] ———, "IDUP and SPKM: Developing public-key-based APIs and mechanisms for communication security services", *Proceedings of the Internet Society Symposium on Network and Distributed System Security*, 128–135, IEEE Computer Society Press, 1996.

[5] C. ADAMS AND H. MEIJER, "Security-related comments regarding McEliece's public-key cryptosystem", *Advances in Cryptology–CRYPTO '87 (LNCS 293)*, 224–228, 1988.

[6] ———, "Security-related comments regarding McEliece's public-key cryptosystem", *IEEE Transactions on Information Theory*, 35 (1989), 454–455. An earlier version appeared in [5].

[7] C. ADAMS AND S.E. TAVARES, "Designing S-boxes for ciphers resistant to differential cryptanalysis", W. Wolfowicz, editor, *Proceedings of the 3rd Symposium on State and Progress of Research in Cryptography, Rome, Italy*, 181–190, 1993.

[8] L.M. ADLEMAN, "A subexponential algorithm for the discrete logarithm problem with applications to cryptography", *Proceedings of the IEEE 20th Annual Symposium on Foundations of Computer Science*, 55–60, 1979.

[9] ———, "The function field sieve", *Algorithmic Number Theory (LNCS 877)*, 108–121, 1994.

[10] ———, "Molecular computation of solutions to combinatorial problems", *Science*, 266 (1994), 1021–1024.

[11] L.M. ADLEMAN AND J. DEMARRAIS, "A subexponential algorithm for discrete logarithms over all finite fields", *Mathematics of Computation*, 61 (1993), 1–15.

[12] L.M. ADLEMAN, J. DEMARRAIS, AND M.-D. HUANG, "A subexponential algorithm for discrete logarithms over the rational subgroup of the Jacobians of large genus hyperelliptic curves over finite fields", *Algorithmic Number Theory (LNCS 877)*, 28–40, 1994.

[13] L.M. ADLEMAN AND M.-D. A. HUANG, *Primality Testing and Abelian Varieties Over Finite Fields*, Springer-Verlag, Berlin, 1992.

[14] L.M. ADLEMAN AND H.W. LENSTRA JR., "Finding irreducible polynomials over finite fields", *Proceedings of the 18th Annual ACM Symposium on Theory of Computing*, 350–355, 1986.

[15] L.M. ADLEMAN AND K.S. MCCURLEY, "Open problems in number theoretic complexity, II", *Algorithmic Number Theory (LNCS 877)*, 291–322, 1994.

[16] L.M. ADLEMAN, C. POMERANCE, AND R.S. RUMELY, "On distinguishing prime numbers from composite numbers", *Annals of Mathematics*, 117 (1983), 173–206.

[17] G.B. AGNEW, "Random sources for cryptographic systems", *Advances in Cryptology–EUROCRYPT '87 (LNCS 304)*, 77–81, 1988.

[18] G.B. AGNEW, R.C. MULLIN, I.M. ONYSZCHUK, AND S.A. VANSTONE, "An implementation for a fast public-key cryptosystem", *Journal of Cryptology*, 3 (1991), 63–79.

[19] G.B. AGNEW, R.C. MULLIN, AND S.A. VANSTONE, "Improved digital signature scheme based on discrete exponentiation", *Electronics Letters*, 26 (July 5, 1990), 1024–1025.

[20] S.G. AKL, "On the security of compressed encodings", *Advances in Cryptology–Proceedings of Crypto 83*, 209–230, 1984.

[21] N. ALEXANDRIS, M. BURMESTER, V. CHRISSIKOPOULOS, AND Y. DESMEDT, "A secure key distribution system", W. Wolfowicz,

editor, *Proceedings of the 3rd Symposium on State and Progress of Research in Cryptography, Rome, Italy*, 30–34, Feb. 1993.

[22] W. ALEXI, B. CHOR, O. GOLDREICH, AND C.P. SCHNORR, "RSA/Rabin bits are $\frac{1}{2} + 1/poly(\log n)$ secure", *Proceedings of the IEEE 25th Annual Symposium on Foundations of Computer Science*, 449–457, 1984.

[23] ——, "RSA and Rabin functions: Certain parts are as hard as the whole", *SIAM Journal on Computing*, 17 (1988), 194–209. An earlier version appeared in [22].

[24] W.R. ALFORD, A. GRANVILLE, AND C. POMERANCE, "There are infinitely many Carmichael numbers", *Annals of Mathematics*, 140 (1994), 703–722.

[25] H. AMIRAZIZI AND M. HELLMAN, "Time-memory-processor trade-offs", *IEEE Transactions on Information Theory*, 34 (1988), 505–512.

[26] R. ANDERSON, "Practical RSA trapdoor", *Electronics Letters*, 29 (May 27, 1993), 995.

[27] ——, "The classification of hash functions", P.G. Farrell, editor, *Codes and Cyphers: Cryptography and Coding IV*, 83–93, Institute of Mathematics & Its Applications (IMA), 1995.

[28] ——, "On Fibonacci keystream generators", B. Preneel, editor, *Fast Software Encryption, Second International Workshop (LNCS 1008)*, 346–352, Springer-Verlag, 1995.

[29] ——, "Searching for the optimum correlation attack", B. Preneel, editor, *Fast Software Encryption, Second International Workshop (LNCS 1008)*, 137–143, Springer-Verlag, 1995.

[30] R. ANDERSON AND E. BIHAM, "Two practical and provably secure block ciphers: BEAR and LION", D. Gollmann, editor, *Fast Software Encryption, Third International Workshop (LNCS 1039)*, 113–120, Springer-Verlag, 1996.

[31] R. ANDERSON AND R. NEEDHAM, "Robustness principles for public key protocols", *Advances in Cryptology–CRYPTO '95 (LNCS 963)*, 236–247, 1995.

[32] N.C. ANKENY, "The least quadratic non residue", *Annals of Mathematics*, 55 (1952), 65–72.

[33] ANSI X3.92, "American National Standard – Data Encryption Algorithm", American National Standards Institute, 1981.

[34] ANSI X3.106, "American National Standard for Information Systems – Data Encryption Algorithm – Modes of Operation", American National Standards Institute, 1983.

[35] ANSI X9.8, "American National Standard for Financial Services – Banking – Personal Identification Number management and security. Part 1: PIN protection principles and techniques; Part 2: Approved algorithms for PIN encipherment", ASC X9 Secretariat – American Bankers Association, 1995.

[36] ANSI X9.9 (REVISED), "American National Standard – Financial institution message authentication (wholesale)", ASC X9 Secretariat – American Bankers Association, 1986 (replaces X9.9–1982).

[37] ANSI X9.17, "American National Standard – Financial institution key management (wholesale)", ASC X9 Secretariat – American Bankers Association, 1985.

[38] ANSI X9.19, "American National Standard – Financial institution retail message authentication", ASC X9 Secretariat – American Bankers Association, 1986.

[39] ANSI X9.23, "American National Standard – Financial institution encryption of wholesale financial messages", ASC X9 Secretariat – American Bankers Association, 1988.

[40] ANSI X9.24, "American National Standard for Financial Services – Financial services retail key management", ASC X9 Secretariat – American Bankers Association, 1992.

[41] ANSI X9.26, "American National Standard – Financial institution sign-on authentication for wholesale financial transactions", ASC X9 Secretariat – American Bankers Association, 1990.

[42] ANSI X9.28, "American National Standard for Financial Services – Financial institution multiple center key management (wholesale)", ASC X9 Secretariat – American Bankers Association, 1991.

[43] ANSI X9.30 (PART 1), "American National Standard for Financial Services – Public key cryptography using irreversible algorithms for the financial services industry – Part 1: The digital signature algorithm (DSA)", ASC X9 Secretariat – American Bankers Association, 1995.

[44] ANSI X9.30 (PART 2), "American National Standard for Financial Services – Public key cryptography using irreversible algorithms for the financial services industry – Part 2: The secure hash algorithm (SHA)", ASC X9 Secretariat – American Bankers Association, 1993.

[45] ANSI X9.31 (PART 1), "American National Standard for Financial Services – Public key cryptography using RSA for the financial services industry – Part 1: The RSA signature algorithm", draft, 1995.

[46] ANSI X9.31 (PART 2), "American National Standard for Financial Services – Public key cryptography using RSA for the financial services industry – Part 2: Hash algorithms for RSA", draft, 1995.

[47] ANSI X9.42, "Public key cryptography for the financial services industry: Management of symmetric algorithm keys using Diffie-Hellman", draft, 1995.

[48] ANSI X9.44, "Public key cryptography using reversible algorithms for the financial services industry: Transport of symmetric algorithm keys using RSA", draft, 1994.

[49] ANSI X9.45, "Public key cryptography for the financial services industry – Enhanced management controls using digital signatures and attribute certificates", draft, 1996.

[50] ANSI X9.52, "Triple data encryption algorithm modes of operation", draft, 1996.

[51] ANSI X9.55, "Public key cryptography for the financial services industry – Extensions to public key certificates and certificate revocation lists", draft, 1995.

[52] ANSI X9.57, "Public key cryptography for the financial services industry – Certificate management", draft, 1995.

[53] K. AOKI AND K. OHTA, "Differential-linear cryptanalysis of FEAL-8", *IEICE Transactions on Fundamentals of Electronics, Communications and Computer Science*, E79-A (1996), 20–27.

[54] B. ARAZI, "Integrating a key distribution procedure into the digital signature standard", *Electronics Letters*, 29 (May 27, 1993), 966–967.

[55] ——, "On primality testing using purely divisionless operations", *The Computer Journal*, 37 (1994), 219–222.

[56] F. ARNAULT, "Rabin-Miller primality test: composite numbers which pass it", *Mathematics of Computation*, 64 (1995), 355–361.

[57] A.O.L. ATKIN AND R.G. LARSON, "On a primality test of Solovay and Strassen", *SIAM Journal on Computing*, 11 (1982), 789–791.

[58] A.O.L. ATKIN AND F. MORAIN, "Elliptic curves and primality proving", *Mathematics of Computation*, 61 (1993), 29–68.

[59] D. ATKINS, M. GRAFF, A.K. LENSTRA, AND P.C. LEYLAND, "The magic words are SQUEAMISH OSSIFRAGE", *Advances in Cryptology–ASIACRYPT '94 (LNCS 917)*, 263–277, 1995.

[60] L. BABAI, "Trading group theory for randomness", *Proceedings of the 17th Annual ACM Symposium on Theory of Computing*, 421–429, 1985.

[61] L. BABAI AND S. MORAN, "Arthur-Merlin games: a randomized proof system, and a hierarchy of complexity classes", *Journal of Computer and System Sciences*, 36 (1988), 254–276.

[62] E. BACH, "Discrete logarithms and factoring", Report No. UCB/CSD 84/186, Computer Science Division (EECS), University of California, Berkeley, California, 1984.

[63] ——, *Analytic Methods in the Analysis and Design of Number-Theoretic Algorithms*, MIT Press, Cambridge, Massachusetts, 1985. An ACM Distinguished Dissertation.

[64] ——, "Explicit bounds for primality testing and related problems", *Mathematics of Computation*, 55 (1990), 355–380.

[65] ——, "Number-theoretic algorithms", *Annual Review of Computer Science*, 4 (1990), 119–172.

[66] ——, "Realistic analysis of some randomized algorithms", *Journal of Computer and System Sciences*, 42 (1991), 30–53.

[67] ——, "Toward a theory of Pollard's rho method", *Information and Computation*, 90 (1991), 139–155.

[68] E. BACH AND J. SHALLIT, "Factoring with cyclotomic polynomials", *Proceedings of the IEEE 26th Annual Symposium on Foundations of Computer Science*, 443–450, 1985.

[69] ——, "Factoring with cyclotomic polynomials", *Mathematics of Computation*, 52 (1989), 201–219. An earlier version appeared in [68].

[70] ——, *Algorithmic Number Theory, Volume I: Efficient Algorithms*, MIT Press, Cambridge, Massachusetts, 1996.

[71] E. BACH AND J. SORENSON, "Sieve algorithms for perfect power testing", *Algorithmica*, 9 (1993), 313–328.

[72] A. BAHREMAN, "PEMToolKit: Building a top-down certification hierarchy", *Proceedings of the Internet Society Symposium on Network and Distributed System Security*, 161–171, IEEE Computer Society Press, 1995.

[73] T. BARITAUD, M. CAMPANA, P. CHAUVAUD, AND H. GILBERT, "On the security of the permuted kernel identification scheme", *Advances in Cryptology–CRYPTO '92 (LNCS 740)*, 305–311, 1993.

[74] W. BARKER, *Cryptanalysis of the Hagelin Cryptograph*, Aegean Park Press, Laguna Hills, California, 1977.

[75] P. BARRETT, "Implementing the Rivest Shamir and Adleman public key encryption algorithm on a standard digital signal processor", *Advances in Cryptology–CRYPTO '86 (LNCS 263)*, 311–323, 1987.

[76] R.K. BAUER, T.A. BERSON, AND R.J. FEIERTAG, "A key distribution protocol using event markers", *ACM Transactions on Computer Systems*, 1 (1983), 249–255.

[77] U. BAUM AND S. BLACKBURN, "Clock-controlled pseudorandom generators on finite groups", B. Preneel, editor, *Fast Software Encryption, Second International Workshop (LNCS 1008)*, 6–21, Springer-Verlag, 1995.

[78] F. BAUSPIESS AND H.-J. KNOBLOCH, "How to keep authenticity alive in a computer network", *Advances in Cryptology–EUROCRYPT '89 (LNCS 434)*, 38–46, 1990.

[79] D. BAYER, S. HABER, AND W.S. STORNETTA, "Improving the efficiency and reliability of digital time-stamping", R. Capocelli, A. De Santis, and U. Vaccaro, editors, *Sequences II: Methods in Communication, Security, and Computer Science*, 329–334, Springer-Verlag, 1993.

[80] P. BEAUCHEMIN AND G. BRASSARD, "A generalization of Hellman's extension to Shannon's approach to cryptography", *Journal of Cryptology*, 1 (1988), 129–131.

[81] P. BEAUCHEMIN, G. BRASSARD, C. CRÉPEAU, C. GOUTIER, AND C. POMERANCE, "The generation of random numbers that are probably prime", *Journal of Cryptology*, 1 (1988), 53–64.

[82] P. BÉGUIN AND J.-J. QUISQUATER, "Secure acceleration of DSS signatures using insecure server", *Advances in Cryptology–ASIACRYPT '94 (LNCS 917)*, 249–259, 1995.

[83] A. BEIMEL AND B. CHOR, "Interaction in key distribution schemes", *Advances in Cryptology–CRYPTO '93 (LNCS 773)*, 444–455, 1994.

[84] H. BEKER AND F. PIPER, *Cipher Systems: The Protection of Communications*, John Wiley & Sons, New York, 1982.

[85] H. BEKER AND M. WALKER, "Key management for secure electronic funds transfer in a retail environment", *Advances in Cryptology–Proceedings of CRYPTO 84 (LNCS 196)*, 401–410, 1985.

[86] M. BELLARE, R. CANETTI, AND H. KRAWCZYK, "Keying hash functions for message authenticaion", *Advances in Cryptology–CRYPTO '96 (LNCS 1109)*, 1–15, 1996.

[87] M. BELLARE AND O. GOLDREICH, "On defining proofs of knowledge", *Advances in Cryptology–CRYPTO '92 (LNCS 740)*, 390–420, 1993.

[88] M. BELLARE, O. GOLDREICH, AND S. GOLDWASSER, "Incremental cryptography: The case of hashing and signing", *Advances in Cryptology–CRYPTO '94 (LNCS 839)*, 216–233, 1994.

[89] ——, "Incremental cryptography and application to virus protection", *Proceedings of the 27th Annual ACM Symposium on Theory of Computing*, 45–56, 1995.

[90] M. BELLARE, R. GUÉRIN, AND P. ROGAWAY, "XOR MACs: New methods for message authentication using finite pseudorandom functions", *Advances in Cryptology–CRYPTO '95 (LNCS 963)*, 15–28, 1995.

[91] M. BELLARE, J. KILIAN, AND P. ROGAWAY, "The security of cipher block chaining", *Advances in Cryptology–CRYPTO '94 (LNCS 839)*, 341–358, 1994.

[92] M. BELLARE AND S. MICALI, "How to sign given any trapdoor function", *Advances in Cryptology–CRYPTO '88 (LNCS 403)*, 200–215, 1990.

[93] M. BELLARE AND P. ROGAWAY, "Random oracles are practical: a paradigm for designing efficient protocols", *1st ACM Conference on Computer and Communications Security*, 62–73, ACM Press, 1993.

[94] ——, "Entity authentication and key distribution", *Advances in Cryptology–CRYPTO '93 (LNCS 773)*, 232–249, 1994.

[95] ——, "Optimal asymmetric encryption", *Advances in Cryptology–EUROCRYPT '94 (LNCS 950)*, 92–111, 1995.

[96] ——, "Provably secure session key distribution – the three party case", *Proceedings of the 27th Annual ACM Symposium on Theory of Computing*, 57–66, 1995.

[97] M.J. BELLER, L.-F. CHANG, AND Y. YACOBI, "Privacy and authentication on a portable communications system", *IEEE Global Telecommunications Conference*, 1922–1927, 1991.

[98] ——, "Security for personal communications services: public-key vs. private key approaches", *The Third IEEE International Symposium on Personal, Indoor and Mobile Radio Communications (PIMRC'92)*, 26–31, 1992.

[99] ——, "Privacy and authentication on a portable communications system", *IEEE Journal on Selected Areas in Communications*, 11 (1993), 821–829.

[100] M.J. BELLER AND Y. YACOBI, "Minimal asymmetric authentication and key agreement schemes", October 1994 unpublished manuscript.

[101] ——, "Fully-fledged two-way public key authentication and key agreement for low-cost terminals", *Electronics Letters*, 29 (May 27, 1993), 999–1001.

[102] S.M. BELLOVIN AND M. MERRITT, "Cryptographic protocol for secure communications", U.S. Patent # 5,241,599, 31 Aug 1993.

[103] ——, "Limitations of the Kerberos authentication system", *Computer Communication Review*, 20 (1990), 119–132.

[104] ——, "Encrypted key exchange: password-based protocols secure against dictionary attacks", *Proceedings of the 1992 IEEE Computer Society Symposium on Research in Security and Privacy*, 72–84, 1992.

[105] ——, "Augmented encrypted key exchange: a password-based protocol secure against dictionary attacks and password file compromise", *1st ACM Conference on Computer and Communications Security*, 244–250, ACM Press, 1993.

[106] ——, "An attack on the Interlock Protocol when used for authentication", *IEEE Transactions on Information Theory*, 40 (1994), 273–275.

[107] I. BEN-AROYA AND E. BIHAM, "Differential cyptanalysis of Lucifer", *Advances in Cryptology–CRYPTO '93 (LNCS 773)*, 187–199, 1994.

[108] ——, "Differential cryptanalysis of Lucifer", *Journal of Cryptology*, 9 (1996), 21–34. An earlier version appeared in [107].

[109] M. BEN-OR, "Probabilistic algorithms in finite fields", *Proceedings of the IEEE 22nd Annual Symposium on Foundations of Computer Science*, 394–398, 1981.

[110] J. BENALOH, "Secret sharing homomorphisms: Keeping shares of a secret secret", *Advances in Cryptology–CRYPTO '86 (LNCS 263)*, 251–260, 1987.

[111] J. BENALOH AND M. DE MARE, "One-way accumulators: A decentralized alternative to digital signatures", *Advances in Cryptology–EUROCRYPT '93 (LNCS 765)*, 274–285, 1994.

[112] J. BENALOH AND J. LEICHTER, "Generalized secret sharing and monotone functions", *Advances in Cryptology–CRYPTO '88 (LNCS 403)*, 27–35, 1990.

[113] S. BENGIO, G. BRASSARD, Y.G. DESMEDT, C. GOUTIER, AND J.-J. QUISQUATER, "Secure implementation of identification systems", *Journal of Cryptology*, 4 (1991), 175–183.

[114] C. BENNETT, G. BRASSARD, S. BREIDBART, AND S. WIESNER, "Quantum cryptography, or unforgeable subway tokens", *Advances in Cryptology–Proceedings of Crypto 82*, 267–275, 1983.

[115] C. BENNETT, G. BRASSARD, AND A. EKERT, "Quantum cryptography", *Scientific American*, special issue (1997), 164–171.

[116] S. BERKOVITS, "How to broadcast a secret", *Advances in Cryptology–EUROCRYPT '91 (LNCS 547)*, 535–541, 1991.

[117] E.R. BERLEKAMP, "Factoring polynomials over finite fields", *Bell System Technical Journal*, 46 (1967), 1853–1859.

[118] ——, *Algebric Coding Theory*, McGraw Hill, New York, 1968.

[119] ——, "Factoring polynomials over large finite fields", *Mathematics of Computation*, 24 (1970), 713–735.

[120] E.R. BERLEKAMP, R.J. MCELIECE, AND H.C.A. VAN TILBORG, "On the inherent intractability of certain coding problems", *IEEE Transactions on Information Theory*, 24 (1978), 384–386.

[121] D.J. BERNSTEIN, "Detecting perfect powers in essentially linear time", preprint, 1995.

[122] D.J. BERNSTEIN AND A.K. LENSTRA, "A general number field sieve implementation", A.K. Lenstra and H.W. Lenstra Jr., editors, *The Development of the Number Field Sieve*, volume 1554 of *Lecture Notes in Mathematics*, 103–126, Springer-Verlag, 1993.

[123] T. BETH, "Efficient zero-knowledge identification scheme for smart cards", *Advances in Cryptology–EUROCRYPT '88 (LNCS 330)*, 77–84, 1988.

[124] T. BETH AND Z.-D. DAI, "On the complexity of pseudo-random sequences – or: If you can describe a sequence it can't be random", *Advances in Cryptology–EUROCRYPT '89 (LNCS 434)*, 533–543, 1990.

[125] T. BETH, H.-J. KNOBLOCH, M. OTTEN, G.J. SIMMONS, AND P. WICHMANN, "Towards acceptable key escrow systems", *2nd ACM Conference on Computer and Communications Security*, 51–58, ACM Press, 1994.

[126] T. BETH AND F.C. PIPER, "The stop-and-go generator", *Advances in Cryptology–Proceedings of EUROCRYPT 84 (LNCS 209)*, 88–92, 1985.

[127] J. BIERBRAUER, T. JOHANSSON, G. KABATIANSKII, AND B. SMEETS, "On families of hash functions via geometric codes and concatenation", *Advances in Cryptology–CRYPTO '93 (LNCS 773)*, 331–342, 1994.

[128] E. BIHAM, "New types of cryptanalytic attacks using related keys", *Advances in Cryptology–EUROCRYPT '93 (LNCS 765)*, 398–409, 1994.

[129] ——, "New types of cryptanalytic attacks using related keys", *Journal of Cryptology*, 7 (1994), 229–246.

[130] ——, "On modes of operation", R. Anderson, editor, *Fast Software Encryption, Cambridge Security Workshop (LNCS 809)*, 116–120, Springer-Verlag, 1994.

[131] ——, "Cryptanalysis of multiple modes of operation", *Advances in Cryptology–ASIACRYPT '94 (LNCS 917)*, 278–292, 1995.

[132] ——, "On Matsui's linear cryptanalysis", *Advances in Cryptology–EUROCRYPT '94 (LNCS 950)*, 341–355, 1995.

[133] E. BIHAM AND A. BIRYUKOV, "How to strengthen DES using existing hardware", *Advances in Cryptology–ASIACRYPT '94 (LNCS 917)*, 398–412, 1995.

[134] E. BIHAM AND A. SHAMIR, "Differential cryptanalysis of DES-like cryptosystems", *Journal of Cryptology*, 4 (1991), 3–72. An earlier version appeared in [135].

[135] ——, "Differential cryptanalysis of DES-like cryptosystems", *Advances in Cryptology–CRYPTO '90 (LNCS 537)*, 2–21, 1991.

[136] ——, "Differential cryptanalysis of Feal and N-Hash", *Advances in Cryptology–EUROCRYPT '91 (LNCS 547)*, 1–16, 1991.

[137] ——, "Differential cryptanalysis of Snefru, Khafre, REDOC-II, LOKI, and Lucifer", *Advances in Cryptology–CRYPTO '91 (LNCS 576)*, 156–171, 1992.

[138] ——, *Differential Cryptanalysis of the Data Encryption Standard*, Springer-Verlag, New York, 1993.

[139] ——, "Differential cryptanalysis of the full 16-round DES", *Advances in Cryptology–CRYPTO '92 (LNCS 740)*, 487–496, 1993.

[140] R. BIRD, I. GOPAL, A. HERZBERG, P. JANSON, S. KUTTEN, R. MOLVA, AND M. YUNG, "Systematic design of two-party authentication protocols", *Advances in Cryptology–CRYPTO '91 (LNCS 576)*, 44–61, 1992.

[141] ——, "Systematic design of a family of attack-resistant authentication protocols", *IEEE Journal on Selected Areas in Communications*, 11 (1993), 679–693.

[142] ——, "The KryptoKnight family of lightweight protocols for authentication and key distribution", *IEEE/ACM Transactions on Networking*, 3 (1995), 31–41.

[143] S. BLACKBURN, S. MURPHY, AND J. STERN, "The cryptanalysis of a public-key implementation of finite group mappings", *Journal of Cryptology*, 8 (1995), 157–166.

[144] R.E. BLAHUT, *Principles and Practice of Information Theory*, Addison-Wesley, Reading, Massachusetts, 1987.

[145] I.F. BLAKE, R. FUJI-HARA, R.C. MULLIN, AND S.A. VANSTONE, "Computing logarithms in finite fields of characteristic two", *SIAM Journal on Algebraic and Discrete Methods*, 5 (1984), 276–285.

[146] I.F. BLAKE, S. GAO, AND R. LAMBERT, "Constructive problems for irreducible polynomials over finite fields", T.A. Gulliver and N.P. Secord, editors, *Information Theory and Applications (LNCS 793)*, 1–23, Springer-Verlag, 1994.

[147] B. BLAKLEY, G.R. BLAKLEY, A.H. CHAN, AND J.L. MASSEY, "Threshold schemes with disenrollment", *Advances in Cryptology–CRYPTO '92 (LNCS 740)*, 540–548, 1993.

[148] G. BLAKLEY, "Safeguarding cryptographic keys", *Proceedings of AFIPS National Computer Conference*, 313–317, 1979.

[149] ——, "A computer algorithm for calculating the product AB modulo M", *IEEE Transactions on Computers*, 32 (1983), 497–500.

[150] G. BLAKLEY AND I. BOROSH, "Rivest-Shamir-Adleman public key cryptosystems do not always conceal messages", *Computers and Mathematics with Applications*, 5:3 (1979), 169–178.

[151] G. BLAKLEY AND C. MEADOWS, "Security of ramp schemes", *Advances in Cryptology–Proceedings of CRYPTO 84 (LNCS 196)*, 242–268, 1985.

[152] M. BLAZE, "Protocol failure in the escrowed encryption standard", *2nd ACM Conference on Computer and Communications Security*, 59–67, ACM Press, 1994.

[153] D. BLEICHENBACHER, "Generating ElGamal signatures without knowing the secret key", *Advances in Cryptology–EUROCRYPT '96 (LNCS 1070)*, 10–18, 1996.

[154] D. BLEICHENBACHER, W. BOSMA, AND A.K. LENSTRA, "Some remarks on Lucas-based cryptosystems", *Advances in Cryptology–CRYPTO '95 (LNCS 963)*, 386–396, 1995.

[155] D. BLEICHENBACHER AND U. MAURER, "Directed acyclic graphs, one-way functions and digital signatures", *Advances in Cryptology–CRYPTO '94 (LNCS 839)*, 75–82, 1994.

[156] U. BLÖCHER AND M. DICHTL, "Fish: A fast software stream cipher", R. Anderson, editor, *Fast Software Encryption, Cambridge Security Workshop (LNCS 809)*, 41–44, Springer-Verlag, 1994.

[157] R. BLOM, "Non-public key distribution", *Advances in Cryptology–Proceedings of Crypto 82*, 231–236, 1983.

[158] ——, "An optimal class of symmetric key generation systems", *Advances in Cryptology–Proceedings of EUROCRYPT 84 (LNCS 209)*, 335–338, 1985.

[159] L. BLUM, M. BLUM, AND M. SHUB, "Comparison of two pseudo-random number generators", *Advances in Cryptology–Proceedings of Crypto 82*, 61–78, 1983.

[160] ——, "A simple unpredictable pseudo-random number generator", *SIAM Journal on Computing*, 15 (1986), 364–383. An earlier version appeared in [159].

[161] M. BLUM, "Independent unbiased coin flips from a correlated biased source: a finite state Markov chain", *Proceedings of the IEEE 25th Annual Symposium on Foundations of Computer Science*, 425–433, 1984.

[162] M. BLUM, A. DE SANTIS, S. MICALI, AND G. PERSIANO, "Noninteractive zero-knowledge", *SIAM Journal on Computing*, 20 (1991), 1084–1118.

[163] M. BLUM, P. FELDMAN, AND S. MICALI, "Non-interactive zero-knowledge and its applications", *Proceedings of the 20th Annual ACM Symposium on Theory of Computing*, 103–112, 1988.

[164] M. BLUM AND S. GOLDWASSER, "An efficient probabilistic public-key encryption scheme which hides all partial information", *Advances in Cryptology–Proceedings of CRYPTO 84 (LNCS 196)*, 289–299, 1985.

[165] M. BLUM AND S. MICALI, "How to generate cryptographically strong sequences of pseudo random bits", *Proceedings of the IEEE 23rd Annual Symposium on Foundations of Computer Science*, 112–117, 1982.

[166] ——, "How to generate cryptographically strong sequences of pseudo-random bits",

SIAM Journal on Computing, 13 (1984), 850–864. An earlier version appeared in [165].

[167] C. BLUNDO AND A. CRESTI, "Space requirements for broadcast encryption", *Advances in Cryptology–EUROCRYPT '94 (LNCS 950)*, 287–298, 1995.

[168] C. BLUNDO, A. CRESTI, A. DE SANTIS, AND U. VACCARO, "Fully dynamic secret sharing schemes", *Advances in Cryptology–CRYPTO '93 (LNCS 773)*, 110–125, 1994.

[169] C. BLUNDO, A. DE SANTIS, A. HERZBERG, S. KUTTEN, U. VACCARO, AND M. YUNG, "Perfectly-secure key distribution for dynamic conferences", *Advances in Cryptology–CRYPTO '92 (LNCS 740)*, 471–486, 1993.

[170] R.V. BOOK AND F. OTTO, "The verifiability of two-party protocols", *Advances in Cryptology–EUROCRYPT '85 (LNCS 219)*, 254–260, 1986.

[171] A. BOOTH, "A signed binary multiplication technique", *The Quarterly Journal of Mechanics and Applied Mathematics*, 4 (1951), 236–240.

[172] J. BOS AND D. CHAUM, "Provably unforgeable signatures", *Advances in Cryptology–CRYPTO '92 (LNCS 740)*, 1–14, 1993.

[173] J. BOS AND M. COSTER, "Additon chain heuristics", *Advances in Cryptology–CRYPTO '89 (LNCS 435)*, 400–407, 1990.

[174] W. BOSMA AND M.-P VAN DER HULST, "Faster primality testing", *Advances in Cryptology–EUROCRYPT '89 (LNCS 434)*, 652–656, 1990.

[175] A. BOSSELAERS, R. GOVAERTS, AND J. VANDEWALLE, "Cryptography within phase I of the EEC-RACE programme", B. Preneel, R. Govaerts, and J. Vandewalle, editors, *Computer Security and Industrial Cryptography: State of the Art and Evolution (LNCS 741)*, 227–234, Springer-Verlag, 1993.

[176] ——, "Comparison of three modular reduction functions", *Advances in Cryptology–CRYPTO '93 (LNCS 773)*, 175–186, 1994.

[177] ——, "Fast hashing on the Pentium", *Advances in Cryptology–CRYPTO '96 (LNCS 1109)*, 298–312, 1996.

[178] A. BOSSELAERS AND B. PRENEEL, editors, *Integrity Primitives for Secure Information Systems: Final Report of RACE Integrity Primitives Evaluation RIPE-RACE 1040*, LNCS 1007, Springer-Verlag, New York, 1995.

[179] J. BOYAR, "Inferring sequences produced by a linear congruential generator missing low-order bits", *Journal of Cryptology*, 1 (1989), 177–184.

[180] ——, "Inferring sequences produced by pseudo-random number generators", *Journal of the Association for Computing Machinery*, 36 (1989), 129–141.

[181] J. BOYAR, D. CHAUM, I.B. DAMGÅRD, AND T. PEDERSEN, "Convertible undeniable signatures", *Advances in Cryptology–CRYPTO '90 (LNCS 537)*, 189–205, 1991.

[182] C. BOYD, "Digital multisignatures", H. Beker and F. Piper, editors, *Cryptography and Coding*, Institute of Mathematics & Its Applications (IMA), 241–246, Clarendon Press, 1989.

[183] C. BOYD AND W. MAO, "On a limitation of BAN logic", *Advances in Cryptology–EUROCRYPT '93 (LNCS 765)*, 240–247, 1994.

[184] B.O. BRACHTL, D. COPPERSMITH, M.M. HYDEN, S.M. MATYAS JR., C.H.W. MEYER, J. OSEAS, S. PILPEL, AND M. SCHILLING, "Data authentication using modification detection codes based on a public one-way encryption function", U.S. Patent # 4,908,861, 13 Mar 1990.

[185] S. BRANDS, "Restrictive blinding of secret-key certificates", *Advances in Cryptology–EUROCRYPT '95 (LNCS 921)*, 231–247, 1995.

[186] J. BRANDT AND I. DAMGÅRD, "On generation of probable primes by incremental search", *Advances in Cryptology–CRYPTO '92 (LNCS 740)*, 358–370, 1993.

[187] J. BRANDT, I. DAMGÅRD, AND P. LANDROCK, "Speeding up prime number generation", *Advances in Cryptology–ASIACRYPT '91 (LNCS 739)*, 440–449, 1993.

[188] J. BRANDT, I. DAMGÅRD, P. LANDROCK, AND T. PEDERSEN, "Zero-knowledge authentication scheme with secret key exchange", *Advances in Cryptology–CRYPTO '88 (LNCS 403)*, 583–588, 1990.

[189] D.K. BRANSTAD, "Encryption protection in computer data communications", *Proceedings of the 4th Data Communications Symposium* (Quebec), 8.1–8.7, IEEE, 1975.

[190] G. BRASSARD, "A note on the complexity of cryptography", *IEEE Transactions on Information Theory*, 25 (1979), 232–233.

[191] ———, "On computationally secure authentication tags requiring short secret shared keys", *Advances in Cryptology–Proceedings of Crypto 82*, 79–86, 1983.

[192] ———, *Modern Cryptology: A Tutorial*, LNCS 325, Springer-Verlag, New York, 1988.

[193] G. BRASSARD, D. CHAUM, AND C. CRÉPEAU, "Minimum disclosure proofs of knowledge", *Journal of Computer and System Sciences*, 37 (1988), 156–189.

[194] G. BRASSARD AND C. CRÉPEAU, "Zero-knowledge simulation of Boolean circuits", *Advances in Cryptology–CRYPTO '86 (LNCS 263)*, 223–233, 1987.

[195] ———, "Sorting out zero-knowledge", *Advances in Cryptology–EUROCRYPT '89 (LNCS 434)*, 181–191, 1990.

[196] R.P. BRENT, "An improved Monte Carlo factorization algorithm", *BIT*, 20 (1980), 176–184.

[197] R.P. BRENT AND J.M. POLLARD, "Factorization of the eighth Fermat number", *Mathematics of Computation*, 36 (1981), 627–630.

[198] D.M. BRESSOUD, *Factorization and Primality Testing*, Springer-Verlag, New York, 1989.

[199] E.F. BRICKELL, "A fast modular multiplication algorithm with applications to two key cryptography", *Advances in Cryptology–Proceedings of Crypto 82*, 51–60, 1983.

[200] ———, "Breaking iterated knapsacks", *Advances in Cryptology–Proceedings of CRYPTO 84 (LNCS 196)*, 342–358, 1985.

[201] ———, "The cryptanalysis of knapsack cryptosystems", R.D. Ringeisen and F.S. Roberts, editors, *Applications of Discrete Mathematics*, 3–23, SIAM, 1988.

[202] E.F. BRICKELL AND J.M. DeLAURENTIS, "An attack on a signature scheme proposed by Okamoto and Shiraishi", *Advances in Cryptology–CRYPTO '85 (LNCS 218)*, 28–32, 1986.

[203] E.F. BRICKELL, D.M. GORDON, AND K.S. MCCURLEY, "Method for exponentiating in cryptographic systems", U.S. Patent # 5,299,262, 29 Mar 1994.

[204] E.F. BRICKELL, D.M. GORDON, K.S. MCCURLEY, AND D.B. WILSON, "Fast exponentiation with precomputation", *Advances in Cryptology–EUROCRYPT '92 (LNCS 658)*, 200–207, 1993.

[205] E.F. BRICKELL, P.J. LEE, AND Y. YACOBI, "Secure audio teleconference", *Advances in Cryptology–CRYPTO '87 (LNCS 293)*, 418–426, 1988.

[206] E.F. BRICKELL AND K.S. MCCURLEY, "An interactive identification scheme based on discrete logarithms and factoring", *Advances in Cryptology–EUROCRYPT '90 (LNCS 473)*, 63–71, 1991.

[207] ———, "An interactive identification scheme based on discrete logarithms and factoring", *Journal of Cryptology*, 5 (1992), 29–39. An earlier version appeared in [206].

[208] E.F. BRICKELL AND A.M. ODLYZKO, "Cryptanalysis: A survey of recent results", *Proceedings of the IEEE*, 76 (1988), 578–593.

[209] ———, "Cryptanalysis: A survey of recent results", G.J. Simmons, editor, *Contemporary Cryptology: The Science of Information Integrity*, 501–540, IEEE Press, 1992. An earlier version appeared in [208].

[210] J. BRILLHART, D. LEHMER, AND J. SELFRIDGE, "New primality criteria and factorizations of $2^m \pm 1$", *Mathematics of Computation*, 29 (1975), 620–647.

[211] J. BRILLHART, D. LEHMER, J. SELFRIDGE, B. TUCKERMAN, AND S. WAGSTAFF JR., *Factorizations of $b^n \pm 1$, $b = 2, 3, 5, 6, 7, 10, 11, 12$ up to High Powers*, volume 22 of *Contemporary Mathematics*, American Mathematical Society, Providence, Rhode Island, 2nd edition, 1988.

[212] J. BRILLHART AND J. SELFRIDGE, "Some factorizations of $2^n \pm 1$ and related results", *Mathematics of Computation*, 21 (1967), 87–96.

[213] D. BRILLINGER, *Time Series: Data Analysis and Theory*, Holden-Day, San Francisco, 1981.

[214] L. BROWN, M. KWAN, J. PIEPRZYK, AND J. SEBERRY, "Improving resistance to differential cryptanalysis and the redesign of LOKI", *Advances in Cryptology–ASIACRYPT '91 (LNCS 739)*, 36–50, 1993.

[215] L. BROWN, J. PIEPRZYK, AND J. SEBERRY, "LOKI – a cryptographic primitive for authentication and secrecy applications", *Advances*

in Cryptology–AUSCRYPT '90 (LNCS 453), 229–236, 1990.

[216] J. BUCHMANN AND S. DÜLLMANN, "On the computation of discrete logarithms in class groups", *Advances in Cryptology–CRYPTO '90 (LNCS 537)*, 134–139, 1991.

[217] J. BUCHMANN, J. LOHO, AND J. ZAYER, "An implementation of the general number field sieve", *Advances in Cryptology–CRYPTO '93 (LNCS 773)*, 159–165, 1994.

[218] J. BUCHMANN AND H.C. WILLIAMS, "A key-exchange system based on imaginary quadratic fields", *Journal of Cryptology*, 1 (1988), 107–118.

[219] J.P. BUHLER, H.W. LENSTRA JR., AND C. POMERANCE, "Factoring integers with the number field sieve", A.K. Lenstra and H.W. Lenstra Jr., editors, *The Development of the Number Field Sieve*, volume 1554 of *Lecture Notes in Mathematics*, 50–94, Springer-Verlag, 1993.

[220] M. BURMESTER, "On the risk of opening distributed keys", *Advances in Cryptology–CRYPTO '94 (LNCS 839)*, 308–317, 1994.

[221] M. BURMESTER AND Y. DESMEDT, "Remarks on soundness of proofs", *Electronics Letters*, 25 (October 26, 1989), 1509–1511.

[222] ———, "A secure and efficient conference key distribution system", *Advances in Cryptology–EUROCRYPT '94 (LNCS 950)*, 275–286, 1995.

[223] M. BURMESTER, Y. DESMEDT, F. PIPER, AND M. WALKER, "A general zero-knowledge scheme", *Advances in Cryptology–EUROCRYPT '89 (LNCS 434)*, 122–133, 1990.

[224] M. BURROWS, M. ABADI, AND R. NEEDHAM, "A logic of authentication", *Proceedings of the Royal Society of London Series A: Mathematical and Physical Sciences*, 246 (1989), 233–271. Preliminary version appeared as 1989 version of [227].

[225] ———, "A logic of authentication", *Proceedings of the 12th Annual ACM Symposium on Operating Systems Principles*, 1–13, 1989.

[226] ———, "A logic of authentication", *ACM Transactions on Computer Systems*, 8 (1990), 18–36.

[227] ———, "A logic of authentication", DEC SRC report #39, Digital Equipment Corporation, Palo Alto, CA, Feb. 1989. Revised Feb. 1990.

[228] J.L. CAMENISCH, J.-M. PIVETEAU, AND M.A. STADLER, "Blind signatures based on the discrete logarithm problem", *Advances in Cryptology–EUROCRYPT '94 (LNCS 950)*, 428–432, 1995.

[229] K.W. CAMPBELL AND M.J. WIENER, "DES is not a group", *Advances in Cryptology–CRYPTO '92 (LNCS 740)*, 512–520, 1993.

[230] C.M. CAMPBELL JR., "Design and specification of cryptographic capabilities", D.K. Branstad, editor, *Computer security and the Data Encryption Standard*, 54–66, NBS Special Publication 500-27, U.S. Department of Commerce, National Bureau of Standards, Washington, D.C., 1977.

[231] E.R. CANFIELD, P. ERDÖS, AND C. POMERANCE, "On a problem of Oppenheim concerning 'Factorisatio Numerorum'", *Journal of Number Theory*, 17 (1983), 1–28.

[232] D.G. CANTOR AND H. ZASSENHAUS, "A new algorithm for factoring polynomials over finite fields", *Mathematics of Computation*, 36 (1981), 587–592.

[233] J.L. CARTER AND M.N. WEGMAN, "Universal classes of hash functions", *Proceedings of the 9th Annual ACM Symposium on Theory of Computing*, 106–112, 1977.

[234] ———, "Universal classes of hash functions", *Journal of Computer and System Sciences*, 18 (1979), 143–154. An earlier version appeared in [233].

[235] F. CHABAUD, "On the security of some cryptosystems based on error-correcting codes", *Advances in Cryptology–EUROCRYPT '94 (LNCS 950)*, 131–139, 1995.

[236] G.J. CHAITIN, "On the length of programs for computing finite binary sequences", *Journal of the Association for Computing Machinery*, 13 (1966), 547–569.

[237] W.G. CHAMBERS, "Clock-controlled shift registers in binary sequence generators", *IEE Proceedings E – Computers and Digital Techniques*, 135 (1988), 17–24.

[238] ———, "Two stream ciphers", R. Anderson, editor, *Fast Software Encryption, Cambridge Security Workshop (LNCS 809)*, 51–55, Springer-Verlag, 1994.

[239] W.G. CHAMBERS AND D. GOLLMANN, "Lock-in effect in cascades of clock-controlled shift-registers", *Advances in Cryptology–EUROCRYPT '88 (LNCS 330)*, 331–343, 1988.

[240] B. CHAR, K. GEDDES, G. GONNET, B. LEONG, M. MONAGAN, AND S. WATT, *Maple V Library Reference Manual*, Springer-Verlag, New York, 1991.

[241] C. CHARNES, L. O'CONNOR, J. PIEPRZYK, R. SAFAVI-NAINI, AND Y. ZHENG, "Comments on Soviet encryption algorithm", *Advances in Cryptology–EUROCRYPT '94 (LNCS 950)*, 433–438, 1995.

[242] D. CHAUM, "Blind signatures for untraceable payments", *Advances in Cryptology–Proceedings of Crypto 82*, 199–203, 1983.

[243] ——, "Security without identification: transaction systems to make big brother obsolete", *Communications of the ACM*, 28 (1985), 1030–1044.

[244] ——, "Demonstrating that a public predicate can be satisfied without revealing any information about how", *Advances in Cryptology–CRYPTO '86 (LNCS 263)*, 195–199, 1987.

[245] ——, "Blinding for unanticipated signatures", *Advances in Cryptology–EUROCRYPT '87 (LNCS 304)*, 227–233, 1988.

[246] ——, "Zero-knowledge undeniable signatures", *Advances in Cryptology–EUROCRYPT '90 (LNCS 473)*, 458–464, 1991.

[247] ——, "Designated confirmer signatures", *Advances in Cryptology–EUROCRYPT '94 (LNCS 950)*, 86–91, 1995.

[248] D. CHAUM, J.-H. EVERTSE, AND J. VAN DE GRAAF, "An improved protocol for demonstrating possession of discrete logarithms and some generalizations", *Advances in Cryptology–EUROCRYPT '87 (LNCS 304)*, 127–141, 1988.

[249] D. CHAUM, J.-H. EVERTSE, J. VAN DE GRAAF, AND R. PERALTA, "Demonstrating possession of a discrete logarithm without revealing it", *Advances in Cryptology–CRYPTO '86 (LNCS 263)*, 200–212, 1987.

[250] D. CHAUM, A. FIAT, AND M. NAOR, "Untraceable electronic cash", *Advances in Cryptology–CRYPTO '88 (LNCS 403)*, 319–327, 1990.

[251] D. CHAUM AND T.P. PEDERSEN, "Wallet databases with observers", *Advances in Cryptology–CRYPTO '92 (LNCS 740)*, 89–105, 1993.

[252] D. CHAUM AND H. VAN ANTWERPEN, "Undeniable signatures", *Advances in Cryptology–CRYPTO '89 (LNCS 435)*, 212–216, 1990.

[253] D. CHAUM, E. VAN HEIJST, AND B. PFITZMANN, "Cryptographically strong undeniable signatures, unconditionally secure for the signer", *Advances in Cryptology–CRYPTO '91 (LNCS 576)*, 470–484, 1992.

[254] D. CHAUM AND E. VAN HEYST, "Group signatures", *Advances in Cryptology–EUROCRYPT '91 (LNCS 547)*, 257–265, 1991.

[255] L. CHEN AND T.P. PEDERSEN, "New group signature schemes", *Advances in Cryptology–EUROCRYPT '94 (LNCS 950)*, 171–181, 1995.

[256] V. CHEPYZHOV AND B. SMEETS, "On a fast correlation attack on certain stream ciphers", *Advances in Cryptology–EUROCRYPT '91 (LNCS 547)*, 176–185, 1991.

[257] B. CHOR AND O. GOLDREICH, "Unbiased bits from sources of weak randomness and probabilistic communication complexity", *Proceedings of the IEEE 26th Annual Symposium on Foundations of Computer Science*, 429–442, 1985.

[258] ——, "Unbiased bits from sources of weak randomness and probabilistic communication complexity", *SIAM Journal on Computing*, 17 (1988), 230–261. An earlier version appeared in [257].

[259] B. CHOR, S. GOLDWASSER, S. MICALI, AND B. AWERBUCH, "Verifiable secret sharing and achieving simultaneity in the presence of faults", *Proceedings of the IEEE 26th Annual Symposium on Foundations of Computer Science*, 383–395, 1985.

[260] B. CHOR AND R.L. RIVEST, "A knapsack type public key cryptosystem based on arithmetic in finite fields", *Advances in Cryptology–Proceedings of CRYPTO 84 (LNCS 196)*, 54–65, 1985.

[261] ——, "A knapsack-type public key cryptosystem based on arithmetic in finite fields", *IEEE Transactions on Information Theory*, 34 (1988), 901–909. An earlier version appeared in [260].

[262] A. CLARK, J. GOLIĆ, AND E. DAWSON, "A comparison of fast correlation attacks", D. Gollmann, editor, *Fast Software Encryption, Third International Workshop (LNCS 1039)*, 145–157, Springer-Verlag, 1996.

[263] H. COHEN, *A Course in Computational Algebraic Number Theory*, Springer-Verlag, Berlin, 1993.

[264] H. COHEN AND A.K. LENSTRA, "Implementation of a new primality test", *Mathematics of Computation*, 48 (1987), 103–121.

[265] H. COHEN AND H.W. LENSTRA JR., "Primality testing and Jacobi sums", *Mathematics of Computation*, 42 (1984), 297–330.

[266] D. COPPERSMITH, "Fast evaluation of logarithms in fields of characteristic two", *IEEE Transactions on Information Theory*, 30 (1984), 587–594.

[267] ——, "Another birthday attack", *Advances in Cryptology–CRYPTO '85 (LNCS 218)*, 14–17, 1986.

[268] ——, "The real reason for Rivest's phenomenon", *Advances in Cryptology–CRYPTO '85 (LNCS 218)*, 535–536, 1986.

[269] ——, "Modifications to the number field sieve", *Journal of Cryptology*, 6 (1993), 169–180.

[270] ——, "Solving linear equations over $GF(2)$: Block Lanczos algorithm", *Linear Algebra and its Applications*, 192 (1993), 33–60.

[271] ——, "The Data Encryption Standard (DES) and its strength against attacks", *IBM Journal of Research and Development*, 38 (1994), 243–250.

[272] ——, "Solving homogeneous linear equations over GF(2) via block Wiedemann algorithm", *Mathematics of Computation*, 62 (1994), 333–350.

[273] ——, "Finding a small root of a bivariate integer equation; factoring with high bits known", *Advances in Cryptology–EUROCRYPT '96 (LNCS 1070)*, 178–189, 1996.

[274] ——, "Finding a small root of a univariate modular equation", *Advances in Cryptology–EUROCRYPT '96 (LNCS 1070)*, 155–165, 1996.

[275] ——, "Analysis of ISO/CCITT Document X.509 Annex D", memorandum, IBM T.J. Watson Research Center, Yorktown Heights, N.Y., 10598, U.S.A., June 11 1989.

[276] ——, "Two broken hash functions", IBM Research Report RC 18397, IBM T.J. Watson Research Center, Yorktown Heights, N.Y., 10598, U.S.A., Oct. 6 1992.

[277] D. COPPERSMITH, M. FRANKLIN, J. PATARIN, AND M. REITER, "Low-exponent RSA with related messages", *Advances in Cryptology–EUROCRYPT '96 (LNCS 1070)*, 1–9, 1996.

[278] D. COPPERSMITH, D.B. JOHNSON, AND S.M. MATYAS, "A proposed mode for triple-DES encryption", *IBM Journal of Research and Development*, 40 (1996), 253–261.

[279] D. COPPERSMITH, H. KRAWCZYK, AND Y. MANSOUR, "The shrinking generator", *Advances in Cryptology–CRYPTO '93 (LNCS 773)*, 22–39, 1994.

[280] D. COPPERSMITH, A.M. ODLZYKO, AND R. SCHROEPPEL, "Discrete logarithms in $GF(p)$", *Algorithmica*, 1 (1986), 1–15.

[281] D. COPPERSMITH AND P. ROGAWAY, "Software-efficient pseudorandom function and the use thereof for encryption", U.S. Patent # 5,454,039, 26 Sep 1995.

[282] T.H. CORMEN, C.E. LEISERSON, AND R.L. RIVEST, *Introduction to Algorithms*, MIT Press, Cambridge, Massachusetts, 1990.

[283] M.J. COSTER, A. JOUX, B.A. LAMACCHIA, A.M. ODLYZKO, C.P. SCHNORR, AND J. STERN, "Improved low-density subset sum algorithms", *Computational Complexity*, 2 (1992), 111–128.

[284] J.-M. COUVEIGNES, "Computing a square root for the number field sieve", A.K. Lenstra and H.W. Lenstra Jr., editors, *The Development of the Number Field Sieve*, volume 1554 of *Lecture Notes in Mathematics*, 95–102, Springer-Verlag, 1993.

[285] T. COVER AND R. KING, "A convergent gambling estimate of the entropy of English", *IEEE Transactions on Information Theory*, 24 (1978), 413–421.

[286] R.E. CRANDALL, "Method and apparatus for public key exchange in a cryptographic system", U.S. Patent # 5,159,632, 27 Oct 1992.

[287] ——, "Method and apparatus for public key exchange in a cryptographic system", U.S. Patent # 5,271,061, 14 Dec 1993 (continuation-in-part of 5,159,632).

[288] R.A. CROFT AND S.P. HARRIS, "Public-key cryptography and re-usable shared secrets", H. Beker and F. Piper, editors, *Cryptography and Coding*, Institute of Mathematics & Its Applications (IMA), 189–201, Clarendon Press, 1989.

[289] J. DAEMEN, *Cipher and hash function design*, PhD thesis, Katholieke Universiteit Leuven (Belgium), 1995.

[290] J. DAEMEN, R. GOVAERTS, AND J. VANDEWALLE, "A new approach to block cipher design", R. Anderson, editor, *Fast Software Encryption, Cambridge Security Workshop (LNCS 809)*, 18–32, Springer-Verlag, 1994.

[291] ——, "Resynchronization weaknesses in synchronous stream ciphers", *Advances in Cryptology–EUROCRYPT '93 (LNCS 765)*, 159–167, 1994.

[292] ——, "Weak keys for IDEA", *Advances in Cryptology–CRYPTO '93 (LNCS 773)*, 224–231, 1994.

[293] Z.-D DAI, "Proof of Rueppel's linear complexity conjecture", *IEEE Transactions on Information Theory*, 32 (1986), 440–443.

[294] Z.-D. DAI AND J.-H. YANG, "Linear complexity of periodically repeated random sequences", *Advances in Cryptology–EUROCRYPT '91 (LNCS 547)*, 168–175, 1991.

[295] I.B. DAMGÅRD, "Collision free hash functions and public key signature schemes", *Advances in Cryptology–EUROCRYPT '87 (LNCS 304)*, 203–216, 1988.

[296] ——, "A design principle for hash functions", *Advances in Cryptology–CRYPTO '89 (LNCS 435)*, 416–427, 1990.

[297] ——, "Towards practical public key systems secure against chosen ciphertext attacks", *Advances in Cryptology–CRYPTO '91 (LNCS 576)*, 445–456, 1992.

[298] ——, "Practical and provably secure release of a secret and exchange of signatures", *Advances in Cryptology–EUROCRYPT '93 (LNCS 765)*, 200–217, 1994.

[299] I.B. DAMGÅRD AND P. LANDROCK, "Improved bounds for the Rabin primality test", M.J. Ganley, editor, *Cryptography and Coding III*, volume 45 of *Institute of Mathematics & Its Applications (IMA)*, 117–128, Clarendon Press, 1993.

[300] I.B. DAMGÅRD, P. LANDROCK, AND C. POMERANCE, "Average case error estimates for the strong probable prime test", *Mathematics of Computation*, 61 (1993), 177–194.

[301] H. DAVENPORT, "Bases for finite fields", *The Journal of the London Mathematical Society*, 43 (1968), 21–39.

[302] G.I. DAVIDA, "Chosen signature cryptanalysis of the RSA (MIT) public key cryptosystem", Technical Report TR-CS-82-2, Department of Electrical Engineering and Computer Science, University of Wisconsin, Milwaukee, WI, 1982.

[303] D.W. DAVIES, "Some regular properties of the 'Data Encryption Standard' algorithm", *Advances in Cryptology–Proceedings of Crypto 82*, 89–96, 1983.

[304] ——, "A message authenticator algorithm suitable for a mainframe computer", *Advances in Cryptology–Proceedings of CRYPTO 84 (LNCS 196)*, 393–400, 1985.

[305] ——, "Schemes for electronic funds transfer at the point of sale", K.M. Jackson and J. Hruska, editors, *Computer Security Reference Book*, 667–689, CRC Press, 1992.

[306] D.W. DAVIES AND D.O. CLAYDEN, "The message authenticator algorithm (MAA) and its implementation", Report DITC 109/88, National Physical Laboratory, U.K., February 1988.

[307] D.W. DAVIES AND G.I.P. PARKIN, "The average cycle size of the key stream in output feedback encipherment", *Advances in Cryptology–Proceedings of Crypto 82*, 97–98, 1983.

[308] D.W. DAVIES AND W.L. PRICE, *Security for Computer Networks*, John Wiley & Sons, New York, 2nd edition, 1989.

[309] D. DAVIS, R. IHAKA, AND P. FENSTERMACHER, "Cryptographic randomness from air turbulence in disk drives", *Advances in Cryptology–CRYPTO '94 (LNCS 839)*, 114–120, 1994.

[310] D. DAVIS AND R. SWICK, "Network security via private-key certificates", *Operating Systems Review*, 24 (1990), 64–67.

[311] J.A. DAVIS, D.B. HOLDRIDGE, AND G.J. SIMMONS, "Status report on factoring (at the Sandia National Labs)", *Advances in Cryptology–Proceedings of EUROCRYPT 84 (LNCS 209)*, 183–215, 1985.

[312] E. DAWSON, "Cryptanalysis of summation generator", *Advances in Cryptology–AUSCRYPT '92 (LNCS 718)*, 209–215, 1993.

[313] W. DE JONGE AND D. CHAUM, "Attacks on some RSA signatures", *Advances in Cryptology–CRYPTO '85 (LNCS 218)*, 18–27, 1986.

[314] P. DE ROOIJ, "On the security of the Schnorr scheme using preprocessing", *Advances in Cryptology–EUROCRYPT '91 (LNCS 547)*, 71–80, 1991.

[315] ——, "On Schnorr's preprocessing for digital signature schemes", *Advances in Cryptology–EUROCRYPT '93 (LNCS 765)*, 435–439, 1994.

[316] ——, "Efficient exponentiation using precomputation and vector addition chains", *Advances in Cryptology–EUROCRYPT '94 (LNCS 950)*, 389–399, 1995.

[317] A. DE SANTIS, S. MICALI, AND G. PERSIANO, "Non-interactive zero-knowledge proof systems", *Advances in Cryptology–CRYPTO '87 (LNCS 293)*, 52–72, 1988.

[318] A. DE SANTIS AND M. YUNG, "On the design of provably secure cryptographic hash functions", *Advances in Cryptology–EUROCRYPT '90 (LNCS 473)*, 412–431, 1991.

[319] D. DE WALEFFE AND J.-J. QUISQUATER, "Better login protocols for computer networks", B. Preneel, R. Govaerts, and J. Vandewalle, editors, *Computer Security and Industrial Cryptography: State of the Art and Evolution (LNCS 741)*, 50–70, Springer-Verlag, 1993.

[320] J.M. DeLAURENTIS, "A further weakness in the common modulus protocol for the RSA cryptoalgorithm", *Cryptologia*, 8 (1984), 253–259.

[321] N. DEMYTKO, "A new elliptic curve based analogue of RSA", *Advances in Cryptology–EUROCRYPT '93 (LNCS 765)*, 40–49, 1994.

[322] B. DEN BOER, "Cryptanalysis of F.E.A.L.", *Advances in Cryptology–EUROCRYPT '88 (LNCS 330)*, 293–299, 1988.

[323] ——, "Diffie-Hellman is as strong as discrete log for certain primes", *Advances in Cryptology–CRYPTO '88 (LNCS 403)*, 530–539, 1990.

[324] B. DEN BOER AND A. BOSSELAERS, "An attack on the last two rounds of MD4", *Advances in Cryptology–CRYPTO '91 (LNCS 576)*, 194–203, 1992.

[325] ——, "Collisions for the compression function of MD5", *Advances in Cryptology–EUROCRYPT '93 (LNCS 765)*, 293–304, 1994.

[326] D.E. DENNING, *Cryptography and Data Security*, Addison-Wesley, Reading, Massachusetts, 1983. Reprinted with corrections.

[327] ——, "Digital signatures with RSA and other public-key cryptosystems", *Communications of the ACM*, 27 (1984), 388–392.

[328] ——, "To tap or not to tap", *Communications of the ACM*, 36 (1993), 24–44.

[329] D.E. DENNING AND D.K. BRANSTAD, "A taxonomy for key escrow encryption systems", *Communications of the ACM*, 39 (1996), 34–40.

[330] D.E. DENNING AND G.M. SACCO, "Timestamps in key distribution protocols", *Communications of the ACM*, 24 (1981), 533–536.

[331] D.E. DENNING AND M. SMID, "Key escrowing today", *IEEE Communications Magazine*, 32 (September 1994), 58–68.

[332] J. B. DENNIS AND E. C. VAN HORN, "Programming semantics for multiprogrammed computations", *Communications of the ACM*, 9 (1966), 143–155.

[333] T. DENNY, B. DODSON, A.K. LENSTRA, AND M.S. MANASSE, "On the factorization of RSA-120", *Advances in Cryptology–CRYPTO '93 (LNCS 773)*, 166–174, 1994.

[334] DEPARTMENT OF DEFENSE (U.S.), "Department of defense password management guideline", CSC-STD-002-85, Department of Defense Computer Security Center, Fort Meade, Maryland, 1985.

[335] Y. DESMEDT, "Unconditionally secure authentication schemes and practical and theoretical consequences", *Advances in Cryptology–CRYPTO '85 (LNCS 218)*, 42–55, 1986.

[336] ——, "Society and group oriented cryptography: A new concept", *Advances in Cryptology–CRYPTO '87 (LNCS 293)*, 120–127, 1988.

[337] ——, "Threshold cryptography", *European Transactions on Telecommunications*, 5 (1994), 449–457.

[338] ——, "Securing traceability of ciphertexts – Towards a secure software key escrow system", *Advances in Cryptology–EUROCRYPT '95 (LNCS 921)*, 147–157, 1995.

[339] Y. DESMEDT AND M. BURMESTER, "Towards practical 'proven secure' authenticated key distribution", *1st ACM Conference on Computer and Communications Security*, 228–231, ACM Press, 1993.

[340] Y. DESMEDT, C. GOUTIER, AND S. BENGIO, "Special uses and abuses of the Fiat-Shamir passport protocol", *Advances in Cryptology–CRYPTO '87 (LNCS 293)*, 21–39, 1988.

[341] Y. DESMEDT AND A.M. ODLYZKO, "A chosen text attack on the RSA cryptosystem and some discrete logarithm schemes", *Advances in Cryptology–CRYPTO '85 (LNCS 218)*, 516–522, 1986.

[342] W. DIFFIE, "The first ten years of public-key cryptography", *Proceedings of the IEEE*, 76 (1988), 560–577.

[343] ——, "The first ten years of public key cryptology", G.J. Simmons, editor, *Contemporary Cryptology: The Science of Information Integrity*, 135–175, IEEE Press, 1992. An earlier version appeared in [342].

[344] W. DIFFIE AND M.E. HELLMAN, "Multiuser cryptographic techniques", *Proceedings of AFIPS National Computer Conference*, 109–112, 1976.

[345] ——, "New directions in cryptography", *IEEE Transactions on Information Theory*, 22 (1976), 644–654.

[346] ——, "Exhaustive cryptanalysis of the NBS Data Encryption Standard", *Computer*, 10 (1977), 74–84.

[347] ——, "Privacy and authentication: An introduction to cryptography", *Proceedings of the IEEE*, 67 (1979), 397–427.

[348] W. DIFFIE, P.C. VAN OORSCHOT, AND M.J. WIENER, "Authentication and authenticated key exchanges", *Designs, Codes and Cryptography*, 2 (1992), 107–125.

[349] C. DING, "The differential cryptanalysis and design of natural stream ciphers", R. Anderson, editor, *Fast Software Encryption, Cambridge Security Workshop (LNCS 809)*, 101–115, Springer-Verlag, 1994.

[350] B. DIXON AND A.K. LENSTRA, "Massively parallel elliptic curve factoring", *Advances in Cryptology–EUROCRYPT '92 (LNCS 658)*, 183–193, 1993.

[351] J.D. DIXON, "Asymptotically fast factorization of integers", *Mathematics of Computation*, 36 (1981), 255–260.

[352] H. DOBBERTIN, "Cryptanalysis of MD4", *Journal of Cryptology*, to appear.

[353] ——, "RIPEMD with two-round compress function is not collision-free", *Journal of Cryptology*, to appear; announced at rump session, Eurocrypt '95.

[354] ——, "Cryptanalysis of MD4", D. Gollmann, editor, *Fast Software Encryption, Third International Workshop (LNCS 1039)*, 53–69, Springer-Verlag, 1996.

[355] H. DOBBERTIN, A. BOSSELAERS, AND B. PRENEEL, "RIPEMD-160: a strengthened version of RIPEMD", D. Gollmann, editor, *Fast Software Encryption, Third International Workshop (LNCS 1039)*, 71–82, Springer-Verlag, 1996.

[356] B. DODSON AND A.K. LENSTRA, "NFS with four large primes: An explosive experiment", *Advances in Cryptology–CRYPTO '95 (LNCS 963)*, 372–385, 1995.

[357] D. DOLEV, C. DWORK, AND M. NAOR, "Non-malleable cryptography", *Proceedings of the 23rd Annual ACM Symposium on Theory of Computing*, 542–552, 1991.

[358] D. DOLEV AND A.C. YAO, "On the security of public key protocols", *Proceedings of the IEEE 22nd Annual Symposium on Foundations of Computer Science*, 350–357, 1981.

[359] ——, "On the security of public key protocols", *IEEE Transactions on Information Theory*, 29 (1983), 198–208. An earlier version appeared in [358].

[360] P. DOWNEY, B. LEONG, AND R. SETHI, "Computing sequences with addition chains", *SIAM Journal on Computing*, 10 (1981), 638–646.

[361] S.R. DUSSÉ AND B.S. KALISKI JR., "A cryptographic library for the Motorola DSP 56000", *Advances in Cryptology–EUROCRYPT '90 (LNCS 473)*, 230–244, 1991.

[362] H. EBERLE, "A high-speed DES implementation for network applications", *Advances in Cryptology–CRYPTO '92 (LNCS 740)*, 521–539, 1993.

[363] W. F. EHRSAM, C.H.W. MEYER, R.L. POWERS, J.L. SMITH, AND W.L. TUCHMAN, "Product block cipher system for data security", U.S. Patent # 3,962,539, 8 Jun 1976.

[364] W.F. EHRSAM, S.M. MATYAS, C.H. MEYER, AND W.L. TUCHMAN, "A cryptographic key management scheme for implementing the Data Encryption Standard", *IBM Systems Journal*, 17 (1978), 106–125.

[365] ELECTRONIC INDUSTRIES ASSOCIATION (EIA), "Dual-mode mobile station – base station compatibility standard", EIA Interim Standard IS-54 Revision B (Rev. B), 1992.

[366] T. ELGAMAL, *Cryptography and logarithms over finite fields*, PhD thesis, Stanford University, 1984.

[367] ——, "A public key cryptosystem and a signature scheme based on discrete logarithms", *Advances in Cryptology–Proceedings of CRYPTO 84 (LNCS 196)*, 10–18, 1985.

[368] ——, "A public key cryptosystem and a signature scheme based on discrete logarithms", *IEEE Transactions on Information Theory*, 31 (1985), 469–472. An earlier version appeared in [367].

[369] ——, "A subexponential-time algorithm for computing discrete logarithms over $GF(p^2)$", *IEEE Transactions on Information Theory*, 31 (1985), 473–481.

[370] P. ELIAS, "The efficient construction of an unbiased random sequence", *The Annals of Mathematical Statistics*, 43 (1972), 865–870.

[371] ——, "Interval and recency rank source encoding: Two on-line adaptive variable-length schemes", *IEEE Transactions on Information Theory*, 33 (1987), 3–10.

[372] E.D. ERDMANN, "Empirical tests of binary keystreams", Master's thesis, Department of Mathematics, Royal Holloway and Bedford New College, University of London, 1992.

[373] P. ERDÖS AND C. POMERANCE, "On the number of false witnesses for a composite number", *Mathematics of Computation*, 46 (1986), 259–279.

[374] D. ESTES, L.M. ADLEMAN, K. KOMPELLA, K.S. MCCURLEY, AND G.L. MILLER, "Breaking the Ong-Schnorr-Shamir signature scheme for quadratic number fields", *Advances in Cryptology–CRYPTO '85 (LNCS 218)*, 3–13, 1986.

[375] A. EVANS JR., W. KANTROWITZ, AND E. WEISS, "A user authentication scheme not requiring secrecy in the computer", *Communications of the ACM*, 17 (1974), 437–442.

[376] S. EVEN AND O. GOLDREICH, "On the power of cascade ciphers", *ACM Transactions on Computer Systems*, 3 (1985), 108–116.

[377] S. EVEN, O. GOLDREICH, AND S. MICALI, "On-line/off-line digital signatures", *Advances in Cryptology–CRYPTO '89 (LNCS 435)*, 263–275, 1990.

[378] ——, "On-line/off-line digital signatures", *Journal of Cryptology*, 9 (1996), 35–67. An earlier version appeared in [377].

[379] S. EVEN AND Y. YACOBI, "Cryptocomplexity and NP-completeness", J.W. de Bakker and J. van Leeuwen, editors, *Automata, Languages, and Programming, 7th Colloquium (LNCS 85)*, 195–207, Springer-Verlag, 1980.

[380] D. EVERETT, "Identity verification and biometrics", K.M. Jackson and J. Hruska, editors, *Computer Security Reference Book*, 37–73, CRC Press, 1992.

[381] J.-H. EVERTSE AND E. VAN HEYST, "Which new RSA-signatures can be computed from certain given RSA-signatures?", *Journal of Cryptology*, 5 (1992), 41–52.

[382] R.C. FAIRFIELD, R.L. MORTENSON, AND K.B. COULTHART, "An LSI random number generator (RNG)", *Advances in Cryptology–Proceedings of CRYPTO 84 (LNCS 196)*, 203–230, 1985.

[383] U. FEIGE, A. FIAT, AND A. SHAMIR, "Zero-knowledge proofs of identity", *Journal of Cryptology*, 1 (1988), 77–94.

[384] U. FEIGE AND A. SHAMIR, "Witness indistinguishable and witness hiding protocols", *Proceedings of the 22nd Annual ACM Symposium on Theory of Computing*, 416–426, 1990.

[385] H. FEISTEL, "Block cipher cryptographic system", U.S. Patent # 3,798,359, 19 Mar 1974.

[386] ——, "Step code ciphering system", U.S. Patent # 3,798,360, 19 Mar 1974.

[387] ——, "Cryptography and computer privacy", *Scientific American*, 228 (May 1973), 15–23.

[388] H. FEISTEL, W.A. NOTZ, AND J.L. SMITH, "Some cryptographic techniques for machine-to-machine data communications", *Proceedings of the IEEE*, 63 (1975), 1545–1554.

[389] F.A. FELDMAN, "Fast spectral tests for measuring nonrandomness and the DES", *Advances in Cryptology–CRYPTO '87 (LNCS 293)*, 243–254, 1988.

[390] P. FELDMAN, "A practical scheme for non-interactive verifiable secret sharing", *Proceedings of the IEEE 28th Annual Symposium on Foundations of Computer Science*, 427–437, 1987.

[391] D.C. FELDMEIER AND P.R. KARN, "UNIX password security – ten years later", *Advances in Cryptology–CRYPTO '89 (LNCS 435)*, 44–63, 1990.

[392] W. FELLER, *An Introduction to Probability Theory and its Applications*, John Wiley & Sons, New York, 3rd edition, 1968.

[393] A. FIAT AND M. NAOR, "Rigorous time/space tradeoffs for inverting functions", *Proceedings of the 23rd Annual ACM Symposium on Theory of Computing*, 534–541, 1991.

[394] ———, "Broadcast encryption", *Advances in Cryptology–CRYPTO '93 (LNCS 773)*, 480–491, 1994.

[395] A. FIAT AND A. SHAMIR, "How to prove yourself: Practical solutions to identification and signature problems", *Advances in Cryptology–CRYPTO '86 (LNCS 263)*, 186–194, 1987.

[396] FIPS 46, "Data encryption standard", Federal Information Processing Standards Publication 46, U.S. Department of Commerce/National Bureau of Standards, National Technical Information Service, Springfield, Virginia, 1977 (revised as FIPS 46-1:1988; FIPS 46-2:1993).

[397] FIPS 74, "Guidelines for implementing and using the NBS data encryption standard", Federal Information Processing Standards Publication 74, U.S. Department of Commerce/National Bureau of Standards, National Technical Information Service, Springfield, Virginia, 1981.

[398] FIPS 81, "DES modes of operation", Federal Information Processing Standards Publication 81, U.S. Department of Commerce/National Bureau of Standards, National Technical Information Service, Springfield, Virginia, 1980.

[399] FIPS 112, "Password usage", Federal Information Processing Standards Publication 112, U.S. Department of Commerce/National Bureau of Standards, National Technical Information Service, Springfield, Virginia, 1985.

[400] FIPS 113, "Computer data authentication", Federal Information Processing Standards Publication 113, U.S. Department of Commerce/National Bureau of Standards, National Technical Information Service, Springfield, Virginia, 1985.

[401] FIPS 140-1, "Security requirements for cryptographic modules", Federal Information Processing Standards Publication 140-1, U.S. Department of Commerce/N.I.S.T., National Technical Information Service, Springfield, Virginia, 1994.

[402] FIPS 171, "Key management using ANSI X9.17", Federal Information Processing Standards Publication 171, U.S. Department of Commerce/N.I.S.T., National Technical Information Service, Springfield, Virginia, 1992.

[403] FIPS 180, "Secure hash standard", Federal Information Processing Standards Publication 180, U.S. Department of Commerce/N.I.S.T., National Technical Information Service, Springfield, Virginia, May 11 1993.

[404] FIPS 180-1, "Secure hash standard", Federal Information Processing Standards Publication 180-1, U.S. Department of Commerce/N.I.S.T., National Technical Information Service, Springfield, Virginia, April 17 1995 (supersedes FIPS PUB 180).

[405] FIPS 185, "Escrowed encryption standard (EES)", Federal Information Processing Standards Publication 185, U.S. Department of Commerce/N.I.S.T., National Technical Information Service, Springfield, Virginia, 1994.

[406] FIPS 186, "Digital signature standard", Federal Information Processing Standards Publication 186, U.S. Department of Commerce/N.I.S.T., National Technical Information Service, Springfield, Virginia, 1994.

[407] FIPS JJJ, "Standard for public key cryptographic entity authentication mechanisms", U.S. Department of Commerce/N.I.S.T., draft (1996 March 29).

[408] A.M. FISCHER, "Public key/signature cryptosystem with enhanced digital signature certification", U.S. Patent # 4,868,877, 19 Sep 1989.

[409] ———, "Public key/signature cryptosystem with enhanced digital signature certification", U.S. Patent # 5,005,200, 2 Apr 1991 (continuation-in-part of 4,868,877).

[410] ———, "Electronic document authorization", *Proceedings of the 13th National Computer*

Security Conference, Washington D.C., sponsored by N.I.S.T. and the National Computer Security Center, USA, 1990.

[411] J.-B. FISCHER AND J. STERN, "An efficient pseudo-random generator provably as secure as syndrome decoding", Advances in Cryptology–EUROCRYPT '96 (LNCS 1070), 245–255, 1996.

[412] M. FISCHER, S. MICALI, AND C. RACKOFF, "A secure protocol for oblivious transfer", unpublished (presented at Eurocrypt'84).

[413] P. FLAJOLET AND A. ODLYZKO, "Random mapping statistics", Advances in Cryptology–EUROCRYPT '89 (LNCS 434), 329–354, 1990.

[414] W. FORD, Computer Communications Security: Principles, Standard Protocols and Techniques, Prentice Hall, Englewood Cliffs, New Jersey, 1994.

[415] ——, "Standardizing information technology security", StandardView, 2 (1994), 64–71.

[416] ——, "Advances in public-key certificate standards", Security, Audit and Control, 13 (1995), ACM Press/SIGSAC, 9–15.

[417] W. FORD AND M. WIENER, "A key distribution method for object-based protection", 2nd ACM Conference on Computer and Communications Security, 193–197, ACM Press, 1994.

[418] R. FORRÉ, "A fast correlation attack on nonlinearly feedforward filtered shift-register sequences", Advances in Cryptology–EUROCRYPT '89 (LNCS 434), 586–595, 1990.

[419] Y. FRANKEL AND M. YUNG, "Cryptanalysis of the immunized LL public key systems", Advances in Cryptology–CRYPTO '95 (LNCS 963), 287–296, 1995.

[420] ——, "Escrow encryption systems visited: Attacks, analysis and designs", Advances in Cryptology–CRYPTO '95 (LNCS 963), 222–235, 1995.

[421] M.K. FRANKLIN AND M.K. REITER, "Verifiable signature sharing", Advances in Cryptology–EUROCRYPT '95 (LNCS 921), 50–63, 1995.

[422] G. FREY AND H.-G. RÜCK, "A remark concerning m-divisibility and the discrete logarithm in the divisor class group of curves", Mathematics of Computation, 62 (1994), 865–874.

[423] W. FRIEDMAN, Military Cryptanalysis, U.S. Government Printing Office, Washington DC, 1944. Volume I – Monoalphabetic substitution systems. Volume II – Simpler varieties of polyalphabetic substitution systems. Volume III – Aperiodic substitutions. Volume IV – Transposition systems.

[424] ——, "Cryptology", Encyclopedia Brittanica, 6 (1967), 844–851.

[425] ——, Elements of Cryptanalysis, Aegean Park Press, Laguna Hills, California, 1976. First published in 1923.

[426] ——, The Index of Coincidence and its Applications in Cryptography, Aegean Park Press, Laguna Hills, California, 1979. First published in 1920.

[427] A.M. FRIEZE, J. HÅSTAD, R. KANNAN, J.C. LAGARIAS, AND A. SHAMIR, "Reconstructing truncated integer variables satisfying linear congruences", SIAM Journal on Computing, 17 (1988), 262–280.

[428] A. FUJIOKA, T. OKAMOTO, AND S. MIYAGUCHI, "ESIGN: An efficient digital signature implementation for smart cards", Advances in Cryptology–EUROCRYPT '91 (LNCS 547), 446–457, 1991.

[429] W. FUMY AND P. LANDROCK, "Principles of key management", IEEE Journal on Selected Areas in Communications, 11 (1993), 785–793.

[430] W. FUMY AND M. LECLERC, "Placement of cryptographic key distribution within OSI: design alternatives and assessment", Computer Networks and ISDN Systems, 26 (1993), 217–225.

[431] W. FUMY AND M. MUNZERT, "A modular approach to key distribution", Advances in Cryptology–CRYPTO '90 (LNCS 537), 274–283, 1991.

[432] W. FUMY AND M. RIETENSPIESS, "Open systems security standards", Encyclopedia of Computer Science and Technology, A. Kent, J.G. Williams, C.M. Hall, eds., Marcel Dekker, New York (to appear, 1996).

[433] K. GAARDER AND E. SNEKKENES, "Applying a formal analysis technique to the CCITT X.509 strong two-way authentication protocol", Journal of Cryptology, 3 (1991), 81–98.

[434] E.M. GABIDULIN, "On public-key cryptosystems based on linear codes: Efficiency and weakness", P.G. Farrell, editor, *Codes and Cyphers: Cryptography and Coding IV*, 17–31, Institute of Mathematics & Its Applications (IMA), 1995.

[435] E.M. GABIDULIN, A.V. PARAMONOV, AND O.V. TRETJAKOV, "Ideals over a non-commutative ring and their application in cryptology", *Advances in Cryptology–EUROCRYPT '91 (LNCS 547)*, 482–489, 1991.

[436] H. GAINES, *Cryptanalysis: A Study of Ciphers and their Solutions*, Dover Publications, New York, 1956.

[437] J. GAIT, "A new nonlinear pseudorandom number generator", *IEEE Transactions on Software Engineering*, 3 (1977), 359–363.

[438] J.M. GALVIN, K. MCCLOGHRIE, AND J.R. DAVIN, "Secure management of SNMP networks", *Integrated Network Management, II*, 703–714, 1991.

[439] R.A. GAMES AND A.H. CHAN, "A fast algorithm for determining the complexity of a binary sequence with period 2^n", *IEEE Transactions on Information Theory*, 29 (1983), 144–146.

[440] M. GARDNER, "A new kind of cipher that would take millions of years to break", *Scientific American*, 237 (Aug 1977), 120–124.

[441] M.R. GAREY AND D.S. JOHNSON, *Computers and Intractability: A Guide to the Theory of NP-completeness*, W.H. Freeman, San Francisco, 1979.

[442] S. GARFINKEL, *PGP: Pretty Good Privacy*, O'Reilly and Associates, Inc., Sebastopol, California, 1995.

[443] H. GARNER, "The residue number system", *IRE Transactions on Electronic Computers*, EC-8 (1959), 140–147.

[444] C.F. GAUSS, *Disquisitiones Arithmeticae*, 1801. English translation by Arthur A. Clarke, Springer-Verlag, New York, 1986.

[445] K. GEDDES, S. CZAPOR, AND G. LABAHN, *Algorithms for Computer Algebra*, Kluwer Academic Publishers, Boston, 1992.

[446] P. GEFFE, "How to protect data with ciphers that are really hard to break", *Electronics*, 46 (1973), 99–101.

[447] J. GEORGIADES, "Some remarks on the security of the identification scheme based on permuted kernels", *Journal of Cryptology*, 5 (1992), 133–137.

[448] J. GERVER, "Factoring large numbers with a quadratic sieve", *Mathematics of Computation*, 41 (1983), 287–294.

[449] P.J. GIBLIN, *Primes and Programming: An Introduction to Number Theory with Computing*, Cambridge University Press, Cambrige, 1993.

[450] J.K. GIBSON, "Some comments on Damgård's hashing principle", *Electronics Letters*, 26 (July 19, 1990), 1178–1179.

[451] ——, "Equivalent Goppa codes and trapdoors to McEliece's public key cryptosystem", *Advances in Cryptology–EUROCRYPT '91 (LNCS 547)*, 517–521, 1991.

[452] ——, "Severely denting the Gabidulin version of the McEliece public key cryptosystem", *Designs, Codes and Cryptography*, 6 (1995), 37–45.

[453] ——, "The security of the Gabidulin public key cryptosystem", *Advances in Cryptology–EUROCRYPT '96 (LNCS 1070)*, 212–223, 1996.

[454] E.N. GILBERT, F.J. MACWILLIAMS, AND N.J.A. SLOANE, "Codes which detect deception", *Bell System Technical Journal*, 53 (1974), 405–424.

[455] H. GILBERT AND G. CHASSÉ, "A statistical attack of the Feal-8 cryptosystem", *Advances in Cryptology–CRYPTO '90 (LNCS 537)*, 22–33, 1991.

[456] H. GILBERT AND P. CHAUVAUD, "A chosen plaintext attack of the 16-round Khufu cryptosystem", *Advances in Cryptology–CRYPTO '94 (LNCS 839)*, 359–368, 1994.

[457] M. GIRAULT, "Hash-functions using modulo-n operations", *Advances in Cryptology–EUROCRYPT '87 (LNCS 304)*, 217–226, 1988.

[458] ——, "An identity-based identification scheme based on discrete logarithms modulo a composite number", *Advances in Cryptology–EUROCRYPT '90 (LNCS 473)*, 481–486, 1991.

[459] ——, "Self-certified public keys", *Advances in Cryptology–EUROCRYPT '91 (LNCS 547)*, 490–497, 1991.

[460] M. GIRAULT, R. COHEN, AND M. CAMPANA, "A generalized birthday attack", *Advances in Cryptology–EUROCRYPT '88 (LNCS 330)*, 129–156, 1988.

[461] M. GIRAULT AND J.C. PAILLÈS, "An identity-based scheme providing zero-knowledge authentication and authenticated key-exchange", *First European Symposium on Research in Computer Security – ESORICS'90*, 173–184, 1990.

[462] M. GIRAULT AND J. STERN, "On the length of cryptographic hash-values used in identification schemes", *Advances in Cryptology–CRYPTO '94 (LNCS 839)*, 202–215, 1994.

[463] V.D. GLIGOR, R. KAILAR, S. STUBBLEBINE, AND L. GONG, "Logics for cryptographic protocols — virtues and limitations", *The Computer Security Foundations Workshop IV*, 219–226, IEEE Computer Security Press, 1991.

[464] C.M. GOLDIE AND R.G.E. PINCH, *Communication Theory*, Cambridge University Press, Cambridge, 1991.

[465] O. GOLDREICH, "Two remarks concerning the Goldwasser-Micali-Rivest signature scheme", *Advances in Cryptology–CRYPTO '86 (LNCS 263)*, 104–110, 1987.

[466] O. GOLDREICH, S. GOLDWASSER, AND S. MICALI, "How to construct random functions", *Proceedings of the IEEE 25th Annual Symposium on Foundations of Computer Science*, 464–479, 1984.

[467] ——, "On the cryptographic applications of random functions", *Advances in Cryptology–Proceedings of CRYPTO 84 (LNCS 196)*, 276–288, 1985.

[468] ——, "How to construct random functions", *Journal of the Association for Computing Machinery*, 33 (1986), 792–807. An earlier version appeared in [466].

[469] O. GOLDREICH AND H. KRAWCZYK, "On the composition of zero-knowledge proof systems", M.S. Paterson, editor, *Automata, Languages and Programming, 17th International Colloquium (LNCS 443)*, 268–282, Springer-Verlag, 1990.

[470] O. GOLDREICH, H. KRAWCZYK, AND M. LUBY, "On the existence of pseudorandom generators", *Proceedings of the IEEE 29th Annual Symposium on Foundations of Computer Science*, 12–24, 1988.

[471] O. GOLDREICH AND L.A. LEVIN, "A hard-core predicate for all one-way functions", *Proceedings of the 21st Annual ACM Symposium on Theory of Computing*, 25–32, 1989.

[472] O. GOLDREICH, S. MICALI, AND A. WIGDERSON, "Proofs that yield nothing but their validity and a methodology of cryptographic protocol design", *Proceedings of the IEEE 27th Annual Symposium on Foundations of Computer Science*, 174–187, 1986.

[473] ——, "How to prove all NP statements in zero-knowledge, and a methodology of cryptographic protocol design", *Advances in Cryptology–CRYPTO '86 (LNCS 263)*, 171–185, 1987.

[474] ——, "Proofs that yield nothing but their validity or all languages in NP have zero-knowledge proof systems", *Journal of the Association for Computing Machinery*, 38 (1991), 691–729. An earlier version appeared in [472].

[475] O. GOLDREICH AND Y. OREN, "Definitions and properties of zero-knowledge proof systems", *Journal of Cryptology*, 7 (1994), 1–32.

[476] S. GOLDWASSER, "The search for provably secure cryptosystems", C. Pomerance, editor, *Cryptology and Computational Number Theory*, volume 42 of *Proceedings of Symposia in Applied Mathematics*, 89–113, American Mathematical Society, 1990.

[477] S. GOLDWASSER AND J. KILIAN, "Almost all primes can be quickly certified", *Proceedings of the 18th Annual ACM Symposium on Theory of Computing*, 316–329, 1986.

[478] S. GOLDWASSER AND S. MICALI, "Probabilistic encryption & how to play mental poker keeping secret all partial information", *Proceedings of the 14th Annual ACM Symposium on Theory of Computing*, 365–377, 1982.

[479] ——, "Probabilistic encryption", *Journal of Computer and System Sciences*, 28 (1984), 270–299. An earlier version appeared in [478].

[480] S. GOLDWASSER, S. MICALI, AND C. RACKOFF, "The knowledge complexity of interactive proof-systems", *Proceedings of the 17th Annual ACM Symposium on Theory of Computing*, 291–304, 1985.

[481] ——, "The knowledge complexity of interactive proof systems", *SIAM Journal on Computing*, 18 (1989), 186–208. An earlier version appeared in [480].

[482] S. GOLDWASSER, S. MICALI, AND R.L. RIVEST, "A "paradoxical" solution to the signature problem", *Proceedings of the IEEE 25th Annual Symposium on Foundations of Computer Science*, 441–448, 1984.

[483] ———, "A "Paradoxical" solution to the signature problem", *Advances in Cryptology–Proceedings of CRYPTO 84 (LNCS 196)*, 467, 1985.

[484] ———, "A digital signature scheme secure against adaptive chosen-message attacks", *SIAM Journal on Computing*, 17 (1988), 281–308. Earlier versions appeared in [482] and [483].

[485] J. GOLIĆ, "Correlation via linear sequential circuit approximation of combiners with memory", *Advances in Cryptology–EUROCRYPT '92 (LNCS 658)*, 113–123, 1993.

[486] ———, "On the security of shift register based keystream generators", R. Anderson, editor, *Fast Software Encryption, Cambridge Security Workshop (LNCS 809)*, 90–100, Springer-Verlag, 1994.

[487] ———, "Intrinsic statistical weakness of keystream generators", *Advances in Cryptology–ASIACRYPT '94 (LNCS 917)*, 91–103, 1995.

[488] ———, "Linear cryptanalysis of stream ciphers", B. Preneel, editor, *Fast Software Encryption, Second International Workshop (LNCS 1008)*, 154–169, Springer-Verlag, 1995.

[489] ———, "Towards fast correlation attacks on irregularly clocked shift registers", *Advances in Cryptology–EUROCRYPT '95 (LNCS 921)*, 248–262, 1995.

[490] ———, "On the security of nonlinear filter generators", D. Gollmann, editor, *Fast Software Encryption, Third International Workshop (LNCS 1039)*, 173–188, Springer-Verlag, 1996.

[491] J. GOLIĆ AND M. MIHALJEVIĆ, "A generalized correlation attack on a class of stream ciphers based on the Levenshtein distance", *Journal of Cryptology*, 3 (1991), 201–212.

[492] J. GOLIĆ AND L. O'CONNOR, "Embedding and probabilistic correlation attacks on clock-controlled shift registers", *Advances in Cryptology–EUROCRYPT '94 (LNCS 950)*, 230–243, 1995.

[493] R.A. GOLLIVER, A.K. LENSTRA, AND K.S. MCCURLEY, "Lattice sieving and trial division", *Algorithmic Number Theory (LNCS 877)*, 18–27, 1994.

[494] D. GOLLMANN, "Pseudo random properties of cascade connections of clock controlled shift registers", *Advances in Cryptology–Proceedings of EUROCRYPT 84 (LNCS 209)*, 93–98, 1985.

[495] ———, "Cryptanalysis of clock controlled shift registers", R. Anderson, editor, *Fast Software Encryption, Cambridge Security Workshop (LNCS 809)*, 121–126, Springer-Verlag, 1994.

[496] D. GOLLMANN AND W.G. CHAMBERS, "Clock-controlled shift registers: a review", *IEEE Journal on Selected Areas in Communications*, 7 (1989), 525–533.

[497] D. GOLLMANN, Y. HAN, AND C. MITCHELL, "Redundant integer representations and fast exponentiation", *Designs, Codes and Cryptography*, 7 (1996), 135–151.

[498] S.W. GOLOMB, *Shift Register Sequences*, Holden-Day, San Francisco, 1967. Reprinted by Aegean Park Press, 1982.

[499] L. GONG, "Using one-way functions for authentication", *Computer Communication Review*, 19 (1989), 8–11.

[500] ———, "A security risk of depending on synchronized clocks", *Operating Systems Review*, 26 (1992), 49–53.

[501] ———, "Variations on the themes of message freshness and replay", *The Computer Security Foundations Workshop VI*, 131–136, IEEE Computer Society Press, 1993.

[502] ———, "New protocols for third-party-based authentication and secure broadcast", *2nd ACM Conference on Computer and Communications Security*, 176–183, ACM Press, 1994.

[503] ———, "Efficient network authentication protocols: lower bounds and optimal implementations", *Distributed Computing*, 9 (1995), 131–145.

[504] L. GONG, T.M.A. LOMAS, R.M. NEEDHAM, AND J.H. SALTZER, "Protecting poorly chosen secrets from guessing attacks", *IEEE Journal on Selected Areas in Communications*, 11 (1993), 648–656.

[505] L. Gong, R. Needham, and R. Ya-halom, "Reasoning about belief in cryptographic protocols", *Proceedings of the IEEE Computer Society Symposium on Research in Security and Privacy*, 234–248, 1990.

[506] L. Gong and D.J. Wheeler, "A matrix key-distribution scheme", *Journal of Cryptology*, 2 (1990), 51–59.

[507] I.J. Good, "The serial test for sampling numbers and other tests for randomness", *Proceedings of the Cambridge Philosophical Society*, 49 (1953), 276–284.

[508] ———, "On the serial test for random sequences", *The Annals of Mathematical Statistics*, 28 (1957), 262–264.

[509] D.M. Gordon, "Designing and detecting trapdoors for discrete log cryptosystems", *Advances in Cryptology–CRYPTO '92 (LNCS 740)*, 66–75, 1993.

[510] ———, "Discrete logarithms in GF(P) using the number field sieve", *SIAM Journal on Discrete Mathematics*, 6 (1993), 124–138.

[511] D.M. Gordon and K.S. McCurley, "Massively parallel computations of discrete logarithms", *Advances in Cryptology–CRYPTO '92 (LNCS 740)*, 312–323, 1993.

[512] J. Gordon, "Very simple method to find the minimum polynomial of an arbitrary nonzero element of a finite field", *Electronics Letters*, 12 (December 9, 1976), 663–664.

[513] ———, "Strong RSA keys", *Electronics Letters*, 20 (June 7, 1984), 514–516.

[514] ———, "Strong primes are easy to find", *Advances in Cryptology–Proceedings of EUROCRYPT 84 (LNCS 209)*, 216–223, 1985.

[515] ———, "How to forge RSA key certificates", *Electronics Letters*, 21 (April 25, 1985), 377–379.

[516] ———, "Fast multiplicative inverse in modular arithmetic", H. Beker and F. Piper, editors, *Cryptography and Coding*, Institute of Mathematics & Its Applications (IMA), 269–279, Clarendon Press, 1989.

[517] J. Gordon and H. Retkin, "Are big *S*-boxes best?", *Cryptography–Proceedings of the Workshop on Cryptography, Burg Feuerstein (LNCS 149)*, 257–262, 1983.

[518] M. Goresky and A. Klapper, "Feedback registers based on ramified extensions of the

2-adic numbers", *Advances in Cryptology–EUROCRYPT '94 (LNCS 950)*, 215–222, 1995.

[519] K.C. Goss, "Cryptographic method and apparatus for public key exchange with authentication", U.S. Patent # 4,956,863, 11 Sep 1990.

[520] R. Graham, D. Knuth, and O. Patashnik, *Concrete Mathematics*, Addison-Wesley, Reading, Massachusetts, 2nd edition, 1994.

[521] A. Granville, "Primality testing and Carmichael numbers", *Notices of the American Mathematical Society*, 39 (1992), 696–700.

[522] E. Grossman, "Group theoretic remarks on cryptographic systems based on two types of addition", IBM Research Report RC 4742, IBM T.J. Watson Research Center, Yorktown Heights, N.Y., 10598, U.S.A., Feb. 26 1974.

[523] L.C. Guillou and J.-J. Quisquater, "Method and apparatus for authenticating accreditations and for authenticating and signing messages", U.S. Patent # 5,140,634, 18 Aug 1992.

[524] ———, "A practical zero-knowledge protocol fitted to security microprocessor minimizing both transmission and memory", *Advances in Cryptology–EUROCRYPT '88 (LNCS 330)*, 123–128, 1988.

[525] L.C. Guillou, J.-J. Quisquater, M. Walker, P. Landrock, and C. Shaer, "Precautions taken against various potential attacks in ISO/IEC DIS 9796", *Advances in Cryptology–EUROCRYPT '90 (LNCS 473)*, 465–473, 1991.

[526] L.C. Guillou and M. Ugon, "Smart card – a highly reliable and portable security device", *Advances in Cryptology–CRYPTO '86 (LNCS 263)*, 464–479, 1987.

[527] L.C. Guillou, M. Ugon, and J.-J. Quisquater, "The smart card: A standardized security device dedicated to public cryptology", G.J. Simmons, editor, *Contemporary Cryptology: The Science of Information Integrity*, 561–613, IEEE Press, 1992.

[528] C.G. Günther, "Alternating step generators controlled by de Bruijn sequences", *Advances in Cryptology–EUROCRYPT '87 (LNCS 304)*, 5–14, 1988.

[529] ———, "A universal algorithm for homophonic coding", *Advances in Cryptology–*

EUROCRYPT '88 (LNCS 330), 405–414, 1988.

[530] ——, "An identity-based key-exchange protocol", *Advances in Cryptology–EUROCRYPT '89 (LNCS 434)*, 29–37, 1990.

[531] H. GUSTAFSON, *Statistical Analysis of Symmetric Ciphers*, PhD thesis, Queensland University of Technology, 1996.

[532] H. GUSTAFSON, E. DAWSON, AND J. GOLIĆ, "Randomness measures related to subset occurrence", E. Dawson and J. Golić, editors, *Cryptography: Policy and Algorithms, International Conference, Brisbane, Queensland, Australia, July 1995 (LNCS 1029)*, 132–143, 1996.

[533] H. GUSTAFSON, E. DAWSON, L. NIELSEN, AND W. CAELLI, "A computer package for measuring the strength of encryption algorithms", *Computers & Security*, 13 (1994), 687–697.

[534] A. GUYOT, "OCAPI: Architecture of a VLSI coprocessor for the gcd and extended gcd of large numbers", *Proceedings of the 10th IEEE Symposium on Computer Arithmetic*, 226–231, IEEE Press, 1991.

[535] S. HABER AND W.S. STORNETTA, "How to time-stamp a digital document", *Journal of Cryptology*, 3 (1991), 99–111.

[536] J.L. HAFNER AND K.S. MCCURLEY, "On the distribution of running times of certain integer factoring algorithms", *Journal of Algorithms*, 10 (1989), 531–556.

[537] ——, "A rigorous subexponential algorithm for computation of class groups", *Journal of the American Mathematical Society*, 2 (1989), 837–850.

[538] T. HANSEN AND G.L. MULLEN, "Primitive polynomials over finite fields", *Mathematics of Computation*, 59 (1992), 639–643.

[539] G.H. HARDY, *A Mathematician's Apology*, Cambridge University Press, London, 1967.

[540] G.H. HARDY AND E.M. WRIGHT, *An Introduction to the Theory of Numbers*, Clarendon Press, Oxford, 5th edition, 1979.

[541] C. HARPES, G.G. KRAMER, AND J.L. MASSEY, "A generalization of linear cryptanalysis and the applicability of Matsui's piling-up lemma", *Advances in Cryptology–EUROCRYPT '95 (LNCS 921)*, 24–38, 1995.

[542] V. HARRIS, "An algorithm for finding the greatest common divisor", *Fibonacci Quarterly*, 8 (1970), 102–103.

[543] J. HÅSTAD, A.W. SCHRIFT, AND A. SHAMIR, "The discrete logarithm modulo a composite hides $O(n)$ bits", *Journal of Computer and System Sciences*, 47 (1993), 376–404.

[544] J. HÅSTAD, "Solving simultaneous modular equations of low degree", *SIAM Journal on Computing*, 17 (1988), 336–341.

[545] ——, "Pseudo-random generators under uniform assumptions", *Proceedings of the 22nd Annual ACM Symposium on Theory of Computing*, 395–404, 1990.

[546] R. HEIMAN, "A note on discrete logarithms with special structure", *Advances in Cryptology–EUROCRYPT '92 (LNCS 658)*, 454–457, 1993.

[547] ——, "Secure audio teleconferencing: A practical solution", *Advances in Cryptology–EUROCRYPT '92 (LNCS 658)*, 437–448, 1993.

[548] M.E. HELLMAN, "An extension of the Shannon theory approach to cryptography", *IEEE Transactions on Information Theory*, 23 (1977), 289–294.

[549] ——, "A cryptanalytic time-memory trade-off", *IEEE Transactions on Information Theory*, 26 (1980), 401–406.

[550] M.E. HELLMAN AND C.E. BACH, "Method and apparatus for use in public-key data encryption system", U.S. Patent # 4,633,036, 30 Dec 1986.

[551] M.E. HELLMAN, B.W. DIFFIE, AND R.C. MERKLE, "Cryptographic apparatus and method", U.S. Patent # 4,200,770, 29 Apr 1980.

[552] M.E. HELLMAN, R. MERKLE, R. SCHROEPPEL, L. WASHINGTON, W. DIFFIE, S. POHLIG, AND P. SCHWEITZER, "Results of an initial attempt to cryptanalyze the NBS Data Encryption Standard", Technical Report SEL 76-042, Information Systems Laboratory, Stanford University, Palo Alto, California, Sept. 9 1976 (revised Nov 10 1976).

[553] M.E. HELLMAN AND R.C. MERKLE, "Public key cryptographic apparatus and method", U.S. Patent # 4,218,582, 19 Aug 1980.

[554] M.E. HELLMAN AND S.C. POHLIG, "Exponentiation cryptographic apparatus and method", U.S. Patent # 4,424,414, 3 Jan 1984.

[555] M.E. HELLMAN AND J.M. REYNERI, "Fast computation of discrete logarithms in GF(q)", *Advances in Cryptology–Proceedings of Crypto 82*, 3–13, 1983.

[556] I.N. HERSTEIN, *Topics in Algebra*, Xerox College Pub., Lexington, Massachusetts, 2nd edition, 1975.

[557] L.S. HILL, "Cryptography in an algebraic alphabet", *American Mathematical Monthly*, 36 (1929), 306–312.

[558] L.J. HOFFMAN, *Modern Methods for Computer Security and Privacy*, Prentice Hall, Englewood Cliffs, New Jersey, 1977.

[559] R.V. HOGG AND E.A. TANIS, *Probability and statistical inference*, Macmillan Publishing Company, New York, 3rd edition, 1988.

[560] W. HOHL, X. LAI, T. MEIER, AND C. WALDVOGEL, "Security of iterated hash functions based on block ciphers", *Advances in Cryptology–CRYPTO '93 (LNCS 773)*, 379–390, 1994.

[561] S.-M. HONG, S.-Y. OH, AND H. YOON, "New modular multiplication algorithms for fast modular exponentiation", *Advances in Cryptology–EUROCRYPT '96 (LNCS 1070)*, 166–177, 1996.

[562] P. HORSTER AND H.-J. KNOBLOCH, "Discrete logarithm based protocols", *Advances in Cryptology–EUROCRYPT '91 (LNCS 547)*, 399–408, 1991.

[563] P. HORSTER, M. MICHELS, AND H. PETERSEN, "Meta-message recovery and meta-blind signature schemes based on the discrete logarithm problem and their applications", *Advances in Cryptology–ASIACRYPT '94 (LNCS 917)*, 224–237, 1995.

[564] P. HORSTER AND H. PETERSEN, "Generalized ElGamal signatures (in German)", *Sicherheit in Informationssystemen, Proceedings der Fachtagung SIS'94*, 89–106, Verlag der Fachvereine Zürich, 1994.

[565] T.W. HUNGERFORD, *Algebra*, Holt, Rinehart and Winston, New York, 1974.

[566] K. HWANG, *Computer Arithmetic, Principles, Architecture and Design*, John Wiley & Sons, New York, 1979.

[567] C. I'ANSON AND C. MITCHELL, "Security defects in CCITT Recommendation X.509 – The directory authentication framework", *Computer Communication Review*, 20 (1990), 30–34.

[568] R. IMPAGLIAZZO, L. LEVIN, AND M. LUBY, "Pseudo-random generation from one-way functions", *Proceedings of the 21st Annual ACM Symposium on Theory of Computing*, 12–24, 1989.

[569] R. IMPAGLIAZZO AND M. NAOR, "Efficient cryptographic schemes provably as secure as subset sum", *Proceedings of the IEEE 30th Annual Symposium on Foundations of Computer Science*, 236–241, 1989.

[570] I. INGEMARSSON AND G.J. SIMMONS, "A protocol to set up shared secret schemes without the assistance of a mutually trusted party", *Advances in Cryptology–EUROCRYPT '90 (LNCS 473)*, 266–282, 1991.

[571] I. INGEMARSSON, D.T. TANG, AND C.K. WONG, "A conference key distribution system", *IEEE Transactions on Information Theory*, 28 (1982), 714–720.

[572] K. IRELAND AND M. ROSEN, *A Classical Introduction to Modern Number Theory*, Springer-Verlag, New York, 2nd edition, 1990.

[573] ISO 7498-2, "Information processing systems – Open Systems Interconnection – Basic reference model – Part 2: Security architecture", International Organization for Standardization, Geneva, Switzerland, 1989 (first edition) (equivalent to ITU-T Rec. X.800).

[574] ISO 8372, "Information processing – Modes of operation for a 64-bit block cipher algorithm", International Organization for Standardization, Geneva, Switzerland, 1987 (first edition; confirmed 1992).

[575] ISO 8730, "Banking – Requirements for message authentication (wholesale)", International Organization for Standardization, Geneva, Switzerland, 1990 (second edition).

[576] ISO 8731-1, "Banking – Approved algorithms for message authentication – Part 1: DEA", International Organization for Standardization, Geneva, Switzerland, 1987 (first edition; confirmed 1992).

[577] ISO 8731-2, "Banking – Approved algorithms for message authentication – Part 2: Message authenticator algorithm", International Organization for Standardization, Geneva, Switzerland, 1992 (second edition).

[578] ISO 8732, "Banking – Key management (wholesale)", International Organization for Standardization, Geneva, Switzerland, 1988 (first edition).

[579] ISO 9564-1, "Banking – Personal Identification Number management and security – Part 1: PIN protection principles and techniques", International Organization for Standardization, Geneva, Switzerland, 1990.

[580] ISO 9564-2, "Banking – Personal Identification Number management and security – Part 2: Approved algorithm(s) for PIN encipherment", International Organization for Standardization, Geneva, Switzerland, 1991.

[581] ISO 9807, "Banking and related financial services – Requirements for message authentication (retail)", International Organization for Standardization, Geneva, Switzerland, 1991.

[582] ISO 10126-1, "Banking – Procedures for message encipherment (wholesale) – Part 1: General principles", International Organization for Standardization, Geneva, Switzerland, 1991.

[583] ISO 10126-2, "Banking – Procedures for message encipherment (wholesale) – Part 2: Algorithms", International Organization for Standardization, Geneva, Switzerland, 1991.

[584] ISO 10202-7, "Financial transaction cards – Security architecture of financial transaction systems using integrated circuit cards – Part 7: Key management", draft (DIS), 1994.

[585] ISO 11131, "Banking – Financial institution sign-on authentication", International Organization for Standardization, Geneva, Switzerland, 1992.

[586] ISO 11166-1, "Banking – Key management by means of asymmetric algorithms – Part 1: Principles, procedures and formats", International Organization for Standardization, Geneva, Switzerland, 1994.

[587] ISO 11166-2, "Banking – Key management by means of asymmetric algorithms – Part 2: Approved algorithms using the RSA cryptosystem", International Organization for Standardization, Geneva, Switzerland, 1995.

[588] ISO 11568-1, "Banking – Key management (retail) – Part 1: Introduction to key management", International Organization for Standardization, Geneva, Switzerland, 1994.

[589] ISO 11568-2, "Banking – Key management (retail) – Part 2: Key management techniques for symmetric ciphers", International Organization for Standardization, Geneva, Switzerland, 1994.

[590] ISO 11568-3, "Banking – Key management (retail) – Part 3: Key life cycle for symmetric ciphers", International Organization for Standardization, Geneva, Switzerland, 1994.

[591] ISO 11568-4, "Banking – Key management (retail) – Part 4: Key management techniques using public key cryptography", draft (DIS), 1996.

[592] ISO 11568-5, "Banking – Key management (retail) – Part 5: Key life cycle for public key cryptosystems", draft (DIS), 1996.

[593] ISO 11568-6, "Banking – Key management (retail) – Part 6: Key management schemes", draft (CD), 1996.

[594] ISO/IEC 9594-1, "Information technology – Open Systems Interconnection – The Directory: Overview of concepts, models, and services", International Organization for Standardization, Geneva, Switzerland, 1995 (equivalent to ITU-T Rec. X.500, 1993).

[595] ISO/IEC 9594-8, "Information technology – Open Systems Interconnection – The Directory: Authentication framework", International Organization for Standardization, Geneva, Switzerland, 1995 (equivalent to ITU-T Rec. X.509, 1993).

[596] ISO/IEC 9796, "Information technology – Security techniques – Digital signature scheme giving message recovery", International Organization for Standardization, Geneva, Switzerland, 1991 (first edition).

[597] ISO/IEC 9797, "Information technology – Security techniques – Data integrity mechanism using a cryptographic check function employing a block cipher algorithm", International Organization for Standardization, Geneva, Switzerland, 1994 (second edition).

[598] ISO/IEC 9798-1, "Information technology – Security techniques – Entity authentication mechanisms – Part 1: General model", International Organization for Standardization, Geneva, Switzerland, 1991 (first edition).

[599] ISO/IEC 9798-2, "Information technology – Security techniques – Entity authentication – Part 2: Mechanisms using symmetric encipherment algorithms", International Organization for Standardization, Geneva, Switzerland, 1994 (first edition).

[600] ISO/IEC 9798-3, "Information technology – Security techniques – Entity authentication mechanisms – Part 3: Entity authen-

tication using a public-key algorithm", International Organization for Standardization, Geneva, Switzerland, 1993 (first edition).

[601] ISO/IEC 9798-4, "Information technology – Security techniques – Entity authentication – Part 4: Mechanisms using a cryptographic check function", International Organization for Standardization, Geneva, Switzerland, 1995 (first edition).

[602] ISO/IEC 9798-5, "Information technology – Security techniques – Entity authentication – Part 5: Mechanisms using zero knowledge techniques", draft (CD), 1996.

[603] ISO/IEC 9979, "Data cryptographic techniques – Procedures for the registration of cryptographic algorithms", International Organization for Standardization, Geneva, Switzerland, 1991 (first edition).

[604] ISO/IEC 10116, "Information processing – Modes of operation for an n-bit block cipher algorithm", International Organization for Standardization, Geneva, Switzerland, 1991 (first edition).

[605] ISO/IEC 10118-1, "Information technology – Security techniques – Hash-functions – Part 1: General", International Organization for Standardization, Geneva, Switzerland, 1994.

[606] ISO/IEC 10118-2, "Information technology – Security techniques – Hash-functions – Part 2: Hash-functions using an n-bit block cipher algorithm", International Organization for Standardization, Geneva, Switzerland, 1994.

[607] ISO/IEC 10118-3, "Information technology – Security techniques – Hash-functions – Part 3: Dedicated hash-functions", draft (CD), 1996.

[608] ISO/IEC 10118-4, "Information technology – Security techniques – Hash-functions – Part 4: Hash-functions using modular arithmetic", draft (CD), 1996.

[609] ISO/IEC 10181-1, "Information technology – Open Systems Interconnection – Security frameworks for open systems – Part 1: Overview", International Organization for Standardization, Geneva, Switzerland, 1995 (equivalent to ITU-T Rec. X.810, 1995).

[610] ISO/IEC 10181-2, "Information technology – Open Systems Interconnection – Security frameworks for open systems – Part 2: Authentication framework", International Organization for Standardization, Geneva, Switzerland, 1995 (equivalent to ITU-T Rec. X.811, 1995).

[611] ISO/IEC 10181-3, "Information technology – Open Systems Interconnection – Security frameworks for open systems – Part 3: Access control framework", draft, 1995.

[612] ISO/IEC 10181-4, "Information technology – Open Systems Interconnection – Security frameworks for open systems – Part 4: Non-repudiation framework", draft, 1995.

[613] ISO/IEC 10181-5, "Information technology – Open Systems Interconnection – Security frameworks for open systems – Part 5: Integrity framework", draft, 1995.

[614] ISO/IEC 10181-6, "Information technology – Open Systems Interconnection – Security frameworks for open systems – Part 6: Confidentiality framework", draft, 1995.

[615] ISO/IEC 10181-7, "Information technology – Open Systems Interconnection – Security frameworks for open systems – Part 7: Security audit framework", draft, 1995.

[616] ISO/IEC 11770-1, "Information technology – Security techniques – Key management – Part 1: Framework", draft (DIS), 1996.

[617] ISO/IEC 11770-2, "Information technology – Security techniques – Key management – Part 2: Mechanisms using symmetric techniques", International Organization for Standardization, Geneva, Switzerland, 1996 (first edition).

[618] ISO/IEC 11770-3, "Information technology – Security techniques – Key management – Part 3: Mechanisms using asymmetric techniques", draft (DIS), 1996.

[619] ISO/IEC 13888-1, "Information technology – Security techniques – Non-repudiation – Part 1: General model", draft (CD), 1996.

[620] ISO/IEC 13888-2, "Information technology – Security techniques – Non-repudiation – Part 2: Using symmetric encipherment algorithms", draft (CD), 1996.

[621] ISO/IEC 13888-3, "Information technology – Security techniques – Non-repudiation – Part 3: Using asymmetric techniques", draft (CD), 1996.

[622] ISO/IEC 14888-1, "Information technology – Security techniques – Digital signatures with appendix – Part 1: General", draft (CD), 1996.

[623] ISO/IEC 14888-2, "Information technology – Security techniques – Digital signatures with appendix – Part 2: Identity-based mechanisms", draft (CD), 1996.

[624] ISO/IEC 14888-3, "Information technology – Security techniques – Digital signatures with appendix – Part 3: Cerificate-based mechanisms", draft (CD), 1996.

[625] M. ITO, A. SAITO, AND T. NISHIZEKI, "Secret sharing scheme realizing general access structure", *IEEE Global Telecommunications Conference*, 99–102, 1987.

[626] ITU-T REC. X.509 (REVISED), "The Directory – Authentication framework", International Telecommunication Union, Geneva, Switzerland, 1993 (equivalent to ISO/IEC 9594-8:1995).

[627] ITU-T REC. X.509 (1988 AND 1993) TECHNICAL CORRIGENDUM 2, "The Directory – Authentication framework", International Telecommunication Union, Geneva, Switzerland, July 1995 (equivalent to Technical Corrigendum 2 to ISO/IEC 9594-8:1990&1995).

[628] ITU-T REC. X.509 (1993) AMENDMENT 1: CERTIFICATE EXTENSIONS, "The Directory – Authentication framework", International Telecommunication Union, Geneva, Switzerland, July 1995 draft for JCT1 letter ballot (equivalent to Ammendment 1 to ISO/IEC 9594-8:1995).

[629] W.-A. JACKSON, K.M. MARTIN, AND C.M. O'KEEFE, "Multisecret threshold schemes", *Advances in Cryptology–CRYPTO '93 (LNCS 773)*, 126–135, 1994.

[630] G. JAESCHKE, "On strong pseudoprimes to several bases", *Mathematics of Computation*, 61 (1993), 915–926.

[631] C.J.A. JANSEN AND D.E. BOEKEE, "On the significance of the directed acyclic word graph in cryptology", *Advances in Cryptology–AUSCRYPT '90 (LNCS 453)*, 318–326, 1990.

[632] ——, "The shortest feedback shift register that can generate a given sequence", *Advances in Cryptology–CRYPTO '89 (LNCS 435)*, 90–99, 1990.

[633] T. JEBELEAN, "Comparing several gcd algorithms", *Proceedings of the 11th Symposium on Computer Arithmetic*, 180–185, IEEE Press, 1993.

[634] J. JEDWAB AND C. MITCHELL, "Minimum weight modified signed-digit representations and fast exponentiation", *Electronics Letters*, 25 (August 17, 1989), 1171–1172.

[635] N. JEFFERIES, C. MITCHELL, AND M. WALKER, "A proposed architecture for trusted third party services", E. Dawson and J. Golić, editors, *Cryptography: Policy and Algorithms, International Conference, Brisbane, Queensland, Australia, July 1995 (LNCS 1029)*, 98–104, 1996.

[636] H.N. JENDAL, Y.J.B. KUHN, AND J.L. MASSEY, "An information-theoretic treatment of homophonic substitution", *Advances in Cryptology–EUROCRYPT '89 (LNCS 434)*, 382–394, 1990.

[637] S.M. JENNINGS, "Multiplexed sequences: Some properties of the minimum polynomial", *Cryptography–Proceedings of the Workshop on Cryptography, Burg Feuerstein (LNCS 149)*, 189–206, 1983.

[638] T. JOHANSSON, G. KABATIANSKII, AND B. SMEETS, "On the relation between A-codes and codes correcting independent errors", *Advances in Cryptology–EUROCRYPT '93 (LNCS 765)*, 1–11, 1994.

[639] D.B. JOHNSON, A. LE, W. MARTIN, S. MATYAS, AND J. WILKINS, "Hybrid key distribution scheme giving key record recovery", *IBM Technical Disclosure Bulletin*, 37 (1994), 5–16.

[640] D.B. JOHNSON AND S.M. MATYAS, "Asymmetric encryption: Evolution and enhancements", *CryptoBytes*, 2 (Spring 1996), 1–6.

[641] D.S. JOHNSON, "The NP-completeness column: an ongoing guide", *Journal of Algorithms*, 9 (1988), 426–444.

[642] R.W. JONES, "Some techniques for handling encipherment keys", *ICL Technical Journal*, 3 (1982), 175–188.

[643] R.R. JUENEMAN, "Analysis of certain aspects of output feedback mode", *Advances in Cryptology–Proceedings of Crypto 82*, 99–127, 1983.

[644] ——, "A high speed manipulation detection code", *Advances in Cryptology–CRYPTO '86 (LNCS 263)*, 327–346, 1987.

[645] R.R. JUENEMAN, S.M. MATYAS, AND C.H. MEYER, "Message authentication with manipulation detection codes", *Proceedings of the 1983 IEEE Symposium on Security and Privacy*, 33–54, 1984.

[646] D. JUNGNICKEL, *Finite Fields: Structure and Arithmetics*, Bibliographisches Institut – Wissenschaftsverlag, Mannheim, 1993.

[647] M. JUST, E. KRANAKIS, D. KRIZANC, AND P. VAN OORSCHOT, "On key distribution via true broadcasting", *2nd ACM Conference on Computer and Communications Security*, 81–88, ACM Press, 1994.

[648] D. KAHN, *The Codebreakers*, Macmillan Publishing Company, New York, 1967.

[649] B.S. KALISKI JR., "A chosen message attack on Demytko's elliptic curve cryptosystem", *Journal of Cryptology*, to appear.

[650] ———, "A pseudo-random bit generator based on elliptic logarithms", *Advances in Cryptology–CRYPTO '86 (LNCS 263)*, 84–103, 1987.

[651] ———, *Elliptic curves and cryptography: a pseudorandom bit generator and other tools*, PhD thesis, MIT Department of Electrical Engineering and Computer Science, 1988.

[652] ———, "Anderson's RSA trapdoor can be broken", *Electronics Letters*, 29 (July 22, 1993), 1387–1388.

[653] ———, "The Montgomery inverse and its applications", *IEEE Transactions on Computers*, 44 (1995), 1064–1065.

[654] B.S. KALISKI JR., R.L. RIVEST, AND A.T. SHERMAN, "Is the Data Encryption Standard a group? (Results of cycling experiments on DES)", *Journal of Cryptology*, 1 (1988), 3–36.

[655] B.S. KALISKI JR. AND M. ROBSHAW, "The secure use of RSA", *CryptoBytes*, 1 (Autumn 1995), 7–13.

[656] B.S. KALISKI JR. AND Y.L. YIN, "On differential and linear cryptanalysis of the RC5 encryption algorithm", *Advances in Cryptology–CRYPTO '95 (LNCS 963)*, 171–184, 1995.

[657] E. KALTOFEN, "Analysis of Coppersmith's block Wiedemann algorithm for the parallel solution of sparse linear systems", *Mathematics of Computation*, 64 (1995), 777–806.

[658] E. KALTOFEN AND V. SHOUP, "Subquadratic-time factoring of polynomials over finite fields", *Proceedings of the 27th Annual ACM Symposium on Theory of Computing*, 398–406, 1995.

[659] J. KAM AND G. DAVIDA, "Structured design of substitution-permutation encryption networks", *IEEE Transactions on Computers*, 28 (1979), 747–753.

[660] N. KAPIDZIC AND A. DAVIDSON, "A certificate management system: structure, functions and protocols", *Proceedings of the Internet Society Symposium on Network and Distributed System Security*, 153–160, IEEE Computer Society Press, 1995.

[661] A. KARATSUBA AND YU. OFMAN, "Multiplication of multidigit numbers on automata", *Soviet Physics – Doklady*, 7 (1963), 595–596.

[662] E.D. KARNIN, J.W. GREENE, AND M.E. HELLMAN, "On secret sharing systems", *IEEE Transactions on Information Theory*, 29 (1983), 35–41.

[663] A. KEHNE, J. SCHÖWÄLDER, AND H. LANGENDÖRFER, "A nonce-based protocol for multiple authentications", *Operating Systems Review*, 26 (1992), 84–89.

[664] R. KEMMERER, C. MEADOWS, AND J. MILLEN, "Three systems for cryptographic protocol analysis", *Journal of Cryptology*, 7 (1994), 79–130.

[665] S. KENT, "Encryption-based protection protocols for interactive user-computer communication", MIT/LCS/TR-162 (M.Sc. thesis), MIT Laboratory for Computer Science, 1976.

[666] ———, "Internet privacy enhanced mail", *Communications of the ACM*, 36 (1993), 48–60.

[667] ———, "Internet security standards: past, present and future", *StandardView*, 2 (1994), 78–85.

[668] A. KERCKHOFFS, "La cryptographie militaire", *Journal des Sciences Militaires*, 9th Series (February 1883), 161–191.

[669] I. KESSLER AND H. KRAWCZYK, "Minimum buffer length and clock rate for the shrinking generator cryptosystem", IBM Research Report RC 19938, IBM T.J. Watson Research Center, Yorktown Heights, N.Y., 10598, U.S.A., 1995.

[670] E. KEY, "An analysis of the structure and complexity of nonlinear binary sequence generators", *IEEE Transactions on Information Theory*, 22 (1976), 732–736.

[671] J. KILIAN AND T. LEIGHTON, "Fair cryptosystems, revisited: A rigorous approach to key-escrow", *Advances in Cryptology–CRYPTO '95 (LNCS 963)*, 208–221, 1995.

[672] J. KILIAN AND P. ROGAWAY, "How to protect DES against exhaustive key search", *Advances in Cryptology–CRYPTO '96 (LNCS 1109)*, 252–267, 1996.

[673] S.-H. KIM AND C. POMERANCE, "The probability that a random probable prime is composite", *Mathematics of Computation*, 53 (1989), 721–741.

[674] M. KIMBERLEY, "Comparison of two statistical tests for keystream sequences", *Electronics Letters*, 23 (April 9, 1987), 365–366.

[675] A. KLAPPER, "The vulnerability of geometric sequences based on fields of odd characteristic", *Journal of Cryptology*, 7 (1994), 33–51.

[676] A. KLAPPER AND M. GORESKY, "Feedback shift registers, combiners with memory, and 2-adic span", *Journal of Cryptology*, to appear.

[677] ———, "2-Adic shift registers", R. Anderson, editor, *Fast Software Encryption, Cambridge Security Workshop (LNCS 809)*, 174–178, Springer-Verlag, 1994.

[678] ———, "Cryptanalysis based on 2-adic rational approximation", *Advances in Cryptology–CRYPTO '95 (LNCS 963)*, 262–273, 1995.

[679] ———, "Large period nearly de Bruijn FCSR sequences", *Advances in Cryptology–EUROCRYPT '95 (LNCS 921)*, 263–273, 1995.

[680] D.V. KLEIN, "Foiling the cracker: a survey of, and improvements to, password security", *Proceedings of the 2nd USENIX UNIX Security Workshop*, 5–14, 1990.

[681] H.-J. KNOBLOCH, "A smart card implementation of the Fiat-Shamir identification scheme", *Advances in Cryptology–EUROCRYPT '88 (LNCS 330)*, 87–95, 1988.

[682] L.R. KNUDSEN, "Cryptanalysis of LOKI", *Advances in Cryptology–ASIACRYPT '91 (LNCS 739)*, 22–35, 1993.

[683] ———, "Cryptanalysis of LOKI91", *Advances in Cryptology–AUSCRYPT '92 (LNCS 718)*, 196–208, 1993.

[684] ———, *Block Ciphers – Analysis, Design and Applications*, PhD thesis, Computer Science Department, Aarhus University (Denmark), 1994.

[685] ———, "A key-schedule weakness in SAFER K-64", *Advances in Cryptology–CRYPTO '95 (LNCS 963)*, 274–286, 1995.

[686] ———, "Truncated and higher order differentials", B. Preneel, editor, *Fast Software Encryption, Second International Workshop (LNCS 1008)*, 196–211, Springer-Verlag, 1995.

[687] L.R. KNUDSEN AND T. BERSON, "Truncated differentials of SAFER", D. Gollmann, editor, *Fast Software Encryption, Third International Workshop (LNCS 1039)*, 15–26, Springer-Verlag, 1996.

[688] L.R. KNUDSEN AND X. LAI, "New attacks on all double block length hash functions of hash rate 1, including the parallel-DM", *Advances in Cryptology–EUROCRYPT '94 (LNCS 950)*, 410–418, 1995.

[689] L.R. KNUDSEN AND W. MEIER, "Improved differential attacks on RC5", *Advances in Cryptology–CRYPTO '96 (LNCS 1109)*, 216–228, 1996.

[690] L.R. KNUDSEN AND T. PEDERSEN, "On the difficulty of software key escrow", *Advances in Cryptology–EUROCRYPT '96 (LNCS 1070)*, 237–244, 1996.

[691] D.E. KNUTH, *The Art of Computer Programming – Fundamental Algorithms*, volume 1, Addison-Wesley, Reading, Massachusetts, 2nd edition, 1973.

[692] ———, *The Art of Computer Programming – Seminumerical Algorithms*, volume 2, Addison-Wesley, Reading, Massachusetts, 2nd edition, 1981.

[693] ———, *The Art of Computer Programming – Sorting and Searching*, volume 3, Addison-Wesley, Reading, Massachusetts, 1973.

[694] D.E. KNUTH AND L. TRABB PARDO, "Analysis of a simple factorization algorithm", *Theoretical Computer Science*, 3 (1976), 321–348.

[695] N. KOBLITZ, "Elliptic curve cryptosystems", *Mathematics of Computation*, 48 (1987), 203–209.

[696] ———, "Hyperelliptic cryptosystems", *Journal of Cryptology*, 1 (1989), 139–150.

[697] ———, *A Course in Number Theory and Cryptography*, Springer-Verlag, New York, 2nd edition, 1994.

[698] C. KOÇ, "High-speed RSA implementation", Technical Report, RSA Laboratories, 1994.

[699] ———, "RSA hardware implementation", Technical Report TR-801, RSA Laboratories, 1996.

[700] C. KOÇ, T. ACAR, AND B.S. KALISKI JR., "Analyzing and comparing Montgomery multiplication algorithms", *IEEE Micro*, 16 (1996), 26–33.

[701] J.T. KOHL, "The use of encryption in Kerberos for network authentication", *Advances in Cryptology–CRYPTO '89 (LNCS 435)*, 35–43, 1990.

[702] L.M. KOHNFELDER, "A method for certification", MIT Laboratory for Computer Science, unpublished (essentially pp.39-43 of [703]), 1978.

[703] ———, "Toward a practical public-key cryptosystem", B.Sc. thesis, MIT Department of Electrical Engineering, 1978.

[704] A. KOLMOGOROV, "Three approaches to the definition of the concept 'quantity of information'", *Problemy Peredachi Informatsii*, 1 (1965), 3–11.

[705] A.G. KONHEIM, *Cryptography, A Primer*, John Wiley & Sons, New York, 1981.

[706] I. KOREN, *Computer Arithmetic Algorithms*, Prentice Hall, Englewood Cliffs, New Jersey, 1993.

[707] V.I. KORZHIK AND A.I. TURKIN, "Cryptanalysis of McEliece's public-key cryptosystem", *Advances in Cryptology–EUROCRYPT '91 (LNCS 547)*, 68–70, 1991.

[708] K. KOYAMA, U. MAURER, T. OKAMOTO, AND S.A. VANSTONE, "New public-key schemes based on elliptic curves over the ring Z_n", *Advances in Cryptology–CRYPTO '91 (LNCS 576)*, 252–266, 1992.

[709] K. KOYAMA AND R. TERADA, "How to strengthen DES-like cryptosystems against differential cryptanalysis", *IEICE Transactions on Fundamentals of Electronics, Communications and Computer Science*, E76-A (1993), 63–69.

[710] E. KRANAKIS, *Primality and Cryptography*, John Wiley & Sons, Stuttgart, 1986.

[711] D.W. KRAVITZ, "Digital signature algorithm", U.S. Patent # 5,231,668, 27 Jul 1993.

[712] H. KRAWCZYK, "How to predict congruential generators", *Advances in Cryptology–CRYPTO '89 (LNCS 435)*, 138–153, 1990.

[713] ———, "How to predict congruential generators", *Journal of Algorithms*, 13 (1992), 527–545. An earlier version appeared in [712].

[714] ———, "LFSR-based hashing and authentication", *Advances in Cryptology–CRYPTO '94 (LNCS 839)*, 129–139, 1994.

[715] ———, "Secret sharing made short", *Advances in Cryptology–CRYPTO '93 (LNCS 773)*, 136–146, 1994.

[716] ———, "The shrinking generator: Some practical considerations", R. Anderson, editor, *Fast Software Encryption, Cambridge Security Workshop (LNCS 809)*, 45–46, Springer-Verlag, 1994.

[717] ———, "New hash functions for message authentication", *Advances in Cryptology–EUROCRYPT '95 (LNCS 921)*, 301–310, 1995.

[718] ———, "SKEME: A versatile secure key exchange mechanism for Internet", *Proceedings of the Internet Society Symposium on Network and Distributed System Security*, 114–127, IEEE Computer Society Press, 1996.

[719] Y. KURITA AND M. MATSUMOTO, "Primitive t-nomials ($t = 3,5$) over GF(2) whose degree is a Mersenne exponent ≤ 44497", *Mathematics of Computation*, 56 (1991), 817–821.

[720] K. KUROSAWA, T. ITO, AND M. TAKEUCHI, "Public key cryptosystem using a reciprocal number with the same intractability as factoring a large number", *Cryptologia*, 12 (1988), 225–233.

[721] K. KUROSAWA, K. OKADA, AND S. TSUJII, "Low exponent attack against elliptic curve RSA", *Advances in Cryptology–ASIACRYPT '94 (LNCS 917)*, 376–383, 1995.

[722] K. KUSUDA AND T. MATSUMOTO, "Optimization of time-memory trade-off cryptanalysis and its application to DES, FEAL-32, and Skipjack", *IEICE Transactions on Fundamentals of Electronics, Communications and Computer Science*, E79-A (1996), 35–48.

[723] J.C. LAGARIAS, "Knapsack public key cryptosystems and diophantine approximation", *Advances in Cryptology–Proceedings of Crypto 83*, 3–23, 1984.

[724] ———, "Pseudorandom number generators in cryptography and number theory", C. Pomerance, editor, *Cryptology and Computational Number Theory*, volume 42 of *Proceedings of Symposia in Applied Mathematics*, 115–143, American Mathematical Society, 1990.

[725] X. LAI, "Condition for the nonsingularity of a feedback shift-register over a general finite field", *IEEE Transactions on Information Theory*, 33 (1987), 747–749.

[726] ——, "On the design and security of block ciphers", ETH Series in Information Processing, J.L. Massey (editor), vol. 1, Hartung-Gorre Verlag Konstanz, Technische Hochschule (Zurich), 1992.

[727] X. LAI AND L.R. KNUDSEN, "Attacks on double block length hash functions", R. Anderson, editor, *Fast Software Encryption, Cambridge Security Workshop (LNCS 809)*, 157–165, Springer-Verlag, 1994.

[728] X. LAI AND J.L. MASSEY, "A proposal for a new block encryption standard", *Advances in Cryptology–EUROCRYPT '90 (LNCS 473)*, 389–404, 1991.

[729] ——, "Hash functions based on block ciphers", *Advances in Cryptology–EUROCRYPT '92 (LNCS 658)*, 55–70, 1993.

[730] X. LAI, J.L. MASSEY, AND S. MURPHY, "Markov ciphers and differential cryptanalysis", *Advances in Cryptology–EUROCRYPT '91 (LNCS 547)*, 17–38, 1991.

[731] X. LAI, R.A. RUEPPEL, AND J. WOOLLVEN, "A fast cryptographic checksum algorithm based on stream ciphers", *Advances in Cryptology–AUSCRYPT '92 (LNCS 718)*, 339–348, 1993.

[732] C.-S. LAIH, L. HARN, J.-Y. LEE, AND T. HWANG, "Dynamic threshold scheme based on the definition of cross-product in an n-dimensional linear space", *Advances in Cryptology–CRYPTO '89 (LNCS 435)*, 286–298, 1990.

[733] C.-S. LAIH, F.-K. TU, AND W.-C TAI, "On the security of the Lucas function", *Information Processing Letters*, 53 (1995), 243–247.

[734] K.-Y. LAM AND T. BETH, "Timely authentication in distributed systems", Y. Deswarte, G. Eizenberg, and J.-J. Quisquater, editors, *Second European Symposium on Research in Computer Security – ESORICS'92 (LNCS 648)*, 293–303, Springer-Verlag, 1992.

[735] K.-Y. LAM AND L.C.K. HUI, "Efficiency of $SS(I)$ square-and-multiply exponentiation algorithms", *Electronics Letters*, 30 (December 8, 1994), 2115–2116.

[736] B.A. LAMACCHIA AND A.M. ODLYZKO, "Computation of discrete logarithms in prime fields", *Designs, Codes and Cryptography*, 1 (1991), 47–62.

[737] ——, "Solving large sparse linear systems over finite fields", *Advances in Cryptology–CRYPTO '90 (LNCS 537)*, 109–133, 1991.

[738] L. LAMPORT, "Constructing digital signatures from a one-way function", Technical report CSL-98, SRI International, Palo Alto, 1979.

[739] ——, "Password authentication with insecure communication", *Communications of the ACM*, 24 (1981), 770–772.

[740] B. LAMPSON, M. ABADI, M. BURROWS, AND E. WOBBER, "Authentication in distributed systems: Theory and practice", *ACM Transactions on Computer Systems*, 10 (1992), 265–310.

[741] S.K. LANGFORD AND M.E. HELLMAN, "Differential-linear cryptanalysis", *Advances in Cryptology–CRYPTO '94 (LNCS 839)*, 17–25, 1994.

[742] P.J. LEE AND E.F. BRICKELL, "An observation on the security of McEliece's public-key cryptosystem", *Advances in Cryptology–EUROCRYPT '88 (LNCS 330)*, 275–280, 1988.

[743] D.H. LEHMER, "Euclid's algorithm for large numbers", *American Mathematical Monthly*, 45 (1938), 227–233.

[744] D.H. LEHMER AND R.E. POWERS, "On factoring large numbers", *Bulletin of the AMS*, 37 (1931), 770–776.

[745] T. LEIGHTON AND S. MICALI, "Secret-key agreement without public-key cryptography", *Advances in Cryptology–CRYPTO '93 (LNCS 773)*, 456–479, 1994.

[746] A.K. LENSTRA, "Posting to sci.crypt", April 11 1996.

[747] ——, "Primality testing", C. Pomerance, editor, *Cryptology and Computational Number Theory*, volume 42 of *Proceedings of Symposia in Applied Mathematics*, 13–25, American Mathematical Society, 1990.

[748] A.K. LENSTRA AND H.W. LENSTRA JR., "Algorithms in number theory", J. van Leeuwen, editor, *Handbook of Theoretical Computer Science*, 674–715, Elsevier Science Publishers, 1990.

[749] ——, *The Development of the Number Field Sieve*, Springer-Verlag, Berlin, 1993.

[750] A.K. LENSTRA, H.W. LENSTRA JR., AND L. LOVÁSZ, "Factoring polynomials with rational coefficients", *Mathematische Annalen*, 261 (1982), 515–534.

[751] A.K. LENSTRA, H.W. LENSTRA JR., M.S. MANASSE, AND J.M. POLLARD, "The factorization of the ninth Fermat number", *Mathematics of Computation*, 61 (1993), 319–349.

[752] ——, "The number field sieve", A.K. Lenstra and H.W. Lenstra Jr., editors, *The Development of the Number Field Sieve*, volume 1554 of *Lecture Notes in Mathematics*, 11–42, Springer-Verlag, 1993.

[753] A.K. LENSTRA AND M.S. MANASSE, "Factoring by electronic mail", *Advances in Cryptology–EUROCRYPT '89 (LNCS 434)*, 355–371, 1990.

[754] ——, "Factoring with two large primes", *Mathematics of Computation*, 63 (1994), 785–798.

[755] A.K. LENSTRA, P. WINKLER, AND Y. YACOBI, "A key escrow system with warrant bounds", *Advances in Cryptology–CRYPTO '95 (LNCS 963)*, 197–207, 1995.

[756] H.W. LENSTRA JR., "Factoring integers with elliptic curves", *Annals of Mathematics*, 126 (1987), 649–673.

[757] ——, "Finding isomorphisms between finite fields", *Mathematics of Computation*, 56 (1991), 329–347.

[758] ——, "On the Chor-Rivest knapsack cryptosystem", *Journal of Cryptology*, 3 (1991), 149–155.

[759] H.W. LENSTRA JR. AND C. POMERANCE, "A rigorous time bound for factoring integers", *Journal of the American Mathematical Society*, 5 (1992), 483–516.

[760] H.W. LENSTRA JR. AND R.J. SCHOOF, "Primitive normal bases for finite fields", *Mathematics of Computation*, 48 (1987), 217–231.

[761] L.A. LEVIN, "One-way functions and pseudorandom generators", *Proceedings of the 17th Annual ACM Symposium on Theory of Computing*, 363–365, 1985.

[762] J. LEVINE, *United States Cryptographic Patents 1861–1981*, Cryptologia, Inc., Terre Haute, Indiana, 1983.

[763] R. LIDL AND W.B. MÜLLER, "Permutation polynomials in RSA-cryptosystems", *Advances in Cryptology–Proceedings of Crypto 83*, 293–301, 1984.

[764] R. LIDL AND H. NIEDERREITER, *Finite Fields*, Cambridge University Press, Cambridge, 1984.

[765] A. LIEBL, "Authentication in distributed systems: A bibliography", *Operating Systems Review*, 27 (1993), 31–41.

[766] C.H. LIM AND P.J. LEE, "Another method for attaining security against adaptively chosen ciphertext attacks", *Advances in Cryptology–CRYPTO '93 (LNCS 773)*, 420–434, 1994.

[767] ——, "More flexible exponentiation with precomputation", *Advances in Cryptology–CRYPTO '94 (LNCS 839)*, 95–107, 1994.

[768] ——, "Server (prover/signer)-aided verification of identity proofs and signatures", *Advances in Cryptology–EUROCRYPT '95 (LNCS 921)*, 64–78, 1995.

[769] S. LIN AND D. COSTELLO, *Error Control Coding: Fundamentals and Applications*, Prentice Hall, Englewood Cliffs, New Jersey, 1983.

[770] J. LIPSON, *Elements of Algebra and Algebraic Computing*, Addison-Wesley, Reading, Massachusetts, 1981.

[771] T.M.A. LOMAS, L. GONG, J.H. SALTZER, AND R.M. NEEDHAM, "Reducing risks from poorly chosen keys", *Operating Systems Review*, 23 (Special issue), 14–18. (Presented at: 12th ACM Symposium on Operating Systems Principles, Litchfield Park, Arizona, Dec. 1989).

[772] D.L. LONG AND A. WIGDERSON, "The discrete logarithm hides $O(\log n)$ bits", *SIAM Journal on Computing*, 17 (1988), 363–372.

[773] R. LOVORN, *Rigorous, subexponential algorithms for discrete logarithms over finite fields*, PhD thesis, University of Georgia, 1992.

[774] M. LUBY, *Pseudorandomness and Cryptographic Applications*, Princeton University Press, Princeton, New Jersey, 1996.

[775] M. LUBY AND C. RACKOFF, "Pseudorandom permutation generators and cryptographic composition", *Proceedings of the 18th Annual ACM Symposium on Theory of Computing*, 356–363, 1986.

[776] ——, "How to construct pseudorandom permutations from pseudorandom functions", *SIAM Journal on Computing*, 17 (1988), 373–386. An earlier version appeared in [775].

[777] S. LUCKS, "Faster Luby-Rackoff ciphers", D. Gollmann, editor, *Fast Software Encryption, Third International Workshop (LNCS 1039)*, 189–203, Springer-Verlag, 1996.

[778] F.J. MACWILLIAMS AND N.J.A. SLOANE, *The Theory of Error-Correcting Codes*, North-Holland, Amsterdam, 1977 (fifth printing: 1986).

[779] W. MADRYGA, "A high performance encryption algorithm", J. Finch and E. Dougall, editors, *Computer Security: A Global Challenge, Proceedings of the Second IFIP International Conference on Computer Security*, 557–570, North-Holland, 1984.

[780] D.P. MAHER, "Crypto backup and key escrow", *Communications of the ACM*, 39 (1996), 48–53.

[781] W. MAO AND C. BOYD, "On the use of encryption in cryptographic protocols", P.G. Farrell, editor, *Codes and Cyphers: Cryptography and Coding IV*, 251–262, Institute of Mathematics & Its Applications (IMA), 1995.

[782] G. MARSAGLIA, "A current view of random number generation", L. Billard, editor, *Computer Science and Statistics: Proceedings of the Sixteenth Symposium on the Interface*, 3–10, North-Holland, 1985.

[783] P. MARTIN-LÖF, "The definition of random sequences", *Information and Control*, 9 (1966), 602–619.

[784] J.L. MASSEY, "Shift-register synthesis and BCH decoding", *IEEE Transactions on Information Theory*, 15 (1969), 122–127.

[785] ——, "An introduction to contemporary cryptology", *Proceedings of the IEEE*, 76 (1988), 533–549.

[786] ——, "Contemporary cryptology: An introduction", G.J. Simmons, editor, *Contemporary Cryptology: The Science of Information Integrity*, 1–39, IEEE Press, 1992. An earlier version appeared in [785].

[787] ——, "SAFER K-64: A byte-oriented block-ciphering algorithm", R. Anderson, editor, *Fast Software Encryption, Cambridge Security Workshop (LNCS 809)*, 1–17, Springer-Verlag, 1994.

[788] ——, "SAFER K-64: One year later", B. Preneel, editor, *Fast Software Encryption, Second International Workshop (LNCS 1008)*, 212–241, Springer-Verlag, 1995.

[789] J.L. MASSEY AND I. INGEMARSSON, "The Rip Van Winkle cipher – A simple and provably computationally secure cipher with a finite key", *IEEE International Symposium on Information Theory (Abstracts)*, p.146, 1985.

[790] J.L. MASSEY AND X. LAI, "Device for converting a digital block and the use thereof", European Patent # 482,154, 29 Apr 1992.

[791] ——, "Device for the conversion of a digital block and use of same", U.S. Patent # 5,214,703, 25 May 1993.

[792] J.L. MASSEY AND J.K. OMURA, "Method and apparatus for maintaining the privacy of digital messages conveyed by public transmission", U.S. Patent # 4,567,600, 28 Jan 1986.

[793] J.L. MASSEY AND R.A. RUEPPEL, "Linear ciphers and random sequence generators with multiple clocks", *Advances in Cryptology–Proceedings of EUROCRYPT 84 (LNCS 209)*, 74–87, 1985.

[794] J.L. MASSEY AND S. SERCONEK, "A Fourier transform approach to the linear complexity of nonlinearly filtered sequences", *Advances in Cryptology–CRYPTO '94 (LNCS 839)*, 332–340, 1994.

[795] M. MATSUI, "The first experimental cryptanalysis of the Data Encryption Standard", *Advances in Cryptology–CRYPTO '94 (LNCS 839)*, 1–11, 1994.

[796] ——, "Linear cryptanalysis method for DES cipher", *Advances in Cryptology–EUROCRYPT '93 (LNCS 765)*, 386–397, 1994.

[797] ——, "On correlation between the order of S-boxes and the strength of DES", *Advances in Cryptology–EUROCRYPT '94 (LNCS 950)*, 366–375, 1995.

[798] M. MATSUI AND A. YAMAGISHI, "A new method for known plaintext attack of FEAL cipher", *Advances in Cryptology–EUROCRYPT '92 (LNCS 658)*, 81–91, 1993.

[799] T. MATSUMOTO AND H. IMAI, "On the key predistribution system: A practical solution to the key distribution problem", *Advances in Cryptology–CRYPTO '87 (LNCS 293)*, 185–193, 1988.

[800] T. MATSUMOTO, Y. TAKASHIMA, AND H. IMAI, "On seeking smart public-key-distribution systems", *The Transactions of the IECE of Japan*, E69 (1986), 99–106.

[801] S.M. MATYAS, "Digital signatures – an overview", *Computer Networks*, 3 (1979), 87–94.

[802] ——, "Key handling with control vectors", *IBM Systems Journal*, 30 (1991), 151–174.

[803] ——, "Key processing with control vectors", *Journal of Cryptology*, 3 (1991), 113–136.

[804] S.M. MATYAS AND C.H. MEYER, "Generation, distribution, and installation of cryptographic keys", *IBM Systems Journal*, 17 (1978), 126–137.

[805] S.M. MATYAS, C.H. MEYER, AND J. OSEAS, "Generating strong one-way functions with cryptographic algorithm", *IBM Technical Disclosure Bulletin*, 27 (1985), 5658–5659.

[806] S.M. MATYAS, C.H.W. MEYER, AND B.O. BRACHTL, "Controlled use of cryptographic keys via generating station established control values", U.S. Patent # 4,850,017, 18 Jul 1989.

[807] U. MAURER, "Fast generation of secure RSA-moduli with almost maximal diversity", *Advances in Cryptology–EUROCRYPT '89 (LNCS 434)*, 636–647, 1990.

[808] ——, "New approaches to the design of self-synchronizing stream ciphers", *Advances in Cryptology–EUROCRYPT '91 (LNCS 547)*, 458–471, 1991.

[809] ——, "A provably-secure strongly-randomized cipher", *Advances in Cryptology–EUROCRYPT '90 (LNCS 473)*, 361–373, 1991.

[810] ——, "A universal statistical test for random bit generators", *Advances in Cryptology–CRYPTO '90 (LNCS 537)*, 409–420, 1991.

[811] ——, "Conditionally-perfect secrecy and a provably-secure randomized cipher", *Journal of Cryptology*, 5 (1992), 53–66. An earlier version appeared in [809].

[812] ——, "Some number-theoretic conjectures and their relation to the generation of cryptographic primes", C. Mitchell, editor, *Cryptography and Coding II*, volume 33 of *Institute of Mathematics & Its Applications (IMA)*, 173–191, Clarendon Press, 1992.

[813] ——, "A universal statistical test for random bit generators", *Journal of Cryptology*, 5 (1992), 89–105. An earlier version appeared in [810].

[814] ——, "Factoring with an oracle", *Advances in Cryptology–EUROCRYPT '92 (LNCS 658)*, 429–436, 1993.

[815] ——, "Secret key agreement by public discussion from common information", *IEEE Transactions on Information Theory*, 39 (1993), 733–742.

[816] ——, "A simplified and generalized treatment of Luby-Rackoff pseudorandom permutation generators", *Advances in Cryptology–EUROCRYPT '92 (LNCS 658)*, 239–255, 1993.

[817] ——, "Towards the equivalence of breaking the Diffie-Hellman protocol and computing discrete logarithms", *Advances in Cryptology–CRYPTO '94 (LNCS 839)*, 271–281, 1994.

[818] ——, "Fast generation of prime numbers and secure public-key cryptographic parameters", *Journal of Cryptology*, 8 (1995), 123–155. An earlier version appeared in [807].

[819] ——, "The role of information theory in cryptography", P.G. Farrell, editor, *Codes and Cyphers: Cryptography and Coding IV*, 49–71, Institute of Mathematics & Its Applications (IMA), 1995.

[820] U. MAURER AND J.L. MASSEY, "Perfect local randomness in pseudo-random sequences", *Advances in Cryptology–CRYPTO '89 (LNCS 435)*, 100–112, 1990.

[821] ——, "Local randomness in pseudorandom sequences", *Journal of Cryptology*, 4 (1991), 135–149. An earlier version appeared in [820].

[822] ——, "Cascade ciphers: The importance of being first", *Journal of Cryptology*, 6 (1993), 55–61.

[823] U. MAURER AND Y. YACOBI, "Non-interactive public-key cryptography", *Advances in Cryptology–EUROCRYPT '91 (LNCS 547)*, 498–507, 1991.

[824] ——, "A remark on a non-interactive public-key distribution system", *Advances in Cryptology–EUROCRYPT '92 (LNCS 658)*, 458–460, 1993.

[825] K.S. MCCURLEY, "A key distribution system equivalent to factoring", *Journal of Cryptology*, 1 (1988), 95–105.

[826] ——, "Cryptographic key distribution and computation in class groups", R.A. Mollin, editor, *Number Theory and Applications*, 459–479, Kluwer Academic Publishers, 1989.

[827] ——, "The discrete logarithm problem", C. Pomerance, editor, *Cryptology and Computational Number Theory*, volume 42 of *Proceedings of Symposia in Applied Mathematics*, 49–74, American Mathematical Society, 1990.

[828] R.J. MCELIECE, "A public-key cryptosystem based on algebraic coding theory", DSN progress report 42-44, Jet Propulsion Laboratory, Pasadena, 1978.

[829] ——, *The Theory of Information and Coding: A Mathematical Framework for Communication*, Cambridge University Press, Cambridge, 1984.

[830] ——, *Finite Fields for Computer Scientists and Engineeers*, Kluwer Academic Publishers, Boston, 1987.

[831] C.A. MEADOWS, "Formal verification of cryptographic protocols: a survey", *Advances in Cryptology–ASIACRYPT '94 (LNCS 917)*, 133–150, 1995.

[832] W. MEIER, "On the security of the IDEA block cipher", *Advances in Cryptology–EUROCRYPT '93 (LNCS 765)*, 371–385, 1994.

[833] W. MEIER AND O. STAFFELBACH, "Fast correlation attacks on stream ciphers", *Advances in Cryptology–EUROCRYPT '88 (LNCS 330)*, 301–314, 1988.

[834] ——, "Fast correlation attacks on certain stream ciphers", *Journal of Cryptology*, 1 (1989), 159–176. An earlier version appeared in [833].

[835] ——, "Analysis of pseudo random sequences generated by cellular automata", *Advances in Cryptology–EUROCRYPT '91 (LNCS 547)*, 186–199, 1991.

[836] ——, "Correlation properties of combiners with memory in stream ciphers", *Advances in Cryptology–EUROCRYPT '90 (LNCS 473)*, 204–213, 1991.

[837] ——, "Correlation properties of combiners with memory in stream ciphers", *Journal of Cryptology*, 5 (1992), 67–86. An earlier version appeared in [836].

[838] ——, "The self-shrinking generator", *Advances in Cryptology–EUROCRYPT '94 (LNCS 950)*, 205–214, 1995.

[839] S. MENDES AND C. HUITEMA, "A new approach to the X.509 framework: allowing a global authentication infrastructure without a global trust model", *Proceedings of the Internet Society Symposium on Network and Distributed System Security*, 172–189, IEEE Computer Society Press, 1995.

[840] A. MENEZES, *Elliptic Curve Public Key Cryptosystems*, Kluwer Academic Publishers, Boston, 1993.

[841] A. MENEZES, I. BLAKE, X. GAO, R. MULLIN, S. VANSTONE, AND T. YAGHOOBIAN, *Applications of Finite Fields*, Kluwer Academic Publishers, Boston, 1993.

[842] A. MENEZES, T. OKAMOTO, AND S. VANSTONE, "Reducing elliptic curve logarithms to logarithms in a finite field", *Proceedings of the 23rd Annual ACM Symposium on Theory of Computing*, 80–89, 1991.

[843] ——, "Reducing elliptic curve logarithms to logarithms in a finite field", *IEEE Transactions on Information Theory*, 39 (1993), 1639–1646. An earlier version appeared in [842].

[844] A. MENEZES, M. QU, AND S. VANSTONE, "Some new key agreement protocols providing implicit authentication", workshop record, 2nd Workshop on Selected Areas in Cryptography (SAC'95), Ottawa, Canada, May 18–19 1995.

[845] R. MENICOCCI, "Cryptanalysis of a two-stage Gollmann cascade generator", W. Wolfowicz, editor, *Proceedings of the 3rd Symposium on State and Progress of Research in Cryptography, Rome, Italy*, 62–69, 1993.

[846] R.C. MERKLE, "Digital signature system and method based on a conventional encryption function", U.S. Patent # 4,881,264, 14 Nov 1989.

[847] ——, "Method and apparatus for data encryption", U.S. Patent # 5,003,597, 26 Mar 1991.

[848] ——, "Method of providing digital signatures", U.S. Patent # 4,309,569, 5 Jan 1982.

[849] ——, "Secure communications over insecure channels", *Communications of the ACM*, 21 (1978), 294–299.

[850] ——, *Secrecy, Authentication, and Public Key Systems*, UMI Research Press, Ann Arbor, Michigan, 1979.

[851] ——, "Secrecy, authentication, and public key systems", Technical Report No.1979-1, Information Systems Laboratory, Stanford University, Palo Alto, California, 1979. Also available as [850].

[852] ——, "Protocols for public key cryptosystems", *Proceedings of the 1980 IEEE Symposium on Security and Privacy*, 122–134, 1980.

[853] ——, "A certified digital signature", *Advances in Cryptology–CRYPTO '89 (LNCS 435)*, 218–238, 1990.

[854] ——, "A fast software one-way hash function", *Journal of Cryptology*, 3 (1990), 43–58.

[855] ——, "One way hash functions and DES", *Advances in Cryptology–CRYPTO '89 (LNCS 435)*, 428–446, 1990.

[856] ——, "Fast software encryption functions", *Advances in Cryptology–CRYPTO '90 (LNCS 537)*, 476–501, 1991.

[857] R.C. MERKLE AND M.E. HELLMAN, "Hiding information and signatures in trapdoor knapsacks", *IEEE Transactions on Information Theory*, 24 (1978), 525–530.

[858] ——, "On the security of multiple encryption", *Communications of the ACM*, 24 (1981), 465–467.

[859] C.H. MEYER AND S.M. MATYAS, *Cryptography: A New Dimension in Computer Data Security*, John Wiley & Sons, New York, 1982 (third printing).

[860] C.H. MEYER AND M. SCHILLING, "Secure program load with manipulation detection code", *Proceedings of the 6th Worldwide Congress on Computer and Communications Security and Protection (SECURICOM'88)*, 111–130, 1988.

[861] S. MICALI, "Fair cryptosystems and methods of use", U.S. Patent # 5,276,737, 4 Jan 1994.

[862] ——, "Fair cryptosystems and methods of use", U.S. Patent # 5,315,658, 24 May 1994 (continuation-in-part of 5,276,737).

[863] ——, "Fair public-key cryptosystems", *Advances in Cryptology–CRYPTO '92 (LNCS 740)*, 113–138, 1993.

[864] S. MICALI, O. GOLDREICH, AND S. EVEN, "On-line/off-line digital signing", U.S. Patent # 5,016,274, 14 May 1991.

[865] S. MICALI, C. RACKOFF, AND B. SLOAN, "The notion of security for probabilistic cryptosystems", *SIAM Journal on Computing*, 17 (1988), 412–426.

[866] S. MICALI AND C.P. SCHNORR, "Efficient, perfect random number generators", *Advances in Cryptology–CRYPTO '88 (LNCS 403)*, 173–198, 1990.

[867] ——, "Efficient, perfect polynomial random number generators", *Journal of Cryptology*, 3 (1991), 157–172. An earlier version appeared in [866].

[868] S. MICALI AND A. SHAMIR, "An improvement of the Fiat-Shamir identification and signature scheme", *Advances in Cryptology–CRYPTO '88 (LNCS 403)*, 244–247, 1990.

[869] S. MICALI AND R. SIDNEY, "A simple method for generating and sharing pseudorandom functions, with applications to Clipper-like key escrow systems", *Advances in Cryptology–CRYPTO '95 (LNCS 963)*, 185–196, 1995.

[870] P. MIHAILESCU, "Fast generation of provable primes using search in arithmetic progressions", *Advances in Cryptology–CRYPTO '94 (LNCS 839)*, 282–293, 1994.

[871] M.J. MIHALJEVIĆ, "A security examination of the self-shrinking generator", presentation at 5th IMA Conference on Cryptography and Coding, Cirencester, U.K., December 1995.

[872] ——, "An approach to the initial state reconstruction of a clock-controlled shift register based on a novel distance measure", *Advances in Cryptology–AUSCRYPT '92 (LNCS 718)*, 349–356, 1993.

[873] ——, "A correlation attack on the binary sequence generators with time-varying output function", *Advances in Cryptology–ASIACRYPT '94 (LNCS 917)*, 67–79, 1995.

[874] M.J. MIHALJEVIĆ AND J.D. GOLIĆ, "A fast iterative algorithm for a shift register initial state reconstruction given the noisy output sequence", *Advances in Cryptology–AUSCRYPT '90 (LNCS 453)*, 165–175, 1990.

[875] ——, "Convergence of a Bayesian iterative error-correction procedure on a noisy shift register sequence", *Advances in Cryptology–EUROCRYPT '92 (LNCS 658)*, 124–137, 1993.

[876] G.L. MILLER, "Riemann's hypothesis and tests for primality", *Journal of Computer and System Sciences*, 13 (1976), 300–317.

[877] S.P. MILLER, B.C. NEUMAN, J.I. SCHILL-ER, AND J.H. SALTZER, "Kerberos authentication and authorization system", Section E.2.1 of Project Athena Technical Plan, MIT, Cambridge, Massachusetts, 1987.

[878] V.S. MILLER, "Use of elliptic curves in cryptography", *Advances in Cryptology–CRYPTO '85 (LNCS 218)*, 417–426, 1986.

[879] C. MITCHELL, "A storage complexity based analogue of Maurer key establishment using public channels", C. Boyd, editor, *Cryptography and Coding, 5th IMA Conference, Proceedings*, 84–93, Institute of Mathematics & Its Applications (IMA), 1995.

[880] ———, "Limitations of challenge-response entity authentication", *Electronics Letters*, 25 (August 17, 1989), 1195–1196.

[881] C. MITCHELL AND F. PIPER, "Key storage in secure networks", *Discrete Applied Mathematics*, 21 (1988), 215–228.

[882] C. MITCHELL, F. PIPER, AND P. WILD, "Digital signatures", G.J. Simmons, editor, *Contemporary Cryptology: The Science of Information Integrity*, 325–378, IEEE Press, 1992.

[883] A. MITROPOULOS AND H. MEIJER, "Zero knowledge proofs – a survey", Technical Report No. 90-IR-05, Queen's University at Kingston, Kingston, Ontario, Canada, 1990.

[884] S. MIYAGUCHI, "The FEAL cipher family", *Advances in Cryptology–CRYPTO '90 (LNCS 537)*, 627–638, 1991.

[885] S. MIYAGUCHI, S. KURIHARA, K. OHTA, AND H. MORITA, "Expansion of FEAL cipher", *NTT Review*, 2 (1990), 117–127.

[886] S. MIYAGUCHI, K. OHTA, AND M. IWATA, "128-bit hash function (N-hash)", *NTT Review*, 2 (1990), 128–132.

[887] S. MIYAGUCHI, A. SHIRAISHI, AND A. SHIMIZU, "Fast data encipherment algorithm FEAL-8", *Review of the Electrical Communications Laboratories*, 36 (1988), 433–437.

[888] A. MIYAJI AND M. TATEBAYASHI, "Public key cryptosystem with an elliptic curve", U.S. Patent # 5,272,755, 21 Dec 1993.

[889] ———, "Method of privacy communication using elliptic curves", U.S. Patent # 5,351,297, 27 Sep 1994 (continuation-in-part of 5,272,755).

[890] S.B. MOHAN AND B.S. ADIGA, "Fast algorithms for implementing RSA public key cryptosystem", *Electronics Letters*, 21 (August 15, 1985), 761.

[891] R. MOLVA, G. TSUDIK, E. VAN HERREWEGHEN, AND S. ZATTI, "KryptoKnight authentication and key distribution system", Y. Deswarte, G. Eizenberg, and J.-J. Quisquater, editors, *Second European Symposium on Research in Computer Security – ESORICS'92 (LNCS 648)*, 155–174, Springer-Verlag, 1992.

[892] L. MONIER, "Evaluation and comparison of two efficient probabilistic primality testing algorithms", *Theoretical Computer Science*, 12 (1980), 97–108.

[893] P. MONTGOMERY, "Modular multiplication without trial division", *Mathematics of Computation*, 44 (1985), 519–521.

[894] ———, "Speeding the Pollard and elliptic curve methods of factorization", *Mathematics of Computation*, 48 (1987), 243–264.

[895] P. MONTGOMERY AND R. SILVERMAN, "An FFT extension to the $P-1$ factoring algorithm", *Mathematics of Computation*, 54 (1990), 839–854.

[896] P.L. MONTGOMERY, "A block Lanczos algorithm for finding dependencies over $GF(2)$", *Advances in Cryptology–EUROCRYPT '95 (LNCS 921)*, 106–120, 1995.

[897] A.M. MOOD, "The distribution theory of runs", *The Annals of Mathematical Statistics*, 11 (1940), 367–392.

[898] J.H. MOORE, "Protocol failures in cryptosystems", *Proceedings of the IEEE*, 76 (1988), 594–602.

[899] ———, "Protocol failures in cryptosystems", G.J. Simmons, editor, *Contemporary Cryptology: The Science of Information Integrity*, 541–558, IEEE Press, 1992. Appeared earlier as [898].

[900] J.H. MOORE AND G.J. SIMMONS, "Cycle structure of the DES for keys having palindromic (or antipalindromic) sequences of round keys", *IEEE Transactions on Software Engineering*, 13 (1987), 262–273. An earlier version appeared in [901].

[901] ———, "Cycle structure of the DES with weak and semi-weak keys", *Advances in Cryptology–CRYPTO '86 (LNCS 263)*, 9–32, 1987.

[902] F. MORAIN, "Distributed primality proving and the primality of $(2^{3539} + 1)/3$", *Advances in Cryptology–EUROCRYPT '90 (LNCS 473)*, 110–123, 1991.

[903] ———, "Prime values of partition numbers and the primality of $p_{1840926}$", LIX Research Report LIX/RR/92/11, Laboratoire d'Informatique de l'Ecole Polytechnique, France, June 1992.

[904] F. MORAIN AND J. OLIVOS, "Speeding up the computations on an elliptic curve using addition-subtraction chains", *Theoretical Informatics and Applications*, 24 (1990), 531–543.

[905] I.H. MORGAN AND G.L. MULLEN, "Primitive normal polynomials over finite fields", *Mathematics of Computation*, 63 (1994), 759–765.

[906] R. MORRIS, "The Hagelin cipher machine (M-209), Reconstruction of the internal settings", *Cryptologia*, 2 (1978), 267–278.

[907] R. MORRIS AND K. THOMPSON, "Password security: a case history", *Communications of the ACM*, 22 (1979), 594–597.

[908] M.A. MORRISON AND J. BRILLHART, "A method of factoring and the factorization of F_7", *Mathematics of Computation*, 29 (1975), 183–205.

[909] W.B. MÜLLER AND R. NÖBAUER, "Cryptanalysis of the Dickson-scheme", *Advances in Cryptology–EUROCRYPT '85 (LNCS 219)*, 50–61, 1986.

[910] W.B. MÜLLER AND W. NÖBAUER, "Some remarks on public-key cryptosystems", *Studia Scientiarum Mathematicarum Hungarica*, 16 (1981), 71–76.

[911] R. MULLIN, I. ONYSZCHUK, S. VANSTONE, AND R. WILSON, "Optimal normal bases in $GF(p^n)$", *Discrete Applied Mathematics*, 22 (1988/89), 149–161.

[912] S. MUND, "Ziv-Lempel complexity for periodic sequences and its cryptographic application", *Advances in Cryptology–EUROCRYPT '91 (LNCS 547)*, 114–126, 1991.

[913] S. MURPHY, "The cryptanalysis of FEAL-4 with 20 chosen plaintexts", *Journal of Cryptology*, 2 (1990), 145–154.

[914] D. NACCACHE, "Can O.S.S. be repaired? – proposal for a new practical signature scheme", *Advances in Cryptology–EUROCRYPT '93 (LNCS 765)*, 233–239, 1994.

[915] D. NACCACHE, D. M'RAÏHI, AND D. RAPHAELI, "Can Montgomery parasites be avoided? A design methodology based on key and cryptosystem modifications", *Designs, Codes and Cryptography*, 5 (1995), 73–80.

[916] D. NACCACHE, D. M'RAÏHI, S. VAUDENAY, AND D. RAPHAELI, "Can D.S.A. be improved? Complexity trade-offs with the digital signature standard", *Advances in Cryptology–EUROCRYPT '94 (LNCS 950)*, 77–85, 1995.

[917] D. NACCACHE AND H. M'SILTI, "A new modulo computation algorithm", *Recherche Opérationnelle – Operations Research (RAIRO-OR)*, 24 (1990), 307–313.

[918] K. NAGASAKA, J.-S. SHIUE, AND C.-W. HO, "A fast algorithm of the Chinese remainder theorem and its application to Fibonacci number", G.E. Bergum, A.N. Philippou, and A.F. Horadam, editors, *Applications of Fibonacci Numbers, Proceedings of the Fourth International Conference on Fibonacci Numbers and their Applications*, 241–246, Kluwer Academic Publishers, 1991.

[919] M. NAOR AND A. SHAMIR, "Visual cryptography", *Advances in Cryptology–EUROCRYPT '94 (LNCS 950)*, 1–12, 1995.

[920] M. NAOR AND M. YUNG, "Universal one-way hash functions and their cryptographic applications", *Proceedings of the 21st Annual ACM Symposium on Theory of Computing*, 33–43, 1989.

[921] ———, "Public-key cryptosystems provably secure against chosen ciphertext attacks", *Proceedings of the 22nd Annual ACM Symposium on Theory of Computing*, 427–437, 1990.

[922] J. NECHVATAL, "Public key cryptography", G.J. Simmons, editor, *Contemporary Cryptology: The Science of Information Integrity*, 177–288, IEEE Press, 1992.

[923] R.M. NEEDHAM AND M.D. SCHROEDER, "Using encryption for authentication in large networks of computers", *Communications of the ACM*, 21 (1978), 993–999.

[924] ———, "Authentication revisited", *Operating Systems Review*, 21 (1987), 7.

[925] B.C. NEUMAN AND S.G. STUBBLEBINE, "A note on the use of timestamps as nonces", *Operating Systems Review*, 27 (1993), 10–14.

[926] B.C. NEUMAN AND T. TS'O, "Kerberos: an authentication service for computer networks", *IEEE Communications Magazine*, 32 (September 1994), 33–38.

[927] H. NIEDERREITER, "The probabilistic theory of linear complexity", *Advances in Cryptology–EUROCRYPT '88 (LNCS 330)*, 191–209, 1988.

[928] ——, "A combinatorial approach to probabilistic results on the linear-complexity profile of random sequences", *Journal of Cryptology*, 2 (1990), 105–112.

[929] ——, "Keystream sequences with a good linear complexity profile for every starting point", *Advances in Cryptology–EUROCRYPT '89 (LNCS 434)*, 523–532, 1990.

[930] ——, "The linear complexity profile and the jump complexity of keystream sequences", *Advances in Cryptology–EUROCRYPT '90 (LNCS 473)*, 174–188, 1991.

[931] K. NISHIMURA AND M. SIBUYA, "Probability to meet in the middle", *Journal of Cryptology*, 2 (1990), 13–22.

[932] I.M. NIVEN AND H.S. ZUCKERMAN, *An Introduction to the Theory of Numbers*, John Wiley & Sons, New York, 4th edition, 1980.

[933] M.J. NORRIS AND G.J. SIMMONS, "Algorithms for high-speed modular arithmetic", *Congressus Numerantium*, 31 (1981), 153–163.

[934] G. NORTON, "Extending the binary gcd algorithm", J. Calmet, editor, *Algebraic Algorithms and Error-Correcting Codes, 3rd International Conference, AAECC-3 (LNCS 229)*, 363–372, Springer-Verlag, 1986.

[935] K. NYBERG, "On one-pass authenticated key establishment schemes", workshop record, 2nd Workshop on Selected Areas in Cryptography (SAC'95), Ottawa, Canada, May 18–19 1995.

[936] K. NYBERG AND R. RUEPPEL, "A new signature scheme based on the DSA giving message recovery", *1st ACM Conference on Computer and Communications Security*, 58–61, ACM Press, 1993.

[937] ——, "Weaknesses in some recent key agreement protocols", *Electronics Letters*, 30 (January 6, 1994), 26–27.

[938] ——, "Message recovery for signature schemes based on the discrete logarithm problem", *Designs, Codes and Cryptography*, 7 (1996), 61–81.

[939] A.M. ODLYZKO, "Cryptanalytic attacks on the multiplicative knapsack cryptosystem and on Shamir's fast signature scheme", *IEEE Transactions on Information Theory*, 30 (1984), 594–601.

[940] ——, "Discrete logarithms in finite fields and their cryptographic significance", *Advances in Cryptology–Proceedings of EUROCRYPT 84 (LNCS 209)*, 224–314, 1985.

[941] ——, "The rise and fall of knapsack cryptosystems", C. Pomerance, editor, *Cryptology and Computational Number Theory*, volume 42 of *Proceedings of Symposia in Applied Mathematics*, 75–88, American Mathematical Society, 1990.

[942] ——, "Discrete logarithms and smooth polynomials", G.L. Mullen and P.J-S. Shiue, editors, *Finite Fields: Theory, Applications, and Algorithms*, volume 168 of *Contemporary Mathematics*, 269–278, American Mathematical Society, 1994.

[943] K. OHTA AND K. AOKI, "Linear cryptanalysis of the Fast Data Encipherment Algorithm", *Advances in Cryptology–CRYPTO '94 (LNCS 839)*, 12–16, 1994.

[944] K. OHTA AND T. OKAMOTO, "Practical extension of Fiat-Shamir scheme", *Electronics Letters*, 24 (July 21, 1988), 955–956.

[945] ——, "A modification of the Fiat-Shamir scheme", *Advances in Cryptology–CRYPTO '88 (LNCS 403)*, 232–243, 1990.

[946] E. OKAMOTO AND K. TANAKA, "Key distribution system based on identification information", *IEEE Journal on Selected Areas in Communications*, 7 (1989), 481–485.

[947] T. OKAMOTO, "A single public-key authentication scheme for multiple users", *Systems and Computers in Japan*, 18 (1987), 14–24. Translated from *Denshi Tsushin Gakkai Ronbunshi* vol. 69-D no.10, October 1986, 1481–1489.

[948] ——, "A fast signature scheme based on congruential polynomial operations", *IEEE Transactions on Information Theory*, 36 (1990), 47–53.

[949] ——, "Provably secure and practical identification schemes and corresponding signature

schemes", *Advances in Cryptology–CRYPTO '92 (LNCS 740)*, 31–53, 1993.

[950] ——, "Designated confirmer signatures and public-key encryption are equivalent", *Advances in Cryptology–CRYPTO '94 (LNCS 839)*, 61–74, 1994.

[951] ——, "An efficient divisible electronic cash scheme", *Advances in Cryptology–CRYPTO '95 (LNCS 963)*, 438–451, 1995.

[952] T. OKAMOTO, S. MIYAGUCHI, A. SHIRAISHI, AND T. KAWAOKA, "Signed document transmission system", U.S. Patent # 4,625,076, 25 Nov 1986.

[953] T. OKAMOTO AND A. SHIRAISHI, "A fast signature scheme based on quadratic inequalities", *Proceedings of the 1985 IEEE Symposium on Security and Privacy*, 123–132, 1985.

[954] T. OKAMOTO, A. SHIRAISHI, AND T. KAWAOKA, "Secure user authentication without password files", Technical Report NI83-92, I.E.C.E., Japan, January 1984. In Japanese.

[955] J. OLIVOS, "On vectorial addition chains", *Journal of Algorithms*, 2 (1981), 13–21.

[956] J.K. OMURA AND J.L. MASSEY, "Computational method and apparatus for finite field arithmetic", U.S. Patent # 4,587,627, 6 May 1986.

[957] H. ONG AND C.P. SCHNORR, "Fast signature generation with a Fiat Shamir-like scheme", *Advances in Cryptology–EUROCRYPT '90 (LNCS 473)*, 432–440, 1991.

[958] H. ONG, C.P. SCHNORR, AND A. SHAMIR, "An efficient signature scheme based on quadratic equations", *Proceedings of the 16th Annual ACM Symposium on Theory of Computing*, 208–216, 1984.

[959] I.M. ONYSZCHUK, R.C. MULLIN, AND S.A. VANSTONE, "Computational method and apparatus for finite field multiplication", U.S. Patent # 4,745,568, 17 May 1988.

[960] G. ORTON, "A multiple-iterated trapdoor for dense compact knapsacks", *Advances in Cryptology–EUROCRYPT '94 (LNCS 950)*, 112–130, 1995.

[961] D. OTWAY AND O. REES, "Efficient and timely mutual authentication", *Operating Systems Review*, 21 (1987), 8–10.

[962] J.C. PAILLÈS AND M. GIRAULT, "CRIPT: A public-key based solution for secure data communications", *Proceedings of the 7th World-wide Congress on Computer and Communications Security and Protection (SECURICOM'89)*, 171–185, 1989.

[963] C.H. PAPADIMITRIOU, *Computational Complexity*, Addison-Wesley, Reading, Massachusetts, 1994.

[964] S.-J. PARK, S.-J. LEE, AND S.-C. GOH, "On the security of the Gollmann cascades", *Advances in Cryptology–CRYPTO '95 (LNCS 963)*, 148–156, 1995.

[965] J. PATARIN, "Hidden fields equations (HFE) and isomorphisms of polynomials (IP): Two new families of asymmetric algorithms", *Advances in Cryptology–EUROCRYPT '96 (LNCS 1070)*, 33–48, 1996.

[966] J. PATARIN AND P. CHAUVAUD, "Improved algorithms for the permuted kernel problem", *Advances in Cryptology–CRYPTO '93 (LNCS 773)*, 391–402, 1994.

[967] W. PENZHORN AND G. KÜHN, "Computation of low-weight parity checks for correlation attacks on stream ciphers", C. Boyd, editor, *Cryptography and Coding, 5th IMA Conference, Proceedings*, 74–83, Institute of Mathematics & Its Applications (IMA), 1995.

[968] R. PERALTA, "Simultaneous security of bits in the discrete log", *Advances in Cryptology–EUROCRYPT '85 (LNCS 219)*, 62–72, 1986.

[969] R. PERALTA AND V. SHOUP, "Primality testing with fewer random bits", *Computational Complexity*, 3 (1993), 355–367.

[970] A. PFITZMANN AND R. ASSMANN, "More efficient software implementations of (generalized) DES", *Computers & Security*, 12 (1993), 477–500.

[971] B. PFITZMANN AND M. WAIDNER, "Fail-stop signatures and their applications", *Proceedings of the 9th Worldwide Congress on Computer and Communications Security and Protection (SECURICOM'91)*, 145–160, 1991.

[972] ——, "Formal aspects of fail-stop signatures", Interner Bericht 22/90, Universität Karlsruhe, Germany, December 1990.

[973] S.J.D. PHOENIX AND P.D. TOWNSEND, "Quantum cryptography: protecting our future networks with quantum mechanics", C. Boyd, editor, *Cryptography and Coding, 5th IMA Conference, Proceedings*, 112–131, Institute of Mathematics & Its Applications (IMA), 1995.

[974] R. PINCH, "The Carmichael numbers up to 10^{15}", *Mathematics of Computation*, 61 (1993), 381–391.

[975] ———, "Some primality testing algorithms", *Notices of the American Mathematical Society*, 40 (1993), 1203–1210.

[976] ———, "Extending the Håstad attack to LUC", *Electronics Letters*, 31 (October 12, 1995), 1827–1828.

[977] ———, "Extending the Wiener attack to RSA-type cryptosystems", *Electronics Letters*, 31 (September 28, 1995), 1736–1738.

[978] V. PLESS, "Encryption schemes for computer confidentiality", *IEEE Transactions on Computers*, 26 (1977), 1133–1136.

[979] J.B. PLUMSTEAD, "Inferring a sequence generated by a linear congruence", *Proceedings of the IEEE 23rd Annual Symposium on Foundations of Computer Science*, 153–159, 1982.

[980] ———, "Inferring a sequence produced by a linear congruence", *Advances in Cryptology–Proceedings of Crypto 82*, 317–319, 1983.

[981] H.C. POCKLINGTON, "The determination of the prime or composite nature of large numbers by Fermat's theorem", *Proceedings of the Cambridge Philosophical Society*, 18 (1914), 29–30.

[982] S.C. POHLIG AND M.E. HELLMAN, "An improved algorithm for computing logarithms over $GF(p)$ and its cryptographic significance", *IEEE Transactions on Information Theory*, 24 (1978), 106–110.

[983] D. POINTCHEVAL, "A new identification scheme based on the perceptrons problem", *Advances in Cryptology–EUROCRYPT '95 (LNCS 921)*, 319–328, 1995.

[984] J.M. POLLARD, "Theorems on factorization and primality testing", *Proceedings of the Cambridge Philosophical Society*, 76 (1974), 521–528.

[985] ———, "A Monte Carlo method for factorization", *BIT*, 15 (1975), 331–334.

[986] ———, "Monte Carlo methods for index computation $(\bmod\ p)$", *Mathematics of Computation*, 32 (1978), 918–924.

[987] ———, "Factoring with cubic integers", A.K. Lenstra and H.W. Lenstra Jr., editors, *The Development of the Number Field Sieve*, volume 1554 of *Lecture Notes in Mathematics*, 4–10, Springer-Verlag, 1993.

[988] J.M. POLLARD AND C. SCHNORR, "An efficient solution of the congruence $x^2 + ky^2 = m \ (\bmod\ n)$", *IEEE Transactions on Information Theory*, 33 (1987), 702–709.

[989] C. POMERANCE, "Analysis and comparison of some integer factoring algorithms", H.W. Lenstra Jr. and R. Tijdeman, editors, *Computational Methods in Number Theory, Part 1*, 89–139, Mathematisch Centrum, 1982.

[990] ———, "The quadratic sieve factoring algorithm", *Advances in Cryptology–Proceedings of EUROCRYPT 84 (LNCS 209)*, 169–182, 1985.

[991] ———, "Fast, rigorous factorization and discrete logarithm algorithms", *Discrete Algorithms and Complexity*, 119–143, Academic Press, 1987.

[992] ———, "Very short primality proofs", *Mathematics of Computation*, 48 (1987), 315–322.

[993] ———, editor, *Cryptology and Computational Number Theory*, American Mathematical Society, Providence, Rhode Island, 1990.

[994] ———, "Factoring", C. Pomerance, editor, *Cryptology and Computational Number Theory*, volume 42 of *Proceedings of Symposia in Applied Mathematics*, 27–47, American Mathematical Society, 1990.

[995] ———, "The number field sieve", W. Gautschi, editor, *Mathematics of Computation, 1943-1993: A Half-Century of Computation Mathematics*, volume 48 of *Proceedings of Symposia in Applied Mathematics*, 465–480, American Mathematical Society, 1994.

[996] C. POMERANCE, J.L. SELFRIDGE, AND S.S. WAGSTAFF JR., "The pseudoprimes to $25 \cdot 10^9$", *Mathematics of Computation*, 35 (1980), 1003–1026.

[997] C. POMERANCE AND J. SORENSON, "Counting the integers factorable via cyclotomic methods", *Journal of Algorithms*, 19 (1995), 250–265.

[998] G.J. POPEK AND C.S. KLINE, "Encryption and secure computer networks", *ACM Computing Surveys*, 11 (1979), 331–356.

[999] E. PRANGE, "An algorithm for factoring $x^n - 1$ over a finite field", AFCRC-TN-59-775, Air Force Cambridge Research Center, 1959.

[1000] V.R. PRATT, "Every prime has a succinct certificate", *SIAM Journal on Computing*, 4 (1975), 214–220.

[1001] B. PRENEEL, "Standardization of cryptographic techniques", B. Preneel, R. Govaerts, and J. Vandewalle, editors, *Computer Security and Industrial Cryptography: State of the Art and Evolution (LNCS 741)*, 162–173, Springer-Verlag, 1993.

[1002] ———, "Cryptographic hash functions", *European Transactions on Telecommunications*, 5 (1994), 431–448.

[1003] ———, *Analysis and design of cryptographic hash functions*, PhD thesis, Katholieke Universiteit Leuven (Belgium), Jan. 1993.

[1004] ———, *Cryptographic Hash Functions*, Kluwer Academic Publishers, Boston, (to appear). Updated and expanded from [1003].

[1005] B. PRENEEL, R. GOVAERTS, AND J. VANDEWALLE, "Differential cryptanalysis of hash functions based on block ciphers", *1st ACM Conference on Computer and Communications Security*, 183–188, ACM Press, 1993.

[1006] ———, "Information authentication: Hash functions and digital signatures", B. Preneel, R. Govaerts, and J. Vandewalle, editors, *Computer Security and Industrial Cryptography: State of the Art and Evolution (LNCS 741)*, 87–131, Springer-Verlag, 1993.

[1007] ———, "Hash functions based on block ciphers: A synthetic approach", *Advances in Cryptology–CRYPTO '93 (LNCS 773)*, 368–378, 1994.

[1008] B. PRENEEL, M. NUTTIN, V. RIJMEN, AND J. BUELENS, "Cryptanalysis of the CFB mode of the DES with a reduced number of rounds", *Advances in Cryptology–CRYPTO '93 (LNCS 773)*, 212–223, 1994.

[1009] B. PRENEEL AND P. VAN OORSCHOT, "MDx-MAC and building fast MACs from hash functions", *Advances in Cryptology–CRYPTO '95 (LNCS 963)*, 1–14, 1995.

[1010] ———, "On the security of two MAC algorithms", *Advances in Cryptology–EUROCRYPT '96 (LNCS 1070)*, 19–32, 1996.

[1011] N. PROCTOR, "A self-synchronizing cascaded cipher system with dynamic control of error propagation", *Advances in Cryptology–Proceedings of CRYPTO 84 (LNCS 196)*, 174–190, 1985.

[1012] G.B. PURDY, "A high security log-in procedure", *Communications of the ACM*, 17 (1974), 442–445.

[1013] M. QU AND S.A. VANSTONE, "The knapsack problem in cryptography", *Contemporary Mathematics*, 168 (1994), 291–308.

[1014] K. QUINN, "Some constructions for key distribution patterns", *Designs, Codes and Cryptography*, 4 (1994), 177–191.

[1015] J.-J. QUISQUATER, "A digital signature scheme with extended recovery", preprint, 1995.

[1016] J.-J. QUISQUATER AND C. COUVREUR, "Fast decipherment algorithm for RSA public-key cryptosystem", *Electronics Letters*, 18 (October 14, 1982), 905–907.

[1017] J.-J. QUISQUATER AND J.-P. DELESCAILLE, "How easy is collision search? Application to DES", *Advances in Cryptology–EUROCRYPT '89 (LNCS 434)*, 429–434, 1990.

[1018] ———, "How easy is collision search. New results and applications to DES", *Advances in Cryptology–CRYPTO '89 (LNCS 435)*, 408–413, 1990.

[1019] J.-J. QUISQUATER AND M. GIRAULT, "2n-bit hash-functions using n-bit symmetric block cipher algorithms", *Advances in Cryptology–EUROCRYPT '89 (LNCS 434)*, 102–109, 1990.

[1020] J.-J. QUISQUATER, L. GUILLOU, AND T. BERSON, "How to explain zero-knowledge protocols to your children", *Advances in Cryptology–CRYPTO '89 (LNCS 435)*, 628–631, 1990.

[1021] M.O. RABIN, "Probabilistic algorithms", J.F. Traub, editor, *Algorithms and Complexity*, 21–40, Academic Press, 1976.

[1022] ———, "Digitalized signatures", R. DeMillo, D. Dobkin, A. Jones, and R. Lipton, editors, *Foundations of Secure Computation*, 155–168, Academic Press, 1978.

[1023] ———, "Digitalized signatures and public-key functions as intractable as factorization", MIT/LCS/TR-212, MIT Laboratory for Computer Science, 1979.

[1024] ———, "Probabilistic algorithm for testing primality", *Journal of Number Theory*, 12 (1980), 128–138.

[1025] ———, "Probabilistic algorithms in finite fields", *SIAM Journal on Computing*, 9 (1980), 273–280.

[1026] ———, "Fingerprinting by random polynomials", TR-15-81, Center for Research in Computing Technology, Harvard University, 1981.

[1027] ——, "Efficient dispersal of information for security, load balancing, and fault tolerance", *Journal of the Association for Computing Machinery*, 36 (1989), 335–348.

[1028] T. RABIN AND M. BEN-OR, "Verifiable secret sharing and multiparty protocols with honest majority", *Proceedings of the 21st Annual ACM Symposium on Theory of Computing*, 73–85, 1989.

[1029] C. RACKOFF AND D.R. SIMON, "Noninteractive zero-knowledge proof of knowledge and chosen ciphertext attack", *Advances in Cryptology–CRYPTO '91 (LNCS 576)*, 433–444, 1992.

[1030] G. RAWLINS, *Compared to What? An Introduction to the Analysis of Algorithms*, Computer Science Press, New York, 1992.

[1031] G. REITWIESNER, "Binary arithmetic", *Advances in Computers*, 1 (1960), 231–308.

[1032] T. RENJI, "On finite automaton one-key cryptosystems", R. Anderson, editor, *Fast Software Encryption, Cambridge Security Workshop (LNCS 809)*, 135–148, Springer-Verlag, 1994.

[1033] RFC 1319, "The MD2 message-digest algorithm", Internet Request for Comments 1319, B. Kaliski, April 1992 (updates RFC 1115, August 1989, J. Linn).

[1034] RFC 1320, "The MD4 message-digest algorithm", Internet Request for Comments 1320, R.L. Rivest, April 1992 (obsoletes RFC 1186, October 1990, R. Rivest).

[1035] RFC 1321, "The MD5 message-digest algorithm", Internet Request for Comments 1321, R.L. Rivest, April 1992 (presented at Rump Session of Crypto'91).

[1036] RFC 1421, "Privacy enhancement for Internet electronic mail – Part I: Message encryption and authentication procedures", Internet Request for Comments 1421, J. Linn, February 1993 (obsoletes RFC 1113 – September 1989; RFC 1040 – January 1988; and RFC 989 – February 1987, J. Linn).

[1037] RFC 1422, "Privacy enhancement for Internet electronic mail – Part II: Certificate-based key management", Internet Request for Comments 1422, S. Kent, February 1993 (obsoletes RFC 1114, August 1989, S. Kent and J. Linn).

[1038] RFC 1423, "Privacy enhancement for Internet electronic mail – Part III: Algorithms, modes, and identifiers", Internet Request for Comments 1423, D. Balenson, February 1993 (obsoletes RFC 1115, September 1989, J. Linn).

[1039] RFC 1424, "Privacy enhancement for Internet electronic mail – Part IV: Key certification and related services", Internet Request for Comments 1424, B. Kaliski, February 1993.

[1040] RFC 1508, "Generic security service application program interface", Internet Request for Comments 1508, J. Linn, September 1993.

[1041] RFC 1510, "The Kerberos network authentication service (V5)", Internet Request for Comments 1510, J. Kohl and C. Neuman, September 1993.

[1042] RFC 1521, "MIME (Multipurpose Internet Mail Extensions) Part One: Mechanisms for specifying and describing the format of Internet message bodies", Internet Request for Comments 1521, N. Borenstein and N. Freed, September 1993 (obsoletes RFC 1341).

[1043] RFC 1750, "Randomness requirements for security", Internet Request for Comments 1750, D. Eastlake, S. Crocker and J. Schiller, December 1994.

[1044] RFC 1828, "IP authentication using keyed MD5", Internet Request for Comments 1828, P. Metzger and W. Simpson, August 1995.

[1045] RFC 1847, "Security multiparts for MIME: Multipart/signed and multipart/encrypted", Internet Request for Comments 1847, J. Galvin, S. Murphy, S. Crocker and N. Freed, October 1995.

[1046] RFC 1848, "MIME object security services", Internet Request for Comments 1848, S. Crocker, N. Freed, J. Galvin and S. Murphy, October 1995.

[1047] RFC 1938, "A one-time password system", Internet Request for Comments 1938, N. Haller and C. Metz, May 1996.

[1048] V. RIJMEN, J. DAEMEN, B. PRENEEL, A. BOSSELAERS, AND E. DE WIN, "The cipher SHARK", D. Gollmann, editor, *Fast Software Encryption, Third International Workshop (LNCS 1039)*, 99–111, Springer-Verlag, 1996.

[1049] V. RIJMEN AND B. PRENEEL, "On weaknesses of non-surjective round functions", presented at the 2nd Workshop on Selected Areas in Cryptography (SAC'95), Ottawa, Canada, May 18–19 1995.

[1050] ——, "Improved characteristics for differential cryptanalysis of hash functions based on block ciphers", B. Preneel, editor, *Fast Software Encryption, Second International Workshop (LNCS 1008)*, 242–248, Springer-Verlag, 1995.

[1051] R.L. RIVEST, "Are 'strong' primes needed for RSA?", Unpublished manuscript, 1991.

[1052] ——, "Remarks on a proposed cryptanalytic attack on the M.I.T. public-key cryptosystem", *Cryptologia*, 2 (1978), 62–65.

[1053] ——, "Statistical analysis of the Hagelin cryptograph", *Cryptologia*, 5 (1981), 27–32.

[1054] ——, "Cryptography", J. van Leeuwen, editor, *Handbook of Theoretical Computer Science*, 719–755, Elsevier Science Publishers, 1990.

[1055] ——, "The MD4 message digest algorithm", *Advances in Cryptology–CRYPTO '90 (LNCS 537)*, 303–311, 1991.

[1056] ——, "The RC5 encryption algorithm", B. Preneel, editor, *Fast Software Encryption, Second International Workshop (LNCS 1008)*, 86–96, Springer-Verlag, 1995.

[1057] R.L. RIVEST AND A. SHAMIR, "How to expose an eavesdropper", *Communications of the ACM*, 27 (1984), 393–395.

[1058] ——, "Efficient factoring based on partial information", *Advances in Cryptology–EUROCRYPT '85 (LNCS 219)*, 31–34, 1986.

[1059] R.L. RIVEST, A. SHAMIR, AND L.M. ADLEMAN, "Cryptographic communications system and method", U.S. Patent # 4,405,829, 20 Sep 1983.

[1060] ——, "A method for obtaining digital signatures and public-key cryptosystems", *Communications of the ACM*, 21 (1978), 120–126.

[1061] R.L. RIVEST AND A.T. SHERMAN, "Randomized encryption techniques", *Advances in Cryptology–Proceedings of Crypto 82*, 145–163, 1983.

[1062] M.J.B. ROBSHAW, "On evaluating the linear complexity of a sequence of least period 2^n", *Designs, Codes and Cryptography*, 4 (1994), 263–269.

[1063] ——, "Stream ciphers", Technical Report TR-701 (version 2.0), RSA Laboratories, 1995.

[1064] M. ROE, "How to reverse engineer an EES device", B. Preneel, editor, *Fast Software Encryption, Second International Workshop (LNCS 1008)*, 305–328, Springer-Verlag, 1995.

[1065] P. ROGAWAY, "Bucket hashing and its application to fast message authentication", *Advances in Cryptology–CRYPTO '95 (LNCS 963)*, 29–42, 1995.

[1066] P. ROGAWAY AND D. COPPERSMITH, "A software-optimized encryption algorithm", R. Anderson, editor, *Fast Software Encryption, Cambridge Security Workshop (LNCS 809)*, 56–63, Springer-Verlag, 1994.

[1067] N. ROGIER AND P. CHAUVAUD, "The compression function of MD2 is not collision free", workshop record, 2nd Workshop on Selected Areas in Cryptography (SAC'95), Ottawa, Canada, May 18–19 1995.

[1068] J. ROMPEL, "One-way functions are necessary and sufficient for secure signatures", *Proceedings of the 22nd Annual ACM Symposium on Theory of Computing*, 387–394, 1990.

[1069] K.H. ROSEN, *Elementary Number Theory and its Applications*, Addison-Wesley, Reading, Massachusetts, 3rd edition, 1992.

[1070] J. ROSSER AND L. SCHOENFELD, "Approximate formulas for some functions of prime numbers", *Illinois Journal of Mathematics*, 6 (1962), 64–94.

[1071] RSA LABORATORIES, "The Public-Key Cryptography Standards – PKCS #11: Cryptographic token interface standard", RSA Data Security Inc., Redwood City, California, April 28 1995.

[1072] ——, "The Public-Key Cryptography Standards (PKCS)", RSA Data Security Inc., Redwood City, California, November 1993 Release.

[1073] A.D. RUBIN AND P. HONEYMAN, "Formal methods for the analysis of authentication protocols", CITI Technical Report 93-7, Information Technology Division, University of Michigan, 1993.

[1074] F. RUBIN, "Decrypting a stream cipher based on J-K flip-flops", *IEEE Transactions on Computers*, 28 (1979), 483–487.

[1075] R.A. RUEPPEL, *Analysis and Design of Stream Ciphers*, Springer-Verlag, Berlin, 1986.

[1076] ——, "Correlation immunity and the summation generator", *Advances in Cryptology–CRYPTO '85 (LNCS 218)*, 260–272, 1986.

[1077] ——, "Linear complexity and random sequences", *Advances in Cryptology–EUROCRYPT '85 (LNCS 219)*, 167–188, 1986.

[1078] ——, "Key agreements based on function composition", *Advances in Cryptology–EUROCRYPT '88 (LNCS 330)*, 3–10, 1988.

[1079] ——, "On the security of Schnorr's pseudo random generator", *Advances in Cryptology–EUROCRYPT '89 (LNCS 434)*, 423–428, 1990.

[1080] ——, "A formal approach to security architectures", *Advances in Cryptology–EUROCRYPT '91 (LNCS 547)*, 387–398, 1991.

[1081] ——, "Stream ciphers", G.J. Simmons, editor, *Contemporary Cryptology: The Science of Information Integrity*, 65–134, IEEE Press, 1992.

[1082] ——, "Criticism of ISO CD 11166 banking — key management by means of asymmetric algorithms", W. Wolfowicz, editor, *Proceedings of the 3rd Symposium on State and Progress of Research in Cryptography, Rome, Italy*, 191–198, 1993.

[1083] R.A. RUEPPEL, A. LENSTRA, M. SMID, K.McCURLEY, Y.DESMEDT, A.ODLYZKO, AND P. LANDROCK, "The Eurocrypt '92 controversial issue: trapdoor primes and moduli", *Advances in Cryptology–EUROCRYPT '92 (LNCS 658)*, 194–199, 1993.

[1084] R.A. RUEPPEL AND J.L. MASSEY, "The knapsack as a non-linear function", *IEEE International Symposium on Information Theory (Abstracts)*, p.46, 1985.

[1085] R.A. RUEPPEL AND O.J. STAFFELBACH, "Products of linear recurring sequences with maximum complexity", *IEEE Transactions on Information Theory*, 33 (1987), 124–131.

[1086] R.A. RUEPPEL AND P.C. VAN OORSCHOT, "Modern key agreement techniques", *Computer Communications*, 17 (1994), 458–465.

[1087] A. RUSSELL, "Necessary and sufficient conditions for collision-free hashing", *Advances in Cryptology–CRYPTO '92 (LNCS 740)*, 433–441, 1993.

[1088] ——, "Necessary and sufficient conditions for collision-free hashing", *Journal of Cryptology*, 8 (1995), 87–99. An earlier version appeared in [1087].

[1089] A. SALOMAA, *Public-key Cryptography*, Springer-Verlag, Berlin, 1990.

[1090] M. SANTHA AND U.V. VAZIRANI, "Generating quasi-random sequences from slightly-random sources", *Proceedings of the IEEE 25th Annual Symposium on Foundations of Computer Science*, 434–440, 1984.

[1091] ——, "Generating quasi-random sequences from semi-random sources", *Journal of Computer and System Sciences*, 33 (1986), 75–87. An earlier version appeared in [1090].

[1092] O. SCHIROKAUER, "Discrete logarithms and local units", *Philosophical Transactions of the Royal Society of London A*, 345 (1993), 409–423.

[1093] B. SCHNEIER, "Description of a new variable-length key, 64-bit block cipher (Blowfish)", R. Anderson, editor, *Fast Software Encryption, Cambridge Security Workshop (LNCS 809)*, 191–204, Springer-Verlag, 1994.

[1094] ——, *Applied Cryptography: Protocols, Algorithms, and Source Code in C*, John Wiley & Sons, New York, 2nd edition, 1996.

[1095] C.P. SCHNORR, "Method for identifying subscribers and for generating and verifying electronic signatures in a data exchange system", U.S. Patent # 4,995,082, 19 Feb 1991.

[1096] ——, "On the construction of random number generators and random function generators", *Advances in Cryptology–EUROCRYPT '88 (LNCS 330)*, 225–232, 1988.

[1097] ——, "Efficient identification and signatures for smart cards", *Advances in Cryptology–CRYPTO '89 (LNCS 435)*, 239–252, 1990.

[1098] ——, "Efficient signature generation by smart cards", *Journal of Cryptology*, 4 (1991), 161–174.

[1099] C.P. SCHNORR AND M. EUCHNER, "Lattice basis reduction: Improved practical algorithms and solving subset sum problems", L. Budach, editor, *Fundamentals of Computation Theory (LNCS 529)*, 68–85, Springer-Verlag, 1991.

[1100] C.P. SCHNORR AND H.H. HÖRNER, "Attacking the Chor-Rivest cryptosystem by improved lattice reduction", *Advances in Cryptology–EUROCRYPT '95 (LNCS 921)*, 1–12, 1995.

[1101] A. SCHÖNAGE, "A lower bound for the length of addition chains", *Theoretical Computer Science*, 1 (1975), 1–12.

[1102] A.W. SCHRIFT AND A. SHAMIR, "On the universality of the next bit test", *Advances in Cryptology–CRYPTO '90 (LNCS 537)*, 394–408, 1991.

[1103] ——, "Universal tests for nonuniform distributions", *Journal of Cryptology*, 6 (1993), 119–133. An earlier version appeared in [1102].

[1104] F. SCHWENK AND J. EISFELD, "Public key encryption and signature schemes based on polynomials over $\mathbb{Z}_n$", *Advances in Cryptology–EUROCRYPT '96 (LNCS 1070)*, 60–71, 1996.

[1105] R. SEDGEWICK, *Algorithms*, Addison-Wesley, Reading, Massachusetts, 2nd edition, 1988.

[1106] R. SEDGEWICK, T.G. SZYMANSKI, AND A.C. YAO, "The complexity of finding cycles in periodic functions", *SIAM Journal on Computing*, 11 (1982), 376–390.

[1107] E.S. SELMER, "Linear recurrence relations over finite fields", Department of Mathematics, University of Bergen, Norway, 1966.

[1108] J. SHALLIT, "On the worst case of three algorithms for computing the Jacobi symbol", *Journal of Symbolic Computation*, 10 (1990), 593–610.

[1109] A. SHAMIR, "A fast signature scheme", MIT/LCS/TM-107, MIT Laboratory for Computer Science, 1978.

[1110] ——, "How to share a secret", *Communications of the ACM*, 22 (1979), 612–613.

[1111] ——, "On the generation of cryptographically strong pseudo-random sequences", S. Even and O. Kariv, editors, *Automata, Languages, and Programming, 8th Colloquium (LNCS 115)*, 544–550, Springer-Verlag, 1981.

[1112] ——, "On the generation of cryptographically strong pseudorandom sequences", *ACM Transactions on Computer Systems*, 1 (1983), 38–44. An earlier version appeared in [1111].

[1113] ——, "A polynomial time algorithm for breaking the basic Merkle-Hellman cryptosystem", *Advances in Cryptology–Proceedings of Crypto 82*, 279–288, 1983.

[1114] ——, "A polynomial-time algorithm for breaking the basic Merkle-Hellman cryptosystem", *IEEE Transactions on Information Theory*, 30 (1984), 699–704. An earlier version appeared in [1113].

[1115] ——, "Identity-based cryptosystems and signature schemes", *Advances in Cryptology–Proceedings of CRYPTO 84 (LNCS 196)*, 47–53, 1985.

[1116] ——, "An efficient identification scheme based on permuted kernels", *Advances in Cryptology–CRYPTO '89 (LNCS 435)*, 606–609, 1990.

[1117] ——, "RSA for paranoids", *CryptoBytes*, 1 (Autumn 1995), 1–4.

[1118] A. SHAMIR AND A. FIAT, "Method, apparatus and article for identification and signature", U.S. Patent # 4,748,668, 31 May 1988.

[1119] M. SHAND AND J. VUILLEMIN, "Fast implementations of RSA cryptography", *Proceedings of the 11th IEEE Symposium on Computer Arithmetic*, 252–259, 1993.

[1120] C.E. SHANNON, "A mathematical theory of communication", *Bell System Technical Journal*, 27 (1948), 379–423, 623–656.

[1121] ——, "Communication theory of secrecy systems", *Bell System Technical Journal*, 28 (1949), 656–715.

[1122] ——, "Prediction and entropy of printed English", *Bell System Technical Journal*, 30 (1951), 50–64.

[1123] J. SHAWE-TAYLOR, "Generating strong primes", *Electronics Letters*, 22 (July 31, 1986), 875–877.

[1124] S. SHEPHERD, "A high speed software implementation of the Data Encryption Standard", *Computers & Security*, 14 (1995), 349–357.

[1125] A. SHIMIZU AND S. MIYAGUCHI, "Data randomization equipment", U.S. Patent # 4,850,019, 18 Jul 1989.

[1126] ——, "Fast data encipherment algorithm FEAL", *Advances in Cryptology–EUROCRYPT '87 (LNCS 304)*, 267–278, 1988.

[1127] Z. SHMUELY, "Composite Diffie-Hellman public-key generating systems are hard to break", Technical Report #356, TECHNION – Israel Institute of Technology, Computer Science Department, 1985.

[1128] P.W. SHOR, "Algorithms for quantum computation: discrete logarithms and factoring", *Proceedings of the IEEE 35th Annual Symposium on Foundations of Computer Science*, 124–134, 1994.

[1129] V. SHOUP, "New algorithms for finding irreducible polynomials over finite fields", *Mathematics of Computation*, 54 (1990), 435–447.

[1130] ——, "Searching for primitive roots in finite fields", *Mathematics of Computation*, 58 (1992), 369–380.

[1131] ——, "Fast construction of irreducible polynomials over finite fields", *Journal of Symbolic Computation*, 17 (1994), 371–391.

[1132] T. SIEGENTHALER, "Correlation-immunity of nonlinear combining functions for cryptographic applications", *IEEE Transactions on Information Theory*, 30 (1984), 776–780.

[1133] ——, "Decrypting a class of stream ciphers using ciphertext only", *IEEE Transactions on Computers*, 34 (1985), 81–85.

[1134] ——, "Cryptanalysts representation of non-linearly filtered ML-sequences", *Advances in Cryptology–EUROCRYPT '85 (LNCS 219)*, 103–110, 1986.

[1135] R.D. SILVERMAN, "The multiple polynomial quadratic sieve", *Mathematics of Computation*, 48 (1987), 329–339.

[1136] R.D. SILVERMAN AND S.S. WAGSTAFF JR., "A practical analysis of the elliptic curve factoring algorithm", *Mathematics of Computation*, 61 (1993), 445–462.

[1137] G.J. SIMMONS, "A "weak" privacy protocol using the RSA crypto algorithm", *Cryptologia*, 7 (1983), 180–182.

[1138] ——, "Authentication theory/coding theory", *Advances in Cryptology–Proceedings of CRYPTO 84 (LNCS 196)*, 411–431, 1985.

[1139] ——, "The subliminal channel and digital signatures", *Advances in Cryptology–Proceedings of EUROCRYPT 84 (LNCS 209)*, 364–378, 1985.

[1140] ——, "A secure subliminal channel (?)", *Advances in Cryptology–CRYPTO '85 (LNCS 218)*, 33–41, 1986.

[1141] ——, "How to (really) share a secret", *Advances in Cryptology–CRYPTO '88 (LNCS 403)*, 390–448, 1990.

[1142] ——, "Prepositioned shared secret and/or shared control schemes", *Advances in Cryptology–EUROCRYPT '89 (LNCS 434)*, 436–467, 1990.

[1143] ——, "Contemporary cryptology: a foreword", G.J. Simmons, editor, *Contemporary Cryptology: The Science of Information Integrity*, vii–xv, IEEE Press, 1992.

[1144] ——, "A survey of information authentication", G.J. Simmons, editor, *Contemporary Cryptology: The Science of Information Integrity*, 379–419, IEEE Press, 1992.

[1145] ——, "An introduction to shared secret and/or shared control schemes and their application", G.J. Simmons, editor, *Contemporary Cryptology: The Science of Information Integrity*, 441–497, IEEE Press, 1992.

[1146] ——, "How to insure that data acquired to verify treaty compliance are trustworthy", G.J. Simmons, editor, *Contemporary Cryptology: The Science of Information Integrity*, 615–630, IEEE Press, 1992.

[1147] ——, "The subliminal channels in the U.S. Digital Signature Algorithm (DSA)", W. Wolfowicz, editor, *Proceedings of the 3rd Symposium on State and Progress of Research in Cryptography, Rome, Italy*, 35–54, 1993.

[1148] ——, "Proof of soundness (integrity) of cryptographic protocols", *Journal of Cryptology*, 7 (1994), 69–77.

[1149] ——, "Subliminal communication is easy using the DSA", *Advances in Cryptology–EUROCRYPT '93 (LNCS 765)*, 218–232, 1994.

[1150] ——, "Protocols that ensure fairness", P.G. Farrell, editor, *Codes and Cyphers: Cryptography and Coding IV*, 383–394, Institute of Mathematics & Its Applications (IMA), 1995.

[1151] G.J. SIMMONS AND M.J. NORRIS, "Preliminary comments on the M.I.T. public-key cryptosystem", *Cryptologia*, 1 (1977), 406–414.

[1152] A. SINKOV, *Elementary Cryptanalysis: A Mathematical Approach*, Random House, New York, 1968.

[1153] M.E. SMID, "Integrating the Data Encryption Standard into computer networks", *IEEE Transactions on Communications*, 29 (1981), 762–772.

[1154] M.E. SMID AND D.K. BRANSTAD, "Cryptographic key notarization methods and apparatus", U.S. Patent # 4,386,233, 31 May 1983.

[1155] ——, "The Data Encryption Standard: Past and future", *Proceedings of the IEEE*, 76 (1988), 550–559.

[1156] ——, "The Data Encryption Standard: Past and future", G.J. Simmons, editor, *Contemporary Cryptology: The Science of Information Integrity*, 43–64, IEEE Press, 1992. Appeared earlier as [1155].

[1157] ——, "Response to comments on the NIST proposed digital signature standard", *Advances in Cryptology–CRYPTO '92 (LNCS 740)*, 76–88, 1993.

[1158] D.R. SMITH AND J.T. PALMER, "Universal fixed messages and the Rivest-Shamir-Adleman cryptosystem", *Mathematika*, 26 (1979), 44–52.

[1159] J.L. SMITH, "Recirculating block cipher cryptographic system", U.S. Patent # 3,796,830, 12 Mar 1974.

[1160] ——, "The design of Lucifer: A cryptographic device for data communications", IBM Research Report RC 3326, IBM T.J. Watson Research Center, Yorktown Heights, N.Y., 10598, U.S.A., Apr. 15 1971.

[1161] P. SMITH AND M. LENNON, "LUC: A new public key system", E. Dougall, editor, *Proceedings of the IFIP TC11 Ninth International Conference on Information Security, IFIP/Sec 93*, 103–117, North-Holland, 1993.

[1162] P. SMITH AND C. SKINNER, "A public-key cryptosystem and a digital signature system based on the Lucas function analogue to discrete logarithms", *Advances in Cryptology–ASIACRYPT '94 (LNCS 917)*, 357–364, 1995.

[1163] R. SOLOVAY AND V. STRASSEN, "A fast Monte-Carlo test for primality", *SIAM Journal on Computing*, 6 (1977), 84–85. Erratum in ibid, 7 (1978), 118.

[1164] J. SORENSON, "Two fast gcd algorithms", *Journal of Algorithms*, 16 (1994), 110–144.

[1165] A. SORKIN, "Lucifer, a cryptographic algorithm", *Cryptologia*, 8 (1984), 22–35.

[1166] M. STADLER, J.-M. PIVETEAU, AND J. CAMENISCH, "Fair blind signatures", *Advances in Cryptology–EUROCRYPT '95 (LNCS 921)*, 209–219, 1995.

[1167] O. STAFFELBACH AND W. MEIER, "Cryptographic significance of the carry for ciphers based on integer addition", *Advances in Cryptology–CRYPTO '90 (LNCS 537)*, 601–614, 1991.

[1168] W. STAHNKE, "Primitive binary polynomials", *Mathematics of Computation*, 27 (1973), 977–980.

[1169] D.G. STEER, L. STRAWCZYNSKI, W. DIFFIE, AND M. WIENER, "A secure audio teleconference system", *Advances in Cryptology–CRYPTO '88 (LNCS 403)*, 520–528, 1990.

[1170] J. STEIN, "Computational problems associated with Racah algebra", *Journal of Computational Physics*, 1 (1967), 397–405.

[1171] J.G. STEINER, C. NEUMAN, AND J.I. SCHILLER, "Kerberos: an authentication service for open network systems", *Proceedings of the Winter 1988 Usenix Conference*, 191–201, 1988.

[1172] M. STEINER, G. TSUDIK, AND M. WAIDNER, "Refinement and extension of encrypted key exchange", *Operating Systems Review*, 29:3 (1995), 22–30.

[1173] J. STERN, "Secret linear congruential generators are not cryptographically secure", *Proceedings of the IEEE 28th Annual Symposium on Foundations of Computer Science*, 421–426, 1987.

[1174] ——, "An alternative to the Fiat-Shamir protocol", *Advances in Cryptology–EUROCRYPT '89 (LNCS 434)*, 173–180, 1990.

[1175] ——, "Designing identification schemes with keys of short size", *Advances in Cryptology–CRYPTO '94 (LNCS 839)*, 164–173, 1994.

[1176] ——, "A new identification scheme based on syndrome decoding", *Advances in Cryptology–CRYPTO '93 (LNCS 773)*, 13–21, 1994.

[1177] D.R. STINSON, "An explication of secret sharing schemes", *Designs, Codes and Cryptography*, 2 (1992), 357–390.

[1178] ——, *Cryptography: Theory and Practice*, CRC Press, Boca Raton, Florida, 1995.

[1179] S.G. STUBBLEBINE AND V.D. GLIGOR, "On message integrity in cryptographic protocols", *Proceedings of the 1992 IEEE Computer Society Symposium on Research in Security and Privacy*, 85–104, 1992.

[1180] D.J. SYKES, "The management of encryption keys", D.K. Branstad, editor, *Computer security and the Data Encryption Standard*, 46–53, NBS Special Publication 500-27, U.S. Department of Commerce, National Bureau of Standards, Washington, D.C., 1977.

[1181] P. SYVERSON, "Knowledge, belief and semantics in the analysis of cryptographic protocols", *Journal of Computer Security*, 1 (1992), 317–334.

[1182] ——, "A taxonomy of replay attacks", *Proceedings of the Computer Security Foundations Workshop VII (CSFW 1994)*, 187–191, IEEE Computer Society Press, 1994.

[1183] P. SYVERSON AND P. VAN OORSCHOT, "On unifying some cryptographic protocol logics", *Proceedings of the 1994 IEEE Computer Society Symposium on Research in Security and Privacy*, 14–28, 1994.

[1184] K. TANAKA AND E. OKAMOTO, "Key distribution using id-related information directory suitable for mail systems", *Proceedings of the 8th Worldwide Congress on Computer and Communications Security and Protection (SECURICOM'90)*, 115–122, 1990.

[1185] A. TARAH AND C. HUITEMA, "Associating metrics to certification paths", Y. Deswarte, G. Eizenberg, and J.-J. Quisquater, editors, *Second European Symposium on Research in Computer Security – ESORICS'92 (LNCS 648)*, 175–189, Springer-Verlag, 1992.

[1186] J.J. TARDO AND K. ALAGAPPAN, "SPX: Global authentication using public key certificates", *Proceedings of the IEEE Symposium on Research in Security and Privacy*, 232–244, 1991.

[1187] A. TARDY-CORFDIR AND H. GILBERT, "A known plaintext attack of FEAL-4 and FEAL-6", *Advances in Cryptology–CRYPTO '91 (LNCS 576)*, 172–182, 1992.

[1188] M. TATEBAYASHI, N. MATSUZAKI, AND D.B. NEWMAN JR., "Key distribution protocol for digital mobile communication systems", *Advances in Cryptology–CRYPTO '89 (LNCS 435)*, 324–334, 1990.

[1189] R. TAYLOR, "An integrity check value algorithm for stream ciphers", *Advances in Cryptology–CRYPTO '93 (LNCS 773)*, 40–48, 1994.

[1190] J.A. THIONG LY, "A serial version of the Pohlig-Hellman algorithm for computing discrete logarithms", *Applicable Algebra in Engineering, Communication and Computing*, 4 (1993), 77–80.

[1191] J. THOMPSON, "S/MIME message specification – PKCS security services for MIME", RSA Data Security Inc., Aug. 29 1995, http://www.rsa.com/.

[1192] T. TOKITA, T. SORIMACHI, AND M. MATSUI, "Linear cryptanalysis of LOKI and s^2DES", *Advances in Cryptology–ASIACRYPT '94 (LNCS 917)*, 293–303, 1995.

[1193] ———, "On applicability of linear cryptanalysis to DES-like cryptosystems – LOKI89, LOKI91 and s^2DES", *IEICE Transactions on Fundamentals of Electronics, Communications and Computer Science*, E78-A (1995), 1148–1153. An earlier version appeared in [1192].

[1194] M. TOMPA AND H. WOLL, "Random self-reducibility and zero-knowledge interactive proofs of possession of information", *Proceedings of the IEEE 28th Annual Symposium on Foundations of Computer Science*, 472–482, 1987.

[1195] ———, "How to share a secret with cheaters", *Journal of Cryptology*, 1 (1988), 133–138.

[1196] G. TSUDIK, "Message authentication with one-way hash functions", *Computer Communication Review*, 22 (1992), 29–38.

[1197] S. TSUJII AND J. CHAO, "A new ID-based key sharing system", *Advances in Cryptology–CRYPTO '91 (LNCS 576)*, 288–299, 1992.

[1198] W. TUCHMAN, "Integrated system design", D.K. Branstad, editor, *Computer security and the Data Encryption Standard*, 94–96, NBS Special Publication 500-27, U.S. Department of Commerce, National Bureau of Standards, Washington, D.C., 1977.

[1199] ———, "Hellman presents no shortcut solutions to the DES", *IEEE Spectrum*, 16 (1979), 40–41.

[1200] J. VAN DE GRAAF AND R. PERALTA, "A simple and secure way to show the validity of your public key", *Advances in Cryptology–CRYPTO '87 (LNCS 293)*, 128–134, 1988.

[1201] E. VAN HEIJST, T.P. PEDERSEN, AND B. PFITZMANN, "New constructions of fail-stop signatures and lower bounds", *Advances in Cryptology–CRYPTO '92 (LNCS 740)*, 15–30, 1993.

[1202] E. VAN HEYST AND T.P. PEDERSEN, "How to make efficient fail-stop signatures", *Advances in Cryptology–EUROCRYPT '92 (LNCS 658)*, 366–377, 1993.

[1203] P. VAN OORSCHOT, "A comparison of practical public key cryptosystems based on integer factorization and discrete logarithms", G.J. Simmons, editor, *Contemporary Cryptology: The Science of Information Integrity*, 289–322, IEEE Press, 1992.

[1204] ———, "Extending cryptographic logics of belief to key agreement protocols", *1st ACM Conference on Computer and Communications Security*, 232–243, ACM Press, 1993.

[1205] ——, "An alternate explanation of two BAN-logic "failures"", *Advances in Cryptology–EUROCRYPT '93 (LNCS 765)*, 443–447, 1994.

[1206] P. VAN OORSCHOT AND M. WIENER, "A known-plaintext attack on two-key triple encryption", *Advances in Cryptology–EUROCRYPT '90 (LNCS 473)*, 318–325, 1991.

[1207] ——, "Parallel collision search with applications to hash functions and discrete logarithms", *2nd ACM Conference on Computer and Communications Security*, 210–218, ACM Press, 1994.

[1208] ——, "Improving implementable meet-in-the-middle attacks by orders of magnitude", *Advances in Cryptology–CRYPTO '96 (LNCS 1109)*, 229–236, 1996.

[1209] ——, "On Diffie-Hellman key agreement with short exponents", *Advances in Cryptology–EUROCRYPT '96 (LNCS 1070)*, 332–343, 1996.

[1210] H.C.A. VAN TILBORG, *An Introduction to Cryptology*, Kluwer Academic Publishers, Boston, 1988.

[1211] ——, "Authentication codes: an area where coding and cryptology meet", C. Boyd, editor, *Cryptography and Coding, 5th IMA Conference, Proceedings*, 169–183, Institute of Mathematics & Its Applications (IMA), 1995.

[1212] J. VAN TILBURG, "On the McEliece public-key cryptosystem", *Advances in Cryptology–CRYPTO '88 (LNCS 403)*, 119–131, 1990.

[1213] S.A. VANSTONE AND R.J. ZUCCHERATO, "Elliptic curve cryptosystems using curves of smooth order over the ring $\mathbb{Z}_n$", *IEEE Transactions on Information Theory*, to appear.

[1214] ——, "Short RSA keys and their generation", *Journal of Cryptology*, 8 (1995), 101–114.

[1215] S. VAUDENAY, "On the need for multipermutations: Cryptanalysis of MD4 and SAFER", B. Preneel, editor, *Fast Software Encryption, Second International Workshop (LNCS 1008)*, 286–297, Springer-Verlag, 1995.

[1216] ——, "On the weak keys of Blowfish", D. Gollmann, editor, *Fast Software Encryption, Third International Workshop (LNCS 1039)*, 27–32, Springer-Verlag, 1996.

[1217] U.V. VAZIRANI, "Towards a strong communication complexity theory, or generating quasi-random sequences from two communicating slightly-random sources", *Proceedings of the 17th Annual ACM Symposium on Theory of Computing*, 366–378, 1985.

[1218] U.V. VAZIRANI AND V.V. VAZIRANI, "Efficient and secure pseudo-random number generation", *Proceedings of the IEEE 25th Annual Symposium on Foundations of Computer Science*, 458–463, 1984. This paper also appeared in [1219].

[1219] ——, "Efficient and secure pseudo-random number generation", *Advances in Cryptology–Proceedings of CRYPTO 84 (LNCS 196)*, 193–202, 1985.

[1220] K. VEDDER, "Security aspects of mobile communications", B. Preneel, R. Govaerts, and J. Vandewalle, editors, *Computer Security and Industrial Cryptography: State of the Art and Evolution (LNCS 741)*, 193–210, Springer-Verlag, 1993.

[1221] G.S. VERNAM, "Secret signaling system", U.S. Patent # 1,310,719, 22 Jul 1919.

[1222] ——, "Cipher printing telegraph systems for secret wire and radio telegraphic communications", *Journal of the American Institute for Electrical Engineers*, 55 (1926), 109–115.

[1223] J. VON NEUMANN, "Various techniques used in connection with random digits", *Applied Mathematics Series, U.S. National Bureau of Standards*, 12 (1951), 36–38.

[1224] J. VON ZUR GATHEN AND V. SHOUP, "Computing Frobenius maps and factoring polynomials", *Computational Complexity*, 2 (1992), 187–224.

[1225] V.L. VOYDOCK AND S.T. KENT, "Security mechanisms in high-level network protocols", *Computing Surveys*, 15 (1983), 135–171.

[1226] D. WACKERLY, W. MENDENHALL III, AND R. SCHEAFFER, *Mathematical Statistics with Applications*, Duxbury Press, Belmont, California, 5th edition, 1996.

[1227] M. WAIDNER AND B. PFITZMANN, "The dining cryptographers in the disco: Unconditional sender and recipient untraceability with computationally secure serviceability", *Advances in Cryptology–EUROCRYPT '89 (LNCS 434)*, 690, 1990.

[1228] C.P. WALDVOGEL AND J.L. MASSEY, "The probability distribution of the Diffie-Hellman key", *Advances in Cryptology–AUSCRYPT '92 (LNCS 718)*, 492–504, 1993.

[1229] S.T. WALKER, S.B. LIPNER, C.M. ELLISON, AND D.M. BALENSON, "Commercial key recovery", *Communications of the ACM*, 39 (1996), 41–47.

[1230] C.D. WALTER, "Faster modular multiplication by operand scaling", *Advances in Cryptology–CRYPTO '91 (LNCS 576)*, 313–323, 1992.

[1231] P.C. WAYNER, "Content-addressable search engines and DES-like systems", *Advances in Cryptology–CRYPTO '92 (LNCS 740)*, 575–586, 1993.

[1232] D. WEBER, "An implementation of the general number field sieve to compute discrete logarithms mod p", *Advances in Cryptology–EUROCRYPT '95 (LNCS 921)*, 95–105, 1995.

[1233] A.F. WEBSTER AND S.E. TAVARES, "On the design of S-boxes", *Advances in Cryptology–CRYPTO '85 (LNCS 218)*, 523–534, 1986.

[1234] M.N. WEGMAN AND J.L. CARTER, "New hash functions and their use in authentication and set equality", *Journal of Computer and System Sciences*, 22 (1981), 265–279.

[1235] D. WELSH, *Codes and Cryptography*, Clarendon Press, Oxford, 1988.

[1236] A.E. WESTERN AND J.C.P. MILLER, *Tables of Indices and Primitive Roots*, volume 9, Royal Society Mathematical Tables, Cambridge University Press, 1968.

[1237] D.J. WHEELER, "A bulk data encryption algorithm", R. Anderson, editor, *Fast Software Encryption, Cambridge Security Workshop (LNCS 809)*, 127–134, Springer-Verlag, 1994.

[1238] D.J. WHEELER AND R.M. NEEDHAM, "TEA, a tiny encryption algorithm", B. Preneel, editor, *Fast Software Encryption, Second International Workshop (LNCS 1008)*, 363–366, Springer-Verlag, 1995.

[1239] D.H. WIEDEMANN, "Solving sparse linear equations over finite fields", *IEEE Transactions on Information Theory*, 32 (1986), 54–62.

[1240] M.J. WIENER, "Cryptanalysis of short RSA secret exponents", *IEEE Transactions on Information Theory*, 36 (1990), 553–558.

[1241] ——, "Efficient DES key search", Technical Report TR-244, School of Computer Science, Carleton University, Ottawa, 1994. Presented at Crypto '93 rump session.

[1242] S. WIESNER, "Conjugate coding", *SIGACT News*, 15 (1983), 78–88. Original manuscript (*cira* 1970).

[1243] H.S. WILF, "Backtrack: An O(1) expected time algorithm for the graph coloring problem", *Information Processing Letters*, 18 (1984), 119–121.

[1244] M.V. WILKES, *Time-Sharing Computer Systems*, American Elsevier Pub. Co., New York, 3rd edition, 1975.

[1245] F. WILLEMS, "Universal data compression and repetition times", *IEEE Transactions on Information Theory*, 35 (1989), 54–58.

[1246] H.C. WILLIAMS, "A modification of the RSA public-key encryption procedure", *IEEE Transactions on Information Theory*, 26 (1980), 726–729.

[1247] ——, "A $p + 1$ method of factoring", *Mathematics of Computation*, 39 (1982), 225–234.

[1248] ——, "Some public-key crypto-functions as intractable as factorization", *Cryptologia*, 9 (1985), 223–237.

[1249] H.C. WILLIAMS AND B. SCHMID, "Some remarks concerning the M.I.T. public-key cryptosystem", *BIT*, 19 (1979), 525–538.

[1250] R.S. WINTERNITZ, "A secure one-way hash function built from DES", *Proceedings of the 1984 IEEE Symposium on Security and Privacy*, 88–90, 1984.

[1251] S. WOLFRAM, "Cryptography with cellular automata", *Advances in Cryptology–CRYPTO '85 (LNCS 218)*, 429–432, 1986.

[1252] ——, "Random sequence generation by cellular automata", *Advances in Applied Mathematics*, 7 (1986), 123–169.

[1253] H. WOLL, "Reductions among number theoretic problems", *Information and Computation*, 72 (1987), 167–179.

[1254] A.D. WYNER, "The wire-tap channel", *Bell System Technical Journal*, 54 (1975), 1355–1387.

[1255] Y. YACOBI, "A key distribution "paradox"", *Advances in Cryptology–CRYPTO '90 (LNCS 537)*, 268–273, 1991.

[1256] Y. YACOBI AND Z. SHMUELY, "On key distribution systems", *Advances in Cryptology–CRYPTO '89 (LNCS 435)*, 344–355, 1990.

[1257] A.C. YAO, "On the evaluation of powers", *SIAM Journal on Computing*, 5 (1976), 100–103.

[1258] ——, "Theory and applications of trapdoor functions", *Proceedings of the IEEE 23rd Annual Symposium on Foundations of Computer Science*, 80–91, 1982.

[1259] S.-M. YEN AND C.-S. LAIH, "New digital signature scheme based on discrete logarithm", *Electronics Letters*, 29 (June 10, 1993), 1120–1121.

[1260] C. YUEN, "Testing random number generators by Walsh transform", *IEEE Transactions on Computers*, 26 (1977), 329–333.

[1261] D. YUN, "Fast algorithm for rational function integration", *Information Processing 77: Proceedings of IFIP Congress 77*, 493–498, 1977.

[1262] G. YUVAL, "How to swindle Rabin", *Cryptologia*, 3 (1979), 187–190.

[1263] K. ZENG AND M. HUANG, "On the linear syndrome method in cryptanalysis", *Advances in Cryptology–CRYPTO '88 (LNCS 403)*, 469–478, 1990.

[1264] K. ZENG, C.-H. YANG, AND T.R.N. RAO, "On the linear consistency test (LCT) in cryptanalysis with applications", *Advances in Cryptology–CRYPTO '89 (LNCS 435)*, 164–174, 1990.

[1265] ——, "An improved linear syndrome algorithm in cryptanalysis with applications", *Advances in Cryptology–CRYPTO '90 (LNCS 537)*, 34–47, 1991.

[1266] K. ZENG, C.-H. YANG, D.-Y WEI, AND T.R.N. RAO, "Pseudorandom bit generators in stream-cipher cryptography", *Computer*, 24 (1991), 8–17.

[1267] C. ZHANG, "An improved binary algorithm for RSA", *Computers and Mathematics with Applications*, 25:6 (1993), 15–24.

[1268] Y. ZHENG, J. PIEPRZYK, AND J. SEBERRY, "HAVAL – a one-way hashing algorithm with variable length of output", *Advances in Cryptology–AUSCRYPT '92 (LNCS 718)*, 83–104, 1993.

[1269] Y. ZHENG AND J. SEBERRY, "Immunizing public key cryptosystems against chosen ciphertext attacks", *IEEE Journal on Selected Areas in Communications*, 11 (1993), 715–724.

[1270] N. ZIERLER, "Primitive trinomials whose degree is a Mersenne exponent", *Information and Control*, 15 (1969), 67–69.

[1271] N. ZIERLER AND J. BRILLHART, "On primitive trinomials (mod 2)", *Information and Control*, 13 (1968), 541–554.

[1272] P.R. ZIMMERMANN, *The Official PGP User's Guide*, MIT Press, Cambridge, Massachusetts, 1995 (second printing).

[1273] J. ZIV AND A. LEMPEL, "On the complexity of finite sequences", *IEEE Transactions on Information Theory*, 22 (1976), 75–81.

[1274] M. ŽIVKOVIĆ, "An algorithm for the initial state reconstruction of the clock-controlled shift register", *IEEE Transactions on Information Theory*, 37 (1991), 1488–1490.

[1275] ——, "A table of primitive binary polynomials", *Mathematics of Computation*, 62 (1994), 385–386.

[1276] ——, "Table of primitive binary polynomials. II", *Mathematics of Computation*, 63 (1994), 301–306.

Index

Symbols

$|S|$ (cardinality of a set S), 49
$\in$ (set member), 49
$\subseteq$ (subset), 49
$\subset$ (proper subset), 49
$\cap$ (set intersection), 49
$\cup$ (set union), 49
$-$ (set difference), 49
$\times$ (Cartesian product), 49
$\emptyset$ (empty set), 50
O-notation (big O), 58
Ω-notation (big Omega), 59
Θ-notation (big Theta), 59
o-notation (little-o), 59
$L_q[\alpha, c]$ (subexponential notation), 60
$\leq_P$ (polytime reduction), 61
$\sim$ (asymptotic equivalence), 134
π (mathematical constant pi), 49
e (base of natural logarithms), 49
$\sum$ (sum), 50
$\prod$ (product), 50
! (factorial), 50
$\lfloor \ \rfloor$ (floor), 49
$\lceil \ \rceil$ (ceiling), 49
ϕ (Euler phi function), 65, 286
$\mu(n)$ (Möbius function), 154
lg (base 2 logarithm), 50
ln (natural logarithm), 50
| (divides relation), 63, 79
$\equiv$ (congruence relation), 67, 79
$\ll$ (much less than), 529
$\gg$ (much greater than), 170
$\binom{n}{k}$ (binomial coefficient), 52
$\left(\frac{a}{p}\right)$ (Legendre symbol), 72
$< \ >$ (inner product), 118
$\|x\|$ (length of a vector x), 118
$a \leftarrow b$ (assignment operator), 66
$a\|b$ (concatenation of strings a,b), 38
$\{0,1\}^k$ (bitstrings of bitlength k), 447
$\{0,1\}^*$ (bitstrings of arbitrary bitlength), 447
$\mathbb{Q}$ (the rational numbers), 49
$\mathbb{R}$ (the real numbers), 49
$\mathbb{Z}$ (the integers), 49
$\mathbb{Z}_n$ (integers modulo n), 68

$\mathbb{Z}_n^*$ (multiplicative group of $\mathbb{Z}_n$), 69
Q_n (quadratic residues modulo n), 70
$\overline{Q}_n$ (quadratic non-residues modulo n), 70
$\mathbb{F}_q$ (finite field of order q), 81
$\mathbb{F}_q^*$ (multiplicative group of $\mathbb{F}_q$), 81
$R[x]$ (polynomial ring), 78
$\vee$ (inclusive-OR), 213
$\oplus$ (exclusive-OR), 20
$\wedge$ (AND), 213
$\boxplus$ (addition mod 2^n), 263
$\boxminus$ (subtraction mod 2^n), 270
$\odot$ (modified multiplication mod $2^n + 1$), 263
$\hookleftarrow$ (left rotation), 213
$\hookrightarrow$ (right rotation), 213
$A \rightarrow B$ (message transfer), 396

A

Abelian group, 75
Abstract Syntax Notation One (ASN.1), 660
Access control, 3
Access control matrix, 387
Access matrix model, 569
Access structure, 526
    monotone, 527
Accredited Standards Committee (ASC), 648
Active adversary, 15, 37
Active attack, 41, 495
Ad hoc security, 43
Adaptive chosen-ciphertext attack, 42
Adaptive chosen-message attack, 433
Adaptive chosen-plaintext attack, 41
Addition chains, 621, 633
Adversary, 13, 495
    active, 15
    insider, 496
        one-time, 496
        permanent, 496
    outsider, 496
    passive, 15
Affine cipher, 239
Algebraic normal form, 205
Algorithm
    definition of, 57
    deterministic, 62
    exponential-time, 59

G